Quasi-Stationary Phenomena in Nonlinearly Perturbed Stochastic Systems

by

Mats Gyllenberg and Dmitrii S. Silvestrov

Walter de Gruyter · Berlin · New York

Authors

Mats Gyllenberg
Department of Mathematics and Statistics
P.O. Box 68
FIN-00014 University of Helsinki
Finland

E-Mail: mats.gyllenberg@helsinki.fi

Dmitrii S. Silvestrov
Division of Applied Mathematics
School of Education, Culture and Communication
Mälardalen University
P.O. Box 883
SE-72123 Västerås
Sweden

E-Mail: dmitrii.silvestrov@mdh.se

Mathematics Subject Classification 2000: Primary 60F05, 60F10, 60K05, 60K15; secondary 60J22, 60J27, 60K20, 60K25, 65C40

Key words: Nonlinear perturbation, quasi-stationary phenomenon, pseudo-stationary phenomenon, stochastic system, renewal equation, asymptotic expansion, ergodic theorem, limit theorem, large deviation, regenerative process, regenerative stopping time, semi-Markov process, Markov chain, absorption time, queueing system, population dynamics, epidemic model, lifetime, risk process, ruin probability, Cramér-Lundberg approximation, diffusion approximation

♾ Printed on acid-free paper which falls within the guidelines
of the ANSI to ensure permanence and durability.

ISSN 0938-6572

ISBN 978-3-11-020437-7

Bibliographic information published by the Deutsche Nationalbibliothek

The Deutsche Nationalbibliothek lists this publication in the Deutsche Nationalbibliografie; detailed bibliographic data are available in the Internet at http://dnb.d-nb.de.

Typeset using the authors' LaTeX Files: Kay Dimler, Müncheberg.
Printing and binding: Hubert & Co. GmbH & Co. KG, Göttingen.
Cover design: Thomas Bonnie, Hamburg.

Preface

Quasi-stationary phenomena in stochastic systems are a subject of intensive studies that started in 60s of 20th century. These phenomena describe the behaviour of stochastic systems with random lifetimes. The core of the quasi-stationary phenomenon is that one can observe something that resembles a stationary behaviour of the system before the lifetime goes to the end.

Examples of stochastic systems, in which quasi-stationary phenomena can be observed, are various queuing systems and reliability models, in which the lifetime is usually considered to be the time in which some kind of a fatal failure occurs in the system. Another class of examples of such stochastic systems is supplied by population dynamics or epidemic models. In population dynamics models, the lifetimes are usually the extinction times for the corresponding populations. In epidemic models, the role of the lifetime is played by the time of extinction of the epidemic in the population.

Usually the behaviour of a stochastic system can be described in terms of some Markov type stochastic process $\eta^{(\varepsilon)}(t)$ and its lifetime defined to be the time $\mu^{(\varepsilon)}$ at which the process $\eta^{(\varepsilon)}(t)$ hits a special absorption subset of the phase space of this process for the first time. The object of interest is the joint distribution of the process $\eta^{(\varepsilon)}(t)$ subject to a condition of non-absorption of the process up to a moment t, i.e., the probabilities $\mathsf{P}\{\eta^{(\varepsilon)}(t) \in A, \mu^{(\varepsilon)} > t\}$. A mathematical description of quasi-stationary phenomena is connected with studies of the asymptotic behaviour of these probabilities for $t \to \infty$.

A typical situation is when the process $\eta^{(\varepsilon)}(t)$ and the absorption time $\mu^{(\varepsilon)}$ depend on a small parameter $\varepsilon \geq 0$ in the sense that some of their local "transition" characteristics depend on the parameter ε. The parameter ε is involved in the model in such a way that the corresponding local characteristics are continuous at the point $\varepsilon = 0$, if regarded as functions of ε. These continuity conditions permit to consider the process $\eta^{(\varepsilon)}(t), t \geq 0$, for $\varepsilon > 0$, as a perturbed version of the process $\eta^{(0)}(t), t \geq 0$.

For example, one can consider queuing systems with servers that could be in a working state or in the state of reparation. The lifetime is the first time when all the servers in the system are in reparation. Then we can consider a system for which the reparation can be quickly completed, so that the expected time of the reparation plays the role of a small perturbation parameter. In population dynamics models, some local extinction probabilities can also be regarded as small perturbation parameters.

In models with perturbations, it is natural to study the asymptotic behaviour of the probabilities $\mathsf{P}\{\eta^{(\varepsilon)}(t) \in A, \mu^{(\varepsilon)} > t\}$ when the time $t \to \infty$ and the perturbation parameter $\varepsilon \to 0$ simultaneously. Here and henceforth in the book symbol $\varepsilon \to 0$ is used to show that $0 < \varepsilon \to 0$, i.e., ε tends to 0 being positive.

The principal novelty of our studies is that we consider the models with nonlinear perturbations. Local transition characteristics that were mentioned above are usually some scalar or vector moment functionals $p^{(\varepsilon)}$ of local transition probabilities for the corresponding processes. By a nonlinear perturbation we mean that these characteristics are nonlinear functions of the perturbation parameter ε and that the assumptions made imply that the characteristics can be expanded in an asymptotic power series with respect to ε up to and including some order k,

$$p^{(\varepsilon)} = p^{(0)} + p[1]\varepsilon + \cdots + p[k]\varepsilon^k + o(\varepsilon^k). \tag{P.1}$$

The case $k = 1$ corresponds to models with usual linear perturbations while the cases $k > 1$ correspond to models with nonlinear perturbations.

It turns out that the relation between the velocities with which ε tends to zero and the time t tends to infinity has a delicate influence on the quasi-stationary asymptotics. Without loss of generality it can be assumed that the time $t = t^{(\varepsilon)}$ is a function of the parameter ε. The balance between the rate of the perturbation and the rate of growth of time can be characterized by the following condition which is assumed to hold for some $1 \le r \le k$:

$$\varepsilon^r t^{(\varepsilon)} \to \lambda_r < \infty \text{ as } \varepsilon \to 0. \tag{P.2}$$

The main result of our studies is the so-called mixed ergodic and limit/large deviation theorems given in a form of *exponential expansions* for the probabilities $\mathsf{P}\{\eta^{(\varepsilon)}(t^{(\varepsilon)}) \in A, \mu^{(\varepsilon)} > t^{(\varepsilon)}\}$. Under the mentioned above assumptions on the perturbations and some additional natural conditions of Cramér type, we obtain exponential asymptotic relations of the following form:

$$\frac{\mathsf{P}\{\eta^{(\varepsilon)}(t^{(\varepsilon)}) \in A, \mu^{(\varepsilon)} > t^{(\varepsilon)}\}}{\exp\{-(\rho^{(0)} + a_1\varepsilon + \cdots + a_{r-1}\varepsilon^{r-1})t^{(\varepsilon)}\}} \to \tilde{\pi}^{(0)}(A)e^{-\lambda_r a_r} \text{ as } \varepsilon \to 0. \tag{P.3}$$

With these relations, we give explicit algorithms for calculating the coefficients $\rho^{(0)}$ and $a_1, \ldots, a_k$ as functions of the coefficients in the expansions of the local transition characteristics that appear in the initial nonlinear perturbation conditions of type (P.1). It is a non-trivial problem due to the nonlinear character of the perturbations involved.

The asymptotic behaviour of $\mathsf{P}\{\eta^{(\varepsilon)}(t) \in A, \mu^{(\varepsilon)} > t\}$ is very much different in the two alternative cases: **(a)** $\rho^{(0)} = 0$ and **(b)** $\rho^{(0)} > 0$. In the first case, the absorption times $\mu^{(\varepsilon)}$ are stochastically unbounded random variables, that is, they tend to ∞ in probability as $\varepsilon \to 0$. In the second case, the absorption times $\mu_0^{(\varepsilon)}$ are stochastically bounded as $\varepsilon \to 0$.

In the literature, the asymptotics related to the second model is known as *quasi-stationary* asymptotics. To distinguish between the asymptotics (P.3) in the first and the second cases, we coin the term *pseudo-stationary* for the first one.

Another class of asymptotic expansions systematically studied in the book concerns the so-called quasi-stationary distributions. Under some natural moment, communication and aperiodicity conditions imposed on the local transition characteristics there exists a so-called *quasi-stationary distribution* for the process $\eta^{(\varepsilon)}(t)$, which is given by the formula

$$\pi^{(\varepsilon)}(A) = \lim_{t \to \infty} \mathsf{P}\{\eta^{(\varepsilon)}(t) \in A \mid \mu^{(\varepsilon)} > t\}. \tag{P.4}$$

Using the asymptotics (P.3) we also obtain *asymptotic expansions* for the quasi-stationary distributions of the nonlinearly perturbed processes $\eta^{(\varepsilon)}(t)$,

$$\pi^{(\varepsilon)}(A) = \pi^{(0)}(A) + g_1(A)\varepsilon + \cdots + g_k(A)\varepsilon^k + o(\varepsilon^k). \tag{P.5}$$

As in the case of asymptotics (P.3), we give an explicit recurrence algorithm for calculating the coefficients $g_1(A), \ldots, g_k(A)$ as functions of the coefficients in the expansions for local transition characteristics of the processes $\eta^{(\varepsilon)}(t)$ in the corresponding initial perturbation conditions.

The classes of processes for which this program is realised include nonlinearly perturbed regenerative processes, semi-Markov processes, and continuous time Markov chains with absorption.

Our approach is based on advanced techniques, developed in the book, of nonlinearly perturbed renewal equations.

Applications to the analysis of quasi-stationary phenomena in models of nonlinearly perturbed stochastic systems considered in the book pertain to models of highly reliable queueing systems, M/G queueing systems with quick service, stochastic systems of birth-death type, including epidemic and population dynamics models, metapopulation dynamic models, and perturbed risk processes.

As was mentioned above, quasi-stationary phenomena were a subject of intensive studies during several decades. An extensive survey of the literature and comments can be found in the bibliographical remarks given in the book.

The results related to the asymptotics given in (P.3) and (P.5) and known in the literature mainly cover the case $k = 1$, which corresponds to the model of linearly perturbed processes. Some known results also relate to the case where the perturbation condition (P.1) has the special form of relation (c) $p^{(\varepsilon)} = p^{(0)} + p[1]\varepsilon$. In context of nonlinearly perturbed models, this is a particular case of nonlinear perturbation condition (P.1) which holds for every $k \geq 1$ with all higher order terms $p[2], p[3], \ldots \equiv 0$. It should be noted that the asymptotic expansions given in (P.3) and (P.5) obviously cover this case but not vise versa. Indeed, the asymptotic expansions obtained for the models with perturbation condition of the form (c) do not give any information about contribution of the terms $p[2], p[3], \ldots$ in nonlinear asymptotic expansions given in (P.3) and (P.5) for the models where these higher order terms take non-zero values.

Throughout the book, we shall make use of several basic types of conditions labeled by letters that make them suggestive. These classes of conditions are the following:

A for **A**bsorption conditions; **B** for **B**alancing conditions which connect and balance the rate of time growth with the rate of the perturbation; **C** for **C**ramér type conditions, that is, the conditions concerning the asymptotic finiteness of the exponential moments of the distribution functions that generate the corresponding renewal equations; **D** is for weak convergence conditions on the **D**istribution functions generating the corresponding renewal equations; **E** for conditions of **E**rgodicity for the corresponding stochastic processes; **F** for continuity conditions on the **F**orcing function of the corresponding renewal equations; **P** for **P**erturbation conditions, that is, the conditions relating the model ingredients with the perturbation parameter ε in terms of expansions of local transition characteristics in asymptotic power series in the parameter ε, etc. Conditions belonging to a specific class have subscripts numbering conditions in the class.

Local conditions used in theorems, lemmas, definitions or remarks are indicated with small Greek letters in parentheses as (α), (β), etc. Local conditions in the text can be indicated as **(a)**, **(b)**, etc. The indicated conditions are referred to always within the *subsection* where these conditions are introduced. Subsections, theorems, lemmas, definitions and remarks have a triple numeration. For example, Theorem 1.2.3 means Theorem 3 of Section 1.2. Formulas also have a triple numeration. For example, label (1.2.3) refers to Formula 3 in Section 1.2.

We finally comment on the structure of the book. The book contains an extended introduction, where we present in informal form the main problems, methods, and algorithms that constitute the content of the book. In Chapters 1 and 2 we present results which deal with a generalisation of the classical renewal theorem to a model of the perturbed renewal equation. These results are interesting by their own and, as we think, can find various applications beyond the areas mentioned in the book. In Chapters 3, 4, and 5 we study asymptotics of the types (P.3) and (P.5) for nonlinearly perturbed regenerative processes, semi-Markov processes, and continuous time Markov chains with absorption. Chapters 6 and 7 are devoted to applications of the theoretical results to studies of quasi-stationary phenomena for various nonlinearly perturbed models of stochastic systems. In Chapter 6, quasi-stationary phenomena are studied for highly reliable queueing systems, M/G queueing systems with quick service, stochastic systems of birth-death type, including epidemic and population dynamics models, and metapopulation dynamic models; Chapter 7 deals with perturbed risk processes. The last Chapter 8 contains three supplements. The first one gives some basic operation formulas for scalar and matrix asymptotic expansions. In the second supplement we discuss some new prospective directions for future research in the area. As was mentioned above, the quasi-stationary phenomena and related problems was a subject of intensive studies during several decades. In the last supplement, we give bibliographical remarks to the bibliography that includes more than 1000 references.

We hope that publication of this new book and a comprehensive bibliography of works related to asymptotic studies of quasi-stationary phenomena for perturbed stochastic processes and systems will be a useful contribution to the continuing intensive studies in this area. In addition to its use for the research and reference purposes, the book can also be used in special courses on the subject and as a complementary reading in general courses on stochastic processes. In this respect, it may be useful for specialists as well as doctoral and advanced undergraduate students.

We would also like to thank all our colleagues at the Department of Mathematics and Statistics at the University of Helsinki and the Division of Applied Mathematics of the School of Education, Culture, and Communication at the Mälardalen University for creating an inspiring research environment and friendly atmosphere which stimulated our work.

Helsinki – Västerås   *Mats Gyllenberg*
March 2008   *Dmitrii Silvestrov*

Contents

Introduction

This book is devoted to studies of quasi-stationary phenomena for nonlinearly perturbed stochastic processes and systems. The introduction intends to present in an informal form the main problems, methods, algorithms, and asymptotic results that constitute the content of the book.

We begin from presentation of a simple example which let us outline the circle of problems and illustrate results presented in the book. Then, we present and comment the content of the book by chapters. We try to show the logic and the ideas that make a basis for new methods and algorithms of asymptotic analysis for nonlinearly perturbed stochastic processes and systems developed in the book, to explain the choice and to comment applications which illustrate the theoretical results as well as to comment new directions for future research that could be a natural continuation of the research studies which results are presented in the book.

A model example and the basic quasi-stationary asymptotics

Let us consider a M/M queueing system that consists of two servers functioning independently of each other with exponential life and repairing times. The failure and the repairing intensities of the server i are $q_{i,-}$ and $q_{i,+}$, respectively.

One can interpret this queueing system as a simple model of a communication system with two channels or a power system with two electric generators.

The state of the server i at instant t is given by a random indicator variable $\eta_i(t)$ that takes the value 1 if the server i works at this instant, and the value 0 if the server i is repaired at this instant. The state of the system is described by the vector process $\eta(t) = (\eta_1(t), \eta_2(t))$, $t \geq 0$, which is, in this case, a continuous time homogeneous Markov chain. It has the phase space $X = \{\bar{i} = (i_1, i_2), i_1, i_2 = 0, 1\}$.

Figure 1 shows the transition graph, the transition probabilities of discrete time imbedded Markov chain, and the parameters of exponential sojourn times for the process $\eta(t)$.

Let us denote $\mu_{\bar{j}}$ the first time at which the process $\eta(t)$ hits into the state $\bar{j} \in X$. We interpret the hitting time $\mu_{\bar{0}}$, where $\bar{0} = (0, 0)$, as a lifetime of the queueing system. By the definition, it is the first time when both servers are out of work.

The main object of our interest is the probability $\mathsf{P}_{\bar{i}}\{\eta(t) = \bar{j}, \mu_{\bar{0}} > t\}$ that the system, which began to function in state $\bar{i} \neq \bar{0}$, will be observed in state $\bar{j} \neq \bar{0}$ at time t and, the lifetime $\mu_{\bar{0}}$ will not yet end up to time t.

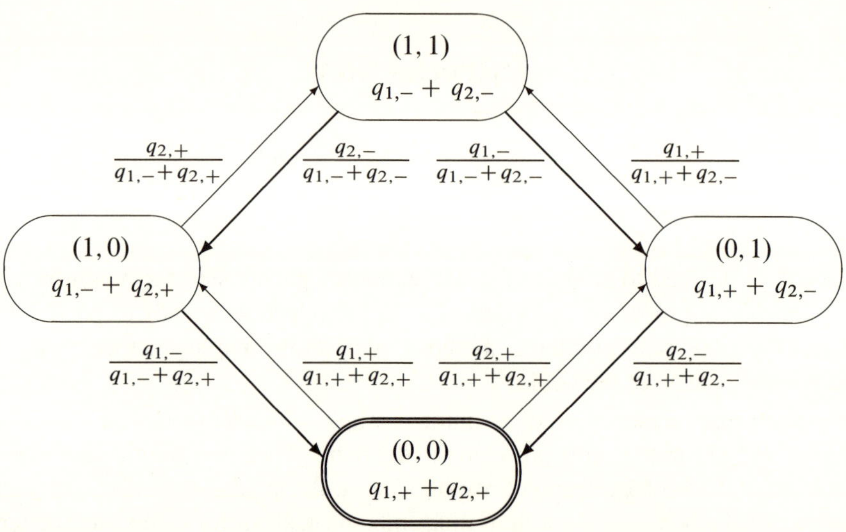

Figure 1 The transition graph for the Markov chain $\eta(t)$.

Let us assume that **(a)** all intensities $q_{i,\pm} > 0, i = 1, 2$. In this case, it is well known that the probability defined above has the following asymptotics

$$\frac{\mathsf{P}_{\bar{i}}\{\eta(t) = \bar{j}, \mu_{\bar{0}} > t\}}{e^{-\rho^{(*)}t}} \to \tilde{\tilde{\pi}}_{\bar{i}\bar{j}}(\rho^{(*)}) \quad \text{as } t \to \infty, \ \bar{i}, \bar{j} \neq \bar{0}. \tag{0.1}$$

The exponential normalising coefficient $\rho^{(*)}$ is the unique root of the following nonlinear characteristic equation **(b)** $\mathsf{E}_{\bar{1}} e^{\rho \mu_{\bar{1}}} \chi(\mu_{\bar{1}} \leq \mu_{\bar{0}}) = 1$, where $\bar{1} = (1, 1)$.

The limits $\tilde{\tilde{\pi}}_{\bar{i}\bar{j}}(\rho^{(*)}), \bar{i}, \bar{j} \neq \bar{0}$ are positive numbers which can depend on the initial state $\bar{i} \neq 0$. They determine so-called quasi-stationary probabilities by the formulas **(c)** $\pi_{\bar{j}}(\rho^{(*)}) = \tilde{\tilde{\pi}}_{\bar{k}\bar{j}}(\rho^{(*)})/\sum_{\bar{r} \neq 0} \tilde{\tilde{\pi}}_{\bar{k}\bar{r}}(\rho^{(*)}), \bar{j} \neq \bar{0}$. The quasi-stationary probabilities do not depend on the choice of the state $\bar{k} \neq \bar{0}$ in these formulas. The following asymptotic relation, which is implied, in an obvious way, by relation (0.1), clarifies this term,

$$\mathsf{P}_{\bar{i}}\{\eta(t) = \bar{j}/\mu_{\bar{0}} > t\} \to \pi_{\bar{j}}(\rho^{(*)}) \quad \text{as } t \to \infty, \ \bar{i}, \bar{j} \neq \bar{0}. \tag{0.2}$$

The asymptotic relation (0.2) shows that the conditional distribution of the process $\eta(t)$ (conditioned by the event that the lifetime will not end before time t) stabilises as time t tends to infinity. Thus, one can observe something that resembles a stationary behaviour of the system before the lifetime goes to the end. This is the sense of quasi-stationary phenomenon.

The asymptotic relations (0.1) and (0.2) are used in the same way as usual stationary type asymptotic relations. These relations let one approximate the quantities on the left hand side in these relations by the corresponding quantities on the right hand side, for large values of t.

As well known, the asymptotic relations of type (0.1) and (0.2) may loose their effectiveness if the parameters of the system take some nearly critical values. Here, it is the case of model with highly reliable servers, where one or both failure intensities $q_{1,-}$ and $q_{2,-}$ take very small values.

An usual formalism here is to assume that the transition intensities $q_{i,\pm} = q_{i,\pm}^{(\varepsilon)}$, $i = 1, 2$ depend on a small perturbation parameter $\varepsilon \geq 0$ in such way that **(d)** $q_{i,\pm}^{(\varepsilon)} \to q_{i,\pm}^{(0)}$ as $\varepsilon \to 0$, $i = 1, 2$. In this case, the process $\eta^{(\varepsilon)}(t)$ and the lifetime $\mu_0^{(\varepsilon)}$ also depend on the perturbation parameter ε. The parameter ε is involved in the model in such a way that the corresponding transition intensities are continuous at the point $\varepsilon = 0$, if regarded as functions of ε. This continuity condition permits to consider the process $\eta^{(\varepsilon)}(t)$, for $\varepsilon > 0$, as a perturbed version of the process $\eta^{(0)}(t)$.

An intuitive hypothesis is that, in this case, the character of asymptotics should also depend somehow of balance between large time values t and small values of failure intensities $q_{1,-}^{(\varepsilon)}$ and $q_{2,-}^{(\varepsilon)}$.

It is natural to assume that the limit repairing intensities **(e)** $q_{i,+}^{(0)} > 0$, $i = 1, 2$. As far as the limit failure intensities $q_{i,-}^{(0)}$, $i = 1, 2$ are concerned, there are three alternatives: **(f)** both limit failure intensities are positive; **(g)** one of limit failure intensities is positive while another equal zero; and **(h)** both limit failure intensities equal zero.

The cases **(g)** and **(h)** correspond to the model with asymptotically reliable servers. The most well investigated is the case of so-called linearly perturbed systems, where the pre-limit intensities can be represented in the following asymptotic form **(i)** $q_{i,\pm}^{(\varepsilon)} = q_{i,\pm}^{(0)} + q_{i,\pm}[1]\varepsilon + o(\varepsilon)$ as $\varepsilon \to 0$, where $|q_{i,\pm}[1]| < \infty$, $i = 1, 2$.

Let us, for example, assume that **(g$_1$)** $q_{1,-}^{(0)} = 0$ while $q_{2,-}^{(0)} > 0$. This means the the first "main" server is asymptotically absolutely reliable while the second "supplementary" server is not.

It turns out that the relation between the velocities with which ε tends to zero and the time t tends to infinity has a delicate influence on the quasi-stationary asymptotics. Without loss of generality it can be assumed that the time $t = t^{(\varepsilon)}$ is a function of the parameter ε. It is well known that in this case and under additional assumption **(j)** $\varepsilon t^{(\varepsilon)} \to \lambda$ as $\varepsilon \to 0$, where $0 \leq \lambda < \infty$, the following asymptotic relation takes place,

$$\mathsf{P}_{\bar{i}}\{\eta^{(\varepsilon)}(t^{(\varepsilon)}) = \bar{j}, \mu_0^{(\varepsilon)} > t^{(\varepsilon)}\} \to \tilde{\tilde{\pi}}_{\bar{i}\bar{j}}^{(0)}(0) e^{-\lambda a_1} \text{ as } \varepsilon \to 0, \ \bar{i}, \bar{j} \neq \bar{0}. \qquad (0.3)$$

Here, coefficient a_1 is explicitly determined by the coefficients penetrating the linear perturbation condition **(i)**. In particular, $a_1 > 0$ if $q_{1,-}[1] > 0$. As far as the coefficients $\tilde{\tilde{\pi}}_{\bar{i}\bar{j}}^{(0)}(0)$ are concerned, they, as expected, take the form $f_{\bar{i}}^{(0)} \tilde{\pi}_{\bar{j}}^{(0)}$, where $\tilde{\pi}_{\bar{j}}^{(0)}$ are the stationary probabilities for the limiting process $\eta^{(0)}(t)$ and $f_{\bar{i}}^{(0)} = \mathsf{P}_{\bar{i}}\{\mu_{\bar{1}}^{(0)} < \mu_{\bar{0}}^{(0)}\}$.

It should be noted that there is a principle difference between the asymptotic relations (0.1) and (0.3).

In the first one, the lifetime $\mu_{\bar{0}}$ is a finite random variable. The asymptotic relation (0.1) provides information about asymptotic behaviour of the mixed tail probabilities $\mathsf{P}_{\bar{i}}\{\eta(t) = \bar{j}, \mu_{\bar{0}} > t\}$, which, according to this relation, can be approximated with the zero asymptotic relative error by the exponential type mixed tail probabilities $\tilde{\tilde{\pi}}_{\bar{i}\bar{j}}(\rho^{(*)}) e^{-\rho^{(*)}t}$ as $t \to \infty$. It is a typical quasi-stationary asymptotic relation.

In the second one, the random variables $\mu_{\bar{0}}^{(\varepsilon)}$ are stochastically unbounded random variables, that is, they tend to ∞ in probability as $\varepsilon \to 0$. The asymptotic relation (0.3) presents some kind of mixed ergodic (for the process $\eta^{(\varepsilon)}(t)$) and limit (for the lifetime $\mu_{\bar{0}}^{(\varepsilon)}$) theorem. According to this relation the state of the perturbed process $\eta^{(\varepsilon)}(t^{(\varepsilon)})$ at the moment $t^{(\varepsilon)}$ and the normalised lifetime $\varepsilon \mu_{\bar{0}}^{(\varepsilon)}$ are asymptotically independent and the joint limit distribution is the product of the stationary distribution $f_{\bar{i}}^{(0)} \tilde{\pi}_{\bar{j}}^{(0)}$ for the limit process $\eta^{(0)}(t)$ and the limit exponential distribution $e^{-\lambda a_1}$ for the normalised lifetimes. The asymptotic relation (0.3) should be classified in some different way. We coin the term "pseudo-stationary" for such kind of asymptotics.

United quasi- and pseudo-stationary asymptotics for nonlinearly perturbed model

The question arises if it is possible to get the asymptotic relations given above in the frame of one united approach? The theory presented in the book gives an affirmative answer to this question.

The linear perturbation condition **(i)** does require some special analysis. In many cases, the perturbation parameter ε has some natural sense. For example, it can be some kind of "reciprocal cost" or "load rate" parameter such that failure intensities decrease if parameter ε tends to zero. In such cases transition intensities may actually be nonlinear functions of this parameter. The linear perturbation condition **(i)** appears as the result of well-known linearization procedure. The question arise, what effect has this linearization on the quality of the corresponding asymptotic formulas?

In this context, it is natural to introduce a nonlinear perturbation condition according to which the transition intensities $q_{i,\pm}^{(\varepsilon)}$ can be expanded in an asymptotic power series with respect to ε up to and including some order $k = 0, 1, \ldots$:

$$\mathbf{P^{(k)}}: \quad q_{i,\pm}^{(\varepsilon)} = q_{i,\pm}^{(0)} + q_{i,\pm}[1]\varepsilon + \cdots + q_{i,\pm}[k]\varepsilon^k + o(\varepsilon^k), \text{ where } |q_{i,\pm}[n]| < \infty \text{ for}$$
$$n = 1, \ldots, k \text{ and } i = 1, 2.$$

It is useful to note that these conditions have a natural subordination in the sense that, for every $k = 0, 1, \ldots$, condition $\mathbf{P^{(k)}}$ is weaker than condition $\mathbf{P^{(k+1)}}$. In the case $k = 0$, condition $\mathbf{P^{(k)}}$ reduces to the weak perturbation condition **(d)**, while, in the case $k = 1$, it coincides with the linear perturbation condition **(i)**.

The methods developed in the book let us improve the asymptotics presented in relations (0.1) and (0.3) for models with nonlinearly perturbed parameters. In particular, we answer the question in what way the higher order terms in the perturbation condition $\mathbf{P^{(k)}}$ influence the corresponding asymptotic relations.

The key role in the asymptotic methods developed in this book is played by the asymptotic expansion for the unique root $\rho^{(\varepsilon)}$ of the perturbed characteristic equation,

$$\mathsf{E}_{\bar{1}} e^{\rho \mu_{\bar{1}}^{(\varepsilon)}} \chi(\mu_{\bar{1}}^{(\varepsilon)} \leq \mu_{\bar{0}}^{(\varepsilon)}) = \int_0^\infty e^{\rho u} \mathsf{P}_{\bar{1}} \{ \mu_{\bar{1}}^{(\varepsilon)} \in du, \mu_{\bar{1}}^{(\varepsilon)} \leq \mu_{\bar{0}}^{(\varepsilon)} \} = 1. \qquad (0.4)$$

We begin our analysis from the case $k = 0$, where the perturbation condition $\mathbf{P^{(0)}}$ is assumed.

This condition implies the asymptotic relation (**k**) $\rho^{(\varepsilon)} \to \rho^{(0)}$ as $\varepsilon \to 0$. Using the methods developed in the book and relation (**k**) we can write down the asymptotic relation that generalises relation (0.1) to the case where the weak perturbation condition (**d**) holds. The following asymptotic relation takes place for any $0 \leq t^{(\varepsilon)} \to \infty$ as $\varepsilon \to 0$,

$$\frac{\mathsf{P}_{\bar{i}} \{ \eta^{(\varepsilon)}(t^{(\varepsilon)}) = \bar{j}, \mu_{\bar{0}}^{(\varepsilon)} > t^{(\varepsilon)} \}}{e^{-\rho^{(\varepsilon)} t^{(\varepsilon)}}} \to \tilde{\tilde{\pi}}_{\bar{i}\bar{j}}^{(0)}(\rho^{(0)}) \quad \text{as} \quad \varepsilon \to 0, \ \bar{i}, \bar{j} \neq \bar{0}. \qquad (0.5)$$

Also, conditions (**d**), (**e**) and (**f**) imply that for every ε small enough all intensities $q_{i,\pm}^{(\varepsilon)} > 0, i = 1, 2$. Thus, the asymptotic relations (0.1) and (0.2) can be written down for the perturbed process $\eta^{(\varepsilon)}(t)$, and the corresponding quasi-stationary distribution $\pi_{\bar{j}}^{(\varepsilon)}(\rho^{(\varepsilon)})$, $\bar{j} \neq \bar{0}$, can be defined by formulas (**c**) applied to this perturbed process.

Relation (0.5) is also supplemented by the asymptotic relation

$$\pi_{\bar{j}}^{(\varepsilon)}(\rho^{(\varepsilon)}) \to \pi_{\bar{j}}^{(0)}(\rho^{(0)}) \quad \text{as} \quad \varepsilon \to 0, \ \bar{j} \neq \bar{0}. \qquad (0.6)$$

The asymptotic behaviour of probabilities $\mathsf{P}_{\bar{i}} \{ \eta^{(\varepsilon)}(t) = \bar{j}, \mu_{\bar{0}}^{(\varepsilon)} > t \}$ is very much different in the two alternative cases: (**1**) $\rho^{(0)} = 0$ and (**2**) $\rho^{(0)} > 0$.

The case (**2**) $\rho^{(0)} > 0$ corresponds to the model where conditions (**d**), (**e**), and (**f**) hold. In this case, the lifetimes $\mu_{\bar{0}}^{(\varepsilon)}$ are stochastically bounded as $\varepsilon \to 0$. The asymptotic relations (0.5) replaces the asymptotic relation (0.1) for the case of model with perturbed parameters. In the literature, the asymptotics similar with (0.1) is referred as quasi-stationary asymptotics.

The case (**1**) $\rho^{(0)} = 0$ corresponds to the model where conditions (**d**), (**e**), and (**g**) or (**h**) hold. In this case, the lifetimes $\mu_{\bar{0}}^{(\varepsilon)}$ are stochastically unbounded random variables, that is, they tend to ∞ in probability as $\varepsilon \to 0$. To distinguish between the asymptotics (0.5) in these two cases, we, as was mentioned above, use the term "pseudo-stationary" in the case (**1**).

Expansions in quasi- and pseudo-stationary asymptotics for nonlinearly perturbed model

The asymptotic relation (0.5) looks nicely but has actually a serious drawback. The exponential normalisation with the coefficient $\rho^{(\varepsilon)}$ is not so effective because of this coefficient is given us only as the root of the nonlinear equation (0.4), for every $\varepsilon \geq 0$.

In order to improve the asymptotics provided by relations (0.5) and (0.6), we should assume that the perturbation condition $\mathbf{P}^{(k)}$, which is stronger than $\mathbf{P}^{(0)}$, holds for some $k \geq 1$. In this case, we show that the following asymptotic expansion holds for the characteristic root $\rho^{(\varepsilon)}$,

$$\rho^{(\varepsilon)} = \rho^{(0)} + a_1\varepsilon + \cdots + a_k\varepsilon^k + o(\varepsilon^k). \tag{0.7}$$

We also give the explicit effective algorithm for calculating the coefficients $a_1, \ldots,$ a_k as functions of the coefficients in the expansions involved in the nonlinear perturbation condition $\mathbf{P}^{(k)}$. It is a non-trivial problem due to the nonlinear character of the perturbations.

Now, we can involve in the analysis balancing conditions for the perturbation parameter ε and time $t^{(\varepsilon)}$. These conditions describe asymptotical time zones of different order. We can assume that the following balancing condition holds for some $1 \leq r \leq k$:

$\mathbf{B}^{(r)}$: $\varepsilon^r t^{(\varepsilon)} \to \lambda_r$ as $\varepsilon \to 0$, where $0 \leq \lambda_r < \infty$.

Using the perturbation condition $\mathbf{P}^{(k)}$ and the balancing condition condition $\mathbf{B}^{(r)}$ we can write down the following asymptotic relation that covers in the united form both cases of pseudo- and quasi-stationary asymptotics,

$$\frac{\mathsf{P}_{\bar{i}}\{\eta^{(\varepsilon)}(t^{(\varepsilon)}) = \bar{j}, \mu_{\bar{0}}^{(\varepsilon)} > t^{(\varepsilon)}\}}{e^{-(\rho^{(0)}+a_1\varepsilon+\cdots+a_{r-1}\varepsilon^{r-1})t^{(\varepsilon)}}} \to \tilde{\tilde{\pi}}_{\bar{i}\bar{j}}^{(0)}(\rho^{(0)})e^{-a_r\lambda_r} \text{ as } \varepsilon \to 0, \bar{i}, \bar{j} \neq \bar{0}. \tag{0.8}$$

This asymptotic relation is also supplemented by the asymptotic expansions for the quasi-stationary probabilities,

$$\pi_{\bar{j}}^{(\varepsilon)}(\rho^{(\varepsilon)}) = \pi_{\bar{j}}^{(0)}(\rho^{(0)}) + g_{\bar{j}}[1]\varepsilon + \cdots + g_{\bar{j}}[k]\varepsilon^k + o(\varepsilon^k), \bar{j} \neq \bar{0}. \tag{0.9}$$

We also give the explicit effective algorithm for calculating the coefficients $g_{\bar{j}}[1], \ldots, g_{\bar{j}}[k]$ as functions of the coefficients in the expansions involved the nonlinear perturbation condition $\mathbf{P}^{(k)}$. This algorithm essentially use the asymptotic expansion (0.7) but also does require several additional non-trivial steps.

Relations (0.8) and (0.9) essentially improve both asymptotic relations (0.5) and (0.6) replacing these simple convergence relations by the corresponding asymptotic expansions.

The exponential normalisation with the coefficient $\rho^{(0)} + a_1\varepsilon + \cdots + a_{r-1}\varepsilon^{r-1}$ involves the root $\rho^{(0)}$. To find it one should solve only one nonlinear equation (0.4), for the case $\varepsilon = 0$. As far as the coefficients $a_1, \ldots, a_{r-1}$ are concerned, they are given in the explicit algebraic recurrence form.

Moreover, the root $\rho^{(0)} = 0$ in the most interesting pseudo-stationary case. Here, the non-linear step connected with finding of the root of equation (0.4) can be omitted.

Relations (0.8) let us get a new type of so-called mixed ergodic (for the process $\eta^{(\varepsilon)}(t)$) and large deviation (for the lifetime $\mu_0^{(\varepsilon)}$) theorems.

Let us, for example, consider the pseudo-stationary case, where **(1)** $\rho^{(0)} = 0$ and conditions **(d)**, **(e)**, and **(g$_1$)** hold.

If $k = 1$, then the only case $r = 1$ is possible for the balancing condition. The asymptotic relation (0.8) reduces to the asymptotic relation (0.3), i.e., it yields the corresponding mixed ergodic and limit theorem for linearly perturbed system.

If $k = 2$, then two cases, $r = 1$ and $r = 2$, are possible for the balancing condition.

In the case $r = 1$, the asymptotic relation (0.8) reduces to (0.3). It show that the mixed tail probabilities $\mathsf{P}_{\bar{i}}\{\eta^{(\varepsilon)}(t^{(\varepsilon)}) = \bar{j}, \varepsilon\mu_0^{(\varepsilon)} > \varepsilon t^{(\varepsilon)}\}$ can be approximated by the exponential type mixed tail probabilities $f_{\bar{i}}^{(0)} \tilde{\pi}_{\bar{j}}^{(0)} e^{-a_1 \varepsilon t^{(\varepsilon)}}$ as $\varepsilon \to 0$, with the zero asymptotic relative error, in every asymptotic time zone which is determined by the relation $\varepsilon t^{(\varepsilon)} \to \lambda_1$ as $\varepsilon \to 0$, where $0 \le \lambda_1 < \infty$.

In the case $r = 2$, the asymptotic relation (0.8) shows that the mixed tail probabilities $\mathsf{P}_{\bar{i}}\{\eta^{(\varepsilon)}(t^{(\varepsilon)}) = \bar{j}, \varepsilon\mu_0^{(\varepsilon)} > \varepsilon t^{(\varepsilon)}\}$ can be approximated by the the exponential type mixed tail probabilities $f_{\bar{i}}^{(0)} \tilde{\pi}_{\bar{j}}^{(0)} e^{-a_1 \varepsilon t^{(\varepsilon)}}$ as $\varepsilon \to 0$, with the asymptotic relative error $1 - e^{-a_2 \lambda_2}$, in every asymptotic time zone which is determined by the relation $\varepsilon^2 t^{(\varepsilon)} \to \lambda_2$ as $\varepsilon \to 0$, where $0 \le \lambda_2 < \infty$.

If $\lambda_2 = 0$, then $\varepsilon t^{(\varepsilon)} = o(\varepsilon^{-1})$, and the asymptotic relative error is 0. Note that this case also covers the situation where $\varepsilon t^{(\varepsilon)}$ is bounded, which corresponds to the asymptotic relation (0.8) with $r = 1$. This is already an extension of this asymptotic relation since it is possible that $\varepsilon t^{(\varepsilon)} \to \infty$.

If $\lambda_2 > 0$, then $\varepsilon t^{(\varepsilon)} = O(\varepsilon^{-1})$, and the asymptotic relative error is $1 - e^{-\lambda_2 a_2}$. It differs from 0. Therefore, $o(\varepsilon^{-1})$ is an asymptotic bound for the large deviation zone with zero asymptotic relative error in the above approximation.

To get the approximation with zero asymptotic relative error in the asymptotic time zone which are determined by the relation $\varepsilon^2 t^{(\varepsilon)} \to \lambda_2$ one should approximate the mixed tail probabilities $\mathsf{P}_{\bar{i}}\{\eta^{(\varepsilon)}(t^{(\varepsilon)}) = \bar{j}, \varepsilon\mu_0^{(\varepsilon)} > \varepsilon t^{(\varepsilon)}\}$ by the exponential type mixed tail probabilities $f_{\bar{i}}^{(0)} \tilde{\pi}_{\bar{j}}^{(0)} e^{-(a_1\varepsilon + a_2\varepsilon^2)t^{(\varepsilon)}} \sim f_{\bar{i}}^{(0)} \tilde{\pi}_{\bar{j}}^{(0)} e^{-a_1 \varepsilon t^{(\varepsilon)} - a_2 \lambda_2}$, i.e., to introduce the corresponding corrections for the parameters in the exponents.

Relation (0.8) can be interpreted as a new type mixed ergodic and large deviation theorem for the nonlinearly perturbed process $\eta^{(\varepsilon)}(t)$ and the lifetime $\mu_0^{(\varepsilon)}$.

A similar interpretation can be made for the asymptotic relation (0.8) if $k > 2$. As above, relation (0.8) can also be interpreted as a mixed ergodic and limit theorem if $r = 1$. On the other hand, this relation is naturally to consider as a mixed ergodic and large deviation theorem for the corresponding asymptotic time large deviation zones if $1 < r \leq k$.

A similar interpretation remains also in the quasi-stationary case, where **(2)** $\rho^{(0)} > 0$.

It is also worth to note that some additional problems should be solved to complete the analysis. First of all, these are solidarity propositions which clarify the role of initial state $\eta^{(\varepsilon)}(0)$ in the corresponding asymptotic relations. The question about pivotal properties of the asymptotic expansions (0.7), i.e., what is the first non-zero coefficient in this expansion, is also of a great interest.

In conclusion, we would like to mention that the full analysis of perturbed queueing systems similar with those investigated above, but with m servers, is given in Section 6.1 of the book.

Contents of the book

The book is devoted to studies of quasi- and pseudo-stationary phenomena for nonlinearly perturbed stochastic processes and systems. The methods used are based on the exponential asymptotics for the nonlinearly perturbed renewal equation. These methods let us get so called mixed ergodic and large deviation theorems and asymptotic expansions in these theorems for nonlinearly perturbed regenerative processes, and then for nonlinearly perturbed semi-Markov processes and Markov chains. Finally, these results are applied to nonlinearly perturbed queueing systems, population dynamics, and epidemic models as well as nonlinearly perturbed risk processes. The book also contains an extended bibliography of works in the area. Let us present in more details the content of the book by chapters.

Chapter 1

In this chapter, we present the theory that can be characterised as a generalisation of classical renewal theory to the model of perturbed renewal equation.

The object of our studies is the classical renewal equation

$$x^{(\varepsilon)}(t) = q^{(\varepsilon)}(t) + \int_0^t x^{(\varepsilon)}(t-s)F^{(\varepsilon)}(ds), \quad t \geq 0. \tag{0.10}$$

The central role in the renewal theory is played by the renewal theorem. This theorem gives conditions for existence of a limit $x^{(\varepsilon)}(\infty)$ for the solution of the renewal equation $x^{(\varepsilon)}(t)$ as $t \to \infty$.

The importance of the renewal equation and the renewal theorem is based on the fact that distributions of regenerative processes, Markov processes, semi-Markov processes, etc., satisfy renewal type equations. This makes the approximation and asymptotic results connected with the renewal equation a very effective tool for getting ergodic and limit theorems for the processes mentioned above.

In Chapter 1, we give a natural generalisation of the classical renewal theorem to the model of perturbed renewal equation, where the forcing function $q^{(\varepsilon)}(t)$ and the distribution function $F^{(\varepsilon)}(s)$ generating this equation depend on a perturbation parameter $\varepsilon \geq 0$ and converge, in some natural sense, to $q^{(0)}(t)$ and $F^{(0)}(s)$ as $\varepsilon \to 0$, correspondingly. These continuity conditions allow to consider equation (0.10), for $\varepsilon > 0$, as a perturbation of equation (0.10) for $\varepsilon = 0$.

In the perturbed model, the solution $x^{(\varepsilon)}(t)$ of the renewal equation (0.10) also depends on the parameter ε, and the problem is to describe the asymptotic behaviour of the solution $x^{(\varepsilon)}(t)$ in the so-called triangular array mode, where $t \to \infty$ and $\varepsilon \to 0$, simultaneously. Without loss of generality, one can assume that the time $t = t^{(\varepsilon)}$ is a function of the perturbation parameter ε.

It is important to generalise the renewal theorem to the model of the perturbed renewal equation in such a way so that, in the case of the non-perturbed equation, the conditions should reduce to the conditions of the classical renewal theorem mentioned above. These results are presented in Chapter 1.

The most interesting asymptotics appears in the case where the distributions $F^{(\varepsilon)}(s)$ may be improper, i.e., $F^{(\varepsilon)}(\infty) \leq 1$ but for the defect of this distribution, $f^{(\varepsilon)} = 1 - F^{(\varepsilon)}(\infty)$, we have $f^{(\varepsilon)} \to 0$ as $\varepsilon \to 0$. In this case, the condition balancing the rate at which the time $t = t^{(\varepsilon)}$ approaches infinity and the convergence rate of $f^{(\varepsilon)}$ to zero as $\varepsilon \to 0$ has a delicate influence upon the asymptotics of the solution $x^{(\varepsilon)}(t)$.

We study this asymptotics under minimal conditions analogous to those used in the classical renewal theorem but also for the case where the distributions $F^{(\varepsilon)}(s)$ satisfy so-called Cramér type conditions for exponential moments of these distributions and the limit distribution $F^{(0)}(s)$ may be either proper or improper.

Asymptotic results presented in Chapter 1 make a basis for obtaining asymptotic exponential expansions for nonlinearly perturbed renewal equations.

Chapter 2

This chapter is devoted to asymptotic exponential expansions for nonlinearly perturbed renewal equations.

We impose some Cramér type condition on the distribution $F^{(\varepsilon)}(s)$ generating the perturbed renewal equation. This condition guarantee existence of the unique root $\rho^{(\varepsilon)}$ of the characteristic equation

$$\int_0^\infty e^{\rho s} F^{(\varepsilon)}(ds) = 1. \tag{0.11}$$

Note that the root $\rho^{(\varepsilon)} = 0$ if $F^{(\varepsilon)}(s)$ is a proper distribution function, i.e., $F^{(\varepsilon)}(\infty) = 1$ while $\rho^{(\varepsilon)} > 0$ if $F^{(\varepsilon)}(s)$ is an improper distribution function, i.e., $F^{(\varepsilon)}(\infty) < 1$.

Asymptotic expansions for this root play the key role in the method introduced in Chapter 2. In order to get such expansions we should impose nonlinear perturbation conditions on the distributions $F^{(\varepsilon)}(s)$. These conditions mean that the power or mixed power-exponential moments of these distributions are nonlinear functions of the perturbation parameter ε which can be expanded in an asymptotic power series with respect to ε up to and including some order k.

The method of asymptotic analysis developed in Chapter 2 is based on the combination of the generalised renewal theorem for perturbed renewal equation obtained in Chapter 1 with the asymptotic expansions for characteristic root $\rho^{(\varepsilon)}$. Using these results and some additional balancing conditions for the perturbation parameter ε and time $t^{(\varepsilon)} \to \infty$ as $\varepsilon \to 0$ we get the asymptotic exponential expansion of the following form:

$$\frac{x^{(\varepsilon)}(t^{(\varepsilon)})}{\exp\{-(\rho^{(0)} + a_1\varepsilon + \cdots + a_{r-1}\varepsilon^{r-1})t^{(\varepsilon)}\}} \to e^{-\lambda_r a_r} \tilde{x}^{(0)}(\infty) \quad \text{as} \quad \varepsilon \to 0. \quad (0.12)$$

This asymptotic expansion is provided by the explicit effective algorithm for calculating the coefficients $a_1, \ldots, a_k$ as functions of the coefficients in the expansions involved in the nonlinear perturbation conditions mentioned above. Despite the nonlinear and complex character of the perturbations involved we achieved our goal and have built the algorithm in an effective form of recurrence algebraic relations.

We also give the asymptotical expansion for the renewal limit of nonlinearly perturbed renewal equation,

$$\tilde{x}^{(\varepsilon)}(\infty) = \frac{\int_0^\infty e^{s\rho^{(\varepsilon)}} q^{(\varepsilon)}(s) m(ds)}{\int_0^\infty s e^{\rho^{(\varepsilon)} s} F^{(\varepsilon)}(ds)}, \quad (0.13)$$

(here $m(ds)$ is the Lebesgue measure) which take the following form

$$\tilde{x}^{(\varepsilon)}(\infty) = \tilde{x}^{(0)}(\infty) + f_1\varepsilon + \cdots + f_k\varepsilon^k + o(\varepsilon^k). \quad (0.14)$$

This expansion is also provided by the explicit recurrence algorithm for calculating the coefficients $f_1, \ldots, f_k$ as functions of the coefficients in the expansions involved in the nonlinear perturbation conditions mentioned above.

The results presented in Chapters 1 and 2, in particular asymptotics of type (0.12) and (0.14) make a basis for our studies of pseudo- and quasi-stationary phenomena for nonlinearly perturbed stochastic processes.

Chapter 3

This chapter is devoted to studies of quasi- and pseudo-stationary asymptotics for nonlinearly perturbed regenerative processes. It contains results concerning mixed ergodic and limit theorems and mixed ergodic and large deviation theorems for nonlinearly perturbed regenerative processes. These theorems are given in a form of exponential asymptotics for joint distributions of positions of regenerative processes and regenerative stopping times. These distributions satisfy some renewal equations, and the corresponding theorems are obtained by applying results of Chapters 1 and 2 to these renewal equations.

We consider a regenerative process $\xi^{(\varepsilon)}(t)$, $t \geq 0$, with regeneration times $\tau_n^{(\varepsilon)}$, $n = 1, 2, \ldots$, and a regenerative stopping time $\mu^{(\varepsilon)}$ that regenerates jointly with the process $\xi^{(\varepsilon)}(t)$, at times $\tau_n^{(\varepsilon)}$.

Both the regenerative process $\xi^{(\varepsilon)}(t)$ and the regenerative stopping time $\mu^{(\varepsilon)}$ are assumed to depend on a small perturbation parameter $\varepsilon \geq 0$. The processes $\xi^{(\varepsilon)}(t)$, for $\varepsilon > 0$ are considered as a perturbation of the process $\xi^{(0)}(t)$, and therefore we assume some weak continuity conditions for certain characteristic quantities of these processes regarded as functions of ε at point $\varepsilon = 0$.

As far as the regenerative stopping times are concerned, we consider two cases. The first one is where the random variables $\mu^{(\varepsilon)}$ are stochastically unbounded, i.e., $\mu^{(\varepsilon)}$ tend to ∞ in probability as $\varepsilon \to 0$. The second one is where the random variables $\mu^{(\varepsilon)}$ are stochastically bounded as $\varepsilon \to 0$.

The object of our study is the probabilities $\mathsf{P}\{\xi^{(\varepsilon)}(t) \in A, \mu^{(\varepsilon)} > t\}$ and their asymptotic behaviour as $t \to \infty$ and $\varepsilon \to 0$. We also present results concerning conditions for convergence and finding asymptotic expansions for quasi-stationary distributions of nonlinearly perturbed regenerative processes.

Two types of characteristic quantities that are related to one regeneration period play a crucial role in our analysis. These are the stopping probability $f^{(\varepsilon)} = \mathsf{P}\{\mu^{(\varepsilon)} < \tau_1^{(\varepsilon)}\}$ and the power moments of the regeneration times, $m_n^{(\varepsilon)} = \mathsf{E}(\tau_1^{(\varepsilon)})^n \chi(\mu^{(\varepsilon)} \geq \tau_1^{(\varepsilon)})$, or their mixed power-exponential analogues $\phi^{(\varepsilon)}(\rho, n) = \mathsf{E}(\tau_1^{(\varepsilon)})^n e^{\rho \tau_1^{(\varepsilon)}} \chi(\mu^{(\varepsilon)} \geq \tau_1^{(\varepsilon)})$ up to the order $n = k$. One of the main new elements in this study is that we consider a model with nonlinear perturbations. This means that the stopping probabilities $f^{(\varepsilon)}$ and the moments $m_n^{(\varepsilon)}$ or $\phi^{(\varepsilon)}(\rho, n)$ are nonlinear functions of ε. However, we restrict our attention to the case where the nonlinear perturbations are smooth and assume that these functions can be expanded in a power series with respect to ε up to and including the order k.

It turns out that the relationship between the rates with which ε tends to zero and the time t tends to infinity is important for the results. The balance between the rate of perturbation and the rate of growth of time is characterised by the following condition controlled by an integer parameter r, $1 \leq r \leq k$ and taking the form of asymptotic relation $\varepsilon^r t^{(\varepsilon)} \to \lambda_r < \infty$ as $\varepsilon \to 0$.

As was mentioned above, the probabilities $x^{(\varepsilon)}(t, A) = \mathsf{P}\{\xi^{(\varepsilon)}(t) \in A, \mu^{(\varepsilon)} > t\}$ satisfy the renewal equation with the forcing function $q^{(\varepsilon)}(t, A) = \mathsf{P}\{\xi^{(\varepsilon)}(t) \in A, \tau_1^{(\varepsilon)} > t, \mu^{(\varepsilon)} > t\}$ and the distribution function $F^{(\varepsilon)}(s) = \mathsf{P}\{\tau_1^{(\varepsilon)} \leq s, \mu^{(\varepsilon)} \geq \tau_1^{(\varepsilon)}\}$ generating the corresponding renewal equation.

This let us write down the asymptotic exponential expansion for the above probabilities that replicates the asymptotic expansion in the relation (0.12) for the probabilities $x^{(\varepsilon)}(t, A)$. We also write down the asymptotic expansion for the the corresponding renewal limits $\tilde{x}^{(\varepsilon)}(\infty, A)$ and then for the quasi-stationary distributions $\pi^{(\varepsilon)}(A) = \tilde{x}^{(\varepsilon)}(\infty, A)/\tilde{x}^{(\varepsilon)}(\infty, X)$ that follows from the asymptotic expansion (0.14).

Finally, we expand this asymptotics to the model of nonlinearly perturbed regenerative processes with transition period.

The asymptotic results obtained in Chapters 1–3 play the key role in further studies. We apply them for getting asymptotic expansions in mixed ergodic and large deviation theorems for nonlinearly perturbed semi-Markov processes with absorption. We make a very detailed analysis of pseudo- and quasi-stationary phenomena for perturbed semi-Markov processes with a finite set of states and absorption in the next two chapters.

Chapter 4

This chapter contains results concerning mixed ergodic theorems (for semi-Markov processes) and limit/large deviation theorems (for absorption times) for nonlinearly perturbed semi-Markov processes. We consider semi-Markov processes with a finite set of states $X = \{0, 1, \dots, N\}$ and the absorption state 0. The first hitting time to this state is the absorption time for the corresponding semi-Markov process.

The results mentioned above are given in the form of an asymptotics for joint distributions of positions of semi-Markov processes and absorption times. They are obtained by applying the corresponding results for regenerative processes given in Chapter 3. It can be done by using the fact that a semi-Markov process can be considered as a regenerative process with regeneration times which are subsequent return moments to any fixed state $i \neq 0$. The first hitting time to the absorption state 0 is, in this case, the regenerative stopping time.

We consider in details not only the generic case where the limiting semi-Markov process has one communication class of recurrent-without absorption states, but also the case, where the limiting semi-Markov process has one communication class of recurrent-without absorption states and, additionally, the class of non-recurrent-without absorption states. The latter model covers a significant part of applications.

We consider a semi-Markov process $\eta^{(\varepsilon)}(t)$, $t \geq 0$, with a phase space $X = \{0, 1, \dots, N\}$ and transition probabilities $Q_{ij}^{(\varepsilon)}(u)$. The semi-Markov process $\eta^{(\varepsilon)}(t)$ is assumed to depend on a perturbation parameter $\varepsilon \geq 0$. The processes $\eta^{(\varepsilon)}(t)$, for $\varepsilon > 0$, are considered as perturbations of the process $\eta^{(0)}(t)$, and therefore we assume some weak continuity conditions satisfied by transition quantities, namely, by the moment functionals of transition probabilities, if regarded as functions of ε, at the point $\varepsilon = 0$.

Our studies are concerned with the random functional $\mu_0^{(\varepsilon)}$ which is the first hitting time for the process $\eta^{(\varepsilon)}(t)$ to hit the absorption state 0. In applications, the hitting times $\mu_0^{(\varepsilon)}$ are often interpreted as transition times for different stochastic systems described by semi-Markov processes; these are occupation times or waiting times in queueing systems, lifetimes in reliability models, extinction times in population dynamic models, etc.

The object of our study are probabilities $P_i\{\eta^{(\varepsilon)}(t) = j, \mu_0^{(\varepsilon)} > t\}$ and their asymptotic behaviour for $t \to \infty$ and $\varepsilon \to 0$.

We give asymptotic relations for these probabilities based on minimal conditions that involve only first moments of transition probabilities $Q_{ij}^{(\varepsilon)}(u)$ as well as asymptotic relations involving conditions imposed on exponential moments of these transition probabilities.

We also formulate and prove a number of auxiliary propositions dealing with convergence of hitting probabilities, distributions, expectations, and exponential moments of hitting times.

Finally, we give solidarity propositions that clarify the role of initial state in the corresponding asymptotic relations as well as solidarity properties of cyclic absorption probabilities, and moment generating functions for hitting times for perturbed semi-Markov processes. They show that the corresponding conditions for convergence formulated in terms of cyclic characteristics of return times, and the form of the corresponding asymptotic relations are invariant with respect to the choice of the recurrent-without-absorption state used to construct the regeneration cycles.

In this case, the the distribution function $_0G_{ii}^{(\varepsilon)}(t)$ of the return-without-absorption time in a state $i \neq 0$ generates the corresponding renewal equation and the corresponding characteristic equation takes the form $\int_0^\infty e^{\rho s}\, _0G_{ii}^{(\varepsilon)}(ds) = 1$. In particular, we show that the characteristic root $\rho^{(\varepsilon)}$ of this equation does not depend on the choice of a recurrent-without-absorption state $i \neq 0$.

The results obtained in Chapter 4 play a preparatory role. The mixed ergodic and large deviation theorems obtained in this chapter, together with the corresponding solidarity propositions for cyclic characteristics deduced for perturbed semi-Markov processes, create a basis for getting exponential expansions in the mixed ergodic and large deviation theorems. Such expansions are given in the next chapter.

Chapter 5

In this chapter, we continue studies began in Chapter 4. We give asymptotic results which improve the mixed ergodic theorems (for semi-Markov processes) and large deviation theorems (for absorption times) to the form of exponential expansions for joint distributions of positions of semi-Markov processes and absorption times. We also give asymptotic expansions for quasi-stationary distributions of nonlinearly perturbed semi-Markov processes.

These asymptotic expansions are obtained by applying the corresponding results, given in Chapter 3, on the exponential expansions for regenerative processes, and mixed ergodic and large deviation theorems obtained in Chapter 4 for nonlinearly perturbed semi-Markov processes.

We consider the same model of perturbed semi-Markov processes as in Chapter 4. However, in Chapter 5, we impose stronger perturbation conditions on the transition probabilities of the imbedded Markov chains $p_{ij}^{(\varepsilon)}$ and the semi-Markov processes

$Q_{ij}^{(\varepsilon)}(u)$. We assume that these transition probabilities $p_{ij}^{(\varepsilon)}$ and the power or the mixed power-exponential moments of transition probabilities $Q_{ij}^{(\varepsilon)}(u)$ (which are improper distribution functions in argument u) can be expanded in a power series with respect to ε up to and including the order k.

As above, the relationship between the rate with which ε tends to zero and the time t tends to infinity has a delicate influence upon the results. The balance between the rate of perturbation and the rate of growth of time is characterised by the following asymptotic relation $\varepsilon^r t^{(\varepsilon)} \to \lambda_r < \infty$ as $\varepsilon \to 0$ that is assumed to hold for some $1 \le r \le k$.

With the above assumptions and under some natural additional non-periodicity and Cramér type conditions for the transition probabilities $Q_{ij}^{(\varepsilon)}(u)$, we obtain the asymptotic relation

$$\frac{\mathsf{P}_i\{\eta^{(\varepsilon)}(t^{(\varepsilon)}) = j, \mu_0^{(\varepsilon)} > t^{(\varepsilon)}\}}{e^{-(\rho^{(0)}+a_1\varepsilon+\cdots+a_{r-1}\varepsilon^{r-1})t^{(\varepsilon)}}} \to \tilde{\tilde{\pi}}_{ij}^{(0)}(\rho^{(0)})e^{-\lambda_r a_r} \quad \text{as } \varepsilon \to 0, \ i, j \neq 0. \quad (0.15)$$

There are to alternatives in this relation, where the characteristic root **(1)** $\rho^{(0)} = 0$ and **(2)** $\rho^{(0)} > 0$.

The first alternative **(1)** holds if an absorption in 0 is impossible for the limiting process $\eta^{(0)}(t)$. In such a case, the absorption times $\mu_0^{(\varepsilon)}$ tend in probability to ∞ as $\varepsilon \to 0$. The asymptotic relation (0.15) describes pseudo-stationary phenomena for perturbed semi-Markov processes.

The second alternative **(2)** holds if an absorption in 0 is possible for the limiting process $\eta^{(0)}(t)$. In this case, the absorption times $\mu_0^{(\varepsilon)}$ are stochastically bounded as $\varepsilon \to 0$. The asymptotic relation (0.15) describes, in this case, quasi-stationary phenomena for perturbed semi-Markov processes.

In the pseudo-stationary case **(1)**, the limiting coefficients $\tilde{\tilde{\pi}}_{ij}^{(0)}(0)$ in the asymptotic relation (0.15), as expected, take the form $f_i^{(0)} \tilde{\pi}_j^{(0)}$, where $\tilde{\pi}_j^{(0)}$ are the stationary probabilities of the limiting process $\eta^{(0)}(t)$ and $f_i^{(0)}$ is the probability of hitting the process $\eta^{(0)}(t)$ to the set of recurrent-without-absorption states before absorption, under condition that $\eta^{(0)}(0) = i \neq 0$.

The situation is not so simple in the quasi-stationary case **(2)**. In this case, under some natural conditions, there exists a so-called quasi-stationary distribution for the semi-Markov process $\eta^{(\varepsilon)}(t)$ given by the formulas

$$\pi_j^{(\varepsilon)}(\rho^{(\varepsilon)}) = \frac{\tilde{\tilde{\pi}}_{kj}^{(\varepsilon)}(\rho^{(\varepsilon)})}{\sum_{r \neq 0} \tilde{\tilde{\pi}}_{kr}^{(\varepsilon)}(\rho^{(\varepsilon)})}, \quad j \neq 0.$$

In this case, the coefficients $\tilde{\tilde{\pi}}_{ij}^{(\varepsilon)}(\rho^{(\varepsilon)})$, which are defined as in (0.15) but for a fixed $\varepsilon \ge 0$, may depend on the initial state $i \neq 0$. But, the quasi-stationary probabilities

$\pi_j^{(\varepsilon)}(\rho^{(\varepsilon)})$, $j \neq 0$, do not depend on the choice of the state $k \neq 0$ in the formulas above.

We also give the following asymptotic expansions for the quasi-stationary distributions for nonlinearly perturbed semi-Markov processes with absorption,

$$\pi_j^{(\varepsilon)}(\rho^{(\varepsilon)}) = \pi_j^{(0)}(\rho^{(0)}) + g_j[1]\varepsilon + \cdots + g_j[k]\varepsilon^k + o(\varepsilon^k), \quad j \neq 0. \qquad (0.16)$$

Both asymptotic expansions (0.15) and (0.16) are provided by the explicit algorithms for calculating the coefficients in these expansions.

We obtain mixed large deviation and ergodic theorems for nonlinearly perturbed semi-Markov processes using exponential expansions in mixed ergodic and large deviation theorems developed in Chapter 3 for nonlinearly perturbed regenerative processes. We use the fact that any semi-Markov processes is a regenerative process with the return times to some fixed state being regeneration times. The time of absorption (the first time of hitting the absorption state 0) is a regenerative stopping time. A semi-Markov process is characterized by prescribing its transition probabilities. It is natural to formulate the perturbation conditions in terms of these transition probabilities. However, the conditions and expansions formulated for regenerative processes are specified in terms of expansions for the moments of regeneration times. Therefore, the corresponding asymptotic expansions for absorption probabilities and the moments of return and hitting times for perturbed semi-Markov processes must be developed.

Thus, as the first step, we construct asymptotic expansions for hitting probabilities, power and mixed power-exponential moments of hitting times using a procedure that is based on recursive systems of linear equations for hitting probabilities and moments of hitting times. These moments satisfy recurrence systems of linear equations with the same perturbed coefficient matrix and the free terms connected by special recurrence systems of relations. In these relations, the free terms for the moments of a given order are given as polynomial functions of moments of lower orders. This permits to build an effective recurrence algorithm for constructing the corresponding asymptotic expansions. Each sub-step in this recurrence algorithm is of a matrix but linear type, where the solution of the system of linear equations with nonlinearly perturbed coefficients and free terms should be expanded in asymptotic series. We also provide these expansions with a detailed analysis of their pivotal properties. These results have, as we think, their own values and possible applications beyond the problems studied in the book.

As soon as asymptotic expansions for moments of return-without-absorption times are constructed, the second nonlinear but scalar step of construction asymptotic expansions expansions for the characteristic root $\rho^{(\varepsilon)}$ can be realised.

The separation of two steps described above, the first one matrix and recurrence but linear and the second one nonlinear but scalar, significantly simplify the whole algorithm.

Asymptotic expansion for the characteristic root $\rho^{(\varepsilon)}$ let us combine it with asymptotic relations for probabilities $\mathsf{P}_i\{\eta^{(\varepsilon)}(t^{(\varepsilon)}) = j, \mu_0^{(\varepsilon)} > t^{(\varepsilon)}\}$ given in Chapter 4 and obtain the exponential expansion (0.15).

The asymptotic expansion (0.16) for quasi-stationary distributions $\pi_j^{(\varepsilon)}(\rho^{(\varepsilon)})$ requires two more steps, which are needed for constructing asymptotic expansions at the point $\rho^{(\varepsilon)}$ for the corresponding moment generation functions, giving expressions for quasi-stationary probabilities in the quotient form, and for transforming the corresponding asymptotic quotient expressions to the form of power asymptotic expansions.

As a result, we get an effective algorithm for a construction of the asymptotic expansions given in relations (0.15) and (0.16). We think that the method used for obtaining the expansions mentioned above has its own value and great potential for future studies.

As a particular but important example, we also consider the model of nonlinearly perturbed continuous time Markov chains with absorption. In this case, it is more natural to formulate the perturbation conditions in terms of generators of the perturbed Markov chains. Here, an additional step in the algorithms is needed, since the initial perturbation conditions for generators must be expressed in terms of the moments for the corresponding semi-Markov transition probabilities. Then, the basic algorithms obtained for nonlinearly perturbed semi-Markov processes can be applied.

Chapters 1–5 present a theory that can be applied in studies of pseudo- and quasi-stationary phenomena in nonlinearly perturbed stochastic systems.

Chapter 6

This chapter deals with applications of the results obtained in Chapters 1–5 to an analysis of pseudo- and quasi-stationary phenomena in nonlinearly perturbed stochastic systems.

Examples of stochastic systems under consideration are queueing systems, epidemic, and population dynamics models with finite lifetimes. In queueing systems, the lifetime is usually the time at which some kind of a fatal failure occurs in the system. In epidemic models, the time of extinction of the epidemic in the population plays the role of the lifetime, while in population dynamics models, the lifetime is usually the extinction time for the corresponding population.

We selected several classical models being the subject of long term research studies. These models serve nowadays mainly as platforms for demonstration of new methods and innovation results. Our goal also is to show what kind of new types results related to quasi-stationary asymptotics can be obtained for such models with nonlinearly perturbed parameters.

We give all types of asymptotic results studied in Chapters 1–5. They include the following: (i) mixed ergodic theorems (for the state of the system) and limit theorems (for the lifetimes) that describe transition phenomena; (ii) mixed ergodic and large deviation theorems that describe pseudo- and quasi-stationary phenomena; (iii) exponential expansions in mixed ergodic and large deviation theorems; (iv) theorems on convergence of quasi-stationary distributions; and (v) asymptotic expansions for quasi-stationary distributions. In all the examples, we try to specify and to describe, in a more explicit form, conditions and algorithms for calculating the limit expressions, the characteristic roots, and the coefficients in the corresponding asymptotic expansions.

As the first example, we consider a M/M queueing system with highly reliable main servers. The simplest variant of such system with nonlinearly perturbed parameters was the subject of discussion in the first three subsections above. This queueing system is our first choice because of its function can be described by some nonlinearly perturbed continuous time Markov chains with absorption. Here, all conditions take a very explicit and clear form.

We also consider M/G queueing system with quick service and a bounded queue buffer. In this case, the perturbed stochastic processes, which describe the dynamics of the queue in the system, belong to the class of so-called stochastic processes with semi-Markov modulation. These processes admit a construction of imbedded semi-Markov processes and are more general than semi-Markov processes. This example was chosen because it shows in which way the main results obtained in the book can be applied to stochastic processes more general than semi-Markov processes, in particular, to stochastic processes with semi-Markov modulation.

Our next example is based on classical semi-Markov and Markov birth-and-death type processes. Some classical models of queueing systems, epidemic or population dynamic models can be described with the use of such processes. We show in which way nonlinear perturbation conditions should be used and what form will take advanced quasi- and pseudo-stationary asymptotics developed in Chapters 1–5.

Finally, we consider an example of nonlinearly perturbed metapopulation model. We think that this example is interesting since it brings, for the first time, the discussion on advanced quasi- and pseudo-stationary asymptotics in this actual area of research in mathematical biology.

Chapter 7

The classical risk processes are still the object of intensive research studies as show, for example, references given in the bibliography. Of course, the purpose of these studies is not any more to derive formulas relevant for field applications. These studies intend to illustrate new methods and types of results that can later be expanded to more complex models. The same approach was used by us when choosing this model. The aim was to show that the innovative methods of analysis for nonlinearly perturbed processes developed in the book can yield new results for this classical models.

This chapter contains results that extend the classical Cramér–Lundberg and diffusion approximations for the ruin probabilities to a model of nonlinearly perturbed risk processes. Both approximations are presented in a unified way using the techniques of perturbed renewal equations developed in Chapters 1 and 2.

The main new element in our results is a high order exponential asymptotic expansion in these approximations for nonlinearly perturbed risk processes. Correction terms are obtained for the Cramér–Lundberg and diffusion type approximations, which provide the right asymptotic behaviour of relative errors in the perturbed model. We study the dependence of these correction terms on the relations between the rate of perturbation and the rate of growth of the initial capital.

We also give various variants of the diffusion type approximation, including the asymptotics for increments and derivatives of the ruin probabilities.

Finally, we give asymptotic expansions in the Cramér–Lundberg and diffusion type approximations for distribution of the capital surplus prior and at ruin for nonlinearly perturbed risk processes.

It seems to us that results presented in Chapters 6 and 7 illustrate well a potential of asymptotic methods developed in the book.

Chapter 8

The last chapter contains three supplements. The first one gives some basic arithmetic operation formulas for scalar and matrix asymptotic expansions.

In the second supplement, we discuss some new prospective directions for future research studies in the area and make comments on some already available results. Let us point out some of these directions.

In the book, we study quasi-stationary phenomena for nonlinearly perturbed stochastic processes and systems with continuous time. However, it is clear that similar results can also be obtained for discrete time models which are important by themselves and in applications.

Another direction for future studies is the exponential asymptotic expansions, which are based on non-polynomial systems of infinitesimals, in mixed ergodic and large-deviation theorems for nonlinearly perturbed regenerative, Markov, and semi-Markov processes.

In the book, we treat models of nonlinearly perturbed Markov chains and semi-Markov processes with a finite phase space. It would be important to realise an analogous program of studies for Markov chains and semi-Markov type processes with countable and general phase spaces.

We have no doubts that the theory developed in the book can be also be successfully applied to more general stochastic processes with semi-Markov modulation (switchings).

Another direction for advancing the studies carried out in the book is connected with models of nonlinearly perturbed Markov chains and semi-Markov processes with asymptotically uncoupled phase spaces.

Explicit estimates for the remainder terms in the exponential asymptotic expansions for nonlinearly perturbed stochastic systems also make a natural object for future studies.

Asymptotic analysis of pseudo- and quasi-stationary phenomena in nonlinearly perturbed queueing systems and networks, which can be described with the use of Markov chains, semi-Markov processes, and stochastic processes with semi-Markov modulation, is an unlimited area for applications of the theory developed in the book. Examples given in the book make just the first step in exploring this unlimited area.

Nonlinearly perturbed epidemic and population dynamics models with different types, e.g., sex, age, health status, etc., of individuals in the population, forms of interactions between individuals, environmental modulation, complex spatial structure, etc., also make interesting objects for asymptotic analysis of pseudo- and quasi-stationary phenomena.

In the book the methods of asymptotic analysis of nonlinearly perturbed renewal equations are also applied to classical risk processes. There are no doubts that these methods can be also applied to more general risk type processes, for example, risk processes with investment components described by Brownian motions, more general Lévy type risk processes, risk processes with modulated claim flows, etc.

Finally, we would like also to mention a possible prospective area of financial applications of asymptotic methods developed in the book that is Markov and semi-Markov processes with absorption describing credit rating dynamics.

We hope that the remarks and the analysis of new directions of research presented in Chapter 8 will be especially useful for young researchers and stimulate their interest to research studies in these areas.

The book contains also extended and carefully gathered bibliography that has more than 1000 references to works in the areas related to the subject of the book. The last supplement in Chapter 8 contains the corresponding brief bibliographical remarks.

Conclusion

Quasi-stationary phenomena and related problems are a subject of intensive studies during several decades. However, the development of theory of quasi-stationary phenomena is still far from its completion. The part of the theory related to conditions of existence of quasi-stationary distributions is comparatively well developed while computational aspects of the theory are underdeveloped. The content of the book is concentrated in this area. The book presents new effective methods for asymptotic analysis of pseudo- and quasi-stationary phenomena for nonlinearly perturbed stochastic processes and systems. Moreover, the results presented in the book unite, for the first time, research studies of pseudo- and quasi-stationary phenomena in the frame of one theory. We also think that methods of asymptotic analysis for nonlinearly perturbed stochastic processes and systems developed in the book have their own values and possible applications beyond the problems studied in the book.

The results presented in the book will be interesting to specialists, who work in such areas of the theory of stochastic processes as ergodic, limit, and large deviation theorems, analytical and computational methods for Markov chains, regenerative, Markov, semi-Markov, risk and other classes of stochastic processes, renewal theory, and their queueing, reliability, population dynamics, and other applications. We hope that the book will also attract attention of those researchers, who are interested in new analytical methods of analysis for nonlinearly perturbed stochastic processes and systems, especially those who like serious analytical work. This gives us hope that the book will find sufficient number of readers interested in stochastic processes and their applications.

Chapter 1

Perturbed renewal equation

Chapter 1 contains results that generalise the classical renewal theorem to the model of the perturbed renewal equation.

The subject of our studies is the following classical *renewal equation* on the half-line:

$$x(t) = q(t) + \int_0^t x(t-s)F(ds), \quad t \geq 0. \tag{1.0.1}$$

The so-called renewal theorem given in its final form by Feller (1966) is one of the main results in the renewal theory. This theorem gives conditions for existence of a limit of $x(t)$ as $t \to \infty$. It states that if (a) the distribution $F(s)$ in non-arithmetic and has a finite mean, and (b) the function $q(t)$ is a directly Riemann integrable on $[0, \infty)$, then

$$x(t) \to x(\infty) \text{ as } t \to \infty, \tag{1.0.2}$$

and it also gives an explicit expression for the limit $x(\infty)$ in terms of $q(s)$ and $F(s)$.

The renewal theorem is a very powerful tool for proving ergodic theorems for regenerative stochastic processes. This class of processes is very broad. It includes Markov processes with discrete phase space. Moreover, any Markov process with a general phase space can be included in a model of regenerative processes with the help of the procedure of artificial regeneration.

Applying the renewal theorem to ergodic theorems for regenerative type processes is based on the well-known fact that the distribution of a regenerative process at a moment t satisfies a renewal equation. This makes it possible to apply the renewal theorem and to describe the asymptotic behaviour of the distribution of the regenerative process as $t \to \infty$.

Some applications, for example when one studies the so-called transition phenomena for stochastic processes, lead to a study of more general ergodic theorems in the so-called triangular array mode. In such models, the characteristics of the stochastic process depend on some perturbation parameter, and one studies the asymptotic behaviour of the distributions of the process as $t \to \infty$ and the perturbation parameter tends to some limiting value, simultaneously.

This problem leads, in a natural way, to generalisations of the renewal theorem to a model with a *perturbed renewal equation*,

$$x^{(\varepsilon)}(t) = q^{(\varepsilon)}(t) + \int_0^t x^{(\varepsilon)}(t-s)F^{(\varepsilon)}(ds), \quad t \geq 0. \tag{1.0.3}$$

In this model, the forcing function $q^{(\varepsilon)}(t)$ and the distribution $F^{(\varepsilon)}(s)$ depend on some *perturbation parameter* $\varepsilon \geq 0$ and converge, in some natural sense, to $q^{(0)}(t)$ and $F^{(0)}(s)$ as $\varepsilon \to 0$, correspondingly. These continuity conditions allow to consider equation (1.0.3) for $\varepsilon > 0$ as perturbations of equation (1.0.3) for $\varepsilon = 0$.

In the perturbed model, the solution $x^{(\varepsilon)}(t)$ of the renewal equation (1.0.1) also depends on the parameter ε, and the problem is to describe the asymptotic behaviour of the solution $x^{(\varepsilon)}(t)$ in the so-called triangular array mode when $t \to \infty$ and $\varepsilon \to 0$, simultaneously. Without loss of generality, one can assume that the time $t = t^{(\varepsilon)}$ is a function of the perturbation parameter ε.

It is also important to generalise the renewal theorem to the model of the perturbed renewal equation in such a way so that, in the case of the non-perturbed equation, the conditions should reduce to the conditions of the classical renewal theorem mentioned above. These results are presented in Chapter 1.

The most interesting asymptotics appears in the case where the distributions $F^{(\varepsilon)}(s)$ may be improper, i.e., $F^{(\varepsilon)}(\infty) \leq 1$ but for the defect of this distribution, $f^{(\varepsilon)} = 1 - F^{(\varepsilon)}(\infty)$, we have $f^{(\varepsilon)} \to 0$ as $\varepsilon \to 0$. In this case, the condition balancing the rate at which the time $t = t^{(\varepsilon)}$ approaches infinity and the convergence rate of $f^{(\varepsilon)}$ to zero as $\varepsilon \to 0$ has a delicate influence upon the asymptotics of the solution $x^{(\varepsilon)}(t)$. This *balancing condition* has the form

$$f^{(\varepsilon)} t^{(\varepsilon)} \to \lambda \ \text{ as } \ \varepsilon \to 0. \tag{1.0.4}$$

Assuming conditions analogous to those in the renewal theorem and the balancing condition (1.0.4) to hold, we obtain an asymptotic relation that is more general than (1.0.2),

$$\frac{x^{(\varepsilon)}(t^{(\varepsilon)})}{\exp\{-\rho^{(\varepsilon)} t^{(\varepsilon)}\}} \to x^{(0)}(\infty) \ \text{ as } \ \varepsilon \to 0, \tag{1.0.5}$$

where ρ_ε is a nonlinear functional of the distributions $F^{(\varepsilon)}(s)$ such that $\rho_\varepsilon \to 0$ as $\varepsilon \to 0$.

Moreover, it can be shown that

$$\rho^{(\varepsilon)} \sim f^{(\varepsilon)} / m_1^{(\varepsilon)} \ \text{ as } \ \varepsilon \to 0, \tag{1.0.6}$$

where $m_1^{(\varepsilon)}$ is the mean of the distribution $F^{(\varepsilon)}(s)$.

It should be noted that this asymptotics also covers the case of proper perturbed renewal equations when $f^{(\varepsilon)} \equiv 0$ for all $\varepsilon \geq 0$. In this case, the balancing condition (1.0.4) holds automatically with $\lambda = 0$ if $t^{(\varepsilon)} \to \infty$ in an arbitrary way. However, in the general case, the balancing condition (1.0.4) imposes a restriction on the rate of growth of time.

This restriction becomes unnecessary if an additional Cramér type condition is imposed on the distributions $F_\varepsilon(s)$. In this case, it can be shown that the asymptotic relation (1.0.5) holds without any restriction on $t^{(\varepsilon)} \to \infty$ with the coefficient $\rho^{(\varepsilon)}$ given as solution of the *characteristic equation*

$$\int_0^\infty e^{s\rho^{(\varepsilon)}} F^{(\varepsilon)}(ds) = 1. \tag{1.0.7}$$

Moreover, in this case the defect $f^{(\varepsilon)}$ can be asymptotically separated from zero, i.e., the limiting renewal equation can also be improper.

The main results of the Chapter 1 are Theorems 1.2.1, 1.3.1, 1.4.1 and 1.4.2.

Theorem 1.2.1 is a direct generalisation of the classical renewal theorem to the model of the perturbed renewal equation. The conditions of this theorem are reduced to the conditions of the classical renewal theorem in the case of the non-perturbed equation, so that it is a generalisation obtained without technical losses in the conditions. Theorems 1.3.1 and 1.4.1 extend the generalisations of the renewal theorem to the model of the improper perturbed renewal equation, where the distributions generating the perturbed renewal equation could have the total variation less than 1 but tending to 1. The deficiency of the distribution of the generating renewal equation causes an appearance of additional exponential normalising functions and additional multiplicative factors in the renewal limits. There is a principal difference between these two theorems. Theorem 1.3.1 uses some condition that balances the rate of growth of time with the rate of convergence of the deficiency of the distributions, which generate the perturbed renewal equation, to zero. In Theorem 1.4.1 there are no restrictions on the rate of growth of time but some Cramér type condition is imposed on the distribution that generates the perturbed renewal equation.

Finally, Theorem 1.4.2 generalises Theorem 1.4.1 to the case where the distributions generating the perturbed renewal equation can have the total variation tending to some limit which may be less, equal or greater than 1.

Theorems 1.2.1, 1.3.1, 1.4.1 and 1.4.2 are basic for both developing exponential asymptotical expansions for the nonlinearly perturbed renewal equation and an analysis and studies of mixed ergodic and large deviation theorems for nonlinearly perturbed stochastic processes and quasi-stationary phenomena in stochastic systems.

A further improvement would be obtaining more explicit representations for the exponential normalising functions in (1.0.5) than those given by the characteristic equation (1.0.7). This problem is discussed in Chapter 2.

1.1 Renewal equation

In this section we formulate some basic facts from the renewal theory. We refer to the classical book by Feller (1966), where one can find a detailed exposition and proofs of these facts.

1.1.1   Renewal equation

Let $F(x)$ be a distribution function on $[0, \infty)$ which is not concentrated at 0. Let also $q(t)$ be a function from the class $\mathcal{L}$ of measurable (Borel) real-valued functions defined on $[0, \infty)$ and bounded on any finite interval.

The following equation, known as the *renewal equation*, plays a fundamental role in the renewal theory and its applications:

$$x(t) = q(t) + \int_0^t x(t - s)F(ds), \quad t \geq 0. \tag{1.1.1}$$

Note that here and in the sequel the integration $\int_s^t$ is performed over the closed interval $[s, t]$ if $0 \leq s \leq t < \infty$ and over the semi-closed interval $[s, \infty)$ if $0 \leq s < t = \infty$.

We will call $F(x)$ the *distribution function generating the renewal equation* (1.1.1), and $q(t)$ the *forcing function* of this equation.

A solution of the renewal equation (1.1.1) is sought for in the same class $\mathcal{L}$ that contains the forcing function $q(s)$.

Let us denote by $F^{(*r)}(t)$ the *r-fold convolution* of the distribution function $F(x)$, i.e., the distribution function defined by recurrence,

$$F^{(*r)}(t) = \begin{cases} \chi_{[0,\infty)}(t),\, t \geq 0 & \text{for } r = 0, \\ F^{(*(r-1))}(t) * F(t),\, t \geq 0 & \text{for } r \geq 1, \end{cases} \tag{1.1.2}$$

where the *convolution* $F(t) = F_1(t) * F_2(t)$ of two proper or improper distribution functions $F_1(t)$ and $F_2(t)$, concentrated on the non-negative half-line, is defined by the following formula:

$$F(t) = \begin{cases} \int_0^t F_1(t - s)F_2(ds) & \text{for } t \geq 0, \\ 0 & \text{for } t < 0. \end{cases}$$

The following function, called a *renewal function*, plays an important role in the theory:

$$U(t) = \sum_{r=0}^{\infty} F^{(*r)}(t), \quad t \geq 0. \tag{1.1.3}$$

It can easily be shown that the series in (1.1.3) converges for all $t \geq 0$ and, moreover, $U(t) \leq K_1 + K_2 t$ for some finite constants K_1 and K_2, so that it has a linear rate of growth in t. Also note that by the definition $U(t)$ is a nondecreasing right-continuous function.

We also consider the *renewal measure* $U(A)$ on the Borel σ-algebra of $[0, \infty)$ generated by the renewal function in the usual way. This measure is defined on intervals by $U((a, b]) = U(b) - U(a),\, 0 \leq a \leq b < \infty.$

Note that this measure always has an atom in zero, since we have the 0-fold convolution $F^{(*0)}(t)$ in (1.1.3). This convolution is, by the definition, a distribution function that has a unit jump in the point zero.

As is well known, a unique *solution of the renewal equation* in the class $\mathfrak{L}$ is of the form

$$x(t) = \int_0^t q(t-s)U(ds), \quad t \geq 0. \tag{1.1.4}$$

1.1.2   Renewal theorem

An important role in probability theory is played by the so-called renewal theorem that describes the behaviour of the solution of (1.1.1) for $t \to \infty$.

There is a long history of developing results related to this theorem. The final form and the proof of the renewal theorem was given by Feller (1966). We follow his presentation. To do this we need in some definitions.

A distribution function $F(x)$ is *arithmetic* if there exists $h > 0$ such that

$$\sum_{r=-\infty}^{\infty} (F(rh) - F(rh - 0)) = 1. \tag{1.1.5}$$

If $F(x)$ is an arithmetic distribution function, then there exists the largest positive h such that (1.1.5) holds. This h is called a *step of the arithmetic distribution function*.

According the above definition, a distribution function $F(x)$ is *non-arithmetic* if relation (1.1.5) does not hold for any positive h, that is,

$$\sum_{r=-\infty}^{\infty} (F(rh) - F(rh - 0)) < 1, \quad h > 0. \tag{1.1.6}$$

Formula (1.1.6) implies that a non-arithmetic distribution function can not be concentrated in zero.

A measurable function $q(t)$ is *directly Riemann integrable* on $[0, \infty)$ if it is continuous almost everywhere with respect to the Lebesgue measure on $[0, \infty)$, bounded on any finite interval, and there exists $h > 0$ such that

$$\lim_{T \to \infty} h \sum_{r \geq T/h} \sup_{rh \leq t \leq (r+1)h} |q(t)| = 0. \tag{1.1.7}$$

Let us remark that a function $q(t)$ that is directly Riemann integrable on $[0, \infty)$ is also absolutely integrable on $[0, \infty)$ in the usual way, and $\int_T^\infty |q(s)|\, ds \to 0$ as $T \to \infty$.

Let us denote the first moment of the distribution function $F(x)$ by

$$m_1 = \int_0^\infty t F(dt).$$

The conditions for the renewal theorem are the following:

$\mathbf{D_1}$: $F(x)$ is a non-arithmetic distribution function;

$\mathbf{M_1}$: $m_1 < \infty$;

and

$\mathbf{F_1}$: $q(t)$ is a directly Riemann integrable function on $[0, \infty)$.

Now we are prepared to formulate the classical variant of the so-called *renewal theorem* in the form as was given by Feller (1966).

Theorem 1.1.1. *Let conditions* $\mathbf{D_1}$, $\mathbf{M_1}$, *and* $\mathbf{F_1}$ *hold. Then*

$$x(t) \to \frac{1}{m_1} \int_0^\infty q(s)\, ds \ \ as \ t \to \infty. \tag{1.1.8}$$

We refer to the book Feller (1966, 1971) for a proof of this theorem. Let us remark that, in the next section, we formulate and give a proof of a theorem that generalises the renewal theorem to the so-called triangular array mode.

1.1.3 Renewal process

We also need to formulate some additional facts concerning the so-called renewal processes.

Let τ_k, $k = 1, 2, \ldots$, be a sequence of nonnegative *independent identically distributed* (*i.i.d.*) *random variables* with a distribution function $F(x)$ that does not have zero in its support. The random variables $\kappa_n = \sum_{k=1}^n \tau_k$, $n = 0, 1, \ldots$, can be interpreted as *renewal times* and a *renewal process* can be defined as $\nu(t) = \max(n : \kappa_n \leq t)$, $t \geq 0$. By definition, $\nu(t)$ is the *number of renewals* in the interval $[0, t]$.

The following well-known formula expresses the renewal functions as the expectation of a renewal process,

$$U(t) = \mathsf{E}\nu(t) + 1, \quad t \geq 0. \tag{1.1.9}$$

We use the following notation:

$$\gamma(t) = \sum_{k=1}^{\nu(t)+1} \tau_k - t, \quad t \geq 0.$$

By definition, $\gamma(t)$ is the time between the moment t and the first after t renewal moment. The processes $\nu(t)$ and $\gamma(t)$ are connected by the following relation:

$$\mathsf{P}\{\nu(t+h) - \nu(t) = n\} = \delta(n, 0)\mathsf{P}\{\gamma(t) > h\}$$

$$+ \int_0^h \mathsf{P}\{\nu(h-s) + 1 = n\}\mathsf{P}\{\gamma(t) \in ds\}, \quad n = 0, 1, \ldots, \tag{1.1.10}$$

where $\delta(r, k)$ is the *Kronecker delta-function* that is $\delta(r, k) = 1$ if $r = k$ and $\delta(r, k) = 0$ if $r \neq k$.

Relation (1.1.10) implies the following formula:

$$\mathsf{E}(v(t + h) - v(t))^r = \int_0^h \mathsf{E}(v(h) + 1)^r \mathsf{P}\{\gamma(t) \in ds\}, \quad r = 1, 2, \ldots . \tag{1.1.11}$$

The following lemma gives a useful upper bound for the moments of the random variables $v(t)$.

Lemma 1.1.1. *If (α) $F(g) \leq G$ for some $g > 0$ and $G < 1$, then for every $r = 1, 2, \ldots$ and $t \geq 0$,*

$$\mathsf{E}v(t)^r \leq (t + g)^r K[r, g, G], \tag{1.1.12}$$

where

$$K[r, g, G] = g^{-r} \sum_{n=0}^{\infty} (n + 1)^r G^n < \infty.$$

Proof. Relation (1.1.11) obviously implies the following inequalities:

$$\mathsf{E}(v(t + h) - v(t))^r \leq \mathsf{E}(v(h) + 1)^r, \quad h, t \geq 0, \quad r = 1, 2, \ldots . \tag{1.1.13}$$

Using inequalities (1.1.13) we get

$$\mathsf{E}v(t)^r = \mathsf{E}\left(\sum_{k=1}^{[t/g]} (v(kg) - v((k-1)g)) + (v(t) - v([t/g]g)) \right)^r$$

$$\leq ([t/g] + 1)^{r-1}\left(\sum_{k=1}^{[t/g]} \mathsf{E}(v(kg) - v((k-1)g)) \right)^r$$

$$+ \ \mathsf{E}(v(t) - v([t/g]g))^r \leq \left(\frac{t + g}{g} \right)^r \mathsf{E}(v(g) + 1)^r. \tag{1.1.14}$$

Under the condition of Lemma 1.1.1,

$$\mathsf{P}\{v(g) \geq n\} = \mathsf{P}\left\{ \sum_{k=1}^{n} \tau_{k-1} \leq g \right\} \leq F(g)^n \leq G^n, \quad n = 0, 1, \ldots . \tag{1.1.15}$$

It follows from (1.1.15) that

$$\mathsf{E}(v(g) + 1)^r \leq L[r, G] = \sum_{n=0}^{\infty} (n + 1)^r G^n < \infty. \tag{1.1.16}$$

Now, inequality (1.1.12) follows from (1.1.14) and (1.1.16). $\qquad\qquad\square$

Lemma 1.1.2. *If condition* **(α)** *given in Lemma* 1.1.1 *holds, then for every* $r = 1,$ 2,... *and* $h \geq 0$,

$$\sup_{t \geq 0} \mathsf{E}(v(t + h) - v(t))^r \leq L[h, r, g, G], \qquad (1.1.17)$$

where

$$L[h, r, g, G] = 2^{r-1}((h + g)^r K[r, g, G] + 1) < \infty.$$

Proof. Using inequality (1.1.13) and Lemma 1.1.1 we get

$$\sup_{t \geq 0} \mathsf{E}(v(t + h) - v(t))^r \leq \mathsf{E}(v(h) + 1)^r \leq 2^{r-1}(\mathsf{E}v(h)^r + 1)$$

$$\leq 2^{r-1}((h + g)^r K[r, g, G] + 1). \qquad (1.1.18)$$

That proves the lemma. $\square$

Let $d(A)$ denote the diameter of the bounded set A.

Lemma 1.1.3. *If condition* **(α)** *given in Lemma* 1.1.1 *holds, then for an arbitrary measurable bounded set* $A \subset [0, \infty)$,

$$\sup_{t \geq 0} U(t + A) \leq (d(A) + 2g)L[h, r, g, G] < \infty. \qquad (1.1.19)$$

Proof. The proof follows in an obvious way from Lemma 1.1.2, the inequalities

$$\mathsf{E}(v(t + h) - v(t)) = U(t + h) - U(t) = U((t, t + h]), \quad t, h \geq 0, \qquad (1.1.20)$$

and the observation that for any bounded subset A there exists an interval of the form $(t, t + h]$ of length $h = d(A) + g$ such that $A \subseteq (t, t + h]$. $\square$

Remark 1.1.1. It is useful to note that inequalities (1.1.19), (1.1.18), and (1.1.20) are uniform with respect to any class $\mathfrak{F}$ of distribution functions $F(x)$ for which the condition $F(g) \leq G$ holds with the same constants $g > 0$ and $G < 1$ for all the distribution functions $F(x)$ that belong to the class $\mathfrak{F}$.

1.2   Renewal theorem for perturbed renewal equation

In this section we formulate and prove a theorem that generalises the classical renewal theorem to a model of the perturbed renewal equation.

1.2.1   Perturbed renewal equation

Consider the family of renewal equations,

$$x^{(\varepsilon)}(t) = q^{(\varepsilon)}(t) + \int_0^t x^{(\varepsilon)}(t - s) F^{(\varepsilon)}(ds), \quad t \geq 0, \tag{1.2.1}$$

where, for every $\varepsilon \geq 0$, we have the following: (a) $q^{(\varepsilon)}(t)$ is a real-valued function on $[0, \infty)$ that is measurable and locally bounded, i.e., bounded on every finite interval, and (b) $F^{(\varepsilon)}(s)$ is a distribution function on $[0, \infty)$ which is not concentrated at 0.

Denote

$$m_1^{(\varepsilon)} = \int_0^\infty s F^{(\varepsilon)}(ds).$$

As usual the symbol $F^{(\varepsilon)}(\cdot) \Rightarrow F^{(0)}(\cdot)$ as $\varepsilon \to 0$ means *weak convergence* of the distribution functions, that is, the pointwise convergence in each point of continuity of the limiting distribution function.

We assume that the functions $q^{(\varepsilon)}(t)$ and the distributions $F^{(\varepsilon)}(s)$ satisfy the following continuity conditions at the point $\varepsilon = 0$, if regarded as functions of ε:

D_2: $F^{(\varepsilon)}(\cdot) \Rightarrow F^{(0)}(\cdot)$ as $\varepsilon \to 0$, where $F^{(0)}(s)$ is a non-arithmetic distribution function;

M_2: $m_1^{(\varepsilon)} \to m_1^{(0)} < \infty$ as $\varepsilon \to 0$;

and

F_2: (a) $\lim_{u \to 0} \overline{\lim}_{0 \leq \varepsilon \to 0} \sup_{|v| \leq u} |q^{(\varepsilon)}(t + v) - q^{(0)}(t)| = 0$ almost everywhere with respect to the Lebesgue measure on $[0, \infty)$;

   (b) $\overline{\lim}_{0 \leq \varepsilon \to 0} \sup_{0 \leq t \leq T} |q^{(\varepsilon)}(t)| < \infty$ for every $T \geq 0$;

   (c) $\lim_{T \to \infty} \overline{\lim}_{0 \leq \varepsilon \to 0} h \sum_{r \geq T/h} \sup_{rh \leq t \leq (r+1)h} |q^{(\varepsilon)}(t)| = 0$ for some $h > 0$.

Conditions D_2, M_2 and F_2 are conditions for convergence of the distribution functions $F^{(\varepsilon)}(x)$ and the forcing functions $q^{(\varepsilon)}(t)$, respectively, to the corresponding limiting functions $F^{(0)}(x)$ and the forcing functions $q^{(0)}(t)$ as $\varepsilon \to 0$. This allows to consider equation (1.2.1) for $\varepsilon > 0$ as a perturbed version of equation (1.2.1) for $\varepsilon = 0$, and to interpret the parameter ε as a perturbation parameter.

Let us make some useful remarks concerning conditions D_2, M_2 and F_2.

Remark 1.2.1. The conditions D_2, M_2 and F_2 reduce to the conditions of the classical renewal theorem, if $F^{(\varepsilon)}(s) \equiv F^{(0)}(s)$ and $q^{(\varepsilon)}(t) \equiv q^{(0)}(t)$ do not depend on the perturbation parameter ε. In particular, condition D_2 reduces to condition D_1, i.e., to the assumption that $F^{(0)}(s)$ is a non-arithmetic distribution function; M_2 to condition M_1, i.e., to the assumption that the expectation $m_1^{(0)}$ is finite; and F_2 to condition F_1, i.e., to the assumption that the function $q^{(0)}(t)$ is directly Riemann integrable on $[0, \infty)$.

Remark 1.2.2. Under condition $\mathbf{D_2}$, condition $\mathbf{M_2}$ is equivalent to the following condition:

$\mathbf{M_3}$: $\lim_{T\to\infty} \overline{\lim}_{\varepsilon\to 0} \int_T^\infty (1 - F^{(\varepsilon)}(v))dv = 0$.

This proposition is a direct corollary of the following well-known formula for expectations of non-negative random variables:

$$m_1^{(\varepsilon)} = \int_0^\infty (1 - F^{(\varepsilon)}(v))dv. \tag{1.2.2}$$

Indeed, let us denote

$$m_+^{(\varepsilon)}(T) = \int_0^T (1 - F^{(\varepsilon)}(v))dv, \quad m_-^{(\varepsilon)}(T) = \int_T^\infty (1 - F^{(\varepsilon)}(v))dv, \quad T > 0.$$

It is clear that $m_1^{(\varepsilon)} = m_+^{(\varepsilon)}(T) + m_-^{(\varepsilon)}(T)$, $T > 0$. Conditions $\mathbf{D_2}$ and $\mathbf{M_2}$ imply, in an obvious way, that **(a)** $m_-^{(\varepsilon)}(T) = m_1^{(\varepsilon)} - m_+^{(\varepsilon)}(T) \to m_1^{(0)} - m_+^{(0)}(T) = m_-^{(0)}(T)$ as $\varepsilon \to 0$, $T > 0$. Also, **(b)** $m_-^{(0)}(T) = m_1^{(0)} - m_+^{(0)}(T) \to 0$ as $T \to \infty$. Relations **(a)** and **(b)** imply condition $\mathbf{M_3}$. On the other hand, condition $\mathbf{D_2}$ implies that **(c)** $\overline{\lim}_{\varepsilon\to 0} m_+^{(\varepsilon)}(T) \geq \overline{\lim}_{\varepsilon\to 0} \int_T^{T'} (1 - F^{(\varepsilon)}(v))dv = \int_T^{T'} (1 - F^{(0)}(v))dv$, for $0 \leq T < T' < \infty$, and, therefore, **(d)** $\overline{\lim}_{\varepsilon\to 0} m_-^{(\varepsilon)}(T) \geq m_-^{(0)}(T)$, for $0 \leq T < \infty$. Relation **(d)** and condition $\mathbf{M_3}$ imply that **(e)** $m_-^{(0)}(T) \to 0$ as $T \to \infty$. Finally, relation **(e)** and conditions $\mathbf{D_2}$ and $\mathbf{M_3}$ imply that **(e)** $\overline{\lim}_{\varepsilon\to 0} |m_1^{(\varepsilon)} - m_1^{(0)}| \leq \overline{\lim}_{\varepsilon\to 0}(|m_+^{(\varepsilon)}(T) - m_+^{(0)}(T)| + m_-^{(\varepsilon)}(T) + m_-^{(0)}(T)) = \overline{\lim}_{\varepsilon\to 0} m_-^{(\varepsilon)}(T) + m_-^{(0)}(T) \to 0$ as $T \to \infty$. Thus, condition $\mathbf{M_2}$ holds.

Note, that, according the proof given above, if condition $\mathbf{D_2}$ holds then condition $\mathbf{M_3}$ can be formulated in the equivalent form that let parameter ε to tend to 0 being non-negative in the corresponding asymptotic relation:

$\mathbf{M_3'}$: $\lim_{T\to\infty} \overline{\lim}_{0\leq\varepsilon\to 0} \int_T^\infty (1 - F^{(\varepsilon)}(v))dv = 0$.

Remark 1.2.3. Denote by $m(A)$ the Lebesgue measure on the real line. Let also C_0 be the set of convergence in the condition $\mathbf{F_2}$. By this condition, $m(\overline{C}_0) = 0$. Condition $\mathbf{F_2}$ implies that every point $t \in C_0$ is a point of continuity of the limiting function $q^{(0)}(t)$. It follows from the assumption that $0 \leq \varepsilon \to 0$, which admit the case where $\varepsilon \equiv 0$. It can be shown that conditions $\mathbf{F_2}$ **(a)**, **(b)** and imply that condition $\mathbf{F_2}$ **(a)** holds for any point of continuity of the limiting function $q^{(0)}(t)$. From these remarks it follows that the limiting function is continuous almost everywhere with respect to the Lebesgue measure on $\mathbb{R}_1$.

Remark 1.2.4. Conditions $\mathbf{F_2}$ **(b)**, **(c)** imply that the functions $q^{(\varepsilon)}(t)$ are Lebesgue integrable on the interval $[0, \infty)$ for all ε small enough. However, there is no guarantee that these functions are directly Riemann integrable on $[0, \infty)$ for $\varepsilon > 0$. As a matter of fact, there is no guarantee that the functions are continuous almost everywhere with respect to the Lebesgue measure on $[0, \infty)$. As far as the limiting function $q^{(0)}(t)$ is concerned, condition $\mathbf{F_2}$ implies that this function is not only Lebesgue integrable but also directly Riemann integrable on $[0, \infty)$. Indeed, as it is mentioned in Remark 1.2.3, this function is continuous almost everywhere with respect to the Lebesgue measure on $[0, \infty)$. Also relation (1.1.7) given in the definition of a directly Riemann integrable function is satisfied for $q^{(0)}(t)$, since the assumption that $0 \leq \varepsilon \to 0$ admits the case where $\varepsilon \equiv 0$.

This remark allows to replace the Lebesgue integral over $[0, \infty)$ for this function with the corresponding Riemann integral in the left-hand side of the following equality:

$$\int_0^\infty q^{(0)}(s)m(ds) = \int_0^\infty q^{(0)}(s)\,ds. \qquad (1.2.3)$$

1.2.2   Renewal theorem for perturbed renewal equation

The main result of this section is the following theorem that generalises the classical renewal theorem to the model of perturbed renewal equation. The conditions of this theorem reduce to to the conditions of the classical renewal Theorem 1.1.1 in the case where of non-perturbed renewal equation.

Theorem 1.2.1. *Let conditions* $\mathbf{D_2}$, $\mathbf{M_2}$, *and* $\mathbf{F_2}$ *hold. Then for any* $0 \leq t^{(\varepsilon)} \to \infty$ *as* $\varepsilon \to 0$,

$$x^{(\varepsilon)}(t^{(\varepsilon)}) \to x^{(0)}(\infty) = \frac{\int_0^\infty q^{(0)}(s)ds}{m_1^{(0)}} \quad as \ \varepsilon \to 0. \qquad (1.2.4)$$

The proof of this Theorem 1.2.1 is based on a generalisation of the so-called Blackwell theorem to the model of the perturbed renewal equation.

Theorem 1.2.2. *Let conditions* $\mathbf{D_2}$ *and* $\mathbf{M_2}$ *hold. Then for any* $0 \leq t^{(\varepsilon)} \to \infty$ *as* $\varepsilon \to 0$ *and* $h \geq 0$,

$$U^{(\varepsilon)}(t^{(\varepsilon)} + h) - U^{(\varepsilon)}(t^{(\varepsilon)}) \to h/m_1^{(0)} \quad as \ \varepsilon \to 0. \qquad (1.2.5)$$

1.2.3   Convergence of Lebesgue integrals in the scheme of series

We postpone the proof of Theorems 1.2.1 and 1.2.2 and, first, prove the following proposition that we shall constantly use in what follows.

Let, for every $\varepsilon \geq 0$, $f^{(\varepsilon)}(s)$ be real-valued bounded Borel functions defined on $\mathbb{R}_1$.

We use the symbol $f^{(\varepsilon)}(s) \xrightarrow{U} f^{(0)}(s)$ as $\varepsilon \to 0$ to indicate that functions $f^{(\varepsilon)}(\cdot)$ converge to a function $f^{(0)}(\cdot)$ *locally uniformly* at point s as $\varepsilon \to 0$. This means that

$$\lim_{0<u\to 0}\ \overline{\lim_{\varepsilon\to 0}}\ \sup_{|v|\leq u}\ |f^{(\varepsilon)}(s+v) - f^{(0)}(s)| = 0. \qquad (1.2.6)$$

Lemma 1.2.1. *Functions* $f^{(\varepsilon)}(s) \overset{U}{\longrightarrow} f^{(\varepsilon)}(s)$ *as* $\varepsilon \to 0$ *if and only if* **(α)** $f^{(\varepsilon)}(s_\varepsilon) \to$ $f^{(0)}(s)$ *as* $\varepsilon \to 0$, *for any* $s_\varepsilon \to s$ *as* $\varepsilon \to 0$.

Proof. Let condition **(α)** holds but relation (1.2.6) does not. This assumption implies that **(a)** there exist $\sigma > 0$ and sequence $0 < u_n \to 0$ as $n \to \infty$ such that $\overline{\lim}_{\varepsilon \to 0} \sup_{|v| \le u_n} |f^{(\varepsilon)}(s + v) - f^{(0)}(s)| \ge \sigma$. Relations **(a)** imply that **(b)** there exist $0 < \varepsilon_n \to 0$, $n = 1, 2, \ldots$, such that $\sup_{|v| \le u_n} |f^{(\varepsilon_n)}(s + v) - f^{(0)}(s)| \ge \sigma/2$. Relations **(b)** imply that **(c)** there exist $|v_n| \le u_n$, $n = 1, 2, \ldots$, such that $|f^{(\varepsilon_n)}(s + v_n) - f^{(0)}(s)| \ge \sigma/4$. Finally, relations **(c)** imply that **(d)** $f^{(\varepsilon_n)}(s_n) \not\to f^{(0)}(s)$ as $n \to \infty$ for sequence $s_n = s + v_n \to s$ as $n \to \infty$. Obviously, **(d)** contradicts to condition **(α)**. Thus condition **(α)** implies relation (1.2.6).

Now, let us assume that relation (1.2.6) holds. Let us take an arbitrary $\sigma > 0$. Relation (1.2.6) implies that **(e)** there exist $u = u(\sigma) > 0$ such that

$$\overline{\lim_{\varepsilon \to 0}} \sup_{|v| \le u(\sigma)} |f^{(\varepsilon)}(s + v) - f^{(0)}(s)| \le \sigma.$$

Let us take an arbitrary $s_\varepsilon \to s$ as $\varepsilon \to 0$. Obviously, there exists $\varepsilon(\delta) > 0$ such that **(f)** $|s_\varepsilon - s| \le u(\sigma)$ for $\varepsilon \le \varepsilon(\delta)$. Relations **(e)** and **(f)** imply that **(g)** $|f^{(\varepsilon)}(s_\varepsilon) - f^{(0)}(s)| \le$ $\sup_{|v| \le u(\sigma)} |f^{(\varepsilon)}(s + v) - f^{(0)}(s)| \le \sigma$ for $\varepsilon \le \varepsilon(\delta)$. Since an arbitrary choice of $\sigma > 0$, relations **(g)** imply that condition **(α)** hold. $\square$

Let $B_{(1)}$ denote the Borel σ-algebra on $\mathbb{R}_1$ and let, for every $\varepsilon \ge 0$, $\mu_\varepsilon(A)$ be a measure on $B_{(1)}$ taking finite values on bounded Borel sets.

We use the symbol $\mu_\varepsilon(A) \Rightarrow \mu_0(A)$ as $\varepsilon \to 0$ to indicate that the measures $\mu_\varepsilon(A)$ weakly converge to a measure $\mu_0(A)$ as $\varepsilon \to 0$. This means that, for all $-\infty < u \le v < \infty$ such that the limiting measure has not atoms in the points u and v,

$$\mu^{(\varepsilon)}((u, v]) \to \mu^{(0)}((u, v]) \quad \text{as} \quad \varepsilon \to 0. \tag{1.2.7}$$

Let now $f^{(\varepsilon)}(s)$ be real-valued bounded Borel functions defined on $\mathbb{R}_1$ and $\mu^{(\varepsilon)}(A)$ be finite measures on $B_{(1)}$ for every $\varepsilon \ge 0$.

Lemma 1.2.2. *Let the following conditions hold:* **(α)** $\mu^{(\varepsilon)}(A) \Rightarrow \mu^{(0)}(A)$ *as* $\varepsilon \to 0$; **(β)** $\mu^{(\varepsilon)}(\mathbb{R}_1) \to \mu^{(0)}(\mathbb{R}_1) < \infty$ *as* $\varepsilon \to 0$; **(γ)** $\overline{\lim}_{\varepsilon \to 0} \sup_{s \in \mathbb{R}_1} |f^{(\varepsilon)}(s)| = f < \infty$; **(δ)** *functions* $f^{(\varepsilon)}(\cdot)$ *converge to* $f^{(0)}(\cdot)$ *as* $\varepsilon \to 0$ *locally uniformly at every point* $s \in S$, *where* S *is some subset of* $B_{(1)}$; **(ε)** $\mu^{(0)}(\overline{S}) = 0$. *Then*

$$\int_{\mathbb{R}_1} f^{(\varepsilon)}(s)\mu^{(\varepsilon)}(ds) \to \int_{\mathbb{R}_1} f^{(0)}(s)\mu^{(0)}(ds) \quad \text{as} \quad \varepsilon \to 0. \tag{1.2.8}$$

Proof. If $\mu^{(0)}(\mathbb{R}_1) = 0$ then the statement of the lemma follows in an obvious way from conditions **(β)** and **(γ)**, and the corresponding limit takes the value zero. If $\mu^{(0)}(\mathbb{R}_1) > 0$ then, condition **(β)** implies that $\mu^{(\varepsilon)}(\mathbb{R}_1) > 0$ for ε small enough, say $\varepsilon \leq \varepsilon_0$. Condition $(\boldsymbol{\beta})$ permits to reduce the proof to the case where $\mu^{(\varepsilon)}(\cdot)$ are probability measures. This follows from the identity

$$\int_{\mathbb{R}_1} f^{(\varepsilon)}(s)\mu^{(\varepsilon)}(ds) = \mu^{(\varepsilon)}(\mathbb{R}_1)\int_{\mathbb{R}_1} f^{(\varepsilon)}(s)\tilde{\mu}^{(\varepsilon)}(ds),$$

where $\tilde{\mu}^{(\varepsilon)}(A) = \mu^{(\varepsilon)}(A)/\mu^{(\varepsilon)}(\mathbb{R}_1)$ is a probability measure and $\varepsilon \leq \varepsilon_0$. Also, taking the definition of a limit in terms of convergent subsequences, one can always reduce the consideration to the case where the parameter $\varepsilon \to \varepsilon_0 = 0$ runs only over some subsequence of positive values, $\varepsilon_n, n = 1, 2, \ldots$ such that $\varepsilon_n \to 0$ as $n \to \infty$.

The Skorokhod (1956) representation theorem (see, for example, Silvestrov (2004a), page 20) implies that, under condition $(\boldsymbol{\alpha})$, it is possible to construct a sequence of random variables $\eta^{(\varepsilon_n)}$ defined on the same probability space $(\Omega, \mathcal{F}, \mathcal{P})$ such that $(\mathbf{h})$ the random variable $\eta^{(\varepsilon_n)}$ has the distribution $\mu^{(\varepsilon_n)}(\cdot)$ for every $n = 0, 1, \ldots$; $(\mathbf{i})$ $\eta^{(\varepsilon_n)} \xrightarrow{a.s.} \eta^{(0)}$ (almost sure) as $n \to \infty$.

Let A be the set of elementary events $\omega \in \Omega$ such that $(\mathbf{j})$ $\eta^{(\varepsilon_n)}(\omega) \to \eta^{(0)}(\omega)$ as $n \to \infty$. Let also B be the set of elementary events ω for which $\eta^{(0)}(\omega) \in S$. By $(\mathbf{i})$ and condition $(\boldsymbol{\epsilon})$ that $(\mathbf{k})$ $\mathsf{P}(A \cap B) = 1$. By Lemma 1.2.1, it follows from $(\mathbf{j})$ and condition $(\boldsymbol{\delta})$ that $(\mathbf{l})$ $f^{(\varepsilon_n)}(\eta^{(\varepsilon_n)}(\omega)) \to f^{(0)}(\eta^{(0)}(\omega))$ as $n \to \infty$ for all $\omega \in A \cap B$. Relations $(\mathbf{k})$ and $(\mathbf{l})$ mean that $(\mathbf{m})$ $f^{(\varepsilon_n)}(\eta^{(\varepsilon_n)}) \xrightarrow{a.s.} f^{(0)}(\eta^{(0)})$ as $n \to \infty$. Note that condition $(\boldsymbol{\gamma})$ implies that $(\mathbf{n})$ there exists $N < \infty$ such that $|f^{(\varepsilon_n)}(\eta^{(\varepsilon_n)})| \leq 2f < \infty$ for $n \geq N$. Also conditions $(\boldsymbol{\gamma})$ and $(\boldsymbol{\delta})$ imply that $(\mathbf{o})$ $\sup_{s \in S}|f^{(0)}(s)| \leq 2f < \infty$, and, therefore, by condition $(\boldsymbol{\epsilon})$, $(\mathbf{p})$ $|f^{(0)}(\eta^{(0)})| \leq 2f < \infty$ with probability 1. Hence, by the Lebesgue theorem, it follows from $(\mathbf{m})$, $(\mathbf{n})$, and $(\mathbf{p})$ that $\mathsf{E}f^{(\varepsilon_n)}(\eta^{(\varepsilon_n)}) \to \mathsf{E}f^{(0)}(\eta^{(0)})$ as $n \to \infty$. By $(\mathbf{h})$, this is equivalent to the assertion of the lemma. $\qquad\square$

1.2.4 Proof of Theorem 1.2.2

Denote by $U^{(\varepsilon)}(t)$ the renewal function generated by the distribution function $F^{(\varepsilon)}(t)$ and defined by formula (1.1.3). It is also convenient to extend the renewal function to the negative half-line by $U^{(\varepsilon)}(t) = 0$ for $t < 0$ and consistently to consider the corresponding renewal measure $U^{(\varepsilon)}(A)$ on the Borel σ-algebra on $\mathbb{R}_1$ defined by the formula $U^{(\varepsilon)}((u, v]) = U^{(\varepsilon)}(v) - U^{(\varepsilon)}(u)$ for all $-\infty < u \leq v < \infty$. By definition, $U^{(\varepsilon)}((-\infty, 0)) = 0$.

The method of the proof follows that in Feller (1966). However, triangular array aspects involved in the model cause many technical complications in the proof.

Let us consider the shifted renewal measures $U^{(\varepsilon)}(t + A), t \geq 0$ (here $t + A$ is the set of the points $x + t$ where $x \in A$). By definition, the measure $U^{(\varepsilon)}(t + A)$ has its support in the interval $[-t, \infty)$.

Since the distribution function $F^{(0)}(t)$ is non-arithmetic, $(\mathbf{q})$ there exists $g > 0$ which is a continuity point for this distribution function such that $F^{(0)}(g) < 1$. Condition $\mathbf{D_2}$ implies that $(\mathbf{r})$ $F^{(\varepsilon)}(g) \to F^{(0)}(g)$ as $\varepsilon \to 0$. Relations $(\mathbf{q})$ and $(\mathbf{r})$ imply that **(s)** there exist $\varepsilon_0 > 0$ such that $F^{(\varepsilon)}(g) \leq G < 1$ for $0 \leq \varepsilon \leq \varepsilon_0$. Due to Lemma 1.1.3 and Remark 1.1.1, relation **(s)** implies that for any measurable bounded set $A \subset \mathbb{R}_1$,

$$\overline{\lim_{0 \leq \varepsilon \to 0}} \sup_{t \geq 0} U^{(\varepsilon)}(t + A) \leq K_1 d(A) + K_2 < \infty, \tag{1.2.9}$$

where K_1 and K_2 are finite constants.

The statement of the theorem is equivalent to the relation

$$U^{(\varepsilon)}(t^{(\varepsilon)} + A) \Rightarrow \frac{1}{m_1^{(0)}} m(A) \quad \text{as } \varepsilon \to 0, \tag{1.2.10}$$

where $m(A)$ is the Lebesgue measure on a Borel σ-algebra $B_{(1)}$.

By a suitable form of the selection theorem (see, for example, Feller (1971), page 270) it follows from (1.2.9) that, given any sequence $0 < \varepsilon = \varepsilon_n \to 0$ as $n \to \infty$, we can choose from this sequence a subsequence $\varepsilon' = \varepsilon_{n_k} \to 0$ as $k \to \infty$ such that

$$U^{(\varepsilon')}(t^{(\varepsilon')} + A) \Rightarrow V(A) \quad \text{as } \varepsilon' \to 0, \tag{1.2.11}$$

where $V(A)$ is a measure defined on the Borel σ-algebra $B_{(1)}$ of subsets of $\mathbb{R}_1$ and taking finite values on bounded Borel sets.

Relation (1.2.10) will be proved if we show that the measure $V(A) = \frac{1}{m_1^{(0)}} m(A)$ is independent of the choice of a sequence $0 < \varepsilon = \varepsilon_n \to 0$ as $n \to \infty$ and the corresponding subsequence $\varepsilon' = e_{n_k} \to 0$ as $k \to \infty$ satisfying (1.2.11).

We observe that by (1.2.9) the measure $V(A)$ satisfies

$$V(A) \leq K_1 d(A) + K_2 \tag{1.2.12}$$

for any bounded set $A \in B_{(1)}$.

The first stage of the proof is to show that the measure $V(A)$ is proportional to the Lebesgue measure. That is, $V(A) = \alpha m(A)$, where α is a constant and $m(A)$ is the Lebesgue measure on $B_{(1)}$.

It is not hard to see that to achieve this it would suffice to show that for any continuous function $q(t)$ on $\mathbb{R}_1$ vanishing outside the interval $[0, a]$ the function

$$h(s) = \int_{s-a}^{s} q(s - v) V(dv), \quad s \in \mathbb{R}_1, \tag{1.2.13}$$

is a constant.

Obviously, any function $h(s)$ defined by (1.2.13) is uniformly continuous in $s \in \mathbb{R}_1$ owing to the uniform continuity of $q(s)$ in s.

Due to (1.2.12), the function $h(s)$ is also bounded, namely,

$$\sup_{s \in \mathbb{R}_1} |h(s)| \leq Q \cdot (K_1 a + K_2), \tag{1.2.14}$$

where

$$Q = \sup_{t \in \mathbb{R}_1} |q(t)| < \infty.$$

We now consider the convolution equation

$$h(s) = \int_0^\infty h(s - v) F^{(0)}(dv), \quad s \in \mathbb{R}_1. \tag{1.2.15}$$

By the Choquet–Deny (1960) lemma (see, for example, Feller (1966), page 364), if $F^{(0)}(\cdot)$ is a non-arithmetic distribution on $[0, \infty)$, with zero not being in the support, equation (1.2.15) has a unique bounded uniformly continuous solution, which is a constant.

Hence, owing to the above remarks, to prove that the measure $V(A)$ is proportional to the Lebesgue measure, it suffices to show that any function given by (1.2.13) satisfies the convolution equation (1.2.15).

So, let $q(s)$ be a continuous function on $\mathbb{R}_1$ vanishing outside the interval $[0, a]$. Let

$$x^{(\varepsilon')}(t) = \int_0^t q(t - s) U^{(\varepsilon')}(ds), \quad t \geq 0,$$

be a solution of the renewal equation

$$x^{(\varepsilon')}(t) = q(t) + \int_0^t x^{(\varepsilon')}(t - s) F^{(\varepsilon')}(ds), \quad t \geq 0. \tag{1.2.16}$$

We consider below this solution in points $t^{(\varepsilon')} + s$ for $s \in \mathbb{R}_1$. It is convenient to define $x^{(\varepsilon)}(s) = 0$ for $s < 0$. In any case, $t^{(\varepsilon')} + s \to \infty$ as $\varepsilon' \to 0$ and therefore, for all ε' small enough, $t^{(\varepsilon')} + s$ is positive.

The function $q(s - v)$ vanishes in 0 and outside the interval $[s - a, s]$. So,

$$\begin{aligned}
x^{(\varepsilon')}(t^{(\varepsilon')} + s) &= \int_0^{t^{(\varepsilon')}+s} q(t^{(\varepsilon')} + s - v) U^{(\varepsilon')}(dv) \\
&= \int_{-\infty}^\infty q(s - v) U^{(\varepsilon')}(t^{(\varepsilon')} + dv) \\
&= \int_{s-a}^s q(s - v) U^{(\varepsilon')}(t^{(\varepsilon')} + dv). \tag{1.2.17}
\end{aligned}$$

We would like to apply Lemma 1.2.2 to the integral in the right-hand side of (1.2.17). The problem is that the limiting measure $V(A)$ in (1.2.11) can have atoms in the endpoints of the interval $[s - a, s]$. We, however, can always extend this interval to a larger interval $[s - a - h_1, s + h_2]$ without changing the value of the integral in the right-hand side of (1.2.17). It is always possible to choose $h_1, h_2 > 0$ in such a way that the measure $V(A)$ has no atoms in the endpoints of the new interval $[s - a - h_1, s + h_2]$.

Then we can replace the function $q(s-v)$ with the function $f(v) = q(s-v)\chi_{[s-a,s]}(v)$ and to define the measures $\mu^{(\varepsilon')}(A) = U^{(\varepsilon')}(t^{(\varepsilon')}+A\cap[s-a-h_1,s+h_2])$. Obviously,

$$\int_{s-a}^{s} q(s-v)U^{(\varepsilon')}(t^{(\varepsilon')}+dv) = \int_{\infty}^{-\infty} f(v)\mu^{(\varepsilon')}(dv). \tag{1.2.18}$$

By (1.2.11) and remarks made above, conditions $(\boldsymbol{\alpha})$ and $(\boldsymbol{\beta})$ of Lemma 1.2.2 are satisfied for the measures $\mu^{(\varepsilon')}(A)$. It is also obvious that conditions $(\boldsymbol{\gamma})$–$(\boldsymbol{\epsilon})$ of this lemma are satisfied for the functions $f(v)$. By applying Lemma 1.2.2 to the integrals in the right-hand side of (1.2.18) and taking into consideration that the actual integration is performed over the interval $[s-a,s]$, we get for all $s \in \mathbb{R}_1$ the following:

$$x^{(\varepsilon')}(t^{(\varepsilon')} + s) = \int_{0}^{t^{(\varepsilon')}+s} q(t^{(\varepsilon')} + s - v)U^{(\varepsilon')}(dv)$$

$$= \int_{\infty}^{-\infty} f(v)\mu^{(\varepsilon')}(dv)$$

$$\to h(s) = \int_{s-a}^{s} q(s-v)V(dv) \ \text{ as } \varepsilon' \to 0. \tag{1.2.19}$$

Now we show that the convergence of $x^{(\varepsilon')}(t^{(\varepsilon')} + s)$ to $h(s)$ is uniform at every point $s \geq 0$.

The function $q(s)$ is uniformly continuous on $\mathbb{R}_1$, since it is continuous and vanishes outside a finite interval. For $c > 0$, denote

$$\Delta_q(c) = \sup_{s\in\mathbb{R}_1} \ \sup_{|h|\leq c} |q(s + h) - q(s)|$$

Using (1.2.19) and (1.2.9) we find that

$$\varlimsup_{\varepsilon'\to 0} \sup_{|h|\leq c} |x^{(\varepsilon')}(t^{(\varepsilon')} + s + h) - h(s)| \leq \varlimsup_{\varepsilon'\to 0} |x^{(\varepsilon')}(t^{(\varepsilon')} + s) - h(s)|$$

$$+ \varlimsup_{\varepsilon'\to 0} \sup_{|h|\leq c} \int_{-\infty}^{+\infty} |g(s + h - v) - g(s - v)|U^{(\varepsilon')}(t^{(\varepsilon')} + dv)$$

$$\leq \Delta_q(c) \cdot \varlimsup_{\varepsilon'\to 0} U^{(\varepsilon')}(t^{(\varepsilon')} + [s - a, s + c])$$

$$\leq \Delta_q(c) \cdot [K_1(a + c) + K_2] \to 0 \ \text{ as } c \to 0. \tag{1.2.20}$$

We observe that the functions $x(t^{(\varepsilon')} + s)$ are bounded uniformly in $s \in \mathbb{R}_1$ and asymptotically uniformly in $\varepsilon' \to 0$, more precisely,

$$\varlimsup_{\varepsilon'\to 0} \sup_{s\in\mathbb{R}_1} |x^{(\varepsilon')}(t^{(\varepsilon')} + s)|$$

$$\leq \varlimsup_{\varepsilon'\to 0} \sup_{s\in\mathbb{R}_1} Q \cdot [U^{(\varepsilon')}(t^{(\varepsilon')} + s) - U^{(\varepsilon')}(t^{(\varepsilon')} + s - a)]$$

$$\leq Q \cdot (K_1 a + K_2). \tag{1.2.21}$$

Since $x^{(\varepsilon')}(t^{(\varepsilon')} + s)$ is a solution of the renewal equation, we find that

$$x^{(\varepsilon')}(t^{(\varepsilon')} + s) = q(t^{(\varepsilon')} + s) + \int_0^{t^{(\varepsilon')}+s} x^{(\varepsilon')}(t^{(\varepsilon')} + s - v)F^{(\varepsilon')}(dv)$$

$$= q(t^{(\varepsilon')} + s) + \int_0^\infty x^{(\varepsilon')}(t^{(\varepsilon')} + s - v)F^{(\varepsilon')}(dv). \qquad (1.2.22)$$

It is obvious that $q(t^{(\varepsilon')} + s) \to 0$ as $\varepsilon' \to 0$, since the function $q(s)$ is positive only in the interval $[0, a]$. Due to (1.2.20), (1.2.21) and condition $\mathbf{D_2}$, Lemma 1.2.2 can be applied to the second term in the right-hand side of (1.2.22). The function $x^{(\varepsilon')}(t^{(\varepsilon')} + s - v)$ plays the role of the function $f^{(\varepsilon')}(v)$ for $\varepsilon' > 0$ and the function $h(s - v)$ is the corresponding limiting function. The measure $F^{(\varepsilon')}(\cdot)$ plays the role of the measure $\mu^{(\varepsilon')}(\cdot)$ for all $\varepsilon' \geq 0$. We find, by applying Lemma 1.2.2, that

$$x^{(\varepsilon')}(t^{(\varepsilon')} + s) \to \int_0^\infty h(s - v)F^{(0)}(dv) \text{ as } \varepsilon' \to 0, \ s \in \mathbb{R}_1. \qquad (1.2.23)$$

It follows from (1.2.19) and (1.2.23) that $h(s)$ satisfies equation (1.2.15).

This proves that $V(A) = \alpha m(A)$, where α is a constant and $m(A)$ is the Lebesgue measure on $B_{(1)}$. Hence, we can rewrite (1.2.11) in the following form:

$$U^{(\varepsilon')}(t^{(\varepsilon')} + A) \Rightarrow \alpha m(A) \text{ as } \varepsilon' \to 0. \qquad (1.2.24)$$

To prove the theorem, it remains to show that $\alpha = 1/m_1^{(0)}$.

If we choose $q^{(\varepsilon')}(t) = 1 - F^{(\varepsilon')}(t)$ in the renewal equation (1.2.16), then its solution is identically equal to 1, that is,

$$1 = \int_0^t (1 - F^{(\varepsilon')}(t - s))U^{(\varepsilon')}(ds). \qquad (1.2.25)$$

Hence, for $t^{(\varepsilon')} > T$,

$$1 = \int_0^{t^{(\varepsilon')}} (1 - F^{(\varepsilon')}(t^{(\varepsilon')} - s))U^{(\varepsilon')}(ds)$$

$$= \int_0^T (1 - F^{(\varepsilon')}(v))U^{(\varepsilon')}(t^{(\varepsilon')} - dv)$$

$$+ \int_T^{t^{(\varepsilon')}} (1 - F^{(\varepsilon')}(v))U^{(\varepsilon')}(t^{(\varepsilon')} - dv). \qquad (1.2.26)$$

Taking into consideration (1.2.9) we find that for any $T, g > 0$,

$$\int_T^{t^{(\varepsilon')}} (1 - F^{(\varepsilon')}(v)) U^{(\varepsilon')}(t^{(\varepsilon')} - dv)$$

$$\leq \varlimsup_{\varepsilon' \to 0} \sum_{r=[T/g]}^{\infty} (1 - F^{(\varepsilon')}(rg))(U^{(\varepsilon')}(t^{(\varepsilon')} + (r+1)g) - U^{(\varepsilon')}(t^{(\varepsilon')} + rg))$$

$$\leq (Kg + K_2) \cdot \varlimsup_{\varepsilon' \to 0} \sum_{r=[T/g]}^{\infty} (1 - F^{(\varepsilon')}(rg))$$

$$\leq (K_1 g + K_2) \cdot \int_{T-2g}^{\infty} (1 - F^{(\varepsilon')}(s))\, ds. \tag{1.2.27}$$

Since conditions $\mathbf{D_2}$ and $\mathbf{M_2}$ imply condition $\mathbf{M_3}$, by using (1.2.27) we get

$$\lim_{T \to \infty} \varlimsup_{\varepsilon' \to 0} \int_T^{t^{(\varepsilon')}} (1 - F^{(\varepsilon')}(v)) U^{(\varepsilon')}(t^{(\varepsilon')} - dv) = 0. \tag{1.2.28}$$

Now we shall verify that the conditions of Lemma 1.2.2 are satisfied for the functions $f^{(\varepsilon')}(v) = 1 - F^{(\varepsilon')}(v)$ and the measures $\mu^{(\varepsilon')}(A) = U^{(\varepsilon')}(t^{(\varepsilon')} - A \cap [0, T])$ (here $-A$ is the set of point $-y$, where $y \in A$).

Indeed, (1.2.24) implies conditions $(\boldsymbol{\alpha})$ and $(\boldsymbol{\beta})$ of this lemma. The functions $1 - F^{(\varepsilon')}(v)$ are bounded by 1 uniformly in ε' and v. Condition $\mathbf{D_2}$ implies that the functions $1 - F^{(\varepsilon')}(v)$ converge to $1 - F^{(0)}(v)$, as $\varepsilon' \to 0$, at any continuity point v of the function $1 - F^{(0)}(v)$. Since the functions $1 - F^{(\varepsilon')}(v)$ are monotone, this convergence is easily seen to be locally uniform at every continuity point of $1 - F^{(0)}(v)$. Also, functions $1 - F^{(\varepsilon')}(v)$ are bounded. Since the set D_0 of discontinuity points of the function $1 - F^{(0)}(v)$ is at most countable, $\alpha m(R_0) = 0$. So, conditions $(\boldsymbol{\gamma})$–$(\boldsymbol{\epsilon})$ of Lemma 1.2.2 are also satisfied.

Applying Lemma 1.2.2 to the functions $1 - F^{(\varepsilon')}(v)$ and to the measures $\mu^{(\varepsilon')}(A)$, we obtain the relation

$$\int_0^T (1 - F^{(\varepsilon')}(v))\, U^{(\varepsilon')}(t^{(\varepsilon')} - dv)$$

$$\to \int_0^T (1 - F^{(0)}(v))\alpha m(dv) \quad \text{as } \varepsilon' \to 0, \tag{1.2.29}$$

which holds for any $T > 0$.

From (1.2.26), (1.2.28) and (1.2.29) it obviously follows that

$$1 = \lim_{T \to \infty} \lim_{\varepsilon' \to 0} \int_0^T (1 - F^{(\varepsilon')}(v)) U^{(\varepsilon')}(t^{(\varepsilon')} - dv)$$

$$= \lim_{T \to \infty} \int_0^T (1 - F^{(0)}(v))\alpha m(dv) = \alpha m_1^{(0)}. \tag{1.2.30}$$

Hence, $\alpha = 1/m_1^{(0)}$ and the proof of the theorem is completed. $\qquad\square$

1.2.5   Proof of Theorem 1.2.1

According to formula (1.1.3), for $t^{(\varepsilon')} > T$ we have

$$x^{(\varepsilon')}(t^{(\varepsilon')}) = \int_0^{t^{(\varepsilon')}} q(t-s)U^{(\varepsilon')}(ds)$$

$$= \int_0^T q^{(\varepsilon')}(v)U^{(\varepsilon')}(t^{(\varepsilon')} - dv)$$

$$+ \int_T^{t^{(\varepsilon')}} q^{(\varepsilon')}(v)U^{(\varepsilon')}(t^{(\varepsilon')} - dv). \qquad (1.2.31)$$

By Lemma 1.1.3, for any interval $I = [a, b]$ of length $h = b - a$,

$$\overline{\lim_{\varepsilon' \to 0}} \sup_{t \geq 0} U^{(\varepsilon')}(t + I) \leq K_1 + K_2 h < \infty,$$

where $K_1, K_2 < \infty$.

Therefore, using of the condition $\mathbf{F_2}$, we find for $h > 0$ in this condition that

$$\overline{\lim_{\varepsilon' \to 0}} \int_T^{t^{(\varepsilon')}} |q^{(\varepsilon')}(v)|U^{(\varepsilon')}(t^{(\varepsilon')} - dv)$$

$$\leq \overline{\lim_{\varepsilon' \to 0}} \sum_{r \geq [T/h]} \sup_{rh \leq t \leq (r+1)h} |q^{(\varepsilon')}(t)|$$

$$\times (U^{(\varepsilon')}(t^{(\varepsilon')} - rh) - U^{(\varepsilon')}(t^{(\varepsilon')} - (r+1)h))$$

$$\leq \frac{K_1 + K_2 h}{h} \cdot \overline{\lim_{\varepsilon' \to 0}} h \sum_{r \geq \frac{T-h}{h}} \sup_{rh \leq t \leq (r+1)h} |q^{(\varepsilon')}(t)| \to 0 \text{ as } T \to 0. \quad (1.2.32)$$

Theorem 1.2.2 and condition $\mathbf{F_2}$ allow to apply Lemma 1.2.2 to the functions $q^{(\varepsilon')}(v)$ and the measures $\mu^{(\varepsilon')}(A) = U^{(\varepsilon')}(t^{(\varepsilon')} - A \cap [0, T])$, which yields the following relation:

$$\int_0^T q^{(\varepsilon')}(t^{(\varepsilon')} - s)U^{(\varepsilon')}(ds) = \int_0^T q^{(\varepsilon')}(v)U^{(\varepsilon')}(t^{(\varepsilon')} - dv)$$

$$\to \frac{1}{m_1^{(0)}} \int_0^T q^{(0)}(v)m(dv) \text{ as } \varepsilon' \to 0. \quad (1.2.33)$$

Relations (1.2.31), (1.2.32) and (1.2.33) imply in an obvious way that

$$\lim_{\varepsilon'\to 0} x^{(\varepsilon')}(t^{(\varepsilon')}) = \lim_{\varepsilon'\to 0} \int_0^{t^{(\varepsilon')}} q^{(\varepsilon')}(t^{(\varepsilon')} - s)U^{(\varepsilon')}(ds)$$

$$= \lim_{T\to\infty}\lim_{\varepsilon'\to 0} \int_0^T q^{(\varepsilon')}(t^{(\varepsilon')} - s)U^{(\varepsilon')}(ds)$$

$$= \lim_{T\to\infty} \frac{1}{m_1^{(0)}} \int_0^T q^{(0)}(v)m(dv)$$

$$= \frac{1}{m_1^{(0)}} \int_0^{\infty} q^{(0)}(v)m(dv). \tag{1.2.34}$$

The limit in (1.2.34) does not depend on the choice of the sequence $0 < \varepsilon = \varepsilon_n \to 0$ as $n \to \infty$ and the corresponding subsequence $\varepsilon' = \varepsilon_{n_k} \to 0$ as $k \to \infty$ for which (1.2.11) is satisfied. Thus relation (1.2.4) in Theorem 1.2.1 holds for any $\varepsilon \to 0$, and the proof is completed. $\square$

As it follows from the proof, Theorem 1.2.1 is in some sense a corollary of Theorem 1.2.2. At the same time, it is obvious that Theorem 1.2.2 can be considered as a particular case of Theorem 1.2.1. Indeed relation (1.2.5) is a particular case of relation (1.2.4), where $q^{(\varepsilon')}(s)$ is the characteristic function of the interval.

1.2.6   Renewal limits

We define, for $\varepsilon \geq 0$,

$$x^{(\varepsilon)}(\infty) = \frac{\int_0^{\infty} q^{(\varepsilon)}(s)m(ds)}{m_1^{(\varepsilon)}}. \tag{1.2.35}$$

Let us assume the following two conditions:

D$_3$: $F^{(\varepsilon)}(s)$ is a non-arithmetic distribution function for all ε small enough, say $\varepsilon \leq \varepsilon_1$;

and

F$_3$: $q^{(\varepsilon)}(s)$ is directly Riemann integrable on $[0, \infty)$ for all ε small enough, say $\varepsilon \leq \varepsilon_2$.

Note that, under condition **F$_2$**, it is enough to assume that $q^{(\varepsilon)}(s)$ is a continuous function almost everywhere with respect to the Lebesgue measure on $[0, \infty)$. In this case, condition **F$_3$** holds and the Lebesgue integration can be replaced with the Riemann integration in the right-hand side of the formula that defines $x^{(\varepsilon)}(\infty)$.

Condition **M$_2$** implies that $m_1^{(\varepsilon)} < \infty$ for all ε small enough, say $\varepsilon \leq \varepsilon_3$.

Define

$$\varepsilon_0 = \min(\varepsilon_1, \varepsilon_2, \varepsilon_3). \tag{1.2.36}$$

The renewal theorem implies the following statement.

Theorem 1.2.3. *Let conditions* **D$_3$**, **M$_2$**, *and* **F$_3$** *hold. Then, for $\varepsilon \leq \varepsilon_0$,*

$$x^{(\varepsilon)}(t) \to x^{(\varepsilon)}(\infty) \ \ as \ t \to \infty. \tag{1.2.37}$$

It should be noted that, under condition **F$_2$**, the expression for $x^{(\varepsilon)}(\infty)$ given by formula (1.2.35) is finite for all ε small enough without the assumption that $q^{(\varepsilon)}(s)$ is a continuous function almost everywhere with respect to the Lebesgue measure on $[0, \infty)$. This is true, since the formula involves the Lebesgue integration.

Moreover, it can be proved that in this case the mean value of the $x^{(\varepsilon)}(\cdot)$ over the interval $[0, t]$ converges to $x^{(\varepsilon)}(\infty)$ as $t \to \infty$.

1.2.7   Convergence of renewal limits

The following theorem supplements the asymptotic relations in (1.2.4) and (1.2.37).

Theorem 1.2.4. *Let conditions* **M$_2$**, *and* **F$_2$** *hold. Then*

$$x^{(\varepsilon)}(\infty) \to x^{(0)}(\infty) \ \ as \ \varepsilon \to 0. \tag{1.2.38}$$

Proof. According to condition **M$_2$**, we have $m_1^{(\varepsilon)} \to m_1^{(0)} \in (0, \infty)$ as $\varepsilon \to 0$. So, it is sufficient to prove that condition **F$_2$** implies the following relation:

$$\int_0^\infty q^{(\varepsilon)}(s)m(ds) \to \int_0^\infty q^{(0)}(s)m(ds) < \infty \ \text{ as } \varepsilon \to 0. \tag{1.2.39}$$

Conditions **F$_2$** (**a**), (**b**) imply that the function $q^{(\varepsilon)}(t)$ is bounded on every finite interval for all ε small enough and that, by the Lebesgue theorem or Lemma 1.2.2, for every $0 < T < \infty$,

$$\int_0^T q^{(\varepsilon)}(s)m(ds) \to \int_0^T q^{(0)}(s)m(ds) \ \text{ as } \varepsilon \to 0. \tag{1.2.40}$$

Also condition **F$_2$** (**c**) implies that, for h in this condition,

$$\lim_{T\to\infty} \overline{\lim_{0\leq\varepsilon\to 0}} \int_T^\infty |q^{(\varepsilon)}(s)|m(ds)$$

$$\leq \lim_{T\to\infty} \overline{\lim_{0\leq\varepsilon\to 0}} \, h \sum_{r\geq T/h} \sup_{rh\leq t\leq (r+1)h} |q^{(\varepsilon)}(t)| = 0. \tag{1.2.41}$$

Relations (1.2.40) and (1.2.41) imply in an obvious way relation (1.2.38), which proves Theorem 1.2.4. $\qquad\square$

Remark 1.2.5. It follows from the proof that conditions $\mathbf{M_2}$ and $\mathbf{F_2}$ can be replaced with the following conditions:

$\mathbf{M_4}$: $m_1^{(\varepsilon)} = \int_0^\infty s F^{(\varepsilon)}(ds) \to m_1^{(0)} \in (0, \infty)$ as $\varepsilon \to 0$;

and

$\mathbf{F_4}$: $q^{(\varepsilon)} = \int_0^\infty q^{(\varepsilon)}(s) m(ds) \to q^{(0)} < \infty$ as $\varepsilon \to 0$.

The limiting constants may or may not be of the form $m_1^{(0)} = \int_0^\infty s F^{(0)}(ds)$ and $q^{(0)} = \int_0^\infty q^{(0)}(s) m(ds)$.

In any case, under conditions $\mathbf{M_4}$ and $\mathbf{F_4}$, the following relation obviously holds:

$$x^{(\varepsilon)}(\infty) \to x^{(0)}(\infty) = q^{(0)} / m_1^{(0)} \quad \text{as } \varepsilon \to 0. \tag{1.2.42}$$

1.2.8 Perturbed renewal processes

Here we give several statements that generalise some useful theorems of renewal theory to the model of a perturbed renewal equation. These theorems will be used in the next section, where Theorems 1.2.1 and 1.2.2 will be generalised to the model of a perturbed improper renewal equation.

Let $\tau_k^{(\varepsilon)}$, $k = 1, 2, \ldots$, be a sequence of i.i.d. (independent identically distributed) random variables with a distribution function $F^{(\varepsilon)}(s)$ the support of which is not concentrated in zero. The random variables $\kappa_n^{(\varepsilon)} = \sum_{k=1}^n \tau_k^{(\varepsilon)}$, $n = 0, 1, \ldots$, can be interpreted as renewal moments and the corresponding renewal process can be defined as $\nu^{(\varepsilon)}(t) = \max(n : \kappa_n^{(\varepsilon)} \leq t), t \geq 0$. By definition, $\nu^{(\varepsilon)}(t)$ is the number of renewals in the interval $[0, t]$.

Let also $t^{(\varepsilon)}$ be a non-random function of $\varepsilon > 0$ such that $0 \leq t^{(\varepsilon)} \to \infty$ as $\varepsilon \to 0$.

As is known (see, for example, Loève (1955)), the conditions

$\mathbf{D_4}$: $\lim_{\varepsilon \to 0} t^{(\varepsilon)}(1 - F^{(\varepsilon)}(ht^{(\varepsilon)})) = 0$ for any $h > 0$;

and

$\mathbf{M_5}$: $\lim_{\varepsilon \to 0} \int_0^{ht^{(\varepsilon)}} x F^{(\varepsilon)}(dx) = m_1 < \infty$ for any $h > 0$;

are necessary and sufficient for the following relation to hold:

$$\frac{1}{t^{(\varepsilon)}} \sum_{k=1}^{[t^{(\varepsilon)}]} \tau_{k-1}^{(\varepsilon)} \overset{P}{\to} m_1 \quad \text{as } \varepsilon \to 0. \tag{1.2.43}$$

Lemma 1.2.3. *Let conditions $\mathbf{D_4}$ and $\mathbf{M_5}$ hold. Then*

$$\frac{\nu^{(\varepsilon)}(t^{(\varepsilon)})}{t^{(\varepsilon)}} \overset{P}{\to} m_1^{-1} \quad \text{as } \varepsilon \to 0. \tag{1.2.44}$$

Proof. According to the definition of the process $\nu^{(\varepsilon)}(t)$ for every $t, x \geq 0$,

$$P\{\nu^{(\varepsilon)}(t) > tx\} = P\left\{ \sum_{k=1}^{[tx]+1} \tau_k^{(\varepsilon)} \leq t \right\}. \tag{1.2.45}$$

Relation (1.2.44) follows in obvious way from (1.2.43) and (1.2.45). $\square$

Consider condition $\mathbf{D_4}$ together with the following condition:

$\mathbf{D_5}$: there exist $g > 0$ and $G < 1$ such that $\overline{\lim}_{\varepsilon \to 0} F^{(\varepsilon)}(g) \leq G$.

Obviously conditions $\mathbf{D_4}$ and $\mathbf{M_5}$ together with condition $\mathbf{D_5}$ imply that the limiting constant satisfies $m_1 > 0$.

Indeed, $\int_0^{ht^{(\varepsilon)}} x F^{(\varepsilon)}(dx) \geq g(1 - F^{(\varepsilon)}(g)$ and, therefore,

$$\lim_{\varepsilon \to 0} \int_0^{ht^{(\varepsilon)}} x F^{(\varepsilon)}(dx) \geq g - g \, \overline{\lim_{\varepsilon \to 0}} \, F^{(\varepsilon)}(g) \geq g(1 - G) > 0.$$

Remark 1.2.6. Condition $\mathbf{D_2}$ implies condition $\mathbf{D_5}$. Also, conditions $\mathbf{D_2}$ and $\mathbf{M_2}$ imply that conditions $\mathbf{D_4}$ and $\mathbf{M_5}$ hold for any $0 \leq t^{(\varepsilon)} \to \infty$ as $\varepsilon \to 0$. In this case, the constant $m_1 = m_1^{(0)}$ is the mean of the limiting distribution function $F^{(0)}(s)$.

1.2.9   Convergence of moments for perturbed renewal processes

In this section we give conditions for convergence of moments for perturbed renewal processes.

Theorem 1.2.5. *Let conditions $\mathbf{D_2}$ and $\mathbf{M_2}$ hold. Then for any $0 \leq t^{(\varepsilon)} \to \infty$ as $\varepsilon \to 0$ and every $r \geq 1$,*

$$\mathsf{E}\left(\frac{\nu^{(\varepsilon)}(t^{(\varepsilon)})}{t^{(\varepsilon)}} \right)^r \to (m_1^{(0)})^{-r} \ \text{as} \ \varepsilon \to 0. \tag{1.2.46}$$

Proof. Condition $\mathbf{D_2}$ implies condition $\mathbf{D_5}$. Now, it follows from Lemma 1.1.1 and Remark 1.1.1 that for every $r \geq 1$,

$$\overline{\lim_{\varepsilon \to 0}} \, \mathsf{E}\left(\frac{\nu^{(\varepsilon)}(t^{(\varepsilon)})}{t^{(\varepsilon)}} \right)^r < \infty. \tag{1.2.47}$$

Now, by (1.2.44) and (1.2.47), the theorem follows from an appropriate variant of the Lebesgue theorem. $\square$

In the case $r = 1$, relation (1.2.46) can be rewritten in an equivalent form, which is actually a triangular array analogue of the elementary renewal theorem.

Theorem 1.2.6. *Let conditions* $\mathbf{D_2}$ *and* $\mathbf{M_2}$ *hold. Then for any* $0 \leq t^{(\varepsilon)} \to \infty$ *as* $\varepsilon \to 0$,

$$\frac{U^{(\varepsilon)}(t^{(\varepsilon)})}{t^{(\varepsilon)}} \to (m_1^{(0)})^{-1} \quad \text{as} \ \ \varepsilon \to 0. \tag{1.2.48}$$

Let us again consider the random variables $\gamma^{(\varepsilon)}(t)$ which is the time that are times between the moment t and the first after t renewal moment defined by

$$\gamma^{(\varepsilon)}(t) = \sum_{k=1}^{\nu^{(\varepsilon)}(t)+1} \tau_k^{(\varepsilon)} - t, \quad t \geq 0.$$

As is known, the probabilities $\mathsf{P}\{\gamma^{(\varepsilon)}(t) > u\}$, as functions of t, satisfy the renewal equation

$$\mathsf{P}\{\gamma^{(\varepsilon)}(t) > u\} = 1 - F^{(\varepsilon)}(t + u)$$

$$+ \int_0^t \mathsf{P}\{\gamma^{(\varepsilon)}(t - s) > u\}F^{(\varepsilon)}(ds), \quad t \geq 0. \tag{1.2.49}$$

Denote

$$G^{(\varepsilon)}(u) = \frac{1}{m_1^{(\varepsilon)}} \int_0^u (1 - F^{(\varepsilon)}(t))\, dt, \quad u \geq 0.$$

Theorem 1.2.7. *Let conditions* $\mathbf{D_2}$ *and* $\mathbf{M_2}$ *hold. Then for any* $0 \leq t^{(\varepsilon)} \to \infty$ *as* $\varepsilon \to 0$ *and* $u \geq 0$,

$$\mathsf{P}\{\gamma^{(\varepsilon)}(t^{(\varepsilon)}) > u\} \to \frac{1}{m_1^{(0)}} \int_0^\infty (1 - F^{(0)}(t + u))\, dt$$

$$= \frac{1}{m_1^{(0)}} \int_u^\infty (1 - F^{(0)}(t))\, dt$$

$$= 1 - G^{(0)}(u) \quad \text{as} \ \ \varepsilon \to 0. \tag{1.2.50}$$

Proof. In order to apply Theorem 1.2.1 to the renewal equation (1.2.49) we have to check that condition $\mathbf{F_2}$ holds for the forcing functions $q^{(\varepsilon)}(t) = 1 - F^{(\varepsilon)}(t + u)$. Condition $\mathbf{D_2}$ implies that the functions $q^{(\varepsilon)}(t) = 1 - F^{(\varepsilon)}(t + u)$ converge to $q^{(0)}(t) = 1 - F^{(0)}(t + u)$, as $\varepsilon \to \infty$, at any continuity point t of $q^{(0)}(t)$. Since the functions $q^{(\varepsilon)}(t)$ are monotone, this convergence is easily seen to be locally uniform at every continuity point of $q^{(0)}(t)$. Since the set C_0 of discontinuity points of the function $q^{(0)}(t) = 1 - F^{(0)}(t + u)$ is at most countable $m(\overline{C}_0) = 0$. So, condition $\mathbf{F_2}$ **(a)** holds. The functions $q^{(\varepsilon)}(t) = 1 - F^{(\varepsilon)}(t + u)$ are bounded by 1 uniformly in ε and

t. Hence, condition $\mathbf{F_2}$ (**b**) also holds. Finally, using condition $\mathbf{M_2}$, Remark 1.2.1, and monotonicity of $q^{(\varepsilon)}(t)$ we get that for every $h > 0$,

$$\lim_{T \to \infty} \overline{\lim_{0 \leq \varepsilon \to 0}} \, h \sum_{r \geq T/h} \sup_{rh \leq t \leq (r+1)h} |q^{(\varepsilon)}(t)|$$

$$= \lim_{T \to \infty} \overline{\lim_{0 \leq \varepsilon \to 0}} \, h \sum_{r \geq T/h} (1 - F^{(\varepsilon)}(rh + u))$$

$$\leq \lim_{T \to \infty} \overline{\lim_{0 \leq \varepsilon \to 0}} \int_{T-h}^{\infty} (1 - F^{(\varepsilon)}(t + u)) \, dt = 0. \tag{1.2.51}$$

Therefore condition $\mathbf{F_2}$ (**c**) holds too. The application of Theorem 1.2.1 to the renewal equation (1.2.49) completes the proof. $\qquad\qquad\square$

By $\xi^{(\varepsilon)}(t), \, t \in T \xrightarrow{\mathrm{d}} \xi^{(0)}(t), \, t \in T$ as $\varepsilon \to 0$ we will denote convergence of stochastic processes in distribution, i.e., weak convergence of finite-dimensional distributions of the stochastic processes $\xi^{(\varepsilon)}(t)$ to the corresponding finite-dimensional distributions of the stochastic process $\xi^{(0)}(t)$ on the set T as $\varepsilon \to 0$.

Also denote by $D_r^{(\varepsilon)}$ the set of discontinuity points of the distribution function of the random variables $\kappa_n^{(\varepsilon)} = \sum_{k=1}^{n} \tau_k^{(\varepsilon)}$. Let also $D^{(\varepsilon)} = \bigcup_{r=0}^{\infty} D_r^{(\varepsilon)}$. The set $D^{(\varepsilon)}$ is at most countable.

We need the following auxiliary proposition.

Lemma 1.2.4. *Let condition $\mathbf{D_2}$ hold. Then*

$$v^{(\varepsilon)}(t), \, t \in \bar{D}^{(0)} \xrightarrow{\mathrm{d}} v^{(0)}(t), \, t \in \bar{D}^{(0)} \ \text{as} \ \varepsilon \to 0. \tag{1.2.52}$$

Proof. By the definition of the renewal process, $v^{(\varepsilon)}(t), \, t \geq 0$, for any positive integer $t_n \geq 0, \, r_n = 0, 1, \ldots,$ and $n, m = 1, 2, \ldots,$ we have

$$\mathsf{P}\{v^{(\varepsilon)}(t_n) \geq r_n, n = 1, \ldots, m\} = \mathsf{P}\left\{ \sum_{k=1}^{r_n} \tau_k^{(\varepsilon)} \leq t_n, n = 1, \ldots, m \right\}. \tag{1.2.53}$$

Relation (1.2.52) easily follows from (1.2.53), the independence of the random variables $\tau_k^{(\varepsilon)}, \, k = 1, 2, \ldots,$ and condition $\mathbf{D_2}$. $\qquad\qquad\square$

Lemma 1.2.5. *If conditions $\mathbf{D_2}$ and $\mathbf{M_2}$ hold, then for any points $t_n \in \bar{D}^{(0)}$ and $n, m \geq 1$,*

$$\mathsf{E} \prod_{n=1}^{m} v^{(\varepsilon)}(t_n) \to \mathsf{E} \prod_{n=1}^{m} v^{(0)}(t_n) \ \text{as} \ \varepsilon \to 0. \tag{1.2.54}$$

Proof. Since conditions $\mathbf{D_2}$ and $\mathbf{M_2}$ imply condition $\mathbf{D_5}$, it follows from Lemma 1.1.1 that there exists a constant $K < \infty$ such that

$$\overline{\lim_{\varepsilon \to 0}} \, \mathsf{E}\nu^{(\varepsilon)}(t)^r \leq Kt^r, \quad t \geq 0, \quad r = 1, 2, \ldots . \tag{1.2.55}$$

Relation (1.2.54) follows at once from (1.2.55) and Lemma 1.2.4. $\qquad\square$

The following relation can be obtained from (1.1.10) by summing over $r \geq n$:

$$\mathsf{P}\{\nu^{(\varepsilon)}(t + h) - \nu^{(\varepsilon)}(t) \geq n\} = \delta(0, n)\mathsf{P}\{\gamma^{(\varepsilon)}(t) > h\}$$

$$+ \int_0^h \mathsf{P}\{\nu^{(\varepsilon)}(h - s) + 1 \geq n\}\mathsf{P}\{\gamma^{(\varepsilon)}(t) \in ds\}, \quad n = 0, 1, \ldots . \tag{1.2.56}$$

Theorem 1.2.8. *Let condition $\mathbf{D_2}$ and $\mathbf{M_2}$ hold. Then, for any $0 \leq t^{(\varepsilon)} \to \infty$ as $\varepsilon \to 0$ and $h \geq 0$ and $n = 0, 1, \ldots,$ the following relation holds:*

$$\mathsf{P}\{\nu^{(\varepsilon)}(t^{(\varepsilon)} + h) - \nu^{(\varepsilon)}(t^{(\varepsilon)}) \geq n\}$$

$$\to \delta(0, n)(1 - G^{(0)}(h)) + \int_0^h \mathsf{P}\{\nu^{(0)}(h - s) + 1 \geq n\}G^{(0)}(ds)$$

$$= 1 - H_h^{(0)}(n) \;\; as \;\; \varepsilon \to 0. \tag{1.2.57}$$

Proof. We would like to apply Lemma 1.2.2 to the integrals in the left-hand side of (1.2.56). By Lemma 1.2.4 we have at every point of continuity of the limiting function $f^{(0)}(s)$,

$$f^{(\varepsilon)}(s) = \mathsf{P}\{\nu^{(\varepsilon)}(h - s) + 1 \geq n\}\chi_{[0,h]}(s)$$

$$\to f^{(0)}(s) = \mathsf{P}\{\nu^{(0)}(h - s) + 1 \geq n\}\chi_{[0,h]}(s) \;\; as \;\; \varepsilon \to 0. \tag{1.2.58}$$

Functions $f^{(\varepsilon)}(s)$ are monotone on the interval $[0, h]$. Hence, this convergence is easily seen to be locally uniform in each point of continuity of the limiting function $f^{(0)}(s)$. The set D of points of discontinuity of the function $f^{(0)}(s)$ is at most countable. Denote $\mu^{(\varepsilon)}(A) = \mathsf{P}\{\gamma^{(\varepsilon)}(t) \in A \cap [0, h]\}$. Weak convergence of these measures is implied by Theorem 1.2.7. Moreover, $\mu^{(\varepsilon)}(\mathbb{R}_1) = \mathsf{P}\{\gamma^{(\varepsilon)}(t^{(\varepsilon)}) \leq h\} \to 1 - G^{(0)}(h)$ as $\varepsilon \to 0$, since $G^{(0)}(h)$ is a distribution function continuous in h. Finally, the limiting measure $\mu^{(0)}(A)$ is absolutely continuous on the interval $[0, h]$, according to the formula (1.2.50). So, $\mu^{(0)}(D) = 0$. Thus, all conditions of Lemma 1.2.2 are satisfied. Applying this lemma in (1.2.56) and using relation (1.2.50) from Theorem 1.2.7 we get relation (1.2.57). $\qquad\square$

1.2.10 Stationary renewal processes

Let $\tilde{\tau}_k^{(0)}$, $k = 1, 2, \ldots$, be a sequence of nonnegative independent random variables such that the random variables $\tilde{\tau}_1^{(0)}$ has a distribution function $G^{(0)}(t)$, while the random variables $\tilde{\tau}_k^{(0)}$, $k > 1$, are identically distributed with a distribution function $F^{(0)}(t)$. Denote $\tilde{\kappa}_k^{(0)} = \sum_{k=1}^{n} \tilde{\tau}_k^{(0)}$, $n = 0, 1, \ldots$, and introduce a renewal process constructed from the random variables $\tilde{\kappa}_k^{(0)}$,

$$\tilde{v}^{(0)}(t) = \max(n : \tilde{\kappa}_n^{(0)} \leq t), \quad t \geq 0. \tag{1.2.59}$$

As is known (see, for example, Feller (1971)) the renewal process $\tilde{v}^{(0)}(t)$, $t \geq 0$, is a stationary renewal process (process with stationary increments). In particular, the distribution of the increments $\tilde{v}^{(0)}(t+h) - \tilde{v}^{(0)}(t)$ of this process depends on $h \geq 0$ but not on $t \geq 0$. They are given by expression in the right-hand side of relation (1.2.57), i.e., for all $h, t \geq 0$ and $n = 0, 1, \ldots$,

$$\mathsf{P}\{\tilde{v}^{(0)}(t + h) - \tilde{v}^{(0)}(t) \geq n\} = \mathsf{P}\{\tilde{v}^{(0)}(h) \geq n\} = 1 - H_h^{(0)}(n). \tag{1.2.60}$$

Since the renewal process $\tilde{v}^{(0)}(t)$ is stationary,

$$\mathsf{E}(\tilde{v}^{(0)}(t + h) - \tilde{v}^{(0)}(t)) = \mathsf{E}\tilde{v}^{(0)}(h)$$
$$= \sum_{n \geq 1}(1 - H_h^{(0)}(n)) = h/m_1^{(0)}, \quad h, t \geq 0. \tag{1.2.61}$$

In principle, it would be possible to show that under conditions of Theorem 1.2.8 we have the following more general relation:

$$v^{(\varepsilon)}(t^{(\varepsilon)} + h) - v^{(\varepsilon)}(t^{(\varepsilon)}), \ h \geq 0 \xrightarrow{\ \mathrm{d}\ } \tilde{v}^{(0)}(h), \ h \geq 0 \ \text{ as } \varepsilon \to 0. \tag{1.2.62}$$

1.2.11 Convergence of moments for increments of perturbed renewal processes

We now return to relation (1.1.11), which has the following form:

$$\mathsf{E}(v^{(\varepsilon)}(t + h) - v^{(\varepsilon)}(t))^r = \int_0^h \mathsf{E}(v^{(\varepsilon)}(h) + 1)^r \mathsf{P}\{\gamma^{(\varepsilon)}(t) \in ds\}, \quad r \geq 1. \tag{1.2.63}$$

Theorem 1.2.9. *Let conditions* $\mathbf{D_2}$ *and* $\mathbf{M_2}$ *hold. Then for any* $0 \leq t^{(\varepsilon)} \to \infty$ *as* $\varepsilon \to 0$ *and* $h \geq 0$ *and* $r = 1, 2, \ldots$, *the following relation holds:*

$$\mathsf{E}(v^{(\varepsilon)}(t^{(\varepsilon)} + h) - v^{(\varepsilon)}(t^{(\varepsilon)}))^r \to \int_0^h \mathsf{E}(v^{(0)}(h) + 1)^r G^{(0)}(ds) \ \text{ as } \varepsilon \to 0. \tag{1.2.64}$$

Proof. The proof is analogous to the proof of Theorem 1.2.8. Lemma 1.2.2 can be applied to the integrals in the left-hand side of (1.2.63). In this case, the functions $\mathsf{P}\{v^{(\varepsilon)}(h-s)+1 \geq n\}\chi_{[0,h]}(s)$ have to be replaced with $f^{(\varepsilon)}(s) = \mathsf{E}(v^{(\varepsilon)}(h-s)+1)^r\,\chi_{[0,h]}(s)$ but we need to use the same measures $\mu^{(\varepsilon)}(A) = \mathsf{P}\{\gamma^{(\varepsilon)}(t) \in A \cap [0,h]\}$. By Lemma 1.2.5 we have at every point of continuity of the limiting function $f^{(0)}(s)$,

$$f^{(\varepsilon)}(s) = \mathsf{E}(v^{(\varepsilon)}(h-s)+1)^r\,\chi_{[0,h]}(s)$$

$$\to f^{(0)}(s) = \mathsf{E}(v^{(0)}(h-s)+1)^r\,\chi_{[0,h]}(s) \quad \text{as } \varepsilon \to 0. \tag{1.2.65}$$

Functions $f^{(\varepsilon)}(s)$ are monotone on the interval $[0, h]$, and $f^{(0)}(h) = \mathsf{E}(v^{(0)}(h)+1)^r < \infty$. Hence, this convergence is easily seen to be locally uniform in each point of continuity of the limiting function $f^{(0)}(s)$, i.e., at every point $s \in \overline{D}$, where D is the same set as in the proof of Theorem 1.2.8. A further reasoning is absolutely analogous to the one given in the proof of Theorem 1.2.8.

One should observe that the proof could also be completed by applying a corresponding variant of the Lebesgue theorem. Theorem 1.2.8 means that

$$v^{(\varepsilon)}(t^{(\varepsilon)} + h) - v^{(\varepsilon)}(t^{(\varepsilon)}) \xrightarrow{\ \text{d}\ } \tilde{v}^{(0)}(h) \quad \text{as } \varepsilon \to 0, \tag{1.2.66}$$

where $\tilde{v}^{(0)}(h)$ is a discrete random variable such that $\mathsf{P}\{\tilde{v}^{(0)}(h) \geq n\} = 1 - H_h^{(0)}(n)$.

At the same time Lemma 1.1.2 imply that there exist constants $K[h, r] < \infty$ such that

$$\overline{\lim_{\varepsilon \to 0}}\, \mathsf{E}(v^{(\varepsilon)}(t^{(\varepsilon)} + h) - v^{(\varepsilon)}(t^{(\varepsilon)}))^r \leq K[h, r]. \tag{1.2.67}$$

Relations (1.2.66) and (1.2.67) imply (1.2.64). $\qquad\qquad\qquad\qquad\qquad\qquad$ $\square$

Note also that the limit expression in the right-hand side of (1.2.64) in Theorem 1.2.9 coincides with $\mathsf{E}\tilde{v}^{(0)}(h)^r$, and relation (1.2.64) could be rewritten in the following form, for $h \geq 0, r = 1, 2, \ldots,$

$$\mathsf{E}(v^{(\varepsilon)}(t^{(\varepsilon)} + h) - v^{(\varepsilon)}(t^{(\varepsilon)}))^r \to \mathsf{E}\tilde{v}^{(0)}(h)^r \quad \text{as } \varepsilon \to 0. \tag{1.2.68}$$

Theorem 1.2.9 is in a sense a generalisation of Theorem 1.2.2. For the case $r = 1$ the assertions of both theorems coincide in view of the relation $U(t) = \mathsf{E}v^{(\varepsilon)}(t) + 1$.

1.3   Improper perturbed renewal equation

In this section we generalise the renewal Theorem 1.2.1 to a model in which the distribution functions $F^{(\varepsilon)}(t)$ that generate the renewal equation (1.2.1) can be improper for $\varepsilon > 0$ but proper for $\varepsilon = 0$, i.e., the total variation satisfies $F^{(\varepsilon)}(\infty) \leq 1$ for $\varepsilon > 0$ but $F^{(0)}(\infty) = 1$. In such a case we say that (1.2.1) is an *improper renewal equation*.

1.3.1   Renewal theorem for improper perturbed renewal equation

All general facts and definitions concerning the renewal equation are translated to the case of an improper renewal equation without any changes. In particular, the renewal function is defined by the same formula and a unique solution of the renewal equation is defined via the renewal function by the same formula as in the case of the proper renewal equation.

We define the first moment of a distribution $F^{(\varepsilon)}(s)$ as usual by the formula

$$m_1^{(\varepsilon)} = \int_0^\infty s F^{(\varepsilon)}(ds).$$

This functional can be finite also in the case where the total variation $F^{(\varepsilon)}(\infty) < 1$. However, in this case, $m_1^{(\varepsilon)}$ can not be interpreted as the expectation of an improper random variable with the distribution $F^{(\varepsilon)}(s)$. Any such random variable has an infinite expectation since it takes the value $+\infty$ with the positive probability $1 - F^{(\varepsilon)}(\infty)$.

Conditions $\mathbf{D_2}$ and $\mathbf{M_2}$ take in this case the following form:

$\mathbf{D_6}$: $F^{(\varepsilon)}(\cdot) \Rightarrow F^{(0)}(\cdot)$ as $\varepsilon \to 0$, where $F^{(0)}(s)$ is a proper non-arithmetic distribution function;

and

$\mathbf{M_6}$: $m_1^{(\varepsilon)} \to m_1^{(0)} < \infty$ as $\varepsilon \to 0$.

Condition $\mathbf{D_6}$ means that $F^{(\varepsilon)}(t) \to F^{(0)}(t)$ as $\varepsilon \to 0$ in every point t that is a point of continuity for the limiting distribution function. This condition also includes the assumption $F^{(0)}(\infty) = 1$. Thus, $\mathbf{D_6}$ also implies the following relation:

$$F^{(\varepsilon)}(\infty) \to F^{(0)}(\infty) = 1 \quad \text{as} \quad \varepsilon \to 0. \tag{1.3.1}$$

Let $f^{(\varepsilon)} = 1 - F^{(\varepsilon)}(\infty)$ denote the defect of the distribution $F^{(\varepsilon)}(t)$.

In this case, we also consider a new type of condition that would balance the rate of growth of time $t^{(\varepsilon)}$, which is supposed to be a non-random nonnegative function of ε, and the rate of deficiency of the renewal equation determined by the defect $f^{(\varepsilon)} = 1 - F^{(\varepsilon)}(\infty)$,

$\mathbf{B_1}$: $0 \le t^{(\varepsilon)} \to \infty$ and $f^{(\varepsilon)} \to 0$ as $\varepsilon \to 0$ in such a way that $f^{(\varepsilon)} t^{(\varepsilon)} \to \lambda$, where $0 \le \lambda \le \infty$.

The main result of this section is the following theorem that generalises the classical renewal theorem to the model of improper perturbed renewal equations.

Theorem 1.3.1. *Let conditions* $\mathbf{D_6}$, $\mathbf{M_6}$, $\mathbf{F_2}$ *hold, and* $0 \le t^{(\varepsilon)} \to \infty$ *as* $\varepsilon \to 0$ *such that condition* $\mathbf{B_1}$ *holds. Then*

$$x^{(\varepsilon)}(t^{(\varepsilon)}) \to e^{-\lambda/m_1^{(0)}} \cdot \frac{\int_0^\infty q^{(0)}(s)\, ds}{m_1^{(0)}} \quad \text{as} \quad \varepsilon \to 0. \tag{1.3.2}$$

Theorem 1.3.1 differs from Theorem 1.2.1. There appears a new exponential factor in the left-hand side of the asymptotic relation (1.3.2). It describes the asymptotical behaviour of a solution of the renewal equation as the time $t^{(\varepsilon)}$ tends to infinity. The appearance of this factor is caused by the deficiency of the distribution functions $F^{(\varepsilon)}(s)$ that generate the renewal equation (1.2.1).

At the same time, Theorem 1.3.1 assumes some restriction on the rate of growth of time $t^{(\varepsilon)}$ in condition $\mathbf{B_1}$, while Theorem 1.2.1 imposes no restrictions on the rate at which $t^{(\varepsilon)}$ tends to infinity.

A proper renewal equation is obtained if $f^{(\varepsilon)} \equiv 0$. In this case, condition $\mathbf{B_1}$ automatically holds, since $f^{(\varepsilon)} t^{(\varepsilon)} \equiv 0$ and $\lambda = 0$. Also note that in this case, $t^{(\varepsilon)}$ can tend to infinity in an arbitrary way. The exponential factor in the left-hand side of (1.3.2) takes the value 1. Theorem 1.3.1 is reduced to Theorem 1.2.1.

Also note that we admit the case where $\lambda = +\infty$. In this case, the exponential factor in the left-hand side of (1.3.2) takes the value 0 and relation (1.3.2) takes the form $x^{(\varepsilon)}(t^{(\varepsilon)}) \to 0$ as $\varepsilon \to 0$.

As in the case of a proper renewal equation, the main part of the proof of Theorem 1.3.1 is based on a generalisation of Blackwel's theorem to the model of an improper perturbed renewal equation.

Theorem 1.3.2. *Let conditions* $\mathbf{D_6}$, $\mathbf{M_6}$ *hold, and* $0 \le t^{(\varepsilon)} \to \infty$ *as* $\varepsilon \to 0$ *such that condition* $\mathbf{B_1}$ *holds. Then for any* $h \ge 0$,

$$U^{(\varepsilon)}(t^{(\varepsilon)} + h) - U^{(\varepsilon)}(t^{(\varepsilon)}) \to e^{-\lambda/m_1^{(0)}} \cdot h/m_1^{(0)} \quad \text{as } \varepsilon \to 0. \qquad (1.3.3)$$

1.3.2 Perturbed renewal processes with improper inter-renewal times

We postpone the proof of Theorems 1.3.1 and 1.3.2. We first formulate and prove analogues of Theorems 1.2.8 and 1.2.9 for the model of perturbed improper renewal processes. These theorems will be used in the proofs of Theorems 1.3.1 and 1.3.2.

Let $\tau_k^{(\varepsilon)}, k = 1, 2, \ldots$, be improper i.i.d. random variables with values in the interval $[0, \infty]$ and a distribution $F^{(\varepsilon)}(t)$. This means that these random variables are improper and take value $+\infty$ with the probability $f^{(\varepsilon)} = 1 - F^{(\varepsilon)}(\infty)$. We also define $\kappa_n^{(\varepsilon)} = \sum_{k=1}^{n} \tau_k^{(\varepsilon)}, n = 0, 1, 2, \ldots$, and construct the corresponding renewal process,

$$\nu^{(\varepsilon)}(t) = \max(n : \kappa_n^{(\varepsilon)} \le t), \quad t \ge 0.$$

Let us define proper distribution functions concentrated on $[0, \infty)$,

$$\bar{F}^{(\varepsilon)}(t) = F^{(\varepsilon)}(t)/F^{(\varepsilon)}(\infty), \quad t \ge 0.$$

Let also $\check{\tau}_k^{(\varepsilon)}, k = 1, 2, \ldots$, be proper i.i.d. random variables with values in the interval $[0, \infty)$ and distribution $\bar{F}^{(\varepsilon)}(t)$. We also define $\check{\kappa}_n^{(\varepsilon)} = \sum_{k=1}^{n} \check{\tau}_k^{(\varepsilon)}, n = 0, 1, 2, \ldots$, and construct the corresponding renewal process,

$$\check{\nu}^{(\varepsilon)}(t) = \max(n : \kappa_n^{(\varepsilon)} \le t), \quad t \ge 0.$$

Finally, let $\breve{\nu}^{(\varepsilon)}$ be a random variable which (a) is defined on the same probability space as the random variables $\breve{\tau}_k^{(\varepsilon)}, k = 1, 2, \ldots$; and (b) independent of these random variables; (c) has a geometric distribution with the parameter $f^{(\varepsilon)}$, i.e.,

$$\mathsf{P}\{\breve{\nu}^{(\varepsilon)} = n\} = f^{(\varepsilon)}(1 - f^{(\varepsilon)})^{n-1}, \quad n = 1, 2, \ldots .$$

The following formula connecting the renewal processes $\nu^{(\varepsilon)}(t)$ and $\breve{\nu}^{(\varepsilon)}(t)$ plays a crucial role in the sequel. It states that for any $0 \leq t_1 \leq t_2 \leq \cdots \leq t_n < \infty$, integer $0 \leq r_1 \leq r_2 \leq \cdots \leq r_n$ and $n \geq 1$,

$$\mathsf{P}\{\nu^{(\varepsilon)}(t_k) \geq r_k, k = 1, \ldots, n\}$$

$$= \mathsf{P}\left\{\sum_{l=1}^{r_k} \tau_l^{(\varepsilon)} \leq t_k, k = 1, \ldots, n\right\}$$

$$= \mathsf{P}\{\tau_l^{(\varepsilon)} < \infty, l = 1, \ldots, r_n\}$$

$$\times \mathsf{P}\left\{\sum_{l=1}^{r_k} \tau_l^{(\varepsilon)} \leq t_k, k = 1, \ldots, n / \tau_l^{(\varepsilon)} < \infty, l = 1, \ldots, r_n\right\}$$

$$= F^{(\varepsilon)}(+\infty)^{r_n} \cdot \mathsf{P}\left\{\sum_{l=1}^{r_k} \breve{\tau}_l^{(\varepsilon)} \leq t_k, k = 1, \ldots, n\right\}$$

$$= \mathsf{P}\{\breve{\nu}^{(\varepsilon)} \geq r_n\} \cdot \mathsf{P}\{\breve{\nu}^{(\varepsilon)}(t_k) \geq r_k, k = 1, \ldots, n\}$$

$$= \mathsf{P}\{\min(\breve{\nu}^{(\varepsilon)}, \breve{\nu}^{(\varepsilon)}(t_k)) \geq r_k, k = 1, \ldots, n\}, \tag{1.3.4}$$

where the random variable $\breve{\nu}^{(\varepsilon)}$ and the renewal process $\breve{\nu}^{(\varepsilon)}(t), t \geq 0$, are independent of the right-hand side of (1.3.4).

Denote by ρ_λ an exponentially distributed random variable with parameter $0 < \lambda < \infty$, i.e., $\mathsf{P}\{\rho_\lambda > t\} = e^{-\lambda t}, t \geq 0$. We also denote by ρ_0 an improper random variable that takes value $+\infty$ with probability 1 and by ρ_∞ a random variable that takes value 0 with probability 1.

Lemma 1.3.1. *Let conditions* $\mathbf{D_6}$, $\mathbf{M_6}$ *hold, and* $0 \leq t^{(\varepsilon)} \to \infty$ *as* $\varepsilon \to 0$ *such that condition* $\mathbf{B_1}$ *holds. Then*

$$\nu^{(\varepsilon)}(t^{(\varepsilon)})/t^{(\varepsilon)} \xrightarrow{\mathrm{d}} \min(\rho_\lambda, 1/\mu_1^{(0)}) \quad \text{as } \varepsilon \to 0. \tag{1.3.5}$$

Theorem 1.3.3. *Let conditions* $\mathbf{D_6}$, $\mathbf{M_6}$ *hold, and* $0 \leq t^{(\varepsilon)} \to \infty$ *as* $\varepsilon \to 0$ *such that condition* $\mathbf{B_1}$ *holds. Then, for all* $r = 1, 2, \ldots$, *the following relation holds:*

$$\mathsf{E}(\nu^{(\varepsilon)}(t^{(\varepsilon)})/t^{(\varepsilon)})^r \to \mathsf{E}(\min(\rho_\lambda, 1/\mu_1^{(0)}))^r \quad \text{as } \varepsilon \to 0. \tag{1.3.6}$$

Theorem 1.3.4. *Let conditions* $\mathbf{D_6}$, $\mathbf{M_6}$ *hold, and* $0 \leq t^{(\varepsilon)} \to \infty$ *as* $\varepsilon \to 0$ *such that condition* $\mathbf{B_1}$ *holds. Then*

$$U^{(\varepsilon)}(t^{(\varepsilon)})/t^{(\varepsilon)} \to \mathsf{E}\min(\rho_\lambda, 1/\mu_1^{(0)}) = \frac{1}{\lambda}(1 - e^{-\lambda/\mu_1^{(0)}}) \ \textit{as} \ \varepsilon \to 0. \qquad (1.3.7)$$

Proof of Lemma 1.3.1 and Theorems 1.3.3 and 1.3.4. It is obvious that under conditions $\mathbf{B_1}$,

$$\breve{\nu}^{(\varepsilon)}/t^{(\varepsilon)} \xrightarrow{\ \mathrm{d}\ } \rho_\lambda \ \text{as} \ \varepsilon \to 0. \qquad (1.3.8)$$

Also, conditions $\mathbf{D_6}$ and $\mathbf{B_1}$ imply that condition $\mathbf{D_2}$ holds for the distribution functions $\bar{F}^{(\varepsilon)}(t)$ with the same limiting mean $\mu_1^{(0)}$. It then follows from Lemma 1.2.3 that

$$\breve{\nu}^{(\varepsilon)}(t^{(\varepsilon)})/t^{(\varepsilon)} \xrightarrow{\ \mathrm{P}\ } 1/\mu_1^{(0)} \ \text{as} \ \varepsilon \to 0. \qquad (1.3.9)$$

As is known, if $\xi^{(\varepsilon)} \xrightarrow{\ \mathrm{d}\ } \xi^{(0)}$ as $\varepsilon \to 0$ and $\eta^{(\varepsilon)} \xrightarrow{\ \mathrm{P}\ } \eta^{(0)}$ as $\varepsilon \to 0$, where $\eta^{(0)} =$ const with probability 1, then random vectors $(\xi^{(\varepsilon)}, \eta^{(\varepsilon)}) \xrightarrow{\ \mathrm{d}\ } (\xi^{(0)}, \eta^{(0)})$ as $\varepsilon \to 0$. Moreover, for any continuous function $g(x, y)$, random variables $g(\xi^{(\varepsilon)}, \eta^{(\varepsilon)}) \xrightarrow{\ \mathrm{d}\ } g(\xi^{(0)}, \eta^{(0)})$ as $\varepsilon \to 0$.

Using this remark and taking into consideration the representation (1.3.4) we get from (1.3.8) and (1.3.9) that

$$\nu^{(\varepsilon)}(t^{(\varepsilon)})/t^{(\varepsilon)} \xrightarrow{\ \mathrm{d}\ } \min(\rho_\lambda, 1/\mu_1^{(0)}) \ \text{as} \ \varepsilon \to 0. \qquad (1.3.10)$$

Since condition $\mathbf{D_2}$ holds for the distribution functions $\bar{F}^{(\varepsilon)}(t)$, condition $\mathbf{D_4}$ also holds, and it follows from Lemma 1.1.1 and Remark 1.1.1 that, for every $r \geq 1$,

$$\varlimsup_{\varepsilon \to 0} \mathsf{E}(\nu^{(\varepsilon)}(t^{(\varepsilon)})/t^{(\varepsilon)})^r \leq \varlimsup_{\varepsilon \to 0} \mathsf{E}(\breve{\nu}^{(\varepsilon)}(t^{(\varepsilon)})/t^{(\varepsilon)})^r < \infty. \qquad (1.3.11)$$

Relation (1.3.6) easily follows from (1.3.10) and (1.3.11). As far as relation (1.3.7) is concerned, it is a particular case of (1.3.6), since $U^{(\varepsilon)}(t) = \mathsf{E}\nu^{(\varepsilon)}(t) + 1$. $\qquad \square$

Now, we formulate and prove analogues of Theorems 1.2.8 and 1.2.9.

Theorem 1.3.5. *Let conditions* $\mathbf{D_6}$, $\mathbf{M_6}$ *hold, and* $0 \leq t^{(\varepsilon)} \to \infty$ *as* $\varepsilon \to 0$ *such that condition* $\mathbf{B_1}$ *holds. Then for all* $h \geq 0$ *and* $n = 1, 2, \ldots$, *the following relation holds:*

$$\mathsf{P}\{\nu^{(\varepsilon)}(t^{(\varepsilon)} + h) - \nu^{(\varepsilon)}(t^{(\varepsilon)}) \geq n\} \to e^{-\lambda/m_1^{(0)}}(1 - H_h^{(0)}(n)) \ \textit{as} \ \varepsilon \to 0. \qquad (1.3.12)$$

Proof. It follows from representation (1.3.7) that for any $t, h \geq 0$ and $n = 1, 2, \ldots$,

$$\mathsf{P}\{v^{(\varepsilon)}(t+h) - v^{(\varepsilon)}(t) \geq n\}$$
$$= \mathsf{P}\{\min(\check{v}^{(\varepsilon)}, \check{v}^{(\varepsilon)}(t+h)) - \min(\check{v}^{(\varepsilon)}, \check{v}^{(\varepsilon)}(t)) \geq n\}, \qquad (1.3.13)$$

and hence for any $t, h \geq 0$ and $n = 1, 2, \ldots$,

$$\mathsf{P}\{v^{(\varepsilon)}(t+h) - v^{(\varepsilon)}(t) \geq n\}$$
$$= \mathsf{P}\{\check{v}^{(\varepsilon)} > \check{v}^{(\varepsilon)}(t+h), \check{v}^{(\varepsilon)}(t+h) - \check{v}^{(\varepsilon)}(t) \geq n\}$$
$$+ \mathsf{P}\{\check{v}^{(\varepsilon)}(t+h) \geq \check{v}^{(\varepsilon)} > \check{v}^{(\varepsilon)}(t), \check{v}^{(\varepsilon)} - \check{v}^{(\varepsilon)}(t) \geq n\}. \qquad (1.3.14)$$

It is obvious that condition $\mathbf{B_1}$ holds for the functions $t^{(\varepsilon)} + h$ if it holds for the functions $t^{(\varepsilon)}$. Hence, similarly to (1.3.9) we have that, for any $h \geq 0$,

$$\check{v}^{(\varepsilon)}(t^{(\varepsilon)} + h)/t^{(\varepsilon)} \xrightarrow{\mathsf{P}} 1/\mu_1^{(0)} \quad \text{as } \varepsilon \to 0. \qquad (1.3.15)$$

From (1.3.15) and (1.3.8) and (1.3.9) it follows that for any $h \geq 0$,

$$(\check{v}^{(\varepsilon)}/t^{(\varepsilon)}, \check{v}^{(\varepsilon)}(t^{(\varepsilon)})/t^{(\varepsilon)}, \check{v}^{(\varepsilon)}(t^{(\varepsilon)} + h)/t^{(\varepsilon)})$$
$$\xrightarrow{\mathrm{d}} (\rho_\lambda, 1/m_1^{(0)}, 1/m_1^{(0)}) \quad \text{as } \varepsilon \to 0. \qquad (1.3.16)$$

Now, (1.3.16) yields for any $h \geq 0$,

$$\mathsf{P}\{\check{v}^{(\varepsilon)}(t^{(\varepsilon)} + h) \geq \check{v}^{(\varepsilon)} > \check{v}^{(\varepsilon)}(t^{(\varepsilon)})\}$$
$$= \mathsf{P}\{\check{v}^{(\varepsilon)}(t^{(\varepsilon)} + h)/t^{(\varepsilon)} \geq \check{v}^{(\varepsilon)}/t^{(\varepsilon)} > \check{v}^{(\varepsilon)}(t^{(\varepsilon)})/t^{(\varepsilon)}\}$$
$$\to 0 \quad \text{as } \varepsilon \to 0. \qquad (1.3.17)$$

Theorem 1.2.8 can be applied to the processes $\check{v}^{(\varepsilon)}(t^{(\varepsilon)})$, since conditions $\mathbf{D_6}$, $\mathbf{M_6}$ and $\mathbf{B_1}$ imply that conditions $\mathbf{D_2}$ and $\mathbf{M_2}$ hold for the distribution functions $\bar{F}^{(\varepsilon)}(t)$. Since $\check{v}^{(\varepsilon)}$ and $\check{v}^{(\varepsilon)}(t^{(\varepsilon)} + h) - \check{v}^{(\varepsilon)}(t^{(\varepsilon)})$ are independent, we get using this remark and (1.3.8) that, for any $h \geq 0$,

$$(\check{v}^{(\varepsilon)}/t^{(\varepsilon)}, \check{v}^{(\varepsilon)}(t^{(\varepsilon)} + h) - \check{v}^{(\varepsilon)}(t^{(\varepsilon)})) \xrightarrow{\mathrm{d}} (\rho_\lambda, \tilde{v}^{(0)}(h)) \quad \text{as } \varepsilon \to 0, \qquad (1.3.18)$$

where the random variables ρ_λ and $\tilde{v}^{(0)}(h)$ are independent and the random variable $\tilde{v}^{(0)}(h)$ is defined in Subsection 1.2.9.

According to the definition of the random variables $\tilde{v}^{(0)}(h)$,

$$\mathsf{P}\{\tilde{v}^{(0)}(h) \geq n\} = 1 - H_h^{(0)}(n), \quad n = 1, 2, \ldots .$$

It follows from (1.3.18) and (1.3.15) that for any $h \geq 0$,

$$(\breve{v}^{(\varepsilon)}/t^{(\varepsilon)} - \breve{v}^{(\varepsilon)}(t^{(\varepsilon)} + h)/t^{(\varepsilon)}, \, \breve{v}^{(\varepsilon)}(t^{(\varepsilon)} + h) - \breve{v}^{(\varepsilon)}(t^{(\varepsilon)}))$$

$$\xrightarrow{\text{d}} (\rho_\lambda - 1/m_1^{(0)}, \tilde{v}^{(0)}(h)) \quad \text{as} \quad \varepsilon \to 0, \tag{1.3.19}$$

where the random variables ρ_λ and $\tilde{v}^{(0)}(h)$ are independent.

Since the distribution function of $\rho_\lambda - 1/m_1^{(0)}$ is continuous and $\tilde{v}^{(0)}(h)$ has a discrete distribution function, from (1.3.19) we obtain for any $h \geq 0$ and $n = 1, 2, \ldots$,

$$\mathsf{P}\{\breve{v}^{(\varepsilon)} - \breve{v}^{(\varepsilon)}(t^{(\varepsilon)} + h) > 0, \, \breve{v}^{(\varepsilon)}(t^{(\varepsilon)} + h) - \breve{v}^{(\varepsilon)}(t^{(\varepsilon)}) \geq n\}$$

$$\to \mathsf{P}\{\rho_\lambda - 1/m_1^{(0)} > 0\}(1 - H_h^{(0)}(n))$$

$$= \exp\{-\lambda/m_1^{(0)}\}(1 - H_h^{(0)}(n)) \quad \text{as} \quad \varepsilon \to 0. \tag{1.3.20}$$

Relation (1.3.12) follows from (1.3.14), (1.3.17) and (1.3.20). $\qquad\qquad\square$

Define the random variables

$$\gamma_h = \tilde{v}^{(0)}(h)\chi(\rho_\lambda \geq 1/m_1^{(0)}),$$

where ρ_λ and $\tilde{v}^{(0)}(h)$ are independent random variables, and $\tilde{v}^{(0)}(h)$ is the number of renewals in the stationary renewal process defined in Subsection 1.2.10.

By definition,

$$\gamma_h = \begin{cases} 0 & \text{with probability } 1 - \exp\{-\lambda/m_1^{(0)}\}[1 - H_h^{(0)}(1)], \\ n & \text{with probability } \exp\{-\lambda/m_1^{(0)}\}(H_h^{(0)}(n) - H_h^{(0)}(n + 1)), \quad n \geq 1. \end{cases}$$

It also follows from the definition of the random variables γ_h that

$$\mathsf{E}\gamma_h^r = \exp\{-\lambda/m_1^{(0)}\}\mathsf{E}(\tilde{v}^{(0)}(h))^r, \quad r = 1, 2, \ldots, \tag{1.3.21}$$

and, in particular,

$$\mathsf{E}\gamma_h = \exp\{-\lambda/m_1^{(0)}\}h/m_1^{(0)}. \tag{1.3.22}$$

Theorem 1.3.6. *Let conditions* $\mathbf{D_6}$, $\mathbf{M_6}$ *hold, and* $0 \leq t^{(\varepsilon)} \to \infty$ *as* $\varepsilon \to 0$ *such that condition* $\mathbf{B_1}$ *holds. Then for all* $h \geq 0$ *and* $r = 1, 2, \ldots$, *the following relation holds:*

$$\mathsf{E}(v^{(\varepsilon)}(t^{(\varepsilon)} + h) - v^{(\varepsilon)}(t^{(\varepsilon)})^r \to e^{-\lambda/m_1^{(0)}}\mathsf{E}(\tilde{v}^{(0)}(h))^r \quad \text{as} \quad \varepsilon \to 0. \tag{1.3.23}$$

Proof. Obviously for all $t, h \geq 0$,

$$v^{(\varepsilon)}(t + h) - v^{(\varepsilon)}(t) = \min(\breve{v}^{(\varepsilon)}, \breve{v}^{(\varepsilon)}(t + h)) - \min(\breve{v}^{(\varepsilon)}, \breve{v}^{(\varepsilon)}(t))$$

$$\leq \breve{v}^{(\varepsilon)}(t + h) - \breve{v}^{(\varepsilon)}(t). \tag{1.3.24}$$

Theorem 1.2.9 can be applied to the processes $\check{v}^{(\varepsilon)}(t^{(\varepsilon)})$, since $\mathbf{D_6}$ and $\mathbf{B_1}$ imply that condition $\mathbf{D_2}$ holds. Using (1.3.24) and the above we have for all $t, h \geq 0$ and $r = 1, 2, \ldots$ that

$$\varlimsup_{\varepsilon \to 0} \mathsf{E}(v^{(\varepsilon)}(t + h) - v^{(\varepsilon)}(t))^r \leq \varlimsup_{\varepsilon \to 0} \mathsf{E}(\check{v}^{(\varepsilon)}(t + h) - \check{v}^{(\varepsilon)}(t))^r < \infty. \qquad (1.3.25)$$

Relation (1.3.23) follows in an obvious way from relation (1.3.12) given in Theorem 1.3.5 and from (1.3.25). $\qquad \square$

1.3.3   Proof of Theorems 1.3.1 and 1.3.2

The renewal function generated by the distribution function $F^{(\varepsilon)}(t)$ is connected with the corresponding renewal function via the same relation as in the model of a proper renewal equation,

$$U^{(\varepsilon)}(t) = \mathsf{E}v^{(\varepsilon)}(t) + 1, \quad t \geq 0. \qquad (1.3.26)$$

Using relation (1.3.26) and formula (1.3.22) we can conclude that Theorem 1.3.2 is a particular case of Theorem 1.3.6 if $r = 1$.

A unique solution of the renewal equation is given by the formulas

$$\begin{aligned}
x^{(\varepsilon)}(t) &= \int_0^t q^{(\varepsilon)}(t - s)U^{(\varepsilon)}(ds) \\
&= \int_0^t q^{(\varepsilon)}(v)U^{(\varepsilon)}(t - dv), \quad t \geq 0.
\end{aligned} \qquad (1.3.27)$$

The proof of Theorem 1.3.1 is same as that given for Theorem 1.2.1 in the previous section. Theorem 1.3.2 and condition $\mathbf{F_2}$ permit to apply Lemma 1.2.2 to the functions $q^{(\varepsilon)}(v)$ and the measures $\mu^{(\varepsilon)}(A) = U^{(\varepsilon)}(t^{(\varepsilon)} - A \cap [0, T])$ for any $T < 0$, which gives the following relation:

$$\int_0^T q^{(\varepsilon)}(v)U^{(\varepsilon)}(t^{(\varepsilon)} - dv) \to \frac{\exp\{-c/m_1^{(0)}\}}{m_1^{(0)}} \int_0^T q^{(0)}(s)m(dv) \quad \text{as } \varepsilon \to 0. \qquad (1.3.28)$$

Denote by $\bar{U}^{(\varepsilon)}(t)$ the renewal function generated by the proper distribution function $\bar{F}^{(\varepsilon)}(t)$. It follows from (1.3.24) and (1.3.26) that for any $t, h \geq 0$,

$$U^{(\varepsilon)}(t + h) - U^{(\varepsilon)}(t) \leq \bar{U}^{(\varepsilon)}(t + h) - \bar{U}^{(\varepsilon)}(t) \qquad (1.3.29)$$

and, therefore,

$$\int_T^{t^{(\varepsilon)}} |q^{(\varepsilon)}(v)|U^{(\varepsilon)}(t^{(\varepsilon)} - dv) \leq \int_T^{t^{(\varepsilon)}} |q^{(\varepsilon)}(v)|\bar{U}^{(\varepsilon)}(t^{(\varepsilon)} - dv). \qquad (1.3.30)$$

Since conditions $\mathbf{D_6}$, $\mathbf{M_6}$ and $\mathbf{B_1}$ imply that conditions $\mathbf{D_2}$ and $\mathbf{M_2}$ hold for the distribution functions $\bar{F}^{(\varepsilon)}(t)$, we can apply Lemma 1.1.3 to the renewal functions $\bar{U}^{(\varepsilon)}(t)$ and get, for any interval $I = [a, b]$ of length $h = b - a$, that

$$\overline{\lim_{\varepsilon \to 0}} \sup_{t \geq 0} \bar{U}^{(\varepsilon')}(t + I) \leq K_1 + K_2 h < \infty, \tag{1.3.31}$$

where $K_1, K_2 < \infty$.

Using (1.3.30), (1.3.31), and condition $\mathbf{F_2}$ we find for $h > 0$ that

$$\overline{\lim_{\varepsilon \to 0}} \int_T^{t^{(\varepsilon)}} |q^{(\varepsilon)}(v)| U^{(\varepsilon)}(t^{(\varepsilon)} - dv)$$

$$\leq \overline{\lim_{\varepsilon \to 0}} \int_T^{t^{(\varepsilon)}} |q^{(\varepsilon)}(v)| \bar{U}^{(\varepsilon)}(t^{(\varepsilon)} - dv)$$

$$\leq \overline{\lim_{\varepsilon \to 0}} \sum_{r \geq [T/h]} \sup_{rh \leq t \leq (r+1)h} |q^{(\varepsilon)}(t)|$$

$$\times (\bar{U}^{(\varepsilon')}(t^{(\varepsilon')} - rh) - \bar{U}^{(\varepsilon')}(t^{(\varepsilon')} - (r+1)h))$$

$$\leq \frac{K_1 + K_2 h}{h} \cdot \overline{\lim_{\varepsilon \to 0}} h \sum_{r \geq \frac{T-h}{h}} \sup_{rh \leq t \leq (r+1)h} |q^{(\varepsilon')}(t)| \to 0 \text{ as } T \to 0. \tag{1.3.32}$$

From relations (1.3.28) and (1.3.32) it follows that

$$\lim_{\varepsilon \to 0} x^{(\varepsilon)}(t^{(\varepsilon)}) = \lim_{\varepsilon \to 0} \int_0^{t^{(\varepsilon')}} q^{(\varepsilon)}(t^{(\varepsilon)} - s) U^{(\varepsilon)}(ds)$$

$$= \lim_{T \to \infty} \lim_{\varepsilon \to 0} \int_0^T q^{(\varepsilon)}(t^{(\varepsilon)} - s) U^{(\varepsilon)}(ds)$$

$$= \lim_{T \to \infty} \frac{\exp\{-c/m_1^{(0)}\}}{m_1^{(0)}} \int_0^T q^{(0)}(v) m(dv)$$

$$= \frac{\exp\{-c/m_1^{(0)}\}}{m_1^{(0)}} \int_0^\infty q^{(0)}(v) m(dv). \tag{1.3.33}$$

This completes the proof. $\square$

1.4   Perturbed renewal equation under Cramér type conditions

In this section, we continue studies of the asymptotic behaviour of the solution $x^{(\varepsilon)}(t^{(\varepsilon)})$ of a perturbed renewal equation in the model with the distribution functions

$F^{(\varepsilon)}(t)$ that generate the renewal equation (1.2.1) that may be improper for $\varepsilon > 0$, i.e., $F^{(\varepsilon)}(\infty) \leq 1$ for $\varepsilon > 0$. In the limiting case $\varepsilon = 0$, we consider both cases when either $F^{(0)}(\infty) = 1$ or $F^{(0)}(\infty) < 1$. However, we additionally impose a Cramér type condition on the distributions $F^{(\varepsilon)}(t)$. As we shall see, this additional assumption allows to omit the restriction on the rate of growth of the time $t^{(\varepsilon)}$ imposed by the balancing condition $\mathbf{B_1}$. We also consider a model in which the perturbed renewal equations are generated by some finite measures $F^{(\varepsilon)}(dt)$ that weakly converge to some limiting finite measure $F^{(0)}(dt)$ with the total variation less, equal or greater than 1.

1.4.1 Perturbed renewal equations generated by asymptotically proper distribution functions

We consider the case when condition $\mathbf{D_6}$ holds, i.e., $F^{(\varepsilon)}(\cdot) \Rightarrow F^{(0)}(\cdot)$ as $\varepsilon \to 0$, where $F^{(0)}(s)$ is a proper non-arithmetic distribution function. As was mentioned in Subsection 1.3.1, this condition also contains the assumption $F^{(0)}(\infty) = 1$ that automatically implies that $F^{(\varepsilon)}(\infty) \to F^{(0)}(\infty) = 1$ as $\varepsilon \to 0$ or, which is the same, that $f^{(\varepsilon)} = 1 - F^{(\varepsilon)}(\infty) \to f^{(0)} = 0$ as $\varepsilon \to 0$.

Now, we assume the following *Cramér type condition*:

$\mathbf{C_1}$: There exists $\delta > 0$ such that $\overline{\lim}_{0 \leq \varepsilon \to 0} \int_0^\infty e^{\delta s} F^{(\varepsilon)}(ds) < \infty$.

Obviously, conditions $\mathbf{D_6}$ and $\mathbf{C_1}$ imply condition $\mathbf{M_6}$, i.e.,

$$m_1^{(\varepsilon)} \to m_1^{(0)} < \infty \quad \text{as} \quad \varepsilon \to 0. \tag{1.4.1}$$

Let us introduce the moment generating function,

$$\phi^{(\varepsilon)}(\rho) = \int_0^\infty e^{\rho s} F^{(\varepsilon)}(ds), \quad \rho \geq 0.$$

By the definition, $\phi^{(\varepsilon)}(\rho)$ is a nonnegative and nondecreasing function. From condition $\mathbf{C_1}$ we have that $\phi^{(\varepsilon)}(\delta) < \infty$ for ε small enough, say $\varepsilon \leq \varepsilon_1$.

The following characteristic equation plays an important role in what follows:

$$\phi^{(\varepsilon)}(\rho) = 1. \tag{1.4.2}$$

Lemma 1.4.1. *Let conditions $\mathbf{D_6}$ and $\mathbf{C_1}$ hold. Then, for any $\beta \in (0, \delta]$, there exists $\varepsilon_0 = \varepsilon_0(\beta) > 0$ (defined below in formula (1.4.4)) such that, for every $0 \leq \varepsilon \leq \varepsilon_0$: ($\alpha$) $\phi^{(\varepsilon)}(\beta) \in (1, \infty)$; ($\beta$) there exists a unique nonnegative root $\rho^{(\varepsilon)}$ of equation (1.4.2); (γ) $\rho^{(\varepsilon)} < \beta$; ($\delta$) $\rho^{(\varepsilon)} \to \rho^{(0)} = 0$ as $\varepsilon \to 0$.*

Proof. By condition $\mathbf{D_6}$, the distribution function $F^{(\varepsilon)}(s)$ is not concentrated in the origin, i.e., $F^{(\varepsilon)}(0) < 1$ for ε small enough, say $\varepsilon \leq \varepsilon_2$.

It is obvious that, for $\varepsilon \leq \min(\varepsilon_1, \varepsilon_2)$, the function $\phi_\varepsilon(\rho)$ is nonnegative, strongly increasing, and continuous on the interval $[0, \delta]$. Note also that $\phi^{(\varepsilon)}(0) = F^{(\varepsilon)}(\infty) \leq 1$.

Since the distribution function $F^{(0)}(s)$ is proper and is not concentrated in 0, equation (1.4.2) for $\varepsilon = 0$ has the unique root $\rho^{(0)} = 0$, i.e., $\phi^{(0)}(0) = 1$, and $\phi^{(0)}(\rho) > 1$ for any $\rho \in (0, \delta]$.

Now, using condition $\mathbf{D_6}$, we have

$$\lim_{\varepsilon \to 0} \phi_\varepsilon(\delta) \geq \lim_{T \to \infty} \lim_{\varepsilon \to \infty} \int_0^T e^{s\delta} F_\varepsilon(ds)$$

$$= \lim_{T \to \infty} \int_0^T e^{s\delta} F_0(ds) = \phi_0(\delta) > 1. \tag{1.4.3}$$

It follows from (1.4.3) that, for any $\beta \in (0, \delta]$, we have $\phi_\varepsilon(\beta) > 1$ for ε small enough, say $\varepsilon \leq \varepsilon_3(\beta)$.

Let us now define

$$\varepsilon_0 = \varepsilon_0(\beta) = \min(\varepsilon_1, \varepsilon_2, \varepsilon_3(\beta)). \tag{1.4.4}$$

It follows from the remarks above that there exists a unique nonnegative root of equation (1.4.2) for $\varepsilon \leq \varepsilon_0$.

This solution satisfies $\rho_\varepsilon \leq \beta$. Since β can be any positive number less or equal δ, we also see that the relation $\rho_\varepsilon \to 0$ holds as $\varepsilon \to 0$. $\square$

Remark 1.4.1. It is worth noting that the assumption that the limiting distribution function $F_0(s)$ is non-arithmetic was not used in the proof of Lemma 1.4.1. This assumption can be replaced in condition $\mathbf{D_6}$ with a weaker assumption, actually used in the proof, that the limiting distribution function $F_0(s)$ is not concentrated in zero.

Now we assume the following condition:

$\mathbf{F_5}$:   (a) $\lim_{u \to 0} \overline{\lim}_{0 \leq \varepsilon \to 0} \sup_{|v| \leq u} |q^{(\varepsilon)}(t + v) - q^{(0)}(t)| = 0$ almost everywhere with respect to the Lebesgue measure on $[0, \infty)$;

    (b) $\overline{\lim}_{0 \leq \varepsilon \to 0} \sup_{0 \leq t \leq T} |q^{(\varepsilon)}(t)| < \infty$ for every $T \geq 0$;

    (c) $\lim_{T \to \infty} \overline{\lim}_{0 \leq \varepsilon \to 0} h \sum_{r \geq T/h} \sup_{rh \leq t \leq (r+1)h} e^{\gamma t} |q^{(\varepsilon)}(t)| = 0$ for some $h > 0$ and some $\gamma > 0$.

Note that condition $\mathbf{F_5}$ can be formulated in the following equivalent form:

$\mathbf{F_5'}$: There exists $\gamma > 0$ such that the function $e^{\gamma t} q^{(\varepsilon)}(t), t \geq 0$ satisfies condition $\mathbf{F_2}$.

Theorem 1.4.1. *Let conditions $\mathbf{D_6}$, $\mathbf{C_1}$, and $\mathbf{F_5}$ hold. Then, for any $0 \leq t^{(\varepsilon)} \to \infty$ as $\varepsilon \to 0$,*

$$\frac{x^{(\varepsilon)}(t^{(\varepsilon)})}{\exp\{-\rho^{(\varepsilon)} t^{(\varepsilon)}\}} \to \frac{\int_0^\infty q^{(0)}(s)\, ds}{m_1^{(0)}} \quad as \ \varepsilon \to 0. \tag{1.4.5}$$

Proof. We use the exponential transformation of renewal equation (1.2.1). Denote

$$\tilde{F}^{(\varepsilon)}(t) = \int_0^t e^{\rho^{(\varepsilon)}t} F^{(\varepsilon)}(ds), \quad \tilde{q}^{(\varepsilon)}(t) = e^{\rho^{(\varepsilon)}t} q^{(\varepsilon)}(t), \quad \tilde{x}^{(\varepsilon)}(t) = e^{\rho^{(\varepsilon)}t} x^{(\varepsilon)}(t), \ t \geq 0.$$

Note first of all that, since $\rho^{(\varepsilon)}$ is a root of equation (1.4.2) by the definition, it implies that $\tilde{F}^{(\varepsilon)}(\infty) = 1$, i.e., $\tilde{F}^{(\varepsilon)}(t)$ is a proper distribution function.

Since $x^{(\varepsilon)}(t)$ is a solution of the renewal equation (1.2.1), the function $\tilde{x}^{(\varepsilon)}(t)$ is a solution of the renewal equation

$$\tilde{x}^{(\varepsilon)}(t) = \tilde{q}^{(\varepsilon)}(t) + \int_0^t \tilde{x}^{(\varepsilon)}(t - s) \tilde{F}^{(\varepsilon)}(ds), \quad t \geq 0. \tag{1.4.6}$$

Since $0 \leq \rho^{(\varepsilon)} \to 0$, conditions $\mathbf{D_6}$ and $\mathbf{C_1}$ imply in an obvious way that the distribution functions $\tilde{F}^{(\varepsilon)}(t)$ satisfy conditions $\mathbf{D_2}$ and $\mathbf{M_2}$. Moreover, for the corresponding limiting distribution, we have $\tilde{F}^{(0)}(t) \equiv F^{(0)}(t)$ and, therefore,

$$\tilde{m}_1^{(0)} = \int_0^\infty s \tilde{F}^{(0)}(ds) = m_1^{(0)} = \int_0^\infty s F^{(0)}(ds). \tag{1.4.7}$$

Also the relation $0 \leq \rho^{(\varepsilon)} \to 0$ and condition $\mathbf{F_5}$ imply that the functions $\tilde{q}^{(\varepsilon)}(t)$ satisfy condition $\mathbf{F_2}$. Moreover, for the corresponding limiting function, $\tilde{q}^{(0)}(t) \equiv q^{(0)}(t)$ and, therefore,

$$\int_0^\infty \tilde{q}^{(0)}(s)\, ds = \int_0^\infty q^{(0)}(s)\, ds. \tag{1.4.8}$$

By applying Theorem (1.2.1) to equation (1.4.6) and taking the remarks made above and relations (1.4.7) and (1.4.8) into consideration we get the relation

$$\tilde{x}^{(\varepsilon)}(t^{(\varepsilon)}) = e^{\rho^{(\varepsilon)}t^{(\varepsilon)}} x^{(\varepsilon)}(t^{(\varepsilon)}) \to \frac{\int_0^\infty q^{(0)}(s)\, ds}{m_1^{(0)}} \quad \text{as } \varepsilon \to 0, \tag{1.4.9}$$

which is equivalent to the relation (1.4.5). $\square$

As was mentioned above, the main difference between Theorem 1.3.1 and Theorem 1.4.1 is that no balancing condition $\mathbf{B_1}$ is assumed in the Theorem 1.4.1. This condition restricts the rate of growth of the time parameter $t^{(\varepsilon)}$ to ∞. However, to be able not to use this condition, the Cramér type condition $\mathbf{C_1}$ should be imposed on the distribution functions $F^{(\varepsilon)}(t)$.

We use the symbol $a^{(\varepsilon)} \sim b^{(\varepsilon)}$ as $\varepsilon \to 0$ to denote that $a^{(\varepsilon)}/b^{(\varepsilon)} \to 1$ as $\varepsilon \to 0$.

The following lemma provides additional information about asymptotics for the coefficients $\rho^{(\varepsilon)}$.

Lemma 1.4.2. *Let conditions* $\mathbf{D_6}$ *and* $\mathbf{C_1}$ *hold. Then, the following asymptotical relation holds*:

$$\rho^{(\varepsilon)} \sim f^{(\varepsilon)}/m_1^{(\varepsilon)} \quad as \quad \varepsilon \to 0. \tag{1.4.10}$$

Proof. It was proved in Lemma 1.4.1 that under conditions $\mathbf{D_6}$ and $\mathbf{C_1}$ there exists a unique nonnegative root $\rho^{(\varepsilon)}$ of equation (1.4.2) for all ε small enough, and that $\rho^{(\varepsilon)} \to 0$ as $\varepsilon \to 0$.

Let us use Taylor's expansion for the function e^t which can be given in the following form (taking onto consideration that this function and its derivatives are increasing)

$$e^{s\rho^{(\varepsilon)}} = 1 + s\rho^{(\varepsilon)} + s^2(\rho^{(\varepsilon)})^2 e^{s\rho^{(\varepsilon)}} \theta^{(\varepsilon)}(s)/2, \tag{1.4.11}$$

where $\theta^{(\varepsilon)}(s), s \geq 0$, is a continuous function such that

$$0 \leq \theta_\varepsilon(s) \leq 1, \quad s \geq 0. \tag{1.4.12}$$

As was shown in the proof of Lemma 1.4.1, for any $\beta \in (0, \delta]$, there exists $\varepsilon_0(\beta)$ such that $\rho_\varepsilon \leq \beta$ for $\varepsilon \leq \varepsilon_0(\beta)$.

By condition $\mathbf{C_1}$,

$$\varlimsup_{0 \leq \varepsilon \to 0} \int_0^\infty s^2 e^{\beta s} F^{(\varepsilon)}(ds) \leq \varlimsup_{0 \leq \varepsilon \to 0} c_2 \int_0^\infty e^{\delta s} F^{(\varepsilon)}(ds) < \infty, \tag{1.4.13}$$

where $c_2 = \sup_{s \geq 0} s^2 e^{-(\delta-\beta)s} < \infty$.

It follows from relation (1.4.13) that there exists $\varepsilon_4 = \varepsilon_4(\beta) > 0$ such that for $\varepsilon \leq \varepsilon_4$,

$$M = \frac{1}{2} \sup_{\varepsilon \leq \varepsilon_4} \int_0^\infty s^2 e^{s\beta} F^{(\varepsilon)}(ds) < \infty. \tag{1.4.14}$$

Define,

$$\varepsilon_5 = \varepsilon_5(\beta) = \min(\varepsilon_0(\beta), \varepsilon_4(\beta)). \tag{1.4.15}$$

By integrating in (1.4.11) and using (1.4.12) and (1.4.14), we get for $\varepsilon \leq \varepsilon_5$ that

$$\int_0^\infty e^{s\rho^{(c)}} F^{(\varepsilon)}(ds) = 1 - f^{(\varepsilon)} + m_1^{(\varepsilon)} \rho^{(\varepsilon)} + (\rho^{(\varepsilon)})^2 M \theta^{(\varepsilon)}, \tag{1.4.16}$$

where

$$\theta^{(\varepsilon)} = \frac{\int_0^\infty s^2 e^{s\rho^{(\varepsilon)}} \theta^{(\varepsilon)}(s) F^{(\varepsilon)}(ds)}{\sup_{\varepsilon \leq \varepsilon_4} \int_0^\infty s^2 e^{s\beta} F^{(\varepsilon)}(ds)} \in [0, 1]. \tag{1.4.17}$$

Therefore, equation (1.4.2) can be rewritten in the form

$$m_1^{(\varepsilon)}\rho^{(\varepsilon)} + (\rho^{(\varepsilon)})^2 M\theta^{(\varepsilon)} = f^{(\varepsilon)}, \tag{1.4.18}$$

or, equivalently,

$$\frac{m_1^{(\varepsilon)}\rho^{(\varepsilon)}}{f^{(\varepsilon)}}(1 + \rho^{(\varepsilon)} M\theta^{(\varepsilon)}/m_1^{(\varepsilon)}) = 1. \tag{1.4.19}$$

For the coefficients, we have $\rho^{(\varepsilon)} \to 0$ and $m_1^{(\varepsilon)} \to m_1^{(0)} \in (0, \infty)$ as $\varepsilon \to 0$. Using these relations we get from (1.4.19), by evaluating the corresponding limits, that

$$m_1^{(\varepsilon)}\rho^{(\varepsilon)}/f^{(\varepsilon)} \to 1 \quad \text{as} \quad \varepsilon \to 0, \tag{1.4.20}$$

which is equivalent to relation (1.4.10). $\square$

Remark 1.4.2. As in Lemma 1.4.1, the assumption that the limiting distribution function $F_0(s)$ is non-arithmetic can be replaced in condition $\mathbf{D_6}$ by the weaker assumption actually used in the proof that the limiting distribution function $F_0(s)$ is not concentrated in zero.

It follows from (1.4.10) that under the balancing condition $\mathbf{B_1}$, relation (1.4.5) in Theorem 1.4.1 is reduced to the relation (1.3.2) given in Theorem 1.3.1.

1.4.2   Perturbed renewal equations generated by finite measures

Consider now, for every $\varepsilon \geq 0$, the renewal equation

$$x^{(\varepsilon)}(t) = q^{(\varepsilon)}(t) + \int_0^t x^{(\varepsilon)}(t - s)F^{(\varepsilon)}(ds), \quad t \geq 0, \tag{1.4.21}$$

where **(a)** $q^{(\varepsilon)}(t)$ is a function in the class $\mathcal{L}$ of (Borel) measurable real-valued functions defined on $[0, \infty)$ and bounded on any finite interval, **(b)** $F^{(\varepsilon)}(ds)$ is a finite measure on $B_{(1)}^+$, the σ-algebra of Borel subsets of $[0, \infty)$, i.e., $F^{(\varepsilon)}([0, \infty)) = F^{(\varepsilon)}(\infty) < \infty$, and **(c)** the measure $F^{(\varepsilon)}(ds)$ is not concentrated in the origin, i.e., $F^{(\varepsilon)}((0, \infty)) > 0$.

As usual, we denote by $F^{(\varepsilon)}(t) = F^{(\varepsilon)}([0, t])$, $t \geq 0$, the corresponding distribution function generated by the measure $F^{(\varepsilon)}(ds)$, and

$$\bar{F}^{(\varepsilon)}(t) = F^{(\varepsilon)}(t)/F^{(\varepsilon)}(\infty), \quad t \geq 0, \quad f^{(\varepsilon)} = 1 - F^{(\varepsilon)}(\infty).$$

All general facts and definitions concerning renewal equations are translated to the case of the renewal equation (1.4.21) without any changes.

In particular, one can define the renewal function as $U^{(\varepsilon)}(t) = \sum_{r=0}^{\infty} F^{(\varepsilon)(*r)}(t)$, $t \geq 0$, where $F^{(\varepsilon)(*r)}(t)$ is the r-fold convolution of the distribution function $F^{(\varepsilon)}(x)$.

The assumption $F^{(\varepsilon)}((0, \infty)) > 0$ implies that $U^{(\varepsilon)}(t) \leq K_1 + K_2 t$ for some finite constants K_1 and K_2, i.e., it has not greater than a linear rate of growth in t. Note also that by the definition, $U^{(\varepsilon)}(t)$ is a function that is nondecreasing and continuous from the right.

The renewal measure $U^{(\varepsilon)}(A)$ on the Borel σ-algebra on $[0, \infty)$ is generated by the renewal function in a usual way. This measure is given by its values on intervals by $U^{(\varepsilon)}((a, b]) = U^{(\varepsilon)}(b) - U^{(\varepsilon)}(a)$, $0 \leq a \leq b < \infty$; it also has an atom in the point zero. This atom appears due to the 0-fold convolution $F^{(\varepsilon)(*0)}(t)$. By the definition, this convolution is a distribution function that has a unit jump in zero.

The unique solution of the renewal equation in class $\mathfrak{L}$ is of the form

$$x^{(\varepsilon)}(t) = \int_0^t q^{(\varepsilon)}(t - s) U^{(\varepsilon)}(ds), \quad t \geq 0.$$

We assume the following perturbation condition:

D$_7$:   **(a)** $\bar{F}^{(\varepsilon)}(\cdot) \Rightarrow \bar{F}^{(0)}(\cdot)$ as $\varepsilon \to 0$, where $\bar{F}^{(0)}(t)$ is a proper non-arithmetic distribution function;

   (b) $f^{(\varepsilon)} \to f^{(0)} \in (-\infty, 1)$ as $\varepsilon \to 0$.

Let us consider the characteristic equation, (1.4.2)

$$\phi^{(\varepsilon)}(\rho) = 1.$$

If $f^{(\varepsilon)} < 0$ then there exists a unique root of this equation, $\rho^{(\varepsilon)} < 0$.

If $f^{(\varepsilon)} = 0$ then there exists a unique root of this equation, $\rho^{(\varepsilon)} = 0$.

If $f^{(\varepsilon)} > 0$, then a Cramér type condition, which requires that $1 < \phi^{(\varepsilon)}(\delta) < \infty$ for some $\delta > 0$, should be imposed on the distribution functions $F^{(\varepsilon)}(t)$ in order to guarantee the existence of a root of the equation (1.4.2). In this case there exists a unique root of this equation, $0 < \rho^{(\varepsilon)} < \delta$.

Denote

$$\mathfrak{N} = \{\varepsilon \geq 0 : f^{(\varepsilon)} > 0\}.$$

We are interested to ensure the existence of a root of the equation (1.4.2) for all ε small enough. Four cases have to be considered:

(i) $f^{(0)} > 0$;

(ii) $f^{(0)} = 0$ and 0 is a limit point in the set $\mathfrak{N}$, i.e., there exists a sequence $0 < \varepsilon_n \to 0$ as $n \to \infty$ such that $f^{(\varepsilon_n)} > 0$;

(iii) $f^{(0)} = 0$ and 0 is not a limit point in the set $\mathfrak{N}$, i.e., $f^{(\varepsilon)} \leq 0$ for all $\varepsilon > 0$ small enough;

(iv) $f^{(0)} < 0$.

In the cases **(iii)** and **(iv)**, no conditions in addition to **D$_7$** are required. In the case **(ii)**, condition **C$_1$** should be additionally imposed. In the case **(i)**, condition **C$_1$** should be replaced with the following stronger condition:

C_2:   There exists $\delta > 0$ such that:

    (a) $\overline{\lim}_{0 \le \varepsilon \to 0} \int_0^\infty e^{\delta s} F^{(\varepsilon)}(ds) < \infty$;

    (b) $\int_0^\infty e^{\delta s} F^{(0)}(ds) > 1$.

Note that condition C_2 **(a)** gives $\int_0^\infty e^{\delta s} F^{(0)}(ds) < \infty$, since this corresponds to the case $\varepsilon = 0$. If $F^{(0)}(s)$ is a proper distribution, then condition C_2 **(b)** automatically holds. However, if $F^{(0)}(s)$ is an improper distribution, then condition C_2 **(b)** plays an essential role.

Let us define $\delta_0 = \sup(\delta \ge 0 : \int_0^\infty e^{\delta s} F^{(0)}(ds) < \infty)$. If $\delta_0 = \infty$, then $\int_0^\infty e^{\delta s} F^{(0)}(ds) \to \infty$ as $\delta \to \infty$. Thus condition C_2 **(b)** holds. If $\delta_0 < \infty$ and $\int_0^\infty e^{\delta_0 s} F^{(0)}(ds) = \infty$, then $\int_0^\infty e^{\delta s} F^{(0)}(ds) \to \infty$ as $\delta < \delta_0, \delta \to \delta_0$. Condition C_2 **(b)** again holds. However, if $\delta_0 < \infty$ and $\int_0^\infty e^{\delta_0 s} F^{(0)}(ds) < \infty$, then condition C_2 **(b)** may not hold.

For example, let

$$F^{(0)}(ds) = \chi_{[1,\infty)}(t)(1 - f)c^{-1}s^{-2}e^{-s}\, ds,$$

where $c = \int_1^\infty s^{-2}e^{-s}\, ds$ and $f \in [0, 1)$. This distribution is improper and $F^{(0)}(\infty) = 1 - f$. It easy to check that, in this case, $\delta_0 = 1$ and $\int_0^\infty e^s F^{(0)}(ds) = \int_1^\infty (1 - f)c^{-1}s^{-2}\, ds = (1 - f)c^{-1}$. If $(1 - f)c^{-1} < 1$, then condition C_2 **(b)** does not hold.

Let us consider a condition weaker than D_7,

D_8:   **(a)** $F^{(\varepsilon)}(\cdot) \Rightarrow F^{(0)}(\cdot)$ as $\varepsilon \to 0$, where $F^{(0)}(t)$ is a distribution function of a finite measure not concentrated in the origin, i.e., such that $F^{(0)}(0) < F^{(0)}(\infty)$;

    (b) $f^{(\varepsilon)} \to f^{(0)} \in (-\infty, 1)$ as $\varepsilon \to 0$.

Lemma 1.4.3. *Let conditions D_8 holds. Then, the following assertions take place:*

(i) *If $f^{(0)} > 0$ and condition C_2 holds, then there exist $\beta \in (0, \delta]$ and $\varepsilon_0' = \varepsilon_0'(\beta) > 0$ (defined below in formula (1.4.24)) such that $\phi^{(\varepsilon)}(\beta) \in (1, \infty)$ and there exists a unique root $0 < \rho^{(\varepsilon)} \le \beta$ of equation (1.4.2) for every $\varepsilon \le \varepsilon_0'$.*

(ii) *If $f^{(0)} = 0$ and 0 is a limit point in the set $\mathfrak{N}$, then for any $\beta \in (0, \delta]$ there exists $\varepsilon_0'' = \varepsilon_0''(\beta) > 0$ (defined below in formula (1.4.26)) such that $\phi^{(\varepsilon)}(\beta) \in (1, \infty)$ and there exists a unique root $\rho^{(\varepsilon)}$ of equation (1.4.2) for every $\varepsilon \le \varepsilon_0''$, moreover, $0 < \rho^{(\varepsilon)} \le \beta$ for $\varepsilon \in [0, \varepsilon_0''] \cap \mathfrak{N}$, $\rho^{(\varepsilon)} \le 0$ for $\varepsilon \in [0, \varepsilon_0''] \cap \overline{\mathfrak{N}}$, and $\rho^{(0)} = 0$.*

(iii) *If $f^{(0)} = 0$ and 0 is not a limit point in the set $\mathfrak{N}$, then there exists $\varepsilon_0''' > 0$ (defined below in formula (1.4.27)) such that there exists a unique root $\rho^{(\varepsilon)} \le 0$ of equation (1.4.2) for every $\varepsilon \le \varepsilon_0'''$, and $\rho^{(0)} = 0$.*

(iv) *If $f^{(0)} < 0$, then there exists ε_0'''' (defined below in formula (1.4.28)) such that there exists a unique root $\rho^{(\varepsilon)} < 0$ of equation (1.4.2) for every $\varepsilon \leq \varepsilon_0''''$.*

Also,

$$\rho^{(\varepsilon)} \to \rho^{(0)} \quad \text{as} \;\; \varepsilon \to 0. \tag{1.4.22}$$

Proof. By condition $\mathbf{D_8}$ **(a)**, the distribution function $F^{(\varepsilon)}(s)$ is not concentrated in the origin for ε small enough, say $\varepsilon \leq \varepsilon_1$. For $\varepsilon \leq \varepsilon_1$, the function $\phi^{(\varepsilon)}(\rho)$ is strongly increasing and continuous on the interval $(-\infty, 0]$. Also, $\phi^{(\varepsilon)}(-\infty) = 0$ and $\phi^{(\varepsilon)}(0) = 1 - f^{(\varepsilon)}$.

Let us consider the case **(i)** with $f^{(0)} > 0$.

In these case, condition $\mathbf{C_2}$ **(a)** implies that $\phi^{(\varepsilon)}(\delta) < \infty$ for all ε small enough, say $\varepsilon \leq \varepsilon_2$. In this case, the function $\phi_\varepsilon(\rho)$ is non-negative, strongly increasing, and continuous on the interval $(-\infty, \delta]$ for every $\varepsilon \leq \min(\varepsilon_1, \varepsilon_2)$.

Also, condition $\mathbf{C_2}$ **(b)** implies that $\phi^{(0)}(\delta) > 1$.

Let us now choose some $\beta \in (0, \delta]$ such that $\phi^{(0)}(\beta) > 1$. Now, using condition $\mathbf{D_8}$, we have

$$\lim_{\varepsilon \to 0} \phi^{(\varepsilon)}(\beta) \geq \lim_{T \to \infty} \lim_{\varepsilon \to \infty} \int_0^T e^{s\beta} F_\varepsilon(ds)$$

$$= \lim_{T \to \infty} \int_0^T e^{s\beta} F_0(ds) = \phi_0(\beta) > 1. \tag{1.4.23}$$

It follows from (1.4.23) that for all ε small enough, say $\varepsilon \leq \varepsilon_3 = \varepsilon_3(\beta)$, we have $\phi_\varepsilon(\beta) > 1$. Also, $\phi_\varepsilon(0) = 1 - f^{(\varepsilon)} < 1$.

The remarks made above imply the existence of a unique positive root of equation (1.4.2) for $\varepsilon \leq \varepsilon_0'$ defined by the following formula:

$$\varepsilon_0' = \varepsilon_0'(\beta) = \min(\varepsilon_1, \varepsilon_2, \varepsilon_3(\beta)). \tag{1.4.24}$$

Note also that $0 < \rho^{(\varepsilon)} \leq \beta$ for $\varepsilon \leq \varepsilon_3$, since $\phi_\varepsilon(\beta) > 1$ and $\phi_\varepsilon(0) = 1 - f^{(\varepsilon)} < 1$ for such ε.

Let us now consider the case **(ii)** with $f^{(0)} = 0$ and 0 being a limit point in the set $\mathfrak{N}$. In this case, the proof is similar to the one given for Lemma 1.4.1.

In these case, condition $\mathbf{C_1}$ implies that $\phi^{(\varepsilon)}(\delta) < \infty$ for all ε small enough, say $\varepsilon \leq \varepsilon_4$. In this case, the function $\phi_\varepsilon(\rho)$ is non-negative, strongly increasing, and continuous on the interval $(-\infty, \delta]$ for every $\varepsilon \leq \min(\varepsilon_1, \varepsilon_4)$.

Because the distribution function $F^{(0)}(s)$ is proper and is not concentrated in 0, equation (1.4.2) for the case $\varepsilon = 0$ has a unique root $\rho^{(0)} = 0$, i.e., $\phi^{(0)}(0) = 1$ and $\phi^{(0)}(\rho) > 1$ for any $\rho \in (0, \delta]$.

Now, using condition $\mathbf{D_8}$ we have

$$\varliminf_{\varepsilon \to 0} \phi_\varepsilon(\delta) \geq \lim_{T \to \infty} \varlimsup_{\varepsilon \to \infty} \int_0^T e^{s\delta} F_\varepsilon(ds)$$

$$= \lim_{T \to \infty} \int_0^T e^{s\delta} F_0(ds) = \phi_0(\delta) > 1. \tag{1.4.25}$$

It follows from (1.4.25) that, for any $\beta \in (0, \delta]$, we have $\phi_\varepsilon(\beta) > 1$ for ε small enough, say $\varepsilon \leq \varepsilon_5(\beta)$.

The remarks made above imply the existence of a unique positive root of equation (1.4.2) for $\varepsilon \leq \varepsilon_0''$ defined by the following formula:

$$\varepsilon_0'' = \varepsilon_0''(\beta) = \min(\varepsilon_1, \varepsilon_4, \varepsilon_5(\beta)). \tag{1.4.26}$$

Note also that $0 < \rho^{(\varepsilon)} \leq \beta$ for $\varepsilon \in [0, \varepsilon_0''(\beta)] \cap \mathfrak{N}$, since $\phi_\varepsilon(\beta) > 1$ and $\phi_\varepsilon(0) = 1 - f^{(\varepsilon)} < 1$ for such ε. Also, $\rho^{(\varepsilon)} \leq 0$ for $\varepsilon \in [0, \varepsilon_0''(\beta)] \cap \overline{\mathfrak{N}}$, since $\phi_\varepsilon(0) = 1 - f^{(\varepsilon)} \geq 1$ for such ε.

Let us consider the case **(iii)** with $f^{(0)} = 0$ and 0 not being a limit point in the set $\mathfrak{N}$.

In this case, $\phi^{(\varepsilon)}(0) = 1 - f^{(\varepsilon)} \geq 1$ for ε small enough, say $\varepsilon \leq \varepsilon_6$. Therefore, there exists a unique non-positive root $\rho^{(\varepsilon)} \leq 0$ of equation (1.4.2) for $\varepsilon \leq \varepsilon_0'''$, where

$$\varepsilon_0''' = \min(\varepsilon_1, \varepsilon_6). \tag{1.4.27}$$

Finally, let us consider the case **(iii)** with $f^{(0)} < 0$.

By condition $\mathbf{D_8}$ **(b)**, $\phi^{(\varepsilon)}(0) = 1 - f^{(\varepsilon)} \to \phi^{(0)}(0) = 1 - f^{(0)} > 1$ and, therefore, $\phi^{(\varepsilon)}(0) = 1 - f^{(\varepsilon)} > 1$ for ε small enough, say $\varepsilon \leq \varepsilon_7$. This implies that there exists a unique negative root $\rho^{(\varepsilon)} < 0$ of equation (1.4.2) for $\varepsilon \leq \varepsilon_0''''$, where

$$\varepsilon_0'''' = \min(\varepsilon_1, \varepsilon_7). \tag{1.4.28}$$

So, the statement **(a)** of the lemma is proved.

To prove the statement **(b)** let us again consider four cases.

Let $f^{(0)} < 0$ or $f^{(0)} = 0$ but assume that there is no a limit point in the set $\mathfrak{N}$. In these cases, the root $\rho^{(\varepsilon)} \leq 0$ for all ε small enough. Therefore, $\mathbf{D_8}$ gives

$$\lim_{T \to \infty} \varlimsup_{\varepsilon \to 0} \int_T^\infty e^{s\rho^{(\varepsilon)}} F^{(\varepsilon)}(ds) \leq \lim_{T \to \infty} \varlimsup_{\varepsilon \to 0} \int_T^\infty F^{(\varepsilon)}(ds)$$

$$= \lim_{T \to \infty} \varlimsup_{\varepsilon \to 0} (F^{(\varepsilon)}(\infty) - F^{(\varepsilon)}(T))$$

$$\leq \varlimsup_{T \to \infty} (F^{(0)}(\infty) - F^{(0)}(T/2)) = 0. \tag{1.4.29}$$

Let $f^{(0)} > 0$ or $f^{(0)} = 0$ but no limit point exists in the set $\mathfrak{N}$. In this case, the fact that $\rho^{(\varepsilon)} < \beta < \delta$ for all ε small enough yields

$$\lim_{T \to \infty} \varlimsup_{\varepsilon \to 0} \int_T^\infty e^{s\rho^{(\varepsilon)}} F^{(\varepsilon)}(ds)$$

$$\leq \lim_{T \to \infty} \varlimsup_{\varepsilon \to 0} \int_T^\infty e^{s\beta} F^{(\varepsilon)}(ds)$$

$$\leq \lim_{T \to \infty} e^{-(\delta-\beta)T} \varlimsup_{\varepsilon \to 0} \int_T^\infty e^{s\delta} F^{(\varepsilon)}(ds) = 0. \qquad (1.4.30)$$

Assume that there exists a subsequence $\varepsilon_k \to 0$ as $k \to \infty$ such that $\rho^{(\varepsilon_k)} \leq \rho^{(0)} - \gamma$, $k = 1, 2, \ldots$, for some $\gamma > 0$. Then, using $\mathbf{D_8}$, we have

$$\lim_{T \to \infty} \varlimsup_{k \to \infty} \int_0^T e^{s\rho^{(\varepsilon_k)}} F^{(\varepsilon_k)}(ds)$$

$$\leq \lim_{T \to \infty} \varlimsup_{k \to \infty} \int_0^T e^{s(\rho^{(0)}-\gamma)} F^{(\varepsilon_k)}(ds)$$

$$\leq \lim_{T \to \infty} \int_0^T e^{s(\rho^{(0)}-\gamma)} F^{(0)}(ds) = \phi^{(0)}(\rho^{(0)} - \gamma) < 1. \qquad (1.4.31)$$

It follows from (1.4.29) or (1.4.30) and (1.4.31) that

$$1 = \varlimsup_{k \to \infty} \phi^{(\varepsilon_k)}(\rho^{(\varepsilon_k)})$$

$$= \lim_{T \to \infty} \varlimsup_{k \to \infty} \left(\int_0^T e^{s\rho^{(\varepsilon_k)}} F^{(\varepsilon_k)}(ds) + \int_T^\infty e^{s\rho^{(\varepsilon_k)}} F^{(\varepsilon_k)}(ds) \right) < 1. \qquad (1.4.32)$$

This contradiction shows that no subsequence ε_k exists, and so

$$\varliminf_{\varepsilon \to 0} \rho^{(\varepsilon)} \geq \rho^{(0)}. \qquad (1.4.33)$$

Assume now that there exists a subsequence $\varepsilon_k \to 0$ as $k \to \infty$ such that $\rho^{(\varepsilon_k)} \geq \rho^{(0)} + \gamma$, $k = 1, 2, \ldots$, for some $\gamma > 0$. Then, using $\mathbf{D_8}$ we have

$$1 \geq \lim_{T \to \infty} \varliminf_{k \to \infty} \int_0^T e^{s\rho^{(\varepsilon_k)}} F^{(\varepsilon_k)}(ds)$$

$$\geq \lim_{T \to \infty} \varliminf_{k \to \infty} \int_0^T e^{s(\rho^{(0)}+\gamma)} F^{(\varepsilon_k)}(ds)$$

$$\geq \lim_{T \to \infty} \int_0^T e^{s(\rho^{(0)}+\gamma)} F^{(0)}(ds) = \phi^{(0)}(\rho^{(0)} + \gamma) > 1. \qquad (1.4.34)$$

This contradiction implies that

$$\varlimsup_{\varepsilon \to 0} \rho^{(\varepsilon)} \leq \rho^{(0)}. \qquad (1.4.35)$$

The proof of Lemma 1.4.3 is complete. $\qquad \qquad \square$

Let us introduce the condition

D$_9$: $F^{(\varepsilon)}(\cdot) \Rightarrow F^{(0)}(\cdot)$ as $\varepsilon \to 0$, where $F^{(0)}(t)$ is a distribution function of a finite measure.

Lemma formulated below shows that condition **D$_8$ (b)** in propositions **(i)** and **(ii)** of Lemma 1.4.3 can also be omitted.

Lemma 1.4.4. *Let conditions* **D$_9$** *and* **C$_1$** *hold. Then* $F^{(\varepsilon)}(\infty) \to F^{(0)}(\infty)$ *as* $\varepsilon \to 0$.

Proof. Choose $0 < \beta < \delta$. Then, using condition **C$_1$**, we have

$$\lim_{T \to \infty} \overline{\lim_{\varepsilon \to 0}} (F^{(\varepsilon)}(\infty) - F^{(\varepsilon)}(T)) = \lim_{T \to \infty} \overline{\lim_{\varepsilon \to 0}} \int_T^\infty F^{(\varepsilon)}(ds)$$

$$\leq \lim_{T \to \infty} \overline{\lim_{\varepsilon \to 0}} \int_T^\infty e^{\beta s} F^{(\varepsilon)}(ds)$$

$$\leq \lim_{T \to \infty} e^{-(\delta - \beta)T} \overline{\lim_{\varepsilon \to 0}} \int_0^\infty e^{\delta s} F^{(\varepsilon)}(ds) = 0. \tag{1.4.36}$$

Take some sequence $0 < T_k \to \infty$ as $k \to \infty$ of continuity points of the distribution function $F^{(0)}(s)$. Condition **D$_9$** implies that $F^{(\varepsilon)}(T_k) \to F^{(0)}(T_k)$ as $\varepsilon \to 0$ for every $k \geq 1$. Obviously, $F^{(0)}(T_k) \to F^{(0)}(\infty)$ as $k \to \infty$. These relations, together with (1.4.36), imply in an obvious way that

$$\overline{\lim_{\varepsilon \to 0}} |F^{(\varepsilon)}(\infty) - F^{(0)}(\infty)|$$

$$\leq \lim_{k \to \infty} \overline{\lim_{\varepsilon \to 0}} (|F^{(\varepsilon)}(\infty) - F^{(\varepsilon)}(T_k)|$$

$$+ |F^{(\varepsilon)}(T_k) - F^{(0)}(T_k)| + |F^{(0)}(T_k) - F^{(0)}(\infty)|) = 0. \tag{1.4.37}$$

The proof is complete. $\qquad\qquad\qquad\qquad\qquad\qquad\qquad\qquad\qquad\qquad\qquad\square$

Taking into consideration that the limiting characteristic root $\rho^{(0)}$ may not be equal to zero, we replace condition **F$_5$** with the following more general condition:

F$_6$: **(a)** $\lim_{u \to 0} \overline{\lim}_{0 \leq \varepsilon \to 0} \sup_{|v| \leq u} |q^{(\varepsilon)}(t + v) - q^{(0)}(t)| = 0$ almost everywhere with respect to the Lebesgue measure on $[0, \infty)$;

  (b) $\overline{\lim}_{0 \leq \varepsilon \to 0} \sup_{0 \leq t \leq T} |q^{(\varepsilon)}(t)| < \infty$ for every $T \geq 0$;

  (c) $\lim_{T \to \infty} \overline{\lim}_{0 \leq \varepsilon \to 0} h \sum_{r \geq T/h} \sup_{rh \leq t \leq (r+1)h} e^{(\rho^{(0)} + \gamma)t} |q^{(\varepsilon)}(t)| = 0$ for some $h > 0$ and $\gamma > 0$.

This condition is used in the cases where **(i)** $f^{(0)} > 0$; or **(ii)** $f^{(0)} = 0$ and 0 is a limit point in the set $\mathfrak{N}$; or **(iv)** $f^{(0)} < 0$. Note that, in case **(ii)**, $\rho^{(0)} = 0$ and condition **F$_6$** coincides with **F$_5$**. In the case where **(iii)** $f^{(0)} = 0$ and 0 is not a limit point in the set $\mathfrak{N}$, the weaker condition $\mathbf{F_2}$ can be used. Obviously, $\mathbf{F_6}$ coincides with condition $\mathbf{F_2}$ if to assign $\gamma = 0$.

Denote

$$\tilde{m}_1^{(\varepsilon)} = \int_0^\infty s e^{\rho^{(\varepsilon)} s} F^{(\varepsilon)}(ds).$$

In order to have the relation

$$\tilde{m}_1^{(\varepsilon)} \to \tilde{m}_1^{(0)} < \infty \quad \text{as } \varepsilon \to 0, \tag{1.4.38}$$

condition $\mathbf{C_2}$ or $\mathbf{C_1}$ must be added to condition $\mathbf{D_7}$, respectively, in the cases **(i)** or **(ii)**. In the case **(iv)**, condition $\mathbf{D_7}$ automatically implies this relation while, in the case **(iii)**, the following condition has to be added to condition $\mathbf{D_7}$:

$\mathbf{M_7}$: $m_1^{(\varepsilon)} \to m_1^{(0)} < \infty$ as $\varepsilon \to 0$.

Let us define, for $\varepsilon \geq 0$,

$$\tilde{x}^{(\varepsilon)}(\infty) = \frac{\int_0^\infty e^{s \rho^{(\varepsilon)}} q^{(\varepsilon)}(s) m(ds)}{\tilde{m}_1^{(\varepsilon)}}. \tag{1.4.39}$$

The following theorem generalises Theorem 1.4.1 to the case where the total variation of the limiting distribution function $F^{(0)}(\infty)$ can be less, equal, or greater than 1.

Theorem 1.4.2. *Let condition $\mathbf{D_7}$ hold together with the following conditions specified for four alternatives*:

  (i) $\mathbf{C_2}$ *and* $\mathbf{F_6}$ *if* $f^{(0)} > 0$;

  (ii) $\mathbf{C_1}$ *and* $\mathbf{F_5}$ *if* $f^{(0)} = 0$ *and 0 is a limit point in the set* $\mathfrak{N}$;

  (iii) $\mathbf{M_7}$ *and* $\mathbf{F_2}$ *if* $f^{(0)} = 0$ *and 0 is not a limit point in the set* $\mathfrak{N}$;

  (iv) $\mathbf{F_6}$ *if* $f^{(0)} < 0$.

Then for any $0 \leq t^{(\varepsilon)} \to \infty$ *as* $\varepsilon \to 0$,

$$\frac{x^{(\varepsilon)}(t^{(\varepsilon)})}{\exp\{-\rho^{(\varepsilon)} t^{(\varepsilon)}\}} \to \tilde{x}^{(0)}(\infty) \quad \text{as } \varepsilon \to 0. \tag{1.4.40}$$

Proof. The proof is similar to the proof of Theorem 1.4.1. Let us use the exponential transformation to the renewal equation (1.4.21). Denote

$$\tilde{F}^{(\varepsilon)}(t) = \int_0^t e^{\rho^{(\varepsilon)} s} F^{(\varepsilon)}(ds), \; \tilde{q}^{(\varepsilon)}(t) = e^{\rho^{(\varepsilon)} t} q^{(\varepsilon)}(t), \; \tilde{x}^{(\varepsilon)}(t) = e^{\rho^{(\varepsilon)} t} x^{(\varepsilon)}(t), \; t \geq 0.$$

Note first of all that by the definition of $\rho^{(\varepsilon)}$ as a root of equation (1.4.2), it implies that $\tilde{F}^{(\varepsilon)}(\infty) = 1$, i.e., $\tilde{F}^{(\varepsilon)}(t)$ is a proper distribution function.

Since $x^{(\varepsilon)}(t)$ is a solution of the renewal equation (1.4.21), the function $\tilde{x}^{(\varepsilon)}(t)$ is a solution of the renewal equation

$$\tilde{x}^{(\varepsilon)}(t) = \tilde{q}^{(\varepsilon)}(t) + \int_0^t \tilde{x}^{(\varepsilon)}(t-s)\tilde{F}^{(\varepsilon)}(ds), \quad t \geq 0. \tag{1.4.41}$$

By Lemma 1.4.3, $\rho^{(\varepsilon)} \to \rho^{(0)}$ as $\varepsilon \to 0$. This relation and condition $\mathbf{D}_7$ imply in an obvious way that

$$\tilde{F}^{(\varepsilon)}(\cdot) \Rightarrow \tilde{F}^{(0)}(\cdot) \quad \text{as } \varepsilon \to 0, \tag{1.4.42}$$

Note also that the limiting distribution function $\tilde{F}^{(0)}(s)$ is non-arithmetic if the limiting distribution function $\bar{F}^{(0)}(s)$ is non-arithmetic.

Thus the distribution functions $\tilde{F}^{(\varepsilon)}(s)$ satisfy condition $\mathbf{D}_2$ with the corresponding limiting distribution function $\tilde{F}^{(0)}(s)$.

Also, as was mentioned above, the conditions of the theorem also imply that relation (1.4.38) holds.

In the cases **(i)**, **(ii)**, and **(iv)**, since $\rho^{(\varepsilon)} \to \rho^{(0)}$ as $\varepsilon \to 0$, we have for any $0 < \gamma_1 < \gamma$ that $\rho^{(\varepsilon)} \leq \rho^{(0)} + \gamma_1$ for all ε small enough. Thus the Cramér type condition $\mathbf{C}_2$ implies that

$$\lim_{T\to\infty} \overline{\lim_{\varepsilon\to 0}} \int_T^\infty se^{s\rho^{(\varepsilon)}} F^{(\varepsilon)}(ds) \leq \lim_{T\to\infty} \overline{\lim_{\varepsilon\to 0}} \int_T^\infty se^{s(\rho^{(0)}+\gamma_1)} F^{(\varepsilon)}(ds)$$

$$\leq \lim_{T\to\infty} \int_T^\infty e^{s(\rho^{(0)}+\gamma)} F^{(0)}(ds) = 0. \tag{1.4.43}$$

In the case **(iii)**, since $\rho^{(\varepsilon)} \leq 0$ for all ε small enough, conditions $\mathbf{D}_7$ and $\mathbf{M}_7$ imply that

$$\lim_{T\to\infty} \overline{\lim_{\varepsilon\to 0}} \int_T^\infty se^{s\rho^{(\varepsilon)}} F^{(\varepsilon)}(ds) \leq \lim_{T\to\infty} \overline{\lim_{\varepsilon\to 0}} \int_T^\infty s F^{(\varepsilon)}(ds)$$

$$= \lim_{T\to\infty} \overline{\lim_{\varepsilon\to 0}} (m_1^{(\varepsilon)} - \int_0^T s F^{(\varepsilon)}(ds))$$

$$\leq \lim_{T\to\infty} (m_1^{(0)} - \int_0^T s F^{(0)}(ds)) = 0. \tag{1.4.44}$$

Relation (1.4.42), together with (1.4.43) or (1.4.44), imply that

$$\lim_{\varepsilon\to 0} \tilde{m}_1^{(\varepsilon)} = \lim_{\varepsilon\to 0} \int_0^\infty se^{s\rho^{(\varepsilon)}} F^{(\varepsilon)}(ds)$$

$$= \lim_{T\to\infty} \lim_{\varepsilon\to 0} \int_0^T se^{s\rho^{(\varepsilon)}} F^{(\varepsilon)}(ds)$$

$$= \lim_{T\to\infty} \int_0^T se^{s\rho^{(0)}} F^{(0)}(ds) = \tilde{m}_1^{(0)}. \tag{1.4.45}$$

Relation (1.4.38) coincides with condition $\mathbf{M_2}$ for the distribution functions $\tilde{F}^{(\varepsilon)}(s)$. The corresponding limiting moment is $\tilde{m}_1^{(0)}$.

As was shown above for the cases (i), (ii), and (iv), for any $0 < \gamma_1 < \gamma$, we have $\rho^{(\varepsilon)} \leq \rho^{(0)} + \gamma_1$ for all ε small enough. So $\rho^{(\varepsilon)} \to \rho^{(0)}$ as $\varepsilon \to 0$, and condition $\mathbf{F_6}$ implies that the functions $\tilde{q}^{(\varepsilon)}(t)$ satisfy condition $\mathbf{F_2}$. Moreover, for the corresponding limiting function we have $\tilde{q}^{(0)}(t) \equiv e^{t\rho^{(0)}} q^{(0)}(t)$ and, therefore, $\int_0^\infty \tilde{q}^{(0)}(s)\,ds = \int_0^\infty e^{s\rho^{(0)}} q^{(0)}(s)\,ds$.

In the case (iii), $\rho^{(\varepsilon)} \leq 0$ for all ε small enough. Hence, the relation $\rho^{(\varepsilon)} \to 0$ as $\varepsilon \to 0$ and condition $\mathbf{F_2}$, assumed to hold for the functions $q^{(\varepsilon)}(t)$, imply that the functions $\tilde{q}^{(\varepsilon)}(t)$ satisfy condition $\mathbf{F_2}$. Moreover, the corresponding limiting function satisfies $\tilde{q}^{(0)}(t) \equiv q^{(0)}(t)$ and, therefore, $\int_0^\infty \tilde{q}^{(0)}(s)\,ds = \int_0^\infty q^{(0)}(s)\,ds$ (note that $\rho^{(0)} = 0$ in this case and, therefore, $e^{s\rho^{(0)}} \equiv 1$).

Now, we can apply Theorem 1.2.1 to equation (1.4.41). Taking into consideration the remarks made above, we get the relation

$$\tilde{x}^{(\varepsilon)}(t^{(\varepsilon)}) = e^{\rho^{(\varepsilon)} t^{(\varepsilon)}} x^{(\varepsilon)}(t^{(\varepsilon)}) \to \frac{\int_0^\infty e^{s\rho^{(0)}} q^{(0)}(s)\,ds}{\tilde{m}_1^{(0)}} \quad \text{as } \varepsilon \to 0, \qquad (1.4.46)$$

which is equivalent to the relation (1.4.40).   □

1.4.3   Renewal limits for improper perturbed renewal equations

Let us see what form will the conditions of Theorem 1.4.2 take if the distribution functions and the forcing function satisfy $F^{(\varepsilon)}(s) \equiv F^{(0)}(s)$ and $q^{(\varepsilon)}(s) \equiv q^{(0)}(s)$ for $\varepsilon \geq 0$, i.e., they do not depend on the parameter ε.

The asymptotic relation (1.4.40) takes in this case the following form:

$$\frac{x^{(0)}(t)}{\exp\{-\rho^{(0)} t\}} \to \tilde{x}^{(0)}(\infty) \quad \text{as } \varepsilon \to 0. \qquad (1.4.47)$$

Condition $\mathbf{D_7}$ reduces to the assumption that $\bar{F}^{(0)}(t)$ is a non-arithmetic distribution function.

In case (i), condition $\mathbf{C_2}$ reduces to the assumption that $\int_0^\infty e^{\delta s} F^{(0)}(ds) \in (1, \infty)$ for some $\delta > 0$, and condition $\mathbf{F_6}$ to the assumption that the function $e^{(\rho^{(0)} + \gamma)s} q^{(0)}(s)$ is directly Riemann integrable on $[0, \infty)$ for some $\gamma > 0$. Note that $\rho^{(0)} > 0$ in this case. These conditions (together with the reduced version of condition $\mathbf{D_7}$) coincide with the conditions of the refined generalisation of the renewal Theorem 1.1.1 to the model of an improper renewal equation given in Feller (1966).

The case (ii) is impossible, which is easily seen, since $\mathfrak{N} = \varnothing$.

In the case (iii), condition $\mathbf{M_7}$ reduces to the assumption $m_1^{(0)} = \int_0^\infty s F^{(0)}(ds) < \infty$, and condition $\mathbf{F_2}$ to the assumption that the function $q^{(0)}(s)$ is directly Riemann integrable on $[0, \infty)$. Together with the reduced version of condition $\mathbf{D_7}$, these conditions are the standard in the renewal Theorem 1.1.1. Note that $\rho^{(0)} = 0$ in this case.

In the case **(iv)**, condition $\mathbf{F_6}$ reduces to the assumption that the function $e^{(\rho^{(0)}+\gamma)s}q^{(0)}(s)$ is directly Riemann integrable on $[0, \infty)$ for some $\gamma > 0$. Note that, in this case, $\rho^{(0)} < 0$.

If the conditions listed above are imposed on characteristics of the perturbed renewal equation by formally replacing $F^{(0)}(s)$ and $q^{(0)}(s)$ with $F^{(\varepsilon)}(s)$ and $q^{(\varepsilon)}(s)$, then the asymptotic relation (1.4.47) would take place if 0 is formally replaced with ε.

The following question arises. Do the conditions of Theorem 1.4.2 guarantee that they remain true for all ε small enough?

The answer is affirmative for moment type conditions. Condition $\mathbf{C_2}$ implies, under condition $\mathbf{D_7}$, that $\int_0^\infty e^{\delta s} F^{(\varepsilon)}(ds) \in (1, \infty)$ for all ε small enough, say $\varepsilon \leq \varepsilon_1$.

Also, condition $\mathbf{M_7}$ implies, under condition $\mathbf{D_7}$, that $\tilde{m}_1^{(\varepsilon)} < \infty$ for all ε small enough, say $\varepsilon \leq \varepsilon_2$, in the case **(iii)**.

The answer is negative for conditions from classes $\mathbf{D}$ and $\mathbf{F}$. Non-arithmetic property of the limiting distribution function $\bar{F}^{(0)}(s)$ does not guarantee that the pre-limiting distribution functions $\bar{F}^{(\varepsilon)}(s)$ also possess this property. Also, the direct Riemann integrability of the limiting forcing function $e^{(\rho^{(0)}+\gamma)s}q^{(0)}(s)$ does not guarantee the same property for the corresponding pre-limiting functions.

The following two conditions should be assumed:

$\mathbf{D_{10}}$: $\bar{F}^{(\varepsilon)}(s)$ is a non-arithmetic distribution function for all ε small enough, say $\varepsilon \leq \varepsilon_3$;

and

   $\mathbf{F_7}$: The function $e^{(\rho^{(\varepsilon)}+\gamma)s}q^{(\varepsilon)}(s)$ is directly Riemann integrable on $[0, \infty)$ for all ε small enough, say $\varepsilon \leq \varepsilon_4$.

Note that, under conditions of Theorem 1.4.2, it is enough to assume that $q^{(\varepsilon)}(s)$ is a continuous function almost everywhere with respect to the Lebesgue measure on $[0, \infty)$. In this case, condition $\mathbf{F_7}$ holds and the Lebesgue integration in the right-hand side of (1.4.39) can be replaced with the Riemann integration.

Define

$$\varepsilon_0 = \min(\varepsilon_1, \varepsilon_2, \varepsilon_3, \varepsilon_4). \tag{1.4.48}$$

The following theorem follows from the above remarks.

Theorem 1.4.3. *Let conditions of Theorem* 1.4.2 *as well as conditions* $\mathbf{D_{10}}$ *and* $\mathbf{F_7}$ *hold. Then, for all* $\varepsilon \leq \varepsilon_0$,

$$\tilde{x}^{(\varepsilon)}(t) = e^{\rho^{(\varepsilon)}t}x^{(\varepsilon)}(t) \to \tilde{x}^{(\varepsilon)}(\infty) \ \ as \ t \to \infty. \tag{1.4.49}$$

Let us remark that, under conditions of Theorem 1.4.2, the expression for $\tilde{x}^{(\varepsilon)}(\infty)$ given by formula (1.4.39) is finite for all $\varepsilon \geq 0$ small enough without the assumption that $q^{(\varepsilon)}(s)$ is a function continuous almost everywhere with respect to the Lebesgue measure on $[0, \infty)$. This is so, since the Lebesgue integration is used in this formula.

Moreover, it can be proved that the mean value of $\tilde{x}^{(\varepsilon)}(\cdot)$ over the interval $[0, t]$ converges to $\tilde{x}^{(\varepsilon)}(\infty)$ as $t \to \infty$.

1.4.4 Convergence of renewal limits for improper perturbed renewal equations

The following theorem supplements the asymptotic relations given in relations (1.4.40) and (1.4.49).

Theorem 1.4.4. *Let conditions of Theorem* 1.4.2 *hold. Then*

$$\tilde{x}^{(\varepsilon)}(\infty) \to \tilde{x}^{(0)}(\infty) \;\; as \;\; \varepsilon \to 0. \tag{1.4.50}$$

Proof. As was shown in the proof of Theorem 1.4.2, the conditions of Theorem 1.4.2 imply that $\rho^{(\varepsilon)} \to \rho^{(0)}$ as $\varepsilon \to 0$; condition $\mathbf{M_2}$ holds for the distributions $\tilde{F}^{(\varepsilon)}(t) = \int_0^t e^{s\rho^{(\varepsilon)}} F^{(\varepsilon)}(ds)$; and condition $\mathbf{F_2}$ holds for the functions $\tilde{q}^{(\varepsilon)}(s) = e^{s\rho^{(\varepsilon)}} q^{(\varepsilon)}(s)$. Hence, relation (1.4.50) follows from Theorem 1.2.4.   □

Chapter 2

Nonlinearly perturbed renewal equation

Chapter 2 contains results concerning exponential expansions for solutions of the perturbed renewal equation for the model in which some Cramér type condition is imposed on the distribution that generates the perturbed renewal equation. Some additional nonlinear perturbation conditions are also imposed on the distributions. These conditions mean that the moments of these distributions are nonlinear functions of the perturbation parameter ε and can be expanded in an asymptotic power series with respect to ε up to and including some order k.

Theorems in Chapter 1 give exponential normalising functions in an implicit form as a solution of the nonlinear characteristic equation (1.0.7). Further improvements should seal with a more explicit form for the exponential normalising functions in the asymptotic relation (1.0.5).

Theorems in Chapter 2 give expressions for the corresponding coefficients in exponential expansions of the corresponding exponential normalising functions in explicit form that makes this asymptotical relation much more useful and effective.

A new improvement is achieved under natural additional perturbation conditions in which we assume that the defect $f^{(\varepsilon)}$ and the corresponding moments of the distribution $F^{(\varepsilon)}(s)$ can be expanded in power series with respect to ε up to and including an order k. Under these perturbation conditions, we give an asymptotic expansion for the characteristic roots $\rho^{(\varepsilon)}$,

$$\rho^{(\varepsilon)} = \rho^{(0)} + a_1\varepsilon + \cdots + a_k\varepsilon^k + o(\varepsilon^k), \tag{2.0.1}$$

and an explicit algorithm for calculating the coefficients $\rho^{(0)}, a_1, a_2, \ldots$, in the expansions of the corresponding perturbation conditions.

Asymptotic expansion (2.0.1) permits to introduce various balancing conditions between the rate of growth of time $t^{(\varepsilon)} \to \infty$ and the perturbation rate determined by the perturbation parameter $\varepsilon \to 0$. These conditions are controlled by an integer parameter r, $1 \le r \le k$, and have the following form:

$$\varepsilon^r t^{(\varepsilon)} \to \lambda_r \quad \text{as} \quad \varepsilon \to 0. \tag{2.0.2}$$

For some of these balancing conditions holds, the asymptotics (1.0.5) can be represented in the following form:

$$\frac{x^{(\varepsilon)}(t^{(\varepsilon)})}{\exp\{-(\rho^{(0)} + a_1\varepsilon + \cdots + a_{r-1}\varepsilon^{r-1})t^{(\varepsilon)}\}} \to e^{-\lambda_r a_r} \tilde{x}^{(0)}(\infty) \quad \text{as} \quad \varepsilon \to 0. \tag{2.0.3}$$

Each quantity $\exp\{-\rho^{(0)}t^{(\varepsilon)}\}$, $\exp\{-a_1\varepsilon t^{(\varepsilon)}\}, \ldots, \exp\{-a_{r-1}\varepsilon t^{(\varepsilon)}\}$ contributes to the normalising function in the left-hand side of (2.0.2) a factor that tends to zero, equal to 1, or tends to ∞ depending on whether the corresponding coefficients $\rho^{(0)}, a_1, \ldots,$ a_{r-1} take positive, zero, or negative values. Also, as was mentioned above, for the asymptotic relation (2.0.2) there is an explicit algorithm for calculating the coefficients $\rho^{(0)}, a_1, \ldots, a_{r-1}$. This makes the asymptotics given in (2.0.2) much more explicit and effective than (1.0.5).

Theorems 2.1.1, 2.1.2, and 2.1.3 yield explicit exponential expansions for solutions of asymptotically proper perturbed renewal equations while Theorems 2.2.1 and 2.2.2 yield similar results for asymptotically improper perturbed renewal equations. They extend the results in Theorems 1.4.1 and 1.4.2.

Finally in Theorem 2.3.2 we give asymptotical expansions for renewal limits of nonlinearly perturbed renewal equations.

The asymptotics of type (2.0.2) is a basis for studies of the so-called mixed ergodic and limit theorems and mixed ergodic and large deviation theorems for perturbed stochastic processes with absorption. These studies are carried out in Chapters 3–5.

2.1 Exponential expansions for perturbed renewal equation

In this section, we give exponential asymptotic expansions for solutions of perturbed asymptotically proper renewal equations.

2.1.1 Exponential expansions based on defects of distributions that generate perturbed renewal equations

Consider, for every $\varepsilon \geq 0$, the renewal equation

$$x^{(\varepsilon)}(t) = q^{(\varepsilon)}(t) + \int_0^t x^{(\varepsilon)}(t-s)F^{(\varepsilon)}(ds), \quad t \geq 0, \tag{2.1.1}$$

where (a) $q^{(\varepsilon)}(t)$ is a measurable and locally bounded real-valued function on $[0, \infty)$, and (b) $F^{(\varepsilon)}(s)$ is a distribution function on $[0, \infty)$ which is not concentrated at 0 but can be improper, i.e., $F^{(\varepsilon)}(\infty) \in (0, 1)$.

As was mentioned in Chapter 1, there is a unique measurable and locally bounded solution $x^{(\varepsilon)}(t)$ of equation (2.1.1).

We use the notations

$$f^{(\varepsilon)} = 1 - F^{(\varepsilon)}(\infty), \quad m_r^{(\varepsilon)} = \int_0^\infty s^r F^{(\varepsilon)}(ds), \quad r \geq 1.$$

Let us recall the condition (introduced in Subsection 1.3.1):

D$_6$: $F^{(\varepsilon)}(\cdot) \Rightarrow F^{(0)}(\cdot)$ as $\varepsilon \to 0$, where $F^{(0)}(s)$ is a proper non-arithmetic distribution function.

Condition $\mathbf{D_6}$ includes the assumption that $F^{(0)}(\infty) = 1$, which automatically implies the following relation:

$$f^{(\varepsilon)} \to f^{(0)} = 0 \text{ as } \varepsilon \to 0. \tag{2.1.2}$$

Now, we assume the following Cramér type condition (introduced in Subsection 1.4.1):

$\mathbf{C_1}$: There exists $\delta > 0$ such that $\overline{\lim}_{0 \leq \varepsilon \to 0} \int_0^\infty e^{\delta s} F^{(\varepsilon)}(ds) < \infty$.

Let us note that conditions $\mathbf{D_6}$ and $\mathbf{C_1}$ imply that

$$m_r^{(\varepsilon)} \to m_r^{(0)} \in (0, \infty) \text{ as } \varepsilon \to 0, \quad r \geq 1. \tag{2.1.3}$$

Denote

$$\phi^{(\varepsilon)}(\rho) = \int_0^\infty e^{\rho s} F^{(\varepsilon)}(ds).$$

It follows from condition $\mathbf{C_1}$ that there exists $\varepsilon_1 > 0$ such that $\phi^{(\varepsilon)}(\delta) < \infty$ for $\varepsilon \leq \varepsilon_1$. Obviously, $m_r^{(\varepsilon)} < \infty, r = 1, 2, \ldots$, for $\varepsilon \leq \varepsilon_1$, as well.

The following plays a crucial role in the further consideration:

$$\phi^{(\varepsilon)}(\rho) = 1. \tag{2.1.4}$$

It follows from Lemma 1.4.1 that, under conditions $\mathbf{D_6}$ and $\mathbf{C_1}$, there exists $\varepsilon_2 > 0$ such that equation (2.1.4) has a unique nonnegative root $\rho^{(\varepsilon)}$ for $\varepsilon \leq \min(\varepsilon_1, \varepsilon_2)$.
Also,

$$\rho^{(\varepsilon)} \to \rho^{(0)} = 0 \text{ as } \varepsilon \to 0. \tag{2.1.5}$$

We also assume the following condition (introduced in Subsection 1.4.1):

$\mathbf{F_5}$: (a) $\lim_{u \to 0} \overline{\lim}_{0 \leq \varepsilon \to 0} \sup_{|v| \leq u} |q^{(\varepsilon)}(t + v) - q^{(0)}(t)| = 0$ almost everywhere with respect to Lebesgue measure on $[0, \infty)$;

    (b) $\overline{\lim}_{0 \leq \varepsilon \to 0} \sup_{0 \leq t \leq T} |q^{(\varepsilon)}(t)| < \infty$ for every $T \geq 0$;

    (c) $\lim_{T \to \infty} \overline{\lim}_{0 \leq \varepsilon \to 0} h \sum_{r \geq T/h} \sup_{rh \leq t \leq (r+1)h} e^{\gamma t} |q^{(\varepsilon)}(t)| = 0$ for some $h > 0$ and some $\gamma > 0$.

Let us also denote

$$x^{(\varepsilon)}(\infty) = \frac{\int_0^\infty q^{(\varepsilon)}(s) m(ds)}{m_1^{(\varepsilon)}}.$$

We shall also use the following balancing condition:

$\mathbf{B_2^{(k)}}$: $0 \leq t^{(\varepsilon)} \to \infty$ and $f^{(\varepsilon)} \to 0$ as $\varepsilon \to 0$ in such a way that $(f^{(\varepsilon)})^k t^{(\varepsilon)} \to \lambda_k$, where $0 \leq \lambda_k < \infty$.

The following theorem generalises Theorem 1.4.1.

Theorem 2.1.1. *Let conditions $\mathbf{D}_6$ and $\mathbf{C}_1$ hold. Then*

(i) *The following asymptotic expansion holds for the root $\rho^{(\varepsilon)}$ of the equation (2.1.4) for every $k \geq 1$:*

$$\rho^{(\varepsilon)} = b_{\varepsilon 1} f^{(\varepsilon)} + \cdots + b_{\varepsilon k}(f^{(\varepsilon)})^k + o((f^{(\varepsilon)})^k), \qquad (2.1.6)$$

where the coefficients $b_{\varepsilon k}$, $k \geq 1$, can be calculated using the recurrence formulas $b_{\varepsilon 1} = (m_1^{(\varepsilon)})^{-1}$, $b_{\varepsilon 2} = -\frac{1}{2}(m_1^{(\varepsilon)})^{-1} m_2^{(\varepsilon)} b_{\varepsilon 1}^2$ and, in general, for $k \geq 2$,

$$b_{\varepsilon k} = -(m_1^{(\varepsilon)})^{-1} \sum_{r=2}^{k} m_r^{(\varepsilon)} \cdot \sum_{n_1,\dots,n_{k-1} \in D_{r,k}} \prod_{i=1}^{k-1} b_{\varepsilon i}^{n_i} / n_i!, \qquad (2.1.7)$$

where $D_{r,k}$, for every $2 \leq r \leq k < \infty$, is the set of all nonnegative, integer solutions of the system

$$n_1 + \cdots + n_{k-1} = r, \quad n_1 + \cdots + (k-1)n_{k-1} = k. \qquad (2.1.8)$$

(ii) *If, additionally, conditions $\mathbf{B}_2^{(k)}$, for some $k \geq 1$, and $\mathbf{F}_5$ hold then the following asymptotic relation takes place:*

$$\frac{x_\varepsilon(t^{(\varepsilon)})}{\exp\{-(b_{\varepsilon 1} f^{(\varepsilon)} + \cdots + b_{\varepsilon k-1}(f^{(\varepsilon)})^{k-1})t^{(\varepsilon)}\}}$$
$$\to e^{-\lambda_k b_{0k}} x_0(\infty) \ \ as \ \ \varepsilon \to 0. \qquad (2.1.9)$$

Proof. As was mentioned above, condition $\mathbf{C}_1$ implies that $\phi^{(\varepsilon)}(\delta) < \infty$ for $\varepsilon \leq \varepsilon_1$, and Lemma 1.4.1 implies that, under conditions $\mathbf{D}_6$ and $\mathbf{C}_1$, there exists a unique nonnegative root $\rho^{(\varepsilon)}$ of equation (2.1.4) for $\varepsilon \leq \min(\varepsilon_1, \varepsilon_2)$ and that $\rho^{(\varepsilon)} \to 0$ as $\varepsilon \to 0$.

Apply Taylor's expansion to the function e^t, which can be given in the following form (using the fact that this function and its derivatives are increasing functions):

$$e^{s\rho^{(\varepsilon)}} = 1 + s\rho^{(\varepsilon)}/1! + \cdots + s^k(\rho^{(\varepsilon)})^k/k!$$
$$+ s^{k+1}(\rho^{(\varepsilon)})^{k+1} e^{s\rho^{(\varepsilon)}} \theta_{k+1}^{(\varepsilon)}(s)/(k+1)!, \qquad (2.1.10)$$

where $\theta_{k+1}^{(\varepsilon)}(s), s \geq 0$, is a continuous function such that

$$0 \leq \theta_{k+1}^{(\varepsilon)}(s) \leq 1, \quad s \geq 0. \qquad (2.1.11)$$

As was shown in the proof of Lemma 1.4.1, for any $\beta \in (0, \delta]$ there exists $\varepsilon_3 = \varepsilon_3(\beta) > 0$ such that $\rho_\varepsilon \leq \beta$ for $\varepsilon \leq \varepsilon_3(\beta)$.

By condition $\mathbf{C}_1$,

$$\overline{\lim_{0 \leq \varepsilon \to 0}} \int_0^\infty s^{k+1} e^{\beta s} F^{(\varepsilon)}(ds) \leq \overline{\lim_{0 \leq \varepsilon \to 0}} c_{k+1} \int_0^\infty e^{\delta s} F^{(\varepsilon)}(ds) < \infty, \qquad (2.1.12)$$

where $c_{k+1} = \sup_{s \geq 0} s^{k+1} e^{-(\delta - \beta)s} < \infty$.

It follows from relation (2.1.12) that there exists $\varepsilon_4 = \varepsilon_4(\beta) > 0$ such that, for $\varepsilon \leq \varepsilon_4$,

$$M_{k+1} = \frac{1}{(k+1)!} \sup_{\varepsilon \leq \varepsilon_4} \int_0^\infty s^{k+1} e^{s\beta} F^{(\varepsilon)}(ds) < \infty. \qquad (2.1.13)$$

Let us define

$$\varepsilon_5 = \varepsilon_5(\beta) = \min(\varepsilon_1, \varepsilon_2, \varepsilon_3(\beta), \varepsilon_4(\beta)). \qquad (2.1.14)$$

Integrating both sides in (2.1.10) and using (2.1.11) and (2.1.13) we get, for $\varepsilon \leq \varepsilon_5$,

$$\int_0^\infty e^{s\rho^{(\varepsilon)}} F^{(\varepsilon)}(ds) = 1 - f^{(\varepsilon)} + m_1^{(\varepsilon)} \rho^{(\varepsilon)}/1! + \cdots$$
$$+ m_k^{(\varepsilon)} (\rho^{(\varepsilon)})^k / k! + (\rho^{(\varepsilon)})^{k+1} M_{k+1} \theta_{k+1}^{(\varepsilon)}, \qquad (2.1.15)$$

where

$$\theta_{k+1}^{(\varepsilon)} = \frac{\int_0^\infty s^{k+1} e^{s\rho^{(\varepsilon)}} \theta_{k+1}^{(\varepsilon)}(s) F^{(\varepsilon)}(ds)}{\sup_{\varepsilon \leq \varepsilon_4} \int_0^\infty s^{k+1} e^{s\beta} F^{(\varepsilon)}(ds)} \in [0, 1]. \qquad (2.1.16)$$

Therefore, equation (2.1.4) can be rewritten in the form

$$m_1^{(\varepsilon)} \rho^{(\varepsilon)}/1! + m_2^{(\varepsilon)} (\rho^{(\varepsilon)})^2/2! + \cdots + m_k^{(\varepsilon)} (\rho^{(\varepsilon)})^k / k!$$
$$+ (\rho^{(\varepsilon)})^{k+1} M_{k+1} \theta_{k+1}^{(\varepsilon)} = f^{(\varepsilon)}. \qquad (2.1.17)$$

The coefficients $\rho^{(\varepsilon)} \to 0$, and the sum of all terms in the left-hand side in (2.1.17), beginning with the second one, is $o(\rho^{(\varepsilon)})$. Also, $m_r^{(\varepsilon)} \to m_r^{(0)} \in (0, \infty)$. Using these facts, we get from (2.1.17), dividing both sides by $f^{(\varepsilon)}$ and evaluating the corresponding limits, that $m_1^{(\varepsilon)} \rho^{(\varepsilon)}/f^{(\varepsilon)} \to 1$ as $\varepsilon \to 0$. This means, in equivalent terms, that the coefficients $\rho^{(\varepsilon)}$ can be represented in the form

$$\rho^{(\varepsilon)} = b_{\varepsilon 1} f^{(\varepsilon)} + \rho_1^{(\varepsilon)}, \qquad (2.1.18)$$

where $b_{\varepsilon 1} = (m_1^{(\varepsilon)})^{-1}$ and $\rho_1^{(\varepsilon)} = o(f^{(\varepsilon)})$.

Relation (2.1.18) becomes (2.1.6) for the case $k = 1$.

Substituting (2.1.18) into (2.1.17) and cancelling the terms $m_1^{(\varepsilon)}\rho^{(\varepsilon)} = f^{(\varepsilon)}$ on the left and the right-hand sides we can rewrite equation (2.1.17) in the form suitable for asymptotic analysis of the residual function $\rho_{\varepsilon 1}$,

$$m_1^{(\varepsilon)}\rho_1^{(\varepsilon)}/1! + m_2^{(\varepsilon)}(b_{\varepsilon 1}f^{(\varepsilon)} + \rho_1^{(\varepsilon)})^2/2! + \cdots$$

$$+ m_k^{(\varepsilon)}(b_{\varepsilon 1}f^{(\varepsilon)} + \rho_1^{(\varepsilon)})^k/k! + (b_{\varepsilon 1}f^{(\varepsilon)} + \rho_1^{(\varepsilon)})^{k+1}M_{k+1}\theta_{k+1}^{(\varepsilon)} = 0. \quad (2.1.19)$$

Using $\rho^{(\varepsilon)} \to 0$ and since $\rho_1^{(\varepsilon)} = o(f^{(\varepsilon)})$ we get from (2.1.19), dividing both sides by $(f^{(\varepsilon)})^2$ and evaluating the corresponding limits, that $m_1^{(\varepsilon)}\rho_1^{(\varepsilon)}/(f^{(\varepsilon)})^2 + b_{\varepsilon 1}^2 m_2^{(\varepsilon)}/2 \to 0$ as $\varepsilon \to 0$. This equivalently means that the coefficients $\rho_1^{(\varepsilon)}$ can be represented in the form

$$\rho_1^{(\varepsilon)} = b_{\varepsilon 2}(f^{(\varepsilon)})^2 + \rho_{\varepsilon 2}, \quad (2.1.20)$$

where $b_{\varepsilon 2} = -(m_1^{(\varepsilon)})^{-1}\frac{1}{2}m_2^{(\varepsilon)}b_{\varepsilon 1}^2$ and $\rho_2^{(\varepsilon)} = o((f^{(\varepsilon)})^2)$.

Relation (2.1.20) is equivalent to relation (2.1.6) for the case $k = 2$.

Continuing in the same way we can obtain relation (2.1.6) for the cases $k > 2$ as well. Note also that the values of the coefficients $b_{\varepsilon r}$, $r \geq 1$, do not depend on the order k chosen in expansion (2.1.6).

Because it is now known that there exists asymptotical expansion (2.1.6), the algorithm described above allows to get the coefficients $b_{\varepsilon k}$ by collecting and setting to 0 the coefficients for f^k, $k \geq 1$, in the formal asymptotic equality that follows from equation (2.1.17),

$$m_1^{(\varepsilon)}(b_{\varepsilon 1}f + b_{\varepsilon 2}f^2 + \cdots)/1!$$

$$+ m_2^{(\varepsilon)}(b_{\varepsilon 1}f + b_{\varepsilon 2}f^2 + \cdots)^2/2! + \cdots = f. \quad (2.1.21)$$

In the case $k = 1$, this equality yields the equation $m_1^{(\varepsilon)}b_{\varepsilon 1} = 1$ and, if $k \geq 2$, the equation

$$m_1^{(\varepsilon)}b_{\varepsilon k} + \sum_{r=2}^{k} m_r^{(\varepsilon)} \cdot \sum_{n_1,\dots,n_{k-1}\in D_{r,k}} \prod_{i=1}^{k-1} b_{\varepsilon i}^{n_i}/n_i! = 0, \quad (2.1.22)$$

which leads to the general recurrence formula (2.1.7) for calculating the coefficients $b_{\varepsilon k}$. The statement (**i**) of Theorem 2.1.1 is proved.

Let us now use Theorem 1.4.1. This theorem implies that, under conditions $\mathbf{D}_6$, $\mathbf{C}_1$, and $\mathbf{F}_5$, for any $0 \leq t^{(\varepsilon)} \to \infty$ the following asymptotic relation holds:

$$\frac{x_\varepsilon(t^{(\varepsilon)})}{\exp\{-\rho^{(\varepsilon)}t^{(\varepsilon)}\}} \to x_0(\infty) \text{ as } \varepsilon \to 0. \quad (2.1.23)$$

The statement **(ii)** is a corollary of relations (2.1.6) and (2.1.23). Under the condition $(f^{(\varepsilon)})^k t_\varepsilon \to \lambda_k \in [0, \infty)$, we have $o((f^{(\varepsilon)})^k)t_\varepsilon \to 0$, and so $e^{o((f^{(\varepsilon)})^k)t_\varepsilon} \to 1$. Also, $b_{\varepsilon k}(f^{(\varepsilon)})^k t_\varepsilon \to b_{0k}\lambda_k$. Using these facts one can easily transform relation (2.1.23) into the form (2.1.9). $\qquad\square$

2.1.2 Asymptotically proper perturbed renewal equations with the defect of order $O(\varepsilon)$

Due to convergence of the moments $m_r^{(\varepsilon)}$, we have

$$b_{\varepsilon r} \to b_{0r} \text{ as } \varepsilon \to 0, \quad r \geq 1.$$

Nevertheless, it is not convenient to have the coefficients in the exponential asymptotic relation (2.1.9) dependent of perturbation parameter ε. This situation can be improved if the probabilities $f^{(\varepsilon)}$ and the moments $m_r^{(\varepsilon)}$ can be expanded in asymptotic power series with respect to the perturbation parameter ε. In this case, expansion (2.1.6) and relation (2.1.9) can be rewritten in the form such that the corresponding coefficients would not depend on ε.

Let us assume that the following perturbation condition holds for some $k \geq 1$:

$\mathbf{P}_1^{(\mathbf{k})}$:   **(a)** $1 - f^{(\varepsilon)} = 1 + b_{1,0}\varepsilon + \cdots + b_{k,0}\varepsilon^k + o(\varepsilon^k)$, where $|b_{r,0}| < \infty$, $r = 1, \ldots, k$;

       (b) $m_r^{(\varepsilon)} = m_r^{(0)} + b_{1,r}\varepsilon + \cdots + b_{k-r,r}\varepsilon^{k-r} + o(\varepsilon^{k-r})$ for $r = 1, \ldots, k$, where $|b_{l,r}| < \infty$, $l = 1, \ldots, k - r$, $r = 1, \ldots, k$.

It is convenient to also set the coefficients $b_{0,0} = 1$ and $b_{0,r} = m_r^{(0)}$, $r \geq 1$.

Let us also remark that the identity $1 - f^{(\varepsilon)} = F^{(\varepsilon)}(\infty) = \int_0^\infty F^{(\varepsilon)}(dx) = m_0^{(\varepsilon)}$ can be interpreted as a moment functional of the same type as the moment functionals $m_r^{(\varepsilon)} = \int_0^\infty x^r F^{(\varepsilon)}(dx), r \geq 1$, but of order $r = 0$. This remark makes the expansions given in conditions $\mathbf{P}_1^{(\mathbf{k})}$ **(a)** and **(b)** consistent.

Condition $\mathbf{P}_1^{(\mathbf{k})}$ should be interpreted as a nonlinear perturbation condition imposed on the moment functionals of the distribution functions that generate the corresponding renewal equations. These conditions mean that the moment functionals $m_r^{(\varepsilon)}$, $r = 0, \ldots, k$, of these distribution functions are nonlinear smooth functions of the perturbation parameter ε such that $m_r^{(\varepsilon)}$ can be expanded in an asymptotic power series with respect to ε up to and including the order $k - r$ for every $r = 0, \ldots, k$.

It follows from condition $\mathbf{D}_6$ that $f^{(\varepsilon)} \to 0$ as $\varepsilon \to 0$, which is consistent with condition $\mathbf{P}_1^{(\mathbf{k})}$ **(a)**. The first non-zero coefficient $b_{r,0}$ in $\mathbf{P}_1^{(\mathbf{k})}$ **(a)** must be negative, since, by the definition, $1 - f^{(\varepsilon)} \leq 1$. Note also that this condition allows for the case where $b_{r,0} = 0$ for all $r = 1, \ldots, k$.

It follows from conditions $\mathbf{D_6}$ and $\mathbf{C_1}$ that the power moments $m_r^{(\varepsilon)}$ are finite for all $r \geq 1$ and all ε small enough, and that

$$m_r^{(\varepsilon)} \to m_r^{(0)} \in (0, \infty) \quad \text{as } \varepsilon \to 0, \quad r \geq 1, \tag{2.1.24}$$

which agrees with condition $\mathbf{P_1^{(k)}}$ (b).

We shall also use the following balancing condition for $1 \leq r \leq k$:

$\mathbf{B_3^{(r)}}$: $0 \leq t^{(\varepsilon)} \to \infty$ as $\varepsilon \to 0$ in such a way that $\varepsilon^r t^{(\varepsilon)} \to \lambda_r$, where $0 \leq \lambda_r < \infty$.

Theorem 2.1.2. *Let conditions $\mathbf{D_6}$, $\mathbf{C_1}$ and $\mathbf{P_1^{(k)}}$ hold. Then*

(i) *For all ε small enough, equation (2.1.4) has a unique nonnegative root $\rho^{(\varepsilon)}$ that has the following asymptotic expansion:*

$$\rho^{(\varepsilon)} = a_1 \varepsilon + \cdots + a_k \varepsilon^k + o(\varepsilon^k), \tag{2.1.25}$$

where the coefficients a_n, $n = 1, \ldots, k$, are given by the recurrence formulas $a_1 = -b_{1,0}/b_{0,1}$ and, in general, for $n = 1, \ldots, k$,

$$a_n = -b_{0,1}^{-1}\left(b_{n,0} + \sum_{q=1}^{n-1} b_{n-q,1} a_q \right.$$
$$\left. + \sum_{m=2}^{n} \sum_{q=m}^{n} b_{n-q,m} \cdot \sum_{n_1,\ldots,n_{q-1} \in D_{m,q}} \prod_{p=1}^{q-1} a_p^{n_p} / n_p! \right), \tag{2.1.26}$$

where $D_{m,q}$, for every $2 \leq m \leq q < \infty$, is the set of all nonnegative, integer solutions of the system

$$n_1 + \cdots + n_{q-1} = m, \quad n_1 + \cdots + (q-1)n_{q-1} = q. \tag{2.1.27}$$

(ii) *If the coefficients satisfy $b_{l,0} = 0$, $l = 1, \ldots, r$, for some $1 \leq r \leq k$, then $a_1, \ldots, a_r = 0$. If $b_{l,0} = 0$, $l = 1, \ldots, r-1$, but $b_{r,0} < 0$ for some $1 \leq r \leq k$, then $a_1, \ldots, a_{r-1} = 0$ but $a_r > 0$.*

(iii) *If, additionally, conditions $\mathbf{B_3^{(r)}}$, for some $1 \leq r \leq k$, and $\mathbf{F_5}$ hold, then the following asymptotic relation holds:*

$$\frac{x^{(\varepsilon)}(t^{(\varepsilon)})}{\exp\{-(a_1\varepsilon + \cdots + a_{r-1}\varepsilon^{r-1})t^{(\varepsilon)}\}} \to e^{-\lambda_r a_r} x^{(0)}(\infty) \quad \text{as } \varepsilon \to 0. \tag{2.1.28}$$

Proof. The proof follows the scheme used in the proof of Theorem 2.1.1. It was proved in Lemma 1.4.1 that under conditions $\mathbf{D_6}$ and $\mathbf{C_1}$ there exists $\varepsilon_5 > 0$, defined in formula (2.1.14), such that equation (2.1.4) has a unique nonnegative root $\rho^{(\varepsilon)}$ for all $\varepsilon \leq \varepsilon_5$, and that $\rho^{(\varepsilon)} \to 0$ as $\varepsilon \to 0$.

Now, we start with an asymptotic analysis for the coefficients $\rho^{(\varepsilon)}$. We use equation (2.1.17) obtained, for $\varepsilon \leq \varepsilon_5$, in the proof of Theorem 2.1.1,

$$m_1^{(\varepsilon)}\rho^{(\varepsilon)}/1! + m_2^{(\varepsilon)}(\rho^{(\varepsilon)})^2/2! + \cdots$$

$$+ m_k^{(\varepsilon)}(\rho^{(\varepsilon)})^k/k! + (\rho^{(\varepsilon)})^{k+1}M_{k+1}\theta_{k+1}^{(\varepsilon)} = f^{(\varepsilon)}, \qquad (2.1.29)$$

where $M_{k+1} < \infty$ and $\theta_{k+1}^{(\varepsilon)} \in [0, 1]$.

The coefficients $\rho^{(\varepsilon)} \to 0$, and the sum of all terms in the left-hand side of (2.1.29), starting with the second one, is $o(\rho^{(\varepsilon)})$. Also, $m_r^{(\varepsilon)} \to m_r^{(0)} \in (0, \infty)$ for $r \geq 1$. Note also that the first term in the asymptotic expansion for the defect $f^{(\varepsilon)}$, as implied by condition $\mathbf{P}_1^{(k)}$ **(a)**, is of the order $O(\varepsilon)$, more precisely, it equals $-b_{1,0}\varepsilon$. Thus, dividing both sides of (2.1.29) by $m_1^{(\varepsilon)}\varepsilon$ and evaluating the corresponding limits we obtain using $\mathbf{P}_1^{(k)}$ that $\rho^{(\varepsilon)}/\varepsilon \to -b_{1,0}/b_{0,1}$. This means that $\rho^{(\varepsilon)}$ can be represented in the form

$$\rho^{(\varepsilon)} = a_1\varepsilon + \rho_1^{(\varepsilon)}, \qquad (2.1.30)$$

where $a_1 = -b_{1,0}/b_{0,1}$ and $\rho_1^{(\varepsilon)} = o(\varepsilon)$.

Relation (2.1.30) reduces to (2.1.25) in the case $k = 1$.

Let $k \geq 2$. Substituting the expansions given in condition $\mathbf{P}_1^{(k)}$ and (2.1.30) into (2.1.29) and taking into account the identity $a_1 = -b_{1,0}/b_{0,1}$, we get

$$(b_{1,1}\varepsilon + o(\varepsilon))a_1\varepsilon + (b_{0,1} + b_{1,1}\varepsilon + o(\varepsilon))\rho_1^{(\varepsilon)}$$

$$+ (b_{0,2} + o(1))(a_1\varepsilon + \rho_1^{(\varepsilon)})^2/2!$$

$$+ \cdots + (b_{0,k} + o(1))(a_1\varepsilon + \rho_1^{(\varepsilon)})^k/k! = -b_{2,0}\varepsilon^2 + o(\varepsilon^2). \qquad (2.1.31)$$

Dividing both sides in (2.1.31) by $b_{0,1}\varepsilon^2$ and evaluating the corresponding limits we obtain that $b_{1,1}a_1/b_{0,1} + \rho_1^{(\varepsilon)}/\varepsilon^2 + b_{0,2}a_1^2/2b_{0,1} \to -b_{2,0}/b_{0,1}$. This is equivalent to the following asymptotical relation:

$$\rho_1^{(\varepsilon)} = a_2\varepsilon^2 + \rho_2^{(\varepsilon)}, \qquad (2.1.32)$$

where $a_2 = -b_{0,1}^{-1}(b_{2,0} + b_{1,1}a_1 + \frac{1}{2}b_{0,2}a_1^2)$ and $\rho_2^{(\varepsilon)} = o(\varepsilon^2)$.

Relations (2.1.30) and (2.1.32) yield relation (2.1.25) for $k = 2$. The expression for a_2 given above is exactly formula (2.1.26) for $k = 2$.

Repeating the above argument in the general case we obtain expansion (2.1.25) and formula (2.1.26) for $k > 2$. However, formula (2.1.26) can be obtained in a simpler way once the asymptotic expansion (2.1.25) has already been obtained. From (2.1.29) we get the following formal equation:

$$(b_{0,1} + b_{1,1}\varepsilon + \cdots)(a_1\varepsilon + a_2\varepsilon^2 + \cdots)/1!$$

$$+ (b_{0,2} + b_{1,2}\varepsilon + \cdots)(a_1\varepsilon + a_2\varepsilon^2 + \cdots)^2/2! + \cdots$$

$$= -(b_{1,0}\varepsilon + b_{2,0}\varepsilon^2 + \cdots). \qquad (2.1.33)$$

Equating the coefficients of ε^n for $n \geq 1$ in the right and the left-hand sides of (2.1.33) we obtain formula (2.1.26) for calculating the coefficients $a_1, \ldots, a_k$.

The first summand in (2.1.26) reflects the contribution of the sum in the right-hand side of (2.1.33). The second and the third terms in the sum in (2.1.26) equal 0 for $n = 1$ and, in this case, formula (2.1.33) yields values of the coefficients $a_1 = -b_{1,0}/b_{0,1}$. Thus the only case where $n \geq 2$ should be considered. The second term in (2.1.26) reflects the contribution of the first product of sums to the left-hand side in (2.1.33). The sum with the index $2 \leq m \leq n$ in (2.1.26) shows the contribution of the product of the sums $(b_{0,m} + \cdots)(a_1\varepsilon + \cdots)^m/m!$ in the right-hand side in (2.1.33). In this case, we collect all terms in this product such that the sum $(b_{0,m} + \cdots)$ contributes ε^{n-q} and $(a_1\varepsilon + \cdots)^m/m!$ contributes ε^q for $m \leq q \leq n$. Note that the cases $1 \leq q < m$ are impossible, since the minimal degree of the power of ε in $(a_1\varepsilon + \cdots)^m/m!$ is m. The coefficient at ε^q, coming from $(a_1\varepsilon + \cdots)^m/m!$, is the sum of the products $a_1^{n_1} \cdots a_q^{n_q}$, where the nonnegative integer coefficients $n_1, \ldots, n_q$ must satisfy the obvious relations (a) $n_1 + \cdots + n_q = m$ and (b) $n_1 + \cdots + qn_q = q$. These relations reduce to (2.1.27), since the coefficient n_q can take the only value 0. Indeed, relation (b) implies that the only values 0 and 1 are possible for the coefficient n_q. However, if $n_q = 1$, then all other coefficients $n_1, \ldots, n_{q-1}$ must equal 0. In this case, $n_1 + \cdots + n_q = 1$, which contradicts the restriction $2 \leq m \leq n$. Therefore, the coefficient n_q must be 0.

Finally, the combinatoric counting of the coefficient at the product $a_1^{n_1} \cdots a_q^{n_q}$ yields the number $m!/n_1! \cdots n_q!$. The coefficient $m!$ cancels since it also appears as the denominator of $(a_1\varepsilon + \cdots)^m/m!$. Also, $n_q = 0$, as was mentioned above. This completes the proof of **(i)**.

The statement **(ii)** is a direct corollary of the recursive formula (2.1.26).

Let us use now Theorem 1.4.1. This theorem implies that under conditions $\mathbf{D_6}$, $\mathbf{C_1}$ and $\mathbf{F_5}$ for any $0 \leq t^{(\varepsilon)} \to \infty$ the following asymptotical relation holds:

$$\frac{x^{(\varepsilon)}(t^{(\varepsilon)})}{\exp\{-\rho^{(\varepsilon)}t^{(\varepsilon)}\}} \to x^{(0)}(\infty) \text{ as } \varepsilon \to 0. \tag{2.1.34}$$

Statement **(iii)** is an obvious corollary of relations (2.1.25) and (2.1.34). Under condition $\varepsilon^r t_\varepsilon \to \lambda_r \in [0, \infty)$, we have $(a_{r+1}\varepsilon^{r+1} + \cdots + a_k\varepsilon^k + o(\varepsilon^k))t_\varepsilon \to 0$ and, so, $\exp\{-(a_{r+1}\varepsilon^{r+1} + \cdots + a_k\varepsilon^k + o(\varepsilon^k))t_\varepsilon\} \to 1$. Also, $a_r\varepsilon^r t_\varepsilon \to a_r\lambda_r$. Using these facts and the asymptotic expansion $\rho^{(\varepsilon)} = a_1\varepsilon + \cdots + a_k\varepsilon^k + o(\varepsilon^k)$ obtained in **(i)**, one can easily transform relation (2.1.34) to the form (2.1.28). The proof is complete. $\qquad\square$

Remark 2.1.1. The coefficient a_n can be calculated by using formula (2.1.26) if $n \leq k$ and the value of coefficient a_n does not depend on the parameter k. Formula (2.1.26) gives the following explicit expressions for first three coefficients in the expansion (2.1.25) in terms of the coefficients that appear in the perturbation condition $\mathbf{P}_1^{(\mathbf{k})}$:

$$a_1 = -\frac{b_{1,0}}{b_{0,1}},$$

$$a_2 = -\frac{1}{b_{0,1}}\left(b_{2,0} + b_{1,1}a_1 + \frac{1}{2}b_{0,2}a_1^2\right) = -\frac{b_{2,0}}{b_{0,1}} + \frac{b_{1,1}b_{1,0}}{b_{0,1}^2} - \frac{b_{0,2}b_{1,0}^2}{2b_{0,1}^3},$$

$$a_3 = -\frac{1}{b_{0,1}}\left(b_{3,0} + b_{2,1}a_1 + b_{1,1}a_2 + \frac{1}{2}b_{1,2}a_1^2 + b_{0,2}a_1a_2 + \frac{1}{6}b_{0,3}a_1^3\right). \tag{2.1.35}$$

2.1.3 Asymptotically proper perturbed renewal equation with the defect of order $O(\varepsilon^h)$

Let us introduce integer parameters $1 \le h \le k_0$ and $k_r \ge 1, r = 1, \ldots,$ and define a vector parameter $\bar{k} = (h, k_0, \ldots, k_{\tilde{N}})$, where $\tilde{N} = \min(n : nh \le k_0)$.

We assume that the following perturbation condition holds:

$\mathbf{P}_2^{(\bar{k})}$:   **(a)** $1 - f^{(\varepsilon)} = 1 + b_{h,0}\varepsilon^h + \cdots + b_{k_0,0}\varepsilon^{k_0} + o(\varepsilon^{k_0})$, where $|b_{l,0}| < \infty$, $l = h, \ldots, k_0$;

       (b) $m_r^{(\varepsilon)} = m_r^{(0)} + b_{1,r}\varepsilon + \cdots + b_{k_r,r}\varepsilon^{k_r} + o(\varepsilon^{k_r})$ for $r = 1, \ldots, \tilde{N}$, where $|b_{l,r}| < \infty, l = 1, \ldots, k_r, r = 1, \ldots, \tilde{N}$.

As before, it is convenient to also define the coefficients $b_{0,0} = 1$ and $b_{0,r} = m_r^{(0)}$, $r \ge 1$.

Condition $\mathbf{P}_2^{(\bar{k})}$ is more general than condition $\mathbf{P}_1^{(\mathbf{k})}$, and the first one reduces to the second one if $h = 1, k_0 = k$ (in this case, $\tilde{N} = k$) and $k_r = k - r, r = 1, \ldots, \tilde{N}$.

The following theorem supplements and generalises Theorem 2.1.2.

Theorem 2.1.3. *Let conditions $\mathbf{D}_6$, $\mathbf{C}_1$, and $\mathbf{P}_2^{(\bar{k})}$ hold. Then*

(i) *For all ε small enough, equation (2.1.4) has a unique nonnegative root $\rho^{(\varepsilon)}$ with the following asymptotic expansion:*

$$\rho^{(\varepsilon)} = a_h\varepsilon^h + \cdots + a_k\varepsilon^k + o(\varepsilon^k), \tag{2.1.36}$$

where $k = \min(k_0, k_1 + h, \ldots, k_{\tilde{N}} + h\tilde{N})$ and the coefficients a_n, $n = h, \ldots, k$, are given by the recurrence formulas $a_h = -b_{h,0}/b_{0,1}$ and, in general, for $n = h, \ldots, k$,

$$a_n = -b_{0,1}^{-1}\left(b_{n,0} + \sum_{i=h}^{n-1} b_{n-i,1}a_i \right.$$

$$\left. + \sum_{r=2}^{[n/h]}\sum_{j=rh}^{n} b_{n-j,r} \cdot \sum_{n_h,\ldots,n_{j-1}\in D_{h,r,j}} \prod_{i=h}^{j-1} a_i^{n_i}/n_i!\right), \tag{2.1.37}$$

where $D_{h,r,j}$, for every $h \geq 1$ and $2 \leq r \leq j < \infty$, is the set of all nonnegative, integer solutions of the system

$$n_h + \cdots + n_{j-1} = r, \quad h n_h + \cdots + (j-1)n_{j-1} = j. \qquad (2.1.38)$$

(ii) *If $b_{l,0} = 0$, $l = h, \ldots, r$, for for some $h \leq r \leq k$, then $a_1, \ldots, a_r = 0$. If $b_{l,0} = 0$, $l = h, \ldots, r-1$ but $b_{r,0} < 0$ for some $h \leq r \leq k$, then $a_h, \ldots, a_{r-1} = 0$ but $a_r > 0$.*

(iii) *If, additionally, conditions $\mathbf{B}_3^{(\mathrm{r})}$, for some $h \leq r \leq k$, and $\mathbf{F_5}$ hold, then the following asymptotic relation holds:*

$$\frac{x^{(\varepsilon)}(t^{(\varepsilon)})}{\exp\{-(a_h\varepsilon + \cdots + a_{r-1}\varepsilon^{r-1})t^{(\varepsilon)}\}} \to e^{-\lambda_r a_r} x^{(0)}(\infty) \quad \text{as } \varepsilon \to 0. \quad (2.1.39)$$

Proof. The first part of the proof that yields equation (2.1.29) is the same as in the proof of Theorem 2.1.2.

We have $\rho^{(\varepsilon)} \to 0$, and the sum of all terms in the left-hand side of (2.1.29), starting with the second one, is $o(\rho^{(\varepsilon)})$. Also, $m_r^{(\varepsilon)} \to m_r^{(0)} \in (0, \infty)$ for $r \geq 1$. However, the first term in the asymptotic expansion for the defect $f^{(\varepsilon)}$ given by condition $\mathbf{P}_2^{(\mathbf{k})}$ (a) is not of order $O(\varepsilon)$, but of the order $O(\varepsilon^h)$, more precisely it equals $-b_{h,0}\varepsilon^h$. Dividing both sides in (2.1.29) by $m_1^{(\varepsilon)}\varepsilon^h$ and evaluating the corresponding limits we obtain using $\mathbf{P}_2^{(\mathbf{k})}$ that $\rho^{(\varepsilon)}/\varepsilon^h \to -b_{h,0}/b_{0,1}$. This means that $\rho^{(\varepsilon)}$ can be represented in the form

$$\rho^{(\varepsilon)} = a_h\varepsilon + \rho_h^{(\varepsilon)}, \qquad (2.1.40)$$

where $a_h = -b_{h,0}/b_{0,1}$ and $\rho_h^{(\varepsilon)} = o(\varepsilon^h)$. Relation (2.1.40) reduces to (2.1.36) if $k = h$.

The following steps in the proof of existence of the corresponding expansion for the root $\rho^{(\varepsilon)}$ are similar to the ones in the proof of Theorem 2.1.2.

Let $k \geq h+1$. This means that $k_0, k_1+h, \ldots, k_{\tilde{N}}+\tilde{N}h \geq h+1$. Thus, substituting the expansions given in condition $\mathbf{P}_2^{(\mathbf{k})}$ and (2.1.40) into (2.1.29) and using the identity $a_h = -b_{h,0}/b_{0,1}$ one can rewrite (2.1.29) in the following form:

$$(b_{1,1}\varepsilon + o(\varepsilon))a_h\varepsilon^h + (b_{0,1} + b_{1,1}\varepsilon + o(\varepsilon))\rho_h^{(\varepsilon)}$$

$$+ (b_{0,2} + o(1))(a_h\varepsilon^h + \rho_h^{(\varepsilon)})^2/2! + \cdots$$

$$+ (b_{0,k} + o(1))(a_h\varepsilon^h + \rho_h^{(\varepsilon)})^k/k! = -b_{h+1,0}\varepsilon^{h+1} + o(\varepsilon^{h+1}). \quad (2.1.41)$$

Then, divide both sides in (2.1.41) by $b_{0,1}\varepsilon^{h+1}$ and evaluate the corresponding limits. The terms $b_{0,r}a_h^r\varepsilon^{rh}/b_{0,1}\varepsilon^{h+1}$ for $r \geq 2$ can contribute with non-zero limits only if $2 \leq r \leq [(h+1)/h]$. In fact, this is possible only if $h = 1$. We thus obtain, by evaluating the corresponding limits, that $b_{1,1}a_h/b_{0,1} + \rho_1^{(\varepsilon)}/\varepsilon^{h+1} + \chi(2 \leq r \leq [(h+1)/h])b_{0,2}a_h^2/2b_{0,1} \to -b_{h+1,0}/b_{0,1}$. This is equivalent to the following asymptotical relation:

$$\rho_1^{(\varepsilon)} = a_{h+1}\varepsilon^{h+1} + \rho_{n+1}^{(\varepsilon)}, \tag{2.1.42}$$

where $a_{h+1} = b_{0,1}^{-1}(-b_{h+1,0} - b_{1,1}a_h - \chi(2 \leq r \leq [(h+1)/h])b_{0,2}a_h^2/2$ and $\rho_{h+1}^{(\varepsilon)} = o(\varepsilon^{h+1})$.

Relations (2.1.40) and (2.1.42) yield the relation (2.1.36) for the case $k = h + 1$. The expression for a_2 given above is exactly formula (2.1.37) with $k = h + 1$.

Repeating the above argument in the general case we obtain expansion (2.1.36) and formula (2.1.37) for $k > h + 1$.

Let us make a comment on the interruption procedure for this algorithm. The algorithm should stop at the point where the limits in (2.1.41) can be evaluated, after the corresponding exclusion transformation and division by $b_{0,1}\varepsilon^n$ were performed, but can not be evaluated for $b_{0,1}\varepsilon^{n+1}$. This will happen when we find for first time (by increasing n) that ε^{n+1} can not be compared with at least one of the remaining terms of the form $o(\varepsilon^{k_0}), o(\varepsilon^{k_1})\varepsilon^h, \ldots, o(\varepsilon^{k_N})\varepsilon^{\tilde{N}h}$, where $o(\varepsilon^{k_0}), o(\varepsilon^{k_1}), \ldots, o(\varepsilon^{k_{\tilde{N}}})$ are the corresponding remaining terms in the perturbation expansions in condition $\mathbf{P}_2^{(k)}$. This will occur for $n = k$.

Formula (2.1.37) can be obtained from the asymptotic expansion (2.1.36) in a way similar to the one used in the proof of Theorem 2.1.2.

The coefficients $a_h, \ldots, a_k$ can be found by equating the coefficients of ε^n for $n \geq 1$ on the right and the left-hand hand sides of the following formal relation obtained from (2.1.29):

$$(b_{0,1} + b_{1,1}\varepsilon + \cdots)(a_h\varepsilon^h + a_{h+1}\varepsilon^{h+1}\cdots)/1!$$
$$+ (b_{0,2} + b_{1,2}\varepsilon + \cdots)(a_h\varepsilon^h + a_{h+1}\varepsilon^{h+1} + \cdots)^2/2! + \cdots$$
$$= -(b_{h,0}\varepsilon^h + b_{h+1,0}\varepsilon^{h+1} + \cdots). \tag{2.1.43}$$

The remaining part of the proof repeats the corresponding part in Theorem 2.1.2. $\square$

Remark 2.1.2. The coefficient a_n can be calculated according to formula (2.1.37) if $h \leq n \leq k$ and the value of the coefficient a_n does not depend on the parameter k. In the case where $h = 1$, formula (2.1.37) gives the same expressions for the coefficients a_n as in formula (2.1.26) in Theorem 2.1.2. The corresponding explicit expressions for the first three coefficients in the expansion (2.1.36) in terms of the coefficients that appear in the perturbation condition $\mathbf{P}_1^{(\bar{k})}$ if $h = 2$ are

$$a_2 = -\frac{b_{2,0}}{b_{0,1}},$$

$$a_3 = -\frac{1}{b_{0,1}}(b_{3,0} + b_{1,1}a_2) = -\frac{b_{3,0}}{b_{0,1}} + \frac{b_{1,1}b_{2,0}}{b_{0,1}^2},$$

$$a_4 = -\frac{1}{b_{0,1}}\left(b_{4,0} + b_{2,1}a_2 + b_{1,1}a_3 + \frac{1}{2}b_{0,2}a_2^2\right). \tag{2.1.44}$$

2.2 Exponential expansions for improper perturbed renewal equation

In this section we give exponential expansions for solutions of the improper perturbed renewal equation.

2.2.1 The asymptotically improper perturbed renewal equation with the perturbation of order $O(\varepsilon)$

Consider again, for every $\varepsilon \geq 0$, the renewal equation (2.1.1).

We use the notations

$$\bar{F}^{(\varepsilon)}(t) = F^{(\varepsilon)}(t)/F^{(\varepsilon)}(\infty), \ t \geq 0, \ f^{(\varepsilon)} = 1 - F^{(\varepsilon)}(\infty).$$

The following condition replaces $\mathbf{D}_6$:

$\mathbf{D}_{11}$: (a) $\bar{F}^{(\varepsilon)}(\cdot) \Rightarrow \bar{F}^{(0)}(\cdot)$ as $\varepsilon \to 0$, where $\bar{F}^{(0)}(t)$ is a proper non-arithmetic distribution function;

(b) $f^{(\varepsilon)} \to f^{(0)} \in [0, 1)$ as $\varepsilon \to 0$.

We also assume the following Cramér type condition (introduced in Subsection 1.4.2):

$\mathbf{C}_2$: There exists $\delta > 0$ such that:

(a) $\overline{\lim}_{0 \leq \varepsilon \to 0} \int_0^\infty e^{\delta s} F^{(\varepsilon)}(ds) < \infty$;

(b) $\int_0^\infty e^{\delta s} F^{(0)}(ds) > 1$.

As follows from Lemma 1.4.4, in the case where condition $\mathbf{C}_2$ (a) holds, condition $\mathbf{D}_{11}$ is implied by the following simpler condition:

$\mathbf{D}_{12}$: $F^{(\varepsilon)}(\cdot) \Rightarrow F^{(0)}(\cdot)$ as $\varepsilon \to 0$, where $F^{(0)}(t)$ is a proper or improper distribution function not concentrated in the origin, i.e., such that $F^{(0)}(0) < F^{(0)}(\infty)$, and $\bar{F}^{(0)}(t) = F^{(0)}(t)/F^{(0)}(\infty)$ is a non-arithmetic distribution function.

Let us again consider the characteristic equation (2.1.4). It follows from Lemma 1.4.3 that under conditions $\mathbf{D}_{11}$ and $\mathbf{C}_2$ there exists a unique nonnegative root $\rho^{(\varepsilon)}$ of equation (2.1.4) for all ε small enough. Note that $\rho^{(\varepsilon)} = 0$ if and only if $f^{(\varepsilon)} = 0$ or, equivalently, $\rho^{(\varepsilon)} > 0$ if and only if $f^{(\varepsilon)} > 0$. Also, $\rho^{(\varepsilon)} \to \rho^{(0)}$ as $\varepsilon \to 0$.

Let us introduce the following mixed moment generating functions:

$$\phi^{(\varepsilon)}(\rho, r) = \int_0^\infty s^r e^{\rho s} F^{(\varepsilon)}(ds), \quad r = 0, 1, \dots .$$

Note that, by the definition, $\phi^{(\varepsilon)}(\rho, 0)$ is the Laplace transform of the distribution function $F^{(\varepsilon)}(x)$. Also, $1 - \phi^{(\varepsilon)}(0, 0) = f^{(\varepsilon)}$ is the defect and $\phi^{(\varepsilon)}(0, r) = m_r^{(\varepsilon)}$, $r \geq 1$, are the corresponding moment functionals for this distribution function. Note also that $\phi^{(\varepsilon)}(\rho, r)$ are non-negative and non-decreasing functions in ρ.

Condition $\mathbf{C}_2$ implies that, for any $\beta < \delta$, the moment generating functions satisfy $\phi^{(\varepsilon)}(\beta, r) < \infty$, $r \geq 1$, for ε small enough.

Indeed, condition $\mathbf{C}_2$ implies that $\phi^{(\varepsilon)}(\delta, 0) < \infty$ for ε small enough, say $\varepsilon \leq \varepsilon_0$. Denote $c_r = \sup_{s \geq 0} s^r e^{-(\delta - \beta)s} < \infty$. Then we have for $\varepsilon \leq \varepsilon_0$ and $r = 0, 1, \dots$ that

$$\phi^{(\varepsilon)}(\beta, r) = \int_0^\infty s^r e^{\beta s} F^{(\varepsilon)}(ds)$$

$$\leq c_r \int_0^\infty e^{\delta s} F^{(\varepsilon)}(ds) = c_r \phi^{(\varepsilon)}(\delta, 0) < \infty.$$

By condition $\mathbf{C}_2$, $\rho^{(0)} < \delta$. Therefore, β can be chosen such that $\rho^{(0)} < \beta < \delta$. Also, $\rho^{(\varepsilon)} \to \rho^{(0)}$ as $\varepsilon \to 0$. Thus, $\rho^{(\varepsilon)} < \beta < \delta$ for all ε small enough. By remarks above, we have $\phi^{(\varepsilon)}(\rho^{(0)}, r), \phi^{(\varepsilon)}(\rho^{(\varepsilon)}, r) < \infty$, $r = 0, 1, \dots$, for all ε small enough.

The following perturbation condition plays a crucial role in the subsequent analysis:

$\mathbf{P}_3^{(\mathbf{k})}$: $\phi^{(\varepsilon)}(\rho^{(0)}, r) = \phi^{(0)}(\rho^{(0)}, r) + b_{1,r}\varepsilon + \cdots + b_{k-r,r}\varepsilon^{k-r} + o(\varepsilon^{k-r})$ for $r = 0, \dots, k$, where $|b_{n,r}| < \infty$, $n = 1, \dots, k-r$, $r = 0, \dots, k$.

It is convenient to define $b_{0,r} = \phi^{(0)}(\rho^{(0)}, r)$, $r = 0, 1, \dots$. From the definition of $\rho^{(0)}$ it is clear that $b_{0,0} = \phi^{(0)}(\rho^{(0)}, 0) = 1$.

Note also that conditions $\mathbf{D}_{11}$ and $\mathbf{P}_3^{(\mathbf{k})}$ imply that

$$\phi^{(\varepsilon)}(\rho^{(0)}, r) \to \phi^{(0)}(\rho^{(0)}, r) \in (0, \infty) \quad \text{as } \varepsilon \to 0, \quad r = 0, \dots, k. \tag{2.2.1}$$

It should be noted that, in the case $\rho^{(0)} = 0$, the perturbation condition $\mathbf{P}_3^{(\mathbf{k})}$ is reduced to the perturbation condition $\mathbf{P}_1^{(\mathbf{k})}$. For this reason, we use the same symbols to denote the coefficients in the corresponding expansions despite that $\mathbf{P}_3^{(\mathbf{k})}$ involves the mixed power-exponential moments while $\mathbf{P}_1^{(\mathbf{k})}$ the power moments.

We also assume that the following condition (introduced in Subsection 1.4.2) holds:

$\mathbf{F_6}$: (a) $\lim_{u \to 0} \overline{\lim}_{0 \le \varepsilon \to 0} \sup_{|v| \le u} |q^{(\varepsilon)}(t+v) - q^{(0)}(t)| = 0$ almost everywhere with respect to Lebesgue measure on $[0, \infty)$;

(b) $\overline{\lim}_{0 \le \varepsilon \to 0} \sup_{0 \le t \le T} |q^{(\varepsilon)}(t)| < \infty$ for every $T \ge 0$;

(c) $\lim_{T \to \infty} \overline{\lim}_{0 \le \varepsilon \to 0} h \sum_{r \ge T/h} \sup_{rh \le t \le (r+1)h} e^{(\rho^{(0)}+\gamma)t} |q^{(\varepsilon)}(t)| = 0$ for some $h > 0$ and $\gamma > 0$.

We denote

$$\tilde{x}^{(\varepsilon)}(\infty) = \frac{\int_0^\infty e^{\rho^{(\varepsilon)}s} q^{(\varepsilon)}(s) m(ds)}{\int_0^\infty s e^{\rho^{(\varepsilon)}s} F^{(\varepsilon)}(ds)}.$$

Conditions $\mathbf{D_{11}}$, $\mathbf{C_2}$, and $\mathbf{F_6}$ imply that $\int_0^\infty e^{\rho^{(\varepsilon)}s} |q^{(\varepsilon)}(s)| m(ds) < \infty$ for all ε small enough. Indeed, $\rho^{(\varepsilon)} \to \rho^{(0)}$ as $\varepsilon \to 0$. Thus, $\rho^{(\varepsilon)} < \rho^{(0)} + \gamma$ for all ε small enough. Therefore, the renewal limit $\tilde{x}^{(\varepsilon)}(\infty)$ is well defined for all ε small enough.

The following theorem generalises Theorem 2.1.2 to the case where the root $\rho^{(0)}$ of the limiting equation (2.1.4) can be equal to 0 or positive. In the first case, the theorem reduces to the case considered in Theorem 2.1.2.

Theorem 2.2.1. *Let conditions $\mathbf{D_{11}}$, $\mathbf{C_2}$, and $\mathbf{P_3^{(k)}}$ hold. Then*

(i) *The root $\rho^{(\varepsilon)}$ of equation (2.1.4) has the asymptotic expansion*

$$\rho^{(\varepsilon)} = \rho^{(0)} + a_1 \varepsilon + \cdots + a_k \varepsilon^k + o(\varepsilon^k), \tag{2.2.2}$$

where the coefficients a_n are given by the recurrence formulas $a_1 = -b_{1,0}/b_{0,1}$ and, in general, for $n = 1, \ldots, k$,

$$a_n = -b_{0,1}^{-1}\left(b_{n,0} + \sum_{q=1}^{n-1} b_{n-q,1} a_q \right.$$

$$\left. + \sum_{2 \le m \le n} \sum_{q=m}^{n} b_{n-q,m} \cdot \sum_{n_1,\ldots,n_{q-1} \in D_{m,q}} \prod_{p=1}^{q-1} a_p^{n_p}/n_p! \right), \tag{2.2.3}$$

where $D_{m,q}$, for every $2 \le m \le q < \infty$, is the set of all nonnegative, integer solutions of the system

$$n_1 + \cdots + n_{q-1} = m, \quad n_1 + \cdots + (q-1)n_{q-1} = q. \tag{2.2.4}$$

(ii) *If $b_{l,0} = 0$, $l = 1, \ldots, r$, for some $1 \le r \le k$, then $a_1, \ldots, a_r = 0$. If $b_{l,0} = 0$, $l = 1, \ldots, r-1$ but $b_{r,0} < 0$, for some $1 \le r \le k$, then $a_1, \ldots, a_{r-1} = 0$ but $a_r > 0$.*

(iii) *If, additionally, conditions* $\mathbf{B}_3^{(r)}$*, for some* $1 \leq r \leq k$*, and* $\mathbf{F}_6$ *hold, then the following asymptotic relation holds:*

$$\frac{x^{(\varepsilon)}(t^{(\varepsilon)})}{\exp\{-(\rho^{(0)} + a_1\varepsilon + \cdots + a_{r-1}\varepsilon^{r-1})t^{(\varepsilon)}\}}$$
$$\to e^{-\lambda_r a_r}\tilde{x}^{(0)}(\infty) \;\; as \;\; \varepsilon \to 0. \tag{2.2.5}$$

Before we prove Theorem 2.2.1 we make a few remarks.

Remark 2.2.1. Formula (2.2.3) has the same form as formula (2.1.26). The only difference is that, in (2.2.3), the coefficients $b_{l,r}$ are taken from perturbation expansions for the mixed power-exponential moments $\phi^{(\varepsilon)}(\rho^{(0)}, r)$ given in condition $\mathbf{P}_3^{(k)}$ while, in (2.1.26), the coefficients $b_{l,r}$ are taken from perturbation expansions for the mixed power moments $m_r^{(\varepsilon)}$ given in condition $\mathbf{P}_1^{(k)}$. In particular,

$$a_1 = -\frac{b_{1,0}}{b_{0,1}},$$

$$a_2 = -\frac{1}{b_{0,1}}(b_{2,0} + b_{1,1}a_1 + \frac{1}{2}b_{0,2}a_1^2) = -\frac{b_{2,0}}{b_{0,1}} + \frac{b_{1,1}b_{1,0}}{b_{0,1}^2} - \frac{b_{0,2}b_{1,0}^2}{2b_{0,1}^3},$$

$$a_3 = -\frac{1}{b_{0,1}}\left(b_{3,0} + b_{2,1}a_1 + b_{1,1}a_2 + \frac{1}{2}b_{1,2}a_1^2 + b_{0,2}a_1a_2 + \frac{1}{6}b_{0,3}a_1^3\right). \tag{2.2.6}$$

Remark 2.2.2. The the case where $F^{(\varepsilon)} = F^{(0)}$ and $q^{(\varepsilon)} = q^{(0)}$ do not depend on ε is also covered by Theorem 2.2.1. In this case, condition $\mathbf{P}_3^{(k)}$ holds for every $k \geq 1$ with the coefficients satisfying $b_{l,r} = 0$ for $r \geq 0, l \geq 1$. Hence, the expansion (2.2.2) takes place for every $k \geq 1$ with the coefficients $a_l = 0$ for $l = 1,\ldots,k$. In this case, the expansion (2.2.2) takes the trivial form $\rho^{(0)} = \rho^{(0)} + o(\varepsilon^k)$ which implies $o(\varepsilon^k) \equiv 0$ and it does not give any information. However, the statement **(iii)** yields the well-known (Feller 1966) exponential asymptotics for solutions of the renewal equation under the Cramér type condition for the distribution $F^{(0)}$. The transformation of conditions $\mathbf{C}_2$ and $\mathbf{F}_6$, in this case, is obvious. One can always take $t^{(\varepsilon)} = \varepsilon^{-1}$, so the condition that balances the rate of perturbation and the rate of growth of time will automatically be satisfied (with $r = 1$). The statement **(iii)** then takes the form $x^{(0)}(\varepsilon^{-1})/e^{-\rho^{(0)}\varepsilon^{-1}} \to \tilde{x}^{(0)}(\infty)$ as $\varepsilon \to 0$. Since $0 < \varepsilon \to 0$ in an arbitrary way, the last relation is equivalent to the relation $x^{(0)}(t)/e^{-\rho^{(0)}t} \to \tilde{x}^{(0)}(\infty)$ as $t \to \infty$.

Proof of Theorem 2.2.1. By Lemma 1.4.3,

$$\Delta^{(\varepsilon)} = \rho^{(\varepsilon)} - \rho^{(0)} \to 0 \;\; as \;\; \varepsilon \to 0. \tag{2.2.7}$$

The Taylor expansion for the function e^s yields

$$e^{s\rho^{(\varepsilon)}} = e^{s\rho^{(0)}}\big(1 + s\Delta^{(\varepsilon)}/1! + \cdots + s^k(\Delta^{(\varepsilon)})^k/k!$$
$$+ s^{k+1}(\Delta^{(\varepsilon)})^{k+1} e^{s|\Delta^{(\varepsilon)}|}\theta_{k+1}^{(\varepsilon)}(s)/(k+1)!\big), \qquad (2.2.8)$$

where $\theta_{k+1}^{(\varepsilon)}(s)$, $s \geq 0$, is a continuous function such that $0 \leq \theta_{k+1}^{(\varepsilon)}(s) \leq 1$, $s \geq 0$.

Recall that $\rho^{(0)} < \delta$ and $\Delta^{(\varepsilon)} \to 0$. Therefore, there exist $\beta < \delta$ and $\varepsilon_1 = \varepsilon_1(\beta)$ such that $\rho^{(\varepsilon)} \leq \rho^{(0)} + |\Delta^{(\varepsilon)}| < \beta$ for ε small enough, say $\varepsilon \leq \varepsilon_1$.

By the remarks above, one can also assume that there exists $\varepsilon_2 = \varepsilon_2(\beta)$ such that $\phi^{(\varepsilon)}(\beta, r) < \infty$, $r = 0, 1, \ldots$, for $\varepsilon \leq \varepsilon_2$.

By condition $\mathbf{C_2}$,

$$\varlimsup_{0 \leq \varepsilon \to 0} \int_0^\infty s^{k+1} e^{\beta s} F^{(\varepsilon)}(ds) \leq \varlimsup_{0 \leq \varepsilon \to 0} c_{k+1} \int_0^\infty e^{\delta s} F^{(\varepsilon)}(ds) < \infty, \qquad (2.2.9)$$

where $c_{k+1} = \sup_{s \geq 0} s^{k+1} e^{-(\delta-\beta)s} < \infty$.

It follows from relation (2.2.9) that there exists $\varepsilon_3 = \varepsilon_3(\beta)$ such that, for $\varepsilon \leq \varepsilon_3$,

$$M_{k+1} = \frac{1}{(k+1)!} \sup_{\varepsilon \leq \varepsilon_3} \int_0^\infty s^{k+1} e^{s\beta} F^{(\varepsilon)}(ds) < \infty. \qquad (2.2.10)$$

Let us define

$$\varepsilon_4 = \varepsilon_4(\beta) = \min(\varepsilon_1(\beta), \varepsilon_2(\beta), \varepsilon_3(\beta)). \qquad (2.2.11)$$

Integrating (2.2.8) and taking condition $\mathbf{C_2}$ into account one obtains for $\varepsilon \leq \varepsilon_4$ that

$$1 = \int_0^\infty e^{s\rho^{(\varepsilon)}} F^{(\varepsilon)}(ds)$$
$$= \phi^{(\varepsilon)}(\rho^{(0)}, 0) + \phi^{(\varepsilon)}(\rho^{(0)}, 1)\Delta^{(\varepsilon)}/1! + \cdots$$
$$+ \phi^{(\varepsilon)}(\rho^{(0)}, k)(\Delta^{(\varepsilon)})^k/k! + (\Delta^{(\varepsilon)})^{k+1} M_{k+1}\theta_{k+1}^{(\varepsilon)}, \qquad (2.2.12)$$

where

$$\theta_{k+1}^{(\varepsilon)} = \frac{\int_0^\infty s^{k+1} e^{s\rho^{(\varepsilon)}} \theta_{k+1}^{(\varepsilon)}(s) F^{(\varepsilon)}(ds)}{\sup_{\varepsilon \leq \varepsilon_3} \int_0^\infty s^{k+1} e^{s\beta} F^{(\varepsilon)}(ds)} \in [0, 1].$$

Formula (2.2.12) can be rewritten in the form

$$\phi^{(\varepsilon)}(\rho^{(0)}, 1)\Delta^{(\varepsilon)}/1! + \phi^{(\varepsilon)}(\rho^{(0)}, 2)(\Delta^{(\varepsilon)})^2/2! + \cdots$$
$$+ \phi^{(\varepsilon)}(\rho^{(0)}, k)(\Delta^{(\varepsilon)})^k/k! + (\Delta^{(\varepsilon)})^{k+1} M_{k+1}\theta_{k+1}^{(\varepsilon)}$$
$$= 1 - \phi^{(\varepsilon)}(\rho^{(0)}, 0). \qquad (2.2.13)$$

The following proof is analogous to that given in Theorem 2.1.2 with the only change that the quantity $\rho^{(\varepsilon)}$ is replaced with $\Delta^{(\varepsilon)}$.

The difference $\Delta^{(\varepsilon)} \to 0$ and the sum of all terms in the left-hand side in (2.2.13), starting with the second one, is $o(\Delta^{(\varepsilon)})$. Also, $\phi^{(\varepsilon)}(\rho^{(0)}, r) \to \phi^{(0)}(\rho^{(0)}, r) \in (0, \infty)$ as $\varepsilon \to 0$ for $r = 0, \ldots, k$. Recall also that $b_{0,0} = 1$ and, therefore, the expression in the right-hand side is of order ε. Dividing both sides in (2.2.13) by $b_{0,1}\varepsilon$ and evaluating the corresponding limits we obtain using $\mathbf{P}_3^{(k)}$ that $\Delta^{(\varepsilon)}/\varepsilon \to -b_{1,0}/b_{0,1}$. This means that $\Delta^{(\varepsilon)} = \rho^{(\varepsilon)} - \rho^{(0)}$ can be represented as

$$\Delta^{(\varepsilon)} = \rho^{(\varepsilon)} - \rho^{(0)} = a_1 \varepsilon + \Delta_1^{(\varepsilon)}, \tag{2.2.14}$$

where $a_1 = -b_{1,0}/b_{0,1}$ and $\Delta_1^{(\varepsilon)} = o(\varepsilon)$.

Relation (2.2.14) reduces to (2.2.2) in the case $k = 1$.

Let $k \geq 2$. Substituting the expansions given in condition $\mathbf{P}_3^{(k)}$ and (2.2.14) into (2.2.13) and using the identity $a_1 = -b_{1,0}/b_{0,1}$ we get

$$(b_{1,1}\varepsilon + o(\varepsilon))a_1\varepsilon + (b_{0,1} + b_{1,1}\varepsilon + o(\varepsilon))\Delta_1^{(\varepsilon)}$$
$$+ (b_{0,2} + o(1))(a_1\varepsilon + \Delta_1^{(\varepsilon)})^2/2! + \cdots$$
$$+ (b_{0,k} + o(1))(a_1\varepsilon + \Delta_1^{(\varepsilon)})^k/k! = -b_{2,0}\varepsilon^2 + o(\varepsilon^2). \tag{2.2.15}$$

Dividing both sides in (2.2.15) by $b_{0,1}\varepsilon^2$ and evaluating the corresponding limits we obtain that $b_{1,1}a_1/b_{0,1} + \Delta_1^{(\varepsilon)}/\varepsilon^2 + b_{0,2}a_1^2/2b_{0,1} \to -b_{2,0}/b_{0,1}$. This is equivalent to the following asymptotic relation:

$$\Delta_1^{(\varepsilon)} = a_2\varepsilon^2 + \Delta_2^{(\varepsilon)}, \tag{2.2.16}$$

where $a_2 = b_{0,1}^{-1}(-b_{2,0} - b_{1,1}a_1 - \frac{1}{2}b_{0,2}a_1^2)$ and $\Delta_2^{(\varepsilon)} = o(\varepsilon^2)$.

Relations (2.2.14) and (2.2.16) yield relation (2.2.2) for $k = 2$. The expression for a_2 given above coincides with formula (2.2.3) for $k = 2$.

Repeating the above argument we obtain expansion (2.2.2) and formula (2.2.3) for $k > 2$. However, formula (2.2.3) can be obtained in a simpler way when the asymptotic expansion (2.2.2) has already been proved. From (2.2.13) we get the following formal equation:

$$(b_{0,1} + b_{1,1}\varepsilon + \cdots)(a_1\varepsilon + a_2\varepsilon^2 + \cdots)/1!$$
$$+ (b_{0,2} + b_{1,2}\varepsilon + \cdots)(a_1\varepsilon + a_2\varepsilon^2 + \cdots)^2/2! + \cdots$$
$$= -(b_{1,0}\varepsilon + b_{2,0}\varepsilon^2 + \cdots). \tag{2.2.17}$$

Equating the coefficients of ε^n for $n \geq 1$ in the right and left-hand sides of (2.2.17) we obtain formula (2.2.3) for calculating the coefficients $a_1, \ldots, a_k$.

The first summand in (2.2.3) reflects the contribution of the sum in the right-hand side of (2.2.17). The second term in the sum in (2.2.3) gives a contribution of the first product of the sums in the left-hand side of (2.2.17). The sum with index $2 \leq m \leq n$ in (2.2.3) reflects the contribution of the product of the sums $(b_{0,m} + \cdots)(a_1\varepsilon + \cdots)^m/m!$ in the right-hand side of (2.2.17). In this case, we collect all terms in this product such that the sum $(b_{0,m} + \cdots)$ contributes ε^{n-q} and $(a_1\varepsilon + \cdots)^m/m!$ the ε^q for $m \leq q \leq n$. Note that the cases $1 \leq q < m$ are impossible, since the minimal power of ε, which $(a_1\varepsilon + \cdots)^m/m!$ can contribute to this product, is m. The coefficient of ε^q, which comes from $(a_1\varepsilon + \cdots)^m/m!$, is the sum of the products $a_1^{n_1} \cdots a_q^{n_q}$, where the nonnegative integer coefficients $n_1, \ldots, n_q$ must satisfy the obvious relations $n_1 + \cdots + n_q = m$ and $n_1 + \cdots + qn_q = q$. Only the values 0 and 1 are suitable for the coefficient n_q. Indeed, if $n_q = 1$, then all other coefficients $n_1, \ldots, n_{q-1}$ must equal 0. In this case, $n_1 + \cdots + n_q = 1$, which contradicts the restriction $2 \leq m \leq n$. Therefore, the coefficient n_q must be 0. Finally, combinatoric counting of the coefficients in the product $a_1^{n_1} \cdots a_q^{n_q}$ yields the number $m!/n_1! \cdots n_q!$. Also, $n_q = 0$, as was mentioned above. The coefficient $m!$ cancels, since it also appears as the denominator of $(a_1\varepsilon + \cdots)^m/m!$. This completes the proof of (i).

Let us use now Theorem 1.4.2. This theorem implies that under conditions $\mathbf{D_{11}}, \mathbf{C_3}$ and $\mathbf{F_6}$, for any $0 \leq t^{(\varepsilon)} \to \infty$, the following asymptotical relation holds:

$$\frac{x^{(\varepsilon)}(t^{(\varepsilon)})}{\exp\{-\rho^{(\varepsilon)}t^{(\varepsilon)}\}} \to \tilde{x}^{(0)}(\infty) \ \text{ as } \ \varepsilon \to 0. \tag{2.2.18}$$

Let us prove (ii). Under the balancing condition $\varepsilon^r t^{(\varepsilon)} \to \lambda_r$, the asymptotics relation (2.2.18) can be rewritten in the following equivalent form:

$$\tilde{x}^{(\varepsilon)}(t^{(\varepsilon)}) = \frac{x^{(\varepsilon)}(t^{(\varepsilon)})}{e^{-(\rho^{(0)}+a_1\varepsilon+\cdots+a_r\varepsilon^r+o(\varepsilon^r))t^{(\varepsilon)}}}$$

$$\sim \frac{x^{(\varepsilon)}(t^{(\varepsilon)})}{e^{-(\rho^{(0)}+a_1\varepsilon+\cdots+a_r\varepsilon^r)t^{(\varepsilon)}}} \to \tilde{x}^{(0)}(\infty) \ \text{ as } \ \varepsilon \to 0. \tag{2.2.19}$$

The latter asymptotic relation is equivalent to (2.2.5). $\square$

Remark 2.2.3. The asymptotic relation (2.2.19) can be written assuming the weaker balancing condition $\overline{\lim}_{0<\varepsilon\to0}\, \varepsilon^r t^{(\varepsilon)} = \lambda_r \in [0, \infty)$. The stronger form of this condition used in condition $\mathbf{B_3^{(r)}}$ allows to transform (2.2.19) into the form given in (2.2.5).

2.2.2 Asymptotically improper perturbed renewal equation with the perturbation of order $O(\varepsilon^h)$

Let us formulate without a proof a theorem which is an analogue of Theorem 2.2.1 for a model of the improper perturbed renewal equation.

As above, let us introduce the integer parameters $1 \leq h \leq k_0$ and $k_r \geq 1$, $r = 1, \ldots$, and define a vector parameter $\bar{k} = (h, k_0, \ldots, k_{\tilde{N}})$, where $\tilde{N} = \min(n : nh \leq k_0)$.

We assume that the following perturbation condition holds:

$\mathbf{P}_4^{(\bar{k})}$:   **(a)** $\phi^{(\varepsilon)}(\rho^{(0)}, 0) = \phi^{(0)}(\rho^{(0)}, 0) + b_{h,0}\varepsilon^h + \cdots + b_{k_0,0}\varepsilon^{k_0} + o(\varepsilon^{k_0})$, where $|b_{n,0}| < \infty$, $n = h, \ldots, k_0$;

        (b) $\phi^{(\varepsilon)}(\rho^{(0)}, r) = \phi^{(0)}(\rho^{(0)}, r) + b_{1,r}\varepsilon + \cdots + b_{k_r,r}\varepsilon^{k_r} + o(\varepsilon^{k_r})$ for $r = 1, \ldots, k$, where $|b_{n,r}| < \infty$, $n = 1, \ldots, k_r$, $r = 1, \ldots, \tilde{N}$.

It is convenient to define $b_{0,r} = \phi^{(0)}(\rho^{(0)}, r)$, $r = 0, 1, \ldots$. Note that $\phi^{(0)}(\rho^{(0)}, 0) = 1$ while $\phi^{(0)}(\rho^{(0)}, r) > 0$, $r = 1, \ldots, \tilde{N}$.

Condition $\mathbf{P}_4^{(\bar{k})}$ is more general than condition $\mathbf{P}_3^{(k)}$, and the first one reduces to the second one if $h = 1$, $k_0 = k$ (in this case, $\tilde{N} = k$) and $k_r = k - r$, $r = 1, \ldots, \tilde{N}$.

The following theorem generalises Theorems 2.1.2. The proof is analogous to the one given in Theorems 2.1.3 and 2.2.1.

Theorem 2.2.2. *Let conditions* $\mathbf{D}_{11}$, $\mathbf{C}_2$, *and* $\mathbf{P}_4^{(\bar{k})}$ *hold. Then*

(i) *For all ε small enough, equation (2.1.4) has a unique nonnegative root $\rho^{(\varepsilon)}$ that has the following asymptotic expansion:*

$$\rho^{(\varepsilon)} = \rho^{(0)} + a_h\varepsilon^h + \cdots + a_k\varepsilon^k + o(\varepsilon^k), \qquad (2.2.20)$$

where $k = \min(k_0, k_1+h, \ldots, k_{\tilde{N}}+h\tilde{N})$ and the coefficients a_n, $n = h, \ldots, k$, are given by the recurrence formulas $a_h = -b_{h,0}/b_{0,1}$ and, in general, for $n = h, \ldots, k$, we have

$$a_n = -b_{0,1}^{-1}\left(b_{n,0} + \sum_{i=h}^{n-1} b_{n-i,1}a_i \right.$$

$$\left. + \sum_{r=2}^{[n/h]} \sum_{j=rh}^{n} b_{n-j,r} \cdot \sum_{n_h,\ldots,n_{j-1}\in D_{h,r,j}} \prod_{i=h}^{j-1} a_i^{n_i}/n_i! \right), \qquad (2.2.21)$$

where $D_{h,r,j}$, for every $h \geq 1$ and $2 \leq r \leq j < \infty$, is the set of all nonnegative, integer solutions of the system

$$n_h + \cdots + n_{j-1} = r, \quad hn_h + \cdots + (j-1)n_{j-1} = j. \qquad (2.2.22)$$

(ii) *If $b_{l,0} = 0$, $l = h, \ldots, r$, for some $h \leq r \leq k$, then $a_1, \ldots, a_r = 0$. If $b_{l,0} = 0$, $l = h, \ldots, r-1$ but $b_{r,0} < 0$ for some $h \leq r \leq k$, then $a_h, \ldots, a_{r-1} = 0$ but $a_r > 0$.*

(iii) *If, additionally, conditions* $\mathbf{B}_3^{(r)}$, *for some* $h \leq r \leq k$, *and* $\mathbf{F}_6$ *hold, then the following asymptotic relation holds*:

$$\frac{x^{(\varepsilon)}(t^{(\varepsilon)})}{\exp\{-(\rho^{(0)} + a_h \varepsilon^h + \cdots + a_{r-1}\varepsilon^{r-1})t^{(\varepsilon)}\}}$$
$$\rightarrow e^{-\lambda_r a_r}\tilde{x}^{(0)}(\infty) \quad as \quad \varepsilon \to 0. \tag{2.2.23}$$

Remark 2.2.4. Formula (2.2.21) has the same form as formula (2.1.37). The only difference is that, in (2.2.21), the coefficients $b_{l,r}$ are taken from the perturbation expansions for the mixed power-exponential moments $\phi^{(\varepsilon)}(\rho^{(0)}, r)$ given in condition $\mathbf{P}_4^{(\bar{k})}$ while, in (2.1.37), the coefficients $b_{l,r}$ are taken from the perturbation expansions for the mixed power moments $m_r^{(\varepsilon)}$ given in condition $\mathbf{P}_2^{(\bar{k})}$.

In the case where $h = 1$, we have the following formulas for the first three coefficients:

$$a_1 = -\frac{b_{1,0}}{b_{0,1}},$$

$$a_2 = -\frac{1}{b_{0,1}}(b_{2,0} + b_{1,1}a_1 + \frac{1}{2}b_{0,2}a_1^2) = -\frac{b_{2,0}}{b_{0,1}} + \frac{b_{1,1}b_{1,0}}{b_{0,1}^2} - \frac{b_{0,2}b_{1,0}^2}{2b_{0,1}^3},$$

$$a_3 = -\frac{1}{b_{0,1}}\Big(b_{3,0} + b_{2,1}a_1 + b_{1,1}a_2 + \frac{1}{2}b_{1,2}a_1^2 + b_{0,2}a_1a_2 + \frac{1}{6}b_{0,3}a_1^3\Big). \tag{2.2.24}$$

In the case where $h = 2$, we have the following formulas for the first three coefficients:

$$a_2 = -\frac{b_{2,0}}{b_{0,1}},$$

$$a_3 = -\frac{1}{b_{0,1}}(b_{3,0} + b_{1,1}a_2) = -\frac{b_{3,0}}{b_{0,1}} + \frac{b_{1,1}b_{2,0}}{b_{0,1}^2},$$

$$a_4 = -\frac{1}{b_{0,1}}\Big(b_{4,0} + b_{2,1}a_2 + b_{1,1}a_3 + \frac{1}{2}b_{0,2}a_2^2\Big). \tag{2.2.25}$$

2.3 Asymptotic expansions for renewal limits

In this section we give asymptotic expansions for renewal limits for proper and improper perturbed renewal equations.

2.3.1 Asymptotic expansions for renewal limits for proper perturbed renewal equations

First we consider the simplest case where **(a)** $F^{(\varepsilon)}(+\infty) = 1$ for all $\varepsilon \geq 0$.

Formula (1.2.35) gives, in this case, an expression for the renewal limit (see, Subsection 1.2.6),

$$x^{(\varepsilon)}(\infty) = \frac{q_0^{(\varepsilon)}}{m_1^{(\varepsilon)}},\tag{2.3.1}$$

where

$$q_0^{(\varepsilon)} = \int_0^\infty q^{(\varepsilon)}(s)m(ds), \quad m_1^{(\varepsilon)} = \int_0^\infty sF^{(\varepsilon)}(ds).$$

Convergence of the renewal limits $x^{(\varepsilon)}(\infty)$ requires a perturbation condition simpler than $\mathbf{P}_1^{(k)}$ and involving only the moment functionals $m_1^{(\varepsilon)}$,

$\mathbf{P}_5^{(k)}$: $m_1^{(\varepsilon)} = m_1^{(0)} + b_{1,1}\varepsilon + \cdots + b_{k,1}\varepsilon^k + o(\varepsilon^k)$, where $m_1^{(0)} \in (0,\infty)$ and $|b_{n,1}| < \infty$, $n = 1,\ldots,k$.

It convenient to also define $b_{0,1} = m_1^{(0)}$.

Also the following perturbation condition imposed on the moment functionals for forcing functions is required:

$\mathbf{P}_6^{(k)}$: $q_0^{(\varepsilon)} = q_0^{(0)} + c_{1,0}\varepsilon + \cdots + c_{k,0}\varepsilon^k + o(\varepsilon^k)$, where $|q_0^{(0)}| < \infty$ and $|c_{n,0}| < \infty$, $n = 1,\ldots,k$.

We define $c_{0,0} = q_0^{(0)}$.

Note that $m_1^{(0)}$ and $q_0^{(0)}$ in the condition above are just some finite constants that may be or not of the form $m_1^{(0)} = \int_0^\infty sF^{(0)}(ds)$ and $q_0^{(0)} = \int_0^\infty q^{(0)}(s)m(ds)$.

The following theorem improves Theorem 1.2.4.

Theorem 2.3.1. *Let* **(a)** $F^{(\varepsilon)}(+\infty) = 1$ *for all* $\varepsilon \geq 0$ *and conditions* $\mathbf{P}_5^{(k)}$ *and* $\mathbf{P}_6^{(k)}$ *hold. Then the functional* $x^{(\varepsilon)}(\infty)$ *has the asymptotic expansions:*

$$x^{(\varepsilon)}(\infty) = \frac{q_0^{(0)} + c_{1,0}\varepsilon + \cdots + c_{k,0}\varepsilon^k + o(\varepsilon^k)}{m_1^{(0)} + b_{1,1}\varepsilon + \cdots + b_{k,1}\varepsilon^k + o(\varepsilon^k)}$$

$$= x^{(0)}(\infty) + f_1\varepsilon + \cdots + f_k\varepsilon^k + o(\varepsilon^k),\tag{2.3.2}$$

where the coefficients f_n *are given by the recurrence formulas* $f_0 = x^{(0)}(\infty) = c_{0,0}/b_{0,1}$, *and in general for* $n = 0,\ldots,k$,

$$f_n = b_{0,1}^{-1}\left(c_{n,0} - \sum_{q=0}^{n-1} b_{n-q,1} f_q\right).\tag{2.3.3}$$

Proof. Using formula (2.3.1) we can write the rational expansion (2.3.2). It always (see, Lemma 8.1.1) can be transformed to a polynomial form, since $b_{0,1} \neq 0$. Hence, we can write the following:

$$(x^{(0)}(\infty) + f_1 \varepsilon + \cdots + f_k \varepsilon^k + o(\varepsilon^k))(m_1^{(0)} + b_{1,1} \varepsilon + \cdots$$
$$+ b_{k,1} \varepsilon^k + o(\varepsilon^k)) = c_{0,0} + c_{1,0} \varepsilon + \cdots + c_{k,0} \varepsilon^k + o(\varepsilon^k). \qquad (2.3.4)$$

The recurrence formulas (2.3.3) can be obtained by grouping the coefficients of ε^n in equation (2.3.4). $\qquad\square$

Remark 2.3.1. The coefficients f_n, $n = 0, \ldots, k$, can be calculated according to formula (2.3.4) and the value of the coefficient f_n does not depend on the parameter k. In particular, the formulas for first three coefficients take the following form:

$$f_0 = \frac{c_{0,0}}{b_{0,1}},$$

$$f_1 = \frac{1}{b_{0,1}}(c_{1,0} - b_{1,1} f_0) = \frac{c_{1,0}}{b_{0,1}} - \frac{c_{0,0} b_{1,1}}{b_{0,1}^2},$$

$$f_2 = \frac{1}{b_{0,1}}(c_{2,0} - b_{2,1} f_0 - b_{1,1} f_1)$$

$$= \frac{c_{2,0}}{b_{0,1}} - \frac{c_{0,0} b_{2,1} + c_{1,0} b_{1,1}}{b_{0,1}^2} + \frac{c_{0,0} b_{1,1}^2}{b_{0,1}^3}. \qquad (2.3.5)$$

2.3.2 Asymptotic expansions for renewal limits for improper perturbed renewal equations

Let us now consider the general case where **(b)** $F^{(\varepsilon)}(+\infty) \leq 1$ for all $\varepsilon \geq 0$.

The following formula, which repeats (1.4.39), gives, in this case, an expression for the renewal limit,

$$\tilde{x}^{(\varepsilon)}(\infty) = \frac{\omega^{(\varepsilon)}(\rho^{(\varepsilon)}, 0)}{\phi^{(\varepsilon)}(\rho^{(\varepsilon)}, 1)}, \qquad (2.3.6)$$

where

$$\phi^{(\varepsilon)}(\rho, r) = \int_0^\infty s^r e^{\rho s} F^{(\varepsilon)}(ds), \qquad \rho \geq 0, \quad r = 0, 1, \ldots.$$

Let us now define for the mixed power-exponential moment functionals for the forcing functions, the following:

$$\omega^{(\varepsilon)}(\rho, r) = \int_0^\infty s^r e^{\rho s} q^{(\varepsilon)}(s) m(ds), \qquad \rho \geq 0, \quad r = 0, 1, \ldots,$$

and

$$\bar{\omega}^{(\varepsilon)}(\rho, r) = \int_0^\infty s^r e^{\rho s} |q^{(\varepsilon)}(s)| m(ds), \quad \rho \geq 0, \quad r = 0, 1, \ldots .$$

Condition $\mathbf{F_6}$ implies that, for any $0 < \beta < \rho^{(0)} + \gamma$, the moment generating functions satisfy $\bar{\omega}^{(\varepsilon)}(\beta, r) < \infty$, $r = 0, 1, \ldots$, for all ε small enough.

Indeed, condition $\mathbf{F_6}$ implies that $\bar{\omega}^{(\varepsilon)}(\rho^{(0)} + \gamma, 0) < \infty$ for ε small enough, say $\varepsilon \leq \varepsilon_0$. Denote $c_r = \sup_{s \geq 0} s^r e^{-(\rho^{(0)} + \gamma - \beta)s} < \infty$. Then, we have for $\varepsilon \leq \varepsilon_0$ and $r = 0, 1, \ldots$ that

$$\bar{\omega}^{(\varepsilon)}(\beta, r) = \int_0^\infty s^r e^{\beta s} |q^{(\varepsilon)}(s)| m(ds)$$

$$\leq c_r \int_0^\infty e^{(\rho^{(0)} + \gamma)s} |q^{(\varepsilon)}(s)| m(ds) = c_r \bar{\omega}^{(\varepsilon)}(\rho^{(0)} + \gamma, 0) < \infty.$$

Note that $\bar{\omega}^{(\varepsilon)}(\rho, r)$ are non-negative and non-decreasing functions of $\rho \geq 0$. Also, $\rho^{(\varepsilon)} < \rho^{(0)} + \gamma$, since $\rho^{(\varepsilon)} \to \rho^{(0)}$ as $\varepsilon \to 0$. Thus, $\bar{\omega}^{(\varepsilon)}(\rho^{(\varepsilon)}, r), \bar{\omega}^{(\varepsilon)}(\rho^{(0)}, r) < \infty$, $r = 0, 1, \ldots$, for all ε small enough.

Therefore, the renewal limit $\tilde{x}^{(\varepsilon)}(\infty)$ is well defined for all ε small enough.

Let us now formulate perturbation conditions for mixed power-exponential moment functionals for the forcing functions,

$\mathbf{P_7^{(k)}}$: $\omega^{(\varepsilon)}(\rho^{(0)}, r) = \omega^{(0)}(\rho^{(0)}, r) + c_{1,r}\varepsilon + \cdots + c_{k-r,r}\varepsilon^{k-r} + o(\varepsilon^{k-r})$ for $r = 0, \ldots, k$, where $|c_{n,r}| < \infty$, $n = 1, \ldots, k - r$, $r = 0, \ldots, k$.

It is convenient to set $c_{0,r} = \omega^{(0)}(\rho^{(0)}, r)$, $r = 0, 1, \ldots .$
The following theorem improves Theorem 1.4.4.

Theorem 2.3.2. *Let conditions* $\mathbf{D_{11}}$, $\mathbf{C_2}$, $\mathbf{F_6}$, $\mathbf{P_3^{(k+1)}}$, *and* $\mathbf{P_7^{(k)}}$ *hold. Then the functional* $\tilde{x}^{(\varepsilon)}(\infty)$ *has the following asymptotic expansions:*

$$\tilde{x}^{(\varepsilon)}(\infty) = \frac{\omega^{(0)}(\rho^{(0)}, 0) + f_1'\varepsilon + \cdots + f_k'\varepsilon^k + o(\varepsilon^k)}{\phi^{(0)}(\rho^{(0)}, 1) + f_1''\varepsilon + \cdots + f_k''\varepsilon^k + o(\varepsilon^k)}$$

$$= \tilde{x}^{(0)}(\infty) + f_1\varepsilon + \cdots + f_k\varepsilon^k + o(\varepsilon^k), \tag{2.3.7}$$

where the coefficients f_n', f_n'' *are given by the formulas* $f_0' = \omega^{(0)}(\rho^{(0)}, 0) = c_{0,0}$, $f_1' = c_{1,0} + c_{0,1}a_1$, $f_0'' = \phi^{(0)}(\rho^{(0)}, 1) = b_{0,1}$, $f_1'' = b_{1,1} + b_{0,2}a_1$, *and in general for* $n = 0, \ldots, k$,

$$f_n' = c_{n,0} + \sum_{q=1}^n c_{n-q,1}a_q$$

$$+ \sum_{2 \leq m \leq n} \sum_{q=m}^n c_{n-q,m} \cdot \sum_{n_1, \ldots, n_{q-1} \in D_{m,q}} \prod_{p=1}^{q-1} a_p^{n_p}/n_p!, \tag{2.3.8}$$

and

$$f_n'' = b_{n,1} + \sum_{q=1}^{n} b_{n-q,2} a_q$$

$$+ \sum_{2 \leq m \leq n} \sum_{q=m}^{n} b_{n-q,m+1} \cdot \sum_{n_1,\ldots,n_{q-1} \in D_{m,q}} \prod_{p=1}^{q-1} a_p^{n_p} / n_p!, \qquad (2.3.9)$$

and the coefficients f_n are given by the recurrence formulas $f_0 = \tilde{x}^{(0)}(\infty) = f_0'/f_0''$ and in general for $n = 0, \ldots, k$,

$$f_n = \left(f_n' - \sum_{q=0}^{n-1} f_{n-q}'' f_q \right) / f_0''. \qquad (2.3.10)$$

Proof. We can write the asymptotic expansion, in powers of the difference $\Delta^{(\varepsilon)} = \rho^{(\varepsilon)} - \rho^{(0)}$, for the numerator and the denominator in the quotient expression for $\tilde{x}^{(\varepsilon)}(\infty)$ given by (2.3.6). Recall that $\rho^{(0)} < \delta$, where δ is defined in $\mathbf{C_2}$, and $\Delta^{(\varepsilon)} \to 0$. Hence, for any $\rho^{(0)} < \beta < \delta$ there exists $\varepsilon_1 = \varepsilon_1(\beta)$ such that $\rho^{(\varepsilon)} \leq \rho^{(0)} + |\Delta^{(\varepsilon)}| \leq \beta$ for $\varepsilon \leq \varepsilon_1$.

By the remarks above, there exists $\varepsilon_2 = \varepsilon_2(\beta) > 0$ such that $\omega^{(\varepsilon)}(\beta, r) < \infty$, $r = 0, 1, \ldots$, and $\phi^{(\varepsilon)}(\beta, r) < \infty, r = 0, 1, \ldots$, for $\varepsilon \leq \varepsilon_2$.

Let us use ones more the Taylor expansion for the function e^s,

$$e^{s\rho^{(\varepsilon)}} = e^{s\rho^{(0)}} \big(1 + s\Delta^{(\varepsilon)}/1! + \cdots + s^k (\Delta^{(\varepsilon)})^k / k!$$

$$+ s^{k+1} (\Delta^{(\varepsilon)})^{k+1} e^{s|\Delta^{(\varepsilon)}|} \theta_{k+1}^{(\varepsilon)}(s) / (k+1)! \big), \qquad (2.3.11)$$

where $\theta_{k+1}^{(\varepsilon)}(s), s \geq 0$, is a continuous function such that $0 \leq \theta_{k+1}^{(\varepsilon)}(s) \leq 1, s \geq 0$.

By condition $\mathbf{C_2}$,

$$\varlimsup_{0 \leq \varepsilon \to 0} \int_0^{\infty} s^{k+2} e^{\beta s} F^{(\varepsilon)}(ds) \leq \varlimsup_{0 \leq \varepsilon \to 0} c_{k+2} \int_0^{\infty} e^{\delta s} F^{(\varepsilon)}(ds) < \infty, \qquad (2.3.12)$$

and, by condition $\mathbf{F_6}$,

$$\varlimsup_{0 \leq \varepsilon \to 0} \int_0^{\infty} s^{k+1} e^{\beta s} |q^{(\varepsilon)}(s)| m(ds)$$

$$\leq \varlimsup_{0 \leq \varepsilon \to 0} c_{k+1} \int_0^{\infty} e^{\delta s} |q^{(\varepsilon)}(s)| m(ds) < \infty, \qquad (2.3.13)$$

where $c_k = \sup_{s \geq 0} s^k e^{-(\delta - \beta)s} < \infty$.

It follows from relations (2.2.9) and (2.3.13) that there exists $\varepsilon_3 = \varepsilon_3(\beta) > 0$ such that, for $\varepsilon \leq \varepsilon_3$,

$$M_{k+2} = \frac{1}{(k+1)!} \sup_{\varepsilon \leq \varepsilon_3} \int_0^{\infty} s^{k+2} e^{s\beta} F^{(\varepsilon)}(ds) < \infty, \qquad (2.3.14)$$

and

$$\hat{M}_{k+1} = \frac{1}{(k+1)!} \sup_{\varepsilon \leq \varepsilon_3} \int_0^\infty s^{k+1} e^{s\beta} |q^{(\varepsilon)}(s)| m(ds) + \frac{1}{(k+1)!} < \infty. \qquad (2.3.15)$$

Define

$$\varepsilon_4 = \varepsilon_4(\beta) = \min(\varepsilon_1(\beta), \varepsilon_2(\beta), \varepsilon_3(\beta)). \qquad (2.3.16)$$

Substituting the Taylor expansion (2.3.11) into the integrals that define $\phi^{(\varepsilon)}(\rho^{(\varepsilon)}, 1)$ and $\omega^{(\varepsilon)}(\rho^{(\varepsilon)}, 0)$, we get for $\varepsilon \leq \varepsilon_4$ that

$$\phi^{(\varepsilon)}(\rho^{(\varepsilon)}, 1) = \phi^{(\varepsilon)}(\rho^{(0)}, 1) + \phi^{(\varepsilon)}(\rho^{(0)}, 2)\Delta^{(\varepsilon)} + \cdots$$
$$+ \phi^{(\varepsilon)}(\rho^{(0)}, k+1)(\Delta^{(\varepsilon)})^k / k! + (\Delta^{(\varepsilon)})^{k+1} M_{k+2}\theta_{k+2}^{(\varepsilon)}, \qquad (2.3.17)$$

where

$$\theta_{k+2}^{(\varepsilon)} = \frac{\int_0^\infty s^{k+2} e^{s\rho^{(\varepsilon)}} \theta_{k+2}^{(\varepsilon)}(s) F^{(\varepsilon)}(ds)}{\sup_{\varepsilon \leq \varepsilon_3} \int_0^\infty s^{k+2} e^{s\beta} F^{(\varepsilon)}(ds)} \in [0, 1],$$

and

$$\omega^{(\varepsilon)}(\rho^{(\varepsilon)}, 0) = \omega^{(\varepsilon)}(\rho^{(0)}, 0) + \omega^{(\varepsilon)}(\rho^{(0)}, 1)\Delta^{(\varepsilon)} + \cdots$$
$$+ \omega^{(\varepsilon)}(\rho^{(0)}, k)(\Delta^{(\varepsilon)})^k / k! + (\Delta^{(\varepsilon)})^{k+1} \hat{M}_{k+1}\hat{\theta}_{k+1}^{(\varepsilon)}, \qquad (2.3.18)$$

where

$$\hat{\theta}_{k+1}^{(\varepsilon)} = \frac{\int_0^\infty s^{k+1} e^{s\rho^{(\varepsilon)}} \theta_{k+1}^{(\varepsilon)}(s) q^{(\varepsilon)}(s) m(ds)}{\sup_{\varepsilon \leq \varepsilon_3} \int_0^\infty s^{k+1} e^{s\beta} |q^{(\varepsilon)}(s)| m(ds) + 1} \in [-1, 1].$$

It was proved in Theorem 2.2.1 that the functions $\Delta^{(\varepsilon)} = \rho^{(\varepsilon)} - \rho^{(0)}$ can be represented in the form of the asymptotic expansion $\Delta^{(\varepsilon)} = a_1\varepsilon + \cdots + a_k\varepsilon^k + o(\varepsilon^k)$. The functions $\omega^{(\varepsilon)}(\rho^{(0)}, n)$ and $\phi^{(\varepsilon)}(\rho^{(0)}, n)$ can also be represented in the form of asymptotic expansions due to conditions $\mathbf{P}_3^{(k+1)}$ and $\mathbf{P}_7^{(k)}$, respectively. Substituting these expansions into (2.3.17) and (2.3.18) we obtain

$$\omega^{(\varepsilon)}(\rho^{(\varepsilon)}, 0) = (c_{0,0} + \cdots + c_{k,0}\varepsilon^k + o(\varepsilon^k))$$
$$+ (c_{0,1} + \cdots + c_{k-1,1}\varepsilon^{k-1} + o(\varepsilon^{k-1}))(a_1\varepsilon + \cdots$$
$$+ a_k\varepsilon^k + o(\varepsilon^k))/1! + \cdots$$
$$+ (c_{0,k} + o(1))(a_1\varepsilon + \cdots + a_k\varepsilon^k + o(\varepsilon^k))^k / k! + o(\varepsilon^k), \qquad (2.3.19)$$

and

$$
\begin{aligned}
\phi^{(\varepsilon)}(\rho^{(0)},1) = &(b_{0,1} + \cdots + b_{k,1}\varepsilon^k + o(\varepsilon^k)) \\
&+ (b_{0,2} + \cdots + b_{k-1,2}\varepsilon^{k-1} + o(\varepsilon^{k-1}))(a_1\varepsilon + \cdots \\
&+ a_k\varepsilon^k + o(\varepsilon^k))/1! + \cdots \\
&+ (b_{0,k+1} + o(1))(a_1\varepsilon + \cdots + a_k\varepsilon^k + o(\varepsilon^k))^k/k! \\
&+ o(\varepsilon^k).
\end{aligned}
\tag{2.3.20}
$$

By grouping the coefficients of powers of ε in these expansions one can transform them to the forms described in (2.3.7)–(2.3.9), i.e.,

$$
\omega^{(\varepsilon)}(\rho^{(\varepsilon)},0) = f_0' + \cdots + f_k'\varepsilon^k + o(\varepsilon^k),
\tag{2.3.21}
$$

and

$$
\phi^{(\varepsilon)}(\rho^{(\varepsilon)},1) = f_0'' + \cdots + f_k''\varepsilon^k + o(\varepsilon^k).
\tag{2.3.22}
$$

The transition of the expansion (2.3.7) to a polynomial form is obvious (see, Lemma 8.1.1), since $f_0'' \neq 0$. The corresponding coefficients can be found by grouping the coefficients of ε^n in the equation

$$
\begin{aligned}
(f_0 + \cdots + f_k\varepsilon^k + o(\varepsilon^k))(f_0'' + \cdots + f_k''\varepsilon^k + o(\varepsilon^k)) \\
= f_0' + \cdots + f_k'\varepsilon^k + o(\varepsilon^k).
\end{aligned}
\tag{2.3.23}
$$

The relations (2.3.10) can be found by grouping the coefficients of ε^n in equation (2.3.23). $\qquad\square$

It is useful to note that in the case where $f^{(0)} = 0$, the characteristic root $\rho^{(0)} = 0$ and $\phi^{(\varepsilon)}(0,r) = m_r^{(\varepsilon)}$, $r \geq 1$, reduce to the usual power moments for the distributions functions $F^{(\varepsilon)}(x)$, and $\omega^{(\varepsilon)}(0,r) = q_r^{(\varepsilon)}$, $r \geq 1$, reduce to the corresponding power moment functionals for the forcing functions $q^{(\varepsilon)}(s)$,

$$
q_r^{(\varepsilon)} = \int_0^\infty s^r q^{(\varepsilon)}(s) m(ds), \quad r \geq 1.
$$

In this case, the perturbation condition $\mathbf{P}_3^{(k+1)}$ reduces to the perturbation condition $\mathbf{P}_1^{(k+1)}$.

The perturbation condition $\mathbf{P}_7^{(k)}$ also reduces, under condition $\mathbf{F_6}$, to the following simpler form:

$\mathbf{P}_8^{(k)}$: $q_r^{(\varepsilon)} = q_r^{(0)} + c_{1,r}\varepsilon + \cdots + c_{k-r,r}\varepsilon^{k-r} + o(\varepsilon^{k-r})$ for $r = 0,\ldots,k$, where $|c_{n,r}| < \infty$, $n = 1,\ldots,k-r$, $r = 0,\ldots,k$.

Remark 2.3.2. The following formulas can be obtained for the first three coefficients given by formulas (2.3.8), (2.3.9), and (2.3.10):

$$f_0' = c_{0,0},$$

$$f_1' = c_{1,0} + c_{0,1}a_1 = c_{1,0} - \frac{c_{0,1}b_{1,0}}{b_{0,1}},$$

$$f_2' = c_{2,0} + c_{0,1}a_2 + c_{1,1}a_1 + \frac{c_{0,2}a_1^2}{2},$$

$$f_0'' = b_{0,1},$$

$$f_1'' = b_{1,1} + b_{0,2}a_1 = b_{1,1} - \frac{b_{0,2}b_{1,0}}{b_{0,1}},$$

$$f_2'' = b_{2,1} + b_{0,2}a_2 + b_{1,2}a_1 + \frac{b_{0,3}a_1^2}{2},$$

$$f_0 = \frac{f_0'}{f_0''} = \frac{c_{0,0}}{b_{0,1}},$$

$$f_1 = \frac{1}{f_0''}(f_1' - f_0 f_1'')$$

$$= \frac{c_{1,0}}{b_{0,1}} - \frac{c_{0,1}b_{1,0}}{b_{0,1}^2} - \frac{c_{0,0}b_{1,1}}{b_{0,1}^2} + \frac{c_{0,0}b_{0,2}b_{1,0}}{b_{0,1}^3},$$

$$f_2 = \frac{1}{f_0''}(f_2' - f_0 f_2'' - f_1 f_1''). \tag{2.3.24}$$

Chapter 3

Nonlinearly perturbed regenerative processes

Chapter 3 contains results concerning mixed ergodic and limit theorems and mixed ergodic and large deviation theorems for nonlinearly perturbed regenerative processes. These theorems are given in a form of exponential asymptotics for joint distributions of positions of regenerative processes and regenerative stopping times. These distributions satisfy some renewal equations and the corresponding theorems are obtained by applying results of Chapters 1 and 2 to these renewal equations.

We consider a regenerative process $\xi^{(\varepsilon)}(t)$, $t \geq 0$, with regeneration times $\tau_n^{(\varepsilon)}$, and a regenerative stopping time $\mu^{(\varepsilon)}$ that regenerates jointly with the process $\xi^{(\varepsilon)}(t)$, $t \geq 0$, at times $\tau_n^{(\varepsilon)}$.

Both the regenerative process $\xi^{(\varepsilon)}(t)$, $t \geq 0$, and the regenerative stopping time $\mu^{(\varepsilon)}$ are assumed to depend on a small perturbation parameter $\varepsilon \geq 0$. The processes $\xi^{(\varepsilon)}(t)$, $t \geq 0$, for $\varepsilon > 0$ are considered as a perturbation of the process $\xi^{(0)}(t)$, $t \geq 0$, and therefore we assume some weak continuity conditions for certain characteristic quantities of these processes regarded as functions of ε at point $\varepsilon = 0$.

As far as the regenerative stopping times are concerned, we consider two cases. The first one is where the random variables $\mu^{(\varepsilon)}$ are stochastically unbounded, i.e., $\mu^{(\varepsilon)}$ tend to ∞ in probability as $\varepsilon \to 0$. The second one is where the random variables $\mu^{(\varepsilon)}$ are stochastically bounded as $\varepsilon \to 0$.

The object of our study is the joint distributions $\mathsf{P}\{\xi^{(\varepsilon)}(t) \in A, \mu^{(\varepsilon)} > t\}$ and their asymptotic behaviour as $t \to \infty$ and $\varepsilon \to 0$. We also present results concerning conditions for convergence and finding asymptotic expansions for quasi-stationary distributions of regenerative processes.

Two types of characteristic quantities that are related to one regeneration period play a crucial role in our analysis. These are the stopping probability $f^{(\varepsilon)} = \mathsf{P}\{\mu^{(\varepsilon)} < \tau_1^{(\varepsilon)}\}$ and the power moments of the regeneration times $m_n^{(\varepsilon)} = \mathsf{E}(\tau_1^{(\varepsilon)})^n \chi(\mu^{(\varepsilon)} \geq \tau_1^{(\varepsilon)})$ or their mixed power-exponential analogues $\phi^{(\varepsilon)}(\rho, n) = \mathsf{E}(\tau_1^{(\varepsilon)})^n e^{\rho \tau_1^{(\varepsilon)}} \chi(\mu^{(\varepsilon)} \geq \tau_1^{(\varepsilon)})$ up to the order $n = k$. One of the main new elements in this study is that we consider a model with *nonlinear* perturbations. This means that the stopping probabilities $f^{(\varepsilon)}$ and the moments $m_n^{(\varepsilon)}$ or $\phi^{(\varepsilon)}(\rho, n)$ are nonlinear functions of ε. However, we restrict our attention to the case where the nonlinear perturbations are smooth and assume that these functions can be expanded in a power series with respect to ε up to and including the order k.

It turns out that the relationship between the rates with which ε tends to zero and the time t tends to infinity is important for the results. Without loss of generality we assume that the time $t = t^{(\varepsilon)}$ is a function of the parameter ε. The balance between the rate of perturbation and the rate of growth of time is characterised by the following condition controlled by an integer parameter r, $1 \leq r \leq k$:

$$\varepsilon^r t^{(\varepsilon)} \to \lambda_r < \infty \quad \text{as} \quad \varepsilon \to 0. \tag{3.0.1}$$

Under the above assumptions and some natural additional conditions of Cramér type for regeneration times, we obtain the asymptotic relation

$$\frac{\mathsf{P}\{\xi^{(\varepsilon)}(t^{(\varepsilon)}) \in A, \mu^{(\varepsilon)} > t^{(\varepsilon)}\}}{\exp\{-(\rho_0 + a_1\varepsilon + \cdots + a_{r-1}\varepsilon^{r-1})t^{(\varepsilon)}\}} \to \pi^{(0)}(A)e^{-\lambda_r a_r} \quad \text{as} \quad \varepsilon \to 0. \tag{3.0.2}$$

This asymptotic relation is provided with an explicit expression for $\pi^{(0)}(A)$, and an explicit recurrence algorithm for calculating the coefficients $a_1, \ldots, a_r$ as rational functions of the coefficients in the expansions for the stopping probabilities $f^{(\varepsilon)}$ and the moments $m_n^{(\varepsilon)}$, $n = 1, \ldots, k$, or $\phi^{(\varepsilon)}(\rho, n)$, $n = 1, \ldots, k$.

The case $\rho_0 = 0$ corresponds to a model with stochastically unbounded random variables $\mu^{(\varepsilon)}$, while the case $\rho_0 > 0$ corresponds to a model with stochastically bounded random variables $\mu^{(\varepsilon)}$. The asymptotic relation (3.0.2) describes in these cases *pseudo-stationary* and *quasi-stationary* phenomena for perturbed regenerative processes, respectively.

To clarify the meaning of the asymptotic relation (3.0.2) for different k, let us consider the case where $\rho_0 = 0$.

If $r = k = 1$, relation (3.0.1) is reduced to $\varepsilon t^{(\varepsilon)} \to \lambda_1$, and the asymptotic relation (3.0.2) can be rewritten in the form

$$\mathsf{P}\{\xi^{(\varepsilon)}(t^{(\varepsilon)}) \in A, \mu^{(\varepsilon)} > t^{(\varepsilon)}\} \to \pi^{(0)}(A)e^{-\lambda_1 a_1} \quad \text{as} \quad \varepsilon \to 0. \tag{3.0.3}$$

Asymptotic relation (3.0.3) shows that the position of the regenerative process $\xi(t^{(\varepsilon)})$ and the normalised regenerative stopping time $\varepsilon\mu^{(\varepsilon)}$ are asymptotically independent and have, in the limit, a stationary distribution and an exponential distribution, respectively. This can be interpreted as a mixed ergodic theorem (for the regenerative processes) and a limit theorem (for regenerative stopping times).

The asymptotic relation (3.0.3) describes the simplest form of pseudo-stationary phenomena for perturbed regenerative processes. In the literature, the term *transient* is also used in connection with such a kind of asymptotic relations.

The simplest form of pseudo-stationary asymptotics given by (3.0.3), as well as the corresponding results on rates of convergence do not give satisfactory information about the asymptotic behaviour of the joint distribution of the normalised regenerative stopping times and the positions of the regenerative process in zones of large deviations for $\varepsilon t^{(\varepsilon)} \to \infty$. In particular, they do not shed light upon the asymptotic behaviour of relative errors of approximation in large deviation zones, which is of great interest. Relation (3.0.2), for the case $k > 1$, provides a means to describe this asymptotics.

In the case $r = k = 2$, the balancing relation (3.0.1) is reduced to $\varepsilon^2 t^{(\varepsilon)} \to \lambda_2$ and the asymptotic relation (3.0.2) can be rewritten in the form

$$\frac{\mathsf{P}\{\xi^{(\varepsilon)}(t^{(\varepsilon)}) \in A, \mu^{(\varepsilon)} > t^{(\varepsilon)}\}}{\pi^{(0)}(A)\exp\{-a_1\varepsilon t^{(\varepsilon)}\}} \to e^{-\lambda_2 a_2} \text{ as } \varepsilon \to 0. \tag{3.0.4}$$

The limit in the right-hand side gives information about the asymptotic behaviour of the relative error of the approximation in (3.0.4) in the large deviation zone of the orders $\varepsilon t^{(\varepsilon)} = o(\varepsilon^{-1})$ or $\varepsilon t^{(\varepsilon)} = O(\varepsilon^{-1})$.

If $\lambda_2 = 0$, then $\varepsilon t^{(\varepsilon)} = o(\varepsilon^{-1})$ and the asymptotic relative error is 0. Note that this case also covers the situation where $\varepsilon t^{(\varepsilon)}$ is bounded, which corresponds to the asymptotic relation (3.0.3) with $r = k = 1$. This is already an extension of the asymptotic result since it is possible that $\varepsilon t^{(\varepsilon)} \to \infty$.

If $\lambda_2 > 0$, then $\varepsilon t^{(\varepsilon)} = O(\varepsilon^{-1})$, and the asymptotic relative error is $1 - e^{-\lambda_2 a_2}$. It differs from 0. Therefore, $o(\varepsilon^{-1})$ is an asymptotic bound for the large deviation zone with the asymptotic relative error 0.

Relation (3.0.3) can be interpreted as a new kind of mixed ergodic and large deviation theorems for nonlinearly perturbed regenerative processes.

A similar interpretation can be made of the asymptotic relation (3.0.2) for $k > 2$. As in the linear case, relation (3.0.2) can also be interpreted as a mixed limit and ergodic theorem if $r = 1$. On the other hand, if $1 < r \le k$ this relation is naturally considered as a mixed large deviation theorem (for stopping times) and an ergodic theorem (for regenerative processes), since $\varepsilon^r t^{(\varepsilon)} \to \infty$ in this case.

Asymptotics of the type (3.0.2) appear in the literature mainly for the case $k = 1$, which corresponds to a model of linearly perturbed processes. Our main result is the asymptotics (3.0.2) for $k > 1$, which corresponds to a model of nonlinearly perturbed processes.

In Theorems 3.2.1–3.3.6 we give an exponential asymptotics which describes pseudo-stationary and quasi-stationary phenomena in perturbed regenerative processes. We improve this asymptotics to the form of exponential expansions in mixed ergodic and large deviation and ergodic theorems for nonlinearly perturbed regenerative processes and regenerative stopping times. This is done in Theorems 3.4.2 and 3.4.4.

Finally, we give asymptotic expansions for stationary and quasi-stationary distributions of nonlinearly perturbed regenerative processes in Theorems 3.5.1, 3.5.2, and 3.5.4, 3.5.5.

The asymptotics of type (3.0.2) makes a basis for studies of the so-called mixed ergodic and limit theorems and mixed ergodic and large deviation theorems for perturbed semi-Markov processes with absorption, which is done in Chapters 4–5.

3.1 Regenerative processes and regenerative stopping times

In this section we introduce an object that plays a key role in our studies. These are objects are regenerative processes and regenerative stopping moments.

3.1.1 Regenerative processes and regenerative stopping times

Let consider a stochastic process $\xi(t), t \geq 0$, with a measurable phase space X (with the corresponding σ-algebra Γ of measurable subsets) defined on some probability space $(\Omega, \mathcal{F}, \mathcal{P})$. We assume that the process $\xi(t), t \geq 0$, is measurable in the sense that $\xi(t, \omega), (t, \omega) \in [0, \infty) \times \Omega$ are measurable functions of (t, ω).

Let also $0 = \tau_0 \leq \tau_1 \leq \ldots$ be a monotonically non-increasing sequence of non-negative random variables defined on the same probability space $(\Omega, \mathcal{F}, \mathcal{P})$.

We denote by $\xi^{(n)}(t) = \xi(\tau_n + t), t \geq 0$, and $\tau_k^{(n)} = \tau_{n+k} - \tau_n, k = 0, 1, \ldots$, for every $n = 0, 1, \ldots$, the shifted to time τ_n versions of the process $\xi(t)$ and the sequence τ_k.

Denote by $\mathcal{F}_n^{(-)}$ the σ-algebra of random events from $\mathcal{F}$ generated by the random variables $\xi(s), 0 \leq s < \tau_n$ (i.e., by random variables $\xi(\tau_n - s \wedge \tau_n), s > 0$), and τ_k, $k = 0, \ldots, n$. This σ-algebra includes random events determined by the trajectory of the process $\xi(t)$ before the moment τ_n and the moments τ_k for $k \leq n$. We also denote by $\mathcal{F}_n^{(+)}$ the σ-algebra of random events from $\mathcal{F}$ generated by the random variables $\xi^{(n)}(t), t \geq 0$, and $\tau_k^{(n)}, k = 0, 1, \ldots$. This σ-algebra includes the random events determined by the trajectory of the process $\xi(t)$ after and including the moment τ_n and the shifted moments $\tau_k^{(n)} = \tau_{n+k} - \tau_n, k = 0, 1, \ldots$.

The stochastic process $\xi(t), t \geq 0$, is a *regenerative process* with *regeneration times* $\tau_n, n = 0, 1, \ldots$, if it possess the following two properties:

$\mathbf{R_1}$: The σ-algebras $\mathcal{F}_n^{(-)}$ and $\mathcal{F}_n^{(+)}$ are independent for every $n = 0, 1, \ldots$;

and

$\mathbf{R_2}$: The joint finite dimensional distributions $\mathsf{P}\{\xi^{(n)}(t_r) \in A_r, \tau_r^{(n)} \leq s_r, r = 1, \ldots, m\} = \mathsf{P}\{\xi(t_r) \in A_r, \tau_r \leq s_r, r = 1, \ldots, m\}, A_r \in \Gamma, t_r, s_r \geq 0, r = 1, \ldots, m, m \geq 1$ are the same for every $n = 0, 1, \ldots$.

According to conditions $\mathbf{R_1}$ and $\mathbf{R_2}$, for every $n = 0, 1, \ldots$, we have $\xi^{(n)}(t) = \xi(\tau_n + t), t \geq 0$, and $\tau_k^{(n)} = \tau_{n+k} - \tau_n, k = 0, 1, \ldots$, is a probability copy of the regenerative process $\xi(t), t \geq 0$, with the regeneration times $\tau_k, k = 0, 1, \ldots$.

We also assume the following condition to hold, which excludes the degenerate case where the regeneration times take zero value with probability 1:

$\mathbf{R_3}$: $\check{f} = \mathsf{P}\{\tau_1 > 0\} > 0$.

By conditions $\mathbf{R_1}$ and $\mathbf{R_2}$, the random variables $\tau_n - \tau_{n-1}, n \geq 1$, are independent and identically distributed. Thus, conditions $\mathbf{R_1}$–$\mathbf{R_3}$ imply that

$$\tau_n \xrightarrow{\ \mathsf{P}\ } \infty \ \text{ as } n \to \infty. \tag{3.1.1}$$

Let now μ be a random variable defined on the same probability space, $(\Omega, \mathcal{F}, \mathcal{P})$, and taking values in the interval $[0, \infty]$. We say that μ is a *regenerative stopping time* if it regenerates together with the process $\xi(t), t \geq 0$, at times $\tau_n, n = 0, 1, \ldots$, in the sense that

$\mathbf{R_4}$: The conditional probabilities $\mathsf{P}\{\xi^{(n)}(t_r) \in A_r, \tau_r^{(n)} \leq s_r, r = 1, \ldots, m, \mu \geq \tau_n + t/B, \mu \geq \tau_n\} = \mathsf{P}\{\xi(t_r) \in B_r, \tau_r \leq s_r, r = 1, \ldots, m, \mu \geq t\}, A_r \in \Gamma, t_r, s_r \geq 0, r = 1, \ldots, m, m \geq 1, t \geq 0$, are the same for any $B \in \mathcal{F}_n^{(-)}$ and for every $n = 0, 1, \ldots$.

We also assume the following condition that excludes the degenerate case where $\mu < \tau_1$ with probability 1,

$\mathbf{R_5}$: $f = \mathsf{P}\{\mu < \tau_1\} \in [0, 1)$.

In what follows we refer to f as a *stopping probability* in one regeneration period.

Let $\xi(t), t \geq 0$, be a regenerative process with regeneration times $\tau_n, n = 0, 1, \ldots$, and μ be a regenerative stopping moment that satisfies conditions $\mathbf{R_1}$–$\mathbf{R_5}$. Then, for any $A \in \Gamma$, the probabilities $P(t, A) = \mathsf{P}\{\xi(t) \in A, \mu > t\}$ and $q(t, A) = \mathsf{P}\{\xi(t) \in A, \tau_1 > t, \mu > t\}$ are measurable functions of t and $P(t, A)$ satisfies the following renewal equation:

$$P(t, A) = q(t, A) + \int_0^t P(t - s, A) F(ds), \ t \geq 0, \tag{3.1.2}$$

where

$$F(t) = \mathsf{P}\{\tau_1 \leq t, \mu \geq \tau_1\}. \tag{3.1.3}$$

Note that $F(t)$ is possibly an improper distribution function such that (a) $F(0) = \mathsf{P}\{\tau_1 = 0\} = 1 - \check{f} < 1$ and (b) $F(\infty) = \mathsf{P}\{\mu \geq \tau_1\} = 1 - f > 0$.

Relation (3.1.2) can be used to simplify the definition of a regenerative process and a regenerative stopping time. One can say that μ is a regenerative stopping time for a regenerative process $\xi(t), t \geq 0$, with regeneration times $\tau_n, n = 0, 1, \ldots$, if relation (3.1.2) holds.

3.1.2   Regenerative processes with transition period

A regenerative process with *transition period* is a useful generalisation of a regenerative process.

The definition differs of the one above only in the formulation of condition $\mathbf{R}_2$. This condition requires that the corresponding joint finite dimensional distributions would have an identity for every $n = 0, 1, \ldots$. A version of this condition adapted to the case of a regenerative process with a transition period does require an identity for the corresponding joint finite dimensional distributions for every $n = 1, 2, \ldots$:

$\mathbf{R}'_2$: The joint finite dimensional distributions $\mathsf{P}\{\xi^{(n)}(t_r) \in A_r, \tau_r^{(n)} \le s_r, r = 1, \ldots, m\} = \mathsf{P}\{\xi^{(1)}(t_r) \in A_r, \tau_r^{(1)} \le s_r, r = 1, \ldots, m\}$, $A_r \in \Gamma, t_r, s_r \ge 0$, $r = 1, \ldots, m, m \ge 1$ are the same for every $n = 1, 2, \ldots$.

According to condition $\mathbf{R}'_2$, the initial process $\xi(t), t \ge 0$, can possess finite dimensional distributions which differ from the corresponding finite dimensional distributions of the shifted processes $\xi(\tau_n + t), t \ge 0$, for $n = 1, 2, \ldots$. Thus one can refer to the period $[0, \tau_1)$ as a transition period.

Condition $\mathbf{R}_3$ should be replaced with the following condition:

$\mathbf{R}'_3$: $\check{f} = \mathsf{P}\{\tau_1^{(1)} = \tau_2 - \tau_1 > 0\} > 0$.

Similar changes should be made in the definition of regenerative stopping moments.

The definition differ from the one above only in the formulation of condition $\mathbf{R}_4$. This condition requires an identity for the corresponding conditional probabilities for every $n = 0, 1, \ldots$. A version of this condition adapted to the case of a regenerative process and regenerative stopping times with a transition period does require an identity for the corresponding conditional probabilities only for every $n = 1, 2, \ldots$:

$\mathbf{R}'_4$: The conditional probabilities $\mathsf{P}\{\xi^{(n)}(t_r) \in A_r, \tau_r^{(n)} \le s_r, r = 1, \ldots, m, \mu \ge \tau_n + t/B, \mu \ge \tau_n\} = \mathsf{P}\{\xi^{(1)}(t_r) \in B_r, \tau_r^{(1)} \le s_r, r = 1, \ldots, m, \mu \ge \tau_1 + t/B, \mu \ge \tau_1\}$, $A_r \in \Gamma, t_r, s_r \ge 0, r = 1, \ldots, m, m \ge 1, t \ge 0$, are the same for any $B \in \mathcal{F}_n^{(-)}$ and for every $n = 1, 2, \ldots$.

Condition $\mathbf{R}_5$ should be replaced with the following condition that excludes the degenerate case (where the regenerative stopping time $\mu < \tau_1$ with probability 1):

$\mathbf{R}'_5$: $\tilde{f} = \mathsf{P}\{\mu < \tau_1\} \in [0, 1)$.

This condition should also be supplemented with the following condition as to exclude the degenerate case (where the regenerative stopping time $\mu < \tau_2$ with probability 1):

$\mathbf{R}''_5$: $f = \mathsf{P}\{\mu - \tau_1 < \tau_2/\mu \ge \tau_1\} \in [0, 1)$.

Let $\xi(t), t \ge 0$, be a regenerative process with a transition period and regeneration times $\tau_n, n = 0, 1, \ldots$, and μ be a regenerative stopping time that satisfies conditions $\mathbf{R}_1$ and $\mathbf{R}'_2$–$\mathbf{R}'_5$, as well as $\mathbf{R}''_5$. Introduce the probabilities $\tilde{P}(t, A) = \mathsf{P}\{\xi(t) \in A, \mu > t\}$ and $\tilde{q}(t, A) = \mathsf{P}\{\xi(t) \in A, \tau_1 > t, \mu > t\}$, and let $P(t, A) =$

$P\{\xi^{(1)}(t) \in A, \mu - \tau_1 > t/\mu \geq \tau_1\}$ and $q(t, A) = P\{\xi^{(1)}(t) \in A, \tau_1^{(1)} > t, \mu - \tau_1 > t/\mu \geq \tau_1\}$. Then, for any $A \in \Gamma$, these probabilities are measurable functions of t, and $\tilde{P}(t, A)$ and $P(t, A)$ satisfy the following renewal relation:

$$\tilde{P}(t, A) = \tilde{q}(t, A) + \int_0^t P(t - s, A)\tilde{F}(ds), \quad t \geq 0, \tag{3.1.4}$$

where

$$\tilde{F}(t) = P\{\tau_1 \leq t, \mu \geq \tau_1\}, \tag{3.1.5}$$

together with the following renewal equation:

$$P(t, A) = q(t, A) + \int_0^t P(t - s, A)F(ds), \quad t \geq 0, \tag{3.1.6}$$

where

$$F(t) = P\{\tau_1^{(1)} \leq t, \mu - \tau_1 \geq \tau_1^{(1)}/\mu \geq \tau_1\}. \tag{3.1.7}$$

Relations (3.1.4) and (3.1.6) can be used to simplify the definition of a regenerative process and a regenerative stopping time with transition period. One can say that μ is a regenerative stopping time for a regenerative process $\xi(t), t \geq 0$, with regeneration times $\tau_n, n = 0, 1, \ldots$, if relations (3.1.4) and (3.1.6) hold.

3.1.3   The structure of regenerative stopping times

The following lemma follows from the definition of regenerative stopping times.

Lemma 3.1.1. *Let conditions* $\mathbf{R_1}$–$\mathbf{R_5}$ *hold. Then*

$$P\{\mu \geq \tau_n\} = (1 - f)^n, \quad n = 0, 1, \ldots . \tag{3.1.8}$$

Proof. Denote $x^{\{h\}} = h[x/h]$. Obviously, $(\tau_{n+1} - \tau_n)^{\{2^{-k}\}}, k \geq 1$, is a sequence of monotonically non-decreasing random variables, and $(\tau_{n+1} - \tau_n)^{\{2^{-k}\}} \xrightarrow{\text{a.s.}} (\tau_{n+1} - \tau_n)$ as $k \to \infty$. Using these facts and condition $\mathbf{R_4}$ we get, for $n = 0, 1, \ldots$, that

$$P\{\mu \geq \tau_{n+1}/\mu \geq \tau_n\} = P\{\mu \geq \tau_n + (\tau_{n+1} - \tau_n)/\mu \geq \tau_n\}$$

$$= \lim_{k \to \infty} P\{\mu \geq \tau_n + (\tau_{n+1} - \tau_n)^{\{2^{-k}\}}/\mu \geq \tau_n\}$$

$$= \lim_{k \to \infty} \sum_{r=0}^{\infty} P\{\mu \geq \tau_n + r2^{-k}, r2^{-k} \leq \tau_{n+1} - \tau_n < (r + 1)2^{-k}/\mu \geq \tau_n\}$$

$$= \lim_{k \to \infty} \sum_{r=0}^{\infty} P\{\mu \geq r2^{-k}, r2^{-k} \leq \tau_1 < (r + 1)2^{-k}\}$$

$$= \lim_{k \to \infty} P\{\mu \geq \tau_1^{\{2^{-k}\}}\} = P\{\mu \geq \tau_1\} = 1 - f. \tag{3.1.9}$$

Using relation (3.1.9) we get, for every $n = 0, 1, \ldots$, that

$$P\{\mu \geq \tau_n\} = \prod_{k=0}^{n-1} P\{\mu \geq \tau_{k+1}/\mu \geq \tau_k\} = (1 - f)^n. \qquad (3.1.10)$$

The proof is complete. $\qquad\qquad\qquad\qquad\qquad\qquad\qquad\qquad\qquad\qquad\qquad\qquad\qquad$ □

Remark 3.1.1. In the case of a regenerative process with a transition period, formula (3.1.8) takes the following form:

$$P\{\mu \geq \tau_n\} = (1 - \tilde{f})(1 - f)^{n-1}, \quad n = 1, 2, \ldots. \qquad (3.1.11)$$

The following lemma is a direct corollary of Lemma 3.1.1.

Lemma 3.1.2. *Let conditions* $\mathbf{R_1}$–$\mathbf{R_5}$ *hold. Then one of the two alternatives holds:* **(a)** $P\{\mu = \infty\} = 1$ *if* $f = 0$, *or* **(b)** $P\{\mu < \infty\} = 1$ *if* $f > 0$.

The only nontrivial case is where $f > 0$. Let us consider it in more details.
We introduce the random variables

$$\nu = \min(n \geq 1 : \tau_{n-1} \leq \mu < \tau_n), \quad \kappa = \mu - \tau_{\nu-1}, \quad \kappa_n = \tau_n - \tau_{n-1}, \quad n = 1, 2, \ldots.$$

It is obvious that the regenerative stopping time μ can be represented in the form of the random sum

$$\mu = \sum_{n=1}^{\nu-1} \kappa_n + \kappa. \qquad (3.1.12)$$

It follows from Lemma 3.1.9 that the random variable ν has a geometric distribution with the parameter f, i.e.,

$$P\{\nu > n\} = (1 - f)^n, \quad n = 0, 1, \ldots. \qquad (3.1.13)$$

Also, the random variables $\kappa_n = \tau_n - \tau_{n-1}, n = 1, 2, \ldots$, are i.i.d. random variables.

Thus μ is a geometric sum of i.i.d. random variables. The representation (3.1.12) has, however, a drawback. The random variables ν, κ, and $\tau_n - \tau_{n-1}, n = 1, 2, \ldots$, employed in this representation, are not independent.

Fortunately, this defect can be overcomed. Let us consider the random variable $\mu \wedge \tau_1$. The distribution function of this random variable can be represented in the form of a linear combination of two conditional distribution functions,

$$\hat{F}(t) = P\{\mu \wedge \tau_1 \leq t\} = \hat{F}_-(t)(1 - f) + \hat{F}_+(t)f, \quad t \geq 0, \qquad (3.1.14)$$

where

$$\hat{F}_-(t) = \mathsf{P}\{\tau_1 \le t / \mu \ge \tau_1\} = F(t)/(1 - f),$$

$$\hat{F}_+(t) = \mathsf{P}\{\mu \le t / \mu < \tau_1\} = (F(t) - \hat{F}(t))/f.$$

In the case where the stopping probability is $f = 0$, the distribution function $\hat{F}_+(t)$ can be taken arbitrarily. This does not affect formula (3.1.14). For simplicity, we take $\hat{F}_+(t) = \chi_{[0,\infty)}(t)$ in this case.

Lemma 3.1.3. *Let conditions* $\mathbf{R_1}$–$\mathbf{R_5}$ *hold, and* $f > 0$. *Then, for every* $0 \le t, t_1, \ldots,$ $t_n < \infty$, $n = 1, 2, \ldots,$

$$\mathsf{P}\{\nu = n, \kappa \le t, \tau_k - \tau_{k-1} \le t_k, k = 1, \ldots, n - 1\}$$

$$= (\hat{F}(t) - F(t)) \prod_{k=1}^{n-1} F(t_k)$$

$$= f(1 - f)^{n-1} \hat{F}_+(t) \prod_{k=1}^{n-1} \hat{F}_-(t_k). \tag{3.1.15}$$

Proof. By conditions of the lemma, $0 < f < 1$. Using the definition of the random variables ν, κ, and $\tau_n - \tau_{n-1}$, $n = 1, 2, \ldots,$ and condition $\mathbf{R_4}$ we get that

$$\mathsf{P}\{\nu = n, \kappa \le t, \tau_k - \tau_{k-1} \le t_k, k = 1, \ldots, n - 1\}$$

$$= \mathsf{P}\{\tau_{n-1} \le \mu < \tau_n, \mu - \tau_{n-1} \le t, \tau_k - \tau_{k-1} \le t_k, k = 1, \ldots, n - 1\}$$

$$= \mathsf{P}\{\mu - \tau_{n-1} \le t, \mu < \tau_n / \tau_{n-1} \le \mu, \tau_k - \tau_{k-1} \le t_k, k = 1, \ldots, n - 1\}$$

$$\times \mathsf{P}\{\tau_{n-1} \le \mu, \tau_k - \tau_{k-1} \le t_k, k = 1, \ldots, n - 1\}$$

$$= \mathsf{P}\{\mu \le t, \mu < \tau_1\}\mathsf{P}\{\tau_{n-1} \le \mu, \tau_k - \tau_{k-1} \le t_k, k = 1, \ldots, n - 1\}$$

$$= (\hat{F}(t) - F(t))\mathsf{P}\{\tau_{n-1} - \tau_{n-2} \le t_{n-1},$$

$$\tau_{n-1} \le \mu / \tau_{n-2} \le \mu, \tau_k - \tau_{k-1} \le t_k, k = 1, \ldots, n - 2\}$$

$$\times \mathsf{P}\{\tau_{n-2} \le \mu, \tau_k - \tau_{k-1} \le t_k, k = 1, \ldots, n - 2\}$$

$$= (\hat{F}(t) - F(t))\mathsf{P}\{\tau_1 \le t_{n-1}, \tau_1 \le \mu\}$$

$$\times \mathsf{P}\{\tau_{n-2} \le \mu, \tau_k - \tau_{k-1} \le t_k, k = 1, \ldots, n - 2\}$$

$$= (\hat{F}(t) - F(t)) F(t_{n-1})\mathsf{P}\{\tau_{n-2} \le \mu, \tau_k - \tau_{k-1} \le t_k, k = 1, \ldots, n - 2\}$$

$$= \ldots$$

$$= (\hat{F}(t) - F(t)) \prod_{k=1}^{n-1} F(t_k) = f(1 - f)^{n-1} \hat{F}_+(t) \prod_{k=1}^{n-1} \hat{F}_-(t_k). \tag{3.1.16}$$

This proves the lemma. $\qquad\qquad\square$

Lemma 3.1.3 permits to represent the regenerative stopping time μ in the following form (the symbol $\overset{\mathrm{d}}{=}$ means that the random variables on the left and the right-hand sides have the same distribution):

$$\mu \overset{\mathrm{d}}{=} \sum_{n=1}^{\hat{\nu}-1} \hat{\kappa}_n + \hat{\kappa}, \tag{3.1.17}$$

where **(a)** $\hat{\nu}$ is a geometrically distributed random variable with the parameter f, **(b)** the random variable $\hat{\kappa}$ has the distribution function $\hat{F}^+(t)$, **(c)** the random variables $\hat{\kappa}_n, n = 1, 2, \ldots$, have the distribution function $\hat{F}^-(t)$, **(d)** the random variables $\hat{\nu}, \hat{\kappa}$ and $\hat{\kappa}_n, n = 1, 2, \ldots$, are mutually independent.

3.2   Mixed ergodic and limit theorems for perturbed regenerative processes

In this section we present mixed ergodic and limit theorems for perturbed regenerative processes which describe pseudo-stationary phenomena for nonlinearly perturbed regenerative processes.

3.2.1   Mixed ergodic and limit theorems for perturbed regenerative processes

Let for every $\varepsilon \geq 0$, $\xi^{(\varepsilon)}(t)$, $t \geq 0$, be a regenerative process on a measurable phase space X (with the corresponding σ-algebra Γ of measurable subsets) with regeneration times $0 = \tau_0^{(\varepsilon)} \leq \tau_1^{(\varepsilon)} \leq \ldots$, and let $\mu^{(\varepsilon)}$ be a random variable defined on the same probability space and taking values in the interval $[0, \infty]$. We assume that the random variable $\mu^{(\varepsilon)}$ is a regenerative stopping time for the regenerative process $\xi^{(\varepsilon)}(t)$, $t \geq 0$, for every $\varepsilon \geq 0$, and thus conditions $\mathbf{R_1}$–$\mathbf{R_5}$ hold for every $\varepsilon \geq 0$.

In this case, according to the definition given above, for any $A \in \Gamma$, the probabilities $P^{(\varepsilon)}(t, A) = \mathsf{P}\{\xi^{(\varepsilon)}(t) \in A, \mu^{(\varepsilon)} > t\}$ and $q^{(\varepsilon)}(t, A) = \mathsf{P}\{\xi^{(\varepsilon)}(t) \in A, \tau_1^{(\varepsilon)} > t, \mu^{(\varepsilon)} > t\}$ are measurable functions of t, and $P^{(\varepsilon)}(t, A)$ satisfies the renewal equation

$$P^{(\varepsilon)}(t, A) = q^{(\varepsilon)}(t, A) + \int_0^t P^{(\varepsilon)}(t - s, A) F^{(\varepsilon)}(ds), \quad t \geq 0, \tag{3.2.1}$$

where

$$F^{(\varepsilon)}(t) = \mathsf{P}\{\tau_1^{(\varepsilon)} \leq t, \mu^{(\varepsilon)} \geq \tau_1^{(\varepsilon)}\}. \tag{3.2.2}$$

An important quantity is the stopping probability in one regeneration period $f^{(\varepsilon)}$, which coincides with the deficiency of the distribution $F^{(\varepsilon)}(t)$,

$$f^{(\varepsilon)} = \mathsf{P}\{\mu^{(\varepsilon)} < \tau_1^{(\varepsilon)}\} = 1 - F^{(\varepsilon)}(\infty).$$

We also use the notations,

$$m_r^{(\varepsilon)} = \mathsf{E}(\tau_1^{(\varepsilon)})^r \chi(\mu^{(\varepsilon)} \geq \tau_1^{(\varepsilon)}) = \int_0^\infty s^r F^{(\varepsilon)}(ds), \quad r \geq 1.$$

To the renewal equation (3.2.1) we apply the results obtained in Chapters 1 and 2, which will yield various variants of asymptotics for the probabilities $P^{(\varepsilon)}(t, A)$.

We impose the following condition:

D_{13}: $F^{(\varepsilon)}(\cdot) \Rightarrow F^{(0)}(\cdot)$ as $\varepsilon \to 0$, where $F^{(0)}(s)$ is a proper non-arithmetic distribution function.

Denote

$$\check{F}^{(\varepsilon)}(t) = \mathsf{P}\{\tau_1^{(\varepsilon)} \leq t\}, \quad \hat{F}^{(\varepsilon)}(t) = \mathsf{P}\{\mu^{(\varepsilon)} \wedge \tau_1^{(\varepsilon)} \leq t\}.$$

By definition, for any $t \geq 0$,

$$\begin{aligned}
1 - f^{(\varepsilon)} - F^{(\varepsilon)}(t) &= \mathsf{P}\{\tau_1^{(\varepsilon)} > t, \mu^{(\varepsilon)} \geq \tau_1^{(\varepsilon)}\} \\
&\leq 1 - \hat{F}^{(\varepsilon)}(t) = \mathsf{P}\{\mu^{(\varepsilon)} \wedge \tau_1^{(\varepsilon)} > t\} \\
&= \mathsf{P}\{\tau_1^{(\varepsilon)} > t, \mu^{(\varepsilon)} \geq \tau_1^{(\varepsilon)}\} + \mathsf{P}\{\mu_1^{(\varepsilon)} > t, \mu^{(\varepsilon)} < \tau_1^{(\varepsilon)}\} \\
&\leq 1 - \check{F}^{(\varepsilon)}(t) = \mathsf{P}\{\tau_1^{(\varepsilon)} > t\}. \qquad (3.2.3)
\end{aligned}$$

Note also that

$$\begin{aligned}
(1 - \check{F}^{(\varepsilon)}(t)) - (1 - f^{(\varepsilon)} - F^{(\varepsilon)}(t)) &= \mathsf{P}\{\tau_1^{(\varepsilon)} > t, \mu^{(\varepsilon)} < \tau_1^{(\varepsilon)}\} \\
&\leq \mathsf{P}\{\mu^{(\varepsilon)} < \tau_1^{(\varepsilon)}\} = f^{(\varepsilon)}. \qquad (3.2.4)
\end{aligned}$$

Condition D_{13} implies that

$$f^{(\varepsilon)} \to f^{(0)} = 0 \text{ as } \varepsilon \to 0. \qquad (3.2.5)$$

Also, condition D_{13} and relations (3.2.3)–(3.2.5) imply that

$$\hat{F}^{(\varepsilon)}(\cdot) \Rightarrow \hat{F}^{(0)}(\cdot) \equiv F^{(0)}(\cdot) \text{ as } \varepsilon \to 0 \qquad (3.2.6)$$

and

$$\check{F}^{(\varepsilon)}(\cdot) \Rightarrow \check{F}^{(0)}(\cdot) \equiv F^{(0)}(\cdot) \text{ as } \varepsilon \to 0. \qquad (3.2.7)$$

We also assume the following condition to hold:

M_8: $\lim_{T \to \infty} \overline{\lim}_{0 \leq \varepsilon \to 0} \int_T^\infty (1 - \hat{F}^{(\varepsilon)}(s))\, ds = 0.$

Condition M_8, due to estimates (3.2.3), implies the following condition:

M_9: $\lim_{T \to \infty} \overline{\lim}_{0 \leq \varepsilon \to 0} \int_T^\infty (1 - f^{(\varepsilon)} - F^{(\varepsilon)}(s))\, ds < \infty;$

and itself is a consequence of the following condition:

$\mathbf{M_{10}}$: $\lim_{T\to\infty} \overline{\lim}_{0\le\varepsilon\to 0} \int_T^\infty (1 - \check{F}^{(\varepsilon)}(s))\,ds < \infty.$

Note also that conditions $\mathbf{D_{11}}$ and $\mathbf{M_9}$ imply that

$$m_1^{(\varepsilon)} \to m_1^{(0)} < \infty \quad \text{as } \varepsilon \to 0. \tag{3.2.8}$$

We also assume the following condition imposed on the forcing function in the renewal equation (3.1.2):

$\mathbf{F_7}$: There exists a class of sets $\Gamma_0 \subseteq \Gamma$ such that, for every $A \in \Gamma_0$, the following relation holds: $\lim_{u\to 0} \overline{\lim}_{0\le\varepsilon\to 0} \sup_{|v|\le u} |q^{(\varepsilon)}(t + v, A) - q^{(0)}(t, A)| = 0$ almost everywhere with respect to the Lebesgue measure on $[0, \infty)$.

Let us also assume the following condition that balances the rate of growth of time $t^{(\varepsilon)} \to \infty$ and the rate of the perturbation determined, in this case, by the speed with which the stopping probability $f^{(\varepsilon)}$ vanish,

$\mathbf{B_4}$: $0 \le t^{(\varepsilon)} \to \infty$ and $f^{(\varepsilon)} \to 0$ as $\varepsilon \to 0$ in such a way that $f^{(\varepsilon)} t^{(\varepsilon)} \to \lambda$, where $0 \le \lambda \le +\infty$.

Note that condition $\mathbf{B_4}$ implies the following: **(a)** if $\lambda > 0$, then automatically we have $f^{(\varepsilon)} > 0$ for all ε small enough, **(b)** if there exists a sequence $0 < \varepsilon_n \to 0$ as $n \to \infty$ such that $f^{(\varepsilon_n)} = 0, n \ge 1$, then $\lambda = 0$.
Denote

$$\pi^{(\varepsilon)}(A) = \frac{\int_0^\infty q^{(\varepsilon)}(s, A)m(ds)}{m_1^{(\varepsilon)}}, \quad A \in \Gamma. \tag{3.2.9}$$

Note that $\pi^{(\varepsilon)}(A)$ is a probability distribution, i.e., $\pi^{(\varepsilon)}(X) = 1$ if and only if $f^{(\varepsilon)} = 0$. In particular, $f^{(0)} = 0$ by condition $\mathbf{D_{13}}$. In this case, $\pi^{(0)}(A)$ is a *stationary distribution* of the regenerative process $\xi^{(0)}(t)$, $t \ge 0$. In the next subsection, we will give a more detailed comment.

Let us formulate a theorem which is similar to Theorem 1.3.1 applied to the renewal equation (3.2.1).

Theorem 3.2.1. *Let conditions $\mathbf{D_{13}}$, $\mathbf{M_8}$, $\mathbf{F_7}$ hold, and $0 \le t^{(\varepsilon)} \to \infty$ as $\varepsilon \to 0$ such that condition $\mathbf{B_4}$ holds. Then, for any $A \in \Gamma_0$,*

$$P^{(\varepsilon)}(t^{(\varepsilon)}, A) = \mathsf{P}\{\xi^{(\varepsilon)}(t^{(\varepsilon)}) \in A, \mu^{(\varepsilon)} > t^{(\varepsilon)}\}$$

$$\to \pi^{(0)}(A)e^{-\lambda/m_1^{(0)}} \quad \text{as} \ \varepsilon \to 0. \tag{3.2.10}$$

Proof. The theorem follows directly from Theorem 1.3.1. Conditions $\mathbf{D_{13}}$ and $\mathbf{B_4}$ coincide, respectively, with conditions $\mathbf{D_6}$ and $\mathbf{B_1}$ for the distribution functions $F^{(\varepsilon)}(t)$ defined above.

Also, conditions $\mathbf{M_8}$ and $\mathbf{F_7}$ imply that condition $\mathbf{F_2}$ holds for the forcing functions $q^{(\varepsilon)}(t, A)$. Condition $\mathbf{F_7}$ coincides in this case with condition $\mathbf{F_2}$ (**a**).

Obviously, for any $A \in \Gamma$ and $t \geq 0$,

$$q^{(\varepsilon)}(t, A) = \mathsf{P}\{\xi^{(\varepsilon)}(t) \in A, \mu^{(\varepsilon)} \wedge \tau_1^{(\varepsilon)} > t\} \leq \mathsf{P}\{\mu^{(\varepsilon)} \wedge \tau_1^{(\varepsilon)} > t\}$$

$$= 1 - \hat{F}^{(\varepsilon)}(t). \tag{3.2.11}$$

It follows from (3.2.11) that condition $\mathbf{M_8}$ implies that conditions $\mathbf{F_2}$ (**b**) and $\mathbf{F_2}$ (**c**) hold. The first implication is obvious and the second one follows from the asymptotic estimates

$$\lim_{T \to \infty} \overline{\lim_{0 \leq \varepsilon \to 0}} \, h \sum_{r \geq T/h} \sup_{rh \leq t \leq (r+1)h} |q^{(\varepsilon)}(t, A)|$$

$$\leq \lim_{T \to \infty} \overline{\lim_{0 \leq \varepsilon \to 0}} \int_{T-h}^{\infty} (1 - \hat{F}^{(\varepsilon)}(s)) \, ds = 0. \tag{3.2.12}$$

So, all the conditions of Theorem 1.3.1 hold. By applying this theorem to the renewal equation (3.1.2) we complete the proof. $\qquad\qquad\square$

3.2.2   Ergodic theorems for perturbed regeneration processes

Let us consider a model for which the regenerative stopping moment $\mu^{(\varepsilon)} = \infty$ with probability 1 for all ε small enough, say $\varepsilon \leq \varepsilon_1$. As follows from Lemma 3.1.2, this is equivalent to the following condition:

$\mathbf{D_{14}}$: $f^{(\varepsilon)} = 0$ for every $\varepsilon \leq \varepsilon_1$.

In this case, condition $\mathbf{B_4}$ automatically holds for any $0 \leq t^{(\varepsilon)} \to \infty$ as $\varepsilon \to 0$ and $\lambda = 0$.

Theorem 3.2.1 yields in this case a variant of the ergodic theorem for perturbed regenerative processes.

Theorem 3.2.2. *Let conditions* $\mathbf{D_{13}}$, $\mathbf{D_{14}}$, $\mathbf{M_8}$, *and* $\mathbf{F_7}$ *hold. Then, for any* $0 \leq t^{(\varepsilon)} \to \infty$ *as* $\varepsilon \to 0$,

$$P^{(\varepsilon)}(t^{(\varepsilon)}, A) = \mathsf{P}\{\xi^{(\varepsilon)}(t^{(\varepsilon)}) \in A\} \to \pi^{(0)}(A) \quad as \quad \varepsilon \to 0, \ A \in \Gamma_0. \tag{3.2.13}$$

It is worth noting that, in the case where the regenerative processes $\xi^{(\varepsilon)}(t) \equiv \xi^{(0)}(t)$, $t \geq 0$, and the regenerative times $\tau_k^{(\varepsilon)} \equiv \tau_k^{(0)}$, $k = 0, 1, \ldots$, for all $\varepsilon \geq 0$, i.e., they do not depend on the parameter ε, the conditions of Theorem 3.2.2 reduce to the standard conditions of the ergodic theorem for the non-perturbed regenerative process $\xi^{(0)}(t)$, $t \geq 0$, which is governed by the classical renewal theorem.

These conditions are: **(a)** the distribution function $\check{F}^{(0)}(t) = \mathsf{P}\{\tau_1^{(0)} \leq t\}$ is non-arithmetic; **(b)** $\check{F}^{(0)}(t)$ has a finite first moment; **(c)** the function $q^{(0)}(t, A) = \mathsf{P}\{\xi^{(0)}(t) \in A, \tau_1^{(0)} > t\}$ is continuous almost everywhere with respect to the Lebesgue measure on $[0, \infty)$ for $A \in \Gamma_0$.

3.2.3 Stationary distributions for perturbed regenerative processes

We continue to consider a model such that condition $\mathbf{D_{14}}$ holds. Let us also assume the following conditions:

$\mathbf{D_{15}}$: $F^{(\varepsilon)}(s)$ is a non-arithmetic distribution function for all ε small enough, say
  $\varepsilon \leq \varepsilon_2$;

and

$\mathbf{F_8}$: $q^{(\varepsilon)}(s, A)$ is directly Riemann integrable on $[0, \infty)$ for $A \in \Gamma_0$ and all ε small
  enough, say $\varepsilon \leq \varepsilon_3$.

Note that, under condition $\mathbf{M_8}$, it is enough to assume that for $A \in \Gamma_0$ the function $q^{(\varepsilon)}(s, A)$ is continuous almost everywhere with respect to the Lebesgue measure on $[0, \infty)$ for all ε small enough. In this case, condition $\mathbf{F_8}$ holds. The Lebesgue integration can be replaced with the Riemann integration in the right-hand side of (3.2.9).

Condition $\mathbf{M_8}$ implies that $m_1^{(\varepsilon)} < \infty$ for all ε small enough, say $\varepsilon \leq \varepsilon_4$.

Let us define,

$$\varepsilon_0 = \min(\varepsilon_1, \varepsilon_2, \varepsilon_3, \varepsilon_4). \tag{3.2.14}$$

The renewal Theorem 1.1.1 implies the following statement.

Theorem 3.2.3. *Let conditions* $\mathbf{D_{14}}$, $\mathbf{D_{15}}$, $\mathbf{M_8}$, *and* $\mathbf{F_8}$ *hold. Then, for* $\varepsilon \leq \varepsilon_0$,

$$\mathsf{P}\{\xi^{(\varepsilon)}(t) \in A\} \to \pi^{(\varepsilon)}(A) \quad as \quad t \to \infty, \quad A \in \Gamma_0. \tag{3.2.15}$$

The expression for $\pi^{(\varepsilon)}(A)$ given by formula (3.2.9) defines a probability distribution, which, due to asymptotic relation (3.2.15), is called a *stationary distribution* for the regenerative process $\xi^{(\varepsilon)}(t)$, $t \geq 0$.

It should be noted that, under condition $\mathbf{M_8}$, the expression for $\pi^{(\varepsilon)}(A)$ given by formula (3.2.9) defines a probability distribution for $\varepsilon \leq \varepsilon_0$ even if condition $\mathbf{F_{10}}$ does not hold. This is so, since we use in this formula the Lebesgue integration.

Moreover it can be proved that the average value of the $\mathsf{P}\{\xi^{(\varepsilon)}(s) \in A\}$ over the interval $[0, t]$ converges to $\pi^{(\varepsilon)}(A)$ as $t \to \infty$.

3.2.4   Convergence of stationary distributions for perturbed regenerative processes

The following theorem supplements the asymptotic relations (3.2.13) and (3.2.15).

Theorem 3.2.4. *Let conditions* $\mathbf{D_{13}}$, $\mathbf{D_{14}}$, $\mathbf{M_8}$, *and* $\mathbf{F_7}$ *hold. Then*

$$\pi^{(\varepsilon)}(A) \to \pi^{(0)}(A) \quad as \quad \varepsilon \to 0, \quad A \in \Gamma_0. \tag{3.2.16}$$

Proof. The theorem follows from Theorem 1.2.4, since conditions $\mathbf{D_{13}}$, $\mathbf{D_{14}}$, $\mathbf{M_8}$, and $\mathbf{F_7}$ imply that conditions $\mathbf{D_2}$, $\mathbf{M_2}$, and $\mathbf{F_2}$ hold for the distribution functions $F^{(\varepsilon)}(t)$ and the forcing functions $q^{(\varepsilon)}(t, A)$. $\square$

Note that convergence in (3.2.16) is guaranteed for $A \in \Gamma_0 \subseteq \Gamma$, while the stationary distribution is defined for $A \in \Gamma$. This is true, because condition $\mathbf{F_7}$ provides convergence of the corresponding forcing functions only for $A \in \Gamma_0$. However, in most applications, $\Gamma_0 = \Gamma$.

3.2.5   Weak convergence of distributions for regenerative stopping times

Let us return to a model for which condition $\mathbf{D_{14}}$ is not assumed to hold. Moreover, we assume a condition that is opposite in some sense:

$\mathbf{D_{16}}$: $f^{(\varepsilon)} > 0$ for every $\varepsilon > 0$.

Note that $f^{(0)} = 0$, according to condition $\mathbf{D_{13}}$.

We consider the case where $A = X$. Obviously, $P^{(\varepsilon)}(t, X) = \mathsf{P}\{\mu^{(\varepsilon)} > t\}$ and $q^{(\varepsilon)}(t, X) = \mathsf{P}\{\tau_1^{(\varepsilon)} > t, \mu^{(\varepsilon)} > t\} = 1 - \hat{F}^{(\varepsilon)}(t)$.

Condition $\mathbf{D_{13}}$ implies in this case that condition $\mathbf{F_7}$ holds. Indeed, relation (3.2.6) implies that $1 - \hat{F}^{(\varepsilon)}(t) \to 1 - F^{(0)}(t)$ as $\varepsilon \to 0$ for every point of continuity of the limit function. Moreover, since the functions $1 - \hat{F}^{(\varepsilon)}(t)$ are monotone, this convergence is locally uniform in every such point. But, the set S_0 of continuity points of the function $1 - F^{(0)}(t)$ is the interval $[0, \infty)$ except for at most a countable set. Thus the Lebesgue measure is $m(\overline{S}_0) = 0$.

Note also that condition $\mathbf{B_4}$ holds for $t^{(\varepsilon)} \sim t / f^{(\varepsilon)}$ as $\varepsilon \to 0$ for every $t > 0$. In this case, $\lambda = t$.

Finally, $\pi^{(0)}(X) = 1$, as was mentioned in Subsection 3.2.1.

Theorem 3.2.1 and the remarks above yield in this case that, under conditions $\mathbf{D_{13}}$, $\mathbf{M_8}$, and $\mathbf{D_{16}}$,

$$P^{(\varepsilon)}(t/f^{(\varepsilon)}, X) = \mathsf{P}\{f^{(\varepsilon)}\mu^{(\varepsilon)} > t\} \to e^{-t/m_1^{(0)}} \quad as \quad \varepsilon \to 0, \ t \geq 0. \tag{3.2.17}$$

Relation (3.2.17) shows that the regenerative stopping moment $f^{(\varepsilon)}\mu^{(\varepsilon)}$, normalised by the quantity $f^{(\varepsilon)}$, has an exponential limit distribution with the parameter $1/m_1^{(0)}$.

Marginal asymptotic relations (3.2.13) and (3.2.17) give a clear interpretation to the mixed asymptotic relation (3.2.10) given in Theorem 3.2.1. This theorem is a *mixed ergodic and limit theorem* for perturbed regenerative processes and regenerative stopping times. The asymptotic relation (3.2.10) shows that the state of the perturbed regenerative process $\xi^{(\varepsilon)}(t/f^{(\varepsilon)})$ at the moment $t/f^{(\varepsilon)}$ and the normalised regenerative stopping time $f^{(\varepsilon)}\mu^{(\varepsilon)}$ are asymptotically independent and the joint limit distribution is the product of the stationary distribution $\pi^{(0)}(A)$ for the limit regenerative process and the limit exponential distribution $e^{-t/m_1^{(0)}}$ for the normalised regenerative stopping times.

As far as the marginal asymptotic relation (3.2.17) is concerned, it can be obtained under conditions weaker than the conditions $\mathbf{D_{13}}$ and $\mathbf{M_8}$ used in Theorem 3.2.1.

To improve these conditions, one can use representation (3.1.17) for the regenerative stopping times μ_ε in the form of the random sum and to apply general theorems for randomly stopped stochastic processes, for example, given in Silvestrov (1974, 2004a). Let us briefly describe this procedure.

Let $\hat{\nu}^{(\varepsilon)}$, $\hat{\kappa}^{(\varepsilon)}$ and $\hat{\kappa}_k^{(\varepsilon)}$, $k \geq 1$, be mutually independent random variables which are used to construct representation (3.1.17) for the regenerative processes $\xi^{(\varepsilon)}(t)$, $t \geq 0$, and the regenerative stopping times $\mu^{(\varepsilon)}$.

Let us first consider the case where conditions $\mathbf{D_{16}}$ and $\mathbf{M_8}$ remain true while condition $\mathbf{D_{13}}$ is weaken by omitting in it the requirement that the limit distribution should be non-arithmetic.

Condition $\mathbf{D_{13}}$, reduced as described above, implies, due to relations (3.2.5) and (3.2.6), that **(a)** $F_-^{(\varepsilon)}(\cdot) \Rightarrow F^{(0)}(\cdot)$ as $\varepsilon \to 0$, where $F_-^{(\varepsilon)}(t) = \mathsf{P}\{\tau_1^{(\varepsilon)} \leq t/\mu^{(\varepsilon)} \geq \tau_1^{(\varepsilon)}\}$. This relation, estimate (3.2.3), and condition $\mathbf{M_8}$ also imply that **(b)** $m_-^{(\varepsilon)} = \int_0^\infty sF_-^{(\varepsilon)}(ds) \to m_1^{(0)} = \int_0^\infty sF^{(0)}(ds)$ as $\varepsilon \to 0$. Relations **(a)** and **(b)** are sufficient for fulfilment of the weak law of large numbers for the sums of i.i.d. random variables $\hat{\kappa}_k^{(\varepsilon)}$, $k \geq 1$, with the distribution function $F_-^{(\varepsilon)}(t)$. Considered in a stochastic processes setting, this limit theorem takes the form of the relation **(c)** $\tau^{(\varepsilon)}(t) = \sum_{k \leq tn^{(\varepsilon)}} \hat{\kappa}_k^{(\varepsilon)}/n^{(\varepsilon)}, t \geq 0 \xrightarrow{\mathrm{d}} \tau^{(0)}(t) = tm_1^{(0)}, t \geq 0$, that holds for any non-random functions $0 < n^{(\varepsilon)} \to \infty$ as $\varepsilon \to 0$. We are interested in the case where $n^{(\varepsilon)} = (f^{(\varepsilon)})^{-1}$ and will take this function in what follows.

The random variable $\hat{\nu}^{(\varepsilon)}$ has a geometric distribution with the parameter $f^{(\varepsilon)}$. Thus, $\mathsf{P}\{f^{(\varepsilon)}\hat{\nu}^{(\varepsilon)} > t\} = (1 - f^{(\varepsilon)})^{[t/f^{(\varepsilon)}]} \to e^{-t}$ as $\varepsilon \to 0$ for $t \geq 0$. In other words, this means that **(e)** $\nu^{(\varepsilon)} = f^{(\varepsilon)}\hat{\nu}^{(\varepsilon)} \xrightarrow{\mathrm{d}} \nu^{(0)}$ as $\varepsilon \to 0$, where $\nu^{(0)}$ is a random variable exponentially distributed with the parameter 1.

The process $\tau^{(0)}(t)$ is a non-random function, relations **(c)** and **(e)**, due to the well-known Slutsky theorem (see, for example, Silvestrov (2004a), page 12), imply also that **(f)** $(\nu_\varepsilon, \tau^{(\varepsilon)}(t)), t \geq 0 \xrightarrow{\mathrm{d}} (\nu^{(0)}, \tau^{(0)}(t)), t \geq 0$, as $\varepsilon \to 0$.

Since the processes $\tau^{(\varepsilon)}(t)$, $t \geq 0$, are non-decreasing and the limit process $\tau^{(0)}(t)$ is also continuous, relation (**f**) implies (see, Silvestrov (2004a), Theorem 2.2.3 and the remarks in Subsection 2.2.5) that (**g**) $\tau^{(\varepsilon)}(\nu^{(\varepsilon)}) \xrightarrow{d} \tau^{(0)}(\nu^{(0)})$ as $\varepsilon \to 0$.

It remains to note that condition $\mathbf{M_8}$ implies, due to estimate (3.2.3), that $1 - \hat{F}_+^{(\varepsilon)}(t/f^{(\varepsilon)}) \leq (1 - \hat{F}^{(\varepsilon)}(t/f^{(\varepsilon)}))/f^{(\varepsilon)} \leq t^{-1} \int_{t/f^{(\varepsilon)}}^{\infty} (1 - \hat{F}^{(\varepsilon)}(s))\, ds \to 0$ as $\varepsilon \to 0$, $t > 0$, where $\hat{F}_+^{(\varepsilon)}(t) = \mathsf{P}\{\mu^{(\varepsilon)} \leq t/\mu^{(\varepsilon)} < \tau_1^{(\varepsilon)}\}$ and $\hat{F}^{(\varepsilon)}(t) = \mathsf{P}\{\mu^{(\varepsilon)} \wedge \tau_1^{(\varepsilon)} \leq t\}$. In other words, this means that (**h**) $\kappa^{(\varepsilon)} = f^{(\varepsilon)}\hat{\kappa}^{(\varepsilon)} \xrightarrow{\mathsf{P}} 0$ as $\varepsilon \to 0$.

Relations (**g**) and (**h**) imply, again due to Slutsky theorem, that $\tau^{(\varepsilon)}(\nu^{(\varepsilon)}) + \kappa^{(\varepsilon)} \xrightarrow{d} \tau^{(0)}(\nu^{(0)})$ as $\varepsilon \to 0$. According to (3.1.17), we have $f^{(\varepsilon)}\mu^{(\varepsilon)} \overset{d}{=} \tau^{(\varepsilon)}(\nu^{(\varepsilon)}) + \kappa^{(\varepsilon)}$. Thus, (**i**) $f^{(\varepsilon)}\mu^{(\varepsilon)} \xrightarrow{d} \tau^{(0)}(\nu^{(0)})$ as $\varepsilon \to 0$. It remains to note that the random variable $\tau^{(0)}(\nu^{(0)}) = m_1^{(0)}\nu^{(0)}$ has an exponential distribution with the parameter $1/m_1^{(0)}$.

The proof presented above suggests a way in which one can achieve further improvements in the conditions that provide asymptotic relation (3.2.17).

First of all, conditions $\mathbf{D_{13}}$ and $\mathbf{M_8}$ can be replaced with the well-known conditions in the main criterion for convergence, which would provide necessary and sufficient conditions for the weak law of large number to hold for sums of i.i.d. random variables in a triangular array mode. Recall that we consider processes of $\tau^{(\varepsilon)}(t) = \sum_{k \leq tn^{(\varepsilon)}} \hat{\kappa}_k^{(\varepsilon)}/n^{(\varepsilon)}$, $t \geq 0$ in the case where $n^{(\varepsilon)} = 1/f^{(\varepsilon)}$. These conditions (see, for example, Loève (1955)) are: (**j**) $(1 - F_-^{(\varepsilon)}(t/f^{(\varepsilon)}))/f^{(\varepsilon)} \to 0$ as $\varepsilon \to 0$, $t > 0$, and (**k**) $\int_0^{1/f^{(\varepsilon)}} s F_-^{(\varepsilon)}(ds) \to m_1^{(0)} \in (0, \infty)$ as $\varepsilon \to 0$. Note that, in this case, $m_1^{(0)}$ may be not the first moment of the distribution function $F^{(0)}(t)$ but just some positive constant.

Conditions (**j**) and (**k**), implied by conditions $\mathbf{D_{13}}$ (the reduced version) and $\mathbf{M_8}$, are necessary and sufficient for relation (**c**) to hold. Thus, by repeating the proof described above, we get relation (**g**). Condition $\mathbf{M_8}$ is also used for getting relation (**h**). As follows from the proof of this relation, it is also implied by the weaker relation (**l**) $1 - \hat{F}_+^{(\varepsilon)}(t/f^{(\varepsilon)}) \to 0$ as $\varepsilon \to 0$.

Summarizing the remarks above we can conclude that conditions (**j**)–(**l**) imply relation (**i**) that is equivalent to the asymptotic relation (3.2.17).

As was pointed out is Subsection 3.2.4, conditions $\mathbf{D_{13}}$ and $\mathbf{M_8}$ used in the mixed ergodic and limit Theorem 3.2.1 serve very well in the marginal ergodic Theorem 3.2.2. In particular, they reduce to the best known ergodic conditions used in the classical renewal theorem for a non-perturbed model. At the same time, the remarks above show that these conditions can be weakened in the marginal limit theorem for regenerative stopping times.

The question arises of why this is so? As a matter of fact, these conditions, combined with the asymptotic relation (3.2.10), should serve in mixed ergodic/limit theorems for both the ergodic and the limit theorems. The ergodic theorems do require conditions stronger than those for the limit theorems. This explains why these conditions can be improved when applying to marginal limit theorems for regenerative stopping times.

We shall see in the next section that the situation is different in the case where the transition phenomena are studied under Cramér type conditions. In this case, we shall get asymptotic relations that are suitable for mixed ergodic theorems for perturbed regenerative processes with large deviation and for theorems with regenerative stopping times. The latter theorems can not be improved in the way described above. In this case, both types of marginal ergodic and large deviation theorems do require the same type of conditions as the corresponding mixed ergodic and large deviation theorems.

3.2.6 Mixed ergodic and limit theorems for perturbed regenerative processes with transition period

In this model, Theorem 3.2.1 can be directly applied to the shifted regenerative processes, which means that the probabilities $P^{(\varepsilon)}(t, A)$ and $q^{(\varepsilon)}(t, A)$, and the distribution functions $F^{(\varepsilon)}(t)$, $\check{F}^{(\varepsilon)}(t)$ and $\hat{F}^{(\varepsilon)}(t)$ introduced in Subsection 3.2.1, should be replaced in the renewal equation (3.2.1), conditions $\mathbf{D_{13}}$, $\mathbf{M_8}$, $\mathbf{F_7}$, $\mathbf{B_4}$, and Theorem 3.2.1 with the probabilities $P^{(\varepsilon)}(t, A) = \mathsf{P}\{\xi^{(\varepsilon)}(\tau_1^{(\varepsilon)} + t) \in A, \mu^{(\varepsilon)} - \tau_1^{(\varepsilon)} > t / \mu^{(\varepsilon)} \geq \tau_1^{(\varepsilon)}\}$, $q^{(\varepsilon)}(t, A) = \mathsf{P}\{\xi^{(\varepsilon)}(\tau_1^{(\varepsilon)} + t) \in A, \tau_2^{(\varepsilon)} - \tau_1^{(\varepsilon)} > t, \mu^{(\varepsilon)} - \tau_1^{(\varepsilon)} > t / \mu^{(\varepsilon)} \geq \tau_1^{(\varepsilon)}\}$, and the distribution functions $F^{(\varepsilon)}(t) = \mathsf{P}\{\tau_2^{(\varepsilon)} - \tau_1^{(\varepsilon)} \leq t, \mu^{(\varepsilon)} - \tau_1^{(\varepsilon)} \geq \tau_2^{(\varepsilon)} - \tau_1^{(\varepsilon)} / \mu^{(\varepsilon)} \geq \tau_1^{(\varepsilon)}\}$, $\check{F}^{(\varepsilon)}(t) = \mathsf{P}\{\tau_2^{(\varepsilon)} - \tau_1^{(\varepsilon)} \leq t\}$, and $\hat{F}^{(\varepsilon)}(t) = \mathsf{P}\{(\mu^{(\varepsilon)} - \tau_1^{(\varepsilon)}) \wedge (\tau_2^{(\varepsilon)} - \tau_1^{(\varepsilon)}) \leq t / \mu^{(\varepsilon)} \geq \tau_1^{(\varepsilon)}\}$.

Theorem 3.2.1, implies in this case that, under conditions $\mathbf{D_{13}}$, $\mathbf{M_8}$, and $\mathbf{F_7}$, and $0 \leq t^{(\varepsilon)} \to \infty$ satisfying condition $\mathbf{B_4}$, for any $A \in \Gamma_0$, we have

$$P^{(\varepsilon)}(t^{(\varepsilon)}, A) \to \pi^{(0)}(A) e^{-\lambda / m_1^{(0)}} \quad \text{as } \varepsilon \to 0, \tag{3.2.18}$$

where

$$\pi^{(\varepsilon)}(A) = \frac{\int_0^\infty q^{(\varepsilon)}(s, A) m(ds)}{m_1^{(\varepsilon)}}, \quad m_1^{(\varepsilon)} = \int_0^\infty s F^{(\varepsilon)}(ds).$$

As far as non-shifted regenerative processes are concerned, the probabilities $P^{(\varepsilon)}(t, A)$ and $q^{(\varepsilon)}(t, A)$, the distribution functions $F^{(\varepsilon)}(t)$, $\check{F}^{(\varepsilon)}(t)$ and $\hat{F}^{(\varepsilon)}(t)$, introduced in Subsection 3.2.1, should be replaced with the probabilities $\tilde{P}^{(\varepsilon)}(t, A) = \mathsf{P}\{\xi^{(\varepsilon)}(t) \in A, \mu^{(\varepsilon)} > t\}$ and $\tilde{q}^{(\varepsilon)}(t, A) = \mathsf{P}\{\xi^{(\varepsilon)}(t) \in A, \tau_1^{(\varepsilon)} > t, \mu^{(\varepsilon)} > t\}$ and the distribution functions $\tilde{F}^{(\varepsilon)}(t) = \mathsf{P}\{\tau_1^{(\varepsilon)} \leq t, \mu^{(\varepsilon)} \geq \tau_1^{(\varepsilon)}\}$, $\check{\tilde{F}}^{(\varepsilon)}(t) = \mathsf{P}\{\tau_1^{(\varepsilon)} \leq t\}$, and $\hat{\tilde{F}}^{(\varepsilon)}(t) = \mathsf{P}\{\mu^{(\varepsilon)} \wedge \tau_1^{(\varepsilon)} \leq t\}$. Let us also denote $\tilde{f}^{(\varepsilon)} = \mathsf{P}\{\mu^{(\varepsilon)} < \tau_1^{(\varepsilon)}\} = \tilde{F}^{(\varepsilon)}(\infty)$.

The probabilities $\tilde{P}^{(\varepsilon)}(t, A)$ are connected with the probabilities $P^{(\varepsilon)}(t, A)$ by the renewal type relation

$$\tilde{P}^{(\varepsilon)}(t, A) = \tilde{q}^{(\varepsilon)}(t, A) + \int_0^t P^{(\varepsilon)}(t - s, A) \tilde{F}^{(\varepsilon)}(ds), \quad t \geq 0. \tag{3.2.19}$$

It is clear that some additional appropriate conditions should be imposed on the functions $\tilde{q}^{(\varepsilon)}(t, A)$ and the distribution functions $\tilde{F}^{(\varepsilon)}(t)$ and $\hat{\tilde{F}}^{(\varepsilon)}(t)$ in order to get, for the probabilities $\tilde{P}^{(\varepsilon)}(t, A)$, asymptotic relation analogous to (3.2.18). These conditions are:

$\mathbf{D_{17}}$: $\tilde{f}^{(\varepsilon)} \to \tilde{f}^{(0)} \in [0, 1]$ as $\varepsilon \to 0$;

and

$\mathbf{D_{18}}$: $\lim_{T \to \infty} \overline{\lim}_{\varepsilon \to 0}(1 - \hat{\tilde{F}}^{(\varepsilon)}(T) = 0.$

The following estimate is analogous to (3.2.3),

$$1 - \tilde{f}^{(\varepsilon)} - \tilde{F}^{(\varepsilon)}(t) = \mathsf{P}\{\tau_1^{(\varepsilon)} > t, \mu^{(\varepsilon)} \geq \tau_1^{(\varepsilon)}\}$$

$$\leq 1 - \hat{\tilde{F}}^{(\varepsilon)}(t) = \mathsf{P}\{\mu^{(\varepsilon)} \wedge \tau_1^{(\varepsilon)} > t\}$$

$$= \mathsf{P}\{\tau_1^{(\varepsilon)} > t, \mu^{(\varepsilon)} \geq \tau_1^{(\varepsilon)}\} + \mathsf{P}\{\mu_1^{(\varepsilon)} > t, \mu^{(\varepsilon)} < \tau_1^{(\varepsilon)}\}$$

$$\leq 1 - \check{\tilde{F}}^{(\varepsilon)}(t) = \mathsf{P}\{\tau_1^{(\varepsilon)} > t\}. \tag{3.2.20}$$

This estimate implies that condition $\mathbf{D_{18}}$ is a consequence of the following condition:

$\mathbf{D_{19}}$: $\lim_{T \to \infty} \overline{\lim}_{\varepsilon \to 0}(1 - \check{\tilde{F}}^{(\varepsilon)}(T) = 0.$

The estimate (3.2.20) also implies that the following condition can be used, together with condition $\mathbf{D_{18}}$, instead of condition $\mathbf{D_{17}}$:

$\mathbf{D_{20}}$: $\tilde{F}^{(\varepsilon)}(\cdot) \Rightarrow \tilde{F}^{(0)}(\cdot)$ as $\varepsilon \to 0$, where $F^{(0)}(s)$ is a proper or improper distribution function.

Note that condition $\mathbf{D_{20}}$ does require convergence of the pre-limit distribution functions $\tilde{F}^{(\varepsilon)}(s)$ to $\tilde{F}^{(0)}(s)$ as $\varepsilon \to 0$ in continuity points s of the limit distribution function, but does not require that the values $\tilde{F}^{(\varepsilon)}(\infty) = 1 - \tilde{f}^{(\varepsilon)}$ would converge to $\tilde{F}^{(0)}(\infty) = 1 - \tilde{f}^{(0)}$ as $\varepsilon \to 0$.

Lemma 3.2.1. *Let condition $\mathbf{D_{18}}$ hold. Then, condition $\mathbf{D_{20}}$ implies condition $\mathbf{D_{17}}$.*

Proof. Let $T_n, n \geq 1$, be a sequence of continuity points of the distribution function $\tilde{F}^{(\varepsilon)}(t)$ such that $T_n \to \infty$ as $n \to \infty$. Using estimate (3.2.20) and conditions $\mathbf{D_{18}}$ and $\mathbf{D_{20}}$, we get

$$\overline{\lim_{\varepsilon \to 0}} |\tilde{f}^{(\varepsilon)} - \tilde{f}^{(0)}|$$

$$\leq \overline{\lim_{\varepsilon \to 0}} \left(|1 - \tilde{f}^{(\varepsilon)} - \tilde{F}^{(\varepsilon)}(T_n)| + |\tilde{F}^{(\varepsilon)}(T_n) - \tilde{F}^{(0)}(T_n)| \right.$$

$$\left. + |1 - \tilde{f}^{(0)} - \tilde{F}^{(0)}(T_n)| \right)$$

$$\leq \overline{\lim_{\varepsilon \to 0}} (1 - \hat{\tilde{F}}^{(\varepsilon)}(T_n)) + |1 - \tilde{f}^{(0)} - \tilde{F}^{(0)}(T_n)| \to 0 \ \text{ as } \ n \to \infty. \tag{3.2.21}$$

This relation proves the lemma. $\qquad\qquad\square$

It is useful to note that the distribution $\tilde{F}^{(0)}(s)$ in condition $\mathbf{D_{20}}$ is proper if and only if the limit constant is $\tilde{f}^{(0)} = 0$ in condition $\mathbf{D_{17}}$.

Note also that we allow for the degenerate case where $\tilde{f}^{(0)} = 1$, i.e., where condition $\mathbf{R'_5}$ does holds for the limit regenerative process.

Theorem 3.2.1 takes the following form.

Theorem 3.2.5. *Let conditions $\mathbf{D_{13}}$, $\mathbf{M_8}$, $\mathbf{F_7}$, $\mathbf{D_{17}}$, $\mathbf{D_{18}}$ hold, and $0 \leq t^{(\varepsilon)} \to \infty$ as $\varepsilon \to 0$ such that condition $\mathbf{B_4}$ holds. Then, for any $A \in \Gamma_0$,*

$$\tilde{\mathsf{P}}(t^{(\varepsilon)}, A) = \mathsf{P}\{\xi^{(\varepsilon)}(t^{(\varepsilon)}) \in A, \mu^{(\varepsilon)} > t^{(\varepsilon)}\}$$

$$\to (1 - \tilde{f}^{(0)})\pi^{(0)}(A)e^{-\lambda/m_1^{(0)}} \quad \text{as } \varepsilon \to 0. \tag{3.2.22}$$

Proof. The proof is based on the use of the asymptotic relation (3.2.10) given in Theorem 3.2.1 and the renewal relation (3.2.19). Note that relation (3.2.10) is implied by conditions $\mathbf{D_{13}}$, $\mathbf{M_8}$, $\mathbf{F_7}$, and by the assumption that $0 \leq t^{(\varepsilon)} \to \infty$ as $\varepsilon \to 0$ such that condition $\mathbf{B_4}$ holds.

Condition $\mathbf{D_{18}}$ implies that, for any $0 \leq t^{(\varepsilon)} \to \infty$ as $\varepsilon \to 0$ and any $A \in \Gamma_0$,

$$\tilde{\mathsf{P}}(t^{(\varepsilon)}, A) - \int_0^{t^{(\varepsilon)}} P^{(\varepsilon)}(t^{(\varepsilon)} - s, A)\tilde{F}^{(\varepsilon)}(ds)$$

$$= \tilde{q}^{(\varepsilon)}(t^{(\varepsilon)}, A) \leq 1 - \hat{\tilde{F}}^{(\varepsilon)}(t^{(\varepsilon)}) \to 0 \quad \text{as } \varepsilon \to 0. \tag{3.2.23}$$

Relation (3.2.23) implies that to get (3.2.22) we should prove that, for any $0 \leq t^{(\varepsilon)} \to \infty$ as $\varepsilon \to 0$ such that condition $\mathbf{B_4}$ holds and any $A \in \Gamma_0$,

$$\int_0^{t^{(\varepsilon)}} P^{(\varepsilon)}(t^{(\varepsilon)} - s, A)\tilde{F}^{(\varepsilon)}(ds) \to (1 - \tilde{f}^{(0)})\pi^{(0)}(A)e^{-\lambda/m_1^{(0)}} \quad \text{as } \varepsilon \to 0. \tag{3.2.24}$$

Two cases should be considered: **(a)** $\tilde{f}^{(0)} = 1$ or **(b)** $\tilde{f}^{(0)} = 1$.

The case **(a)** is trivial. Indeed in this case we get, for any $A \in \Gamma_0$,

$$\int_0^{t^{(\varepsilon)}} P^{(\varepsilon)}(t^{(\varepsilon)} - s, A)\tilde{F}^{(\varepsilon)}(ds) \leq 1 - \tilde{f}^{(\varepsilon)} \to 0 \quad \text{as } \varepsilon \to 0. \tag{3.2.25}$$

Let us consider the case **(b)**. Let us use Lemma 1.2.2 for asymptotic analysis of the integral in the right-hand side of (3.2.24).

Note first that this integral can be rewritten in the form where the limits of integration do not depend on ε,

$$\int_0^{t^{(\varepsilon)}} P^{(\varepsilon)}(t^{(\varepsilon)} - s, A)\tilde{F}^{(\varepsilon)}(ds) = \int_0^{\infty} P^{(\varepsilon)}(t^{(\varepsilon)} - s, A)\tilde{F}^{(\varepsilon)}(ds), \tag{3.2.26}$$

where one should take $P^{(\varepsilon)}(t - s, A) = 0$ for $s > t$.

For every $A \in \Gamma_0$, the functions $P^{(\varepsilon)}(t^{(\varepsilon)} - s, A)$ are asymptotically bounded, i.e.,

$$\overline{\lim_{\varepsilon \to 0}} \sup_{0 \le s \le t^{(\varepsilon)}} P^{(\varepsilon)}(t^{(\varepsilon)} - s, A) \le 1. \tag{3.2.27}$$

Let us show that, for every $A \in \Gamma_0$ and any $s \ge 0$ and $v > 0$,

$$\sup_{|u| \le v} |P^{(\varepsilon)}(t^{(\varepsilon)} - s + u, A) - \pi^{(0)}(A)e^{-\lambda/m_1^{(0)}}| \to 0 \text{ as } \varepsilon \to 0. \tag{3.2.28}$$

Suppose that asymptotic relation (3.2.28) does not take place. This means that there exists a sequence $0 < \varepsilon_k \to 0$ as $k \to \infty$ and u_k with $|u_k| \le v$ such that

$$|P^{(\varepsilon_k)}(t^{(\varepsilon_k)} - s + u_k, A) - \pi^{(0)}(A)e^{-\lambda/m_1^{(0)}}| \nrightarrow 0 \text{ as } \varepsilon \to 0. \tag{3.2.29}$$

Since $|u_k| \le v$ and $f^{(\varepsilon_k)} \to 0$ as $k \to \infty$, condition $\mathbf{B_4}$ implies that $f^{(\varepsilon)}(t^{(\varepsilon_k)} - s + u_k) \to \lambda$ as $k \to \infty$ for any $s \ge 0$ and $v > 0$. Therefore, by (3.2.19),

$$|P^{(\varepsilon_k)}(t^{(\varepsilon_k)} - s + u_k, A) - \pi^{(0)}(A)e^{-\lambda/m_1^{(0)}}| \to 0 \text{ as } \varepsilon \to 0. \tag{3.2.30}$$

Relation (3.2.29) contradicts (3.2.30). Thus, relation (3.2.28) holds.
Let us now assume that condition $\mathbf{D_{20}}$ holds, instead of condition $\mathbf{D_{17}}$.
Conditions $\mathbf{D_{18}}$ and $\mathbf{D_{20}}$ imply, by Lemma 3.2.1, that

$$1 - \tilde{f}^{(\varepsilon)} = \tilde{F}^{(\varepsilon)}(\infty) \to 1 - \tilde{f}^{(0)} = \tilde{F}^{(0)}(\infty) \text{ as } \varepsilon \to 0. \tag{3.2.31}$$

Condition $\mathbf{D_{20}}$ and relations (3.2.27), (3.2.28) and (3.2.31) imply that for every $A \in \Gamma_0$, the functions $P^{(\varepsilon)}(t^{(\varepsilon)} - s, A)$ and the measures $\tilde{F}^{(\varepsilon)}(ds)$ satisfy conditions of Lemma 1.2.2. Thus, we get by applying Lemma 1.2.2 that, for any $A \in \Gamma_0$,

$$\int_0^\infty P^{(\varepsilon)}(t^{(\varepsilon)} - s, A)\tilde{F}^{(\varepsilon)}(ds) \to \int_0^\infty \pi^{(0)}(A)e^{-\lambda/m_1^{(0)}} \tilde{F}^{(0)}(ds)$$

$$= (1 - \tilde{f}^{(0)})\pi^{(0)}(A)e^{-\lambda/m_1^{(0)}} \text{ as } \varepsilon \to 0. \tag{3.2.32}$$

Let us consider now the case where condition $\mathbf{D_{17}}$ holds. We show that the proof can be reduced to the case considered above, i.e., to the case where condition $\mathbf{D_{20}}$ holds.

Indeed, let us take an arbitrary sequence $0 < \varepsilon_n \to 0$ as $n \to \infty$. It is always possible to select from this sequence a subsequence $\varepsilon_{n_k} \to 0$ as $k \to \infty$ such that

$$\tilde{F}^{(\varepsilon_{n_k})}(\cdot) \Rightarrow \tilde{F}(\cdot) \text{ as } \varepsilon \to 0, \tag{3.2.33}$$

where $\tilde{F}(s)$ is a proper or improper distribution function.

The proof given above for the case where condition $\mathbf{D_{20}}$ holds can be applied to a subsequence of models with the perturbation parameters ε_{n_k}, $k \geq 1$, which and to yields, for any $A \in \Gamma_0$, the following relation:

$$\int_0^\infty P^{(\varepsilon_{n_k})}(t^{(\varepsilon_{n_k})} - s, A)\tilde{F}^{(\varepsilon_{n_k})}(ds)$$

$$\to (1 - \tilde{f})\pi^{(0)}(A)e^{-\lambda/m_1^{(0)}} \quad \text{as } \varepsilon \to 0, \tag{3.2.34}$$

where $\tilde{f} = 1 - \tilde{F}(\infty)$.

The limit distribution $\tilde{F}(s)$ does depend on the choice of the sequence ε_n and the subsequence ε_{n_k}, but the limit $\tilde{f}$ does not.

Indeed, conditions $\mathbf{D_{18}}$ and relation (3.2.33) imply, by Lemma 3.2.1, that

$$1 - \tilde{F}^{(\varepsilon_{n_k})}(\infty) = \tilde{f}^{(\varepsilon_{n_k})} \to 1 - \tilde{F}(\infty) = \tilde{f} \quad \text{as } \varepsilon \to 0. \tag{3.2.35}$$

On the other hand, condition $\mathbf{D_{17}}$ implies that

$$1 - \tilde{F}^{(\varepsilon_{n_k})}(\infty) = \tilde{f}^{(\varepsilon_{n_k})} \to \tilde{f}^{(0)} \quad \text{as } \varepsilon \to 0. \tag{3.2.36}$$

Thus, $\tilde{f} = \tilde{f}^{(0)}$ and, therefore, the limit in the right hand-side of (3.2.34) does not depend on the choice of the sequence ε_n and the subsequence ε_{n_k}. Moreover, it coincides with the limit in the right-hand side of relation (3.2.24). This implies that relation (3.2.24) holds.

The proof of the theorem is complete. $\qquad\qquad\qquad\qquad\qquad\qquad\qquad\square$

3.2.7 Ergodic theorems for perturbed regenerative processes with transition period

Let us again consider a model in which the regenerative stopping time is $\mu^{(\varepsilon)} = \infty$ with probability 1 for every $\varepsilon \geq 0$. In this case, $f^{(\varepsilon)} \equiv 0$, $F^{(\varepsilon)}(s) = \hat{F}^{(\varepsilon)}(s) = \check{F}^{(\varepsilon)}(s) = \mathsf{P}\{\tau_2^{(\varepsilon)} - \tau_1^{(\varepsilon)} \leq s\}$ and $m_1^{(\varepsilon)} = \mathsf{E}(\tau_2^{(\varepsilon)} - \tau_1^{(\varepsilon)})$. Also, $q^{(\varepsilon)}(s, A) = \mathsf{P}\{\xi^{(\varepsilon)}(\tau_1^{(\varepsilon)} + t) \in A, \tau_2^{(\varepsilon)} - \tau_1^{(\varepsilon)} > t\}$. Condition $\mathbf{B_4}$ automatically holds for any $0 \leq t^{(\varepsilon)} \to \infty$ as $\varepsilon \to 0$ and $\lambda = 0$. As far as characteristics of the transition period are concerned, in this case, $\tilde{F}^{(\varepsilon)}(s) = \hat{\tilde{F}}^{(\varepsilon)}(s) = \check{\tilde{F}}^{(\varepsilon)}(s) = \mathsf{P}\{\tau_1^{(\varepsilon)} \leq s\}$, $\tilde{q}^{(\varepsilon)}(s, A) = \mathsf{P}\{\xi^{(\varepsilon)}(t) \in A, \tau_1^{(\varepsilon)} > t\}$ and $\tilde{f}^{(\varepsilon)} = 0$ for every $\varepsilon \geq 0$.

Note that, in this case, condition $\mathbf{D_{17}}$ automatically holds with the limit constant $\tilde{f}^{(0)} = 0$, conditions $\mathbf{D_{18}}$ and $\mathbf{D_{19}}$ are equivalent, and condition $\mathbf{D_{20}}$ implies condition $\mathbf{D_{18}}$.

Theorem 3.2.5 yields in this case a variant of the ergodic theorem for perturbed regenerative processes.

Theorem 3.2.6. *Let conditions $\mathbf{D_{13}}$, $\mathbf{D_{14}}$, $\mathbf{M_8}$, $\mathbf{F_7}$, $\mathbf{D_{19}}$ hold. Then, for any $0 \leq t^{(\varepsilon)} \to \infty$ as $\varepsilon \to 0$,*

$$P^{(\varepsilon)}(t^{(\varepsilon)}, A) = \mathsf{P}\{\xi^{(\varepsilon)}(t^{(\varepsilon)}) \in A\} \to \pi^{(0)}(A) \quad as \quad \varepsilon \to 0, \quad A \in \Gamma_0. \tag{3.2.37}$$

Conditions $\mathbf{D_{17}}$ and $\mathbf{D_{18}}$ hold automatically for any fixed $\varepsilon \geq 0$. Thus, convergence $\mathsf{P}\{\xi^{(\varepsilon)}(t) \in A\}$ to a stationary distribution $\pi^{(\varepsilon)}(A)$ is also preserved for regenerative processes with transition period. Let ε_0 be the parameter defined in Theorem 3.2.3.

Theorem 3.2.7. *Let conditions* $\mathbf{D_{14}}$, $\mathbf{D_{15}}$, $\mathbf{M_8}$, *and* $\mathbf{F_8}$ *hold. Then, for any all* $\varepsilon \leq \varepsilon_0$,

$$\mathsf{P}\{\xi^{(\varepsilon)}(t) \in A\} \to \pi^{(\varepsilon)}(A) \quad as \quad t \to \infty, \quad A \in \Gamma_0. \tag{3.2.38}$$

3.3   Mixed ergodic and large deviation theorems for perturbed regenerative processes

In this section we give mixed ergodic and large deviation theorems for perturbed regenerative processes and regenerative stopping times. These theorems describe pseudo-stationary and quasi-stationary phenomena for perturbed regenerative processes.

3.3.1   Pseudo-stationary exponential asymptotics for perturbed regenerative processes

Let us now assume the condition:

$\mathbf{C_3}$:   There exists $\delta > 0$ such that $\overline{\lim}_{0 \leq \varepsilon \to 0} \int_0^\infty e^{\delta s} \hat{F}^{(\varepsilon)}(ds) < \infty$.

Obviously,

$$\mathsf{E}e^{\delta \tau_1^{(\varepsilon)}} \chi(\mu^{(\varepsilon)} \geq \tau_1^{(\varepsilon)}) = \mathsf{E} \int_0^\infty e^{\delta s} F^{(\varepsilon)}(ds) \leq \mathsf{E}e^{\delta(\mu^{(\varepsilon)} \wedge \tau_1^{(\varepsilon)})}$$

$$= \mathsf{E}e^{\delta \tau_1^{(\varepsilon)}} \chi(\mu^{(\varepsilon)} \geq \tau_1^{(\varepsilon)}) + \mathsf{E}e^{\delta \mu_1^{(\varepsilon)}} \chi(\mu^{(\varepsilon)} < \tau_1^{(\varepsilon)})$$

$$= \int_0^\infty e^{\delta s} \hat{F}^{(\varepsilon)}(ds) \leq \mathsf{E}e^{\delta \tau_1^{(\varepsilon)}} = \int_0^\infty e^{\delta s} \check{F}^{(\varepsilon)}(ds). \tag{3.3.1}$$

Therefore, condition $\mathbf{C_3}$ implies the following condition:

$\mathbf{C_4}$:   There exists $\delta > 0$ such that $\overline{\lim}_{0 \leq \varepsilon \to 0} \int_0^\infty e^{\delta s} F^{(\varepsilon)}(ds) < \infty$.

At the same time, condition $\mathbf{C_3}$ is implied by the following condition:

$\mathbf{C_5}$:   There exists $\delta > 0$ such that $\overline{\lim}_{0 \leq \varepsilon \to 0} \int_0^\infty e^{\delta s} \check{F}^{(\varepsilon)}(ds) < \infty$.

Denote

$$\phi^{(\varepsilon)}(\rho) = \int_0^\infty e^{\rho s} F^{(\varepsilon)}(ds).$$

The following equation plays an important role in what follows:

$$\phi^{(\varepsilon)}(\rho) = 1. \tag{3.3.2}$$

As follows from Lemma 1.4.1, conditions $\mathbf{D}_{13}$ and $\mathbf{C}_3$ imply that there exists a unique nonnegative root $\rho^{(\varepsilon)}$ of equation (3.3.2) for all ε small enough, and $\rho^{(\varepsilon)} \to 0$ as $\varepsilon \to 0$.

Note also that conditions $\mathbf{D}_{13}$ and $\mathbf{C}_3$ imply that $m_1^{(\varepsilon)} \to m_1^{(0)} < \infty$ as $\varepsilon \to 0$.

Let us formulate a theorem which is an analogue of Theorem 1.4.1 applied to renewal equation (3.2.1).

Theorem 3.3.1. *Let conditions* $\mathbf{D}_{13}$, $\mathbf{C}_3$, *and* $\mathbf{F}_7$ *hold. Then, for any* $0 \leq t^{(\varepsilon)} \to \infty$ *as* $\varepsilon \to 0$ *and any* $A \in \Gamma_0$,

$$\frac{\mathsf{P}\{\xi^{(\varepsilon)}(t^{(\varepsilon)}) \in A, \mu^{(\varepsilon)} > t^{(\varepsilon)}\}}{\exp\{-\rho^{(\varepsilon)}t^{(\varepsilon)}\}} \to \pi^{(0)}(A) \quad as \quad \varepsilon \to 0. \tag{3.3.3}$$

Proof. Condition $\mathbf{D}_{13}$ implies that condition $\mathbf{D}_6$ holds for the distribution functions $F^{(\varepsilon)}(s)$. Also, condition $\mathbf{C}_3$ implies condition $\mathbf{C}_4$, which in this case coincides with condition $\mathbf{C}_1$.

Condition $\mathbf{F}_7$ implies that, for every $A \in \Gamma_0$, condition $\mathbf{F}_5$ **(a)** holds for the function $q^{(\varepsilon)}(s, A)$. Since $q^{(\varepsilon)}(s, A) \leq 1$, condition $\mathbf{F}_5$ **(b)** obviously holds.

Let us show now that condition $\mathbf{C}_3$ implies that for every $A \in \Gamma_0$ condition $\mathbf{F}_5$ **(c)** holds for the functions $q^{(\varepsilon)}(s, A)$ for any $0 < \gamma < \delta$. Indeed, it obviously follows from condition $\mathbf{C}_3$ that for any ε small enough,

$$e^{\delta T}(1 - \hat{F}^{(\varepsilon)}(T)) \leq \int_T^\infty e^{\delta s}\hat{F}^{(\varepsilon)}(ds) \to 0 \quad as \quad T \to \infty. \tag{3.3.4}$$

Using this relation, integration by parts and condition $\mathbf{C}_3$ we have

$$\lim_{T \to \infty} \overline{\lim_{0 \leq \varepsilon \to 0}} \, h \sum_{r \geq T/h} \sup_{rh \leq t \leq (r+1)h} e^{\gamma t}|q^{(\varepsilon)}(t, A)|$$

$$\leq \lim_{T \to \infty} \overline{\lim_{0 \leq \varepsilon \to 0}} \int_{T-h}^\infty e^{\gamma s}(1 - \hat{F}^{(\varepsilon)}(s))ds$$

$$\leq \lim_{T \to \infty} \overline{\lim_{0 \leq \varepsilon \to 0}} \, e^{-(\delta-\gamma)(T-h)} \int_{T-h}^\infty e^{\delta s}(1 - \hat{F}^{(\varepsilon)}(s))ds$$

$$\leq \lim_{T \to \infty} e^{-(\delta-\gamma)(T-h)} \overline{\lim_{0 \leq \varepsilon \to 0}} \int_0^\infty e^{\delta s}(1 - \hat{F}^{(\varepsilon)}(s))ds$$

$$= \lim_{T \to \infty} e^{-(\delta-\gamma)(T-h)} \overline{\lim_{0 < \varepsilon \to 0}} \left(\delta^{-1}(1 - \hat{F}^{(\varepsilon)}(0))\right.$$

$$\left. + \delta^{-1} \int_0^\infty e^{\delta s}\hat{F}^{(\varepsilon)}(ds)\right)$$

$$\leq \lim_{T \to \infty} e^{-(\delta-\gamma)(T-h)}(\delta^{-1} + \delta^{-1} \overline{\lim_{0 \leq \varepsilon \to 0}} \int_0^\infty e^{\delta s}\hat{F}^{(\varepsilon)}(ds)) = 0. \tag{3.3.5}$$

Now we can finish the proof of the theorem by applying Theorem 1.4.1. $\qquad\square$

3.3.2   Quasi-stationary exponential asymptotics for perturbed regenerative processes

Let us now consider the case where the limit distribution $F^{(0)}(t)$ can be either proper or improper.

We use the representation

$$\hat{F}_{-}^{(\varepsilon)}(t) = \mathsf{P}\{\tau_1^{(\varepsilon)} \leq t / \mu^{(\varepsilon)} \geq \tau_1^{(\varepsilon)}\} = F^{(\varepsilon)}(t)/(1 - f^{(\varepsilon)}), \quad t \geq 0.$$

The following condition replaces $\mathbf{D_{13}}$:

$\mathbf{D_{21}}$:   (a) $\hat{F}_{-}^{(\varepsilon)}(\cdot) \Rightarrow \hat{F}_{-}^{(0)}(\cdot)$ as $\varepsilon \to 0$, where $\hat{F}_{-}^{(0)}(t)$ is a non-arithmetic distribution function;

   (b) $f^{(\varepsilon)} \to f^{(0)} \in [0, 1)$ as $\varepsilon \to 0$.

We assume also the following version of the Cramér type condition:

$\mathbf{C_6}$: There exists $\delta > 0$ such that:

   (a) $\overline{\lim}_{0 \leq \varepsilon \to 0} \int_0^\infty e^{\delta s} \hat{F}^{(\varepsilon)}(ds) < \infty$;

   (b) $\int_0^\infty e^{\delta s} F^{(0)}(ds) > 1$.

Let us consider the condition:

$\mathbf{D_{22}}$: $F^{(\varepsilon)}(\cdot) \Rightarrow F^{(0)}(\cdot)$ as $\varepsilon \to 0$, where $F^{(0)}(s)$ is a proper or improper distribution not concentrated in zero, i.e., such that $F^{(0)}(0) < F^{(0)}(\infty)$, and $\hat{F}_{-}^{(0)}(t) = F^{(0)}(t)/F^{(0)}(\infty)$ is a non-arithmetic distribution function.

As follows from Lemma 1.4.4, conditions $\mathbf{D_{22}}$ and $\mathbf{C_4}$, which are weaker than $\mathbf{C_6}$, imply that $f^{(\varepsilon)} \to f^{(0)} \in [0, 1)$ as $\varepsilon \to 0$.

Therefore, under condition $\mathbf{C_6}$, condition $\mathbf{D_{21}}$ is implied by the simpler condition $\mathbf{D_{22}}$.

Let us again consider the characteristic equation (3.3.2). It follows from Lemma 1.4.3 that under conditions $\mathbf{D_{21}}$ and $\mathbf{C_6}$ there exists a unique nonnegative root $\rho^{(\varepsilon)}$ of equation (3.3.2) for all ε small enough. Note that $\rho^{(\varepsilon)} = 0$ if and only if $f^{(\varepsilon)} = 0$ or, equivalently, $\rho^{(\varepsilon)} > 0$ if and only if $f^{(\varepsilon)} > 0$. Also, $\rho^{(\varepsilon)} \to \rho^{(0)}$ as $\varepsilon \to 0$. Note also that $\rho^{(0)} < \delta$.

It is also useful to note that, in the case $f^{(0)} = 0$, conditions $\mathbf{D_{21}}$ and $\mathbf{C_6}$ (a) imply condition $\mathbf{C_6}$ (b).

Denote

$$\tilde{\pi}^{(\varepsilon)}(A) = \frac{\int_0^\infty e^{\rho^{(\varepsilon)} s} q^{(\varepsilon)}(s, A) m(ds)}{\int_0^\infty s e^{\rho^{(\varepsilon)} s} F^{(\varepsilon)}(ds)}, \quad A \in \Gamma. \tag{3.3.6}$$

Let us formulate a theorem which is analogous to Theorem 1.4.2 applied to the renewal equation (3.2.1). This theorem covers both cases, $f^{(0)} > 0$ and $f^{(0)} = 0$.

Recall that $f^{(0)} > 0$ if and only if $\rho^{(0)} > 0$ or, equivalently, $f^{(0)} = 0$ if and only if $\rho^{(0)} = 0$. Theorem formulated below generalises Theorem 3.3.1 and reduces to this theorem in the case where $f^{(0)} = 0$.

Theorem 3.3.2. *Let conditions* $\mathbf{D_{21}}$, $\mathbf{C_6}$, *and* $\mathbf{F_7}$ *hold. Then, for any* $0 \leq t^{(\varepsilon)} \to \infty$ *as* $\varepsilon \to 0$ *and any* $A \in \Gamma_0$,

$$\frac{\mathsf{P}\{\xi^{(\varepsilon)}(t^{(\varepsilon)}) \in A, \mu^{(\varepsilon)} > t^{(\varepsilon)}\}}{\exp\{-\rho^{(\varepsilon)}t^{(\varepsilon)}\}} \to \tilde{\pi}^{(0)}(A) \quad as \quad \varepsilon \to 0. \tag{3.3.7}$$

Proof. Condition $\mathbf{D_{21}}$ implies that condition $\mathbf{D_{11}}$ holds for the distribution functions $F^{(\varepsilon)}(s)$. Also, condition $\mathbf{C_6}$ implies condition $\mathbf{C_4}$ that, in this case, coincides with condition $\mathbf{C_2}$.

Condition $\mathbf{F_7}$ implies that for every $A \in \Gamma_0$, condition $\mathbf{F_6}$ (**a**) holds for the function $q^{(\varepsilon)}(s, A)$. Since $0 \leq q^{(\varepsilon)}(s, A) \leq 1$, condition $\mathbf{F_6}$ (**b**) obviously holds.

Let us show now that condition $\mathbf{C_6}$ implies that, for every $A \in \Gamma_0$, condition $\mathbf{F_6}$ (**c**) holds for the functions $q^{(\varepsilon)}(s, A)$ if $0 < \rho^{(0)} + \gamma < \delta$. Indeed, we have similarly to (3.3.5) that

$$\lim_{T \to \infty} \overline{\lim_{0 \leq \varepsilon \to 0}} \, h \sum_{r \geq T/h} \sup_{rh \leq t \leq (r+1)h} e^{(\rho^{(0)}+\gamma)t}|q^{(\varepsilon)}(t, A)|$$

$$\leq \lim_{T \to \infty} \overline{\lim_{0 \leq \varepsilon \to 0}} \int_{T-h}^{\infty} e^{(\rho^{(0)}+\gamma)s}(1 - \hat{F}^{(\varepsilon)}(s))ds$$

$$\leq \lim_{T \to \infty} e^{-(\delta-(\rho^{(0)}+\gamma))(T-h)} \overline{\lim_{0 \leq \varepsilon \to 0}} \int_0^{\infty} e^{\delta s}(1 - \hat{F}^{(\varepsilon)}(s))ds$$

$$\leq \lim_{T \to \infty} e^{-(\delta-(\rho^{(0)}+\gamma))(T-h)}(\delta^{-1} + \delta^{-1} \overline{\lim_{0 \leq \varepsilon \to 0}} \int_0^{\infty} e^{\delta s} \hat{F}^{(\varepsilon)}(ds)) = 0. \tag{3.3.8}$$

Application of Theorem 1.4.2 finishes the proof. $\qquad\square$

3.3.3 Quasi-stationary distributions for perturbed regenerative processes

Condition $\mathbf{D_{21}}$ does not imply that the distribution function $\hat{F}_-^{(\varepsilon)}(t)$ is non-arithmetic for all ε small enough. Let us assume the following condition:

$\mathbf{D_{23}}$: $\hat{F}_-^{(\varepsilon)}(t)$ is a non-arithmetic distribution function for all ε small enough, say $\varepsilon \leq \varepsilon_1$.

Let us also assume the following condition:

$\mathbf{F_9}$: There exists $\gamma > 0$ such that $\rho^{(0)} + \gamma < \delta$ and, for every $A \in \Gamma_0$, the function $e^{(\rho^{(0)}+\gamma)s}q^{(\varepsilon)}(s, A)$ is directly Riemann integrable on $[0, \infty)$ for all ε small enough, say $\varepsilon \leq \varepsilon_2$.

Note that under condition $\mathbf{C_6}$ it is sufficient to assume that, for $A \in \Gamma_0$, the function $q^{(\varepsilon)}(s, A)$ is continuous almost everywhere with respect to Lebesgue measure on $[0, \infty)$ for all ε small enough. In this case, condition $\mathbf{F_9}$ holds and Lebesgue integration can be replaced with Riemann integration in the right-hand side of (3.3.6).

Condition $\mathbf{C_6}$ implies that $\int_0^\infty e^{\delta s} \hat{F}^{(\varepsilon)}(ds) \in (1, \infty)$ for all ε small enough, say $\varepsilon \leq \varepsilon_3$.

Let

$$\varepsilon_0 = \min(\varepsilon_1, \varepsilon_2, \varepsilon_3). \tag{3.3.9}$$

The following theorem is a direct corollary of Theorem 3.3.2.

Theorem 3.3.3. *Let conditions* $\mathbf{D_{23}}$*,* $\mathbf{C_6}$*, and* $\mathbf{F_9}$ *hold. Then, for* $\varepsilon \leq \varepsilon_0$*,*

$$e^{\rho^{(\varepsilon)}t}\mathsf{P}\{\xi^{(\varepsilon)}(t) \in A, \mu^{(\varepsilon)} > t\} \to \tilde{\pi}^{(\varepsilon)}(A) \quad as \quad t \to \infty, \quad A \in \Gamma_0. \tag{3.3.10}$$

Note that, under conditions $\mathbf{D_{21}}$ and $\mathbf{C_6}$, the expression for $\tilde{\pi}^{(\varepsilon)}(A)$ given by formula (3.3.6) is finite for all $\varepsilon \geq 0$ even if condition $\mathbf{F_9}$ does not hold. This is so, since this formula uses Lebesgue integration.

It should be noted that, in general, $\tilde{\pi}^{(\varepsilon)}(A)$ may not be a probability distribution, i.e., it may be such that $0 < \tilde{\pi}^{(\varepsilon)}(X) \neq 1$. Let us now define

$$\pi^{(\varepsilon)}(A) = \frac{\tilde{\pi}^{(\varepsilon)}(A)}{\tilde{\pi}^{(\varepsilon)}(X)} = \frac{\int_0^\infty e^{\rho^{(\varepsilon)}s} q^{(\varepsilon)}(s, A) m(ds)}{\int_0^\infty e^{\rho^{(\varepsilon)}s}(1 - \hat{F}^{(\varepsilon)}(s)) ds}, \quad A \in \Gamma. \tag{3.3.11}$$

Note that our notations are consistent with those introduced in formula (3.2.9). Indeed, in the case considered in (3.2.9), $\rho^{(\varepsilon)} = 0$ and $1 - \hat{F}^{(\varepsilon)}(s) \equiv 1 - F^{(\varepsilon)}(s)$. In this case, the denominator in (3.3.11) is equal to $m_1^{(\varepsilon)}$.

By the definition, $\pi^{(\varepsilon)}(A)$ is a probability distribution. It is called a *quasi-stationary distribution* for the regenerative process $\xi^{(\varepsilon)}(t)$, $t \geq 0$, with the regenerative stopping time $\mu^{(\varepsilon)}$.

Theorem 3.3.4. *Let conditions* $\mathbf{D_{21}}$*,* $\mathbf{C_6}$*, and* $\mathbf{F_9}$ *hold. Then, for* $\varepsilon \leq \varepsilon_0$*,*

$$\mathsf{P}\{\xi^{(\varepsilon)}(t) \in A / \mu^{(\varepsilon)} > t\} = \frac{e^{\rho^{(\varepsilon)}t}\mathsf{P}\{\xi^{(\varepsilon)}(t) \in A, \mu^{(\varepsilon)} > t\}}{e^{\rho^{(\varepsilon)}t}\mathsf{P}\{\mu^{(\varepsilon)} > t\}}$$

$$\to \pi^{(\varepsilon)}(A) \quad as \quad t \to \infty, \quad A \in \Gamma_0. \tag{3.3.12}$$

Relation (3.3.12) clarifies why it is natural to call the distribution $\pi^{(\varepsilon)}(A)$ a *quasi-stationary* distribution for the regenerative process $\xi^{(\varepsilon)}(t)$ with the regenerative stopping time $\mu^{(\varepsilon)}$.

3.3.4   Convergence of quasi-stationary distributions for perturbed regenerative processes

The following theorem supplements the asymptotic relations (3.3.7) and (3.3.10).

Theorem 3.3.5. *Let conditions* $\mathbf{D_{21}}$, $\mathbf{C_6}$, *and* $\mathbf{F_7}$ *hold. Then*

$$\tilde{\pi}^{(\varepsilon)}(A) \to \tilde{\pi}^{(0)}(A) \quad as \quad \varepsilon \to 0, \quad A \in \Gamma_0. \tag{3.3.13}$$

Proof. As was shown in the proof of Theorem 3.3.2, conditions $\mathbf{D_{21}}$, $\mathbf{C_6}$ and $\mathbf{F_7}$ imply that conditions $\mathbf{D_{11}}$, $\mathbf{C_2}$ and $\mathbf{F_6}$ of Theorem 1.4.1 hold for the distribution functions $F^{(\varepsilon)}(s)$ and the functions $q^{(\varepsilon)}(s, A)$ for every $A \in \Gamma_0$. So we can prove the theorem by applying Theorem 1.4.4. $\square$

The following theorem is a direct corollary of formula (3.3.11) and Theorem 3.3.5.

Theorem 3.3.6. *Let conditions* $\mathbf{D_{21}}$, $\mathbf{C_6}$, *and* $\mathbf{F_7}$ *hold. Then*

$$\pi^{(\varepsilon)}(A) \to \pi^{(0)}(A) \quad as \quad \varepsilon \to 0, \quad A \in \Gamma_0. \tag{3.3.14}$$

Remark 3.3.1. It is useful to note that formula (3.3.6) that defines the quantities $\tilde{\pi}^{(\varepsilon)}(A)$ and formula (3.3.11) that defines the quasi-stationary probabilities $\pi^{(\varepsilon)}(A)$ do not require for the distribution $F^{(\varepsilon)}(t)$ to be non-arithmetic. This remark allows to drop the requirement that the limit distribution $F^{(0)}(t)$ be non-arithmetic in condition $\mathbf{D_{21}}$ used in Theorems 3.3.5 and 3.3.6.

3.3.5   Large deviation asymptotics for distributions of regenerative stopping moments

We assume that condition $\mathbf{D_{16}}$ holds and consider the case where $A = X$. As was pointed out in Subsection 3.2.4, $P^{(\varepsilon)}(t, X) = \mathsf{P}\{\mu^{(\varepsilon)} > t\}$ and $q^{(\varepsilon)}(t, X) = \mathsf{P}\{\mu^{(\varepsilon)} \wedge \tau_1^{(\varepsilon)} > t\} = 1 - \hat{F}^{(\varepsilon)}(t)$. Recall also that $F^{(\varepsilon)}(t) = \mathsf{P}\{\tau_1^{(\varepsilon)} \le t, \tau_1^{(\varepsilon)} \le \mu^{(\varepsilon)}\}$.

Condition $\mathbf{F_7}$ is equivalent in this case to the following condition:

$\mathbf{D_{24}}$: $\hat{F}^{(\varepsilon)}(\cdot) \Rightarrow \hat{F}^{(0)}(\cdot)$ as $\varepsilon \to 0$, where $\hat{F}^{(0)}(t)$ is a proper distribution function.

Indeed, condition $\mathbf{F_7}$ obviously implies that $\hat{F}^{(\varepsilon)}(t) \to \hat{F}^{(0)}(t)$ as $\varepsilon \to 0$ almost everywhere with respect to Lebesgue measure on $[0, \infty)$. Let S_0' be the corresponding set of convergence. Obviously, S_0' is dense in the interval $[0, \infty)$. Therefore, condition $\mathbf{F_{10}}$ holds, since weak convergence of distribution functions concentrated on the interval $[0, \infty)$ is implied by their pointwise convergence in points of any set dense in this interval.

On the other hand, condition $\mathbf{D_{24}}$ implies that $1 - \hat{F}^{(\varepsilon)}(t) \to 1 - \hat{F}^{(0)}(t)$ as $\varepsilon \to 0$ for every point of continuity of the limit function. Moreover, since the functions $1 - \hat{F}^{(\varepsilon)}(t)$ are monotone, this convergence is locally uniform for every such point.

But the set S_0 of continuity points of the function $1 - F^{(0)}(t)$ is the interval $[0, \infty)$ except for at most a countable set. Thus the Lebesgue measure $m(\overline{S}_0) = 0$. Therefore, condition $\mathbf{F_7}$ holds.

Note also that, in the case where $f^{(0)} = 0$, condition $\mathbf{D_{21}}$ implies, due to relation (3.2.6), that condition $\mathbf{D_{24}}$ holds. Also, in this case, $\hat{F}_-^{(0)}(t) \equiv \hat{F}^{(0)}(t) \equiv F^{(0)}(t)$.

Define

$$\tilde{\pi}^{(\varepsilon)}(X) = \frac{\int_0^\infty e^{\rho^{(\varepsilon)}s}(1 - \hat{F}^{(\varepsilon)}(s))ds}{\int_0^\infty se^{\rho^{(\varepsilon)}s}F^{(\varepsilon)}(ds)}. \tag{3.3.15}$$

Formula (3.3.15) is a particular case of formula (3.3.6) taken for $A = X$. Note that Lebesgue integration, used in the integral in the numerator of (3.3.6), is replaced with Riemann integration in (3.3.15). As a matter of fact, the function $e^{\rho^{(\varepsilon)}s}(1 - \hat{F}^{(\varepsilon)}(s))$ has at most countable set of discontinuity points.

As was mentioned above, it may happen that $\tilde{\pi}^{(\varepsilon)}(X) \neq 1$. However, if $f^{(\varepsilon)} = 0$, then $\rho^{(\varepsilon)} = 0$ and $\hat{F}^{(\varepsilon)}(t) \equiv F^{(\varepsilon)}(t)$. In this case,

$$\tilde{\pi}^{(\varepsilon)}(X) = \frac{\int_0^\infty (1 - F^{(\varepsilon)}(s))ds}{\int_0^\infty sF^{(\varepsilon)}(ds)} = 1.$$

Theorem 3.3.2 takes in this case the following form.

Theorem 3.3.7. *Let conditions $\mathbf{D_{21}}$, $\mathbf{C_6}$, and $\mathbf{D_{24}}$ hold. Then for any $0 \leq t^{(\varepsilon)} \to \infty$ as $\varepsilon \to 0$,*

$$\frac{\mathsf{P}\{\mu^{(\varepsilon)} > t^{(\varepsilon)}\}}{\exp\{-\rho^{(\varepsilon)}t^{(\varepsilon)}\}} \to \tilde{\pi}^{(0)}(X) \quad as \quad \varepsilon \to 0. \tag{3.3.16}$$

Asymptotic relation (3.3.16) becomes trivial if condition $\mathbf{D_{14}}$ holds, i.e., $f^{(\varepsilon)} = 0$ for ε small enough, say $\varepsilon \leq \varepsilon_0$. Indeed, in this case, $\mu^{(\varepsilon)} = \infty$ with probability 1 for every $\varepsilon \leq \varepsilon_0$. In this case, both expressions in the left and the right hand-sides of (3.3.16) equal to 1.

Let us consider the case where condition $\mathbf{D_{16}}$ holds, i.e., $f^{(\varepsilon)} > 0$ for $\varepsilon > 0$.

There are two alternative cases: **(i)** $f^{(0)} = 0$ and **(ii)** $f^{(0)} > 0$.

Let us first consider case **(i)**. Condition $\mathbf{D_{21}}$ implies that $f^{(\varepsilon)} \to 0$ as $\varepsilon \to 0$. Therefore, $\rho^{(\varepsilon)} \to 0$ as $\varepsilon \to 0$. Also, in this case, $\tilde{\pi}^{(0)}(X) = 1$.

Choose $t^{(\varepsilon)} \to \infty$ such that $\rho^{(\varepsilon)}t^{(\varepsilon)} \to 0$ as $\varepsilon \to 0$. Now, the asymptotic relation (3.3.16) takes the following form:

$$\mathsf{P}\{\mu^{(\varepsilon)} > t^{(\varepsilon)}\} \to 1 \quad as \quad \varepsilon \to 0. \tag{3.3.17}$$

Relation (3.3.17) implies that, in case **(i)**, the stopping times $\mu^{(\varepsilon)}$ are asymptotically stochastically unbounded random variables as $\varepsilon \to 0$, i.e.,

$$\mu^{(\varepsilon)} \xrightarrow{\mathsf{P}} \infty \quad as \quad \varepsilon \to 0. \tag{3.3.18}$$

Let us take $t^{(\varepsilon)} = t/\rho^{(\varepsilon)} \to \infty$ as $\varepsilon \to 0$, where $t > 0$. Asymptotic relation (3.3.16) takes the following form:

$$\frac{\mathsf{P}\{\rho^{(\varepsilon)}\mu^{(\varepsilon)} > t\}}{\exp\{-t\}} \to 1 \quad \text{as } \varepsilon \to 0, \quad t > 0. \tag{3.3.19}$$

This is a weak convergence asymptotic relation. Indeed, relation (3.3.19) means that the random variables converge, $\rho^{(\varepsilon)}\mu^{(\varepsilon)} \xrightarrow{\mathrm{d}} \nu_0$ as $\varepsilon \to 0$, where ν_0 is an exponentially distributed random variable with parameter 1.

Note that (3.3.19) is consistent with the asymptotic relation (3.2.17). According to Lemma 1.4.2, in this case, $\rho^{(\varepsilon)} \sim f^{(\varepsilon)}/m_1^{(\varepsilon)}$ as $\varepsilon \to 0$. Also, $m_1^{(\varepsilon)} \to m_1^{(0)}$ as $\varepsilon \to 0$. Thus, (3.3.19) can be rewritten in the equivalent form $\mathsf{P}\{f^{(\varepsilon)}\mu^{(\varepsilon)} > t\} \to \exp\{-t/m_1^{(0)}\}$ as $\varepsilon \to 0$, $t > 0$, identical to (3.2.17). It should be noted however that asymptotic relation (3.2.17) was obtained under weaker conditions than (3.3.19).

We can, however, take $t^{(\varepsilon)} = \tilde{t}^{(\varepsilon)}/\rho^{(\varepsilon)}$, where $\tilde{t}^{(\varepsilon)} \to \infty$ as $\varepsilon \to 0$. In this case (3.3.16) takes the form of a large deviation asymptotic relation,

$$\frac{\mathsf{P}\{\rho^{(\varepsilon)}\mu^{(\varepsilon)} > \tilde{t}^{(\varepsilon)}\}}{\exp\{-\tilde{t}^{(\varepsilon)}\}} \to 1 \quad \text{as } \varepsilon \to 0. \tag{3.3.20}$$

Asymptotic relation (3.3.20) is much stronger than (3.3.19). This relation yields that the asymptotic relative error for approximation of the tail probabilities $P\{\rho^{(\varepsilon)}\mu^{(\varepsilon)} > \tilde{t}^{(\varepsilon)}\}$ with the corresponding tail probabilities of the limit exponential distribution equals 0 for any $\tilde{t}^{(\varepsilon)} \to \infty$. This is achieved because of the choice of the very well-fitted normalising coefficients $\rho^{(\varepsilon)}$. The payment for this result is a need to use the Cramér type condition $\mathbf{C_6}$.

The drawback of (3.3.20) is that the normalising coefficient $\rho^{(\varepsilon)}$ is given as a root of the nonlinear equation (3.3.2). The question arises whether it is possible to replace $\rho^{(\varepsilon)}$ in (3.3.20), for example, with a simpler coefficient $f^{(\varepsilon)}/m_1^{(\varepsilon)}$ and to rewrite this relation in the form $\mathsf{P}\{f^{(\varepsilon)}\mu^{(\varepsilon)} > \tilde{t}^{(\varepsilon)}m_1^{(\varepsilon)}\}/\exp\{-\tilde{t}^{(\varepsilon)}m_1^{(\varepsilon)}\} \to 1$ as $\varepsilon \to 0$? The answer is partly affirmative. Relation $\rho^{(\varepsilon)} \sim f^{(\varepsilon)}/m_1^{(\varepsilon)}$ means that $\rho^{(\varepsilon)} = f^{(\varepsilon)}/m_1^{(\varepsilon)} + o(f^{(\varepsilon)})$. Thus, (3.3.20) implies that $\mathsf{P}\{f^{(\varepsilon)}\mu^{(\varepsilon)} > \tilde{t}^{(\varepsilon)}m_1^{(\varepsilon)}\} \sim \exp\{-\rho^{(\varepsilon)}\tilde{t}^{(\varepsilon)}m_1^{(\varepsilon)}/f^{(\varepsilon)}\} \sim \exp\{-(f^{(\varepsilon)}/m_1^{(\varepsilon)} + o(f^{(\varepsilon)}))\tilde{t}^{(\varepsilon)}m_1^{(\varepsilon)}/f^{(\varepsilon)}\}$. Therefore, the suggested replacement would additionally require to assume that $\tilde{t}^{(\varepsilon)}o(f^{(\varepsilon)})/f^{(\varepsilon)} \to 0$, i.e., to restrict the rate of convergence of $\tilde{t}^{(\varepsilon)}$ to ∞.

The questions related to the use of normalising coefficients given in a more explicit form will be investigated in the next section where we shall get asymptotic expansions for the roots $\rho^{(\varepsilon)}$ in asymptotic power series.

Let us now consider case (ii). Condition $\mathbf{D_{21}}$ implies in this case that $f^{(\varepsilon)} \to f^{(0)} > 0$ as $\varepsilon \to 0$. Therefore, $\rho^{(\varepsilon)} \to \rho^{(0)} > 0$ as $\varepsilon \to 0$. Also, in this case, it is possible that $\tilde{\pi}^{(0)}(X) \neq 1$.

Let us also write the representation formula (3.1.14) for the regeneration processes $\xi^{(\varepsilon)}(t)$, $t \geq 0$, and the regenerative stopping times $\mu^{(\varepsilon)}$,

$$\hat{F}^{(\varepsilon)}(t) = \hat{F}_-^{(\varepsilon)}(t)(1 - f^{(\varepsilon)}) + \hat{F}_+^{(\varepsilon)}(t) f^{(\varepsilon)}, \quad t \geq 0, \tag{3.3.21}$$

where

$$\hat{F}_+^{(\varepsilon)}(t) = \mathsf{P}\{\mu^{(\varepsilon)} \leq t / \mu^{(\varepsilon)} < \tau_1^{(\varepsilon)}\}, \quad t \geq 0.$$

Representation (3.3.21) implies that, under conditions $\mathbf{D_{21}}$ and $\mathbf{D_{24}}$,

$$\hat{F}_+^{(\varepsilon)}(\cdot) \Rightarrow \hat{F}_+^{(0)}(\cdot) \quad \text{as } \varepsilon \to 0. \tag{3.3.22}$$

Let also $\hat{v}^{(\varepsilon)}$, $\hat{\kappa}^{(\varepsilon)}$ and $\hat{\kappa}_k^{(\varepsilon)}$, $k \geq 1$, be mutually independent random variables that are used to construct representation (3.1.17) for the regenerative processes $\xi^{(\varepsilon)}(t)$, $t \geq 0$, and the regenerative stopping times $\mu^{(\varepsilon)}$.

Denote by $\phi_-^{(\varepsilon)}(s) = \mathsf{E} \exp\{-s\hat{\kappa}_1^{(\varepsilon)}\} = \int_0^\infty e^{-st} \hat{F}_-^{(\varepsilon)}(dt)$ and $\phi_+^{(\varepsilon)}(s) = \mathsf{E} \exp\{-s\hat{\kappa}^{(\varepsilon)}\} = \int_0^\infty e^{-st} \hat{F}_+^{(\varepsilon)}(dt)$ the Laplace transforms for, respectively, the random variables $\hat{\kappa}_1^{(\varepsilon)}$ and $\hat{\kappa}^{(\varepsilon)}$.

In terms of the Laplace transforms, representation (3.1.17) takes, for the regenerative processes $\xi^{(\varepsilon)}(t)$, $t \geq 0$, and the regenerative stopping times $\mu^{(\varepsilon)}$, the following form:

$$\mathsf{E} \exp\{-s\mu^{(\varepsilon)}\} = \frac{\phi_+^{(\varepsilon)}(s) f^{(\varepsilon)}}{1 - (1 - f^{(\varepsilon)})\phi_-^{(\varepsilon)}(s)}, \quad s \geq 0. \tag{3.3.23}$$

Conditions $\mathbf{D_{21}}$, $\mathbf{D_{24}}$, and relation (3.3.22) imply that $\phi_+^{(\varepsilon)}(s) \to \phi_+^{(0)}(s)$ and $\phi_-^{(\varepsilon)}(s) \to \phi_-^{(0)}(s)$ as $\varepsilon \to 0, s \geq 0$. Also, as was mentioned above, $f^{(\varepsilon)} \to f^{(0)} > 0$ as $\varepsilon \to 0$. Thus, it follows from (3.3.23) that $\mathsf{E} \exp\{-s\mu^{(\varepsilon)}\} \to \mathsf{E} \exp\{-s\mu^{(0)}\}$ as $\varepsilon \to 0, s \geq 0$. Therefore,

$$\mu^{(\varepsilon)} \xrightarrow{\text{d}} \mu^{(0)} \quad \text{as } \varepsilon \to 0. \tag{3.3.24}$$

Relation (3.3.24) implies that, in case **(ii)**, the stopping times $\mu^{(\varepsilon)}$ are asymptotically stochastically bounded random variables as $\varepsilon \to 0$, i.e.,

$$\lim_{T \to \infty} \overline{\lim_{\varepsilon \to 0}} \, \mathsf{P}\{\mu^{(\varepsilon)} > T\} = 0. \tag{3.3.25}$$

Under conditions $\mathbf{D_{21}}$, $\mathbf{C_6}$, and $\mathbf{D_{24}}$, relation (3.3.16) yields, in case **(ii)**, that for any $0 \leq t^{(\varepsilon)} \to \infty$ as $\varepsilon \to 0$,

$$\frac{\mathsf{P}\{\mu^{(\varepsilon)} > t^{(\varepsilon)}\}}{\exp\{-\rho^{(\varepsilon)} t^{(\varepsilon)}\}} \to \tilde{\pi}^{(0)}(X) \quad \text{as } \varepsilon \to 0. \tag{3.3.26}$$

Note that in case **(ii)**, $\rho^{(\varepsilon)} \to \rho^{(0)} > 0$, thus the tail probabilities for stopping times have exponential decay with parameter $\rho^{(\varepsilon)}$, asymptotically, as $\varepsilon \to 0$, separated from zero.

In conclusion, let us see what form will the corresponding conditions take in the case where the processes $\xi^{(\varepsilon)}(t) \equiv \xi^{(0)}(t)$, $t \geq 0$, the regeneration times $\tau_k^{(\varepsilon)} \equiv \tau_k^{(0)}$, $k = 0, 1, \ldots$, and the regenerative stopping times $\mu^{(\varepsilon)} \equiv \mu^{(0)}$ for all $\varepsilon \geq 0$, i.e., if they do not depend on the parameter ε.

Condition $\mathbf{D_{24}}$ reduces to the assumptions that **(a)** $F_{-}^{(0)}(t)$ is non-arithmetic, and **(b)** $f^{(0)} > 0$; condition $\mathbf{D_{24}}$ automatically holds; and condition $\mathbf{C_6}$ reduces to the assumption that **(c)** there exists $\delta > 0$ such that $\int_0^\infty e^{\delta s} F^{(0)}(ds) \in (1, \infty)$. Relation (3.3.16) takes the following form:

$$\frac{\mathsf{P}\{\mu^{(0)} > t\}}{\tilde{\pi}^{(0)}(X) \exp\{-\rho^{(0)}t\}} \to 1 \quad \text{as } t \to \infty. \tag{3.3.27}$$

Asymptotic relation (3.3.27) means that the tail probabilities $P\{\mu^{(0)} > t\}$ have exponential decay and can be approximated by the corresponding tail probabilities for an exponential type distribution with the corresponding tail probabilities $\tilde{\pi}^{(0)}(X)e^{-\rho^{(0)}t}$. The asymptotic relative error of such an approximation, as $t \to \infty$, equals 0.

The remarks made in Subsections 3.3.3 and 3.3.4 permit to interpret the mixed asymptotic relation (3.3.7) given in Theorem 3.3.2 as a *mixed ergodic and large deviation theorem* for perturbed regenerative processes and regenerative stopping times.

3.3.6 Pseudo-stationary exponential asymptotics for perturbed regenerative processes with transition period

Let us consider the model of perturbed regenerative processes with transition period introduced in Subsection 3.2.5. We use all the notations introduced in this section.

We assume the following two conditions:

$\mathbf{C_7}$: There exists $0 < \gamma_1 < \delta$ such that $\overline{\lim}_{0 \leq \varepsilon \to 0} \int_0^\infty e^{\gamma_1 s} \tilde{F}^{(\varepsilon)}(ds) < \infty$;

and

$\mathbf{D_{25}}$: There exists $0 < \gamma_2 < \delta$ such that $\lim_{T \to \infty} \overline{\lim}_{\varepsilon \to 0} e^{\gamma_2 T}(1 - \hat{\tilde{F}}^{(\varepsilon)}(T)) = 0$.

Note that estimate (3.3.1) implies that both conditions $\mathbf{C_7}$ and $\mathbf{D_{25}}$ can be replaced with the following condition:

$\mathbf{C_8}$: There exists $0 < \gamma < \delta$ such that $\overline{\lim}_{0 \leq \varepsilon \to 0} \int_0^\infty e^{\gamma s} \hat{\tilde{F}}^{(\varepsilon)}(ds) < \infty$.

More precisely, $\mathbf{C_7}$ and $\mathbf{D_{25}}$ hold for any $0 < \gamma_1, \gamma_2 < \gamma$.

Let us formulate a theorem that generalises Theorem 3.3.1 to the case of perturbed regeneration processes with transition period and asymptotically proper distribution functions $F^{(\varepsilon)}(t)$.

Theorem 3.3.8. *Let conditions* $\mathbf{D_{13}}$, $\mathbf{C_3}$, $\mathbf{F_7}$, $\mathbf{D_{17}}$, $\mathbf{C_7}$, *and* $\mathbf{D_{25}}$ *hold. Then, for any* $0 \leq t^{(\varepsilon)} \to \infty$ *as* $\varepsilon \to 0$ *and any* $A \in \Gamma_0$,

$$\frac{\mathsf{P}\{\xi^{(\varepsilon)}(t^{(\varepsilon)}) \in A, \mu^{(\varepsilon)} > t^{(\varepsilon)}\}}{\exp\{-\rho^{(\varepsilon)}t^{(\varepsilon)}\}} \to (1 - f^{(0)})\pi^{(0)}(A) \quad as \quad \varepsilon \to 0. \qquad (3.3.28)$$

Proof. The proof is based on the use of the renewal relation (3.2.19) and the asymptotic relation (3.3.3) given in Theorem 3.3.1. Note that relation (1.4.9) is provided by conditions $\mathbf{D_{13}}$, $\mathbf{C_3}$, $\mathbf{F_7}$, and that $0 \leq t^{(\varepsilon)} \to \infty$ as $\varepsilon \to 0$ in an arbitrary way in this relation.

Renewal relation (3.2.19) can be rewritten in the following form:

$$e^{\rho^{(\varepsilon)}t}\tilde{\mathsf{P}}(t^{(\varepsilon)}, A) = e^{\rho^{(\varepsilon)}t}\tilde{q}^{(\varepsilon)}(t^{(\varepsilon)}, A)$$

$$+ \int_0^{t^{(\varepsilon)}} e^{\rho^{(\varepsilon)}(t-s)} P^{(\varepsilon)}(t^{(\varepsilon)} - s, A)e^{\rho^{(\varepsilon)}s}\tilde{F}^{(\varepsilon)}(ds), \ t \geq 0. \quad (3.3.29)$$

Conditions $\mathbf{D_{13}}$ and $\mathbf{C_3}$ imply that $\rho^{(\varepsilon)} \to 0$ as $\varepsilon \to 0$. Therefore, $\rho^{(\varepsilon)} \leq \gamma_1 \wedge \gamma_2$ for all ε small enough. Thus we get by $\mathbf{D_{25}}$ that, for any $0 \leq t^{(\varepsilon)} \to \infty$ as $\varepsilon \to 0$ and any $A \in \Gamma_0$,

$$e^{\rho^{(\varepsilon)}t^{(\varepsilon)}}\tilde{\mathsf{P}}(t^{(\varepsilon)}, A) - \int_0^{t^{(\varepsilon)}} e^{\rho^{(\varepsilon)}(t^{(\varepsilon)}-s)} P^{(\varepsilon)}(t^{(\varepsilon)} - s, A)e^{\rho^{(\varepsilon)}s}\tilde{F}^{(\varepsilon)}(ds)$$

$$= e^{\rho^{(\varepsilon)}t^{(\varepsilon)}}\tilde{q}^{(\varepsilon)}(t^{(\varepsilon)}, A) \leq e^{\gamma_2 t^{(\varepsilon)}}(1 - \hat{\tilde{F}}^{(\varepsilon)}(t^{(\varepsilon)})) \to 0 \quad as \quad \varepsilon \to 0. \quad (3.3.30)$$

Relation (3.3.30) implies that, in order to get (3.3.28), we should prove that, for any $0 \leq t^{(\varepsilon)} \to \infty$ as $\varepsilon \to 0$ and any $A \in \Gamma_0$,

$$\int_0^{t^{(\varepsilon)}} e^{\rho^{(\varepsilon)}(t^{(\varepsilon)}-s)} P^{(\varepsilon)}(t^{(\varepsilon)} - s, A)e^{\rho^{(\varepsilon)}s}\tilde{F}^{(\varepsilon)}(ds)$$

$$\to (1 - f^{(0)})\pi^{(0)}(A) \quad as \quad \varepsilon \to 0. \qquad (3.3.31)$$

Let us also show that, for every $A \in \Gamma_0$, the functions $e^{\rho^{(\varepsilon)}(t^{(\varepsilon)}-s)} P^{(\varepsilon)}(t^{(\varepsilon)} - s, A)$ are asymptotically bounded, i.e.,

$$\overline{\lim_{\varepsilon \to 0}} \sup_{0 \leq s \leq t^{(\varepsilon)}} e^{\rho^{(\varepsilon)}(t^{(\varepsilon)}-s)} P^{(\varepsilon)}(t^{(\varepsilon)} - s, A) = R < \infty. \qquad (3.3.32)$$

Let us suppose, conversely, that there exists $0 \leq s^{(\varepsilon)} \leq t^{(\varepsilon)}$ such that

$$e^{\rho^{(\varepsilon)}(t^{(\varepsilon)}-s^{(\varepsilon)})} P^{(\varepsilon)}(t^{(\varepsilon)} - s^{(\varepsilon)}, A) \to \infty \quad as \quad \varepsilon \to 0. \qquad (3.3.33)$$

We can suppose that $t^{(\varepsilon)} - s^{(\varepsilon)} \to w$ as $\varepsilon \to 0$, where $0 \leq w \leq \infty$ (if not we can use subsequences).

The limit constant w can not be finite, since in this case, $\rho^{(\varepsilon)}(t^{(\varepsilon)} - s^{(\varepsilon)}) \to 0 \cdot w = 0$ as $\varepsilon \to 0$ and, therefore, in contradiction with (3.3.33),

$$\overline{\lim_{\varepsilon \to 0}} \, e^{\rho^{(\varepsilon)}(t^{(\varepsilon)} - s^{(\varepsilon)})} P^{(\varepsilon)}(t^{(\varepsilon)} - s^{(\varepsilon)}, A) \leq \overline{\lim_{\varepsilon \to 0}} \, e^{\rho^{(\varepsilon)}(t^{(\varepsilon)} - s^{(\varepsilon)})} = 1. \tag{3.3.34}$$

Neither can w be equal to ∞. Indeed, in this case, $u^{(\varepsilon)} = t^{(\varepsilon)} - s^{(\varepsilon)} \to \infty$ as $\varepsilon \to 0$. Then it would follow from (3.3.3) that

$$e^{\rho^{(\varepsilon)} u^{(\varepsilon)}} P^{(\varepsilon)}(u^{(\varepsilon)}, A) \to \pi^{(0)}(A) \quad \text{as } \varepsilon \to 0, \tag{3.3.35}$$

which would again contradict (3.3.33).

Therefore, relation (3.3.33) can not hold but, on the contrary, relation (3.3.32) holds. Two following cases should be considered: **(a)** $\tilde{f}^{(0)} = 1$ and **(b)** $\tilde{f}^{(0)} = 1$.

Let us consider case **(a)**. Using that $\rho^{(\varepsilon)} \leq \gamma_1 \wedge \gamma_2$ for all ε small enough, relation (3.3.32), and inequality $\tilde{F}^{(\varepsilon)}(T) \leq \tilde{F}^{(\varepsilon)}(\infty) = 1 - \tilde{f}^{(\varepsilon)}$, we get that

$$\overline{\lim_{\varepsilon \to 0}} \int_0^\infty e^{\rho^{(\varepsilon)}(t^{(\varepsilon)} - s)} P^{(\varepsilon)}(t^{(\varepsilon)} - s, A) e^{\rho^{(\varepsilon)} s} \tilde{F}^{(\varepsilon)}(ds)$$

$$\leq \overline{\lim_{\varepsilon \to 0}} R \int_0^\infty e^{\rho^{(\varepsilon)} s} \tilde{F}^{(\varepsilon)}(ds)$$

$$\leq \overline{\lim_{\varepsilon \to 0}} R(e^{\gamma_1 T} \tilde{F}^{(\varepsilon)}(T) + e^{-(\delta - \gamma_1)T} \int_T^\infty e^{\delta s} \tilde{F}^{(\varepsilon)}(ds))$$

$$= \overline{\lim_{\varepsilon \to 0}} R e^{-(\delta - \gamma_1)T} \int_0^\infty e^{\delta s} \tilde{F}^{(\varepsilon)}(ds) \to 0 \quad \text{as } T \to \infty. \tag{3.3.36}$$

Consider case **(b)**. Let us apply Lemma 1.2.2 for an asymptotic analysis of the integral in the right-hand side of (3.3.30). To do this, note first that the integral in (3.3.31) can be rewritten in the form where the limits of integration do not depend on ε,

$$\int_0^{t^{(\varepsilon)}} e^{\rho^{(\varepsilon)}(t^{(\varepsilon)} - s)} P^{(\varepsilon)}(t^{(\varepsilon)} - s, A) e^{\rho^{(\varepsilon)} s} \tilde{F}^{(\varepsilon)}(ds)$$

$$= \int_0^\infty e^{\rho^{(\varepsilon)}(t^{(\varepsilon)} - s)} P^{(\varepsilon)}(t^{(\varepsilon)} - s, A) e^{\rho^{(\varepsilon)} s} \tilde{F}^{(\varepsilon)}(ds), \tag{3.3.37}$$

where one should take $P^{(\varepsilon)}(t - s, A) = 0$ for $s > t$.

Let us show that, for every $A \in \Gamma_0$ and any $s \geq 0$ and $v > 0$,

$$\sup_{|u| \leq v} |e^{\rho^{(\varepsilon)}(t^{(\varepsilon)} - s + u)} P^{(\varepsilon)}(t^{(\varepsilon)} - s + u, A) - \pi^{(0)}(A)| \to 0 \quad \text{as } \varepsilon \to 0. \tag{3.3.38}$$

Suppose that asymptotic relation (3.2.26) does not take place. This means that there exists a sequence $0 < \varepsilon_k \to 0$ as $k \to \infty$ and u_k with $|u_k| \leq v$ such that

$$|e^{\rho^{(\varepsilon)}(t^{(\varepsilon)} - s + u)} P^{(\varepsilon_k)}(t^{(\varepsilon_k)} - s + u_k, A) - \pi^{(0)}(A)| \not\to 0 \quad \text{as } \varepsilon \to 0. \tag{3.3.39}$$

Since the sequence $|u_k| \leq v$ is bounded, $(t^{(\varepsilon_k)} - s + u_k) \to \infty$ as $k \to \infty$ for any $s \geq 0$ and $v > 0$. Therefore, by (3.3.3),

$$|e^{\rho^{(\varepsilon)}(t^{(\varepsilon)} - s + u)} P^{(\varepsilon_k)}(t^{(\varepsilon_k)} - s + u_k, A) - \pi^{(0)}(A)| \to 0 \text{ as } \varepsilon \to 0. \qquad (3.3.40)$$

Relation (3.3.39) contradicts (3.3.40). Thus, relation (3.3.38) holds.
Introduce the distribution functions

$$\tilde{\tilde{F}}^{(\varepsilon)}(t) = \int_0^t e^{\rho^{(\varepsilon)}s} \tilde{F}^{(\varepsilon)}(ds), \quad t \geq 0.$$

Condition $\mathbf{C_7}$ and the relation $\rho^{(\varepsilon)} \to 0$ as $\varepsilon \to 0$ imply that there exists $0 < \gamma_1' < \gamma_1$ such that $\rho^{(\varepsilon)} + \gamma_1' \leq \gamma_1$ for all ε small enough, and thus,

$$\overline{\lim_{0 \leq \varepsilon \to 0}} \int_0^\infty e^{\gamma_1' s} \tilde{\tilde{F}}^{(\varepsilon)}(ds)$$

$$\leq \overline{\lim_{0 \leq \varepsilon \to 0}} \int_0^\infty e^{\gamma_1 s} \tilde{F}^{(\varepsilon)}(ds) < \infty. \qquad (3.3.41)$$

Relation (3.3.41) implies that $\tilde{\tilde{F}}^{(\varepsilon)}(ds)$ is a finite measure for ε small enough.
Let us now assume that condition $\mathbf{D_{20}}$ holds instead of condition $\mathbf{D_{17}}$.
Condition $\mathbf{D_{20}}$ and the relation $\rho^{(\varepsilon)} \to 0$ as $\varepsilon \to 0$ imply that

$$\tilde{\tilde{F}}^{(\varepsilon)}(\cdot) \Rightarrow \tilde{F}^{(0)}(\cdot) \text{ as } \varepsilon \to 0. \qquad (3.3.42)$$

Relations (3.3.41) and (3.3.42) imply, by Lemma 1.4.4, that

$$\tilde{\tilde{F}}^{(\varepsilon)}(\infty) \to \tilde{F}^{(0)}(\infty) \text{ as } \varepsilon \to 0. \qquad (3.3.43)$$

Condition $\mathbf{D_{20}}$ and relations (3.3.32), (3.3.38), and (3.3.42) imply that, for any $0 \leq t^{(\varepsilon)} \to \infty$ as $\varepsilon \to 0$ and any $A \in \Gamma_0$, conditions of Lemma 1.2.2 hold for the functions $e^{\rho^{(\varepsilon)}(t^{(\varepsilon)} - s)} P^{(\varepsilon)}(t^{(\varepsilon)} - s, A)$ and the measures $\tilde{\tilde{F}}^{(\varepsilon)}(ds)$. Thus we get, by applying Lemma 1.2.2, that for any $0 \leq t^{(\varepsilon)} \to \infty$ as $\varepsilon \to 0$ and any $A \in \Gamma_0$,

$$\int_0^\infty e^{\rho^{(\varepsilon)}(t^{(\varepsilon)} - s)} P^{(\varepsilon)}(t^{(\varepsilon)} - s, A) e^{\rho^{(\varepsilon)}s} \tilde{F}^{(\varepsilon)}(ds)$$

$$= \int_0^\infty e^{\rho^{(\varepsilon)}(t^{(\varepsilon)} - s)} P^{(\varepsilon)}(t^{(\varepsilon)} - s, A) \tilde{\tilde{F}}^{(\varepsilon)}(ds)$$

$$\to \int_0^\infty \pi^{(0)}(A) \tilde{F}^{(0)}(ds) = (1 - f^{(0)})\pi^{(0)}(A) \text{ as } \varepsilon \to 0. \qquad (3.3.44)$$

Let us consider now the case where condition $\mathbf{D_{17}}$ holds. We show that the proof can be reduced to the case considered above, i.e., to the case where condition $\mathbf{D_{20}}$ holds.

Indeed, let us take an arbitrary sequence $0 < \varepsilon_n \to 0$ as $n \to \infty$. It is always possible to choose from this sequence a subsequence $\varepsilon_{n_k} \to 0$ as $k \to \infty$ such that

$$\tilde{F}^{(\varepsilon_{n_k})}(\cdot) \Rightarrow \tilde{F}(\cdot) \quad \text{as } k \to \infty, \tag{3.3.45}$$

where $\tilde{F}(s)$ is a proper or improper distribution function.

Relation (3.3.45) and the relation $\rho^{(\varepsilon)} \to 0$ as $\varepsilon \to 0$ imply that

$$\tilde{\tilde{F}}^{(\varepsilon_{n_k})}(\cdot) \Rightarrow \tilde{F}(\cdot) \quad \text{as } k \to \infty. \tag{3.3.46}$$

The proof given above for the case where condition $\mathbf{D_{20}}$ holds can be applied to a subsequence of models with the perturbation parameters ε_{n_k}, $k \geq 1$, to yield, for any $A \in \Gamma_0$, the following relation:

$$\int_0^\infty P^{(\varepsilon_{n_k})}(t^{(\varepsilon_{n_k})} - s, A)\tilde{F}^{(\varepsilon_{n_k})}(ds) \to (1 - \tilde{f})\pi^{(0)}(A) \quad \text{as } k \to \infty, \tag{3.3.47}$$

where $\tilde{f} = 1 - \tilde{F}(\infty)$.

The limit distribution $\tilde{F}(s)$ does depend on the choice of the sequence ε_n and the subsequence ε_{n_k}, but the limit $\tilde{f}$ does not.

Indeed, relations (3.3.41) and (3.3.46), by Lemma 1.4.4, imply that

$$1 - \tilde{\tilde{F}}^{(\varepsilon_{n_k})}(\infty) = \tilde{\tilde{f}}^{(\varepsilon_{n_k})} \to 1 - \tilde{F}(\infty) = \tilde{f} \quad \text{as } k \to \infty. \tag{3.3.48}$$

On the other hand, by condition $\mathbf{D_{17}}$,

$$1 - \tilde{\tilde{F}}^{(\varepsilon_{n_k})}(\infty) = \tilde{\tilde{f}}^{(\varepsilon_{n_k})} \to \tilde{f}^{(0)} \quad \text{as } k \to \infty. \tag{3.3.49}$$

Thus, $\tilde{f} = \tilde{f}^{(0)}$ and, therefore, the limit in the right-hand side of (3.3.47) does not depend on the choice of the sequence ε_n and the subsequence ε_{n_k}. Moreover, it coincides with the limit in the right-hand side of relation (3.3.44). This implies that relation (3.3.44) holds.

The proof of the theorem is complete.   □

3.3.7 Quasi-stationary exponential asymptotics for perturbed regenerative processes with transition period

Let us continue to consider the model of perturbed regenerative processes with transition period, introduced in Subsection 3.2.5. We use all the notations introduced there. However, we consider now the case where $\tilde{F}^{(\varepsilon)}(t)$ are asymptotically proper or improper distribution functions.

In this case we impose, on the characteristics of the transition period, the following Cramér type condition:

C9: There exists $0 < \gamma' < \delta - \rho^{(0)}$ such that $\overline{\lim}_{0 \le \varepsilon \to 0} \int_0^\infty e^{(\rho^{(0)} + \gamma')s} \tilde{F}^{(\varepsilon)}(ds) < \infty$;

and the condition:

D26: There exists $0 < \gamma'' < \delta - \rho^{(0)}$ such that $\lim_{T \to \infty} \overline{\lim}_{\varepsilon \to 0} e^{(\rho^{(0)} + \gamma'')T}(1 - \hat{\tilde{F}}^{(\varepsilon)}(T)) = 0$.

Note that estimate (3.3.1) implies that both conditions **C9** and **D26** can be replaced with the following condition:

C10: There exists $0 < \gamma < \delta - \rho^{(0)}$ such that $\overline{\lim}_{0 \le \varepsilon \to 0} \int_0^\infty e^{(\rho^{(0)} + \gamma)s} \hat{\tilde{F}}^{(\varepsilon)}(ds) < \infty$.

More precisely, **C9** and **D26** hold for any $0 < \gamma', \gamma'' < \gamma$.

Let us introduce the moment generating function

$$\tilde{\phi}^{(\varepsilon)}(\rho) = \int_0^\infty e^{\rho s} \tilde{F}^{(\varepsilon)}(ds), \quad \rho \in \mathbb{R}_1.$$

Condition **C9** guarantees that $\tilde{\phi}^{(\varepsilon)}(\rho^{(0)} + \gamma') < \infty$ for ε small enough.

Condition **D17** should be replaced in this case with the following condition:

C11: $\tilde{\phi}^{(\varepsilon)}(\rho^{(0)}) \to \tilde{\phi}^{(0)} \in [0, \infty)$ as $\varepsilon \to 0$.

Lemma 3.3.1. *Let condition* **C9** *hold. Then condition* **D20** *implies that condition* **C11** *holds, and, moreover, in this case the limiting constant is* $\tilde{\phi}^{(0)} = \tilde{\phi}^{(0)}(\rho^{(0)})$.

Proof. Take $0 < \beta < \rho^{(0)} + \gamma'$. Then, using condition **C9**, we have

$$\lim_{T \to \infty} \overline{\lim_{\varepsilon \to 0}} (\tilde{\phi}^{(\varepsilon)}(\rho^{(0)}) - \int_0^T e^{\rho^{(0)} s} F^{(\varepsilon)}(ds))$$

$$= \lim_{T \to \infty} \overline{\lim_{\varepsilon \to 0}} \int_T^\infty e^{\rho^{(0)} s} F^{(\varepsilon)}(ds)$$

$$\le \lim_{T \to \infty} e^{-\gamma' T} \overline{\lim_{\varepsilon \to 0}} \int_0^\infty e^{(\rho^{(0)} + \gamma')s} F^{(\varepsilon)}(ds) = 0. \qquad (3.3.50)$$

Let us take some sequence $0 < T_k \to \infty$ as $k \to \infty$ of continuity points of the distribution function $F^{(0)}(s)$. Condition **D20** implies that $F^{(\varepsilon)}(T_k) \to F^{(0)}(T_k)$ as

$\varepsilon \to 0$ for every $k \geq 1$. Obviously, $\int_0^{T_k} e^{\rho^{(0)}s} F^{(\varepsilon)}(ds) \to \int_0^{T_k} e^{\rho^{(0)}s} F^{(0)}(ds)$ as $k \to \infty$. These relations, together with (3.3.50), imply in an obvious way that

$$\overline{\lim_{\varepsilon \to 0}} \mid \tilde{\phi}^{(\varepsilon)}(\rho^{(0)}) - \tilde{\phi}^{(0)}(\rho^{(0)}) \mid$$

$$\leq \lim_{k \to \infty} \overline{\lim_{\varepsilon \to 0}} (\tilde{\phi}^{(\varepsilon)}(\rho^{(0)}) - \int_0^{T_k} e^{\rho^{(0)}s} F^{(\varepsilon)}(ds))$$

$$+ \lim_{k \to \infty} \overline{\lim_{\varepsilon \to 0}} \mid \int_0^{T_k} e^{\rho^{(0)}s} F^{(\varepsilon)}(ds) - \int_0^{T_k} e^{\rho^{(0)}s} F^{(0)}(ds) \mid$$

$$+ \lim_{k \to \infty} (\tilde{\phi}^{(0)}(\rho^{(0)}) - \int_0^{T_k} e^{\rho^{(0)}s} F^{(0)}(ds)) = 0. \tag{3.3.51}$$

The proof is complete. $\qquad\square$

Therefore, under condition $\mathbf{C_9}$, condition $\mathbf{C_{11}}$ can be replaced with condition $\mathbf{D_{20}}$.

Let us formulate a theorem which is an analogue of Theorem 3.3.2 in the case of perturbed regenerative processes with transition period and distribution functions $F^{(\varepsilon)}(t)$ and $\tilde{F}^{(\varepsilon)}(t)$ which could be asymptotically proper or improper.

Theorem 3.3.9. *Let conditions $\mathbf{D_{21}}$, $\mathbf{C_6}$, $\mathbf{F_7}$, $\mathbf{C_9}$, $\mathbf{D_{26}}$, and $\mathbf{C_{11}}$ hold. Then for any $0 \leq t^{(\varepsilon)} \to \infty$ as $\varepsilon \to 0$ and any $A \in \Gamma_0$,*

$$\frac{\mathsf{P}\{\xi^{(\varepsilon)}(t^{(\varepsilon)}) \in A, \mu^{(\varepsilon)} > t^{(\varepsilon)}\}}{\exp\{-\rho^{(\varepsilon)}t^{(\varepsilon)}\}} \to \tilde{\phi}^{(0)}\tilde{\pi}^{(0)}(A) \quad as \quad \varepsilon \to 0. \tag{3.3.52}$$

Proof. The proof is similar to the proof of Theorem 3.3.8 and is based on the use of the renewal relation (3.2.19) and the asymptotic relation (3.3.3) given in Theorem 3.3.1. Note that relation (3.3.3) is implied by conditions $\mathbf{D_{21}}$, $\mathbf{C_6}$, $\mathbf{F_7}$, and that $0 \leq t^{(\varepsilon)} \to \infty$ as $\varepsilon \to 0$ in an arbitrary way in this relation.

Let us again rewrite the renewal relation (3.2.19) in the following form:

$$e^{\rho^{(\varepsilon)}t}\tilde{\mathsf{P}}(t^{(\varepsilon)}, A) = e^{\rho^{(\varepsilon)}t}\tilde{q}^{(\varepsilon)}(t^{(\varepsilon)}, A)$$

$$+ \int_0^{t^{(\varepsilon)}} e^{\rho^{(\varepsilon)}(t-s)} P^{(\varepsilon)}(t^{(\varepsilon)} - s, A)e^{\rho^{(\varepsilon)}s}\tilde{F}^{(\varepsilon)}(ds), \ t \geq 0. \tag{3.3.53}$$

Conditions $\mathbf{D_{21}}$ and $\mathbf{C_6}$ imply that $\rho^{(\varepsilon)} \to \rho^{(0)}$ as $\varepsilon \to 0$. Therefore, $\rho^{(\varepsilon)} \leq \rho^{(0)} + \gamma' \wedge \gamma''$ for all ε small enough. Thus we get, by $\mathbf{D_{26}}$, that, for any $0 \leq t^{(\varepsilon)} \to \infty$ as $\varepsilon \to 0$ and any $A \in \Gamma_0$,

$$e^{\rho^{(\varepsilon)}t^{(\varepsilon)}}\tilde{\mathsf{P}}(t^{(\varepsilon)}, A) - \int_0^{t^{(\varepsilon)}} e^{\rho^{(\varepsilon)}(t^{(\varepsilon)}-s)} P^{(\varepsilon)}(t^{(\varepsilon)} - s, A)e^{\rho^{(\varepsilon)}s}\tilde{F}^{(\varepsilon)}(ds)$$

$$= e^{\rho^{(\varepsilon)}t^{(\varepsilon)}}\tilde{q}^{(\varepsilon)}(t^{(\varepsilon)}, A) \leq e^{(\rho^{(0)}+\gamma'')t^{(\varepsilon)}}(1 - \hat{\tilde{F}}^{(\varepsilon)}(t^{(\varepsilon)})) \to 0 \text{ as } \varepsilon \to 0. \tag{3.3.54}$$

Relation (3.3.54) implies that, in order to get (3.3.52), we should prove that, for any $0 \le t^{(\varepsilon)} \to \infty$ as $\varepsilon \to 0$ and any $A \in \Gamma_0$,

$$\int_0^{t^{(\varepsilon)}} e^{\rho^{(\varepsilon)}(t^{(\varepsilon)}-s)} P^{(\varepsilon)}(t^{(\varepsilon)} - s, A) e^{\rho^{(\varepsilon)}s} \tilde{F}^{(\varepsilon)}(ds) \to \tilde{\phi}^{(0)} \pi^{(0)}(A) \quad \text{as } \varepsilon \to 0. \tag{3.3.55}$$

Let us also show that, for every $A \in \Gamma_0$, the functions $e^{\rho^{(\varepsilon)}(t^{(\varepsilon)}-s)} P^{(\varepsilon)}(t^{(\varepsilon)} - s, A)$ are asymptotically bounded, i.e.,

$$\varlimsup_{\varepsilon \to 0} \sup_{0 \le s \le t^{(\varepsilon)}} e^{\rho^{(\varepsilon)}(t^{(\varepsilon)}-s)} P^{(\varepsilon)}(t^{(\varepsilon)} - s, A) = R < \infty. \tag{3.3.56}$$

Assume to the contrary that there exists $0 \le s^{(\varepsilon)} \le t^{(\varepsilon)}$ such that

$$e^{\rho^{(\varepsilon)}(t^{(\varepsilon)}-s^{(\varepsilon)})} P^{(\varepsilon)}(t^{(\varepsilon)} - s^{(\varepsilon)}, A) \to \infty \quad \text{as } \varepsilon \to 0. \tag{3.3.57}$$

We can suppose that $t^{(\varepsilon)} - s^{(\varepsilon)} \to w$ as $\varepsilon \to 0$, where $0 \le w \le \infty$ (if not we can go to subsequences).

The limit constant w can not be finite, since in this case, $\rho^{(\varepsilon)}(t^{(\varepsilon)} - s^{(\varepsilon)}) \to \rho^{(0)} w$ as $\varepsilon \to 0$ and, therefore, in contradiction with (3.3.57),

$$\varlimsup_{\varepsilon \to 0} e^{\rho^{(\varepsilon)}(t^{(\varepsilon)}-s^{(\varepsilon)})} P^{(\varepsilon)}(t^{(\varepsilon)} - s^{(\varepsilon)}, A) \le \varlimsup_{\varepsilon \to 0} e^{\rho^{(\varepsilon)}(t^{(\varepsilon)}-s^{(\varepsilon)})} = e^{\rho^{(0)} w}. \tag{3.3.58}$$

Neither can w be equal to ∞. Indeed, in this case $u^{(\varepsilon)} = t^{(\varepsilon)} - s^{(\varepsilon)} \to \infty$ as $\varepsilon \to 0$. Then, it would follow from (3.3.3) that

$$e^{\rho^{(\varepsilon)}u^{(\varepsilon)}} P^{(\varepsilon)}(u^{(\varepsilon)}, A) \to \pi^{(0)}(A) \quad \text{as } \varepsilon \to 0, \tag{3.3.59}$$

which would again contradict (3.3.57).

Therefore, relation (3.3.57) can not hold but, on the contrary, relation (3.3.56) holds.

The following cases should be considered: **(a)** $\tilde{\phi}^{(0)} = 0$ and **(b)** $\tilde{\phi}^{(0)} \in (0, \infty)$.

Let us consider case **(a)**. Since $\rho^{(\varepsilon)} \to \rho^{(0)}$ as $\varepsilon \to 0$, there exists $0 < \gamma_0 < \gamma' \wedge \gamma''$ such that $\rho^{(\varepsilon)} \le \rho^{(0)} + \gamma_0$ for all ε small enough. This proposition, relation (3.3.56), and inequality $\int_0^T e^{\rho^{(0)}s} \tilde{F}^{(\varepsilon)}(ds) \le \tilde{\phi}^{(\varepsilon)}(\rho^{(0)})$ imply that

$$\varlimsup_{\varepsilon \to 0} \int_0^\infty e^{\rho^{(\varepsilon)}(t^{(\varepsilon)}-s)} P^{(\varepsilon)}(t^{(\varepsilon)} - s, A) e^{\rho^{(\varepsilon)}s} \tilde{F}^{(\varepsilon)}(ds)$$

$$\le \varlimsup_{\varepsilon \to 0} R \int_0^\infty e^{\rho^{(\varepsilon)}s} \tilde{F}^{(\varepsilon)}(ds)$$

$$\le \varlimsup_{\varepsilon \to 0} R\left(e^{\gamma' T} \int_0^T e^{\rho^{(0)}s} \tilde{F}^{(\varepsilon)}(ds) \right.$$

$$\left. + e^{-(\gamma'-\gamma_0)T} \int_T^\infty e^{(\rho^{(0)}+\gamma')s} \tilde{F}^{(\varepsilon)}(ds) \right)$$

$$= \varlimsup_{\varepsilon \to 0} R e^{-(\gamma'-\gamma_0)T} \int_0^\infty e^{(\rho^{(0)}+\gamma')s} \tilde{F}^{(\varepsilon)}(ds) \to 0 \quad \text{as } T \to \infty. \tag{3.3.60}$$

Consider now case (**b**). We apply Lemma 1.2.2 for asymptotic analysis of the integral in the right-hand side of (3.3.54). To do this, note first that the integral in (3.3.55) can be rewritten in a form where the limits of integration do not depend on ε,

$$\int_0^{t^{(\varepsilon)}} e^{\rho^{(\varepsilon)}(t^{(\varepsilon)}-s)} P^{(\varepsilon)}(t^{(\varepsilon)} - s, A) e^{\rho^{(\varepsilon)}s} \tilde{F}^{(\varepsilon)}(ds)$$

$$= \int_0^{\infty} e^{\rho^{(\varepsilon)}(t^{(\varepsilon)}-s)} P^{(\varepsilon)}(t^{(\varepsilon)} - s, A) e^{\rho^{(\varepsilon)}s} \tilde{F}^{(\varepsilon)}(ds), \qquad (3.3.61)$$

where we take $P^{(\varepsilon)}(t - s, A) = 0$ for $s > t$.

Let us show that, for every $A \in \Gamma_0$ and any $s \geq 0$ and $v > 0$,

$$\sup_{|u| \leq v} |e^{\rho^{(\varepsilon)}(t^{(\varepsilon)}-s+u)} P^{(\varepsilon)}(t^{(\varepsilon)} - s + u, A) - \pi^{(0)}(A)| \to 0 \text{ as } \varepsilon \to 0. \quad (3.3.62)$$

Suppose that asymptotic relation (3.3.62) does not take place. This means that there exist sequences $0 < \varepsilon_k \to 0$ as $k \to \infty$ and u_k with $|u_k| \leq v$ such that

$$|e^{\rho^{(\varepsilon)}(t^{(\varepsilon)}-s+u)} P^{(\varepsilon_k)}(t^{(\varepsilon_k)} - s + u_k, A) - \pi^{(0)}(A)| \nrightarrow 0 \text{ as } \varepsilon \to 0. \quad (3.3.63)$$

Since the sequence $|u_k| \leq v$ is bounded, $(t^{(\varepsilon_k)} - s + u_k) \to \infty$ as $k \to \infty$ for any $s \geq 0$ and $v > 0$. Therefore, by (3.3.3),

$$|e^{\rho^{(\varepsilon)}(t^{(\varepsilon)}-s+u)} P^{(\varepsilon_k)}(t^{(\varepsilon_k)} - s + u_k, A) - \pi^{(0)}(A)| \to 0 \text{ as } \varepsilon \to 0. \quad (3.3.64)$$

Relation (3.3.63) contradicts (3.3.64). Thus, relation (3.3.62) holds.
Now introduce the distribution functions

$$\tilde{\tilde{F}}^{(\varepsilon)}(t) = \int_0^t e^{\rho^{(\varepsilon)}s} \tilde{F}^{(\varepsilon)}(ds), \quad t \geq 0.$$

Condition $\mathbf{C_9}$ and relation $\rho^{(\varepsilon)} \to \rho^{(0)}$ as $\varepsilon \to 0$ imply that there exists $0 < \gamma_0' < \gamma'$ such that $\rho^{(\varepsilon)} + \gamma_0' \leq \rho^{(0)} + \gamma'$ for all ε small enough, and thus

$$\varlimsup_{0 \leq \varepsilon \to 0} \int_0^{\infty} e^{\gamma_0's} \tilde{F}^{(\varepsilon)}(ds) \leq \varlimsup_{0 \leq \varepsilon \to 0} \int_0^{\infty} e^{(\rho^{(0)}+\gamma')s} \tilde{F}^{(\varepsilon)}(ds) < \infty. \quad (3.3.65)$$

Relation (3.3.65) implies that $\tilde{\tilde{F}}^{(\varepsilon)}(ds)$ is a finite measure for ε small enough.
Let us now assume that condition $\mathbf{D_{20}}$ holds, instead of condition $\mathbf{C_{11}}$.
Condition $\mathbf{D_{20}}$ and relation $\rho^{(\varepsilon)} \to \rho^{(0)}$ as $\varepsilon \to 0$ imply that

$$\tilde{\tilde{F}}^{(\varepsilon)}(\cdot) \Rightarrow \tilde{\tilde{F}}^{(0)}(\cdot) \text{ as } \varepsilon \to 0. \quad (3.3.66)$$

Relations (3.3.65) and (3.3.66) imply, by Lemma 1.4.4, that

$$\tilde{\tilde{F}}^{(\varepsilon)}(\infty) = \tilde{\phi}^{(\varepsilon)}(\rho^{(\varepsilon)}) \to \tilde{\tilde{F}}^{(0)}(\infty) = \tilde{\phi}^{(0)}(\rho^{(0)}) \text{ as } \varepsilon \to 0. \quad (3.3.67)$$

Condition $\mathbf{D_{20}}$ and relations (3.3.56), (3.3.62), and (3.3.66) imply that, for any $0 \leq t^{(\varepsilon)} \to \infty$ as $\varepsilon \to 0$ and any $A \in \Gamma_0$, conditions of Lemma 1.2.2 hold for the functions $e^{\rho^{(\varepsilon)}(t^{(\varepsilon)}-s)} P^{(\varepsilon)}(t^{(\varepsilon)} - s, A)$ and the measures $\tilde{F}^{(\varepsilon)}(ds)$. We thus get, by applying Lemma 1.2.2, that for any $0 \leq t^{(\varepsilon)} \to \infty$ as $\varepsilon \to 0$ and any $A \in \Gamma_0$,

$$\int_0^\infty e^{\rho^{(\varepsilon)}(t^{(\varepsilon)}-s)} P^{(\varepsilon)}(t^{(\varepsilon)} - s, A) e^{\rho^{(\varepsilon)}s} \tilde{F}^{(\varepsilon)}(ds)$$

$$= \int_0^\infty e^{\rho^{(\varepsilon)}(t^{(\varepsilon)}-s)} P^{(\varepsilon)}(t^{(\varepsilon)} - s, A) \tilde{\tilde{F}}^{(\varepsilon)}(ds)$$

$$\to \int_0^\infty \pi^{(0)}(A) \tilde{\tilde{F}}^{(0)}(ds) = \tilde{\phi}^{(0)}(\rho^{(0)}) \pi^{(0)}(A) \quad \text{as } \varepsilon \to 0. \qquad (3.3.68)$$

Let us consider now the case where condition $\mathbf{C_{11}}$ holds. We show that the proof can be reduced to the case considered above, i.e., to the case where condition $\mathbf{D_{20}}$ holds.

Indeed, let us take an arbitrary sequence $0 < \varepsilon_n \to 0$ as $n \to \infty$. It is always possible to select from this sequence a subsequence $\varepsilon_{n_k} \to 0$ as $k \to \infty$ such that

$$\tilde{F}^{(\varepsilon_{n_k})}(\cdot) \Rightarrow \tilde{F}(\cdot) \quad \text{as } k \to \infty, \qquad (3.3.69)$$

where $\tilde{F}(s)$ is a proper or improper distribution function.

Relation (3.3.69) and the relation $\rho^{(\varepsilon)} \to \rho^{(0)}$ as $\varepsilon \to 0$ imply that

$$\tilde{\tilde{F}}^{(\varepsilon_{n_k})}(\cdot) \Rightarrow \tilde{\tilde{F}}(\cdot) \quad \text{as } k \to \infty, \qquad (3.3.70)$$

where

$$\tilde{\tilde{F}}(t) = \int_0^t e^{\rho^{(0)}s} \tilde{F}(ds), \quad t \geq 0.$$

The proof given above for the case where condition $\mathbf{D_{20}}$ holds can be applied to a subsequence of models with perturbation parameters ε_{n_k}, $k \geq 1$, and to yield, for any $A \in \Gamma_0$, the following relation:

$$\int_0^\infty P^{(\varepsilon_{n_k})}(t^{(\varepsilon_{n_k})} - s, A) \tilde{\tilde{F}}^{(\varepsilon_{n_k})}(ds) \to \tilde{\phi}(\rho^{(0)}) \pi^{(0)}(A) \quad \text{as } k \to \infty, \qquad (3.3.71)$$

where

$$\tilde{\phi}(\rho^{(0)}) = \int_0^t e^{\rho^{(0)}s} \tilde{F}(ds).$$

The limit distribution $\tilde{F}(s)$ does depend on the choice of the sequence ε_n and the subsequence ε_{n_k}, but the limit $\tilde{\phi}(\rho^{(0)})$ does not.

Indeed, relations (3.3.65) and (3.3.70), by Lemma 3.3.1, imply that

$$\tilde{\phi}^{(\varepsilon_{n_k})}(\rho^{(\varepsilon_{n_k})}) \to \tilde{\phi}(\rho^{(0)}) \quad \text{as } k \to \infty. \qquad (3.3.72)$$

On the other hand, by condition $\mathbf{C_{11}}$,

$$\tilde{\phi}^{(\varepsilon_{n_k})}(\rho^{(0)}) \to \tilde{\phi}^{(0)} \quad \text{as} \ k \to \infty. \tag{3.3.73}$$

Let us now show that

$$\tilde{\phi}^{(\varepsilon_{n_k})}(\rho^{(\varepsilon_{n_k})}) - \tilde{\phi}^{(\varepsilon_{n_k})}(\rho^{(0)}) \to 0 \quad \text{as} \ k \to \infty. \tag{3.3.74}$$

Indeed, using condition $\mathbf{C_9}$, the relation $\rho^{(\varepsilon_{n_k})} \to \rho^{(0)}$ as $k \to \infty$, and the inequality $\rho^{(\varepsilon)} + \gamma_0' \le \rho^{(0)} + \gamma'$, which holds for all ε small enough, we get

$$\overline{\lim_{k \to \infty}} \ | \ \tilde{\phi}^{(\varepsilon_{n_k})}(\rho^{(\varepsilon_{n_k})}) - \tilde{\phi}^{(\varepsilon_{n_k})}(\rho^{(0)}) \ |$$

$$\le \overline{\lim_{k \to \infty}} \ (\sup_{0 \le s \le T} \ | \ e^{\rho^{(\varepsilon_{n_k})}s} - e^{\rho^{(0)}s} \ |$$

$$+ \int_T^\infty e^{\rho^{(\varepsilon_{n_k})}s} \tilde{F}^{(\varepsilon_{n_k})}(ds) + \int_T^\infty e^{\rho^{(0)}s} \tilde{F}^{(\varepsilon_{n_k})}(ds))$$

$$\le (e^{-\gamma_0'T} + e^{-\gamma'T}) \, \overline{\lim_{k \to \infty}} \int_0^\infty e^{(\rho^{(0)}+\gamma')s} \tilde{F}^{(\varepsilon_{n_k})}(ds) \to 0 \quad \text{as} \ T \to \infty. \tag{3.3.75}$$

Relations (3.3.72), (3.3.73), and (3.3.74) imply that $\tilde{\phi}(\rho^{(0)}) = \tilde{\phi}^{(0)}$ and, therefore, the limit in the right-hand side of (3.3.71) does not depend on the choice of the sequence ε_n and the subsequence ε_{n_k}. Moreover, it coincides with the limit in the right-hand side of relation (3.3.68). This implies that relation (3.3.68) holds.

The proof of the theorem is complete. $\square$

There is a difference between Theorems 3.3.8 and 3.3.9.

In Theorem 3.3.8, the limit in the asymptotic relation (3.3.28) depends on characteristics of the transition period via the quantity $1 - \tilde{f}^{(0)} = \lim_{\varepsilon \to 0} \tilde{F}^{(\varepsilon)}(\infty)$ while, in Theorem 3.3.9, the limit in the asymptotic relation (3.3.52) depends on characteristics of the transition period via the quantity $\tilde{\phi}^{(0)} = \lim_{\varepsilon \to 0} \int_0^\infty e^{\rho^{(0)}s} \tilde{F}^{(\varepsilon)}(ds)$. Obviously, $1 - \tilde{f}^{(0)} \le \tilde{\phi}^{(0)}$.

Let us consider the case where condition $\mathbf{D_{20}}$ holds, i.e., $\tilde{F}^{(\varepsilon)}(\cdot) \Rightarrow \tilde{F}^{(0)}(\cdot)$ as $\varepsilon \to 0$. In this case, under an additional assumption that condition $\mathbf{C_9}$ holds, we have $1 - \tilde{f}^{(0)} = \tilde{F}^{(0)}(\infty)$ and $\tilde{\phi}^{(0)} = \tilde{\phi}^{(0)}(\rho^{(0)}) = \int_0^\infty e^{\rho^{(0)}s} \tilde{F}^{(0)}(ds)$.

If $\rho^{(0)} = 0$, then $\tilde{\phi}^{(0)} = 1 - \tilde{f}^{(0)}$. In this case, Theorem 3.3.9 just reduces to Theorem 3.3.8. The limits in (3.3.28) and (3.3.52) do not depend on the characteristics of the transition period if $1 - \tilde{f}^{(0)} = 1$. If $\rho^{(0)} > 0$ and $\tilde{F}^{(0)}(0) < 1$, i.e., the distribution function $\tilde{F}^{(0)}(t)$ is not concentrated in zero, then $\tilde{F}^{(0)}(\infty) < \tilde{\phi}^{(0)}(\rho^{(0)})$. The limit in (3.3.52) does not depend on the characteristics of the transition period if $\tilde{\phi}^{(0)}(\rho^{(0)}) = 1$.

The following theorem, which can be referred as a quasi-ergodic theorem, is a direct corollary of Theorem 3.3.9.

Theorem 3.3.10. *Let conditions* $\mathbf{D_{21}}$*,* $\mathbf{C_6}$*,* $\mathbf{F_7}$*,* $\mathbf{C_9}$*,* $\mathbf{D_{26}}$*, and* $\mathbf{C_{11}}$*, together with the additional assumption that* $\tilde{\phi}^{(0)} > 0$*, hold. Then, for any* $0 \leq t^{(\varepsilon)} \to \infty$ *as* $\varepsilon \to 0$ *and any* $A \in \Gamma_0$*,*

$$\mathsf{P}\{\xi^{(\varepsilon)}(t^{(\varepsilon)}) \in A / \mu^{(\varepsilon)} > t^{(\varepsilon)}\} = \frac{\mathsf{P}\{\xi^{(\varepsilon)}(t^{(\varepsilon)}) \in A, \mu^{(\varepsilon)} > t^{(\varepsilon)}\}}{\mathsf{P}\{\mu^{(\varepsilon)} > t^{(\varepsilon)}\}}$$

$$\to \frac{\tilde{\pi}^{(0)}(A)}{\tilde{\pi}^{(0)}(X)} = \pi^{(0)}(A) \ \ as \ \ \varepsilon \to 0. \tag{3.3.76}$$

Thus, convergence of the conditional distribution of a perturbed regenerative process to the corresponding quasi-stationary distribution takes place also in a model with transition period.

Note that Theorem 3.3.9 yields directly that the conditional probabilities in (3.3.76) converge to the fraction $\tilde{\phi}^{(0)}\tilde{\pi}^{(0)}(A)/\tilde{\phi}^{(0)}\tilde{\pi}^{(0)}(X)$ as $\varepsilon \to 0$. In order to be able to reduce this expression to $\tilde{\pi}^{(0)}(A)/\tilde{\pi}^{(0)}(X) = \pi^{(0)}(A)$, one should require that $\tilde{\phi}^{(0)} > 0$. Without this assumption, the asymptotic relation (3.3.52) would still take place, but the asymptotic relation (3.3.76) possibly would not.

3.3.8   Quasi-stationary distributions for perturbed regenerative processes with transition period

Let ε_1, ε_2, and ε_3 be parameters used, respectively, in conditions $\mathbf{D_{23}}$, $\mathbf{C_6}$, and $\mathbf{F_9}$.

Condition $\mathbf{C_{10}}$ implies that $\int_0^\infty e^{(\rho^{(0)}+\gamma)s}\hat{\tilde{F}}^{(\varepsilon)}(ds) < \infty$ for all ε small enough, say $\varepsilon \leq \varepsilon_4$.

Define

$$\varepsilon_0 = \min(\varepsilon_1, \varepsilon_2, \varepsilon_3, \varepsilon_4), \tag{3.3.77}$$

and denote

$$\tilde{\tilde{\pi}}^{(\varepsilon)}(A) = \tilde{\phi}^{(\varepsilon)}(\rho^{(\varepsilon)})\tilde{\pi}^{(\varepsilon)}(A), \quad A \in \Gamma. \tag{3.3.78}$$

The following theorem is a direct corollary of Theorems 3.3.3 and 3.3.9.

Theorem 3.3.11. *Let conditions* $\mathbf{D_{23}}$*,* $\mathbf{C_6}$*,* $\mathbf{F_9}$*, and* $\mathbf{C_{10}}$ *hold. Then, for* $\varepsilon \leq \varepsilon_0$*,*

$$e^{\rho^{(\varepsilon)}t}\mathsf{P}\{\xi^{(\varepsilon)}(t) \in A, \mu^{(\varepsilon)} > t\} \to \tilde{\tilde{\pi}}^{(\varepsilon)}(A) \ \ as \ \ t \to \infty, \ A \in \Gamma_0. \tag{3.3.79}$$

Let us assume the following condition:

$\mathbf{D_{27}}$: $\tilde{F}^{(\varepsilon)}(\infty) > 0$ for $\varepsilon \leq \varepsilon_5$.

Condition $\mathbf{D_{27}}$ obviously implies that $\tilde{\phi}^{(\varepsilon)}(\rho^{(\varepsilon)}) > 0$ for $\varepsilon \leq \varepsilon_5$.
Let us also define

$$\varepsilon_0' = \min(\varepsilon_0, \varepsilon_5). \tag{3.3.80}$$

Theorem 3.3.11 implies the following statement.

Theorem 3.3.12. *Let conditions* $\mathbf{D_{23}}$, $\mathbf{C_6}$, $\mathbf{F_9}$, $\mathbf{C_{10}}$, *and* $\mathbf{D_{27}}$ *hold. Then, for* $\varepsilon \leq \varepsilon_0'$,

$$
\mathsf{P}\{\xi^{(\varepsilon)}(t) \in A / \mu^{(\varepsilon)} > t\} = \frac{\mathsf{P}\{\xi^{(\varepsilon)}(t) \in A, \mu^{(\varepsilon)} > t\}}{\mathsf{P}\{\mu^{(\varepsilon)} > t\}}
$$

$$
\to \frac{\tilde{\pi}^{(\varepsilon)}(A)}{\tilde{\pi}^{(\varepsilon)}(X)} = \pi^{(\varepsilon)}(A) \ \text{as} \ t \to \infty, \ A \in \Gamma_0. \quad (3.3.81)
$$

Finally, we also give conditions for convergence of the functionals $\tilde{\tilde{\pi}}^{(\varepsilon)}(A)$. In this case, it is natural to assume also that condition $\mathbf{D_{20}}$ holds.

Theorem 3.3.13. *Let conditions* $\mathbf{D_{21}}$, $\mathbf{C_6}$, $\mathbf{F_7}$, *and* $\mathbf{D_{20}}$, $\mathbf{C_9}$ *hold. Then*

$$
\tilde{\tilde{\pi}}^{(\varepsilon)}(A) \to \tilde{\tilde{\pi}}^{(0)}(A) \ \text{as} \ \varepsilon \to 0, \quad A \in \Gamma_0. \quad (3.3.82)
$$

Proof. According to (3.3.13), conditions $\mathbf{D_{21}}$, $\mathbf{C_6}$, and $\mathbf{F_7}$ imply that $\tilde{\pi}^{(\varepsilon)}(A) \to \tilde{\pi}^{(0)}(A)$ as $\varepsilon \to 0$ for $A \in \Gamma_0$. Thus, it is sufficient to show that conditions $\mathbf{D_{24}}$ and $\mathbf{C_9}$ imply that

$$
\tilde{\phi}^{(\varepsilon)}(\rho^{(\varepsilon)}) \to \tilde{\phi}^{(0)}(\rho^{(0)}) \ \text{as} \ \varepsilon \to 0. \quad (3.3.83)
$$

Since $\rho^{(\varepsilon)} \to \rho^{(0)}$ as $\varepsilon \to 0$, we have $\rho^{(\varepsilon)} < \beta < \rho^{(0)} + \gamma'$ for all ε small enough. Thus,

$$
\lim_{T \to \infty} \overline{\lim_{\varepsilon \to 0}} \int_T^\infty e^{s\rho^{(\varepsilon)}} \tilde{F}^{(\varepsilon)}(ds)
$$

$$
\leq \lim_{T \to \infty} \overline{\lim_{\varepsilon \to 0}} \int_T^\infty e^{s\beta} \tilde{F}^{(\varepsilon)}(ds)
$$

$$
\leq \lim_{T \to \infty} e^{-T(\rho^{(0)}+\gamma'-\beta)} \overline{\lim_{\varepsilon \to 0}} \int_0^\infty e^{s(\rho^{(0)}+\gamma')} \tilde{F}^{(\varepsilon)}(ds) = 0. \quad (3.3.84)
$$

Condition $\mathbf{D_{24}}$ and the relation $\rho^{(\varepsilon)} \to \rho^{(0)}$ as $\varepsilon \to 0$ imply that, for any T which is a point of continuity of the distribution $\tilde{F}^{(0)}(s)$,

$$
\lim_{\varepsilon \to 0} \int_0^T e^{s\rho^{(\varepsilon)}} \tilde{F}^{(\varepsilon)}(ds) = \int_0^T e^{s\rho^{(0)}} \tilde{F}^{(0)}(ds). \quad (3.3.85)
$$

Relations (3.3.84) and (3.3.85) imply relation (3.3.83). $\qquad\square$

3.4 Exponential expansions for nonlinearly perturbed regenerative processes

In this section, we present exponential asymptotic expansions in mixed ergodic and large deviation theorems for nonlinearly perturbed regenerative processes and regenerative stopping times.

3.4.1   Pseudo-stationary exponential expansions for perturbed regenerative processes based on stopping probabilities

Let us now formulate an analogue of Theorem 2.1.1. We assume the following balancing condition:

$\mathbf{B}_4^{(k)}$: $0 \leq t^{(\varepsilon)} \to \infty$ and $f^{(\varepsilon)} \to 0$ as $\varepsilon \to 0$ in such a way that $(f^{(\varepsilon)})^k t^{(\varepsilon)} \to \lambda_k$, where $0 \leq \lambda_k < \infty$.

Obviously, conditions $\mathbf{D}_{13}$ and $\mathbf{C}_3$ imply that the power moments $m_r^{(\varepsilon)}$ are finite for all $r \geq 1$ and all ε small enough, and that

$$m_r^{(\varepsilon)} \to m_r^{(0)} < \infty \text{ as } \varepsilon \to 0, \quad r \geq 1. \tag{3.4.1}$$

Theorem 3.4.1. *Let conditions $\mathbf{D}_{13}$ and $\mathbf{C}_3$ hold. Then*

(i) *The following asymptotic Taylor's expansion holds for the root $\rho^{(\varepsilon)}$ of the equation (3.3.2) for every $k \geq 1$:*

$$\rho^{(\varepsilon)} = b_{\varepsilon 1} f^{(\varepsilon)} + \cdots + b_{\varepsilon k}(f^{(\varepsilon)})^k + o((f^{(\varepsilon)})^k), \tag{3.4.2}$$

where the coefficients $b_{\varepsilon k}$, $k \geq 1$, can be calculated using the recurrence formulas $b_{\varepsilon 1} = (m_1^{(\varepsilon)})^{-1}$, $b_{\varepsilon 2} = -\frac{1}{2}(m_1^{(\varepsilon)})^{-1} m_2^{(\varepsilon)} b_{\varepsilon 1}^2$ and, in general, for $k \geq 2$,

$$b_{\varepsilon k} = -(m_1^{(\varepsilon)})^{-1} \sum_{r=2}^{k} m_r^{(\varepsilon)} \cdot \sum_{n_1,\ldots,n_{k-1} \in D_{r,k}} \prod_{i=1}^{k-1} b_{\varepsilon i}^{n_i} / n_i!, \tag{3.4.3}$$

where $D_{r,k}$, for every $2 \leq r \leq k < \infty$, is the set of all nonnegative, integer solutions of the system

$$n_1 + \cdots + n_{k-1} = r, \quad n_1 + \cdots + (k-1)n_{k-1} = k. \tag{3.4.4}$$

(ii) *If, additionally, conditions $\mathbf{B}_4^{(k)}$, for some $k \geq 1$, and $\mathbf{F}_7$ hold, then the following asymptotic relation holds for any $A \in \Gamma_0$:*

$$\frac{\mathsf{P}\{\xi^{(\varepsilon)}(t^{(\varepsilon)}) \in A, \mu^{(\varepsilon)} > t^{(\varepsilon)}\}}{\exp\{-(b_{\varepsilon 1} f^{(\varepsilon)} + \cdots + b_{\varepsilon k-1}(f^{(\varepsilon)})^{k-1})t^{(\varepsilon)}\}}$$
$$\to \pi^{(0)}(A)e^{-\lambda_k b_{0k}} \text{ as } \varepsilon \to 0. \tag{3.4.5}$$

Proof. The proof can be obtained by directly applying Theorem 2.1.1 to the renewal equation (3.2.1).

Condition $\mathbf{D}_{13}$ implies that condition $\mathbf{D}_6$ holds for the distribution functions $F^{(\varepsilon)}(s)$. Also, condition $\mathbf{C}_3$ implies condition $\mathbf{C}_1$. Condition $\mathbf{F}_7$ (**a**) implies that, for every $A \in \Gamma_0$, condition $\mathbf{F}_4$ (**a**) holds for the function $q^{(\varepsilon)}(s, A)$. Since $0 \leq q^{(\varepsilon)}(s, A) \leq 1$, condition $\mathbf{F}_5$ (**b**) obviously holds. As was shown in the proof of Theorem 3.3.1, condition $\mathbf{C}_3$ implies that, for every $A \in \Gamma_0$, condition $\mathbf{F}_5$ (**c**) holds. Finally, condition $\mathbf{B}_4^{(k)}$ coincides in this case with condition $\mathbf{B}_2^{(k)}$. Thus, all conditions of Theorem 2.1.1 hold. $\qquad\square$

3.4.2 Pseudo-stationary exponential expansions for perturbed regenerative processes in the case where the stopping probabilities are of order $O(\varepsilon)$

Let us now assume the following perturbation condition:

$\mathbf{P_9^{(k)}}$: (a) $1 - f^{(\varepsilon)} = 1 + b_{1,0}\varepsilon + \cdots + b_{k,0}\varepsilon^k + o(\varepsilon^k)$, where $|b_{r,0}| < \infty, r = 1, \ldots, k$;

      (b) $m_r^{(\varepsilon)} = m_r^{(0)} + b_{1,r}\varepsilon + \cdots + b_{k-r,r}\varepsilon^{k-r} + o(\varepsilon^{k-r})$ for $r = 1, \ldots, k$, where $|b_{l,r}| < \infty, l = 1, \ldots, k - r, r = 1, \ldots, k$.

It is convenient to also set $b_{0,0} = 1$ and $b_{0,r} = m_r^{(0)}, r \geq 1$.

We shall also use the following balancing condition for $1 \leq r \leq k$:

$\mathbf{B_5^{(r)}}$: $0 \leq t^{(\varepsilon)} \to \infty$ as $\varepsilon \to 0$ such that $\varepsilon^r t^{(\varepsilon)} \to \lambda_r$, where $0 \leq \lambda_r < \infty$.

Theorem 3.4.2. *Let conditions* $\mathbf{D_{13}}$, $\mathbf{C_3}$ *and* $\mathbf{P_9^{(k)}}$ *hold. Then*

(i) *For all ε small enough there exists a unique nonnegative root $\rho^{(\varepsilon)}$ of the equation (3.3.2) with the following asymptotic Taylor's expansion:*

$$\rho^{(\varepsilon)} = a_1\varepsilon + \cdots + a_k\varepsilon^k + o(\varepsilon^k), \tag{3.4.6}$$

where the coefficients $a_n, n = 1, \ldots, k$, are given by the recurrence formulas $a_1 = -b_{1,0}/b_{0,1}$ and, in general, for $n = 1, \ldots, k$,

$$a_n = -b_{0,1}^{-1}\left(b_{n,0} + \sum_{q=1}^{n-1} b_{n-q,1}a_q \right.$$

$$\left. + \sum_{m=2}^{n}\sum_{q=m}^{n} b_{n-q,m} \cdot \sum_{n_1,\ldots,n_{q-1}\in D_{m,q}} \prod_{p=1}^{q-1} a_p^{n_p}/n_p! \right), \tag{3.4.7}$$

where $D_{m,q}$, for every $2 \leq m \leq q < \infty$, is the set of all nonnegative, integer solutions of the system

$$n_1 + \cdots + n_{q-1} = m, \quad n_1 + \cdots + (q-1)n_{q-1} = q. \tag{3.4.8}$$

(ii) *If $b_{l,0} = 0, l = 1, \ldots, r$, for some $1 \leq r \leq k$, then $a_1, \ldots, a_r = 0$. If $b_{l,0} = 0, l = 1, \ldots, r - 1$, but $b_{r,0} < 0$ for some $1 \leq r \leq k$, then $a_1, \ldots, a_{r-1} = 0$ but $a_r > 0$.*

(iii) *If, additionally, conditions $\mathbf{B_5^{(r)}}$ for some $1 \leq r \leq k$, and $\mathbf{F_7}$ hold, then the following asymptotic relation holds for any $A \in \Gamma_0$:*

$$\frac{\mathsf{P}\{\xi^{(\varepsilon)}(t^{(\varepsilon)}) \in A, \mu^{(\varepsilon)} > t^{(\varepsilon)}\}}{\exp\{-(a_1\varepsilon + \cdots + a_{r-1}\varepsilon^{r-1})t^{(\varepsilon)}\}} \to \pi^{(0)}(A)e^{-\lambda_r a_r} \quad as \ \varepsilon \to 0. \tag{3.4.9}$$

Proof. The proof can be obtained by a direct application of Theorem 2.1.2 to the renewal equation (3.2.1).

Condition D_{13} implies that condition D_6 holds for the distribution functions $F^{(\varepsilon)}(s)$. Also, condition C_3 implies condition C_1. Condition F_7 (a) implies that, for every $A \in \Gamma_0$, condition F_5 (a) holds for the function $q^{(\varepsilon)}(s, A)$. Since $0 \leq q^{(\varepsilon)}(s, A) \leq 1$, condition F_5 (b) obviously holds. As was shown in the proof of Theorem 3.3.1, condition C_3 implies that for every $A \in \Gamma_0$ condition F_5 (c) is fulfilled. Finally, condition $B_5^{(r)}$ coincides in this case with condition $B_3^{(r)}$, while condition $P_9^{(k)}$ coincides with condition $P_1^{(k)}$. Thus, all conditions of Theorem 2.1.2 hold. $\qquad\square$

3.4.3 Pseudo-stationary exponential expansions for perturbed regenerative processes in the case where the stopping probabilities are of order $O(\varepsilon^h)$

Let us introduce integer parameters $1 \leq h \leq k_0$ and $k_r \geq 1, r = 1, \ldots$, and define a vector parameter $\bar{k} = (h, k_0, \ldots, k_{\tilde{N}})$, where $\tilde{N} = \min(n : nh \leq k_0)$.

We assume to hold the following perturbation condition:

$P_{10}^{(\bar{k})}$: (a) $1 - f^{(\varepsilon)} = 1 + b_{h,0}\varepsilon^h + \cdots + b_{k_0,0}\varepsilon^{k_0} + o(\varepsilon^{k_0})$, where $|b_{l,0}| < \infty$, $l = h, \ldots, k_0$;

  (b) $m_r^{(\varepsilon)} = m_r^{(0)} + b_{1,r}\varepsilon + \cdots + b_{k_r,r}\varepsilon^{k_r} + o(\varepsilon^{k_r})$ for $r = 1, \ldots, \tilde{N}$, where $|b_{l,r}| < \infty, l = 1, \ldots, k_r, r = 1, \ldots, \tilde{N}$.

As before, it is convenient to also define $b_{0,0} = 1$ and $b_{0,r} = m_r^{(0)}, r \geq 1$.

Condition $P_{10}^{(\bar{k})}$ is more general than condition $P_9^{(k)}$, and the first one reduces to the second one if $h = 1, k_0 = k$ (in this case, $\tilde{N} = k$) and $k_r = k - r, r = 1, \ldots, \tilde{N}$.

The following theorem supplements and generalises Theorem 3.4.2.

Theorem 3.4.3. *Let conditions D_{13}, C_3 and $P_{10}^{(\bar{k})}$ hold. Then*

(i) *For all ε small enough there exists a unique nonnegative root $\rho^{(\varepsilon)}$ of the equation (3.3.2) with the following asymptotic expansion:*

$$\rho^{(\varepsilon)} = a_h \varepsilon^h + \cdots + a_k \varepsilon^k + o(\varepsilon^k), \tag{3.4.10}$$

where $k = \min(k_0, k_1 + h, \ldots, k_{\tilde{N}} + h\tilde{N})$ and the coefficients $a_n, n = h, \ldots, k$, are given by the recurrence formulas $a_h = -b_{h,0}/b_{0,1}$ and for $n = h, \ldots, k$,

$$a_n = -b_{0,1}^{-1}\left(b_{n,0} + \sum_{i=h}^{n-1} b_{n-i,1} a_i \right.$$

$$\left. + \sum_{r=2}^{[n/h]} \sum_{j=rh}^{n} b_{n-j,r} \cdot \sum_{n_h,\ldots,n_{j-1} \in D_{h,r,j}} \prod_{i=h}^{j-1} a_i^{n_i}/n_i! \right), \tag{3.4.11}$$

where $D_{h,r,j}$, for every $h \geq 1$ and $2 \leq r \leq j < \infty$, is the set of all nonnegative, integer solutions of the system

$$n_h + \cdots + n_{j-1} = r, \quad hn_h + \cdots + (j-1)n_{j-1} = j. \qquad (3.4.12)$$

(ii) *If $b_{l,0} = 0$, $l = h, \ldots, r$, for some $h \leq r \leq k$, then $a_1, \ldots, a_r = 0$. If $b_{l,0} = 0$, $l = h, \ldots, r - 1$ but $b_{r,0} < 0$ for some $h \leq r \leq k$, then $a_h, \ldots, a_{r-1} = 0$ but $a_r > 0$.*

(iii) *If, additionally, conditions $\mathbf{B}_5^{(r)}$, for some $h \leq r \leq k$, and $\mathbf{F}_7$ hold, then the following asymptotic relation holds:*

$$\frac{\mathsf{P}\{\xi^{(\varepsilon)}(t^{(\varepsilon)}) \in A, \mu^{(\varepsilon)} > t^{(\varepsilon)}\}}{\exp\{-(a_h\varepsilon + \cdots + a_{r-1}\varepsilon^{r-1})t^{(\varepsilon)}\}} \to \pi^{(0)}(A)e^{-\lambda_r a_r} \quad \text{as } \varepsilon \to 0. \quad (3.4.13)$$

Proof. The proof can be obtained by a direct application of Theorem 2.1.3 to the renewal equation (3.2.1). Condition $\mathbf{D}_{13}$ implies that condition $\mathbf{D}_6$ holds for the distribution functions $F^{(\varepsilon)}(s)$. Also, condition $\mathbf{C}_3$ implies condition $\mathbf{C}_1$. Condition $\mathbf{F}_7$ **(a)** implies that, for every $A \in \Gamma_0$, condition $\mathbf{F}_5$ **(a)** holds for the function $q^{(\varepsilon)}(s, A)$. Since $0 \leq q^{(\varepsilon)}(s, A) \leq 1$, condition $\mathbf{F}_5$ **(b)** obviously holds. As was shown in the proof of Theorem 3.3.1, condition $\mathbf{C}_3$ implies that for every $A \in \Gamma_0$ condition $\mathbf{F}_5$ **(c)** holds true. Finally, condition $\mathbf{B}_5^{(r)}$ coincides in this case with condition $\mathbf{B}_3^{(r)}$, while condition $\mathbf{P}_9^{(k)}$ coincides with condition $\mathbf{P}_1^{(k)}$. Thus, all conditions of Theorem 2.1.3 are satisfied. $\qquad\square$

3.4.4 Quasi-stationary exponential expansions for perturbed regenerative processes in the case where the perturbation is of order $O(\varepsilon)$

We finally consider the most general case where the limit distribution $F^{(0)}(s)$ can be proper or improper.

Let us introduce the following moment generating functions:

$$\phi^{(\varepsilon)}(\rho, r) = \int_0^\infty s^r e^{\rho s} F^{(\varepsilon)}(ds), \quad r \geq 1.$$

Condition $\mathbf{C}_3$ implies that for any $\beta < \delta$ the mixed moment generating functions satisfy $\phi^{(\varepsilon)}(\beta, r) < \infty, r = 0, 1, \ldots$, for all ε small enough.

Let us introduce the following perturbation condition:

$\mathbf{P}_{11}^{(k)}$: $\phi^{(\varepsilon)}(\rho^{(0)}, r) = \phi^{(0)}(\rho^{(0)}, r) + b_{1,r}\varepsilon + \cdots + b_{k-r,r}\varepsilon^{k-r} + o(\varepsilon^{k-r})$ for $r = 0, \ldots, k$, where $|b_{n,r}| < \infty, n = 1, \ldots, k - r, r = 0, \ldots, k$.

It is convenient to define $b_{0,r} = \phi^{(0)}(\rho^{(0)}, r), r = 0, 1, \ldots$. It clearly follows from the definition of $\rho^{(0)}$ that $b_{0,0} = \phi^{(0)}(\rho^{(0)}, 0) = 1$.

Let us formulate an analogue of Theorem 2.2.1.

Theorem 3.4.4. *Let conditions* $\mathbf{D_{21}}$, $\mathbf{C_6}$ *and* $\mathbf{P_{11}^{(k)}}$ *hold. Then*

(i) *The root* $\rho^{(\varepsilon)}$ *of equation (3.3.2) has the asymptotic expansion*

$$\rho^{(\varepsilon)} = \rho^{(0)} + a_1\varepsilon + \cdots + a_k\varepsilon^k + o(\varepsilon^k), \qquad (3.4.14)$$

where the coefficients a_n *are given by the recurrence formulas* $a_1 = b_{1,0}/b_{0,1}$
and, in general for $n = 1,\ldots,k$,

$$a_n = -b_{0,1}^{-1}\left(b_{n,0} + \sum_{q=1}^{n-1} b_{n-q,1}a_q \right.$$

$$\left. + \sum_{2 \le m \le n}\sum_{q=m}^{n} b_{n-q,m} \cdot \sum_{n_1,\ldots,n_{q-1}\in D_{m,q}} \prod_{p=1}^{q-1} a_p^{n_p}/n_p! \right), \qquad (3.4.15)$$

were $D_{m,q}$, *for every* $2 \le m \le q < \infty$, *is the set of all nonnegative, integer*
solutions of the system

$$n_1 + \cdots + n_{q-1} = m, \quad n_1 + \cdots + (q-1)n_{q-1} = q. \qquad (3.4.16)$$

(ii) *If the coefficients satisfy* $b_{l,0} = 0$, $l = 1,\ldots,r$, *for some* $1 \le r \le k$, *then*
$a_1,\ldots,a_r = 0$. *If* $b_{l,0} = 0$, $l = 1,\ldots,r-1$, *but* $b_{r,0} < 0$ *for some* $1 \le r \le k$,
then $a_1,\ldots,a_{r-1} = 0$ *but* $a_r > 0$.

(iii) *If, additionally, conditions* $\mathbf{B_5^{(r)}}$, *for some* $1 \le r \le k$, *and* $\mathbf{F_7}$ *hold, then for any*
$A \in \Gamma_0$, *the following asymptotic relation holds:*

$$\frac{\mathsf{P}\{\xi^{(\varepsilon)}(t^{(\varepsilon)}) \in A, \mu^{(\varepsilon)} > t^{(\varepsilon)}\}}{\exp\{-(\rho^{(0)} + a_1\varepsilon + \cdots + a_{r-1}\varepsilon^{r-1})t^{(\varepsilon)}\}}$$

$$\to \tilde{\pi}^{(0)}(A)e^{-\lambda_r a_r} \quad as \ \varepsilon \to 0. \qquad (3.4.17)$$

Proof. The proof can be obtained by directly applying Theorem 2.2.1 to the renewal
equation (3.2.1).

   Condition $\mathbf{D_{21}}$ implies that condition $\mathbf{D_{11}}$ holds for the distribution functions
$F^{(\varepsilon)}(s)$. Also, condition $\mathbf{C_6}$ implies condition $\mathbf{C_2}$. Condition $\mathbf{F_7}$ **(a)** implies that
for every $A \in \Gamma_0$ condition $\mathbf{F_6}$ **(a)** holds for the function $q^{(\varepsilon)}(s, A)$. Since $0 \le$
$q^{(\varepsilon)}(s, A) \le 1$, condition $\mathbf{F_6}$ **(b)** obviously holds. As was shown in the proof of Theorem 3.3.1, condition $\mathbf{C_6}$ implies that for every $A \in \Gamma_0$, condition $\mathbf{F_6}$ **(c)** is verified.
Finally, condition $\mathbf{B_5^{(r)}}$ coincides in this case with condition $\mathbf{B_3^{(r)}}$, while condition $\mathbf{P_{11}^{(k)}}$
coincides with condition $\mathbf{P_3^{(k)}}$. Thus, all conditions of Theorem 2.2.1 hold. $\qquad\square$

3.4.5 Quasi-stationary exponential expansions for perturbed regenerative processes in the case where the perturbation is of order $O(\varepsilon^h)$

Let us also formulate a theorem which is an analogue of Theorem 3.4.3 for the model of improper perturbed regenerative processes.

As above, let us introduce the integer parameters $1 \le h \le k_0$ and $k_r \ge 1$, $r = 1,\ldots$, and define a vector parameter $\bar{k} = (h, k_0, \ldots, k_{\tilde{N}})$, where $\tilde{N} = \min(n : nh \le k_0)$.

We assume that the following perturbation condition holds:

$\mathbf{P}_{12}^{(\bar{k})}$:   (a) $\phi^{(\varepsilon)}(\rho^{(0)}, 0) = \phi^{(0)}(\rho^{(0)}, 0) + b_{h,0}\varepsilon^h + \cdots + b_{k_0,0}\varepsilon^{k_0} + o(\varepsilon^{k_0})$, where $|b_{n,0}| < \infty$, $n = h, \ldots, k_0$;

   (b) $\phi^{(\varepsilon)}(\rho^{(0)}, r) = \phi^{(0)}(\rho^{(0)}, r) + b_{1,r}\varepsilon + \cdots + b_{k_r,r}\varepsilon^{k_r} + o(\varepsilon^{k_r})$ for $r = 1, \ldots, k$, where $|b_{n,r}| < \infty$, $n = 1, \ldots, k_r$, $r = 1, \ldots, \tilde{N}$.

It is convenient to define $b_{0,r} = \phi^{(0)}(\rho^{(0)}, r)$, $r = 0, 1, \ldots$. Note that $\phi^{(0)}(\rho^{(0)}, 0) = 1$, while $\phi^{(0)}(\rho^{(0)}, r) > 0$, $r = 1, \ldots, \tilde{N}$.

Condition $\mathbf{P}_{12}^{(\bar{k})}$ is more general than condition $\mathbf{P}_{11}^{(k)}$, and the first one reduces to the second one if $h = 1$, $k_0 = k$ (in this case, $\tilde{N} = k$) and $k_r = k - r$, $r = 1, \ldots, N$.

The following theorem generalises Theorems 3.4.4.

Theorem 3.4.5. *Let conditions* $\mathbf{D}_{21}$, $\mathbf{C}_6$ *and* $\mathbf{P}_{12}^{(\bar{k})}$ *hold. Then*

(i) *For all ε small enough there exists a unique nonnegative root $\rho^{(\varepsilon)}$ of equation (2.1.4) with the following asymptotic expansion:*

$$\rho^{(\varepsilon)} = \rho^{(0)} + a_h\varepsilon^h + \cdots + a_k\varepsilon^k + o(\varepsilon^k), \tag{3.4.18}$$

where $k = \min(k_0, k_1+h, \ldots, k_{\tilde{N}}+h\tilde{N})$ and the coefficients a_n, $n = h, \ldots, k$, are given by the recurrence formulas $a_h = -b_{h,0}/b_{0,1}$ and for $n = h, \ldots, k$,

$$a_n = -b_{0,1}^{-1}\left(b_{n,0} + \sum_{i=h}^{n-1} b_{n-i,1}a_i \right.$$

$$\left. + \sum_{r=2}^{[n/h]} \sum_{j=rh}^{n} b_{n-j,r} \cdot \sum_{n_h,\ldots,n_{j-1} \in D_{h,r,j}} \prod_{i=h}^{j-1} a_i^{n_i}/n_i! \right), \tag{3.4.19}$$

where $D_{h,r,j}$, for every $h \ge 1$ and $2 \le r \le j < \infty$, is the set of all nonnegative, integer solutions of the system

$$n_h + \cdots + n_{j-1} = r, \quad hn_h + \cdots + (j-1)n_{j-1} = j. \tag{3.4.20}$$

(ii) *If $b_{l,0} = 0$, $l = h, \ldots, r$, for some $h \le r \le k$, then $a_1, \ldots, a_r = 0$. If $b_{l,0} = 0$, $l = h, \ldots, r-1$ but $b_{r,0} < 0$ for some $h \le r \le k$, then $a_h, \ldots, a_{r-1} = 0$ but $a_r > 0$.*

(iii) *If, additionally, conditions* $\mathbf{B}_5^{(r)}$*, for some* $h \leq r \leq k$*, and* $\mathbf{F}_7$ *hold, then the following asymptotic relation holds:*

$$\frac{\mathsf{P}\{\xi^{(\varepsilon)}(t^{(\varepsilon)}) \in A, \mu^{(\varepsilon)} > t^{(\varepsilon)}\}}{\exp\{-(\rho^{(0)} + a_h\varepsilon^h + \cdots + a_{r-1}\varepsilon^{r-1})t^{(\varepsilon)}\}}$$
$$\to \tilde{\pi}^{(0)}(A)e^{-\lambda_r a_r} \quad \text{as } \varepsilon \to 0. \tag{3.4.21}$$

Proof. The proof can be obtained by applying Theorem 2.2.2 to the renewal equation (3.2.1).

Condition $\mathbf{D}_{21}$ implies that condition $\mathbf{D}_{11}$ holds for the distribution functions $F^{(\varepsilon)}(s)$. Also, condition $\mathbf{C}_6$ implies condition $\mathbf{C}_2$. Condition $\mathbf{F}_7$ **(a)** implies that, for every $A \in \Gamma_0$, condition $\mathbf{F}_6$ **(a)** holds for the function $q^{(\varepsilon)}(s, A)$. Since $0 \leq q^{(\varepsilon)}(s, A) \leq 1$, condition $\mathbf{F}_6$ **(b)** obviously holds. As was shown in the proof of Theorem 3.3.1, condition $\mathbf{C}_6$ implies that, for every $A \in \Gamma_0$, condition $\mathbf{F}_6$ **(c)** also holds. Finally, condition $\mathbf{B}_5^{(r)}$ coincides with condition $\mathbf{B}_3^{(r)}$, and condition $\mathbf{P}_{12}^{(k)}$ with condition $\mathbf{P}_4^{(k)}$. This shows that all conditions of Theorem 2.2.2 hold. $\square$

3.4.6 Pseudo-stationary and quasi-stationary exponential expansions for perturbed regenerative processes with transition period

Let us also give the corresponding exponential expansions for perturbed regenerative processes with transition period.

The exponential expansions presented above in Theorems 3.4.1, 3.4.2, 3.4.3, 3.4.4, and 3.4.5 are based on combination of two asymptotic results.

The first one is a mixed ergodic and large deviation asymptotic relation of the form

$$\frac{\mathsf{P}\{\xi^{(\varepsilon)}(t^{(\varepsilon)}) \in A, \mu^{(\varepsilon)} > t^{(\varepsilon)}\}}{\exp\{-\rho^{(\varepsilon)}t^{(\varepsilon)}\}} \to \tilde{\pi}^{(0)}(A) \quad \text{as } \varepsilon \to 0. \tag{3.4.22}$$

The second one is a representation of the characteristic root $\rho^{(\varepsilon)}$ of equation (2.1.4) in the form of the asymptotic expansion

$$\rho^{(\varepsilon)} = \rho^{(0)} + a_h\varepsilon^h + \cdots + a_k\varepsilon^k + o(\varepsilon^k). \tag{3.4.23}$$

Representation (3.4.23) and the additional balancing condition, $\varepsilon^r t^{(\varepsilon)} \to \lambda_r$ as $\varepsilon \to 0$, permit to transform asymptotic relation (3.4.22) to an equivalent but more explicit form,

$$\frac{\mathsf{P}\{\xi^{(\varepsilon)}(t^{(\varepsilon)}) \in A, \mu^{(\varepsilon)} > t^{(\varepsilon)}\}}{\exp\{-(\rho^{(0)} + a_h\varepsilon^h + \cdots + a_{r-1}\varepsilon^{r-1})t^{(\varepsilon)}\}} \to \tilde{\pi}^{(0)}(A)e^{-\lambda_r a_r} \quad \text{as } \varepsilon \to 0. \tag{3.4.24}$$

In the model of asymptotically proper perturbed regenerative processes ($f^{(\varepsilon)} \to f^{(0)} = 0$ as $\varepsilon \to 0$), the mixed ergodic and large deviation asymptotic relation (3.4.22) takes the same form for the perturbed regenerative processes with transition period as for the corresponding standard regenerative processes. In this case, the asymptotic relation (3.4.24) is also preserved, for processes with transition period, in the same form as for the corresponding standard regenerative processes. However, the corresponding conditions imposed on characteristics of the transition period should be augmented.

The next three theorems are related to the model of asymptotically proper perturbed regenerative processes with transition period.

Theorem 3.4.6. *Let conditions* $\mathbf{D_{13}}$, $\mathbf{C_3}$, $\mathbf{F_7}$, $\mathbf{D_{17}}$, $\mathbf{C_7}$, $\mathbf{D_{25}}$, *and* $\mathbf{B_4^{(k)}}$, *for some* $k \geq 1$, *hold. Then, for any* $A \in \Gamma_0$,

$$\frac{\mathsf{P}\{\xi^{(\varepsilon)}(t^{(\varepsilon)}) \in A, \mu^{(\varepsilon)} > t^{(\varepsilon)}\}}{\exp\{-(b_{\varepsilon 1} f^{(\varepsilon)} + \cdots + b_{\varepsilon k-1}(f^{(\varepsilon)})^{k-1})t^{(\varepsilon)}\}}$$
$$\to (1 - f^{(0)})\pi^{(0)}(A)e^{-\lambda_k b_{0k}} \quad \text{as } \varepsilon \to 0. \tag{3.4.25}$$

Theorem 3.4.7. *Let conditions* $\mathbf{D_{13}}$, $\mathbf{C_3}$, $\mathbf{F_7}$, $\mathbf{D_{17}}$, $\mathbf{C_7}$, $\mathbf{D_{25}}$, $\mathbf{P_9^{(k)}}$, *and* $\mathbf{B_5^{(r)}}$, *for some* $1 \leq r \leq k$, *hold. Then, for any* $A \in \Gamma_0$,

$$\frac{\mathsf{P}\{\xi^{(\varepsilon)}(t^{(\varepsilon)}) \in A, \mu^{(\varepsilon)} > t^{(\varepsilon)}\}}{\exp\{-(a_1\varepsilon + \cdots + a_{r-1}\varepsilon^{r-1})t^{(\varepsilon)}\}}$$
$$\to (1 - f^{(0)})\pi^{(0)}(A)e^{-\lambda_r a_r} \quad \text{as } \varepsilon \to 0. \tag{3.4.26}$$

Theorem 3.4.8. *Let conditions* $\mathbf{D_{13}}$, $\mathbf{C_3}$, $\mathbf{F_7}$, $\mathbf{D_{17}}$, $\mathbf{C_7}$, $\mathbf{D_{25}}$, $\mathbf{P_{10}^{(k)}}$, *and* $\mathbf{B_5^{(r)}}$, *for some* $1 \leq r \leq k$, *hold. Then, for any* $A \in \Gamma_0$,

$$\frac{\mathsf{P}\{\xi^{(\varepsilon)}(t^{(\varepsilon)}) \in A, \mu^{(\varepsilon)} > t^{(\varepsilon)}\}}{\exp\{-(a_h\varepsilon + \cdots + a_{r-1}\varepsilon^{r-1})t^{(\varepsilon)}\}}$$
$$\to (1 - f^{(0)})\pi^{(0)}(A)e^{-\lambda_r a_r} \quad \text{as } \varepsilon \to 0. \tag{3.4.27}$$

Slightly different is the situation for the model of asymptotically improper perturbed regenerative processes ($f^{(\varepsilon)} \to f^{(0)} \geq 0$ as $\varepsilon \to 0$). The mixed ergodic and large deviation asymptotic relation takes the form (3.4.22) for standard regenerative processes, while, for the perturbed regenerative processes with transition period, it takes a similar but different form,

$$\frac{\mathsf{P}\{\xi^{(\varepsilon)}(t^{(\varepsilon)}) \in A, \mu^{(\varepsilon)} > t^{(\varepsilon)}\}}{\exp\{-\rho^{(\varepsilon)}t^{(\varepsilon)}\}} \to \tilde{\phi}^{(0)}(\rho^{(0)})\tilde{\pi}^{(0)}(A) \quad \text{as } \varepsilon \to 0. \tag{3.4.28}$$

Representation (3.4.23) plus an additional balancing condition, $\varepsilon^r t^{(\varepsilon)} \to \lambda_r$ as $\varepsilon \to 0$, allow to transform the asymptotic relation (3.4.28) to an equivalent but more explicit form,

$$\frac{\mathsf{P}\{\xi^{(\varepsilon)}(t^{(\varepsilon)}) \in A, \mu^{(\varepsilon)} > t^{(\varepsilon)}\}}{\exp\{-(\rho^{(0)} + a_h \varepsilon^h + \cdots + a_{r-1}\varepsilon^{r-1})t^{(\varepsilon)}\}}$$
$$\to \tilde{\phi}^{(0)}(\rho^{(0)})\tilde{\pi}^{(0)}(A)e^{-\lambda_r a_r} \quad \text{as } \varepsilon \to 0. \tag{3.4.29}$$

Again, the corresponding conditions imposed on characteristics of the transition period should augmented.

The next two theorems relate to the model of asymptotically improper perturbed regenerative processes with transition period. It should be noted that these theorems cover, in fact, both the cases where $f^{(0)} > 0$ and $f^{(0)} = 0$. In the latter case, these theorems reduce, respectively, to Theorems 3.4.7 and 3.4.8.

Theorem 3.4.9. *Let conditions* $\mathbf{D}_{21}$, $\mathbf{C}_6$, $\mathbf{F}_7$, $\mathbf{C}_9$, $\mathbf{D}_{26}$, $\mathbf{C}_{11}$, $\mathbf{P}_{11}^{(k)}$ *and* $\mathbf{B}_5^{(r)}$, *for some* $h \le r \le k$, *hold. Then, for any* $A \in \Gamma_0$,

$$\frac{\mathsf{P}\{\xi^{(\varepsilon)}(t^{(\varepsilon)}) \in A, \mu^{(\varepsilon)} > t^{(\varepsilon)}\}}{\exp\{-(\rho^{(0)} + a_1 \varepsilon + \cdots + a_{r-1}\varepsilon^{r-1})t^{(\varepsilon)}\}}$$
$$\to \tilde{\phi}^{(0)}\tilde{\pi}^{(0)}(A)e^{-\lambda_r a_r} \quad \text{as } \varepsilon \to 0. \tag{3.4.30}$$

Theorem 3.4.10. *Let conditions* $\mathbf{D}_{21}$, $\mathbf{C}_6$, $\mathbf{F}_7$, $\mathbf{C}_9$, $\mathbf{D}_{26}$, $\mathbf{C}_{11}$, $\mathbf{P}_{12}^{(k)}$ *and* $\mathbf{B}_5^{(r)}$, *for some* $h \le r \le k$, *hold. Then, for any* $A \in \Gamma_0$,

$$\frac{\mathsf{P}\{\xi^{(\varepsilon)}(t^{(\varepsilon)}) \in A, \mu^{(\varepsilon)} > t^{(\varepsilon)}\}}{\exp\{-(\rho^{(0)} + a_h \varepsilon^h + \cdots + a_{r-1}\varepsilon^{r-1})t^{(\varepsilon)}\}}$$
$$\to \tilde{\phi}^{(0)}\tilde{\pi}^{(0)}(A)e^{-\lambda_r a_r} \quad \text{as } \varepsilon \to 0. \tag{3.4.31}$$

3.5   Asymptotic expansions for quasi-stationary distributions

In this section, we give asymptotic expansions for stationary and quasi-stationary distributions of perturbed regenerative processes.

3.5.1   Asymptotic expansions for stationary distributions of perturbed regenerative processes

Let us define, for $r = 0, 1, \ldots$, moment functionals by

$$q_r^{(\varepsilon)}(A) = \int_0^\infty s^r q^{(\varepsilon)}(s, A)m(ds), \quad A \in \Gamma.$$

Let us first consider the simplest case where condition $\mathbf{D}_{14}$ holds, i.e., $f^{(\varepsilon)} = 0$ for every $\varepsilon \leq \varepsilon_0$. As was shown in Lemma 3.1.2, this condition holds if and only if the regenerative stopping time $\mu^{(\varepsilon)} = \infty$ with probability 1 for all $\varepsilon \leq \varepsilon_0$.

In this case, obviously, $F^{(\varepsilon)}(s) = \hat{F}^{(\varepsilon)}(s) = \mathsf{P}\{\tau_1^{(\varepsilon)} \leq s\}$ and $m_1^{(\varepsilon)} = \mathsf{E}\tau_1^{(\varepsilon)}$. Also, $q^{(\varepsilon)}(s, A) = \mathsf{P}\{\xi^{(\varepsilon)}(t) \in A, \tau_1^{(\varepsilon)} > t\}$.

The following formula defines a *stationary distribution* for the regenerative process $\xi^{(\varepsilon)}(t), t \geq 0$:

$$\pi^{(\varepsilon)}(A) = \frac{\int_0^\infty q^{(\varepsilon)}(s, A) m(ds)}{m_1^{(\varepsilon)}}, \quad A \in \Gamma. \tag{3.5.1}$$

Let us assume the following two perturbation conditions:

$\mathbf{P}_{13}^{(k)}$: $m_1^{(\varepsilon)} = m_1^{(0)} + b_{1,1}\varepsilon + \cdots + b_{k,1}\varepsilon^k + o(\varepsilon^k)$, where $m_1^{(0)} \in (0, \infty)$ and $|b_{n,1}| < \infty, n = 1, \ldots, k$;

and

$\mathbf{P}_{14}^{(k)}$: $q_0^{(\varepsilon)}(A) = q_0^{(0)}(A) + c_{1,0}(A)\varepsilon + \cdots + c_{k,0}(A)\varepsilon^k + o(\varepsilon^k)$, where $|c_{n,0}(A)| < \infty$, $n = 1, \ldots, k, A \in \Gamma_0 \subseteq \Gamma$.

Note that $m_1^{(0)}$ and $q_0^{(0)}(A)$ in the conditions above are just some constants that may or may not be the corresponding characteristics of the limit regenerative process. The condition $m_1^{(0)} \in (0, \infty)$ is essential. As far as the functional $q_0^{(0)}(A)$ is concerned, it automatically takes values in the interval $[0, 1]$, since the pre-limit functional $q_0^{(\varepsilon)}(A)$ takes values in this interval.

It is also convenient to define $b_{0,1} = m_1^{(0)}$ and $c_{0,0}(A) = q_0^{(0)}(A)$.

Theorem 3.5.1. *Let conditions $\mathbf{P}_{13}^{(k)}$ and $\mathbf{P}_{14}^{(k)}$ hold. Then, for any $A \in \Gamma_0$, the following asymptotic expansions hold*:

$$\pi^{(\varepsilon)}(A) = \frac{q_0^{(0)}(A) + c_{1,0}(A)\varepsilon + \cdots + c_{k,0}(A)\varepsilon^k + o(\varepsilon^k)}{m_1^{(0)} + b_{1,1}\varepsilon + \cdots + b_{k,1}\varepsilon^k + o(\varepsilon^k)}$$

$$= \pi^{(0)}(A) + f_1(A)\varepsilon + \cdots + f_k(A)\varepsilon^k + o(\varepsilon^k), \tag{3.5.2}$$

where the coefficients $f_n(A)$ are given by the recurrence formulas $f_0(A) = \pi^{(0)}(A) = c_{0,0}(A)/b_{0,1}$ and, for $n = 0, \ldots, k$,

$$f_n(A) = \left(c_{n,0}(A) - \sum_{q=0}^{n-1} b_{n-q,1} f_q(A)\right)/b_{0,1}. \tag{3.5.3}$$

Proof. The theorem directly follows from Theorem 2.3.1 that should be applied to any $A \in \Gamma_0$ with $q^{(\varepsilon)} = q_0^{(\varepsilon)}(A)$, $m_1^{(\varepsilon)} = \mathsf{E}\tau_1^{(\varepsilon)}$ and $\tilde{x}^{(\varepsilon)}(\infty) = \pi^{(\varepsilon)}(A) = q_0^{(\varepsilon)}(A)/m_1^{(\varepsilon)}$. In this case, conditions $\mathbf{P}_{13}^{(k)}$ and $\mathbf{P}_{14}^{(k)}$ reduce, respectively, to conditions $\mathbf{P}_3^{(k)}$ and $\mathbf{P}_4^{(k)}$. Alternatively, one can directly transform the rational expansion (3.5.2) given by formula (3.5.1) and conditions $\mathbf{P}_{13}^{(k)}$ and $\mathbf{P}_{14}^{(k)}$ to a polynomial form. Note that $b_{0,1} \neq 0$. $\square$

3.5.2 Asymptotic expansions for quasi-stationary distributions of perturbed regenerative processes

Let us consider the general case. We define, for $r = 0, 1, \ldots$, the moment functionals

$$\omega^{(\varepsilon)}(\rho, r, A) = \int_0^\infty s^r e^{\rho s} q^{(\varepsilon)}(s, A) m(ds), \quad \rho \in \mathbb{R}_1, \quad A \in \Gamma.$$

It follows from conditions $\mathbf{D}_{21}$, $\mathbf{C}_6$, and $\mathbf{F}_7$ that for any $0 \leq \beta < \delta$ we have $\omega^{(\varepsilon)}(\beta, r, A) < \infty$ with $r = 0, 1, \ldots$ for all ε small enough. Indeed, let $c_r = \sup_{s \geq 0} s^r e^{-(\delta - \beta)s}$. Then, for $r = 0, 1, \ldots$,

$$\omega^{(\varepsilon)}(\beta, r, A) = \int_0^\infty s^r e^{\beta s} q^{(\varepsilon)}(s, A) m(ds)$$

$$\leq \int_0^\infty s^r e^{\beta s} (1 - \hat{F}^{(\varepsilon)}(s))\, ds \leq c_r \int_0^\infty e^{\delta s} (1 - \hat{F}^{(\varepsilon)}(s))\, ds$$

$$\leq c_r \delta^{-1} \Big(1 + \int_0^\infty e^{\delta s} \hat{F}^{(\varepsilon)}(ds)\Big) < \infty. \tag{3.5.4}$$

We now formulate a perturbation condition for the forcing functions $q^{(\varepsilon)}(t, A)$:

$\mathbf{P}_{15}^{(k)}$: $\omega^{(\varepsilon)}(\rho^{(0)}, r, A) = \omega^{(0)}(\rho^{(0)}, r, A) + c_{1,r}(A)\varepsilon + \cdots + c_{k-r,r}(A)\varepsilon^{k-r} + o(\varepsilon^{k-r})$
for $r = 0, \ldots, k$, where $|c_{n,r}| < \infty$, $n = 1, \ldots, k-r$, $r = 0, \ldots, k$, $A \in \Gamma_0$.

It is convenient to define $c_{0,r}(A) = \omega^{(0)}(\rho^{(0)}, r, A)$, $r = 0, 1, \ldots$.

It is useful to note that in the case where $f^{(0)} = 0$, the characteristic root is $\rho^{(0)} = 0$ and, therefore, the mixed moment functionals $\omega^{(\varepsilon)}(0, r, A) = q_r^{(\varepsilon)}(A) = \int_0^\infty s^r q^{(\varepsilon)}(s, A) m(ds)$ and $\phi^{(\varepsilon)}(0, r) = m_r^{(\varepsilon)} = \int_0^\infty s^r F^{(\varepsilon)}(ds)$ are the usual moment functionals.

In this case, perturbation condition $\mathbf{P}_{11}^{(k)}$ reduces to the perturbation condition $\mathbf{P}_9^{k}$.

The perturbation condition $\mathbf{P}_{15}^{(k)}$ also reduces to a simpler form:

$\mathbf{P}_{16}^{(k)}$: $q_r^{(\varepsilon)}(A) = q_r^{(0)}(A) + c_{1,r}(A)\varepsilon + \cdots + c_{k-r,r}(A)\varepsilon^{k-r} + o(\varepsilon^{k-r})$ for $r = 0, \ldots, k$,
where $|c_{n,r}(A)| < \infty$, $n = 1, \ldots, k-r$, $r = 0, \ldots, k$, $A \in \Gamma_0$.

Recall that

$$\tilde{\pi}^{(\varepsilon)}(A) = \frac{\omega^{(\varepsilon)}(\rho^{(\varepsilon)}, 0, A)}{\phi^{(\varepsilon)}(\rho^{(\varepsilon)}, 1)}, \quad A \in \Gamma. \tag{3.5.5}$$

Theorem 3.5.2. *Let conditions* $\mathbf{D}_{21}$, $\mathbf{C}_6$, $\mathbf{F}_7$, $\mathbf{P}_{11}^{(k+1)}$, *and* $\mathbf{P}_{15}^{(k)}$ *hold. Then, for any* $A \in \Gamma_0$, *the following asymptotic expansions hold:*

$$\tilde{\pi}^{(\varepsilon)}(A) = \frac{\omega^{(0)}(\rho^{(0)}, 0, A) + f_1'(A)\varepsilon + \cdots + f_k'(A)\varepsilon^k + o(\varepsilon^k)}{\phi^{(0)}(\rho^{(0)}, 1) + f_1''\varepsilon + \cdots + f_k''\varepsilon^k + o(\varepsilon^k)}$$

$$= \tilde{\pi}^{(0)}(A) + f_1(A)\varepsilon + \cdots + f_k(A)\varepsilon^k + o(\varepsilon^k), \tag{3.5.6}$$

where the coefficients $f_n'(A)$, f_n'' *are given by the formulas* $f_0'(A) = \omega^{(0)}(\rho^{(0)}, 0, A) = c_{0,0}(A)$, $f_1'(A) = c_{1,0}(A) + c_{0,1}(A)a_1$, $f_0'' = \phi^{(0)}(\rho^{(0)}, 1) = b_{0,1}$, $f_1'' = b_{1,1} + b_{0,2}a_1$, *and, for* $n = 0, \ldots, k$,

$$f_n'(A) = c_{n,0}(A) + \sum_{q=1}^{n} c_{n-q,1}(A)a_q$$

$$+ \sum_{2 \le m \le n} \sum_{q=m}^{n} c_{n-q,m}(A) \cdot \sum_{n_1,\ldots,n_{q-1} \in D_{m,q}} \prod_{p=1}^{q-1} a_p^{n_p}/n_p!, \tag{3.5.7}$$

and

$$f_n'' = b_{n1} + \sum_{q=1}^{n} b_{n-q,2}a_q$$

$$+ \sum_{2 \le m \le n} \sum_{q=m}^{n} b_{n-q,m+1} \cdot \sum_{n_1,\ldots,n_{q-1} \in D_{m,q}} \prod_{p=1}^{q-1} a_p^{n_p}/n_p!, \tag{3.5.8}$$

and the coefficients $f_n(A)$ *are given by the recurrence formulas* $f_0(A) = \tilde{\pi}^{(0)}(A) = f_0'(A)/f_0''$ *and, for* $n = 0, \ldots, k$,

$$f_n(A) = \left(f_n'(A) - \sum_{q=0}^{n-1} f_{n-q}'' f_q(A) \right)/f_0''. \tag{3.5.9}$$

Proof. The theorem directly follows from Theorem 2.3.2 that should be applied, for any $A \in \Gamma_0$, to a model in which $q^{(\varepsilon)}(s) = q^{(\varepsilon)}(s, A)$, $F^{(\varepsilon)}(s) = \mathsf{P}\{\tau_1^{(\varepsilon)} \le s, \mu^{(\varepsilon)} > \tau_1^{(\varepsilon)}\}$ and $\tilde{x}^{(\varepsilon)}(\infty) = \tilde{\pi}^{(\varepsilon)}(A) = \omega^{(\varepsilon)}(\rho^{(\varepsilon)}, 0, A)/\phi^{(\varepsilon)}(\rho^{(\varepsilon)}, 1)$.

In this case, as was shown in the proof of Theorem 3.4.4, conditions $\mathbf{D}_{21}$, $\mathbf{C}_6$ and $\mathbf{F}_7$ imply that conditions $\mathbf{D}_{13}$, $\mathbf{C}_2$ and $\mathbf{F}_6$ hold. Also, conditions $\mathbf{P}_{11}^{(k+1)}$ and $\mathbf{P}_{15}^{(k)}$ coincide with conditions $\mathbf{P}_3^{(k+1)}$ and $\mathbf{P}_7^{(k)}$, respectively. $\qquad\square$

Recall that

$$\tilde{\tilde{\pi}}^{(\varepsilon)}(A) = \tilde{\phi}^{(\varepsilon)}(\rho^{(\varepsilon)}, 0)\tilde{\pi}^{(\varepsilon)}(A), \quad A \in \Gamma, \tag{3.5.10}$$

where

$$\tilde{\phi}^{(\varepsilon)}(\rho, r) = \int_0^\infty s^r e^{\rho s} \tilde{F}^{(\varepsilon)}(ds), \quad \rho \in \mathbb{R}_1, \quad r = 0, 1, \dots .$$

Note that, by the definition, $\tilde{\phi}^{(\varepsilon)}(\rho, 0) = \tilde{\phi}^{(\varepsilon)}(\rho)$.

Condition $\mathbf{C_9}$ implies that, for any $0 < \beta < \rho^{(0)} + \gamma'$, the moment generating functions satisfy $\tilde{\phi}^{(\varepsilon)}(\beta, r) < \infty, r = 0, 1, \dots$, for all ε small enough.

Indeed, condition $\mathbf{C_9}$ implies that $\tilde{\phi}^{(\varepsilon)}(\rho^{(0)} + \gamma', 0) < \infty$ for ε small enough, say $\varepsilon \leq \varepsilon_1$. Let us denote $c_r = \sup_{s \geq 0} s^r e^{-(\rho^{(0)} + \gamma' - \beta)s} < \infty$. Then, we have for $\varepsilon \leq \varepsilon_1$ and $r = 0, 1, \dots$ that

$$\tilde{\phi}^{(\varepsilon)}(\beta, r) = \int_0^\infty s^r e^{\beta s} \tilde{F}^{(\varepsilon)}(ds)$$

$$\leq c_r \int_0^\infty e^{(\rho^{(0)} + \gamma')s} \tilde{F}^{(\varepsilon)}(ds) = c_r \tilde{\phi}^{(\varepsilon)}(\rho^{(0)} + \gamma', r) < \infty.$$

Note that $\tilde{\phi}^{(\varepsilon)}(\rho, r)$ are non-negative and non-decreasing functions of $\rho \geq 0$. Also, $\rho^{(\varepsilon)} < \rho^{(0)} + \gamma$, since $\rho^{(\varepsilon)} \to \rho^{(0)}$ as $\varepsilon \to 0$. Thus, $\tilde{\phi}^{(\varepsilon)}(\rho^{(0)}, r), \tilde{\phi}^{(\varepsilon)}(\rho^{(\varepsilon)}, r) < \infty$, $r = 0, 1, \dots$, for all ε small enough, say $\varepsilon \leq \varepsilon_2$.

Let us now formulate perturbation conditions for mixed power-exponential moment functionals for forcing functions:

$\mathbf{P_{17}^{(k)}}$: $\tilde{\phi}^{(\varepsilon)}(\rho^{(0)}, r) = \tilde{\phi}^{(0)}(\rho^{(0)}, r) + d_{1,r}\varepsilon + \cdots + d_{k-r,r}\varepsilon^{k-r} + o(\varepsilon^{k-r})$ for $r = 0, \dots, k$, where $|d_{n,r}| < \infty, n = 1, \dots, k - r, r = 0, \dots, k$.

It is convenient to define $d_{0,r} = \tilde{\phi}^{(0)}(\rho^{(0)}, r), r = 0, 1, \dots$.

Theorem 3.5.3. *Let conditions* $\mathbf{D_{21}}$, $\mathbf{C_6}$, $\mathbf{F_7}$, $\mathbf{P_{11}^{(k+1)}}$, $\mathbf{P_{15}^{(k)}}$, $\mathbf{D_{20}}$, $\mathbf{C_9}$, *and* $\mathbf{P_{17}^{(k)}}$ *hold. Then, for any* $A \in \Gamma_0$, *the following asymptotic expansions hold:*

$$\tilde{\tilde{\pi}}^{(\varepsilon)}(A) = \tilde{\tilde{\pi}}^{(0)}(A) + h_1(A)\varepsilon + \cdots + h_k(A)\varepsilon^k + o(\varepsilon^k), \tag{3.5.11}$$

where the coefficients $h_n(A)$ *are given by the formulas* $h_0(A) = \tilde{\tilde{\pi}}^{(0)}(\rho^{(0)}, 0) = f_0''' f_0(A), h_1(A) = f_0''' f_1(A) + f_1''' f_0(A)$, *and, for* $n = 0, \dots, k$,

$$h_n(A) = \sum_{q=0}^{n} f_q''' f_{n-q}(A) \tag{3.5.12}$$

and

$$f_n''' = d_{n,0} + \sum_{q=1}^{n} d_{n-q,1} a_q$$

$$+ \sum_{2 \leq m \leq n} \sum_{q=m}^{n} d_{n-q,m} \cdot \sum_{n_1, \dots, n_{q-1} \in D_{m,q}} \prod_{p=1}^{q-1} a_p^{n_p} / n_p!. \tag{3.5.13}$$

Proof. We can write the asymptotic expansion of the first factor $\tilde{\phi}^{(\varepsilon)}(\rho^{(\varepsilon)}, 0)$ in the product expression for $\tilde{\tilde{\pi}}^{(\varepsilon)}(A)$ given by (3.5.10) in powers of the difference $\Delta^{(\varepsilon)} = \rho^{(\varepsilon)} - \rho^{(0)}$. Recall that $\Delta^{(\varepsilon)} \to 0$. Hence, for any $\rho^{(0)} < \beta < \rho^{(0)} + \gamma'$, the sum $\rho^{(0)} + |\Delta^{(\varepsilon)}| \leq \beta$ for ε small enough, say $\varepsilon \leq \varepsilon_3$.

Condition $\mathbf{C_9}$ implies that, for any $0 < \beta < \rho^{(0)} + \gamma'$, the moment generating functions satisfy $\tilde{\phi}^{(\varepsilon)}(\beta, r) < \infty, r = 0, 1, \ldots$, for ε small enough, say $\varepsilon \leq \varepsilon_4$.

By condition $\mathbf{C_9}$,

$$\varlimsup_{0 \leq \varepsilon \to 0} \int_0^\infty s^{k+1} e^{\beta s} \tilde{F}^{(\varepsilon)}(ds) \leq \varlimsup_{0 \leq \varepsilon \to 0} c_{k+1} \int_0^\infty e^{\delta s} \tilde{F}^{(\varepsilon)}(ds) < \infty, \qquad (3.5.14)$$

where $c_{k+1} = \sup_{s \geq 0} s^{k+1} e^{-(\delta - \beta)s} < \infty$.

It follows from relation (3.5.14) that there exists $\varepsilon_5 > 0$ such that, for $\varepsilon \leq \varepsilon_5$,

$$\tilde{M}_{k+1} = \frac{1}{(k+1)!} \sup_{\varepsilon \leq \varepsilon_5} \int_0^\infty s^{k+1} e^{s\beta} \tilde{F}^{(\varepsilon)}(ds) < \infty. \qquad (3.5.15)$$

Let us once more use Taylor expansion (2.2.8) for the function e^s,

$$e^{s\rho^{(\varepsilon)}} = e^{s\rho^{(0)}} \left(1 + s\Delta^{(\varepsilon)}/1! + \cdots + s^k (\Delta^{(\varepsilon)})^k / k! \right.$$

$$\left. + s^{k+1} (\Delta^{(\varepsilon)})^{k+1} e^{s|\Delta^{(\varepsilon)}|} \theta_{k+1}^{(\varepsilon)}(s)/(k+1)!\right), \qquad (3.5.16)$$

where $\theta_{k+1}^{(\varepsilon)}(s), s \geq 0$ is a continuous function such that $0 \leq \theta_{k+1}^{(\varepsilon)}(s) \leq 1, s \geq 0$.

Define

$$\varepsilon_6 = \min(\varepsilon_3, \varepsilon_4, \varepsilon_5). \qquad (3.5.17)$$

Substituting the Taylor expansion (3.5.16) into the integrals that define $\tilde{\phi}^{(\varepsilon)}(\rho^{(\varepsilon)}, 0)$, we obtain for $\varepsilon \leq \varepsilon_6$ that

$$\tilde{\phi}^{(\varepsilon)}(\rho^{(\varepsilon)}, 0) = \tilde{\phi}^{(\varepsilon)}(\rho^{(0)}, 0) + \tilde{\phi}^{(\varepsilon)}(\rho^{(0)}, 1)\Delta^{(\varepsilon)} + \cdots$$

$$+ \tilde{\phi}^{(\varepsilon)}(\rho^{(0)}, k)(\Delta^{(\varepsilon)})^k / k! + (\Delta^{(\varepsilon)})^{k+1} \tilde{M}_{k+1} \tilde{\theta}_{k+1}^{(\varepsilon)}, \qquad (3.5.18)$$

where

$$\tilde{\theta}_{k+1}^{(\varepsilon)} = \frac{\int_0^\infty s^{k+1} e^{s\rho^{(\varepsilon)}} \theta_{k+1}^{(\varepsilon)}(s) \tilde{F}^{(\varepsilon)}(ds)}{\sup_{\varepsilon \leq \varepsilon_3} \int_0^\infty s^{k+1} e^{s\beta} \tilde{F}^{(\varepsilon)}(ds)} \in [0, 1].$$

Let us use the asymptotic expansion $\rho^{(\varepsilon)} = \rho^{(0)} = a_1 \varepsilon + \cdots + a_k \varepsilon^k + o(\varepsilon^k)$ given in Theorem 3.4.4. This expansion can be rewritten as an expansion for $\Delta^{(\varepsilon)} = \rho^{(\varepsilon)} - \rho^{(0)}$. The functions $\tilde{\phi}^{(\varepsilon)}(\rho^{(0)}, n)$ can be represented in the form of asymptotic expansions, due to condition $\mathbf{P_{17}^{(k)}}$. Substituting these expansions into (3.5.18) we obtain

$$\tilde{\phi}^{(\varepsilon)}(\rho^{(\varepsilon)}, 0) = (d_{0,0} + \cdots + d_{k,0}\varepsilon^k + o(\varepsilon^k))$$

$$+ (d_{0,1} + \cdots + d_{k-1,1}\varepsilon^{k-1} + o(\varepsilon^{k-1}))$$

$$\times (a_1 \varepsilon + \cdots + a_k \varepsilon^k + o(\varepsilon^k))/1! + \cdots$$

$$+ (d_{0,k} + o(1))(a_1 \varepsilon + \cdots + a_k \varepsilon^k + o(\varepsilon^k))^k / k! + o(\varepsilon^k). \qquad (3.5.19)$$

By grouping the coefficients at powers of ε in this expansion, one can transform it to the forms described in (3.5.13),

$$\tilde{\phi}^{(\varepsilon)}(\rho^{(\varepsilon)}, 0) = f_0''' + \cdots + f_k''' \varepsilon^k + o(\varepsilon^k). \tag{3.5.20}$$

The asymptotic expansion (3.5.6) for $\tilde{\pi}^{(\varepsilon)}(A)$ given in Theorem 3.5.2 has the following form:

$$\tilde{\pi}^{(\varepsilon)}(A) = f_0(A) + f_1(A)\varepsilon + \cdots + f_k(A)\varepsilon^k + o(\varepsilon^k). \tag{3.5.21}$$

Formulas and (3.5.20) and (3.5.21), in an obvious way, yield the asymptotic expansion (3.5.12) for the product $\tilde{\tilde{\pi}}^{(\varepsilon)}(A) = \tilde{\phi}^{(\varepsilon)}(\rho^{(\varepsilon)}, 0)\tilde{\pi}^{(\varepsilon)}(A)$. The corresponding coefficients can be found by grouping the coefficients of ε^n in the equation

$$h_0 + \cdots + h_k \varepsilon^k + o(\varepsilon^k)$$
$$= (f_0''' + \cdots + f_k''' \varepsilon^k + o(\varepsilon^k))(f_0(A) + \cdots + f_k(A)\varepsilon^k + o(\varepsilon^k)). \tag{3.5.22}$$

The proof is complete. $\square$

As a corollary, we give asymptotic expansions for the quasi-stationary distribution $\pi^{(\varepsilon)}(A)$ given by formula (3.3.11).

Recall that

$$\pi^{(\varepsilon)}(A) = \frac{\tilde{\pi}^{(\varepsilon)}(A)}{\tilde{\pi}^{(\varepsilon)}(X)}, \quad A \in \Gamma. \tag{3.5.23}$$

Note that $\tilde{\pi}^{(\varepsilon)}(A) = \omega^{(\varepsilon)}(\rho^{(\varepsilon)}, 0, A)/\phi^{(\varepsilon)}(\rho^{(\varepsilon)}, 1)$ and, therefore,

$$\tilde{\pi}^{(\varepsilon)}(X) = \int_0^\infty e^{\rho^{(\varepsilon)}s}(1 - \hat{F}^{(\varepsilon)}(s))\, ds / \int_0^\infty s e^{\rho^{(\varepsilon)}s} F^{(\varepsilon)}(ds).$$

Due to condition $\mathbf{D_{21}}$, both distribution functions $\hat{F}^{(\varepsilon)}(s)$ and $F^{(\varepsilon)}(s)$ are not concentrated in zero for ε small enough. Thus, $\tilde{\pi}^{(\varepsilon)}(X) \neq 0$ for such ε.

Theorem 3.5.4. *Let conditions* $\mathbf{D_{21}}$, $\mathbf{C_6}$, $\mathbf{F_7}$, $\mathbf{P_{11}^{(k+1)}}$, *and* $\mathbf{P_{15}^{(k)}}$ *hold. Then for any* $A \in \Gamma_0$, *the following asymptotic expansions hold:*

$$\pi^{(\varepsilon)}(A) = \frac{\tilde{\pi}^{(0)}(A) + f_1(A)\varepsilon + \cdots + f_k(A)\varepsilon^k + o(\varepsilon^k)}{\tilde{\pi}^{(0)}(X) + f_1(X)\varepsilon + \cdots + f_k(X)\varepsilon^k + o(\varepsilon^k)}$$
$$= \pi^{(0)}(A) + g_1(A)\varepsilon + \cdots + g_k(A)\varepsilon^k + o(\varepsilon^k), \tag{3.5.24}$$

where the coefficients $g_n(A)$ *are given by the recurrence formulas* $g_0(A) = \pi^{(0)}(A) = f_0(A)/f_0(X)$, $f_0(A) = \tilde{\pi}^{(0)}(A)$, $f_0(X) = \tilde{\pi}^{(0)}(X)$ *and, for* $n = 0, \ldots, k$,

$$g_n(A) = \left(f_n(A) - \sum_{q=0}^{n-1} f_{n-q}(X) g_q(A) \right) / f_0(X). \tag{3.5.25}$$

Proof. In this case according formula (3.3.11), $\pi^{(\varepsilon)}(A) = \tilde{\pi}^{(\varepsilon)}(A)/\tilde{\pi}^{(\varepsilon)}(X)$. It follows from this formula that we can use the rational expansion (3.5.24) that can be transformed into a polynomial form. Note that $\tilde{\pi}^{(0)}(X) \neq 0$. We have

$$\left(\pi^{(0)}(A) + \sum_{r=1}^{k} g_r(A)\varepsilon^r + o(\varepsilon^k)\right)\left(\tilde{\pi}^{(0)}(X) + \sum_{r=1}^{k} f_r(X)\varepsilon^r + o(\varepsilon^k)\right)$$

$$= \tilde{\pi}^{(0)}(A) + \sum_{r=1}^{k} f_r(A)\varepsilon^r + o(\varepsilon^k). \tag{3.5.26}$$

Relations (3.5.25) can be obtained by grouping the coefficients of ε^n in equation (3.5.26). $\qquad\square$

Formulas (3.5.5) and (3.5.23) also give the following representation for the *quasi-stationary distribution* $\pi^{(\varepsilon)}(A)$, alternative to (3.5.23):

$$\pi^{(\varepsilon)}(A) = \frac{\omega^{(\varepsilon)}(\rho^{(\varepsilon)}, 0, A)}{\omega^{(\varepsilon)}(\rho^{(\varepsilon)}, 0, X)}, \quad A \in \Gamma. \tag{3.5.27}$$

Representation (3.5.27) does not contain the functions $\phi^{(\varepsilon)}(\rho^{(\varepsilon)}, 1)$. Using this, we can write an asymptotic expansion for quasi-stationary distributions, which is alternative to (3.5.24) and does not involve condition $\mathbf{P}_{11}^{(k+1)}$.

Theorem 3.5.5. *Let conditions* $\mathbf{D}_{21}$, $\mathbf{C}_6$, $\mathbf{F}_7$, *and* $\mathbf{P}_{15}^{(k)}$ *hold. Then, the following asymptotic expansions hold for any* $A \in \Gamma_0$:

$$\pi^{(\varepsilon)}(A) = \frac{\omega^{(0)}(\rho^{(0)}, 0, A) + f_1'(A)\varepsilon + \cdots + f_k'(A)\varepsilon^k + o(\varepsilon^k)}{\omega^{(0)}(\rho^{(0)}, 0, X) + f_1'(X)\varepsilon + \cdots + f_k'(X)\varepsilon^k + o(\varepsilon^k)}$$

$$= \pi^{(0)}(A) + g_1(A)\varepsilon + \cdots + g_k(A)\varepsilon^k + o(\varepsilon^k), \tag{3.5.28}$$

where the coefficients $f_n'(A)$ *are given by the formulas* $f_0'(A) = \omega^{(0)}(\rho^{(0)}, 0, A) = c_{0,0}(A)$, $f_1'(A) = c_{1,0}(A) + c_{0,1}(A)a_1$, *and, for* $n = 0, \ldots, k$,

$$f_n'(A) = c_{n,0}(A) + \sum_{q=1}^{n} c_{n-q,1}(A)a_q$$

$$+ \sum_{2 \leq m \leq n} \sum_{q=m}^{n} c_{n-q,m}(A) \cdot \sum_{n_1,\ldots,n_{q-1} \in D_{m,q}} \prod_{p=1}^{q-1} a_p^{n_p}/n_p!, \tag{3.5.29}$$

and the coefficients $g_n(A)$ *are given by the recurrence formulas* $g_0(A) = \pi^{(0)}(A) = f_0'(A)/f_0'(X)$ *and, for* $n = 0, \ldots, k$,

$$g_n(A) = \left(f_n'(A) - \sum_{q=0}^{n-1} f_{n-q}'(X)g_q(A)\right)/f_0'(X). \tag{3.5.30}$$

Proof. In this case according formula (3.5.27), $\pi^{(\varepsilon)}(A) = \omega^{(\varepsilon)}(\rho^{(\varepsilon)}, 0, A)/\omega^{(\varepsilon)}(\rho^{(\varepsilon)}, 0, X)$. This shows that we can write the rational expansion (3.5.28) that can be transformed to a polynomial form. Note that $\omega^{(0)}(\rho^{(0)}, 0, X) \neq 0$. We have

$$\left(\pi^{(0)}(A) + \sum_{r=1}^{k} g_r(A)\varepsilon^r + o(\varepsilon^k)\right)\left(\omega^{(0)}(\rho^{(0)}, 0, X) + \sum_{r=1}^{k} f_r(X)\varepsilon^r + o(\varepsilon^k)\right)$$

$$= \omega^{(0)}(\rho^{(0)}, 0, A) + \sum_{r=1}^{k} f_r'(A)\varepsilon^r + o(\varepsilon^k). \tag{3.5.31}$$

Relations (3.5.30) can be obtained by grouping the coefficients of ε^n in equation (3.5.31). $\qquad\square$

It is useful to note that despite the difference in the algorithms, the coefficients $g_n(A)$, $n = 1, \ldots, k$, in both asymptotic expansions (3.5.24) and (3.5.28) coincide.

Chapter 4

Perturbed semi-Markov processes

Chapter 4 contains results concerning mixed ergodic theorems (for semi-Markov processes) and limit/large deviation theorems (for absorption times) for nonlinearly perturbed semi-Markov processes. We consider semi-Markov processes with a finite set of states $X = \{0, 1, \dots, N\}$ and the absorption state 0. The first hitting time to this state is the absorption time for the corresponding semi-Markov process.

The asymptotic results mentioned above are given in the form of an asymptotics for joint distributions of positions of semi-Markov processes and absorption times. They are obtained by applying the corresponding results to regenerative processes given in Chapter 3. It can be done by using the fact that a semi-Markov process can be considered as a regenerative process with regeneration times which are subsequent return moments to any fixed state $i \neq 0$. The first hitting time to the absorption state 0 is, in this case, the regenerative stopping time.

We consider in details not only the generic case where the limiting semi-Markov process has one communication class of recurrent-without absorption states, but also the case, where the limiting semi-Markov process has one communication class of recurrent-without absorption states and, additionally, the class of non-recurrent-without absorption states. The latter model covers a significant part of applications.

We also formulate and prove a number of auxiliary propositions dealing with convergence of hitting probabilities, distributions, expectations, and exponential moments. We give propositions on so-called solidarity properties of cyclic absorption probabilities, and moment generating functions for hitting times for perturbed semi-Markov processes. These propositions, as we think, are useful by themselves.

We consider a semi-Markov process $\eta^{(\varepsilon)}(t)$, $t \geq 0$, with a phase space $X = \{0, 1, \dots, N\}$ and transition probabilities $Q_{ij}^{(\varepsilon)}(u)$. The semi-Markov process $\eta^{(\varepsilon)}(t)$, $t \geq 0$, is assumed to depend on a small perturbation parameter $\varepsilon \geq 0$. The processes $\eta^{(\varepsilon)}(t)$, $t \geq 0$, for $\varepsilon > 0$ are considered as perturbations of the process $\eta^{(0)}(t)$, $t \geq 0$, and therefore we assume some weak continuity conditions satisfied by transition quantities, namely, by the moment functionals on transition probabilities if regarded as as functions of ε at the point $\varepsilon = 0$.

Our studies are concerned with the random functional $\mu_0^{(\varepsilon)}$ which is the first hitting time for the process $\eta^{(\varepsilon)}(t)$, $t \geq 0$, to hit the absorption state 0. In applications, the hitting times $\mu_0^{(\varepsilon)}$ are often interpreted as transition times for different stochastic systems described by semi-Markov processes; these are occupation times or waiting times in queuing systems, lifetimes in reliability models, extinction times in population dynamic models, etc.

The object of our study is the joint distributions $P_i\{\eta^{(\varepsilon)}(t) = j, \mu_0^{(\varepsilon)} > t\}$ and their asymptotic behaviour for $t \to \infty$ and $\varepsilon \to 0$.

The mixed ergodic and limit theorems mentioned above describe the asymptotic behaviour of these joint distributions in the model with the one-step probability of absorption, $f^{(\varepsilon)}$, which is averaged by the stationary distribution of the corresponding limiting embedded Markov chain, tending to 0 as $\varepsilon \to 0$, and satisfying the condition that balances the rates for $f^{(\varepsilon)} \to 0$ and time $t^{(\varepsilon)} \to \infty$ as $\varepsilon \to 0$,

$$f^{(\varepsilon)} t^{(\varepsilon)} \to \lambda \quad \text{as} \quad \varepsilon \to 0, \tag{4.0.1}$$

while the corresponding mixed ergodic and limit theorems take the form of the following asymptotic relation for the generic case with one class of communication non-absorption states for the limiting semi-Markov process:

$$P_i\{\eta^{(\varepsilon)}(t^{(\varepsilon)}) = j, \mu_0^{(\varepsilon)} > t^{(\varepsilon)}\} \to \pi_j^{(0)} e^{-\lambda/m^{(0)}} \quad \text{as} \quad \varepsilon \to 0, \quad i, j \neq 0, \tag{4.0.2}$$

where $\pi_j^{(0)}$, $j \neq 0$, are stationary probabilities for the limiting semi-Markov process $\eta^{(\varepsilon)}(t)$ and $m^{(0)}$ is the mean value of the sojourn time averaged by the stationary distribution of the corresponding limiting embedded Markov chain.

The asymptotic relation (4.0.2) means that the position in the semi-Markov process $\eta^{(\varepsilon)}(t^{(\varepsilon)})$ and the normalised absorption time $f^{(\varepsilon)} \mu_0^{(\varepsilon)}$ are asymptotically independent and have, in the limit, a stationary distribution and an exponential distribution, respectively. This can be interpreted as a mixed ergodic theorem (for semi-Markov processes) and a limit theorem (for absorption times). In the literature, the term *transient* is also used in connection with such a kind of asymptotic relations.

For the asymptotic relation (4.0.2) to hold, we need to assume only the conditions of asymptotic ergodicity imposed on the semi-Markov processes $\eta^{(\varepsilon)}(t)$. These conditions, are based on the first moments for the distribution functions $_0G_{jj}^{(\varepsilon)}(t)$ of the return-without-absorption times for some state $j \neq 0$.

A much stronger asymptotic relation can be obtained under an additional Cramér type condition imposed on the exponential moments for distribution functions $_0G_{jj}^{(\varepsilon)}(t)$. In this case, the asymptotic relation (4.0.2) can be replaced with the following much stronger asymptotic relation:

$$\frac{P_i\{\eta^{(\varepsilon)}(t^{(\varepsilon)}) = j, \mu_0^{(\varepsilon)} > t^{(\varepsilon)}\}}{\exp\{-\rho^{(\varepsilon)} t^{(\varepsilon)}\}} \to \tilde{\tilde{\pi}}_{ij}^{(0)}(\rho^{(0)}) \quad \text{as} \quad \varepsilon \to 0, \quad i, j \neq 0, \tag{4.0.3}$$

where $\rho^{(\varepsilon)}$ is given as a unique solution of the characteristic equation

$$\int_0^\infty e^{s\rho^{(\varepsilon)}} {}_0G_{jj}^{(\varepsilon)}(ds) = 1, \tag{4.0.4}$$

and the limiting coefficients $\tilde{\tilde{\pi}}_{ir}^{(0)}(\rho^{(0)})$ are explicitly defined via moment generating functions for the distributions of the first hitting times (without absorption) ${}_0G_{ij}^{(\varepsilon)}(t)$ and ${}_0G_{jj}^{(\varepsilon)}(t)$ at the point $\rho^{(0)}$.

There is a principal difference between the asymptotic relations (4.0.2) and (4.0.3).

Firstly, there is no restriction on the rate with which $t^{(\varepsilon)} \to \infty$ as $\varepsilon \to 0$ in the asymptotic relation (4.0.4). This relation allows for the case where $f^{(\varepsilon)}t^{(\varepsilon)} \to \infty$, so that it can be interpreted as a mixed ergodic theorem (for semi-Markov processes) and a large deviation theorem (for absorption times).

Secondly, the latter relation covers not only the case where **(a)** $f^{(\varepsilon)} \to f^{(0)} = 0$ but also **(b)** $f^{(\varepsilon)} \to f^{(0)} > 0$ as $\varepsilon \to 0$.

The first alternative **(a)** means that an absorption in 0 is impossible for the limiting process $\eta^{(0)}(t)$, $t \geq 0$, and it holds if and only if $\rho^{(0)} = 0$. In such a case, the absorption times $\mu_0^{(\varepsilon)} \xrightarrow{P} \infty$ as $\varepsilon \to 0$. The asymptotic relation (4.0.2) describes pseudo-stationary phenomena for perturbed semi-Markov processes.

The second alternative **(b)** means that an absorption in 0 is possible for the limiting process $\eta^{(0)}(t)$, $t \geq 0$, and it holds if and only if $\rho^{(0)} > 0$. In the latter case, the absorption times $\mu_0^{(\varepsilon)}$ are stochastically bounded as $\varepsilon \to 0$. The asymptotic relation (4.0.3) describes quasi-stationary phenomena for perturbed semi-Markov processes.

The sequential return times at any recurrent state $i \neq 0$ are regeneration times for a semi-Markov process. The proofs of the asymptotic relations (4.0.2) and (4.0.3) is based on a use of this regeneration property for semi-Markov processes.

The proofs mentioned above include, as parts, the so-called solidarity propositions. They show that the corresponding conditions for convergence formulated in terms of cyclic characteristics of return times, and the form of the corresponding asymptotic relations are invariant with respect to the choice of the recurrent state used to construct the regeneration cycles. In particular, we show that the characteristic root of $\rho^{(\varepsilon)}$ does not depend on the choice of a recurrent state $i \neq 0$.

Another important part of these proofs clarifies the role of the initial state in the corresponding asymptotic relations.

In the pseudo-stationary case **(a)**, the limiting coefficients $\tilde{\tilde{\pi}}_{ij}^{(0)}(\rho^{(0)})$ in the asymptotic relation (4.0.3) do not depend on the initial state $i \neq 0$. Moreover, $\tilde{\tilde{\pi}}_{ij}^{(0)}(\rho^{(0)}) = \pi_j^{(0)}$, $i \neq 0$, i.e., these coefficients, as expected, coincide with the corresponding stationary probabilities of the limiting semi-Markov process $\eta^{(0)}(t)$.

The situation is not so simple in the quasi-stationary case **(b)**. In this case, under some natural conditions, there exists a so-called quasi-stationary distribution for the semi-Markov process $\eta^{(\varepsilon)}(t)$, $t \geq 0$, and it is given by the formula

$$\mathsf{P}_i\{\eta^{(\varepsilon)}(t) = j/\mu_0^{(\varepsilon)} > t\} \to \pi_j^{(\varepsilon)}(\rho^{(\varepsilon)}) \text{ as } t \to \infty, \quad i, j \neq 0, \tag{4.0.5}$$

where

$$\pi_j^{(\varepsilon)}(\rho^{(\varepsilon)}) = \frac{\tilde{\tilde{\pi}}_{kj}^{(0)}(\rho^{(0)})}{\sum_{k \neq 0} \tilde{\tilde{\pi}}_{kr}^{(0)}(\rho^{(0)})}, \quad j \neq 0.$$

In this case, the limiting coefficients $\tilde{\tilde{\pi}}_{ij}^{(0)}(\rho^{(0)})$ may depend on the initial state $i \neq 0$, but the quasi-stationary probabilities $\pi_j^{(\varepsilon)}(\rho^{(\varepsilon)})$, $j \neq 0$, do not depend on the choice of the state $k \neq 0$ in the formula above.

We also give conditions for convergence of the quasi-stationary distributions of nonlinearly perturbed semi-Markov processes,

$$\pi_j^{(\varepsilon)}(\rho^{(\varepsilon)}) \to \pi_j^{(0)}(\rho^{(0)}) \text{ as } \varepsilon \to 0, \quad j \neq 0. \tag{4.0.6}$$

Theorems 4.4.1 and 4.4.2 are mixed ergodic and limit theorems. Theorems 4.6.1 and 4.6.2 are mixed ergodic and large deviation theorems which cover the case of the pseudo-stationary asymptotics if the absorption probabilities are asymptotically small. Theorems 4.6.3, 4.6.5, 4.6.8, 4.6.7, and 4.6.11 cover the case of the so-called quasi-stationary asymptotics if the absorption probabilities can be asymptotically separated from zero.

Theorems 4.4.5 and 4.6.6 give conditions for convergence of stationary and quasi-stationary distributions for perturbed semi-Markov processes.

The mixed ergodic and large deviation theorems obtained in Chapter 4, together with the corresponding solidarity propositions for cyclic characteristics deduced for perturbed semi-Markov processes, create a basis for getting exponential expansions in the mixed ergodic and large deviation theorems. Such expansions are given in Chapter 5.

4.1 Semi-Markov processes with absorption

In this section, we give a definition for semi-Markov processes and semi-Markov processes with absorption.

4.1.1 Semi-Markov processes

Let (η_n, κ_n), $n = 0, 1, \ldots$ be a homogeneous *Markov chain*, with a *phase space* $X \times [0, \infty)$, where $X = \{0, 1, \ldots, N\}$, an *initial distribution* $\mathsf{P}\{\eta_0 = i, \kappa_0 = 0\} = p_i$, $i \in X$, and *transition probabilities*

$$\mathsf{P}\{\eta_{n+1} = j, \kappa_{n+1} \leq t / \eta_n = i, \kappa_n = s, \eta_k = i_k, \kappa_k = s_k, k = 0, \ldots, n-1\}$$

$$= \mathsf{P}\{\eta_{n+1} = j, \kappa_{n+1} \leq t / \eta_n = i, \kappa_n = s\}$$

$$= Q_{ij}(t), \quad i, j \in X, \quad s, t \in [0, \infty). \tag{4.1.1}$$

A characteristic property of the Markov chain (η_n, κ_n), $n = 0, 1, \ldots$, is that its transition probabilities do not depend on the current position of the second component. Such a Markov chain is called a Markov renewal process.

In what follows, we use the notations $\mathsf{P}_i(A) = \mathsf{P}\{A/\eta_0 = i\}$ and $\mathsf{E}_i \xi = \mathsf{E}\{\xi/\eta_0 = i\}$, respectively, to denote the conditional probabilities of events and expectations of the random variables determined by the Markov chain (η_n, κ_n), $n = 0, 1, \ldots$.

Let us now define the random variables $\tau(0) = 0$ and $\tau(n) = \kappa_1 + \cdots + \kappa_n$ for $n \geq 1$, and then consider the stochastic processes

$$\eta(t) = \eta_{v(t)}, \quad v(t) = \max(n : \tau(n) \leq t), \quad t \geq 0.$$

A process $\eta(t)$, $t \geq 0$, defined in this way is called a *semi-Markov process*.

The space X is referred to as a *phase space* of the semi-Markov process $\eta(t)$. The probabilities $Q_{ij}(t)$, $i, j \in X$, $t \geq 0$, will be called *transition probabilities* of the semi-Markov process $\eta(t)$, and the distribution p_i, $i \in X$, an *initial distribution* of this process.

By the definition, the random variables τ_n, κ_n and η_n are, respectively, the successive moments of jumps, the times between the successive moments of jumps, and the positions of the process $\eta(t)$, $t \geq 0$, at successive moments of jumps. The random variable $v(t)$ counts the number of jumps of the process $\eta(t)$, $t \geq 0$, in the interval $[0, t]$.

As was mentioned, the transition probabilities of the Markov renewal process (η_n, κ_n) do not depend on a current position of the second component. This implies that the first component η_n, $n = 0, 1, \ldots$, itself is a homogeneous Markov chain with the phase space X, the initial distribution $\mathsf{P}\{\eta_0 = i\} = p_i$, $i \in X$, and the transition probabilities

$$\mathsf{P}\{\eta_{n+1} = j/\eta_n = i\} = p_{ij} = Q_{ij}(\infty), \quad i, j \in X. \tag{4.1.2}$$

This Markov chain is called an *imbedded Markov chain* for the semi-Markov process $\eta(t)$, $t \geq 0$.

It is always possible to represent the transition probabilities $Q_{ij}(t)$ in the following form:

$$Q_{ij}(t) = p_{ij} F_{ij}(t), \quad i, j \in X, \quad t \geq 0, \tag{4.1.3}$$

where

$$F_{ij}(t) = \mathsf{P}\{\kappa_{n+1} \leq t/\eta_n = i, \eta_{n+1} = j\}, \quad i, j \in X, \quad t \in [0, \infty). \tag{4.1.4}$$

Note that the distribution function $F_{ij}(t)$ can be chosen in an arbitrary way if $p_{ij} = 0$. This choice does not affect the transition probability $Q_{ij}(t)$.

Let us also introduce the distribution functions of sojourn times,

$$F_i(t) = \sum_{j \in X} Q_{ij}(t) = \sum_{j \in X} p_{ij} F_{ij}(t), \quad i \in X.$$

We exclude from the consideration the case where the process $\eta(t)$, $t \geq 0$, has instant jumps by assuming the following condition:

$\mathbf{I_1}$: $F_i(0) = 0, i \in X$.

Under this condition, the sequence of random variables $\tau(n)$, $n = 0, 1, \ldots$, is strictly increasing with probability 1. Moreover, it is easy to show that, in this case, $\tau(n) \overset{P}{\to} \infty$ as $n \to \infty$. This implies that $\nu(t) < \infty$ with probability 1 for every $t \geq 0$, i.e., the semi-Markov process $\eta(t)$, $t \geq 0$, has a finite number of jumps in any finite interval.

4.1.2 Semi-Markov processes with absorption

The following random variables play an important role in what follows. These random variables are $\nu_j = \min\{n \geq 1 : \eta_n = j\}$, which is the *first hitting time* for the imbedded Markov chain η_n in the state j, and $\mu_j = \tau(\nu_j)$, which is the *first hitting time* for the semi-Markov process $\eta(t)$ in the state j.

By the definition, the random variable ν_j can take values $1, 2, \ldots$ and ∞, i.e., it may be a proper or improper integer-valued random variable. The random variable μ_j can take values i the interval $[0, \infty]$, i.e., it may be a proper or improper integer-valued random variable.

We assume that the sate 0 is an *absorbing state* if the following condition holds:

$\mathbf{A_1}$: $p_{0j} = 0, j \neq 0$.

Under condition $\mathbf{A_1}$, we have $\mathsf{P}_i\{\eta(t) = 0, t \geq \mu_0 / \mu_0 < \infty\} = 1, i \in X$. This explains the term *absorption time* applied to the random variable μ_0.

In this case it is natural to call $\eta(t)$, $t \geq 0$, a semi-Markov process with absorption.

Let us introduce the *hitting-without-absorption probabilities*

$$_0 f_{ij} = \mathsf{P}_i\{\nu_j < \nu_0\}, \quad i, j \neq 0.$$

It is easy to show that $_0 f_{ij} > 0$ if and only if there exists a chain of states $i = j_0, j_1, \ldots, j_{k_{ij}} = j$ such that $j_0, j_1, \ldots, j_{k_{ij}} \neq 0$ and $\prod_{n=0}^{k_{ij}-1} p_{j_n j_{n+1}} > 0$.

We say that a state $j \neq 0$ is a *recurrent-without-absorption state* if the hitting non-absorption probability satisfies $_0 f_{ij} > 0$ for all $i \neq 0$, i.e., the imbedded Markov chain η_n can reach, with positive probability, a state j from any state i before being absorbed in the state 0.

Let us also introduce a class of recurrent-without-absorption states,

$$X_1 = \{j \neq 0 : {}_0 f_{ij} > 0, i \neq 0\}.$$

We say that a state $j \neq 0$ is a *non-recurrent-without-absorption state* if the hitting non-absorption probability satisfies ${}_0 f_{ij} = 0$ for some $i \neq 0$, i.e., the imbedded Markov chain η_n can not reach with positive probability the state j from some state $i \neq 0$ before being absorbed in the state 0.

Let us also introduce a class of non-recurrent-without-absorption states,

$$X_2 = \{j \neq 0 : {}_0 f_{ij} = 0, \text{ for some } i \neq 0\}.$$

By the definition,

$$X_1 \cup X_2 = X_0 = \{1, \ldots, N\}.$$

Let also define *absorption probabilities* by

$$_j f_{i0} = \mathsf{P}_i\{v_0 < v_j\}, \quad i, j \neq 0.$$

The following conditions that guarantee existence of recurrent-without-absorption states will be used in the sequel:

$\mathbf{E}_1$: $X_1 \neq \varnothing$, i.e., there exists a state $j \neq 0$ such that the hitting-without-absorption probability satisfies ${}_0 f_{ij} > 0$ for any state $i \neq 0$.

The most important is the case where the following condition holds:

$\mathbf{E}_1'$: $X_1 = \{1, \ldots, N\}, X_2 = \varnothing$, i.e., ${}_0 f_{ij} > 0$ for all $i, j \neq 0$.

However, we will also consider an intermediate case where the following condition holds:

$\mathbf{E}_1''$: $X_1, X_2 \neq \varnothing$, i.e., there exist $j, r \neq 0$ such that ${}_0 f_{ij} > 0$ for every $i \neq 0$, and ${}_0 f_{ir} = 0$ for some $i \neq 0$.

4.1.3   Classification of states

Let us now classify communicative properties of a semi-Markov process $\eta(t)$, $t \geq 0$, and the corresponding imbedded Markov chain η_n, $n = 0, 1, \ldots$, in traditional terms (see, for example, Feller (1950, 1968)).

We introduce non-conditional *hitting probabilities* by

$$f_{ij} = \mathsf{P}_i\{v_j < \infty\}, \quad i, j \in X.$$

By the definition, the hitting probability satisfies $f_{ij} > 0$ if and only if there exists a chain of states $i = j_0, j_1, \ldots, j_{k_{ij}} = j$ such that $\prod_{n=0}^{k_{ij}-1} p_{j_n j_{n+1}} > 0$.

The following four cases have to be considered:

(i) Condition $\mathbf{E}_1'$ holds and $p_{i0} > 0$ for some state $i \in X_1$. In this case, X_1 is a non-closed communicative class of non-essential states, while $\{0\}$ is a closed communicative class with one essential state 0. The hitting probabilities are: **(a)** $f_{ij} > 0$, $i, j \in X_1$, and $f_{i0} > 0, i \in X_1$; **(b)** $f_{00} = 1$, $f_{0j} = 0$, $j \in X_1$.

(ii) Condition $\mathbf{E}'_1$ holds and $p_{i0} = 0$ for every state $i \in X_1$. In this case, X_1 and $\{0\}$ are two closed communicative classes of essential states. The hitting probabilities are: **(c)** $f_{ij} > 0$, $i, j \in X_1$, while $f_{i0} = 0$, $i \in X_1$; **(d)** $f_{00} = 1$, $f_{0j} = 0$, $j \in X_1$.

(iii) Condition $\mathbf{E}''_1$ holds and $p_{i0} > 0$ for some state $i \in X_1$. In this case, X_1 is a non-closed communicative class of non-essential states, X_2 is a non-closed class of non-essential states, while $\{0\}$ is a closed communicative class with one essential state 0. The hitting probabilities are: **(e)** $f_{ij} > 0$, $i, j \in X_1$ and $f_{i0} > 0$, $i \in X_1$, while $f_{ij} = 0$, $i \in X_1$, $j \in X_2$; **(f)** $f_{ij} > 0$, $i \in X_2$, $j \in X_1$, while $f_{i0} \in [0, 1)$, $i \in X_2$; **(g)** $f_{00} = 1$, $f_{0j} = 0$, $j \in X_1 \cup X_2$.

(iv) Condition $\mathbf{E}''_1$ holds and $p_{i0} = 0$ for every state $i \in X_1$. In this case, X_1 and $\{0\}$ are two closed communicative classes of essential states, while X_2 is a non-closed class of non-essential states. The hitting probabilities are: **(h)** $f_{ij} > 0$, $i, j \in X_1$, while $f_{i0} = 0$, $i \in X_1$ and $f_{ij} = 0$, $i \in X_1$, $j \in X_2$; **(i)** $f_{ij} > 0$, $i \in X_2$, $j \in X_1$, while $f_{i0} \in [0, 1)$, $i \in X_2$; **(j)** $f_{00} = 1$, $f_{0j} = 0$, $j \in X_1 \cup X_2$.

Note that alternatives **(i)** and **(iii)** correspond to a model in which absorption is possible from recurrent states, while alternatives **(ii)** ad **(iv)** correspond to a model in which absorption from recurrent states is not possible.

Let us consider the random variables $v_j \wedge v_0$, $j \neq 0$. By the definition of these random variables,

$$\mathsf{P}_i\{v_j \wedge v_0 < \infty\} = {}_0 f_{ij} + {}_j f_{i0}, \quad i, j \neq 0. \tag{4.1.5}$$

The following useful lemma follows directly from the classification of the states described above.

Lemma 4.1.1. *If $j \neq 0$ is a recurrent-without-absorption state, then the random variable $v_i \wedge v_0$ is finite with probability 1, i.e.,*

$$_0 f_{ij} + {}_j f_{i0} = 1. \tag{4.1.6}$$

4.1.4   Removing the absorption

Let $\mathfrak{F}_t = \sigma(\eta(s), s \in [0, t])$, $t \geq 0$, be a natural filtration of the σ-algebras of the random events generated by the semi-Markov process $\eta(t)$.

Suppose that we are interested in finding the probabilities $\mathsf{P}_i\{A_t, \mu_0 > t\}$, $t \geq 0$, for some random events $A_t \in \mathfrak{F}_t$, $t \geq 0$, and an initial state $i \neq 0$.

It is easy to see that this probability does not depend on the transition probabilities $Q_{0k}(u)$, $u \geq 0$, $k \in X$. So, we can change these transition probabilities without changing values of the probabilities $\mathsf{P}_i\{A_t, \mu_0 > t\}$, $t \geq 0$.

In particular, this remarks can be applied to the probabilities $\mathsf{P}_i\{\eta(t) = j, \mu_0 > t\}$, $t \geq 0$, where $i, j \neq 0$. These probabilities make a subject of our studies in Chapter 4.

One of natural ways to change transition probabilities is for the absorption state to take new values, $Q_{0k}(u) = \frac{1}{N+1}(1 - e^{-u})$, $u \geq 0$, $k \in X$. For the new modified version of the semi-Markov process $\eta(t)$, the state 0 is not an absorbing state any more.

Note that this change of transition probabilities in the state 0 also changes the classification of states described in Subsection 4.1.3.

In the cases **(i)** and **(iii)**, the whole phase space $X = X_1 \cup \{0\}$ becomes a single communicative class of essential states. In the case **(ii)**, X_1 becomes a closed communicative class of essential states, while the state 0 becomes non-essential. In case **(iv)**, X_1 becomes a closed communicative class of essential states, while the class $X_2 \cup \{0\}$ becomes a non-closed class of non-essential states.

4.1.5   Instant transitions

Condition $\mathbf{I_1}$, which prohibits existence of instant states, can be weakened in the case where condition $\mathbf{E_1}$ holds.

Condition $\mathbf{I_1}$ can be replaced with any condition that would imply $\mathsf{P}_i(A) = 1$, $i \in X$, where the random event is $A = \{\lim_{n\to\infty} \tau(n) = \infty\}$. Indeed, in this case the random variable satisfies $\mathsf{P}_i\{\nu(t) < \infty, t \geq 0\} = 1$, $i \in X$, and, therefore, the semi-Markov process $\eta(t) = \eta_{\nu(t)}$, $t \geq 0$, will be well defined for $t \geq 0$.

Lemma 4.1.1 implies that, under condition $\mathbf{E_1}$, condition $\mathbf{I_1}$ can be replaced with the following much weaker condition that allows for instant transitions:

$\mathbf{I_2}$: **(a)** $F_i(0) < 1$ for some state $i \in X_1$; **(b)** $F_0(0) < 1$.

Let us define return times by $\nu_r(n) = \min(n > \nu_r(n-1) : \eta_n = r)$, $n = 1, 2, \ldots$, $\nu_r(0) = 0$. By the definition, $\nu_j(n)$ is the time of n-th return into the state r for the imbedded Markov chain η_n. Let also condition $\mathbf{E_1}$ hold and $j \in X_1$.

If $\eta_0 = 0$, then κ_n, $n = 1, 2, \ldots$, are i.i.d. random variables with the distribution function $F_0(t)$. Thus, condition $\mathbf{I_2}$ **(b)** implies that $\mathsf{P}_0(A) = 1$.

In the cases **(i)** and **(iii)**, the random variable satisfies $f_{i0} = \mathsf{P}_i\{\nu_0(1) < \infty\} = 1$, $i \in X$. Therefore, $\mathsf{P}_i(A) = f_{i0}\mathsf{P}_0(A) = 1$ for any $i \in X$.

In the cases **(ii)** and **(vi)**, if $j \in X_1$, then $\mathsf{P}_j\{\nu_j(n) < \infty, n = 1, 2, \ldots\} = 1$, and the random variables $\kappa_{\nu_j(n)+1}$, $n = 1, 2, \ldots$, are i.i.d. random variables with the distribution function $F_j(t)$. Thus, condition $\mathbf{I_2}$ **(a)** implies that $\mathsf{P}_j(A) = 1$. In this case, using Lemma 4.1.1, we get $\mathsf{P}_i(A) = {}_jf_{i0}\mathsf{P}_0(A) + {}_0f_{ij}\mathsf{P}_j(A) = {}_jf_{i0} + {}_0f_{ij} = 1$ for any $i \in X$.

4.2   Perturbed semi-Markov processes

In this section, we consider a model of perturbed semi-Markov processes, and study convergence of some basic characteristics of return times for such processes.

4.2.1   Perturbed semi-Markov processes

Let, for every $\varepsilon \geq 0$, $(\eta_n^{(\varepsilon)}, \kappa_n^{(\varepsilon)})$, $n = 0, 1, \ldots$ be a Markov renewal process with the phase space $X = \{0, 1, \ldots, N\}$ and transition probabilities $Q_{ij}^{(\varepsilon)}(u)$.

Let $0 = \tau^{(\varepsilon)}(0), \tau^{(\varepsilon)}(n) = \kappa_1^{(\varepsilon)} + \cdots + \kappa_n^{(\varepsilon)}, n \geq 1$, and $\nu^{(\varepsilon)}(t) = \max(n : \tau^{(\varepsilon)}(n) \leq t), t \geq 0$, and $\eta^{(\varepsilon)}(t) = \eta_{\nu^{(\varepsilon)}(t)}^{(\varepsilon)}, t \geq 0$, be the corresponding semi-Markov process associated with the Markov renewal process $(\eta_n^{(\varepsilon)}, \kappa_n^{(\varepsilon)})$.

We write the transition probabilities as $Q_{ij}^{(\varepsilon)}(t) = p_{ij}^{(\varepsilon)} F_{ij}^{(\varepsilon)}(t)$, where $p_{ij}^{(\varepsilon)} = Q_{ij}^{(\varepsilon)}(\infty)$ are the transition probabilities of the corresponding imbedded Markov chain and $F_{ij}^{(\varepsilon)}(t)$ are the distribution functions of transition times.

Let also $\nu_j^{(\varepsilon)} = \min\{n \geq 1 : \eta_n^{(\varepsilon)} = j\}$ and $\mu_j^{(\varepsilon)} = \tau^{(\varepsilon)}(\nu_j^{(\varepsilon)})$ be the first hitting time at which the imbedded Markov chain $\eta_n^{(\varepsilon)}$ and the semi-Markov process $\eta^{(\varepsilon)}(t)$ hit the state j, respectively. The first hitting time $\mu_0^{(\varepsilon)}$ is the absorption time for the semi-Markov process $\eta^{(\varepsilon)}(t), t \geq 0$.

It is also useful to define return times by $\nu_j^{(\varepsilon)}(n) = \min(n > \nu_j^{(\varepsilon)}(n-1) : \eta_n^{(\varepsilon)} = j)$, $n = 1, 2, \ldots, \nu_j^{(\varepsilon)}(0) = 0$ and $\mu_j^{(\varepsilon)}(n) = \tau^{(\varepsilon)}(\nu_j^{(\varepsilon)}(n))$, $n = 0, 1, \ldots$. By the definition, $\nu_j^{(\varepsilon)}(n)$ and $\mu_j^{(\varepsilon)}(n)$ are the times of the n-th return into the state j for the imbedded Markov chain $\eta_n^{(\varepsilon)}$ and the semi-Markov process $\eta^{(\varepsilon)}(t)$, respectively.

Denote

$$F_i^{(\varepsilon)}(t) = \sum_{j \in X} Q_{ij}^{(\varepsilon)}(t) = \sum_{j \in X} p_{ij}^{(\varepsilon)} F_{ij}^{(\varepsilon)}(t), \quad i \in X.$$

For simplicity we assume that the following condition preventing instant jumps holds:

I₃: $F_i^{(\varepsilon)}(0) = 0, i \in X$, for all $\varepsilon \geq 0$.

We consider a model in which 0 is an absorbing state, that is, the following condition holds:

A₂: $p_{0j}^{(\varepsilon)} = 0, j \neq 0$, for all $\varepsilon \geq 0$.

We also assume the following condition imposed on the transition probabilities of the semi-Markov process $\eta^{(\varepsilon)}(t), t \geq 0$, which allows to consider these processes for $\varepsilon > 0$ as perturbed versions of the semi-Markov process $\eta^{(0)}(t), t \geq 0$,

T₁: (a) $p_{ij}^{(\varepsilon)} \to p_{ij}^{(0)}$ as $\varepsilon \to 0, i \neq 0, j \in X$;

(b) $Q_{ij}^{(\varepsilon)}(\cdot) \Rightarrow Q_{ij}^{(0)}(\cdot)$ as $\varepsilon \to 0, i \neq 0, j \in X$.

The following condition obviously implies condition **T₁** to hold:

T₁′: (a) $p_{ij}^{(\varepsilon)} \to p_{ij}^{(0)}$ as $\varepsilon \to 0, i \neq 0, j \in X$;

(b) $F_{ij}^{(\varepsilon)}(\cdot) \Rightarrow F_{ij}^{(0)}(\cdot)$ as $\varepsilon \to 0, i \neq 0, j \in X$.

It is useful to note that the asymptotic relations in conditions $\mathbf{T_1}$ and $\mathbf{T'_1}$ are equivalent for $i \neq 0$, $j \in X$ such that $p_{ij}^{(0)} > 0$. However, it can be that the asymptotic relation **(b)** in condition $\mathbf{T_1}$ holds while the corresponding asymptotic relation **(b)** in condition $\mathbf{T'_1}$ does not hold, if $p_{ij}^{(0)} = 0$. It is so, if $p_{ij}^{(\varepsilon)} \to p_{ij}^{(0)} = 0$ as $\varepsilon \to 0$ but $F_{ij}^{(\varepsilon)}(\cdot) \not\Rightarrow F_{ij}^{(0)}(\cdot)$ as $\varepsilon \to 0$. Indeed, in this case, $Q_{ij}^{(\varepsilon)}(\cdot) \Rightarrow Q_{ij}^{(0)}(\cdot)$ as $\varepsilon \to 0$, where $Q_{ij}^{(0)}(t) = 0, t \geq 0$.

We also assume the following recurrence condition for the limit semi-Markov process:

$\mathbf{E_2}$: The set of recurrent-without-absorption states for the semi-Markov process $\eta^{(0)}(t)$ is not empty, $X_1^{(0)} \neq \varnothing$, i.e. there exists a state $j \neq 0$ such that the hitting-without-absorption probability satisfies $_0 f_{ij}^{(0)} > 0$ for any state $i \neq 0$.

The most important is the case where the following condition holds:

$\mathbf{E'_2}$: $X_1^{(0)} = \{1, \ldots, N\}$, $X_2^{(0)} = \varnothing$, i.e., $_0 f_{ij}^{(0)} > 0$ for all $i, j \neq 0$.

However, we also consider an intermediate case where the following condition holds:

$\mathbf{E''_2}$: $X_1^{(0)}, X_2^{(0)} \neq \varnothing$, i.e., there exist $j, r \neq 0$ such that $_0 f_{ij}^{(0)} > 0$ for every $i \neq 0$, and $_0 f_{ir}^{(0)} = 0$ for some $i \neq 0$.

4.2.2   Reduction to the model of perturbed regenerative processes

We are interested in an asymptotic analysis for the probabilities

$$P_{ij}^{(\varepsilon)}(t) = \mathsf{P}_i\{\eta^{(\varepsilon)}(t) = j, \mu_0^{(\varepsilon)} > t\}, \quad t \geq 0, \quad i, j \neq 0.$$

These probabilities are measurable functions of $t \geq 0$ with respect to the Borel σ-algebra of subsets of $[0, \infty)$ for every $i, j \neq 0$.

The semi-Markov process $\eta^{(\varepsilon)}(t)$, $t \geq 0$, can be considered as a particular case of a regenerative process with regeneration times $\mu_j^{(\varepsilon)}(n)$, which are times of subsequent return of the process $\eta^{(\varepsilon)}(t)$ into some fixed recurrent state j.

The absorption time $\mu_0^{(\varepsilon)}$ plays a role of the regenerative stopping time.

However, we have to be careful. Under the absorption condition $\mathbf{A_2}$, the process $\eta^{(\varepsilon)}(t)$ is a regenerative process with improper regeneration times that can take the value $+\infty$ (in the case of absorption). In Chapter 3, we have considered regenerative processes with proper regeneration times. Thus, the procedure of removing the absorption, described in Subsection 4.1.4, should be applied first.

For example, transition probabilities in the absorption state can be changed and take new values such as $Q_{0k}^{(\varepsilon)}(u) = \frac{1}{N+1}(1 - e^{-u})$, $u \geq 0$, $k \in X$. For the new modified version of the semi-Markov process $\eta^{(\varepsilon)}(t)$, the state 0 is not absorbing any more. As was mentioned in Subsection 4.1.3, the initial and the modified semi-Markov processes have the same probabilities, $P_{ij}^{(\varepsilon)}(t) = \mathsf{P}_i\{\eta^{(\varepsilon)}(t) = j, \mu_0^{(\varepsilon)} > t\}, t \geq 0$, $i, j \neq 0$.

As we will show below, conditions $\mathbf{A_2}$, $\mathbf{T_1}$ and $\mathbf{E_2}$ imply that condition $\mathbf{E_2}$ holds for the semi-Markov process $\eta^{(\varepsilon)}(t), t \geq 0$, moreover, all the states $j \in X_1$ are recurrent-without-absorption states for this process, if ε is small enough.

If condition $\mathbf{E_2}$ holds for the semi-Markov process $\eta^{(\varepsilon)}(t), t \geq 0$, then there exists a unique closed communicative class of essential states for the corresponding modified version of this semi-Markov process, and any recurrent state $j \in X_1$ belongs to this class.

This implies that the return times $\mu_j^{(\varepsilon)}(n), n = 0, 1, \ldots$, are, with probability 1, finite random variables for the modified version of the semi-Markov process $\eta^{(\varepsilon)}(t)$.

Let us fix some state $j \in X_1^{(0)}$. If the initial state is $i = j$, then the process $\eta^{(\varepsilon)}(t)$, $t \geq 0$, is a standard regenerative process. However, if the initial state is $i \neq j$, the process $\eta^{(\varepsilon)}(t), t \geq 0$, is a regenerative process with a transition period.

Let us introduce the distribution functions for the first hitting times,

$$_0G_{ij}^{(\varepsilon)}(t) = \mathsf{P}_i\{\mu_j^{(\varepsilon)} \leq t, v_j^{(\varepsilon)} \leq v_0^{(\varepsilon)}\}, \quad t \geq 0, \quad i, j \neq 0,$$

and the distribution functions for the absorption times,

$$_jG_{i0}^{(\varepsilon)}(t) = \mathsf{P}_i\{\mu_j^{(\varepsilon)} \leq t, v_0^{(\varepsilon)} \leq v_j^{(\varepsilon)}\}, \quad t \geq 0, \quad i, j \neq 0.$$

Using the regeneration property, which the semi-Markov process $\eta^{(\varepsilon)}(t), t \geq 0$, possesses at the time of the first hitting into any state, we can write the following renewal equation for the case where the initial state is $i = j \neq 0$:

$$P_{jj}^{(\varepsilon)}(t) = 1 - F_j^{(\varepsilon)}(t) + \int_0^t P_{jj}^{(\varepsilon)}(t - s)\,_0G_{jj}^{(\varepsilon)}(ds), \quad t \geq 0. \tag{4.2.1}$$

It is worth to note a very simple form the forcing function $1 - F_j^{(\varepsilon)}(t)$ takes in equation (4.2.1). As a matter of fact, if the initial state is j, then $\{\eta^{(\varepsilon)}(t) = j, \mu_j^{(\varepsilon)} \wedge \mu_0^{(\varepsilon)} > t\} = \{\tau^{(\varepsilon)}(1) > t\}$, that is, the process $\eta^{(\varepsilon)}(t)$ can be observed in the state j at the moment t before the time it hits this state for the first time or being absorbed, only if no jumps occur before the moment t.

As was mentioned above, the semi-Markov process $\eta^{(\varepsilon)}(t), t \geq 0$, considered as a regenerative process, has a transition period in the case where the initial state is $i \neq j, 0$. The renewal equation (4.2.1) should be supplemented, in this case, with the following renewal type relation that can be written for every initial state $i \neq j$, $i, j \neq 0$:

$$P_{ij}^{(\varepsilon)}(t) = \int_0^t P_{jj}^{(\varepsilon)}(t - s)\,_0G_{ij}^{(\varepsilon)}(ds), \quad t \geq 0. \tag{4.2.2}$$

Note that there is no free forcing function in this renewal type relation. As a matter of fact, if the initial state is $i \neq j, 0$, then $\{\eta^{(\varepsilon)}(t) = j, \mu_j^{(\varepsilon)} \wedge \mu_0^{(\varepsilon)} > t\} = \varnothing$.

The renewal relations (4.2.1) and (4.2.2) can be used to make an asymptotic analysis of the probabilities $P_{ij}^{(\varepsilon)}(t)$ in the case where $j \in X_1^{(\varepsilon)}$, i.e., it is a recurrent-without-absorption state.

In a general case, for any state $r \neq 0$ and $j \in X_1^{(\varepsilon)}$, the following renewal equation can be used:

$$P_{jr}^{(\varepsilon)}(t) = q_{jr}^{(\varepsilon)}(t) + \int_0^t P_{jr}^{(\varepsilon)}(t - s)_0 G_{jj}^{(\varepsilon)}(ds), \quad t \geq 0, \tag{4.2.3}$$

where

$$q_{jr}^{(\varepsilon)}(t) = \mathsf{P}_j\{\eta^{(\varepsilon)}(t) = r, \mu_j^{(\varepsilon)} > t, \mu_0^{(\varepsilon)} > t\}, \quad t \geq 0.$$

It is obvious that

$$G_{ij}^{(\varepsilon)}(t) = {}_0G_{ij}^{(\varepsilon)}(t) + {}_jG_{i0}^{(\varepsilon)}(t) = \mathsf{P}_i\{\mu_j^{(\varepsilon)} \wedge \mu_0^{(\varepsilon)} \leq t\}, \quad t \geq 0, \ i, j \neq 0, \tag{4.2.4}$$

and that the following estimate takes place:

$$q_{jr}^{(\varepsilon)}(t) \leq 1 - G_{jj}(t), \quad t \geq 0, \quad j, r \neq 0. \tag{4.2.5}$$

The renewal equation (4.2.3) should be supplemented with the following renewal type relation that can be written for every initial state $i \neq j, 0$ as

$$P_{ir}^{(\varepsilon)}(t) = q_{ijr}^{(\varepsilon)}(t) + \int_0^t P_{jr}^{(\varepsilon)}(t - s)_0 G_{ij}^{(\varepsilon)}(ds), \quad t \geq 0, \tag{4.2.6}$$

where

$$q_{ijr}^{(\varepsilon)}(t) = \mathsf{P}_i\{\eta^{(\varepsilon)}(t) = r, \mu_j^{(\varepsilon)} > t, \mu_0^{(\varepsilon)} > t\}, \quad t \geq 0.$$

It is obvious that the following estimate takes place:

$$q_{ijr}^{(\varepsilon)}(t) \leq 1 - G_{ij}(t), \quad t \geq 0, \quad i, j, r \neq 0. \tag{4.2.7}$$

The renewal relations (4.2.1) and (4.2.2) are simpler than the renewal relations (4.2.3) and (4.2.6), because they have simpler forcing functions. The first two renewal relations can be used in the asymptotic analysis of the probabilities $P_{ij}^{(\varepsilon)}(t)$ in the case where $j \in X_1^{(\varepsilon)}$, i.e., j is a recurrent-without-absorption state for the Markov chain $\eta_n^{(\varepsilon)}$. The latter two renewal relations can be used, in any case, for an asymptotic analysis of the probabilities $P_{ir}^{(\varepsilon)}(t)$. However, it makes sense to use them in the case where $X_2^{(\varepsilon)} \neq \varnothing$ and $r \in X_2^{(\varepsilon)}$.

We can now apply results of Chapters 1–3 to these equations. To do this, we should translate the corresponding perturbation conditions imposed on the transition probabilities $Q_{ij}^{(\varepsilon)}(t)$ of perturbed semi-Markov processes to the distribution functions ${}_0G_{ij}^{(\varepsilon)}(t)$ and the forcing functions $\delta(i, j)(1 - F_j^{(\varepsilon)}(t))$ used in (4.2.1) and (4.2.2).

It should be noted that the renewal equation (4.2.1) and the renewal relations (4.2.2) are the same for the modified and the initial versions of the semi-Markov process $\eta^{(\varepsilon)}(t)$, and all asymptotic results can be related to the initial version of the semi-Markov process $\eta^{(\varepsilon)}(t)$.

4.2.3   Convergence of hitting and absorption probabilities

Let us introduce the hitting and absorption probabilities by

$$_0 f_{ij}^{(\varepsilon)} = \mathsf{P}_i\{v_j^{(\varepsilon)} \le v_0^{(\varepsilon)}\} = {}_0 G_{ij}^{(\varepsilon)}(\infty), \quad i, j \ne 0,$$

and

$$_j f_{i0}^{(\varepsilon)} = \mathsf{P}_i\{v_0^{(\varepsilon)} \le v_j^{(\varepsilon)}\} = {}_j G_{i0}^{(\varepsilon)}(\infty), \quad i, j \ne 0.$$

Lemma 4.2.1. *Let conditions* $\mathbf{A_2}$, $\mathbf{T_1}$ **(a)**, *and* $\mathbf{E_2}$ *hold. Then,* **(α)** $_0 f_{ij}^{(\varepsilon)} \to {}_0 f_{ij}^{(0)}$ *as* $\varepsilon \to 0$, $i \ne 0$, $j \in X_1^{(0)}$; **(β)** $_j f_{i0}^{(\varepsilon)} \to {}_j f_{i0}^{(0)}$ *as* $\varepsilon \to 0$, $i \ne 0$, $j \in X_1^{(0)}$.

Proof. The probabilities $_0 f_{ij}^{(\varepsilon)}$, $i \ne 0$, satisfy the following system of linear equations for every $j \ne 0$:

$$_0 f_{ij}^{(\varepsilon)} = p_{ij}^{(\varepsilon)} + \sum_{k \ne j, 0} p_{ik}^{(\varepsilon)} \, _0 f_{kj}^{(\varepsilon)}, \quad i \ne 0. \tag{4.2.8}$$

Systems (4.2.8) can be rewritten, for every $j \ne 0$, in the following matrix form:

$$_0\mathbf{f}_j^{(\varepsilon)} = \mathbf{p}_j^{(\varepsilon)} + {}_j\mathbf{P}^{(\varepsilon)}{}_0\mathbf{f}_j^{(\varepsilon)}, \tag{4.2.9}$$

where

$$_0\mathbf{f}_j^{(\varepsilon)} = \begin{bmatrix} _0 f_{1j}^{(\varepsilon)} \\ \vdots \\ _0 f_{Nj}^{(\varepsilon)} \end{bmatrix}, \quad \mathbf{p}_j^{(\varepsilon)} = \begin{bmatrix} p_{1j}^{(\varepsilon)} \\ \vdots \\ p_{Nj}^{(\varepsilon)} \end{bmatrix},$$

and

$$_j\mathbf{P}^{(\varepsilon)} = \begin{bmatrix} p_{11}^{(\varepsilon)} & \cdots & p_{1(j-1)}^{(\varepsilon)} & 0 & p_{1(j+1)}^{(\varepsilon)} & \cdots & p_{1N}^{(\varepsilon)} \\ \vdots & & \vdots & \vdots & \vdots & & \vdots \\ p_{N1}^{(\varepsilon)} & \cdots & p_{N(j-1)}^{(\varepsilon)} & 0 & p_{N(j+1)}^{(\varepsilon)} & \cdots & p_{NN}^{(\varepsilon)} \end{bmatrix}.$$

Let us show that the coefficient matrix of this system, $\mathbf{I} - {}_j\mathbf{P}^{(\varepsilon)}$, for every $j \in X_1$ has an inverse for all ε small enough, and, therefore, a unique solution of the systems (4.2.9) has the following form:

$$_0\mathbf{f}_j^{(\varepsilon)} = [\mathbf{I} - {}_j\mathbf{P}^{(\varepsilon)}]^{-1}\mathbf{p}_j^{(\varepsilon)}. \tag{4.2.10}$$

We introduce the probabilities

$$_{0j}f_i^{(\varepsilon)}(n) = \mathsf{P}_i\{v_0^{(\varepsilon)} \wedge v_j^{(\varepsilon)} > n\}, \quad i,j \neq 0, \quad n = 0,1,\dots.$$

Condition $\mathbf{T_1}$ **(a)** implies that, for $i,j \neq 0$, $n = 0,1,\dots$,

$$_{0j}f_i^{(\varepsilon)}(n) = \sum_{i=i_0,\dots,i_n \neq 0,j} \prod_{k=1}^{n} p_{i_{k-1}i_k}^{(\varepsilon)}$$

$$\rightarrow \ _{0j}f_i^{(0)}(n) = \sum_{i=i_0,\dots,i_n \neq 0,j} \prod_{k=1}^{n} p_{i_{k-1}i_k}^{(0)} \quad \text{as } \varepsilon \to 0. \tag{4.2.11}$$

Condition $\mathbf{E_2}$ implies, by Lemma 4.1.1, that for $j \in X_1$,

$$_{0j}f_i^{(0)}(n) \to 0 \quad \text{as } n \to 0, \quad i \neq 0, \tag{4.2.12}$$

and thus there exists m such that

$$\max_{i \neq 0} {}_{0j}f_i^{(0)}(m) < L < 1. \tag{4.2.13}$$

It follows from (4.2.11) and (4.2.13) that there exists $\varepsilon_0 > 0$ such that

$$\sup_{\varepsilon \leq \varepsilon_0} \max_{i \neq 0} {}_{0j}f_i^{(\varepsilon)}(m) \leq L. \tag{4.2.14}$$

Obviously,

$$_{0j}f_i^{(\varepsilon)}(2m) = \sum_{k \neq 0,j} \mathsf{P}_i\{v_j^{(\varepsilon)} \wedge v_0^{(\varepsilon)} > m, \eta_m^{(\varepsilon)} = k\} \ {}_{0j}f_k^{(\varepsilon)}(m). \tag{4.2.15}$$

Using estimate (4.2.14) and relation (4.2.15) we get

$$\sup_{\varepsilon \leq \varepsilon_0} \max_{i \neq 0} {}_{0j}f_i^{(\varepsilon)}(nm) < L^n, \quad n = 1,2,\dots. \tag{4.2.16}$$

Now introduce the probabilities

$$_{0j}f_{ik}^{(\varepsilon)}(n) = \mathsf{P}_i\{\eta_n^{(\varepsilon)} = k, v_j^{(\varepsilon)} \wedge v_0^{(\varepsilon)} > n\}, \quad i,k,j \neq 0, \quad n = 0,1,\dots.$$

It is easily seen that the n-th power of the matrix $_j\mathbf{P}^{(\varepsilon)}$ takes the following form:

$$(_j\mathbf{P}^{(\varepsilon)})^n = \|_{0j}f_{ik}^{(\varepsilon)}(n)\|, \quad n = 0,1,\dots. \tag{4.2.17}$$

For every $j \in X_1$, estimate (4.2.16) yields the following estimate for the norm of the matrix $({}_j\mathbf{P}^{(\varepsilon)})^n$, and the estimate is uniform for $\varepsilon \leq \varepsilon_0$:

$$| ({}_j\mathbf{P}^{(\varepsilon)})^n | = \max_{i \neq 0} \sum_{k=1}^{N} {}_{0j}f_{ik}^{(\varepsilon)}(n)$$

$$= \max_{i \neq 0} {}_{0j}f_i^{(\varepsilon)}(n) \leq L^{[n/m]} \to 0 \ \text{ as } \ n \to 0. \tag{4.2.18}$$

Relation (4.2.18) implies that for every $j \in X_1$ and every $\varepsilon \leq \varepsilon_0$ there exists an inverse matrix, $[\mathbf{I} - {}_j\mathbf{P}^{(\varepsilon)}]^{-1}$, given by the following matrix series, which is norm convergent:

$$[\mathbf{I} - {}_j\mathbf{P}^{(\varepsilon)}]^{-1} = \mathbf{I} + {}_j\mathbf{P}^{(\varepsilon)} + ({}_j\mathbf{P}^{(\varepsilon)})^2 + \cdots . \tag{4.2.19}$$

For every $j, k \neq 0$, let us introduce a random variable $\delta_{jk}^{(\varepsilon)}$ as the number of visits of the imbedded Markov chain $\eta_n^{(\varepsilon)}$ in the state k before the first hitting of one of the sates j or 0,

$$\delta_{jk}^{(\varepsilon)} = \sum_{n=1}^{v_j^{(\varepsilon)} \wedge v_0^{(\varepsilon)}} \chi(\eta_{n-1}^{(\varepsilon)} = k).$$

By the definition,

$$\mathsf{E}_i \delta_{jk}^{(\varepsilon)} = \sum_{n=1}^{\infty} \mathsf{P}_i\{\eta_{n-1}^{(\varepsilon)} = k, v_j^{(\varepsilon)} \wedge v_0^{(\varepsilon)} > n - 1\}. \tag{4.2.20}$$

Finally, comparing formulas (4.2.19) and (4.2.20) we get that $\mathsf{E}_i \delta_{jk}^{(\varepsilon)} < \infty, i, k \neq 0$, $j \in X_1^{(0)}$, and that for every $j \in X_1^{(0)}$ and $\varepsilon \leq \varepsilon_0$,

$$[\mathbf{I} - {}_j\mathsf{P}^{(\varepsilon)}]^{-1} = \| \mathsf{E}_i \delta_{jk}^{(\varepsilon)} \|. \tag{4.2.21}$$

Condition $\mathbf{T}_1$ implies in an obvious way that, for every $n = 0, 1, \ldots,$

$$({}_j\mathbf{P}^{(\varepsilon)})^n \to ({}_j\mathbf{P}^{(0)})^n \ \text{ as } \ \varepsilon \to 0. \tag{4.2.22}$$

Relations (4.2.18) and (4.2.22) imply that

$$[\mathbf{I} - {}_j\mathbf{P}^{(\varepsilon)}]^{-1} \to [\mathbf{I} - {}_j\mathbf{P}^{(0)}]^{-1} \ \text{ as } \ \varepsilon \to 0. \tag{4.2.23}$$

One can prove in the following simple way that the inverse matrix $[\mathbf{I} - {}_j\mathbf{P}^{(\varepsilon)}]^{-1}$ exists for all ε small enough and that convergence relation (4.2.25) holds true. As was mentioned above, the inverse matrix $[\mathbf{I} - {}_j\mathbf{P}^{(0)}]^{-1}$ does exist. This means that $\det(\mathbf{I} - {}_j\mathbf{P}^{(0)}) \neq 0$. Condition $\mathbf{T}_1$ obviously implies that ${}_j\mathbf{P}^{(\varepsilon)} \to {}_j\mathbf{P}^{(0)}$ as $\varepsilon \to 0$. Thus,

$\det(\mathbf{I} - {}_j\mathbf{P}^{(\varepsilon)}) \neq 0$ for all ε small enough. Moreover, the elements of the inverse matrix $[\mathbf{I} - {}_j\mathbf{P}^{(\varepsilon)}]^{-1}$ are continuous rational functions of elements of the matrix $\mathbf{I} - {}_j\mathbf{P}^{(\varepsilon)}$. This rational function has the non-zero denominator $\det(\mathbf{I} - {}_j\mathbf{P}^{(\varepsilon)})$. This implies existence of $[\mathbf{I} - {}_j\mathbf{P}^{(\varepsilon)}]^{-1}$ for ε small enough and relation (4.2.25).

Formulas (4.2.10) and (4.2.21) imply that, for every $j \in X_1^{(0)}$ and $\varepsilon \leq \varepsilon_0$,

$$
{}_0 f_{ij}^{(\varepsilon)} = \sum_{k=1}^{N} \mathsf{E}_i \delta_{jk}^{(\varepsilon)} p_{kj}^{(\varepsilon)}, \quad i \neq 0.
\tag{4.2.24}
$$

According to relation (4.2.23),

$$
\mathsf{E}_i \delta_{jk}^{(\varepsilon)} \to \mathsf{E}_i \delta_{jk}^{(0)} \quad \text{as} \quad \varepsilon \to 0, \quad i, j, k \neq 0.
\tag{4.2.25}
$$

Relation (4.2.25), condition $\mathbf{T_1}$ **(a)** and formula (4.2.24) imply the convergence relation $(\boldsymbol{\alpha})$ in Lemma 4.2.1.

For every $j \neq 0$, the probabilities ${}_j f_{i0}^{(\varepsilon)}$, $i \neq 0$, satisfy the following system of linear equations that are similar to (4.2.8):

$$
{}_j f_{i0}^{(\varepsilon)} = p_{i0}^{(\varepsilon)} + \sum_{k \neq j, 0} p_{ik}^{(\varepsilon)} \, {}_j f_{k0}^{(\varepsilon)}, \quad i \neq 0.
\tag{4.2.26}
$$

System (4.2.26) can be rewritten, for every $j \neq 0$, in the following matrix form:

$$
{}_j\mathbf{f}_0^{(\varepsilon)} = \mathbf{p}_0^{(\varepsilon)} + {}_j\mathbf{P}^{(\varepsilon)}{}_j\mathbf{f}_0^{(\varepsilon)},
\tag{4.2.27}
$$

where

$$
{}_j\mathbf{f}_0^{(\varepsilon)} = \begin{bmatrix} {}_j f_{10}^{(\varepsilon)} \\ \vdots \\ {}_j f_{N0}^{(\varepsilon)} \end{bmatrix}, \quad
\mathbf{p}_0^{(\varepsilon)} = \begin{bmatrix} p_{10}^{(\varepsilon)} \\ \vdots \\ p_{N0}^{(\varepsilon)} \end{bmatrix}.
$$

This system has the same coefficient matrix $\mathbf{I} - {}_j\mathbf{P}^{(\varepsilon)}$ as system (4.2.9) for every $j \in X_1^{(0)}$. As was shown in the proof of Lemma 4.2.10, this matrix has an inverse for $\varepsilon \leq \varepsilon_0$. Therefore, a unique solution of the system (4.2.27) has the following form:

$$
{}_j\mathbf{f}_0^{(\varepsilon)} = [\mathbf{I} - {}_j\mathbf{P}^{(\varepsilon)}]^{-1}\mathbf{p}_0^{(\varepsilon)}.
\tag{4.2.28}
$$

Taking into account formula (4.2.21) and (4.2.28) we get the following formula for every $j \in X_1^{(0)}$ and $\varepsilon \leq \varepsilon_0$ as above:

$$
{}_j f_{i0}^{(\varepsilon)} = \sum_{k=1}^{N} \mathsf{E}_i \delta_{jk}^{(\varepsilon)} p_{k0}^{(\varepsilon)}, \quad i \neq 0.
\tag{4.2.29}
$$

Relation (4.2.25), condition $\mathbf{T_1}$ **(a)**, and formula (4.2.29) imply the convergence relation $(\boldsymbol{\beta})$ in Lemma 4.2.1. $\qquad\square$

The following lemma is a direct corollary of Lemma 4.2.1.

Lemma 4.2.2. *Let conditions* $\mathbf{A_2}$, $\mathbf{T_1}$ **(a)**, *and* $\mathbf{E_2}$ *hold. Then, any state* $j \in X_1^{(0)}$ *is also a recurrent-without-absorption state for the perturbed semi-Markov process* $\eta^{(\varepsilon)}(t)$, $t \geq 0$, *for all* $\varepsilon \leq \varepsilon_0$ *(defined in formula (4.2.14)), i.e.,* $X_1^{(0)} \subseteq X_1^{(\varepsilon)}$ *for all* $\varepsilon \leq \varepsilon_0$.

4.2.4 Weak convergence of distribution functions of the first hitting times

The following lemma gives a condition for weak convergence of the distribution functions of the first hitting times.

Lemma 4.2.3. *Let conditions* $\mathbf{A_2}$, $\mathbf{T_1}$, *and* $\mathbf{E_2}$ *hold. Then* **(α)** $_0G_{ij}^{(\varepsilon)}(\cdot) \Rightarrow {}_0G_{ij}^{(0)}(\cdot)$ *as* $\varepsilon \to 0$, $i \neq 0$, $j \in X_1^{(0)}$; **(β)** $_jG_{i0}^{(\varepsilon)}(\cdot) \Rightarrow {}_jG_{i0}^{(0)}(\cdot)$ *as* $\varepsilon \to 0$, $i \neq 0$, $j \in X_1^{(0)}$; **(γ)** $G_{ij}^{(\varepsilon)}(\cdot) \Rightarrow G_{ij}^{(0)}(\cdot)$ *as* $\varepsilon \to 0$, $i \neq 0$, $j \in X_1^{(0)}$.

Proof. Let us use the Laplace transform,

$$\bar{p}_{ij}^{(\varepsilon)}(\rho) = \int_0^\infty e^{-\rho s} Q_{ij}^{(\varepsilon)}(ds), \quad _0\bar{\phi}_{ij}^{(\varepsilon)}(\rho) = \int_0^\infty e^{-\rho s}\, {}_0G_{ij}^{(\varepsilon)}(ds), \quad \rho \geq 0, \ i, j \neq 0.$$

Note that condition $\mathbf{T_1}$ can be expressed in the following equivalent form:

$\mathbf{T_1''}$: $\bar{p}_{ij}^{(\varepsilon)}(\rho) \to \bar{p}_{ij}^{(0)}(\rho)$ as $\varepsilon \to 0$ for $\rho \geq 0$, $i \neq 0$, $j \in X$.

The quantities $_0\bar{\phi}_{ij}^{(\varepsilon)}(\rho)$, $i \neq 0$, for every $j \neq 0$ and $\rho \geq 0$, satisfy the following system of linear equations:

$$_0\bar{\phi}_{ij}^{(\varepsilon)}(\rho) = \bar{p}_{ij}^{(\varepsilon)}(\rho) + \sum_{k \neq j,0} \bar{p}_{ik}^{(\varepsilon)}(\rho)\, {}_0\bar{\phi}_{kj}^{(\varepsilon)}(\rho), \quad i \neq 0. \tag{4.2.30}$$

System (4.2.30), for every $j \neq 0$, can be rewritten in the following matrix form:

$$_0\bar{\boldsymbol{\phi}}_j^{(\varepsilon)}(\rho) = \bar{\mathbf{p}}_j^{(\varepsilon)}(\rho) + {}_j\bar{\mathbf{P}}^{(\varepsilon)}(\rho)\, {}_0\bar{\boldsymbol{\phi}}_j^{(\varepsilon)}(\rho), \tag{4.2.31}$$

where

$$_0\bar{\boldsymbol{\phi}}_j^{(\varepsilon)}(\rho) = \begin{bmatrix} _0\bar{\phi}_{1j}^{(\varepsilon)}(\rho) \\ \vdots \\ _0\bar{\phi}_{Nj}^{(\varepsilon)}(\rho) \end{bmatrix}, \quad \bar{\mathbf{p}}_j^{(\varepsilon)}(\rho) = \begin{bmatrix} \bar{p}_{1j}^{(\varepsilon)}(\rho) \\ \vdots \\ \bar{p}_{Nj}^{(\varepsilon)}(\rho) \end{bmatrix},$$

and

$$_j\bar{\mathbf{P}}^{(\varepsilon)}(\rho) = \begin{bmatrix} \bar{p}_{11}^{(\varepsilon)}(\rho) & \cdots & \bar{p}_{1(j-1)}^{(\varepsilon)}(\rho) & 0 & \bar{p}_{1(j+1)}^{(\varepsilon)}(\rho) & \cdots & \bar{p}_{1N}^{(\varepsilon)}(\rho) \\ \vdots & & \vdots & \vdots & \vdots & & \vdots \\ \bar{p}_{N1}^{(\varepsilon)}(\rho) & \cdots & \bar{p}_{N(j-1)}^{(\varepsilon)}(\rho) & 0 & \bar{p}_{N(j+1)}^{(\varepsilon)}(\rho) & \cdots & \bar{p}_{NN}^{(\varepsilon)}(\rho) \end{bmatrix}.$$

Let us show that the coefficient matrix of this system, $\mathbf{I} - {}_{j}\bar{\mathbf{P}}^{(\varepsilon)}(\rho)$, for every $j \in X_1^{(0)}$ and $\rho \geq 0$, has an inverse for all ε small enough, and, therefore, a unique solution of system (4.2.31) has the following form:

$$_{0}\bar{\phi}_{j}^{(\varepsilon)}(\rho) = [\mathbf{I} - {}_{j}\bar{\mathbf{P}}^{(\varepsilon)}(\rho)]^{-1}\bar{\mathbf{p}}_{j}^{(\varepsilon)}(\rho). \tag{4.2.32}$$

Indeed, by the definition, the elements of the matrices ${}_{j}\bar{\mathbf{P}}^{(\varepsilon)}(\rho)$ and ${}_{j}\mathbf{P}^{(\varepsilon)}$ satisfy the following inequalities for every $\rho \geq 0$:

$$0 \leq \chi(k \neq j)\bar{p}_{ik}^{(\varepsilon)}(\rho) \leq \chi(k \neq j)p_{ik}^{(\varepsilon)}, \quad i,k \neq 0, \tag{4.2.33}$$

and thus the corresponding norms satisfy the following relation for every $j \in X_1^{(0)}$, $\rho \geq 0$ and $\varepsilon \leq \varepsilon_0$:

$$| ({}_{j}\bar{\mathbf{P}}^{(\varepsilon)}(\rho))^n | \leq | ({}_{j}\bar{\mathbf{P}}^{(\varepsilon)})^n | \leq L^{[n/m]} \to 0 \text{ as } n \to 0. \tag{4.2.34}$$

Relation (4.2.34) implies that for every $j \in X_1^{(0)}, \rho \geq 0$, and every $\varepsilon \leq \varepsilon_0$ there exists an inverse matrix, $[\mathbf{I} - {}_{j}\mathbf{P}^{(\varepsilon)}(\rho)]^{-1}$, given by the following (convergent in the norm) matrix series,

$$[\mathbf{I} - {}_{j}\bar{\mathbf{P}}^{(\varepsilon)}(\rho)]^{-1} = \mathbf{I} + {}_{j}\bar{\mathbf{P}}^{(\varepsilon)}(\rho) + ({}_{j}\bar{\mathbf{P}}^{(\varepsilon)}(\rho))^2 + \cdots . \tag{4.2.35}$$

For every $j, k \neq 0, \rho \geq 0$, let us introduce random variables by

$$\bar{\delta}_{jk}^{(\varepsilon)}(\rho) = \sum_{n=1}^{\infty} e^{-\rho \tau^{(\varepsilon)}(n-1)} \chi(\eta_{n-1}^{(\varepsilon)} = k, v_j^{(\varepsilon)} \wedge v_0^{(\varepsilon)} > n - 1).$$

By the Lebesgue theorem,

$$\mathsf{E}_i \bar{\delta}_{jk}^{(\varepsilon)}(\rho) = \sum_{n=1}^{\infty} \mathsf{E}_i e^{-\rho \tau^{(\varepsilon)}(n-1)} \chi(\eta_{n-1}^{(\varepsilon)} = k, v_j^{(\varepsilon)} \wedge v_0^{(\varepsilon)} > n - 1\}. \tag{4.2.36}$$

Finally, by comparing formulas (4.2.35) and (4.2.36), we get that $\mathsf{E}_i \bar{\delta}_{jk}^{(\varepsilon)}(\rho) < \infty$, $i, k \neq 0, j \in X_1^{(0)}, \rho \geq 0$, and that for every $j \in X_1^{(0)}, \rho \geq 0$, and $\varepsilon \leq \varepsilon_0$,

$$[\mathbf{I} - {}_{j}\bar{\mathbf{P}}^{(\varepsilon)}(\rho)]^{-1} = \| \mathsf{E}_i \bar{\delta}_{jk}^{(\varepsilon)}(\rho) \|. \tag{4.2.37}$$

Condition $\mathbf{T_1}$ implies in an obvious way that, for every $j \in X_1, \rho \geq 0$,

$$_{j}\bar{\mathbf{P}}^{(\varepsilon)}(\rho) \to {}_{j}\bar{\mathbf{P}}^{(0)}(\rho) \text{ as } \varepsilon \to 0. \tag{4.2.38}$$

As was mentioned above, the inverse matrix $[\mathbf{I} - {}_{j}\bar{\mathbf{P}}^{(\varepsilon)}(\rho)]^{-1}$ exists for all ε small enough. Moreover, elements of the inverse matrix $[\mathbf{I} - {}_{j}\bar{\mathbf{P}}^{(\varepsilon)}(\rho)]^{-1}$ are continuous rational functions of elements of the matrix $\mathbf{I} - {}_j\bar{\mathbf{P}}^{(\varepsilon)}(\rho)$. This rational function has the non-zero denominator $\det(\mathbf{I} - {}_j\bar{\mathbf{P}}^{(\varepsilon)}(\rho))$. This implies that, for every $j \in X_1^{(0)}, \rho \geq 0$,

$$[\mathbf{I} - {}_j\bar{\mathbf{P}}^{(\varepsilon)}(\rho)]^{-1} \to [\mathbf{I} - {}_j\bar{\mathbf{P}}^{(0)}(\rho)]^{-1} \quad \text{as } \varepsilon \to 0. \tag{4.2.39}$$

Formulas (4.2.32) and (4.2.37) imply that, for every $j \in X_1^{(0)}, \rho \geq 0$, and $\varepsilon \leq \varepsilon_0$,

$$_0\bar{\phi}_{ij}^{(\varepsilon)}(\rho) = \sum_{k=1}^{N} \mathsf{E}_i \bar{\delta}_{jk}^{(\varepsilon)}(\rho) \bar{p}_{kj}^{(\varepsilon)}(\rho), \quad i \neq 0. \tag{4.2.40}$$

According to relation (4.2.39), for every $\rho \geq 0$,

$$\mathsf{E}_i \bar{\delta}_{jk}^{(\varepsilon)}(\rho) \to \mathsf{E}_i \bar{\delta}_{jk}^{(0)}(\rho) \quad \text{as } \varepsilon \to 0, \quad i, j, k \neq 0. \tag{4.2.41}$$

Relation (4.2.41), condition $\mathbf{T}_1''$, and formula (4.2.40) imply that, for every $j \in X_1^{(0)}$ and $\rho \geq 0$,

$$_0\bar{\phi}_{ij}^{(\varepsilon)}(\rho) \to {}_0\bar{\phi}_{ij}^{(\varepsilon)}(\rho) \quad \text{as } \varepsilon \to 0, \quad i \neq 0. \tag{4.2.42}$$

This relation is equivalent to the weak convergence relation $(\boldsymbol{\alpha})$ in Lemma 4.2.3.

The proof of the weak convergence relation $(\boldsymbol{\beta})$ given in Lemma 4.2.3 repeats in main the proof given above. Let us now use the Laplace transforms

$$\bar{p}_{i0}^{(\varepsilon)}(\rho) = \int_0^\infty e^{-\rho s} Q_{i0}^{(\varepsilon)}(ds), \quad {}_j\bar{\phi}_{i0}^{(\varepsilon)}(\rho) = \int_0^\infty e^{-\rho s} {}_j G_{i0}^{(\varepsilon)}(ds), \quad \rho \geq 0, \ i, j \neq 0.$$

For every $j \neq 0$ and $\rho \geq 0$, the quantities ${}_j\bar{\phi}_{i0}^{(\varepsilon)}(\rho), i \neq 0$, satisfy the following system of linear equations:

$$_j\bar{\phi}_{i0}^{(\varepsilon)}(\rho) = \bar{p}_{i0}^{(\varepsilon)}(\rho) + \sum_{k \neq j, 0} \bar{p}_{ik}^{(\varepsilon)}(\rho) \, {}_j\bar{\phi}_{k0}^{(\varepsilon)}(\rho), \quad i \neq 0. \tag{4.2.43}$$

For every $j \neq 0$ and $\rho \geq 0$, system (4.2.43) can be rewritten in the following matrix form:

$$_j\bar{\boldsymbol{\phi}}_0^{(\varepsilon)}(\rho) = \bar{\mathbf{p}}_0^{(\varepsilon)}(\rho) + {}_j\bar{\mathbf{P}}^{(\varepsilon)}(\rho) \, {}_j\bar{\boldsymbol{\phi}}_0^{(\varepsilon)}(\rho), \tag{4.2.44}$$

where

$$_j\bar{\boldsymbol{\phi}}_0^{(\varepsilon)}(\rho) = \begin{bmatrix} {}_j\bar{\phi}_{10}^{(\varepsilon)}(\rho) \\ \vdots \\ {}_j\bar{\phi}_{N0}^{(\varepsilon)}(\rho) \end{bmatrix}, \qquad \bar{\mathbf{p}}_0^{(\varepsilon)}(\rho) = \begin{bmatrix} \bar{p}_{10}^{(\varepsilon)}(\rho) \\ \vdots \\ \bar{p}_{N0}^{(\varepsilon)}(\rho) \end{bmatrix}.$$

For every $j \in X_1$ and $\rho \geq 0$, this system has the same coefficient matrix $\mathbf{I} - {}_j\bar{\mathbf{P}}^{(\varepsilon)}(\rho)$ as system (4.2.31). As was shown in the proof of Lemma 4.2.10, this matrix has an inverse for $\varepsilon \le \varepsilon_0$. Therefore, a unique solution of system (4.2.44) has the following form:

$$_j\bar{\boldsymbol{\phi}}_0^{(\varepsilon)}(\rho) = [\mathbf{I} - {}_j\bar{\mathbf{P}}^{(\varepsilon)}(\rho)]^{-1}\bar{\mathbf{p}}_0^{(\varepsilon)}(\rho). \qquad (4.2.45)$$

Taking into account formulas (4.2.32) and (4.2.45) we get the following formula for every $j \in X_1^{(0)}$, $\rho \ge 0$, and $\varepsilon \le \varepsilon_0$ defined in the proof of Lemma 4.2.1:

$$_j\bar{\phi}_{i0}^{(\varepsilon)}(\rho) = \sum_{k=1}^{N} \mathsf{E}_i \bar{\delta}_{jk}^{(\varepsilon)}(\rho)\bar{p}_{k0}^{(\varepsilon)}(\rho), \quad i \ne 0. \qquad (4.2.46)$$

Relation (4.2.41), condition $\mathbf{T}_1''$, and formula (4.2.46) imply that, for every $j \in X_1^{(0)}$ and $\rho \ge 0$,

$$_0\bar{\phi}_{ij}^{(\varepsilon)}(\rho) \to {}_0\bar{\phi}_{ij}^{(0)}(\rho) \quad \text{as } \varepsilon \to 0, \quad i \ne 0. \qquad (4.2.47)$$

This relation is equivalent to the weak convergence relation $(\boldsymbol{\beta})$ given in Lemma 4.2.3.

The weak convergence relation $(\boldsymbol{\gamma})$ in Lemma 4.2.3 follows from relation (4.2.4) and the weak convergence relations $(\boldsymbol{\alpha})$ and $(\boldsymbol{\beta})$ in this lemma. $\qquad \square$

Remark 4.2.1. It follows from the proof of Lemma 4.2.3 that the assumption of the weak convergence $F_{i0}^{(\varepsilon)}(\cdot) \Rightarrow F_{i0}^{(0)}(\cdot)$ as $\varepsilon \to 0$, $i \ne 0$, is not required in claim $(\boldsymbol{\alpha})$ of this lemma.

Lemmas 4.2.1–4.2.3 allow us to also weaken condition $\mathbf{I}_3$ and replace it with the following conditions:

$\mathbf{I}_4$: $F_j^{(0)}(0) < 1$ for some state $j \in X_1^{(0)}$;

and

$\mathbf{I}_5$: $F_0^{(0)}(0) < 1$.

Condition $\mathbf{T}_1$ implies that $F_i^{(\varepsilon)}(\cdot) \Rightarrow F_i^{(0)}(\cdot)$ as $\varepsilon \to 0$, $i \in X$, and thus there exists $\varepsilon_1 > 0$ such that $F_i^{(\varepsilon)}(0) < 1$ and $F_0^{(\varepsilon)}(0) < 1$ for $\varepsilon \le \varepsilon_1$.

As was pointed out in Lemma 4.2.2, conditions $\mathbf{A}_2$, $\mathbf{T}_1$, $\mathbf{E}_2$ imply that any state $j \in X_1^{(0)}$, for $\varepsilon \le \varepsilon_0$, is a recurrent-without-absorption state for the semi-Markov processes $\eta^{(\varepsilon)}(t)$, $t \ge 0$.

Therefore, conditions $\mathbf{A}_2$, $\mathbf{T}_1$, $\mathbf{E}_2$, and $\mathbf{I}_4$, $\mathbf{I}_5$ imply that condition $\mathbf{I}_2$ holds for the semi-Markov processes $\eta^{(\varepsilon)}(t)$, $t \ge 0$, for $\varepsilon \le \min(\varepsilon_0, \varepsilon_1)$.

Note also that condition $\mathbf{I}_5$ is not restrictive at all. It automatically holds for the modified semi-Markov process obtained by removing the absorption state, as described in Subsection 4.2.3. Thus, the use of this condition can be omitted in the corresponding asymptotic relations for non-absorption events.

4.2.5 Convergence of expectations for the first hitting times

We denote

$$m_{ij}^{(\varepsilon)} = \int_0^\infty s F_{ij}^{(\varepsilon)}(ds), \quad p_{ij}^{(\varepsilon)}[1] = \int_0^\infty s Q_{ij}^{(\varepsilon)}(ds) = m_{ij}^{(\varepsilon)} p_{ij}^{(\varepsilon)}, \quad i \neq 0, \quad j \in X.$$

Asymptotic results that describe transition phenomena in perturbed semi-Markov processes also involve the following moment perturbation condition:

$\mathbf{M_{11}}$: $p_{ij}^{(\varepsilon)}[1] \to p_{ij}^{(0)}[1] < \infty$ as $\varepsilon \to 0$, $i \neq 0, j \in X$.

Condition $\mathbf{T_1}$ **(a)** and the following condition obviously imply condition $\mathbf{M_{11}}$ to hold:

$\mathbf{M'_{11}}$: $m_{ij}^{(\varepsilon)} \to m_{ij}^{(0)} < \infty$ as $\varepsilon \to 0, i \neq 0, j \in X$.

It is useful to note that the asymptotic relations in conditions $\mathbf{M_{11}}$ and $\mathbf{M'_{11}}$ taken for fixed $i \neq 0, j \in X$ are equivalent if the the asymptotic relation in condition $\mathbf{T_1}$ **(a)** holds for the same i and j and $p_{ij}^{(0)} > 0$. However, if $p_{ij}^{(0)} = 0$ then the asymptotic relation in condition $\mathbf{M_{11}}$ may hold while the corresponding relation in condition $\mathbf{M'_{11}}$ does not.

We introduce the following notations:

$$_0M_{ij}^{(\varepsilon)} = \int_0^\infty s\,_0G_{ij}^{(\varepsilon)}(ds) = \mathsf{E}_i \mu_j^{(\varepsilon)} \chi(\nu_j^{(\varepsilon)} < \nu_0^{(\varepsilon)}), \quad i, j \neq 0,$$

$$_jM_{i0}^{(\varepsilon)} = \int_0^\infty s_j G_{i0}^{(\varepsilon)}(ds) = \mathsf{E}_i \mu_j^{(\varepsilon)} \chi(\nu_0^{(\varepsilon)} < \nu_j^{(\varepsilon)}), \quad i, j \neq 0,$$

and

$$M_{ij}^{(\varepsilon)} = \int_0^\infty s G_{ij}^{(\varepsilon)}(ds) = {}_0M_{ij}^{(\varepsilon)} + {}_jM_{i0}^{(\varepsilon)} = \mathsf{E}_i(\mu_j^{(\varepsilon)} \wedge \mu_0^{(\varepsilon)}), \quad i, j \neq 0.$$

Lemma 4.2.4. *Let conditions* $\mathbf{A_2}$, $\mathbf{T_1}$, $\mathbf{M_{11}}$, *and* $\mathbf{E_2}$ *hold. Then* **(α)** $_0M_{ij}^{(\varepsilon)} \to {}_0M_{ij}^{(0)} < \infty$ *as* $\varepsilon \to 0, i \neq 0, j \in X_1^{(0)}$; **($\beta$)** $_jM_{i0}^{(\varepsilon)} \to {}_jM_{i0}^{(0)} < \infty$ *as* $\varepsilon \to 0, i \neq 0, j \in X_1^{(0)}$; **($\gamma$)** $M_{ij}^{(\varepsilon)} \to M_{ij}^{(0)} < \infty$ *as* $\varepsilon \to 0, i \neq 0, j \in X_1^{(0)}$.

Proof. Let us first note that condition $\mathbf{M_{11}}$ implies that there exists $\varepsilon_2 > 0$ such that

$$\sup_{\varepsilon \leq \varepsilon_3} \max_{i \neq 0} \sum_{j \in X} p_{ij}^{(\varepsilon)}[1] \leq M < \infty. \tag{4.2.48}$$

Using this estimate we get for every $i \neq 0$, $j \in X_1^{(0)}$, and $\varepsilon \leq \varepsilon_2$ that

$$M_{ij}^{(\varepsilon)} = \mathsf{E}_i \sum_{n=1}^{\nu_j^{(\varepsilon)} \wedge \nu_0^{(\varepsilon)}} \kappa_n^{(\varepsilon)} = \mathsf{E}_i \sum_{n=1}^{\infty} \kappa_n^{(\varepsilon)} \chi(\nu_j^{(\varepsilon)} \wedge \nu_0^{(\varepsilon)} > n - 1)$$

$$= \sum_{n=1}^{\infty} \sum_{k \neq j,0} \mathsf{E}_k \kappa_1^{(\varepsilon)} \mathsf{E}_i \chi(\nu_j^{(\varepsilon)} \wedge \nu_0^{(\varepsilon)} > n - 1, \eta_{n-1}^{(\varepsilon)} = k)$$

$$\leq M \sum_{n=1}^{\infty} \mathsf{P}_i \{\nu_j^{(\varepsilon)} \wedge \nu_0^{(\varepsilon)} > n - 1\} \leq M \sum_{n=0}^{\infty} L^{[n/m]} = \frac{Mm}{1-L} < \infty. \quad (4.2.49)$$

The expectations $_0M_{ij}^{(\varepsilon)}$, $i \neq 0$, satisfy the following system of linear equations analogous to (4.2.8) for every $j \neq 0$:

$$_0M_{ij}^{(\varepsilon)} = L_{ij}^{(\varepsilon)} + \sum_{k \neq j,0} p_{ik}^{(\varepsilon)} {}_0M_{kj}^{(\varepsilon)}, \quad i \neq 0, \quad (4.2.50)$$

where

$$L_{ij}^{(\varepsilon)} = p_{ij}^{(\varepsilon)}[1] + \sum_{k \neq j,0} p_{ik}^{(\varepsilon)}[1] {}_0f_{kj}^{(\varepsilon)}, \quad i, j \neq 0.$$

Note that there are the hitting probabilities $_0f_{kj}^{(\varepsilon)}$ appear in the expression for the free terms $L_{ij}^{(\varepsilon)}$, because we operate with the expectations $_0M_{ij}^{(\varepsilon)} = \mathsf{E}_i \mu_j^{(\varepsilon)} \chi(\nu_j^{(\varepsilon)} \leq \nu_0^{(\varepsilon)})$ of the first hitting times $\mu_j^{(\varepsilon)}$ multiplied by the indicators of the hitting events $\chi(\nu_j^{(\varepsilon)} \leq \nu_0^{(\varepsilon)})$.

For every $j \neq 0$, system (4.2.50) can be rewritten in the following matrix form:

$$_0\mathbf{M}_j^{(\varepsilon)} = \mathbf{L}_j^{(\varepsilon)} + {}_j\mathbf{P}^{(\varepsilon)} {}_0\mathbf{M}_j^{(\varepsilon)}, \quad (4.2.51)$$

where

$$_0\mathbf{M}_j^{(\varepsilon)} = \begin{bmatrix} _0M_{1j}^{(\varepsilon)} \\ \vdots \\ _0M_{Nj}^{(\varepsilon)} \end{bmatrix}, \qquad \mathbf{L}_j^{(\varepsilon)} = \begin{bmatrix} L_{1j}^{(\varepsilon)} \\ \vdots \\ L_{Nj}^{(\varepsilon)} \end{bmatrix}.$$

This system has the same coefficient matrix $\mathbf{I} - {}_j\mathbf{P}^{(\varepsilon)}$ for every $j \in X_1^{(0)}$ as system (4.2.9). As was shown in the proof of Lemma 4.2.10, this matrix has an inverse for $\varepsilon \leq \varepsilon_0$, defined in formula (4.2.14).

Define

$$\varepsilon_3 = \min(\varepsilon_0, \varepsilon_2). \quad (4.2.52)$$

A unique solution of system (4.2.51) has the following form for $j \in X_1^{(0)}$ and $\varepsilon \leq \varepsilon_3$:

$$_0\mathbf{M}_j^{(\varepsilon)} = [\mathbf{I} - {}_j\mathbf{P}^{(\varepsilon)}]^{-1}\mathbf{L}_j^{(\varepsilon)}. \tag{4.2.53}$$

Taking into account (4.2.21) and (4.2.53) we get the following formula for every $j \in X_1^{(0)}$ and $\varepsilon \leq \varepsilon_3$:

$$_0M_{ij}^{(\varepsilon)} = \sum_{k=1}^{N} \mathsf{E}_i \delta_{jk}^{(\varepsilon)} L_{kj}^{(\varepsilon)}, \quad i \neq 0. \tag{4.2.54}$$

By Lemma 4.2.1 and relation (4.2.25),

$$L_{kj}^{(\varepsilon)} \to L_{kj}^{(0)} \quad \text{as } \varepsilon \to 0, \quad k, j \neq 0. \tag{4.2.55}$$

Relation (4.2.55), conditions $\mathbf{T_1}$ **(a)**, $\mathbf{M_{11}}$, and formula (4.2.54) imply the convergence relation **(α)** in Lemma 4.2.4.

The expectations ${}_jM_{i0}^{(\varepsilon)}$, $i \neq 0$, for every $j \neq 0$, satisfy the following system of linear equations that are similar to (4.2.50):

$$_jM_{i0}^{(\varepsilon)} = L_{j0}^{(\varepsilon)} + \sum_{k \neq j, 0} p_{ik}^{(\varepsilon)} {}_jM_{k0}^{(\varepsilon)}, \quad i \neq 0, \tag{4.2.56}$$

where

$$L_{i0}^{(\varepsilon)} = p_{i0}^{(\varepsilon)}[1] + \sum_{k \neq j, 0} p_{ik}^{(\varepsilon)}[1] {}_jf_{k0}^{(\varepsilon)}, \quad i, j \neq 0.$$

For every $j \neq 0$, system (4.2.56) can be rewritten in a matrix form,

$$_j\mathbf{M}_0^{(\varepsilon)} = \mathbf{L}_0^{(\varepsilon)} + {}_j\mathbf{P}^{(\varepsilon)} {}_j\mathbf{M}_0^{(\varepsilon)}, \tag{4.2.57}$$

where

$$_j\mathbf{M}_0^{(\varepsilon)} = \begin{bmatrix} {}_jM_{10}^{(\varepsilon)} \\ \vdots \\ {}_jM_{Nj0}^{(\varepsilon)} \end{bmatrix}, \qquad \mathbf{L}_0^{(\varepsilon)} = \begin{bmatrix} L_{10}^{(\varepsilon)} \\ \vdots \\ L_{N0}^{(\varepsilon)} \end{bmatrix}.$$

For every $j \in X_1^{(0)}$, this system has the same coefficient matrix $\mathbf{I} - {}_j\mathbf{P}^{(\varepsilon)}$ as system (4.2.9). As was shown in the proof of Lemma 4.2.10, this matrix has an inverse for $\varepsilon \leq \varepsilon_0$. Therefore, a unique solution of system (4.2.51) has the following form for $j \in X_1^{(0)}$ and $\varepsilon \leq \varepsilon_3$ that are defined in formula (4.2.52):

$$_j\mathbf{M}_0^{(\varepsilon)} = [\mathbf{I} - {}_j\mathbf{P}^{(\varepsilon)}]^{-1}\mathbf{L}_0^{(\varepsilon)}. \tag{4.2.58}$$

Taking into account formulas (4.2.21) and (4.2.58) we get for every $j \in X_1^{(0)}$ and $\varepsilon \leq \varepsilon_3$ that

$$_jM_{i0}^{(\varepsilon)} = \sum_{k=1}^{N} \mathsf{E}_i \delta_{jk}^{(\varepsilon)} L_{k0}^{(\varepsilon)}, \quad i \neq 0. \qquad (4.2.59)$$

By Lemma 4.2.1 and relation (4.2.25), we have

$$L_{k0}^{(\varepsilon)} \to L_{k0}^{(0)} \text{ as } \varepsilon \to 0, \quad k \neq 0. \qquad (4.2.60)$$

Relation (4.2.60), conditions $\mathbf{T_1}$ (a), $\mathbf{M_{11}}$, and formula (4.2.59) imply the convergence relation $(\boldsymbol{\beta})$ in Lemma 4.2.4.

Relation $(\boldsymbol{\gamma})$ in Lemma 4.2.4 is a direct corollary of relations $(\boldsymbol{\alpha})$ and $(\boldsymbol{\beta})$ given in this lemma. $\qquad\qquad\qquad\qquad\qquad\qquad\qquad\qquad\qquad\qquad\qquad\qquad\qquad\square$

Remark 4.2.2. It follows from the proof of Lemma 4.2.4 that the assumption that $m_{i0}^{(\varepsilon)} \to m_{i0}^{(0)}$ as $\varepsilon \to 0, i \neq 0$, is not required in claim $(\boldsymbol{\alpha})$ of this lemma.

4.2.6 Convergence of generalised hitting and absorption probabilities

In what follows, we will also need to consider more general hitting and absorption probabilities.

Let us set consider the set $Z \subseteq \{1, \ldots, N\}$, and introduce the first hitting time in the set Z for the imbedded Markov chain $\eta_n^{(\varepsilon)}$ by $\nu_Z^{(\varepsilon)} = \min(n \geq 1 : \eta_n^{(\varepsilon)} \in Z)$, and then define the hitting probabilities

$$_0f_{iZr}^{(\varepsilon)} = \mathsf{P}_i\{\nu_Z^{(\varepsilon)} < \nu_0^{(\varepsilon)}, \eta_{\nu^{(\varepsilon)}(Z)}^{(\varepsilon)} = j\}, \quad i \neq 0, \quad r \in Z,$$

and the absorption probabilities

$$_Zf_{i0}^{(\varepsilon)} = \mathsf{P}_i\{\nu_0^{(\varepsilon)} < \nu_Z^{(\varepsilon)}\}, \ i \neq 0.$$

Lemma 4.2.5. *Let conditions* $\mathbf{A_2}$, $\mathbf{T_1}$ **(a)**, $\mathbf{E_2}$ *hold, and* $Z \cap X_1^{(0)} \neq \varnothing$. *Then* $(\boldsymbol{\alpha})$ $_0f_{iZr}^{(\varepsilon)} \to {_0f_{iZr}^{(0)}}$ *as* $\varepsilon \to 0, i \neq 0, r \in Z$; $(\boldsymbol{\beta})$ $_Zf_{i0}^{(\varepsilon)} \to {_Zf_{i0}^{(0)}}$ *as* $\varepsilon \to 0, i \neq 0$.

Proof. Without loss of generality, we can assume that $Z = \{1, \ldots, m\}$.

The case $m = N$ is trivial. Indeed, by the definition, $\nu_Z^{(\varepsilon)} \wedge \nu_0^{(\varepsilon)} = 1$ with probability 1 in this case and, therefore, $_0f_{iZr}^{(\varepsilon)} = p_{ir}^{(\varepsilon)}$, $_Zf_{i0}^{(\varepsilon)} = p_{i0}^{(\varepsilon)}, i \neq 0, r \in Z$.

Thus, we assume that $m < N$. The probabilities $_0f_{iZr}^{(\varepsilon)}, i \neq 0$, satisfy the following system of linear equations for Z such that $Z \cap X_1 \neq \varnothing$ and every $r \in Z$:

$$_0f_{iZr}^{(\varepsilon)} = p_{ir}^{(\varepsilon)} + \sum_{k=m+1}^{N} p_{ik}^{(\varepsilon)} {_0f_{kZr}^{(\varepsilon)}}, \quad i \neq 0. \qquad (4.2.61)$$

It is important that the first m equations in this system are just formulas that express the quantities $_0f_{iZj}^{(\varepsilon)}$, $i = 1, \ldots, m$, as linear combinations of the quantities $_0f_{iZj}^{(\varepsilon)}$, $i = m + 1, \ldots, N$. The last $N - m$ equations make a determined system of linear equations for the latter quantities. Thus, $N - m$ is the real dimension of system (4.2.61).

Systems (4.2.61) can be rewritten in the following matrix form:

$$_0\mathbf{f}_{Zr}^{(\varepsilon)} = \mathbf{p}_r^{(\varepsilon)} + {}_Z\mathbf{P}^{(\varepsilon)}{}_0\mathbf{f}_{Zr}^{(\varepsilon)}, \tag{4.2.62}$$

where

$$_0\mathbf{f}_{Zr}^{(\varepsilon)} = \begin{bmatrix} _0f_{1Zr}^{(\varepsilon)} \\ \vdots \\ _0f_{NZr}^{(\varepsilon)} \end{bmatrix}, \qquad \mathbf{p}_r^{(\varepsilon)} = \begin{bmatrix} p_{1r}^{(\varepsilon)} \\ \vdots \\ p_{Nr}^{(\varepsilon)} \end{bmatrix},$$

and

$$_Z\mathbf{P}^{(\varepsilon)} = \begin{bmatrix} 0 & \cdots & 0 & p_{1(m+1)}^{(\varepsilon)} & \cdots & p_{1N}^{(\varepsilon)} \\ \vdots & & \vdots & \vdots & & \vdots \\ 0 & \cdots & 0 & p_{N(m+1)}^{(\varepsilon)} & \cdots & p_{NN}^{(\varepsilon)} \end{bmatrix}.$$

Let us show that the coefficient matrix of this system, $\mathbf{I} - {}_Z\mathbf{P}^{(\varepsilon)}$, has an inverse for every $r \in Z, Z \cap X_1^{(0)} \neq \varnothing$, and all ε small enough, and, hence, a unique solution of systems (4.2.62) takes the following form:

$$_0\mathbf{f}_{Zr}^{(\varepsilon)} = [\mathbf{I} - {}_Z\mathbf{P}^{(\varepsilon)}]^{-1}\mathbf{p}_r^{(\varepsilon)}. \tag{4.2.63}$$

Indeed, take some recurrent-without-absorption state $j \in Z \cap X_1^{(0)}$. By the definition, the elements of the matrices $_Z\mathbf{P}^{(\varepsilon)}$ and $_j\mathbf{P}^{(\varepsilon)}$ satisfy the inequalities

$$0 \le \chi(k \ge m + 1)p_{ik}^{(\varepsilon)} \le \chi(k \neq j)p_{ik}^{(\varepsilon)}, \quad i, k \neq 0, \tag{4.2.64}$$

and, thus, the corresponding norms satisfy the following relation for every $j \in Z \cap X_1^{(0)}$ and $\varepsilon \le \varepsilon_0$ that are defined in formula (4.2.14):

$$\mid ({}_Z\mathbf{P}^{(\varepsilon)})^n \mid \le \mid ({}_j\mathbf{P}^{(\varepsilon)})^n \mid \le L^{[n/m]} \to 0 \text{ as } n \to 0. \tag{4.2.65}$$

Relation (4.2.65) implies that for every $r \in Z, Z \cap X_1^{(0)} \neq \varnothing$, there exists the inverse $[\mathbf{I} - {}_Z\mathbf{P}^{(\varepsilon)}]^{-1}$ given by the following matrix series that converges in the norm:

$$[\mathbf{I} - {}_Z\mathbf{P}^{(\varepsilon)}]^{-1} = \mathbf{I} + {}_Z\mathbf{P}^{(\varepsilon)} + ({}_Z\mathbf{P}^{(\varepsilon)})^2 + \cdots. \tag{4.2.66}$$

Let us introduce the random variables

$$\delta_{Zk}^{(\varepsilon)} = \sum_{n=1}^{v_Z^{(\varepsilon)} \wedge v_0^{(\varepsilon)}} \chi(\eta_{n-1}^{(\varepsilon)} = k), \quad k \neq 0.$$

By the definition,

$$\mathsf{E}_i \delta_{Zk}^{(\varepsilon)} = \sum_{n=1}^{\infty} \mathsf{E}_i \chi(\eta_{n-1}^{(\varepsilon)} = k, v_Z^{(\varepsilon)} \wedge v_0^{(\varepsilon)} > n - 1). \qquad (4.2.67)$$

Finally, comparing formulas (4.2.66) and (4.2.67) we get that $\mathsf{E}_i \delta_{Zk}^{(\varepsilon)} < \infty, i, k \neq 0, Z \cap X_1 \neq \varnothing$, for every $\varepsilon \leq \varepsilon_0$, and that

$$[\mathbf{I} - {}_Z\mathbf{P}^{(\varepsilon)}]^{-1} = \| \mathsf{E}_i \delta_{Zk}^{(\varepsilon)} \|. \qquad (4.2.68)$$

It is useful to note that, by the definition,

$$\mathsf{E}_i \delta_{Zk}^{(\varepsilon)} = \chi(i = k), \quad i, k \in Z. \qquad (4.2.69)$$

Condition $\mathbf{T}_1$ **(a)** implies in an obvious way that

$$_Z\mathbf{P}^{(\varepsilon)} \to {}_Z\mathbf{P}^{(0)} \quad \text{as } \varepsilon \to 0. \qquad (4.2.70)$$

As was mentioned above, the inverse matrix $[\mathbf{I} - {}_Z\mathbf{P}^{(\varepsilon)}]^{-1}$ exists for all ε small enough. Moreover, the elements of the inverse matrix $[\mathbf{I} - {}_Z\mathbf{P}^{(\varepsilon)}]^{-1}$ are continuous rational functions of elements of the matrix $\mathbf{I} - {}_Z\mathbf{P}^{(\varepsilon)}$. These rational functions have the non-zero denominator $\det(\mathbf{I} - {}_Z\mathbf{P}^{(\varepsilon)})$. This implies that

$$[\mathbf{I} - {}_Z\mathbf{P}^{(\varepsilon)}]^{-1} \to [\mathbf{I} - {}_Z\mathbf{P}^{(0)}]^{-1} \quad \text{as } \varepsilon \to 0. \qquad (4.2.71)$$

Formulas (4.2.63) and (4.2.68) imply that, for every $r \in Z, Z \cap X_1^{(0)} \neq \varnothing$, and $\varepsilon \leq \varepsilon_0$,

$$_0 f_{iZr}^{(\varepsilon)} = \sum_{k=1}^{N} \mathsf{E}_i \delta_{Zk}^{(\varepsilon)} p_{kr}^{(\varepsilon)}, \quad i \neq 0. \qquad (4.2.72)$$

Using relations (4.2.68) and (4.2.71) we see that, if $Z \cap X_1^{(0)} \neq \varnothing$, then

$$\mathsf{E}_i \delta_{Zk}^{(\varepsilon)} \to \mathsf{E}_i \delta_{Zk}^{(0)} \quad \text{as } \varepsilon \to 0, \quad i, k \neq 0. \qquad (4.2.73)$$

Relation (4.2.73), condition $\mathbf{T}_1$ **(a)**, and formula (4.2.72) imply the convergence relation **(α)** given in Lemma 4.2.5.

For every Z such that $Z \cap X_1^{(0)} \neq \emptyset$, the probabilities ${}_Z f_{i0}^{(\varepsilon)}$, $i \neq 0$, satisfy the following system of linear equations that are similar to (4.2.61):

$$
{}_Z f_{i0}^{(\varepsilon)} = p_{i0}^{(\varepsilon)} + \sum_{k=m+1}^{N} p_{ik}^{(\varepsilon)} \, {}_Z f_{k0}^{(\varepsilon)}, \quad i \neq 0. \tag{4.2.74}
$$

System (4.2.74) can be rewritten in the following matrix form:

$$
{}_Z \mathbf{f}_0^{(\varepsilon)} = \mathbf{p}_0^{(\varepsilon)} + {}_Z \mathbf{P}^{(\varepsilon)} {}_Z \mathbf{f}_0^{(\varepsilon)}, \tag{4.2.75}
$$

where

$$
{}_Z \mathbf{f}_0^{(\varepsilon)} = \begin{bmatrix} {}_Z f_{10}^{(\varepsilon)} \\ \vdots \\ {}_Z f_{N0}^{(\varepsilon)} \end{bmatrix}, \qquad \mathbf{p}_0^{(\varepsilon)} = \begin{bmatrix} p_{10}^{(\varepsilon)} \\ \vdots \\ p_{N0}^{(\varepsilon)} \end{bmatrix}.
$$

This system has the same coefficient matrix $\mathbf{I} - {}_Z \mathbf{P}^{(\varepsilon)}$ as system (4.2.62). As was shown above, this matrix has an inverse for every Z such that $Z \cap X_1^{(0)} \neq \emptyset$ and $\varepsilon \leq \varepsilon_0$. Therefore, a unique solution of system (4.2.75) has the following form:

$$
{}_Z \mathbf{f}_0^{(\varepsilon)} = [\mathbf{I} - {}_Z \mathbf{P}^{(\varepsilon)}]^{-1} \mathbf{p}_0^{(\varepsilon)}. \tag{4.2.76}
$$

Taking into account formulas (4.2.68) and (4.2.76) we get the following formula for every Z such that $Z \cap X_1^{(0)} \neq \emptyset$ and $\varepsilon \leq \varepsilon_0$ defined in (4.2.14):

$$
{}_Z f_{i0}^{(\varepsilon)} = \sum_{k=1}^{N} \mathsf{E}_i \delta_{Zk}^{(\varepsilon)} p_{k0}^{(\varepsilon)}, \quad i \neq 0. \tag{4.2.77}
$$

Relation (4.2.73), condition $\mathbf{T_1}$ **(a)**, and formula (4.2.77) imply the convergence relation $(\boldsymbol{\beta})$ in Lemma 4.2.5. $\qquad\qquad\qquad\qquad\qquad\qquad\qquad\qquad\qquad\quad\;\square$

4.3 Asymptotics of cyclic absorption probabilities

In this section we give a description of asymptotics for cyclic absorption probabilities for perturbed semi-Markov processes.

4.3.1 Stationary distribution for the imbedded Markov chain

Let us assume that the following two conditions hold for the limit imbedded Markov chain $\eta_n^{(0)}$:

$\mathbf{A_3}$: $\sum_{i \in X_1^{(0)}} p_{i0}^{(0)} = 0.$

Conditions $\mathbf{A_3}$ and $\mathbf{E_2'}$ are equivalent to the following condition:

$\mathbf{E_2'''}$: $X_1^{(0)} = \{1, \ldots, N\}$ is a closed communicative class of states for the limit imbedded Markov chain $\eta_n^{(0)}$, $n = 0, 1, \ldots$.

Let introduce a random variable $\varsigma_i^{(0)}(n)$ which is the number of visits of the Markov chain $\eta_n^{(0)}$ in the state i during the first n transitions,

$$\varsigma_j^{(0)}(n) = \sum_{r=1}^{n} \chi(\eta_{r-1}^{(0)} = j), \quad n = 0, 1, \ldots, \quad i \in X_1^{(0)}.$$

The following formula follows directly from the definition of the random variables $\varsigma_i^{(0)}(n)$:

$$\frac{\mathsf{E}_i \varsigma_i^{(0)}(n)}{n} = \bar{p}_{ij}^{(0)}(n-1)$$

$$= \frac{\sum_{r=1}^{n} p_{ij}^{(0)}(r-1)}{n}, \quad n = 0, 1, \ldots, \quad i, j \in X_1^{(0)}, \tag{4.3.1}$$

where
$$p_{ij}^{(0)}(n) = \mathsf{P}_i\{\eta_n^{(0)} = j\}, \quad n = 0, 1, \ldots, \quad i, j \in X_1^{(0)}.$$

The statement in the lemmas given below can serve to define a stationary distribution $\varrho_i^{(0)}$, $i \in X_1^{(0)}$, for the closed communicative class of states $X_1^{(0)}$.

Lemma 4.3.1. *Let condition $\mathbf{E_2'''}$ hold. Then* $(\boldsymbol{\alpha})$ $\mathsf{P}_i\{\varsigma_j^{(0)}(n)/n \to \varrho_j^{(0)}\} = 1$, $i, j \in X_1^{(0)}$, *where* $(\mathbf{a})$ $\varrho_j^{(0)} > 0$, $i \in X_1^{(0)}$, $(\mathbf{b})$ $\sum_{j \in X_1^{(0)}} \varrho_j^{(0)} = 1$; $(\boldsymbol{\beta})$ $\bar{p}_{ij}^{(0)}(n-1))/n \to \varrho_j^{(0)}$ *as* $n \to \infty$, $i, j \in X_1^{(0)}$.

The *stationary distribution* $\varrho_i^{(0)}$, $i \in X_1^{(0)}$ can also be uniquely characterised as a solution of the following system of linear equations:

$$\varrho_i^{(0)} = \sum_{k \in X_1^{(0)}} \varrho_k^{(0)} p_{ki}^{(0)}, \quad i \neq 0, \quad \sum_{k \in X_1^{(0)}} \varrho_k^{(0)} = 1. \tag{4.3.2}$$

It should be noted that the first N equations of system (4.3.2) are linearly dependent, since the sum of the quantities in the left and the right hand sides equals 1. Thus, in fact, there are only N linearly independent equations in this system.

Lemma 4.3.2. *Let condition $\mathbf{E_2'''}$ hold. Then* $(\boldsymbol{\alpha})$ *the stationary distribution* $\varrho_i^{(0)}$, $i \in X_1^{(0)}$, *is a unique solution of the system of linear equations* (4.3.2).

The following lemma supplies two useful formulas for the stationary distribution $\varrho_i^{(0)}, i \in X_1^{(0)}$, that will be used in the sequel.

Lemma 4.3.3. *Let condition* $\mathbf{E}_2'''$ *hold. Then* **(α)** $\varrho_i^{(0)} = (\mathsf{E}_i v_i^{(0)})^{-1}, i \in X_1^{(0)}$; **($\beta$)** $\mathsf{E}_i \delta_{ik}^{(0)} = \varrho_k^{(0)} / \varrho_i^{(0)}, i, k \in X_1^{(0)}$.

Proof of Lemmas 4.3.1–4.3.3. If one chooses the distribution functions $F_{ij}^{(0)}(t) = \chi_{[1,\infty)}(t)$, then the semi-Markov process $\eta^{(0)}(t)$ coincides with the Markov chain $\eta_n^{(0)}$ in the sense that $\eta^{(0)}(t) = \eta_{[t]}^{(0)}, t \geq 0$.

By applying Lemma 4.2.4 in this case, we get that **(c)** $\mathsf{E}_i v_j^{(0)} < \infty, i, j \in X_1^{(0)}$. Let us introduce the random variable $v_j^{(0)}(n) = \min(k : \varsigma_i^{(0)}(k) = n)$ that is the moment of the n-th hitting of the Markov chain $\eta_n^{(0)}$ into the state j. As is known, the moments $v_j^{(0)}(n), n = 0, 1, \ldots$, are regeneration times for the Markov chain $\eta_n^{(0)}$, and, thus, the random variables $v_j^{(0)}(n) - v_j^{(0)}(n-1), n = 1, 2, \ldots$, are independent, moreover, these random variables have the same distribution $\mathsf{P}_j\{v_j^{(0)} \leq t\}$ for $n \geq 2$. Thus, by the standard strong law of large numbers, we have **(d)** $\mathsf{P}_i\{v_j^{(0)}(n)/n \to \mathsf{E}_j v_j^{(0)}\} = 1, i, j \in X_1^{(0)}$. By the definition, $\varsigma_i^{(0)}(n) = \max(k : v_j^{(0)}(k) \leq n)$, and, thus, relation **(c)** implies, in an obvious way, that **(e)** $\mathsf{P}_i\{\varsigma_j^{(0)}(n)/n \to (\mathsf{E}_j v_j^{(0)})^{-1}\} = 1, i, j \in X_1^{(0)}$. It follows from **(c)** that $(\mathsf{E}_j v_j^{(0)})^{-1} > 0, j \in X_1^{(0)}$. Since, by the definition, $\sum_{j \in X_1^{(0)}} \varsigma_j^{(0)}(n)/n = 1, n = 0, 1, \ldots$, relation **(e)** also implies that $\sum_{j \in X_1^{(0)}} (\mathsf{E}_j v_j^{(0)})^{-1} = 1$. Thus, relation **($\alpha$)** given in Lemma 4.3.1 is proved. Also we have proved that the stationary probabilities satisfy $\varrho_j^{(0)} = (\mathsf{E}_j v_j^{(0)})^{-1}, i \in X_1^{(0)}$, that is relation **($\delta$)** in Lemma 4.3.3.

Taking into account that, by the definition, the random variables $|\varsigma_j^{(0)}(n)/n| \leq 1$ are bounded, we get **(α)**, and by the Lebesgue theorem, **(f)** $\mathsf{E}_i \varsigma_j^{(0)}(n)/n \to \varrho_j^{(0)}$ as $n \to \infty, i, j \in X_1^{(0)}$. Relation **(f)** is equivalent, due to (4.3.1), to the relation **(β)** in Lemma 4.3.1. Now, let us use the relation **(g)** $\sum_{r=1}^{n} p_{ij}^{(0)}(r-1) = \delta(i, j) + \sum_{k \in X_1^{(0)}} (\sum_{r=1}^{n-1} p_{ik}^{(0)}) p_{kj}^{(0)}, n \geq 1, i, j \in X_1^{(0)}$. Dividing the expressions in the left and the right hand-side of **(g)** by n and then making n tend to ∞ we get the identities **(h)** $\varrho_j^{(0)} = \sum_{k \in X_1^{(0)}} \varrho_k^{(0)} p_{kj}^{(0)}, j \in X_1^{(0)}$. Thus, the stationary distribution is a solution of system (4.3.2). Let us assume that $\tilde{\varrho}_j^{(0)}, j \in X_1^{(0)}$, is another solution of this system. Then, by iterating the equations in this system we get, in an obvious way, that $\tilde{\varrho}_j^{(0)}, j \in X_1^{(0)}$, should also satisfy the relations **(k)** $\tilde{\varrho}_j^{(0)} = \sum_{k \in X_1^{(0)}} \tilde{\varrho}_k^{(0)} p_{kj}^{(0)}(n)$, $j \in X_1^{(0)}, n \geq 1$. By summing up these relations, we get also the relations **(l)** $n\tilde{\varrho}_j^{(0)} =$

$\sum_{k \in X_1^{(0)}} \tilde{\varrho}_k^{(0)} \sum_{r=1}^{n} p_{kj}^{(0)}(r)$, $j \in X_1^{(0)}$, $n \geq 1$. Dividing the expressions in the left and the right hand sides of **(l)** by n and then approach n to ∞ we get the identities **(m)** $\tilde{\varrho}_j^{(0)} = \sum_{k \in X_1^{(0)}} \tilde{\varrho}_k^{(0)} \varrho_j^{(0)} = \varrho_j^{(0)}$, $j \in X_1^{(0)}$. Here we also used that, by the assumption made above, $\tilde{\varrho}_j^{(0)}$, $j \in X_1^{(0)}$, is a solution of system (4.3.2), and, therefore, $\sum_{k \in X_1^{(0)}} \tilde{\varrho}_k^{(0)} = 1$. Identities **(m)** prove that any solution of system (4.3.2) coincides with the stationary distribution. Thus, this distribution is a unique solution of system (4.3.2).

It remains to prove the formulas given in Lemma 4.3.3. Let us introduce the random variable $\delta_{jk}^{(0)}(n) = \sum_{v_j^{(0)}(n-1) \leq r < v_j^{(0)}(n)} \chi(\eta_r^{(0)} = k)$, $n \geq 1$, $j, k \in X_1^{(0)}$. By the definition, $\delta_{jk}^{(0)}(n)$ is the number of times the Markov chain $\eta_n^{(0)}$ hits the the state k between the $(n-1)$-th and n-th visit to the state j. Let us also denote $\Delta_{jk}^{(0)}(n) = \sum_{r=1}^{n} \delta_{jk}^{(0)}(r)$, $n \geq 1$, $j, k \in X_1^{(0)}$. The moments $v_j^{(0)}(n)$, $n = 0, 1, \ldots$, are regeneration times for the Markov chain $\eta_n^{(0)}$, and, thus, the random variables $\delta_{jk}^{(0)}(n)$, $n \geq 1$, are independent, moreover, these random variables have the same distribution $\mathsf{P}_j\{\delta_{jk}^{(0)}(1) \leq t\}$ for $n \geq 2$. Thus, by the standard strong law of large numbers, we have **(n)** $\mathsf{P}_i\{\Delta_{jk}^{(0)}(n))/n \to \mathsf{E}_j \delta_{jk}^{(0)}(1)\} = 1$, $i, j \in X_1^{(0)}$. Relations **(e)** and **(n)** imply, in an obvious way, that **(o)** $\mathsf{P}_i\{\Delta_{jk}^{(0)}(\varsigma_j^{(0)}(n))/n \to \varrho_j^{(0)} \mathsf{E}_j \delta_{jk}^{(0)}(1)\} = 1$ and $\mathsf{P}_i\{\Delta_{jk}^{(0)}(\varsigma_j^{(0)}(n) + 1)/n \to \varrho_j^{(0)} \mathsf{E}_j \delta_{jk}^{(0)}(1)\} = 1$, $i, j \in X_1^{(0)}$. On the other hand, the following two-sided estimates hold true: **(p)** $\Delta_{jk}^{(0)}(\varsigma_j^{(0)}(n)) \leq \varsigma_k^{(0)}(n) \leq \Delta_{jk}^{(0)}(\varsigma_j^{(0)}(n) + 1)$, $n \geq 1$, $k, j \in X_1^{(0)}$. Relation **(o)** and estimates **(p)** imply that **(r)** $\mathsf{P}_i\{\varsigma_k^{(0)}(n)/n \to \varrho_j^{(0)} \mathsf{E}_j \delta_{jk}^{(0)}(1)\} = 1$, $k, j \in X_1^{(0)}$. Finally, relations **(e)** and **(r)** imply that $\varrho_k^{(0)} = \varrho_j^{(0)} \mathsf{E}_j \delta_{jk}^{(0)}\} = 1$, $k, j \in X_1^{(0)}$, since, by the definition, $\delta_{jk}^{(0)}(1) = \delta_{jk}^{(0)}$. $\square$

4.3.2 Asymptotics of cyclic absorption probabilities for semi-Markov processes without non-recurrent-without-absorption states

The following formula, which is a particular case of (4.2.29), is a starting point in our further analysis:

$$_j f_{j0}^{(\varepsilon)} = \sum_{k \neq 0} \mathsf{E}_j \, \delta_{jk}^{(\varepsilon)} p_{k0}^{(\varepsilon)}, \quad j \neq 0. \tag{4.3.3}$$

The asymptotics of the cyclic absorption probabilities $_j f_{j0}^{(\varepsilon)}$ very much depends on the structure of the set of recurrent-without-absorption states. The basic and the most important case is where the conditions $\mathbf{T}_1$ **(a)**, $\mathbf{A}_3$, and $\mathbf{E}_2'$ hold.

Conditions $\mathbf{T}_1$ **(a)** and $\mathbf{A}_3$ imply that local absorption probabilities satisfy

$$p_{i0}^{(\varepsilon)} \to 0 \quad \text{as} \quad \varepsilon \to 0, \quad i \neq 0. \tag{4.3.4}$$

Condition $\mathbf{A_3}$ and relation (4.3.4) mean that the absorption is "asymptotically" impossible for the pre-limit ($\varepsilon > 0$) processes $\eta^{(\varepsilon)}(t)$ and it is impossible for the limit ($\varepsilon = 0$) semi-Markov process $\eta^{(0)}(t)$.

Let us now consider the absorption probability averaged by this stationary distribution for the limit semi-Markov process

$$f^{(\varepsilon)} = \sum_{i \in X_1^{(0)}} \varrho_i^{(0)} p_{i0}^{(\varepsilon)}. \tag{4.3.5}$$

Conditions $\mathbf{A_3}$, $\mathbf{T_1}$ **(a)**, and $\mathbf{E_2'}$ obviously imply that

$$f^{(\varepsilon)} \to f^{(0)} = 0 \text{ as } \varepsilon \to 0. \tag{4.3.6}$$

We are interested to give a more precise description of the *cyclic absorption probabilities* $_jf_{j0}^{(\varepsilon)}$ for recurrent-without-absorption states.

Let us also assume the following condition that balances the rate of growth of time $t^{(\varepsilon)} \to \infty$ and the rate of perturbation determined, in this case, by the speed with which the averaged absorption probability vanishes, $f^{(\varepsilon)} \to 0$:

$\mathbf{B_6}$: $0 \leq t^{(\varepsilon)} \to \infty$ and $f^{(\varepsilon)} \to 0$ as $\varepsilon \to 0$ in such a way that $f^{(\varepsilon)}t^{(\varepsilon)} \to \lambda$, where $0 \leq \lambda \leq \infty$.

It should be noted that condition $\mathbf{B_6}$ implies the following: **(a)** if $\lambda > 0$, then $f^{(\varepsilon)} > 0$ automatically for all ε small enough, **(b)** if there exists a sequence $0 < \varepsilon_n \to 0$ as $n \to \infty$ such that $f^{(\varepsilon_n)} = 0, n \geq 1$, then $\lambda = 0$.

Lemma 4.3.4. *Let conditions* $\mathbf{A_2}$, $\mathbf{A_3}$, $\mathbf{T_1}$ **(a)**, *and* $\mathbf{E_2'}$ *hold. Then,* **(α)** $_jf_{j0}^{(\varepsilon)} \to 0$ *as* $\varepsilon \to 0, j \in X_1^{(0)}$. *Let also condition* $\mathbf{B_6}$ *holds. Then,* **(β)** $_jf_{j0}^{(\varepsilon)}t^{(\varepsilon)} \to \lambda/\varrho_j^{(0)}$ *as* $\varepsilon \to 0, j \in X_1^{(0)}$.

Proof. Statement **(α)** follows directly from Lemma 4.2.1 and condition $\mathbf{A_3}$, which imply that

$$_jf_{j0}^{(\varepsilon)} \to {}_jf_{j0}^{(0)} = 0 \text{ as } \varepsilon \to 0. \tag{4.3.7}$$

Let us prove statement **(β)**. Obviously, condition $\mathbf{A_3}$ holds for the first hitting time $v_0^{(0)} = \infty$ a.s., and, thus, the following identity holds with probability 1:

$$\delta_{jk}^{(0)} = \sum_{n=1}^{v_j^{(0)} \wedge v_0^{(0)}} \chi(\eta_{n-1}^{(0)} = k) = \sum_{n=1}^{v_j^{(0)}} \chi(\eta_{n-1}^{(0)} = k). \tag{4.3.8}$$

Let us now use the well-known formula that connects the expected value of the number of visits of the imbedded Markov chain $\eta_n^{(0)}$ into the state k, up to the first hitting into the state j, with the stationary probabilities

$$\mathsf{E}_j\, \delta_{jk}^{(0)} = \varrho_k^{(0)}/\varrho_j^{(0)}, \quad j,k \in X_1^{(0)}. \tag{4.3.9}$$

Relations (4.2.21) and (4.2.22) imply that

$$\mathsf{E}_j \delta_{jk}^{(\varepsilon)} \to \mathsf{E}_j \delta_{jk}^{(0)} \quad \text{as } \varepsilon \to 0, \quad j,k \in X_1^{(0)}. \tag{4.3.10}$$

Let us first consider the case where $\lambda = 0$ in condition $\mathbf{B_6}$.

Obviously, $p_{k0}^{(\varepsilon)} \leq f^{(\varepsilon)}/\varrho_k^{(0)}, k \in X_1^{(0)}$. Using these inequalities, relations (4.2.21), (4.2.22), and formula (4.3.3) we have

$$\overline{\lim_{\varepsilon \to 0}} \, _j f_{j0}^{(\varepsilon)} \cdot t^{(\varepsilon)} \leq \overline{\lim_{\varepsilon \to 0}} \sum_{k \in X_1^{(0)}} \mathsf{E}_j \, \delta_{jk}^{(\varepsilon)} \, p_{k0}^{(\varepsilon)} t^{(\varepsilon)}$$

$$\leq \overline{\lim_{\varepsilon \to 0}} \Big(\sum_{k \in X_1^{(0)}} \mathsf{E}_j \, \delta_{jk}^{(\varepsilon)}/\varrho_k^{(0)} \Big) \cdot \overline{\lim_{\varepsilon \to 0}} \, f^{(\varepsilon)} t^{(\varepsilon)}$$

$$\leq (N/\varrho_j^{(0)}) \cdot \overline{\lim_{\varepsilon \to 0}} \, f^{(\varepsilon)} t^{(\varepsilon)} = 0. \tag{4.3.11}$$

Let us now consider the case where $\lambda > 0$ in condition $\mathbf{B_6}$. As was mentioned above, $f^{(\varepsilon)} > 0$ in this case for all ε small enough, say $0 < \varepsilon \leq \varepsilon_4$.

Using relations (4.2.21), (4.2.25), and formulas (4.3.3), (4.3.9) we get for $0 < \varepsilon \leq \varepsilon_4$ that

$$\frac{|\, _j f_{j0}^{(\varepsilon)} - f^{(\varepsilon)}/\varrho_j^{(0)} \,|}{f^{(\varepsilon)}/\varrho_j^{(0)}} \leq \sum_{k=1}^{N} |\, \mathsf{E}_j \, \delta_{jk}^{(\varepsilon)} - \varrho_k^{(0)}/\varrho_j^{(0)} \,| \cdot \frac{\varrho_j^{(0)} p_{k0}^{(\varepsilon)}}{\sum_{i=1}^{N} \varrho_i^{(0)} p_{i0}^{(\varepsilon)}}$$

$$\leq \sum_{k=1}^{N} |\, \mathsf{E}_j \, \delta_{jk}^{(\varepsilon)} - \varrho_k^{(0)}/\varrho_j^{(0)} \,| \frac{\varrho_j^{(0)}}{\varrho_k^{(0)}}$$

$$\to 0 \quad \text{as } \varepsilon \to 0. \tag{4.3.12}$$

Relation (4.3.12) implies that

$$\frac{_j f_{j0}^{(\varepsilon)} \varrho_j^{(0)}}{f^{(\varepsilon)}} \to 1 \quad \text{as } \varepsilon \to 0, \tag{4.3.13}$$

and, thus,

$$_j f_{j0}^{(\varepsilon)} \cdot t^{(\varepsilon)} = \frac{_j f_{j0}^{(\varepsilon)} \varrho_j^{(0)}}{f^{(\varepsilon)}} \cdot \frac{f^{(\varepsilon)} t^{(\varepsilon)}}{\varrho_j^{(0)}} \to \lambda/\varrho_j^{(0)} \quad \text{as } \varepsilon \to 0. \tag{4.3.14}$$

The proof is complete. $\qquad\qquad\qquad\qquad\qquad\qquad\qquad\qquad\qquad\quad \square$

4.3.3   Asymptotics of cyclic absorption probabilities for semi-Markov processes with non-recurrent-without-absorption states

An asymptotic analysis of cyclic absorption probabilities becomes more complicated in the case where condition $\mathbf{E}_2''$ holds.

Let conditions $\mathbf{A}_2$, $\mathbf{A}_3$, $\mathbf{T}_1$ (a), and $\mathbf{E}_2''$ hold, and $\eta_0^{(\varepsilon)} = i \in X_1^{(0)}$. We define subsequent return times by $\hat{v}^{(\varepsilon)}(n) = \min(k > v^{(\varepsilon)}(n-1) : \eta_k^{(\varepsilon)} \in Y)$, $n = 1, 2, \ldots, \hat{v}^{(\varepsilon)}(0) = 0$, where $Y = X_1^{(0)} \cup \{0\}$. In this case, as follows from Lemma 4.2.2, the random variables $\hat{v}^{(\varepsilon)}(n)$, $n = 0, 1, \ldots$ are finite with probability 1 for $\varepsilon \le \varepsilon_0$. Thus, for $\varepsilon \le \varepsilon_0$ we can define the random sequence $\hat{\eta}^{(\varepsilon)}(n) = \eta_{\hat{v}^{(\varepsilon)}(n)}^{(\varepsilon)}$, $n = 0, 1, \ldots$.

This sequence is a homogeneous Markov chain with the phase space Y. Obviously, 0 is an absorption state while $X_1^{(0)}$ is one class of communicative states for this Markov chain. The transition probabilities of this Markov chain for non-absorbing states are given by the following formulas:

$$\hat{p}_{ij}^{(\varepsilon)} = p_{ij}^{(\varepsilon)} + \sum_{k \in X_2^{(0)}} p_{ik}^{(\varepsilon)} {}_0 f_{kX_1^{(0)}j}^{(\varepsilon)}, \quad i, j \in X_1^{(0)}, \tag{4.3.15}$$

and

$$\hat{p}_{i0}^{(\varepsilon)} = p_{i0}^{(\varepsilon)} + \sum_{k \in X_2^{(0)}} p_{ik}^{(\varepsilon)} {}_{X_1^{(0)}} f_{k0}^{(\varepsilon)}, \quad i \in X_1^{(0)}. \tag{4.3.16}$$

It follows from conditions $\mathbf{A}_3$ and $\mathbf{T}_1$ (a) that $p_{ik}^{(\varepsilon)} \to 0$ as $\varepsilon \to 0$ for the states $i \in X_1^{(0)}, k \in X_2^{(0)}$. Thus,

$$\hat{p}_{ij}^{(\varepsilon)} \to \hat{p}_{ij}^{(0)} = p_{ij}^{(0)} \text{ as } \varepsilon \to 0, \quad i, j \in X_1^{(0)}, \tag{4.3.17}$$

and

$$\hat{p}_{i0}^{(\varepsilon)} \to \hat{p}_{i0}^{(0)} = 0 \text{ as } \varepsilon \to 0, \quad i \in X_1^{(0)}. \tag{4.3.18}$$

It is obvious that the following identity holds:

$$_j\hat{f}_{j0}^{(\varepsilon)} = \mathbf{P}_j\{\hat{\eta}_1^{(\varepsilon)} = 0\} = {}_j f_{j0}^{(\varepsilon)}, \quad j \in X_1^{(0)}. \tag{4.3.19}$$

Thus the cyclic absorption probabilities coincide for the Markov chains $\eta_n^{(\varepsilon)}$ and $\hat{\eta}_n^{(\varepsilon)}$.

Also, relations (4.3.17) and (4.3.18) imply that conditions $\mathbf{A}_2$, $\mathbf{T}_1$ (a), and $\mathbf{E}_2'$ hold for the Markov chain $\hat{\eta}_n^{(\varepsilon)}$.

Denote

$$\hat{f}^{(\varepsilon)} = \sum_{i \in X_1^{(0)}} \varrho_i^{(0)} \hat{p}_{i0}^{(\varepsilon)}.$$

Conditions $\mathbf{A_3}$, $\mathbf{T_1}$ (a), and $\mathbf{E_2''}$ evidently imply that

$$\hat{f}^{(\varepsilon)} \to \hat{f}^{(0)} = 0 \ \text{ as } \ \varepsilon \to 0. \tag{4.3.20}$$

Let us also assume the following condition that relates the rate of growth of time, $t^{(\varepsilon)} \to \infty$, and the speed of the perturbation determined, in this case, by the rate with which the averaged absorption probability vanishes, $\hat{f}^{(\varepsilon)} \to 0$:

$\mathbf{B_7}$: $0 \le t^{(\varepsilon)} \to \infty$ and $\hat{f}^{(\varepsilon)} \to 0$ as $\varepsilon \to 0$ in such a way that $\hat{f}^{(\varepsilon)} t^{(\varepsilon)} \to \lambda$, where $0 \le \lambda \le \infty$.

Let us remark that condition $\mathbf{B_7}$ implies the following: **(c)** if $\lambda > 0$ then automatically $\hat{f}^{(\varepsilon)} > 0$ for all ε small enough, **(d)** if there exists a sequence $0 < \varepsilon_n \to 0$ as $n \to \infty$ such that $\hat{f}^{(\varepsilon_n)} = 0, n \ge 1$, then $\lambda = 0$.

Finally, we can get an asymptotics for the cyclic absorption probabilities ${}_j f_{j0}^{(\varepsilon)}$, $j \in X_1^{(0)}$ by applying Lemma 4.3.4 to the Markov chains $\hat{\eta}_n^{(\varepsilon)}$.

Lemma 4.3.5. *Let conditions* $\mathbf{A_2}$, $\mathbf{A_3}$, $\mathbf{T_1}$ (a), *and* $\mathbf{E_2''}$ *hold. Then,* $(\boldsymbol{\alpha})$ ${}_j f_{j0}^{(\varepsilon)} \to 0$ *as* $\varepsilon \to 0$, $j \in X_1^{(0)}$. *Let also condition* $\mathbf{B_7}$ *hold. Then* $(\boldsymbol{\beta})$ ${}_j f_{j0}^{(\varepsilon)} t^{(\varepsilon)} \to \lambda / \varrho_j^{(0)}$ *as* $\varepsilon \to 0$, $j \in X_1^{(0)}$.

Lemmas 4.3.4 and 4.3.5 play an important role in what follows. They make it possible to formulate the corresponding balancing conditions between the time and the absorption probabilities in a form invariant with respect to the choice of the state used to build the regeneration cycles.

4.3.4   Balancing conditions for cyclic absorption probabilities

In conclusion, let us formulate a condition under which $\mathbf{B_6}$ implies that $\mathbf{B_7}$ holds with the same function $t^{(\varepsilon)}$ and the limit λ.

By using formulas (4.2.70), (4.2.77) in the case where $Z = X_1^{(0)}$ and then formula (4.3.16), we can represent the quantity $\hat{f}^{(\varepsilon)}$ in the following form:

$$\hat{f}^{(\varepsilon)} = \sum_{i \in X_1^{(0)}} \varrho_i^{(0)} \Big(p_{i0}^{(\varepsilon)} + \sum_{k \in X_2^{(0)}} p_{ik}^{(\varepsilon)} {}_{X_1^{(0)}} f_{k0}^{(\varepsilon)} \Big) = f^{(\varepsilon)} + f_1^{(\varepsilon)}, \tag{4.3.21}$$

where

$$f_1^{(\varepsilon)} = \sum_{i \in X_1^{(0)}} \varrho_i^{(0)} \sum_{k \in X_2^{(0)}} p_{ik}^{(\varepsilon)} \sum_{r \in X_2^{(0)}} \mathsf{E}_k \delta_{X_1^{(0)}, r}^{(\varepsilon)} \, p_{r0}^{(\varepsilon)}.$$

Conditions $\mathbf{T_1}$ (a) and $\mathbf{E_2''}$ imply that (e) $p_{ik}^{(\varepsilon)} \to 0$ as $\varepsilon \to 0$, $i \in X_1^{(0)}$, $k \in X_2$, and (f) $\mathsf{E}_k \delta_{X_1^{(0)} r}^{(\varepsilon)} \to \mathsf{E}_k \delta_{X_1^{(0)} r}^{(\varepsilon)} < \infty$ as $\varepsilon \to 0$, $k, r \in X_2^{(0)}$.

It is useful to note that relations (e) and (f) imply that (g) $f_1^{(\varepsilon)} \to 0$ as $\varepsilon \to 0$, and, therefore, (h) $\hat{f}^{(\varepsilon)} \to 0$ as $\varepsilon \to 0$ even if (i) $p_{k0}^{(\varepsilon)} \not\to 0$ as $\varepsilon \to 0$ for $k \in X_2^{(0)}$.

Let us assume the following condition:

$\mathbf{B_8}$: $\overline{\lim}_{\varepsilon \to 0}\, p_{k0}^{(\varepsilon)} t^{(\varepsilon)} < \infty$, $k \in X_2^{(0)}$.

If conditions $\mathbf{B_6}$ and $\mathbf{B_8}$ hold, then formula (4.3.21) and relations (e) and (f) above easily give that (h) $f_1^{(\varepsilon)} t^{(\varepsilon)} \to 0$ as $\varepsilon \to 0$. Therefore, condition $\mathbf{B_7}$ holds with the same function $t^{(\varepsilon)}$ and the limit λ as in condition $\mathbf{B_6}$, i.e.,

$$\lim_{\varepsilon \to 0} \hat{f}^{(\varepsilon)} t^{(\varepsilon)} = \lim_{\varepsilon \to 0} f^{(\varepsilon)} t^{(\varepsilon)} = \lambda. \tag{4.3.22}$$

If condition $\mathbf{B_8}$ does not hold, in particular, if $f^{(\varepsilon)} = 0$ for $\varepsilon > 0$, then the asymptotics for the quantities $\hat{f}^{(\varepsilon)}$ and, as a consequence, the cyclic absorption probabilities ${}_j f_{j0}^{(\varepsilon)}$ can be more complicated and require an additional careful analysis. We will consider examples of such models in Chapter 5.

4.4   Mixed ergodic and limit theorems for perturbed semi-Markov processes

In this section we present mixed ergodic and limit theorems for perturbed regenerative processes. These theorems describe pseudo-stationary phenomena for nonlinearly perturbed semi-Markov processes.

4.4.1   Non-arithmetic properties of distributions of return times

Let us formulate now a condition that would guarantee that the limit distribution functions of the return times $G_{ii}^{(0)}(t)$ are arithmetic.

A distribution function $F(x)$ is called *generalised arithmetic* if there exists a positive h and a real constant d such that the following relation holds:

$$\sum_{n=-\infty}^{\infty} (F(nh - d) - F(nh - d - 0)) = 1. \tag{4.4.1}$$

Denote by $U_h = \{0, \pm h, \pm 2h, \ldots\}$ the lattice of points with step $h > 0$.

Let also $I_i^{(m)}$ be, for $i \in X_1^{(0)}$, $m \geq 1$, the set of all cyclic chains of states $(i_0, \ldots, i_m)$ such that (a) $i_0 = i, i_1, \ldots, i_{m-1} \neq i, 0$ and $i_1, \ldots, i_{m-1} \in X_1^{(0)}$, $i_m = i$; (b) $\prod_{n=1}^{m} p_{i_{n-1} i_n}^{(0)} > 0$.

Let us introduce the following condition:

$\bar{\mathbf{N}}_1$: There exist positive h and real constants d_{rk}, $r, k \in X_1^{(0)}$, satisfying the following:

(a) $\sum_{n=-\infty}^{\infty}(F_{rk}^{(0)}(nh - d_{rk}) - F_{rk}^{(0)}(nh - d_{rk} - 0)) = 1$ for any $r, k \in X_1^{(0)}$, such that $p_{rk}^{(0)} > 0$;

(b) $\sum_{n=1}^{m} d_{i_{n-1}i_n} \in U_h$ for any cyclic chain of states $(i_0, \ldots, i_m) \in I_i^{(m)}$ for every $i \in X_1^{(0)}$ and $m \geq 1$.

Denote

$$_0\bar{G}_{ii}^{(0)}(t) = {_0}G_{ii}^{(0)}(t)/{_0}G_{ii}^{(0)}(\infty), \quad t \geq 0, \quad i \in X_1^{(0)}.$$

Note that condition $\mathbf{E}_2$ implies that $_0G_{ii}^{(0)}(\infty) = {_0}f_{ii}^{(0)} > 0$ for $i \in X_1^{(0)}$.

Lemma 4.4.1. *Let conditions $\mathbf{A}_2$ and $\mathbf{E}_2$ hold. Then (α) if $_0\bar{G}_{ii}^{(0)}(t)$ is an arithmetic distribution function for some $i \in X_1^{(0)}$, then it is an arithmetic distribution function for every $i \in X_1^{(0)}$; (β) condition $\bar{\mathbf{N}}_1$ is necessary and sufficient for the distribution functions $_0\bar{G}_{ii}^{(0)}(t)$, $i \in X_1^{(0)}$, to be arithmetic.*

Proof. Denote

$$\psi_{ij}^{(0)}(\lambda) = \int_0^{\infty} e^{i\lambda s} F_{ij}^{(0)}(ds), \quad _0\varphi_{ij}^{(0)}(\lambda) = \int_0^{\infty} e^{i\lambda s} {_0}\bar{G}_{ij}^{(0)}(ds), \quad \lambda \in \mathbb{R}_1, \ i, j \in X.$$

Let us assume that the distribution function $_0\bar{G}_{ii}^{(0)}(t)$ is arithmetic. As is known (see, for example, Feller (1971)) this is so if and only if (a) $_0\varphi_{ii}^{(0)}(\lambda_h) = 1$ for some $\lambda_h = 2\pi/h > 0$ or, equivalently, (b) $\sum_{n=-\infty}^{\infty}({_0}\bar{G}_{ii}^{(0)}(nh) - {_0}\bar{G}_{ii}^{(0)}(nh - 0)) = 1$.

Obviously,

$$_0\varphi_{ii}^{(0)}(\lambda) = ({_0}f_{ii}^{(0)})^{-1} \sum_{m=1}^{\infty} \sum_{(i_0,\ldots,i_m)\in I_i^{(m)}} \prod_{n=1}^{m} \psi_{i_{n-1}i_n}^{(0)}(\lambda) p_{i_{n-1}i_n}^{(0)}, \quad \lambda \in \mathbb{R}_1, \quad (4.4.2)$$

and, thus,

$$|_0\varphi_{ii}^{(0)}(\lambda)| \leq ({_0}f_{ii}^{(0)})^{-1} \sum_{m=1}^{\infty} \sum_{(i_0,\ldots,i_m)\in I_i^{(m)}} \prod_{n=1}^{m} |\psi_{i_{n-1}i_n}^{(0)}(\lambda)| p_{i_{n-1}i_n}^{(0)}, \quad \lambda \in \mathbb{R}_1. \quad (4.4.3)$$

Note that, by the definition,

$$_0f_{ii}^{(0)} = \sum_{m=1}^{\infty} \sum_{(i_0,\ldots,i_m)\in I_i^{(m)}} \prod_{n=1}^{m} p_{i_{n-1}i_n}^{(0)}. \quad (4.4.4)$$

It follows from relations (4.4.3) and (4.4.4) that **(c)** $|\psi_{rk}^{(0)}(\lambda_h)| = 1$ if $p_{rk}^{(0)} > 0$.

Indeed, since $X_1^{(0)}$ is a communicative class of states, for $r, k \in X_1^{(0)}$ such that $p_{rk}^{(0)} > 0$ there exists a chain of states $(i_0, \dots, i_m) \in I_i^{(m)}$ such that $\sum_{n=1}^m \chi((i_{n-1}, i_n) = (r, k)) > 0$. Thus, if **(c)** would not hold, i.e., $|\psi_{rk}^{(0)}(\lambda_h)| < 1$ for some $r, k \in X_1^{(0)}$ such that $p_{rk}^{(0)} > 0$, then relations (4.4.3) and (4.4.4) would imply that $|_0\varphi_{ii}^{(0)}(\lambda_h)| < 1$.

Relation **(d)** $|\psi_{rk}^{(0)}(\lambda_h)| = 1$ means that there exists real d_{rk} such that $\psi_{rk}^{(0)}(\lambda_h) = e^{i\lambda_h d_{rk}}$. In this case, the characteristic function $\psi_{rk}^{(0)}(\lambda)e^{-i\lambda d_{rk}}$ takes value 1 in point λ_h. Thus, by the same result as mentioned above, relation **(d)** is equivalent to the following relation: **(e)** $\sum_{n=-\infty}^{\infty}(F_{rk}^{(0)}(nh-d_{rk}) - F_{rk}^{(0)}(nh-d_{rk}-0)) = 1$. Therefore, condition $\bar{\mathbf{N}}_1$ **(a)** holds.

Relation (4.4.2), taken for $\lambda = \lambda_h$, can be re-written in the following form:

$$1 = ({}_0 f_{ii}^{(0)})^{-1} \sum_{m=1}^{\infty} \sum_{(i_0,\dots,i_m)\in I_i^{(m)}} \exp\{i\lambda_h \sum_{n=1}^m d_{i_{n-1}i_n}\} \prod_{n=1}^m p_{i_{n-1}i_n}^{(0)}. \qquad (4.4.5)$$

It follows from (4.4.4) that this equality holds if and only if $\lambda_h \sum_{n=1}^m d_{i_{n-1}i_n} \in U_{2\pi}$ or, equivalently, **(f)** $\sum_{n=1}^m d_{i_{n-1}i_n} \in U_h$ for any chain of states $(i_0, \dots, i_m) \in I_i^{(m)}$ for every $m \geq 1$.

Let us now prove that **(f)** implies that **(g)** $\sum_{n=1}^m d_{i_{n-1}i_n} \in U_h$ for any chain of states $(i_0, \dots, i_m) \in I_j^{(m)}$ for every $m \geq 1$ and any $j \neq i$, $j \in X_1^{(0)}$.

Take an arbitrary chain of states $(i_0, \dots, i_m) \in I_j^{(m)}$. Two cases are possible for this chain: **(h$'$)** $i_1, \dots, i_{m-1} \neq i$, and **(h$''$)** there exists a subsequence $0 < n_1 < \cdots < n_k < m$ such that $i_{n_1}, \dots, i_{n_k} = i$ while $i_n \neq i, 0 < n < m, n \neq n_1, \dots, n_k$.

Let us first consider the case **(h$'$)**. Since $X_1^{(0)}$ is a class of communicative states, **(i)** there exist two sequences of states $r_0 = i, r_1 \neq i, j, r_1 \in X_1^{(0)}, \dots, r_{q-1} \neq i$, $j, j_{q-1} \in X_1^{(0)}, j_q = j$ and $j_0 = j, j_1 \neq j, i, j_1 \in X_1^{(0)}, \dots, j_{l-1} \neq j, i, j_{l-1} \in X_1^{(0)}, j_l = i$ such that $\prod_{n=1}^q p_{r_{n-1}r_n} > 0$ and $\prod_{n=1}^l p_{j_{n-1}j_n} > 0$. Then **(h)** and **(i)** imply that **(j)** two chains of states $(r_0, \dots, r_{q-1}, i_0, \dots, i_{m-1}, j_0, \dots, j_l)$ and $(r_0, \dots, r_{q-1}, i_0, \dots, i_{m-1}, i_0, \dots, i_{m-1}, j_0, \dots, j_l)$ belong, respectively, to the sets $I_i^{(q+m+l)}$ and $I_i^{(q+2m+l)}$. Thus, **(k)** $\sum_{n=1}^q d_{r_{n-1}r_n} + \sum_{n=1}^m d_{i_{n-1}i_n} + \sum_{n=1}^l d_{j_{n-1}j_n} \in U_h$ and $\sum_{n=1}^q d_{r_{n-1}r_n} + 2\sum_{n=1}^m d_{i_{n-1}i_n} + \sum_{n=1}^l d_{j_{n-1}j_n} \in U_h$. Obviously, **(k)** implies that **(l)** the difference of these sums is $\sum_{n=1}^m d_{i_{n-1}i_n} \in U_h$, i.e., **(g)** holds for the chain of states $(i_0, \dots, i_m)$.

Let us now consider the case **(h$''$)**. In this case, the chains of states satisfy $(i_{n_1}, \dots, i_{n_2}) \in I_i^{(n_2-n_1)}, \dots, (i_{n_{k-1}}, \dots, i_{n_k}) \in I_i^{(n_k-n_{k-1})}$ and $(i_{n_k}, \dots, i_{m-1}, i_0, \dots, i_{n_1}) \in I_i^{(m-n_k+n_1)}$. Thus, **(m)** $\sum_{n=n_1+1}^{n_2} d_{i_{n-1}i_n} + \cdots + \sum_{n=n_{k-1}+1}^{n_k} d_{i_{n-1}i_n}$

$+ \sum_{n=n_k+1}^{m} d_{i_{n-1}i_n} = \sum_{n=1}^{m} d_{i_{n-1}i_n} > 0$, i.e., again **(g)** holds for the chain of states $(i_0, \ldots, i_m)$.

Proposition **(g)** implies that condition $\bar{\mathbf{N}}_1$ **(b)** also holds.

Thus, **(n)** the assumption that the distribution $_0\bar{G}_{ii}^{(0)}(t)$ is arithmetic for some $i \in X_1^{(0)}$ implies that condition $\bar{\mathbf{N}}_1$ holds.

On the other hand, **(o)** condition $\bar{\mathbf{N}}_1$ **(a)** implies that $\psi_{rk}^{(0)}(\lambda_h)e^{-i\lambda_h d_{rk}} = 1$ if $p_{rk}^{(0)} > 0$, while **(p)** condition $\bar{\mathbf{N}}_1$ **(b)** implies that relation (4.4.5) holds. Using (4.4.4), **(o)** and **(p)** we get for any $i \in X_1^{(0)}$ that

$$_0\varphi_{ii}^{(0)}(\lambda_h)$$

$$= (_0f_{ii}^{(0)})^{-1} \sum_{m=1}^{\infty} \sum_{(i_0,\ldots,i_m)\in I_i^{(m)}} \prod_{n=1}^{m} \psi_{i_{n-1}i_n}^{(0)}(\lambda_h)e^{-i\lambda_h d_{i_{n-1}i_n}} e^{i\lambda_h d_{i_{n-1}i_n}} p_{i_{n-1}i_n}^{(0)}$$

$$= (_0f_{ii}^{(0)})^{-1} \sum_{m=1}^{\infty} \sum_{(i_0,\ldots,i_m)\in I_i^{(m)}} \exp\{i\lambda_h \sum_{n=1}^{m} d_{i_{n-1}i_n}\} \prod_{n=1}^{m} p_{i_{n-1}i_n}^{(0)}$$

$$= (_0f_{ii}^{(0)})^{-1} \sum_{m=1}^{\infty} \sum_{(i_0,\ldots,i_m)\in I_i^{(m)}} \prod_{n=1}^{m} p_{i_{n-1}i_n}^{(0)} = 1. \tag{4.4.6}$$

As was mentioned above, the equality $_0\varphi_{ii}^{(0)}(\lambda_h) = 1$ for some $\lambda_h > 0$ implies that the distribution function $\bar{G}_{ii}^{(0)}(t)$ is arithmetic.

Thus we have proved that **(q)** condition $\bar{\mathbf{N}}_1$ implies that distribution $_0\bar{G}_{ii}^{(0)}(t)$ is arithmetic for any $i \in X_1^{(0)}$.

Statements **(n)** and **(q)** prove both propositions of Lemma 4.4.1. $\square$

In fact, we will use the following condition that is opposite to condition $\bar{\mathbf{N}}_1$:

$\mathbf{N}_1$:   There does not exist positive h and real constants $d_{rk}, r, k \in X_1^{(0)}$ such that both conditions $\bar{\mathbf{N}}_1$ **(a)** and $\bar{\mathbf{N}}_1$ **(b)** hold.

According to Lemma 4.4.3, condition $\mathbf{N}_1$ is necessary and sufficient for the distribution functions of return times, $_0G_{ii}^{(0)}(t), i \in X_1^{(0)}$, to be non-arithmetic.

Condition $\mathbf{N}_1$ is not restrictive. It holds, for example, if the distribution function $F_{rk}^{(0)}(t)$ has an absolutely continuous component for some $r, k \in X_1^{(0)}$ such that $p_{rk}^{(0)} > 0$.

It is also useful to mention that condition $\mathbf{N}_1$ automatically implies condition $\mathbf{I}_4$ **(a)**. Indeed, if this condition would not hold, i.e., $F_i^{(0)}(0) = 1, i \in X_1^{(0)}$, then it is obvious that $_0G_{ii}^{(0)}(0) = 1, i \in X_1^{(0)}$, i.e., the distribution functions $G_{ii}^{(0)}(t), i \in X_1^{(0)}$, would be concentrated in zero. This would contradict to the non-arithmetic property of these distribution functions.

4.4.2 Mixed ergodic and limit theorems for perturbed semi-Markov processes without non-recurrent-without-absorption states

Now we are in a position to formulate basic results concerning pseudo-stationary phenomena for perturbed semi-Markov processes. We assume that conditions $\mathbf{A_2}$, $\mathbf{T_1}$, $\mathbf{M_{11}}$, and $\mathbf{A_3}$ hold.

Let us first consider the case where condition $\mathbf{E'_2}$ holds.

Denote

$$\pi_j^{(0)} = \frac{\varrho_j^{(0)} m_j^{(0)}}{\sum_{i \in X_1^{(0)}} \varrho_i^{(0)} m_i^{(0)}}, \quad j \in X_1^{(0)}, \tag{4.4.7}$$

where

$$m_i^{(0)} = \sum_{k \in X} p_{ik}^{(0)}[1] = \sum_{k \in X} m_{ik}^{(0)} p_{ik}^{(0)}, \quad i \neq 0.$$

Note that here and henceforth the product $p_{ik}^{(0)}[1] = m_{ik}^{(0)} p_{ik}^{(0)}$ should be counted as 0 if $m_{ik}^{(0)} = \infty$ but $p_{ik}^{(0)} = 0$. In particular, condition $\mathbf{A_3}$ implies that $p_{i0}^{(0)}[1] = m_{i0}^{(0)} p_{i0}^{(0)} = 0$.

As is known, $\pi_j^{(0)}$, $j \in X_1^{(0)}$, is a *stationary distribution* for the limit semi-Markov process $\eta^{(0)}(t)$, $t \geq 0$, for the class of states $X_1^{(0)}$.

Let us also denote

$$m^{(0)} = \sum_{i \in X_1^{(0)}} \varrho_i^{(0)} m_i^{(0)}.$$

Theorem 4.4.1. *Let conditions* $\mathbf{A_2}$, $\mathbf{T_1}$, $\mathbf{E'_2}$, $\mathbf{M_{11}}$, $\mathbf{A_3}$, $\mathbf{N_1}$ *hold, and* $0 \leq t^{(\varepsilon)} \to \infty$ *as* $\varepsilon \to 0$ *such that condition* $\mathbf{B_6}$ *holds. Then*

$$P_{ij}^{(\varepsilon)}(t^{(\varepsilon)}) = \mathsf{P}_i\{\eta^{(\varepsilon)}(t^{(\varepsilon)}) = j, \mu_0^{(\varepsilon)} > t^{(\varepsilon)}\}$$

$$\to \pi_j^{(0)} e^{-\lambda/m^{(0)}} \quad \text{as} \ \varepsilon \to 0, \quad i, j \neq 0. \tag{4.4.8}$$

Proof. According to the remarks made in Subsection 4.2.1, the semi-Markov process $\eta^{(\varepsilon)}(t)$, more precisely, its modified version with the absorption state removed according to the procedure described in that subsection, can be considered as a perturbed regenerative process. The absorption time $\mu_0^{(\varepsilon)}$ is regarded as the regenerative stopping time.

Let us chose some state $j \in X_1^{(0)}$. Note that, according to condition $\mathbf{E'_2}$, we have $X_1^{(0)} = X_0^{(0)} = \{1, \ldots, N\}$. We can apply Theorem 3.2.1 to prove the asymptotic relation (4.4.8) in the case where $i = j$, since, in this case, $\eta^{(\varepsilon)}(t)$ is a standard regenerative process with regenerative times $\mu_j^{(\varepsilon)}(n)$ that are subsequent times of return of the process $\eta^{(\varepsilon)}(t)$ into the state j.

The renewal equation (3.2.1) now takes the form of renewal equation (4.2.1). In this case, $A = \{j\}$ and the probability $P_{jj}^{(\varepsilon)}(t)$ plays the role of the probability $P^{(\varepsilon)}(t, A)$.

The distribution function $_0G_{jj}^{(\varepsilon)}(t)$ is regarded as the distribution function $F^{(\varepsilon)}(t)$ that generates this renewal equation. Conditions $\mathbf{A_2}$, $\mathbf{T_1}$, $\mathbf{E_2'}$, $\mathbf{A_3}$, and $\mathbf{N_1}$ imply, by Lemmas 4.2.1, 4.2.3, 4.3.4, and 4.4.1, respectively, that (**k**) the distribution functions $_0G_{jj}^{(\varepsilon)}(t) \Rightarrow {_0G_{jj}^{(0)}}(t)$ as $\varepsilon \to 0$, (**l**) the limit distribution function $_0G_{jj}^{(0)}(t)$ is proper, and (**m**) it is also non-arithmetic. It follows from (**k**)–(**m**) that the condition of weak convergence $\mathbf{D_{13}}$ holds for the distribution functions $_0G_{jj}^{(\varepsilon)}(t)$.

The distribution function $G_{jj}^{(\varepsilon)}(t)$ is regarded here as the distribution function $\hat{F}^{(\varepsilon)}(t)$. Conditions $\mathbf{A_2}$, $\mathbf{T_1}$, $\mathbf{E_2'}$, and $\mathbf{M_{11}}$ imply, by Lemmas 4.2.3 and 4.2.4, (**n**) the distribution functions $_0G_{jj}^{(\varepsilon)}(t) \Rightarrow {_0G_{jj}^{(0)}}(t)$ as $\varepsilon \to 0$, and (**o**) convergence of the first moments of these distribution functions, $M_{jj}^{(\varepsilon)} \to M_{jj}^{(0)}$ as $\varepsilon \to 0$. The convergence relations (**k**) and (**o**) imply, by Remark 1.2.2, that (**p**) condition $\mathbf{M_8}$ holds for the distribution functions $G_{jj}^{(\varepsilon)}(t)$.

The function $1 - F_j^{(\varepsilon)}(t)$ is a forcing function for the renewal equation (4.2.1). The distribution functions $F_j^{(\varepsilon)}(t)$ are non-decreasing, and, by condition $\mathbf{T_1}$, weakly converge. Thus, (**q**) the functions $1 - F_j^{(\varepsilon)}(t)$ converge locally uniformly in every continuity point of the limit function $1 - F_j^{(0)}(t)$.

Indeed, denote by C_j the set of positive points of continuity of the distribution function $F_j^{(0)}(s)$. Take a point $t \in C_j$ and let s_n, $n = 0, 1, \ldots$, be a sequence of positive numbers such that (**r**) $s_n \to 0$ as $n \to \infty$ and (**s**) the points $t \pm s_n$ are points of continuity of the distribution function $F_j^{(0)}(s)$ for all $n = 0, 1, \ldots$. Using (**r**), (**s**), and weak convergence of $F_j^{(\varepsilon)}(t)$ we have that for any $n = 0, 1, \ldots$,

$$
\lim_{u \to 0} \varlimsup_{0 \leq \varepsilon \to 0} \sup_{|v| \leq u} |(1 - F_j^{(\varepsilon)}(t + v)) - (1 - F_j^{(0)}(t))|
$$

$$
= \varlimsup_{0 \leq \varepsilon \to 0} (|(1 - F_j^{(\varepsilon)}(t - s_n)) - (1 - F_j^{(0)}(t + s_n))|
$$

$$
+ |(1 - F_j^{(\varepsilon)}(t + s_n)) - (1 - F_j^{(0)}(t - s_n))|)
$$

$$
\leq 2|(1 - F_j^{(0)}(t - s_n)) - (1 - F_j^{(0)}(t + s_n))| \to 0 \text{ as } \varepsilon \to 0. \tag{4.4.9}
$$

It remains to note that the set $\overline{C}_j$ is at most countable and, therefore, (**t**) $\overline{C}_j$ has Lebesgue measure zero, $m(\overline{C}_j) = 0$. Thus condition $\mathbf{F_7}$ holds for the functions $1 - F_j^{(\varepsilon)}(t)$.

Finally, the probability $_0f_{jj}^{(\varepsilon)}$ can be regarded as the stopping probability $f^{(\varepsilon)}$ in one regeneration cycle. The balancing condition $\mathbf{B_6}$ implies, by Lemma 4.3.4, that

$_0 f_{jj}^{(\varepsilon)} t^{(\varepsilon)} \to \lambda/\varrho_j^{(0)}$ as $\varepsilon \to 0$. Thus, the balancing condition $\mathbf{B_4}$ holds with the limit constant $\lambda' = \lambda/\varrho_j^{(0)}$.

Thus, all conditions of Theorem 3.2.1 are satisfied. It remains to calculate the parameters in the corresponding limit expression $\pi^{(0)}(A)e^{-\lambda'/m_1^{(0)}}$ in the asymptotic relation (3.2.10) in this theorem.

Due to conditions $\mathbf{E_2'}$ and $\mathbf{A_3}$, we have $_0 f_{kj}^{(0)} = 1$, $k, j \in X_1^{(0)}$ and $p_{i0}^{(0)} = 0$, $i \in X_1^{(0)}$.

Therefore, the coefficients $L_{ij}^{(0)}$ take, in this case, the following form:

$$L_{ij}^{(0)} = m_{ij}^{(0)} p_{ij}^{(0)} + \sum_{k \neq j,0} m_{ik}^{(0)} p_{ik}^{(0)}$$

$$= \sum_{k \in X_1^{(0)}} m_{ik}^{(0)} p_{ik}^{(0)} = m_i^{(0)}, \quad i, j \in X_1^{(0)}. \tag{4.4.10}$$

Thus, according to formulas (4.2.54), (4.3.9), and (4.4.10),

$$_0 M_{jj}^{(0)} = \sum_{k \in X_1^{(0)}} \mathsf{E}_j \, \delta_{jk}^{(0)} L_{kj}^{(0)}$$

$$= \sum_{k \in X_1^{(0)}} (\varrho_k^{(0)}/\varrho_j^{(0)}) m_k^{(0)} = m^{(0)}/\varrho_j^{(0)}, \quad j \in X_1^{(0)}, \tag{4.4.11}$$

and, therefore,

$$\lambda'/_0 M_{jj}^{(0)} = \frac{\lambda/\varrho_j^{(0)}}{m^{(0)}/\varrho_j^{(0)}} = \lambda/m^{(0)}, \quad j \in X_1^{(0)}. \tag{4.4.12}$$

The corresponding limit stationary probability calculated with the use of formula (3.2.9) for the set $A = \{j\}$ yields

$$\pi^{(0)}(A) = \frac{\int_0^\infty (1 - F_j^{(0)}(s))ds}{_0 M_{jj}^{(0)}}$$

$$= \frac{m_j^{(0)}}{m^{(0)}/\varrho_j^{(0)}} = \frac{\varrho_j^{(0)} m_j^{(0)}}{m^{(0)}} = \pi_j^{(0)}, \quad j \in X_1^{(0)}. \tag{4.4.13}$$

Now, by applying the asymptotic relation (3.2.10) given in Theorem 3.2.1, we get the asymptotic relation (4.4.8) for the case where $i = j$.

We can apply Theorem 3.2.5 to prove the asymptotic relation (4.4.8) for $i \neq j$. In this case, $\eta^{(\varepsilon)}(t)$ is again a regenerative process with regeneration times $\mu_j^{(\varepsilon)}(n)$ that are subsequent times of return of the process $\eta^{(\varepsilon)}(t)$ into the state j. This process has a transition period, because the initial state differs from j but the shifted process $\eta^{(\varepsilon)}(\mu_j^{(\varepsilon)}(1) + t)$ is the standard regenerative process considered above.

The renewal type relation (3.2.19) takes in this case a form of the renewal type relation (4.2.2). In this case again, $A = \{j\}$, and the probability $P_{ij}^{(\varepsilon)}(t)$ plays the role of the probability $\tilde{P}^{(\varepsilon)}(t, A)$. The distribution function $_0 G_{ij}^{(\varepsilon)}(t)$ is regarded as the distribution function $\tilde{F}^{(\varepsilon)}(t)$ that generates this renewal type relation, and the probability $_j f_{i0}^{(\varepsilon)} = 1 - {_0 f_{ij}^{(\varepsilon)}} = 1 - {_0 G_{ij}^{(\varepsilon)}(\infty)}$ replaces the stopping probability $\tilde{f}^{(\varepsilon)}$ in the transition period. Conditions $\mathbf{A_2}$, $\mathbf{T_1}$, $\mathbf{E'_2}$, and $\mathbf{A_3}$ imply, by Lemma 4.2.1, that (u) $_j f_{i0}^{(\varepsilon)} \to {_j f_{i0}^{(0)}} = 0$ as $\varepsilon \to 0$. Thus, condition $\mathbf{D_{17}}$ holds with the limit constant $\tilde{f}^{(0)} = 0$.

The distribution function $G_{ij}^{(\varepsilon)}(t)$ replaces the distribution function $\hat{\tilde{F}}^{(\varepsilon)}(t)$. Conditions $\mathbf{A_2}$, $\mathbf{T_1}$, $\mathbf{E'_2}$ and $\mathbf{M_{11}}$ imply, by Lemma 4.2.4, that (v) $\lim_{T \to \infty} \overline{\lim}_{\varepsilon \to 0}(1 - G_{ij}^{(\varepsilon)}(T)) \leq \lim_{T \to \infty} \overline{\lim}_{\varepsilon \to 0} M_{ij}^{(\varepsilon)}/T = 0$. Thus, condition $\mathbf{D_{18}}$ holds.

Finally, by applying the asymptotic relation (3.2.22) given in Theorem 3.2.5, we get the asymptotic relation (4.4.8) for the case where $i \neq j$. $\square$

4.4.3   Mixed ergodic and limit theorems for perturbed semi-Markov processes with non-recurrent-without absorption states

Let us formulate the basic results on transition phenomena for perturbed semi-Markov processes in the case where condition $\mathbf{E''_2}$ holds.

In this case, define

$$
\pi_r^{(0)} =
\begin{cases}
\dfrac{\varrho_r^{(0)} m_r^{(0)}}{\sum_{i \in X_1^{(0)}} \varrho_i^{(0)} m_i^{(0)}} & \text{for } r \in X_1^{(0)}, \\[2ex]
0 & \text{for } r \in X_2^{(0)},
\end{cases}
\tag{4.4.14}
$$

where $\varrho_r^{(0)}$, $r \in X_1^{(0)}$, are as above the stationary probabilities for the states from the class $X_1^{(0)}$ given by a unique solution of the system of linear equation (4.3.2), and the coefficients $m_r^{(0)}$ are the same as in formula (4.4.7).

As is known, $\pi_r^{(0)}$, $r \in X_1^{(0)} \cup X_2^{(0)}$, is a stationary distribution for the limit semi-Markov process $\eta^{(0)}(t)$, $t \geq 0$, in the class of states $X_0^{(0)} = X_1^{(0)} \cup X_2^{(0)}$.

Let us use the absorption probabilities $_{X_1^{(0)}} f_{i0}^{(0)} = \mathsf{P}_i\{v_0^{(0)} < v_{X_1^{(0)}}^{(0)}\}$ introduced in Subsection 4.2.6 and given by formula (4.2.77) with the set $Z = X_1^{(0)}$.

It useful to note that, according to condition $\mathbf{E''_2}$, the states $i \in X_1^{(0)}$ satisfy $_{X_1^{(0)}} f_{i0}^{(0)} = 0$ while $_{X_1^{(0)}} f_{i0}^{(0)} \in [0, 1)$ for the states $i \in X_2^{(0)}$.

Theorem 4.4.2. *Let conditions* $\mathbf{A_2}$, $\mathbf{T_1}$, $\mathbf{E_2''}$, $\mathbf{M_{11}}$, $\mathbf{A_3}$, $\mathbf{N_1}$ *hold, and* $0 \le t^{(\varepsilon)} \to \infty$ *as* $\varepsilon \to 0$ *such that condition* $\mathbf{B_7}$ *holds. Then*

$$P_{ir}^{(\varepsilon)}(t^{(\varepsilon)}) = \mathsf{P}_i\{\eta^{(\varepsilon)}(t^{(\varepsilon)}) = r, \mu_0^{(\varepsilon)} > t^{(\varepsilon)}\}$$

$$\to (1 - {}_{X_1^{(0)}}f_{i0}^{(0)})\pi_r^{(0)}e^{-\lambda/m^{(0)}} \quad as \ \varepsilon \to 0, \quad i, r \ne 0. \qquad (4.4.15)$$

Proof. Let us first consider that case where $i = r$ and $r = j \in X_1^{(0)}$.

The first part of the proof in which we verify conditions $\mathbf{D_{13}}$, $\mathbf{M_8}$, $\mathbf{F_7}$ and $\mathbf{B_4}$ repeats that in the proof of Theorem 4.4.1. The only difference is that the reference to the conditions $\mathbf{E_2'}$ and $\mathbf{B_6}$ should be replaced with the equivalent reference to conditions $\mathbf{E_2''}$ and $\mathbf{B_7}$.

Thus again, all conditions of Theorem 3.2.1 hold and it only remains to calculate the parameters in the corresponding limit expression $\pi^{(0)}(A)e^{-\lambda'/m_1^{(0)}}$ in the asymptotic relation (3.2.10) given in this theorem.

However, a care should be taken. As a matter of fact, under conditions $\mathbf{E_2''}$ and $\mathbf{A_3}$, we have ${}_0f_{kj}^{(0)} = 1$, $k, j \in X_1^{(0)}$, but, due to condition $\mathbf{E_2''}$, it is only known that ${}_0f_{kj}^{(0)} \in (0, 1]$, $k \in X_2^{(0)}$, $j \in X_1^{(0)}$.

Taking also into account that $p_{ik}^{(0)} = 0$, $i \in X_1^{(0)}$, $k \in X_2^{(0)} \cup \{0\}$, we can represent the coefficients $L_{ij}^{(0)}$ in the following form, for $i, j \in X_1^{(0)}$:

$$L_{ij}^{(0)} = m_{ij}^{(0)} p_{ij}^{(0)} + \sum_{k \ne j,0} m_{ik}^{(0)} p_{ik}^{(0)}{}_0 f_{kj}^{(0)} = \sum_{k \in X_1^{(0)}} m_{ik}^{(0)} p_{ik}^{(0)} = m_i^{(0)}. \qquad (4.4.16)$$

The coefficients $L_{ij}^{(0)}$ take a more complicated form for $i \in X_2^{(0)}$, $j \in X_1^{(0)}$,

$$L_{ij}^{(0)} = m_{ij}^{(0)} p_{ij}^{(0)} + \sum_{k \ne j,0} m_{ik}^{(0)} p_{ik}^{(0)}{}_0 f_{kj}^{(0)}$$

$$= \sum_{k \in X_1^{(0)}} m_{ik}^{(0)} p_{ik}^{(0)} + \sum_{k \in X_2^{(0)}} m_{ik}^{(0)} p_{ik}^{(0)}{}_0 f_{kj}^{(0)}. \qquad (4.4.17)$$

Formula (4.3.9) takes, under condition $\mathbf{E_2''}$, the following form:

$$\mathsf{E}_j\,\delta_{jk}^{(0)} = \begin{cases} \varrho_k^{(0)}/\varrho_j^{(0)} & \text{for } j, k \in X_1^{(0)}, \\ 0 & \text{for } j \in X_1^{(0)}, k \in X_2^{(0)}. \end{cases} \qquad (4.4.18)$$

Using formulas (4.2.54), (4.4.16), and (4.4.18) we get

$$
\begin{aligned}
{}_0 M_{jj}^{(0)} &= \sum_{k \in X_1^{(0)} \cup X_2^{(0)}} \mathsf{E}_j \, \delta_{jk}^{(0)} L_{kj}^{(0)} \\
&= \sum_{k \in X_1^{(0)}} (\varrho_k^{(0)} / \varrho_j^{(0)}) m_k^{(0)} = m^{(0)} / \varrho_j^{(0)}, \quad j \in X_1^{(0)}.
\end{aligned}
\tag{4.4.19}
$$

Thus, formula (4.4.19) has the same form as formula (4.4.11). Formulas (4.4.12) and (4.4.13) also do no change.

Let us now consider the case where $i \neq r, 0$ and $r = j \in X_1^{(0)}$.

As above, the renewal type relation (3.2.19) takes in this case the form of the equation (4.2.2), the distribution function ${}_0 G_{ij}^{(\varepsilon)}(t)$ is regarded as the distribution function $\tilde{F}^{(\varepsilon)}(t)$ that generates this renewal type relation, and the probability ${}_j f_{i0}^{(\varepsilon)} = 1 - {}_0 f_{ij}^{(\varepsilon)} = 1 - {}_0 G_{ij}^{(\varepsilon)}(\infty)$ replaces the stopping probability $\tilde{f}^{(\varepsilon)}$ in the transition period. Conditions $\mathbf{A_2}$, $\mathbf{T_1}$, $\mathbf{E_2''}$, and $\mathbf{A_3}$ imply, by Lemma 4.2.1, that $(\mathbf{w})$ ${}_j f_{i0}^{(\varepsilon)} \to {}_j f_{i0}^{(0)}$ as $\varepsilon \to 0$. Thus, condition $\mathbf{D_{17}}$ holds with the limit constant $\tilde{f}^{(0)} = {}_j f_{i0}^{(0)}$.

Conditions $\mathbf{E_2''}$ and $\mathbf{A_3}$ imply that, for $i, j \in X_1^{(0)}$,

$$
{}_j f_{i0}^{(0)} = {}_{X_1^{(0)}} f_{i0}^{(0)} = 0.
\tag{4.4.20}
$$

Using (4.4.20) we also get for $i \in X_2^{(0)}$, $j \in X_1^{(0)}$ that

$$
\begin{aligned}
{}_j f_{i0}^{(0)} &= 1 - {}_0 f_{ij}^{(0)} \\
&= 1 - {}_0 f_{i X_1^{(0)} j}^{(0)} - \sum_{k \in X_1^{(0)}, k \neq j} {}_0 f_{i X_1^{(0)} k}^{(0)} \, {}_j f_{k0}^{(0)} \\
&= 1 - \sum_{k \in X_1^{(0)}} {}_0 f_{i X_1^{(0)} k}^{(0)} = {}_{X_1^{(0)}} f_{i0}^{(0)}.
\end{aligned}
\tag{4.4.21}
$$

As above, the distribution function $G_{ij}^{(\varepsilon)}(t)$ is regarded as the distribution function $\hat{\tilde{F}}^{(\varepsilon)}(t)$. Conditions $\mathbf{A_2}$, $\mathbf{T_1}$, $\mathbf{E_2''}$ imply, by Lemma 4.2.4, that $(\mathbf{x})$ $\lim_{T \to \infty} \overline{\lim}_{\varepsilon \to 0}(1 - G_{ij}^{(\varepsilon)}(T)) \leq \lim_{T \to \infty} \overline{\lim}_{\varepsilon \to 0} M_{ij}^{(\varepsilon)} / T = 0$. Thus, condition $\mathbf{D_{18}}$ holds.

By applying the asymptotic relation (3.2.22) given in Theorem 3.2.5 we get the asymptotic relation (4.4.15).

It still remains to prove that the asymptotic relation (4.4.15) holds if $r \in X_2^{(0)}$. In this case, we can choose some state $j \in X_1^{(0)}$ and use the renewal relations (4.2.3) and (4.2.6), instead of the renewal relations (4.2.1) and (4.2.2).

Let us first consider the case where $i = j \in X_1^{(0)}$, $r \in X_2^{(0)}$.

The renewal equations (4.2.3) and (4.2.1) have the same distribution function $_0G_{jj}^{(\varepsilon)}(t)$ that generates these equations, and the distribution function $G_{jj}^{(\varepsilon)}(t)$ replaces the distribution function $\hat{F}^{(\varepsilon)}(t)$. Thus, conditions $\mathbf{D_{13}}$ and $\mathbf{M_8}$ do not require a new verification.

However, the renewal equation (4.2.3) has a forcing function that differs from the forcing function for the renewal equation (4.2.1). Thus, condition $\mathbf{F_7}$ needs to be checked again.

Consider $q_{jr}^{(\varepsilon)}(t)$, $t \geq 0$. The following estimate can be applied to the forcing function for the renewal equation (4.2.3):

$$q_{jr}^{(\varepsilon)}(t) = \mathsf{P}_j\{\eta^{(\varepsilon)}(t) = r, \mu_j^{(\varepsilon)} > t, \mu_0^{(\varepsilon)} > t\}$$

$$\leq \mathsf{P}_j\{v_r^{(\varepsilon)} < v_j^{(\varepsilon)} \wedge v_0^{(\varepsilon)}\} = {}_0f_{jZr}^{(\varepsilon)}, \quad t \geq 0, \tag{4.4.22}$$

where $Z = \{j, r\}$.

Conditions $\mathbf{A_2}$, $\mathbf{T_1}$, $\mathbf{E_2''}$, and $\mathbf{A_3}$, by Lemma 4.2.5, imply that $(\mathbf{y})$ $_0f_{jZr}^{(\varepsilon)} \to {}_0f_{jZr}^{(0)} = 0$ as $\varepsilon \to 0$ for $j \in X_1^{(0)}$, $r \in X_2^{(0)}$.

Estimate (4.4.22) and relation $(\mathbf{y})$ imply that the limit forcing function is zero, $q_{jr}^{(0)}(t) = 0$, $t \geq 0$, and the following relation holds for $j \in X_1^{(0)}$, $r \in X_2^{(0)}$:

$$\sup_{t \geq 0} q_{jr}^{(\varepsilon)}(t) \to 0 \quad \text{as} \quad \varepsilon \to 0. \tag{4.4.23}$$

Relation (4.4.23) implies that the functions $q_{jr}^{(\varepsilon)}(t)$ converge locally uniformly to $q_{jr}^{(0)}(t)$ as $\varepsilon \to 0$ at every point $t \geq 0$. Thus, condition $\mathbf{F_7}$ holds.

In this case, for the set $A = \{r\}$, $r \in X_2^{(0)}$, we have

$$\pi^{(0)}(A) = \frac{\int_0^\infty q_{jr}^{(0)}(s)\, ds}{{}_0M_{jj}^{(0)}} = \pi_r^{(0)} = 0. \tag{4.4.24}$$

By applying the asymptotic relation (3.2.22) given in Theorem 3.2.5, we get the asymptotic relation (4.4.15) that takes the following form for $i = j \in X_1^{(0)}$, $r \in X_2^{(0)}$:

$$p_{jr}^{(\varepsilon)}(t^{(\varepsilon)}) \to 0 \quad \text{as} \quad \varepsilon \to 0. \tag{4.4.25}$$

The last case to be considered is where $i \neq j, 0$, $j \in X_1^{(0)}$ and $r \in X_2^{(0)}$.

The renewal relation (3.2.19) takes in this case the form of renewal relation (4.2.6) that should be used instead of the renewal relation (4.2.2). The renewal relations (4.2.6) and (4.2.2) have the same distribution function $_0G_{ij}^{(\varepsilon)}(t)$ that generates these equations and the same distribution function $G_{ij}^{(\varepsilon)}(t)$ which plays the role of the distribution function $\hat{\hat{F}}^{(\varepsilon)}(t)$. Thus, conditions $\mathbf{D_{17}}$ and $\mathbf{D_{18}}$ do not require a new verification.

Finally, by applying the asymptotic relation (3.2.22) given in Theorem 3.2.5, we get the asymptotic relation (4.4.15) which again takes a simple form of the asymptotic relation (4.4.25). $\qquad\square$

Remark 4.4.1. It is useful to note that, according to the remarks made in Subsection 4.2.2, condition $\mathbf{A_2}$ can be omitted in both Theorems 4.4.1 and 4.4.2.

4.4.4   Ergodic theorems for perturbed semi-Markov processes

Let us consider a model in which absorption is impossible for all ε small enough, say $\varepsilon \leq \varepsilon_5$, i.e., a condition stronger than $\mathbf{A_3}$ holds,

$\mathbf{A_4}$: $\sum_{i \neq 0} p_{i0}^{(\varepsilon)} = 0$ for every $\varepsilon \leq \varepsilon_5$.

Note that, under condition $\mathbf{A_4}$, condition $\mathbf{T_1}$ can be slightly simplified. Namely, the convergence relations in this condition can only be required for $i, j \neq 0$, so it can be re-formulated in the following form:

$\mathbf{T_2}$:   **(a)** $p_{ij}^{(\varepsilon)} \to p_{ij}^{(0)}$ as $\varepsilon \to 0$, $i, j \neq 0$;

     (b) $Q_{ij}^{(\varepsilon)}(\cdot) \Rightarrow Q_{ij}^{(0)}(\cdot)$ as $\varepsilon \to 0$, $i, j \neq 0$.

The following condition obviously implies condition $\mathbf{T_2}$ to hold:

$\mathbf{T_2'}$:   **(a)** $p_{ij}^{(\varepsilon)} \to p_{ij}^{(0)}$ as $\varepsilon \to 0$, $i, j \neq 0$;

     (b) $F_{ij}^{(\varepsilon)}(\cdot) \Rightarrow F_{ij}^{(0)}(\cdot)$ as $\varepsilon \to 0$, $i, j \neq 0$.

Let us first consider the case where condition $\mathbf{E_2'}$ holds. In this case the averaged absorption probability is zero, $f^{(\varepsilon)} = 0$, for $\varepsilon \leq \varepsilon_5$ and, therefore, for any function $0 \leq t^{(\varepsilon)} \to \infty$ as $\varepsilon \to 0$, we have $f^{(\varepsilon)} t^{(\varepsilon)} \to 0$ as $\varepsilon \to 0$, i.e., the balancing condition $\mathbf{B_6}$ automatically holds with the limit constant $\lambda = 0$.

Let us now consider the case where condition $\mathbf{E_2''}$ holds. In this case, the averaged absorption probability satisfies $\hat{f}^{(\varepsilon)} = 0$ for $\varepsilon \leq \varepsilon_5$ and, therefore, for any function $0 \leq t^{(\varepsilon)} \to \infty$ as $\varepsilon \to 0$ the product $\hat{f}^{(\varepsilon)} t^{(\varepsilon)} \to 0$ as $\varepsilon \to 0$, i.e., the balancing condition $\mathbf{B_7}$ automatically holds with the limit constant $\lambda = 0$. Also, condition $\mathbf{A_4}$ implies that $_{X_1^{(0)}} f_{i0}^{(0)} = 0$, $i \neq 0$.

In this case, formulations of Theorems 4.4.1 and 4.4.2 coincide. The only difference is that the corresponding stationary distribution is given by formula (4.4.7) if condition $\mathbf{E_2'}$ is satisfied, or by formula (4.4.14) if condition $\mathbf{E_2''}$ holds. Note that these conditions are variants of the condition $\mathbf{E_2}$.

Theorem 4.4.1 and 4.4.2 take the following form of an ergodic theorem for perturbed semi-Markov processes.

Theorem 4.4.3. *Let conditions* $\mathbf{A_4}$, $\mathbf{T_2}$, $\mathbf{E_2}$, $\mathbf{M_{11}}$, *and* $\mathbf{N_1}$ *hold. Then, for any* $0 \le t^{(\varepsilon)}$ $\to \infty$ *as* $\varepsilon \to 0$,

$$P_{ir}^{(\varepsilon)}(t^{(\varepsilon)}) = \mathsf{P}_i\{\eta^{(\varepsilon)}(t^{(\varepsilon)}) = r\} \to \pi_r^{(0)} \quad as \quad \varepsilon \to 0, \quad i, r \ne 0. \qquad (4.4.26)$$

It should be noted that if $\eta^{(\varepsilon)}(t) \equiv \eta^{(0)}(t)$, $t \ge 0$, for all $\varepsilon \ge 0$, i.e., the processes do not depend on the parameter ε, the conditions of Theorem 4.4.3 reduce to standard conditions of the ergodic theorem for the non-perturbed semi-Markov process $\eta^{(0)}(t)$, $t \ge 0$, given by the renewal theorem.

Condition $\mathbf{A_4}$ reduces to the form $p_{i0}^{(0)} = 0$, $i \ne 0$. In this case, the state 0 can be excluded from the phase set X when analysing asymptotics of the probabilities $p_{ir}^{(0)}(t)$ for the states $i, r \ne 0$.

Condition $\mathbf{E_2}$ assumes that there is at least one recurrent state. This assumption is equivalent to the requirement that the set $X_0 = \{1, \ldots, N\} = X_1^{(0)} \cup X_2^{(0)}$ be a union of the closed class of communicative states, $X_1^{(0)} \ne \varnothing$, and the class of non-essential states, $X_2^{(0)}$, that communicate with the set $X_1^{(0)}$. Note that the class $X_2^{(0)}$ is empty if condition $\mathbf{E_2'}$ holds and non-empty if condition $\mathbf{E_2''}$ holds.

Condition $\mathbf{M_{11}}$ reduces to the requirement that the moments satisfy $m_{ir}^{(0)} < \infty$, $i, r \ne 0$. Note that the condition $m_{i0}^{(0)} < \infty$, $i \ne 0$, can be omitted since, according to condition $\mathbf{A_4}$, the probabilities satisfy $p_{i0}^{(0)} = 0$, $i \ne 0$.

Condition $\mathbf{N_1}$ remains. This is a necessary and sufficient condition for the distribution functions of return times $G_{jj}^{(0)}(t)$, $j \in X_1^{(0)}$, to be non-arithmetic.

4.4.5 Stationary distributions for perturbed semi-Markov processes

We continue to consider a model in which condition $\mathbf{A_4}$ holds. Condition $\mathbf{A_4}$ means that absorption is impossible for the semi-Markov process $\eta^{(\varepsilon)}(t)$ if $\varepsilon \le \varepsilon_5$.

Let $X_1^{(\varepsilon)}$ and $X_2^{(\varepsilon)}$ be the corresponding classes of recurrent and non-recurrent (without absorption) states for the semi-Markov process $\eta^{(\varepsilon)}(t)$. Note that, by the definition of these sets given in Subsection 4.1.2, $X_0 = \{1, \ldots, N\} = X_1^{(\varepsilon)} \cup X_2^{(\varepsilon)}$. Note also that, if set $X_1^{(\varepsilon)} \ne \varnothing$ then all distribution functions $G_{jj}^{(\varepsilon)}(t)$, $j \in X_1^{(\varepsilon)}$ are proper.

Let us also assume the following conditions:

$\mathbf{E_3}$: $X_1^{(\varepsilon)} \ne \varnothing$ for the semi-Markov process $\eta^{(\varepsilon)}(t)$ for all ε small enough, say $\varepsilon \le \varepsilon_6$;

$\mathbf{N_2}$: $G_{jj}^{(\varepsilon)}(t)$, $j \in X_1^{(\varepsilon)}$ are non-arithmetic distribution functions for all ε small enough, say $\varepsilon \le \varepsilon_7$;

$\mathbf{M_{12}}$: $p_{ij}^{(\varepsilon)}[1] < \infty$, $i, j \ne 0$, for all ε small enough, say $\varepsilon \le \varepsilon_8$.

The following condition obviously implies condition $\mathbf{M_{12}}$ to hold:

$\mathbf{M'_{12}}$: $m_{ij}^{(\varepsilon)} < \infty$, $i, j \neq 0$, for all ε small enough, say $\varepsilon \leq \varepsilon_8$.

Let $\varrho_i^{(\varepsilon)}$, $i \in X_1^{(\varepsilon)}$, be stationary probabilities for the states from the class $X_1^{(\varepsilon)}$. As known, they are are unique solutions of the following system of linear equation:

$$\varrho_i^{(\varepsilon)} = \sum_{k \in X_1^{(\varepsilon)}} \varrho_k^{(\varepsilon)} p_{ki}^{(\varepsilon)}, \quad i \neq 0, \quad \sum_{k \in X_1^{(\varepsilon)}} \varrho_k^{(\varepsilon)} = 1. \tag{4.4.27}$$

Define

$$\pi_r^{(\varepsilon)} = \begin{cases} \dfrac{\varrho_r^{(\varepsilon)} m_r^{(\varepsilon)}}{m^{(\varepsilon)}} & \text{for } r \in X_1^{(\varepsilon)}, \\ 0 & \text{for } r \in X_2^{(\varepsilon)}, \end{cases} \tag{4.4.28}$$

where

$$m_i^{(\varepsilon)} = \sum_{k \in X} p_{ik}^{(\varepsilon)}[1] = \sum_{k \in X} m_{ik}^{(\varepsilon)} p_{ik}^{(\varepsilon)}, \quad i \neq 0, \quad m^{(\varepsilon)} = \sum_{i \in X_1^{(\varepsilon)}} \varrho_i^{(\varepsilon)} m_i^{(\varepsilon)}. \tag{4.4.29}$$

Note that the product $p_{ik}^{(\varepsilon)}[1] = m_{ik}^{(\varepsilon)} p_{ik}^{(\varepsilon)}$ should be counted az 0 if $m_{ik}^{(\varepsilon)} = \infty$ but $p_{ik}^{(\varepsilon)} = 0$. In particular, condition $\mathbf{A_3}$ implies that $p_{i0}^{(\varepsilon)}[1] = m_{i0}^{(\varepsilon)} p_{i0}^{(\varepsilon)} = 0$.

As follows from the remarks made in Subsection 4.3.5, Theorem 4.4.3 can be applied to every semi-Markov process $\eta^{(\varepsilon)}(t)$, $t \geq 0$, for $\varepsilon \leq \varepsilon_9$, where

$$\varepsilon_9 = \min(\varepsilon_5, \varepsilon_6, \varepsilon_7, \varepsilon_8). \tag{4.4.30}$$

Theorem 4.4.4. *Let conditions* $\mathbf{A_2}$, $\mathbf{A_4}$, $\mathbf{E_3}$, $\mathbf{M_{12}}$, *and* $\mathbf{N_2}$ *hold. Then, for* $\varepsilon \leq \varepsilon_9$,

$$P_{ij}^{(\varepsilon)}(t) = \mathsf{P}_i\{\eta^{(\varepsilon)}(t) = j\} \to \pi_j^{(\varepsilon)} \quad \text{as } t \to \infty, \quad i, j \neq 0. \tag{4.4.31}$$

As is known, $\pi_r^{(\varepsilon)}$, $r \neq 0$, given by formula (4.4.28), define a stationary distribution for the semi-Markov process $\eta^{(\varepsilon)}(t)$, $t \geq 0$.

Note that, under conditions $\mathbf{E_3}$ and $\mathbf{M_{12}}$, the expression for $\pi^{(\varepsilon)}(A)$, given by formula (4.4.28), defines a probability distribution for $\varepsilon \leq \varepsilon_9$, even if condition $\mathbf{N_2}$ does not hold.

Moreover, it can be proved that the average value of $\mathsf{P}_i\{\eta^{(\varepsilon)}(s) = r\}$ over the interval $[0, t]$ converges to $\pi_r^{(\varepsilon)}$ as $t \to \infty$, $r \neq 0$.

4.4.6　Convergence of stationary distributions for perturbed semi-Markov processes

Let us first formulate conditions that would provide convergence of stationary distributions for imbedded Markov chains.

Let us assume that conditions $\mathbf{T_2}$ **(a)**, $\mathbf{A_4}$, and $\mathbf{E_2}$ hold. In this case, according to Lemma 4.2.1, for the hitting probabilities we have

$$_0 f_{ij}^{(\varepsilon)} = f_{ij}^{(\varepsilon)} = \mathsf{P}_i\{v_j^{(\varepsilon)} < \infty\}$$

$$\to {}_0 f_{ij}^{(0)} = f_{ij}^{(0)} = \mathsf{P}_i\{v_j^{(0)} < \infty\} \text{ as } \varepsilon \to 0, \quad i, j \neq 0. \quad (4.4.32)$$

Thus, for recurrent states $j \in X_1^{(0)}$, we have $f_{ij}^{(\varepsilon)} > 0, i \neq 0, j \in X_1^{(0)}$, for $\varepsilon \leq \varepsilon_9$. Therefore, condition $\mathbf{E_3}$ holds and $X_1^{(0)} \subseteq X_1^{(\varepsilon)}$ for $\varepsilon \leq \varepsilon_9$.

Let $\varrho^{(\varepsilon)}, j \in X_1^{(\varepsilon)}$, be a stationary distribution of the Markov chain $\eta_n^{(\varepsilon)}$ in the closed communicative class $X_1^{(\varepsilon)}$. Taking in account that the set of non-recurrent states $X_2^{(\varepsilon)}$ can be non empty, we define the stationary distribution in the class $X_0^{(0)} = \{1, \dots, N\} = X_1^{(\varepsilon)} \cup X_2^{(\varepsilon)}$ by the following formula:

$$\varrho_r^{(\varepsilon)} = \begin{cases} \varrho_r^{(\varepsilon)} & \text{for } r \in X_1^{(\varepsilon)}, \\ 0 & \text{for } r \in X_2^{(\varepsilon)}. \end{cases} \quad (4.4.33)$$

Lemma 4.4.2. *Let conditions* $\mathbf{T_2}$ **(a)**, $\mathbf{A_4}$, *and* $\mathbf{E_2}$ *hold. Then*

$$\varrho_r^{(\varepsilon)} \to \varrho_r^{(0)} \text{ as } \varepsilon \to 0, \quad r \neq 0. \quad (4.4.34)$$

Proof. Let us consider the semi-Markov process with the phase space X and transition probabilities $Q_{ij}^{(\varepsilon)}(t) = p_{ij}^{(\varepsilon)} \chi(t \geq 1)$.

Let $\varepsilon \leq \varepsilon_9$ and, therefore, $X_1^{(0)} \subseteq X_1^{(\varepsilon)}$. In this case,

$$\mathsf{E}_r v_r^{(\varepsilon)} = M_{rr}^{(\varepsilon)} < \infty, \quad r \in X_1^{(0)}. \quad (4.4.35)$$

Using relation (4.4.35) and Lemmas 4.3.3 and 4.2.4, we get

$$\varrho_r^{(\varepsilon)} = (M_{rr}^{(\varepsilon)})^{-1} \to \varrho_r^{(0)} = (M_{rr}^{(0)})^{-1} \text{ as } \varepsilon \to 0, \quad r \in X_1^{(0)}. \quad (4.4.36)$$

Relation (4.4.36) also implies that

$$\sum_{r \in X_2^{(0)}} \varrho_r^{(\varepsilon)} = 1 - \sum_{r \in X_1^{(0)}} \varrho_r^{(\varepsilon)}$$

$$\to 1 - \sum_{r \in X_1^{(0)}} \varrho_r^{(0)} = 0 \text{ as } \varepsilon \to 0. \quad (4.4.37)$$

Relations (4.4.36) and (4.4.37) imply relation (4.4.34). $\qquad\qquad\square$

Let us now give a condition that would imply convergence of stationary distributions of perturbed semi-Markov processes. The following theorem supplements the asymptotic relations (4.4.26) and (4.4.31).

Theorem 4.4.5. *Let conditions* $\mathbf{T_2}$, $\mathbf{A_4}$, $\mathbf{E_2}$, *and* $\mathbf{M_{11}}$ *hold. Then*

$$\pi_r^{(\varepsilon)} \to \pi_r^{(0)} \quad as \ \varepsilon \to 0, \quad r \neq 0. \tag{4.4.38}$$

Proof. Conditions $\mathbf{T_2}$, $\mathbf{A_4}$, and $\mathbf{M_{11}}$ imply that

$$m_r^{(\varepsilon)} \to m_r^{(0)} \quad as \ \varepsilon \to 0, \quad r \neq 0. \tag{4.4.39}$$

Formula (4.4.28), relation (4.4.39) and Lemma 4.4.2 imply relation (4.4.38).   □

4.4.7   Weak convergence of distributions for regenerative stopping moments

Let us go back to a model in which conditions $\mathbf{A_2}$, $\mathbf{T_1}$, $\mathbf{A_3}$, $\mathbf{E_2}$, and $\mathbf{M_{11}}$ hold.

We will describe an asymptotic behaviour of the probabilities

$$P_i^{(\varepsilon)}(t) = \sum_{r \neq 0} P_{ir}^{(\varepsilon)}(t) = \mathsf{P}_i\{\mu_0^{(\varepsilon)} > t\}, \quad t \geq 0, \quad i \neq 0.$$

First, consider the case where conditions $\mathbf{E_2'}$, $\mathbf{A_3}$ hold. Also assume that condition $\mathbf{B_6}$ holds for some function $0 \leq t^{(\varepsilon)} \to \infty$ as $\varepsilon \to 0$, i.e., $f^{(\varepsilon)} t^{(\varepsilon)} \to \lambda$ as $\varepsilon \to 0$. In this case, condition $\mathbf{B_6}$ also holds for the function $t t^{(\varepsilon)}$ for every $t > 0$ with the limit $t\lambda$.

Theorem 4.4.6. *Let conditions* $\mathbf{A_2}$, $\mathbf{T_1}$, $\mathbf{A_3}$, $\mathbf{E_2'}$, $\mathbf{M_{11}}$ *hold, and* $0 \leq t^{(\varepsilon)} \to \infty$ *as* $\varepsilon \to 0$ *such that condition* $\mathbf{B_6}$ *holds. Then*

$$P_i^{(\varepsilon)}(t t^{(\varepsilon)}) = \mathsf{P}_i\{\mu_0^{(\varepsilon)} > t t^{(\varepsilon)}\} \to e^{-t\lambda/m^{(0)}} \quad as \ \varepsilon \to 0, \quad t > 0. \tag{4.4.40}$$

Proof. Relation (4.4.40) is a direct corollary of asymptotic relation (4.4.8) given in Theorem 4.4.1. It can be obtained by using the function $t^{(\varepsilon)} = t t^{(\varepsilon)}$ in relation (4.4.8) and summing up the pre-limit and the limit expressions over $j \in X_1^{(0)}$ in the left and the right-hand sides of (4.4.8).   □

Relation (4.4.40) shows that the properly normalised absorption time $\mu_0^{(\varepsilon)}/t^{(\varepsilon)}$ has an exponential limit distribution with the parameter $\lambda/m^{(0)}$.

Marginal asymptotic relations (4.4.26) and (4.4.40) give a clear interpretation to the mixed asymptotic relation (4.4.8) in Theorem 4.4.1. This theorem is a *mixed ergodic and limit theorem* for perturbed semi-Markov processes and absorption times.

Let us now consider the case where conditions $\mathbf{E_2''}$, $\mathbf{A_3}$ hold. Let us also assume that condition $\mathbf{B_7}$ holds for some function $0 \leq t^{(\varepsilon)} \to \infty$ as $\varepsilon \to 0$, i.e., $\hat{f}^{(\varepsilon)} t^{(\varepsilon)} \to \lambda$ as $\varepsilon \to 0$. In this case, condition $\mathbf{B_7}$ also holds for the function $t t^{(\varepsilon)}$ for every $t > 0$ with the limit $t\lambda$.

Theorem 4.4.7. *Let conditions* $\mathbf{A}_2$, $\mathbf{T}_1$, $\mathbf{A}_3$, $\mathbf{E}_2''$, $\mathbf{M}_{11}$ *hold, and* $0 \leq t^{(\varepsilon)} \to \infty$ *as* $\varepsilon \to 0$ *such that condition* $\mathbf{B}_7$ *holds. Then*

$$P_i^{(\varepsilon)}(tt^{(\varepsilon)}) = \mathsf{P}_i\{\mu_0^{(\varepsilon)} > tt^{(\varepsilon)}\}$$

$$\to (1 - {}_{X_1}f_{i0}^{(0)})e^{-t\lambda/m^{(0)}} \quad as \ \varepsilon \to 0, \ t > 0. \tag{4.4.41}$$

Proof. Relation (4.4.41) is a direct corollary of the asymptotic relation (4.4.15) given in Theorem 4.4.2. It can be obtained by using in (4.4.15) the function $tt^{(\varepsilon)}$ instead of $t^{(\varepsilon)}$ and summing up over $r \neq 0$ the pre-limit and the limit expressions in the left and the right-hand sides of (4.4.15). □

4.5 Asymptotic solidarity properties for hitting times

In this section, we will study asymptotic solidarity properties for moment generating functions of hitting times.

4.5.1 Moment generating functions for hitting times

Let us introduce the moment generating functions, for $\rho \in \mathbb{R}_1, i \neq 0, j \in X$,

$$\psi_{ij}^{(\varepsilon)}(\rho) = \int_0^\infty e^{\rho s} F_{ij}^{(\varepsilon)}(ds), \quad p_{ij}^{(\varepsilon)}(\rho) = \int_0^\infty e^{\rho s} Q_{ij}^{(\varepsilon)}(ds) = \psi_{ij}^{(\varepsilon)}(\rho) p_{ij}^{(\varepsilon)},$$

and, for $\rho \in \mathbb{R}_1, i, j \neq 0$,

$${}_0\phi_{ij}^{(\varepsilon)}(\rho) = \int_0^\infty e^{\rho s} \, {}_0G_{ij}^{(\varepsilon)}(ds) = \mathsf{E}_i e^{\rho \mu_j^{(\varepsilon)}} \chi(\nu_j^{(\varepsilon)} < \nu_0^{(\varepsilon)}).$$

Note that, by the definition, $p_{ij}^{(\varepsilon)}(\rho) = 0$ if $p_{ij}^{(\varepsilon)} = 0$.

By the definition, these moment generating functions are non-negative and non-decreasing of the argument $\rho \in \mathbb{R}_1$ taking values in the interval $[0, +\infty]$. However, it is obvious that these functions take finite values at least on the interval $(-\infty, 0]$. Moreover, these generating functions are continuous in points ρ at which they take finite values.

The following *characteristic equation* plays an important role in the sequel:

$${}_0\phi_{jj}^{(\varepsilon)}(\rho) = \int_0^\infty e^{\rho s} \, {}_0G_{jj}^{(\varepsilon)}(ds) = 1. \tag{4.5.1}$$

The following system of linear equations can be written for every $\rho \in \mathbb{R}_1, i, j \neq 0$, by using the Markov property of the semi-Markov process $\eta^{(\varepsilon)}(t)$ at the moment of the first jump and then calculating the corresponding expectation that depends on the position of this process at the moment of the first jump:

$${}_0\phi_{ij}^{(\varepsilon)}(\rho) = p_{ij}^{(\varepsilon)}(\rho) + \sum_{k \neq j,0} p_{ik}^{(\varepsilon)}(\rho) \, {}_0\phi_{kj}^{(\varepsilon)}(\rho), \quad i \neq 0. \tag{4.5.2}$$

Relations that constitute system (4.5.2) hold for every $\rho \in \mathbb{R}_1$, $i, j \neq 0$, even in the cases where the moment generating functions ${}_0\phi_{ij}^{(\varepsilon)}(\rho)$ or $p_{ik}^{(\varepsilon)}(\rho)$ take the value $+\infty$. In these cases, one needs to take the products $p_{ij}^{(\varepsilon)}(\rho) = \psi_{ij}^{(\varepsilon)}(\rho)p_{ij}^{(\varepsilon)}$ and $p_{ik}^{(\varepsilon)}(\rho){}_0\phi_{kj}^{(\varepsilon)}(\rho) = \psi_{ik}^{(\varepsilon)}(\rho)p_{ik}^{(\varepsilon)}{}_0\phi_{kj}^{(\varepsilon)}(\rho)$ as zeros if $p_{ij}^{(\varepsilon)} = 0$ or $p_{ik}^{(\varepsilon)} = 0$, respectively.

The system (4.5.2) can be rewritten in the following matrix form:

$$
{}_0\boldsymbol{\phi}_j^{(\varepsilon)}(\rho) = \mathbf{p}_j^{(\varepsilon)}(\rho) + {}_j\mathbf{P}^{(\varepsilon)}(\rho){}_0\boldsymbol{\phi}_j^{(\varepsilon)}(\rho),
\tag{4.5.3}
$$

where

$$
{}_0\boldsymbol{\phi}_j^{(\varepsilon)}(\rho) = \begin{bmatrix} {}_0\phi_{1j}^{(\varepsilon)}(\rho) \\ \vdots \\ {}_0\phi_{Nj}^{(\varepsilon)}(\rho) \end{bmatrix}, \qquad
\mathbf{p}_j^{(\varepsilon)}(\rho) = \begin{bmatrix} p_{1j}^{(\varepsilon)}(\rho) \\ \vdots \\ p_{Nj}^{(\varepsilon)}(\rho) \end{bmatrix},
$$

and

$$
{}_j\mathbf{P}^{(\varepsilon)}(\rho) = \begin{bmatrix} p_{11}^{(\varepsilon)}(\rho) & \cdots & p_{1(j-1)}^{(\varepsilon)}(\rho) & 0 & p_{1(j+1)}^{(\varepsilon)}(\rho) & \cdots & p_{1N}^{(\varepsilon)}(\rho) \\ \vdots & & \vdots & \vdots & \vdots & & \vdots \\ p_{N1}^{(\varepsilon)}(\rho) & \cdots & p_{N(j-1)}^{(\varepsilon)}(\rho) & 0 & p_{N(j+1)}^{(\varepsilon)}(\rho) & \cdots & p_{NN}^{(\varepsilon)}(\rho) \end{bmatrix}.
$$

Note that we admit the case where some elements of the vectors $\mathbf{p}_j^{(\varepsilon)}(\rho)$, ${}_0\boldsymbol{\phi}_j^{(\varepsilon)}(\rho)$, and the matrix ${}_j\mathbf{P}^{(\varepsilon)}(\rho)$ can be equal to $+\infty$. One should also set the product $0 \cdot \infty$ to be 0 when calculating the products that represent elements of these vectors and matrices and the elements of the vectors in the right-hand side of equality (4.5.3).

The n-the power of this matrix ${}_j\mathbf{P}^{(\varepsilon)}(\rho)$ is given by the following formula:

$$
({}_j\mathbf{P}^{(\varepsilon)}(\rho))^n = \| \, \mathsf{E}_i e^{\rho \tau_n^{(\varepsilon)}} \chi(\nu_j^{(\varepsilon)} \wedge \nu_0^{(\varepsilon)} > n, \eta_n^{(\varepsilon)} = k) \, \| \,.
\tag{4.5.4}
$$

This equality is also valid and becomes $+\infty = +\infty$ for elements of the matrices in the left and the right-hand sides if the corresponding expectation takes the infinite value. Also note that one should take the corresponding expectation to be zero if $\mathsf{P}_i\{\nu_j^{(\varepsilon)} \wedge \nu_0^{(\varepsilon)} > n, \eta_n^{(\varepsilon)} = k\} = 0$.

Let us now consider the matrix series

$$
{}_j\mathbf{A}^{(\varepsilon)}(\rho) = \mathbf{I} + {}_j\mathbf{P}^{(\varepsilon)}(\rho) + ({}_j\mathbf{P}^{(\varepsilon)}(\rho))^2 + \cdots .
\tag{4.5.5}
$$

Note that we admit for some elements of the matrix ${}_j\mathbf{A}(\rho)$ to take the value $+\infty$ in the case where the corresponding series in the right-hand side diverges, which also includes the case where one of the summands in this series equals to $+\infty$.

Let us introduce, for every $\rho \in \mathbb{R}_1$, $j, k \neq 0$, the random variables

$$
\delta_{jk}^{(\varepsilon)}(\rho) = \sum_{n=1}^{\infty} e^{\rho \tau^{(\varepsilon)}(n-1)} \chi(\nu_j^{(\varepsilon)} \wedge \nu_0^{(\varepsilon)} > n-1, \eta_{n-1}^{(\varepsilon)} = k).
$$

The random variable $\delta_{jk}^{(\varepsilon)}(\rho)$ may be improper, i.e., take the value $+\infty$ with a positive probability. In any case, by the Lebesgue theorem,

$$\mathsf{E}_i \delta_{jk}^{(\varepsilon)}(\rho) = \sum_{n=1}^{\infty} \mathsf{E}_i e^{\rho \tau^{(\varepsilon)}(n-1)} \chi(\nu_j^{(\varepsilon)} \wedge \nu_0^{(\varepsilon)} > n-1, \eta_{n-1}^{(\varepsilon)} = k). \qquad (4.5.6)$$

This identity remains true and takes the form $+\infty = +\infty$ in the case where at least one of the expectations under the summation sign takes the value $+\infty$ or the corresponding series diverges. Also note that one should take the expectations under the summation sign to be zero if $\mathsf{P}_i\{\nu_j^{(\varepsilon)} \wedge \nu_0^{(\varepsilon)} > n-1, \eta_{n-1}^{(\varepsilon)} = k\} = 0$.

Comparing (4.5.4) and (4.5.6) yields a formula that gives a probabilistic interpretation to elements of the matrix $_j\mathbf{A}^{(\varepsilon)}(\rho)$,

$$_j\mathbf{A}^{(\varepsilon)}(\rho) = \mathbf{I} + {}_j\mathbf{P}^{(\varepsilon)}(\rho) + ({}_j\mathbf{P}^{(\varepsilon)}(\rho))^2 + \cdots = \| \mathsf{E}_i \delta_{jk}^{(\varepsilon)}(\rho) \|. \qquad (4.5.7)$$

Identity (4.5.7) is also valid if the matrix series in the left hand-side diverges in the sense that at least one of the series that represents elements of this matrix diverges. This includes the case where at least one element of the corresponding series is equal to $+\infty$ or the expectation that represents the corresponding element of the matrix in the right-hand side equals $+\infty$. In this case, the equalities of the corresponding elements of these matrices just take the form $+\infty = +\infty$.

In the lemmas formulated below and their proofs, we adopt similar conventions, i.e., corresponding equalities or inequalities for components of the vectors can take the form $+\infty = +\infty$ or $+\infty \leq +\infty$. Also, the product $\infty \cdot 0$ is set to be 0.

Lemma 4.5.1. *Let condition* $\mathbf{E}_1$ *hold for the Markov chain* $\eta_n^{(\varepsilon)}$, $n = 0, 1, \ldots$, *and* $j \in X_1^{(\varepsilon)}$. *Then the following claims hold true for any given value of* $\rho \in \mathbb{R}_1$: **(α)** $_0\boldsymbol{\phi}_j^{(\varepsilon)}(\rho) = {}_j\mathbf{A}^{(\varepsilon)}(\rho)\mathbf{p}_j^{(\varepsilon)}(\rho)$; **($\beta$)** $_0\boldsymbol{\phi}_j^{(\varepsilon)}(\rho)$ *is a minimal non-negative solution of the system of linear equations* (4.5.3); **(γ)** $_0\boldsymbol{\phi}_j^{(\varepsilon)}(\rho)$ *is a vector with all positive components*; **(δ)** *the vector* $_0\boldsymbol{\phi}_j^{(\varepsilon)}(\rho)$ *is finite if and only if the vector* $\mathbf{p}_j^{(\varepsilon)}(\rho)$ *and the matrices* $_j\mathbf{P}^{(\varepsilon)}(\rho)$ *and* $_j\mathbf{A}^{(\varepsilon)}(\rho)$ *are finite*; **(ϵ)** *if the matrix* $_j\mathbf{P}^{(\varepsilon)}(\rho)$ *is finite then the inverse matrix* $[\mathbf{I} - {}_j\mathbf{P}^{(\varepsilon)}(\rho)]^{-1}$ *exists if and only if* $_j\mathbf{A}^{(\varepsilon)}(\rho)$ *is finite and, in this case,* $_j\mathbf{A}^{(\varepsilon)}(\rho) = [\mathbf{I} - {}_j\mathbf{P}^{(\varepsilon)}(\rho)]^{-1}$; **($\varepsilon$)** $_0\boldsymbol{\phi}_j^{(\varepsilon)}(\rho)$ *is finite if and only if the vector* $\mathbf{p}_j^{(\varepsilon)}(\rho)$ *and the matrix* $_j\mathbf{P}^{(\varepsilon)}(\rho)$ *are finite, and the inverse matrix* $[\mathbf{I} - {}_j\mathbf{P}^{(\varepsilon)}(\rho)]^{-1}$ *exists, moreover, in this case,* $_0\boldsymbol{\phi}_j^{(\varepsilon)}(\rho)$ *is a unique solution of the system of linear equations* (4.5.3); **(ζ)** *if the vector* $_0\boldsymbol{\phi}_j^{(\varepsilon)}(\rho)$ *is finite and the vector* $\mathbf{p}_j^{(\varepsilon)}(\rho')$ *and the matrix* $_j\mathbf{P}(\rho')$ *are finite for some* $\rho' > \rho$ *then there exists* $\rho < \rho'' \leq \rho'$ *such that the vector* $_0\boldsymbol{\phi}_j^{(\varepsilon)}(\rho'')$ *is finite.*

Proof. Direct calculations using relation (4.5.6) yield the following formula:

$$_0\phi_{ij}^{(\varepsilon)}(\rho) = \mathsf{E}_i e^{\rho\mu_j^{(\varepsilon)}} \chi(v_j^{(\varepsilon)} < v_0^{(\varepsilon)})$$

$$= \mathsf{E}_i \sum_{n=1}^{\infty} \sum_{k \neq 0} e^{\rho\tau_{n-1}^{(\varepsilon)}} \chi(\eta_{n-1}^{(\varepsilon)} = k, v_j^{(\varepsilon)} \wedge v_0^{(\varepsilon)} > n-1)$$

$$\times e^{\rho\kappa_n^{(\varepsilon)}} \chi(\eta_n^{(\varepsilon)} = j)$$

$$= \sum_{n=1}^{\infty} \sum_{k \neq 0} \mathsf{E}_i e^{\rho\tau_{n-1}^{(\varepsilon)}} \chi(\eta_{n-1}^{(\varepsilon)} = k, v_j^{(\varepsilon)} \wedge v_0^{(\varepsilon)} > n-1) p_{kj}^{(\varepsilon)}(\rho)$$

$$= \sum_{k \neq 0} p_{kj}^{(\varepsilon)}(\rho) \sum_{n=1}^{\infty} \mathsf{E}_i e^{\rho\tau_{n-1}^{(\varepsilon)}} \chi(\eta_{n-1}^{(\varepsilon)} = k, v_j^{(\varepsilon)} \wedge v_0^{(\varepsilon)} > n-1)$$

$$= \sum_{k \neq 0} p_{kj}^{(\varepsilon)}(\rho) \mathsf{E}_i \delta_{jk}^{(\varepsilon)}(\rho), \quad \rho \in \mathbb{R}_1, \ i, j \neq 0. \tag{4.5.8}$$

Note that identities in (4.5.8) also hold if the corresponding series diverge, which includes the case where at least one expectation representing the corresponding summand equals $+\infty$. In this case, the equalities take the form $+\infty = +\infty$. One also should set the expectations under the summation sign to zero if $\mathsf{P}_i\{v_j^{(\varepsilon)} \wedge v_0^{(\varepsilon)} > n - 1, \eta_{n-1}^{(\varepsilon)} = k\} = 0$; the product $\infty \cdot 0$ is also equal to 0.

Relations (4.5.7) and (4.5.8) yield the formula given in the proposition **(α)**.

By the definition, $_0\phi_j^{(\varepsilon)}(\rho)$ is a vector with components taking values in the interval $[0, \infty]$. As was pointed above, the vector $_0\phi_j^{(\varepsilon)}(\rho)$ is a solution of system (4.5.3).

Let $\mathbf{v}(\rho)$ be a vector components of which take values in the interval $[0, \infty]$ and such that it is also a solution of the system **(a)** $\mathbf{v}(\rho) = \mathbf{p}_j^{(\varepsilon)}(\rho) +_j \mathbf{P}^{(\varepsilon)}(\rho)\mathbf{v}(\rho)$. Iterating relation **(a)** yields that $\mathbf{v}(\rho)$ is also a solution of the system **(b)** $\mathbf{v}(\rho) = \mathbf{p}_j^{(\varepsilon)}(\rho) + \mathbf{p}_j^{(\varepsilon)}(\rho)_j\mathbf{P}^{(\varepsilon)}(\rho) + \cdots + \mathbf{p}_j^{(\varepsilon)}(\rho)\,(_j\mathbf{P}^{(\varepsilon)}(\rho))^n + (_j\mathbf{P}^{(\varepsilon)}(\rho))^{n+1}\mathbf{v}(\rho)$ for every $n \geq 1$ and, thus, **(c)** $\mathbf{v}(\rho) \geq \mathbf{p}_j^{(\varepsilon)}(\rho) + \mathbf{p}_j^{(\varepsilon)}(\rho)_j\mathbf{P}^{(\varepsilon)}(\rho) + \cdots + \mathbf{p}_j^{(\varepsilon)}(\rho)\,(_j\mathbf{P}^{(\varepsilon)}(\rho))^n$ for $n \geq 1$. By letting n tend to ∞ in **(c)**, we get the following inequality: **(d)** $\mathbf{v}(\rho) \geq \mathbf{p}_j^{(\varepsilon)}(\rho) + \mathbf{p}_j^{(\varepsilon)}(\rho)_j\mathbf{P}^{(\varepsilon)}(\rho) + \cdots = \mathbf{p}_j^{(\varepsilon)}(\rho)_j\mathbf{A}^{(\varepsilon)}(\rho) = {}_0\phi_j^{(\varepsilon)}(\rho)$. Inequality **(d)** proves that $_0\phi_j^{(\varepsilon)}(\rho)$ is a minimal non-negative solution of system (4.5.3), i.e., the claim **(β)** is true.

Since j is a recurrent-without-absorption state, **(e)** for any $i \neq 0$ there exists a natural number n_{ij} such that $\mathsf{P}_i\{v_j^{(\varepsilon)} \wedge v_0^{(\varepsilon)} > n_{ij}, \eta_{n_{ij}+1}^{(\varepsilon)} = j\} > 0$. Denote $C_j = \{k \neq 0 : p_{kj}^{(\varepsilon)} > 0\}$. Obviously **(e)** implies that $\mathsf{P}_i\{v_j^{(\varepsilon)} \wedge v_0^{(\varepsilon)} > n_{ij}, \eta_{n_{ij}+1}^{(\varepsilon)} = j\} = \sum_{k \in C_j} \mathsf{P}_i\{v_j^{(\varepsilon)} \wedge v_0^{(\varepsilon)} > n_{ij}, \eta_{n_{ij}}^{(\varepsilon)} = k\} p_{kj}^{(\varepsilon)} > 0$, and, therefore, **(f)** $\sum_{k \in C_j} \mathsf{P}_i\{v_j^{(\varepsilon)} \wedge v_0^{(\varepsilon)} > n_{ij}, \eta_{n_{ij}}^{(\varepsilon)} = k\} > 0$. Let also $n = \max_{i \neq 0} n_{ij}$. Since $p_{kj}^{(\varepsilon)}(\rho) > 0$ if

$p_{kj}^{(\varepsilon)} > 0$, relation **(d)** implies that **(g)** ${}_0\boldsymbol{\phi}_j^{(\varepsilon)}(\rho) \geq \mathbf{p}_j^{(\varepsilon)}(\rho) + \mathbf{p}_j^{(\varepsilon)}(\rho){}_j\mathbf{P}^{(\varepsilon)}(\rho) + \cdots + \mathbf{p}_j^{(\varepsilon)}(\rho)({}_j\mathbf{P}^{(\varepsilon)}(\rho))^n$, while **(f)** implies that the vector in the right-hand side of inequality **(g)** has all positive components, i.e., the claim $(\boldsymbol{\gamma})$ is also proved.

Since the vector ${}_0\boldsymbol{\phi}_j^{(\varepsilon)}(\rho)$ satisfies the relation **(h)** ${}_0\boldsymbol{\phi}_j^{(\varepsilon)}(\rho) = \mathbf{p}_j^{(\varepsilon)}(\rho) + {}_j\mathbf{P}^{(\varepsilon)}(\rho){}_0\boldsymbol{\phi}_j^{(\varepsilon)}(\rho)$, it also satisfies the relation **(i)** ${}_0\boldsymbol{\phi}_j^{(\varepsilon)}(\rho) = \mathbf{p}_j^{(\varepsilon)}(\rho) + \mathbf{p}_j^{(\varepsilon)}(\rho){}_j\mathbf{P}^{(\varepsilon)}(\rho) + \cdots + \mathbf{p}_j^{(\varepsilon)}(\rho)({}_j\mathbf{P}^{(\varepsilon)}(\rho))^n + ({}_j\mathbf{P}^{(\varepsilon)}(\rho))^{n+1}{}_0\boldsymbol{\phi}_j^{(\varepsilon)}(\rho)$ for $n \geq 1$.

Let us assume that ${}_0\boldsymbol{\phi}_j^{(\varepsilon)}(\rho)$ is a finite vector, which, as was proved above, also has in this case all positive components. Then **(i)** implies that all the quantities $p_{ik}^{(\varepsilon)}(\rho)$, $i, k \neq 0$, are finite, i.e., **(j)** the vector $\mathbf{p}_j^{(\varepsilon)}(\rho)$ and the matrix ${}_j\mathbf{P}^{(\varepsilon)}(\rho)$ are finite. Also, by making n increase to ∞ in relation **(k)** and taking into account that $\mathbf{p}_j^{(\varepsilon)}(\rho) + \mathbf{p}_j^{(\varepsilon)}(\rho){}_j\mathbf{P}^{(\varepsilon)}(\rho) + \cdots + \mathbf{p}_j^{(\varepsilon)}(\rho)({}_j\mathbf{P}^{(\varepsilon)}(\rho))^n \to {}_0\boldsymbol{\phi}_j^{(\varepsilon)}(\rho)$ as $n \to \infty$, and **(i)**, we get that **(l)** $({}_j\mathbf{P}^{(\varepsilon)}(\rho))^{n+1}{}_0\boldsymbol{\phi}_j^{(\varepsilon)}(\rho) \to 0$ as $n \to \infty$, where the last relations means that all components of the pre-limit vectors tend to 0. Since ${}_0\boldsymbol{\phi}_j^{(\varepsilon)}(\rho)$ is a finite vector with all positive components, **(m)** implies that **(n)** $({}_j\mathbf{P}^{(\varepsilon)}(\rho))^{n+1} \to 0$ as $n \to \infty$, where the last relations means that all components of the pre-limit matrices tend to 0. As is known, relation **(n)** holds if and only if the matrix series ${}_j\mathbf{A}^{(\varepsilon)}(\rho) = \mathbf{I} + {}_j\mathbf{P}^{(\varepsilon)}(\rho) + {}_j\mathbf{P}^{(\varepsilon)}(\rho))^2 + \cdots$ converges in norms, i.e., the matrix ${}_j\mathbf{A}^{(\varepsilon)}(\rho)$ is finite. Indeed, relation **(n)** implies that there exists m such that $|({}_j\mathbf{P}^{(\varepsilon)}(\rho))^m| \leq L_m < 1$. Let us also denote $K_m = \max_{1 \leq r \leq m-1}|({}_j\mathbf{P}^{(\varepsilon)}(\rho))^r|$. Then **(n)** implies the estimates **(o)** $|({}_j\mathbf{P}^{(\varepsilon)}(\rho))^n| \leq K_m(L_m)^{[n/m]}, n \geq 1$. Estimates **(o)** implies convergence of the matrix series ${}_j\mathbf{A}^{(\varepsilon)}(\rho)$ in norms. Conversely, if the vector $\mathbf{p}_j^{(\varepsilon)}(\rho)$ and the matrix ${}_j\mathbf{A}^{(\varepsilon)}(\rho)$ are finite, then the vector ${}_0\boldsymbol{\phi}_j^{(\varepsilon)}(\rho)$ is finite by the formula given in $(\boldsymbol{\alpha})$. This proves claim $(\boldsymbol{\delta})$.

The matrix ${}_j\mathbf{A}^{(\varepsilon)}(\rho)$ defined in formula (4.5.5) satisfies the following relations: **(p)** ${}_j\mathbf{A}^{(\varepsilon)}(\rho) = \mathbf{I} + {}_j\mathbf{P}^{(\varepsilon)}(\rho) + {}_j\mathbf{P}^{(\varepsilon)}(\rho))^2 + \cdots = \mathbf{I} + {}_j\mathbf{P}^{(\varepsilon)}(\rho)(\mathbf{I} + {}_j\mathbf{P}^{(\varepsilon)}(\rho) + {}_j\mathbf{P}^{(\varepsilon)}(\rho))^2 + \cdots) = \mathbf{I} + {}_j\mathbf{P}^{(\varepsilon)}(\rho){}_j\mathbf{A}^{(\varepsilon)}(\rho)$.

Let us now assume that the matrix ${}_j\mathbf{P}^{(\varepsilon)}(\rho)$ is finite. In this case, if ${}_j\mathbf{A}^{(\varepsilon)}(\rho)$ is also finite, then relation **(p)** can be rewritten as **(r)** $\mathbf{I} = {}_j\mathbf{A}^{(\varepsilon)}(\rho) \cdot [\mathbf{I} - {}_j\mathbf{P}^{(\varepsilon)}(\rho)]$. Relation **(r)** means that there exists the inverse matrix $[\mathbf{I} - {}_j\mathbf{P}^{(\varepsilon)}(\rho)]^{-1} = {}_j\mathbf{A}^{(\varepsilon)}(\rho)$. Conversely, if the matrix ${}_j\mathbf{P}^{(\varepsilon)}(\rho)$ is finite and the inverse matrix $[\mathbf{I} - {}_j\mathbf{P}^{(\varepsilon)}(\rho)]^{-1}$ exists, it, by the definition, satisfies the relation **(s)** $\mathbf{I} = [\mathbf{I} - {}_j\mathbf{P}^{(\varepsilon)}(\rho)][\mathbf{I} - {}_j\mathbf{P}^{(\varepsilon)}(\rho)]^{-1}$ that can be rewritten in an equivalent form, $[\mathbf{I} - {}_j\mathbf{P}^{(\varepsilon)}(\rho)]^{-1} = \mathbf{I} + {}_j\mathbf{P}^{(\varepsilon)}(\rho)[\mathbf{I} - {}_j\mathbf{P}^{(\varepsilon)}(\rho)]^{-1}$. An iteration of this relation implies that **(t)** $[\mathbf{I} - {}_j\mathbf{P}^{(\varepsilon)}(\rho)]^{-1} = \mathbf{I} + {}_j\mathbf{P}^{(\varepsilon)}(\rho) + \cdots + ({}_j\mathbf{P}^{(\varepsilon)}(\rho))^n + ({}_j\mathbf{P}^{(\varepsilon)}(\rho))^{n+1}[\mathbf{I} - {}_j\mathbf{P}^{(\varepsilon)}(\rho)]^{-1}, n \geq 1$. Relation **(t)** implies that the matrix series $\mathbf{I} + {}_j\mathbf{P}^{(\varepsilon)}(\rho) + \cdots$ converges in norm, and, therefore, **(u)** $({}_j\mathbf{P}^{(\varepsilon)}(\rho))^{n+1}[\mathbf{I} - {}_j\mathbf{P}^{(\varepsilon)}(\rho)]^{-1} \to 0$ as $n \to \infty$, where the last relation means that all components of the pre-limit matrices tend to 0. Relations **(t)** and **(u)** imply, in an obvious way, that **(v)** $[\mathbf{I} - {}_j\mathbf{P}^{(\varepsilon)}(\rho)]^{-1} = \mathbf{I} + {}_j\mathbf{P}^{(\varepsilon)}(\rho) + \cdots = {}_j\mathbf{A}^{(\varepsilon)}(\rho)$.

Claim (ε) follows from the claims (α) and (ϵ).

It remains to prove the claim (ζ). Since the functions $p_{ik}^{(\varepsilon)}(\rho)$, $i, k \neq 0$, (which should be 0 if $p_{ik}^{(\varepsilon)} = 0$) are non-decreasing in ρ, we have that $(\mathbf{w})$ $p_{ik}^{(\varepsilon)}(\rho) < \infty, i, k \neq 0$, for $\rho < \rho'$ and, moreover, these function are continuous for $\rho < \rho'$. As was proved above, if the vector $_0\boldsymbol{\phi}_j^{(\varepsilon)}(\rho)$ is finite then the inverse matrix $[\mathbf{I} - {}_j\mathbf{P}^{(\varepsilon)}(\rho)]^{-1}$ exists, and, therefore, $(\mathbf{t})$ $\det(\mathbf{I} - {}_j\mathbf{P}^{(\varepsilon)}(\rho)) \neq 0$. But $\det(\mathbf{I} - {}_j\mathbf{P}^{(\varepsilon)}(\rho))$ is a continuous function of elements of the matrix $_j\mathbf{P}^{(\varepsilon)}(\rho)$. Thus, due to $(\mathbf{w})$, we also have that $(\mathbf{x})$ $\det(\mathbf{I} - {}_j\mathbf{P}^{(\varepsilon)}(\rho'')) \neq 0$ for some $\rho < \rho'' < \rho'$. Thus, the inverse matrix $[\mathbf{I} - {}_j\mathbf{P}^{(\varepsilon)}(\rho'')]^{-1}$ exists. Therefore, by $(\boldsymbol{\theta})$, the vector $_0\boldsymbol{\phi}_j^{(\varepsilon)}(\rho'')$ is finite. $\qquad\square$

Remark 4.5.1. It follows from the proof of claim (ζ) in Lemma 4.5.1 that the parameter ρ'' depends on ε. In fact, define $\rho''' = \sup(\rho'' > \rho : \det(\mathbf{I} - {}_j\mathbf{P}^{(\varepsilon)}(\rho'')) \neq 0)$. Then any value $\rho < \rho'' < \min(\rho', \rho''')$ can be used in this proposition.

Let us also consider the moment generating functions,

$$_j\phi_{i0}^{(\varepsilon)}(\rho) = \int_0^\infty e^{\rho s} \, {}_j G_{i0}^{(\varepsilon)}(ds)$$

$$= \mathsf{E}_i e^{\rho \mu_0^{(\varepsilon)}} \chi(v_0^{(\varepsilon)} < v_j^{(\varepsilon)}), \quad \rho \in \mathbb{R}_1, \quad i, j \neq 0,$$

and

$$\phi_{ij}^{(\varepsilon)}(\rho) = {}_0\phi_{ij}^{(\varepsilon)}(\rho) + {}_j\phi_{i0}^{(\varepsilon)}(\rho)$$

$$= \int_0^\infty e^{\rho s} G_{ij}^{(\varepsilon)}(ds) = \mathsf{E}_i e^{\rho(\mu_j^{(\varepsilon)} \wedge \mu_0^{(\varepsilon)})}, \quad \rho \in \mathbb{R}_1, \quad i, j \neq 0.$$

By the definition, these moment generating functions are non-negative and non-decreasing in $\rho \in \mathbb{R}_1$ that takes values in the interval $[0, +\infty]$. However, it is obvious that these functions take finite values at least in the interval $(-\infty, 0]$. Moreover, these generating functions are continuous in points ρ in which they take finite values.

For every $\rho \in \mathbb{R}_1$, $i, j \neq 0$, by using the Markov property of the semi-Markov process $\eta^{(\varepsilon)}(t)$ at the moment of the first jump and calculating the corresponding expectation conditioned by the position of this process at the moment of the first jump, we can write the following system of linear equations:

$$_j\phi_{i0}^{(\varepsilon)}(\rho) = p_{i0}^{(\varepsilon)}(\rho) + \sum_{k \neq j, 0} p_{ik}^{(\varepsilon)}(\rho) \, {}_0\phi_{kj}^{(\varepsilon)}(\rho), \quad i \neq 0. \qquad (4.5.9)$$

Relations that constitute the system (4.5.9) are satisfied for every $\rho \in \mathbb{R}_1$, $i, j \neq 0$, even if the moment generating functions $_0\phi_{i0}^{(\varepsilon)}(\rho)$ or $\psi_{ik}^{(\varepsilon)}(\rho)$ take the value $+\infty$. In such a case, we take the products $p_{i0}^{(\varepsilon)}(\rho) = \psi_{i0}^{(\varepsilon)}(\rho) p_{i0}^{(\varepsilon)}$ and $p_{i0}^{(\varepsilon)}(\rho) {}_0\phi_{kj}^{(\varepsilon)}(\rho) = \psi_{ik}^{(\varepsilon)}(\rho) p_{ik}^{(\varepsilon)} \, {}_0\phi_{kj}^{(\varepsilon)}(\rho)$ to be zero if $p_{ij}^{(\varepsilon)} = 0$ or $p_{ik}^{(\varepsilon)} = 0$, respectively.

System (4.5.9) can be rewritten in the following matrix form:

$$_0\boldsymbol{\phi}_j^{(\varepsilon)}(\rho) = \mathbf{p}_0^{(\varepsilon)}(\rho) + {}_j\mathbf{P}^{(\varepsilon)}(\rho)\,{}_0\boldsymbol{\phi}_j^{(\varepsilon)}(\rho), \qquad (4.5.10)$$

where

$$_j\boldsymbol{\phi}_0^{(\varepsilon)}(\rho) = \begin{bmatrix} {}_j\phi_{10}^{(\varepsilon)}(\rho) \\ \vdots \\ {}_j\phi_{N0}^{(\varepsilon)}(\rho) \end{bmatrix}, \qquad \mathbf{p}_0^{(\varepsilon)}(\rho) = \begin{bmatrix} p_{10}^{(\varepsilon)}(\rho) \\ \vdots \\ p_{N0}^{(\varepsilon)}(\rho) \end{bmatrix}.$$

Note that we admit the case in which some elements of the the vectors $\mathbf{p}_0^{(\varepsilon)}(\rho)$ and $_j\boldsymbol{\phi}_0^{(\varepsilon)}(\rho)$ could be equal to $+\infty$. The product $0 \cdot \infty$ is taken to be 0 when calculating the products of elements of these vectors and matrices and the elements of the vectors in the right-hand side of (4.5.10).

Let us also define a vector $\boldsymbol{\phi}_j^{(\varepsilon)}(\rho)$, the components of which are moment generating functions $\phi_{ij}^{(\varepsilon)}(\rho)$, $i \neq 0$, i.e.,

$$\boldsymbol{\phi}_j^{(\varepsilon)}(\rho) = {}_0\boldsymbol{\phi}_j^{(\varepsilon)}(\rho) + {}_j\boldsymbol{\phi}_0^{(\varepsilon)}(\rho). \qquad (4.5.11)$$

The following lemma supplements Lemma 4.5.1.

Lemma 4.5.2. *Let condition* $\mathbf{E_1}$ *hold for the Markov chain* $\eta_n^{(\varepsilon)}$, $n = 0, 1, \ldots$, *and let* $j \in X_1^{(\varepsilon)}$. *Then we have the following for any given value of* $\rho \in \mathbb{R}_1$: $(\boldsymbol{\alpha})$ $_j\boldsymbol{\phi}_0^{(\varepsilon)}(\rho) = {}_j\mathbf{A}^{(\varepsilon)}(\rho)\mathbf{p}_0^{(\varepsilon)}(\rho)$; $(\boldsymbol{\beta})$ $_j\boldsymbol{\phi}_0^{(\varepsilon)}(\rho)$ *is a minimal non-negative solution of the system of linear equations* (4.5.10); $(\boldsymbol{\gamma})$ *the vector* $_j\boldsymbol{\phi}_0^{(\varepsilon)}(\rho)$ *is finite if the vector* $\mathbf{p}_0^{(\varepsilon)}(\rho)$ *and the matrices* $_j\mathbf{P}^{(\varepsilon)}(\rho)$ *and* $_j\mathbf{A}^{(\varepsilon)}(\rho) = [\mathbf{I} - {}_j\mathbf{P}^{(\varepsilon)}(\rho)]^{-1}$ *are finite and, in this case,* $_j\boldsymbol{\phi}_0^{(\varepsilon)}(\rho)$ *is a unique solution of the system of linear equations* (4.5.10); $(\boldsymbol{\delta})$ *in this case,* $\boldsymbol{\phi}_j^{(\varepsilon)}(\rho)$ *is also a finite vector.*

Proof. The proof of claims $(\boldsymbol{\alpha})$ and $(\boldsymbol{\beta})$ in Lemma 4.5.2 are absolutely similar to the proofs of $(\boldsymbol{\alpha})$ and $(\boldsymbol{\beta})$ in Lemma 4.5.1. Claim $(\boldsymbol{\gamma})$ is a direct corollary of $(\boldsymbol{\delta})$ in Lemma 4.5.1 and $(\boldsymbol{\alpha})$ in Lemma 4.5.2. Claim $(\boldsymbol{\delta})$ directly follows from (4.5.11) and $(\boldsymbol{\epsilon})$ in Lemma 4.5.1 and $(\boldsymbol{\gamma})$ in Lemma 4.5.2. $\qquad \square$

4.5.2 Solidarity properties of moment generating functions of hitting times

As above, we let $X_1^{(\varepsilon)}$ and $X_2^{(\varepsilon)}$ to be the classes of all recurrent-without-absorption and non-recurrent-without-absorption states $i \neq 0$, respectively, for the Markov chain $\eta_n^{(\varepsilon)}$, $n = 0, 1, \ldots$.

Let begin from the remark that solidarity properties for exponential moments of hitting times differ of solidarity properties for power moments.

As is known if the cyclic power moment $\mathsf{E}_i(\mu_i^{(\varepsilon)})^n \chi(\nu_i^{(\varepsilon)} < \nu_0^{(\varepsilon)}) < \infty$ for some recurrent-without-absorption state $i \in X_1^{(\varepsilon)}$ then $\mathsf{E}_j(\mu_j^{(\varepsilon)})^n \chi(\nu_j^{(\varepsilon)} < \nu_0^{(\varepsilon)}) < \infty$ for all states $j \in X_1^{(\varepsilon)}$.

However, inequality $_0\phi_{ii}^{(\varepsilon)}(\rho) < \infty$ for some $i \in X_1^{(\varepsilon)}$ and $\rho > 0$ does not imply that $_0\phi_{jj}^{(\varepsilon)}(\rho) < \infty$ for other states $j \in X_1^{(\varepsilon)}$, $j \neq i$.

The following simple example illustrates this. Let the set $X = \{0, 1, 2\}$ and the transition probabilities $p_{ij}^{(\varepsilon)} \in (0, 1)$, $i, j = 1, 2$. In this case, the set $X_1^{(\varepsilon)} = \{1, 2\}$. Let also the distribution functions $F_{ij}^{(\varepsilon)}(t) = 1 - e^{-a_i^{(\varepsilon)} t}$, $t \geq 0$, where $a_i^{(\varepsilon)} > 0$, $i = 1, 2$.

Simple calculations show that the moment generating function $_0\phi_{ii}^{(\varepsilon)}(\rho) < \infty$ if and only if $\rho < a_1^{(\varepsilon)} \wedge a_2^{(\varepsilon)}$ and $\frac{a_j^{(\varepsilon)}}{a_j^{(\varepsilon)} - \rho} p_{jj}^{(\varepsilon)} < 1$, where $i, j = 1, 2$, $i \neq j$, and, in this case,

$$
_0\phi_{ii}^{(\varepsilon)}(\rho) = \frac{a_1^{(\varepsilon)}}{\rho - a_i^{(\varepsilon)}} p_{ii}^{(\varepsilon)} + \frac{a_i^{(\varepsilon)}}{\rho - a_i^{(\varepsilon)}} p_{ij}^{(\varepsilon)} \left(1 + \frac{a_j^{(\varepsilon)}}{\rho - a_j^{(\varepsilon)}} p_{jj}^{(\varepsilon)} \right.
$$

$$
\left. + \left(\frac{a_j^{(\varepsilon)}}{\rho - a_j^{(\varepsilon)}} p_{jj}^{(\varepsilon)} \right)^2 + \cdots \right) \frac{a_j^{(\varepsilon)}}{\rho - a_j^{(\varepsilon)}} p_{ji}^{(\varepsilon)}
$$

$$
= \frac{a_1^{(\varepsilon)}}{\rho - a_i^{(\varepsilon)}} p_{ii}^{(\varepsilon)} + \frac{a_1^{(\varepsilon)}}{\rho - a_i^{(\varepsilon)}} p_{ij}^{(\varepsilon)}
$$

$$
\times \left(1 - \frac{a_j^{(\varepsilon)}}{a_j^{(\varepsilon)} - \rho} p_{jj}^{(\varepsilon)} \right)^{-1} \frac{a_i^{(\varepsilon)}}{\rho - a_j^{(\varepsilon)}} p_{ji}^{(\varepsilon)}. \tag{4.5.12}
$$

The inequality $\frac{a_j^{(\varepsilon)}}{a_j^{(\varepsilon)} - \rho} p_{jj}^{(\varepsilon)} < 1$ can be rewritten in the equivalent form as $\rho < a_j^{(\varepsilon)} p_{ji}^{(\varepsilon)}$. Let, for example, $a_1^{(\varepsilon)} p_{12}^{(\varepsilon)} < a_2^{(\varepsilon)} p_{21}^{(\varepsilon)}$. Then, as follows from the remarks above, $_0\phi_{11}^{(\varepsilon)}(\rho) < \infty$ but $_0\phi_{22}^{(\varepsilon)}(\rho) = \infty$ for any $a_1^{(\varepsilon)} p_{12}^{(\varepsilon)} < \rho < a_2^{(\varepsilon)} p_{21}^{(\varepsilon)}$.

This example clarifies specific forms of cyclic solidarity properties for the moment generating functions as they presented below.

Lemma 4.5.3. *Let* **(i)** *condition* $\mathbf{E_1}$ *hold for the Markov chain* $\eta_n^{(\varepsilon)}$, *and* **(ii)** *there exists a state* $i \in X_1^{(\varepsilon)}$ *such that* $_0\phi_{ii}^{(\varepsilon)}(\beta_i) \in (1, \infty)$ *for some* $\beta_i > 0$. *Then we have the following:* **(α)** *there exists* $0 < \beta \leq \beta_i$ *such that* $_0\phi_{jj}^{(\varepsilon)}(\beta) \in (1, \infty)$ *for every state* $j \in X_1^{(\varepsilon)}$; **($\beta$)** $_0\phi_{kj}^{(\varepsilon)}(\beta) < \infty$ *for all states* $k, j \in X_1^{(\varepsilon)}$; **($\gamma$)** *there exists a unique root* $\rho^{(\varepsilon)}$ *of the equation* $_0\phi_{jj}^{(\varepsilon)}(\rho) = 1$ *for every state* $j \in X_1^{(\varepsilon)}$; **($\delta$)** *the root* $\rho^{(\varepsilon)}$ *is the same for all states* $j \in X_1^{(\varepsilon)}$; **($\epsilon$)** $0 \leq \rho^{(\varepsilon)} < \beta$; **($\epsilon$)** *if in addition to the assumptions* **(i)** *and*

(ii), *the assumption* **(iii)** $_0\phi_{ki}^{(\varepsilon)}(\beta_i) < \infty, k \in X_2^{(\varepsilon)}$, *also holds, then* **($\zeta$)** $_0\phi_{kj}^{(\varepsilon)}(\beta) < \infty$, $j \in X_1^{(\varepsilon)}, k \in X_2^{(\varepsilon)}$, *for β defined in* **(α)**.

Proof. The lemma is trivial if the set $X_1^{(\varepsilon)}$ contains only one state. Thus, we assume that this set contains at least two states.

Using the Markov property possessed by the semi-Markov process $\eta^{(\varepsilon)}(t), t \geq 0$, at the first hitting time, we can write the following renewal relations for two states $i \neq j$, $i, j \neq 0$:

$$_0G_{ii}^{(\varepsilon)}(t) = {}_{0j}G_{ii}^{(\varepsilon)}(t) + {}_{0i}G_{ij}^{(\varepsilon)}(t) * {}_0G_{ji}^{(\varepsilon)}(t), \quad t \geq 0, \tag{4.5.13}$$

and

$$_0G_{ji}^{(\varepsilon)}(t) = {}_{0j}G_{ji}^{(\varepsilon)}(t) + {}_{0i}G_{jj}^{(\varepsilon)}(t) * {}_0G_{ji}^{(\varepsilon)}(t), \quad t \geq 0. \tag{4.5.14}$$

These relation can be rewritten in terms of the corresponding moment generating functions for $i \neq j, i, j \neq 0$ in the following equivalent form:

$$_0\phi_{ii}^{(\varepsilon)}(\rho) = {}_{0j}\phi_{ii}^{(\varepsilon)}(\rho) + {}_{0i}\phi_{ij}^{(\varepsilon)}(\rho) {}_0\phi_{ji}^{(\varepsilon)}(\rho), \quad \rho \in \mathbb{R}_1, \tag{4.5.15}$$

and

$$_0\phi_{ji}^{(\varepsilon)}(\rho) = {}_{0j}\phi_{ji}^{(\varepsilon)}(\rho) + {}_{0i}\phi_{jj}^{(\varepsilon)}(\rho) {}_0\phi_{ji}^{(\varepsilon)}(\rho), \quad \rho \in \mathbb{R}_1. \tag{4.5.16}$$

Similar two relations can be obtained by interchanging the states i and j in (4.5.15) and (4.5.16),

$$_0\phi_{jj}^{(\varepsilon)}(\rho) = {}_{0i}\phi_{jj}^{(\varepsilon)}(\rho) + {}_{0j}\phi_{ji}^{(\varepsilon)}(\rho) {}_0\phi_{ij}^{(\varepsilon)}(\rho), \quad \rho \in \mathbb{R}_1, \tag{4.5.17}$$

and

$$_0\phi_{ij}^{(\varepsilon)}(\rho) = {}_{0i}\phi_{ij}^{(\varepsilon)}(\rho) + {}_{0j}\phi_{ii}^{(\varepsilon)}(\rho) {}_0\phi_{ij}^{(\varepsilon)}(\rho), \quad \rho \in \mathbb{R}_1. \tag{4.5.18}$$

Relations (4.5.15)–(4.5.18) can take the form $+\infty = +\infty$ if the corresponding moment generating functions take the value $+\infty$. Here, we take $\infty \cdot 0$ to be 0.

Let us now assume that **(a)** $i, j \in X_1^{(\varepsilon)}$, i.e., both are recurrent-without-absorption states. This assumption imply that, for any $\rho \in \mathbb{R}_1$,

$$_0\phi_{ii}^{(\varepsilon)}(\rho), \ {}_0\phi_{jj}^{(\varepsilon)}(\rho), \ {}_0\phi_{ij}^{(\varepsilon)}(\rho), \ {}_0\phi_{ji}^{(\varepsilon)}(\rho), \ {}_{0i}\phi_{ij}^{(\varepsilon)}(\rho), \ {}_{0j}\phi_{ji}^{(\varepsilon)}(\rho) \in (0, \infty]. \tag{4.5.19}$$

Let us also assume that **(b)** $_0\phi_{ii}^{(\varepsilon)}(\rho) < \infty$ for some $\rho > 0$.

Assumptions **(a)** and **(b)**, and relations (4.5.15) and (4.5.19) imply that, for the value of ρ satisfying assumption **(b)**,

$$_{0j}\phi_{ii}^{(\varepsilon)}(\rho), \ {}_{0i}\phi_{ij}^{(\varepsilon)}(\rho), \ {}_0\phi_{ji}^{(\varepsilon)}(\rho) < \infty. \tag{4.5.20}$$

Then, assumptions **(a)** and **(b)**, and relations (4.5.16), (4.5.19), and (4.5.20) imply that if ρ satisfies the assumption **(b)**, then

$$_{0j}\phi_{ji}^{(\varepsilon)}(\rho), \;\; _{0i}\phi_{jj}^{(\varepsilon)}(\rho) < \infty. \tag{4.5.21}$$

Moreover, relation (4.5.16) can be rewritten as $1 = {}_{0j}\phi_{ji}^{(\varepsilon)}(\rho)/{}_{0}\phi_{ji}^{(\varepsilon)}(\rho) + {}_{0i}\phi_{jj}^{(\varepsilon)}(\rho)$. Therefore, the assumptions **(a)** and **(b)**, and relations (4.5.16), (4.5.19), and (4.5.20) imply that

$$_{0i}\phi_{jj}^{(\varepsilon)}(\rho) < 1, \tag{4.5.22}$$

if ρ satisfies **(b)**.

However, the assumptions **(a)** and **(b)** and relations (4.5.16)–(4.5.22) do not guarantee that $_{0}\phi_{jj}^{(\varepsilon)}(\rho)$ and $_{0}\phi_{ij}^{(\varepsilon)}(\rho)$ are finite.

Let us now assume that **(c)** $_{0}\phi_{ii}^{(\varepsilon)}(\rho) \in [0, 1]$ for some $\rho > 0$. Note that the assumption **(c)** is stronger than the assumption **(b)**.

In this case, relations (4.5.15), (4.5.19), and (4.5.20) imply that

$$_{0j}\phi_{ii}^{(\varepsilon)}(\rho) \in [0, 1). \tag{4.5.23}$$

Let us again consider the return times $v_j^{(\varepsilon)}(n) = \min(n > v_j^{(\varepsilon)}(n - 1) : \eta_n^{(\varepsilon)} = j)$, $n = 1, 2, \ldots$, $v_j^{(\varepsilon)}(0) = 0$ and $\mu_j^{(\varepsilon)}(n) = \tau^{(\varepsilon)}(v_j^{(\varepsilon)}(n))$, $n = 0, 1, \ldots$. By the definition, $v_j^{(\varepsilon)}(n)$ and $\mu_j^{(\varepsilon)}(n)$ are the times of the n-th return into the state j for the imbedded Markov chain $\eta_n^{(\varepsilon)}$ and for the semi-Markov process $\eta^{(\varepsilon)}(t)$, respectively.

A direct calculation yields the following formula for every $\rho \in \mathbb{R}_1$:

$$
\begin{aligned}
_{0}\phi_{ij}^{(\varepsilon)}(\rho) &= \mathsf{E}_i e^{\rho\mu_j^{(\varepsilon)}} \chi(v_j^{(\varepsilon)} < v_0^{(\varepsilon)}) \\
&= \mathsf{E}_i e^{\rho\mu_i^{(\varepsilon)}} \chi(v_j^{(\varepsilon)} < v_0^{(\varepsilon)} \wedge v_i^{(\varepsilon)}) + \mathsf{E}_i \sum_{n=1}^{\infty} e^{\rho(\mu_i^{(\varepsilon)}(n) + \mu_j^{(\varepsilon)} - \mu_i^{(\varepsilon)}(n))} \\
&\qquad \times \chi(v_i^{(\varepsilon)}(n) < v_0^{(\varepsilon)} \wedge v_j^{(\varepsilon)}, v_j^{(\varepsilon)} < v_0^{(\varepsilon)} \wedge v_i^{(\varepsilon)}(n + 1)) \\
&= {}_{0i}\phi_{ij}^{(\varepsilon)}(\rho) + \sum_{n=1}^{\infty} ({}_{0j}\phi_{ii}^{(\varepsilon)}(\rho))^n \; _{0i}\phi_{ij}^{(\varepsilon)}(\rho).
\end{aligned}
\tag{4.5.24}
$$

Relation (4.5.24) holds for any value $\rho \in \mathbb{R}_1$ and, hence, for any possible value of the quantity $_{0j}\phi_{ii}^{(\varepsilon)}(\rho)$. In the case, if $_{0j}\phi_{ii}^{(\varepsilon)}(\rho) \in [1, \infty]$, this relation yields that $_{0}\phi_{ij}^{(\varepsilon)}(\rho) = +\infty$. However, in the case where the assumption **(c)** is verified relations (4.5.24) and (4.5.21) imply for ρ satisfying **(c)** that

$$_{0}\phi_{ij}^{(\varepsilon)}(\rho) = {}_{0j}\phi_{ii}^{(\varepsilon)}(\rho)/(1 - {}_{0j}\phi_{ii}^{(\varepsilon)}(\rho)) < \infty. \tag{4.5.25}$$

Relation (4.5.25) is consistent with relation (4.5.18). Relations (4.5.19) and (4.5.17) yield in this case for ρ satisfying **(c)** that

$$_0\phi_{jj}^{(\varepsilon)}(\rho) < \infty. \qquad (4.5.26)$$

Thus, if the assumptions **(a)** and **(c)** hold, all quantities in relations (4.5.15)–(4.5.18) are finite for any ρ satisfying assumption **(c)**.

For any ρ satisfying assumption **(c)**, we can perform the following transformations. Relation (4.5.16) can be rewritten in the form **(d)** $_0\phi_{ji}^{(\varepsilon)}(\rho)(1-{}_{0i}\phi_{jj}^{(\varepsilon)}(\rho)) = {}_{0j}\phi_{ji}^{(\varepsilon)}(\rho)$. Multiplying relation (4.5.15) by the quantity $1 - {}_{0i}\phi_{jj}^{(\varepsilon)}(\rho)$ and using **(d)** we can transform this relation to the following form: **(e)** $_0\phi_{ii}^{(\varepsilon)}(\rho)(1 - {}_{0i}\phi_{jj}^{(\varepsilon)}(\rho)) = {}_{0j}\phi_{ii}^{(\varepsilon)}(\rho)(1 - {}_{0i}\phi_{jj}^{(\varepsilon)}(\rho)) + {}_{0i}\phi_{ij}^{(\varepsilon)}(\rho)\,{}_0\phi_{ji}^{(\varepsilon)}(\rho)(1 - {}_{0i}\phi_{jj}^{(\varepsilon)}(\rho)) = {}_{0j}\phi_{ii}^{(\varepsilon)}(\rho)(1 - {}_{0i}\phi_{jj}^{(\varepsilon)}(\rho)) + {}_{0i}\phi_{ij}^{(\varepsilon)}(\rho)\,{}_{0j}\phi_{ji}^{(\varepsilon)}(\rho)$. Now, by subtracting $1 - {}_{0i}\phi_{jj}^{(\varepsilon)}(\rho)$ from the left and right hand-side of **(e)**, transform this relation to the form **(f)** $_0\phi_{ii}^{(\varepsilon)}(\rho)(1 - {}_{0i}\phi_{jj}^{(\varepsilon)}(\rho)) - (1 - {}_{0i}\phi_{jj}^{(\varepsilon)}(\rho)) = -(1 - {}_0\phi_{ii}^{(\varepsilon)}(\rho))(1 - {}_{0i}\phi_{jj}^{(\varepsilon)}(\rho)) = -(1 - {}_{0j}\phi_{ii}^{(\varepsilon)}(\rho))(1 - {}_{0i}\phi_{jj}^{(\varepsilon)}(\rho)) + {}_{0i}\phi_{ij}^{(\varepsilon)}(\rho)\,{}_{0j}\phi_{ji}^{(\varepsilon)}(\rho)$. Using in the same manner relations (4.5.17) and (4.5.18), we can get the following relation: **(g)** $-(1 - {}_0\phi_{jj}^{(\varepsilon)}(\rho))(1 - {}_{0j}\phi_{ii}^{(\varepsilon)}(\rho)) = -(1 - {}_{0i}\phi_{jj}^{(\varepsilon)}(\rho))(1 - {}_{0j}\phi_{ii}^{(\varepsilon)}(\rho)) + {}_{0j}\phi_{ji}^{(\varepsilon)}(\rho)\,{}_{0i}\phi_{ij}^{(\varepsilon)}(\rho)$. Note that the expressions in right-hand side of relations **(f)** and **(g)** are completely symmetric with respect to the indices i and j and, moreover, they coincide. Thus, the quantities in the left-hand side of relations **(f)** and **(g)** are the same.

Finally, using assumptions **(a)** and **(c)**, for any value of ρ satisfying **(c)** we get

$$(1 - {}_0\phi_{ii}^{(\varepsilon)}(\rho))(1 - {}_{0i}\phi_{jj}^{(\varepsilon)}(\rho)) = (1 - {}_0\phi_{jj}^{(\varepsilon)}(\rho))(1 - {}_{0j}\phi_{ii}^{(\varepsilon)}(\rho)). \qquad (4.5.27)$$

An identity similar with (4.5.27) was first set down for semi-Markov processes without absorption in Teugels (1968b). This identity plays a key role in proofs of various solidarity results. We use it to prove claim (δ) of the lemma.

Assume that **(h)** $_0\phi_{ii}^{(\varepsilon)}(\beta_i) \in (1, \infty)$ for some $\beta_i > 0$.

First of all note that assumption **(h)** implies that **(i)** the distribution function $_0G_{ii}^{(\varepsilon)}(t)$ is not concentrated in zero. Indeed, if this would be so, then the identity $_0\phi_{ii}^{(\varepsilon)}(\rho) = {}_0f_{ii}^{(\varepsilon)} \le 1$, $\rho \in \mathbb{R}_1$ would hold.

It follows from **(h)** and **(i)** that the moment generating function $_0\phi_{ii}^{(\varepsilon)}(\rho)$ is a continuous and strictly increasing function for $\rho \le \beta_i$. Note also that $_0\phi_{ii}^{(\varepsilon)}(0) = {}_0f_{ii}^{(\varepsilon)} \le 1$. Therefore, **(j)** there exists a unique root $\rho_i^{(\varepsilon)}$ of the equation $_0\phi_{ii}^{(\varepsilon)}(\rho) = 1$, and, moreover, **(k)** $\rho_i^{(\varepsilon)} \in [0, \beta_i)$.

Note that according to relation (4.5.15), $_{0j}\phi_{ii}^{(\varepsilon)}(\rho) < {}_0\phi_{ii}^{(\varepsilon)}(\rho)$ for any ρ in the assumption **(c)**. Thus, $_{0j}\phi_{ii}^{(\varepsilon)}(\rho_i^{(\varepsilon)}) < 1$ and, therefore, the equality (4.5.27) can hold for $\rho = \rho_i^{(\varepsilon)}$ if and only if $_0\phi_{jj}^{(\varepsilon)}(\rho_i^{(\varepsilon)}) = 1$. Thus, **(l)** $\rho_i^{(\varepsilon)}$ is also a root of the equation $_0\phi_{jj}^{(\varepsilon)}(\rho) = 1$ for every $j \in X_1^{(\varepsilon)}$. By **(l)**, we can omit the index i and denote this root by $\rho^{(\varepsilon)}$.

All statements in Lemma 4.5.3 deal with the case where the initial state j belongs to the class of all recurrent-without-absorption states $X_1^{(\varepsilon)}$ for the Markov chain $\eta_n^{(\varepsilon)}$. Since, by the definition, the states in set of non-recurrent-without-absorption states $X_2^{(\varepsilon)}$ can not be reached by the Markov chain $\eta_n^{(\varepsilon)}$ that starts in an arbitrary initial state $j \in X_1^{(\varepsilon)}$, we can, therefore, reduce the consideration to the case where the set $X_1^{(\varepsilon)} = \{1, \ldots, N\}$, and this allows to apply Lemma 4.5.1.

As was proved above (see, relation (4.5.20)), the assumption **(h)** implies that **(m)** $_0\phi_{ki}^{(\varepsilon)}(\beta_i) < \infty$, $k \in X_1^{(\varepsilon)}$. Thus, we can apply Lemma 4.5.1. According to this lemma, relations **(m)** imply that the vector $\mathbf{p}_i^{(\varepsilon)}(\beta_i)$ and the matrix $_i\mathbf{P}(\beta_i)$ are finite. In fact, this means that **(m')** $p_{rk}^{(\varepsilon)}(\beta_i) < \infty$, $r, k \in X_1^{(\varepsilon)}$. Thus, **(m'')** the vector $\mathbf{p}_j^{(\varepsilon)}(\beta_i)$ and the matrix $_j\mathbf{P}(\beta_i)$ are finite. As was proved above (again see relation (4.5.20)), the relation $_0\phi_{jj}^{(\varepsilon)}(\rho^{(\varepsilon)}) = 1$ implies that **(n)** $_0\phi_{kj}^{(\varepsilon)}(\rho^{(\varepsilon)}) < \infty$, $k \in X_1^{(\varepsilon)}$. Using the last statement of Lemma 4.5.1 we get that **(o)** relations **(n)** and **(m'')** imply that **(p)** there exist $\rho^{(\varepsilon)} < \beta_j < \beta_i$ such that $_0\phi_{kj}^{(\varepsilon)}(\beta_j) < \infty, k \in X_1^{(\varepsilon)}$.

Note also that the proposition **(i)** implies that **(r)** the distribution function $_0G_{jj}^{(\varepsilon)}(t)$ is not concentrated in zero for every $j \in X_1^{(\varepsilon)}$.

It follows from **(p)** and **(r)** that the moment generating function $_0\phi_{jj}^{(\varepsilon)}(\rho)$ is continuous and strictly increasing for $\rho \le \beta_j$. Thus **(s)** $\rho^{(\varepsilon)}$ is the unique root of the equation $_0\phi_{jj}^{(\varepsilon)}(\rho) = 1$, and **(t)** $_0\phi_{jj}^{(\varepsilon)}(\beta_j) \in (1, \infty)$. Let now $\beta = \min(\beta_j, j \in X_1^{(\varepsilon)})$. By the definition, **(u)** $_0\phi_{rk}^{(\varepsilon)}(\beta) < \infty$ for $r, k \in X_1^{(\varepsilon)}$, and **(v)** $\rho^{(\varepsilon)} < \beta$. Therefore, **(w)** $_0\phi_{jj}^{(\varepsilon)}(\beta) \in (1, \infty)$, for $j \in X_1^{(\varepsilon)}$.

Let us now go back and consider the general case where, in general, $X_2^{(\varepsilon)} \ne \varnothing$ and, in addition to **(h)**, assume that **(x)** $_0\phi_{ki}^{(\varepsilon)}(\beta_i) < \infty, k \in X_2^{(\varepsilon)}$.

For $Z \subseteq X$, introduce the distribution functions

$$_Z G_{ij}^{(\varepsilon)}(t) = \mathsf{P}_i\{\mu_j^{(\varepsilon)} \le t, \nu_j^{(\varepsilon)} \le \nu_Z^{(\varepsilon)}\}, \quad t \ge 0, \quad i, j, k \ne 0,$$

and the corresponding moment generating functions

$$_Z \phi_{ij}^{(\varepsilon)}(t) = \mathsf{E}_i e^{\rho \mu_j^{(\varepsilon)}} \chi(\nu_j^{(\varepsilon)} \le \nu_Z^{(\varepsilon)}) = \int_0^\infty e^{\rho t} {}_Z G_{ij}^{(\varepsilon)}(dt), \quad \rho \in \mathbb{R}_1, \quad i, j \ne 0.$$

Using the Markov property of the semi-Markov process $\eta^{(\varepsilon)}(t)$, $t \ge 0$, at the first hitting time, we can write the following renewal relation for three states $j \in X_1^{(\varepsilon)}$,

$k \in X_2^{(\varepsilon)}$, and the set $Z^{(\varepsilon)} = X_1^{(\varepsilon)} \cup \{0\}$:

$$_0G_{kj}^{(\varepsilon)}(t) = \sum_{r \in X_1^{(\varepsilon)}} {}_{Z^{(\varepsilon)}}G_{kr}^{(\varepsilon)}(t) * {}_0G_{rj}^{(\varepsilon)}(t), \quad t \geq 0. \tag{4.5.28}$$

This relation can be rewritten in an equivalent form in terms of the corresponding moment generating functions,

$$_0\phi_{kj}^{(\varepsilon)}(\rho) = \sum_{r \in X_1^{(\varepsilon)}} {}_{Z^{(\varepsilon)}}\phi_{kr}^{(\varepsilon)}(\rho){}_0\phi_{rj}^{(\varepsilon)}(\rho), \quad \rho \in \mathbb{R}_1. \tag{4.5.29}$$

Relation (4.5.29) may take the form $+\infty = +\infty$ if the corresponding moment generating functions take the value $+\infty$. In such a case, we set $\infty \cdot 0$ to 0.

Relations (4.5.19) and **(u)** imply that **(y)** $_0\phi_{rj}^{(\varepsilon)}(\beta) \in (0, \infty)$, $r, j \in X_1^{(\varepsilon)}$. Thus, assumption **(x)** and relation (4.5.29), used for $j = i$, imply that **(z)** $_{Z^{(\varepsilon)}}\phi_{kr}^{(\varepsilon)}(\beta) < \infty$, $r \in X_1^{(\varepsilon)}$, $k \in X_2^{(\varepsilon)}$. Finally, relations **(y)**, **(z)**, and (4.5.29) imply that $_0\phi_{kj}^{(\varepsilon)}(\beta) < \infty$ for any $j \in X_1^{(\varepsilon)}$, $k \in X_2^{(\varepsilon)}$. $\square$

Remark 4.5.2. It follows from the proof of Lemma 4.5.3 that the parameter β depends on ε. In fact, define $\beta_j' = \sup(\beta_j > \rho^{(\varepsilon)} : \det(\mathbf{I} - {}_j\mathbf{P}^{(\varepsilon)}(\beta_j)) \neq 0)$, $j \in X_1^{(\varepsilon)}$. Then any values $\rho^{(\varepsilon)} < \beta_j < \min(\beta_j', \beta_i)$, $j \in X_1^{(\varepsilon)}$ can be used to define $\beta = \min(\beta_j, j \in X_1^{(\varepsilon)})$.

The following lemma supplements Lemma 4.5.3.

Lemma 4.5.4. *Let* **(i)** *condition* $\mathbf{E}_1$ *hold for the Markov chain* $\eta_n^{(\varepsilon)}$; **(ii)** *there exist a state* $i \in X_1^{(\varepsilon)}$ *such that* $_0\phi_{ii}^{(\varepsilon)}(\beta_i) \in (1, \infty)$ *for some* $\beta_i > 0$; **(iii)** $_0\phi_{ki}^{(\varepsilon)}(\beta_i) < \infty$, $k \in X_2^{(\varepsilon)}$; *and* **(iv)** $p_{k0}^{(\varepsilon)}(\beta) < \infty$, $k \neq 0$. *Then* **(α)** $_j\phi_{k0}^{(\varepsilon)}(\beta) < \infty$ *and* $\phi_{kj}^{(\varepsilon)}(\beta) < \infty$ *for all states* $k \neq 0$, $j \in X_1^{(\varepsilon)}$, *and* β *defined in statement* **(ζ)** *of Lemma* 4.5.3.

Proof. As was shown in the proof of Lemma 4.5.3 **(a)**, the assumptions **(i)**–**(iii)** imply that $_0\phi_{kj}^{(\varepsilon)}(\beta) < \infty$ for all $k \neq 0$, $j \in X_1^{(\varepsilon)}$. By Lemma 4.5.1, **(a)** implies that **(b)** the matrices $_j\mathbf{A}^{(\varepsilon)}(\beta) = [\mathbf{I} - {}_j\mathbf{P}^{(\varepsilon)}(\beta)]^{-1}$ are finite for all $j \in X_1^{(\varepsilon)}$. Claim **(b)**, assumption **(iv)** and Lemma 4.5.2 yield the statement of Lemma 4.5.4 in an obvious way. $\square$

4.5.3 Test functions and upper and lower bounds for exponential moments of hitting times

Let us formulate simple necessary and sufficient conditions in terms of the so-called test functions for the relation $_0\phi_{jj}^{(\varepsilon)}(\rho) \in (1, \infty)$ to hold.

Lemma 4.5.5. *Let condition* $\mathbf{E_1}$ *hold for the Markov chain* $\eta_n^{(\varepsilon)}$, $n = 0, 1, \ldots,$ *and* $j \neq 0$ *be a recurrent-without-absorption state. Then the following takes place for any given* $\rho \in \mathbb{R}_1$: **(α)** $_0\phi_j^{(\varepsilon)}(\rho)$ *is a finite vector if and only if there exists a finite vector* $\mathbf{v}(\rho)$ *with non-negative components such that* $\mathbf{v}(\rho) \geq \mathbf{p}_j^{(\varepsilon)}(\rho) + {}_j\mathbf{P}^{(\varepsilon)}(\rho)\mathbf{v}(\rho)$ *and, in this case,* $_0\phi_j^{(\varepsilon)}(\rho) \leq \mathbf{v}(\rho)$; **($\beta$)** *if also there exists a vector* $\mathbf{u}(\rho) \leq \mathbf{v}(\rho)$ *with non-negative components such that* $\mathbf{u}(\rho) \leq \mathbf{p}_j^{(\varepsilon)}(\rho) + {}_j\mathbf{P}^{(\varepsilon)}(\rho)\mathbf{u}(\rho)$, *then* $_0\phi_j^{(\varepsilon)}(\rho) \geq \mathbf{u}(\rho)$; **($\gamma$)** $_0\phi_{jj}^{(\varepsilon)}(\rho) \in (1, \infty)$ *if and only if there exists a vector* $\mathbf{u}(\rho)$ *described in* **(β)** *such that the j-th component of this vector is greater than* 1.

Proof. Let $\mathbf{v}(\rho)$ be a finite vector with non-negative components and satisfying the system of test inequalities, **(a)** $\mathbf{v}(\rho) \geq \mathbf{p}_j^{(\varepsilon)}(\rho) + {}_j\mathbf{P}^{(\varepsilon)}(\rho)\mathbf{v}(\rho)$. Iterating the relation **(a)** yields the inequalities **(b)** $\mathbf{v}(\rho) \geq \mathbf{p}_j^{(\varepsilon)}(\rho) + \mathbf{p}_j^{(\varepsilon)}(\rho){}_j\mathbf{P}^{(\varepsilon)}(\rho) + \cdots + \mathbf{p}_j^{(\varepsilon)}(\rho)\, ({}_j\mathbf{P}^{(\varepsilon)}(\rho))^n + ({}_j\mathbf{P}^{(\varepsilon)}(\rho))^{n+1}\mathbf{v}(\rho) \geq \mathbf{p}_j^{(\varepsilon)}(\rho) + \mathbf{p}_j^{(\varepsilon)}(\rho){}_j\mathbf{P}^{(\varepsilon)}(\rho) + \cdots + \mathbf{p}_j^{(\varepsilon)}(\rho)\, ({}_j\mathbf{P}^{(\varepsilon)}(\rho))^n$ for $n \geq 1$. By letting n tend to ∞ in **(b)**, we get the following inequality: **(c)** $\mathbf{v}(\rho) \geq \mathbf{p}_j^{(\varepsilon)}(\rho) + \mathbf{p}_j^{(\varepsilon)}(\rho){}_j\mathbf{P}^{(\varepsilon)}(\rho) + \cdots = \mathbf{p}_j^{(\varepsilon)}(\rho){}_j\mathbf{A}^{(\varepsilon)}(\rho) = {}_0\phi_j^{(\varepsilon)}(\rho)$. Inequality **(c)** proves that $_0\phi_j^{(\varepsilon)}(\rho)$ is a finite vector if there exist a finite vector with non-negative components $\mathbf{v}(\rho)$ that satisfies the system of inequalities **(a)**. Moreover, in this case, the inequality $_0\phi_j^{(\varepsilon)}(\rho) \leq \mathbf{v}(\rho)$ also holds.

On the other hand, the vector $_0\phi_j^{(\varepsilon)}(\rho)$, by the definition, has non-negative components and satisfies the system of linear equations (4.5.3), which is a particular case of the system **(a)**. Thus, in the case where $_0\phi_j^{(\varepsilon)}(\rho)$ is a finite vector, it can serve as the vector $\mathbf{v}(\rho)$ satisfying system **(a)**.

Let $\mathbf{u}(\rho)$ be a vector with non-negative components such that $\mathbf{u}(\rho) \leq \mathbf{v}(\rho)$ and satisfying the system of test inequalities **(d)** $\mathbf{u}(\rho) \leq \mathbf{p}_j^{(\varepsilon)}(\rho) + {}_j\mathbf{P}^{(\varepsilon)}(\rho)\mathbf{u}(\rho)$. By iterating the relation **(d)**, we get yields inequalities **(e)** $\mathbf{u}(\rho) \leq \mathbf{p}_j^{(\varepsilon)}(\rho) + \mathbf{p}_j^{(\varepsilon)}(\rho){}_j\mathbf{P}^{(\varepsilon)}(\rho) + \cdots + \mathbf{p}_j^{(\varepsilon)}(\rho)\, ({}_j\mathbf{P}^{(\varepsilon)}(\rho))^n + ({}_j\mathbf{P}^{(\varepsilon)}(\rho))^{n+1}\mathbf{u}(\rho)$ for $n \geq 1$. Since $_0\phi_j^{(\varepsilon)}(\rho)$ is a finite vector, using what was proved in Lemma 4.5.1 we get **(f)** $({}_j\mathbf{P}^{(\varepsilon)}(\rho))^n \to 0$ as $n \to 0$, where the last relations means that all components of the pre-limit vectors tend to 0. Relation **(f)** implies that **(g)** $({}_j\mathbf{P}^{(\varepsilon)}(\rho))^{n+1}\mathbf{u}(\rho) \to 0$ as $n \to 0$, where the last relation means that all components of the pre-limit vectors tend to 0. By making n increase to ∞ in **(e)** and using **(g)** we get the inequalities **(h)** $\mathbf{u}(\rho) \leq \mathbf{p}_j^{(\varepsilon)}(\rho) + \mathbf{p}_j^{(\varepsilon)}(\rho){}_j\mathbf{P}^{(\varepsilon)}(\rho) + \cdots = {}_0\phi_j^{(\varepsilon)}(\rho)$. If the j-th component of the vector $\mathbf{u}(\rho)$ is greater than 1, then the j-th component of the vector $_0\phi_j^{(\varepsilon)}(\rho)$ is also greater than 1.

On the other hand, the vector $_0\phi_j^{(\varepsilon)}(\rho)$, by the definition, has non-negative components, and it satisfies the system of linear equations (4.5.3) that are also a particular case of system **(d)**. Thus, in the case where $_0\phi_j^{(\varepsilon)}(\rho)$ is a finite vector, we can take it as the vector $\mathbf{u}(\rho)$ that satisfies system **(d)**. $\square$

4.5.4 Asymptotic Cramér type conditions for hitting times

In what follows, we assume that conditions $\mathbf{A_2}$, $\mathbf{T_1}$ and $\mathbf{E_2}$ hold.

Consider the sets $X_1^{(0)}$ and $X_2^{(0)}$ of recurrent-without-absorption and non-recurrent-without-absorption states for the limiting Markov chain $\eta_n^{(0)}$, respectively.

Recall that, according to condition $\mathbf{E_2}$, $X_1^{(0)} \neq \varnothing$, while $X_2^{(0)} = \varnothing$ if condition $\mathbf{E_2'}$ holds and $X_2^{(0)} \neq \varnothing$ if condition $\mathbf{E_2''}$ holds.

We will assume the following Cramér type condition:

$\mathbf{C_{12}}$: There exists $\delta > 0$ such that $\overline{\lim}_{0 \leq \varepsilon \to 0} \, p_{ij}^{(\varepsilon)}(\delta) < \infty, i \neq 0, j \in X$.

The following condition obviously implies condition $\mathbf{C_{12}}$ to hold:

$\mathbf{C_{12}'}$: There exists $\delta > 0$ such that $\overline{\lim}_{0 \leq \varepsilon \to 0} \, \psi_{ij}^{(\varepsilon)}(\delta) < \infty, i \neq 0, j \in X$.

Condition $\mathbf{C_{12}}$ implies that there exists $\varepsilon_1' > 0$ such that $\psi_{ij}^{(\varepsilon)}(\delta) < \infty, i \neq 0,$ $j \in X_1^{(0)}$ for $\varepsilon \leq \varepsilon_1'$.

The following condition plays a key role in our considerations:

$\mathbf{C_{13}}$: There exist $i \in X_1^{(0)}$ and $\beta_i \in (0, \delta]$, for δ defined in condition $\mathbf{C_{12}}$, such that

$\quad$ **(a)** $_0\phi_{ii}^{(0)}(\beta_i) \in (1, \infty)$;

$\quad$ **(b)** $_0\phi_{ki}^{(0)}(\beta_i) < \infty, k \in X_2^{(0)}$.

Note that condition $\mathbf{C_{13}}$ **(b)** is automatically satisfied if condition $\mathbf{E_2'}$ holds, i.e., if $X_2^{(0)} = \varnothing$.

Lemma 4.5.6. *Let conditions $\mathbf{A_2}$, $\mathbf{T_1}$, $\mathbf{E_2}$, $\mathbf{C_{12}}$, and $\mathbf{C_{13}}$ hold. Then, (α) there exist $0 < \beta \leq \beta_i$ and $\varepsilon_0' > 0$ (defined by (4.5.30)) such that $_0\phi_{jj}^{(\varepsilon)}(\beta) \in (1, \infty)$ for every state $j \in X_1^{(0)}$ and $\varepsilon \leq \varepsilon_0'$; (β) $_0\phi_{kj}^{(\varepsilon)}(\beta) < \infty$ for all states $k \neq 0, j \in X_1^{(0)}$, and $\varepsilon \leq \varepsilon_0'$; (γ) there exists a unique root $\rho^{(\varepsilon)}$ of the equation $_0\phi_{jj}^{(\varepsilon)}(\rho) = 1$ for every state $j \in X_1^{(0)}$ and $\varepsilon \leq \varepsilon_0'$; (δ) the root $\rho^{(\varepsilon)}$ is the same for all states $j \in X_1^{(0)}$; (ϵ) $0 \leq \rho^{(\varepsilon)} < \beta$ for $\varepsilon \leq \varepsilon_0'$; (ϵ) $\rho^{(\varepsilon)} \to \rho^{(0)}$ as $\varepsilon \to 0$; (ζ) $_0\phi_{kj}^{(\varepsilon)}(\rho) \to {_0\phi_{kj}^{(0)}}(\rho)$ as $\varepsilon \to 0$ for all $k \neq 0, j \in X_1^{(0)}$, and $\rho \leq \beta$.*

Proof. Let us first apply Lemma 4.5.3 to the case where $\varepsilon = 0$. According to condition $\mathbf{C_{13}}$, this lemma implies that **(a)** there exists $0 < \beta \leq \beta_i$ such that $_0\phi_{jj}^{(0)}(\beta) \in (1, \infty)$, $j \in X_1^{(0)}$; **(a')** $_0\phi_{kj}^{(0)}(\beta) < \infty$ for $k \neq 0, j \in X_1^{(0)}$; **(a'')** there exists a unique root $\rho^{(0)}$ of the equation $_0\phi_{jj}^{(0)}(\rho) = 1$ for every state $j \in X_1^{(0)}$; **(a''')** the root $\rho^{(0)}$ is the same for all states $j \in X_1^{(0)}$; **(a'''')** $0 \leq \rho^{(0)} < \beta$.

Lemma 4.5.1 and (**a′**) implies that, in this case, (**b**) there exists the inverse matrix $[\mathbf{I} - {}_j\mathbf{P}^{(\varepsilon)}(\beta)]^{-1}$ for any $j \in X_1^{(0)}$, i.e., (**b′**) $\det(\mathbf{I} - {}_j\mathbf{P}^{(0)}(\beta)) \neq 0$ for $j \in X_1^{(0)}$.

Conditions $\mathbf{T_1}$ and $\mathbf{C_{12}}$ imply that (**c**) the moment generating functions converge, $p_{kr}^{(\varepsilon)}(\rho) \to p_{kr}^{(0)}(\rho)$ as $\varepsilon \to 0$ for $k, r \neq 0$, and $\rho \leq \beta$.

Since the determinant of a matrix is a continuous function of elements of the matrix, relations (**b′**) and (**c**) imply that there exists $\varepsilon_2' > 0$ such that (**d**) $\det(\mathbf{I} - {}_j\mathbf{P}^{(\varepsilon)}(\beta)) \neq 0$ for $j \in X_1^{(0)}$ and $\varepsilon \leq \min(\varepsilon_1', \varepsilon_2')$. Moreover, (**e**) ${}_0\boldsymbol{\phi}_j^{(\varepsilon)}(\beta) = [\mathbf{I} - {}_j\mathbf{P}^{(\varepsilon)}(\beta)]^{-1}\mathbf{p}_j^{(\varepsilon)}(\beta)$, $j \in X_1^{(0)}$.

Since, the elements of the inverse matrix are continuous functions of elements of the matrix itself, (**c**) implies that (**c′**) $[\mathbf{I} - {}_j\mathbf{P}^{(\varepsilon)}(\beta)]^{-1} \to [\mathbf{I} - {}_j\mathbf{P}^{(0)}(\beta)]^{-1}$ as $\varepsilon \to 0$, $j \in X_1^{(0)}$. Now, (**c**), (**c′**), and (**e**) imply that (**f**) ${}_0\phi_{kj}^{(\varepsilon)}(\beta) \to {}_0\phi_{kj}^{(0)}(\beta) < \infty$ as $\varepsilon \to 0$, for $k \neq 0$, $j \in X_1^{(0)}$. Relations (**a′**) and (**f**) imply that there exists $\varepsilon_3 > 0$ such that (**g**) ${}_0\phi_{jj}^{(\varepsilon)}(\beta) \in (1, \infty)$ for $j \in X_1^{(0)}$ and $\varepsilon \leq \varepsilon_0'$ and (**h**) ${}_0\phi_{kj}^{(\varepsilon)}(\beta) < \infty$ for $k \neq 0$, $j \in X_1^{(0)}$ and $\varepsilon \leq \varepsilon_0'$, where

$$\varepsilon_0' = \min(\varepsilon_1', \varepsilon_2', \varepsilon_3'). \tag{4.5.30}$$

Lemma 4.5.3 and relations (**g**) and (**h**) now give statements (**α**)–(**ε**) of Lemma 4.5.6.

According to Lemma 4.2.3, conditions $\mathbf{A_2}$, $\mathbf{T_1}$, and $\mathbf{E_2}$ imply that (**i**) ${}_0G_{kj}^{(\varepsilon)}(\cdot) \Rightarrow {}_0G_{kj}^{(0)}(\cdot)$ as $\varepsilon \to 0$, $k \neq 0$, $j \in X_1^{(0)}$. Relations (**f**) and (**i**) easily give that (**f′**) ${}_0\phi_{kj}^{(\varepsilon)}(\rho) \to {}_0\phi_{kj}^{(0)}(\rho)$ as $\varepsilon \to 0$ for all $k \neq 0$, $j \in X_1^{(0)}$, and $\rho \leq \beta$, so that we see that claim (**ζ**) holds.

Note that relation (**a**) implies that (**j**) the distribution function ${}_0G_{jj}^{(0)}(t)$ is not concentrated in zero for every $j \in X_1^{(0)}$. Finally, relations (**i**) and (**f′**), both taken for $k = j$, and proposition (**j**) yield, by Lemma 1.4.3, that statement (**ε**) of Lemma 4.5.6 holds.   □

The following lemma supplements Lemma 4.5.6.

Lemma 4.5.7. *Let conditions* $\mathbf{A_2}$, $\mathbf{T_1}$, $\mathbf{E_2}$, $\mathbf{C_{12}}$, *and* $\mathbf{C_{13}}$ *hold. Then,* (**α**) *there exists* $\varepsilon_4' > 0$ *(defined in formula (4.5.31)) such that* ${}_j\phi_{k0}^{(\varepsilon)}(\beta) < \infty$ *and* $\phi_{kj}^{(\varepsilon)}(\beta) < \infty$ *for* $0 < \beta \leq \delta$ *defined in* (**α**) *of Lemma 4.5.6,* $k \neq 0$, $j \in X_1^{(0)}$, *and* $\varepsilon \leq \varepsilon_4'$; (**β**) ${}_j\phi_{k0}^{(\varepsilon)}(\rho) \to {}_j\phi_{k0}^{(0)}(\rho) < \infty$ *as* $\varepsilon \to 0$ *and* $\phi_{kj}^{(\varepsilon)}(\rho) \to \phi_{kj}^{(0)}(\rho) < \infty$ *as* $\varepsilon \to 0$ *for* $\rho \leq \beta$ *and* $k \neq 0$, $j \in X_1^{(0)}$.

Proof. As was shown in the proof of Lemma 4.5.6, (**k**) the inverse matrices $[\mathbf{I} - {}_j\mathbf{P}^{(\varepsilon)}(\beta)]^{-1}$ exist for $j \in X_1^{(0)}$ and $\varepsilon \leq \varepsilon_0'$, and (**k′**) $[\mathbf{I} - {}_j\mathbf{P}^{(\varepsilon)}(\beta)]^{-1} \to [\mathbf{I} - {}_j\mathbf{P}^{(0)}(\beta)]^{-1}$ as $\varepsilon \to 0$, $j \in X_1^{(0)}$.

Condition $\mathbf{C_{12}}$ implies that there exists $\varepsilon_5' > 0$ such that $(\mathbf{l})$ $p_{k0}^{(\varepsilon)}(\beta) < \infty$, $k \neq 0$, for $\varepsilon \leq \varepsilon_5'$. According to Lemma 4.5.2, $(\mathbf{k})$ and $(\mathbf{l})$ imply that $(\mathbf{m})$ $_j\boldsymbol{\phi}_0^{(\varepsilon)}(\beta) = [\mathbf{I} - {_j}\mathbf{P}^{(\varepsilon)}(\beta)]^{-1}\mathbf{p}_0^{(\varepsilon)}(\beta)$, for $j \in X_1^{(0)}$ and $\varepsilon \leq \varepsilon_4'$, where

$$\varepsilon_4' = \min(\varepsilon_0', \varepsilon_5'). \tag{4.5.31}$$

Conditions $\mathbf{T_1}$ $(\mathbf{b})$ and $\mathbf{C_{12}}$ imply that $(\mathbf{n})$ $p_{k0}^{(\varepsilon)}(\rho) \to p_{k0}^{(0)}(\rho)$ as $\varepsilon \to 0$ for $k \neq 0$ and $\rho \leq \beta$. Relations $(\mathbf{k'})$, $(\mathbf{m})$, and $(\mathbf{n'})$ imply that $(\mathbf{o})$ $_j\phi_{k0}^{(\varepsilon)}(\beta) \to {_j}\phi_{k0}^{(0)}(\beta) < \infty$ as $\varepsilon \to 0$ for $k \neq 0$, $j \in X_1^{(0)}$.

According to Lemma 4.2.3, conditions $\mathbf{A_2}$, $\mathbf{T_1}$, and $\mathbf{E_2}$ imply that $(\mathbf{p})$ $_jG_{k0}^{(\varepsilon)}(\cdot) \Rightarrow {_j}G_{k0}^{(0)}(\cdot)$ as $\varepsilon \to 0$, $k \neq 0$, $j \in X_1^{(0)}$. Relations $(\mathbf{o})$ and $(\mathbf{p})$ easily give that $(\mathbf{o'})$ $_j\phi_{k0}^{(\varepsilon)}(\rho) \to {_j}\phi_{k0}^{(0)}(\rho)$ as $\varepsilon \to 0$ for $k \neq 0$, $j \in X_1^{(0)}$ and $\rho \leq \beta$. Finally, $(\mathbf{o'})$ and statement $(\boldsymbol{\zeta})$ of Lemma 4.5.6 imply that $(\mathbf{o''})$ $\phi_{kj}^{(\varepsilon)}(\rho) \to \phi_{kj}^{(0)}(\rho)$ as $\varepsilon \to 0$ for $k \neq 0$, $j \in X_1^{(0)}$ and $\rho \leq \beta$. $\qquad\square$

Remark 4.5.3. It follows from the proof of Lemmas 4.5.6 and 4.5.7 that the assumption $\overline{\lim}_{0 \leq \varepsilon \to 0}\, p_{i0}^{(\varepsilon)}(\delta) < \infty$, $i \neq 0$, is not needed in Lemma 4.5.6, but it is required in Lemma 4.5.7.

4.5.5 Sufficient asymptotic Cramér type conditions for hitting times

Let us recall formula (4.2.29),

$$1 - {_0}G_{jj}^{(0)}(\infty) = {_j}f_{j0}^{(0)} = \sum_{k=1}^{N} \mathsf{E}_j\, \delta_{jk}^{(0)}\, p_{k0}^{(0)}, \quad j \in X_1^{(0)}. \tag{4.5.32}$$

It follows from formula (4.5.32) that, $(\mathbf{a})$ under conditions $\mathbf{A_2}$ and $\mathbf{E_2}$, the distribution functions $_0G_{jj}^{(0)}(t)$ are proper or improper simultaneously for all $j \in X_1^{(0)}$, and that $(\mathbf{b})$ condition $\mathbf{A_3}$ is necessary and sufficient for these distribution functions to be proper.

Also recall that, $(\mathbf{c})$ under conditions $\mathbf{A_2}$ and $\mathbf{E_2}$, the distribution functions $_0G_{jj}^{(0)}(t)$ are concentrated or not concentrated in zero simultaneously for all $j \in X_1^{(0)}$, and that $(\mathbf{d})$ condition $\mathbf{I_4}$ $(\mathbf{a})$ is necessary and sufficient for these distribution functions to be not concentrated in zero.

If $_0G_{jj}^{(0)}(t)$ is a proper distribution function and not concentrated, then $_0\phi_{jj}^{(0)}(0) = 1$, while $_0\phi_{jj}^{(0)}(\rho) \in (0, 1)$ for $\rho < 0$, and $_0\phi_{jj}^{(0)}(\rho) \in (1, \infty]$ for any $\rho > 0$. Therefore, $(\mathbf{e})$ 0 in this case, is a unique root for the characteristic equation (4.5.1).

If $_0G_{jj}^{(0)}(t)$ is an improper distribution function, i.e., condition $\mathbf{A_2}$ does not hold and $_0\phi_{jj}^{(0)}(\beta) < \infty$ for some $\beta > 0$ and $_0G_{jj}^{(0)}(t)$ is not concentrated in zero, then

$_0\phi_{jj}^{(0)}(\rho)$ is a continuous and strictly increasing function for $\rho \leq \beta$. Moreover, in this case, $_0\phi_{jj}^{(0)}(0) = {}_0G_{jj}^{(0)}(\infty) < 1$. Thus, **(f)** if $_0\phi_{jj}^{(0)}(\beta) \in (1, \infty)$ there exists a unique root $0 < \rho^{(0)} < \beta$ of the characteristic equation (4.5.1).

The following lemma gives an additional useful information about the characteristic roots $\rho^{(\varepsilon)}$.

Lemma 4.5.8. *Let conditions* $\mathbf{A_2}$, $\mathbf{T_1}$, $\mathbf{I_4}$, $\mathbf{E_2}$, $\mathbf{C_{12}}$, *and* $\mathbf{A_3}$ *hold. Then,* **(α)** *condition* $\mathbf{C_{13}}$ *holds for any* $i \in X_1^{(0)}$ *and* $0 < \beta_i \leq \beta^{(0)}$ *for some* $0 < \beta^{(0)} \leq \delta$; **($\beta$)** $\rho^{(\varepsilon)} \to \rho^{(0)} = 0$ *as* $\varepsilon \to 0$.

Proof. As was shown in the proof of Lemma 4.2.1, **(g)** conditions $\mathbf{A_2}$ and $\mathbf{E_2}$ imply that the inverse matrix $[\mathbf{I} - {}_i\mathbf{P}^{(0)}]^{-1}$ exists for every $i \in X_1^{(0)}$, i.e., **(h)** $\det(\mathbf{I} - {}_i\mathbf{P}^{(0)}) \neq 0$. By condition $\mathbf{C_{12}}$, **(i)** $p_{kr}^{(0)}(\rho) \to p_{kr}^{(0)}(0) = p_{kr}^{(0)}$ as $\rho \to 0$ for all $k, r \neq 0$, and thus **(j)** $_i\mathbf{P}^{(0)}(\rho) \to {}_i\mathbf{P}^{(0)}$ as $\rho \to 0$. Relations **(h)** and **(j)** imply that **(k)** there exists small enough $0 < \beta^{(0)} \leq \delta$ such that $\det(\mathbf{I} - {}_i\mathbf{P}^{(0)}(\beta^{(0)})) \neq 0$ for every $i \in X_1^{(0)}$. Also, by condition $\mathbf{C_{12}}$, **(l)** $p_{kr}^{(0)}(\beta^{(0)}) < \infty$, $k, r \in X_1^{(0)}$. Now, Lemma 4.5.1, **(k)** and **(l)** imply that **(m)** $_0\phi_{ii}^{(0)}(\beta^{(0)}) < \infty$, $i \in X_1^{(0)}$. Note that **(n)** $_0\phi_{ii}^{(0)}(\rho) > 1$ for $\rho > 0$ since, due to conditions $\mathbf{I_4}$ and $\mathbf{A_3}$, the distribution function $_0G_{ii}^{(0)}(t)$ is proper and not concentrated in zero. Relations **(m)** and **(n)** prove **(α)**. The statement **(β)** follows from Lemma 4.5.6 and the remarks above which imply that **(o)** $\rho^{(0)} = 0$. $\square$

Let us introduce moment generating functions

$$p_i^{(\varepsilon)}(\rho) = \sum_{j \neq 0} p_{ij}^{(\varepsilon)}(\rho), \quad \rho \geq 0, \quad i \neq 0,$$

and define for $i \neq 0$,

$$\rho_i^* = \sup(\rho \geq 0 : p_i^{(0)}(\rho) < \infty). \tag{4.5.33}$$

By the definition, **(z′)** $\rho_i^* \in [0, \infty]$; **(z″)** $p_i^{(0)}(\rho) < \infty$ for $\rho < \rho_i^*$ and $p_i^{(0)}(\rho_i^*) \leq \infty$; **(z‴)** $p_i^{(0)}(\rho) \to p_i^{(0)}(\rho_i^*)$ as $\rho < \rho_i^*, \rho \to \rho_i^*$.

Let us also define

$$\rho^* = \min_{i \in X_1^{(0)}} \rho_i^*. \tag{4.5.34}$$

Let us now introduce the following condition:

$\mathbf{G_1}$: **(a)** $\rho^* > 0$;

    (b) $p_i^{(0)}(\rho^*) = \infty$ for some $i \in X_1^{(0)}$.

Lemma 4.5.9. *Let conditions* $\mathbf{A}_2$, $\mathbf{E}_2$, *and* $\mathbf{G}_1$ *hold. Then,* $(\boldsymbol{\alpha})$ *there exist* $i \in X_1^{(0)}$ *and* $0 < \beta_i < \rho^*$ *such that* $_0\phi_{ii}^{(0)}(\beta_i) \in (1, \infty)$.

Proof. Let $i \in X_1^{(0)}$ be a state such that condition $\mathbf{G}_1$ **(b)** holds. Then, **(p)** there exists $k \in X_1^{(0)}$ such that $p_{ik}^{(0)} > 0$ and $p_{ik}^{(0)}(\rho^*) = \infty$. This obviously imply that $_0\phi_{ii}^{(0)}(\rho^*) = \infty$. By repeating the first part of the proof of Lemma 4.5.8 we get that **(r)** there exists $0 < \bar{\rho}^* < \rho^*$ such that $\det(\mathbf{I} - {}_i\mathbf{P}^{(0)}(\bar{\rho}_i^*)) \neq 0$. Since $0 < \bar{\rho}_i^* < \rho^*$ we also have **(s)** $p_{kr}^{(0)}(\bar{\rho}_i^*) < \infty$, $k, r \in X_1^{(0)}$. Relations **(r)** and **(s)** imply that **(t)** $_0\phi_{ii}^{(0)}(\bar{\rho}_i^*) < \infty$. Relations **(p)** and **(t)** imply that **(u)** there exist $\bar{\rho}_i^* < \hat{\rho}_i^* \leq \rho^*$ such that $_0\phi_{ii}^{(0)}(\rho) < \infty$ for $\rho < \hat{\rho}_i^*$ and $_0\phi_{ii}^{(0)}(\rho) \to \infty$ as $\rho < \hat{\rho}_i^*, \rho \to \hat{\rho}_i^*$. Since function $_0\phi_{ii}^{(0)}(\rho)$ is continuous and strictly monotone for $\rho < \hat{\rho}_i^*$, relations **(u)** imply that there exist $\bar{\rho}_i^* \leq \beta_i < \hat{\rho}_i^*$ such that $_0\phi_{ii}^{(0)}(\beta_i) \in (1, \infty)$. $\qquad\square$

Let us also introduce condition:

$\mathbf{G}_2$: **(a)**: $\overline{\lim}_{0 \leq \varepsilon \to 0}\, p_{ij}^{(\varepsilon)}(\rho) < \infty$ for $\rho < \rho^*, i \neq 0, j \in X$;

  (b): $_0\phi_{ki}^{(0)}(\rho) < \infty$ for $\rho < \rho^*$, $k \in X_2^{(0)}$ for some $i \in X_1^{(0)}$.

Lemma 4.5.10. *Let conditions* $\mathbf{A}_2$, $\mathbf{E}_2$, $\mathbf{G}_1$ *and* $\mathbf{G}_2$ *hold. Then,* $(\boldsymbol{\alpha})$ *there exist* $i \in X_1^{(0)}$ *and* $0 < \beta_i < \delta < \rho^*$ *such that conditions* $\mathbf{C}_{12}$ *and* $\mathbf{C}_{13}$ *hold.*

Proof. By condition $\mathbf{G}_2$ **(a)** and the definition of ρ^*, **(v)** $p_{ij}^{(\varepsilon)}(\rho) < \infty$ for $i \neq 0$, $j \in X$. Also, as follows from Lemma 4.5.9, **(w)** conditions $\mathbf{A}_2$, $\mathbf{E}_2$, and $\mathbf{G}_1$ imply that there exist $i \in X_1^{(0)}$ and $\beta_i < \rho^*$ such that $_0\phi_{ii}^{(0)}(\beta_i) \in (1, \infty)$. Relations **(v)**, **(w)**, and condition $\mathbf{G}_2$ imply that conditions $\mathbf{C}_{12}$ and $\mathbf{C}_{13}$ holds with any $0 < \beta_i < \delta < \rho^*$. $\quad\square$

Remark 4.5.4. Lemmas 4.5.9 and 4.5.10 let one to replace conditions and $\mathbf{C}_{13}$ by conditions $\mathbf{G}_1$ and $\mathbf{G}_2$ in Lemmas 4.5.6 and 4.5.7. The assumption $\overline{\lim}_{0 \leq \varepsilon \to 0}\, p_{i0}^{(\varepsilon)}(\rho) < \infty, \rho < \rho^*, i \neq 0$, is not needed in the corresponding variant of Lemma 4.5.6 but it is not required in the corresponding variant of Lemma 4.5.7.

In conclusion, we would like to note that two cases are characteristic for most applications. Either we have a pseudo-stationary model, where the root $\rho^{(0)} = 0$ or the quasi-stationary model, where the root $\rho^{(0)} > 0$ and condition $\mathbf{E}_2'$ holds, i.e., the set of non-recurrent-without-absorption states $X_2^{(0)}$ is empty. In the first case, Lemma 4.5.8 let one escape checking of the "unpleasant" condition $\mathbf{C}_{13}$ **(b)**. In the second case, the "unpleasant" condition $\mathbf{G}_2$, which is an analogue of $\mathbf{C}_{13}$ **(b)**, can obviously be omitted in Lemma 4.5.10. This let one effectively use this lemma for checking condition $\mathbf{C}_{13}$.

4.6 Mixed ergodic and large deviation theorems for perturbed semi-Markov processes

In this section we formulate a so-called mixed ergodic and large deviation theorem for perturbed semi-Markov processes.

4.6.1 Pseudo-stationary exponential asymptotics for perturbed semi-Markov processes without non-recurrent-without-absorption states

The following two theorems require Cramér type conditions on transition times for semi-Markov processes but do not impose any restrictions on the rate of the time growth as it was in Theorems 4.4.1 and 4.4.2.

Note that, in the theorem formulated below, the stationary probabilities $\pi_j^{(0)}$, $j \neq 0$, are given by formula (4.4.7).

Theorem 4.6.1. *Let conditions* $\mathbf{A_2}$, $\mathbf{T_1}$, $\mathbf{E'_2}$, $\mathbf{C_{12}}$, $\mathbf{A_3}$, *and* $\mathbf{N_1}$ *hold. Then, for any* $0 \leq t^{(\varepsilon)} \to \infty$ *as* $\varepsilon \to 0$,

$$\frac{\mathsf{P}_i\{\eta^{(\varepsilon)}(t^{(\varepsilon)}) = j, \mu_0^{(\varepsilon)} > t^{(\varepsilon)}\}}{\exp\{-\rho^{(\varepsilon)} t^{(\varepsilon)}\}} \to \pi_j^{(0)} \quad as \quad \varepsilon \to 0, \quad i, j \neq 0. \tag{4.6.1}$$

Proof. The proof mainly repeats the proof of Theorem 4.4.1. According to the remarks made in Subsection 4.2.1, the semi-Markov process $\eta^{(\varepsilon)}(t)$, more precisely its modified version with the absorption state removed using the procedure described in that subsection, can be considered as a perturbed regenerative process. The absorption time $\mu_0^{(\varepsilon)}$ plays the role of the regenerative stopping moment.

Let us chose some state $j \in X_1^{(0)}$. Note that, by condition $\mathbf{E'_2}$, $X_1^{(0)} = X_0^{(0)} = \{1, \ldots, N\}$. We can apply Theorem 3.3.1 to prove the asymptotic relation (4.6.1) for the case where the initial state $i = j$, since $\eta^{(\varepsilon)}(t)$ is a standard regenerative process with the regeneration times $\mu_j^{(\varepsilon)}(n)$ that are subsequent times of return of the process $\eta^{(\varepsilon)}(t)$ into the state j.

The renewal equation (3.2.1) takes now the form of the renewal equation (4.2.1). In this case, we have $A = \{j\}$ and the probability $P_{jj}^{(\varepsilon)}(t)$ can be regarded as the probability $P^{(\varepsilon)}(t, A)$.

The distribution function $_0G_{jj}^{(\varepsilon)}(t)$ replaces the distribution function $F^{(\varepsilon)}(t)$ that generates this renewal equation. As was pointed out in the proof of Theorem 4.4.1, conditions $\mathbf{A_2}$, $\mathbf{T_1}$, $\mathbf{E'_2}$, $\mathbf{A_3}$, and $\mathbf{N_1}$ imply, by Lemmas 4.2.1, 4.2.3, 4.3.4, and 4.4.1, respectively, that **(a)** the distribution functions $_0G_{jj}^{(\varepsilon)}(t)$ weakly converge to $_0G_{jj}^{(0)}(t)$ as $\varepsilon \to 0$, **(b)** the limit distribution function $_0G_{jj}^{(0)}(t)$ is proper, and **(c)** it is also non-arithmetic. It follows from **(a)**–**(c)** that the condition of weak convergence $\mathbf{D_{13}}$ holds for the distribution functions $_0G_{jj}^{(\varepsilon)}(t)$.

The distribution function $G_{jj}^{(\varepsilon)}(t)$ is regarded as the distribution function $\hat{F}^{(\varepsilon)}(t)$. Conditions $\mathbf{A_2}$, $\mathbf{T_1}$, $\mathbf{E_2'}$, and $\mathbf{C_{12}}$ yield, by Lemmas 4.5.6, 4.5.7 and 4.5.8, **(d)** convergence of the corresponding exponential moments, $\phi_{jj}^{(\varepsilon)}(\beta) \to \phi_{jj}^{(0)}(\beta) < \infty$ as $\varepsilon \to 0$. Lemmas 4.5.7 – 4.5.8 and the convergence relation **(d)** imply that condition $\mathbf{C_3}$ holds for the distribution functions $G_{jj}^{(\varepsilon)}(t)$.

The function $1 - F_j^{(\varepsilon)}(t)$ is a forcing function for the renewal equation (4.2.1). These functions are **(e)** non-decreasing, and, **(f)** by condition $\mathbf{T_1}$, weakly converge. As was shown in the proof of Theorem 4.4.1, **(e)** and **(f)** imply that the functions $1 - F_j^{(\varepsilon)}(t)$ converge locally uniformly in every point of the set C_j of positive continuity points of the limit function $1 - F_j^{(0)}(t)$. Note that the set C_j has the value of Lebesgue measure $m(\overline{C}_j) = 0$. Thus condition $\mathbf{F_7}$ holds for the functions $1 - F_j^{(\varepsilon)}(t)$.

All conditions of Theorem 3.3.1 hold. The corresponding evaluation of the stationary probabilities $\pi_j^{(0)}$ has been made in the proof of Theorem 4.4.1.

By applying the asymptotic relation (1.4.9), given in Theorem 3.3.1, we get the asymptotic relation (4.6.1) for $i = j$.

We can apply Theorem 3.3.8 to prove the asymptotic relation (4.6.1) for the case where $i \neq j$. In this case, $\eta^{(\varepsilon)}(t)$ is again a regenerative process with regeneration times $\mu_j^{(\varepsilon)}(n)$ that are subsequent times of return of the process $\eta^{(\varepsilon)}(t)$ into the state j. This process has a transition period, since the initial state differs from j but the shifted process $\eta^{(\varepsilon)}(\mu_j^{(\varepsilon)}(1) + t)$ is the standard regenerative process considered above.

The renewal type relation (3.2.19) takes in this case the form of the renewal type relation (4.2.2). In this case again, $A = \{j\}$ and the probability $P_{ij}^{(\varepsilon)}(t)$ plays the role of the probability $\tilde{P}^{(\varepsilon)}(t, A)$. The distribution function $_0G_{ij}^{(\varepsilon)}(t)$ can be regarded as the distribution function $\tilde{F}^{(\varepsilon)}(t)$ that generates this renewal type relation.

The probability $_jf_{i0}^{(\varepsilon)} = 1 - {_0f_{ij}^{(\varepsilon)}} = 1 - {_0G_{ij}^{(\varepsilon)}}(\infty)$ becomes the stopping probability $\tilde{f}^{(\varepsilon)}$ in the transition period. Conditions $\mathbf{A_2}$, $\mathbf{T_1}$, $\mathbf{E_2'}$, and $\mathbf{A_3}$ imply, by Lemma 4.2.1, that **(g)** $_jf_{i0}^{(\varepsilon)} \to {_jf_{i0}^{(0)}} = 0$ as $\varepsilon \to 0$. Thus, condition $\mathbf{D_{17}}$ holds with the limit constant $\tilde{f}^{(0)} = 0$.

The distribution function $G_{ij}^{(\varepsilon)}(t)$ is now the distribution function $\hat{\tilde{F}}^{(\varepsilon)}(t)$. Conditions $\mathbf{A_2}$, $\mathbf{T_1}$, $\mathbf{E_2'}$ and $\mathbf{C_{12}}$ imply, by Lemmas 4.5.6, 4.5.7 and 4.5.8, that **(h)** condition $\mathbf{C_8}$ holds for these distribution functions. This condition implies conditions $\mathbf{C_7}$ and $\mathbf{D_{25}}$.

Finally, by applying the asymptotic relation (3.3.28) given in Theorem 3.3.8, we get the asymptotic relation (4.6.1) for the case where $i \neq j$. $\square$

4.6.2 Pseudo-stationary exponential asymptotics for perturbed semi-Markov processes with non-recurrent-without-absorption states

Let us now formulate a theorem which is an analogue of Theorem 4.6.1 for the case where the limit set of non-recurrent-without-absorption states is $X_2^{(0)} \neq 0$.

Note that the stationary probabilities $\pi_j^{(0)}$, $j \neq 0$, and the absorption probabilities $_{X_1^{(0)}} f_{i0}^{(0)}$ are given in the theorem formulated below by formulas (4.4.28) and (4.2.77), respectively, with $Z = X_1^{(0)}$.

Theorem 4.6.2. *Let conditions* $\mathbf{A_2}$, $\mathbf{T_1}$, $\mathbf{E_2''}$, $\mathbf{C_{12}}$, $\mathbf{A_3}$, *and* $\mathbf{N_1}$ *hold. Then, for any* $0 \leq t^{(\varepsilon)} \to \infty$ *as* $\varepsilon \to 0$,

$$\frac{\mathsf{P}_i\{\eta^{(\varepsilon)}(t^{(\varepsilon)}) = r, \mu_0^{(\varepsilon)} > t^{(\varepsilon)}\}}{\exp\{-\rho^{(\varepsilon)} t^{(\varepsilon)}\}} \to (1 - _{X_1^{(0)}} f_{i0}^{(0)})\pi_r^{(0)} \quad \text{as } \varepsilon \to 0, \ i, r \neq 0. \quad (4.6.2)$$

Proof. Let us first consider that case where $i = r$ and $r = j \in X_1^{(0)}$.

The first part of the proof in which conditions $\mathbf{D_{13}}$, $\mathbf{C_3}$, $\mathbf{F_7}$ are verified repeats the corresponding part in the proof of Theorem 4.6.1. We only need to replace the reference to the condition $\mathbf{E_2'}$ with a reference to condition $\mathbf{E_2''}$.

Thus again, all conditions of Theorem 3.3.1 hold true. An evaluation of the corresponding stationary probabilities $\pi_j^{(0)}$ was made in the proof of Theorem 4.4.1. There was also pointed out a proof that $_{X_1^{(0)}} f_{i0}^{(0)} = 0$ for $i \in X_1^{(0)}$.

By applying the asymptotic relation (1.4.9) given in Theorem 3.3.1, we get the asymptotic relation (4.6.1) for the case where $i = r$ and $j = r \in X_1^{(0)}$.

Let us now consider the case where the initial state $i \neq r, 0$ and $r = j \in X_1^{(0)}$.

As above, the renewal type relation (3.2.19) takes the form of the renewal equation (4.2.2), the distribution function $_0G_{ij}^{(\varepsilon)}(t)$ becomes the distribution function $\tilde{F}^{(\varepsilon)}(t)$ that generates this renewal type relation, and the probability $_jf_{i0}^{(\varepsilon)} = 1 - _0f_{ij}^{(\varepsilon)} = 1 - _0G_{ij}^{(\varepsilon)}(\infty)$ plays the role of the stopping probability $\tilde{f}^{(\varepsilon)}$ in the transition period. Conditions $\mathbf{A_2}$, $\mathbf{T_1}$, $\mathbf{E_2''}$, and $\mathbf{A_3}$ imply, by Lemma 4.2.1, that **(a)** $_jf_{i0}^{(\varepsilon)} \to _jf_{i0}^{(0)}$ as $\varepsilon \to 0$. Thus, condition $\mathbf{D_{17}}$ holds with the limit constant $\tilde{f}^{(0)} = _jf_{i0}^{(0)}$. As was shown in the proof of Theorem 4.4.2, $_jf_{i0}^{(0)} = _{X_1^{(0)}} f_{i0}^{(0)}$ in this case.

Similarly to the above, the distribution function $G_{ij}^{(\varepsilon)}(t)$ can be regarded as the distribution function $\hat{\tilde{F}}^{(\varepsilon)}(t)$. Conditions $\mathbf{A_2}$, $\mathbf{T_1}$, $\mathbf{E_2'}$ and $\mathbf{C_{12}}$ imply, by Lemmas 4.5.6, 4.5.7 and 4.5.8, that **(b)** condition $\mathbf{C_8}$ holds for these distribution functions, which implies that conditions $\mathbf{C_7}$ and $\mathbf{D_{25}}$ are also verified.

By applying the asymptotic relation (3.3.28) given in Theorem 3.3.8, we get the asymptotic relation (4.6.2) for the case where $i \neq r, 0$ and $r = j \in X_1^{(0)}$.

It still remains to prove asymptotic relation (4.6.2) for $r \in X_2^{(0)}$. In this case, we can choose some state $j \in X_1^{(0)}$ and use the renewal relations (4.2.3) and (4.2.6) instead of the renewal relations (4.2.1) and (4.2.2).

Let us first consider the case where $i = j \in X_1^{(0)}, r \in X_2^{(0)}$.

The renewal equations (4.2.3) and (4.2.1) have the same distribution function $_0G_{jj}^{(\varepsilon)}(t)$ that generates these equations, and the distribution function $G_{jj}^{(\varepsilon)}(t)$ plays the role of the distribution function $\hat{F}^{(\varepsilon)}(t)$. Thus, conditions $\mathbf{D}_{13}$ and $\mathbf{C}_3$ do not require a new proof.

However, the renewal equation (4.2.3) has a forcing function that differs from the forcing function for the renewal equation (4.2.1). Condition $\mathbf{F}_7$ was checked in this case, in the proof of Theorem 4.4.2. The corresponding stationary probabilities $\pi_r^{(0)}$ were also given in the proof of this theorem.

By applying the asymptotic relation (1.4.9) given in Theorem 3.3.1, we get the asymptotic relation (4.6.2) for the case where $i = j \in X_1^{(0)}, r \in X_2^{(0)}$.

The last case to be considered is the one where the initial state $i \neq j, 0, j \in X_1^{(0)}$, and $r \in X_2^{(0)}$.

The renewal relation (3.2.19) takes now the form of the renewal relation (4.2.6) that should be used instead of the renewal relation (4.2.2). The renewal relations (4.2.6) and (4.2.2) have the same distribution function $_0G_{ij}^{(\varepsilon)}(t)$ generating these equations and the same distribution function $G_{ij}^{(\varepsilon)}(t)$ that plays the role of the distribution function $\hat{\hat{F}}^{(\varepsilon)}(t)$. Thus, condition $\mathbf{C}_8$ and, consequently, conditions $\mathbf{C}_7$ and $\mathbf{D}_{25}$ do not need a new proof.

Finally, by applying the asymptotic relation (3.3.28) given in Theorem 3.3.8, we get the asymptotic relation (4.6.2) for the case where $i \neq j, 0, j \in X_1^{(0)}$, and $r \in X_2^{(0)}$. $\square$

Remark 4.6.1. By Lemma 1.4.2, under conditions $\mathbf{A}_2$, $\mathbf{T}_1$, $\mathbf{E}_2$ $\mathbf{C}_{12}$, $\mathbf{A}_3$, we have $\rho^{(\varepsilon)} \sim {}_jf_{j0}^{(\varepsilon)}/{}_0M_{jj}^{(\varepsilon)}$ as $\varepsilon \to 0$. Thus, if conditions $\mathbf{E}_2'$ and $\mathbf{B}_6$ or $\mathbf{E}_2''$ and $\mathbf{B}_7$ hold with a limit constant $\lambda > 0$, then $\rho^{(\varepsilon)}t^{(\varepsilon)} \to \lambda/m^{(0)}$ as $\varepsilon \to 0$. In these cases, the asymptotic relations given in Theorems 4.6.1 and 4.6.2 reduce to the asymptotic relations given, respectively, in Theorems 4.4.1 and 4.4.2.

4.6.3   Quasi-stationary exponential asymptotics for perturbed semi-Markov processes without non-recurrent-without-absorption states

Let us now consider the general case where the asymptotic non-absorption condition $\mathbf{A}_3$ may or may not hold.

Note that the characteristic root is $\rho^{(0)} > 0$ if and only if the condition $\mathbf{A}_3$ does not hold or, equivalently, $\rho^{(0)} = 0$ if and only if condition $\mathbf{A}_3$ holds.

Denote

$$\tilde{\pi}_j^{(\varepsilon)}(\rho^{(\varepsilon)}) = \frac{\int_0^\infty e^{\rho^{(\varepsilon)}s}(1 - F_j^{(\varepsilon)}(s))\,ds}{\int_0^\infty se^{\rho^{(\varepsilon)}s}\,{}_0G_{jj}^{(\varepsilon)}(ds)}, \qquad j \in X_1^{(\varepsilon)}. \tag{4.6.3}$$

Note that, under conditions $\mathbf{A}_2$, $\mathbf{T}_1$, and $\mathbf{E}_2'$, $X_1^{(\varepsilon)} = X_0 = \{1,\dots,N\}$ for all ε small enough. This follows from Lemma 4.2.2.

Let us also denote

$$\tilde{\tilde{\pi}}_{ij}^{(\varepsilon)}(\rho^{(\varepsilon)}) = {}_0\phi_{ij}^{(\varepsilon)}(\rho^{(\varepsilon)})\tilde{\pi}_j^{(\varepsilon)}(\rho^{(\varepsilon)}), \quad i,j \in X_1^{(\varepsilon)}. \tag{4.6.4}$$

Note that, under conditions $\mathbf{A_2}$, $\mathbf{T_1}$, $\mathbf{E_2'}$, $\mathbf{C_{12}}$, and $\mathbf{C_{13}}$, the integrals in the expression that defines the quantities $\tilde{\pi}_j^{(\varepsilon)}(\rho^{(\varepsilon)})$ and $\tilde{\tilde{\pi}}_{ij}^{(\varepsilon)}(\rho^{(\varepsilon)})$ converge for all ε small enough. This follows from Lemma 4.5.7.

Theorem 4.6.3. *Let conditions* $\mathbf{A_2}$, $\mathbf{T_1}$, $\mathbf{E_2'}$, $\mathbf{C_{12}}$, $\mathbf{C_{13}}$, *and* $\mathbf{N_1}$ *hold. Then, for any* $0 \le t^{(\varepsilon)} \to \infty$ *as* $\varepsilon \to 0$,

$$\frac{\mathsf{P}_i\{\eta^{(\varepsilon)}(t^{(\varepsilon)}) = j, \mu_0^{(\varepsilon)} > t^{(\varepsilon)}\}}{\exp\{-\rho^{(\varepsilon)}t^{(\varepsilon)}\}} \to \tilde{\tilde{\pi}}_{ij}^{(0)}(\rho^{(0)}) \text{ as } \varepsilon \to 0, \quad i,j \neq 0. \tag{4.6.5}$$

Proof. Using the remarks made in the Subsection 4.2.1 we see that the semi-Markov process $\eta^{(\varepsilon)}(t)$, more precisely its modified version with the absorption state removed according to the procedure described in that subsection, can be considered as a perturbed regenerative process. The absorption time $\mu_0^{(\varepsilon)}$ plays the role of the regenerative stopping moment.

Let us chose some state $j \in X_1^{(0)}$. Note that, according to condition $\mathbf{E_2'}$, $X_1^{(0)} = X_0 = \{1,\dots,N\}$. In this case, we can apply Theorem 3.3.2 to prove the asymptotic relation (4.6.5) for the case where $i = j$, since in this case, $\eta^{(\varepsilon)}(t)$ is a standard regenerative process with regeneration times $\mu_j^{(\varepsilon)}(n)$ that are subsequent times of return of the process $\eta^{(\varepsilon)}(t)$ into the state j.

The renewal equation (3.2.1) takes now the form of renewal equation (4.2.1). In this case, $A = \{j\}$ and the probability $P_{jj}^{(\varepsilon)}(t)$ is regarded as the probability $P^{(\varepsilon)}(t, A)$.

The distribution function ${}_0G_{jj}^{(\varepsilon)}(t)$ is now the distribution function $F^{(\varepsilon)}(t)$ that generates this renewal equation. As was pointed out in the proof of Theorem 4.4.1, conditions $\mathbf{A_2}$, $\mathbf{T_1}$, $\mathbf{E_2'}$, $\mathbf{A_3}$, and $\mathbf{N_1}$ imply, by Lemmas 4.2.1, 4.2.3, 4.3.4, and 4.4.1, respectively, that **(a)** the distribution functions ${}_0G_{jj}^{(\varepsilon)}(t)$ weakly converge to ${}_0G_{jj}^{(0)}(t)$ as $\varepsilon \to 0$, **(b)** the quantities ${}_0G_{jj}^{(\varepsilon)}(\infty)$ converge to ${}_0G_{jj}^{(0)}(\infty)$ as $\varepsilon \to 0$, **(c)** ${}_0G_{jj}^{(0)}(\infty) > 0$ and the normalised limit distribution function ${}_0G_{jj}^{(0)}(t)/{}_0G_{jj}^{(0)}(\infty)$ is non-arithmetic. It follows from **(a)**–**(d)** that the condition of weak convergence $\mathbf{D_{22}}$ holds for the distribution functions ${}_0G_{jj}^{(\varepsilon)}(t)$.

The distribution function $G_{jj}^{(\varepsilon)}(t)$ replaces the distribution function $\hat{F}^{(\varepsilon)}(t)$. Conditions $\mathbf{A_2}$, $\mathbf{T_1}$, $\mathbf{E_2'}$, $\mathbf{C_{12}}$, and $\mathbf{C_{13}}$ imply, by Lemmas 4.5.6 and 4.5.7, that **(d)** the corresponding exponential moments converge, $\phi_{jj}^{(\varepsilon)}(\beta) \to \phi_{jj}^{(0)}(\beta) < \infty$ as $\varepsilon \to 0$, and also give the relation **(e)** $\phi_{jj}^{(0)}(\beta) \in (1, \infty)$. Relations **(d)** and **(e)** imply that condition $\mathbf{C_6}$ holds for the distribution functions $G_{jj}^{(\varepsilon)}(t)$.

Note that conditions $\mathbf{D_{22}}$ and $\mathbf{C_6}$ imply condition $\mathbf{D_{21}}$.

The function $1 - F_j^{(\varepsilon)}(t)$ is a forcing function for the renewal equation (4.2.1). Such functions are **(f)** non-decreasing, and **(g)** weakly converge by condition $\mathbf{T_1}$. As was shown in the proof of Theorem 4.4.1, **(f)** and **(g)** imply that the functions $1 - F_j^{(\varepsilon)}(t)$ converge locally uniformly in every point of the set C_j of positive continuity points of the limit function $1 - F_j^{(0)}(t)$. Note that, for the Lebesgue measure, $m(\overline{C}_j) = 0$. Thus condition $\mathbf{F_7}$ holds for the functions $1 - F_j^{(\varepsilon)}(t)$.

This shows that all conditions of Theorem 3.3.2 hold. By applying the asymptotic relation (2.2.18) in Theorem 3.3.2, we get the asymptotic relation (4.6.5) for $i = j$.

We can apply Theorem 3.3.7 to prove the asymptotic relation (4.6.5) for the case where the initial state is $i \neq j$. In this case, $\eta^{(\varepsilon)}(t)$ is again a regenerative process with regeneration times $\mu_j^{(\varepsilon)}(n)$ that are subsequent times of return of the process $\eta^{(\varepsilon)}(t)$ into the state j. This process has a transition period, since the initial state differs from j but the shifted process $\eta^{(\varepsilon)}(\mu_j^{(\varepsilon)}(1) + t)$ is the standard regenerative process considered above.

The renewal type relation (3.2.19) takes in this case the form of the renewal type relation (4.2.2). In this case again, $A = \{j\}$ and the probability $P_{ij}^{(\varepsilon)}(t)$ plays the role of the probability $\tilde{P}^{(\varepsilon)}(t, A)$. The distribution function $_0G_{ij}^{(\varepsilon)}(t)$ replaces the function $\tilde{F}^{(\varepsilon)}(t)$ that generates this renewal type relation.

The moment generating function $_0\phi_{ij}^{(\varepsilon)}(\rho)$ coincides in this case with the moment generating function $\tilde{\phi}^{(\varepsilon)}(\rho)$. Conditions $\mathbf{A_2}, \mathbf{T_1}, \mathbf{E_2'}$ imply, by Lemmas 4.5.6 and 4.5.7, that **(h)** $_0\phi_{ij}^{(\varepsilon)}(\rho^{(0)}) \to {}_0\phi_{ij}^{(0)}(\rho^{(0)}) < \infty$ as $\varepsilon \to 0$. Thus, condition $\mathbf{C_{11}}$ holds with the limit constant $\tilde{\phi}^{(0)} = {}_0\phi_{ij}^{(0)}(\rho^{(0)})$.

The distribution function $G_{ij}^{(\varepsilon)}(t)$ is regarded as the function $\hat{\tilde{F}}^{(\varepsilon)}(t)$. Conditions $\mathbf{A_2}, \mathbf{T_1}, \mathbf{E_2'}, \mathbf{C_{12}}$, and $\mathbf{C_{13}}$ imply, by Lemma 4.5.7, that **(i)** condition $\mathbf{C_{10}}$ holds for these distribution functions, which implies that conditions $\mathbf{C_9}$ and $\mathbf{D_{26}}$ also hold.

Finally, by applying the asymptotic relation (3.3.52) given in Theorem 3.3.7, we get the asymptotic relation (4.6.5) for the case where $i \neq j$. $\qquad\square$

4.6.4 Quasi-stationary distributions for perturbed semi-Markov processes without non-recurrent-without-absorption states

Let us assume that the following conditions hold:

$\mathbf{E_3'}$: $X_1^{(\varepsilon)} = X_0 = \{1, \ldots, N\}$ for all ε small enough, say $\varepsilon \leq \varepsilon_6'$;

$\mathbf{C_{14}}$: $p_{ij}^{(\varepsilon)}(\delta) < \infty, i \neq 0, j \in X$, for all ε small enough, say $\varepsilon \leq \varepsilon_7'$;

Note that the following condition implies condition $\mathbf{C_{14}}$:

$\mathbf{C_{14}'}$: $\psi_{ij}^{(\varepsilon)}(\delta) < \infty, i \neq 0, j \in X$, for all ε small enough, say $\varepsilon \leq \varepsilon_7'$;

Let us also assume that the following conditions hold:

$\mathbf{C_{15}}$: There exists $i \neq 0$ and $\beta_i \in (0, \delta]$, for δ defined in condition $\mathbf{C_{14}}$, such that: $\varphi_{ii}^{(\varepsilon)}(\beta_i) \in (1, \infty)$, for all ε small enough, say $\varepsilon \leq \varepsilon_8'$;

$\mathbf{N_3}$: $_0\bar{G}_{jj}^{(\varepsilon)}(t)$, $j \neq 0$, are non-arithmetic distribution functions for ε small enough, say $\varepsilon \leq \varepsilon_9'$.

Set

$$\varepsilon_{10}' = \min(\varepsilon_6', \varepsilon_7', \varepsilon_8', \varepsilon_9'). \tag{4.6.6}$$

The following theorem is a direct corollary of Theorem 4.6.3.

Theorem 4.6.4. *Let conditions* $\mathbf{E_3'}$, $\mathbf{C_{14}}$, $\mathbf{C_{15}}$, *and* $\mathbf{N_3}$ *hold. Then, for every* $\varepsilon \leq \varepsilon_{10}'$,

$$e^{\rho^{(\varepsilon)}t}\mathsf{P}_i\{\eta^{(\varepsilon)}(t^{(\varepsilon)}) = j, \mu_0^{(\varepsilon)} > t\} \to \tilde{\tilde{\pi}}_{ij}^{(\varepsilon)}(\rho^{(\varepsilon)}) \ \text{as} \ t \to \infty, \quad i, j \neq 0. \tag{4.6.7}$$

Proof. Choose some $\varepsilon' \leq \varepsilon_{10}'$ and consider a model in which the transition probabilities $Q_{ij}^{(\varepsilon)}(t)$ of the semi-Markov processes $\eta^{(\varepsilon)}(t)$, $t \geq 0$, do not depend on the parameter ε and coincide with the transition probabilities $Q_{ij}^{(\varepsilon')}(t)$. In this case, conditions of Theorem 4.6.3 just reduce to conditions $\mathbf{E_3'}$, $\mathbf{C_{14}}$, $\mathbf{C_{15}}$, and $\mathbf{N_3}$. The asymptotic relation (4.6.5) takes in this case the form of the asymptotic relation (4.6.7). In the same way one can prove relation (4.6.7) for any $\varepsilon \leq \varepsilon_{10}'$. $\qquad\square$

Let

$$\pi_j^{(\varepsilon)}(\rho^{(\varepsilon)}) = \frac{\tilde{\tilde{\pi}}_{ij}^{(\varepsilon)}(\rho^{(\varepsilon)})}{\sum_{r \neq 0} \tilde{\tilde{\pi}}_{ir}^{(\varepsilon)}(\rho^{(\varepsilon)})}, \quad j \in X_1^{(\varepsilon)}. \tag{4.6.8}$$

Conditions $\mathbf{A_2}$, $\mathbf{T_1}$, $\mathbf{E_3'}$, $\mathbf{C_{12}}$, and $\mathbf{C_{13}}$, imply that $\tilde{\tilde{\pi}}_r^{(\varepsilon)}(\rho^{(\varepsilon)}) \in (0, \infty)$ for all ε small enough, namely, $\varepsilon \leq \min(\varepsilon_6', \varepsilon_7', \varepsilon_8')$. This follows from Lemma 4.5.7.

Now, **(a)** $\pi_j^{(\varepsilon)}(\rho^{(\varepsilon)}) > 0$, $j \neq 0$, and **(b)** $\sum_{j \in X_1^{(\varepsilon)}} \pi_j^{(\varepsilon)}(\rho^{(\varepsilon)}) = 1$. Thus, formula (4.6.8) defines a discrete probability distribution.

This distribution is called a *quasi-stationary distribution* for the semi-Markov process $\eta^{(\varepsilon)}(t)$, $t \geq 0$, with absorption times $\mu_0^{(\varepsilon)}$.

The following quasi-ergodic theorem is a direct corollary of Theorem 4.6.3.

Theorem 4.6.5. *Let conditions* $\mathbf{E_3'}$, $\mathbf{C_{14}}$, $\mathbf{C_{15}}$, *and* $\mathbf{N_3}$ *hold. Then, for* $\varepsilon \leq \varepsilon_{10}'$,

$$\mathsf{P}_i\{\eta^{(\varepsilon)}(t) = j/\mu_0^{(\varepsilon)} > t\} = \frac{e^{\rho^{(\varepsilon)}t}\mathsf{P}_i\{\eta^{(\varepsilon)}(t) = j, \mu_0^{(\varepsilon)} > t\}}{e^{\rho^{(\varepsilon)}t}\mathsf{P}_i\{\mu_0^{(\varepsilon)} > t\}}$$

$$\to \pi_j^{(\varepsilon)}(\rho^{(\varepsilon)}) \ \text{as} \ t \to \infty, \quad j \neq 0. \tag{4.6.9}$$

Relation (4.6.9) clarifies why it is natural to call the distribution $\pi_j^{(\varepsilon)}(\rho^{(\varepsilon)})$ a *quasi-stationary* distribution for the regenerative process $\eta^{(\varepsilon)}(t)$ with absorption moments $\mu_0^{(\varepsilon)}$.

Formula (4.6.8) gives the quasi-stationary distribution in a form that seems to be dependent on the choice of the "initial" state $i \neq 0$. The following lemma shows that the quasi-stationary probabilities $\pi_j^{(\varepsilon)}(\rho^{(\varepsilon)})$ defined in (4.6.8) are, actually, invariant with respect to the choice of the state $i \neq 0$.

Lemma 4.6.1. *Let conditions* $\mathbf{E}_3'$, $\mathbf{C}_{14}$, $\mathbf{C}_{15}$, *and* $\mathbf{N}_3$ *hold. Then, for every* $\varepsilon \leq \varepsilon_{10}'$,

$$\pi_j^{(\varepsilon)}(\rho^{(\varepsilon)}) = \frac{\tilde{\tilde{\pi}}_{ij}^{(\varepsilon)}(\rho^{(\varepsilon)})}{\sum_{r \neq 0} \tilde{\tilde{\pi}}_{ir}^{(\varepsilon)}(\rho^{(\varepsilon)})}, \quad i, j \neq 0. \tag{4.6.10}$$

Proof. Let us return to the proof of Theorem 4.6.3 for the case where the initial state $i \neq j, i, j \in X_1^{(0)}$.

The semi-Markov process $\eta^{(\varepsilon)}(t)$, more precisely, its modified version with the absorption state removed following the procedure described in Subsection 4.2.1, can be considered as a regenerative process.

But the first regeneration time can be defined as the first time at which the semi-Markov process $\eta^{(\varepsilon)}(t)$ hits the state j after first hitting some state $k \neq 0$, and then the following regeneration times are defined to be the subsequent times of hitting the state j.

More precisely, let us define the corresponding hitting times for the imbedded Markov chain $\eta_n^{(\varepsilon)}$ and the semi-Markov process $\eta^{(\varepsilon)}(t)$, respectively, by $v_{kj}^{(\varepsilon)}(1) = \min(n \geq v_k^{(\varepsilon)} : \eta_n^{(\varepsilon)} = j)$ and $\mu_{kj}^{(\varepsilon)}(1) = \tau^{(\varepsilon)}(v_{kj}^{(\varepsilon)})$, and then $v_{kj}^{(\varepsilon)}(n) = \min(n > v_{kj}^{(\varepsilon)}(n-1) : \eta_n^{(\varepsilon)} = j)$ and $\mu_{kj}^{(\varepsilon)}(n) = \tau^{(\varepsilon)}(v_{kj}^{(\varepsilon)}(n))$ for $n = 2, 3, \ldots$.

The semi-Markov process $\eta^{(\varepsilon)}(t)$ can be considered as a regenerative process with regeneration times $\mu_{kj}^{(\varepsilon)}(n)$. This process has a transition period, since the initial state differs from j but the shifted process $\eta^{(\varepsilon)}(\mu_{kj}^{(\varepsilon)}(1) + t)$ is the standard regenerative process considered above with regeneration times that are subsequent return times into the state j.

Denote

$$_0G_{ikj}^{(\varepsilon)}(t) = \mathsf{P}_i\{\mu_{kj}^{(\varepsilon)}(1) \leq t, v_{kj}^{(\varepsilon)}(1) < v_0^{(\varepsilon)}\}, \quad t \geq 0, \quad i, j, k \neq 0.$$

Let us apply Theorem 3.3.7. The renewal type relation (3.2.19) takes now the form of a renewal type relation different from (4.2.2). In this case again, $A = \{j\}$ and the probability $P_{ij}^{(\varepsilon)}(t)$ is regarded as the probability $\tilde{P}^{(\varepsilon)}(t, A)$. But the distribution function $_0G_{ikj}^{(\varepsilon)}(t)$ replaces the distribution function $\tilde{F}^{(\varepsilon)}(t)$ that generates the following renewal type relation replacing (4.2.2):

$$P_{ij}^{(\varepsilon)}(t) = q_{ikjj}^{(\varepsilon)}(t) + \int_0^t P_{jj}^{(\varepsilon)}(t-s)\,{}_0G_{ikj}^{(\varepsilon)}(ds), \quad t \geq 0, \tag{4.6.11}$$

where

$$q_{ikjr}^{(\varepsilon)}(t) = \mathsf{P}_i\{\eta^{(\varepsilon)}(t) = r, \mu_{kj}^{(\varepsilon)}(1) > t, \mu_0^{(\varepsilon)} > t\}, \quad t \geq 0.$$

Using the regeneration property of the semi-Markov process $\eta^{(\varepsilon)}(t)$ at hitting times, we can write the following formula for $i, j, k \neq 0$:

$$_0G_{ikj}^{(\varepsilon)}(t) = {}_0G_{ik}^{(\varepsilon)}(t) * {}_0G_{kj}^{(\varepsilon)}(t), \quad t \geq 0. \tag{4.6.12}$$

It follows from this formula that the distribution function ${}_0G_{ikj}^{(\varepsilon)}(t)$ has the following moment generating function for $i, j, k \neq 0$:

$$_0\phi_{ikj}^{(\varepsilon)}(\rho) = {}_0\phi_{ik}^{(\varepsilon)}(\rho) \cdot {}_0\phi_{kj}^{(\varepsilon)}(\rho), \quad \rho \in \mathbb{R}_1. \tag{4.6.13}$$

The moment generating function ${}_0\phi_{ikj}^{(\varepsilon)}(\rho)$ plays the role of the moment generating function $\tilde{\phi}^{(\varepsilon)}(\rho)$. Conditions $\mathbf{A_2}$, $\mathbf{T_1}$, $\mathbf{E_2'}$ imply, by Lemmas 4.5.6 and 4.5.7, that **(a)** ${}_0\phi_{ikj}^{(\varepsilon)}(\rho^{(0)}) \to {}_0\phi_{ikj}^{(0)}(\rho^{(0)}) < \infty$ as $\varepsilon \to 0$. Thus, condition $\mathbf{C_{11}}$ holds with the limit constant $\tilde{\phi}^{(0)} = {}_0\phi_{ikj}^{(0)}(\rho^{(0)}) = {}_0\phi_{ik}^{(0)}(\rho^{(0)}) \cdot {}_0\phi_{kj}^{(0)}(\rho^{(0)})$.

Denote

$$G_{ikj}^{(\varepsilon)}(t) = \mathsf{P}_i\{\mu_{kj}^{(\varepsilon)}(1) \wedge \mu_0^{(\varepsilon)} \leq t\}, \quad t \geq 0, \quad i, j, k \neq 0.$$

Using the regeneration property of the semi-Markov process $\eta^{(\varepsilon)}(t)$ at hitting times, we obtain the following formula for $i, j, k \neq 0$:

$$G_{ikj}^{(\varepsilon)}(t) = {}_0G_{ikj}^{(\varepsilon)}(t) + {}_kG_{i0}^{(\varepsilon)}(t) + {}_0G_{ik}^{(\varepsilon)}(t) * {}_jG_{k0}^{(\varepsilon)}(t), \quad t \geq 0, \; i, j, k \neq 0. \tag{4.6.14}$$

According to formulas (4.6.13) and (4.6.14), the moment generating functions $G_{ikj}^{(\varepsilon)}(t)$ have the following moment generating function for $i, j, k \neq 0$:

$$\phi_{ikj}^{(\varepsilon)}(\rho) = {}_0\phi_{ik}^{(\varepsilon)}(\rho) \cdot {}_0\phi_{kj}^{(\varepsilon)}(\rho) + {}_k\phi_{i0}^{(\varepsilon)}(\rho) + {}_0\phi_{ik}^{(\varepsilon)}(\rho) \cdot {}_j\phi_{k0}^{(\varepsilon)}(\rho), \quad \rho \in \mathbb{R}_1. \tag{4.6.15}$$

The distribution function $G_{ikj}^{(\varepsilon)}(t)$ here is viewed as the distribution function $\hat{\tilde{F}}^{(\varepsilon)}(t)$. Conditions $\mathbf{A_2}$, $\mathbf{T_1}$, $\mathbf{E_2'}$, $\mathbf{C_{12}}$, $\mathbf{C_{13}}$ imply, by Lemmas 4.5.6 and 4.5.7, that **(b)** these distribution functions satisfy condition $\mathbf{C_{10}}$. This implies that conditions $\mathbf{C_9}$ and $\mathbf{D_{26}}$ hold.

Finally, by applying the asymptotic relation (3.3.52) given in Theorem 3.3.7 and assuming conditions $\mathbf{E}_3'$, $\mathbf{C}_{14}$, $\mathbf{C}_{15}$, and $\mathbf{N}_2$ to hold, we get the following asymptotic relation that is different from (4.6.5):

$$\frac{\mathsf{P}_i\{\eta^{(\varepsilon)}(t^{(\varepsilon)}) = j, \mu_0^{(\varepsilon)} > t^{(\varepsilon)}\}}{\exp\{-\rho^{(\varepsilon)} t^{(\varepsilon)}\}}$$

$$\to {}_0\phi_{ik}^{(0)}(\rho^{(0)}) {}_0\phi_{kj}^{(0)}(\rho^{(0)})\tilde{\pi}_j^{(0)}(\rho^{(0)}) \quad \text{as } \varepsilon \to 0, \quad i,k,j \neq 0. \qquad (4.6.16)$$

Comparing the asymptotic relations (4.6.5) and (4.6.16) we get an interesting relation,

$$ {}_0\phi_{ik}^{(0)}(\rho^{(0)}) {}_0\phi_{kj}^{(0)}(\rho^{(0)}) = {}_0\phi_{ij}^{(0)}(\rho^{(0)}), \quad i,k,j \neq 0. \qquad (4.6.17)$$

Let us consider a model in which the transition probabilities $Q_{ij}^{(\varepsilon)}(t)$ of the semi-Markov processes $\eta^{(\varepsilon)}(t)$, $t \geq 0$, do not depend on the parameter ε, more precisely, their values coincide for some particular value of the parameter $\varepsilon \leq \varepsilon_{10}'$. In this case, conditions of Theorem 4.6.3 just reduce to conditions $\mathbf{E}_3'$, $\mathbf{C}_{14}$, $\mathbf{C}_{15}$, and $\mathbf{N}_2$. Relation (4.6.17) takes in this case the following form for every $\varepsilon \leq \varepsilon_{10}'$:

$$ {}_0\phi_{ik}^{(\varepsilon)}(\rho^{(\varepsilon)}) {}_0\phi_{kj}^{(\varepsilon)}(\rho^{(\varepsilon)}) = {}_0\phi_{ij}^{(\varepsilon)}(\rho^{(\varepsilon)}), \quad i,k,j \neq 0. \qquad (4.6.18)$$

Relation (4.6.18) permits to complete the proof of Lemma 4.6.1. Using this formula we get for $i,k,j \neq 0$ that

$$\pi_j^{(\varepsilon)}(\rho^{(\varepsilon)}) = \frac{\tilde{\tilde{\pi}}_{ij}^{(\varepsilon)}(\rho^{(\varepsilon)})}{\sum_{r \neq 0} \tilde{\tilde{\pi}}_{ir}^{(\varepsilon)}(\rho^{(\varepsilon)})} = \frac{{}_0\phi_{ij}^{(\varepsilon)}(\rho^{(\varepsilon)})\tilde{\pi}_j^{(\varepsilon)}(\rho^{(\varepsilon)})}{\sum_{r \neq 0} {}_0\phi_{ir}^{(\varepsilon)}(\rho^{(\varepsilon)})\tilde{\pi}_r^{(\varepsilon)}(\rho^{(\varepsilon)})}$$

$$= \frac{{}_0\phi_{ik}^{(\varepsilon)}(\rho^{(\varepsilon)}) {}_0\phi_{kj}^{(\varepsilon)}(\rho^{(\varepsilon)})\tilde{\pi}_j^{(\varepsilon)}(\rho^{(\varepsilon)})}{\sum_{r \neq 0} {}_0\phi_{ik}^{(\varepsilon)}(\rho^{(\varepsilon)}) {}_0\phi_{kr}^{(\varepsilon)}(\rho^{(\varepsilon)})\tilde{\pi}_r^{(\varepsilon)}(\rho^{(\varepsilon)})}$$

$$= \frac{{}_0\phi_{kj}^{(\varepsilon)}(\rho^{(\varepsilon)})\tilde{\pi}_j^{(\varepsilon)}(\rho^{(\varepsilon)})}{\sum_{r \neq 0} {}_0\phi_{kr}^{(\varepsilon)}(\rho^{(\varepsilon)})\tilde{\pi}_r^{(\varepsilon)}(\rho^{(\varepsilon)})} = \frac{\tilde{\tilde{\pi}}_{kj}^{(\varepsilon)}(\rho^{(\varepsilon)})}{\sum_{r \neq 0} \tilde{\tilde{\pi}}_{kr}^{(\varepsilon)}(\rho^{(\varepsilon)})}. \qquad (4.6.19)$$

The proof is complete.   □

It is useful to note that formulas (4.6.18) and (4.6.19) hold without assuming that condition $\mathbf{N}_3$ holds.

To see this, let us consider a semi-Markov process $\eta^{(0)}(t)$ with the transition probabilities $Q_{ij}^{(0)}(t)$ such that (a) $X_1^{(0)} = X_0 = \{1,\ldots,N\}$; (b) $\psi_{ij}^{(0)}(\delta) < \infty$, $i \neq 0$, $j \in X$; (c) there exist $i \neq 0$ and $\beta_i \in (0, \delta]$ for δ defined in (a) such that $\varphi_{ii}^{(0)}(\beta_i) \in (1, \infty)$. Condition $\mathbf{N}_3$ is not assumed to hold for the process $\eta^{(0)}(t)$.

Now, let us choose and fix some state $i \in X_0$ and perturb the transition probabilities $Q_{ij}^{(0)}(t)$, $j \in X$, by replacing them with the convolutions $Q_{ij}^{(\varepsilon)}(t) = Q^{(\varepsilon)}(t) * Q_{ij}^{(0)}(t)$, $j \in X$, where $Q^{(\varepsilon)}(t) = (t/\varepsilon) \wedge 1$, $t \geq 0$, is the uniform distribution function on the interval $[0, \varepsilon]$. It is obvious that, for any $\varepsilon > 0$, conditions **(a)**–**(c)** also hold for the semi-Markov processes $\eta^{(\varepsilon)}(t)$, $t \geq 0$, with the transition probabilities $Q_{ij}^{(\varepsilon)}(t)$. Obviously, condition $\mathbf{N_3}$ holds for the process $\eta^{(\varepsilon)}(t)$. Thus, by Lemma 4.6.1, the process $\eta^{(\varepsilon)}(t)$ satisfies formulas (4.6.18) and (4.6.19) for every $\varepsilon > 0$. Then, these formulas can be verified for the process $\eta^{(0)}(t)$ by letting ε tend to 0 and using Lemma 4.5.6.

Lemma 4.6.1 permits to vary the choice of the state $i \neq 0$ in formula (4.6.10) as to simplify this formula. For example, one can choose the state $i = j$ for every $j \neq 0$ and get the following representation for quasi-stationary probabilities:

$$\pi_j^{(\varepsilon)}(\rho^{(\varepsilon)}) = \frac{\tilde{\pi}_j^{(\varepsilon)}(\rho^{(\varepsilon)})}{\tilde{\pi}_j^{(\varepsilon)}(\rho^{(\varepsilon)}) + \sum_{r \neq 0, j} \tilde{\pi}_{jr}^{(\varepsilon)}(\rho^{(\varepsilon)})}, \qquad j \neq 0. \qquad (4.6.20)$$

Here we used the identities $_0\phi_{jj}^{(\varepsilon)}(\rho^{(\varepsilon)}) = 1$, $j \neq 0$, and $\tilde{\tilde{\pi}}_{jj}^{(\varepsilon)}(\rho^{(\varepsilon)}) = \tilde{\pi}_j^{(\varepsilon)}(\rho^{(\varepsilon)})$, $j \neq 0$.

4.6.5   Convergence of quasi-stationary distributions for perturbed semi-Markov processes without non-recurrent-without-absorption states

The following lemma gives asymptotic relations that permit to formulate conditions for convergence of quasi-stationary distributions of perturbed semi-Markov processes.

Lemma 4.6.2. *Let conditions* $\mathbf{A_2}$, $\mathbf{T_1}$, $\mathbf{E_2'}$, $\mathbf{C_{12}}$, *and* $\mathbf{C_{13}}$ *hold. Then* $(\boldsymbol{\alpha})$ $\tilde{\pi}_j^{(\varepsilon)}(\rho^{(\varepsilon)}) \to \tilde{\pi}_j^{(0)}(\rho^{(0)})$ *as* $\varepsilon \to 0$, $j \neq 0$; $(\boldsymbol{\beta})$ $\tilde{\pi}_{ij}^{(\varepsilon)}(\rho^{(\varepsilon)}) \to \tilde{\pi}_{ij}^{(0)}(\rho^{(0)})$ *as* $\varepsilon \to 0$, $i, j \neq 0$.

Proof. It follows from Lemma 4.5.6 that, **(a)** under conditions of Lemma 4.6.2, the characteristic roots converge, $\rho^{(\varepsilon)} \to \rho^{(0)}$ as $\varepsilon \to 0$, and that **(b)** $0 \leq \rho^{(0)} < \beta \leq \delta$. Relations **(a)** and **(b)** imply that **(c)** there exist $0 < \rho^{(0)} < \beta' < \beta$ and $\varepsilon_{11}' = \varepsilon_{11}'(\beta')$ such that $0 \leq \rho^{(\varepsilon)} \leq \beta'$ for $\varepsilon \leq \varepsilon_{11}'$.

Denote

$$\bar{\psi}_j^{(\varepsilon)}(\rho) = \int_0^\infty e^{\rho s}(1 - F_j^{(\varepsilon)}(s))\, ds, \qquad \rho \in \mathbb{R}_1, \quad j \neq 0.$$

Conditions $\mathbf{T_1}$ and $\mathbf{C_{12}}$ imply that **(d)** $\bar{\psi}_j^{(\varepsilon)}(\rho) \to \bar{\psi}_j^{(0)}(\rho) < \infty$ as $\varepsilon \to 0$ for $0 \leq \rho < \delta$, $j \neq 0$. Relation **(d)** implies that **(e)** there exists $\varepsilon_{12}' > 0$ such that $\bar{\psi}_j^{(\varepsilon)}(\beta') < \infty$ for $j \neq 0$ and $\varepsilon \leq \min(\varepsilon_{11}', \varepsilon_{12}')$. Consequently, relation **(e)** implies that **(f)** $\bar{\psi}_j^{(\varepsilon)}(\rho)$ is a finite, non-negative, non-decreasing, and continuous function on the closed interval $[0, \beta']$ for $j \neq 0$ and $\varepsilon \leq \min(\varepsilon'_{11}, \varepsilon'_{12})$. It follows from **(d)** and **(f)** that **(g)** $\sup_{0 \leq \rho \leq \beta'} |\bar{\psi}_j^{(\varepsilon)}(\rho) - \bar{\psi}_j^{(0)}(\rho)| \to 0$ as $\varepsilon \to 0$ for $j \neq 0$.

Relations **(a)**, **(c)**, **(f)**, and **(g)** show that, for $j \neq 0$,

$$\varlimsup_{\varepsilon \to 0} |\bar{\psi}_j^{(\varepsilon)}(\rho^{(\varepsilon)}) - \bar{\psi}_j^{(0)}(\rho^{(0)})|$$

$$\leq \varlimsup_{\varepsilon \to 0} \left(\sup_{0 \leq \rho \leq \beta'} |\bar{\psi}_j^{(\varepsilon)}(\rho) - \bar{\psi}_j^{(0)}(\rho)| + |\bar{\psi}_j^{(0)}(\rho^{(\varepsilon)}) - \bar{\psi}_j^{(0)}(\rho^{(0)})| \right) = 0. \quad (4.6.21)$$

Conditions $\mathbf{T_1}$, $\mathbf{C_{12}}$, and $\mathbf{C_{13}}$ also imply, by Lemma 4.5.6, that **(h)** ${}_0\phi_{ij}^{(\varepsilon)}(\rho) \to {}_0\phi_{ij}^{(0)}(\rho) < \infty$ as $\varepsilon \to 0$ for $0 \leq \rho < \beta$, $i, j \neq 0$. Relation **(h)** implies that **(i)** there exists $\varepsilon'_{13} > 0$ such that ${}_0\phi_{ij}^{(\varepsilon)}(\beta') < \infty$ for $i, j \neq 0$ and $\varepsilon \leq \min(\varepsilon'_{11}, \varepsilon'_{12}, \varepsilon'_{13})$. Hence, relation **(i)** implies that **(j)** the functions ${}_0\phi_{ij}^{(\varepsilon)}(\rho)$ are finite, non-negative, non-decreasing, and continuous on the closed interval $[0, \beta']$ for $i, j \neq 0$ and $\varepsilon \leq \min(\varepsilon'_{11}, \varepsilon'_{12}, \varepsilon'_{13})$. It follows from **(h)** and **(j)** that **(k)** $\sup_{0 \leq \rho \leq \beta'} |{}_0\phi_{ij}^{(\varepsilon)}(\rho) - {}_0\phi_{ij}^{(0)}(\rho)| \to 0$ as $\varepsilon \to 0$ for $i, j \neq 0$.

Relations **(a)**, **(c)**, **(j)**, and **(k)** imply that, for $i, j \neq 0$,

$$\varlimsup_{\varepsilon \to 0} |{}_0\phi_{ij}^{(\varepsilon)}(\rho^{(\varepsilon)}) - {}_0\phi_{ij}^{(\varepsilon)}(\rho^{(0)})|$$

$$\leq \varlimsup_{\varepsilon \to 0} \left(\sup_{0 \leq \rho \leq \beta'} |{}_0\phi_{ij}^{(\varepsilon)}(\rho) - {}_0\phi_{ij}^{(\varepsilon)}(\rho)| \right.$$

$$\left. + |{}_0\phi_{ij}^{(0)}(\rho^{(\varepsilon)}) - {}_0\phi_{ij}^{(0)}(\rho^{(0)})| \right) = 0. \quad (4.6.22)$$

Formula (4.6.3), which defines the coefficients $\tilde{\pi}_j^{(\varepsilon)}(\rho^{(\varepsilon)})$, and relations (4.6.21) and (4.6.22) imply that, for $j \neq 0$,

$$\tilde{\pi}_j^{(\varepsilon)}(\rho^{(\varepsilon)}) = \frac{\bar{\psi}_j^{(\varepsilon)}(\rho^{(\varepsilon)})}{{}_0\phi_{jj}^{(\varepsilon)}(\rho^{(\varepsilon)})} \to \tilde{\pi}_j^{(0)}(\rho^{(0)}) = \frac{\bar{\psi}_j^{(0)}(\rho^{(0)})}{{}_0\phi_{jj}^{(0)}(\rho^{(0)})} \quad \text{as } \varepsilon \to 0, \quad (4.6.23)$$

and, for $i, j \neq 0$,

$$\tilde{\tilde{\pi}}_{ij}^{(\varepsilon)}(\rho^{(\varepsilon)}) = {}_0\phi_{ij}^{(\varepsilon)}(\rho^{(\varepsilon)}) \frac{\bar{\psi}_j^{(\varepsilon)}(\rho^{(\varepsilon)})}{{}_0\phi_{jj}^{(\varepsilon)}(\rho^{(\varepsilon)})}$$

$$\to \tilde{\tilde{\pi}}_j^{(0)}(\rho^{(0)}) = {}_0\phi_{ij}^{(0)}(\rho^{(0)}) \frac{\bar{\psi}_j^{(0)}(\rho^{(0)})}{{}_0\phi_{jj}^{(0)}(\rho^{(0)})} \quad \text{as } \varepsilon \to 0. \quad (4.6.24)$$

The proof is complete. $\qquad\qquad\qquad\qquad\qquad\qquad\qquad\qquad\qquad\qquad\qquad\square$

The following theorem is a direct corollary of formula (4.6.8) and Lemma 4.6.2. It gives an asymptotic relation that supplements the asymptotic relations (4.6.5) and (4.6.7).

Theorem 4.6.6. *Let conditions* $\mathbf{A}_2$, $\mathbf{T}_1$, $\mathbf{E}'_2$, $\mathbf{C}_{12}$, *and* $\mathbf{C}_{13}$ *hold. Then*

$$\pi_j^{(\varepsilon)}(\rho^{(\varepsilon)}) \to \pi_j^{(0)}(\rho^{(0)}) \ \ as \ \varepsilon \to 0, \ j \neq 0. \tag{4.6.25}$$

4.6.6   Quasi-stationary exponential asymptotics for perturbed semi-Markov processes with non-recurrent-without-absorption states

Let us now consider a model for which condition $\mathbf{E}''_2$ holds.

Define

$$\tilde{\pi}_r^{(\varepsilon)}(\rho^{(\varepsilon)}) = \begin{cases} \tilde{\pi}_r^{(\varepsilon)}(\rho^{(\varepsilon)}) & \text{for } r \in X_1^{(\varepsilon)}, \\[2mm] 0 & \text{for } r \in X_2^{(\varepsilon)}, \end{cases} \tag{4.6.26}$$

and

$$\tilde{\tilde{\pi}}_{ir}^{(\varepsilon)}(\rho^{(\varepsilon)}) = \begin{cases} {}_0\phi_{ir}^{(\varepsilon)}(\rho^{(\varepsilon)})\tilde{\pi}_r^{(\varepsilon)}(\rho^{(\varepsilon)}) & \text{for } i \neq 0,\, r \in X_1^{(\varepsilon)}, \\[2mm] 0 & \text{for } i \neq 0,\, r \in X_2^{(\varepsilon)}, \end{cases} \tag{4.6.27}$$

where the coefficients $\tilde{\pi}_r^{(\varepsilon)}(\rho^{(\varepsilon)})$, $r \in X_1^{(\varepsilon)}$ are defined by formula (4.6.3).

Theorem 4.6.7. *Let conditions* $\mathbf{A}_2$, $\mathbf{T}_1$, $\mathbf{E}''_2$, $\mathbf{C}_{12}$, $\mathbf{C}_{13}$, *and* $\mathbf{N}_1$ *hold. Then, for any* $0 \leq t^{(\varepsilon)} \to \infty$ *as* $\varepsilon \to 0$,

$$\frac{\mathsf{P}_i\{\eta^{(\varepsilon)}(t^{(\varepsilon)}) = j, \mu_0^{(\varepsilon)} > t^{(\varepsilon)}\}}{\exp\{-\rho^{(\varepsilon)}t^{(\varepsilon)}\}} \to \tilde{\tilde{\pi}}_{ij}^{(0)}(\rho^{(0)}) \ \ as \ \varepsilon \to 0, \ \ i, j \neq 0. \tag{4.6.28}$$

Proof. Let us first consider that case where the initial state is $i = r$ and $r = j \in X_1^{(0)}$.

The first part of the proof, where conditions $\mathbf{D}_{22}$, $\mathbf{C}_6$, and $\mathbf{F}_7$ are verified, repeats the same part of the proof of Theorem 4.6.5. The only difference is that the reference to condition $\mathbf{E}'_2$ should be replaced with an equivalent reference to condition $\mathbf{E}''_2$.

Thus, all conditions of Theorem 3.3.2 hold. By applying the asymptotic relation (2.2.18) given in Theorem 3.3.2, we get the asymptotic relation (4.6.28) for the case where $i = r$ and $r = j \in X_1^{(0)}$.

Let us now consider the case where the initial state $i \neq r, 0$ and $r = j \in X_1^{(0)}$.

As above, the renewal type relation (3.2.19) takes in this case the form of the renewal equation (4.2.2) with the distribution function ${}_0 G_{ij}^{(\varepsilon)}(t)$, being regarded as the distribution function $\tilde{F}^{(\varepsilon)}(t)$ that generates this renewal type relation, and the moment generating function ${}_0\phi_{ij}^{(\varepsilon)}(\rho)$ replacing the moment generating function, $\tilde{\phi}^{(\varepsilon)}(\rho)$.

Conditions $\mathbf{A_2}$, $\mathbf{T_1}$, $\mathbf{E_2''}$ imply, by Lemmas 4.5.6 and 4.5.7, that **(a)** $_0\phi_{ij}^{(\varepsilon)}(\rho^{(0)}) \to$ $_0\phi_{ij}^{(0)}(\rho^{(0)}) < \infty$ as $\varepsilon \to 0$. Thus, condition $\mathbf{C_{11}}$ holds with the limit constant $\tilde{\phi}^{(0)} = {}_0\phi_{ij}^{(0)}(\rho^{(0)})$.

The distribution function $G_{ij}^{(\varepsilon)}(t)$ plays the role of the distribution function $\hat{\tilde{F}}^{(\varepsilon)}(t)$. Conditions $\mathbf{A_2}$, $\mathbf{T_1}$, $\mathbf{E_2'}$ and $\mathbf{C_{12}}$ and $\mathbf{C_{13}}$ imply, by Lemmas 4.5.7, that **(b)** condition $\mathbf{C_{10}}$ holds for these distribution functions. This condition implies that conditions $\mathbf{C_9}$ and $\mathbf{D_{26}}$ are satisfied.

Finally, by applying the asymptotic relation (3.3.52) given in Theorem 3.3.7, we get the asymptotic relation (4.6.28) for the case where $i \neq r, 0$ and $r = j \in X_1^{(0)}$.

It still remains to prove validity of the asymptotic relation (4.6.28) for $r \in X_2^{(0)}$. In this case, we can choose some state $j \in X_1^{(0)}$ and use the renewal relations (4.2.3) and (4.2.6), instead of the renewal relations (4.2.1) and (4.2.2).

To this end, let us first consider the case where the initial state is $i = j \in X_1^{(0)}$, $r \in X_2^{(0)}$.

The renewal equations (4.2.3) and (4.2.1) have the same distribution function $_0G_{jj}^{(\varepsilon)}(t)$ that generates these equations, and the distribution function $G_{jj}^{(\varepsilon)}(t)$ takes place of the distribution function $\hat{F}^{(\varepsilon)}(t)$. This shows that conditions $\mathbf{D_{22}}$ and $\mathbf{C_6}$ hold.

However, the renewal equation (4.2.3) has a forcing function that differs from the forcing function for renewal equation (4.2.1). Condition $\mathbf{F_7}$ was verified, in this case, in the proof of Theorem 4.4.2. In particular, it was shown that the limit forcing function is $q_{jr}^{(0)}(t) = 0, t \geq 0$. Thus, for the set $A = \{r\}, r \in X_2^{(0)}$, we have

$$\tilde{\pi}^{(0)}(A) = \frac{\int_0^\infty e^{\rho^{(0)}s} q_{jr}^{(0)}(s)\, ds}{\int_0^\infty s e^{\rho^{(0)}s}\, {}_0G_{jj}^{(0)}(ds)} = \tilde{\pi}_r^{(0)}(\rho^{(0)}) = 0. \tag{4.6.29}$$

Finally, by applying the asymptotic relation (3.3.52) given in Theorem 3.3.7, we get the asymptotic relation (4.6.28) for the case where $i = j \in X_1^{(0)}, r \in X_2^{(0)}$.

The last case to be considered is where the initial state $i \neq j, 0, j \in X_1^{(0)}$, and $r \in X_2^{(0)}$.

The renewal relation (3.2.19) takes now the form of the renewal relation (4.2.6) that should be used instead of the renewal relation (4.2.2). The renewal relations (4.2.6) and (4.2.2) have the same distribution function $_0G_{ij}^{(\varepsilon)}(t)$ that generates these equations and the same distribution function $G_{ij}^{(\varepsilon)}(t)$ which replaces the distribution function $\hat{\tilde{F}}^{(\varepsilon)}(t)$.

The moment generating function $_0\phi_{ij}^{(\varepsilon)}(\rho)$ coincides in this case with the moment generating function $\tilde{\phi}^{(\varepsilon)}(\rho)$. Conditions $\mathbf{A_2}$, $\mathbf{T_1}$, $\mathbf{E_2''}$ imply, by Lemmas 4.5.6 and 4.5.7, that **(c)** $_0\phi_{ij}^{(\varepsilon)}(\rho^{(0)}) \to {}_0\phi_{ij}^{(0)}(\rho^{(0)}) < \infty$ as $\varepsilon \to 0$. Thus, condition $\mathbf{C_{11}}$ holds with the limit constant $\tilde{\phi}^{(0)} = {}_0\phi_{ij}^{(0)}(\rho^{(0)})$.

The distribution function $G_{ij}^{(\varepsilon)}(t)$ is regarded as the distribution function $\hat{\tilde{F}}^{(\varepsilon)}(t)$. Conditions $\mathbf{A_2}$, $\mathbf{T_1}$, $\mathbf{E_2''}$, and $\mathbf{C_{12}}$ and $\mathbf{C_{13}}$ imply, by Lemmas 4.5.7, that $(\mathbf{d})$ condition $\mathbf{C_{10}}$ holds for the distribution function $G_{ij}^{(\varepsilon)}(t)$ that plays the role of the distribution function $\hat{\tilde{F}}^{(\varepsilon)}(t)$. This condition implies that conditions $\mathbf{C_9}$ and $\mathbf{D_{26}}$ are verified.

Finally, by applying the asymptotic relation (3.3.52) given in Theorem 3.3.7, we get the asymptotic relation (4.6.28) for the case where $i \neq j, 0, j \in X_1^{(0)}$, and $r \in X_2^{(0)}$. $\square$

4.6.7   Quasi-stationary distributions for perturbed semi-Markov processes with non-recurrent-without-absorption states

Let us consider the general case where the following condition (introduced in Subsection 4.4.5) holds:

$\mathbf{E_3}$: $X_1^{(\varepsilon)} \neq \varnothing$ for the semi-Markov process $\eta^{(\varepsilon)}(t)$ for ε small enough, say $\varepsilon \leq \varepsilon_6$.

We also assume to hold the following condition that should replace condition $\mathbf{C_{15}}$:

$\mathbf{C_{16}}$: There exists $i \in X_1^{(\varepsilon)}$ and $\beta_i \in (0, \delta]$ for δ defined in condition $\mathbf{C_{14}}$ such that

$\quad$ (a): $_0\phi_{ii}^{(\varepsilon)}(\beta_i) \in (1, \infty)$ for all ε small enough, say $\varepsilon \leq \varepsilon_{14}'$;

$\quad$ (b): $_0\phi_{ki}^{(\varepsilon)}(\beta_i) < \infty, k \in X_2^{(0)}$ for all $\varepsilon \leq \varepsilon_{14}'$.

Let

$$\varepsilon_{15}' = \min(\varepsilon_6, \varepsilon_7', \varepsilon_9', \varepsilon_{14}'). \tag{4.6.30}$$

The following theorem is an analogue of Theorem 4.6.5.

Theorem 4.6.8. *Let conditions* $\mathbf{E_3}$, $\mathbf{C_{14}}$, $\mathbf{C_{15}}$, *and* $\mathbf{N_2}$ *hold. Then, for every* $\varepsilon \leq \varepsilon_{15}'$,

$$e^{\rho^{(\varepsilon)}t} \mathsf{P}_i\{\eta^{(\varepsilon)}(t^{(\varepsilon)}) = j, \mu_0^{(\varepsilon)} > t\} \to \tilde{\tilde{\pi}}_{ij}^{(\varepsilon)}(\rho^{(\varepsilon)}) \text{ as } t \to \infty, \quad i, j \neq 0. \tag{4.6.31}$$

Proof. Choose some $\varepsilon' \leq \varepsilon_{15}'$ and consider a model in which the transition probabilities $Q_{ij}^{(\varepsilon)}(t)$ of semi-Markov processes $\eta^{(\varepsilon)}(t), t \geq 0$, do not depend on the parameter ε and coincide with the transition probabilities $Q_{ij}^{(\varepsilon')}(t)$. In this case, conditions of Theorem 4.6.7 just reduce to conditions $\mathbf{E_3}$, $\mathbf{C_{14}}$, $\mathbf{C_{15}}$, and $\mathbf{N_2}$. The asymptotic relation (4.6.28) take the form of the asymptotic relation (4.6.31). Relation (4.6.31) can be proved in the same way for any $\varepsilon \leq \varepsilon_{15}'$. $\square$

Denote

$$\pi_j^{(\varepsilon)}(\rho^{(\varepsilon)}) = \frac{\tilde{\tilde{\pi}}_{ij}^{(\varepsilon)}(\rho^{(\varepsilon)})}{\sum_{r \neq 0} \tilde{\tilde{\pi}}_{ir}^{(\varepsilon)}(\rho^{(\varepsilon)})}, \quad j \neq 0, \tag{4.6.32}$$

where $\tilde{\tilde{\pi}}_{ij}^{(\varepsilon)}(\rho^{(\varepsilon)}), i, j \neq 0$, are defined by (4.6.27).

Conditions $\mathbf{A_2}$, $\mathbf{T_1}$, and $\mathbf{E_3}$, $\mathbf{C_{12}}$, and $\mathbf{C_{13}}$, imply that $\tilde{\tilde{\pi}}_r^{(\varepsilon)}(\rho^{(\varepsilon)}) \in (0, \infty)$ for all ε small enough, namely, $\varepsilon \leq \min(\varepsilon_6, \varepsilon_7', \varepsilon_{14}')$. This follows from Lemma 4.5.7.

In this case, **(a)** $\pi_j^{(\varepsilon)}(\rho^{(\varepsilon)}) > 0$, $j \in X_1^{(\varepsilon)}$; **(b)** $\pi_j^{(\varepsilon)}(\rho^{(\varepsilon)}) = 0$, $j \in X_2^{(\varepsilon)}$; **(c)** $\sum_{j \in X_1^{(\varepsilon)}} \pi_j^{(\varepsilon)}(\rho^{(\varepsilon)}) = 1$. Thus, formula (4.6.32) defines a discrete probability distribution.

This distribution is called a *quasi-stationary distribution* for the semi-Markov process $\eta^{(\varepsilon)}(t)$, $t \geq 0$, with absorption times $\mu_0^{(\varepsilon)}$.

The following quasi-stationary theorem is a direct corollary of Theorem 4.6.8.

Theorem 4.6.9. *Let conditions* $\mathbf{E_3}$, $\mathbf{C_{14}}$, $\mathbf{C_{15}}$, *and* $\mathbf{N_2}$ *hold. Then, for* $\varepsilon \leq \varepsilon_{15}'$,

$$\mathsf{P}_i\{\eta^{(\varepsilon)}(t) = j/\mu_0^{(\varepsilon)} > t\} = \frac{e^{\rho^{(\varepsilon)}t}\mathsf{P}_i\{\eta^{(\varepsilon)}(t) = j, \mu_0^{(\varepsilon)} > t\}}{e^{\rho^{(\varepsilon)}t}\mathsf{P}_i\{\mu_0^{(\varepsilon)} > t\}}$$

$$\rightarrow \pi_j^{(\varepsilon)}(\rho^{(\varepsilon)}) \quad as \ t \rightarrow \infty, \quad j \neq 0. \qquad (4.6.33)$$

Relation (4.6.33) clarifies why it is natural to call the distribution $\pi_j^{(\varepsilon)}(\rho^{(\varepsilon)})$ a *quasi-stationary* distribution for the semi-Markov process $\eta^{(\varepsilon)}(t)$ with the absorption moments $\mu_0^{(\varepsilon)}$.

Formula (4.6.32) defines the quasi-stationary distribution in the form that seems to be dependent on the choice of the "initial" state $i \neq 0$. The following lemma permits to prove that the quasi-stationary probabilities $\pi_j^{(\varepsilon)}(\rho^{(\varepsilon)})$ defined in (4.6.32) are, actually, independent of a choice of the state $i \neq 0$.

Lemma 4.6.3. *Let conditions* $\mathbf{E_3}$, $\mathbf{C_{14}}$, $\mathbf{C_{15}}$, *and* $\mathbf{N_2}$ *hold. Then, for every* $\varepsilon \leq \varepsilon_{15}'$,

$$\pi_j^{(\varepsilon)}(\rho^{(\varepsilon)}) = \frac{\tilde{\tilde{\pi}}_{ij}^{(\varepsilon)}(\rho^{(\varepsilon)})}{\sum_{r \neq 0} \tilde{\tilde{\pi}}_{ir}^{(\varepsilon)}(\rho^{(\varepsilon)})}, \quad i, j \neq 0. \qquad (4.6.34)$$

Proof. The case $j \in X_2^{(0)}$ is trivial. Indeed, formula (4.6.26) implies in this case that $\tilde{\tilde{\pi}}_{ij}^{(\varepsilon)}(\rho^{(\varepsilon)}) = 0$ for such j. In the case $j \in X_1^{(0)}$, the proof repeats the proof of Lemma 4.6.1. The only difference is that now one should take the states $i \neq j, 0, j, k \in X_1^{(0)}$, and then repeat the proof of Lemma 4.6.1. In particular, one can refer to conditions $\mathbf{A_2}$, $\mathbf{T_1}$, $\mathbf{E_2''}$, instead of conditions $\mathbf{A_2}$, $\mathbf{T_1}$, $\mathbf{E_2'}$, when applying Lemmas 4.5.6 and 4.5.7 to get, with the use conditions $\mathbf{A_2}$, $\mathbf{T_1}$, $\mathbf{E_2''}$, and $\mathbf{N_2}$, the following formula for every $i \neq j, 0, j, k \in X_1^{(0)}$:

$$_0\phi_{ikj}^{(\varepsilon)}(\rho) = {}_0\phi_{ik}^{(\varepsilon)}(\rho) \cdot {}_0\phi_{kj}^{(\varepsilon)}(\rho), \quad \rho \in \mathbb{R}_1. \qquad (4.6.35)$$

Then one considers a model in which the transition probabilities $Q_{ij}^{(\varepsilon)}(t)$ of the semi-Markov processes $\eta^{(\varepsilon)}(t)$, $t \geq 0$, do not depend on the parameter ε, more precisely, their values coincide for some particular value of the parameter $\varepsilon \leq \varepsilon_{15}'$. In this case, condition $\mathbf{E}_3$ coincides with either condition $\mathbf{E}_2'$ or $\mathbf{E}_2''$. Formula (4.6.17) in the first case and formula (4.6.35) in the second case yield, for every $i \neq j, 0$, $j, k \in X_1^{(0)}$,

$$_0\phi_{ik}^{(\varepsilon)}(\rho^{(\varepsilon)})\,_0\phi_{kj}^{(\varepsilon)}(\rho^{(\varepsilon)}) = \,_0\phi_{ij}^{(\varepsilon)}(\rho^{(\varepsilon)}). \tag{4.6.36}$$

Using relation (4.6.36) we prove formula (4.6.34) repeating transformations given in relation (4.6.19). $\qquad\square$

4.6.8   Convergence of quasi-stationary distributions of perturbed semi-Markov processes with non-recurrent-without-absorption states

The following lemma is an analogue of Lemma 4.6.2 for the model of perturbed semi-Markov processes with non-recurrent-without-absorption states.

Lemma 4.6.4. *Let conditions* $\mathbf{A}_2$, $\mathbf{T}_1$, $\mathbf{E}_2''$, $\mathbf{C}_{12}$, *and* $\mathbf{C}_{13}$ *hold. Then* ($\boldsymbol{\alpha}$) $\tilde{\pi}_j^{(\varepsilon)}(\rho^{(\varepsilon)}) \to \tilde{\pi}_j^{(0)}(\rho^{(0)})$ *as* $\varepsilon \to 0$, $j \neq 0$; ($\boldsymbol{\beta}$) $\tilde{\tilde{\pi}}_{ij}^{(\varepsilon)}(\rho^{(\varepsilon)}) \to \tilde{\tilde{\pi}}_{ij}^{(0)}(\rho^{(0)})$ *as* $\varepsilon \to 0$, $i, j \neq 0$.

Proof. Note that, under conditions $\mathbf{A}_2$, $\mathbf{T}_1$, and $\mathbf{E}_2''$, we have $X_1^{(0)} \subseteq X_1^{(\varepsilon)}$ for all ε small enough. Taking into account this fact and repeating the first part of the proof of Lemma 4.6.2 we see that, for $j \in X_1^{(0)}$,

$$\tilde{\pi}_j^{(\varepsilon)}(\rho^{(\varepsilon)}) \to \tilde{\pi}_j^{(0)}(\rho^{(0)}) \quad \text{as} \quad \varepsilon \to 0. \tag{4.6.37}$$

Then, using relation (4.6.37) we get

$$\sum_{j \in X_2^{(0)}} \tilde{\pi}_j^{(\varepsilon)}(\rho^{(\varepsilon)}) = 1 - \sum_{j \in X_1^{(0)}} \tilde{\pi}_j^{(\varepsilon)}(\rho^{(\varepsilon)})$$

$$\to 1 - \sum_{j \in X_1^{(0)}} \tilde{\pi}_j^{(0)}(\rho^{(0)}) = 0 \quad \text{as} \quad \varepsilon \to 0. \tag{4.6.38}$$

The proof of the following relation repeats the corresponding part in the proof of Lemma 4.6.2:

$$_0\phi_{ij}^{(\varepsilon)}(\rho^{(\varepsilon)}) \to \,_0\phi_{ij}^{(0)}(\rho^{(0)}) \quad \text{as} \quad \varepsilon \to 0, \quad i \neq 0, \quad j \in X_1^{(0)}. \tag{4.6.39}$$

Relations (4.6.37), (4.6.38), and (4.6.39) prove the lemma. $\qquad\square$

The following theorem is a direct corollary of Lemma 4.6.4.

Theorem 4.6.10. *Let conditions* $\mathbf{A}_2$, $\mathbf{T}_1$, $\mathbf{E}_2''$, $\mathbf{C}_{12}$, *and* $\mathbf{C}_{13}$ *hold. Then*

$$\pi_j^{(\varepsilon)}(\rho^{(\varepsilon)}) \to \pi_j^{(0)}(\rho^{(0)}) \quad \text{as} \quad \varepsilon \to 0, \quad j \neq 0. \tag{4.6.40}$$

4.6.9   Large deviation asymptotics for distributions of absorption times

Denote

$$\tilde{\tilde{\pi}}_i^{(\varepsilon)}(\rho^{(\varepsilon)}) = \sum_{r \in X_1^{(\varepsilon)}} \tilde{\tilde{\pi}}_{ir}^{(\varepsilon)}(\rho^{(\varepsilon)}), \quad i \neq 0.$$

The following theorem is an obvious corollary of Theorems 4.6.3 and 4.6.7.

Theorem 4.6.11. *Let conditions* $\mathbf{A_2}$, $\mathbf{T_1}$, $\mathbf{E_2}$, $\mathbf{C_{12}}$, $\mathbf{C_{13}}$, *and* $\mathbf{N_1}$ *hold. Then, for any* $0 \leq t^{(\varepsilon)} \to \infty$ *as* $\varepsilon \to 0$,

$$\frac{\mathsf{P}_i\{\mu_0^{(\varepsilon)} > t^{(\varepsilon)}\}}{\exp\{-\rho^{(\varepsilon)}t^{(\varepsilon)}\}} \to \tilde{\tilde{\pi}}_i^{(0)}(\rho^{(0)}) \quad \text{as } \varepsilon \to 0, \quad i \neq 0. \tag{4.6.41}$$

Proof. Relation (4.6.41) is a direct corollary of asymptotic relations (4.6.5) and (4.6.28) given, respectively, in Theorems 4.6.3 and 4.6.7. It can be obtained by summing over $r \neq 0$ the pre-limit and limit expressions in the left and the right-hand sides of (4.6.5) or (4.6.28) depending on whether condition $\mathbf{E_2'}$ or $\mathbf{E_2''}$ holds. $\quad\square$

Let us comment on the asymptotic relation (4.6.41) given in Theorem 4.6.11. We consider the main case where condition $\mathbf{E_2}$ is realised in the form of condition $\mathbf{E_2'}$.

Asymptotic relation (4.6.41) becomes trivial if condition $\mathbf{A_4}$ holds, i.e., if $\sum_{i \neq 0} p_{i0}^{(\varepsilon)} = 0$ for all ε small enough. Indeed, in this case, $\mathsf{P}_i\{\mu^{(\varepsilon)} = \infty\} = 1$, $i \neq 0$, for all such ε and both expressions in the left and the right-hand sides of (4.6.41) are equal to 1.

Let us assume, additionally to conditions of Theorem 4.6.11, the following condition that guarantees that $\mathsf{P}_i\{\mu^{(\varepsilon)} < \infty\} = 1, i \neq 0$ for every $\varepsilon > 0$:

$\mathbf{A_5}$: $\sum_{i \in X_1^{(\varepsilon)}} p_{i0}^{(\varepsilon)} > 0$ for $\varepsilon > 0$.

Condition $\mathbf{A_5}$ is equivalent to the assumption that **(a)** $\rho^{(\varepsilon)} > 0$ for $\varepsilon > 0$.

Also, as follows from Lemma 4.5.6, conditions $\mathbf{A_2}$, $\mathbf{T_1}$, $\mathbf{E_2'}$, $\mathbf{C_{12}}$, and $\mathbf{C_{13}}$ imply that **(b)** $\rho^{(\varepsilon)} \to \rho^{(0)}$ as $\varepsilon \to 0$.

Note also that, according to Lemma 4.2.2, conditions $\mathbf{T_1}$ and $\mathbf{E_2'}$ imply that $X_1^{(\varepsilon)} = X_0 = \{1, \ldots, N\}$ for all ε small enough, say $\varepsilon \leq \varepsilon_{16}'$.

There are two alternative cases. One is where **(i)** condition $\mathbf{A_3}$ holds or, equivalently, $\rho^{(0)} = 0$ and another one is where **(ii)** condition $\mathbf{A_3}$ does not hold or, equivalently, $\rho^{(0)} > 0$.

Let us first consider case **(i)**. By Lemma 4.5.8, condition $\mathbf{C_{13}}$ can be omitted.

Relation **(b)** in this case takes the form **(b′)** $\rho^{(\varepsilon)} \to \rho^{(0)} = 0$ as $\varepsilon \to 0$.

Also, in this case, **(c)** $\tilde{\tilde{\pi}}_i^{(0)}(0) = 1, i \neq 0$. Indeed, **(d)** $_0\phi_{ij}^{(0)}(0) = 1, i, j \neq 0$. Also, **(e)** $\tilde{\pi}_j^{(0)}(0) = \pi_j^{(0)}(0), j \neq 0$, where $\pi_j^{(0)}, j \neq 0$, are stationary probabilities given by formula (4.4.7). Thus, using **(d)** and **(e)** we get

$$\tilde{\tilde{\pi}}_i^{(0)}(0) = \sum_{r \in X_1^{(0)}} {}_0\phi_{ir}^{(0)}(0)\,\tilde{\pi}_r^{(0)}(0) = \sum_{r \in X_1^{(0)}} \pi_r^{(0)}(0) = 1. \tag{4.6.42}$$

Let us first choose $t^{(\varepsilon)} \to \infty$ such that $\rho^{(\varepsilon)} t^{(\varepsilon)} \to 0$ as $\varepsilon \to 0$. Then the asymptotic relation (4.6.41) takes the following form for $i \neq 0$:

$$\mathsf{P}_i\{\mu^{(\varepsilon)} > t^{(\varepsilon)}\} \to 1 \quad \text{as } \varepsilon \to 0. \tag{4.6.43}$$

Relation (4.6.43) implies that, in case **(f)**, the stopping moments $\mu^{(\varepsilon)}$ are asymptotically stochastically unbounded random variables as $\varepsilon \to 0$, i.e., for $i \neq 0$,

$$\lim_{\varepsilon \to 0} \mathsf{P}_i\{\mu^{(\varepsilon)} > T\} = 1, \quad T > 0. \tag{4.6.44}$$

Let us take $t^{(\varepsilon)} = t/\rho^{(\varepsilon)} \to \infty$ as $\varepsilon \to 0$, where $t > 0$. Asymptotic relation (4.6.41) takes the following form for $i \neq 0$:

$$\frac{\mathsf{P}_i\{\rho^{(\varepsilon)} \mu^{(\varepsilon)} > t\}}{\exp\{-t\}} \to 1 \quad \text{as } \varepsilon \to 0, \quad t > 0. \tag{4.6.45}$$

This is a weak convergence asymptotic relation. Indeed, relation (4.6.45) means that, for any initial state $i \neq 0$, the random variables converge, $\rho^{(\varepsilon)} \mu^{(\varepsilon)} \xrightarrow{\text{d}} \nu_0$ as $\varepsilon \to 0$, where ν_0 is an exponentially distributed random variable with parameter 1.

Note that (4.6.45) is consistent with the asymptotic relation (4.4.40) given in Theorem 4.4.6. Indeed, as was mentioned in Remark 4.6.1, by Lemma 1.4.2 with conditions $\mathbf{A_2}$, $\mathbf{T_1}$, $\mathbf{E_2}$ $\mathbf{C_{12}}$, $\mathbf{A_3}$ fulfilled, **(g)** the characteristic root satisfies $\rho^{(\varepsilon)} \sim {}_jf_{j0}^{(\varepsilon)}/{}_0M_{jj}^{(\varepsilon)}$ as $\varepsilon \to 0$. According to Lemmas 4.2.4 and 4.3.4, and formula (4.4.11), we have in this case that **(h)** ${}_jf_{j0}^{(\varepsilon)}/{}_0M_{jj}^{(\varepsilon)} \sim f^{(\varepsilon)}/m^{(0)}$ as $\varepsilon \to 0$, where $f^{(\varepsilon)}$ and $m^{(0)}$ are defined in formulas (4.3.5) and (4.4.29). If $t^{(\varepsilon)} = t/\rho^{(\varepsilon)}$, then **(g)** and **(h)** imply that $\rho^{(\varepsilon)} t^{(\varepsilon)} \to \lambda/m^{(0)}$ as $\varepsilon \to 0$. In these cases, asymptotic relation (4.6.41) given in Theorem 4.6.11 reduces to the asymptotic relation **(i)** $\mathsf{P}_i\{f^{(\varepsilon)} \mu_0^{(\varepsilon)} > t\} \to \exp\{-t/m^{(0)}\}$ as $\varepsilon \to 0, t > 0, i \neq 0$. This relation is equivalent to the asymptotic relation (4.4.40) given in Theorem 4.4.6.

It should be noted that the asymptotic relation (4.4.40) was obtained under conditions weaker than the asymptotic relation (4.6.41).

We can, however, take $t^{(\varepsilon)} = \tilde{t}^{(\varepsilon)}/\rho^{(\varepsilon)}$ where $\tilde{t}^{(\varepsilon)} \to \infty$ as $\varepsilon \to 0$. In this case, (4.6.41) takes the form of a large deviation asymptotic relation for $i \neq 0$,

$$\frac{\mathsf{P}_i\{\rho^{(\varepsilon)} \mu^{(\varepsilon)} > \tilde{t}^{(\varepsilon)}\}}{\exp\{-\tilde{t}^{(\varepsilon)}\}} \to 1 \quad \text{as } \varepsilon \to 0. \tag{4.6.46}$$

Asymptotic relation (4.6.46) is much stronger than asymptotic relation **(i)**. This relation implies that the asymptotic relative error, if the tail probabilities

$\mathsf{P}_i\{\rho^{(\varepsilon)}\mu^{(\varepsilon)} > \tilde{t}^{(\varepsilon)}\}$ are approximated with the corresponding tail probabilities for limit exponential distribution, equals 0 for any $\tilde{t}^{(\varepsilon)} \to \infty$. This is achieved by choosing very well-fitted normalising coefficients $\rho^{(\varepsilon)}$. A payment for this is the need to use the Cramér type condition $\mathbf{C}_{12}$.

A drawback of (4.6.46) is that the normalising coefficient $\rho^{(\varepsilon)}$ is given as a root of the nonlinear equation (4.5.1). The question arises if it is possible to replace $\rho^{(\varepsilon)}$ in (4.6.46), for example, by a simpler coefficient $f^{(\varepsilon)}/m^{(0)}$ and to rewrite this relation in the form (**j**) $\mathsf{P}_i\{f^{(\varepsilon)}\mu^{(\varepsilon)} > \tilde{t}^{(\varepsilon)}m^{(0)}\}/\exp\{-\tilde{t}^{(\varepsilon)}m^{(0)}\} \to 1$ as $\varepsilon \to 0, t > 0$, $i \neq 0$? The answer is partly in the affirmative. Relation $\rho^{(\varepsilon)} \sim f^{(\varepsilon)}/m^{(0)}$ means that $\rho^{(\varepsilon)} = f^{(\varepsilon)}/m^{(0)} + o(f^{(\varepsilon)})$. Thus, (4.6.46) implies that (**j**) $\mathsf{P}_i\{f^{(\varepsilon)}\mu^{(\varepsilon)} > \tilde{t}^{(\varepsilon)}m^{(0)}\} \sim \exp\{-\rho^{(\varepsilon)}\tilde{t}^{(\varepsilon)}m^{(0)}/f^{(\varepsilon)}\} \sim \exp\{-(f^{(\varepsilon)}/m^{(0)} + o(f^{(\varepsilon)}))\tilde{t}^{(\varepsilon)}m^{(0)}/f^{(\varepsilon)}\}$. Therefore, the suggested replacement would additionally require to assume that (**k**) $\tilde{t}^{(\varepsilon)}o(f^{(\varepsilon)})/f^{(\varepsilon)} \to 0$, i.e., to restrict the rate of $\tilde{t}^{(\varepsilon)}$ approaching ∞.

The questions related to the use of normalising coefficients that are given in a more explicit form will be investigated in the next chapter where we shall get asymptotic expansions for the roots $\rho^{(\varepsilon)}$ in asymptotic power series.

Let us now consider the case (**ii**).

Relation (**b**) takes in this case the form (**b″**) $\rho^{(\varepsilon)} \to \rho^{(0)} > 0$ as $\varepsilon \to 0$.

Also, (**l**) it is possible that $\tilde{\tilde{\pi}}_i^{(0)}(\rho^{(0)}) \neq 1$, at least for some $i \neq 0$. Indeed, it can happen that (**m**) $_0\phi_{ij}^{(0)}(\rho^{(0)}) \neq 1$ at least for some $i, j \neq 0$, and (**e**) $\tilde{\pi}_j^{(0)}(\rho^{(0)}) \neq \pi_j^{(0)}$ at least for some $j \neq 0$. Moreover, the coefficients $\tilde{\tilde{\pi}}_i^{(0)}(\rho^{(0)})$ can depend on i, i.e., the limit expression in the asymptotic relation (4.6.41) can depend on the choice of the initial state $i \neq 0$ used in the left-hand side of this relation. However, we have the following formula: (**m**) $\tilde{\tilde{\pi}}_i^{(0)}(\rho^{(0)})/\tilde{\tilde{\pi}}_j^{(0)}(\rho^{(0)}) = {}_0\phi_{ij}^{(0)}(\rho^{(0)}), i, j \neq 0$. Indeed, using formula (4.6.18) we get for $i, j \neq 0$, that

$$\frac{\tilde{\tilde{\pi}}_i^{(0)}(\rho^{(0)})}{\tilde{\tilde{\pi}}_j^{(0)}(\rho^{(0)})} = \frac{\sum_{r \neq 0} {}_0\phi_{ij}^{(0)}(\rho^{(0)}){}_0\phi_{jr}^{(0)}(\rho^{(0)})\tilde{\pi}_r^{(0)}(\rho^{(0)})}{\sum_{r \neq 0} {}_0\phi_{jr}^{(\varepsilon)}(\rho^{(\varepsilon)})\tilde{\pi}_r^{(\varepsilon)}(\rho^{(\varepsilon)})} = {}_0\phi_{ij}^{(0)}(\rho^{(0)}). \quad (4.6.47)$$

Let us introduce a distribution function for absorption times by

$$G_i^{(\varepsilon)}(t) = \mathsf{P}_i\{\mu_0^{(\varepsilon)} \leq t\}, \quad t \geq 0, \quad i \neq 0,$$

and the corresponding Laplace transforms

$$\bar{\phi}_i^{(\varepsilon)}(\rho) = \int_0^\infty e^{-\rho s} G_i^{(\varepsilon)}(ds), \quad \rho \geq 0, \quad i \neq 0.$$

By using the regeneration property possessed by the semi-Markov process $\eta^{(\varepsilon)}(t)$, $t \geq 0$, at the time of first hitting into any state, we can write the following renewal type relation for $i \neq 0$:

$$G_i^{(\varepsilon)}(t) = {}_iG_{i0}^{(\varepsilon)}(t) + {}_0G_{ii}^{(\varepsilon)}(t) * G_i^{(\varepsilon)}(t), \quad t \geq 0. \quad (4.6.48)$$

In terms of the Laplace transform, this formula takes the following form for $i \neq 0$:

$$\bar{\phi}_i^{(\varepsilon)}(\rho) = {}_i\bar{\phi}_{i0}^{(\varepsilon)}(\rho) + {}_0\bar{\phi}_{ii}^{(\varepsilon)}(\rho)\bar{\phi}_i^{(\varepsilon)}(\rho), \quad \rho \geq 0. \tag{4.6.49}$$

Note that if condition $\mathbf{E}_2'$ is verified, the assumption $\rho^{(0)} > 0$ is equivalent to the assumption that $\mathbf{(n)}$ ${}_0 f_{ii}^{(0)} < 1$, $i \neq 0$. Under condition $\mathbf{T}_1$, we have $\mathbf{(0)}$ ${}_0\bar{\phi}_{ii}^{(\varepsilon)}(0) \to {}_0 f_{ii}^{(0)}$ as $\varepsilon \to 0$, $i \neq 0$ and, thus, $\mathbf{(p)}$ ${}_0\bar{\phi}_{ii}^{(\varepsilon)}(0) = {}_0 f_{ii}^{(\varepsilon)} < 1$, $i \neq 0$, for ε small enough, say $\varepsilon \leq \varepsilon_{17}'$.

Taking into account relation $\mathbf{(p)}$ we can re-write relation (4.6.49) in the following form for $i \neq 0$ and $\varepsilon \leq \min(\varepsilon_{16}', \varepsilon_{17}')$:

$$\bar{\phi}_i^{(\varepsilon)}(\rho) = \frac{{}_i\bar{\phi}_{i0}^{(\varepsilon)}(\rho)}{1 - {}_0\bar{\phi}_{ii}^{(\varepsilon)}(\rho)}, \quad \rho \geq 0. \tag{4.6.50}$$

Using Lemmas 4.2.1 and 4.2.3, we have $\mathbf{(q)}$ ${}_i\bar{\phi}_{i0}^{(\varepsilon)}(\rho) \to {}_i\bar{\phi}_{i0}^{(0)}(\rho)$ as $\varepsilon \to 0$ for $\rho \geq 0$ and $i \neq 0$, and $\mathbf{(r)}$ ${}_0\bar{\phi}_{ii}^{(\varepsilon)}(\rho) \to {}_0\bar{\phi}_{ii}^{(0)}(\rho)$ as $\varepsilon \to 0$ for $\rho \geq 0$ and $i \neq 0$. Relations $\mathbf{(q)}$ and $\mathbf{(r)}$ imply that $\mathbf{(s)}$ $\bar{\phi}_i^{(\varepsilon)}(\rho) \to \bar{\phi}_i^{(0)}(\rho)$ as $\varepsilon \to 0$ for $\rho \geq 0$ and $i \neq 0$. Thus, for $i \neq 0$,

$$G_i^{(\varepsilon)}(\cdot) \Rightarrow G_i^{(\varepsilon)}(\cdot) \quad \text{as } \varepsilon \to 0. \tag{4.6.51}$$

Relation (4.6.51) implies that, in the case $\mathbf{(ii)}$, the stopping moments $\mu_0^{(\varepsilon)}$ are asymptotically stochastically bounded random variables as $\varepsilon \to 0$, i.e.,

$$\lim_{T\to\infty} \overline{\lim_{\varepsilon\to 0}} \, \mathsf{P}_i\{\mu_0^{(\varepsilon)} > T\} = 0, \quad i \neq 0. \tag{4.6.52}$$

Under conditions of Theorem 4.6.11, relation (4.6.41) yields, in case $\mathbf{(ii)}$, that for any $0 \leq t^{(\varepsilon)} \to \infty$ as $\varepsilon \to 0$,

$$\frac{\mathsf{P}_i\{\mu_0^{(\varepsilon)} > t^{(\varepsilon)}\}}{\exp\{-\rho^{(\varepsilon)}t^{(\varepsilon)}\}} \to \tilde{\pi}_i^{(0)}(\rho^{(0)}) \quad \text{as } \varepsilon \to 0, \quad i \neq 0. \tag{4.6.53}$$

Note that in case $\mathbf{(ii)}$, we also have that $\rho^{(\varepsilon)} \to \rho^{(0)} > 0$, thus the tail probabilities for stopping moments have exponential decay with parameter $\rho^{(\varepsilon)}$ that is separated from zero asymptotically as $\varepsilon \to 0$.

In conclusion, let us show what form will the corresponding conditions take in the case where $\eta^{(\varepsilon)}(t) \equiv \eta^{(0)}(t)$, $t \geq 0$, and, therefore, $\mu_0^{(\varepsilon)} \equiv \mu_0^{(0)}$ for all $\varepsilon \geq 0$, i.e., the processes and the absorption moments do not depend on the parameter ε.

Conditions of Theorem 4.6.11 reduce to the following assumptions: $\mathbf{(t)}$ condition $\mathbf{E}_2'$ holds, i.e., the class of recurrent-without-absorption states is $X_1^{(0)} = X_0 = \{1, \ldots, N\}$; $\mathbf{(u)}$ $\psi_{ij}^{(0)}(\delta) < \infty$, $i \neq 0$, $j \in X$; $\mathbf{(v)}$ ${}_0\phi_{ii}^{(0)}(\beta_i) \in (1, \infty)$ for some

$0 < \beta_i \leq \delta$ and $i \neq 0$; (**w**) $\sum_{i \neq 0} p_{i0}^{(0)} > 0$, i.e., $\rho^{(0)} > 0$; (**x**) $_0 G_{ii}^{(0)}(t)$, $i \neq 0$, are non-arithmetic distribution functions, i.e., condition $\mathbf{N_1}$ holds.

Relation (4.6.41) takes the following form:

$$\frac{\mathsf{P}_i\{\mu_0^{(0)} > t\}}{\tilde{\tilde{\pi}}_i^{(0)}(\rho^{(0)}) \exp\{-\rho^{(0)}t\}} \to 1 \text{ as } t \to \infty, \quad i \neq 0. \tag{4.6.54}$$

Asymptotic relation (4.6.54) means that the tail probabilities $\mathsf{P}_i\{\mu_0^{(0)} > t\}$ have exponential decay and can be approximated with the corresponding tail probabilities for an exponential type distribution with the tail probabilities $\tilde{\tilde{\pi}}_i^{(0)}(\rho^{(0)})e^{-\rho^{(0)}t}$. The asymptotic relative error of such an approximation equals 0 as $t \to \infty$.

The asymptotic relation (4.6.41) given in Theorem 4.6.11 can be commented on in a similar way when condition $\mathbf{E_2}$ is realised in the form of condition $\mathbf{E_2''}$, i.e., when the limit semi-Markov process has a non-empty class of non-recurrent-without-absorption states.

The remarks above allow us to interpret the mixed asymptotic relations given in Theorems 4.6.1–4.6.3 and 4.6.7 as *mixed ergodic and large deviation theorems* for perturbed semi-Markov processes with absorption.

These theorems, together with Theorems 4.6.4–4.6.6, 4.6.8–4.6.11, describe the so-called pseudo-stationary and quasi-stationary phenomena for perturbed semi-Markov processes.

We would like to mention once more that there is an essential difference between the asymptotic relations given in Theorems 4.6.1–4.6.11 for the cases where (**i**) $\rho^{(0)} = 0$, or (**ii**) where $\rho^{(0)} > 0$.

In the first case, absorption is impossible for the limit semi-Markov process $\eta^{(0)}(t)$, which implies asymptotic stochastic unboundedness of absorption times $\mu_0^{(\varepsilon)}$ as $\varepsilon \to 0$. The corresponding asymptotic relations describe the pseudo-stationary phenomenon for perturbed semi-Markov processes.

In the second case, absorption is possible for the limit semi-Markov process $\eta^{(0)}(t)$, which implies asymptotic stochastic boundedness of the absorption times $\mu_0^{(\varepsilon)}$ as $\varepsilon \to 0$. The corresponding asymptotic relations describe the quasi-stationary phenomenon for perturbed semi-Markov processes.

Chapter 5

Nonlinearly perturbed semi-Markov processes

Chapter 5 contains asymptotic results which improve the mixed ergodic theorems (for semi-Markov processes) and large deviation theorems (for absorption times), obtained in Chapter 4, to the case of nonlinearly perturbed semi-Markov processes with a finite set of states. As in the preceding chapter, we consider semi-Markov processes with a finite set of states, $X = \{0, 1, \ldots, N\}$, and the absorption state 0. The first time of hitting this state is the absorption time for the corresponding semi-Markov process.

The above asymptotic results are given in the form of exponential expansions for joint distributions of positions of semi-Markov processes and absorption times. These expansions reduce, in the simplest zero-order case, to the mixed ergodic and large deviation theorems given in Chapter 4.

They are obtained by applying the corresponding results, discussed in Chapter 3, on the exponential expansions for regenerative processes and mixed ergodic and large deviation theorems obtained in Chapter 4 for nonlinearly perturbed semi-Markov processes.

It also contains results related to asymptotic expansions for quasi-stationary distributions and asymptotic expansions and their pivotal properties for potential matrices.

We consider the same model of perturbed semi-Markov processes as in Chapter 4. However, in Chapter 5, we impose stronger perturbation conditions on the transition probabilities of the imbedded Markov chains $p_{ij}^{(\varepsilon)}$ and the semi-Markov processes $Q_{ij}^{(\varepsilon)}(t)$. We assume that these transition probabilities $p_{ij}^{(\varepsilon)}$ and the power or the mixed power-exponential moments of transition probabilities $Q_{ij}^{(\varepsilon)}(t)$ (which are improper distribution functions in argument t) can be expanded in a power series with respect to ε up to and including the order k.

As above, the relationship between the rate with which ε tends to zero and the time t tends to infinity has a delicate influence upon the results. Without loss of generality we assume that the time $t = t^{(\varepsilon)}$ is a function of the parameter ε. The balance between the rate of perturbation and the rate of growth of time is characterised by the following condition that is assumed to hold for some $1 \le r \le k$,

$$\varepsilon^r t^{(\varepsilon)} \to \lambda_r < \infty \quad \text{as } \varepsilon \to 0. \tag{5.0.1}$$

With the above assumptions and under some natural additional conditions of Cramér type for the transition probabilities $Q_{ij}^{(\varepsilon)}(u)$, we obtain the asymptotic relation

$$\frac{\mathsf{P}_i\{\eta^{(\varepsilon)}(t^{(\varepsilon)}) = j, \mu_0^{(\varepsilon)} > t^{(\varepsilon)}\}}{\exp\{-(\rho^{(0)} + a_1\varepsilon + \cdots + a_{r-1}\varepsilon^{r-1})t^{(\varepsilon)}\}} \to \tilde{\tilde{\pi}}_{ij}^{(0)}(\rho^{(0)})e^{-\lambda_r a_r} \quad \text{as} \quad \varepsilon \to 0. \quad (5.0.2)$$

We derive an equation for finding the coefficient $\rho^{(0)}$, systems of linear equations to determine $\tilde{\tilde{\pi}}_{ij}^{(0)}(\rho^{(0)})$, $i, j \neq 0$, and finally an explicit recurrence algorithm for calculating the coefficients and $a_1, \ldots, a_k$ as rational functions of the coefficients in the expansions for the moments of the transition probabilities.

The results known in the literature and related to the asymptotics given in (5.0.2) mainly cover the case $k = 1$, which corresponds to a model of linearly perturbed processes. Relation (5.0.2) for $k > 1$ corresponds to a model of nonlinearly perturbed processes. This is one of the main new elements in our results.

One more new element is a unified treatment of different types of asymptotics. Both the pseudo- and the quasi-stationary cases are covered by the same analytical approach.

Relation (5.0.2) can be interpreted as a new kind of ergodic and mixed large deviation theorems for nonlinearly perturbed semi-Markov processes.

We obtain mixed large deviation and ergodic theorems for nonlinearly perturbed semi-Markov processes using exponential expansions in mixed ergodic and large deviation theorems developed in Chapter 3 for nonlinearly perturbed regenerative processes. We use the fact that any semi-Markov processes is a regenerative process with the times of returning to some fixed state being regeneration times. The time of absorption (the first time of hitting the absorbing state) is a stopping time. A semi-Markov process is characterized by prescribing its transition probabilities. It is natural to formulate the perturbation conditions in terms of these transition probabilities. However, the conditions and expansions formulated for regenerative processes are specified in terms of expansions for the moments of regeneration times. Therefore, the corresponding asymptotic expansions for absorption probabilities and the moments of return and hitting times for perturbed semi-Markov processes must be developed.

We construct asymptotic expansions for hitting probabilities, and for power moments and mixed power-exponential moments of hitting times using a procedure that is based on recursive systems of linear equations for hitting probabilities and moments of hitting times as well as provide these expansions with a detailed analysis of their pivotal properties. These results seem to be important by themselves.

We also obtain nonlinear asymptotic expansions for quasi-stationary distributions of nonlinearly perturbed semi-Markov processes,

$$\pi_j^{(\varepsilon)}(\rho^{(\varepsilon)}) = \pi_j^{(0)}(\rho^{(0)}) + g_j[1]\varepsilon + \cdots + g_j[k]\varepsilon^k + o(\varepsilon^k). \quad (5.0.3)$$

We give an explicit recurrence algorithm for calculating the coefficients and $g_j[1]$, $\ldots, g_j[k]$ as rational functions of the coefficients in the expansions for the moment functionals of transition probabilities.

In the case of the asymptotic relation (5.0.3), the results known in the literature are related to the model of linearly perturbed Markov chains and the first order asymptotics.

New elements in our studies are nonlinear perturbations and nonlinear higher order asymptotic expansions of type (5.0.3) for quasi-stationary and usual stationary distributions of the perturbed semi-Markov processes.

As a particular but important example, we consider the model of continuous time Markov chains with absorption. In this case, it is more natural to formulate the perturbation conditions in terms of generators of the Markov chains. Here, an additional step in the algorithm is needed, since the initial perturbation conditions for generators must be expressed in terms of the moments for the corresponding semi-Markov transition probabilities before the basic algorithm can be applied.

The first three sections are devoted to studies of asymptotic expansions for hitting probabilities, and for power moments and mixed power-exponential moments of hitting times for perturbed semi-Markov processes. Lemmas 5.1.1, 5.1.2, 5.1.3, and 5.1.4 present asymptotic expansions and describe pivotal properties of solutions of nonlinearly perturbed "hitting type" systems of linear equations. Lemma 5.2.1 gives a recursion algorithm for constructing asymptotic expansions for power moments of hitting times while Lemmas 5.3.5 and 5.3.6 present similar results for mixed power-exponential moments of hitting times.

Theorems 5.4.1, 5.4.2, 5.4.3, and 5.4.4 give exponential expansions in mixed ergodic and large deviation theorem for perturbed semi-Markov processes. The first two theorems cover the case of the pseudo-stationary asymptotics, where the absorption probabilities are asymptotically small. The latter two theorems deal with the case of the quasi-stationary asymptotics for the absorption probabilities that are asymptotically separated from zero.

Theorem 5.5.1 gives higher order asymptotic expansions for stationary distributions of ergodic nonlinearly perturbed semi-Markov processes. Theorems 5.5.4 and 5.5.5 give higher order asymptotic expansions for quasi-stationary distributions in the case of nonlinearly perturbed semi-Markov processes.

The mixed ergodic and limit/large deviation theorems obtained in Chapters 4 and 5 as well as the exponential expansions provided by these theorems are used in Chapter 6 as a tool for an analysis of quasi-stationary phenomena in models of population dynamics, epidemic models, and highly reliable queueing systems.

5.1   Asymptotic expansions for absorption and hitting probabilities

In this section, we describe a recursion algorithm for the asymptotic analysis of hitting and absorption probabilities for nonlinearly perturbed semi-Markov processes. The algorithm is important in its own right and makes an essential part of the proof of our main results concerning the semi-Markov processes.

5.1.1 Asymptotic expansions for potential matrices

We assume that the absorption condition $\mathbf{A_2}$ holds. Let us also assume that the following perturbation conditions hold for some integer $k \geq 0$:

$\mathbf{P_{18}^{(k)}}$: $p_{ij}^{(\varepsilon)} = p_{ij}^{(0)} + e_{ij}[1,0]\varepsilon + \cdots + e_{ij}[k,0]\varepsilon^k + o(\varepsilon^k)$ for $i, j \neq 0$, where $|e_{ij}[r,0]| < \infty, r = 1, \ldots, k$.

Condition $\mathbf{P_{18}^{(k)}}$ obviously implies that condition $\mathbf{T_1}$ **(a)** holds, i.e.,

$$p_{ij}^{(\varepsilon)} \to p_{ij}^{(0)} \quad \text{as } \varepsilon \to 0, \quad i \neq 0, \quad j \in X. \tag{5.1.1}$$

Moreover, in the case where $k = 0$, condition $\mathbf{P_{18}^{(0)}}$ is equivalent to the convergence relation (5.1.1).

We will rewrite condition $\mathbf{P_{18}^{(k)}}$ in a matrix form.

For $r = 1, \ldots, k$, introduce the matrices

$$\mathbf{P}^{(\varepsilon)} = \begin{bmatrix} p_{11}^{(\varepsilon)} & \cdots & p_{1N}^{(\varepsilon)} \\ \vdots & \vdots & \vdots \\ p_{N1}^{(\varepsilon)} & \cdots & p_{NN}^{(\varepsilon)} \end{bmatrix}, \qquad \mathbf{E}[r,0] = \begin{bmatrix} e_{11}[r,0] & \cdots & e_{1N}[r,0] \\ \vdots & \vdots & \vdots \\ e_{N1}[r,0] & \cdots & e_{NN}[r,0] \end{bmatrix}.$$

Here and in the sequel, $\mathbf{o}(\varepsilon^k)$ denotes a matrix or a vector such that $|\mathbf{o}(\varepsilon^k)| = o(\varepsilon^k)$ as $\varepsilon \to 0$.

Condition $\mathbf{P_{18}^{(k)}}$ can be rewritten in the following equivalent matrix form:

$\mathbf{P_{18}^{(k)}}$: $\mathbf{P}^{(\varepsilon)} = \mathbf{P}^{(0)} + \mathbf{E}[1,0]\varepsilon + \cdots + \mathbf{E}[k,0]\varepsilon^k + \mathbf{o}(\varepsilon^k)$, where $\mathbf{E}[r,0], r = 1, \ldots, k$, are finite matrices.

For $j \neq 0$, consider the matrices

$$_j\mathbf{P}^{(\varepsilon)} = \begin{bmatrix} p_{11}^{(\varepsilon)} & \cdots & p_{1(j-1)}^{(\varepsilon)} & 0 & p_{1(j+1)}^{(\varepsilon)} & \cdots & p_{1N}^{(\varepsilon)} \\ \vdots & & \vdots & \vdots & \vdots & & \vdots \\ p_{N1}^{(\varepsilon)} & \cdots & p_{N(j-1)}^{(\varepsilon)} & 0 & p_{N(j+1)}^{(\varepsilon)} & \cdots & p_{NN}^{(\varepsilon)} \end{bmatrix},$$

and also, for $r = 1, \ldots, k$ and $j \neq 0$, the matrices

$$_j\mathbf{E}[r,0] = \begin{bmatrix} e_{11}[r,0] & \cdots & e_{1(j-1)}[r,0] & 0 & e_{1(j+1)}[r,0] & \cdots & e_{1N}[r,0] \\ \vdots & & \vdots & \vdots & \vdots & & \vdots \\ e_{N1}[r,0] & \cdots & e_{N(j-1)}[r,0] & 0 & e_{N(j+1)}[r,0] & \cdots & e_{NN}[r,0] \end{bmatrix}.$$

Condition $\mathbf{P}_{18}^{(k)}$ yields the following matrix asymptotic expansion:

$$_{j}\mathbf{P}^{(\varepsilon)} = {}_{j}\mathbf{P}^{(0)} + {}_{j}\mathbf{E}[1,0]\varepsilon + \cdots + {}_{j}\mathbf{E}[k,0]\varepsilon^{k} + \mathbf{o}(\varepsilon^{k}). \tag{5.1.2}$$

It is convenient to set $e_{ij}[0,0] = p_{ij}^{(0)}$, $i,j \neq 0$, and also $\mathbf{E}[0,0] = \mathbf{P}^{(0)}$ and $_{j}\mathbf{E}[0,0] = {}_{j}\mathbf{P}^{(0)}$, $j \neq 0$.

As was shown in the proof of Lemma 4.2.1, conditions $\mathbf{A}_2$, $\mathbf{T}_1$ (a) and $\mathbf{E}_2$ imply that for $\varepsilon_0 > 0$ defined in relation (4.2.14) and $j \in X_1^{(0)}$,

$$\sup_{\varepsilon \leq \varepsilon_0} |\,(_{j}\mathbf{P}^{(\varepsilon)})^{n}\,| \to 0 \text{ as } n \to \infty. \tag{5.1.3}$$

Relation (5.1.3) implies that for $\varepsilon \leq \varepsilon_0$ and $j \in X_1^{(0)}$ there exists the inverse matrix $[\mathbf{I} - {}_{j}\mathbf{P}^{(\varepsilon)}]^{-1}$. Denote this matrix by

$$_{j}\mathbf{U}^{(\varepsilon)} = \|\,_{j}u_{il}^{(\varepsilon)}\,\| = [\mathbf{I} - {}_{j}\mathbf{P}^{(\varepsilon)}]^{-1}.$$

Elements of the matrix $_{j}\mathbf{U}^{(\varepsilon)}$ have a clear probability meaning, namely, for $i,l \neq 0$, $j \in X_1^{(0)}$,

$$_{j}u_{il}^{(\varepsilon)} = \mathsf{E}_{i}\delta_{jl}^{(\varepsilon)}, \tag{5.1.4}$$

where

$$\delta_{jl}^{(\varepsilon)} = \sum_{n=1}^{\nu_{j}^{(\varepsilon)} \wedge \nu_{0}^{(\varepsilon)}} \delta(\eta_{n-1}^{(\varepsilon)}, l) = \sum_{n=1}^{\infty} \chi(\nu_{j}^{(\varepsilon)} \wedge \nu_{0}^{(\varepsilon)} > n-1, \eta_{n-1}^{(\varepsilon)} = l).$$

is the number of times the imbedded Markov chain $\eta_{n}^{(\varepsilon)}$ visits the state l until it first hits one of the states j or 0.

The matrices $_{j}\mathbf{U}^{(\varepsilon)}$, $j \in X_1^{(0)}$ can be considered as some kind of potential matrices associated with the matrix $\mathbf{P}^{(\varepsilon)}$ of transition probabilities.

The following lemma, which is a direct corollary of Lemma 8.1.2, gives an asymptotic matrix expansion for the potential matrices $_{j}\mathbf{U}^{(\varepsilon)}$, $j \in X_1^{(0)}$.

Lemma 5.1.1. *Let conditions $\mathbf{A}_2$, $\mathbf{E}_2$, and $\mathbf{P}_{18}^{(k)}$ hold. Then*

(i) *The potential matrices $_{j}\mathbf{U}^{(\varepsilon)}$ have, for every $j \in X_1^{(0)}$, the following asymptotic matrix expansions:*

$$_{j}\mathbf{U}^{(\varepsilon)} = {}_{j}\mathbf{U}^{(0)} + {}_{j}\mathbf{U}[1]\varepsilon + \cdots + {}_{j}\mathbf{U}[k]\varepsilon^{k} + \mathbf{o}(\varepsilon^{k}), \tag{5.1.5}$$

where $_{j}\mathbf{U}[r]$, $r = 1,\ldots,k$, are finite matrices;

(ii) *The matrix coefficients $_j\mathbf{U}[r]$ are given by the following recurrence formulas:*
$_j\mathbf{U}[0] = {}_j\mathbf{U}^{(0)} = [\mathbf{I} - {}_j\mathbf{P}^{(0)}]^{-1}$ *and, for $r = 1,\ldots,k$,*

$$_j\mathbf{U}[r] = {}_j\mathbf{U}[0] \sum_{q=1}^{r} {}_j\mathbf{E}[q,0]_j\mathbf{U}[r-q]. \tag{5.1.6}$$

Proof. Relations (5.1.2) and (5.1.3) imply that conditions of Lemma 8.1.2 are satisfied by the nonlinearly perturbed matrices $\mathbf{I} - {}_j\mathbf{P}^{(\varepsilon)}$ for every $j \in X_1^{(0)}$. By applying the statement **(iv)** of Lemma 8.1.2 to these matrices we prove the claim of Lemma 5.1.1. $\square$

5.1.2   Pivotal properties of asymptotic expansions for potential matrices

Let us use the following notations for the matrices $_j\mathbf{U}[r], r = 0,\ldots,k$:

$$_j\mathbf{U}[r] = \begin{bmatrix} {}_ju_{11}[r] & \cdots & {}_ju_{1N}[r] \\ \vdots & \vdots & \vdots \\ {}_ju_{N1}[r] & \cdots & {}_ju_{NN}[r] \end{bmatrix}.$$

Explicit expressions for elements of the matrices $_j\mathbf{U}[r], r = 0,\ldots,k$, can be easily derived from the recurrence matrix formulas (5.1.6).

We are interested in clarifying pivotal properties of the asymptotic expansions for the components $_ju_{il}^{(\varepsilon)}$ of elements of the potential matrices $_j\mathbf{U}^{(\varepsilon)}$. A description of such properties is especially important for *cyclic elements* of these matrices; these are elements $_ju_{jl}^{(\varepsilon)}$ for the recurrent-without-absorption states $j \in X_1^{(0)}$.

For a pair of states (i, j), where $i, j \neq 0$, we define a parameter, called a *local asymptotic order*, via the coefficients in the asymptotic expansion for the transition probabilities $p_{ij}^{(\varepsilon)}$ given in condition $\mathbf{P}_{18}^{(k)}$,

$$r(i, j) = \begin{cases} r & \text{if } e_{ij}[n,0] = 0, n < r \text{ but } e_{ij}[r,0] > 0, \text{ for } r = 0,\ldots,k, \\ k+1 & \text{if } e_{ij}[n,0] = 0, n \leq k. \end{cases}$$

Let $i, j \neq 0$ and assume that a chain of states $(i_0, i_1, \ldots, i_m)$ is an (i, j)-chain if $i_0 = i, i_m = j$ and $i_1, \ldots, i_{m-1} \neq 0, i, j$.

For an (i, j)-chain $(i_0, i_1, \ldots, i_m)$, define a parameter, called a *chain asymptotic order*, via the corresponding local asymptotic order of subsequent pairs of states in this chain,

$$r(i_0, i_1, \ldots, i_m) = \begin{cases} r & \text{if } \sum_{n=1}^{m} r(i_{n-1}, i_n) = r \text{ for } r = 0,\ldots,k, \\ k+1 & \text{if } \sum_{n=1}^{m} r(i_{n-1}, i_n) \geq k+1. \end{cases}$$

We say that an (i, j)-chain $(i_0, i_1, \ldots, i_m)$ is an *irreducible chain of states* if all the states $i_1, \ldots, i_{m-1}$ are distinct and $i_1, \ldots, i_{m-1} \neq 0, i, j$.

Let $(i_0, \ldots, i_m)$ be an (i, j)-chain. If there exists $1 \le n' < n'' \le m - 1$ such that $i_{n'} = i_{n''}$, then we say that the (i, j)-chain $(i_0, \ldots, i_m)$ contains a *loop of states* $i_{n'}, \ldots, i_{n''}$ that starts at the state $i_{n'} = i_{n''}$. By removing this loop of states, i.e., replacing the subsequence $i_{n'}, \ldots, i_{n''}$ in the initial (i, j)-chain $(i_0, \ldots, i_m)$ with the state $i_{n'}$, this chain can be replaced with the reduced (i, j)-chain $(i_0, \ldots, i_{n'}, i_{n''+1}, \ldots, i_m)$.

It is obvious that **(a)** the chain asymptotic orders of the initial and the reduced (i, j)-chains are connected by the following inequality:

$$r(i_0, \ldots, i_{n'}, i_{n''+1}, \ldots, i_m) \le r(i_0, \ldots, i_m). \tag{5.1.7}$$

Let now $(i'_1, \ldots, i'_{m'})$ be an (i, j)-chain. We now describe a multi-step reduction algorithm that reduces this (i, j)-chain to an irreducible (i, j)-chain $(i_1, \ldots, i_m)$ by removing all loops from the initial (i, j)-chain.

At step 0 one should check whether $(i'_1, \ldots, i'_{m'})$ is an irreducible (i, j)-chain. If this is so, then the (i, j)-chain $(i_1, \ldots, i_m)$ coincides with the (i, j)-chain $(i'_1, \ldots, i'_{m'})$ and the algorithm stops. Otherwise we go step 1.

Step 1 starts by finding the index $n_1 = \max(1 \le n \le m' - 1 : i'_n = i'_1)$. If $n_1 = 1$, then the (i, j)-chain $(i'_1, \ldots, i'_{m'})$ has no loops of states that start at the state i'_1. In this case, the (i, j)-chain $(i_1, \ldots, i_m)$ coincides with the (i, j)-chain $(i'_1, \ldots, i'_{m'})$, step 1 of the algorithm stops, and we go to step 2. If $n_1 > 1$, then the loop of states $i'_1, \ldots, i'_{n_1}$ that starts at the state $i'_1 = i'_{n_1}$ should be removed from the (i, j)-chain $(i'_1, \ldots, i'_{m'})$. This means that this (i, j)-chain should be replaced with the reduced (i, j)-chain $(i'_0, i'_1, i'_{n_1+1}, \ldots, i'_{m'})$. If the (i, j)-chain is irreducible, then the (i, j)-chain $(i_1, \ldots, i_m)$ coincides with the (i, j)-chain $(i'_0, i'_1, i'_{n_1+1}, \ldots, i'_{m'})$ and the algorithm stops. Otherwise we start step 2.

Step 2 starts by finding the index $n_2 = \max(n_1 \le n \le m' - 1 : i'_n = i'_{n_1+1})$. If $n_2 = n_1 + 1$, then the (i, j)-chain $(i'_1, \ldots, i'_{m'})$ has no loops of states that start at the state i'_{n_1+1}. In this case, the (i, j)-chain $(i_1, \ldots, i_m)$ coincides with the (i, j)-chain $(i'_0, i'_1, i'_{n_1+1}, \ldots, i'_{m'})$ step 2 of the algorithm ends and we pass to step 3. If $n_2 > n_1 + 1$, then the loop of states $i'_{n_1+1}, \ldots, i'_{n_1}$ that starts at the state $i'_{n_1+1}, \ldots, i'_{n_2}$ should be removed from the (i, j)-chain $(i'_0, i'_1, i'_{n_1+1}, \ldots, i'_{m'})$. This means that this (i, j)-chain is replaced with the reduced (i, j)-chain $(i'_0, i'_1, i'_{n_1+1}, i'_{n_2+1}, \ldots, i'_{m'})$. If this reduced (i, j)-chain is irreducible, then the (i, j)-chain $(i_1, \ldots, i_m)$ coincides with the (i, j)-chain $(i'_0, i'_1, i'_{n_1+1}, i'_{n_2+1}, \ldots, i'_{m'})$ and the algorithm stops. Otherwise, the algorithm continues.

The algorithm stops when the subsequent removing of loops as described above results in an irreducible (i, j)-chain $(i_1, \ldots, i_m)$.

It is obvious that the final irreducible (i, j)-chain $(i_1, \ldots, i_m)$ is uniquely determined by the initial (i, j)-chain $(i'_1, \ldots, i'_{m'})$.

Also, inequality (5.1.7) implies that

$$r(i_0, \ldots, i_m) \le r(i'_0, \ldots, i'_{m'}). \tag{5.1.8}$$

Denote by R_{ij}^* the set of all (i, j)-chains and by R_{ij} the set of all irreducible (i, j)-chains.

The multi-step reduction algorithm described above defines a mapping of the set R_{ij}^* to the set R_{ij}. Let $R_{ij}(i_0, \ldots, i_m)$ be the set of inverse images of an irreducible (i, j)-chain $(i_0, \ldots, i_m)$ with respect to this mapping.

By the definition, **(c)** the sets $R_{ij}(i_0, \ldots, i_m)$ have empty intersections for different irreducible (i, j)-chains and

$$R_{ij}^* = \bigcup_{(i_0, \ldots, i_m) \in R_{ij}} R_{ij}(i_0, \ldots, i_m). \tag{5.1.9}$$

We will refer to the (i, j)-chain $(i_0, \ldots, i_m)$ as the *core (i, j)-chain* for the set of (i, j)-chains $R_{ij}(i_0, \ldots, i_m)$.

The multi-step reduction algorithm described above allows to also define, for the set of (i, j)-chains $R_{ij}(i_0, \ldots, i_m)$, intermediate sets of (i, j)-chains obtained after performing a fixed number of steps of this algorithm.

Denote by $R_{ij}^{(1)}(i_0, \ldots, i_m)$ the set of all (i, j)-chains obtained after the first step of the multi-step reduction algorithm applied to the (i, j)-chains from the set $R_{ij}(i_0, \ldots, i_m)$, by $R_{ij}^{(2)}(i_0, \ldots, i_m)$ the set of all (i, j)-chains obtained after the second step of the multi-step reduction algorithm applied to the (i, j)-chains from the set $R_{ij}^{(1)}(i_0, \ldots, i_m)$, etc. Finally, denote by $R_{ij}^{(m-1)}(i_0, \ldots, i_m)$ the set of all (i, j)-chains obtained after step $(m - 1)$ of the multi-step reduction algorithm applied to (i, j)-chains in the set $R_{ij}^{(m-2)}(i_0, \ldots, i_m)$. It is readily seen that the set $R_{ij}^{(m-1)}(i_0, \ldots, i_m) = \{(i_0, \ldots, i_m)\}$ contains only the core (i, j)-chain $(i_0, \ldots, i_m)$.

By the definition, the sets $R_{ij}^{(r)}(i_0, \ldots, i_m)$ satisfy the following inclusion relations:

$$R_{ij}(i_0, \ldots, i_m) \supseteq R_{ij}^{(1)}(i_0, \ldots, i_m) \supseteq \cdots \supseteq R_{ij}^{(m-1)}(i_0, \ldots, i_m). \tag{5.1.10}$$

Finally, let us define a *global asymptotic order* of a pair of states (i, j), where $i, j \neq 0$, via the chain asymptotic orders of the (i, j)-chains,

$$\hat{r}(i, j) = \min_{(i_0', 1, \ldots, i_{m'}') \in R_{ij}^*} r(i_0', \ldots, i_{m'}').$$

It follows from the remarks above that **(d)** the global asymptotic order can also be defined in a simpler way,

$$\hat{r}(i, j) = \min_{(i_0, i_1, \ldots, i_m) \in R_{ij}} r(i_0, i_1, \ldots, i_m). \tag{5.1.11}$$

Indeed, the inequalities for the chain orders of (i, j)-chains and their reduced counterparts given in (5.1.7) and (5.1.8) imply that **(e)** the following relation holds for any irreducible (i, j)-chain $(i_0, i_1, \ldots, i_m)$:

$$r(i_0, i_1, \ldots, i_m) = \min_{(i_0', \ldots, i_{m'}') \in R_{ij}(i_0, i_1, \ldots, i_m)} r(i_0', \ldots, i_{m'}'). \tag{5.1.12}$$

Relation (5.1.12) implies that

$$\hat{r}(i,j) = \min_{(i'_0,1,\dots,i'_{m'})\in R^*_{ij}} r(i'_0,\dots,i'_{m'})$$

$$= \min_{(i_0,i_1,\dots,i_m)\in R_{ij}} \; \min_{(i'_0,\dots,i'_{m'})\in R_{ij}(i_0,i_1,\dots,i_m)} r(i'_0,\dots,i'_{m'})$$

$$= \min_{(i_0,i_1,\dots,i_m)\in R_{ij}} r(i_0,i_1,\dots,i_m). \tag{5.1.13}$$

Formula (5.1.11) has an obvious advantage as compared with formula (5.1.2). Indeed, the set R^*_{ij} is infinite, while R_{ij} is a finite set, since any irreducible (i,j)-chain $(i_0,i_1,\dots,i_m)$ has the length $m \le N$. Thus, a calculation of the global asymptotic order r_{ij} according to formula (5.1.11) requires a finite number of comparison operations.

It useful to note that in application models the transition matrices $\mathbf{P}^{(\varepsilon)}$ usually have a number of zero elements, **(f)** $p^{(\varepsilon)}_{l'l''} = 0$ for $\varepsilon \ge 0$.

A pair (l',l'') such that **(f)** holds obviously has the local asymptotic order equal to $k+1$.

This implies that **(g)** an irreducible (i,j)-chain $(i_0,i_1,\dots,i_m)$ has the chain asymptotic order $r(i_0,i_1,\dots,i_m) = k+1$ if it includes a pair of states $(i_{n-1},i_n) = (l',l'')$ which satisfies **(a)**.

Denote by R'_{ij} the set of all irreducible (i,j)-chains $(i_0,i_1,\dots,i_m)$ that do not include pairs of states satisfying **(a)**.

In follows from the remarks above that, **(h)** in order to calculate the global asymptotic order of a pair of states (i,j), it is actually necessary to calculate and to compare the chain asymptotic orders only for irreducible (i,j)-chains from set R'_{ij}, i.e.,

$$\hat{r}(i,j) = \min_{(i_0,i_1,\dots,i_m)\in R'_{ij}} r(i_0,i_1,\dots,i_m). \tag{5.1.14}$$

In many cases, this simplifies the calculation of the global asymptotic order of a pair of states (i,j).

Another useful remark is that **(i)** a (i,j)-chain $(i_0,i_1,\dots,i_m)$ has $r(i_0,i_1,\dots,i_m) = 0$ if and only if $p^{(0)}_{i_0 i_1} \cdots p^{(0)}_{i_{m-1} i_m} > 0$.

It is also useful to note that, according to inequality (5.1.7), **(j)** if an (i,j)-chain has the chain asymptotic order equal to 0 and belongs to the class $R_{ij}(i_0,i_1,\dots,i_m)$ with a core irreducible chain $(i_0,i_1,\dots,i_m)$, then this core irreducible chain also has the chain asymptotic order $r(i_0,i_1,\dots,i_m) = 0$.

These remarks show that **(k)** a pair (i,j) has the global asymptotic order $\hat{r}(i,j) = 0$ if and only if there exists an irreducible (i,j)-chain $(i_0,i_1,\dots,i_m)$ with the chain asymptotic order $r(i_0,i_1,\dots,i_m) = 0$ or, equivalently, with $p^{(0)}_{i_0 i_1} \cdots p^{(0)}_{i_{m-1} i_m} > 0$.

The notion of the global asymptotic order allows to give a constructive description of pivotal properties of the asymptotic expansion of the cyclic elements ${}_j u^{(\varepsilon)}_{jl}$ for $j \in X_1^{(0)}, l \ne 0$.

Introduce the sets

$$Z_{j,r} = \begin{cases} \{l \in X_0 : {}_ju_{jl}[n] = 0, n = 0,\ldots,r-1, {}_ju_{jl}[r] \neq 0\} & \text{for } r = 1,\ldots,k, \\ \{l \in X_0 : {}_ju_{jl}[n] = 0, n = 0,\ldots,k\} & \text{for } r = k+1. \end{cases}$$

We shall show below that the sets $Z_{j,r} = Z_r, r = 0,\ldots,k+1$, do not depend on the choice of the state $j \in X_1^{(0)}$.

Also, by the definition, the sets $Z_r, r = 0,\ldots,k+1$, are disjoint, i.e.,

$$Z_{r'} \cap Z_{r''} = \emptyset, \quad r', r'' \neq 0, \quad r' \neq r''. \tag{5.1.15}$$

Note also that, by the definition, some of the sets $Z_r, r = 1,\ldots,k+1$, can be empty, but, as we shall show, $Z_0 = X_1^{(0)}$ and $\bigcup_{r=1}^{k+1} Z_r = X_2^{(0)}$.

The following lemma describes properties of the sets $Z_r, r = 0,\ldots,k+1$, mentioned above, which permits to not only prove invariance properties but also to give a constructive description of these sets in terms of the corresponding global asymptotic orders.

Lemma 5.1.2. *Let conditions* $\mathbf{A}_2$, $\mathbf{E}_2$, *and* $\mathbf{P}_{18}^{(k)}$ *hold. Then*

(i) *The global asymptotic order* $\hat{r}(j,l) = \hat{r}(l)$ *does not depend on the choice of the state* $j \in X_1^{(0)}$ *for every state* $l \neq 0$.

(ii) $Z_{j,r} = Z_r, r = 0,\ldots,k+1$, *do not depend on the choice of the state* $j \in X_1^{(0)}$.

(iii) $Z_0 = X_1^{(0)}$, $\bigcup_{r=1}^{k+1} Z_r = X_2^{(0)}$.

(iv) $Z_r = \{l \neq 0 : \hat{r}(l) = r\}, r = 0,\ldots,k+1$.

Proof. Let us prove statement **(i)** of the lemma. Note first of all that $X_1^{(0)} \neq \emptyset$ by condition $\mathbf{E}_2$. Statement **(i)** is trivial if the set $X_1^{(0)}$ contains only one state. So, assume that it contains at least two states.

Let $j', j'' \in X_1^{(0)}$, $j' \neq j''$ and $l \neq 0$. Let also $(i_0', \ldots, i_{m'}')$ and $(i_0'', \ldots, i_{m''}'')$ be irreducible (j',l)-chain and (j'',l)-chain which have the chain asymptotic orders equal to the global asymptotic orders $\hat{r}(j',l)$ and $\hat{r}(j'',l)$, respectively.

Assume that $(\mathbf{l}_1)$ $\hat{r}(j',l) < \hat{r}(j'',l)$ and show that this assumption implies that $(\mathbf{m})$ one can find an irreducible (j'',l)-chain with the chain asymptotic order less than or equal to $\hat{r}(j',l)$.

There are two cases to be considered. The first case is where the state j'' coincides with some state i_n' in the (j',l)-chain $(i_0', \ldots, i_{m'}')$. Then, $(i_n', \ldots, i_{m'}')$ is an irreducible (j'',l)-chain and $r(i_n', \ldots, i_{m'}') \leq r(i_0', \ldots, i_{m'}') = \hat{r}(j',l)$.

The second case is where the state j'' does not coincide with any state i_n' in the (j',l)-chain $(i_0', \ldots, i_{m'}')$. Denote by V the set of all states i_n' from the (j',l)-chain $(i_0', \ldots, i_{m'}')$ that belong to the set $X_1^{(0)}$. We use the notations introduced in Section

4.2 and consider the hitting probabilities $_0f^{(0)}_{j''Zi'_n} = P_{j''}\{v^{(0)}_V < v^{(0)}_0, \eta^{(0)}_{v^{(0)}(V)} = i'_n\}$, $i'_n \in V$. Since $X^{(0)}_1$ is the set of all recurrent-without-absorption states, there exists at least one state $i'_n \in V$ such that $_0f^{(0)}_{j''Vi'_n} > 0$. Obviously, this probability is positive if and only if $(\mathbf{n_1})$ there exists an irreducible (j'', i_n)-chain $(i_0, \ldots, i_{m-1})$ such that $i_1, \ldots, i_m \notin V \cup \{0\}$ and $p^{(0)}_{i_0 i_1} \cdots p^{(0)}_{i_{m-1} i_m} > 0$. The transition probabilities are $p^{(0)}_{ij} = 0$ for all states $i \in X^{(0)}_1$, $j \in X^{(0)}_2$ (if $X^{(0)}_2 \neq \emptyset$). Thus, $(\mathbf{n_2})$ all states satisfy $i_0, \ldots, i_{m-1} \in X^{(0)}_1$. It follows from $(\mathbf{n_1})$ and $(\mathbf{n_2})$ that the chain $(i_1, \ldots, i_m, i'_n, \ldots, i'_{m'})$ is irreducible. Obviously $r(i_1, \ldots, i_m) = 0$ and the chain order $r(i'_n, \ldots, i'_{m'}) \leq r(i'_1, \ldots, i'_{m'})$. Therefore, $r(i_1, \ldots, i_m, i'_n, \ldots i'_{m'}) \leq r(i_1, \ldots, i_m) + r(i'_n, \ldots, i'_{m'}) \leq r(i'_1, \ldots, i'_{m'}) = \hat{r}(j', l)$.

Statement $(\mathbf{m})$ contradicts $(\mathbf{l_1})$, since the global asymptotic order $\hat{r}(j'', l)$, by the definition, is greater than or equal to the chain asymptotic order for any irreducible (j'', l)-chain.

Thus the inequality $(\mathbf{l_1})$ is impossible and, therefore, $(\mathbf{h_2})$ $\hat{r}(j'', l) \geq \hat{r}(j', l)$. In the same way, one can prove that $(\mathbf{l_3})$ $\hat{r}(j', l) \geq \hat{r}(j'', l)$. Inequalities $(\mathbf{l_2})$ and $(\mathbf{l_3})$ imply that $\hat{r}(j', l) = \hat{r}(j'', l)$, which proves $(\mathbf{m})$ in the lemma.

In what follows, we will use two formulas. The first one is the obvious formula

$$_ju^{(\varepsilon)}_{ij} = \mathsf{E}_i\delta^{(\varepsilon)}_{jj} = \delta(i, j), \quad i, j \neq 0, \tag{5.1.16}$$

where $\delta(i, j)$ is the Kronecker delta-function.

The second one is a formula that plays a key role in the further proof,

$$_ju^{(\varepsilon)}_{il} = \mathsf{E}_i\delta^{(\varepsilon)}_{jl} = {}_{0j}f^{(\varepsilon)}_{il}\mathsf{E}_l\delta^{(\varepsilon)}_{jl} = \frac{_{0j}f^{(\varepsilon)}_{il}}{1 - {}_{0j}f^{(\varepsilon)}_{ll}}, \quad i, j, l \neq 0, \ j \neq l, \tag{5.1.17}$$

We prove statements $(\mathbf{ii})$–$(\mathbf{iii})$ in a simple case where condition $\mathbf{E'_2}$ holds. Recall that condition $\mathbf{E'_2}$ means that the set of recurrent-without-absorption states is $X^{(0)}_1 = X_0$, where $X_0 = \{1, \ldots, N\}$.

The case $N = 1$ is trivial. Indeed, by formula (5.1.16), the expectation $(\mathbf{o_1})$ satisfies $_ju^{(0)}_{jj} = 1$. Thus, $Z_0 = \{j\}$. This automatically implies that $Z_1, \ldots, Z_{k+1} = \emptyset$. In this case, the only irreducible (j, j)-chain is the chain (j, j) of length 1. The state j is recurrent-without absorption, which implies that $p^{(0)}_{jj} > 0$. Thus, $(\mathbf{p_1})$ the chain asymptotic order is $r(j, j) = 0$ and, therefore, the global asymptotic order satisfies $\hat{r}(j) = \hat{r}(j, j) = 0$. The remarks above imply that $Z_0 = \{j\} = \{j : \hat{r}(j) = 0\}$.

Let us now assume that $N > 1$. Formula (5.1.16) implies that $(\mathbf{o_2})$ $_ju^{(0)}_{jj} = 1$. Also, in this case, $_{0j}f^{(0)}_{jl} > 0$ and $_{0j}f^{(0)}_{ll} < 1$ for all $j, l \in X^{(0)}_1$, $j \neq l$. Therefore, $(\mathbf{o_3})$ the expectations satisfy $\mathsf{E}_j\delta^{(0)}_{jl} > 0$ for all $j, l \in X^{(0)}_1$, $j \neq l$. Thus the set $Z_{j,0} = Z_0$ does not depend on $j \in X^{(0)}_1$ and, moreover, it coincides with the set $X^{(0)}_1 = X_0$.

This automatically implies that $Z_1, \ldots, Z_{k+1} = \varnothing$. Since all $j \in X_1^{(0)}$ are recurrent-without-absorption states, the probabilities are positive, ${}_{0j}f_{jl}^{(0)} > 0$ for all $j, l \in X_1^{(0)}$. We have ${}_{0j}f_{jl}^{(0)} > 0$ if and only if there exists an irreducible (i, l)-chain $(i_0, \ldots, i_m)$ such that $p_{i_0 i_1}^{(0)} \cdots p_{i_{m-1} i_m}^{(0)} > 0$ or, equivalently, $r(i_0, \ldots, i_m) = 0$. This implies that $(\mathbf{p_2})$ the global asymptotic order is zero, $\hat{r}(j, l) = \hat{r}(l) = 0$. This holds for all $l \in X_1^{(0)}$. Thus the set $\{l \neq 0 : \hat{r}(l) = 0\}$ coincides with the set $X_1^{(0)} = Z_0$.

Let us now consider the case where condition $\mathbf{E_2''}$ is satisfied. Recall that condition $\mathbf{E_2'}$ means that $X_1^{(0)} \neq \varnothing$ but also the set of non-recurrent-without-absorption is nonempty, $X_2^{(0)} \neq \varnothing$.

The proofs of the claims that $(\mathbf{o_4})$ all states $l \in X_1^{(0)}$ belong to the set $Z_{j,0}$ for every $j \in X_1^{(0)}$, and that $(\mathbf{p_3})$ the global asymptotic order is zero, $\hat{r}(j, l) = \hat{r}(l) = 0$, for all states $l \in X_1^{(0)}$ repeats the above. Let us prove that $(\mathbf{o_5})$ all states $l \in X_2^{(0)}$ do not belong to the set $Z_{j,0}$ for every $j \in X_1^{(0)}$. Indeed, by the definition of the set $X_1^{(0)}$, the probabilities $p_{jl}^{(0)} = 0$ for all states $j \in X_1^{(0)}, l \in X_2^{(0)}$. This implies, in an obvious way, that ${}_{0j}f_{jl}^{(0)} = 0$. Also, by condition $\mathbf{E_2''}$, we have that ${}_{0j}f_{ll}^{(0)} < 1$ for all $j \in X_1^{(0)}, l \in X_2^{(0)}$. Therefore, by formula (5.1.17), the expectations are zero, $\mathsf{E}_j \delta_{jl}^{(0)} = 0$, for all $j \in X_1^{(0)}, l \in X_2^{(0)}$. Obviously, $(\mathbf{o_4})$ and $(\mathbf{o_5})$ imply that the set $Z_{j,0} = Z_0$ does not depend on $j \in X_1^{(0)}$ and, moreover, this set coincides with the set $X_1^{(0)}$. Let us also prove that $(\mathbf{p_4})$ the global asymptotic order is positive, $\hat{r}(j, l) = \hat{r}(l) > 0$, for all states $l \in X_2^{(0)}$. Indeed, as was mentioned above, in this case, ${}_{0j}f_{jl}^{(0)} = 0$ for all $j \in X_1^{(0)}, l \in X_2^{(0)}$. Now, ${}_{0j}f_{jl}^{(0)} = 0$ if and only if there are no irreducible (j, l)-chains with the chain asymptotic order 0. This implies that the global asymptotic order satisfies $\hat{r}(j, l) = \hat{r}(l) > 0$. This holds for all $j \in X_1^{(0)}, l \in X_2^{(0)}$. Obviously, $(\mathbf{p_3})$ and $(\mathbf{p_4})$ imply that the set $\{l \neq 0 : \hat{r}(l) = 0\}$ coincides with the set $X_1^{(0)} = Z_0$. Note that this also implies that $\bigcup_{r=1}^{k+1} Z_r = X_2^{(0)}$.

Let us now prove that $(\mathbf{q})$ $Z_{j,r} = \{l \neq 0 : \hat{r}(l) = r\}$ for all states $j \in X_1^{(0)}$ and $r = 1, \ldots, k + 1$. This will complete the proof of the lemma.

As was mentioned above, condition $\mathbf{P_{18}^{(k)}}$ implies that condition $\mathbf{T_1}$ $(\mathbf{a})$ is satisfied, and, as was shown in the proof of Lemma 4.2.1, conditions $\mathbf{A_2}$, $\mathbf{E_2}$, and $\mathbf{T_1}$ $(\mathbf{a})$ imply that

$$\mathsf{E}_i \delta_{jl}^{(\varepsilon)} \to \mathsf{E}_i \delta_{jl}^{(0)} < \infty \text{ as } \varepsilon \to 0, \ i, l \neq 0, \ j \in X_1^{(0)}. \tag{5.1.18}$$

As was mentioned above, ${}_{0j}f_{ll}^{(0)} < 1$ for all $j \in X_1^{(0)}, l \in X_2^{(0)}$. Thus,

$$\mathsf{E}_l \delta_{jl}^{(0)} \geq 1, \quad j \in X_1^{(0)}, \quad l \in X_2^{(0)}. \tag{5.1.19}$$

According to Lemma 5.1.1, the expectations ${}_j u_{jl}^{(\varepsilon)} = \mathsf{E}_j \delta_{jl}^{(\varepsilon)}$, $j \in X_1^{(0)}, l \in X_2^{(0)}$ admit asymptotic expansions of order k. Hence, formula (5.1.17) used in the case

$i = j \in X_1^{(0)}, l \in X_2^{(0)}$, relations (5.1.18) and (5.1.19) imply that, **(r)** for every $j \in X_1^{(0)}$ and $r = 1,\ldots,k+1$, we will have $l \in Z_{j,r}$ if and only if there exists a coefficient $f_{jl}[r] > 0$ such that the probability $_{0j}f_{jl}^{(\varepsilon)}$ has the following asymptotic representation:

$$_{0j}f_{jl}^{(\varepsilon)} = \begin{cases} \varepsilon^r f_{jl}[r] + o(\varepsilon^r) & \text{if } r = 1,\ldots,k, \\ o(\varepsilon^k) & \text{if } r = k+1. \end{cases} \tag{5.1.20}$$

The probability $_{0j}f_{jl}^{(\varepsilon)}$ can be represented as the following sum:

$$_{0j}f_{jl}^{(\varepsilon)} = \sum_{(i_0',\ldots,i_{m'}') \in R_{il}^*} p_{i_0'i_1'}^{(\varepsilon)} \cdots p_{i_{m'-1}'i_{m'}'}^{(\varepsilon)}$$

$$= \sum_{(i_0,\ldots,i_m) \in R_{il}} \sum_{(i_0',\ldots,i_{m'}') \in R_{il}(i_0,\ldots,i_m)} p_{i_0'i_1'}^{(\varepsilon)} \cdots p_{i_{m'-1}'i_{m'}'}^{(\varepsilon)}. \tag{5.1.21}$$

Denote by $N_q(i_0',\ldots,i_{m'}')$ the number of states i_q in the chain $(i_0',\ldots,i_{m'}')$ from the set $R_{jl}^{(q-1)}(i_0,\ldots,i_m)$ for $q = 1,\ldots,m-1$. Here we set, for convenience, $R_{jl}^{(0)}(i_0,\ldots,i_m) = R_{jl}(i_0,\ldots,i_m)$.

Let us also denote $V_q = V_q(i_0,\ldots,i_{q-1},i_m) = \{0,j,l,i_1,\ldots,i_{q-1}\}$ for $q = 1,\ldots,m-1$, and $f_q^{(\varepsilon)}(i_0,\ldots,i_{q-1},i_m) = {}_{V_q}f_{i_q i_q}^{(\varepsilon)} = \mathsf{P}_{i_q}\{v_{i_q}^{(\varepsilon)} < v_{V_q}^{(\varepsilon)}\}$ for $q = 1,\ldots,m-1$.

A key role in our calculations plays the following formulas which hold for every $q = 1,\ldots,m-1$:

$$\sum_{(i_0',\ldots,i_{m'}') \in R_{jl}^{(q-1)}(i_0,\ldots,i_m)} p_{i_0'i_1'}^{(\varepsilon)} \cdots p_{i_{m'-1}'i_{m'}'}^{(\varepsilon)}$$

$$= \sum_{n=1}^{\infty} \sum_{(i_0',\ldots,i_{m'}') \in R_{jl}^{(q-1)}(i_0,\ldots,i_m), \, N_q(i_0',\ldots,i_{m'}')=n} p_{i_0'i_1'}^{(\varepsilon)} \cdots p_{i_{m'-1}'i_{m'}'}^{(\varepsilon)}$$

$$= \sum_{n=1}^{\infty} (f_q^{(\varepsilon)}(i_0,\ldots,i_{q-1},i_m))^{n-1}$$

$$\times \sum_{(i_0'',\ldots,i_{m''}'') \in R_{jl}^{(q)}(i_0,\ldots,i_m)} p_{i_0''i_1''}^{(\varepsilon)} \cdots p_{i_{m''-1}''i_{m''}''}^{(\varepsilon)}$$

$$= \frac{1}{1 - f_q^{(\varepsilon)}(i_0,\ldots,i_{q-1},i_m)}$$

$$\times \sum_{(i_0'',\ldots,i_{m''}'') \in R_{jl}^{(q)}(i_0,\ldots,i_m)} p_{i_0''i_1''}^{(\varepsilon)} \cdots p_{i_{m''-1}''i_{m''}''}^{(\varepsilon)}. \tag{5.1.22}$$

Combining formulas (5.1.21)–(5.1.22) together and taking into account that the set $R_{jl}^{(m-1)}(i_0, \ldots, i_m)$ contains the only core irreducible (j, l)-chain $(i_0, \ldots, i_m)$, we get the following important formula:

$$_{0j}f_{jl}^{(\varepsilon)} = \sum_{(i_0,\ldots,i_m)\in R_{jl}} \prod_{q=1}^{m-1} \frac{1}{1 - f_q^{(\varepsilon)}(i_0, \ldots, i_{q-1}, i_m)} \cdot p_{i_0 i_1}^{(\varepsilon)} \cdots p_{i_{m-1} i_m}^{(\varepsilon)}. \qquad (5.1.23)$$

By Lemma 4.2.5, for every $q = 1, \ldots, m - 1$,

$$f_q^{(\varepsilon)}(i_0, \ldots, i_{q-1}, i_m) = {}_{V_q}f_{i_q i_q}^{(\varepsilon)}$$

$$\rightarrow f_q^{(0)}(i_0, \ldots, i_{q-1}, i_m) = {}_{V_q}f_{i_q i_q}^{(0)} \quad \text{as } \varepsilon \rightarrow 0. \qquad (5.1.24)$$

Since j is a recurrent-without-absorption state, $f_q^{(0)}(i_0, \ldots, i_{q-1}, i_m) = {}_{V_q}f_{i_q i_q}^{(0)} < 1$, $q = 1, \ldots, m - 1$, and, therefore, the following relation holds:

$$\prod_{q=1}^{m-1} \frac{1}{1 - f_q^{(0)}(i_0, \ldots, i_{q-1}, i_m)} > 0. \qquad (5.1.25)$$

It follows from relations (5.1.23), (5.1.24), and (5.1.25) that **(s)** the asymptotic relation (5.1.20) holds if and only if the global asymptotic order satisfies the following relation $\min_{(i_0,\ldots,i_m)\in R_{jl}} r(i_0, \ldots, i_m) = \hat{r}(j, l) = r$ for some $r = 1, \ldots, k + 1$.

Moreover, in the cases $r = 1, \ldots, k$, the coefficients $f_{jl}[r]$ can be calculated with the use of the following formula:

$$f_{jl}[r] = \sum_{(i_0,\ldots,i_m)\in R_{jl}, r_{jl}(i_0,\ldots,i_m)=r} \prod_{q=1}^{m-1} \frac{1}{1 - f_q^{(0)}(i_0, \ldots, i_{q-1}, i_m)}$$

$$\times \prod_{n=1}^{n} e_{i_{n-1} i_n}[r(i_{n-1}, i_n), 0]. \qquad (5.1.26)$$

It follows from propositions **(r)** and **(s)** that $Z_{j,r} = \{l \neq 0 : \hat{r}(j, l) = r\}$ for every $r = 1, \ldots, k+1$. According to claim **(i)** in the lemma, the asymptotic orders $\hat{r}(j, l) = \hat{r}(l), l \neq 0$, do not depend on $j \in X_1^{(0)}$. Thus, **(t)** the set $Z_{j,r} = \{l \neq 0 : \hat{r}(j, l) = r\}$ coincides with the set $Z_r = \{l \neq 0 : \hat{r}(l) = r\}$ for every $r = 1, \ldots, k + 1$.

The proof of the lemma is complete. $\qquad \square$

5.1.3 Asymptotic expansions for solutions of hitting type systems of linear equations

Let us consider functions $y_i^{(\varepsilon)}$, $i \neq 0$, and assume that they satisfy the following non-negativity condition:

$\mathbf{S_1}$: $y_i^{(\varepsilon)} \geq 0$ for $\varepsilon \geq 0$ and $i \neq 0$;

and that the following perturbation condition holds:

$\mathbf{P}_{19}^{(k)}$: $y_i^{(\varepsilon)} = y_i^{(0)} + y_i[1]\varepsilon + \cdots + y_i[k]\varepsilon^k + o(\varepsilon^k)$ for $i \neq 0$, where $|y_i[r]| < \infty$, $r = 1,\ldots,k$.

For $r = 1,\ldots,k$, consider the vectors

$$\mathbf{y}^{(\varepsilon)} = \begin{bmatrix} y_1^{(\varepsilon)} \\ \vdots \\ y_N^{(\varepsilon)} \end{bmatrix}, \qquad \mathbf{y}[r] = \begin{bmatrix} y_1[r] \\ \vdots \\ y_N[r] \end{bmatrix}.$$

Condition $\mathbf{P}_{19}^{(k)}$ can be rewritten in the following equivalent vector form:

$\mathbf{P}_{19}^{(k)}$: $\mathbf{y}^{(\varepsilon)} = \mathbf{y}^{(0)} + \mathbf{y}[1]\varepsilon + \cdots + \mathbf{y}[k]\varepsilon^k + \mathbf{o}(\varepsilon^k)$, where $\mathbf{y}[r]$, $r = 1,\ldots,k$, are finite vectors.

It is convenient to define $y_i[0] = y_i^{(0)}$, $i \neq 0$, and $\mathbf{y}[0] = \mathbf{y}^{(0)}$.

For every $j \in X_1^{(0)}$, consider the system of linear equations

$$\mathbf{x}_j^{(\varepsilon)} = \mathbf{y}^{(\varepsilon)} + {}_j\mathbf{P}^{(\varepsilon)}\mathbf{x}_j^{(\varepsilon)}. \tag{5.1.27}$$

Under conditions $\mathbf{A}_2$, $\mathbf{E}_2$, and $\mathbf{T}_1$ (a), system (5.1.27) has a unique solution for every $\varepsilon \leq \varepsilon_0$ and $j \in X_1^{(0)}$,

$$\mathbf{x}_j^{(\varepsilon)} = [\mathbf{I} - {}_j\mathbf{P}^{(\varepsilon)}]^{-1}\mathbf{y}^{(\varepsilon)}. \tag{5.1.28}$$

It is also useful to note that if condition $\mathbf{S}_1$ is satisfied, the vector $\mathbf{x}_j^{(\varepsilon)}$ has all components nonnegative.

The following lemma gives vector asymptotic expansions for solutions of these systems.

Lemma 5.1.3. *Let conditions* $\mathbf{A}_2$, $\mathbf{E}_2$, $\mathbf{P}_{18}^{(k)}$, $\mathbf{S}_1$, *and* $\mathbf{P}_{19}^{(k)}$ *hold. Then*

(i) *The vector* $\mathbf{x}_j^{(\varepsilon)}$ *has the following asymptotic expansion for every* $j \in X_1^{(0)}$:

$$\mathbf{x}_j^{(\varepsilon)} = \mathbf{x}_j^{(0)} + \mathbf{x}_j[1]\varepsilon + \cdots + \mathbf{x}_j[k]\varepsilon^k + \mathbf{o}(\varepsilon^k), \tag{5.1.29}$$

where $\mathbf{x}_j[r]$, $r = 1,\ldots,k$, *are finite vectors.*

(ii) *The vector coefficients* $\mathbf{x}_j[r]$ *are given by the recurrence formulas* $\mathbf{x}_j[0] = \mathbf{x}_j^{(0)} = {}_j\mathbf{U}[0]\mathbf{y}^{(0)} = [\mathbf{I} - {}_j\mathbf{P}^{(0)}]^{-1}\mathbf{y}^{(0)}$ *and, for* $r = 0,\ldots,k$,

$$\mathbf{x}_j[r] = {}_j\mathbf{U}[0]\Big(\mathbf{y}[r] + \sum_{q=1}^{r} {}_j\mathbf{E}[q,0]\,\mathbf{x}_j[r-q]\Big). \tag{5.1.30}$$

(iii) *The vector coefficients* $\mathbf{x}_j[r]$ *can also be calculated by the following formulas for* $r = 0, \ldots, k$:

$$\mathbf{x}_j[r] = \sum_{p=0}^{r} {}_j\mathbf{U}[p]\,\mathbf{y}[r - p].\tag{5.1.31}$$

Proof. Relations (5.1.2) and (5.1.3) and condition $\mathbf{P}_{19}^{(k)}$, expressed in the vector form, imply that conditions of Lemma 8.1.4 are satisfied for the system of nonlinearly perturbed linear equations (5.1.27) for every $j \in X_1^{(0)}$. By applying this lemma to systems (5.1.27), we get the claims of Lemma 5.1.1. $\qquad\square$

For $r = 1, \ldots, k$ and $j \in X_1^{(0)}$ denote

$$\mathbf{x}_j^{(\varepsilon)} = \begin{bmatrix} x_{1j}^{(\varepsilon)} \\ \vdots \\ x_{Nj}^{(\varepsilon)} \end{bmatrix}, \qquad \mathbf{x}_j[r] = \begin{bmatrix} x_{1j}[r] \\ \vdots \\ x_{Nj}[r] \end{bmatrix}.$$

Explicit expressions for components of the vectors $\mathbf{x}_j[r]$, $r = 0, \ldots, k$, can be derived from recurrence formulas (5.1.30) or formulas (5.1.31).

There is a simple probabilistic interpretation of solutions of system (5.1.27). Indeed, using the representation for elements of the matrix $[\mathbf{I} - {}_j\mathbf{P}^{(\varepsilon)}]^{-1}$ given by formula (5.1.4), we can write for every $i \neq 0$, $j \in X_1^{(0)}$ the following representation for components of the vector $\mathbf{x}_j^{(\varepsilon)}$,

$$x_{ij}^{(\varepsilon)} = \sum_{l \neq 0} {}_ju_{il}^{(\varepsilon)}y_l^{(\varepsilon)} = \sum_{l \neq 0}\mathsf{E}_i\delta_{jl}^{(\varepsilon)}y_l^{(\varepsilon)} = \mathsf{E}_i\sum_{n=1}^{v_j^{(\varepsilon)}\wedge v_0^{(\varepsilon)}} y_{\eta_n^{(\varepsilon)}}^{(\varepsilon)}.\tag{5.1.32}$$

Formula (5.1.32) permits to refer to systems of type (5.1.27) as a *hitting system of linear equations*.

Hitting systems of type (5.1.27) play an important role in our analysis. Hitting and absorption probabilities and power type moments for hitting times are solutions of such systems.

We are especially interested in a study of pivotal properties of asymptotic expansions for the components $x_{ij}^{(\varepsilon)}$ of the solutions $\mathbf{x}_j^{(\varepsilon)}$ of system (5.1.32). A description of such properties is especially important for the *cyclic components* $x_{jj}^{(\varepsilon)}$ of the vectors $\mathbf{x}_j^{(\varepsilon)}$.

Let us use perturbation condition $\mathbf{P}_{19}^{(k)}$ to classify the states $i \neq 0$ by the order of the first nonzero coefficient in the corresponding asymptotic expansion.

Consider the sets

$$
Y_r = \begin{cases} \{i \in X_0 : y_i[n] = 0, \ n = 0, \ldots, r-1, \ y_i[r] \neq 0\} & \text{for } r = 1, \ldots, k, \\ \{i \in X_0 : y_i[n] = 0, \ n = 0, \ldots, k\} & \text{for } r = k+1, \end{cases}
$$

where $X_0 = \{1, \ldots, N\}$.

Note that, by the definition, **(a)** some of the sets Y_r, $r = 0, \ldots, k+1$, could be empty, and **(b)** $Y_{r'} \cap Y_{r''} = \varnothing$, $r' \neq r''$, but **(c)** $\bigcup_{r=0}^{k+1} Y_r = X_0$.

We are now in a position to give a clear description of pivotal properties of asymptotic expansions for the cyclic components $x_{jj}^{(\varepsilon)}$, $j \in X_1^{(0)}$.

Define the sets

$$
W_r = \bigcup_{r'+r''=r} (Y_{r'} \cap Z_{r''}), \ r = 0, \ldots, k, \qquad W_{k+1} = \bigcup_{r'+r'' \geq k+1} (Y_{r'} \cap Z_{r''}).
$$

By the definition, **(d)** some of the sets W_r, $r = 0, \ldots, k+1$, could be empty, **(e)** $W_{r'} \cap W_{r''} = \varnothing$, $r' \neq r''$, but **(f)** $\bigcup_{r=0}^{k+1} W_r = X_0$.

The case where condition $\mathbf{E}_2'$ holds, is a simple one. Indeed, here $Z_0 = X_1^{(0)} = X_0$, and, therefore, $Z_1, \ldots, Z_{k+1} = \varnothing$. Thus,

$$
W_r = Y_r, \quad r = 0, \ldots, k+1.
$$

For an integer $0 \leq h \leq k$, consider the following two conditions:

$\mathbf{O}_1^{(h)}$: $W_r = \varnothing$, for $r \leq h$;

and

$\mathbf{O}_2^{(h)}$: $W_r = \varnothing$, for $r < h$, but $W_h \neq \varnothing$.

Lemma 5.1.4. *Let conditions* $\mathbf{A}_2$, $\mathbf{E}_2$, $\mathbf{P}_{18}^{(k)}$, $\mathbf{S}_1$, *and* $\mathbf{P}_{19}^{(k)}$ *hold. Then*

(i) *If condition* $\mathbf{O}_1^{(h)}$ *holds for some* $0 \leq h \leq k$, *then* $x_{jj}[n] = 0$, $n = 0, \ldots, h$, $j \in X_1^{(0)}$.

(ii) *If condition* $\mathbf{O}_2^{(h)}$ *holds for some* $0 \leq h \leq k$, *then* $x_{jj}[n] = 0$, $n = 0, \ldots, h-1$, $j \in X_1^{(0)}$, *but* $x_{jj}[h] > 0$, $j \in X_1^{(0)}$.

Proof. By using asymptotic expansions (5.1.5) for the potential matrices $_j\mathbf{U}^{(\varepsilon)}$, and condition $\mathbf{P}_{19}^{(k)}$, relation (5.1.32) can be rewritten for the cyclic components $x_{jj}^{(\varepsilon)}$, $j \in$

$X_1^{(0)}$, in the following form:

$$x_{jj}^{(\varepsilon)} = \sum_{l \neq 0} {}_j u_{il}^{(\varepsilon)} y_l^{(\varepsilon)}$$

$$= \sum_{0 \le r \le k} \sum_{r'+r''=r} \sum_{l \in Y_{r'} \cap Z_{r''}} {}_j u_{il}^{(\varepsilon)} y_l^{(\varepsilon)} + \sum_{r'+r'' \ge k+1} \sum_{l \in Y_{r'} \cap Z_{r''}} {}_j u_{il}^{(\varepsilon)} y_l^{(\varepsilon)}$$

$$= \sum_{0 \le r \le k} \sum_{r'+r''=r} \sum_{l \in Y_{r'} \cap Z_{r''}} \left(\sum_{r'' \le p'' \le k} \varepsilon^{p''} {}_j u_{jl}[p''] \right)$$

$$\times \left(\sum_{r' \le p' \le k} \varepsilon^{p'} y_l[p'] \right) + o(\varepsilon^k)$$

$$= \sum_{0 \le r \le k} \left(\varepsilon^r L_j[r,r] + \cdots + \varepsilon^k L_j[k,r] \right) + o(\varepsilon^k)$$

$$= \sum_{0 \le r \le k} \varepsilon^r \left(L_j[r,0] + \cdots + L_j[r,r] \right) + o(\varepsilon^k), \tag{5.1.33}$$

where the coefficients $L_j[p,r]$, $p = r, \ldots, k$, $r = 0, \ldots, k$, are readily calculated by multiplying the inner sums in (5.1.33) and collecting the coefficients at ε^p.

For every $r = 0, \ldots, k$, (g) the coefficients $L_j[p,r]$, $p = r, \ldots, k$, are sums of numbers from the set W_r. In particular, (h) the coefficients $L_j[r,r]$ have the following form for $r = 0, \ldots, k$:

$$L_j[r,r] = \sum_{r'+r''=r} \sum_{l \in Y_{r'} \cap Z_{r''}} {}_j u_{jl}[r''] y_l[r']. \tag{5.1.34}$$

Formula (5.1.33) implies that (i) the coefficients $x_{jj}[r]$ can be represented in the following form for $r = 0, \ldots, k$:

$$x_{jj}[r] = \sum_{0 \le p \le r} L_j[r,p]. \tag{5.1.35}$$

If condition $\mathbf{O}_1^{(h)}$ holds for some $0 \le h \le k$, then $W_r = \emptyset$, $r \le h$. Thus, (j) the sum that defines the coefficients $L[p,r]$, $p = r, \ldots, k$, are actually sums over the empty set for $r \le h$. Therefore, (k) $L[p,r] = 0$, $p = r, \ldots, k$, for $r \le h$. The claim (k) and formula (5.1.35) imply that (l) $x_{jj}[r] = 0$, $r \le h$. This proves the first statement of the lemma.

If condition $\mathbf{O}_2^{(h)}$ holds for some $0 \le h \le k$, then $W_r = \emptyset$, $r \le h - 1$, but $W_h \neq \emptyset$. This means that condition $\mathbf{O}_1^{(h-1)}$ holds and, therefore, (m) the coefficients satisfy $x_{jj}[r] = 0$, $r \le h - 1$, as was proved above. Moreover, similarly to the above, (n) $L[p,r] = 0$, $p = r, \ldots, k$ for $r \le h - 1$. Obviously, (n) and formula (5.1.35) imply that, in the case under consideration, (o) $x_{jj}[h] = L_j[r,r]$.

If $W_h \neq \varnothing$, then $Y_{r'} \cap Z_{r''} \neq \varnothing$ for some non-negative integer r' and r'' such that $r' + r'' = h$. By the definition of the set $Z_{r''}$, we have $_j u_{jl}[r''] > 0$, $l \in Z_{r''}$. Also, by the definition of the set $Y_{r'}$ and condition $\mathbf{S_1}$, the coefficients satisfy $y_l[r'] > 0$, $l \in Y_{r'}$. Thus, $\sum_{l \in Y_{r'} \cap Z_{r''}} {}_j u_{jl}[r''] y_l[r'] > 0$. Therefore, $x_{jj}[r] = L_j[r, r] > 0$ according to formula (5.1.35). This proves the second statement of the lemma. $\square$

5.1.4 Asymptotic expansions for absorption and hitting probabilities

It is useful to note that condition $\mathbf{P}_{18}^{(k)}$ automatically implies that a similar asymptotic expansion takes place for the local absorption probabilities $p_{i0}^{(\varepsilon)}$, $i \neq 0$,

$$
\begin{aligned}
p_{i0}^{(\varepsilon)} &= 1 - \sum_{j \neq 0} p_{ij}^{(\varepsilon)} \\
&= 1 - \sum_{j \neq 0} (p_{ij}^{(0)} + e_{ij}[1, 0]\varepsilon + \cdots + e_{ij}[k, 0]\varepsilon^k + o(\varepsilon^k)) \\
&= p_{i0}^{(0)} + e_{i0}[1, 0]\varepsilon + \cdots + e_{i0}[k, 0]\varepsilon^k + o(\varepsilon^k),
\end{aligned}
\tag{5.1.36}
$$

where, for $r = 1, \ldots, k$,

$$
e_{i0}[r, 0] = - \sum_{j \neq 0} e_{ij}[r, 0].
$$

For $r = 1, \ldots, k$ and $j \in X$, we introduce the vectors,

$$
\mathbf{p}_j^{(\varepsilon)} = \begin{bmatrix} p_{1j}^{(\varepsilon)} \\ \vdots \\ p_{Nj}^{(\varepsilon)} \end{bmatrix}, \qquad
\mathbf{e}_j[r, 0] = \begin{bmatrix} e_{1j}[r, 0] \\ \vdots \\ e_{Nj}[r, 0] \end{bmatrix}.
$$

Condition $\mathbf{P}_{18}^{(k)}$ and asymptotic relation (5.1.36) give the following vector asymptotic expansion for $j \in X$:

$$
\mathbf{p}_j^{(\varepsilon)} = \mathbf{p}_j^{(0)} + \mathbf{e}_j[1, 0]\varepsilon + \cdots + \mathbf{e}_j[r, 0]\varepsilon^k + o(\varepsilon^k).
\tag{5.1.37}
$$

Recall that $e_{ij}[0, 0] = p_{ij}^{(0)}$, $i, j \neq 0$. It will be convenient in the sequel to set $\mathbf{e}_j[0, 0] = \mathbf{p}_j^{(0)}$, $j \neq 0$.

Now, let us recall that the absorption probabilities satisfy

$$
_j f_{i0}^{(\varepsilon)} = \mathsf{P}_i\{v_0^{(\varepsilon)} < v_j^{(\varepsilon)}\}, \quad i, j \neq 0,
$$

and, for the hitting probabilities, we have

$$
_0 f_{ij}^{(\varepsilon)} = \mathsf{P}_i\{v_j^{(\varepsilon)} < v_0^{(\varepsilon)}\}, \quad i, j \neq 0.
$$

For $j \neq 0$, consider the vectors

$$_j\mathbf{f}_0^{(\varepsilon)} = \begin{bmatrix} _jf_{10}^{(\varepsilon)} \\ \vdots \\ _jf_{N0}^{(\varepsilon)} \end{bmatrix}, \qquad _0\mathbf{f}_j^{(\varepsilon)} = \begin{bmatrix} _0f_{1j}^{(\varepsilon)} \\ \vdots \\ _0f_{Nj}^{(\varepsilon)} \end{bmatrix}.$$

As was shown in Lemma 4.2.1, conditions $\mathbf{A}_2$, $\mathbf{T}_1$ **(a)**, and $\mathbf{E}_2$ imply for every $\varepsilon \leq \varepsilon_0$ and $j \in X_1^{(0)}$ that the absorption probabilities uniquely solve the following system of linear equations:

$$_j\mathbf{f}_0^{(\varepsilon)} = \mathbf{p}_0^{(\varepsilon)} + _j\mathbf{P}^{(\varepsilon)}{}_j\mathbf{f}_0^{(\varepsilon)}, \tag{5.1.38}$$

and that for every $\varepsilon \leq \varepsilon_0$ and $j \in X_1^{(0)}$, the hitting probabilities make a unique solution the system

$$_0\mathbf{f}_j^{(\varepsilon)} = \mathbf{p}_j^{(\varepsilon)} + _j\mathbf{P}^{(\varepsilon)}{}_0\mathbf{f}_j^{(\varepsilon)}. \tag{5.1.39}$$

The following lemma is a direct corollary of Lemma 5.1.3 applied to the absorption probability vectors $_j\mathbf{f}_0^{(\varepsilon)}$, $j \in X_1^{(0)}$.

Lemma 5.1.5. *Let conditions $\mathbf{A}_2$, $\mathbf{E}_2$, and $\mathbf{P}_{18}^{(k)}$, hold. Then*

(i) *For every $j \in X_1^{(0)}$, the vector $_j\mathbf{f}_0^{(\varepsilon)}$ has the following asymptotic expansion:*

$$_j\mathbf{f}_0^{(\varepsilon)} = _j\mathbf{f}_0^{(0)} + _j\mathbf{b}_0[1,0]\varepsilon + \cdots + _j\mathbf{b}_0[k,0]\varepsilon^k + o(\varepsilon^k), \tag{5.1.40}$$

where $_j\mathbf{b}_0[r,0]$, $r = 1, \ldots, k$, are finite vectors.

(ii) *The vector coefficients $_j\mathbf{b}_0[r,0]$ are given by the recurrence formulas $_j\mathbf{b}_0[0,0] = {}_j\mathbf{f}_0^{(0)} = {}_j\mathbf{U}[0]\mathbf{p}_0^{(0)} = [\mathbf{I} - {}_j\mathbf{P}^{(0)}]^{-1}\mathbf{p}_0^{(0)}$ and, for $r = 0, \ldots, k$,*

$$_j\mathbf{b}_0[r,0] = {}_j\mathbf{U}[0]\left(\mathbf{e}_0[r,0] + \sum_{q=1}^{r} {}_j\mathbf{E}[q,0]{}_j\mathbf{b}_0[r-q,0]\right). \tag{5.1.41}$$

(iii) *The vector coefficients $_j\mathbf{b}_0[r,0]$ can also be calculated by the following formulas for $r = 0, \ldots, k$:*

$$_j\mathbf{b}_0[r,0] = \sum_{p=0}^{r} {}_j\mathbf{U}[p]\,\mathbf{e}_0[r-p,0]. \tag{5.1.42}$$

Proof. Lemma 5.1.3 can be applied to the system of linear equations (5.1.38) for every $j \in X_1^{(0)}$. In this case, the vector $\mathbf{p}_0^{(\varepsilon)}$ plays the role of the vector $\mathbf{y}^{(\varepsilon)}$. The non-negativity condition $\mathbf{S}_1$ obviously holds. Also, the asymptotic relation (5.1.37) implies that condition $\mathbf{P}_{19}^{(k)}$ holds for $j = 0$. By applying Lemma 5.1.3 to the system of linear equations (5.1.38), we finish the proof of Lemma 5.1.5. $\square$

The following lemma is a direct corollary of Lemma 5.1.3 if applied to the hitting probability vectors $_0\mathbf{f}_j^{(\varepsilon)}$, $j \in X_1^{(0)}$.

Lemma 5.1.6. *Let conditions* $\mathbf{A}_2$, $\mathbf{E}_2$, *and* $\mathbf{P}_{18}^{(k)}$ *hold. Then*

(i) *The vector* $_0\mathbf{f}_j^{(\varepsilon)}$ *has the following asymptotic expansion for every* $j \in X_1^{(0)}$:

$$_0\mathbf{f}_j^{(\varepsilon)} = {}_0\mathbf{f}_j^{(0)} + {}_0\mathbf{b}_j[1,0]\varepsilon + \cdots + {}_0\mathbf{b}_j[k,0]\varepsilon^k + \mathbf{o}(\varepsilon^k), \tag{5.1.43}$$

where $_0\mathbf{b}_j[r,0]$, $r = 1,\ldots,k$, *are finite vectors.*

(ii) *The vector coefficients* $_0\mathbf{b}_j[r,0]$ *are given by the recurrence formulas* $_0\mathbf{b}_j[0,0] = {}_0\mathbf{f}_j^{(0)} = {}_j\mathbf{U}[0]\mathbf{p}_j^{(0)} = [\mathbf{I} - {}_j\mathbf{P}^{(0)}]^{-1}\mathbf{p}_j^{(0)}$, *for* $r = 0,\ldots,k$,

$$_0\mathbf{b}_j[r,0] = {}_j\mathbf{U}[0]\left(\mathbf{e}_j[r,0] + \sum_{q=1}^{r} {}_j\mathbf{E}[q,0]_0\mathbf{b}_j[r-q,0]\right). \tag{5.1.44}$$

(iii) *The vector coefficients* $_0\mathbf{b}_j[r,0]$ *can also be calculated for* $r = 0,\ldots,k$ *by the following formulas:*

$$_0\mathbf{b}_j[r,0] = \sum_{p=0}^{r} {}_j\mathbf{U}[p]\,\mathbf{e}_j[r-p,0]. \tag{5.1.45}$$

Proof. Apply Lemma 5.1.3 to the system of linear equations (5.1.39) for every $j \in X_1^{(0)}$. In this case, the vector $\mathbf{p}_j^{(\varepsilon)}$ is regarded as the vector $\mathbf{y}^{(\varepsilon)}$. The non-negativity condition $\mathbf{S}_1$ obviously holds. Also, the asymptotic relations (5.1.37) imply that condition $\mathbf{P}_{19}^{(k)}$ holds for $j \in X_1^{(0)}$. Applying Lemma 5.1.3 to the system of linear equations (5.1.39) yields the claims of Lemma 5.1.6.   □

5.1.5   Pivotal properties of asymptotic expansions for absorption and hitting probabilities

We use the following notations for the vectors $_j\mathbf{b}_0[r,0]$ and $_0\mathbf{b}_j[r,0]$, $r = 0,\ldots,k$:

$$_j\mathbf{b}_0[r,0] = \begin{bmatrix} {}_jb_{10}[r,0] \\ \vdots \\ {}_jb_{N0}[r,0] \end{bmatrix}, \qquad _0\mathbf{b}_j[r,0] = \begin{bmatrix} {}_0b_{1j}[r,0] \\ \vdots \\ {}_0b_{Nj}[r,0] \end{bmatrix}.$$

Explicit expressions for components of the vectors $_j\mathbf{b}_0[r,0]$, $r = 0,\ldots,k$, and $_0\mathbf{b}_j[r,0]$, $r = 0,\ldots,k$, can be derived from the recurrence formulas (5.1.41) and (5.1.42) or (5.1.44) and (5.1.45), respectively.

Note that if $k = 0$, Lemmas 5.1.6 and 5.1.5 reduce to Lemma 4.2.1, i.e., it yields convergence of the absorption and hitting probabilities. Thus, $_jb_{i0}[0, 0] = {}_jf_{i0}^{(0)}$ and $_0b_{ij}[0, 0] = {}_0f_{ij}^{(0)}$ for $i \neq 0$, $j \in X_1^{(0)}$.

We are interested to clarify pivotal properties of asymptotic expansions for the cyclic absorption probabilities $_jf_{j0}^{(\varepsilon)}$, $j \in X_1^{(0)}$, and the cyclic hitting probabilities $_0f_{ij}^{(\varepsilon)}$, $i \neq 0$, $j \in X_1^{(0)}$.

For $j \in X$, let

$$
Y_{j,r} = \begin{cases} \{i \in X_0 : e_{ij}[n, 0] = 0, n = 0, \ldots, r - 1, e_{ij}[r, 0] \neq 0\} & \text{for } r = 1, \ldots, k, \\ \{i \in X_0 : e_{ij}[n, 0] = 0, n = 0, \ldots, k\} & \text{for } r = k + 1. \end{cases}
$$

The set $Y_{j,r}$ is regarded as the set Y_r if the vectors $\mathbf{p}_j^{(0)}$ are taken to be $\mathbf{y}^{(\varepsilon)}$.

By the definition, **(a)** some of the sets $Y_{j,r}$, $r = 0, \ldots, k + 1$, could be empty, **(b)** $Y_{j,r'} \cap Y_{j,r''} = \varnothing$, $r' \neq r''$, but **(c)** $\bigcup_{r=0}^{k+1} Y_{j,r} = X_0$.

For $j \in X_1^{(0)}$, consider the sets

$$
W_{j,r} = \bigcup_{r'+r''=r} (Y_{j,r'} \cap Z_{r''}), \quad r = 0, \ldots, k,
$$

$$
W_{j,k+1} = \bigcup_{r'+r'' \geq k+1} (Y_{j,r'} \cap Z_{r''}).
$$

The case where condition $\mathbf{E}_2'$ holds is simple. Indeed $Z_0 = X_1^{(0)} = X_0$, and, therefore, $Z_1, \ldots, Z_{k+1} = \varnothing$. Thus, for $j \in X_1^{(0)}$,

$$
W_{j,r} = Y_{j,r}, \quad r = 0, \ldots, k + 1. \tag{5.1.46}
$$

For integer $0 \leq h \leq k$, introduce two conditions,

$\mathbf{O}_3^{(h)}$: $W_{0,r} = \varnothing$, for $r \leq h$;

and

$\mathbf{O}_4^{(h)}$: $W_{0,r} = \varnothing$, for $r < h$, but $W_{0,h} \neq \varnothing$.

The following lemma gives a complete description of pivotal properties of asymptotic expansions for the cyclic absorption probabilities $_jf_{j0}^{(\varepsilon)}$, $j \in X_1^{(0)}$.

Lemma 5.1.7. *Let conditions* $\mathbf{A}_2$, $\mathbf{E}_2$, *and* $\mathbf{P}_{18}^{(k)}$ *hold. Then*

    (i) *If condition* $\mathbf{O}_3^{(h)}$ *holds for some* $0 \leq h \leq k$, *then* $_jb_{j0}[n] = 0$, $n = 0, \ldots, h$, $j \in X_1^{(0)}$.

(ii) *If condition* $\mathbf{O}_4^{(h)}$ *holds for some* $0 \le h \le k$, *then* $_jb_{j0}[n] = 0$, $n = 0,\ldots,h-1$, $j \in X_1^{(0)}$, *but* $_jb_{j0}[h] > 0$, $j \in X_1^{(0)}$.

Proof. We will again use Lemma 5.1.4 with the vector $\mathbf{p}_0^{(\varepsilon)}$ replacing the vectors $\mathbf{y}^{(\varepsilon)}$ and the absorption probabilities $_jf_{j0}^{(\varepsilon)}$ replacing the cyclic components $x_{jj}^{(\varepsilon)}$. The non-negativity condition $\mathbf{S}_1$ obviously holds. Also, the asymptotic relation (5.1.37) implies that condition $\mathbf{P}_{19}^{(k)}$ holds for the vector $\mathbf{p}_0^{(\varepsilon)}$. Conditions $\mathbf{O}_1^{(h)}$ and $\mathbf{O}_2^{(h)}$ take, respectively, the form of conditions $\mathbf{O}_3^{(h)}$ and $\mathbf{O}_4^{(h)}$. Thus, Lemma 5.1.4 just takes the form of Lemma 5.1.7. $\qquad\square$

The following simple lemma gives a useful information about pivotal properties of the non-cyclic hitting probabilities $_0f_{ij}^{(\varepsilon)}$, $i \ne 0$, $j \in X_1^{(0)}$.

Lemma 5.1.8. *Let conditions* $\mathbf{A}_2$, $\mathbf{E}_2$, *and* $\mathbf{P}_{18}^{(k)}$ *hold. Then the asymptotic expansions for the hitting probabilities* $_0f_{ij}^{(\varepsilon)}$, $i \ne 0$, $j \in X_1^{(0)}$, *are all 0-pivotal, i.e., the corresponding zero coefficients are positive,* $_0f_{ij}^{(0)} > 0$ *for all* $i \ne 0$, $j \in X_1^{(0)}$.

Proof. The statement of the lemma immediately follows from the definition of the set of recurrent-without-absorption states $X_1^{(0)}$, according to which we have $_0f_{ij}^{(0)} > 0$ for all $i \ne 0$, $j \in X_1^{(0)}$. $\qquad\square$

It is useful to recall that if conditions $\mathbf{A}_2$, $\mathbf{T}_1$, **(a)**, and $\mathbf{E}_2$ are satisfied, then there exists $0 < \varepsilon_1 \le \varepsilon_0$ such that $X_1^{(0)} \subseteq X_1^{(\varepsilon)}$ for $\varepsilon \le \varepsilon_1$. For $\varepsilon \le \varepsilon_1$, the hitting and absorption probabilities are connected by the following relation:

$$_0f_{ij}^{(\varepsilon)} = 1 - {}_jf_{i0}^{(\varepsilon)}, \quad i \ne 0, \quad j \in X_1^{(0)}. \tag{5.1.47}$$

Thus, the asymptotic expansion for the vectors of hitting and absorption probabilities satisfy the relation

$$
\begin{aligned}
_0\mathbf{f}_j^{(\varepsilon)} &= {}_0\mathbf{f}_j^{(0)} + {}_0\mathbf{b}_j[1,0]\varepsilon + \cdots + {}_0\mathbf{b}_j[k,0]\varepsilon^k + o(\varepsilon^k) \\
&= \mathbf{1} - {}_j\mathbf{f}_0^{(0)} - {}_j\mathbf{b}_0[1,0]\varepsilon - \cdots - {}_j\mathbf{b}_0[k,0]\varepsilon^k + o(\varepsilon^k),
\end{aligned}
\tag{5.1.48}
$$

where $\mathbf{1}$ is a column vector with N components all equal 1.

This means that the coefficients in the corresponding asymptotic expansions for components of these vectors satisfy for $i \ne 0$, $j \in X_1^{(0)}$ the following:

$$_0b_{ij}[r,0] = \begin{cases} 1 - {}_jb_{i0}[0,0] & \text{for } r = 0, \\ -{}_jb_{i0}[r,0] & \text{for } r = 1,\ldots,k. \end{cases} \tag{5.1.49}$$

This formula gives a much more detailed description of "higher order" pivotal properties of the hitting probabilities.

Lemma 5.1.9. *Let conditions* $\mathbf{A}_2$, $\mathbf{E}_2$, *and* $\mathbf{P}_{18}^{(k)}$ *hold. Then*

(i) *If condition* $\mathbf{O}_3^{(0)}$ *holds, then* ${}_0 b_{jj}[0, 0] = 1 - {}_j b_{j0}[0, 0] = 1$, $j \in X_1^{(0)}$.

(ii) *If condition* $\mathbf{O}_3^{(h)}$ *holds for some* $1 \leq h \leq k$, *then* ${}_0 b_{jj}[0, 0] = 1$, $j \in X_1^{(0)}$ *and*
$${}_0 b_{jj}[n, 0] = 0, \; 1 \leq n \leq h, \; j \in X_1^{(0)}.$$

(iii) *If condition* $\mathbf{O}_4^{(0)}$ *holds, then* ${}_0 b_{jj}[0, 0] = {}_0 f_{jj}^{(0)} = 1 - {}_j b_{j0}[0, 0] > 0$, $j \in X_1^{(0)}$.

(iv) *If condition* $\mathbf{O}_4^{(h)}$ *holds for some* $1 \leq h \leq k$, *then* ${}_0 b_{jj}[0, 0] = 1$, $j \in X_1^{(0)}$ *and*
${}_0 b_{jj}[n, 0] = 0, \; 1 \leq n \leq h - 1, \; j \in X_1^{(0)}$ *but* ${}_0 b_{jj}[h, 0] = -{}_j b_{j0}[h, 0] < 0$,
$j \in X_1^{(0)}$.

Proof. The lemma is a direct corollary of relation (5.1.49) and Lemmas 5.1.7 and 5.1.8. $\qquad\Box$

5.2 Asymptotic expansions for power moments of hitting times

In this section we describe a recursive algorithm used in the asymptotic analysis of power moments of the absorption and hitting times for nonlinearly perturbed semi-Markov processes. The algorithm is important in its own right and makes an essential part of the proof of our main results concerning semi-Markov processes.

5.2.1 Power moment perturbation conditions for transition times

Consider the power moments of transition times, for $n = 0, 1, \ldots, k$, $i \neq 0$, $j \in X$,

$$p_{ij}^{(\varepsilon)}[n] = \int_0^\infty s^n Q_{ij}^{(\varepsilon)}(ds) = m_{ij}^{(\varepsilon)}[n] p_{ij}^{(\varepsilon)}, \quad m_{ij}^{(\varepsilon)}[n] = \int_0^\infty s^n F_{ij}^{(\varepsilon)}(ds).$$

Note that, by the definition, $m_{ij}^{(\varepsilon)}[0] = 1$, $i, j \neq 0$.

Note also that the power moments defined above are connected by the following formulas, for $n = 0, 1, \ldots, k$, $i \neq 0$, $j \in X$,

$$p_{ij}^{(\varepsilon)}[n] = p_{ij}^{(\varepsilon)} m_{ij}^{(\varepsilon)}[n].$$

Also note that $m_{ij}^{(\varepsilon)}[1] = m_{ij}^{(\varepsilon)}$, $i, j \neq 0$, are expectations for the transition times.

We further assume that conditions $\mathbf{A}_2$ and $\mathbf{E}_2$ hold, and also that the following condition holds:

$\mathbf{M}_{13}^{(k)}$: $p_{ij}^{(\varepsilon)}[n] \to p_{ij}^{(0)}[n] < \infty$ as $\varepsilon \to 0$, for $n = 0, \ldots, k$, $i, j \neq 0$.

The following condition and condition $\mathbf{T_1}$ obviously imply condition $\mathbf{M}_{13}^{(k)}$ to hold:

$\mathbf{M}_{13}^{'(k)}$: $m_{ij}^{(\varepsilon)}[n] \to m_{ij}^{(0)}[n] < \infty$ as $\varepsilon \to 0$, for $n = 0, \ldots, k, i, j \neq 0$.

Condition $\mathbf{M}_{13}^{(k)}$ guarantees that there exists $0 < \varepsilon_2 \leq \varepsilon_1$ such that, for $i, j \neq 0$,

$$\sup_{\varepsilon \leq \varepsilon_2} \max_{i,j \neq 0} p_{ij}^{(\varepsilon)}[k] = M_k < \infty. \tag{5.2.1}$$

By the definition, $p_{ij}^{(\varepsilon)}[0] = p_{ij}^{(\varepsilon)}$, $i, j \neq 0$.

We assume that the following perturbation condition, which is more general than $\mathbf{P}_{18}^{(k)}$, to hold:

$\mathbf{P}_{20}^{(k)}$: $p_{ij}^{(\varepsilon)}[n] = p_{ij}^{(0)}[n] + e_{ij}[1, n]\varepsilon + \cdots + e_{ij}[k - n, n]\varepsilon^{k-n} + o(\varepsilon^{k-n})$ for $n = 0, \ldots, k$, $i, j \neq 0$, where $|e_{ij}[r, n]| < \infty$, $r = 1, \ldots, k - n$, $n = 0, \ldots, k$, $i, j \neq 0$.

It is convenient to define $e_{ij}[0, n] = p_{ij}^{(0)}[n]$ for $n = 0, \ldots, k$, $i, j \neq 0$.

Note that condition $\mathbf{P}_{20}^{(k)}$ implies condition $\mathbf{M}_{13}^{(k)}$.

It is useful to note that moments $p_{i0}^{(\varepsilon)}[n]$ are not involved in condition $\mathbf{P}_{20}^{(k)}$. It is because of these moments do not penetrate formulas for power moments of hitting without absorption times.

In some cases, it can be more convenient to use the perturbation condition $\mathbf{P}_{18}^{(k)}$ for the transition probabilities $p_{ij}^{(\varepsilon)}$, $i, j \neq 0$, and, additionally, the following perturbation condition imposed on the moments of transition times, $m_{ij}^{(\varepsilon)}[n]$, $n = 1, \ldots, k$, $i, j \neq 0$:

$\mathbf{P}_{21}^{(k)}$: $m_{ij}^{(\varepsilon)}[n] = m_{ij}^{(0)}[n] + v_{ij}[1, n]\varepsilon + \cdots + v_{ij}[k - n, n]\varepsilon^{k-n} + o(\varepsilon^{k-n})$ for $n = 1, \ldots, k$, $i, j \neq 0$ where (a) $m_{ij}^{(0)}[n] < \infty$, $n = 1, \ldots, k$, $i, j \neq 0$, and (b) $|v_{ij}[r, n]| < \infty$, $r = 1, \ldots, k - n$, $n = 1, \ldots, k$, $i, j \neq 0$.

Define $v_{ij}[0, n] = m_{ij}^{(0)}[n]$ for $i, j \neq 0$, $n = 1, \ldots, k$.

For $k = 0$, condition $\mathbf{P}_{21}^{(0)}$ should be interpreted as an "empty" condition that holds automatically.

If conditions $\mathbf{P}_{18}^{(k)}$ and $\mathbf{P}_{21}^{(k)}$ are satisfied, then conditions $\mathbf{M}_{13}^{(k)}$ and $\mathbf{P}_{21}^{(k)}$ also hold. Moreover, in such a case, the corresponding asymptotic expansions are connected with the asymptotic expansion in condition $\mathbf{P}_{20}^{(k)}$ by the following formulas for $i, j \neq 0$:

$$e_{ij}[r, n] = \begin{cases} e_{ij}[r, 0] & \text{if } r = 0, \ldots, k, n = 0, \\ \sum_{m=0}^{r} e_{ij}[m, 0]v_{ij}[r - m, n] & \text{if } r = 0, \ldots, k - n, n = 1, \ldots, k. \end{cases} \tag{5.2.2}$$

We will rewrite condition $\mathbf{P}_{20}^{(k)}$ in a matrix form.

For $n = 0, 1, \ldots, k$, introduce the matrices

$$\mathbf{P}^{(\varepsilon)}[n] = \begin{bmatrix} p_{11}^{(\varepsilon)}[n] & \cdots & p_{1N}^{(\varepsilon)}[n] \\ \vdots & \vdots & \vdots \\ p_{N1}^{(\varepsilon)}[n] & \cdots & p_{NN}^{(\varepsilon)}[n] \end{bmatrix},$$

and, for $r = 0, \ldots, k - n$, $n = 1, \ldots, k$,

$$\mathbf{E}[r, n] = \begin{bmatrix} e_{11}[r, n] & \cdots & e_{1N}[r, n] \\ \vdots & \vdots & \vdots \\ e_{N1}[r, n] & \cdots & e_{NN}[r, n] \end{bmatrix}.$$

Condition $\mathbf{P}_{20}^{(k)}$ can be rewritten in the following equivalent matrix form:

$\mathbf{P}_{20}^{(k)}$: $\mathbf{P}^{(\varepsilon)}[n] = \mathbf{P}^{(0)}[n] + \mathbf{E}[1, n]\varepsilon + \cdots + \mathbf{E}[k - n, n]\varepsilon^k + \mathbf{o}(\varepsilon^{k-n})$, where $\mathbf{P}^{(0)}[n]$ and $\mathbf{E}[r, n]$, $r = 1, \ldots, k - n$, $n = 1, \ldots, k$ are finite matrices.

For $n = 0, \ldots, k$, $j \neq 0$, consider the matrices

$$_j\mathbf{P}^{(\varepsilon)}[n] = \begin{bmatrix} p_{11}^{(\varepsilon)}[n] & \cdots & p_{1(j-1)}^{(\varepsilon)}[n] & 0 & p_{1(j+1)}^{(\varepsilon)}[n] & \cdots & p_{1N}^{(\varepsilon)}[n] \\ \vdots & & \vdots & \vdots & \vdots & & \vdots \\ p_{N1}^{(\varepsilon)}[n] & \cdots & p_{N(j-1)}^{(\varepsilon)}[n] & 0 & p_{N(j+1)}^{(\varepsilon)}[n] & \cdots & p_{NN}^{(\varepsilon)}[n] \end{bmatrix},$$

and also, for $r = 0, \ldots, k - n$, $n = 1, \ldots, k$, and $j \neq 0$,

$$_j\mathbf{E}[r, n] = \begin{bmatrix} e_{11}[r, n] & \cdots & e_{1(j-1)}[r, n] & 0 & e_{1(j+1)}[r, n] & \cdots & e_{1N}[r, n] \\ \vdots & & \vdots & \vdots & \vdots & & \vdots \\ e_{N1}[r, n] & \cdots & e_{N(j-1)}[r, n] & 0 & e_{N(j+1)}[r, n] & \cdots & e_{NN}[r, n] \end{bmatrix}.$$

If $j \neq 0$, condition $\mathbf{P}_{20}^{(k)}$ gives the following matrix asymptotic expansions for $n = 0, \ldots, k$:

$$_j\mathbf{P}^{(\varepsilon)}[n] = {}_j\mathbf{P}^{(0)}[n] + {}_j\mathbf{E}[1, n]\varepsilon + \cdots + {}_j\mathbf{E}[k - n, n]\varepsilon^{k-n} + \mathbf{o}(\varepsilon^{k-n}). \qquad (5.2.3)$$

It is convenient to define $e_{ij}[0, n] = p_{ij}^{(0)}[n]$, $n = 0, \ldots, k$, $i, j \neq 0$, and the matrices $\mathbf{E}[0, n] = \mathbf{P}^{(0)}[n]$, $_j\mathbf{E}[0, n] = {}_j\mathbf{P}^{(0)}[n]$, $n = 0, \ldots, k$, $j \neq 0$.

For $r = k - n$, $n = 0, \ldots, k$, and $j \neq 0$, consider the vectors

$$\mathbf{p}_j^{(\varepsilon)}[n] = \begin{bmatrix} p_{1j}^{(\varepsilon)}[n] \\ \vdots \\ p_{Nj}^{(\varepsilon)}[n] \end{bmatrix}, \qquad \mathbf{e}_j[r, n] = \begin{bmatrix} e_{1j}[r, n] \\ \vdots \\ e_{Nj}[r, n] \end{bmatrix}.$$

Condition $\mathbf{P}_{20}^{(k)}$ gives the following vector asymptotic expansion for $n = 0,\ldots,k$ and $j \neq 0$:

$$\mathbf{p}_j^{(\varepsilon)}[n] = \mathbf{p}_j^{(0)}[n] + \mathbf{e}_j[1,n]\varepsilon + \cdots \mathbf{e}_j[k-n,n]\varepsilon^{k-n} + o(\varepsilon^{k-n}). \qquad (5.2.4)$$

Recall that $e_{ij}[0,n] = p_{ij}^{(0)}[n]$, $n = 0,\ldots,k$, $i,j \neq 0$. In the sequel, it will be convenient to define the vectors $\mathbf{e}_j[0,n] = \mathbf{p}_j^{(0)}[n]$, $n = 0,\ldots,k$, $j \neq 0$.

5.2.2   Recursive systems of linear equations for power moments of hitting times

Let us also introduce power moments for the hitting times,

$$_0M_{ij}^{(\varepsilon)}[n] = \mathsf{E}_i(\mu_j^{(\varepsilon)})^n \chi(v_j^{(\varepsilon)} \leq v_0^{(\varepsilon)}), \quad n = 0,1,\ldots,k, \quad i,j \neq 0.$$

By the definition we have

$$_0M_{ij}^{(\varepsilon)}[0] = \mathsf{P}_i\{v_j^{(\varepsilon)} \leq v_0^{(\varepsilon)}\} = {}_0f_{ij}^{(\varepsilon)}, \quad i,j \neq 0.$$

Recall again that if conditions $\mathbf{A}_2$, $\mathbf{T}_1$, (a), and $\mathbf{E}_2$ hold, then there exists $\varepsilon_1 > 0$ such that $X_1^{(0)} \subseteq X_1^{(\varepsilon)}$ for $\varepsilon \leq \varepsilon_1$. Therefore, (a) the hitting probabilities satisfy $_0f_{ij}^{(\varepsilon)} > 0$, $i \neq 0$, $j \in X_1^{(0)}$. Note also that condition $\mathbf{M}_{13}^{(k)}$ implies that (c) there exists $0 < \varepsilon_2 \leq \varepsilon_1$ such that $p_{ij}^{(\varepsilon)}[k] \leq M_k < \infty$ for $\varepsilon \leq \varepsilon_2$ and $i,j \neq 0$.

As was shown in Subsection 4.2.3, (c) there exist $0 < L < 1$ and $0 < \varepsilon_3 \leq \varepsilon_2$ such that $\max_{l \neq 0} \mathsf{P}_l\{v_j^{(\varepsilon)} \wedge v_0^{(\varepsilon)} > n\} \leq L^{[n/m]}$ for $n \geq 1$ and $\varepsilon \leq \varepsilon_3$.

These relations imply that for $\varepsilon \leq \varepsilon_3$

$$\sup_{\varepsilon \leq \varepsilon_3} {}_0M_{ij}^{(\varepsilon)}[k] < \infty, \quad i \neq 0, \quad j \in X_1^{(0)}. \qquad (5.2.5)$$

Indeed, using relation (4.2.17), we get for $\varepsilon \leq \varepsilon_3$,

$$_0M_{ij}^{(\varepsilon)}[k] \leq \sum_{n=1}^{\infty} n^k \sum_{i_0=i,i_1,\ldots,i_{n-1}\neq 0,j,\,i_n=j} \left(\frac{1}{n}\sum_{r=1}^{n} m_{i_{r-1}i_r}^{(\varepsilon)}[k]\right) \prod_{r=1}^{n} p_{i_{r-1}i_r}^{(\varepsilon)}$$

$$\leq \sum_{n=1}^{\infty} n^k M_k \max_{l \neq 0} \mathsf{P}_l\left\{v_j^{(\varepsilon)} \wedge v_0^{(\varepsilon)} > \left[\frac{n-1}{2} - 1\right]\right\}$$

$$\leq M_k \sum_{n=1}^{\infty} n^k L^{[\frac{n-1}{2}-1]/m} < \infty. \qquad (5.2.6)$$

The moments $M_{ij}^{(\varepsilon)}[n]$, $i \neq 0$, satisfy the following system of linear equations for every $\varepsilon \leq \varepsilon_3$, $j \in X_1^{(0)}$, $n = 0,1,\ldots,k$:

$$_0M_{ij}^{(\varepsilon)}[n] = q_{ij}^{(\varepsilon)}[n] + \sum_{l \neq 0,j} p_{il}^{(\varepsilon)} {}_0M_{lj}^{(\varepsilon)}[n], \quad i \neq 0, \qquad (5.2.7)$$

where

$$q_{ij}^{(\varepsilon)}[n] = p_{ij}^{(\varepsilon)}[n] + \sum_{m=1}^{n} C_n^m \sum_{l \neq 0, j} p_{il}^{(\varepsilon)}[m]\, _0M_{lj}^{(\varepsilon)}[n-m], \quad i \neq 0, \qquad (5.2.8)$$

and

$$C_n^m = \frac{n!}{m!(n-m)!}, \quad 0 \le m \le n < \infty, \quad 0! = 1.$$

These systems can be rewritten in a matrix form. To do this, introduce for $j \in X_1^{(0)}$ and $n = 0, \ldots, k$, the vectors

$$_0\mathbf{m}_j^{(\varepsilon)}[n] = \begin{bmatrix} _0M_{1j}^{(\varepsilon)}[n] \\ \vdots \\ _0M_{Nj}^{(\varepsilon)}[n] \end{bmatrix}, \qquad \mathbf{q}_j^{(\varepsilon)}[n] = \begin{bmatrix} q_{1j}^{(\varepsilon)}[n] \\ \vdots \\ q_{Nj}^{(\varepsilon)}[n] \end{bmatrix}.$$

Then, the system of linear equations (5.2.7) can be rewritten in the following form for every $j \in X_1^{(0)}$ and $n = 0, 1, \ldots, k$:

$$_0\mathbf{m}_j^{(\varepsilon)}[n] = \mathbf{q}_j^{(\varepsilon)}[n] + \,_j\mathbf{P}^{(\varepsilon)} \,_0\mathbf{m}_j^{(\varepsilon)}[n]. \qquad (5.2.9)$$

These are systems of hitting type considered in Subsection 5.1.3, with the same coefficient matrix $_j\mathbf{P}^{(\varepsilon)}$.

As was pointed out in this subsection, conditions $\mathbf{A}_2$, $\mathbf{E}_2$, $\mathbf{T}_1$ **(a)**, and $\mathbf{S}_1$ imply that there exists $\varepsilon_0 > 0$ such that for every $\varepsilon \le \varepsilon_0$ any hitting type system of linear equation with the coefficient matrix $_j\mathbf{P}^{(\varepsilon)}$ has a unique solution.

Condition $\mathbf{M}_{13}^{(k)}$ should be added as to provide finiteness of the k-order moments of the transition times. Obviously, the vectors $\mathbf{q}_j^{(\varepsilon)}[n]$, $n = 0, \ldots, k$, $j \in X_1^{(0)}$, have non-negative components.

A solution of system (5.2.9) has the following form for $\varepsilon \le \varepsilon_3$, $j \in X_1^{(0)}$, and $n = 0, \ldots, k$:

$$_0\mathbf{m}_j^{(\varepsilon)}[n] = [\mathbf{I} - \,_j\mathbf{P}^{(\varepsilon)}]^{-1} \mathbf{q}_j^{(\varepsilon)}[n]. \qquad (5.2.10)$$

As was mentioned above, systems (5.2.9) have the same coefficient matrix $_j\mathbf{P}^{(\varepsilon)}$ for a given $j \in X_1^{(0)}$ but different inhomogeneous terms $\mathbf{q}_j^{(\varepsilon)}[n]$ for $n = 0, 1, \ldots, k$. These systems should be solved recursively.

First, system (5.2.7) should be solved for $n = 0$. Note that it is actually a system of linear equations for the hitting probabilities $_0f_{ij}^{(\varepsilon)} = \,_0M_{ij}^{(\varepsilon)}$, $i \neq 0$.

Second, system (5.2.7) should be solved for $n = 1$. Note that the expressions for the inhomogeneous terms $q_{ij}^{(\varepsilon)}[1] = p_{ij}^{(\varepsilon)}[1] + \sum_{l \neq 0, j} p_{il}^{(\varepsilon)}[1] M_{lj}^{(\varepsilon)}[0]$, $i \neq 0$, in (5.2.8) include the solutions $M_{ij}^{(\varepsilon)}[0]$, $i \neq 0$, of systems (5.2.7) for $n = 0$.

This recursive procedure should be repeated for $n = 1, \ldots, k$. The expressions for the inhomogeneous terms $q_{ij}^{(\varepsilon)}[n]$ given in (5.2.8) include the solutions $M_{ij}^{(\varepsilon)}[m]$, $i \neq 0$, of systems (5.2.7) for $m = 0, 1, \ldots, n - 1$.

Formula (5.2.8) can be rewritten in the following vector form for every $j \in X_1^{(0)}$ and $n = 0, 1, \ldots, k$:

$$\mathbf{q}_j^{(\varepsilon)}[n] = \mathbf{p}_j^{(\varepsilon)}[n] + \sum_{m=1}^{n} C_n^m {}_j\mathbf{P}^{(\varepsilon)}[m]\, {}_0\mathbf{m}_j^{(\varepsilon)}[n - m]. \tag{5.2.11}$$

Using formula (5.2.11) we can rewrite formula (5.2.10) in a form that uses the recurrence procedure for calculating solutions of the system of linear equations (5.2.7) for every $j \in X_1^{(0)}$ and $n = 0, 1, \ldots, k$,

$$\begin{aligned}
{}_0\mathbf{m}_j^{(\varepsilon)}[n] &= [\mathbf{I} - {}_j\mathbf{P}^{(\varepsilon)}]^{-1}\mathbf{q}_j^{(\varepsilon)}[n] \\
&= [\mathbf{I} - {}_j\mathbf{P}^{(\varepsilon)}]^{-1}\mathbf{p}_j^{(\varepsilon)}[n] \\
&\quad + \sum_{m=1}^{n} C_n^m [\mathbf{I} - {}_j\mathbf{P}^{(\varepsilon)}]^{-1} {}_j\mathbf{P}^{(\varepsilon)}[m]\, {}_0\mathbf{m}_j^{(\varepsilon)}[n - m]. \tag{5.2.12}
\end{aligned}$$

5.2.3   Asymptotic expansions for power moments of hitting times

The following lemma is a corollary of Lemma 5.1.3 applied to the vectors of power moments of the hitting times, ${}_0\mathbf{m}_j^{(\varepsilon)}[n]$, for $j \in X_1^{(0)}$ and $n = 0, 1, \ldots, k$.

Lemma 5.2.1. *Let conditions* $\mathbf{A}_2$, $\mathbf{E}_2$, $\mathbf{M}_{13}^{(k)}$, *and* $\mathbf{P}_{20}^{(k)}$, *hold. Then*

(i) *The vector* ${}_0\mathbf{m}_j^{(\varepsilon)}[n]$ *has the following asymptotic expansion for every* $j \in X_1^{(0)}$ *and* $n = 0, \ldots, k$:

$${}_0\mathbf{m}_j^{(\varepsilon)}[n] = {}_0\mathbf{m}_j^{(0)}[n] + {}_0\mathbf{b}_j[1, n]\varepsilon + \cdots + {}_0\mathbf{b}_j[k - n, n]\varepsilon^k + o(\varepsilon^{k-n}), \tag{5.2.13}$$

where ${}_0\mathbf{b}_j[r, n]$, $r = 1, \ldots, k - n$, $n = 0, \ldots, k$, *are finite vectors.*

(ii) *The vector coefficients* ${}_0\mathbf{b}_j[r, n]$ *are given by the recurrence formulas* ${}_0\mathbf{b}_j[0, 0] = {}_0\mathbf{m}_j^{(0)}[0] = {}_j\mathbf{U}[0]\mathbf{q}_j^{(0)}[0] = [\mathbf{I} - {}_j\mathbf{P}^{(0)}]^{-1}\mathbf{p}_j^{(0)}$ *and for* $r = 0, \ldots, k - n$ *(for given* n*) and sequentially for every* $n = 0, \ldots, k$,

$${}_0\mathbf{b}_j[r, n] = {}_j\mathbf{U}[0]\Big({}_j\mathbf{e}[r, n] + \sum_{m=1}^{n} C_n^m \sum_{q=0}^{r} {}_j\mathbf{E}[q, m]\, {}_0\mathbf{b}_j[r - q, n - m]$$

$$+ \sum_{p=1}^{r} \mathbf{E}_j[p, 0]\, {}_0\mathbf{b}_j[r - p, n]\Big). \tag{5.2.14}$$

(iii) *The vector coefficients $_0\mathbf{b}_j[r, n]$ also can be calculated by the recurrence formulas for $r = 0, \ldots, k - n$ (for given n) and sequentially for every $n = 0, \ldots, k$,*

$$_0\mathbf{b}_j[r, n] = \sum_{q=0}^{r} {}_j\mathbf{U}[q]\mathbf{e}_j[r - q, n]$$

$$+ \sum_{m=1}^{n} C_n^m \sum_{q=0}^{r} \left(\sum_{p=0}^{q} {}_j\mathbf{U}[p]_j\mathbf{E}[q - p, m] \right) {}_0\mathbf{b}_j[r - q, n - m]. \quad (5.2.15)$$

Proof. Fix some $j \in X_1^{(0)}$. The asymptotic expansions (5.2.13) are constructed recursively for the vectors $_0\mathbf{m}_j^{(\varepsilon)}[n]$ where $n = 0, \ldots, k$.

If $n = 0$, then $_0\mathbf{m}_j^{(\varepsilon)}[0] = {}_0\mathbf{f}_j^{(\varepsilon)}$. The asymptotic expansion given for $n = 0$ in condition $\mathbf{P}_{20}^{(k)}$ coincides with the asymptotic expansion given in condition $\mathbf{P}_{18}^{(k)}$. The asymptotic expansion (5.2.13) for the vectors $_0\mathbf{m}_j^{(\varepsilon)}[0]$ and formulas (5.2.14) and (5.2.15) for the corresponding vector coefficients just coincide, respectively, with the asymptotic expansion (5.1.43) for the vectors $_0\mathbf{f}_j^{(\varepsilon)}$ and formulas (5.1.44) and (5.1.45) for the corresponding vector coefficients. Note that this is an $(N, 1)$-matrix $(0, k)$-expansion.

Now we can write an asymptotic expansion for the vectors $\mathbf{q}_j^{(\varepsilon)}[1] = \mathbf{p}_j^{(\varepsilon)}[1] + {}_j\mathbf{P}^{(\varepsilon)}[1] \cdot {}_0\mathbf{m}_j^{(\varepsilon)}[0]$ by using Lemma 8.1.2. This will be an $(N, 1)$-matrix $(0, k - 1)$-expansion. Then Lemma 5.1.3 can be applied to the vectors $_0\mathbf{m}_j^{(\varepsilon)}[1] = [\mathbf{I} - {}_j\mathbf{P}^{(\varepsilon)}]^{-1} \cdot \mathbf{q}_j^{(\varepsilon)}[1]$. The resulted asymptotic expansion is an $(N, 1)$-matrix $(0, k - 1)$-expansion.

The recursive procedure can be repeated for $n = 1, \ldots, k$. At the step n, the corresponding expansion for the vectors $\mathbf{q}_j^{(\varepsilon)}[n]$ given by formula (5.2.11) make an $(N, 1)$-matrix $(0, k - n)$-expansion that takes the following form:

$$\mathbf{q}_j^{(\varepsilon)}[n] = \mathbf{q}_j[0, n] + \mathbf{q}_j[1, n]\varepsilon + \cdots + \mathbf{q}_j[k - n, n]\varepsilon^{k-n} + o(\varepsilon^{k-n}), \quad (5.2.16)$$

where the coefficients $\mathbf{q}_j[r, n]$ for $r = 0, \ldots, k - n$ are given by the formulas

$$\mathbf{q}_j[r, n] = \mathbf{e}_j[r, n] + \sum_{m=1}^{n} C_n^m \sum_{q=0}^{r} {}_j\mathbf{E}[q, m]_0\mathbf{b}_j[r - q, n - m]. \quad (5.2.17)$$

Then, we apply Lemma 5.1.3 to the vectors $_0\mathbf{m}_j^{(\varepsilon)}[n] = [\mathbf{I} - {}_j\mathbf{P}^{(\varepsilon)}]^{-1}\mathbf{q}_j^{(\varepsilon)}[n]$. The obtained asymptotic expansion makes an $(N, 1)$-matrix $(0, k - n)$-expansion given by the asymptotic expansion (5.2.13) and formulas (5.2.14) and (5.2.15) for the corresponding vector coefficients. $\square$

Remark 5.2.1. It should be noted that both formulas (5.2.14) and (5.2.15) for the corresponding vector coefficients in the asymptotic expansions (5.2.13) have a recursive character. However, they differ in their structure. Formulas (5.2.14) are used for calculating the coefficients $_0\mathbf{b}_j[r,n]$ and $_0\mathbf{b}_j[l,m]$, $l = 0,\ldots,r$, $m = 0,\ldots,n-1$. While formulas (5.2.15) give the coefficients $_0\mathbf{b}_j[r,n]$ and $_0\mathbf{b}_j[l,m]$, $l = 0,\ldots,r$, $m = 0,\ldots,n-1$, and also the coefficients $_0\mathbf{b}_j[l,n]$, $l = 0,\ldots,r-1$.

Introducing the integer parameters $\tilde{n} \geq 1$ and $\tilde{k}_n \geq 0$, $n = 0,\ldots,\tilde{n}$, define the vector parameter $\tilde{k} = (\tilde{n},\tilde{k}_0,\ldots,\tilde{k}_{\tilde{n}})$.

Let us now assume that the following perturbation condition, which is more general than $\mathbf{P}_{20}^{(\mathbf{k})}$, is satisfied:

$\mathbf{P}_{22}^{(\tilde{\mathbf{k}})}$: $p_{ij}^{(\varepsilon)}[n] = p_{ij}^{(0)}[n] + e_{ij}[1,n]\varepsilon + \cdots + e_{ij}[\tilde{k}_n,n]\varepsilon^{\tilde{k}_n} + o(\varepsilon^{\tilde{k}_n})$ for $n = 0,\ldots,\tilde{n}$,
  $i,j \neq 0$, where $p_{ij}^{(0)}[n] < \infty$, $n = 0,\ldots,\tilde{n}$, $i,j \neq 0$ and $|e_{ij}[r,n]| < \infty$,
  $r = 1,\ldots,\tilde{k}_n$, $n = 0,\ldots,\tilde{n}$, $i,j \neq 0$.

Condition $\mathbf{P}_{22}^{(\tilde{\mathbf{k}})}$ reduces to condition $\mathbf{P}_{20}^{(\mathbf{k})}$ if $\tilde{n} = k$, $\tilde{k}_n = k - n$, $n = 0,\ldots,k$.

The following lemma generalises Lemma 5.2.1 and reduces to this lemma in the case mentioned above.

Lemma 5.2.2. *Let conditions* $\mathbf{A}_2$, $\mathbf{E}_2$, $\mathbf{M}_{13}^{(\tilde{n})}$, *and* $\mathbf{P}_{22}^{(\tilde{\mathbf{k}})}$ *hold. Then*

(i) *The vector* $_0\mathbf{m}_j^{(\varepsilon)}[n]$, *for every* $j \in X_1^{(0)}$ *and* $n = 0,\ldots,\tilde{n}$, *has the following asymptotic expansion:*

$$_0\mathbf{m}_j^{(\varepsilon)}[n] = {}_0\mathbf{m}_j^{(0)}[n] + {}_0\mathbf{b}_j[1,n]\varepsilon + \cdots + {}_0\mathbf{b}_j[k_n,n]\varepsilon^{k_n} + \mathbf{o}(\varepsilon^{k_n}), \quad (5.2.18)$$

where $k_n = \min(\tilde{k}_0,\ldots,\tilde{k}_n)$, $n = 0,\ldots,\tilde{n}$, *and* $_0\mathbf{b}_j[r,n]$, $r = 1,\ldots,k_n$, $n = 0,\ldots,\tilde{n}$, *are finite vectors.*

(ii) *The vector coefficients* $_0\mathbf{b}_j[r,n]$ *are given by the recurrence formulas* $_0\mathbf{b}_j[0,0] = {}_0\mathbf{m}_j^{(0)}[0] = {}_j\mathbf{U}[0]\mathbf{q}_j^{(0)}[0] = [\mathbf{I} - {}_j\mathbf{P}^{(0)}]^{-1}\mathbf{p}_j^{(0)}$ *and for* $r = 0,\ldots,k_n$ *(for given* n*) and sequentially for every* $n = 0,\ldots,\tilde{n}$,

$$_0\mathbf{b}_j[r,n] = {}_j\mathbf{U}[0]\Big(\mathbf{e}_j[r,n] + \sum_{m=1}^{n} C_n^m \sum_{q=0}^{r} {}_j\mathbf{E}[q,m]{}_0\mathbf{b}_j[r-q,n-m]$$

$$+ \sum_{p=1}^{r} \mathbf{E}_j[p,0]{}_0\mathbf{b}_j[r-p,n]\Big). \qquad (5.2.19)$$

(iii) *The vector coefficients* $_0\mathbf{b}_j[r,n]$ *can also be calculated by the recurrence formulas for* $r = 0,\ldots,k_n$ *(for given* n*) and sequentially for every* $n = 0,\ldots,\tilde{n}$,

$$_0\mathbf{b}_j[r,n] = \sum_{q=0}^{r} {}_j\mathbf{U}[q]\mathbf{e}_j[r-q,n]$$

$$+ \sum_{m=1}^{n} C_n^m \sum_{q=0}^{r} \Big(\sum_{p=0}^{q} {}_j\mathbf{U}[p]{}_j\mathbf{E}[q-p,m]\Big) {}_0\mathbf{b}_j[r-q,n-m]. \quad (5.2.20)$$

Proof. The proof repeats the proof of Lemma 5.2.1. Lemma 5.2.2 differs from Lemma 5.2.1 in the orders of the corresponding asymptotic expansions for the power moments of hitting times, $_0m_j^{(\varepsilon)}[n]$. These orders also should be calculated recursively.

As was mentioned above, $_0m_j^{(\varepsilon)}[0] = _0f_j^{(\varepsilon)}$ if $n = 0$. The asymptotic expansion given for $n = 0$ in condition $\mathbf{P}_{22}^{(k)}$ coincides with the asymptotic expansion given in condition $\mathbf{P}_{18}^{(k)}$ if we choose $k = \tilde{k}_0$. The asymptotic expansion (5.2.18) for the vectors $_0m_j^{(\varepsilon)}[0] = [\mathbf{I} - _j\mathbf{P}^{(\varepsilon)}]^{-1}\mathbf{q}_j^{(\varepsilon)}[0]$ and formulas (5.2.19) and (5.2.20) for the corresponding vector coefficients just coincide, respectively, with the asymptotic expansion (5.1.43) for the vectors $_0f_j^{(\varepsilon)}$ and formulas (5.1.44) and (5.1.45) for the corresponding vector coefficients. Note that this is an $(N, 1)$-matrix $(0, k_0)$-expansion, where $k_0 = \tilde{k}_0$. Indeed, the order of this expansion is the minimum of orders of the asymptotic expansions for the matrices $[\mathbf{I} - _j\mathbf{P}^{(\varepsilon)}]^{-1}$ and the vectors $\mathbf{q}_j^{(\varepsilon)}[0]$. Both these expansions have, according to condition $\mathbf{P}_{22}^{(k)}$, the order $\tilde{k}_0$.

Now we can write an asymptotic expansion for the vectors $\mathbf{q}_j^{(\varepsilon)}[1] = \mathbf{p}_j^{(\varepsilon)}[1] + _j\mathbf{P}^{(\varepsilon)}[1] \cdot _0m_j^{(\varepsilon)}[0]$ by using Lemma 8.1.2. This will be an $(N, 1)$-matrix $(0, k_1)$-expansion, where $k_1 = \min(\tilde{k}_0, \tilde{k}_1)$. Indeed, according to Lemma 5.1.3, the order of this expansion is the minimum of the orders of the asymptotic expansions for the vectors $\mathbf{p}_j^{(\varepsilon)}[1]$, the matrices $_j\mathbf{P}^{(\varepsilon)}[1]$, and the vectors $_0m_j^{(\varepsilon)}[0]$ that are, respectively, $\tilde{k}_1$, $\tilde{k}_1$, and $\tilde{k}_0$. This minimum is $k_1 = \min(\tilde{k}_0, \tilde{k}_1)$. Then, Lemma 5.1.3 can be applied to the vectors $_0m_j^{(\varepsilon)}[1] = [\mathbf{I} - _j\mathbf{P}^{(\varepsilon)}]^{-1}\mathbf{q}_j^{(\varepsilon)}[1]$. The resulted asymptotic expansion is an $(N, 1)$-matrix $(0, k_1)$-expansion given in formula (5.2.18) for $n = 1$. According to Lemma 5.1.3, the order of this expansion is the minimum of orders of the asymptotic expansions for the matrices $[\mathbf{I} - _j\mathbf{P}^{(\varepsilon)}]^{-1}$ and the vectors $\mathbf{q}_j^{(\varepsilon)}[1]$. These expansions have, respectively, the order $\tilde{k}_1$, where $k_1 = \min(\tilde{k}_0, \tilde{k}_1)$. This minimum is k_1.

The recurrence procedure can be repeated for $n = 1, \ldots, \tilde{n}$, which will yield the corresponding asymptotic expansions given in (5.2.18). $\qquad\square$

5.2.4   Pivotal properties of asymptotic expansions for power moments of hitting times

We will use the following notations for the vectors $_0\mathbf{b}_j[r, n]$, $r = 0, \ldots, k_n$, $n = 0, \ldots, \tilde{n}$:

$$_0\mathbf{b}_j[r, n] = \begin{bmatrix} _0b_{1j}[r, n] \\ \vdots \\ _0b_{Nj}[r, n] \end{bmatrix}.$$

Explicit expressions for components of the vectors $_0\mathbf{b}_j[r, n]$, $r = 0, \ldots, k_n$, $n = 0, \ldots, \tilde{n}$, can be derived from the recurrence formulas (5.2.14) and (5.2.15), respectively.

We are interested in finding the cyclic power moments $_0M_{jj}^{(\varepsilon)}[n]$, $n = 0,\dots,\tilde{n}$, $j \in X_1^{(0)}$.

We will limit our consideration to the case where condition $\mathbf{I_4}$ is satisfied, that is, there exists a state $i \in X_1^{(0)}$ such that $F_i^{(0)}(0) < 1$. This condition guarantees that the corresponding semi-Markov process $\eta^{(\varepsilon)}(t)$, $t \geq 0$, has a finite number of transitions in every finite interval for ε small enough.

Lemma 5.2.3. *Let conditions* $\mathbf{A_2}$, $\mathbf{E_2}$, $\mathbf{I_4}$, $\mathbf{M_{13}^{(\tilde{n})}}$, *and* $\mathbf{P_{22}^{(\tilde{k})}}$ *hold. Then the asymptotic expansions for the power moments of the hitting probabilities* $_0M_{jj}^{(\varepsilon)}[n]$, $n = 0,\dots,\tilde{n}$, $j \in X_1^{(0)}$ *are all* 0-*pivotal, i.e., the corresponding zero coefficients satisfy* $_0M_{jj}^{(0)}[n] = \,_0b_{jj}[0,n] > 0$ *for all* $n = 0,\dots,\tilde{n}$ *and* $j \in X_1^{(0)}$.

Proof. Condition $\mathbf{I_4}$ implies that **(c)** there exists a state $i' \in X_1^{(0)}$, and $j' \neq 0$ such that **(d)** $p_{i'j'}^{(0)} > 0$ and **(e)** $1 - F_{i'j'}^{(0)}(0) > 0$. Let us choose an arbitrary state $j \in X_1^{(0)}$. There are two cases to be considered. The first case is where $i' = j$. Here, **(d)** and **(e)** imply that **(f)** $1 - \,_0G_{jj}(0) > p_{i'j'}^{(0)}(1 - F_{i'j'}^{(0)}(0)) > 0$. The second case is where $i' \neq j$. In this case, **(g)** the cyclic hitting probability satisfies $_0{}_jf_{ji'}^{(0)} > 0$, since i' is a recurrent-without-absorption state. Using **(d)**, **(e)**, and **(g)** we get that **(h)** $1 - \,_0G_{jj}(0) > \,_0{}_jf_{ji'}^{(0)} p_{i'j'}^{(0)}(1 - F_{i'j'}^{(0)}(0)) > 0$. Clearly, **(f)** as well as **(h)** implies that **(i)** $_0M_{jj}^{(0)}[n] > 0$, $n = 1,\dots,\tilde{n}$. The relation $_0M_{jj}^{(0)}[0] > 0$, $j \in X_1^{(0)}$, was proved in Lemma 5.1.8. $\square$

5.3   Asymptotic expansions for power-exponential moments of hitting times

In this section, we describe a recurrence algorithm for a use in the asymptotic analysis of mixed power-exponential moments of hitting times for nonlinearly perturbed semi-Markov processes. The algorithm is also important by itself and is an essential in the proof of our main results concerning semi-Markov processes.

5.3.1   Asymptotic expansions for ρ-potential matrices

Let us recall some notations introduced in Section 4.4 for moment generating functions,

$$p_{ij}^{(\varepsilon)}(\rho) = \int_0^\infty e^{\rho s} Q_{ij}^{(\varepsilon)}(ds), \quad \psi_{ij}^{(\varepsilon)}(\rho) = \int_0^\infty e^{\rho s} F_{ij}^{(\varepsilon)}(ds), \quad \rho \in \mathbb{R}_1, \quad i,j \neq 0.$$

The moment generating functions introduced above are connected by the following formulas, for $\rho \in \mathbb{R}_1$ and $i,j \neq 0$,

$$p_{ij}^{(\varepsilon)}(\rho) = p_{ij}^{(\varepsilon)} \psi_{ij}^{(\varepsilon)}(\rho).$$

We assume that the following condition holds:

$\mathbf{C_{17}}$: There exists $\delta > 0$ such that $\overline{\lim}_{0 \leq \varepsilon \to 0} \, p_{ij}^{(\varepsilon)}(\delta) < \infty$, $i, j \neq 0$.

Condition $\mathbf{C_{17}}$ is a reduced form of condition $\mathbf{C_{12}}$ where the corresponding asymptotic relation is also required for $i \neq 0$, $j = 0$. According to Remark 4.5.3, these additional asymptotic relations are not required when considering moment generating functions of hitting times without absorption.

The following condition obviously implies condition $\mathbf{C_{17}}$ be satisfied:

$\mathbf{C'_{17}}$: There exists $\delta > 0$ such that $\overline{\lim}_{0 \leq \varepsilon \to 0} \, \psi_{ij}^{(\varepsilon)}(\delta) < \infty$, $i, j \neq 0$.

Condition $\mathbf{C_{17}}$ guarantees that for every $0 \leq \beta < \delta$ there exists $\varepsilon'_0 = \varepsilon'_0(\beta) > 0$ such that, for $\rho \leq \beta$,

$$\sup_{\varepsilon \leq \varepsilon'_0} \max_{i,j \neq 0} \, p_{ij}^{(\varepsilon)}(\rho) < \infty. \tag{5.3.1}$$

Let us also assume the following condition (introduced in Subsection 4.4.4):

$\mathbf{T_2}$: **(a)** $p_{ij}^{(\varepsilon)} \to p_{ij}^{(0)}$ as $\varepsilon \to 0$, $i, j \neq 0$;

   (b) $Q_{ij}^{(\varepsilon)}(\cdot) \Rightarrow Q_{ij}^{(0)}(\cdot)$ as $\varepsilon \to 0$, $i, j \neq 0$.

The following condition obviously implies condition $\mathbf{T_2}$ be satisfied:

$\mathbf{T'_2}$: **(a)** $p_{ij}^{(\varepsilon)} \to p_{ij}^{(0)}$ as $\varepsilon \to 0$, $i, j \neq 0$;

   (b) $F_{ij}^{(\varepsilon)}(\cdot) \Rightarrow F_{ij}^{(0)}(\cdot)$ as $\varepsilon \to 0$, $i, j \neq 0$.

Condition $\mathbf{T_2}$ is a reduced form of condition $\mathbf{T_1}$, where the corresponding asymptotic relation is also required for $i \neq 0, j = 0$. These asymptotic relations are not required when considering moment generating functions of hitting times without absorption.

Conditions $\mathbf{T_2}$ and $\mathbf{C_{17}}$ imply that, for $i, j \neq 0$ and $\rho \leq \beta$,

$$p_{ij}^{(\varepsilon)}(\rho) \to p_{ij}^{(0)}(\rho) \text{ as } \varepsilon \to 0. \tag{5.3.2}$$

We assume that the following perturbation condition is satisfied for some $\rho \leq \beta$:

$\mathbf{P_{23}^{(\rho,k)}}$: $p_{ij}^{(\varepsilon)}(\rho) = p_{ij}^{(0)}(\rho) + e_{ij}[\rho, 1, 0]\varepsilon + \cdots + e_{ij}[\rho, k, 0]\varepsilon^k + o(\varepsilon^k)$ for $i, j \neq 0$, where $|e_{ij}[\rho, r, 0]| < \infty$, $r = 1, \ldots, k$, $i, j \neq 0$.

It is convenient to define $e_{ij}[\rho, 0, 0] = p_{ij}^{(0)}[\rho, 0]$ for $i, j \neq 0, \rho \leq \beta$.

It is useful to note that moment generating functions $p_{i0}^{(\varepsilon)}(\rho)$ are not involved in condition $\mathbf{P_{23}^{(\rho,k)}}$. It is because of they do not penetrate formulas for power moment generating functions for hitting without absorption times.

In some cases, it can be more convenient to use the perturbation condition $\mathbf{P_{18}^{(k)}}$ for the transition probabilities $p_{ij}^{(\varepsilon)}$, $i, j \neq 0$, and, additionally, the following perturbation condition imposed on moments of the transition times $\psi_{ij}^{(\varepsilon)}(\rho)$, $i, j \neq 0$:

$\mathbf{P}_{24}^{(\rho,k)}$: $\psi_{ij}^{(\varepsilon)}(\rho) = \psi_{ij}^{(0)}(\rho) + v_{ij}[\rho, 1, 0]\varepsilon + \cdots + v_{ij}[\rho, k, 0]\varepsilon^k + o(\varepsilon^k)$ for $i, j \neq 0$,
where $|v_{ij}[\rho, r, 0]| < \infty$, $r = 1, \ldots, k$, $i, j \neq 0$.

Define $v_{ij}[\rho, 0, 0] = \psi_{ij}^{(0)}(\rho)$ for $i, j \neq 0$, $\rho \leq \beta$.

It can be proved that conditions $\mathbf{P}_{18}^{(k)}$ and $\mathbf{P}_{24}^{(\rho,k)}$ imply that condition $\mathbf{P}_{23}^{(\rho,k)}$ holds. Moreover, the corresponding asymptotic expansions, assuming that conditions $\mathbf{P}_{18}^{(k)}$ and $\mathbf{P}_{24}^{(\rho,k)}$ hold, are connected for $i, j \neq 0$ with the asymptotic expansion in condition $\mathbf{P}_{23}^{(\rho,k)}$ by the following formulas:

$$e_{ij}[\rho, r, 0] = \sum_{m=0}^{r} e_{ij}[r, 0]v_{ij}[\rho, r - m, 0], \quad r = 0, \ldots, k. \tag{5.3.3}$$

We will rewrite condition $\mathbf{P}_{23}^{(\rho,k)}$ in a matrix form.
For $r = 0, 1, \ldots, k$, consider the matrices

$$\mathbf{P}^{(\varepsilon)}(\rho) = \begin{bmatrix} p_{11}^{(\varepsilon)}(\rho) & \cdots & p_{1N}^{(\varepsilon)}(\rho) \\ \vdots & \vdots & \vdots \\ p_{N1}^{(\varepsilon)}(\rho) & \cdots & p_{NN}^{(\varepsilon)}(\rho) \end{bmatrix},$$

and

$$\mathbf{E}[\rho, r, 0] = \begin{bmatrix} e_{11}[\rho, r, 0] & \cdots & e_{1N}[\rho, r, 0] \\ \vdots & \vdots & \vdots \\ e_{N1}[\rho, r, 0] & \cdots & e_{NN}[\rho, r, 0] \end{bmatrix}.$$

Condition $\mathbf{P}_{23}^{(\rho,k)}$ can be rewritten in the following equivalent matrix form:

$\mathbf{P}_{23}^{(\rho,k)}$: $\mathbf{P}^{(\varepsilon)}(\rho) = \mathbf{P}^{(0)}(\rho) + \mathbf{E}[\rho, 1, 0]\varepsilon + \cdots + \mathbf{E}[\rho, k, 0]\varepsilon^k + o(\varepsilon^k)$, where $\mathbf{E}[\rho, r, 0]$,
$r = 0, \ldots, k$ are finite matrices.

For $j \neq 0$, set

$$_j\mathbf{P}^{(\varepsilon)}(\rho) = \begin{bmatrix} p_{11}^{(\varepsilon)}(\rho) & \cdots & p_{1(j-1)}^{(\varepsilon)}(\rho) & 0 & p_{1(j+1)}^{(\varepsilon)}(\rho) & \cdots & p_{1N}^{(\varepsilon)}(\rho) \\ \vdots & & \vdots & \vdots & \vdots & & \vdots \\ p_{N1}^{(\varepsilon)}(\rho) & \cdots & p_{N(j-1)}^{(\varepsilon)}(\rho) & 0 & p_{N(j+1)}^{(\varepsilon)}(\rho) & \cdots & p_{NN}^{(\varepsilon)}(\rho) \end{bmatrix},$$

and also, for $r = 1, \ldots, k$ and $j \neq 0$,

$_j\mathbf{E}[\rho, r, 0]$

$$= \begin{bmatrix} e_{11}[\rho, r, 0] & \cdots & e_{1(j-1)}[\rho, r, 0] & 0 & e_{1(j+1)}[\rho, r, n] & \cdots & e_{1N}[\rho, r, 0] \\ \vdots & & \vdots & \vdots & \vdots & & \vdots \\ e_{N1}[\rho, r, 0] & \cdots & e_{N(j-1)}[\rho, r, 0] & 0 & e_{N(j+1)}[\rho, r, 0] & \cdots & e_{NN}[\rho, r, 0] \end{bmatrix}.$$

Condition $\mathbf{P}_{23}^{(\rho,k)}$, for $j \neq 0$, gives the following matrix asymptotic expansions:

$$_j\mathbf{P}^{(\varepsilon)}[\rho,0] = {}_j\mathbf{P}^{(0)}[\rho,0] + {}_j\mathbf{E}[\rho,1,0]\varepsilon + \cdots + {}_j\mathbf{E}[\rho,k,0]\varepsilon^k + \mathbf{o}(\varepsilon^k). \qquad (5.3.4)$$

It is convenient to define $e_{ij}[\rho,0,0] = p_{ij}^{(0)}(\rho)$, $i, j \neq 0$, and the matrices $\mathbf{E}[\rho,0,0] = \mathbf{P}^{(0)}(\rho)$ and $_j\mathbf{E}[\rho,0,0] = {}_j\mathbf{P}^{(0)}(\rho)$, $j \neq 0$.

We also assume that the following condition holds:

$\mathbf{C_{18}}$: There exist $i \in X_1^{(0)}$ and $\beta_i \in (0,\delta]$, for δ defined in condition $\mathbf{C_{17}}$, such that:

> **(a)**: $_0\phi_{ii}^{(0)}(\beta_i) \in (1,\infty)$;
>
> **(b)**: $_0\phi_{ki}^{(0)}(\beta_i) < \infty$, $k \in X_2^{(0)}$.

Condition $\mathbf{C_{18}}$ is a variant of condition $\mathbf{C_{13}}$ in which δ defined in condition $\mathbf{C_{12}}$ is replaced with δ defined in condition $\mathbf{C_{17}}$.

It should be mentioned that necessary and sufficient conditions for condition $\mathbf{C_{18}}$ to hold, formulated in terms of so-called test functions, are given in Lemma 4.5.5.

In what follows we assume that conditions $\mathbf{A_2}$, $\mathbf{T_2}$, $\mathbf{E_2}$, $\mathbf{C_{17}}$, and $\mathbf{C_{18}}$ hold.

As was shown in Lemmas 4.5.6 and 8.1.4, these conditions imply that **(a)** there exist $0 < \beta < \beta_i$, where β_i is from condition $\mathbf{C_{18}}$, and $0 < \varepsilon_1' = \varepsilon_1(\beta) \leq \varepsilon_0'(\beta)$ such that, for $j \in X_1^{(0)}$,

$$\sup_{\varepsilon \leq \varepsilon_1'} |\,(_j\mathbf{P}^{(\varepsilon)}(\beta))^n\,| \to 0 \text{ as } n \to \infty. \qquad (5.3.5)$$

Relation (5.3.5) implies that **(b)** there exists the inverse matrix $[\mathbf{I} - {}_j\mathbf{P}^{(\varepsilon)}(\rho)]^{-1}$ for all $\rho \leq \beta$, $j \in X_1^{(0)}$, and $\varepsilon \leq \varepsilon_1'$. Introduce a notation for this matrix,

$$_j\mathbf{U}^{(\varepsilon)}(\rho) = \|_j u_{il}^{(\varepsilon)}(\rho)\| = [\mathbf{I} - {}_j\mathbf{P}^{(\varepsilon)}(\rho)]^{-1}.$$

Elements of the matrix $_j\mathbf{U}^{(\varepsilon)}(\rho)$ have a clear probability meaning, namely, for $i, l \neq 0$, $j \in X_1^{(0)}$,

$$_j u_{il}^{(\varepsilon)}(\rho) = \mathsf{E}_i \delta_{jl}^{(\varepsilon)}(\rho), \qquad (5.3.6)$$

where

$$\delta_{jl}^{(\varepsilon)}(\rho) = \sum_{n=1}^{v_j^{(\varepsilon)} \wedge v_0^{(\varepsilon)}} e^{\rho\tau^{(\varepsilon)}(n-1)} \delta(\eta_{n-1}^{(\varepsilon)}, l)$$

$$= \sum_{n=1}^{\infty} e^{\rho\tau^{(\varepsilon)}(n-1)} \chi(v_j^{(\varepsilon)} \wedge v_0^{(\varepsilon)} > n - 1, \eta_{n-1}^{(\varepsilon)} = l).$$

We refer to the matrix $_j\mathbf{U}^{(\varepsilon)}(\rho)$ as a ρ-potential matrix associated with the matrix $\mathbf{P}^{(\varepsilon)}(\rho)$.

It is useful to note that $_j\mathbf{U}^{(\varepsilon)}(0) = {}_j\mathbf{U}^{(\varepsilon)}$, i.e., 0-potential matrices coincide with the corresponding usual potential matrices studied in Section 5.1.

The following lemma, which is a direct corollary of Lemma 8.1.2, gives an asymptotic matrix expansion for the potential matrices $_j\mathbf{U}^{(\varepsilon)}(\rho)$, $j \in X_1^{(0)}$.

In Lemmas 5.3.1–5.3.4, β is defined in (5.3.5).

Lemma 5.3.1. *Let conditions* $\mathbf{A}_2$, $\mathbf{T}_2$, $\mathbf{E}_2$, $\mathbf{C}_{17}$, *and* $\mathbf{C}_{18}$ *hold, and also condition* $\mathbf{P}_{23}^{(\rho,k)}$ *be satisfied for some* $\rho \leq \beta$. *Then*

(i) *The* ρ-*potential matrices* $_j\mathbf{U}^{(\varepsilon)}(\rho)$, *for every* $j \in X_1^{(0)}$, *have the following asymptotic matrix expansions:*

$$_j\mathbf{U}^{(\varepsilon)}(\rho) = {}_j\mathbf{U}^{(0)}(\rho) + {}_j\mathbf{U}[\rho, 1]\varepsilon + \cdots + {}_j\mathbf{U}[\rho, k]\varepsilon^k + \mathbf{o}(\varepsilon^k), \qquad (5.3.7)$$

where $_j\mathbf{U}[\rho, r]$, $r = 1, \ldots, k$, *are finite matrices.*

(ii) *The matrix coefficients* $_j\mathbf{U}[\rho, r]$ *are given by the following recurrence formulas* $_j\mathbf{U}[\rho, 0] = {}_j\mathbf{U}^{(0)}(\rho) = [\mathbf{I} - {}_j\mathbf{P}^{(0)}(\rho)]^{-1}$ *and, for* $r = 1, \ldots, k$,

$$_j\mathbf{U}[\rho, r] = {}_j\mathbf{U}[\rho, 0] \sum_{q=1}^{r} {}_j\mathbf{E}[\rho, q, 0]{}_j\mathbf{U}[\rho, r - q]. \qquad (5.3.8)$$

Proof. Relations (5.3.4) and (5.3.5) imply that conditions of Lemma 8.1.2 are satisfied by the nonlinearly perturbed matrices $\mathbf{I} - {}_j\mathbf{P}^{(\varepsilon)}(\rho)$ for every $j \in X_1^{(0)}$. By applying statement (iv) of Lemma 8.1.2 to these matrices, we get the claims of Lemma 5.3.1. $\square$

5.3.2 Pivotal properties of asymptotic expansions for ρ-potential matrices

Let us use the following notations for the matrices $_j\mathbf{U}[\rho, r]$, $r = 0, \ldots, k$:

$$_j\mathbf{U}[\rho, r] = \begin{bmatrix} _ju_{11}[\rho, r] & \cdots & _ju_{1N}[\rho, r] \\ \vdots & \vdots & \vdots \\ _ju_{N1}[\rho, r] & \cdots & _ju_{NN}[\rho, r] \end{bmatrix}.$$

Explicit expressions for elements of the matrices $_j\mathbf{U}[\rho, r]$, $r = 0, \ldots, k$, can be easily derived from recurrence formulas (5.3.8).

We would like to clarify pivotal properties of asymptotic expansions for the components $_ju_{il}^{(\varepsilon)}(\rho)$ of elements of the potential matrices $_j\mathbf{U}^{(\varepsilon)}(\rho)$. A description of such properties is especially important for *cyclic elements* of these matrices; these are elements $_ju_{jl}^{(\varepsilon)}(\rho)$ for the recurrent-without-absorption states $j \in X_1^{(0)}$.

We are going to show that, under a natural assumption that conditions of Lemma 5.3.1 and the perturbation condition $\mathbf{P}_{18}^{(k)}$ hold, $_j u_{jl}^{(\varepsilon)}(\rho)$ and $_j u_{jl}^{(\varepsilon)}$ have the same pivotal properties.

For a pair of states (i, j), where $i, j \neq 0$, define a parameter, called a *local asymptotic ρ-order*, via coefficients in the asymptotic expansion for the quantities $p_{ij}^{(\varepsilon)}(\rho)$ given in condition $\mathbf{P}_{23}^{(\rho,k)}$,

$$r_\rho(i, j) = \begin{cases} r & \text{if } e_{ij}[\rho, n, 0] = 0,\ n < r \text{ but } e_{ij}[\rho, r, 0] > 0,\ \text{for } r = 0, \ldots, k, \\ k+1 & \text{if } e_{ij}[\rho, n, 0] = 0,\ n \le k. \end{cases}$$

Consider an (i, j)-chain $(i_0, i_1, \ldots, i_m)$ and define a parameter, called a *chain asymptotic ρ-order*, via the corresponding local asymptotic ρ-order of sequential pairs of states in this chain,

$$r_\rho(i_0, i_1, \ldots, i_m) = \begin{cases} r & \text{if } \sum_{n=1}^m r_\rho(i_{n-1}, i_n) = r \text{ for } r = 0, \ldots, k, \\ k+1 & \text{if } \sum_{n=1}^m r_\rho(i_{n-1}, i_n) \ge k+1. \end{cases}$$

Finally, we define a *global asymptotic ρ-order* for a pair of states (i, j), where $i, j \neq 0$, via the chain asymptotic ρ-orders of the (i, j)-chains,

$$\hat{r}_\rho(i, j) = \min_{(i_0', 1, \ldots, i_{m'}') \in R_{ij}^*} r_\rho(i_0', \ldots, i_{m'}').$$

Consider the sets

$$Z_{j,r,\rho} =$$

$$\begin{cases} \{l \in X_0 : {}_j u_{jl}[\rho, n] = 0,\ n = 0, \ldots, r-1,\ {}_j u_{jl}[r] \neq 0\} & \text{for } r = 1, \ldots, k, \\ \{l \in X_0 : {}_j u_{jl}[\rho, n] = 0,\ n = 0, \ldots, k\} & \text{for } r = k+1. \end{cases}$$

Lemma 5.3.2. *Let conditions* $\mathbf{A}_2$, $\mathbf{T}_2$, $\mathbf{E}_2$, $\mathbf{C}_{17}$, $\mathbf{C}_{18}$, $\mathbf{P}_{18}^{(k)}$ *hold, and condition* $\mathbf{P}_{23}^{(\rho,k)}$ *be satisfied for some $\rho \le \beta$. Then*

(i) $r_\rho(i, j) = r(i, j)$, $i, j \neq 0$, *and* $r_\rho(i_0, i_1, \ldots, i_m) = r(i_0, i_1, \ldots, i_m)$, $(i_0, i_1, \ldots, i_m) \in R_{ij}^*$, $i, j \neq 0$.

(ii) $\hat{r}_\rho(i, j) = \hat{r}(j, l) = \hat{r}(l)$ *does not depend on the choice of the state $j \in X_1^{(0)}$ for every state $l \neq 0$.*

(iii) $Z_{j,r,\rho} = Z_r$, $j \neq 0$, $r = 0, \ldots, k+1$.

Proof. Take an arbitrary $i, j \neq 0$. By the definition, **(a)** $0 < p_{ij}^{(0)}(\rho) < \infty$ and, by relation (5.3.2), **(b)** $p_{ij}^{(\varepsilon)}(\rho) \to p_{ij}^{(0)}(\rho)$ as $\varepsilon \to 0$. It follows from statements **(a)** and **(b)** and conditions $\mathbf{P}_{18}^{(k)}$, $\mathbf{P}_{23}^{(\rho,k)}$ that either **(c)** there exists $0 \le r \le k$ such that

$\lim_{\varepsilon \to 0} \varepsilon^{-r} p_{ij}^{(\varepsilon)} = e_{ij}[r, 0] > 0$ and $\lim_{\varepsilon \to 0} \varepsilon^{-r} p_{ij}^{(\varepsilon)}(\rho) = e_{ij}[\rho, r, 0] > 0$, i.e., both expansions given in these conditions are r-pivotal, or **(d)** $p_{ij}^{(\varepsilon)} = o(\varepsilon^k)$ and $p_{ij}^{(\varepsilon)}(\rho) = o(\varepsilon^k)$. Thus, claim $(\boldsymbol{\alpha})$ is verified. Statements $(\boldsymbol{\beta})$ and $(\boldsymbol{\gamma})$ are obvious corollaries of $(\boldsymbol{\alpha})$.

Both $_j\mathbf{P}^{(\varepsilon)}(\rho)$ and $_j\mathbf{P}^{(\varepsilon)}$ are matrices with non-negative elements and with zero element in the column j. The matrices $_j\mathbf{P}^{(\varepsilon)}(\rho)$ can be considered as usual potential matrices $_j\mathbf{P}^{(\varepsilon)}$ but with elements $p_{ij}^{(\varepsilon)}(\rho)$ instead of elements $p_{ij}^{(\varepsilon)}$. Note also that the elements $p_{ij}^{(\varepsilon)}(\rho)$ and $p_{ij}^{(\varepsilon)}$ take positive or zero values simultaneously. The asymptotic ρ-orders $r_\rho(i, j), r_\rho(i_0, i_1, \ldots, i_m)$ and $\hat{r}_\rho(i, j)$ replace, respectively, the usual asymptotic orders $r(i, j), r(i_0, i_1, \ldots, i_m)$, and $\hat{r}(i, j)$.

Conditions $\mathbf{A}_2$, $\mathbf{T}_2$, $\mathbf{E}_2$, $\mathbf{C}_{17}$, $\mathbf{C}_{18}$, $\mathbf{P}_{23}^{(\rho,k)}$ imply that the conditions of Lemma 5.1.2 are satisfied for these matrices. Therefore, claims **(i)**–**(iv)** of Lemma 5.1.2 are verified for the matrices $_j\mathbf{P}^{(\varepsilon)}(\rho)$.

In particular, **(e)** the global asymptotic orders $\hat{r}_\rho(i, j) = \hat{r}_\rho(j), i \in X_1^{(0)}, j \neq 0$, do not depend on the choice of $i \in X_1^{(0)}$, and **(f)** the sets $Z_{j,r,\rho} = Z_{r,\rho}, r = 0, \ldots, k + 1$, do not depend on the choice of $j \in X_1^{(0)}$. By the definition, $Z_{r,\rho} = \{j \neq 0 : \hat{r}_\rho(j) = r\}$. But statement $(\boldsymbol{\gamma})$ implies that **(g)** $\hat{r}_\rho(j) = \hat{r}(j), j \neq 0$. Thus, $Z_{r,\rho} = Z_r$, $r = 0, \ldots, k + 1$, since $Z_r = \{j \neq 0 : \hat{r}(j) = r\}, r = 0, \ldots, k + 1$. $\qquad\square$

5.3.3   Asymptotic expansions for solutions of ρ-hitting type systems of linear equations

Let us again consider the vector function $\mathbf{y}^{(\varepsilon)}$ and assume that it satisfies the non-negativity condition $\mathbf{S}_1$. For every $j \in X_1^{(0)}$ consider the system of linear equations,

$$\mathbf{x}_j^{(\varepsilon)}(\rho) = \mathbf{y}^{(\varepsilon)} + {}_j\mathbf{P}^{(\varepsilon)}(\rho)\mathbf{x}_j^{(\varepsilon)}(\rho). \tag{5.3.9}$$

If conditions $\mathbf{A}_2$, $\mathbf{E}_2$, $\mathbf{T}_2$, $\mathbf{C}_{17}$, $\mathbf{C}_{18}$, and $\mathbf{S}_1$ hold, then system (5.3.9) has a unique solution for every $\varepsilon \leq \varepsilon_1'$ and $j \in X_1^{(0)}$,

$$\mathbf{x}_j^{(\varepsilon)}(\rho) = [\mathbf{I} - {}_j\mathbf{P}^{(\varepsilon)}(\rho)]^{-1}\mathbf{y}^{(\varepsilon)}. \tag{5.3.10}$$

Also note that, under condition $\mathbf{S}_1$, the components of the vector $\mathbf{x}_j^{(\varepsilon)}(\rho)$ are all non-negative.

The following lemma gives vector asymptotic expansions for solutions of these systems.

Lemma 5.3.3. *Let conditions* $\mathbf{A}_2$, $\mathbf{E}_2$, $\mathbf{T}_2$, $\mathbf{C}_{17}$, $\mathbf{C}_{18}$, *and* $\mathbf{S}_1$ *hold, and also conditions* $\mathbf{P}_{23}^{(\rho,k)}$ *and* $\mathbf{P}_{19}^{(k)}$ *be satisfied for some* $\rho \leq \beta$. *Then*

(i) *The vector* $\mathbf{x}_j^{(\varepsilon)}(\rho)$ *has the following asymptotic expansion for every* $j \in X_1^{(0)}$:

$$\mathbf{x}_j^{(\varepsilon)}(\rho) = \mathbf{x}_j^{(0)}(\rho) + \mathbf{x}_j[\rho, 1]\varepsilon + \cdots + \mathbf{x}_j[\rho, k]\varepsilon^k + \mathbf{o}(\varepsilon^k), \tag{5.3.11}$$

where $\mathbf{x}_j[\rho, r]$, $r = 1, \ldots, k$, *are finite vectors.*

(ii) *The vector coefficients* $\mathbf{x}_j[\rho, r]$ *are given by the recurrence formulas* $\mathbf{x}_j[\rho, 0] = \mathbf{x}_j^{(0)}(\rho) = {}_j\mathbf{U}[\rho, 0]\mathbf{y}^{(0)} = [\mathbf{I} - {}_j\mathbf{P}^{(0)}(\rho)]^{-1}\mathbf{y}^{(0)}$ *and, for* $r = 0, \ldots, k$,

$$\mathbf{x}_j[\rho, r] = {}_j\mathbf{U}[\rho, 0]\left(\mathbf{y}[r] + \sum_{q=1}^{r} {}_j\mathbf{E}[\rho, q, 0]\,\mathbf{x}_j[\rho, r - q]\right). \tag{5.3.12}$$

(iii) *The vector coefficients* $\mathbf{x}_j[\rho, r]$ *can also be calculated by the following formulas for* $r = 0, \ldots, k$:

$$\mathbf{x}_j[\rho, r] = \sum_{p=0}^{r} {}_j\mathbf{U}[\rho, p]\,\mathbf{y}[r - p]. \tag{5.3.13}$$

Proof. Relations (5.3.4) and (5.3.5) and condition $\mathbf{P}_{19}^{(k)}$, expressed in a vector form, imply that conditions of Lemma 8.1.4 are satisfied by the system of nonlinearly perturbed linear equations (5.3.9) for every $j \in X_1^{(0)}$. By applying this lemma to systems (5.3.9) we get the claims of Lemma 5.1.1. $\qquad\square$

For $r = 1, \ldots, k$ and $j \in X_1^{(0)}$, let

$$\mathbf{x}_j^{(\varepsilon)}(\rho) = \begin{bmatrix} x_{1j}^{(\varepsilon)}(\rho) \\ \vdots \\ x_{Nj}^{(\varepsilon)}(\rho) \end{bmatrix}, \qquad \mathbf{x}_j[\rho, r] = \begin{bmatrix} x_{1j}[\rho, r] \\ \vdots \\ x_{Nj}[\rho, r] \end{bmatrix}.$$

Explicit expressions for components of the vectors $\mathbf{x}[\rho, r]$, $r = 0, \ldots, k$, can be derived from the recurrence formulas (5.3.12) or formulas (5.3.13).

There is a simple probabilistic interpretation of solution of system (5.3.9). In fact, using the probability representation for elements of the matrix $[\mathbf{I} - {}_j\mathbf{P}^{(\varepsilon)}(\rho)]^{-1}$ given by formula (5.3.6), we can write the following representation for components of the vector $\mathbf{x}_j^{(\varepsilon)}(\rho)$ for $i \neq 0$, $j \in X_1^{(0)}$:

$$x_{ij}^{(\varepsilon)}(\rho) = \sum_{l \neq 0} {}_j u_{il}^{(\varepsilon)}(\rho) y_l^{(\varepsilon)}$$

$$= \sum_{l \neq 0} \mathsf{E}_i \delta_{jl}^{(\varepsilon)}(\rho) y_l^{(\varepsilon)} = \mathsf{E}_i \sum_{n=1}^{\nu_j^{(\varepsilon)} \wedge \nu_0^{(\varepsilon)}} e^{\rho \tau^{(\varepsilon)}(n-1)} y_{\eta_{n-1}^{(\varepsilon)}}^{(\varepsilon)}. \tag{5.3.14}$$

Formula (5.3.14) suggests to refer to systems of type (5.3.9) as *ρ-hitting system of linear equations*.

Hitting systems of type (5.3.9) play an important role in our considerations. Mixed power-exponential moments of hitting times are solutions of such systems.

We are especially interested in a study of pivotal properties of asymptotic expansions for the components $x_{ij}^{(\varepsilon)}(\rho)$ of the solutions $\mathbf{x}_j^{(\varepsilon)}(\rho)$ of system (5.3.9). A description of such properties is especially important for *cyclic components* of the vectors $\mathbf{x}_j^{(\varepsilon)}(\rho)$, which are the components $x_{jj}^{(\varepsilon)}(\rho)$.

We will use the sets $Y_r, W_r, r = 0, \ldots, k$, and conditions $\mathbf{O}_1^{(h)}, \mathbf{O}_2^{(h)}$ introduced in Subsection 5.1.3.

Lemma 5.3.4. *Let conditions $\mathbf{A}_2$, $\mathbf{E}_2$, $\mathbf{T}_2$, $\mathbf{C}_{17}$, $\mathbf{C}_{18}$, and $\mathbf{S}_1$ hold, and conditions $\mathbf{P}_{23}^{(\rho,k)}$ and $\mathbf{P}_{19}^{(k)}$ be satisfied for some $\rho \leq \beta$. Then*

   (i) *If condition $\mathbf{O}_1^{(h)}$ holds for some $0 \leq h \leq k$, then the first h coefficients are zeros, i.e., $x_{jj}[\rho, n] = 0$, $j \in X_1^{(0)}$, $n = 0, \ldots, h$.*

   (ii) *If condition $\mathbf{O}_2^{(h)}$ holds for some $0 \leq h \leq k$, then the first $h - 1$ coefficients are zeros, i.e., $x_{jj}[\rho, n] = 0$, $j \in X_1^{(0)}$, $n = 0, \ldots, h - 1$, but $x_{jj}[\rho, h] > 0$, $j \in X_1^{(0)}$.*

Proof. The proof is analogous to the proof of Lemma 5.1.4. Using asymptotic expansions (5.3.7) for the ρ-potential matrices ${}_j\mathbf{U}^{(\varepsilon)}(\rho)$, condition $\mathbf{P}_{19}^{(k)}$, and Lemma 5.3.2, we rewrite relation (5.3.14) for the cyclic components $x_{jj}^{(\varepsilon)}(\rho)$, $j \in X_1^{(0)}$, in the following form:

$$x_{jj}^{(\varepsilon)}(\rho) = \sum_{l \neq 0} {}_j u_{jl}^{(\varepsilon)}(\rho) y_l^{(\varepsilon)}$$

$$= \sum_{0 \leq r \leq k} \sum_{r'+r''=r} \sum_{l \in Y_{r'} \cap Z_{r''}} {}_j u_{jl}^{(\varepsilon)} y_l^{(\varepsilon)} + \sum_{r'+r'' \geq k+1} \sum_{l \in Y_{r'} \cap Z_{r''}} {}_j u_{jl}^{(\varepsilon)}(\rho) y_l^{(\varepsilon)}$$

$$= \sum_{0 \leq r \leq k} \sum_{r'+r''=r} \sum_{l \in Y_{r'} \cap Z_{r''}} \left(\sum_{r'' \leq p'' \leq k} \varepsilon^{p''} {}_j u_{jl}[\rho, p''] \right)$$

$$\times \left(\sum_{r' \leq p' \leq k} \varepsilon^{p'} y_l[p'] \right) + o(\varepsilon^k)$$

$$= \sum_{0 \leq r \leq k} \left(\varepsilon^r L_j[\rho, r, r] + \cdots + \varepsilon^k L_j[\rho, k, r] \right) + o(\varepsilon^k)$$

$$= \sum_{0 \leq r \leq k} \varepsilon^r \left(L_j[\rho, r, 0] + \cdots + L_j[\rho, r, r] \right) + o(\varepsilon^k), \tag{5.3.15}$$

where the coefficients $L_j[\rho, p, r]$, $p = r, \ldots, k$, $r = 0, \ldots, k$, are readily calculated by multiplying the inner sums in (5.3.15) and collecting the coefficients at ε^p.

For every $r = 0, \ldots, k$, (g) the coefficients $L_j[\rho, p, r]$, $p = r, \ldots, k$, are sums of numbers over the set W_r. In particular, (h) the coefficients $L_j[\rho, r, r]$ have the following form for $r = 0, \ldots, k$:

$$L_j[\rho, r, r] = \sum_{r'+r''=r} \sum_{l \in Y_{r'} \cap Z_{r''}} {}_j u_{jl}[\rho, r''] \, y_l[r']. \tag{5.3.16}$$

Formula (5.3.15) implies that (i) the coefficients $x_{jj}[\rho, r]$ can be represented for $r = 0, \ldots, k$ as

$$x_{jj}[\rho, r] = \sum_{0 \le p \le r} L_j[\rho, r, p]. \tag{5.3.17}$$

If condition $\mathbf{O}_1^{(h)}$ holds for some $0 \le h \le k$, then $W_r = \varnothing, r \le h$. Thus, (j) the sums that define the coefficients $L[\rho, p, r], p = r, \ldots, k$, are actually the sums over the empty set for $r \le h$. Therefore, (k) $L[\rho, p, r] = 0, p = r, \ldots, k$, for $r \le h$. The claim (k) and formula (5.1.35) imply that (l) $x_{jj}[\rho, r] = 0, r \le h$. This proves the first statement of the lemma.

If condition $\mathbf{O}_2^{(h)}$ holds for some $0 \le h \le k$, then $W_r = \varnothing, r \le h-1$, but $W_h \ne \varnothing$. This means that condition $\mathbf{O}_1^{(h-1)}$ holds and, therefore, (m) $x_{jj}[\rho, r] = 0, r \le h - 1$, as was proved above. Moreover, as was shown, (n) $L[\rho, p, r] = 0, p = r, \ldots, k$ for $r \le h - 1$. It is clear that (n) and formula (5.1.35) imply in this case that (o) $x_{jj}[\rho, h] = L_j[\rho, r, r]$.

If $W_h \ne \varnothing$, then $Y_{r'} \cap Z_{r''} \ne \varnothing$ for some non-negative integers r' and r'' such that $r' + r'' = h$. By Lemma 5.3.2 and the definition of the set $Z_{r''}$, we have ${}_j u_{jl}[\rho, r''] > 0, l \in Z_{r''}$. Also, by the definition of the set $Y_{r'}$, and condition $\mathbf{S}_1$, $y_l[r'] > 0, l \in Y_{r'}$. Thus, $\sum_{l \in Y_{r'} \cap Z_{r''}} {}_j u_{jl}[\rho, r''] y_l[r'] > 0$. Therefore, $x_{jj}[\rho, r] = L_j[\rho, r, r] > 0$ according formula (5.3.16). This proves the second statement of the lemma. $\qquad \square$

5.3.4 Perturbation conditions for mixed power-exponential moments of hitting times

We introduce mixed power-exponential moments of transition times, for $n = 0, 1, \ldots$, $i, j \ne 0$, and $\rho \in \mathbb{R}_1$,

$$p_{ij}^{(\varepsilon)}[\rho, n] = \int_0^\infty s^n e^{\rho s} Q_{ij}^{(\varepsilon)}(ds), \quad \psi_{ij}^{(\varepsilon)}[\rho, n] = \int_0^\infty s^n e^{\rho s} F_{ij}^{(\varepsilon)}(ds).$$

The mixed power-exponential moments introduced above are connected by the following formulas, for $\rho \in \mathbb{R}_1$ and $i, j \ne 0, n = 0, 1, \ldots$,

$$p_{ij}^{(\varepsilon)}[\rho, n] = p_{ij}^{(\varepsilon)} \psi_{ij}^{(\varepsilon)}[\rho, n].$$

Choose an arbitrary $0 < \beta < \delta$. The following inequality holds for every $n = 1, \ldots$, $i, j \neq 0$:

$$p_{ij}^{(\varepsilon)}[\beta, n] = \int_0^\infty s^n e^{\beta s} Q_{ij}^{(\varepsilon)}(ds) \leq c_n \int_0^\infty e^{\beta s} Q_{ij}^{(\varepsilon)}(ds) = c_n p_{ij}^{(\varepsilon)}(\beta), \quad (5.3.18)$$

where $c_n = c_n(\delta, \beta) = \sup_{s \geq 0} s^n e^{-(\delta - \beta)s} < \infty$.

Thus, condition $\mathbf{C}_{17}$ guarantees that for every $0 \leq \beta < \delta$ there exists $\varepsilon_0' = \varepsilon_0'(\beta) > 0$ such that, for $n = 0, 1, \ldots$, $i, j \neq 0$, $\rho \leq \beta$, and $\varepsilon \leq \varepsilon_0$,

$$\sup_{\varepsilon \leq \varepsilon_0'} p_{ij}^{(\varepsilon)}[\beta, n] < \infty. \quad (5.3.19)$$

Conditions $\mathbf{T}_2$ and $\mathbf{C}_{17}$ imply that, for $n = 0, 1, \ldots$, $j \neq 0$ and $\rho \leq \beta$, we have

$$p_{ij}^{(\varepsilon)}[\rho, n] \to p_{ij}^{(0)}[\rho, n] \quad \text{as } \varepsilon \to 0. \quad (5.3.20)$$

We assume that the following perturbation condition holds for some $\rho \leq \beta$:

$\mathbf{P}_{25}^{(\rho, k)}$: $\quad p_{ij}^{(\varepsilon)}[\rho, n] = p_{ij}^{(0)}[\rho, n] + e_{ij}[\rho, 1, n]\varepsilon + \cdots + e_{ij}[\rho, k - n, n]\varepsilon^{k-n} + o(\varepsilon^{k-n})$
for $n = 0, \ldots, k$, $i, j \neq 0$, where $|e_{ij}[\rho, r, n]| < \infty$, $r = 1, \ldots, k - n$,
$n = 0, \ldots, k$, $i, j \neq 0$.

It is convenient to define $e_{ij}[\rho, 0, n] = p_{ij}^{(0)}[\rho, n]$ for $n = 0, \ldots, k$, $i, j \neq 0$.

Note that moments $p_{i0}^{(\varepsilon)}[\rho, n]$ are not involved in condition $\mathbf{P}_{25}^{(\rho, k)}$. It is because of these moments do not penetrate formulas for mixed power-exponential moments of hitting without absorption times.

In some cases, it will be more convenient to use the perturbation condition $\mathbf{P}_{18}^{(k)}$ for the transition probabilities $p_{ij}^{(\varepsilon)}$, $i, j \neq 0$, and to additionally use the following perturbation condition imposed on the mixed power-exponential moments of transition times, $\psi_{ij}^{(\varepsilon)}[\rho, n]$, $n = 1, \ldots, k$, $i, j \neq 0$:

$\mathbf{P}_{26}^{(\rho, k)}$: $\quad \psi_{ij}^{(\varepsilon)}[\rho, n] = \psi_{ij}^{(0)}[\rho, n] + v_{ij}[\rho, 1, n]\varepsilon + \cdots + v_{ij}[\rho, k - n, n]\varepsilon^{k-n} + o(\varepsilon^{k-n})$
for $n = 0, \ldots, k$, $i, j \neq 0$, where $|v_{ij}[\rho, r, n]| < \infty$, $r = 1, \ldots, k - n$,
$n = 0, \ldots, k$, $i, j \neq 0$.

Define $v_{ij}[\rho, 0, n] = \psi_{ij}^{(0)}[\rho, n]$ for $i, j \neq 0$, $n = 0, \ldots, k$.

It can be proved that conditions $\mathbf{P}_{18}^{(k)}$ and $\mathbf{P}_{26}^{(\rho, k)}$ imply that condition $\mathbf{P}_{25}^{(\rho, k)}$ is verified. Moreover, the corresponding asymptotic expansions in conditions $\mathbf{P}_{18}^{(k)}$ and $\mathbf{P}_{26}^{(\rho, k)}$ are connected with the asymptotic expansion in condition $\mathbf{P}_{25}^{(\rho, k)}$ by the following formulas for $i, j \neq 0$ and $r = 0, \ldots, k - n$, $n = 0, \ldots, k$:

$$e_{ij}[\rho, r, n] = \sum_{m=0}^r e_{ij}[r, 0]v_{ij}[\rho, r - m, n]. \quad (5.3.21)$$

300   5 Nonlinearly perturbed semi-Markov processes

For $n = 0, 1, \ldots, k$, consider the matrices

$$\mathbf{P}^{(\varepsilon)}[\rho, n] = \begin{bmatrix} p_{11}^{(\varepsilon)}[\rho, n] & \cdots & p_{1N}^{(\varepsilon)}[\rho, n] \\ \vdots & \vdots & \vdots \\ p_{N1}^{(\varepsilon)}[\rho, n] & \cdots & p_{NN}^{(\varepsilon)}[\rho, n] \end{bmatrix},$$

and, for $r = 0, \ldots, k - n$, $n = 1, \ldots, k$,

$$\mathbf{E}[\rho, r, n] = \begin{bmatrix} e_{11}[\rho, r, n] & \cdots & e_{1N}[\rho, r, n] \\ \vdots & \vdots & \vdots \\ e_{N1}[\rho, r, n] & \cdots & e_{NN}[\rho, r, n] \end{bmatrix}.$$

Condition $\mathbf{P}_{25}^{(\rho,k)}$ can be rewritten in the following equivalent matrix form:

$\mathbf{P}_{25}^{(\rho,k)}$: $\mathbf{P}^{(\varepsilon)}[\rho, n] = \mathbf{P}^{(0)}[\rho, n] + \mathbf{E}[\rho, 1, n]\varepsilon + \cdots + \mathbf{E}[\rho, k - n, n]\varepsilon^{k-n} + \mathbf{o}(\varepsilon^{k-n})$,
where $\mathbf{P}^{(0)}[\rho, n]$ and $\mathbf{E}[\rho, r, n]$, $r = 0, \ldots, k - n$, $n = 0, \ldots, k$ are finite matrices.

For $n = 0, \ldots, k$, $j \neq 0$, consider the matrices

$${}_j\mathbf{P}^{(\varepsilon)}[\rho, n] =$$

$$\begin{bmatrix} p_{11}^{(\varepsilon)}[\rho, n] & \cdots & p_{1(j-1)}^{(\varepsilon)}[\rho, n] & 0 & p_{1(j+1)}^{(\varepsilon)}[\rho, n] & \cdots & p_{1N}^{(\varepsilon)}[\rho, n] \\ \vdots & & \vdots & \vdots & \vdots & & \vdots \\ p_{N1}^{(\varepsilon)}[\rho, n] & \cdots & p_{N(j-1)}^{(\varepsilon)}[\rho, n] & 0 & p_{N(j+1)}^{(\varepsilon)}[\rho, n] & \cdots & p_{NN}^{(\varepsilon)}[\rho, n] \end{bmatrix},$$

and also, for $r = 0, \ldots, k - n$, $n = 1, \ldots, k$ and $j \neq 0$,

$${}_j\mathbf{E}[\rho, r, n] =$$

$$\begin{bmatrix} e_{11}[\rho, r, n] & \cdots & e_{1(j-1)}[\rho, r, n] & 0 & e_{1(j+1)}[\rho, r, n] & \cdots & e_{1N}[\rho, r, n] \\ \vdots & & \vdots & \vdots & \vdots & & \vdots \\ e_{N1}[\rho, r, n] & \cdots & e_{N(j-1)}[\rho, r, n] & 0 & e_{N(j+1)}[\rho, r, n] & \cdots & e_{NN}[\rho, r, n] \end{bmatrix}.$$

Condition $\mathbf{P}_{23}^{(\rho,k)}$ implies that for $j \neq 0$, the following matrix asymptotic expansions take place for $n = 0, \ldots, k$:

$${}_j\mathbf{P}^{(\varepsilon)}[\rho, n] = {}_j\mathbf{P}^{(0)}[\rho, n] + {}_j\mathbf{E}[\rho, 1, n]\varepsilon + \cdots$$

$$+ {}_j\mathbf{E}[\rho, k - n, n]\varepsilon^{k-n} + \mathbf{o}(\varepsilon^{k-n}). \tag{5.3.22}$$

Let $e_{ij}[\rho, 0, n] = p_{ij}^{(0)}[\rho, n]$, $n = 0, \ldots, k$, $i, j \neq 0$, and, consequently, $\mathbf{E}[\rho, 0, n] = \mathbf{P}^{(0)}[\rho, n]$ and ${}_j\mathbf{E}[\rho, 0, n] = {}_j\mathbf{P}^{(0)}[\rho, n]$, $n = 0, \ldots, k$, $j \neq 0$. We shall also use the notation ${}_j\mathbf{P}^{(\varepsilon)}(\rho) = {}_j\mathbf{P}^{(\varepsilon)}[\rho, 0]$.

For $r = 0, \ldots, k - n, n = 0, \ldots, k$ and $j \neq 0$, consider the vectors

$$\mathbf{p}_j^{(\varepsilon)}[\rho, n] = \begin{bmatrix} p_{1j}^{(\varepsilon)}[\rho, n] \\ \vdots \\ p_{Nj}^{(\varepsilon)}[\rho, n] \end{bmatrix}, \quad \mathbf{e}_j[\rho, r, n] = \begin{bmatrix} e_{1j}[\rho, r, n] \\ \vdots \\ e_{Nj}[\rho, r, n] \end{bmatrix}.$$

Condition $\mathbf{P}_{23}^{(\rho, k)}$ gives the following vector asymptotic expansion for $j \neq 0$:

$$\mathbf{p}_j^{(\varepsilon)}(\rho) = \mathbf{p}_j^{(0)}(\rho) + \mathbf{e}_j[\rho, 1, 0]\varepsilon + \cdots + \mathbf{e}_j[\rho, k, 0]\varepsilon^k + \mathbf{o}(\varepsilon^k). \tag{5.3.23}$$

Condition $\mathbf{P}_{25}^{(\rho, k)}$, for $j \neq 0$ and $n = 0, \ldots, k$, yields

$$\mathbf{p}_j^{(\varepsilon)}[\rho, n] = \mathbf{p}_j^{(0)}[\rho, n] + \mathbf{e}_j[\rho, 1, n]\varepsilon + \cdots$$
$$+ \mathbf{e}_j[\rho, k - n, n]\varepsilon^{k-n} + \mathbf{o}(\varepsilon^{k-n}). \tag{5.3.24}$$

Recall that $e_{ij}[\rho, 0, n] = p_{ij}^{(0)}[\rho, n], n = 0, \ldots, k, i, j \neq 0$. Hence, $\mathbf{e}_j[\rho, 0, n] = \mathbf{p}_j^{(0)}[\rho, n], n = 0, \ldots, k, j \neq 0$. We shall also use the notation $\mathbf{p}_j^{(\varepsilon)}(\rho) = \mathbf{p}_j^{(\varepsilon)}[\rho, 0]$.

It is useful to note that the perturbation conditions $\mathbf{P}_{23}^{(0, k)}$ and $\mathbf{P}_{25}^{(0, k)}$, in the case where $\rho = 0$, reduce to more simple conditions $\mathbf{P}_{18}^{(k)}$ and $\mathbf{P}_{20}^{(k)}$, respectively,

5.3.5   Recursive systems of linear equations for mixed power-exponential moments of hitting times

Recall that the moment generating functions for hitting times, introduced in Section 4.5, are given by

$$_0\phi_{ij}^{(\varepsilon)}(\rho) = \int_0^\infty e^{\rho s}\, _0G_{ij}^{(\varepsilon)}(ds) = \mathsf{E}_i e^{\rho \mu_j^{(\varepsilon)}} \chi\{v_j^{(\varepsilon)} < v_0^{(\varepsilon)}\}, \quad \rho \in \mathbb{R}_1, \quad i, j \neq 0,$$

and the mixed power-exponential moments for hitting times for $n = 0, 1, \ldots$ are

$$_0\phi_{ij}^{(\varepsilon)}[\rho, n] = \int_0^\infty s^n e^{\rho s}\, _0G_{ij}^{(\varepsilon)}(ds) = \mathsf{E}_i e^{\rho \mu_j^{(\varepsilon)}} \chi\{v_j^{(\varepsilon)} < v_0^{(\varepsilon)}\}, \quad \rho \in \mathbb{R}_1, \quad i, j \neq 0.$$

By the definition, $_0\phi_{ij}^{(\varepsilon)}[\rho, 0] = {}_0\phi_{ij}^{(\varepsilon)}(\rho), i, j \neq 0$.

Assume that conditions $\mathbf{A}_2$, $\mathbf{E}_2$, $\mathbf{T}_2$, $\mathbf{C}_{17}$, and $\mathbf{C}_{18}$ hold.

As was mentioned above, Lemma 4.5.6 implies that **(a)** there exist $\beta < \beta_i$, where β_i are given in condition $\mathbf{C}_{18}$, and $0 < \varepsilon_1' = \varepsilon_1(\beta) \leq \varepsilon_0'(\beta)$ such that relation (5.3.5) holds and the inverse matrix $[\mathbf{I} - {}_j\mathbf{P}^{(\varepsilon)}(\rho)]^{-1}$ exists for every $\rho \leq \beta, j \in X_1^{(0)}$ and $\varepsilon \leq \varepsilon_1'$.

Lemma 4.5.6 also implies that **(b)** $0 < \beta < \beta_i$ can be chosen so that, for every $\rho \leq \beta$, $i \neq 0$, $j \in X_1^{(0)}$, and $\varepsilon \leq \varepsilon_1'$, we have

$$_0\phi_{ij}^{(\varepsilon)}(\rho) \to {}_0\phi_{ij}^{(0)}(\rho) < \infty \text{ as } \varepsilon \to 0. \tag{5.3.25}$$

Moreover, what is also very important for further consideration is that **(c)** $0 < \beta < \beta_i$ can be chosen in such way that, for every $\varepsilon \leq \varepsilon_1'$, there exists a unique root $\rho^{(\varepsilon)} \geq 0$ of the characteristic equation $_0\phi_{ij}^{(\varepsilon)}(\rho) = 1$, which **(c$_1$)** does not depend on the choice of the state $j \in X_1^{(0)}$, **(c$_2$)** satisfies the inequality $\rho^{(\varepsilon)} < \beta$, and is such that **(c$_3$)** $\rho^{(\varepsilon)} \to \rho^{(0)}$ as $\varepsilon \to 0$.

Relation (5.3.25) implies that **(d)** there exists $0 < \varepsilon_2' = \varepsilon_2'(\beta) < \varepsilon_1'(\beta)$ such that, for $i \neq 0$, $j \in X_1^{(0)}$,

$$\sup_{\varepsilon \leq \varepsilon_2} {}_0\phi_{ij}^{(\varepsilon)}(\beta) < \infty. \tag{5.3.26}$$

Choose an arbitrary $\rho^{(0)} < \beta' < \beta$. Then, for every $n = 0, 1, \ldots, i \neq 0$, $j \in X_1^{(0)}$,

$$_0\phi_{ij}^{(\varepsilon)}[\beta', n] = \int_0^\infty s^n e^{\beta' s} {}_0G_{ij}^{(\varepsilon)}(ds)$$

$$\leq c_n' \int_0^\infty e^{\beta s} {}_0G_{ij}^{(\varepsilon)}(ds) = c_n' {}_0\phi_{ij}^{(\varepsilon)}(\beta), \tag{5.3.27}$$

where $c_n' = c_n'(\beta, \beta') = \sup_{s \geq 0} s^n e^{-(\beta - \beta')s} < \infty$.

Relation (5.3.27) implies that **(e)** for every $n = 0, 1, \ldots, i \neq 0$, $j \in X_1^{(0)}$,

$$\sup_{\varepsilon \leq \varepsilon_2} {}_0\phi_{ij}^{(\varepsilon)}[\beta, n] < \infty. \tag{5.3.28}$$

The properties **(c$_1$)**–**(c$_3$)** imply that **(f)** there exists $0 < \varepsilon_3' = \varepsilon_3'(\beta, \beta') \leq \varepsilon_2(\beta)$ such that

$$\sup_{\varepsilon \leq \varepsilon_3} \rho^{(\varepsilon)} < \beta'. \tag{5.3.29}$$

Relation (5.3.29) allows us to work with perturbation conditions for mixed power-exponential moments of hitting times for $\rho = \rho^{(0)}$ and to get corresponding asymptotic expansions for the roots $\rho^{(\varepsilon)}$. These expansions will be obtained in the next section. To do this, however, there is no need in obtaining asymptotic expansions for the mixed power-exponential moments $_0\phi_{jj}^{(\varepsilon)}[\rho, n]$.

By using the Markov property of the semi-Markov process $\eta^{(\varepsilon)}(t)$ at the moment of the first jump and calculating the corresponding expectation, which depends on the

position of this process at the moment of first jump, **(f)** we obtain the following system
of linear equations for every $\rho \leq \beta'$, $j \in X_1^{(0)}$ and $\varepsilon \leq \varepsilon_3$:

$$_0\phi_{ij}^{(\varepsilon)}(\rho) = p_{ij}^{(\varepsilon)}(\rho) + \sum_{l \neq j,0} p_{il}^{(\varepsilon)}(\rho)_0\phi_{lj}^{(\varepsilon)}(\rho), \quad i \neq 0. \tag{5.3.30}$$

It follows from relation (5.3.19) that **(g)** for every $n \geq 1$ there exists a derivative of
order n of the moment generating function $\psi_{ij}^{(\varepsilon)}(\rho)$ and this derivative is the function
$\psi_{ij}^{(\varepsilon)}[\rho, n]$. This is true for every $i, j \neq 0$, $\rho \leq \beta$, and $\varepsilon \leq \varepsilon_3$.

It easily follows from relation (5.3.28) that **(h)** for every $n = 1, \ldots$ the moment
generating function $\varphi_{ij}^{(\varepsilon)}(\rho)$ has a derivative of order n, and it is the function $\varphi_{ij}^{(\varepsilon)}[\rho, n]$.
This is true for every $i \neq 0$, $j \in X_1^{(0)}$, $\rho \leq \beta'$, and $\varepsilon \leq \varepsilon_3$.

Using statements **(f)**–**(h)**, we can differentiate equations (5.3.30) and get the follow-
ing system of linear equation for every $j \in X_1^{(0)}$, $\rho \leq \beta'$, $\varepsilon \leq \varepsilon_3$, and $n = 0, 1, \ldots$:

$$\phi_{ij}^{(\varepsilon)}[\rho, n] = \lambda_{ij}^{(\varepsilon)}[\rho, n] + \sum_{l \neq 0,j} p_{il}^{(\varepsilon)}(\rho)\phi_{lj}^{(\varepsilon)}[\rho, n], \quad i \neq 0, \tag{5.3.31}$$

where

$$\lambda_{ij}^{(\varepsilon)}[\rho, n] = p_{ij}^{(\varepsilon)}[\rho, n] + \sum_{l \neq 0,j} \sum_{m=1}^{n} C_n^m p_{il}^{(\varepsilon)}[\rho, m]\phi_{lj}^{(\varepsilon)}[\rho, n - m]. \tag{5.3.32}$$

Note that $\lambda_{ij}^{(\varepsilon)}[\rho, 0] = p_{ij}^{(\varepsilon)}[\rho, 0] = p_{ij}^{(\varepsilon)}(\rho)$, $i \neq 0$, $j \in X_1^{(0)}$.
Thus, system (5.3.31) coincides with system (5.3.30) for $n = 0$.

These systems can be rewritten in a matrix form. For $j \in X_1^{(0)}$ and $n = 0, \ldots$,
consider the vectors

$$_0\boldsymbol{\phi}_j^{(\varepsilon)}[\rho, n] = \begin{bmatrix} _0\phi_{1j}^{(\varepsilon)}[\rho, n] \\ \vdots \\ _0\phi_{Nj}^{(\varepsilon)}[\rho, n] \end{bmatrix}, \quad \boldsymbol{\lambda}_j^{(\varepsilon)}[\rho, n] = \begin{bmatrix} \lambda_{1j}^{(\varepsilon)}[\rho, n] \\ \vdots \\ \lambda_{Nj}^{(\varepsilon)}[\rho, n] \end{bmatrix}.$$

Then, systems (5.3.30) (see, also, (4.5.3)) can be rewritten for every $j \in X_1^{(0)}$,
$\rho \leq \beta'$, $\varepsilon \leq \varepsilon_3$, as

$$_0\boldsymbol{\phi}_j^{(\varepsilon)}(\rho) = \mathbf{p}_j^{(\varepsilon)}(\rho) + {}_j\mathbf{P}^{(\varepsilon)}(\rho)\,_0\boldsymbol{\phi}_j^{(\varepsilon)}(\rho), \tag{5.3.33}$$

and systems (5.3.31) for every $j \in X_1^{(0)}$, $\rho \leq \beta'$, $\varepsilon \leq \varepsilon_3$, and $n = 0, 1, \ldots$ become

$$_0\boldsymbol{\phi}_j^{(\varepsilon)}[\rho, n] = \boldsymbol{\lambda}_j^{(\varepsilon)}[\rho, n] + {}_j\mathbf{P}^{(\varepsilon)}(\rho)\,_0\boldsymbol{\phi}_j^{(\varepsilon)}[\rho, n]. \tag{5.3.34}$$

304   5 Nonlinearly perturbed semi-Markov processes

If $n = 0$, system (5.3.34) coincides with system (5.3.33), since $\lambda_j^{(\varepsilon)}[\rho, 0] = \mathbf{p}_j^{(\varepsilon)}(\rho)$, $j \in X_1^{(0)}$.

The systems of linear equations (5.3.34) are systems of ρ-hitting type with the same coefficient matrix $_j\mathbf{P}^{(\varepsilon)}(\rho)$.

As was pointed above, conditions $\mathbf{A_2}$, $\mathbf{T_1}$, $\mathbf{E_2}$, $\mathbf{C_{17}}$, $\mathbf{C_{18}}$, and $\mathbf{S_1}$ imply that there exists $\varepsilon_0 > 0$ such that for every $\varepsilon \leq \varepsilon_0$ any ρ-hitting type system of linear equation with the coefficient matrix $_j\mathbf{P}^{(\varepsilon)}(\rho)$ has a unique solution for $j \in X_1^{(0)}$, $\rho \leq \beta'$, and $\varepsilon \leq \varepsilon_3$.

Obviously, the vectors $\lambda_j^{(\varepsilon)}[\rho, n]$ have non-negative components for every $j \in X_1^{(0)}$, $\rho \leq \beta'$, $\varepsilon \leq \varepsilon_3$, and $n = 0, \dots$.

In the case of the mixed power-exponential moments of hitting times, this solution has the following form for $j \in X_1^{(0)}$, $\rho \leq \beta'$, $\varepsilon \leq \varepsilon_3'$, and $n = 0, \dots$:

$$_0\boldsymbol{\phi}_j^{(\varepsilon)}[\rho, n] = [\mathbf{I} - {}_j\mathbf{P}^{(\varepsilon)}(\rho)]^{-1} \lambda_j^{(\varepsilon)}[\rho, n]. \tag{5.3.35}$$

As was mentioned above, systems (5.3.34), for a given $j \in X_1^{(0)}$, have the same coefficient matrix $_j\mathbf{P}^{(\varepsilon)}(\rho)$ but different inhomogeneous terms $\lambda_j^{(\varepsilon)}[\rho, n]$ for $n = 0, 1, \dots$. These systems should be solved recursively.

First, the system (5.3.34) should be solved for $n = 0$. Note that it actually becomes a system of linear equations for the moment generating functions $_0\phi_{ij}^{(\varepsilon)}(\rho)$, $i \neq 0$.

Second, we solve system (5.3.34) for $n = 1$. Note that expressions for the inhomogeneous terms $\lambda_{ij}^{(\varepsilon)}[\rho, 1] = p_{ij}^{(\varepsilon)}[\rho, 1] + \sum_{l \neq 0, j} p_{il}^{(\varepsilon)}[\rho, 1]{}_0\phi_{lj}^{(\varepsilon)}[\rho, 0]$, $i \neq 0$, given in (5.3.32) include the solutions $_0\phi_{ij}^{(\varepsilon)}[\rho, 0]$, $i \neq 0$, of systems (5.3.34) for $n = 0$.

This recursive procedure should be repeated for $n = 1, \dots$. The expressions for the inhomogeneous terms $\lambda_{ij}^{(\varepsilon)}[\rho, n]$ given in (5.3.32) include the solutions $\phi_{ij}^{(\varepsilon)}[\rho, m]$, $i \neq 0$, of systems (5.3.34) for $m = 0, 1, \dots, n - 1$.

Formula (5.3.32) can be rewritten for every $j \in X_1^{(0)}$, $\rho \leq \beta'$, $\varepsilon \leq \varepsilon_3'$, and $n = 0, 1, \dots$ in the following vector form:

$$\lambda_j^{(\varepsilon)}[\rho, n] = \mathbf{p}_j^{(\varepsilon)}[\rho, n] + \sum_{m=1}^{n} C_n^m {}_j\mathbf{P}^{(\varepsilon)}[\rho, m] {}_0\boldsymbol{\phi}_j^{(\varepsilon)}[\rho, n - m]. \tag{5.3.36}$$

Using formula (5.3.36) we can rewrite formula (5.2.10) in a form representing the recursive procedure for calculating solutions of the system of linear equations (5.3.34) for $j \in X_1^{(0)}$, $\rho \leq \beta'$, $\varepsilon \leq \varepsilon_3'$, and $n = 0, 1, \dots$,

$$_0\boldsymbol{\phi}_j^{(\varepsilon)}[\rho, n] = [\mathbf{I} - {}_j\mathbf{P}^{(\varepsilon)}(\rho)]^{-1} \lambda_j^{(\varepsilon)}[\rho, n]$$

$$= [\mathbf{I} - {}_j\mathbf{P}^{(\varepsilon)}(\rho)]^{-1} \mathbf{p}_j^{(\varepsilon)}[\rho, n]$$

$$+ \sum_{m=1}^{n} C_n^m [\mathbf{I} - {}_j\mathbf{P}^{(\varepsilon)}(\rho)]^{-1} {}_j\mathbf{P}^{(\varepsilon)}[\rho, m] {}_0\boldsymbol{\phi}_j^{(\varepsilon)}[\rho, n - m]. \tag{5.3.37}$$

5.3.6   Asymptotic expansions for power-exponential moments of hitting times

The following lemmas are corollaries of Lemma 5.3.4 applied to the vectors of power moments of the hitting times, $_0\phi_j^{(\varepsilon)}[\rho, n]$, for $j \in X_1^{(0)}$ and $n = 0, 1, \ldots$.

The next lemma is a direct corollary of Lemma 5.3.4 applied to the vectors of the moment generating functions $_0\phi_j^{(\varepsilon)}(\rho) = {}_0\phi_j^{(\varepsilon)}[\rho, 0]$ for $j \in X_1^{(0)}$.

In Lemmas 5.3.5–5.3.8, $\rho^{(0)} < \beta' < \beta$, where β is defined in (5.3.25).

Lemma 5.3.5. *Let conditions* $\mathbf{A_2}$, $\mathbf{T_2}$, $\mathbf{E_2}$, $\mathbf{C_{17}}$, $\mathbf{C_{18}}$ *hold, and condition* $\mathbf{P}_{23}^{(\rho,k)}$ *be satisfied for some* $\rho \leq \beta'$. *Then*

(i) *For every* $j \in X_1^{(0)}$, *the vector* $_0\phi_j^{(\varepsilon)}(\rho)$ *has the following asymptotic expansion:*

$$_0\phi_j^{(\varepsilon)}(\rho) = {}_0\phi_j^{(0)}(\rho) + {}_0\mathbf{b}_j[\rho, 1, 0]\varepsilon + \cdots + {}_0\mathbf{b}_j[\rho, k, 0]\varepsilon^k + o(\varepsilon^k), \quad (5.3.38)$$

where $_0\mathbf{b}_j[\rho, r, 0]$, $r = 1, \ldots, k$, *are finite vectors.*

(ii) *The vector coefficients* $_0\mathbf{b}_j[\rho, r, 0]$ *are given by the recurrence formulas* $_0\mathbf{b}_j[\rho, 0, 0] = {}_0\phi_j^{(0)}(\rho) = {}_j\mathbf{U}[\rho, 0]\mathbf{p}_j^{(0)}(\rho) = [\mathbf{I} - {}_j\mathbf{P}^{(0)}(\rho)]^{-1}\mathbf{p}_j^{(0)}(\rho)$ *and, consequently, for* $r = 0, \ldots, k$,

$$_0\mathbf{b}_j[\rho, r, 0] = {}_j\mathbf{U}[\rho, 0]\Big(\mathbf{e}_j[\rho, r, 0] + \sum_{q=1}^{r} {}_j\mathbf{E}[\rho, q, 0]{}_0\mathbf{b}_j[\rho, r - q, 0]\Big). \quad (5.3.39)$$

(iii) *The vector coefficients* $_0\mathbf{b}_j[\rho, r, 0]$ *can also be calculated by the following formulas for* $r = 0, \ldots, k$:

$$_0\mathbf{b}_j[\rho, r, 0] = \sum_{p=0}^{r} {}_j\mathbf{U}[\rho, p]\,\mathbf{e}_j[\rho, r - p, 0]. \quad (5.3.40)$$

Proof. We apply Lemma 5.3.4 to the system of linear equations (5.3.34) for every $j \in X_1^{(0)}$ and $n = 0$. Here, the vector $\mathbf{p}_j^{(\varepsilon)}(\rho)$ is regarded as the vector $\mathbf{y}^{(\varepsilon)}$. The non-negativity condition $\mathbf{S_1}$ obviously holds. Also, the asymptotic relations (5.3.23) imply that, for every $j \in X_1^{(0)}$, condition $\mathbf{P}_{19}^{(k)}$ holds. Applying Lemma 5.3.4 to the system of linear equations (5.3.34) we finish the proof of Lemma 5.3.5. $\square$

The following lemma is a direct corollary of Lemma 5.3.4 applied to the vectors of mixed power-exponential moments $_0\phi_j^{(\varepsilon)}[\rho, n]$, $n = 0, 1, \ldots, k$ for $j \in X_1^{(0)}$.

Lemma 5.3.6. *Let conditions* $\mathbf{A_2}$, $\mathbf{T_2}$, $\mathbf{E_2}$, $\mathbf{C_{17}}$, $\mathbf{C_{18}}$ *hold, and condition* $\mathbf{P}_{25}^{(\rho,k)}$ *be satisfied for some* $\rho \leq \beta'$. *Then*

(i) *For every $j \in X_1^{(0)}$ and $n = 0, \ldots, k$, the vector $_0\boldsymbol{\phi}_j^{(\varepsilon)}[\rho, n]$ has the following asymptotic expansion:*

$$_0\boldsymbol{\phi}_j^{(\varepsilon)}[\rho, n] = {}_0\boldsymbol{\phi}_j^{(0)}[\rho, n] + {}_0\mathbf{b}_j[\rho, 1, n]\varepsilon + \cdots$$

$$+ {}_0\mathbf{b}_j[\rho, k - n, n]\varepsilon^{k-n} + \mathbf{o}(\varepsilon^{k-n}), \qquad (5.3.41)$$

where $_0\mathbf{b}_j[\rho, r, n]$, $r = 1, \ldots, k - n$, $n = 0, \ldots, k$, are finite vectors.

(ii) *The vector coefficients $_0\mathbf{b}_j[\rho, r, n]$ are given by the recurrence formulas $_0\mathbf{b}_j[\rho, 0, 0] = {}_0\boldsymbol{\phi}_j^{(0)}[\rho, 0] = {}_j\mathbf{U}[\rho, 0]\boldsymbol{\lambda}_j^{(0)}[\rho, 0] = [\mathbf{I} - {}_j\mathbf{P}^{(0)}(\rho)]^{-1}\mathbf{p}_j^{(0)}(\rho)$ for $r = 0, \ldots, k - n$ and fixed n and sequentially for, $n = 0, \ldots, k$,*

$$_0\mathbf{b}_j[\rho, r, n] = {}_j\mathbf{U}[\rho, 0](\mathbf{e}_j[\rho, r, n]$$

$$+ \sum_{m=1}^{n} C_n^m \sum_{q=0}^{r} {}_j\mathbf{E}[\rho, q, m]{}_0\mathbf{b}_j[\rho, r - q, n - m]$$

$$+ \sum_{p=1}^{r} \mathbf{E}_j[\rho, p, 0]{}_0\mathbf{b}_j[\rho, r - p, n]). \qquad (5.3.42)$$

(iii) *The vector coefficients $_0\mathbf{b}_j[\rho, r, n]$ can also be calculated by the recurrence formulas for $r = 0, \ldots, k - n$ for given and fixed n and sequentially for, $n = 0, \ldots, k$,*

$$_0\mathbf{b}_j[\rho, r, n] = \sum_{q=0}^{r} {}_j\mathbf{U}[\rho, q]\mathbf{e}_j[\rho, r - q, n]$$

$$+ \sum_{m=1}^{n} C_n^m \sum_{q=0}^{r} \left(\sum_{p=0}^{q} {}_j\mathbf{U}[\rho, p]{}_j\mathbf{E}[\rho, q - p, m] \right)$$

$$\times {}_0\mathbf{b}_j[\rho, r - q, n - m]. \qquad (5.3.43)$$

Proof. Fix some $j \in X_1^{(0)}$. The asymptotic expansions (5.3.41) constructed for the vectors $_0\boldsymbol{\phi}_j^{(\varepsilon)}[\rho, n]$ are obtained recursively for $n = 0, \ldots, k$.

For $n = 0$, we set $_0\boldsymbol{\lambda}_j^{(\varepsilon)}[0] = {}_0\mathbf{p}_j^{(\varepsilon)}(\rho)$. The asymptotic expansion for $n = 0$ in condition $\mathbf{P}_{25}^{(\rho,k)}$ coincides with the asymptotic expansion given in condition $\mathbf{P}_{23}^{(\rho,k)}$. The asymptotic expansion (5.3.41) for the vectors $_0\boldsymbol{\phi}_j^{(\varepsilon)}[\rho, 0]$ and formulas (5.3.42) and (5.3.43) for the corresponding vector coefficients coincide, respectively, with the asymptotic expansion (5.3.38) for the vectors $_0\boldsymbol{\phi}_j^{(\varepsilon)}(\rho)$ and formulas (5.3.39) and (5.3.40) for the corresponding vector coefficients. Note that this is an $(N, 1)$-matrix $(0, k)$-expansion.

Now we can write an asymptotic expansion for the vectors $\lambda_j^{(\varepsilon)}[\rho, 1] = \mathbf{p}_j^{(\varepsilon)}[\rho, 1] + {}_j\mathbf{P}^{(\varepsilon)}[\rho, 1]_0\phi_j^{(\varepsilon)}[\rho, 0]$ using Lemma 8.1.2. This will be an $(N, 1)$-matrix $(0, k - 1)$-expansion. Then, Lemma 5.1.3 can be applied to the vectors ${}_0\phi_j^{(\varepsilon)}[1] = [\mathbf{I} - {}_j\mathbf{P}^{(\varepsilon)}(\rho)]^{-1} \cdot \lambda_j^{(\varepsilon)}[\rho, 1]$. The resulted asymptotic expansion is an $(N, 1)$-matrix $(0, k - 1)$-expansion.

The recurrence procedure is repeated for $n = 1, \ldots, k$. At step n, the expansion for the vectors $\lambda_j^{(\varepsilon)}[\rho, n]$ given by formula (5.2.11) is an $(N, 1)$-matrix $(0, k-n)$-expansion that takes the following form:

$$\lambda_j^{(\varepsilon)}[\rho, n] = \lambda_j^{(0)}[\rho, n] + \lambda_j[\rho, 1, n]\varepsilon + \cdots$$
$$+ \lambda_j[\rho, k - n, n]\varepsilon^{k-n} + o(\varepsilon^{k-n}), \tag{5.3.44}$$

where $\lambda_j^{(0)}[\rho, n] = \lambda_j[\rho, 0, n]$ and the coefficients $\lambda_j[\rho, r, n]$, $r = 0, \ldots, k - n$, are given by the formulas

$$\lambda_j[\rho, r, n] = \mathbf{e}_j[\rho, r, n] + \sum_{m=1}^{n} C_n^m \sum_{q=0}^{r} {}_j\mathbf{E}[\rho, q, m]_0\mathbf{b}_j[\rho, r - q, n - m]. \tag{5.3.45}$$

Now, apply Lemma 5.3.4 to the vectors ${}_0\phi_j^{(\varepsilon)}[\rho, n] = [\mathbf{I} - {}_j\mathbf{P}^{(\varepsilon)}(\rho)]^{-1}\lambda_j^{(\varepsilon)}[\rho, n]$. The resulted asymptotic expansion is an $(N, 1)$-matrix $(0, k - n)$-expansion given by asymptotic expansion (5.3.41) and formulas (5.3.42) and (5.3.43) for the corresponding vector coefficients. $\qquad\square$

Remark 5.3.1. It should be noted that both formulas (5.3.42) and (5.3.43) for the corresponding vector coefficients in the asymptotic expansions (5.3.41) have a recursive character. However, they differ in their structure. For calculating the coefficients ${}_0\mathbf{b}_j[\rho, r, n]$ in (5.3.42) we use the coefficients ${}_0\mathbf{b}_j[\rho, l, m]$, $l = 0, \ldots, r$, $m = 0, \ldots, n - 1$. However, in formulas (5.3.43) the coefficients ${}_0\mathbf{b}_j[\rho, r, n]$ are calculated from the coefficients ${}_0\mathbf{b}_j[\rho, l, m]$, $l = 0, \ldots, r$, $m = 0, \ldots, n - 1$, and also the coefficients ${}_0\mathbf{b}_j[\rho, l, n]$, $l = 0, \ldots, r - 1$.

Take integer parameters $\tilde{n} \geq 1$ and $\tilde{k}_n \geq 0, n = 0, \ldots, \tilde{n}$, and define a vector parameter $\tilde{k} = (\tilde{n}, \tilde{k}_0, \ldots, \tilde{k}_{\tilde{n}})$.

Assume that the following perturbation condition, which is more general than $\mathbf{P}_{25}^{(\rho, \mathbf{k})}$, holds for some $\rho \leq \beta'$:

$\mathbf{P}_{27}^{(\rho, \tilde{\mathbf{k}})}$: $\quad p_{ij}^{(\varepsilon)}[\rho, n] = p_{ij}^{(0)}[\rho, n] + e_{ij}[\rho, 1, n]\varepsilon + \cdots + e_{ij}[\rho, \tilde{k}_n, n]\varepsilon^{\tilde{k}_n} + o(\varepsilon^{\tilde{k}_n})$ for

$\quad n = 0, \ldots, \tilde{n}, \, i, j \neq 0$, where $p_{ij}^{(0)}[\rho, n] < \infty, n = 0, \ldots, \tilde{n}, \, i, j \neq 0$, and

$\quad |e_{ij}[\rho, r, n]| < \infty, \, r = 1, \ldots, \tilde{k}_n, n = 0, \ldots, \tilde{n}, i, j \neq 0$.

Condition $\mathbf{P}_{27}^{(\rho, \tilde{\mathbf{k}})}$ reduces to condition $\mathbf{P}_{25}^{(\rho, \mathbf{k})}$ if $\tilde{n} = k, \tilde{k}_n = k - n, n = 0, \ldots, k$.

The following lemma generalises Lemma 5.2.1 and reduces to this lemma in the case mentioned above.

Lemma 5.3.7. *Let conditions* $\mathbf{A}_2$, $\mathbf{T}_2$, $\mathbf{E}_2$, $\mathbf{C}_{17}$, $\mathbf{C}_{18}$ *hold, and condition* $\mathbf{P}_{27}^{(\rho,\mathbf{k})}$ *be satisfied for some* $\rho \leq \beta'$. *Then*

(i) *The vector* $_0\boldsymbol{\phi}_j^{(\varepsilon)}[\rho, n]$ *has the following asymptotic expansion for every* $j \in X_1^{(0)}$ *and* $n = 0, \ldots, \tilde{n}$:

$$_0\boldsymbol{\phi}_j^{(\varepsilon)}[\rho, n] = {}_0\boldsymbol{\phi}_j^{(0)}[\rho, n] + {}_0\mathbf{b}_j[\rho, 1, n]\varepsilon + \cdots$$

$$+ {}_0\mathbf{b}_j[\rho, k_n, n]\varepsilon^{k_n} + o(\varepsilon^{k_n}), \tag{5.3.46}$$

where $k_n = \min(\tilde{k}_0, \ldots, \tilde{k}_n)$, $n = 0, \ldots, \tilde{n}$, *and* $_0\mathbf{b}_j[\rho, r, n]$, $r = 1, \ldots, k_n$, $n = 0, \ldots, \tilde{n}$, *are finite vectors.*

(ii) *The vector coefficients* $_0\mathbf{b}_j[\rho, r, n]$ *are given by the recurrence formulas* $_0\mathbf{b}_j[\rho, 0, 0] = {}_0\boldsymbol{\phi}_j^{(0)}[\rho, 0] = {}_j\mathbf{U}[\rho, 0]\boldsymbol{\lambda}_j^{(0)}[\rho, 0] = [\mathbf{I} - {}_j\mathbf{P}^{(0)}(\rho)]^{-1}\mathbf{p}_j^{(0)}(\rho)$ *for* $r = 0, \ldots, k_n$ *and sequentially for* $n = 0, \ldots, \tilde{n}$,

$$_0\mathbf{b}_j[\rho, r, n] = {}_j\mathbf{U}[\rho, 0](\mathbf{e}_j[\rho, r, n]$$

$$+ \sum_{m=1}^{n} C_n^m \sum_{q=0}^{r} {}_j\mathbf{E}[\rho, q, m] {}_0\mathbf{b}_j[\rho, r - q, n - m]$$

$$+ \sum_{p=1}^{r} \mathbf{E}_j[\rho, p, 0] {}_0\mathbf{b}_j[\rho, r - p, n]). \tag{5.3.47}$$

(iii) *The vector coefficients* $_0\mathbf{b}_j[\rho, r, n]$ *are also calculated by the recurrence formulas for* $r = 0, \ldots, k_n$ *and sequentially for* $n = 0, \ldots, \tilde{n}$,

$$_0\mathbf{b}_j[\rho, r, n] = \sum_{q=0}^{r} {}_j\mathbf{U}[\rho, q]\mathbf{e}_j[\rho, r - q, n]$$

$$+ \sum_{m=1}^{n} C_n^m \sum_{q=0}^{r} \left(\sum_{p=0}^{q} {}_j\mathbf{U}[\rho, p] {}_j\mathbf{E}[\rho, q - p, m] \right)$$

$$\times {}_0\mathbf{b}_j[\rho, r - q, n - m]. \tag{5.3.48}$$

Proof. The proof repeats the proof of Lemma 5.3.6. Lemma 5.3.7 differs from Lemma 5.3.6 in the orders of the corresponding asymptotic expansions for the power moments of the hitting times $_0\boldsymbol{\phi}_j^{(\varepsilon)}[\rho, n]$. These orders also should be calculated recursively.

As was mentioned above, $_0\boldsymbol{\phi}_j^{(\varepsilon)}[\rho, 0] = {}_0\boldsymbol{\phi}_j^{(\varepsilon)}(\rho)$ for $n = 0$. The asymptotic expansion in condition $\mathbf{P}_{27}^{(\rho,\tilde{\mathbf{k}})}$ given for $n = 0$ coincides with the asymptotic expansion in condition $\mathbf{P}_{23}^{(\rho,\mathbf{k})}$ if we set $k = \tilde{k}_0$. The asymptotic expansion (5.3.46) for the vectors $_0\mathbf{m}_j^{(\varepsilon)}[\rho, 0] = [\mathbf{I} - {}_j\mathbf{P}^{(\varepsilon)}(\rho)]^{-1}\boldsymbol{\lambda}_j^{(\varepsilon)}[\rho, 0]$ and formulas (5.3.47) and (5.3.48) for the corresponding vector coefficients just coincide, respectively, with the asymptotic expansion (5.3.38) for the vectors $_0\boldsymbol{\phi}_j^{(\varepsilon)}(\rho) = {}_0\boldsymbol{\phi}_j^{(\varepsilon)}[\rho, 0]$ and formulas (5.3.39) and (5.3.40) for the corresponding vector coefficients. Note that this is an $(N, 1)$-matrix $(0, k_0)$-expansion, where $k_0 = \tilde{k}_0$. Indeed, according to Lemma 8.1.4, the order of this expansion is the minimum of orders of the asymptotic expansions for the matrices $_j\mathbf{P}^{(\varepsilon)}(\rho)$ and the vectors $\boldsymbol{\lambda}_j^{(\varepsilon)}[\rho, 0]$. Both these expansions, by condition $\mathbf{P}_{23}^{(\rho, \mathbf{k})}$, have order $k = \tilde{k}_0$.

Now we can write an asymptotic expansion for the vectors $\boldsymbol{\lambda}_j^{(\varepsilon)}[\rho, 1] = \mathbf{p}_j^{(\varepsilon)}[\rho, 1] + {}_j\mathbf{P}^{(\varepsilon)}[\rho, 1]\,{}_0\boldsymbol{\phi}_j^{(\varepsilon)}[\rho, 0]$ using Lemma 8.1.2. This is an $(N, 1)$-matrix $(0, k_1)$-expansion, where $k_1 = \min(\tilde{k}_0, \tilde{k}_1)$. Indeed, by Lemma 5.1.3, the order of this expansion is the minimum of orders of the asymptotic expansions for the vectors $\mathbf{p}_j^{(\varepsilon)}[\rho, 1]$, matrices $_j\mathbf{P}^{(\varepsilon)}[\rho, 1]$, and the vectors $_0\boldsymbol{\phi}_j^{(\varepsilon)}[\rho, 0]$ that are, respectively, $\tilde{k}_1$, $\tilde{k}_1$, and $\tilde{k}_0$. This minimum is $k_1 = \min(\tilde{k}_0, \tilde{k}_1)$. Then, apply Lemma 5.1.3 to the vectors $_0\boldsymbol{\phi}_j^{(\varepsilon)}[\rho, 1] = [\mathbf{I} - {}_j\mathbf{P}^{(\varepsilon)}(\rho)]^{-1}\boldsymbol{\lambda}_j^{(\varepsilon)}[\rho, 1]$. The obtained asymptotic expansion is an $(N, 1)$-matrix $(0, k_1)$-expansion given in formula (5.3.46) for $n = 1$. According to Lemma 8.1.4, the order of this expansion is the minimum of orders of the asymptotic expansions for the matrices $_j\mathbf{P}^{(\varepsilon)}(\rho)$ and the vectors $\boldsymbol{\lambda}_j^{(\varepsilon)}[\rho, 1]$. These expansions have, respectively, the order $\tilde{k}_1$ and $k_1 = \min(\tilde{k}_0, \tilde{k}_1)$. Their minimum is k_1.

This recursive procedure can be repeated for $n = 1, \ldots, \tilde{n}$ that will yield the corresponding asymptotic expansions given in (5.3.46). $\square$

5.3.7   Pivotal properties of asymptotic expansions for mixed power-exponential moments of hitting times

For vectors $_0\mathbf{b}_j[r, n]$, $r = 0, \ldots, k - n$, $n = 0, \ldots, k$, we denote

$$_0\mathbf{b}_j[\rho, r, n] = \begin{bmatrix} _0b_{1j}[\rho, r, n] \\ \vdots \\ _0b_{Nj}[\rho, r, n] \end{bmatrix}.$$

Explicit expressions for components of the vectors $_0\mathbf{b}_j[r, n]$, $r = 0, \ldots, k - n$, $n = 0, \ldots, k$, can be derived from recurrence formulas (5.3.47) and (5.3.48), respectively.

Pivotal properties of asymptotic cyclic mixed power-exponential moments $_0\phi_{jj}^{(\varepsilon)}[\rho, n]$, $n = 0, \ldots, \tilde{n}$, $j \in X_1^{(0)}$ are simple.

Lemma 5.3.8. *Let conditions* $\mathbf{A}_2$, $\mathbf{T}_2$, $\mathbf{E}_2$, $\mathbf{I}_4$, $\mathbf{C}_{17}$, $\mathbf{C}_{18}$ *hold, and condition* $\mathbf{P}_{27}^{(\rho, \mathbf{k})}$ *be satisfied for some* $\rho \leq \beta'$. *Then the asymptotic expansions for the power moments of hitting probabilities* $_0\phi_{jj}^{(\varepsilon)}[\rho, n]$, $n = 0, \ldots, k$, $j \in X_1^{(0)}$, *are all 0-pivotal, i.e.,* *the corresponding zero coefficients are* $_0\phi_{jj}^{(0)}[\rho, n] = {}_0b_{jj}[\rho, 0, n] > 0$ *for all* $n = 0, \ldots, \tilde{n}$ *and* $j \in X_1^{(0)}$.

Proof. Condition $\mathbf{I_4}$ implies that **(a)** there exists a state $i' \in X_1^{(0)}$ and $j' \neq 0$ such that **(b)** $p_{i'j'}^{(0)} > 0$ and **(c)** $1 - F_{i'j'}^{(0)}(0) > 0$. Choose an arbitrary state $j \in X_1^{(0)}$. There are two cases to be considered. The first case is where $i' = j$. Here, **(b)** and **(c)** imply that **(d)** $1 - {}_0G_{jj}(0) > p_{i'j'}^{(0)}(1 - F_{i'j'}^{(0)}(0)) > 0$. The second case is where $i' \neq j$. There, **(e)** the cyclic hitting probability is $_0{}_jf_{ji'}^{(0)} > 0$, since i' is a recurrent-without-absorption state. Using **(b)**, **(c)**, and **(e)** we get that **(f)** $1 - {}_0G_{jj}(0) > {}_0{}_jf_{ji'}^{(0)} p_{i'j'}^{(0)}(1 - F_{i'j'}^{(0)}(0)) > 0$. Obviously, **(d)** as well as **(f)** implies that **(i)** $_0\phi_{jj}^{(0)}[\rho, n] > 0$, $n = 0, \ldots, \tilde{n}$. $\square$

Remark 5.3.2. Condition $\mathbf{I_4}$ is not required to be imposed on the moment generating functions $_0\phi_{jj}^{(\varepsilon)}[\rho, 0] = {}_0\phi_{jj}^{(\varepsilon)}(\rho)$ in the corresponding proposition, since $_0\phi_{jj}^{(0)}[\rho, 0] > 0$ by the definition.

The question about "higher" order pivotal properties of mixed power-exponential moments of hitting times is not simple. We are especially interested in such properties for the moment generating functions $_0\phi_{jj}^{(\varepsilon)}(\rho) = {}_0\phi_{jj}^{(\varepsilon)}[\rho, 0]$, since these assumptions are involved in the conditions of theorems about asymptotic exponential expansions in mixed ergodic and large deviation theorems for perturbed semi-Markov processes.

It would be nice to be able to formulate, for the moment generating functions $_0\phi_{jj}^{(\varepsilon)}(\rho)$, a statement about higher order coefficients in the corresponding asymptotic expansions similar to those formulated for the hitting probabilities $_0f_{jj}^{(\varepsilon)}$ in Lemma 5.1.9.

The proof of Lemma 5.1.9 is based on the use of the formula **(a)** $_0f_{jj}^{(\varepsilon)} = 1 - {}_jf_{j0}^{(\varepsilon)}$ which connects the hitting probabilities $_0f_{jj}^{(\varepsilon)}$ and the absorption probabilities $_jf_{j0}^{(\varepsilon)}$. This formula shows that the higher order coefficients in the asymptotic expansions for the hitting probabilities $_0f_{jj}^{(\varepsilon)}$ equal zero if the corresponding higher order coefficients in the asymptotic expansions for the absorption probabilities $_jf_{j0}^{(\varepsilon)}$ equal zero.

Unfortunately, an analogous formula that connects the moment generating functions $_0\phi_{jj}^{(\varepsilon)}(\rho)$ and $_j\phi_{j0}^{(\varepsilon)}(\rho)$ has not such a simple form, **(b)** $_0\phi_{jj}^{(\varepsilon)}(\rho) = \phi_{jj}^{(\varepsilon)}(\rho) - {}_j\phi_{j0}^{(\varepsilon)}(\rho)$, where $\phi_{jj}^{(\varepsilon)}(\rho) = \mathsf{E}_j e^{\rho(\mu_j^{(\varepsilon)} \wedge \mu_0^{(\varepsilon)})}$ and $_j\phi_{j0}^{(\varepsilon)}(\rho) = \mathsf{E}_j e^{\rho\mu_0^{(\varepsilon)}} \chi(v_0^{(\varepsilon)} < v_j^{(\varepsilon)})$.

It is possible to prove, for the moment generation functions $_j\phi_{j0}^{(\varepsilon)}(\rho)$, statements similar to those formulated in Lemma 5.1.7 on the absorption probabilities $_jf_{j0}^{(\varepsilon)}$, if we extend the perturbation condition $\mathbf{P_{23}^{(\rho,k)}}$ by assuming that the asymptotic expansion given in this condition also holds for $i \neq 0$, $j = 0$. However, this would not yield a desirable form of an asymptotic expansion for the moment generating functions $_0\phi_{jj}^{(\varepsilon)}(\rho)$ similar to those given for the hitting probabilities $_0f_{jj}^{(\varepsilon)}$ in Lemma 5.1.9.

This is so because the moment generating functions $\phi_{jj}^{(\varepsilon)}(\rho) \neq 1$ and the coefficients of asymptotic expansions for these functions can also contribute to higher order coefficients in the asymptotic expansions for $_0\phi_{jj}^{(\varepsilon)}(\rho)$ even if the corresponding higher order coefficients of $_j\phi_{j0}^{(\varepsilon)}(\rho)$ would vanish.

For this reason, we can not formulate conditions which would guarantee vanishing the higher order coefficients in the asymptotic expansions for the moment generating functions $_0\phi_{jj}^{(\varepsilon)}(\rho)$ in such an explicit form as for the hitting probabilities $_0f_{jj}^{(\varepsilon)}$ in Lemma 5.1.9. Recall that these conditions are $\mathbf{O}_3^{(h)}$ and $\mathbf{O}_4^{(h)}$ that are given in terms of the coefficients of the initial perturbation condition for the transition probabilities $p_{ij}^{(\varepsilon)}$.

We will replace these condition with a less explicit condition that imposes a "zero" assumption directly on the coefficients in the corresponding expansions for the moment generating functions $_0\phi_{jj}^{(\varepsilon)}(\rho)$. These conditions for $1 \leq h \leq k$ are the following:

$\mathbf{O}_5^{(\rho,\mathbf{h})}$: $b_{jj}[\rho, r, 0] = 0, r = 1,\ldots,h$ for some $j \in X_1^{(0)}$;

and

$\mathbf{O}_6^{(\rho,\mathbf{h})}$: $b_{jj}[\rho, r, 0] = 0, r = 1,\ldots,h - 1$ but $b_{jj}[\rho, h, 0] > 0$ for some $j \in X_1^{(0)}$.

The following lemma describes a corresponding solidarity property connected with conditions $\mathbf{O}_5^{(\rho,\mathbf{h})}$ and $\mathbf{O}_6^{(\rho,\mathbf{h})}$.

Note that the perturbation condition $\mathbf{P}_{27}^{(\rho,k)}$ is used in this lemma for the value $\rho = \rho^{(0)}$ which is the root of the equation $_0\phi_{jj}^{(0)}(\rho) = 1$.

Lemma 5.3.9. *Let conditions* $\mathbf{A}_2$, $\mathbf{T}_2$, $\mathbf{E}_2$, $\mathbf{C}_{17}$, $\mathbf{C}_{18}$ *hold, and condition* $\mathbf{P}_{27}^{(\rho^{(0)},k)}$ *be satisfied. Then*

(i) *If condition* $\mathbf{O}_5^{(\rho^{(0)},\mathbf{h})}$ *holds for some* $1 \leq h \leq k$, *then* $b_{jj}[\rho^{(0)}, n, 0] = 0$, $n = 1,\ldots,h$ *for every* $j \in X_1^{(0)}$.

(ii) *If condition* $\mathbf{O}_6^{(\rho^{(0)},\mathbf{h})}$ *holds for some* $1 \leq h \leq k$, *then* $b_{jj}[\rho^{(0)}, n, 0] = 0$, $n = 0,\ldots,h - 1$ *but* $b_{jj}[\rho^{(0)}, h, 0] > 0$ *for every* $j \in X_1^{(0)}$.

Proof. The statement of the lemma is trivial if the set $X_1^{(0)}$ contains only one state. Thus, we assume that this set contains at least two states. Let $j \in X_1^{(0)}$ be a state for which condition $\mathbf{O}_5^{(\rho^{(0)},\mathbf{h})}$ or $\mathbf{O}_6^{(\rho^{(0)},\mathbf{h})}$ holds, and $i \neq j, i \in X_1^{(0)}$.

It follows from relation (4.5.27) that **(g)** $(1 - _0\phi_{ii}^{(\varepsilon)}(\rho))(1 - _{0i}\phi_{jj}^{(\varepsilon)}(\rho)) = (1 - _0\phi_{jj}^{(\varepsilon)}(\rho))(1 - _{0j}\phi_{ii}^{(\varepsilon)}(\rho))$. As was shown in the proof of Lemma 4.5.3, this formula is valid for any $i \neq j, i, j \neq 0, \rho \in \mathbb{R}_1$.

Apply this formula to $\rho = \rho^{(0)}$. As, as was mentioned above, (**h**) $_0\phi_{jj}^{(0)}(\rho^{(0)})) = 1$, $j \in X_1^{(0)}$. Also, $_{0i}\phi_{jj}^{(\varepsilon)}(\rho^{(0)}))$, $_{0j}\phi_{ii}^{(\varepsilon)}(\rho^{(0)}) < 1$ for any $i \neq j, i, j \in X_1^{(\varepsilon)}$. It follows from Lemma 4.2.2 that (**i**) there exists $\varepsilon_0 > 0$ such that $X_1^{(0)} \subseteq X_1^{(\varepsilon)}$ for $\varepsilon \leq \varepsilon_0$.

Using (**g**)–(**i**) we can rewrite formula (4.5.27) for $\varepsilon \leq \varepsilon_0$ in the following form:

$$(_0\phi_{ii}^{(0)}(\rho^{(0)}) - _0\phi_{ii}^{(\varepsilon)}(\rho^{(0)}))$$

$$= (_0\phi_{jj}^{(0)}(\rho^{(0)}) - _0\phi_{jj}^{(\varepsilon)}(\rho^{(0)}))\frac{(1 - _{0j}\phi_{ii}^{(\varepsilon)}(\rho^{(0)}))}{(1 - _{0i}\phi_{jj}^{(\varepsilon)}(\rho^{(0)}))}. \tag{5.3.49}$$

As follows from Lemma 4.2.5, (**j**) $_{0i}\phi_{jj}^{(\varepsilon)}(\rho^{(0)}) \rightarrow _{0i}\phi_{jj}^{(0)}(\rho^{(0)})$ as $\varepsilon \rightarrow 0$ and $_{0j}\phi_{ii}^{(\varepsilon)}(\rho^{(0)}) \rightarrow _{0j}\phi_{ii}^{(0)}(\rho^{(0)})$ as $\varepsilon \rightarrow 0$ for $i \neq j$, $i, j \in X_1^{(0)}$. Thus,

$$\frac{(1 - _{0j}\phi_{ii}^{(\varepsilon)}(\rho^{(0)}))}{(1 - _{0i}\phi_{jj}^{(\varepsilon)}(\rho^{(0)}))} \rightarrow \frac{(1 - _{0j}\phi_{ii}^{(0)}(\rho^{(0)}))}{(1 - _{0i}\phi_{jj}^{(0)}(\rho^{(0)}))} > 0 \text{ as } \varepsilon \rightarrow 0. \tag{5.3.50}$$

If condition $\mathbf{O}_5^{(\rho^{(0)},\mathbf{h})}$ holds then, by the Lemma 5.3.7,

$$\varepsilon^{-h}(_0\phi_{jj}^{(0)}(\rho^{(0)}) - _0\phi_{jj}^{(\varepsilon)}(\rho^{(0)})) \rightarrow 0 \text{ as } \varepsilon \rightarrow 0. \tag{5.3.51}$$

Relations (5.3.49), (5.3.50), and (5.3.51) imply that, for $i \neq j, i, j \in X_1^{(0)}$,

$$\varepsilon^{-h}(_0\phi_{ii}^{(0)}(\rho^{(0)}) - _0\phi_{ii}^{(\varepsilon)}(\rho^{(0)})) \rightarrow 0 \text{ as } \varepsilon \rightarrow 0. \tag{5.3.52}$$

Relation (5.3.52) obviously implies that (**k**) $_0b_{ii}[\rho^{(0)}, n, 0] = 0, n = 1, \ldots, h$, for $i \neq j, i, j \in X_1^{(0)}$. Claim (**k**) proves the first statement of the lemma.

If condition $\mathbf{O}_6^{(\rho^{(0)},\mathbf{h})}$ holds, then by Lemma 5.3.7,

$$\varepsilon^{-h}(_0\phi_{jj}^{(0)}(\rho^{(0)}) - _0\phi_{jj}^{(\varepsilon)}(\rho^{(0)})) \rightarrow _0b_{jj}[\rho^{(0)}, h, 0] > 0 \text{ as } \varepsilon \rightarrow 0. \tag{5.3.53}$$

Relations (5.3.49), (5.3.53), and (5.3.51) imply that, for $i \neq j, i, j \in X_1^{(0)}$,

$$\varepsilon^{-h}(_0\phi_{ii}^{(0)}(\rho^{(0)}) - _0\phi_{ii}^{(\varepsilon)}(\rho^{(0)}))$$

$$\rightarrow _0b_{jj}[\rho^{(0)}, h, 0]\frac{(1 - _{0j}\phi_{ii}^{(0)}(\rho^{(0)}))}{(1 - _{0i}\phi_{jj}^{(0)}(\rho^{(0)}))} > 0 \text{ as } \varepsilon \rightarrow 0. \tag{5.3.54}$$

Relation (5.3.54) obviously implies that (**l**) $_0b_{ii}[\rho^{(0)}, n, 0] = 0, n = 1, \ldots, h - 1$, for $i \neq j, i, j \in X_1^{(0)}$ and also that (**m**) $_0b_{ii}[\rho^{(0)}, h, 0] > 0$ for $i \neq j, i, j \in X_1^{(0)}$. Now we use claims (**l**) and (**m**) to prove the second statement of the lemma. $\qquad\square$

5.4   Exponential expansions for nonlinearly perturbed semi-Markov processes

In this section we give pseudo-stationary exponential asymptotic expansions in mixed ergodic and large deviation theorems for perturbed semi-Markov processes.

5.4.1   Imbedding in the model of perturbed regenerative processes

We will be interested in obtaining mixed limit and large deviation theorems for the transition probabilities $P_{il}^{(\varepsilon)}(t) = \mathsf{P}_i\{\eta^{(\varepsilon)}(t) = l, \mu_0^{(\varepsilon)} > t\}$.

The process $\eta^{(\varepsilon)}(t), t \geq 0$, is a regenerative process with absorption, and the times of return to the initial state are the corresponding regeneration times. Also, $\mu_0^{(\varepsilon)}$ is a regenerative stopping time which regenerates together with the process $\eta^{(\varepsilon)}(t), t \geq 0$, at the times of return. Therefore, we can use theorems given in Chapter 3. The corresponding imbedding procedure is described in Section 4.2.

The basic renewal equation for the case where the corresponding imbedded regenerative process has not a transition period is given in relation (4.2.3), and it takes the following form for $j \in X_1^{(0)}, l \neq 0$:

$$P_{jl}^{(\varepsilon)}(t) = q_{jl}^{(\varepsilon)}(t) + \int_0^t P_{jl}^{(\varepsilon)}(t - s)\, {}_0G_{jj}^{(\varepsilon)}(ds), \quad t \geq 0, \tag{5.4.1}$$

where

$$q_{jl}^{(\varepsilon)}(t) = \mathsf{P}_j\{\eta^{(\varepsilon)}(t) = l, \mu_j^{(\varepsilon)} > t, \mu_0^{(\varepsilon)} > t\}, \quad t \geq 0.$$

The distribution of the hitting (return) time ${}_0G_{jj}^{(\varepsilon)}(t)$ plays the role of the distribution of regeneration times that generates this renewal equation.

A crucial role is played by the root of the characteristic equation,

$$\int_0^\infty e^{\rho s}\, {}_0G_{jj}^{(\varepsilon)}(ds) = 1. \tag{5.4.2}$$

Theorems that we obtain below improve the corresponding mixed limit and large deviation theorems given in Chapter 4, since we construct asymptotic expansions for the roots $\rho^{(\varepsilon)}$ of equation (5.4.2). These asymptotic expansions are based on corresponding asymptotic expansions for power and mixed power-exponential moments of the distributions of regeneration times. Here, we will use the asymptotic expansions for such moments constructed for the hitting times in Sections 5.1–5.3.

We shall also use the following condition (introduced in Subsection 3.4.2) that balances the rate of growth of time and the rate of the perturbation:

$\mathbf{B}_5^{(\mathbf{r})}$: $0 \leq t^{(\varepsilon)} \to \infty$ as $\varepsilon \to 0$ such that $\varepsilon^r t^{(\varepsilon)} \to \lambda_r$, where $0 \leq \lambda_r < \infty$.

5.4.2 Pseudo-stationary exponential expansions in mixed ergodic and large deviation theorems

In this section we present pseudo-stationary exponential asymptotic expansions in mixed ergodic and large deviation theorems for perturbed semi-Markov processes.

First of all note that the pseudo-stationary asymptotics appear in the case where the absorption from the states $i \in X_1^{(0)}$ is asymptotically impossible. This is the case where condition $\mathbf{O}_3^{(0)}$ holds, i.e., $p_{i0}^{(0)} = 0$ for every $i \in X_1^{(0)}$.

Theorem 5.4.1. *Let conditions* $\mathbf{A}_2$, $\mathbf{T}_1$, $\mathbf{E}_2$, $\mathbf{C}_{12}$, $\mathbf{N}_1$, $\mathbf{O}_3^{(0)}$, *and* $\mathbf{P}_{20}^{(k)}$ *hold. Then*

(i) *For all ε small enough there exists a unique nonnegative root $\rho^{(\varepsilon)}$ of equation (5.4.2), it does not depend on the choice of the state $j \in X_1^{(0)}$, and has the following asymptotic Taylor's expansion:*

$$\rho^{(\varepsilon)} = a_1 \varepsilon + \cdots + a_k \varepsilon^k + o(\varepsilon^k), \tag{5.4.3}$$

where the coefficients a_n, $n = 1, \ldots, k$, do not depend on the choice of state $j \in X_1^{(0)}$ and are given by the recurrence formulas $a_1 = -b_{jj}[1,0]/b_{jj}[0,1]$ for $n = 1, \ldots, k$,

$$a_n = -b_{jj}[0,1]^{-1}\left(b_{jj}[n,0] + \sum_{q=1}^{n-1} b_{jj}[n-q,1]a_q \right.$$

$$\left. + \sum_{m=2}^{n} \sum_{q=m}^{n} b_{jj}[n-q,m] \cdot \sum_{n_1,\ldots,n_{q-1} \in D_{m,q}} \prod_{p=1}^{q-1} a_p^{n_p}/n_p! \right), \tag{5.4.4}$$

where $D_{m,q}$, for every $2 \le m \le q < \infty$, is the set of all nonnegative, integer solutions of the system

$$n_1 + \cdots + n_{q-1} = m, \quad n_1 + \cdots + (q-1)n_{q-1} = q \tag{5.4.5}$$

and the coefficients $b_{jj}[r,n]$, $r = 0, \ldots, k-n$, $n = 0, \ldots, k$, for every $j \in X_1^{(0)}$, are given by the recurrence formulas (5.2.14) or (5.2.15) in Lemma 5.2.1.

(ii) *If condition $\mathbf{O}_3^{(h)}$ holds for some $1 \le h \le k$, then $a_1, \ldots, a_h = 0$. If $\mathbf{O}_4^{(h)}$ holds for some $1 \le h \le k$, then $a_1, \ldots, a_{h-1} = 0$ but $a_h > 0$.*

(iii) *If condition $\mathbf{B}_5^{(r)}$ holds for some $1 \le r \le k$, then the following asymptotic relation holds for $i, l \ne 0$:*

$$\frac{\mathsf{P}_i\{\eta^{(\varepsilon)}(t^{(\varepsilon)}) = l, \mu_0^{(\varepsilon)} > t^{(\varepsilon)}\}}{\exp\{-(a_1\varepsilon + \cdots + a_{r-1}\varepsilon^{r-1})t^{(\varepsilon)}\}}$$

$$\to (1 - {}_{X_1^{(0)}}f_{i0}^{(0)})\pi_l^{(0)} e^{-\lambda_r a_r} \quad as \ \varepsilon \to 0. \tag{5.4.6}$$

Remark 5.4.1. In asymptotic relation (5.4.6), $\pi_l^{(0)}$, $l \neq 0$, is the stationary distribution for the semi-Markov process $\eta^{(0)}(t)$, $t \geq 0$, given in formula (4.4.14), and $_{X_1^{(0)}} f_{i0}^{(0)}$ is the absorption probability given in formula (4.2.77) with the set $Z = X_1^{(0)}$.

Remark 5.4.2. If condition $\mathbf{E_2}$ is realised in the form of condition $\mathbf{E_2'}$, then $X_1^{(0)} = X_0 = \{1, \ldots, N\}$, and the stationary distribution $\pi_l^{(0)}$, $l \neq 0$, is given by a simpler formula (4.4.7). In this case, we also have that $_{X_1^{(0)}} f_{i0}^{(0)} = 0$, $i \neq 0$.

Proof. The proof is based on a use of the corresponding Theorems 3.4.2 and 3.4.7 for perturbed regenerative processes or on the improvements of the corresponding asymptotic relations in Theorems 4.6.1 and 4.6.2 for perturbed semi-Markov processes. The latter theorems yield the following asymptotic relation for $i, l \neq 0$ under conditions $\mathbf{A_2}, \mathbf{T_1}, \mathbf{E_2}, \mathbf{C_{13}}, \mathbf{N_1}$, and $\mathbf{O_3^{(0)}}$:

$$\frac{\mathsf{P}_i\{\eta^{(\varepsilon)}(t^{(\varepsilon)}) = l, \mu_0^{(\varepsilon)} > t^{(\varepsilon)}\}}{\exp\{-\rho^{(\varepsilon)} t^{(\varepsilon)}\}} \to (1 - {}_{X_1^{(0)}} f_{i0}^{(0)}) \pi_l^{(0)} \quad \text{as } \varepsilon \to 0. \tag{5.4.7}$$

We should add to this asymptotic relation the asymptotic expansion (5.4.3) for the root $\rho^{(\varepsilon)}$ of the characteristic equation (5.4.2). This can be done by using Theorem 2.1.2. The distribution function $_0 G_{jj}^{(\varepsilon)}(t)$ plays the role of the distribution function $F^{(\varepsilon)}(t)$ in the characteristic equation (2.1.4). Note that, according to Lemma 4.5.6, the root $\rho^{(\varepsilon)}$ does not depend on the choice of the state $j \in X_1^{(0)}$.

Lemmas 4.2.3, 4.5.6, and 4.5.8 imply that, if conditions $\mathbf{A_2}, \mathbf{T_1}, \mathbf{E_2}, \mathbf{C_{13}}, \mathbf{N_1}$, and $\mathbf{O_3^{(0)}}$ hold, then the distribution function $_0 G_{jj}^{(\varepsilon)}(t)$ satisfies conditions $\mathbf{D_6}$ and $\mathbf{C_1}$. Lemmas 5.1.6, 5.1.9, and 5.2.1 also imply that, if conditions $\mathbf{A_2}, \mathbf{T_1}, \mathbf{E_2}, \mathbf{C_{12}}, \mathbf{O_3^{(0)}}$, and $\mathbf{P_{20}^{(k)}}$ hold, the perturbation condition $\mathbf{P_1^{(k)}}$ is also satisfied. The hitting probabilities $_0 f_{jj}^{(\varepsilon)}$ and the power moments of hitting times $_0 M_{jj}^{(\varepsilon)}[r]$, $r = 1, \ldots, k$, replace in this case the probabilities $1 - f^{(\varepsilon)}$ and the moments $m_r^{(\varepsilon)}$, $r = 1, \ldots, k$, in condition $\mathbf{P_1^{(k)}}$.

The asymptotic expansion (2.1.25) takes in this case the form of the asymptotic expansion (5.4.3). This proves statement (**i**) of the theorem.

Lemmas 5.1.7 and 5.1.9 imply claim (**ii**).

By substituting the asymptotic expansion (5.4.3) in the asymptotic relation (5.4.7), we transform this relation into the following form:

$$\frac{\mathsf{P}_i\{\eta^{(\varepsilon)}(t^{(\varepsilon)}) = l, \mu_0^{(\varepsilon)} > t^{(\varepsilon)}\}}{\exp\{-(a_1\varepsilon + \cdots + a_k\varepsilon^k + o(\varepsilon^k))t^{(\varepsilon)}\}} \to (1 - {}_{X_1^{(0)}} f_{i0}^{(0)}) \pi_l^{(0)} \quad \text{as } \varepsilon \to 0. \tag{5.4.8}$$

If the balancing condition $\mathbf{B_5^{(r)}}$ holds for some $1 \leq r \leq k$, then (**a**) $\exp\{-a_r\varepsilon^r t^{(\varepsilon)}\} \to \exp\{-a_r\lambda_r\}$ as $\varepsilon \to 0$ and (**b**) $\exp\{-(a_{r+1}\varepsilon^{r+1} + \cdots + a_k\varepsilon^k + o(\varepsilon^k))t^{(\varepsilon)}\} \to 1$ as $\varepsilon \to 0$. Using (**a**) and (**b**) we can rewrite relation (5.4.8) in the desirable form (5.4.6). This proves assertion (**iii**) of the theorem. $\qquad\square$

Introduce the integer parameters $h \geq 1$, $k_0 \geq h$, $\tilde{n} = \tilde{n}(h) = \max(n : nh \leq k_0)$, and $k_0 \geq k_1 \geq \cdots \geq k_{\tilde{n}} \geq 1$. Let us also denote $\tilde{k}(h) = (\tilde{n}(h), k_0, \ldots, k_{\tilde{n}(h)})$.

Theorem 5.4.2. *Let conditions* $\mathbf{A_2}$, $\mathbf{T_1}$, $\mathbf{E_2}$, $\mathbf{C_{12}}$, $\mathbf{N_1}$, $\mathbf{O_3^{(h-1)}}$, *and* $\mathbf{P_{20}^{(\tilde{k}(h))}}$ *hold. Then*

(i) *For all ε small enough, equation (5.4.2) has a unique nonnegative root $\rho^{(\varepsilon)}$, which does not depend on the choice of the state $j \in X_1^{(0)}$ and has the following asymptotic expansion:*

$$\rho^{(\varepsilon)} = a_h \varepsilon + \cdots + a_k \varepsilon^k + o(\varepsilon^k), \tag{5.4.9}$$

where $k = \min(k_0, k_1+h, \ldots, k_{\tilde{n}(h)}+h\tilde{n}(h))$, the coefficients a_n, $n = h, \ldots, k$ do not depend on the choice of state $j \in X_1^{(0)}$ and are given by the recurrence formulas $a_h = -b_{jj}[h, 0]/b_{jj}[0, 1]$ and, for $n = h, \ldots, k$,

$$a_n = -b_{jj}[0, 1]^{-1} \Bigg(b_{jj}[n, 0] + \sum_{q=h}^{n-1} b_{jj}[n-q, 1]a_q$$

$$+ \sum_{m=2}^{[n/h]} \sum_{q=mh}^{n} b_{jj}[n-q, m] \cdot \sum_{n_1, \ldots, n_{q-1} \in D_{h,m,q}} \prod_{p=1}^{q-1} a_p^{n_p}/n_p! \Bigg), \tag{5.4.10}$$

where $D_{h,m,q}$, for every $h \geq 1$ and $2 \leq m \leq q < \infty$, is the set of all nonnegative, integer solutions of the system

$$n_h + \cdots + n_{q-1} = m, \quad hn_h + \cdots + (q-1)n_{q-1} = q, \tag{5.4.11}$$

and the coefficients $b_{jj}[r, n]$, $r = 0, \ldots, k_n$, $n = 0, \ldots, \tilde{n}(h)$ are given by the recurrence formulas (5.2.19) or (5.2.20) in Lemma 5.2.2 for every $j \in X_1^{(0)}$.

(ii) *If condition $\mathbf{O_3^{(\tilde{h})}}$ holds for some $h \leq \tilde{h} \leq k$, then $a_h, \ldots, a_{\tilde{h}} = 0$. If condition $\mathbf{O_4^{(\tilde{h})}}$ holds for some $h \leq \tilde{h} \leq k$, then $a_h, \ldots, a_{\tilde{h}-1} = 0$ but $a_{\tilde{h}} > 0$.*

(iii) *If condition $\mathbf{B_5^{(r)}}$ holds for some $1 \leq r \leq k$, then the following asymptotic relation holds for $i, l \neq 0$:*

$$\frac{\mathsf{P}_i\{\eta^{(\varepsilon)}(t^{(\varepsilon)}) = l, \mu_0^{(\varepsilon)} > t^{(\varepsilon)}\}}{\exp\{-(a_h \varepsilon^h + \cdots + a_{r-1} \varepsilon^{r-1})t^{(\varepsilon)}\}}$$

$$\to (1 - {}_{X_1^{(0)}} f_{i0}^{(0)}) \pi_l^{(0)} e^{-\lambda_r a_r} \quad \text{as } \varepsilon \to 0. \tag{5.4.12}$$

Proof. The proof is based on using of the corresponding Theorems 3.4.3 and 3.4.8 for perturbed regenerative processes and on the improvement of the corresponding asymptotic relations given in Theorems 4.6.1 and 4.6.2 for perturbed semi-Markov processes.

Again, the distribution function $_0G_{jj}^{(\varepsilon)}(t)$ plays in this case the role of the distribution function $F^{(\varepsilon)}(t)$ in the characteristic equation (2.1.4). The only difference with the proof of Theorem 5.4.1 is that the corresponding asymptotic expansion for the characteristic root $\rho^{(\varepsilon)}$ of the characteristic equation (5.4.2) takes another form with the first $h-1$ coefficients equal zero. This can be done by using Theorem 2.1.3 instead of Theorem 2.1.2. Lemmas 5.1.6, 5.1.9, and 5.2.1 imply that, under conditions $\mathbf{A_2}, \mathbf{T_1},$ $\mathbf{E_2}, \mathbf{C_{12}}, \mathbf{O_3^{(0)}}$, and $\mathbf{P_{20}^{(\tilde{k}(h))}}$, the perturbation condition $\mathbf{P_2^{(\tilde{k}(h))}}$ holds. The asymptotic expansion (2.1.36) takes in this case the form of the asymptotic expansion (5.4.9). This proves statement (**i**) of the theorem. The proof of statements (**ii**) and (**iii**) is analogous to those given in Theorem 5.4.1. $\qquad\Box$

5.4.3   Quasi-stationary exponential expansions in mixed ergodic and large deviation theorems

In this section, we obtain quasi-stationary exponential asymptotic expansions in mixed ergodic and large deviation theorems for perturbed semi-Markov processes.

Note that quasi-stationary asymptotics appear in the case where the absorption from the states $i \in X_1^{(0)}$ is asymptotically possible. This is the case where condition $\mathbf{O_4^{(0)}}$ holds, that is, $p_{i0}^{(0)} \neq 0$ for some $i \in X_1^{(0)}$.

It should be noted that the theorems formulated below do not require condition $\mathbf{O_4^{(0)}}$ to hold and cover both pseudo- and quasi-stationary cases. However, in the pseudo-stationary case, i.e., if condition $\mathbf{O_3^{(0)}}$ holds, Theorems 5.4.1 and 5.4.2 give more effective conditions for obtaining the corresponding asymptotic relations, in particular, those related to pivotal properties of these expansions.

In the quasi-stationary case, the pivotal properties of the asymptotic expansions for the moment generating functions $_0\phi_{jj}^{(\varepsilon)}$ can not be expressed so explicitly via the coefficients in the corresponding asymptotic expansion given in the initial perturbation conditions as it is done for the pseudo-stationary case. Conditions $\mathbf{O_3^{(h)}}$ and $\mathbf{O_4^{(h)}}$ should be replaced with the less explicit but more general conditions $\mathbf{O_5^{(\rho^{(0)},h)}}$ and $\mathbf{O_6^{(\rho^{(0)},h)}}$.

Theorem 5.4.3. *Let conditions* $\mathbf{A_2}, \mathbf{T_1}, \mathbf{E_2}, \mathbf{C_{12}}, \mathbf{C_{13}}, \mathbf{N_1},$ *and* $\mathbf{P_{25}^{(\rho^{(0)},k)}}$ *hold. Then*

(**i**) *For all ε small enough, equation (5.4.2) has a unique nonnegative root $\rho^{(\varepsilon)}$ which does not depend on the choice of the state $j \in X_1^{(0)}$ and has the following* *asymptotic expansion*:

$$\rho^{(\varepsilon)} = \rho^{(0)} + a_1 \varepsilon + \cdots + a_k \varepsilon^k + o(\varepsilon^k), \qquad (5.4.13)$$

where the coefficients a_n, $n = 1, \ldots, k$ do not depend on the choice of state $j \in X_1^{(0)}$ and are given by the recurrence formulas $a_1 = -b_{jj}[\rho^{(0)}, 1, 0]/ b_{jj}[\rho^{(0)}, 0, 1]$ and, for $n = 1, \ldots, k$,

$$a_n = - b_{jj}[\rho^{(0)}, 0, 1]^{-1} \left(b_{jj}[\rho^{(0)}, n, 0] + \sum_{q=1}^{n-1} b_{jj}[\rho^{(0)}, n - q, 1] a_q \right.$$

$$\left. + \sum_{m=2}^{n} \sum_{q=m}^{n} b_{jj}[\rho^{(0)}, n - q, m] \cdot \sum_{n_1, \ldots, n_{q-1} \in D_{m,q}} \prod_{p=1}^{q-1} a_p^{n_p} / n_p! \right), \qquad (5.4.14)$$

where $D_{m,q}$, for every $2 \leq m \leq q < \infty$, is the set of all nonnegative, integer solutions of the system

$$n_1 + \cdots + n_{q-1} = m, \quad n_1 + \cdots + (q - 1)n_{q-1} = q, \qquad (5.4.15)$$

and the coefficients $b_{jj}[\rho^{(0)}, r, n]$, $r = 0, \ldots, k - n$, $n = 0, \ldots, k$, are given by the recurrence formulas (5.3.39) or (5.3.40) in Lemma 5.3.5 for every $j \in X_1^{(0)}$.

(ii) *If condition $\mathbf{O}_5^{(\rho^{(0)}, \mathbf{h})}$ holds for some $1 \leq h \leq k$, then $a_1, \ldots, a_h = 0$. If condition $\mathbf{O}_6^{(\rho^{(0)}, \mathbf{h})}$ holds for some $1 \leq h \leq k$, then $a_1, \ldots, a_{h-1} = 0$ but $a_h > 0$.*

(iii) *If condition $\mathbf{B}_5^{(\mathbf{r})}$ holds for some $1 \leq r \leq k$, then the following asymptotic relation holds for $i, l \neq 0$:*

$$\frac{\mathsf{P}_i\{\eta^{(\varepsilon)}(t^{(\varepsilon)}) = l, \mu_0^{(\varepsilon)} > t^{(\varepsilon)}\}}{\exp\{-(\rho^{(0)} + a_1 \varepsilon + \cdots + a_{r-1}\varepsilon^{r-1})t^{(\varepsilon)}\}}$$

$$\to \tilde{\tilde{\pi}}_{il}^{(0)}(\rho^{(0)}) e^{-\lambda_r a_r} \quad as \ \varepsilon \to 0. \qquad (5.4.16)$$

Remark 5.4.3. In the asymptotic relation (5.4.16), the quantities $\tilde{\pi}_l^{(0)}(\rho^{(0)})$ and $\tilde{\tilde{\pi}}_{il}^{(0)}(\rho^{(0)}) = {}_0\phi_{il}^{(0)}(\rho^{(0)})\tilde{\pi}_l^{(0)}(\rho^{(0)})$ are given by formulas (4.6.27) and (4.6.26).

Proof. The proof is based on a use of the corresponding Theorems 3.4.4 and 3.4.9 for perturbed regenerative processes or on the improvement of the corresponding asymptotic relations given in Theorems 4.6.3 and 4.6.7 for perturbed semi-Markov processes. The latter theorems yield, under conditions $\mathbf{A}_2$, $\mathbf{T}_1$, $\mathbf{E}_2$, $\mathbf{C}_{12}$, $\mathbf{C}_{13}$, and $\mathbf{N}_1$, the following asymptotic relation for $i, l \neq 0$:

$$\frac{\mathsf{P}_i\{\eta^{(\varepsilon)}(t^{(\varepsilon)}) = l, \mu_0^{(\varepsilon)} > t^{(\varepsilon)}\}}{\exp\{-\rho^{(\varepsilon)}t^{(\varepsilon)}\}} \to \tilde{\tilde{\pi}}_{il}^{(0)}(\rho^{(0)}) \quad as \ \varepsilon \to 0. \qquad (5.4.17)$$

The asymptotic relation (5.4.17) should be supplemented with the asymptotic expansion (5.4.13) for the root $\rho^{(\varepsilon)}$ of the characteristic equation (5.4.2). This can be done by using Theorem 2.2.1, where the distribution function $_0G_{jj}^{(\varepsilon)}(t)$ replaces the distribution function $F^{(\varepsilon)}(t)$ in the characteristic equation (2.1.4). Note that, according to Lemma 4.5.6, the root $\rho^{(\varepsilon)}$ does not depend on the choice of the state $j \in X_1^{(0)}$.

Lemmas 4.2.3, 4.5.6, and 4.5.8 imply that, under conditions $\mathbf{A_2}$, $\mathbf{T_1}$, $\mathbf{E_2}$, $\mathbf{C_{12}}$, $\mathbf{C_{13}}$, and $\mathbf{N_1}$, the distribution function $_0G_{jj}^{(\varepsilon)}(t)$ satisfies conditions $\mathbf{D_{11}}$ and $\mathbf{C_2}$. Lemmas 5.3.5 and 5.3.6 also imply that the perturbation condition $\mathbf{P_3^{(k)}}$ holds under conditions $\mathbf{A_2}$, $\mathbf{T_1}$, $\mathbf{E_2}$, $\mathbf{C_{12}}$, $\mathbf{C_{13}}$, and $\mathbf{P_{25}^{(\rho^{(0)},k)}}$. The mixed power-exponential moments of hitting times, $_0\phi_{jj}^{(\varepsilon)}[\rho^{(0)},r]$, $r = 0,\ldots,k$, replace mixed moment generating functions $\phi^{(\varepsilon)}(\rho^{(0)},r)$, $r = 0,\ldots,k$, in condition $\mathbf{P_3^{(k)}}$.

The asymptotic expansion (2.2.2) takes in this case a form of the asymptotic expansion (5.4.13). This proves statement **(i)** of the theorem.

Lemma 5.3.9 implies the statement **(ii)** of the theorem.

By substituting the asymptotic expansion (5.4.13) in the asymptotic relation (5.4.17), we transform this relation to the following form:

$$\frac{\mathsf{P}_i\{\eta^{(\varepsilon)}(t^{(\varepsilon)}) = l, \mu_0^{(\varepsilon)} > t^{(\varepsilon)}\}}{\exp\{-(\rho^{(0)} + a_1\varepsilon + \cdots + a_k\varepsilon^k + o(\varepsilon^k))t^{(\varepsilon)}\}} \to \tilde{\tilde{\pi}}_{il}^{(0)}(\rho^{(0)}) \text{ as } \varepsilon \to 0. \quad (5.4.18)$$

If the balancing condition $\mathbf{B_5^{(r)}}$ holds for some $1 \le r \le k$, then **(a)** $\exp\{-a_r\varepsilon^r t^{(\varepsilon)}\} \to \exp\{-a_r\lambda_r\}$ as $\varepsilon \to 0$, and **(b)** $\exp\{-(a_{r+1}\varepsilon^{r+1} + \cdots + a_k\varepsilon^k + o(\varepsilon^k))t^{(\varepsilon)}\} \to 1$ as $\varepsilon \to 0$. Using **(a)** and **(b)** we can transform relation (5.4.18) into the desirable form (5.4.16). This proves the statement **(ii)** of the theorem. $\square$

Let us introduce the integer parameters $h \ge 1$, $k_0 \ge h$, $\tilde{n} = \tilde{n}(h) = \max(n : nh \le k_0)$, and $k_0 \ge k_1 \ge \cdots \ge k_{\tilde{n}} \ge 1$, and denote $\bar{k}(h) = (\tilde{n}(h), k_0,\ldots,k_{\tilde{n}(h)})$.

Theorem 5.4.4. *Let conditions* $\mathbf{A_2}$, $\mathbf{T_1}$, $\mathbf{E_2}$, $\mathbf{C_{12}}$, $\mathbf{C_{13}}$, $\mathbf{N_1}$, $\mathbf{O_5^{(\rho^{(0)},h-1)}}$, *and* $\mathbf{P_{27}^{(\rho^{(0)},\bar{k}(h))}}$ *hold. Then*

(i) *For all ε small enough there exists a unique nonnegative root $\rho^{(\varepsilon)}$ of equation (5.4.2), it does not depend on the choice of the state $j \in X_1^{(0)}$, and has the following asymptotic expansion:*

$$\rho^{(\varepsilon)} = \rho^{(0)} + a_h\varepsilon + \cdots + a_k\varepsilon^k + o(\varepsilon^k), \quad (5.4.19)$$

where $k = \min(k_0, k_1 + h,\ldots,k_{\tilde{n}(h)} + h\tilde{n}(h))$, the coefficients a_n, $n = h,\ldots,k$, do not depend on the choice of the state $j \in X_1^{(0)}$, and are given by the *recurrence formulas* $a_h = -b_{jj}[\rho^{(0)}, h, 0]/b_{jj}[\rho^{(0)}, 0, 1]$ *and, for* $n = h, \ldots, k,$

$$a_n = -b_{jj}[\rho^{(0)}, 0, 1]^{-1} \left(b_{jj}[\rho^{(0)}, n, 0] + \sum_{q=h}^{n-1} b_{jj}[\rho^{(0)}, n-q, 1]a_q \right.$$

$$\left. + \sum_{m=2}^{[n/h]} \sum_{q=mh}^{n} b_{jj}[\rho^{(0)}, n-q, m] \cdot \sum_{n_1,\ldots,n_{q-1} \in D_{h,m,q}} \prod_{p=1}^{q-1} a_p^{n_p}/n_p! \right), \quad (5.4.20)$$

where $D_{h,m,q}$, *for every* $h \geq 1$ *and* $2 \leq m \leq q < \infty$, *is the set of all nonnegative, integer solutions of the system*

$$n_h + \cdots + n_{q-1} = m, \quad hn_h + \cdots + (q-1)n_{q-1} = q, \quad (5.4.21)$$

and the coefficients $b_{jj}[\rho^{(0)}, r, n]$, $r = 0, \ldots, k_n$, $n = 0, \ldots, \tilde{n}(h)$, *are given, for every* $j \in X_1^{(0)}$, *by the recurrence formulas* (5.3.39) *or* (5.3.40) *in Lemma 5.3.5.*

(ii) *If condition* $\mathbf{O}_5^{(\rho^{(0)}, \tilde{h})}$ *holds for some* $h \leq \tilde{h} \leq k$, *then* $a_h, \ldots, a_{\tilde{h}} = 0$. *If condition* $\mathbf{O}_6^{(\rho^{(0)}, \tilde{h})}$ *holds for some* $h \leq \tilde{h} \leq k$, *then* $a_h, \ldots, a_{\tilde{h}-1} = 0$ *but* $a_{\tilde{h}} > 0$.

(iii) *If condition* $\mathbf{B}_5^{(r)}$ *holds for some* $1 \leq r \leq k$, *then the following asymptotic relation holds for* $i, l \neq 0$:

$$\frac{\mathsf{P}_i\{\eta^{(\varepsilon)}(t^{(\varepsilon)}) = l, \mu_0^{(\varepsilon)} > t^{(\varepsilon)}\}}{\exp\{-(\rho^{(0)} + a_h \varepsilon^h + \cdots + a_{r-1}\varepsilon^{r-1})t^{(\varepsilon)}\}}$$

$$\to \tilde{\tilde{\pi}}_{il}^{(0)}(\rho^{(0)})e^{-\lambda_r a_r} \quad as \ \varepsilon \to 0. \quad (5.4.22)$$

Proof. The proof uses the corresponding Theorems 3.4.5 and 3.4.31 for perturbed regenerative processes and the improvement of the corresponding asymptotic relations given in Theorems 4.6.1 and 4.6.2 for perturbed semi-Markov processes.

Again, the distribution function $_0G_{jj}^{(\varepsilon)}(t)$ plays the role of the distribution function $F^{(\varepsilon)}(t)$ in the characteristic equation (2.1.4). The only difference with the proof of Theorem 5.4.3 is that the corresponding asymptotic expansion for the characteristic root $\rho^{(\varepsilon)}$ of the characteristic equation (5.4.2) takes another form with the coefficients $a_1, \ldots, a-h-1$ equal zero. This can be done by using Theorem 2.2.2 instead of Theorem 2.2.1. Lemmas 5.1.6, 5.1.9, and 5.2.1 imply that, under conditions $\mathbf{A}_2$, $\mathbf{T}_1$, $\mathbf{E}_2$, $\mathbf{C}_{12}$, $\mathbf{C}_{13}$, and $\mathbf{P}_{27}^{(\rho^{(0)}, \tilde{k}(h))}$, the perturbation condition $\mathbf{P}_4^{(\tilde{k}(h))}$ holds. The asymptotic expansion (2.2.20) takes the form of the asymptotic expansion (5.4.19). This proves statement **(i)** of the theorem. The proof of statements **(ii)** and **(iii)** is analogous to those given for Theorem 5.4.3. $\qquad\square$

5.5 Asymptotics of quasi-stationary distributions for nonlinearly perturbed semi-Markov processes

In this section, we give asymptotic expansions for stationary and quasi-stationary distributions of perturbed semi-Markov processes.

5.5.1 Cyclic representation for stationary distribution of ergodic semi-Markov process

In this section we consider a model without absorption, where condition $\mathbf{A_4}$ holds, i.e., **(a)** there exists $\varepsilon_1 > 0$ such that $\sum_{i \neq 0} p_{i0}^{(\varepsilon)} = 0$ for $\varepsilon \leq \varepsilon_1$.

We also assume that condition $\mathbf{T_1}$ **(a)** holds, i.e., **(b)** $p_{ij}^{(\varepsilon)} \to p_{ij}^{(0)}$ as $\varepsilon \to 0$ for $i, j \neq 0$.

Condition $\mathbf{E_2}$ is also assumed to hold. According to Lemma 4.2.2, conditions $\mathbf{T_1}$ **(a)**, $\mathbf{A_4}$, and $\mathbf{E_2}$ imply that **(c)** there exists $\varepsilon_0 > 0$ such that the corresponding sets of recurrent states $X_1^{(\varepsilon)}$ for the imbedded Markov chains $\eta_n^{(\varepsilon)}, n = 0, 1, \ldots$, are connected by the relation $X_1^{(0)} \subseteq X_1^{(\varepsilon)}$ for every $\varepsilon \leq \varepsilon_0$.

We assume that condition $\mathbf{M_{13}^{(1)}}$ holds, i.e., **(d)** $p_{ij}^{(\varepsilon)}[1] \to p_{ij}^{(0)}[1] < \infty$ as $\varepsilon \to 0$ for $i, j \neq 0$.

This condition implies that **(e)** there exists $\varepsilon_2 > 0$ such $p_{ij}^{(\varepsilon)}[1] = p_{ij}^{(\varepsilon)} m_{ij}^{(\varepsilon)} < \infty$ for $i, j \neq 0$ and $\varepsilon \leq \varepsilon_2$.

Let us now introduce the expectations of sojourn times,

$$m_i^{(\varepsilon)} = \sum_{j \in X} p_{ij}^{(\varepsilon)}[1] = \sum_{j \in X} p_{ij}^{(\varepsilon)} m_{ij}^{(\varepsilon)}, \quad i \neq 0.$$

Note that, due to condition $\mathbf{A_4}$, moments $p_{i0}^{(\varepsilon)}[1] = p_{i0}^{(\varepsilon)} m_{i0}^{(\varepsilon)} = 0, i \neq 0$ for $\varepsilon \leq \varepsilon_1$.

Relations **(b)**–**(e)** imply that **(f)** $m_i^{(\varepsilon)} \to m_i^{(0)}$ as $\varepsilon \to 0$ for $i \neq 0$.

Finally, we assume that condition $\mathbf{I_4}$ holds. Obviously, statement **(e)** implies that **(g)** $m_i^{(0)} < \infty, i \neq 0$. In this case, condition $\mathbf{I_4}$ is equivalent to the relation **(h)** $\sum_{i \in X_1^{(0)}} m_i^{(0)} > 0$.

Relations **(f)**–**(h)** imply that **(i)** there exists $\varepsilon_3 > 0$ such that **(j)** $\sum_{i \in X_1^{(0)}} m_i^{(\varepsilon)} > 0$ for $\varepsilon \leq \varepsilon_3$.

If conditions $\mathbf{A_4}$, $\mathbf{T_1}$ **(a)**, $\mathbf{E_2}$, $\mathbf{M_{13}^{(1)}}$, and $\mathbf{I_4}$ hold, then the semi-Markov process $\eta^{(\varepsilon)}(t), t \geq 0$, is ergodic for every $\varepsilon \leq \varepsilon_4$, where $\varepsilon_4 = \min(\varepsilon_0, \varepsilon_1, \varepsilon_2, \varepsilon_3)$.

The semi-Markov process $\eta^{(\varepsilon)}(t), t \geq 0$, can be characterized as an ergodic process since, under the above conditions, the following relation holds for every $\varepsilon \leq \varepsilon_4$ and $i, l \neq 0$:

$$\mathsf{P}_i \left\{ \lim_{t \to \infty} \frac{1}{t} \int_0^t \chi(\eta^{(\varepsilon)}(s) = l)\, ds = \pi_l^{(\varepsilon)} \right\} = 1. \tag{5.5.1}$$

Here $\pi_l^{(\varepsilon)}$, $l \neq 0$, is a *stationary distribution* of the semi-Markov process $\eta^{(\varepsilon)}(t)$, $t \geq 0$. This distribution is given by the following formula for $j \in X_1^{(\varepsilon)}$:

$$\pi_l^{(\varepsilon)} = \frac{M_{jjl}^{(\varepsilon)}}{M_{jj}^{(\varepsilon)}}, \quad l \neq 0, \tag{5.5.2}$$

where

$$M_{ij}^{(\varepsilon)} = \mathsf{E}_i \mu_j^{(\varepsilon)}, \quad M_{ijl}^{(\varepsilon)} = \mathsf{E}_i \int_0^{\mu_j^{(\varepsilon)}} \chi(\eta^{(\varepsilon)}(s) = l)\, ds, \quad i, j, l \neq 0.$$

It is clear that the stationary distribution satisfies the relations (**k**) $\pi_l^{(\varepsilon)} \geq 0$, $l \neq 0$, and (**l**) $\sum_{l \neq 0} \pi_l^{(\varepsilon)} = 1$.

Also, (**m**) the stationary distribution does not depend on the choice of the state $j \in X_1^{(\varepsilon)}$ used to calculate the stationary distribution by formula (5.5.2).

Note also that (**n**) the limits in the relation (5.5.1) do not depend on the choice of the initial state $i \neq 0$.

A proof of the ergodic theorem, which uses the asymptotic relation (5.5.1), can be found, for example, in Silvestrov (1980a). Here we give a sketch of this proof.

Choose and fix a recurrent state $j \in X_1^{(\varepsilon)}$. For this state, we will carry out a cyclic construction as described below. For every $t \geq 0$, we have

$$\sum_{n=1}^{\upsilon_j^{(\varepsilon)}(t)} \kappa_{jl}^{(\varepsilon)}(n) \leq \int_0^t \chi(\eta^{(\varepsilon)}(s) = l)\, ds \leq \sum_{n=1}^{\upsilon_j^{(\varepsilon)}(t)+1} \kappa_{jl}^{(\varepsilon)}(n), \tag{5.5.3}$$

where (**o$_1$**) $\mu_j^{(\varepsilon)}(n) = \tau_{\upsilon_j^{(\varepsilon)}(n)}^{(\varepsilon)}$, $n = 0, 1, \ldots$ are the subsequent moments the semi-Markov process $\eta^{(\varepsilon)}(s)$ hits the state j; (**o$_2$**) $\upsilon_j^{(\varepsilon)}(0) = 0$, $\upsilon_j^{(\varepsilon)}(n) = \min(k > \upsilon_j^{(\varepsilon)}(n-1), \eta_k^{(\varepsilon)} = j)$, $n = 1, 2, \ldots$, are subsequent moments at which the imbedded Markov chain $\eta_n^{(\varepsilon)}$ hits the state j; (**o$_3$**) $\upsilon_j^{(\varepsilon)}(t) = \max(n : \mu_j^{(\varepsilon)}(n) \leq t)$ is, for given $t \geq 0$, the number of times the process $\eta^{(\varepsilon)}(s)$ hits the state j in the time interval $[0, t]$; (**o$_4$**) $\kappa_j^{(\varepsilon)}(n) = \mu_j^{(\varepsilon)}(n) - \mu_j^{(\varepsilon)}(n-1)$, $n = 1, 2, \ldots$, are times between subsequent moments at which the process $\eta^{(\varepsilon)}(s)$ hits the state j; and (**o$_5$**) $\kappa_{jl}^{(\varepsilon)}(n) = \int_{\mu_j^{(\varepsilon)}(n-1)}^{\mu_j^{(\varepsilon)}(n)} \chi(\eta^{(\varepsilon)}(s) = l)\, ds$, $n = 1, 2, \ldots$, are times spent by the process $\eta^{(\varepsilon)}(s)$ in the state l between subsequent moments the process hits the state j.

By the definition, the random variables $\kappa_j^{(\varepsilon)}(n)$, $n = 1, 2, \ldots$, are independent, identically distributed at least for $n \geq 2$, and $\mathsf{E}_j \kappa_j^{(\varepsilon)}(2) = M_{jj}^{(\varepsilon)}$. Thus, (**p**) $\mathsf{P}_i\{\lim_{n \to \infty} n^{-1} \mu_j^{(\varepsilon)}(n) = M_{jj}^{(\varepsilon)}\} = 1$ for $i \neq 0$, $j \in X_1^{(\varepsilon)}$. It easily follows from (**p**) and the definition of the random variables $v_j^{(\varepsilon)}(t)$ that **(q)** $\mathsf{P}_i\{\lim_{t\to\infty} t^{-1} v_j^{(\varepsilon)}(t) = (M_{jj}^{(\varepsilon)})^{-1}\} = 1$ for $i \neq 0$, $j \in X_1^{(\varepsilon)}$. By the definition, the random variables $\kappa_{jl}^{(\varepsilon)}(n)$, $n = 1, 2, \ldots$, are independent, identically distributed at least for $n \geq 2$, and $\mathsf{E}_j \kappa_{jl}^{(\varepsilon)}(2) = M_{jjl}^{(\varepsilon)}$. Thus, **(r)** $\mathsf{P}_i\{\lim_{n\to\infty} n^{-1} \sum_{r=1}^n \kappa_{jl}^{(\varepsilon)}(r) = M_{jjl}^{(\varepsilon)}\} = 1$ for $i \neq 0$, $j \in X_1^{(\varepsilon)}$. Relations **(q)**–**(r)** imply that **(s)** $\mathsf{P}_i\{\lim_{t\to\infty} t^{-1} \sum_{n=1}^{v_j^{(\varepsilon)}(t)} \kappa_{jl}^{(\varepsilon)}(n) = M_{jjl}^{(\varepsilon)}(M_{jj}^{(\varepsilon)})^{-1}\} = 1$ for $i \neq 0$, $j \in X_1^{(\varepsilon)}$, and also that **(t)** $\mathsf{P}_i\{\lim_{t\to\infty} t^{-1} \cdot \sum_{n=1}^{v_j^{(\varepsilon)}(t)+1} \kappa_{jl}^{(\varepsilon)}(n) = M_{jjl}^{(\varepsilon)}(M_{jj}^{(\varepsilon)})^{-1}\} = 1$ for $i \neq 0$, $j \in X_1^{(\varepsilon)}$. Relations **(s)**, **(t)** and (5.5.3) imply that **(u)** $\mathsf{P}_i\{\lim_{t\to\infty} t^{-1} \int_0^\infty \chi(\eta^{(\varepsilon)}(s) = l)\, ds = M_{jjl}^{(\varepsilon)}(M_{jj}^{(\varepsilon)})^{-1}\} = 1$ for $i \neq 0$, $j \in X_1^{(\varepsilon)}$.

Note also that claim **(m)** holds, since any recurrent state $j \in X_1^{(\varepsilon)}$ can be chosen as to realise the cyclic construction described above. Thus, almost sure convergence of the random variables $\frac{1}{t} \int_0^t \chi(\eta^{(\varepsilon)}(s) = l)\, ds$ in (5.5.1) to the quantity $M_{jjl}^{(\varepsilon)}(M_{jj}^{(\varepsilon)})^{-1}$ can be proved for every $j \in X_1^{(\varepsilon)}$. Of course, this is possible only if these quantities take the same value for every $j \in X_1^{(\varepsilon)}$.

Claim **(n)** follows from the above proof. As a matter of fact, the choice of the initial state $i \neq 0$ influences the random variables introduced in $(\mathbf{o}_1)$–$(\mathbf{o}_5)$ only via the random variables $\kappa_j^{(\varepsilon)}(1)$ and $\kappa_{jl}^{(\varepsilon)}(1)$. These random variables have distributions different from the random variables $\kappa_j^{(\varepsilon)}(2)$ and $\kappa_{jl}^{(\varepsilon)}(2)$, respectively, if $i \neq j$. However, this does not effect the corresponding asymptotic relations **(p)**–**(u)**, since these random variables, normalised in an appropriate way, converge almost surely to zero.

Relation (5.5.1) can be connected with the "individual" ergodic theorem given in Theorem 4.4.3.

Note that the variables that enter this relation are bounded, more precisely, we have

$$0 \leq \frac{1}{t} \int_0^t \chi(\eta^{(\varepsilon)}(s) = l)\, ds \leq 1, \quad t \geq 0, \quad l \neq 0. \tag{5.5.4}$$

Relations (5.5.1) and (5.5.4) imply, by the Lebesgue and Fubini theorems, that, for every $\varepsilon \leq \varepsilon_4$ and $i, l \neq 0$,

$$\mathsf{E}_i \frac{1}{t} \int_0^t \chi(\eta^{(\varepsilon)}(s) = l)\, ds = \frac{1}{t} \int_0^t \mathsf{P}_i\{\eta^{(\varepsilon)}(s) = l\}\, ds \to \pi_l^{(\varepsilon)} \quad \text{as } t \to \infty. \tag{5.5.5}$$

Theorem 4.4.3 also needs to assume the additional non-arithmetic condition $\mathbf{N_2}$ which guarantees the existence of $\varepsilon_5 > 0$ such that the distributions $_0G_{jj}^{(\varepsilon)}(t), j \in X_1^{(\varepsilon)}$, are non-arithmetic for $\varepsilon \leq \varepsilon_5$.

If conditions $\mathbf{A_4}$, $\mathbf{T_1}$ **(a)**, $\mathbf{E_2}$, $\mathbf{M_{13}^{(1)}}$, $\mathbf{I_4}$, and $\mathbf{N_2}$ hold, Theorem 4.4.3 yields the following asymptotic relation for every $\varepsilon \leq \min(\varepsilon_4, \varepsilon_5)$ and $i, l \neq 0$:

$$\mathsf{P}_i\{\eta^{(\varepsilon)}(t) = l\} \to \pi_l^{(\varepsilon)} \quad \text{as } t \to \infty. \tag{5.5.6}$$

This relation is obviously stronger than relation (5.5.5), since (5.5.6) implies (5.5.5) but not vice versa.

In fact, Theorem 4.4.3 gives a stationary distribution in the form given in formula (4.4.28).

However, the semi-Markov process $\eta^{(\varepsilon)}(t)$, $t \geq 0$, is a regenerative process with times of returns to any fixed recurrent state considered as regeneration times. Theorem 4.4.3 is just a particular case of the "individual" ergodic Theorems 3.2.3 and 3.2.7 for regenerative processes. These theorems, applied to the case where a state $j \in X_1^{(\varepsilon)}$ is taken for constructing regeneration cycles and a one-state set $A = \{l\}$ is chosen in the asymptotic relation (3.2.15), give the following formula for a stationary distribution of the ergodic semi-Markov process $\eta^{(\varepsilon)}(t)$, $t \geq 0$:

$$\pi_l^{(\varepsilon)} = \frac{\int_0^\infty \mathsf{P}_j\{\eta^{(\varepsilon)}(s) = l, \mu_j^{(\varepsilon)} > s\}\,ds}{M_{jj}^{(\varepsilon)}}, \quad l \neq 0. \tag{5.5.7}$$

Both forms given in (4.4.28) and (5.5.7) coincide, since they give limits of the same probabilities $\mathsf{P}_i\{\eta^{(\varepsilon)}(t) = l\}$. Also, the form of the stationary distribution given in formula (5.5.2) coincides with that given in (5.5.7), since the Cesàro limit in (5.5.5) must coincide with the limit in (5.5.6). The same is implied by the following relation obtained with the use of the Fubini theorem:

$$M_{jjl}^{(\varepsilon)} = \mathsf{E}_j \int_0^{\mu_j^{(\varepsilon)}} \chi(\eta^{(\varepsilon)}(s) = l)\,ds = \mathsf{E}_j \int_0^\infty \chi(\eta^{(\varepsilon)}(s) = l, \mu_j^{(\varepsilon)} > s)\,ds$$

$$= \int_0^\infty \mathsf{E}_j \chi(\eta^{(\varepsilon)}(s) = l, \mu_j^{(\varepsilon)} > s)\,ds$$

$$= \int_0^\infty \mathsf{P}_j\{\eta^{(\varepsilon)}(s) = l, \mu_j^{(\varepsilon)} > s\}\,ds. \tag{5.5.8}$$

It also useful to note that formula (5.5.7) is a variant of the corresponding formula (3.2.9) for regenerative processes, where the Lebesgue integration is used. But the function $\mathsf{P}_i\{\eta^{(\varepsilon)}(s) = l, \mu_j^{(\varepsilon)} > s\}$ is directly Riemann integrable on $[0, \infty)$. Indeed, this function has at most a countable set of discontinuity points, since the process $\eta^{(\varepsilon)}(s)$, $s \geq 0$ has continuous from the right stepwise trajectories. Also, this function is non-negative and bounded from above by the Riemann integrable monotone function $\mathsf{P}_i\{\mu_j^{(\varepsilon)} > s\}$. Thus, the Lebesgue integration can be replaced with the Riemann integration in the integral in the numerator of the fraction in (5.5.7) and in the integrals that enter (5.5.8).

5.5.2　Asymptotic expansions for expectations of times spend in a given state before hitting a given state

The cyclic representation formula (5.5.2) for the stationary distribution plays a key role in what follows.

We assume that the perturbation condition $\mathbf{P}_{18}^{(k)}$ (introduced in Subsection 5.1.1) holds for the transition probabilities of the imbedded Markov chains,

$\mathbf{P}_{18}^{(k)}$: $p_{ij}^{(\varepsilon)} = p_{ij}^{(0)} + e_{ij}[1,0]\varepsilon + \cdots + e_{ij}[k,0]\varepsilon^k + o(\varepsilon^k)$ for $i, j \neq 0$, where $|e_{ij}[r,0]| < \infty$, $r = 1, \ldots, k$.

We also assume that the following perturbation condition holds for the expectation of transition times:

$\mathbf{P}_{28}^{(k)}$: $p_{ij}^{(\varepsilon)}[1] = p_{ij}^{(0)}[1] + e_{ij}[1,1]\varepsilon + \cdots + e_{ij}[k,1]\varepsilon^k + o(\varepsilon^k)$ for $i, j \neq 0$, where $p_{ij}^{(0)}[1] < \infty$, $i, j \neq 0$, and $|e_{ij}[r,1]| < \infty$, $r = 1, \ldots, k$, $i, j \neq 0$.

Obviously condition $\mathbf{P}_{28}^{(k)}$ is implied by the condition $\mathbf{P}_{18}^{(k)}$ and the following condition:

$\mathbf{P'}_{28}^{(k)}$: $m_{ij}^{(\varepsilon)} = m_{ij}^{(0)} + v_{ij}[1,1]\varepsilon + \cdots + v_{ij}[k,1]\varepsilon^k + o(\varepsilon^k)$ for $i, j \neq 0$, where $m_{ij}^{(0)} < \infty$, $i, j \neq 0$, and $|v_{ij}[r,1]| < \infty$, $r = 1, \ldots, k$, $i, j \neq 0$.

Moreover, the following formula connect the corresponding asymptotic expansions, for $i, j \neq 0$ and $r = 0, \ldots, k$,

$$e_{ij}[r,1] = \sum_{m=0}^{r} e_{ij}[m,0]v_{ij}[r-m,1], \qquad (5.5.9)$$

where $e_{ij}[0,1] = p_{ij}^{(0)}[1]$, $e_{ij}[0,0] = p_{ij}^{(0)}$, and $v_{ij}[0,1] = m_{ij}^{(0)}$.

Condition $\mathbf{P}_{18}^{(k)}$ obviously implies that condition $\mathbf{T}_1$ (a) holds and condition $\mathbf{P}_{28}^{(k)}$ implies condition $\mathbf{M}_{13}^{(1)}$ to hold.

The following lemmas, which are direct corollaries of statements (i) and (ii) in Lemma 8.1.1, gives asymptotic expansions for the expectations $m_i^{(\varepsilon)}$, $i \neq 0$.

Lemma 5.5.1. *Let conditions $\mathbf{P}_{18}^{(k)}$ and $\mathbf{P}_{28}^{(k)}$ hold. Then*

(i) *The following expansion holds for $i \neq 0$:*

$$m_i^{(\varepsilon)} = m_i^{(0)} + v_i[1,1]\varepsilon + \cdots + v_i[k,1]\varepsilon^k + o(\varepsilon^k), \qquad (5.5.10)$$

where the coefficients $v_i[r,1]$, $r = 0, \ldots, k$, are given for $i \neq 0$ by the formulas $v_i[0,1] = m_i^{(0)}$ and, for $r = 0, \ldots, k$,

$$v_i[r,1] = \sum_{j \neq 0} e_{ij}[r,1]. \qquad (5.5.11)$$

(ii) *If condition $\mathbf{P}_{28}^{'(k)}$ holds instead of condition $\mathbf{P}_{28}^{(k)}$, then the formulas (5.5.11) for the coefficients in the asymptotic expansion (5.5.10) can be rewritten in the following form*:

$$v_i[r, 1] = \sum_{j \neq 0} e_{ij}[r, 1] = \sum_{j \neq 0} \sum_{q=0}^{r} e_{ij}[q, 0] v_{ij}[r - q, 1]. \qquad (5.5.12)$$

The asymptotic expansion (5.5.10) can be rewritten in a vector form. Let us introduce the following vectors for $l \neq 0$ and $r = 1, \ldots, k$:

$$\mathbf{v}^{(\varepsilon)} = \begin{bmatrix} m_1^{(\varepsilon)} \\ \vdots \\ m_N^{(\varepsilon)} \end{bmatrix}, \qquad \mathbf{v}[r, 1] = \begin{bmatrix} v_1[r, 1] \\ \vdots \\ v_N[r, 1] \end{bmatrix}.$$

Then the perturbation condition $\mathbf{P}_{28}^{(k)}$ gives the asymptotic vector expansion

$$\mathbf{v}^{(\varepsilon)} = \mathbf{v}^{(0)} + \mathbf{v}[1, 1]\varepsilon + \cdots + \mathbf{v}[k, 1]\varepsilon^k + \mathbf{o}(\varepsilon^k). \qquad (5.5.13)$$

Note that, under condition $\mathbf{A}_4$, $M_{jj}^{(\varepsilon)} = {}_0 M_{jj}^{(\varepsilon)}$, $j \in X_1^{(\varepsilon)}$. The asymptotic expansions for these expectations are given in Lemma 5.2.2. Note that it is necessary to set $\tilde{n} = 1$, $\tilde{k}_0 = \tilde{k}_1 = k$ in this lemma in order to get asymptotic expansions of the order k for these expectations.

Similar asymptotic expansions can be given for the expectations $M_{jjl}^{(\varepsilon)}$.

The following representation can be written for $l \neq 0$:

$$\int_0^{\mu_j^{(\varepsilon)}} \chi(\eta^{(\varepsilon)}(s) = l)\, ds = \sum_{n=1}^{v_j^{(\varepsilon)}} \kappa_n^{(\varepsilon)} \chi(\eta_{n-1}^{(\varepsilon)} = l). \qquad (5.5.14)$$

This representation immediately shows that the expectations $M_{ijl}^{(\varepsilon)}$, $i \neq 0$, satisfy the following system of linear equations for every $j \in X_1^{(0)}$ and $l \neq 0$:

$$M_{ijl}^{(\varepsilon)} = m_i^{(\varepsilon)} \delta(i, l) + \sum_{i' \neq 0, j} p_{ii'}^{(\varepsilon)} M_{i'jl}^{(\varepsilon)}, \quad i \neq 0. \qquad (5.5.15)$$

These systems can be rewritten in a matrix form. For $j \in X_1^{(0)}$, $l \neq 0$, and $r = 1, \ldots, k$, introduce the vectors

$$\mathbf{m}_{jl}^{(\varepsilon)} = \begin{bmatrix} M_{1jl}^{(\varepsilon)} \\ \vdots \\ M_{Njl}^{(\varepsilon)} \end{bmatrix}, \qquad \mathbf{v}_1^{(\varepsilon)} = \begin{bmatrix} m_1^{(\varepsilon)} \delta(1, l) \\ \vdots \\ m_N^{(\varepsilon)} \delta(N, l) \end{bmatrix}, \qquad \mathbf{v}_1[r, 1] = \begin{bmatrix} v_1[r, 1] \delta(1, l) \\ \vdots \\ v_N[r, 1] \delta(N, l) \end{bmatrix}.$$

Then the perturbation condition $\mathbf{P}_{28}^{(k)}$ implies that, for $l \neq 0$,

$$\mathbf{v}_1^{(\varepsilon)} = \mathbf{v}_1^{(0)} + \mathbf{v}_1[1,1]\varepsilon + \cdots + \mathbf{v}_1[k,1]\varepsilon^k + \mathbf{o}(\varepsilon^k). \tag{5.5.16}$$

It is also convenient to set $v_i[0,1] = m_i^{(0)}$, $i \neq 0$, and $\mathbf{v}[0,1] = \mathbf{v}^{(0)}$, and $\mathbf{v}_1[0,1] = \mathbf{v}_1^{(0)}$, $l \neq 0$.

Then the system of linear equations (5.5.15) can be rewritten in the following form for every $j \in X_1^{(0)}$ and $l \neq 0$:

$$\mathbf{m}_{jl}^{(\varepsilon)} = \mathbf{v}_1^{(\varepsilon)} + {}_j\mathbf{P}^{(\varepsilon)}\,\mathbf{m}_{jl}^{(\varepsilon)}. \tag{5.5.17}$$

These linear systems are systems of the hitting type, considered in Section 5.1, with the same coefficient matrix ${}_j\mathbf{P}^{(\varepsilon)}$.

Conditions $\mathbf{A}_4$, $\mathbf{E}_2$, $\mathbf{T}_1$ (a), and $\mathbf{M}_{12}^{(1)}$ imply that for every $\varepsilon \leq \varepsilon_4$ there exists a unique solution of the systems of linear equations (5.5.17), which has the following form for $\varepsilon \leq \varepsilon_4$ and $j \in X_1^{(0)}$ and $l \neq 0$:

$$\mathbf{m}_{jl}^{(\varepsilon)} = [\mathbf{I} - {}_j\mathbf{P}^{(\varepsilon)}]^{-1}\mathbf{v}_1^{(\varepsilon)}. \tag{5.5.18}$$

The following lemma is a corollary of Lemma 5.1.3 applied to the vectors of the power moments of hitting times, $\mathbf{m}_{jl}^{(\varepsilon)}$, for $j \in X_1^{(0)}$ and $l \neq 0$.

Lemma 5.5.2. *Let conditions* $\mathbf{A}_4$, $\mathbf{E}_2$, $\mathbf{P}_{18}^{(k)}$, *and* $\mathbf{P}_{28}^{(k)}$ *hold. Then*

(i) *The vector* $\mathbf{m}_{jl}^{(\varepsilon)}$ *has the following asymptotic expansion for every* $j \in X_1^{(0)}$ *and* $l \neq 0$:

$$\mathbf{m}_{jl}^{(\varepsilon)} = \mathbf{m}_{jl}^{(0)} + \mathbf{c}_{jl}[1,0]\varepsilon + \cdots + \mathbf{c}_{jl}[k,0]\varepsilon^k + \mathbf{o}(\varepsilon^k), \tag{5.5.19}$$

where $\mathbf{c}_{jl}[r,0]$, $r = 1,\ldots,k$, *are finite vectors.*

(ii) *The vector coefficients* $\mathbf{c}_{jl}[r,0]$ *are given by the recurrence formulas* $\mathbf{c}_{jl}[0,0] = \mathbf{m}_{jl}^{(0)} = {}_j\mathbf{U}[0]\mathbf{v}_1[0,1] = [\mathbf{I} - {}_j\mathbf{P}^{(0)}]^{-1}\mathbf{v}_1^{(0)}$ *and, for* $r = 0,\ldots,k$,

$$\mathbf{c}_{jl}[r,0] = {}_j\mathbf{U}[0]\big(\mathbf{v}_1[r,1] + \sum_{q=1}^{r} {}_j\mathbf{E}[q,0]\mathbf{c}_{jl}[r-q,0]\big). \tag{5.5.20}$$

(iii) *The vector coefficients* $\mathbf{c}_j[r,0]$ *can also be calculated from the following formulas for* $r = 0,\ldots,k$:

$$\mathbf{c}_{jl}[r,0] = \sum_{p=0}^{r} {}_j\mathbf{U}[p]\,\mathbf{v}_1[r-p,1]. \tag{5.5.21}$$

It is useful to note that the vectors $\mathbf{m}_j^{(\varepsilon)}$ and $\mathbf{m}_{jl}^{(\varepsilon)}, l \neq 0$, for $j \in X_1^{(0)}$ are connected by the formula

$$\mathbf{m}_j^{(\varepsilon)} = \sum_{l \neq 0} \mathbf{m}_{jl}^{(\varepsilon)}. \qquad (5.5.22)$$

The corresponding asymptotic expansions for the vectors $\mathbf{m}_j^{(\varepsilon)} = {}_0\mathbf{m}_j^{(\varepsilon)}$ are given in Lemma 5.2.2, where we must take $\tilde{n} = 1, \tilde{k}_0 = \tilde{k}_1 = k$. The coefficients in these expansions are given in formulas (5.2.19) and (5.2.20), where $r = 1$.

Formula (5.5.22) implies that the corresponding coefficients in the asymptotic expansions for the vectors $\mathbf{m}_j^{(\varepsilon)} = {}_0\mathbf{m}_j^{(\varepsilon)}$ and $\mathbf{m}_{jl}^{(\varepsilon)}, l \neq 0$, are connected for $j \in X_1^{(0)}$ and $r = 0, \ldots, k$ by the formula

$$_0\mathbf{b}_j[r, 1] = \sum_{l \neq 0} \mathbf{c}_{jl}[r, 0]. \qquad (5.5.23)$$

5.5.3 Pivotal properties of asymptotic expansions for expectations of times spend in a given state before hitting a given state

We will clarify pivotal properties of asymptotic expansions for the cyclic expectations $M_{jjl}^{(\varepsilon)}, l \neq 0, j \in X_1^{(0)}$.

For the vectors $\mathbf{c}_{jl}[r, 0], r = 0, \ldots, k, l \neq 0, j \in X_1^{(0)}$, we denote

$$\mathbf{c}_{jl}[r, 0] = \begin{bmatrix} c_{1jl}[r, 0] \\ \vdots \\ c_{Njl}[r, 0] \end{bmatrix}.$$

Explicit expressions for components of these vectors can be derived from recurrence formulas (5.5.20) and (5.5.21), respectively.

Consider the sets

$$\tilde{Y}_r = \begin{cases} \{l \neq 0 : v_l[n, 1] = 0, \, n = 0, \ldots, r-1, \, v_l[r, 1] \neq 0\} & \text{for } r = 1, \ldots, k, \\ \{l \neq 0 : v_l[n, 1] = 0, \, n = 0, \ldots, k\} & \text{for } r = k+1. \end{cases}$$

By the definition, **(a)** some of the sets $\tilde{Y}_r, r = 0, \ldots, k+1$, could be empty, **(b)** $\tilde{Y}_{r'} \cap \tilde{Y}_{r''} = \varnothing, r' \neq r''$, but **(c)** $\bigcup_{r=0}^{k+1} \tilde{Y}_r = X_0$.

For $j \in X_1^{(0)}$, define the sets

$$\tilde{W}_r = \bigcup_{r'+r''=r} (\tilde{Y}_{r'} \cap Z_{r''}), \quad r = 0, \ldots, k, \quad \tilde{W}_{k+1} = \bigcup_{r'+r'' \geq k+1} (\tilde{Y}_{r'} \cap Z_{r''}).$$

The case where condition $\mathbf{E}_2'$ holds is simple. Indeed, in this case, $Z_0 = X_1^{(0)} = X_0$, and, therefore, $Z_1, \ldots, Z_{k+1} = \varnothing$. Thus, for $j \in X_1^{(0)}$, we have

$$\tilde{W}_r = \tilde{Y}_r, \quad r = 0, \ldots, k+1. \qquad (5.5.24)$$

For an integer $0 \leq h \leq k$, introduce two conditions,

$\mathbf{O}_7^{(l,h)}$: $l \notin \tilde{W}_r$, for $r \leq h$;

and

$\mathbf{O}_8^{(l,h)}$: $l \notin \tilde{W}_r$, for $r < h$, but $l \in \tilde{W}_h$.

The following lemma gives a complete description of pivotal properties of asymptotic expansions for the cyclic expectations $M_{jjl}^{(\varepsilon)}$, $l \neq 0$, $j \in X_1^{(0)}$.

Lemma 5.5.3. *Let conditions* $\mathbf{A}_4$, $\mathbf{E}_2$, $\mathbf{P}_{18}^{(k)}$, *and* $\mathbf{P}_{28}^{(k)}$ *hold. Then*

(i) *If condition* $\mathbf{O}_7^{(l,h)}$ *holds for some* $l \neq 0$ *and* $0 \leq h \leq k$, *then* $c_{jjl}[n,0] = 0$, $n = 0, \dots, h$, $j \in X_1^{(0)}$.

(ii) *If condition* $\mathbf{O}_8^{(l,h)}$ *holds for some* $l \neq 0$ *and* $0 \leq h \leq k$, *then* $c_{jjl}[n,0] = 0$, $n = 0, \dots, h-1$, $j \in X_1^{(0)}$, *but* $c_{jjl}[h,0] > 0$, $j \in X_1^{(0)}$.

Proof. Recall that $X_1^{(0)} \subseteq X_1^{(\varepsilon)}$ for $\varepsilon \leq \varepsilon_4$. Also, according to formula (5.5.18), the cyclic components $M_{jjl}^{(\varepsilon)}$ can be calculated by the following formula for every $l \neq 0$, $j \in X_1^{(0)}$:

$$M_{jjl}^{(\varepsilon)} = {}_j u_{jl}^{(\varepsilon)} m_l^{(\varepsilon)}. \tag{5.5.25}$$

Since $\bigcup_{r=0}^{k+1} \tilde{W}_r = X_0$, there exists $0 \leq r \leq k+1$ such that $l \in \tilde{W}_r$.

Let us first consider the case where $r \leq k$. Then **(a)** there exist $0 \leq r', r'' \leq r$ such that $r' + r'' = r$ and $l \in \tilde{Y}_{r'} \cap Z_{r''}$.

In this case, the cyclic components $M_{jjl}^{(\varepsilon)}$ are products of two functions ${}_j u_{jl}^{(\varepsilon)}$ and $m_l^{(\varepsilon)}$ which have, respectively, a pivotal (r'',k)-expansion and a pivotal (r',k)-expansion. According to claim **(ii)** in Lemma 8.1.1, **(b)** $M_{jjl}^{(\varepsilon)}$ is a pivotal $(r' + r'', (r' + k) \wedge (r'' + k))$-expansion. Obviously such an expansion can be considered as a pivotal (r,k)-expansion.

Let us now consider the case where $r = k + 1$. In this case, either **(c)** there exist $0 \leq r', r'' \leq k$ such that $r' + r'' \geq k + 1$ and $l \in \tilde{Y}_{r'} \cap Z_{r''}$, or **(d)** there exist $0 \leq r', r'' \leq k + 1$, $\max(r', r'') \geq k + 1$ and $l \in \tilde{Y}_{r'} \cap Z_{r''}$.

In the case **(c)**, the cyclic components $M_{jjl}^{(\varepsilon)}$ are products of two functions ${}_j u_{jl}^{(\varepsilon)}$ and $m_l^{(\varepsilon)}$ which have, respectively, a pivotal (r'',k)-expansion and a pivotal (r',k)-expansion if $r', r'' \leq k$. Using claim **(ii)** in Lemma 8.1.1, we see that $M_{jjl}^{(\varepsilon)}$ is a pivotal $(r' + r'', (r' + k) \wedge (r'' + k))$-expansion. Since $r' + r'' \geq k + 1$, this means that **(e)** $M_{jjl}^{(\varepsilon)}$ is $o(\varepsilon^k)$.

In the case **(d)**, at least one of the functions ${}_j u_{jl}^{(\varepsilon)}$ or $m_l^{(\varepsilon)}$ is $o(\varepsilon^k)$ and, therefore, **(f)** their product $M_{jjl}^{(\varepsilon)}$ is $o(\varepsilon^k)$.

If condition $\mathbf{O}_7^{(l,h)}$ holds for some $0 \leq h \leq k$, then (g) $l \in \tilde{W}_r$ for some $h + 1 \leq r \leq k$ or (h) $l \in \tilde{W}_{k+1}$. In the case (g), $M_{jjl}^{(\varepsilon)}$ is, by (b), a pivotal (r, k)-expansion and, therefore, $c_{jjl}[n, 0] = 0, n = 0, \ldots, h$. In the case (h), $M_{jjl}^{(\varepsilon)}$ is $o(\varepsilon^k)$ by (e) or (f) and, again, $c_{jjl}[n, 0] = 0, n = 0, \ldots, h$. This proves the first statement of the lemma.

If condition $\mathbf{O}_8^{(l,h)}$ holds for some $0 \leq h \leq k$, then $M_{jjl}^{(\varepsilon)}$ is, by (b), a pivotal (h, k)-expansion. This proves the second statement of the lemma.   □

Remark 5.5.1. Note that condition $\mathbf{I}_4$ is not required either in Lemma 5.2.2 or in Lemma 5.5.2. However, this condition is necessary in ergodic theorems. If condition $\mathbf{I}_4$ holds, then there exists at least one state $l \in \tilde{W}_0 \subseteq X_1^{(0)}$. The corresponding coefficients are $c_{jjl}[0, 0] > 0, j \in X_1^{(0)}$.

5.5.4 Asymptotic expansions for stationary distributions of perturbed ergodic semi-Markov processes

Now we are in a position to get asymptotic expansions for stationary distributions of perturbed ergodic semi-Markov processes. We will use the asymptotic expansions for the cyclic expectations $M_{jj}^{(\varepsilon)} = {}_0M_{jj}^{(\varepsilon)}$ and $M_{jjl}^{(\varepsilon)}, l \neq 0$, for $j \in X_1^{(0)}$, given in Lemmas 5.2.2 and 5.5.2, and the cyclic representation for stationary distribution given in formula (5.5.2).

Theorem 5.5.1. *Let conditions* $\mathbf{A}_4$, $\mathbf{E}_2$, $\mathbf{I}_4$, $\mathbf{P}_{18}^{(k)}$, *and* $\mathbf{P}_{28}^{(k)}$ *hold. Then*

(i) *The stationary probabilities* $\pi_l^{(\varepsilon)} = M_{jjl}^{(\varepsilon)} / M_{jj}^{(\varepsilon)}, l \neq 0$, *which do not depend on the choice of the state* $j \in X_1^{(0)}$, *have the following asymptotic expansion for every* $l \neq 0$ *and* $j \in X_1^{(0)}$:

$$\pi_l^{(\varepsilon)} = \frac{M_{jjl}^{(0)} + c_{jjl}[1, 0]\varepsilon + \cdots + c_{jjl}[k, 0]\varepsilon^k + o(\varepsilon^k)}{M_{jj}^{(0)} + {}_0b_{jj}[1, 1]\varepsilon + \cdots + {}_0b_{jj}[k, 1]\varepsilon^k + o(\varepsilon^k)}$$

$$= \pi_l^{(0)} + f_l[1]\varepsilon + \cdots + f_l[k]\varepsilon^k + o(\varepsilon^k), \qquad (5.5.26)$$

where the coefficients $f_l[r]$ *are given by the recurrence formulas* $f_l[0] = \pi_l^{(0)} = M_{jjl}^{(0)} / M_{jj}^{(0)} = c_{jjl}[0, 0]/{}_0b_{jj}[0, 1]$ *and, for* $n = 0, \ldots, k$,

$$f_l[n] = (c_{jjl}[n, 0] - \sum_{q=0}^{n-1} {}_0b_{jj}[n - q, 1] f_l[q]) / {}_0b_{jj}[0, 1]. \qquad (5.5.27)$$

(ii) *If condition* $\mathbf{O}_7^{(l,h)}$ *holds for given* $l \neq 0$ *and* $0 \leq h \leq k$, *then* $c_{jjl}[n, 0] = 0$, $n = 0, \ldots, h, j \in X_1^{(0)}$, *and* $f_l[n] = 0, n = 0, \ldots, h$.

(iii) *If condition* $\mathbf{O}_8^{(l,h)}$ *holds for given* $l \neq 0$ *and* $0 \leq h \leq k$, *then* $c_{jjl}[n, 0] = 0$,
$n = 0, \ldots, h - 1$, $j \in X_1^{(0)}$, *but* $c_{jjl}[h, 0] > 0$, $j \in X_1^{(0)}$, *and* $f_l[n] = 0$,
$n = 0, \ldots, h - 1$, *but* $f_l[h] > 0$.

Proof. Lemmas 5.2.2 and 5.5.2 yield asymptotic expansions for the functions $M_{jjl}^{(\varepsilon)}$
and $M_{jj}^{(\varepsilon)}$, which permits to imply statement **(iii)** of Lemma 8.1.1 to the quotient of
these functions for every $l \neq 0$ and $j \in X_1^{(0)}$. This yields asymptotic expansions for
the stationary probabilities $\pi_l^{(\varepsilon)}$, $i \neq 0$, given in Theorem 5.5.1. The pivotal properties
described in statements **(ii)** and **(iii)** directly follow from Lemma 5.5.3 and statement
(iv) in Lemma 8.1.1. $\qquad\qquad\qquad\qquad\qquad\qquad\qquad\qquad\qquad\qquad\qquad\qquad\qquad$ $\square$

5.5.5   Cyclic representation for quasi-stationary distribution of semi-Markov process

In this section we consider a model in which the absorption is possible both for perturbed and for the corresponding limit semi-Markov processes.

   We assume that condition $\mathbf{T}_1$ holds, i.e., **(a)** $p_{ij}^{(\varepsilon)} \to p_{ij}^{(0)}$ as $\varepsilon \to 0$ for $i \neq 0$,
$j \in X$, and **(b)** $Q_{ij}^{(\varepsilon)}(\cdot) \Rightarrow Q_{ij}^{(0)}(\cdot)$ as $\varepsilon \to 0$ for $i \neq 0$, $j \in X$.

   We also assume that condition $\mathbf{E}_2$ holds, i.e., the set of recurrent-without-absorption
states is not empty, and that condition $\mathbf{A}_2$ is satisfied, i.e., 0 is an absorption state.

   According to Lemma 4.2.2, relation **(a)** and conditions $\mathbf{A}_2$, $\mathbf{T}_1$ **(a)**, and $\mathbf{E}_2$ imply
that **(c)** there exists $\varepsilon_0' > 0$ such that the corresponding sets of recurrent states, $X_1^{(\varepsilon)}$,
for the imbedded Markov chains $\eta_n^{(\varepsilon)}$, $n = 0, 1, \ldots$, are connected by the relation
$X_1^{(0)} \subseteq X_1^{(\varepsilon)}$ for every $\varepsilon \leq \varepsilon_0'$.

   We assume that condition $\mathbf{C}_{12}$ is satisfied, i.e., **(d)** $\overline{\lim}_{\varepsilon \to 0} \, p_{ij}^{(\varepsilon)}(\delta) < \infty, i \neq 0$,
$j \in X$ for some $\delta > 0$.

   Obviously condition $\mathbf{C}_{12}$ implies that **(e$_1$)** there exists $\varepsilon_1' > 0$ such that $p_{ij}^{(\varepsilon)}(\delta) < \infty$
for $i \neq 0$, $j \in X$ and $\varepsilon \leq \varepsilon_1'$. As was shown in Subsection 5.3.5, **(e$_2$)** $p_{ij}^{(\varepsilon)}[\beta, n] \leq$
$c_n(\delta, \beta) p_{ij}^{(\varepsilon)}(\beta)$, $n = 0, 1, \ldots$, for any $\beta < \delta$. Thus, **(e$_3$)** $p_{ij}^{(\varepsilon)}[\beta, n] < \infty$ for $\varepsilon \leq \varepsilon_1'$,
$i \neq 0$, $j \in X$. Note also that **(e$_4$)** the function $p_{ij}^{(\varepsilon)}[\rho, n]$ is a derivative (in the variable
ρ) of order n of the function $p_{ij}^{(\varepsilon)}[\rho, 0]$ for $\rho \leq \beta$, $\varepsilon \leq \varepsilon_1'$, $n = 1, 2, \ldots, i \neq 0$, $j \in X$,

   For $i \neq 0, n = 0, 1, \ldots$, introduce the mixed power-exponential moment generating
functions for sojourn times by

$$\psi_i^{(\varepsilon)}(\rho) = \sum_{j \in X} p_{ij}^{(\varepsilon)}(\rho) = \sum_{j \in X} p_{ij}^{(\varepsilon)} \psi_{ij}^{(\varepsilon)}(\rho), \quad \rho \in \mathbb{R}_1,$$

and

$$\psi_i^{(\varepsilon)}[\rho, n] = \sum_{j \in X} p_{ij}^{(\varepsilon)}[\rho, n] = \sum_{j \in X} p_{ij}^{(\varepsilon)} \psi_{ij}^{(\varepsilon)}[\rho, n], \quad \rho \in \mathbb{R}_1.$$

Note that, by the definition, $\psi_i^{(\varepsilon)}[\rho, 0] = \psi_i^{(\varepsilon)}(\rho), i \neq 0$.

Relations **(a)** and **(e$_3$)** imply that **(f$_1$)** $p_i^{(\varepsilon)}[\beta, n] < \infty$ for $\varepsilon \leq \varepsilon_1', n = 0, 1, \ldots$, $i \neq 0$. Note also that **(f$_2$)** for every $n = 1, 2, \ldots$ and $i \neq 0$, the function $p_i^{(\varepsilon)}[\rho, n]$ is a derivative (in the variable ρ) of order n of the function $p_i^{(\varepsilon)}[\rho, 0]$ for $\rho \leq \beta, \varepsilon \leq \varepsilon_1'$, $i \neq 0$.

Assume that condition **C$_{13}$** holds, i.e., **(g)** there exist $i \in X_1^{(0)}$ and $\beta_i \in (0, \delta]$ for δ defined in condition **C$_{12}$** such that $_0\phi_{ii}^{(0)}(\beta_i) \in (1, \infty)$ and $_0\phi_{ki}^{(0)}(\beta_i) < \infty, k \in X_2^{(0)}$.

As follows from Lemmas 4.5.3 and 4.5.6, conditions **A$_2$, T$_1$, E$_2$, C$_{12}$**, and **C$_{13}$** imply that **(h$_1$)** there exist $0 < \beta < \delta$ and $\varepsilon_2' > 0$ such that $_0\phi_{ij}^{(\varepsilon)}(\beta) < \infty$ for $\varepsilon \leq \varepsilon_2'$, $i \neq 0, j \in X_1^{(\varepsilon)}$.

As follows from Lemmas 4.5.3 and 4.5.6, conditions **A$_2$, T$_1$, E$_2$, C$_{12}$**, and **C$_{13}$** also imply that **(i$_1$)** there exists a unique root $\rho^{(\varepsilon)}$ of the equation $_0\phi_{jj}^{(\varepsilon)}(\rho) = 1$ for $j \in X_1^{(\varepsilon)}$ and $\varepsilon \leq \varepsilon_2'$, **(i$_2$)** the root $\rho^{(\varepsilon)}$ does not depend on the choice of $j \in X_1^{(\varepsilon)}$, **(i$_3$)** the root $\rho^{(\varepsilon)}$ satisfies the inequality $\rho^{(\varepsilon)} < \beta$ for $\varepsilon \leq \varepsilon_2'$, **(i$_4$)** $\rho^{(\varepsilon)} \to \rho^{(0)}$ as $\varepsilon \to 0$.

Choose an arbitrary $\rho^{(0)} < \beta' < \beta$.

As was shown in Subsection 5.3.5, **(j$_1$)** $_0\phi_{ij}^{(\varepsilon)}[\beta, n] \leq c_n(\beta, \beta')\, _0\phi_{ij}^{(\varepsilon)}(\beta), n = 0, 1, \ldots$. Thus, **(j$_2$)** $_0\phi_{ij}^{(\varepsilon)}[\beta', n] < \infty$ for $\varepsilon \leq \varepsilon_2', n = 0, 1, \ldots, i \neq 0, j \in X_1^{(\varepsilon)}$. Note also that **(j$_3$)** the function $_0\phi_{ij}^{(\varepsilon)}[\rho, n]$ is a derivative (in the variable ρ) of order n of the function $_0\phi_{ij}^{(\varepsilon)}[\rho, 0]$ for $\rho \leq \beta, \varepsilon \leq \varepsilon_2', n = 1, 2, \ldots, i \neq 0$.

We will use the following notations introduced in Subsection 4.4.1 for the moment generating functions, $\rho \in \mathbb{R}_1, i, j \neq 0$:

$$\phi_{ij}^{(\varepsilon)}(\rho) = \mathsf{E}_i e^{\rho(\mu_j^{(\varepsilon)} \wedge \mu_0^{(\varepsilon)})} = \int_0^\infty e^{\rho s} G_{ij}^{(\varepsilon)}(ds).$$

Also, consider the corresponding mixed power-exponential generating functions for $\rho \in \mathbb{R}_1, i, j, l \neq 0$ and $n = 0, 1, \ldots$,

$$\phi_{ij}^{(\varepsilon)}[\rho, n] = \int_0^\infty s^n e^{\rho s} G_{ij}^{(\varepsilon)}(ds).$$

By the definition, $\phi_{ij}^{(\varepsilon)}[\rho, 0] = \phi_{ij}^{(\varepsilon)}(\rho), i, j \neq 0$.

As follows from Lemmas 4.5.4 and 4.5.7, conditions **A$_2$, T$_1$, E$_2$, C$_{12}$**, and **C$_{13}$** imply that **(k$_1$)** there exists $\varepsilon_3' > 0$ such that $\phi_{ij}^{(\varepsilon)}(\beta) < \infty$ for $\varepsilon \leq \varepsilon_3'$ and $i \neq 0, j \in X_1^{(\varepsilon)}$. Similarly to inequalities **(j$_2$)**, it can be proved that **(k$_2$)** $\phi_{ij}^{(\varepsilon)}[\beta', n] \leq c_n'(\beta, \beta')\phi_{ij}^{(\varepsilon)}(\beta')$, $n = 0, 1, \ldots$. Thus, **(k$_3$)** $\phi_{ij}^{(\varepsilon)}[\beta', n] < \infty$ for $\varepsilon \leq \varepsilon_3', n = 0, 1, \ldots, i \neq 0, j \in X_1^{(\varepsilon)}$. Note also that **(k$_4$)** the function $\phi_{ij}^{(\varepsilon)}[\rho, n]$ is the derivative (in the variable ρ) of order n of the function $\phi_{ij}^{(\varepsilon)}[\rho, 0]$ for $\rho \leq \beta, \varepsilon \leq \varepsilon_3', n = 1, 2, \ldots, i \neq 0$.

Let us introduce the following moment generating functions for $\rho \in \mathbb{R}_1, i, j, l \neq 0$:

$$\omega_{ijl}^{(\varepsilon)}(\rho) = \int_0^\infty e^{\rho s} \mathsf{P}_i\{\eta^{(\varepsilon)}(s) = l, \mu_j^{(\varepsilon)} \wedge \mu_0^{(\varepsilon)} > s\}\, ds.$$

Consider the following mixed power-exponential moment generating functions for $\rho \in \mathbb{R}_1, i, j, l \neq 0$ and $n = 0, 1, \ldots$:

$$\omega_{ijl}^{(\varepsilon)}[\rho, n] = \int_0^\infty s^n e^{\rho s} \mathsf{P}_i\{\eta^{(\varepsilon)}(s) = l, \mu_j^{(\varepsilon)} \wedge \mu_0^{(\varepsilon)} > s\}\, ds.$$

By the definition, $\omega_{ijl}^{(\varepsilon)}[\rho, 0] = \omega_{ijl}^{(\varepsilon)}(\rho)$, $i, j, l \neq 0$.

For $\rho \in \mathbb{R}_1, i, j \neq 0$, consider the moment generating functions

$$\omega_{ij}^{(\varepsilon)}(\rho) = \int_0^\infty e^{\rho s} \mathsf{P}_i\{\mu_j^{(\varepsilon)} \wedge \mu_0^{(\varepsilon)} > s\}\, ds,$$

Also consider the following mixed power-exponential moment generating functions for $\rho \in \mathbb{R}_1, i, j, l \neq 0$ and $n = 0, 1, \ldots$:

$$\omega_{ij}^{(\varepsilon)}[\rho, n] = \int_0^\infty s^n e^{\rho s} \mathsf{P}_i\{\mu_j^{(\varepsilon)} \wedge \mu_0^{(\varepsilon)} > s\}\, ds.$$

By the definition, $\omega_{ij}^{(\varepsilon)}[\rho, 0] = \omega_{ij}^{(\varepsilon)}(\rho)$, $i, j, \neq 0$.
It is clear that, for $\rho \in \mathbb{R}_1$ and $i, j \neq 0$,

$$\omega_{ij}^{(\varepsilon)}(\rho) = \sum_{l \neq 0} \omega_{ijl}^{(\varepsilon)}(\rho). \tag{5.5.28}$$

The following formula plays a key role in what follows:

$$\begin{aligned}
\omega_{ij}^{(\varepsilon)}(\rho) &= \int_0^\infty e^{\rho s} \mathsf{P}_i\{\mu_j^{(\varepsilon)} \wedge \mu_0^{(\varepsilon)} > s\}\, ds \\
&= \mathsf{E}_i \int_0^\infty e^{\rho s} \chi(\mu_j^{(\varepsilon)} \wedge \mu_0^{(\varepsilon)} > s)\, ds \\
&= \mathsf{E}_i \int_0^{\mu_j^{(\varepsilon)} \wedge \mu_0^{(\varepsilon)}} e^{\rho s}\, ds = \begin{cases} (\phi_{ij}^{(\varepsilon)}[\rho, 0] - 1)/\rho, & \text{if } \rho > 0, \\ \phi_{ij}^{(\varepsilon)}[0, 1], & \text{if } \rho = 0. \end{cases}
\end{aligned} \tag{5.5.29}$$

We have the following estimate for $i, j \neq 0$ and $n = 0, 1, \ldots$:

$$\omega_{ij}^{(\varepsilon)}[\beta' n] \leq c_n' \int_0^\infty e^{\beta s} \mathsf{P}_i\{\mu_j^{(\varepsilon)} \wedge \mu_0^{(\varepsilon)} > s\}\, ds = c_n' \omega_{ij}^{(\varepsilon)}(\beta), \tag{5.5.30}$$

where $c_n' = c_n(\beta, \beta') = \sup_{s \geq 0} s^n e^{-(\beta - \beta')s} < \infty$.

It follows from $(\mathbf{k}_1)$ and relations (5.5.28), (5.5.29), (5.5.30) that $(\mathbf{l}_1)$ $\omega_{ijl}^{(\varepsilon)}[\beta, n] \leq \omega_{ij}^{(\varepsilon)}[\beta, n] < \infty$ for $\varepsilon \leq \varepsilon_3'$, $n = 0, 1, \ldots$, and $i, l \neq 0$, $j \in X_1^{(\varepsilon)}$. Also note that $(\mathbf{k}_2)$ the functions $\omega_{ijl}^{(\varepsilon)}[\rho, n]$ and $\omega_{ij}^{(\varepsilon)}[\rho, n]$ are derivatives (in the variable ρ) of order n, respectively, of the functions $\omega_{ij}^{(\varepsilon)}[\rho, 0]$ and $\omega_{ij}^{(\varepsilon)}[\rho, 0]$ for $\rho \leq \beta'$, $\varepsilon \leq \varepsilon_3'$, $n = 1, 2, \ldots, i \neq 0, j \in X_1^{(\varepsilon)}$.

Relation ($\mathbf{i_4}$) also implies that ($\mathbf{l}$) there exists $\varepsilon'_4 > 0$ such that $\rho^{(\varepsilon)} < \beta'$ for $\varepsilon \leq \varepsilon'_4$.

We assume that condition $\mathbf{I_4}$ holds. Together with conditions $\mathbf{T_1}$ and $\mathbf{C_{12}}$, condition $\mathbf{I_4}$ implies that ($\mathbf{m_1}$) there exists $\varepsilon'_5 > 0$ such that $\max_{i \in X_1^{(\varepsilon)}} \psi_i^{(\varepsilon)}(\rho) > 1$ for $0 < \rho \leq \beta'$, $\varepsilon \leq \varepsilon'_5$.

Also, conditions $\mathbf{A_2}$, $\mathbf{T_1}$, $\mathbf{E_2}$, $\mathbf{C_{12}}$, $\mathbf{C_{13}}$ and $\mathbf{I_4}$ imply that ($\mathbf{n_1}$) there exists $\varepsilon'_6 > 0$ such that $\phi_{jj}^{(\varepsilon)}(\rho) > 1$ for $0 < \rho \leq \beta'$, $\varepsilon \leq \varepsilon'_6$, $j \in X_1^{(\varepsilon)}$.

Finally, we require the non-arithmetic condition $\mathbf{N_2}$, which implies ($\mathbf{o_1}$) existence of $\varepsilon'_8 > 0$ such that the distributions $_0G_{jj}^{(\varepsilon)}(t)$, $j \in X_1^{(\varepsilon)}$, are non-arithmetic for $\varepsilon \leq \varepsilon'_6$.

According to Theorem 4.6.8, if conditions $\mathbf{A_2}$, $\mathbf{T_1}$, $\mathbf{E_2}$, $\mathbf{C_{12}}$, $\mathbf{C_{13}}$, $\mathbf{I_4}$ and $\mathbf{N_2}$ hold, then ($\mathbf{p_1}$) the semi-Markov process $\eta^{(\varepsilon)}(t)$, $t \geq 0$, is quasi-ergodic for every $\varepsilon \leq \min(\varepsilon'_7, \varepsilon'_8)$, where $\varepsilon'_7 = \min(\varepsilon'_0, \varepsilon'_1, \varepsilon'_2, \varepsilon'_3, \varepsilon'_4, \varepsilon'_5, \varepsilon'_6)$, in the sense that the following asymptotic quasi-ergodic relation holds for $i, l \neq 0$:

$$e^{\rho^{(\varepsilon)}t} \mathsf{P}_i\{\eta^{(\varepsilon)}(t^{(\varepsilon)}) = l, \mu_0^{(\varepsilon)} > t\} \to \tilde{\tilde{\pi}}_{il}^{(\varepsilon)}(\rho^{(\varepsilon)}) \text{ as } t \to \infty, \tag{5.5.31}$$

where the limit quantities $\tilde{\tilde{\pi}}_{il}^{(\varepsilon)}(\rho^{(\varepsilon)})$ are given by formulas (4.6.3), (4.6.26) and (4.6.27).

We prefer however to use an alternative cyclic representation formula for the limits $\tilde{\tilde{\pi}}_{il}^{(\varepsilon)}(\rho^{(\varepsilon)})$. As was mentioned above, the semi-Markov process $\eta^{(\varepsilon)}(t)$, $t \geq 0$, is a regenerative process with the times of returns to any fixed recurrent state considered as regeneration times. Theorem 4.6.8 is just a particular case of the "individual" quasi-ergodic Theorems 3.3.3 and 3.3.11 for regenerative processes. These theorems, applied to the case where a state $j \in X_1^{(\varepsilon)}$ is chosen to construct regeneration cycles and a one-state set $A = \{l\}$ is chosen in the asymptotic relation (3.3.79), give the following formula for the limits $\tilde{\tilde{\pi}}_{il}^{(\varepsilon)}(\rho^{(\varepsilon)})$ for $i, l \neq 0$ and $j \in X_1^{(\varepsilon)}$:

$$\tilde{\tilde{\pi}}_{il}^{(\varepsilon)}(\rho^{(\varepsilon)}) = {}_0\phi_{ij}^{(\varepsilon)}(\rho^{(\varepsilon)})\tilde{\pi}_{jl}^{(\varepsilon)}(\rho^{(\varepsilon)}), \tag{5.5.32}$$

where

$$\tilde{\pi}_{jl}^{(\varepsilon)}(\rho^{(\varepsilon)}) = \frac{\omega_{jjl}^{(\varepsilon)}(\rho^{(\varepsilon)})}{_0\phi_{jj}^{(\varepsilon)}(\rho^{(\varepsilon)})} = \frac{\int_0^\infty e^{\rho^{(\varepsilon)}s}\mathsf{P}_i\{\eta^{(\varepsilon)}(s) = l, \mu_j^{(\varepsilon)} \wedge \mu_0^{(\varepsilon)} > s\}\,ds}{\int_0^\infty s e^{\rho^{(\varepsilon)}s}\,_0G_{jj}^{(\varepsilon)}(ds)}. \tag{5.5.33}$$

Both forms of the quasi-stationary limits $\tilde{\tilde{\pi}}_{il}^{(\varepsilon)}(\rho^{(\varepsilon)})$ given in formulas (5.5.32) and (5.5.33) and in formulas (4.6.3), (4.6.26), and (4.6.27) coincide, since they give limits of the same quantities $e^{\rho^{(\varepsilon)}t}\mathsf{P}_i\{\eta^{(\varepsilon)}(t) = l, \mu_0^{(\varepsilon)} > t\}$.

Formulas (5.5.32) and (5.5.33) are variants of the corresponding formulas for regenerative processes (3.3.78) and (3.3.6), respectively. These formulas use the Lebesgue integration. But the function $e^{\rho^{(\varepsilon)}s}\mathsf{P}_i\{\eta^{(\varepsilon)}(s) = l, \mu_j^{(\varepsilon)} \wedge \mu_0^{(\varepsilon)} > s\}$ is directly Riemann integrable on $[0, \infty)$. Indeed, it has at most a countable set of discontinuity points, since the process $\eta^{(\varepsilon)}(s)$, $s \geq 0$ has continuous from the right stepwise trajectories. Also this function is non-negative and bounded from above by the directly

Riemann integrable function $e^{\rho^{(\varepsilon)}s}\mathsf{P}_i\{\mu_j^{(\varepsilon)} \wedge \mu_0^{(\varepsilon)} > s\}$. Thus, the Lebesgue integration can be replaced in the numerator of the fraction in (5.5.33) with Riemann integration.

Note also that the limits $\tilde{\tilde{\pi}}_{il}^{(\varepsilon)}(\rho^{(\varepsilon)})$ do not depend on the choice of the state $j \in X_1^{(\varepsilon)}$, because we can choose any recurrent state $j \in X_1^{(\varepsilon)}$ for constructing a corresponding regenerative process. Thus, convergence of the quantities $e^{\rho^{(\varepsilon)}t}\mathsf{P}_i\{\eta^{(\varepsilon)}(t) = l\}$ in (5.5.31) to the quantity $\tilde{\tilde{\pi}}_{il}^{(\varepsilon)}(\rho^{(\varepsilon)})$ can be proved for every $j \in X_1^{(\varepsilon)}$. Obviously, this is possible only if these quantities take the same value for every $j \in X_1^{(\varepsilon)}$.

A dependence on the initial state $i \neq 0$ is possible in the quasi-stationary case for $\rho^{(\varepsilon)} > 0$.

This dependence disappears in the pseudo-stationary case, where $\rho^{(\varepsilon)} = 0$. In this case, $_0\phi_{ij}^{(\varepsilon)}(0) = 1$, $i \neq 0$, $j \in X_1^{(\varepsilon)}$ and $\tilde{\pi}_{jl}^{(\varepsilon)}(0) = \pi_l^{(\varepsilon)}$, $l \neq 0$, coincide with the corresponding stationary probabilities given in formula (5.5.7) for every $j \in X_1^{(\varepsilon)}$.

Finally, it is useful to note that formulas (5.5.32) and (5.5.33) define the corresponding quasi-stationary quantities $\tilde{\tilde{\pi}}_{il}^{(\varepsilon)}(\rho^{(\varepsilon)})$ for $\varepsilon \leq \varepsilon_7'$ without the assumption that the non-arithmetic condition $\mathbf{N_2}$ holds.

It can be proved that $\tilde{\tilde{\pi}}_{il}^{(\varepsilon)}(\rho^{(\varepsilon)})$ do not depend on the choice of the state $j \in X_1^{(\varepsilon)}$ by using the same method as in the proof of Lemma 4.6.1.

Let us prove it for the process $\eta^{(0)}(t)$, $t \geq 0$. The same method can be used for the process $\eta^{(\varepsilon)}(t)$, $t \geq 0$, corresponding to any other value of the parameter ε.

Thus, we assume that conditions $\mathbf{A_2}$, $\mathbf{E_2}$, $\mathbf{C_{12}}$, $\mathbf{C_{13}}$, and $\mathbf{I_4}$ hold for this process. Let us choose and fix some state $i \in X_1^{(\varepsilon)}$ and perturb the transition probabilities $Q_{ij}^{(0)}(t)$, $j \in X$, by replacing them with the convolutions $Q_{ij}^{(\varepsilon)}(t) = Q^{(\varepsilon)}(t) * Q_{ij}^{(0)}(t)$, $j \in X$, where $Q^{(\varepsilon)}(t) = (t/\varepsilon) \wedge 1$, $t \geq 0$, is a uniform distribution function on the interval $[0, \varepsilon]$. Fix some $\varepsilon' > 0$ and, for $0 \leq \varepsilon \leq \varepsilon'$, consider the family of perturbed semi-Markov processes $\eta^{(\varepsilon)}(t)$, $t \geq 0$, with the transition probabilities $Q_{ij}^{(\varepsilon)}(t)$. Obviously, conditions $\mathbf{T_1}$ and $\mathbf{N_2}$ are satisfied by the perturbed semi-Markov processes $\eta^{(\varepsilon)}(t)$, $t \geq 0$. Thus there exists $\varepsilon'' > 0$ such that the asymptotic relation (5.5.31) takes place for these processes for $\varepsilon \leq \min(\varepsilon', \varepsilon'')$. The corresponding limits $\tilde{\tilde{\pi}}_{il}^{(\varepsilon)}(\rho^{(\varepsilon)})$ do not depend on the choice of the state $j \in X_1^{(\varepsilon)}$. Here $X_1^{(\varepsilon)}$ is the set of recurrent states for the process $\eta^{(\varepsilon)}(t)$, $t \geq 0$, introduced above. Note that this set is the same for all $0 \leq \varepsilon \leq \varepsilon'$, since all processes $\eta^{(\varepsilon)}(t)$, $t \geq 0$, have the same transition matrix of imbedded Markov chains. By making $\varepsilon \to 0$ and performing the corresponding limit transition using Lemma 4.6.2, we can get the same formula for the process $\eta^{(0)}(t)$ as well as to prove that the quantities $\tilde{\tilde{\pi}}_{il}^{(0)}(\rho^{(0)})$ do not depend on the choice of the state $j \in X_1^{(0)}$.

Considering the quantities $\tilde{\pi}_{jl}^{(\varepsilon)}(\rho^{(\varepsilon)})$, $l \neq 0$, $j \in X_1^{(\varepsilon)}$, we see that $\phi_{jj}^{(\varepsilon)}(\rho^{(\varepsilon)}) = 1$, $j \in X_1^{(\varepsilon)}$, and, therefore, $\tilde{\tilde{\pi}}_{jl}^{(\varepsilon)}(\rho^{(\varepsilon)}) = \tilde{\pi}_{jl}^{(\varepsilon)}(\rho^{(\varepsilon)})$, $l \neq 0$, $j \in X_1^{(\varepsilon)}$.

Thus, it is possible that $\tilde{\pi}_{jl}^{(\varepsilon)}(\rho^{(\varepsilon)})$, $l \neq 0$, depend on the choice of $j \in X_1^{(0)}$ in the quasi-stationary case where $\rho^{(\varepsilon)} > 0$.

This dependence disappears in the pseudo-stationary case where $\rho^{(\varepsilon)} = 0$. In this case, $\tilde{\pi}_{jl}^{(\varepsilon)}(0) = \pi_l^{(\varepsilon)}$, $l \neq 0$, coincide with the corresponding stationary probabilities given in formula (5.5.7) for every $j \in X_1^{(\varepsilon)}$.

According to Theorem 4.6.5, if conditions $\mathbf{A}_2$, $\mathbf{T}_1$, $\mathbf{E}_2$, $\mathbf{C}_{12}$, $\mathbf{C}_{13}$, $\mathbf{I}_4$ and $\mathbf{N}_2$ hold, then the following asymptotic quasi-ergodic relation also holds for every $\varepsilon \leq \min(\varepsilon_7', \varepsilon_8')$ and $i, l \neq 0$:

$$\mathsf{P}_i\{\eta^{(\varepsilon)}(t) = l / \mu_0^{(\varepsilon)} > t\} \to \pi_l^{(\varepsilon)}(\rho^{(\varepsilon)}) \quad \text{as } t \to \infty, \tag{5.5.34}$$

where

$$\pi_l^{(\varepsilon)}(\rho^{(\varepsilon)}) = \frac{\tilde{\tilde{\pi}}_{il}^{(\varepsilon)}(\rho^{(\varepsilon)})}{\sum_{r \neq 0} \tilde{\tilde{\pi}}_{ir}^{(\varepsilon)}(\rho^{(\varepsilon)})}, \quad l \neq 0. \tag{5.5.35}$$

Formula (5.5.34) defines a *quasi-stationary distribution* for the semi-Markov process $\eta^{(\varepsilon)}(t)$, $t \geq 0$. The quasi-stationary distribution $\pi_l^{(\varepsilon)}(\rho^{(\varepsilon)})$, $l \neq 0$, is defined via the quantities $\tilde{\tilde{\pi}}_{il}^{(\varepsilon)}(\rho^{(\varepsilon)})$ that are dependent on the choice of the initial state $i \neq 0$. However, Lemma 4.6.1 shows that the quasi-stationary probabilities $\pi_l^{(\varepsilon)}(\rho^{(\varepsilon)})$, $l \neq 0$, do not depend on the choice of the initial state $i \neq 0$.

Substituting the expressions for $\tilde{\tilde{\pi}}_{il}^{(\varepsilon)}(\rho^{(\varepsilon)})$, $i, l \neq 0$, given in formulas (5.5.32) and (5.5.33) into formula (5.5.35), canceling the factors $_0\phi_{ij}^{(\varepsilon)}(\rho^{(\varepsilon)})$ and $_0\phi_{jj}^{(\varepsilon)}(\rho^{(\varepsilon)})$ in the numerator and the denominator we obtain the following expressions for the quasi-stationary probabilities for $i \neq 0$, $j \in X_1^{(\varepsilon)}$:

$$\begin{aligned}
\pi_l^{(\varepsilon)}(\rho^{(\varepsilon)}) &= \frac{\tilde{\tilde{\pi}}_{jl}^{(\varepsilon)}(\rho^{(\varepsilon)})}{\sum_{r \neq 0} \tilde{\tilde{\pi}}_{jr}^{(\varepsilon)}(\rho^{(\varepsilon)})} = \frac{\tilde{\pi}_{jl}^{(\varepsilon)}(\rho^{(\varepsilon)})}{\sum_{r \neq 0} \tilde{\pi}_{jr}^{(\varepsilon)}(\rho^{(\varepsilon)})} \\[2mm]
&= \frac{\omega_{jjl}^{(\varepsilon)}(\rho^{(\varepsilon)})}{\sum_{r \neq 0} \omega_{jjr}^{(\varepsilon)}(\rho^{(\varepsilon)})} = \frac{\omega_{jjl}^{(\varepsilon)}(\rho^{(\varepsilon)})}{\omega_{jj}^{(\varepsilon)}(\rho^{(\varepsilon)})}, \quad l \neq 0.
\end{aligned} \tag{5.5.36}$$

As follows from the remarks above, these probabilities do not depend on the choice of the initial state $i \neq 0$ or the state $j \in X_1^{(\varepsilon)}$.

5.5.6   Perturbation conditions used in asymptotic expansions for quasi-stationary distributions

The cyclic representation formula (5.5.32) for the quasi-stationary limits $\tilde{\tilde{\pi}}_{il}^{(\varepsilon)}(\rho^{(\varepsilon)})$ and formula (5.5.34) for the quasi-stationary probabilities $\pi_l^{(\varepsilon)}(\rho^{(\varepsilon)})$ will play a key role in what follows.

We assume that conditions $\mathbf{A_2}$, $\mathbf{T_1}$, $\mathbf{E_2}$, $\mathbf{C_{12}}$, and $\mathbf{C_{13}}$ hold.

We will also assume that the following perturbation condition $\mathbf{P_{18}^{(k)}}$ (introduced in Subsection 5.1.1) for the transition probabilities of imbedded Markov chains holds:

$\mathbf{P_{18}^{(k)}}$: $p_{ij}^{(\varepsilon)} = p_{ij}^{(0)} + e_{ij}[1,0]\varepsilon + \cdots + e_{ij}[k,0]\varepsilon^k + o(\varepsilon^k)$ for $i, j \neq 0$, where $|e_{ij}[r,0]| < \infty, r = 1,\ldots,k$.

Note that this condition involves an asymptotic perturbation expansion for transition probabilities up to the degree k. Also, this condition automatically yields an asymptotic expansion for the local absorption probabilities $p_{i0}^{(\varepsilon)}$, $i \neq 0$, given in the asymptotic relation (5.1.37).

Consider the following perturbation condition for mixed power-exponential moment of transition times for $\rho \leq \beta'$:

$\mathbf{P_{29}^{(\rho,k)}}$: $p_{ij}^{(\varepsilon)}[\rho,n] = p_{ij}^{(0)}[\rho,n] + e_{ij}[\rho,1,n]\varepsilon + \cdots + e_{ij}[\rho,k,n]\varepsilon^{k-n} + o(\varepsilon^{k-n})$ for $n = 0,\ldots,k, i \neq 0, j \in X$, where $|e_{ij}[\rho,r,n]| < \infty, r = 1,\ldots,k-n$, $n = 0,\ldots,k, i \neq 0, j \in X$.

This condition resembles condition $\mathbf{P_{25}^{(\rho,k)}}$. However, condition $\mathbf{P_{29}^{(\rho,k)}}$ includes also the asymptotic expansions for mixed power-exponential moments $p_{i0}^{(\varepsilon)}[\rho,n]$, $n = 0,\ldots,k, i \neq 0$, while condition $\mathbf{P_{25}^{(\rho,k)}}$ does not. This is because of the moments $p_{i0}^{(\varepsilon)}[\rho,n]$ penetrate formulas for quasi-stationary distributions.

Recall the formulas that connect the mixed power-exponential moments $p_{ij}^{(\varepsilon)}[\rho,n]$ and $\psi_{ij}^{(\varepsilon)}[\rho,n]$, for every $n = 0,1,\ldots, i \neq 0, j \in X$, and $\rho \in \mathbb{R}_1$,

$$p_{ij}^{(\varepsilon)}[\rho,n] = p_{ij}^{(\varepsilon)}\psi_{ij}^{(\varepsilon)}[\rho,n].$$

These formulas imply that condition $\mathbf{P_{29}^{(\rho,k)}}$ is implied by condition $\mathbf{P_{18}^{(k)}}$ and the following condition:

$\mathbf{P_{29}^{'(\rho,k)}}$: $\psi_{ij}^{(\varepsilon)}[\rho,n] = \psi_{ij}^{(0)}[\rho,n] + v_{ij}[\rho,1,n]\varepsilon + \cdots + v_{ij}[\rho,k,n]\varepsilon^{k-n} + o(\varepsilon^{k-n})$ for $n = 0,\ldots,k, i \neq 0, j \in X$, where $|v_{ij}[\rho,r,n]| < \infty, r = 1,\ldots,k-n$, $n = 0,\ldots,k, i \neq 0, j \in X$.

Moreover, the corresponding asymptotic expansions in conditions $\mathbf{P_{18}^{(k)}}$ and $\mathbf{P_{29}^{'(\rho,k)}}$ are connected with the asymptotic expansion in condition $\mathbf{P_{29}^{(\rho,k)}}$ by the following formulas for $i \neq 0, j \in X$ and $r = 0,\ldots,k-n$, $n = 0,\ldots,k$:

$$e_{ij}[\rho,r,n] = \sum_{m=0}^{r} e_{ij}[r,0]v_{ij}[\rho,r-m,n]. \tag{5.5.37}$$

Consider the mixed power-exponential moments of sojourn times for $\rho \in \mathbb{R}_1, n = 0,1,\ldots,$ and $i \neq 0$,

$$\psi_i^{(\varepsilon)}[\rho,n] = \sum_{i\in X} p_{ij}^{(\varepsilon)}[\rho,n] = \sum_{i\in X} p_{ij}^{(\varepsilon)}\psi_{ij}^{(\varepsilon)}[\rho,n].$$

The following lemma, which is the direct corollary of statements **(i)** and **(ii)** in Lemma 8.1.1, gives asymptotic expansions for the expectations $\psi_i^{(\varepsilon)}[\rho, n]$, $i \neq 0$.

Lemma 5.5.4. *Let condition* $\mathbf{P}_{18}^{(k)}$, *together with condition* $\mathbf{P}_{29}^{(\rho, k)}$ *for some* $\rho \leq \beta'$, *hold. Then*

(i) *The following asymptotic expansion takes place for* $n = 0, 1, \ldots, k$ *and* $i \neq 0$:

$$\psi_i^{(\varepsilon)}[\rho, n] = \psi_i^{(0)}[\rho, n] + v_i[\rho, 1, n]\varepsilon + \cdots$$
$$+ v_i[\rho, k - n, n]\varepsilon^{k-n} + o(\varepsilon^{k-n}), \qquad (5.5.38)$$

where the coefficients $v_i[\rho, r, n]$, $r = 0, \ldots, k - n$, $n = 0, \ldots, k$, $i \neq 0$, *are given by the formulas* $v_i[\rho, 0, n] = \psi_i^{(0)}[\rho, n]$, $n = 0, \ldots, k$, $i \neq 0$, *and, for* $r = 0, \ldots, k - n$, $n = 0, \ldots, k$, $i \neq 0$,

$$v_i[\rho, r, n] = \sum_{j \in X} e_{ij}[\rho, r, n]. \qquad (5.5.39)$$

(ii) *If condition* $\mathbf{P}_{29}^{'(\rho, k)}$ *holds instead of condition* $\mathbf{P}_{29}^{(\rho, k)}$, *then the formulas* (5.5.39) *for the coefficients in the asymptotic expansion* (5.5.38) *can be rewritten in the following form:*

$$v_i[\rho, r, n] = \sum_{j \in X} e_{ij}[\rho, r, n] = \sum_{j \in X} \sum_{q=0}^{r} e_{ij}[q, 0] v_{ij}[\rho, r - q, n]. \qquad (5.5.40)$$

The asymptotic expansions (5.5.38) can be rewritten in a vector form. Let us introduce the following vectors for $r = 0, \ldots, k + 1 - n$ and $n = 0, \ldots, k + 1$:

$$\boldsymbol{\psi}^{(\varepsilon)}[\rho, n] = \begin{bmatrix} \psi_1^{(\varepsilon)}[\rho, n] \\ \vdots \\ \psi_N^{(\varepsilon)}[\rho, n] \end{bmatrix}, \qquad \mathbf{v}[\rho, r, n] = \begin{bmatrix} v_1[\rho, r, n] \\ \vdots \\ v_N[\rho, r, n] \end{bmatrix}.$$

The asymptotic expansions (5.5.38) can be rewritten in the following vector form for $r = 0, \ldots, k - n$ and $n = 0, \ldots, k$:

$$\boldsymbol{\psi}^{(\varepsilon)}[\rho, n] = \boldsymbol{\psi}^{(0)}[\rho, n] + \mathbf{v}[\rho, 1, n]\varepsilon + \cdots + \mathbf{v}[\rho, k - n, n]\varepsilon^{k-n} + \mathbf{o}(\varepsilon^{k-n}). \qquad (5.5.41)$$

For $\varepsilon \leq \varepsilon_6'$ and $i \neq 0$, define the functions

$$\varphi_i^{(\varepsilon)}[\rho, 0] = \begin{cases} \frac{1}{\rho}(\psi_i^{(\varepsilon)}[\rho, 0] - 1) & \text{for } \rho \neq 0, \rho \leq \beta', \\ \psi_i^{(\varepsilon)}[\rho, 1] & \text{for } \rho = 0, \end{cases}$$
$$= \begin{cases} \int_0^\infty \frac{1}{\rho}(e^{s\rho} - 1) F_i^{(\varepsilon)}(ds) & \text{for } \rho \neq 0, \rho \leq \beta', \\ \int_0^\infty s F_i^{(\varepsilon)}(ds) & \text{for } \rho = 0. \end{cases} \qquad (5.5.42)$$

As was pointed out in Subsection 5.5.4, the functions $\psi_i^{(\varepsilon)}[\rho, 0]$, $i \neq 0$, have derivatives of any order $n = 1, 2, \ldots$ for $\varepsilon \leq \varepsilon_7'$ and $\rho \leq \beta'$. This implies that the functions $\varphi_i^{(\varepsilon)}[\rho, 0]$, $i \neq 0$, have derivatives of any order $n = 1, 2, \ldots$ for $\varepsilon \leq \varepsilon_7'$ and $\rho \leq \beta'$.

We can obtain a relationship between these derivatives for $\varepsilon \leq \varepsilon_7'$ by differentiating the relation (a) $\psi_i^{(\varepsilon)}[\rho, 0] = \varphi_i^{(\varepsilon)}[\rho, 0]\rho + 1$, $\rho \leq \beta'$, $i \neq 0$, that follows from formula (5.5.42). Differentiating (a) yields the following relation: (b) $\psi_i^{(\varepsilon)}[\rho, n] = \varphi_i^{(\varepsilon)}[\rho, n]\rho + n\varphi_i^{(\varepsilon)}[\rho, n-1]$, $n = 1, 2, \ldots$, $\rho \leq \beta'$, $i \neq 0$. In particular, (b) yields the following relation for $\rho = 0$: (c) $\psi_i^{(\varepsilon)}[0, n] = n\varphi_i^{(\varepsilon)}[0, n-1]$, $n = 1, 2, \ldots$, $i \neq 0$.

Relations (a)–(c) yield the following recurrence relations for derivatives of the functions $\varphi_i^{(\varepsilon)}[\rho, n]$ for $n = 1, 2, \ldots$, $\varepsilon \leq \varepsilon_7'$ and $\rho \leq \beta'$:

$$\varphi_i^{(\varepsilon)}[\rho, n] = \begin{cases} (\psi_i^{(\varepsilon)}[\rho, n] - n\varphi_i^{(\varepsilon)}[\rho, n-1])/\rho & \text{for } \rho \neq 0, \rho \leq \beta', \\ \psi_i^{(\varepsilon)}[0, n+1]/(n+1) & \text{for } \rho = 0. \end{cases} \tag{5.5.43}$$

Lemmas 5.5.4 and 8.1.1 and formulas (5.5.42) and (5.5.43) easily yield asymptotic expansions for the functions $\varphi_i^{(\varepsilon)}[\rho, n]$, $n = 0, 1, \ldots$, $i \neq 0$ for the cases $\rho \neq 0$ and $\rho = 0$.

Lemma 5.5.5. *Let condition* $\mathbf{P}_{18}^{(k)}$ *hold condition* $\mathbf{P}_{29}^{(\rho, k)}$ *for some* $\rho \leq \beta'$, $\rho \neq 0$. *Then the following asymptotic expansion takes place for* $n = 0, \ldots, k$ *and* $i \neq 0$:

$$\varphi_i^{(\varepsilon)}[\rho, n] = \varphi_i^{(0)}[\rho, n] + w_i[\rho, 1, n]\varepsilon + \cdots$$
$$+ w_i[\rho, k-n, n]\varepsilon^{k-n} + o(\varepsilon^{k-n}), \tag{5.5.44}$$

where the coefficients $w_i[\rho, r, n]$, $r = 0, \ldots, k-n$, $n = 0, \ldots, k$ *are given for every* $i \neq 0$ *by the formulas*

$$w_i[\rho, r, n]$$
$$= \begin{cases} \frac{1}{\rho}(v_i[\rho, r, 0] - \delta(r, 0)) & \text{for } r = 0, \ldots, k, \ n = 0, \\ \frac{1}{\rho}(v_i[\rho, r, n] - nw_i[\rho, r, n-1]) & \text{for } r = 0, \ldots, k-n, \ n = 1, \ldots, k. \end{cases} \tag{5.5.45}$$

Lemma 5.5.6. *Let conditions* $\mathbf{P}_{18}^{(k)}$ *and* $\mathbf{P}_{29}^{(0,k+1)}$ *hold. Then the following asymptotic expansion takes place for* $n = 0, \ldots, k$ *and* $i \neq 0$:

$$\varphi_i^{(\varepsilon)}[0, n] = \varphi_i^{(0)}[0, n] + w_i[0, 1, n]\varepsilon + \cdots$$
$$+ w_i[0, k-n, n]\varepsilon^{k-n} + o(\varepsilon^{k-n}), \tag{5.5.46}$$

where the coefficients $w_i[0, r, n]$, $r = 0, \ldots, k-n$, $n = 0, \ldots, k$, *are given for every* *i $\neq$ 0 by the formulas*

$$w_i[\rho, r, n] = \begin{cases} v_i[0, r, 1] & \textit{for } r = 0, \ldots, k, \, n = 0, \\ \frac{1}{n+1} v_i[0, r, n+1] & \textit{for } r = 0, \ldots, k-n, \, n = 1, \ldots, k. \end{cases} \tag{5.5.47}$$

Note that the asymptotic expansions of the order k for functions $\varphi_i^{(\varepsilon)}[\rho, n]$ require condition $\mathbf{P}_{29}^{(\rho,k)}$, if $\rho \neq 0$, and condition $\mathbf{P}_{29}^{(0,k+1)}$, if $\rho = 0$.

The asymptotic expansions (5.5.46) and (5.5.44) for functions $\varphi_i^{(\varepsilon)}[\rho, n]$ can be rewritten in a vector form. For $r = 0, \ldots, k-n$ and $n = 0, \ldots, k$, consider the vectors

$$\boldsymbol{\varphi}^{(\varepsilon)}[\rho, n] = \begin{bmatrix} \varphi_1^{(\varepsilon)}[\rho, n] \\ \vdots \\ \varphi_N^{(\varepsilon)}[\rho, n] \end{bmatrix}, \qquad \mathbf{w}[\rho, r, n] = \begin{bmatrix} w_1[\rho, r, n] \\ \vdots \\ w_N[\rho, r, n] \end{bmatrix}.$$

The asymptotic expansion (5.5.46) can be rewritten in the following vector form for $n = 0, \ldots, k$:

$$\boldsymbol{\varphi}^{(\varepsilon)}[\rho, n] = \boldsymbol{\varphi}^{(0)}[\rho, n] + \mathbf{w}[\rho, 1, n]\varepsilon + \cdots$$
$$+ \mathbf{w}[\rho, k-n, n]\varepsilon^{k-n} + \mathbf{o}(\varepsilon^{k-n}). \tag{5.5.48}$$

For $r = 0, \ldots, k-n$ and $n = 0, \ldots, k, \, l \neq 0$, introduce the vectors

$$\boldsymbol{\varphi}_l^{(\varepsilon)}[\rho, n] = \begin{bmatrix} \varphi_1^{(\varepsilon)}[\rho, n]\delta(1, l) \\ \vdots \\ \varphi_N^{(\varepsilon)}[\rho, n]\delta(N, l) \end{bmatrix}, \qquad \mathbf{w}_l[\rho, r, n] = \begin{bmatrix} w_1[\rho, r, n]\delta(1, l) \\ \vdots \\ w_N[\rho, r, n]\delta(N, l) \end{bmatrix}.$$

The asymptotic expansion (5.5.48) also gives the following asymptotic expansions for $n = 0, \ldots, k$, and $l \neq 0$:

$$\boldsymbol{\varphi}_l^{(\varepsilon)}[\rho, n] = \boldsymbol{\varphi}_l^{(0)}[\rho, n] + \mathbf{w}_l[\rho, 1, n]\varepsilon + \cdots$$
$$+ \mathbf{w}_l[\rho, k-n, n]\varepsilon^{k-n} + \mathbf{o}(\varepsilon^{k-n}). \tag{5.5.49}$$

Also, it is convenient to define the vectors $\mathbf{w}^{(0)}[\rho, 0, n] = \boldsymbol{\varphi}^{(0)}[\rho, n]$ and $\mathbf{w}_l^{(0)}[\rho, 0, n] = \boldsymbol{\varphi}_l^{(0)}[\rho, n]$ for $l \neq 0$ and $n = 0, \ldots, k$.

5.5.7 Asymptotic expansions for mixed power-exponential moments for the time spend in a given state before hitting a given state or an absorption

Note that, under conditions $\mathbf{A}_2$, $\mathbf{T}_1$, $\mathbf{E}_2$, $\mathbf{C}_{12}$, and $\mathbf{C}_{13}$, asymptotic expansions for the mixed power-exponential moments $_0\phi_{ij}^{(\varepsilon)}[\rho, n]$ are given in Lemmas 5.3.5 and 5.3.6.

Similar asymptotic expansions can be given for the expectations $\omega_{ij}^{(\varepsilon)}[\rho, n]$. The following representation can be written for $l, j \neq 0$:

$$\int_0^{\mu_j^{(\varepsilon)} \wedge \mu_0^{(\varepsilon)}} e^{s\rho} \chi(\eta^{(\varepsilon)}(s) = l)\, ds$$

$$= \int_0^{\tau_1^{(\varepsilon)}} e^{s\rho} \chi(\eta^{(\varepsilon)}(s) = l)\, ds + \sum_{i' \neq 0, j} \chi(\eta^{(\varepsilon)}(\tau_1^{(\varepsilon)}) = i')$$

$$\times\, e^{\tau_1^{(\varepsilon)}\rho} \int_0^{\mu_j^{(\varepsilon)} \wedge \mu_0^{(\varepsilon)} - \tau_1^{(\varepsilon)}} e^{s\rho} \chi(\eta^{(\varepsilon)}(\tau_1^{(\varepsilon)} + s) = l)\, ds. \qquad (5.5.50)$$

We will use the following simple relation for $\varepsilon \leq \varepsilon_6'$, $\rho \leq \beta'$ and $i, l \neq 0$:

$$\mathsf{E}_i \int_0^{\tau_1^{(\varepsilon)}} e^{s\rho} \chi(\eta^{(\varepsilon)}(s) = l)\, ds = \delta(i, l)\phi_i^{(\varepsilon)}[\rho, 0]$$

$$= \begin{cases} \delta(i, l)(\mathsf{E}_i e^{\rho\tau_1^{(\varepsilon)}} - 1)/\rho & \text{for } \rho \neq 0, \\ \delta(i, l)\mathsf{E}_i \tau_1^{(\varepsilon)} & \text{for } \rho = 0. \end{cases} \qquad (5.5.51)$$

Note again that $X_1^{(0)} \subseteq X_1^{(\varepsilon)}$ for $\varepsilon \leq \varepsilon_7'$.

By taking the expectation in the left and the right-hand sides of (5.5.50) and using the Markov property of the semi-Markov process $\eta^{(\varepsilon)}(t)$ at the time of the first jump, as well as relation (5.5.51), this representation permits to readily show that the moment generating functions $\omega_{ijl}^{(\varepsilon)}[\rho, 0]$, $i \neq 0$, satisfy the following system of linear equations for every $\varepsilon \leq \varepsilon_6'$, $\rho \leq \beta'$, $l \neq 0$, and $j \in X_1^{(0)}$:

$$\omega_{ijl}^{(\varepsilon)}[\rho, 0] = \varphi_i^{(\varepsilon)}[\rho, 0]\delta(i, l) + \sum_{i' \neq 0, j} p_{ii'}^{(\varepsilon)}(\rho)\, \omega_{i'jl}^{(\varepsilon)}[\rho, 0], \quad i \neq 0. \qquad (5.5.52)$$

In virtue of the remarks above, we can differentiate equations (5.5.52) with respect to the variable ρ and get the following system of linear equations for every $\varepsilon \leq \varepsilon_7'$, $\rho \leq \beta'$, and $n = 0, 1, \ldots, j \in X_1^{(0)}$:

$$\omega_{ijl}^{(\varepsilon)}[\rho, n] = \vartheta_{ijl}^{(\varepsilon)}[\rho, n] + \sum_{i' \neq 0, j} p_{ii'}^{(\varepsilon)}(\rho)\omega_{i'jl}^{(\varepsilon)}[\rho, n], \quad i \neq 0, \qquad (5.5.53)$$

where

$$\vartheta_{ijl}^{(\varepsilon)}[\rho, n] = \delta(i, l)\varphi_i^{(\varepsilon)}[\rho, n] + \sum_{i' \neq 0, j} \sum_{m=1}^{n} C_n^m\, p_{ii'}^{(\varepsilon)}[\rho, m]\omega_{i'jl}^{(\varepsilon)}[\rho, n - m]. \qquad (5.5.54)$$

These systems can be rewritten in a matrix form. For $\rho \leq \beta'$, $\varepsilon \leq \varepsilon_7'$, $l \neq 0$ and $n = 0, 1, \ldots, l \neq 0$, and $j \in X_1^{(0)}$, consider the vectors

$$\boldsymbol{\omega}_{jl}^{(\varepsilon)}[\rho, n] = \begin{bmatrix} \omega_{1jl}^{(\varepsilon)}[\rho, n] \\ \vdots \\ \omega_{Njl}^{(\varepsilon)}[\rho, n] \end{bmatrix}, \qquad \boldsymbol{\vartheta}_{jl}^{(\varepsilon)}[\rho, n] = \begin{bmatrix} \vartheta_{1jl}^{(\varepsilon)}[\rho, n] \\ \vdots \\ \vartheta_{Njl}^{(\varepsilon)}[\rho, n] \end{bmatrix}.$$

Then, the linear systems (5.5.52) can be rewritten in the following form for every $\rho \leq \beta'$, $\varepsilon \leq \varepsilon_7'$, $l \neq 0$ and $j \in X_1^{(0)}$:

$$\boldsymbol{\omega}_{jl}^{(\varepsilon)}[\rho, 0] = \boldsymbol{\varphi}_l^{(\varepsilon)}[\rho, 0] + {}_j\mathbf{P}^{(\varepsilon)}[\rho, 0]\,\boldsymbol{\omega}_{jl}^{(\varepsilon)}[\rho, 0], \qquad (5.5.55)$$

and for every $\varepsilon \leq \varepsilon_7'$, $\rho \leq \beta'$, $n = 0, 1, \ldots, l \neq 0$, and $j \in X_1^{(0)}$, (5.5.53) becomes

$$\boldsymbol{\omega}_{jl}^{(\varepsilon)}[\rho, n] = \boldsymbol{\vartheta}_{jl}^{(\varepsilon)}[\rho, n] + {}_j\mathbf{P}^{(\varepsilon)}(\rho)\,\boldsymbol{\omega}_{jl}^{(\varepsilon)}[\rho, n]. \qquad (5.5.56)$$

It is convenient to also define $\vartheta_{ijl}^{(\varepsilon)}[\rho, 0] = \delta(i, l)\varphi_l^{(\varepsilon)}(\rho)$, $i, l \neq 0$, $j \in X_1^{(0)}$ and, respectively, $\boldsymbol{\vartheta}_{jl}^{(\varepsilon)}[\rho, 0] = \boldsymbol{\varphi}_l^{(\varepsilon)}[\rho, 0]$, $l \neq 0$, $j \in X_1^{(0)}$.

In this case, systems (5.5.53) and (5.5.56) coincide with systems (5.5.52) and (5.5.55), respectively, for $n = 0$.

Recall also that $\omega_{ijl}^{(\varepsilon)}(\rho) = \omega_{ijl}^{(\varepsilon)}[\rho, 0]$, $l \neq 0$, $j \in X_1^{(0)}$. Define the vectors $\boldsymbol{\omega}_{jl}^{(\varepsilon)}(\rho) = \boldsymbol{\omega}_{jl}^{(\varepsilon)}[\rho, 0]$, $l \neq 0$, $j \in X_1^{(0)}$.

The systems of linear equations (5.5.56) are systems of ρ-hitting type with the same coefficient matrix ${}_j\mathbf{P}^{(\varepsilon)}(\rho)$.

Under conditions $\mathbf{A}_2$, $\mathbf{T}_1$, $\mathbf{E}_2$, $\mathbf{C}_{12}$, $\mathbf{C}_{13}$, these systems have unique solutions for every $\varepsilon \leq \varepsilon_7'$, $\rho \leq \beta'$, $n = 0, 1, \ldots, l \neq 0$, and $j \in X_1^{(0)}$. These solutions have the following forms:

$$\boldsymbol{\omega}_{jl}^{(\varepsilon)}[\rho, n] = [\mathbf{I} - {}_j\mathbf{P}^{(\varepsilon)}(\rho)]^{-1}\boldsymbol{\vartheta}_{jl}^{(\varepsilon)}[\rho, n]. \qquad (5.5.57)$$

For given $l \neq 0$, $j \in X_1^{(0)}$, systems (5.5.56) have the same coefficient matrix ${}_j\mathbf{P}^{(\varepsilon)}(\rho)$ but different inhomogeneous terms $\boldsymbol{\vartheta}_{jl}^{(\varepsilon)}[\rho, n]$ for $n = 0, 1, \ldots$. These systems should be solved recursively.

First, system (5.5.56) should be solved for $n = 0$. Second, we solve system (5.5.56) for $n = 1$. Note that the expressions for inhomogeneous terms $\vartheta_{ijl}^{(\varepsilon)}[\rho, 1] = \delta(i, l)\varphi_i^{(\varepsilon)}[\rho, 1] + \sum_{l \neq 0, j} p_{ii'}^{(\varepsilon)}[\rho, 1]\omega_{i'jl}^{(\varepsilon)}[\rho, 0]$, $i \neq 0$, given in (5.5.54), include the solutions $\omega_{ijl}^{(\varepsilon)}[\rho, 0]$, $i \neq 0$, of systems (5.5.56) for $n = 0$.

This recursive procedure is to be repeated for $n = 1, \ldots$. The expressions for the inhomogeneous terms $\vartheta_{ijl}^{(\varepsilon)}[\rho, n]$, given in (5.5.54), include the solutions $\omega_{ijl}^{(\varepsilon)}[\rho, m]$, $i \neq 0$, of systems (5.5.56) for $m = 0, 1, \ldots, n-1$.

Formula (5.5.54) can be rewritten in the following vector form for every $\rho \leq \beta'$, $\varepsilon \leq \varepsilon_7'$, and $n = 0, 1, \ldots, l \neq 0, j \in X_1^{(0)}$:

$$\boldsymbol{\vartheta}_{jl}^{(\varepsilon)}[\rho, n] = \boldsymbol{\varphi}_1^{(\varepsilon)}[\rho, n] + \sum_{m=1}^{n} C_n^m {}_j \mathbf{P}^{(\varepsilon)}[\rho, m] \boldsymbol{\omega}_{jl}^{(\varepsilon)}[\rho, n-m]. \tag{5.5.58}$$

Using formula (5.5.58) we can rewrite formula (5.5.56) in a form that indicates the recurrence procedure used for calculating solutions of the linear system (5.5.56) for $\rho \leq \beta'$, $\varepsilon \leq \varepsilon_7'$, and $n = 0, 1, \ldots, l \neq 0, j \in X_1^{(0)}$,

$$\begin{aligned}
\boldsymbol{\omega}_{jl}^{(\varepsilon)}[\rho, n] &= [\mathbf{I} - {}_j\mathbf{P}^{(\varepsilon)}(\rho)]^{-1} \boldsymbol{\vartheta}_{jl}^{(\varepsilon)}[\rho, n] \\
&= [\mathbf{I} - {}_j\mathbf{P}^{(\varepsilon)}(\rho)]^{-1} \boldsymbol{\varphi}_1^{(\varepsilon)}[\rho, n] \\
&\quad + \sum_{m=1}^{n} C_n^m [\mathbf{I} - {}_j\mathbf{P}^{(\varepsilon)}(\rho)]^{-1} {}_j\mathbf{P}^{(\varepsilon)}[\rho, m] \boldsymbol{\omega}_{jl}^{(\varepsilon)}[\rho, n-m].
\end{aligned} \tag{5.5.59}$$

We are now in a position to derive asymptotic expansions for the functions $\omega_{ijl}^{(\varepsilon)}[\rho, n], n = 0, 1, \ldots, k, l \neq 0, j \in X_1^{(0)}$.

The following lemmas are corollaries of Lemma 5.3.4 applied to the vector-valued functions $\boldsymbol{\omega}_j^{(\varepsilon)}[\rho, n]$ for $l \neq 0, j \in X_1^{(0)}$ and $n = 0, 1, \ldots, k$. Their proofs are similar to those given for Lemmas 5.3.5 and 5.3.6.

Lemma 5.5.7. *Let conditions* $\mathbf{A}_2$, $\mathbf{T}_2$, $\mathbf{E}_2$, $\mathbf{C}_{12}$, $\mathbf{C}_{13}$, $\mathbf{P}_{18}^{(k)}$ *hold, and, for some* $\rho \leq \beta'$, *either condition* $\mathbf{P}_{29}^{(\rho,k)}$, *if* $\rho \leq \beta', \rho \neq 0$, *or* $\mathbf{P}_{29}^{(0,k+1)}$, *if* $\rho = 0$, *be satisfied. Then*

(i) *For every* $l \neq 0, j \in X_1^{(0)}$, *the vector* $\boldsymbol{\omega}_{jl}^{(\varepsilon)}[\rho, 0]$ *admits an asymptotic expansion,*

$$\boldsymbol{\omega}_{jl}^{(\varepsilon)}[\rho, 0] = \boldsymbol{\omega}_{jl}^{(0)}[\rho, 0] + \mathbf{c}_{jl}[\rho, 1, 0]\varepsilon + \cdots + \mathbf{c}_{jl}[\rho, k, 0]\varepsilon^k + o(\varepsilon^k), \tag{5.5.60}$$

where $\mathbf{c}_{jl}[\rho, r, 0]$, $r = 1, \ldots, k$, *are finite vectors.*

(ii) *The vector coefficients* $\mathbf{c}_{jl}[\rho, r, 0]$ *are given by the recurrence formulas* $\mathbf{c}_{jl}[\rho, 0, 0] = \boldsymbol{\omega}_{jl}^{(0)}[\rho, 0] = [\mathbf{I} - {}_j\mathbf{P}^{(0)}(\rho)]^{-1} \boldsymbol{\varphi}_1^{(0)}[\rho, 0]$ *and, for* $r = 0, \ldots, k$,

$$\mathbf{c}_{jl}[\rho, r, 0] = {}_j\mathbf{U}[\rho, 0]\Big(\mathbf{w}_1[\rho, r, 0] + \sum_{q=1}^{r} {}_j\mathbf{E}[\rho, q, 0]\mathbf{c}_{jl}[\rho, r-q, 0]\Big). \tag{5.5.61}$$

(iii) *The vector coefficients $\mathbf{c}_{jl}[\rho, r, 0]$ can also be calculated by the following formulas for $r = 0, \ldots, k$:*

$$\mathbf{c}_{jl}[\rho, r, 0] = \sum_{p=0}^{r} {}_j\mathbf{U}[\rho, p]\, \mathbf{w}_1[\rho, r - p, 0]. \qquad (5.5.62)$$

(iv) *For every $l \neq 0$, $j \in X_1^{(0)}$, and $n = 0, \ldots, k$, the vector $\boldsymbol{\omega}_{jl}^{(\varepsilon)}[\rho, n]$ admits the following asymptotic expansion:*

$$\boldsymbol{\omega}_{jl}^{(\varepsilon)}[\rho, n] = \boldsymbol{\omega}_{jl}^{(0)}[\rho, n] + \mathbf{c}_{j1}[\rho, 1, n]\varepsilon + \cdots$$
$$+ \mathbf{c}_{jl}[\rho, k - n, n]\varepsilon^{k-n} + \mathbf{o}(\varepsilon^{k-n}), \qquad (5.5.63)$$

where $\mathbf{c}_{jl}[\rho, r, n]$, $r = 1, \ldots, k - n$, $n = 0, \ldots, k$, are finite vectors.

(v) *The vector coefficients $\mathbf{c}_{jl}[\rho, r, n]$ are given by the recurrence formulas $\mathbf{c}_{jl}[\rho, 0, 0] = \boldsymbol{\omega}_{jl}^{(0)}[\rho, 0] = {}_j\mathbf{U}[\rho, 0]\boldsymbol{\vartheta}_{jl}^{(0)}[\rho, 0] = [\mathbf{I} - {}_j\mathbf{P}^{(0)}(\rho)]^{-1}\boldsymbol{\phi}_1^{(0)}[\rho, 0]$ and, for $r = 0, \ldots, k - n$, and and sequentially for $n = 0, \ldots, k$,*

$$\mathbf{c}_{jl}[\rho, r, n] = {}_j\mathbf{U}[\rho, 0](\mathbf{w}_1[\rho, r, n]$$
$$+ \sum_{m=1}^{n} C_n^m \sum_{q=0}^{r} {}_j\mathbf{E}[\rho, q, m]\mathbf{c}_{jl}[\rho, r - q, n - m]$$
$$+ \sum_{p=1}^{r} \mathbf{E}_j[\rho, p, 0]\mathbf{c}_{jl}[\rho, r - p, n]). \qquad (5.5.64)$$

(vi) *The vector coefficients $\mathbf{c}_{jl}[\rho, r, n]$ can also be calculated by the recurrence formulas for $r = 0, \ldots, k - n$ and sequentially for $n = 0, \ldots, k$,*

$$\mathbf{c}_{jl}[\rho, r, n] = \sum_{q=0}^{r} {}_j\mathbf{U}[\rho, q]\mathbf{w}_1[\rho, r - q, n]$$
$$+ \sum_{m=1}^{n} C_n^m \sum_{q=0}^{r} \left(\sum_{p=0}^{q} {}_j\mathbf{U}[\rho, p]_j\mathbf{E}[\rho, q - p, m] \right)$$
$$\times \mathbf{c}_{jl}[\rho, r - q, n - m]. \qquad (5.5.65)$$

Remark 5.5.2. Note that the asymptotic expansions of the order k for vectors $\boldsymbol{\omega}_{jl}^{(\varepsilon)}[\rho, n]$ require condition $\mathbf{P}_{29}^{(\rho,k)}$, if $\rho \neq 0$, and condition $\mathbf{P}_{29}^{(0,k+1)}$, if $\rho = 0$.

Remark 5.5.3. Note that condition $\mathbf{I}_4$ is not required in Lemma 5.5.7.

Formula (5.5.28) implies that, for $n = 0, 1, \ldots$, $i, j, l \neq 0$,

$$\omega_{ij}^{(\varepsilon)}[\rho, n] = \sum_{l \neq 0} \omega_{ijl}^{(\varepsilon)}[\rho, n]. \tag{5.5.66}$$

For $\rho \leq \beta'$, $\varepsilon \leq \varepsilon_7'$, $l \neq 0$, and $n = 0, 1, \ldots, k$, $l \neq 0$, and $j \in X_1^{(0)}$, introduce the vectors

$$\mathbf{c}_j[\rho, r, n] = \sum_{l \neq 0} \mathbf{c}_{jl}[\rho, r, n]. \tag{5.5.67}$$

For $r = 0, \ldots, k - n$, $n = 0, \ldots, k$, also define the vectors

$$\boldsymbol{\omega}_j^{(\varepsilon)}[\rho, n] = \begin{bmatrix} \omega_{1j}^{(\varepsilon)}[\rho, n] \\ \vdots \\ \omega_{Nj}^{(\varepsilon)}[\rho, n] \end{bmatrix},$$

and

$$\mathbf{c}_j[\rho, r, n] = \begin{bmatrix} c_{1j}[\rho, r, n] \\ \vdots \\ c_{Nj}[\rho, r, n] \end{bmatrix}, \qquad \mathbf{c}_{jl}[\rho, r, n] = \begin{bmatrix} c_{1jl}[\rho, r, n] \\ \vdots \\ c_{Njl}[\rho, r, n] \end{bmatrix}.$$

Relation (5.5.66) permits to obtain asymptotic expansions for the vectors $\boldsymbol{\omega}_j^{(\varepsilon)}[\rho, n]$ by using asymptotic expansions for the vectors $\boldsymbol{\omega}_{jl}^{(\varepsilon)}[\rho, n]$ given in Lemma 5.5.7. These expansions take the following form for $\varepsilon \leq \varepsilon_6'$, $\rho \leq \beta'$, $j \in X_1^{(0)}$, and $n = 0, 1, \ldots, k$:

$$\boldsymbol{\omega}_j^{(\varepsilon)}[\rho, n] = \boldsymbol{\omega}_j^{(0)}[\rho, n] + \mathbf{c}_j[\rho, 1, n]\varepsilon + \cdots + \mathbf{c}_j[\rho, k, n]\varepsilon^{k-n} + \mathbf{o}(\varepsilon^{k-n}). \tag{5.5.68}$$

Usually, it is also convenient to define the vectors $\mathbf{c}_j[\rho, 0, n] = \boldsymbol{\omega}_j^{(0)}[\rho, n]$, $n = 0, \ldots, k$.

5.5.8 Asymptotic expansions for quasi-stationary distributions of perturbed semi-Markov processes

We are now in a position to give asymptotic expansions for quasi-stationary distributions and related quasi-stationary limits of perturbed semi-Markov processes.

The process $\eta^{(\varepsilon)}(t)$, $t \geq 0$, is a regenerative process with absorption, and the times of return to the initial state are the corresponding regeneration times. Also, $\mu_0^{(\varepsilon)}$ is a regenerative stopping time which regenerate together with the process $\eta^{(\varepsilon)}(t)$, $t \geq 0$, at the times of return. Therefore, we can use theorems given in Section 3.5.

The corresponding imbedding procedure is described in Section 4.2. In all "imbedding" calculations a one-state set $A = \{l\}$ and a recurrent-without-absorption state $j \in X_1^{(0)}$ are used for constructing the corresponding regeneration cycles.

The moment generating functions $\phi^{(\varepsilon)}(\rho)$, $\tilde{\phi}^{(\varepsilon)}(\rho)$, and $\omega^{(\varepsilon)}(\rho, r, A)$ are regarded as the moment generating functions ${}_0\phi_{jj}^{(\varepsilon)}(\rho)$, ${}_0\phi_{ij}^{(\varepsilon)}(\rho)$, and $\omega_{jjl}^{(\varepsilon)}(\rho)$, respectively.

The respective asymptotic expansions for the moment generating functions ${}_0\phi_{jj}^{(\varepsilon)}(\rho)$, ${}_0\phi_{ij}^{(\varepsilon)}(\rho)$, and $\omega_{jjl}^{(\varepsilon)}(\rho)$ are given in terms of the transition probabilities $p_{ir}^{(\varepsilon)}$ of imbedded Markov chains, and the moment generation functions $p_{ir}^{(\varepsilon)}[\rho, n]$ of transition times. These conditions are used in Lemmas 5.3.5, 5.3.6, and 5.5.7, where the corresponding perturbation expansions for the moment generation functions ${}_0\phi_{jj}^{(\varepsilon)}(\rho)$, ${}_0\phi_{ij}^{(\varepsilon)}(\rho)$, and $\omega_{jjl}^{(\varepsilon)}(\rho)$ are obtained.

All other conditions used in Theorems 3.5.2–3.5.4 are checked in the proofs of Theorems 5.4.1–5.4.4.

Since any recurrent-without-absorption state $j \in X_1^{(0)}$ can be used in the construction of corresponding regeneration cycles, the corresponding quasi-stationary distributions do not depend on the choice of the state $j \in X_1^{(0)}$.

Taking into account the above remarks, we formulate the following four theorems without explicit proofs.

The next theorem is a variant of Theorem 5.4.1.

Theorem 5.5.2. *Let conditions* $\mathbf{A_2}$, $\mathbf{I_4}$ $\mathbf{T_2}$, $\mathbf{E_2}$, $\mathbf{P_{18}^{(k)}}$ *hold, and either condition* $\mathbf{P_{29}^{(\rho^{(0)},k)}}$, *if* $\rho^{(0)} > 0$ *or* $\mathbf{P_{29}^{(0,k+1)}}$, *if* $\rho^{(0)} = 0$, *be satisfied. Then, the quasi-stationary limits* $\tilde{\pi}_{jl}^{(\varepsilon)}(\rho^{(\varepsilon)}) = \omega_{jjl}^{(\varepsilon)}(\rho^{(\varepsilon)})/{}_0\phi_{jj}^{(\varepsilon)}(\rho^{(\varepsilon)})$, $l \neq 0$, $j \in X_1^{(0)}$, *have the following asymptotic expansion for every* $l \neq 0$ *and* $j \in X_1^{(0)}$:

$$
\begin{aligned}
\tilde{\pi}_{jl}^{(\varepsilon)}(\rho^{(\varepsilon)}) &= \frac{\omega_{jjl}^{(0)}(\rho^{(0)}) + f'_{jjl}[\rho^{(0)}, 1]\varepsilon + \cdots + f'_{jjl}[\rho^{(0)}, k]\varepsilon^k + o(\varepsilon^k)}{{}_0\phi_{jj}^{(0)}(\rho^{(0)}) + f''_{jj}[\rho^{(0)}, 1]\varepsilon + \cdots + f''_{jj}[\rho^{(0)}, k]\varepsilon^k + o(\varepsilon^k)} \\
&= \tilde{\pi}_{jl}^{(0)}(\rho^{(0)}) + f_{jl}[\rho^{(0)}, 1]\varepsilon + \cdots + f_{jl}[\rho^{(0)}, k]\varepsilon^k + o(\varepsilon^k), \quad (5.5.69)
\end{aligned}
$$

where the coefficients $f'_{jjl}[\rho^{(0)}, n]$ *and* $f''_{jj}[\rho^{(0)}, n]$ *are given by the formulas*

$$
f'_{jjl}[\rho^{(0)}, 0] = \omega_{jjl}^{(0)}(\rho^{(0)}) = c_{jjl}[\rho^{(0)}, 0, 0],
$$

$$
f'_{jjl}[\rho^{(0)}, 1] = c_{jjl}[\rho^{(0)}, 1, 0] + c_{jjl}[\rho^{(0)}, 0, 1]a_1,
$$

$$
f''_{jj}[\rho^{(0)}, 0] = {}_0\phi_{jj}^{(0)}(\rho^{(0)}) = {}_0b_{jj}[\rho^{(0)}, 0, 1],
$$

$$
f''_{jj}[\rho^{(0)}, 1] = {}_0b_{jj}[\rho^{(0)}, 1, 1] + {}_0b_{jj}[\rho^{(0)}, 0, 2]a_1,
$$

and, for $n = 0, \ldots, k$,

$$f'_{jjl}[\rho^{(0)}, n]$$

$$= c_{jjl}[\rho^{(0)}, n, 0] + \sum_{q=1}^{n} c_{jjl}[\rho^{(0)}, n-q, 1]a_q$$

$$+ \sum_{2 \leq m \leq n} \sum_{q=m}^{n} c_{jjl}[\rho^{(0)}, n-q, m] \cdot \sum_{n_1, \ldots, n_{q-1} \in D_{m,q}} \prod_{p=1}^{q-1} a_p^{n_p} / n_p! \qquad (5.5.70)$$

and

$$f'_{jj}[\rho^{(0)}, n]$$

$$= {}_0 b_{jj}[\rho^{(0)}, n, 1] + \sum_{q=1}^{n} {}_0 b_{jj}[\rho^{(0)}, n-q, 2]a_q$$

$$+ \sum_{2 \leq m \leq n} \sum_{q=m}^{n} {}_0 b_{jj}[\rho^{(0)}, n-q, m+1] \cdot \sum_{n_1, \ldots, n_{q-1} \in D_{m,q}} \prod_{p=1}^{q-1} a_p^{n_p} / n_p!, \qquad (5.5.71)$$

and the coefficients $f_{jl}[\rho^{(0)}, n]$ are given by the recurrence formulas $f_{jl}[\rho^{(0)}, 0] = \tilde{\pi}_{jl}^{(0)}(\rho^{(0)}) = f'_{jjl}[\rho^{(0)}, 0]/f''_{jj}[\rho^{(0)}, n]$ and, for $n = 0, \ldots, k$,

$$f_{jl}[\rho^{(0)}, n] = \left(f'_{jjl}[\rho^{(0)}, n] - \sum_{q=0}^{n-1} f''_{jj}[\rho^{(0)}, n-q] f_{jl}[\rho^{(0)}, q] \right) / f''_{jj}[\rho^{(0)}, 0]. \qquad (5.5.72)$$

For $j \in X_1^{(0)}$, consider the quasi-stationary quantities

$$\tilde{\pi}_j^{(\varepsilon)}(\rho^{(\varepsilon)}) = \sum_{r \neq 0} \tilde{\pi}_{jr}^{(\varepsilon)}(\rho^{(\varepsilon)}).$$

Theorem 5.5.2 gives the following asymptotic expansion for these quantities; it is a direct corollary of the asymptotic expansion (5.5.69),

$$\tilde{\pi}_j^{(\varepsilon)}(\rho^{(\varepsilon)}) = \tilde{\pi}_j^{(0)}(\rho^{(0)}) + f_j[\rho^{(0)}, 1]\varepsilon + \cdots + f_j[\rho^{(0)}, k]\varepsilon^k + o(\varepsilon^k), \qquad (5.5.73)$$

where the coefficients $f_j[\rho^{(0)}, n]$ are given by the recurrence formulas $f_j[\rho^{(0)}, 0] = \tilde{\pi}_j^{(0)}(\rho^{(0)})$ and, for $n = 0, \ldots, k$,

$$f_j[\rho^{(0)}, n] = \sum_{r \neq 0} f_{jr}[\rho^{(0)}, n]. \qquad (5.5.74)$$

The following theorem is a modification of Theorem 5.4.2.

Theorem 5.5.3. *Let conditions* $\mathbf{A}_2$, $\mathbf{I}_4$ $\mathbf{T}_2$, $\mathbf{E}_2$, $\mathbf{P}_{18}^{(k)}$ *hold, and either condition* $\mathbf{P}_{29}^{(\rho^{(0)},k)}$, *if* $\rho^{(0)} > 0$ *or* $\mathbf{P}_{29}^{(0,k+1)}$, *if* $\rho^{(0)} = 0$, *be satisfied. Then, the quasi-stationary limits* $\tilde{\tilde{\pi}}_{il}^{(\varepsilon)}(\rho^{(\varepsilon)}) = {}_0\phi_{ij}^{(\varepsilon)}(\rho^{(\varepsilon)})\omega_{jjl}^{(\varepsilon)}(\rho^{(\varepsilon)})/{}_0\phi_{jj}^{(\varepsilon)}(\rho^{(\varepsilon)})$, $i,l \neq 0$, *which do not depend on the choice of the state* $j \in X_1^{(0)}$, *have the following asymptotic expansion for every* $i,l \neq 0$ *and* $j \in X_1^{(0)}$:

$$\tilde{\tilde{\pi}}_{il}^{(\varepsilon)}(\rho^{(\varepsilon)}) = \tilde{\tilde{\pi}}_{il}^{(0)}(\rho^{(0)}) + h_{il}[\rho^{(0)},1]\varepsilon + \cdots + h_{il}[\rho^{(0)},k]\varepsilon^k + o(\varepsilon^k), \quad (5.5.75)$$

where the coefficients $h_{il}[\rho^{(0)},n]$ *are given by the formulas*

$$h_{il}[\rho^{(0)},0] = \tilde{\tilde{\pi}}_{il}^{(0)}(\rho^{(0)}) = f_{ij}'''[\rho^{(0)},0]f_{jl}[\rho^{(0)},0],$$

$$h_{il}[\rho^{(0)},1] = f_{ij}'''[\rho^{(0)},0]f_{jl}[\rho^{(0)},1] + f_{ij}'''[\rho^{(0)},1]f_{jl}[\rho^{(0)},0],$$

and, for $n = 0,\ldots,k$,

$$h_{il}[\rho^{(0)},n] = \sum_{q=0}^{n} f_{ij}'''[\rho^{(0)},q]f_{jl}[\rho^{(0)},n-q] \qquad (5.5.76)$$

and

$$f_{ij}'''[\rho^{(0)},n]$$

$$= {}_0b_{ij}[\rho^{(0)},n,0] + \sum_{q=1}^{n} {}_0b_{ij}[\rho^{(0)},n-q,0]a_q$$

$$+ \sum_{2 \leq m \leq n}\sum_{q=m}^{n} {}_0b_{ij}[\rho^{(0)},n-q,m] \cdot \sum_{n_1,\ldots,n_{q-1}\in D_{m,q}}\prod_{p=1}^{q-1} a_p^{n_p}/n_p!. \quad (5.5.77)$$

The following theorem is a version of Theorem 5.4.3.

Theorem 5.5.4. *Let conditions* $\mathbf{A}_2$, $\mathbf{I}_4$ $\mathbf{T}_2$, $\mathbf{E}_2$, $\mathbf{P}_{18}^{(k)}$ *hold, and either condition* $\mathbf{P}_{29}^{(\rho^{(0)},k)}$, *if* $\rho^{(0)} > 0$ *or* $\mathbf{P}_{29}^{(0,k+1)}$, *if* $\rho^{(0)} = 0$, *be satisfied. Then, the quasi-stationary probabilities* $\pi_l^{(\varepsilon)}(\rho^{(\varepsilon)}) = \tilde{\pi}_{jl}^{(\varepsilon)}(\rho^{(\varepsilon)})/\tilde{\pi}_j^{(\varepsilon)}(\rho^{(\varepsilon)})$, $l \neq 0$, $j \in X_1^{(0)}$, *which do not depend on the choice of the state* $j \in X_1^{(0)}$, *have the following asymptotic expansion for every* $l \neq 0$ *and* $j \in X_1^{(0)}$:

$$\pi_l^{(\varepsilon)}(\rho^{(\varepsilon)}) = \frac{\tilde{\pi}_{jl}^{(0)}(\rho^{(0)}) + f_{jl}[\rho^{(0)},1]\varepsilon + \cdots + f_{jl}[\rho^{(0)},k]\varepsilon^k + o(\varepsilon^k)}{\tilde{\pi}_j^{(0)}(\rho^{(0)}) + f_j[\rho^{(0)},1]\varepsilon + \cdots + f_j[\rho^{(0)},k]\varepsilon^k + o(\varepsilon^k)}$$

$$= \pi^{(0)}(\rho^{(0)}) + g_l[\rho^{(0)},1]\varepsilon + \cdots + g_l[\rho^{(0)},k]\varepsilon^k + o(\varepsilon^k), \quad (5.5.78)$$

where the coefficients $g_l[\rho^{(0)}, n]$ are given by the recurrence formulas

$$g_l[\rho^{(0)}, 0] = \pi^{(0)}(\rho^{(0)}) = f_{jl}[\rho^{(0)}, 0]/f_j[\rho^{(0)}, 1],$$

$$f_{jl}[\rho^{(0)}, 0] = \tilde{\pi}_{jl}^{(0)}(\rho^{(0)}),$$

$$f_j[\rho^{(0)}, 0] = \tilde{\pi}_j^{(0)}(\rho^{(0)}),$$

and, for $n = 0, \ldots, k$,

$$g_l[\rho^{(0)}, n] = (f_{jl}[\rho^{(0)}, n] - \sum_{q=0}^{n-1} f_j[\rho^{(0)}, n - q]g_l[\rho^{(0)}, q])/f_j[\rho^{(0)}, 0]. \qquad (5.5.79)$$

The next theorem is a variant of Theorem 5.4.4.

Theorem 5.5.5. *Let conditions $\mathbf{A_2}$, $\mathbf{I_4}$ $\mathbf{T_2}$, $\mathbf{E_2}$, $\mathbf{P_{18}^{(k)}}$ hold, and either condition $\mathbf{P_{29}^{(\rho^{(0)}, k)}}$, if $\rho^{(0)} > 0$ or $\mathbf{P_{29}^{(0, k+1)}}$, if $\rho^{(0)} = 0$, be satisfied. Then, the quasi-stationary probabilities $\pi_l^{(\varepsilon)}(\rho^{(\varepsilon)}) = \omega_{jjl}^{(\varepsilon)}(\rho^{(\varepsilon)})/\omega_{jj}^{(\varepsilon)}(\rho^{(\varepsilon)}), l \neq 0, j \in X_1^{(0)}$, which do not depend on the choice of the state $j \in X_1^{(0)}$, have the following asymptotic expansion for every $l \neq 0$ and $j \in X_1^{(0)}$:*

$$\pi_l^{(\varepsilon)}(\rho^{(\varepsilon)}) = \frac{\omega_{jjl}^{(0)}(\rho^{(0)}) + f_{jjl}'[\rho^{(0)}, 1]\varepsilon + \cdots + f_{jjl}'[\rho^{(0)}, k]\varepsilon^k + o(\varepsilon^k)}{\omega_{jj}^{(0)}(\rho^{(0)}) + f_{jj}'[\rho^{(0)}, 1]\varepsilon + \cdots + f_{jj}'[\rho^{(0)}, k]\varepsilon^k + o(\varepsilon^k)}$$

$$= \pi^{(0)}(\rho^{(0)}) + g_l[\rho^{(0)}, 1]\varepsilon + \cdots + g_l[\rho^{(0)}, k]\varepsilon^k + o(\varepsilon^k), \qquad (5.5.80)$$

where the coefficients $f_{jjl}'[\rho^{(0)}, n]$ and $f_{jj}'[\rho^{(0)}, n]$ are given by the formulas

$$f_{jjl}'[\rho^{(0)}, 0] = \omega_{jjl}^{(0)}(\rho^{(0)}) = c_{jjl}[\rho^{(0)}, 0, 0],$$

$$f_{jjl}'[\rho^{(0)}, 1] = c_{jjl}[\rho^{(0)}, 1, 0] + c_{jjl}[\rho^{(0)}, 0, 1]a_1,$$

$$f_{jj}'[\rho^{(0)}, 0] = \omega_{jj}^{(0)}(\rho^{(0)}) = c_{jj}[\rho^{(0)}, 0, 0],$$

$$f_{jj}'[\rho^{(0)}, 1] = c_{jj}[\rho^{(0)}, 1, 0] + c_{jj}[\rho^{(0)}, 0, 1]a_1,$$

and, for $n = 0, \ldots, k$,

$$f_{jjl}'[\rho^{(0)}, n]$$

$$= c_{jjl}[\rho^{(0)}, n, 0] + \sum_{q=1}^{n} c_{jjl}[\rho^{(0)}, n - q, 1]a_q$$

$$+ \sum_{2 \leq m \leq n} \sum_{q=m}^{n} c_{jjl}[\rho^{(0)}, n - q, m] \cdot \sum_{n_1, \ldots, n_{q-1} \in D_{m,q}} \prod_{p=1}^{q-1} a_p^{n_p}/n_p!, \qquad (5.5.81)$$

and

$$f'_{jj}[\rho^{(0)}, n]$$

$$= c_{jj}[\rho^{(0)}, n, 0] + \sum_{q=1}^{n} c_{jj}[\rho^{(0)}, n-q, 1]a_q$$

$$+ \sum_{2 \le m \le n} \sum_{q=m}^{n} c_{jj}[\rho^{(0)}, n-q, m] \cdot \sum_{n_1,\dots,n_{q-1} \in D_{m,q}} \prod_{p=1}^{q-1} a_p^{n_p}/n_p!, \qquad (5.5.82)$$

and coefficients $g_l[\rho^{(0)}, n]$ are given by the recurrence formulas $g_l[\rho^{(0)}, 0] = \pi^{(0)}(\rho^{(0)}) = f'_{jjl}[\rho^{(0)}, 0]/f'_{jj}[\rho^{(0)}, 0]$, and, for $n = 0, \dots, k$,

$$g_l[\rho^{(0)}, n] = (f'_{jjl}[\rho^{(0)}, n] - \sum_{q=0}^{n-1} f'_{jj}[\rho^{(0)}, n-q]g_l[\rho^{(0)}, q])/f'_{jj}[\rho^{(0)}, 0]. \quad (5.5.83)$$

It is useful to note that despite the difference in the algorithms, coefficients $g_l[\rho^{(0)}, n]$, $n = 1, \dots, k$, in both asymptotic expansions (5.5.78) and (5.5.80) coincide.

In conclusion we would like to remark that the theorems formulated above cover both the pseudo- and quasi-stationary models for $\rho^{(0)} = 0$ and $\rho^{(0)} > 0$, respectively.

However, even in the pseudo-stationary case, Theorems 5.5.2, 5.5.3, 5.5.4, and 5.5.5 differ from Theorem 5.5.1. Indeed, the former four theorems allow for an absorption of pre-limit perturbed processes, while this is impossible for pre-limit perturbed processes in the Theorem 5.5.1.

5.6   Nonlinearly perturbed continuous-time Markov chains

In this section we consider a particular but important case where nonlinearly perturbed continuous-time Markov chains with absorption.

5.6.1   Perturbed continuous-time Markov chains

Let $\eta^{(\varepsilon)}(t)$, $t \ge 0$ be a continuous-time homogeneous Markov chain for every $\varepsilon \ge 0$, with the phase space $X = \{0, \dots, N\}$ and a generator $\mathbf{Q}^{(\varepsilon)}$ that has the following form:

$$\mathbf{Q}^{(\varepsilon)} = \begin{bmatrix} -q_0^{(\varepsilon)} & q_{01}^{(\varepsilon)} & \cdots & q_{0N}^{(\varepsilon)} \\ q_{10}^{(\varepsilon)} & -q_1^{(\varepsilon)} & \cdots & q_{1N}^{(\varepsilon)} \\ \vdots & \vdots & \vdots & \vdots \\ q_{N0}^{(\varepsilon)} & q_{N1}^{(\varepsilon)} & \cdots & -q_N^{(\varepsilon)} \end{bmatrix},$$

where **(a)** $0 \le q_{ij}^{(\varepsilon)} < \infty$, $i \ne j$, $i, j \in X$, **(b)** $q_i^{(\varepsilon)} = \sum_{j \ne i} q_{ij}^{(\varepsilon)}$, $i \in X$.

We consider the case where 0 is an absorption state, so that the following condition holds:

$\mathbf{A_6}$: $q_0^{(\varepsilon)} = 0$ for any $\varepsilon \geq 0$.

We also assume that the following condition is satisfied:

$\mathbf{T_3}$: $q_{ij}^{(\varepsilon)} \to q_{ij}^{(0)}$ as $\varepsilon \to 0$ for $i \neq 0, j \in X$.

Condition $\mathbf{T_3}$ allows to consider the Markov chain $\eta^{(\varepsilon)}(t), t \geq 0$, for $\varepsilon > 0$ as a perturbed version of the Markov chain $\eta^{(0)}(t), t \geq 0$.

If condition $\mathbf{T_3}$ holds, then, for $i \neq 0$,

$$q_i^{(\varepsilon)} \to q_i^{(0)} \quad \text{as } \varepsilon \to 0. \tag{5.6.1}$$

In order to exclude the existence of absorption states $j \neq 0$, we also assume that the following condition is satisfied:

$\mathbf{I_6}$: $q_j^{(0)} > 0, j \neq 0$.

Conditions $\mathbf{I_6}$ and $\mathbf{T_3}$ imply that **(a)** there exists $\varepsilon_0 > 0$ such that $q_j^{(\varepsilon)} > 0, j \neq 0$, for $\varepsilon \leq \varepsilon_0$.

The Markov chain $\eta^{(\varepsilon)}(t), t \geq 0$ is a particular case of the semi-Markov process with absorption, for $\varepsilon \leq \varepsilon_0$. It has the transition probabilities

$$Q_{ij}^{(\varepsilon)}(t) = \begin{cases} p_{ij}^{(\varepsilon)}(1 - e^{-q_i^{(\varepsilon)}t}) & \text{if } i \neq 0, j \neq i, j \in X, \\ 0 & \text{if } i \neq 0, j = i, \end{cases} \tag{5.6.2}$$

where

$$p_{ij}^{(\varepsilon)} = \begin{cases} q_{ij}^{(\varepsilon)}/q_i^{(\varepsilon)} & \text{if } i \neq 0, j \neq i, j \in X, \\ 0 & \text{if } i \neq 0, j = i. \end{cases} \tag{5.6.3}$$

In this case, the distributions of transition times are exponential, with the parameters that do not depend on $j \in X$. They have the following form for $i \neq 0, j \in X$:

$$F_{ij}^{(\varepsilon)}(t) = F_i^{(\varepsilon)}(t) = (1 - e^{-q_i^{(\varepsilon)}t}). \tag{5.6.4}$$

Conditions $\mathbf{A_6}$, $\mathbf{T_3}$, and $\mathbf{I_6}$ imply and replace, respectively, conditions $\mathbf{A_2}$, $\mathbf{T_1}$, and $\mathbf{I_4}$.

Conditions $\mathbf{E_2}$, $\mathbf{E_2'}$ and $\mathbf{E_2''}$ do not change their formulations.

5.6.2   Regularisation of continuous-time Markov chains

The standard definition of continuous-time Markov chain excludes possibility for a Markov chain to have virtual transitions $i \rightarrow i$. In some cases, it is convenient to allow such transitions.

This can be achieved by adding some additional intensities $q_i \geq 0$, $i \neq 0$ for such transitions and by replacing the Markov chain $\eta^{(\varepsilon)}(t)$, $t \geq 0$, by a semi-Markov process $\tilde{\eta}^{(\varepsilon)}(t)$, $t \geq 0$, with the transition probabilities

$$\tilde{Q}_{ij}^{(\varepsilon)}(t) = \begin{cases} \tilde{p}_{ij}^{(\varepsilon)}(1 - e^{-\tilde{q}_i^{(\varepsilon)}t}) & \text{if } i \neq 0, j \neq i, j \in X, \\ 0 & \text{if } i \neq 0, j = i, \end{cases} \tag{5.6.5}$$

where

$$\tilde{q}_i^{(\varepsilon)} = q_i + q_i^{(\varepsilon)}, \quad i \neq 0. \tag{5.6.6}$$

and

$$\tilde{p}_{ij}^{(\varepsilon)} = \begin{cases} q_{ij}^{(\varepsilon)}/\tilde{q}_i^{(\varepsilon)} & \text{if } i \neq 0, j \neq i, j \in X, \\ q_i/\tilde{q}_i^{(\varepsilon)} & \text{if } i \neq 0, j = i. \end{cases} \tag{5.6.7}$$

Simple calculation shows that the processes $\eta^{(\varepsilon)}(t)$, $t \geq 0$, by a semi-Markov process $\tilde{\eta}^{(\varepsilon)}(t)$, $t \geq 0$, have the same finite-dimensional distributions, i.e., for any $i, i_1, \ldots, i_n \neq 0$ and $0 \leq t_1 \leq \cdots \leq t_n < \infty$, $n = 1, 2, \ldots$,

$$\mathsf{P}_i\{\eta^{(\varepsilon)}(t_r) = i_r, r = 1, \ldots, n\} = \mathsf{P}_i\{\tilde{\eta}^{(\varepsilon)}(t_r) = i_r, r = 1, \ldots, n\}. \tag{5.6.8}$$

Relation (5.6.8) obviously implies that the semi-Markov process $\tilde{\eta}^{(\varepsilon)}(t)$, $t \geq 0$ is, in fact, a homogeneous Markov process. This process can replace the Markov chain $\eta^{(\varepsilon)}(t)$, $t \geq 0$ in any quasi-stationary asymptotic relations.

5.6.3   Moment conditions

The distributions of transition times have finite power moments of all orders, do not depend on $j \in X$, and are given by the following explicit formula for $n = 0, 1, \ldots$, $i \neq 0$, $j \in X$ and $\varepsilon \leq \varepsilon_0$:

$$m_{ij}^{(\varepsilon)}[n] = m_i^{(\varepsilon)}[n] = \int_0^\infty s^n q_i^{(\varepsilon)} e^{-q_i^{(\varepsilon)}s}, ds = n!(q_i^{(\varepsilon)})^{-n} < \infty. \tag{5.6.9}$$

The moment generating functions of random variables, $\tau_1^{(\varepsilon)}$, also do not depend on $j \in X$ and are given by the following explicit formula for $n = 0, 1, \ldots, i \neq 0, j \in X$:

$$\psi_{ij}^{(\varepsilon)}[\rho, n] = \psi_i^{(\varepsilon)}[\rho, n] = \int_0^\infty s^n e^{\rho s} q_i^{(\varepsilon)} e^{-q_i^{(\varepsilon)}s} \, ds$$

$$= \begin{cases} \dfrac{n! q_i^{(\varepsilon)}}{(q_i^{(\varepsilon)} - \rho)^{n+1}} < \infty & \text{if } \rho < q_i^{(\varepsilon)}, \\ \infty & \text{if } \rho \geq q_i^{(\varepsilon)}. \end{cases} \tag{5.6.10}$$

Condition $\mathbf{C_{12}}$ holds for any $0 < \delta < \rho_*$, where

$$\rho_* = \min(q_1^{(0)}, \ldots, q_N^{(0)}). \qquad (5.6.11)$$

Condition $\mathbf{C_{13}}$ remains the same.

The pseudo-stationary case, where the characteristic root $\rho^{(0)} = 0$, corresponds with condition $\mathbf{A_3}$ that takes the form $\sum_{i \in X_1^{(0)}} q_{i0}^{(0)} = 0$.

In this case, as was shown in Lemma 4.5.8, condition $\mathbf{C_{13}}$ automatically holds.

The quasi-stationary case, where the characteristic root $\rho^{(0)} > 0$, corresponds with the assumption that condition $\mathbf{A_3}$ does not hold, i.e., $\sum_{i \in X_1^{(0)}} q_{i0}^{(0)} > 0$.

In this case, condition $\mathbf{G_1}$ holds with the parameter

$$\rho^* = \min(q_i^{(0)} : i \in X_1^{(0)}). \qquad (5.6.12)$$

Condition $\mathbf{G_2}$ **(a)** holds if $\rho^* = \rho_*$. Indeed, in this case, for $\rho < \rho^*$ and $i \neq 0$,

$$\overline{\lim_{\varepsilon \to 0}} \, \psi_i^{(\varepsilon)}[\rho, 0] = \overline{\lim_{\varepsilon \to 0}} \, \frac{q_i^{(\varepsilon)}}{q_i^{(\varepsilon)} - \rho} = \psi_i^{(0)}[\rho, 0] = \frac{q_i^{(0)}}{q_i^{(0)} - \rho} < \infty. \qquad (5.6.13)$$

Condition $\mathbf{G_2}$ **(b)** remains.

As was shown in Lemma 4.5.10, condition $\mathbf{C_{13}}$ is implied by conditions $\mathbf{G_1}$ and $\mathbf{G_2}$. In this case, the characteristic root $0 < \rho^{(0)} < \rho^*$.

Fortunately, as was mentioned in Subsection 4.5.5, in the quasi-stationary case, condition $\mathbf{E_2'}$ usually holds, i.e. the set of non-recurrent-without-absorption states $X_2^{(0)} = \varnothing$. In this case, conditions $\rho^* = \rho_*$ and $\mathbf{G_2}$, which is an analogue of $\mathbf{C_{13}}$ **(b)**, automatically holds.

5.6.4   Perturbation conditions

The perturbation condition that implies and replaces the corresponding perturbation conditions on the transition probabilities of imbedded Markov chain, the power moments, and the mixed power-exponential moments of distributions of transition times, and was used in Chapter 5, takes the following form:

$$\mathbf{P_{30}^{(k)}}: \ q_{ij}^{(\varepsilon)} = q_{ij}^{(0)} + q_{ij}[1]\varepsilon + \cdots + q_{ij}[k]\varepsilon^k + o(\varepsilon^k) \text{ for } i \neq 0, \, j \in X, \text{ where } |q_{ij}[r]| < \infty, r = 0, \ldots, k, i \neq 0, j \in X.$$

It is convenient to also set $q_{ij}[0] = q_{ij}^{(0)}$, $i \neq 0$, $j \in X$. The parameter k in condition $\mathbf{P_{30}^{(k)}}$ can take any non-negative integer value.

Note also that condition $\mathbf{P_{30}^{(k)}}$ implies that condition $\mathbf{T_3}$ holds.

The following two lemmas show that condition $\mathbf{P_{30}^{(k)}}$ implies the perturbation condition $\mathbf{P_{18}^{(k)}}$ for transition probabilities of an imbedded Markov chain.

Lemma 5.6.1. *Let conditions* $\mathbf{I_6}$, *and* $\mathbf{P}_{30}^{(k)}$ *hold. Then the following expansion holds for every* $i \neq 0$, $j \in X$:

$$q_i^{(\varepsilon)} = q_i[0] + q_i[1]\varepsilon + \cdots + q_i[k]\varepsilon^k + o(\varepsilon^k), \tag{5.6.14}$$

where $q_i[0] = q_i^{(0)}$ *and, for* $r = 0,\ldots,k$,

$$q_i[r] = \sum_{j \neq i} q_{ij}[r], \quad r = 0,\ldots,k. \tag{5.6.15}$$

Proof. The lemma is a direct corollary of formula (5.6.3), relation **(b)**, and Lemma 8.1.1. According to relation **(b)**, the intensities $q_i^{(\varepsilon)}$, $i \neq 0$, are obtained as sums of the intensities $q_{ij}^{(\varepsilon)}$. Thus, the asymptotic expansion (5.6.14) is a corollary of statement **(ii)** of Lemma 8.1.1. $\qquad\qquad\square$

Lemma 5.6.2. *Let conditions* $\mathbf{A_5}$, $\mathbf{I_6}$, *and* $\mathbf{P}_{30}^{(k)}$ *hold. Then we have the following expansion for every* $i \neq 0$, $j \in X$:

$$p_{ij}^{(\varepsilon)} = p_{ij}^{(0)} + e_{ij}[1,0]\varepsilon + \cdots + e_{ij}[k,0]\varepsilon^k + o(\varepsilon^k), \tag{5.6.16}$$

where the coefficients $e_{ij}[r,0]$, $r = 0,\ldots,k$, *are given by the formulas* $e_{ij}[0,0] = p_{ij}^{(0)} = q_{ij}[0]/q_i[0]$, $e_{ij}[1,0] = (q_{ij}[1] - e_{ij}[0,0]q_i[1])/q_i[0]$ *and, in general, for* $r = 0,\ldots,k$,

$$e_{ij}[r,0] = \left(q_{ij}[r] - \sum_{m=0}^{r-1} e_{ij}[m,0]q_i[r-m]\right)/q_i[0]. \tag{5.6.17}$$

Proof. The lemma immediately follows from formula (5.6.3), relation **(b)**, and Lemma 8.1.1. The transition probabilities $p_{ij}^{(\varepsilon)}$ are given by formula (5.6.3) as a quotient of the intensity functions $q_{ij}^{(\varepsilon)}$ and $q_i^{(\varepsilon)}$. Therefore, asymptotic expansion (5.6.16) can be used for these transition probabilities, due to statement **(v)** of Lemma 8.1.1. $\qquad\square$

The explicit formula (5.6.9) yields expansions for the power moments of transition times and permits to prove that condition $\mathbf{P}_{30}^{(k)}$ implies to hold the perturbation conditions $\mathbf{P}_{20}^{(k)}$, $\mathbf{P}_{21}^{(k)}$, and $\mathbf{P}_{22}^{(\tilde{k})}$ for the power moments of transition times.

Lemma 5.6.3. *Let conditions* $\mathbf{A_5}$, $\mathbf{I_6}$, *and* $\mathbf{P}_{30}^{(k)}$ *be satisfied. Then the following expansion takes place for every* $i \neq 0$, $j \in X$ *and* $n \geq 1$:

$$m_i^{(\varepsilon)}[n] = m_i^{(0)}[n] + v_i[1,n]\varepsilon + \cdots + v_i[k,n]\varepsilon^k + o(\varepsilon^k), \tag{5.6.18}$$

where the coefficients $v_i[r,n]$, $r = 0,\ldots,k$, $n \geq 1$, are given by the formulas $v_i[0,n] = m_{ij}^{(0)}[n] = n!/(q_i[0])^n$, $n \geq 1$, and, for $r = 1,\ldots,k$ and $n \geq 1$,

$$v_i[r,n] = -\frac{n!}{(q_i[0])^n}\left(\sum_{l=1}^{r} v_i[r-l,n] \cdot \sum_{n_0,\ldots,n_l \in D^*_{n,l}} \prod_{p=0}^{l} q_i[p]^{n_p}/n_p!\right), \quad (5.6.19)$$

where $D^*_{n,l}$ is the set of all nonnegative, integer solutions of the system

$$n_0 + n_1 + \cdots + n_l = n, \quad n_1 + \cdots + ln_l = l. \quad (5.6.20)$$

Proof. It follows from the explicit formula (5.6.9) and Lemma 8.1.1 that the power moments $m_i^{(\varepsilon)}[n]$, $n \geq 1$, can be expanded in $(0,k)$-asymptotic expansions. To see this, one must first apply statement **(iv)** of this lemma to get an $(0,k)$-asymptotic expansion for the function $(q_i^{(\varepsilon)})^{-1}$ and then apply $n-1$ times the product rule, given in statement **(iii)**, to this function, in order to get the $(0,k)$-asymptotic expansion for function $(q_i^{(\varepsilon)})^{-n}$.

However, it is better to find the coefficients in expansion (5.6.18) by equating the coefficients of ε^r, $r = 0,1,\ldots,k$, in the left and right hand-sides of the formal asymptotic identity

$$(v_i[0,n] + v_i[1,n]\varepsilon + \cdots)(q_i[0] + q_i[1]\varepsilon + \cdots)^n = n!, \quad (5.6.21)$$

which immediately yields formulas (5.6.19). $\qquad\square$

Consider the functions used in the perturbation conditions $\mathbf{P}_{20}^{(k)}$ and $\mathbf{P}_{22}^{(\tilde{k})}$,

$$p_{ij}^{(\varepsilon)}[n] = p_{ij}^{(\varepsilon)} m_i^{(\varepsilon)}[n], \quad n \geq 1, \quad i \neq 0, \quad j \in X.$$

Lemma 5.6.4. *Let conditions $\mathbf{A}_5$, $\mathbf{I}_6$, and $\mathbf{P}_{30}^{(k)}$ hold. Then there is an expansion that holds for every $i \neq 0$, $j \in X$ and $n \geq 1$,*

$$p_{ij}^{(\varepsilon)}[n] = p_{ij}^{(0)}[n] + e_{ij}[1,n]\varepsilon + \cdots + e_{ij}[k,n]\varepsilon^k + o(\varepsilon^k), \quad (5.6.22)$$

where the coefficients $e_{ij}[r,n]$, $r = 0,\ldots,k$, $n \geq 1$, are given by the formulas $e_{ij}[0,n] = p_{ij}^{(0)}[n] = p_{ij}^{(0)} m_i^{(0)}[n] = e_{ij}[0,0]v_i[0,n]$, $n \geq 1$, and, for $r = 0,\ldots,k$ and $n \geq 1$,

$$e_{ij}[r,n] = \begin{cases} e_{ij}[r,0] & \text{for } r = 0,\ldots,k, \ n = 0, \\ \sum_{m=0}^{r} e_{ij}[m,0]v_i[r-m,n] & \text{for } r = 0,\ldots,k, \ n \geq 1. \end{cases} \quad (5.6.23)$$

Proof. The proof follows from statement **(iii)** of Lemma 8.1.1 applied to the product $p_{ij}^{(\varepsilon)} m_i^{(\varepsilon)}[n]$. $\qquad\square$

Remark 5.6.1. According to Lemma 5.6.4, condition $\mathbf{P}_{30}^{(k)}$ implies that the perturbation condition $\mathbf{P}_{22}^{(\tilde{k})}$ holds for any $\tilde{n} \geq 1$ and $0 \leq \tilde{k}_n \leq k$, $n = 0, \ldots, \tilde{n}$.

The explicit formula (5.6.10) yields expansions for the mixed power-exponential moments of transition times and permits to prove that if condition $\mathbf{P}_{30}^{(k)}$ holds, then implies the perturbation conditions $\mathbf{P}_{23}^{(\rho,k)} - \mathbf{P}_{26}^{(\rho,k)}$, $\mathbf{P}_{27}^{(\rho,\tilde{k})}$, $\mathbf{P}_{28}^{(\rho,k)}$, and $\mathbf{P}_{29}^{(\rho,k)}$ also hold for the mixed power-exponential moments of transition times.

Lemma 5.6.5. *Let conditions $\mathbf{A}_5$, $\mathbf{I}_6$, and $\mathbf{P}_{30}^{(k)}$ hold. Then we have the following expansion for every $i \neq 0$, $j \in X$, $n \geq 1$, and $\rho < q_i^{(0)} = q_i[0]$:*

$$\psi_i^{(\varepsilon)}[\rho, n] = \psi_i^{(0)}[\rho, n] + v_i[\rho, 1, n]\varepsilon + \cdots + v_i[\rho, k, n]\varepsilon^k + o(\varepsilon^k), \qquad (5.6.24)$$

where the coefficients $v_i[\rho, r, n]$, $r = 0, \ldots, k$, $n \geq 1$, are given by the formulas $v_i[\rho, 0, n] = \psi_{ij}^{(0)}[\rho, n] = n!/(q_i[0] - \rho)^{n+1}$, $n \geq 1$, and, for $r = 1, \ldots, k$ and $n \geq 1$,

$$v_i[\rho, r, n] = \frac{n!}{(q_i[0] - \rho)^{n+1}} - \frac{(n+1)!}{(q_i[0] - \rho)^{n+1}} \left(\sum_{l=1}^{r} v_i[\rho, r - l, n] \right.$$

$$\left. \times \sum_{n_0, \ldots, n_l \in D_{n+1,l}^*} \frac{(q_i[0] - \rho)^{n_0}}{n_0!} \prod_{p=1}^{l} \frac{q_i[p]^{n_p}}{n_p!} \right), \qquad (5.6.25)$$

where $D_{n+1,l}^*$ is the set of all nonnegative, integer solutions of the system

$$n_0 + n_1 + \cdots + n_l = n + 1, \quad n_1 + \cdots + l n_l = l. \qquad (5.6.26)$$

Proof. It follows from the explicit formula (5.6.10) and Lemma 8.1.1 that the power moments $\psi_i^{(\varepsilon)}[\rho, n]$, $n \geq 1$, can be expanded in $(0, k)$-asymptotic expansions. To see this, one must first apply statement **(iv)** of this lemma to get an $(0, k)$-asymptotic expansion for the function $(q_i^{(\varepsilon)} - \rho)^{-1}$, and then apply $n + 1$ times the product rule in statement **(iii)** to this function for obtaining an $(0, k)$-asymptotic expansion for the function $q_i^{(\varepsilon)}(q_i^{(\varepsilon)} - \rho)^{-n-1}$.

However, it is easier to find the coefficients in expansion (5.6.24) by equating the coefficients of ε^r, $r = 0, 1, \ldots, k$, in the left and right hand-sides of the formal asymptotic identity

$$\left(v_i[\rho, 0, n] + v_i[\rho, 1, n]\varepsilon + \cdots \right) \left((q_i[0] - \rho) + q_i[1]\varepsilon + \cdots \right)^{n+1}$$

$$= n! \left(q_i[0] + q_i[1]\varepsilon + \cdots \right), \qquad (5.6.27)$$

which at once yields formulas (5.6.25). $\qquad\qquad \square$

Remark 5.6.2. Condition $\mathbf{P}_{30}^{(k)}$ implies that the perturbation condition $\mathbf{P}_{27}^{(\rho,\tilde{k})}$ holds for any $0 \leq \tilde{k}_n \leq k, n = 0, \ldots, \tilde{n}$ and $\tilde{n} \geq 1$.

Now, the theorems given in Section 5.5 can be applied to construct corresponding exponential asymptotic expansions. Here, however, an additional step is needed, because the conditions are now given in terms of the intensities $q_{ij}^{(\varepsilon)}$ instead of the transition probabilities $p_{ij}^{(\varepsilon)}$, the power moment of transition times $m_{ij}^{(\varepsilon)}[n]$, and the mixed power-exponential moments $\psi_{ij}^{(\varepsilon)}[\rho, n]$.

Applications of the results obtained in Chapters 3–5 to an analysis of pseudo- and quasi-stationary phenomena in nonlinearly perturbed stochastic systems and nonlinearly perturbed risk processes are given in Chapters 6 and 7.

Quasi-stationary phenomena in stochastic systems

Chapter 6 deals with applications of the results obtained in Chapters 3–5 to an analysis of pseudo- and quasi-stationary phenomena in nonlinearly perturbed stochastic systems. Examples of stochastic systems under consideration are queueing systems, epidemic, and population dynamics models with finite lifetimes. In queueing systems, the lifetime $\mu^{(\varepsilon)}$ is usually the time at which some kind of a fatal failure occurs in the system. In epidemic models, the time of extinction of the epidemic in the population plays the role of the lifetime, while in population dynamics models, the lifetime is usually the extinction time for the corresponding population. In all examples, we study asymptotic behaviour of the probabilities $\mathsf{P}\{\eta^{(\varepsilon)}(t) = j, \mu^{(\varepsilon)} > t\}$ for the corresponding perturbed Markov/semi-Markov type process $\eta^{(\varepsilon)}(t)$, which represents the state of the system at time t, and the lifetime $\mu^{(\varepsilon)}$.

We give all types of asymptotic results studied in Chapters 3–5. They include the following: (i) mixed ergodic theorems (for the state of the system) and limit theorems (for the lifetimes) that describe transition phenomena; (ii) mixed ergodic and large deviation theorems that describe pseudo- and quasi-stationary phenomena; (iii) exponential expansions in mixed ergodic and large deviation theorems; (iv) theorems on convergence of quasi-stationary distributions; and (v) asymptotic expansions for quasi-stationary distributions.

In all the examples, we try to specify and to describe, in a more explicit form, conditions and algorithms for calculating the limit expressions, the characteristic roots, and the coefficients in the corresponding asymptotic expansions.

As the first application, we consider a M/M queueing system that consists of N servers functioning independently of each other with exponential life and repairing times. The failure and the repairing intensities of the server i are $q_{i,-}^{(\varepsilon)}$ and $q_{i,+}^{(\varepsilon)}$, respectively. The state of the server i at an instant t is given by a random indicator variable $\eta_i^{(\varepsilon)}(t)$ that takes the value 1 if the server i works at this instant, and the value 0 if the server i is being repaired at this moment. The state of the system is described by the vector $\eta^{(\varepsilon)}(t) = (\eta_1^{(\varepsilon)}(t), \ldots, \eta_N^{(\varepsilon)}(t))$, $t \geq 0$, which is, in this case, a continuous time homogeneous Markov chain. In this case, the lifetime $\mu^{(\varepsilon)}$ is the first time at which the process $\eta^{(\varepsilon)}(t)$ gets into the state $\bar{0} = (0, \ldots, 0)$. The perturbation conditions take the form of asymptotic Taylor type expansions for the intensities $q_{i,\pm}^{(\varepsilon)}$.

We consider the cases where the limiting intensities satisfy (a) $q_{i,+}^{(0)} > 0$ for all i, while

$q_{i,-}^{(0)} = 0$ for some i, or **(b)** $q_{i,\pm}^{(0)} > 0$ for all i. The cases **(a)** and **(b)** correspond to pseudo- and quasi-stationary models, respectively.

As the second application, we consider a M/G queueing system with one server and a bounded queue buffer of size N. We assume that the input flow is a standard Poisson flow with parameter λ and the service distribution function $G^{(\varepsilon)}(t)$ depends on a perturbation parameter $\varepsilon \geq 0$. Functioning of the system is described by a random variable $\eta^{(\varepsilon)}(t)$ that is the number of free places in the queue buffer at the moment t. This process is not semi-Markov but belongs to a more general class of the so-called stochastic processes with semi-Markov modulation (switchings). These processes admit a construction of imbedded semi-Markov processes. This example shows that the main results obtained in the book can be applied to more general stochastic processes, in particular, to stochastic processes with semi-Markov modulation. The lifetime $\mu^{(\varepsilon)}$ is the first time at which the input item can not be placed in the queue buffer, i.e., the first time at which the process $\eta^{(\varepsilon)}(t)$ hits the state N. We assume that the distribution functions $G^{(\varepsilon)}(t)$ weakly converge to $G^{(0)}(t)$ as $\varepsilon \to 0$ and that a Cramér type condition holds for the service distribution function $G^{(\varepsilon)}(t)$ asymptotically uniformly for small ε. The perturbation conditions take the form of asymptotic Taylor type expansions for power or mixed power-exponential moments the service distribution functions $G^{(\varepsilon)}(t)$. We consider two cases where the limiting distribution $G^{(0)}(t)$ **(a)** is concentrated in zero, and **(b)** is not concentrated in zero. The cases **(a)** and **(b)** correspond to the pseudo- and quasi-stationary models, respectively.

The next application deals with birth-and-death type models described by a semi-Markov chain, where the phase space is $X = \{0, 1, \ldots, N\}$ and the transition probabilities from a state i to the neighbouring states $i - 1$ and $i + 1$ for $0 < i < N$ ($N - 1$ and N for $i = N$) have the form $Q_{i\,i\pm1}^{(\varepsilon)}(t) = p_{i,\pm}^{(\varepsilon)} F_i^{(\varepsilon)}(t)$. The state 0 is an absorption state and the first time the process $\eta^{(\varepsilon)}(t)$ hits it is interpreted as the lifetime of the corresponding stochastic system. The perturbation conditions take the form of asymptotic Taylor type expansions for the transition probabilities $p_{i,\pm}^{(\varepsilon)}$ and the power or mixed power-exponential moments for the distribution functions of the sojourn times $F_i^{(\varepsilon)}(t)$. The limit transition probabilities satisfy the condition **(a)** $p_{i,+}^{(0)} > 0$ for all $i \neq 0$, while $p_{i,-}^{(0)} = 0$ for some $i \neq 0$, or **(b)** $p_{i,\pm}^{(0)} > 0$ for all $i \neq 0$. The cases **(a)** and **(b)** correspond to pseudo- and quasi-stationary models, respectively.

A standard birth-and-death type model corresponds to the case where all the distribution functions $F_i^{(\varepsilon)}(t)$ are exponential and, therefore, $\eta^{(\varepsilon)}(t)$ is a continuous time Markov chain. In this case the transition characteristics are usually defined via the transition intensities $q_{i,\pm}^{(\varepsilon)}$ and have the form $p_{i,\pm}^{(\varepsilon)} = q_{i,\pm}^{(\varepsilon)}/q_i^{(\varepsilon)}$ and $F_i^{(\varepsilon)}(t) = 1 - \exp\{-q_i^{(\varepsilon)}t\}$, where $q_i^{(\varepsilon)} = q_{i,-}^{(\varepsilon)} + q_{i,+}^{(\varepsilon)}$. In this case, the perturbation conditions take the form of asymptotic Taylor type expansions for the intensities $q_{i,\pm}^{(\varepsilon)}$. As well known, standard birth-and-death processes can be used for a description of M/M type queueing systems, epidemic or population dynamic models. We consider several examples. Let us mention two of them. The model with $q_{i,-}^{(\varepsilon)} = iq_-^{(\varepsilon)}, q_{i,+}^{(\varepsilon)} = (N-i)q_+^{(\varepsilon)}$ can be used to describe a M/M queueing system which consists of N servers functioning independently of each other with failure and repairing intensities $q_-^{(\varepsilon)}$ and $q_+^{(\varepsilon)}$, respectively. In this case, the process $\eta^{(\varepsilon)}(t)$ represents the number of working servers at the time t. Here, the lifetime of the system is the first moment at which the number of working servers becomes zero. The model with $q_{i,-}^{(\varepsilon)} = i q_-^{(\varepsilon)}$, $q_{i,+}^{(\varepsilon)} = i(N-i)q_+^{(\varepsilon)}$ can be used to describe an epidemic in a population with N individuals recovering independently with intensity $q_-^{(\varepsilon)}$ and infecting each other pairwise independently with intensity $q_+^{(\varepsilon)}$. In this case, the process $\eta^{(\varepsilon)}(t)$ represents the number of ill individuals at the time t. The time of extinction of the epidemic is the first moment at which the number of ill individuals is zero.

In the last example, we consider a metapopulation system consisting of N patches each of which at any time $t \geq 0$ can be either occupied or empty. The state of the metapopulation at a time t is described by the vector $\eta^{(\varepsilon)}(t) = (\eta_1^{(\varepsilon)}(t), \ldots, \eta_N^{(\varepsilon)}(t))$, $t \geq 0$, where $\eta_i^{(\varepsilon)}(t)$ is either 1 or 0, according to whether the patch i is occupied or empty at the time t. The first time the process $\eta^{(\varepsilon)}(t)$ hits the state $\bar{0} = (0, \ldots, 0)$ is the extinction time for the metapopulation. We assume the following: (i) in the absence of migration, the local population inhibiting a patch i will extinct in the time interval of length Δt with probability $p_{ii}^{(\varepsilon)}(\Delta t) = g_{ii}^{(\varepsilon)} \Delta t + o(\Delta t)$; (ii) the probability that an empty patch j will be colonised in a time interval of length Δt by migrants originating from a patch i is $p_{ij}^{(\varepsilon)}(\Delta t) = g_{ij}^{(\varepsilon)} \Delta t + o(\Delta t)$; (iii) all interaction processes (extinction and colonisation) are independent. Under the assumptions made for the local patch dynamics, we deduce that a natural time evolution model for the process $\eta^{(\varepsilon)}(t)$ is a homogeneous Markov chain with absorption. The perturbation conditions take the form of asymptotic Taylor type expansions for the intensities $g_{ij}^{(\varepsilon)}$. We consider the cases where the limit intensities $g_{ij}^{(0)} > 0$, for $i \neq j$, while **(a)** $g_{ii}^{(0)} = 0$ for some i, or **(b)** $g_{ij}^{(0)} > 0$ for all i, j. The cases **(a)** and **(b)** correspond, respectively, to pseudo- and quasi-stationary models.

Chapter 6 gives typical examples of applying results obtained from an analysis of quasi-stationary phenomena for regenerative and semi-Markov processes to a study of quasi-stationary phenomena in nonlinearly perturbed stochastic systems. In Section 6.1, we consider queueing systems with highly reliable main servers. In Sections 6.2 and 6.3, we consider M/G queueing system with one server and a bounded queue buffer. In Section 6.4, semi-Markov and Markov birth-and-death type processes are considered. Some classical models of M/M type queueing systems, epidemic or population dynamic models can be described with the use of such processes. In Section 6.5, an example of metapopulation model is considered.

6.1   Queueing systems with highly reliable main servers

In this section, we study conditions on pseudo- and quasi-stationary asymptotics for queueing systems with asymptotically highly reliable servers.

6.1.1 Queueing systems with highly reliable main servers

As our first application, we will consider a queueing system which consists of m servers functioning independently of each other with exponential life and repairing times. The failure and repairing intensities of the servers are $q_{i,-}^{(\varepsilon)}$ and $q_{i,+}^{(\varepsilon)}$, respectively.

The state of a server i at an instant t is given by a random indicator variable $\eta_i^{(\varepsilon)}(t)$ that takes the value 1 if the server i is functioning at this instant, and the value 0 if the server i is being repaired at this moment.

The state of the system is described by the vector $\eta^{(\varepsilon)}(t) = (\eta_1^{(\varepsilon)}(t),\ldots,\eta_m^{(\varepsilon)}(t))$, $t \geq 0$. The phase space of the stochastic process $\eta^{(\varepsilon)}(t)$ is the m-dimensional binary hypercube $X = \{\bar{x} = (x_1,\ldots,x_m) : x_i = 0, 1,\ i = 1,\ldots,m\}$.

We assume that the continuous-time process $\eta^{(\varepsilon)}(t)$ is a homogeneous Markov chain, continuous from the right, with the transition intensities

$$q_{\bar{x}\bar{y}}^{(\varepsilon)} = \begin{cases} q_{i,-}^{(\varepsilon)} & \text{if } x_i = 1,\ y_i = 0,\ x_j = y_j,\ j \neq i, \\ q_{i,+}^{(\varepsilon)} & \text{if } x_i = 0,\ y_i = 1,\ x_j = y_j,\ j \neq i, \\ 0 & \text{otherwise.} \end{cases} \tag{6.1.1}$$

In this case, the parameter of the corresponding exponential distribution of the sojourn time in a state $\bar{x} = (x_1,\ldots,x_m) \in X$ is given by the following formula:

$$q_{\bar{x}}^{(\varepsilon)} = \sum_{\bar{y} \neq \bar{x}} q_{\bar{x}\bar{y}}^{(\varepsilon)} = \sum_{i:x_i=1} q_{i,-}^{(\varepsilon)} + \sum_{i:x_i=0} q_{i,+}^{(\varepsilon)}. \tag{6.1.2}$$

Let $\tau^{(\varepsilon)}(n)$, $n = 0, 1, \ldots$, be subsequent moments of jumps of the Markov chain $\eta^{(\varepsilon)}(t)$, $t \geq 0$, and $\eta_n^{(\varepsilon)} = \eta^{(\varepsilon)}(\tau^{(\varepsilon)}(n))$, $n = 0, 1, \ldots$, be the states of the Markov chain $\eta^{(\varepsilon)}(t)$ at subsequent moments of jumps. By the definition, the random sequence $\eta_n^{(\varepsilon)}$, $n = 0, 1, \ldots$ is an imbedded discrete time Markov chain for the continuous time Markov chain $\eta^{(\varepsilon)}(t)$, $t \geq 0$.

Let us also define the hitting times in a state $\bar{y} \in X$ for the imbedded Markov chain $\eta_n^{(\varepsilon)}$ and the Markov chain $\eta^{(\varepsilon)}(t)$ to be $\nu_{\bar{y}}^{(\varepsilon)} = \min(n \geq 1 : \eta_n^{(\varepsilon)} = \bar{y})$ and $\mu_{\bar{y}}^{(\varepsilon)} = \tau^{(\varepsilon)}(\nu_{\bar{y}}^{(\varepsilon)})$, respectively.

The first hitting time $\mu_{\bar{0}}^{(\varepsilon)}$ is the lifetime of the system. By the definition, $\mu_{\bar{0}}^{(\varepsilon)}$ is the first time when all servers occur in a failure state.

The model described above is a particular case of the model considered in Section 5.6. The notations used for the states of the Markov chain above and for the ones in Section 5.6 can easily be agreed by a natural re-numeration of states, $(0, 0, \ldots, 0) \leftrightarrow 0$, $(0, 0, \ldots, 1) \leftrightarrow 1, \ldots, (1, 1, \ldots, 1) \leftrightarrow 2^m - 1$. Note that the number of non-absorption states in this model is $N = 2^m - 1$.

We will use all the notations introduced in Chapter 4 for the transition and hitting probabilities and the moment generation functions with the only change in the notations of indices used for the states. Instead of the indices $i, j, \ldots$ we shall use the indices $\bar{x}, \bar{y}, \ldots$, etc.

As was pointed out in Subsection 4.3.1, we need not assume that $\bar{0} = (0, \ldots, 0)$ is an absorption state, since the probabilities $\mathsf{P}_{\bar{x}}\{\eta^{(\varepsilon)}(t) = \bar{y}, \mu_{\bar{0}}^{(\varepsilon)} > t\}$ do not depend on the probabilities of transition from the state $\bar{0}$.

Let us therefore see what form will the corresponding conditions formulated in Chapters 4 and 5 take. It is natural to re-formulate these conditions in terms of the intensities $q_{i,\pm}^{(\varepsilon)}, i = 1, \ldots, m$.

We assume the following condition, which would allow to interpret the models with $\varepsilon > 0$ as perturbed versions of the model with $\varepsilon = 0$:

$\mathbf{T_4}$: $q_{i,\pm}^{(\varepsilon)} \to q_{i,\pm}^{(0)}$ as $\varepsilon \to 0, i = 1, \ldots, m$.

As far as the limit model is concerned, we assume that the following condition holds:

$\mathbf{E_4}$: $q_{i,+}^{(0)} > 0, i = 1, \ldots, m$.

Condition $\mathbf{E_4}$ can be realised in two variants. Condition $\mathbf{T_4}$ obviously implies that the limiting intensities satisfy $q_{i,-}^{(0)} \geq 0, i = 1, \ldots, m$. Let us introduce the set

$$E = \{1 \leq i \leq m : q_{i,-}^{(0)} = 0\}.$$

The following condition is the first variant for realisation of condition $\mathbf{E_4}$:

$\mathbf{E_4'}$: $q_{i,+}^{(0)} > 0, i = 1, \ldots, m$, and $E \neq \varnothing$.

In this case we assume, without loss of generality, that there exists $1 \leq l \leq m$ such that $E = \{1, \ldots, l\}$. We shall refer to servers with the indices $i \in E$ as "main" servers.

It is obvious that conditions $\mathbf{T_4}$ and $\mathbf{E_4'}$ imply that the following condition holds:

$\mathbf{A_7}$: $q_{i,+}^{(\varepsilon)} > 0, i = 1, \ldots, m$, and $q_{i,-}^{(\varepsilon)} > 0, i = l + 1, \ldots, m, \varepsilon \leq \varepsilon_1$ for some $\varepsilon_1 > 0$.

Condition $\mathbf{A_7}$ means that, for every model with the parameter ε small enough, every server with index $i = 1, \ldots, m$ can be repaired, i.e., changed from the failure state 0 to the functioning state 1 with a positive reparation intensity $q_{i,+}^{(\varepsilon)}$, and that every server with index $i = l + 1, \ldots, m$ can break, i.e., change from the functioning state 1 to the failure state 0 with a positive failure intensity $q_{i,-}^{(\varepsilon)}$. As far as servers with the indices $i = 1, \ldots, l$ are concerned, conditions $\mathbf{T_4}$ and $\mathbf{E_4'}$ imply that the corresponding failure intensities are small for ε small enough and that these servers can not break, i.e., they are absolutely reliable for the unperturbed model with $\varepsilon = 0$.

This case can be interpreted as a model with asymptotically reliable main servers.

As was mentioned above, conditions $\mathbf{T_4}$ and $\mathbf{E_4'}$ do not guarantee that the failure intensities satisfy $q_{i,-}^{(\varepsilon)} = 0, i = 1, \ldots, l$, for ε small enough. In order to guarantee this property, one should impose, additionally to $\mathbf{T_4}$ and $\mathbf{E'_4}$, the following condition:

$\mathbf{A_8}$: $q_{i,-}^{(\varepsilon)} = 0, i \in E, \varepsilon \leq \varepsilon_2$ for some $\varepsilon_2 > 0$.

Conditions $\mathbf{A_7}$ and $\mathbf{A_8}$ mean that, for every model with the parameter ε that is small enough, every server with the index $i = 1, \ldots, m$ can be repaired, every server with the index $i = 1, \ldots, l$ can not break, i.e., it is absolutely reliable, while every server with the index $i = l + 1, \ldots, m$ may break.

The following condition is the second variant for realisation of condition $\mathbf{E_4}$:

$\mathbf{E''_4}$: $q_{i,+}^{(0)} > 0, i = 1, \ldots, m$, and $E = \varnothing$.

It is obvious that conditions $\mathbf{T_4}$ and $\mathbf{E''_4}$ imply that the following condition holds:

$\mathbf{A_9}$: $q_{i,\pm}^{(\varepsilon)} > 0, i = 1, \ldots, m, \varepsilon \leq \varepsilon_3$ for some $\varepsilon_3 > 0$.

Condition $\mathbf{A_9}$ means that, for every model with the parameter ε small enough, every server can break, i.e., to change from the functioning state 1 to the failure state 0 and can be repaired, i.e., changed from the failure state 0 to the functioning state 1.

Let us note that only the first case, where condition $\mathbf{E'_4}$ holds, actually corresponds to queueing systems with asymptotically high reliable servers. Indeed, in this case the failure intensities satisfy $q_{i,-}^{(\varepsilon)} \to 0$ as $\varepsilon \to 0$ for $i \in E$. However, condition $\mathbf{E''_4}$ can also be relevant to an analysis of queueing systems with highly reliable servers, if one would like to get the corresponding asymptotical results in the case where the failure intensities $q_{i,-}^{(\varepsilon)}, i \in E$, actually take some small values but are separated from zero as $\varepsilon \to 0$.

Condition $\mathbf{E_4}$ guarantees that $q_{\bar{x}}^{(0)} > 0$ for every $\bar{x} \neq \bar{1}$, where $\bar{1} = (1, 1, \ldots, 1)$. The state $\bar{1}$ has a special status. For this state, we have that $q_{\bar{1}}^{(0)} = \sum_{i=1}^{m} q_{i,-}^{(0)}$. If all the intensities satisfy $q_{i,-}^{(0)} = 0, i = 1, \ldots, m$, i.e., all the servers for the unperturbed system are absolutely reliable, then $q_{\bar{1}}^{(0)} = 0$ and, therefore, the state $\bar{1}$ is an absorption state. Not to deal with this situation we shall use a regularisation transformation by adding an additional intensity $q > 0$ to this state for the virtual transition $\bar{1} \to \bar{1}$.

Thus, let $\tilde{\eta}^{(\varepsilon)}(t) = (\tilde{\eta}_1^{(\varepsilon)}(t), \ldots, \tilde{\eta}_m^{(\varepsilon)}(t)), t \geq 0$, be, for every $\varepsilon \geq 0$, a semi-Markov process with a phase space X and the transition probabilities

$$Q_{\bar{x}\bar{y}}^{(\varepsilon)}(t) = \frac{\tilde{q}_{\bar{x}\bar{y}}^{(\varepsilon)}}{\tilde{q}_{\bar{x}}^{(\varepsilon)}} \left(1 - \exp\{-\tilde{q}_{\bar{x}}^{(\varepsilon)}t\}\right), \quad \bar{x}, \bar{y} \in X, \quad t \geq 0, \tag{6.1.3}$$

where

$$\tilde{q}_{\bar{x}\bar{y}}^{(\varepsilon)} = \begin{cases} q_{i,-}^{(\varepsilon)} & \text{if } \bar{x} \neq \bar{1}, x_i = 1, y_i = 0, x_j = y_j, j \neq i, \\ q & \text{if } \bar{x} = \bar{1}, \bar{x} = \bar{y}, \\ q_{i,+}^{(\varepsilon)} & \text{if } x_i = 0, y_i = 1, x_j = y_j, j \neq i, \\ 0 & \text{otherwise,} \end{cases} \tag{6.1.4}$$

and

$$\tilde{q}_{\bar{x}}^{(\varepsilon)} = \sum_{\bar{y} \neq \bar{x}} q_{\bar{x}\bar{y}}^{(\varepsilon)} + q\delta(\bar{x},\bar{1}) = \sum_{i:x_i=1} q_{i,-}^{(\varepsilon)} + \sum_{i:x_i=0} q_{i,+}^{(\varepsilon)} + q\delta(\bar{x},\bar{1}). \qquad (6.1.5)$$

As was mentioned in Section 5.6, the semi-Markov process $\tilde{\eta}^{(\varepsilon)}(t), t \geq 0$, and the Markov chain $\eta^{(\varepsilon)}(t), t \geq 0$, have the same finite-dimensional distributions. Consequently, the probabilities $\mathsf{P}_{\bar{x}}\{\eta^{(\varepsilon)}(t) = \bar{y}, \mu_{\bar{0}}^{(\varepsilon)} > t\}$ and $\mathsf{P}_{\bar{x}}\{\tilde{\eta}^{(\varepsilon)}(t) = \bar{y}, \tilde{\mu}_{\bar{0}}^{(\varepsilon)} > t\}$ coincide for any $\bar{x}, \bar{y} \neq \bar{0}$, and $t \geq 0$.

Thus, we can always consider a regularised semi-Markov process $\tilde{\eta}^{(\varepsilon)}(t)$ instead of the Markov chain $\eta^{(\varepsilon)}(t)$. To simplify the notations we interpret $\tilde{\eta}^{(\varepsilon)}(t)$ as a regularised version of the Markov chain $\eta^{(\varepsilon)}(t)$ and use the initial notation $\eta^{(\varepsilon)}(t), t \geq 0$, for the regularised version as well as the notations $q_{\bar{x}\bar{y}}^{(\varepsilon)}$ and $q_{\bar{x}}^{(\varepsilon)}$ for the corresponding transition intensities.

6.1.2   Conditions for mixed ergodic and limit/large deviation theorems

Let us study relations between conditions $\mathbf{T_4}$, $\mathbf{E_4}$ and the conditions introduced in Chapter 4 for mixed ergodic and limit/large deviation theorems.

Denote

$$X_0 = \{\bar{x} \in X : \bar{x} \neq \bar{0}\},$$

$$X_1^{(0)} = \{\bar{x} \in X_0 : x_1, \ldots, x_l = 1\},$$

$$X_2^{(0)} = X_0 \setminus X_1^{(0)}, \qquad (6.1.6)$$

and

$$U_0 = X_1^{(0)}, \quad U_r = \{\bar{x} \in X_2^{(0)} : \sum_{i=1}^{l} \delta(x_i,0) = r\}, \quad r = 1, \ldots, l. \qquad (6.1.7)$$

By the definition,

$$\bigcup_{r=1}^{l} U_r = X_2^{(0)}.$$

As was mentioned above, conditions $\mathbf{T_4}$, $\mathbf{E_4'}$ imply that condition $\mathbf{A_7}$ holds. According to this condition, for every $0 \leq \varepsilon \leq \varepsilon_1$, the hitting probabilities are positive, $_0 f_{\bar{x}\bar{y}}^{(\varepsilon)} > 0$, for any $\bar{x} \in X_0, \bar{y} \in X_1^{(0)}$.

Indeed, in this case, all reparation and failure intensities that can cause a transition from a state $\bar{x}$ to a state $\bar{y}$ without visiting the state $\bar{0}$ are positive.

The property of hitting probabilities described above means that the Markov chain $\eta^{(\varepsilon)}(t)$ has one class of recurrent-without-absorption states $X_1^{(\varepsilon)} \supseteq X_1^{(0)}$.

With regard to the absorption probabilities, condition $\mathbf{A_7}$ guarantees only that, for every $0 \leq \varepsilon \leq \varepsilon_1$, the absorption probabilities satisfy $_{\bar{y}} f_{\bar{x}\bar{0}}^{(\varepsilon)} > 0$ for any $\bar{x} \in U_l$, $\bar{y} \in X_1^{(0)}$.

Indeed, in this case, the class U_l consists of states of the form $(0, \ldots, 0, x_{l+1}, \ldots, x_m)$ such that $\sum_{k=l+1}^{m} \delta(x_k, 1) \geq 1$. All reparation and failure intensities, which can provide any transition within the class U_l, are positive. The one-step transition to the absorption state $\bar{0} = (0, \ldots, 0)$ is possible from the states in this class satisfying $\sum_{k=l+1}^{m} \delta(x_k, 1) = 1$.

Also, in the prelimit cases, i.e., if $0 < \varepsilon \leq \varepsilon_1$, the absorption probabilities satisfy $_{\bar{y}} f_{\bar{x}\bar{0}}^{(\varepsilon)} > 0$ for the states $\bar{x} \in X_0 \setminus U_l$, $\bar{y} \in X_1^{(0)}$ such that $x_i = 1$ if the corresponding failure intensity are $q_{i,-}^{(\varepsilon)} > 0$, and $x_i = 0$ if the corresponding failure intensity are zero, $q_{i,-}^{(\varepsilon)} = 0$.

Indeed, in this case, it is possible that a transition to some state $\bar{z}$ from the class U_l occurs without visiting the class $X_1^{(0)}$, and then a transition from the state $\bar{z}$ to the state $\bar{0}$ occurs without leaving the class U_l.

The situation is different in the limiting case where $\varepsilon = 0$. In this case, as was pointed out above, $_{\bar{z}} f_{\bar{x}\bar{y}}^{(0)} = 0$ for any $\bar{x}, \bar{z} \in X_1^{(0)}$, $\bar{y} \in X_2^{(0)}$.

This is true because all the failure intensities equal zero for the servers with the indices $i \in E$.

Thus, in this case, $X_1^{(0)}$ is a closed set of recurrent-without absorption states, while $X_2^{(0)}$ is a class of non-recurrent-without absorption states.

The absorption probabilities are $_{\bar{y}} f_{\bar{x}\bar{0}}^{(0)} = 0$ for any $\bar{x} \in X_0 \setminus U_l$, $\bar{y} \in X_1^{(0)}$, but $_{\bar{y}} f_{\bar{x}\bar{0}}^{(0)} \in (0, 1)$ for $\bar{x} \in U_l$, $\bar{y} \in X_1^{(0)}$.

This is so also because all the failure intensities equal zero for the servers with the indices $i \in E$.

Thus, the case where conditions $\mathbf{T_4}$ and $\mathbf{E_4'}$ hold corresponds to a pseudo-stationary model. Moreover, in this case, the class of recurrent-without-absorption states is $X_1^{(0)}$ and the class of non-recurrent-without-absorption states is $X_2^{(0)}$.

As was mentioned above, conditions $\mathbf{T_4}$ and $\mathbf{E_4''}$ imply that condition $\mathbf{A_9}$ holds. According to this condition, for every $0 \leq \varepsilon \leq \varepsilon_3$, the hitting probabilities are positive, $_{\bar{0}} f_{\bar{x}\bar{y}}^{(\varepsilon)} > 0$, for any $\bar{x}, \bar{y} \in X_0$. This means that the Markov chain $\eta^{(\varepsilon)}(t)$ has one class of recurrent-without-absorption states, $X_1^{(\varepsilon)} = X_0$, Moreover, in this case, it can readily be seen that also all the absorption probabilities are positive, $_{\bar{y}} f_{\bar{x}\bar{0}}^{(\varepsilon)} > 0$, for any $\bar{x}, \bar{y} \in X_0$.

This is true, since the reparation and failure intensities are positive for all servers.

Thus, the case where conditions $\mathbf{T_4}$ and $\mathbf{E_4''}$ hold corresponds to a quasi-stationary model. Moreover, in this case, the class of recurrent-without-absorption states is $X_1^{(0)} = X_0$.

Let us discuss how the conditions introduced in Chapters 4 are realised for the model described above. We assume that conditions $\mathbf{T_4}$ and $\mathbf{E_4}$ hold.

Relations (6.1.1), (6.1.2), and condition $\mathbf{T_4}$ obviously imply that, for $\bar{x}, \bar{y} \in X$,

$$q_{\bar{x}\bar{y}}^{(\varepsilon)} \to q_{\bar{x}\bar{y}}^{(0)} \quad \text{as} \quad \varepsilon \to 0, \tag{6.1.8}$$

and

$$q_{\bar{x}}^{(\varepsilon)} \to q_{\bar{x}}^{(0)} \quad \text{as} \quad \varepsilon \to 0. \tag{6.1.9}$$

By formula (6.1.3), the distributions of sojourn times are exponential and, for $\bar{x}, \bar{y} \in X$, we have

$$F_{\bar{x}\bar{y}}^{(\varepsilon)}(t) = F_{\bar{x}}^{(\varepsilon)}(t) = 1 - \exp\{-q_{\bar{x}}^{(\varepsilon)}t\}, \quad t \geq 0. \tag{6.1.10}$$

Relations (6.1.3) and (6.1.9), and formula (6.1.10) imply that condition $\mathbf{T_3}$ holds. Thus, condition $\mathbf{T_1}$ holds. Formulas (6.1.5) and (6.1.10) imply that condition $\mathbf{I_3}$ holds.

Formula (6.1.3) and relation (6.1.9) also imply that conditions $\mathbf{M_{11}}$ and $\mathbf{M_{12}}$ are satisfied.

The non-arithmetic conditions $\mathbf{N_1}$ and $\mathbf{N_2}$ for return times to the states $\bar{x} \in X_1^{(0)}$ obviously hold since the distributions of sojourn times are exponential.

Let us define

$$\rho^* = \min(q_{\bar{x}}^{(0)}, \bar{x} \neq \bar{0}). \tag{6.1.11}$$

By formula (6.1.5), $\rho^* > 0$.

As was mentioned in Subsection 5.6.3, condition $\mathbf{C_{12}}$ holds for any $0 < \delta < \rho^*$.

Conditions $\mathbf{T_4}$ and $\mathbf{E_4'}$ correspond to the pseudo-stationary case. In this case, as follows from Lemma 4.5.8, condition $\mathbf{C_{13}}$ also holds.

Conditions $\mathbf{T_4}$ and $\mathbf{E_4''}$ correspond to the quasi-stationary case. In this case, condition $\mathbf{G_1}$ holds with the parameter ρ^* given in formula (6.1.11). Condition $\mathbf{G_2}$ also holds since the set $X_2^{(0)} = \varnothing$. Therefore, as follows from Lemma 4.5.10, condition $\mathbf{C_{13}}$ also holds.

Conditions $\mathbf{C_{14}}$ and $\mathbf{C_{15}}$ also hold as follows from the remarks made in Section 5.6. Condition $\mathbf{C_{16}}$ is not needed, since $X_2^{(0)} = \varnothing$ in the quasi-stationary case.

So, all the conditions introduced in Chapter 4 are verified. All results given in Chapter 4 can also be applied to the Markov chain $\eta^{(\varepsilon)}(t) = (\eta_1^{(\varepsilon)}(t), \ldots, \eta_m^{(\varepsilon)}(t)), t \geq 0$ that describes functioning of queueing systems with highly reliable main servers.

6.1.3   Mixed ergodic and limit/large deviation theorems

Conditions $\mathbf{T_4}$ and $\mathbf{E_4'}$ correspond to the pseudo-stationary model with the absorption probabilities asymptotically vanishing to zero. Under these conditions, we have analogues for mixed ergodic and limit Theorems 4.4.1 and 4.4.2 and mixed ergodic and large deviation Theorems 4.6.1 and 4.6.2, covering the case of the pseudo-stationary asymptotics. Also, analogues of Theorems 4.4.3 and 4.4.5 give conditions for convergence of the corresponding stationary distributions under the additional non-absorption condition $\mathbf{A_8}$.

Conditions $\mathbf{T_4}$ and $\mathbf{E_4''}$ correspond to a quasi-stationary model with absorption probabilities asymptotically separated from zero. Under these conditions, we can obtain analogues of mixed ergodic and large deviation Theorems 4.6.3, 4.6.5, 4.6.7, and 4.6.11, covering the case quasi-stationary asymptotics, as well as Theorem 4.6.6 that gives conditions for convergence of the corresponding quasi-stationary distributions. It is also useful to note that the theorems listed above can be applied in the pseudo-stationary case as well.

The limit parameters, including the stationary and quasi-stationary distributions, can be calculated by using the corresponding algorithms and formulas given in Chapter 4. They can be used for the semi-Markov processes $\eta^{(\varepsilon)}(t)$, $t \geq 0$, introduced in the current section.

6.1.4   Conditions for exponential expansions in mixed ergodic and large deviation theorems

Let us formulate conditions that replace the corresponding perturbation conditions introduced in Chapter 5 for the corresponding theorems representing the exponential asymptotic expansions in mixed ergodic and limit/large deviation theorems.

The perturbation condition $\mathbf{P_{30}^{(k)}}$ can be replaced with the following condition:

$\mathbf{P_{31}^{(k)}}$: $q_{i,\pm}^{(\varepsilon)} = q_{i,\pm}^{(0)} + q_{i,\pm}[1]\varepsilon + \cdots + q_{i,\pm}[k]\varepsilon^k + o(\varepsilon^k)$, for $i = 1, \ldots, m$, where $|q_{i,\pm}[r]| < \infty$, $r = 0, \ldots, k$, $i = 1, \ldots, m$.

It is also convenient to define $q_{i,\pm}[0] = q_{i,\pm}^{(0)}$, $i = 1, \ldots, m$.

Note that condition $\mathbf{P_{31}^{(k)}}$ obviously implies that condition $\mathbf{T_4}$ holds.

Condition $\mathbf{P_{31}^{(k)}}$ also implies that condition $\mathbf{P_{30}^{(k)}}$ and the corresponding expansions for transition intensities $q_{\bar{x}\bar{y}}^{(\varepsilon)}$ and $q_{\bar{x}}^{(\varepsilon)}$ take the following form for $\bar{x} \neq \bar{0}$, $\bar{y} \in X$:

$$q_{\bar{x}\bar{y}}^{(\varepsilon)} = q_{\bar{x}\bar{y}}^{(0)} + q_{\bar{x}\bar{y}}[1]\varepsilon + \cdots + q_{\bar{x}\bar{y}}[k]\varepsilon^k + o(\varepsilon^k), \tag{6.1.12}$$

where, for $r = 1, \ldots, k$,

$$q_{\bar{x}\bar{y}}[r] = \begin{cases} q_{i,-}[r] & \text{if } \bar{x} \neq \bar{1}, x_i = 1, y_i = 0, x_j = y_j, j \neq i, \\ q_{i,+}[r] & \text{if } x_i = 0, y_i = 1, x_j = y_j, j \neq i, \\ 0 & \text{otherwise,} \end{cases} \tag{6.1.13}$$

and

$$q_{\bar{x}}^{(\varepsilon)} = q_{\bar{x}}^{(0)} + q_{\bar{x}}[1]\varepsilon + \cdots + q_{\bar{x}}[k]\varepsilon^k + o(\varepsilon^k), \tag{6.1.14}$$

where, for $r = 1, \ldots, k$,

$$q_{\bar{x}}[r] = \sum_{i:x_i=1} q_{i,-}[r] + \sum_{i:x_i=0} q_{i,+}[r]. \tag{6.1.15}$$

It is also convenient to define $q_{\bar{x}\bar{y}}[0] = q_{\bar{x}\bar{y}}^{(0)}$ and $q_{\bar{x}}[0] = q_{\bar{x}}^{(0)}$ for $\bar{x} \neq \bar{0}$, $\bar{y} \in X$.

As soon as the asymptotic expansions for transition intensities $q_{\bar{x}\bar{y}}^{(\varepsilon)}$ and $q_{\bar{x}}^{(\varepsilon)}$ are obtained, one can get, for every $\bar{x} \in X_0$ and $n = 0, 1, \ldots$, asymptotic expansions for power moments

$$m_{\bar{x}\bar{y}}^{(\varepsilon)}[n] = m_{\bar{x}}^{(\varepsilon)}[n] = \int_0^\infty s^n q_{\bar{x}}^{(\varepsilon)} \exp\{-q_{\bar{x}}^{(\varepsilon)} s\}\, ds = n!(q_{\bar{x}}^{(\varepsilon)})^{-n} < \infty, \tag{6.1.16}$$

and moment generating functions

$$\psi_{\bar{x}\bar{y}}^{(\varepsilon)}[\rho, n] = \psi_{\bar{x}}^{(\varepsilon)}[\rho, n] = \int_0^\infty s^n e^{\rho s} q_{\bar{x}}^{(\varepsilon)} \exp\{-q_{\bar{x}}^{(\varepsilon)} s\}\, ds$$

$$= \frac{n! q_{\bar{x}}^{(\varepsilon)}}{(q_{\bar{x}}^{(\varepsilon)} - \rho)^{n+1}} < \infty, \quad \rho < q_{\bar{x}}^{(\varepsilon)}. \tag{6.1.17}$$

These expansions can be obtained with the use of Lemmas 5.6.3 and 5.6.5. Note that here the only condition $\mathbf{P}_{31}^{(k)}$ is required.

Therefore, condition $\mathbf{P}_{31}^{(k)}$ implies that the perturbation conditions $\mathbf{P}_{18}^{(k)}$ and $\mathbf{P}_{20}^{(k)}$ hold and that the perturbation conditions $\mathbf{P}_{24}^{(\rho,k)}$ and $\mathbf{P}_{26}^{(\rho,k)}$ hold for $\rho < \rho^*$.

The only conditions that do require an additional analysis are the conditions $\mathbf{O}_3^{(h)}$, $\mathbf{O}_4^{(h)}$.

Recall that these conditions determine pivotal properties of asymptotic expansions for the characteristic root $\rho^{(\varepsilon)}$ of the equation $_0\phi_{\bar{x}\bar{x}}^{(\varepsilon)}(\rho) = 1$ for $\bar{x} \in X_1^{(0)}$ in the most interesting pseudo-stationary case.

Let us first consider the case where the following condition, additional to conditions $\mathbf{E}_4'$ and $\mathbf{P}_{31}^{(k)}$, holds:

$\mathbf{O}_9$: $q_{i,-}[1] > 0$, $i \in E$.

Condition $\mathbf{O}_9$ means that the failure intensities for the main (asymptotically reliable) servers are of order $O(\varepsilon)$.

Let us first clarify the pivotal properties for cyclic potential quantities,

$$_{\bar{x}} u_{\bar{x}\bar{y}}^{(\varepsilon)} = \mathsf{E}_{\bar{x}} \sum_{n=1}^{\nu_{\bar{x}}^{(\varepsilon)} \wedge \nu_{\bar{0}}^{(\varepsilon)}} \delta(\eta_{n-1}^{(\varepsilon)}, \bar{y}), \quad \bar{x} \in X_1^{(0)}, \quad \bar{y} \in X_0.$$

The following two cases should be considered: **(1)** $k \geq l$ and **(2)** $k < l$. Recall that $k \geq 0$ is the order of the asymptotic expansion in the perturbation condition $\mathbf{P}_{31}^{(k)}$, while $l \geq 1$ is the number of indices in the set E, i.e., the number of main servers.

As shown in Lemma 5.1.1, the potential quantities $_{\bar{x}}u_{\bar{x}\bar{y}}^{(\varepsilon)}$ admit $(0, k)$-expansions,

$$_{\bar{x}}u_{\bar{x}\bar{y}}^{(\varepsilon)} = {}_{\bar{x}}u_{\bar{x}\bar{y}}[0] + {}_{\bar{x}}u_{\bar{x}\bar{y}}[1]\varepsilon + \cdots + {}_{\bar{x}}u_{\bar{x}\bar{y}}[k]\varepsilon^k + o(\varepsilon^k). \qquad (6.1.18)$$

Let us show that **(a)** in case **(1)**, where $k \geq l$, conditions $\mathbf{E}'_4$, $\mathbf{P}_{31}^{(k)}$, and $\mathbf{O}_9$ imply that the coefficients satisfy $_{\bar{x}}u_{\bar{x}\bar{y}}[n] = 0$, $n < r$, while $_{\bar{x}}u_{\bar{x}\bar{y}}[r] > 0$ for $\bar{x} \in X_1^{(0)} = Z_0$, $\bar{y} \in U_r, 0 \leq r \leq l$ (the sets U_r were defined in relations (6.1.6) and (6.1.7)).

According to Lemma 5.1.2, statement **(a)** is equivalent to that **(b)** the global asymptotic order $\hat{r}(\bar{x}, \bar{y}) = r$ for the states $\bar{x} \in U_0, \bar{y} \in U_r, 0 \leq r \leq l$.

Statement **(b)** follows from an obvious observation that **(c)** the local asymptotic order $r(\bar{x}', \bar{x}'')$ equals 1 if $\bar{x}' \in U_n$, $\bar{x}'' \in U_{n+1}$ for some $n = 0, \ldots, l - 1$, or 0 otherwise. Let $(\bar{x}_0, \ldots, \bar{x}_q)$ be an irreducible $(\bar{x}', \bar{x}'')$-chain of states from X_0, where $\bar{x}' = \bar{x}_0 \in U_0, \bar{x}'' = \bar{x}_q \in U_r$. Obviously, **(d)** the minimal number of pairs $(\bar{x}_n, \bar{x}_{n+1})$ with the local asymptotic order $r(\bar{x}_n, \bar{x}_{n+1}) = 1$ in such an irreducible $(\bar{x}', \bar{x}'')$-chain is r. These pairs correspond to transitions from some state in the class U_n to some state in the class U_{n+1} for $n = 0, \ldots, r - 1$. This implies statement **(b)** and, as a consequence, statement **(a)**.

Note that **(a)** implies that **(e)** $Z_r = U_r, r = 0, \ldots, l$ (the sets Z_r were defined in Subsection 5.1). Also, $Z_r = \varnothing, r = l + 1, \ldots, k + 1$, since $\bigcup_{n=0}^{l} U_n = X_0$.

Denote by U'_{l-1} the set of states $\bar{x} = (x_1, \ldots, x_m)$ such that $x_i = 1$ for some $1 \leq i \leq l$ and $x_j = 0$ for $j \neq i$, and by U''_l the set of states $\bar{x} = (x_1, \ldots, x_m)$ such that $x_i = 1$ for some $l + 1 \leq i \leq m$ and $x_j = 0$ for $j \neq i$. By the definition, $U'_{l-1} \subseteq U_{l-1}$ and $U''_l \subseteq U_l$.

Formula (6.1.3) and conditions $\mathbf{E}'_4$ and $\mathbf{O}_9$ imply that **(f)** the transition intensities satisfy $q_{\bar{x}\bar{0}}^{(\varepsilon)} = 0$ if $\bar{x} \in \bigcup_{n=0}^{l-2} U_n$; **(g)** the transition intensities satisfy $q_{\bar{x}\bar{0}}^{(\varepsilon)} \sim q_{i,-}[1]\varepsilon$ as $\varepsilon \to 0$ (note that $q_{i,-}[1] > 0$) if $x_i = 1, \bar{x} \in U'_{l-1}$, for $i = 1, \ldots, l$, and $q_{\bar{x}\bar{0}}^{(\varepsilon)} = 0$ if $\bar{x} \in U_{l-1} \setminus U'_{l-1}$; **(h)** the transition intensities satisfy $q_{\bar{x}\bar{0}}^{(\varepsilon)} \sim q_{i,-}^{(0)}$ as $\varepsilon \to 0$ (note that $q_{i,-}^{(0)} > 0$) if $x_i = 1, \bar{x} \in U''_l$, for $i = l + 1, \ldots, m$, and $q_{\bar{x}\bar{0}}^{(\varepsilon)} = 0$ if $\bar{x} \in U_l \setminus U''_l$.

Using relations **(f)**–**(h)**, asymptotic expansions (6.1.12) and (6.1.14) and the quotient rule for the asymptotic expansions given in statement **(iv)** of Lemma 8.1.1, one can readily get properties similar to **(f)**–**(h)** for the one-step absorption transition probabilities $p_{\bar{x}\bar{0}}^{(\varepsilon)} = q_{\bar{x}\bar{0}}^{(\varepsilon)}/q_{\bar{x}}^{(\varepsilon)}$. These properties are the following: **(i)** the transition probabilities are zero, $p_{\bar{x}\bar{0}}^{(\varepsilon)} = 0$ if $\bar{x} \in \bigcup_{n=0}^{l-2} U_n$; **(j)** the transition probabilities satisfy $p_{\bar{x}\bar{0}}^{(\varepsilon)} \sim q_{i,-}[1]\varepsilon/q_{\bar{x}}^{(0)}$ as $\varepsilon \to 0$ if $x_i = 1, \bar{x} \in Z'_{l-1}$ for $i = 1, \ldots, l$, and $p_{\bar{x}\bar{0}}^{(\varepsilon)} = 0$ if $\bar{x} \in U_{l-1} \setminus U'_{l-1}$; **(k)** the transition probabilities satisfy $p_{\bar{x}\bar{0}}^{(\varepsilon)} \sim q_{i,-}^{(0)}/q_{\bar{x}}^{(0)}$ as $\varepsilon \to 0$ if $x_i = 1, \bar{x} \in Z''_l$, for $i = l + 1, \ldots, m$, and $p_{\bar{x}\bar{0}}^{(\varepsilon)} = 0$ if $\bar{x} \in U_l \setminus U''_l$.

Statements **(i)**–**(k)** imply that **(l)** the set $Y_{\bar{0},r}$ (defined in Subsection 5.1.5) coincides with U''_l for $r = 0$, with U'_{l-1} for $r = 1$, $\varnothing$ for $r = 2, \ldots, k$, and with $X_0 \setminus (U'_{l-1} \cup U''_l)$ for $r = k + 1$.

Statements **(a)** and **(l)** imply in an obvious way that **(m)** the set $W_{\bar{0},r}$ (defined in Subsection 5.1.5) coincides with $\varnothing$ for $r = 0,\ldots,l-1$; $U'_{l-1} \cup U''_l$ for $r = l$; $\varnothing$ for $r = l+1,\ldots,k$; and $X_0 \setminus (U'_{l-1} \cup U''_l)$ for $r = k+1$.

Therefore, **(n)** if conditions $\mathbf{E}'_4$ and $\mathbf{P}^{(k)}_{31}$, and $\mathbf{O}_9$ hold and the additional assumption **(1)** $k \geq l$ is satisfied, then condition $\mathbf{O}^{(l)}_4$ holds.

In this case, by Lemmas 5.1.5 and 5.1.7, the cyclic absorption probabilities $_{\bar{x}}f^{(\varepsilon)}_{\bar{x}\bar{0}}$, $\bar{x} \in X^{(0)}_1$, have (l,k)-asymptotic pivotal expansions (where $_{\bar{x}}\bar{b}_{\bar{x}\bar{0}}[l] > 0$),

$$_{\bar{x}}f^{(\varepsilon)}_{\bar{x}\bar{0}} = {}_{\bar{x}}\bar{b}_{\bar{x}\bar{0}}[l]\varepsilon^l + \cdots + {}_{\bar{x}}\bar{b}_{\bar{x}\bar{0}}[k]\varepsilon^k + o(\varepsilon^k). \tag{6.1.19}$$

Also, by Theorem 5.4.1, the characteristic root $\rho^{(\varepsilon)}$ of the equation $_{\bar{0}}\phi^{(\varepsilon)}_{\bar{x}\bar{x}}(\rho) = 1$ has the following (l,k)-asymptotic pivotal expansion:

$$\rho^{(\varepsilon)} = a_l\varepsilon^l + \cdots + a_k\varepsilon^k + o(\varepsilon^k). \tag{6.1.20}$$

Similarly it can be shown that **(o)** in case **(2)** with $k < l$, conditions $\mathbf{E}'_4$, $\mathbf{P}^{(k)}_{31}$, and $\mathbf{O}_9$ imply that $_{\bar{x}}u_{\bar{x}\bar{y}}[n] = 0$, $n < r$, while $_{\bar{x}}u_{\bar{x}\bar{y}}[r] > 0$ for $\bar{x} \in X^{(0)}_1 = U_0$, $\bar{y} \in U_r, 0 \leq r \leq k$ (these sets were defined in relations (6.1.6) and (6.1.7)), and **(p)** $_{\bar{x}}u_{\bar{x}\bar{y}}[n] = 0$, $n < k+1$, for $\bar{x} \in X^{(0)}_1 = U_0$, $\bar{y} \in \bigcup^l_{r=k+1} U_r$.

The proposition **(o)** implies that **(q)** $Z_r = U_r$, $r = 0,\ldots,k$, and $Z_{k+1} = \bigcup^l_{r=k+1} U_r$. The sets $Y_{\bar{0},r}$ were defined in statement **(l)**.

It follows from statements **(o)** and **(l)** that **(r)** the set $W_{\bar{0},r}$ coincides with $\varnothing$ for $r = 0,\ldots,k$ and $W_{\bar{0},k+1} = X_0$.

Therefore, **(s)** if conditions $\mathbf{E}'_4$ and $\mathbf{P}^{(k)}_{31}$, and $\mathbf{O}_9$ hold and the additional assumption **(1)** $k \leq l$ is satisfied, then condition $\mathbf{O}^{(k)}_3$ holds.

By Lemmas 5.1.5 and 5.1.7, the cyclic absorption probabilities $_{\bar{x}}f^{(\varepsilon)}_{\bar{x}\bar{0}}$, $\bar{x} \in X^{(0)}_1$ satisfy the following relations: **(t)** $_{\bar{x}}f^{(\varepsilon)}_{\bar{x}\bar{0}} = o(\varepsilon^k)$, $\bar{x} \in X^{(0)}_1$. Also, by Theorem 5.4.1, the characteristic root $\rho^{(\varepsilon)}$ of the equation $_{\bar{0}}\phi^{(\varepsilon)}_{\bar{x}\bar{x}}(\rho) = 1$ satisfies the relation **(u)** $\rho^{(\varepsilon)} = o(\varepsilon^k)$.

The asymptotic relations and **(t)** and **(u)** are not surprising. One should not expect that the lemmas and theorem mentioned above would give more precise asymptotics under the perturbation conditions providing exact asymptotics for transition intensities only up to terms of order ε^k.

Let us also consider a more complicated case where the following condition, additional to conditions $\mathbf{E}'_4$ and $\mathbf{P}^{(k)}_{31}$, holds:

$\mathbf{O}_{10}$: $q_{i,-}[n] = 0$, $n < k_i$ and $q_{i,-}[k_i] > 0$ for some $1 \leq k_i \leq k$, for every $i \in E$.

Condition $\mathbf{O}_{10}$ means that the failure intensity for main (asymptotically reliable) servers $i \in E$ are of a different orders $O(\varepsilon^{k_i})$.

As above, two cases should be considered, **(3)** $k \geq k_1 + \cdots + k_l = \bar{k}_l$ and **(4)** $k < \bar{k}_l$.

By repeating the asymptotic analysis presented above, it can be shown that, under conditions $\mathbf{E}'_4$, $\mathbf{P}^{(k)}_{31}$, and $\mathbf{O}_{10}$, condition $\mathbf{O}^{(\bar{k}_l)}_3$ holds in case **(3)**, or condition $\mathbf{O}^{(k)}_4$ holds in case **(4)**. Conditions $\mathbf{O}^{(\rho,h)}_5$, $\mathbf{O}^{(\rho,h)}_6$, as was mentioned in Subsection 5.3.7, can not be expressed so explicitly in terms of coefficients in the asymptotic expansions used in the perturbation condition $\mathbf{P}^{(k)}_{31}$. Here, we need to use the coefficients of the corresponding expansions for the moment generating functions $_{\bar{0}}\phi^{(\varepsilon)}_{\bar{x}\bar{x}}(\rho)$, $x \in X^{(0)}_1$, given in Lemma 5.3.5 and specified for the semi-Markov processes considered in the current section.

6.1.5 Exponential expansions in mixed ergodic and large deviation theorems

All results given in Chapter 5 can be specified for semi-Markov processes $\eta^{(\varepsilon)}(t)$, $t \geq 0$ that describe functioning of queueing systems with highly reliable main servers.

Conditions $\mathbf{T}_4$, $\mathbf{E}'_4$, $\mathbf{P}^{(k)}_{31}$, and $\mathbf{O}_9$ or $\mathbf{O}_{10}$ correspond to a pseudo-stationary model with the absorption probabilities asymptotically vanishing to zero. Under these conditions, we can formulate analogues of Theorems 5.4.1 and 5.4.2, which would give exponential asymptotic expansion in mixed ergodic and large deviation theorems. Also an analogue of Theorem 5.5.1 gives higher order asymptotic expansions for stationary distributions under the additional non-absorption condition $\mathbf{A}_8$.

Conditions $\mathbf{T}_4$, $\mathbf{E}''_4$, and $\mathbf{P}^{(k)}_{31}$ correspond to a quasi-stationary model with the absorption probabilities asymptotically separated from zero. Under these conditions, analogues of Theorems 5.4.3 and 5.4.4, which give exponential asymptotic expansion in mixed ergodic and large deviation theorems, can be obtained. Also, the analogues of Theorems 5.5.4 and 5.5.5 give higher order asymptotic expansions for the quasi-stationary distributions. It is useful to note that the theorems given above also cover the pseudo-stationary case, i.e., they can be applied if conditions $\mathbf{T}_4$, $\mathbf{E}'_4$, and $\mathbf{P}^{(k)}_{31}$ hold as well.

The limit parameters, including the stationary and quasi-stationary distributions, and the coefficients in the corresponding asymptotic expansions, can be calculated by using the corresponding algorithms and formulas given in Chapters 4 and 5. They should be specified for the semi-Markov processes $\eta^{(\varepsilon)}(t)$, $t \geq 0$, introduced in the current section.

6.2 M/G queueing systems with quick service

In this section, we investigate conditions for quasi- and pseudo-stationary asymptotics for M/G queueing systems with quick service and a bounded queue buffer. In this case, the perturbed stochastic processes, which describe the dynamics of the queue in the system, belong to the class of so-called stochastic processes with semi-Markov modulation. They admit a construction of imbedded semi-Markov processes and are more general than semi-Markov processes. An example presented in this section shows how the main results obtained in the book can be applied to stochastic processes more general than semi-Markov processes, in particular, to stochastic processes with semi-Markov modulation.

6.2.1 M/G queueing systems with quick service

Let us consider a standard M/G queueing system with a bounded queue buffer. This system has one server and a bounded queue buffer of size $N \geq 2$ (including the place in the server). We make the following assumptions: **(a)** the input flow of customers is a standard Poisson flow with parameter $\lambda > 0$; **(b)** the customers are placed in the queue buffer in the order they come in the system; **(c)** a new customer coming in the system is immediately served if the queue buffer is empty, is placed in the queue buffer if the number of customers in the system is less than N, or leaves the system if the queue buffer is full, i.e., the number of customers in the system equals to N; **(d)** the service times for different customers are independent random variables with the distribution function $G^{(\varepsilon)}(t)$; **(f)** the processes of a customer coming in the system and the service processes are independent.

We assume that the distribution of the service time depends on the perturbation parameter $\varepsilon \geq 0$ and that $G^{(\varepsilon)}(t)$ is a distribution concentrated on the interval $[0, \infty)$ and is not concentrated in zero for every $\varepsilon > 0$.

The state of the system is determined by a random variable $\xi^{(\varepsilon)}(t)$, which is the number of empty places in the queue buffer at the moment t. The stochastic process $\xi^{(\varepsilon)}(t), t \geq 0$ has a space of states, $X = \{0, 1, \ldots, N\}$. We assume that the process $\xi^{(\varepsilon)}(t), t \geq 0$ is right continuous. We also assume that the initial queue is $\xi^{(\varepsilon)}(0) = \text{const} > 0$.

The lifetime of the system $\mu^{(\varepsilon)} = \inf(t > 0 : \xi^{(\varepsilon)}(t) = 0)$ is defined as the first moment when the number of empty places in the queue buffer becomes equal 0.

Instead of the stochastic process $\xi^{(\varepsilon)}(t), t \geq 0$, one can also use the process $\tilde{\xi}^{(\varepsilon)}(t) = N - \xi^{(\varepsilon)}(t), t \geq 0$. Obviously, the random variable $\tilde{\xi}^{(\varepsilon)}(t)$ is the number of customers in the queue buffer at the moment t. In this case, the lifetime of the system $\mu^{(\varepsilon)} = \inf(t > 0 : \tilde{\xi}^{(\varepsilon)}(t) = N)$ is defined to be the first moment at which the number of customers in the queue buffer becomes equal N.

Note that the definition of the lifetime given above agrees with the conventions used in Chapters 4 and 5, where the phase space for the corresponding processes is always the set $X = \{1, \ldots, N\}$ and the absorption times are always the first hitting time in the state 0. In the example considered in the current section, the lifetime $\mu^{(\varepsilon)}$, in fact, can be interpreted as the first time at which the customer, coming in the system with the actual queue buffer size $N - 1$, leaves the system because the number of customers in the system is $N - 1$, i.e., the queue buffer is full.

We assume the following condition, which allows us to interpret the models with $\varepsilon > 0$ as perturbed versions of the model with $\varepsilon = 0$,

T$_5$: $G^{(\varepsilon)}(\cdot) \Rightarrow G^{(0)}(\cdot)$ as $\varepsilon \to 0$, where $G^{(0)}(t)$ is a proper distribution function concentrated on the interval $[0, \infty)$.

There are two cases possible. The first one corresponds to a model with asymptotically quick service subject to the following condition:

E$_5'$: $G^{(0)}(t)$ is a distribution function concentrated in 0, i.e., $G^{(0)}(0) = 1$.

This is the case with pseudo-stationary asymptotics for the corresponding probabilities conditioned by non-absorption events.

For the second case, which corresponds to a model with the usual service, the following condition holds:

E$_5''$: $G^{(0)}(t)$ is a distribution function not concentrated in 0, i.e., $G^{(0)}(0) < 1$.

This is the case for quasi-stationary asymptotics of the corresponding probabilities conditioned by non-absorption events.

6.2.2   An imbedded semi-Markov process

The process $\xi^{(\varepsilon)}(t), t \geq 0$ is a regenerative process. The subsequent moments at which this process hits the state N, which results in the end of the service for some customer, $\tau_n^{(\varepsilon)}, n = 1, 2 \ldots$, can be taken as subsequent regeneration moments for this process.

In fact, the subsequent moments at which this process hits any other fixed state $i > 1$ can be taken to be subsequent regeneration moments as well. Note that the state 1 can not be used to construct appropriate regeneration cycles. In this case, the corresponding stopping probability for one regeneration cycle is $f^{(\varepsilon)} = \mathsf{P}\{\mu^{(\varepsilon)} < \tau_1^{(\varepsilon)}\} = 1$.

Since $\xi^{(\varepsilon)}(t)$ is a regenerative process, the results obtained in Chapter 3 can be applied. However, one should take into account that conditions and results in this chapter are formulated in terms of the characteristics related to one regeneration cycle.

We are going to show how these conditions and the results can be formulated in terms of characteristics of the corresponding queueing systems, which are the parameter λ, the distribution functions $G^{(\varepsilon)}(t)$, and the queue buffer size N.

We shall use the special property of the process $\xi^{(\varepsilon)}(t), t \geq 0$, which is connected to the possibility of constructing, for this process, a special imbedded semi-Markov process.

Let $\kappa_n^{(\varepsilon)}$ be the times between subsequent moments at which customers come in the system. Because of the assumption that the input flow is a Poisson flow with parameter λ, the random variables $\kappa_n^{(\varepsilon)}, n = 1, 2, \ldots$, are i.i.d. random variables with an exponential distribution function with parameter λ.

Let us also denote by $\zeta^{(\varepsilon)}(t) = \max(n : \sum_{k=1}^{n} \kappa_k^{(\varepsilon)} \le t), t \ge 0$, the corresponding Poisson process that counts the number of customers came in the system in the interval $[0, t]$.

Let also $\varsigma_n^{(\varepsilon)}$ be the service time for the n-th customer. These random variables, $\varsigma_n^{(\varepsilon)}, n = 1, 2, \ldots$, are i.i.d. random variables with the distribution function $G^{(\varepsilon)}(t)$.

According to the description of the queueing system given above, the sequences of random variables $\kappa_n^{(\varepsilon)}, n = 1, 2, \ldots$, and $\varsigma_n^{(\varepsilon)}, n = 1, 2, \ldots$, are independent.

Let $\alpha_0^{(\varepsilon)} = 0$ and $\alpha_n^{(\varepsilon)}, n = 1, 2, \ldots$, be subsequent moments at which one of the following events occurs: **(a)** the end of the service for some customer; **(b)** the end of the waiting time in the case where the number of empty places in the queue buffer equals N (there is no customers in the system); **(c)** the process $\xi^{(\varepsilon)}(t)$ hits the state 0 from the state 1 (there are N customers in the system).

Let us also define the random variable $v^{(\varepsilon)} = \min(n \ge 1 : \xi^{(\varepsilon)}(\alpha_n^{(\varepsilon)}) = 0)$.

Now define the imbedded semi-Markov process with absorption for the process $\xi^{(\varepsilon)}(t), t \ge 0$,

$$\eta^{(\varepsilon)}(t) = \begin{cases} \xi^{(\varepsilon)}(\alpha_n^{(\varepsilon)}) & \text{if } \alpha_n^{(\varepsilon)} \le t < \alpha_{n+1}^{(\varepsilon)}, n < v^{(\varepsilon)}, \\ 0 & \text{if } t \ge \alpha_n^{(\varepsilon)}, n = v^{(\varepsilon)}. \end{cases} \tag{6.2.1}$$

The semi-Markov process $\eta^{(\varepsilon)}(t), t \ge 0$, has the following transition probabilities:

$$Q_{ij}^{(\varepsilon)}(t)$$

$$= \begin{cases} 0 & \text{if } i = 0, \\ 0 & \text{if } 1 \le i \le N - 1, j > i + 1, \\ \int_0^t \frac{(\lambda s)^{i-j+1}}{(i-j+1)!} e^{-\lambda s} G^{(\varepsilon)}(ds) & \text{if } 1 \le i \le N - 1, 2 \le j \le i + 1, \\ 0 & \text{if } 1 \le i \le N - 1, j = 1, \\ \int_0^t \frac{\lambda^i s^{i-1}}{(i-1)!} e^{-\lambda s}(1 - G^{(\varepsilon)}(s))ds & \text{if } 1 \le i \le N - 1, j = 0, \\ 1 - e^{-\lambda t} & \text{if } i = N, j = N - 1, \\ 0 & \text{if } i = N, j \ne N - 1. \end{cases} \tag{6.2.2}$$

The zero value of the transition probability in the first line in the right-hand side of (6.2.2) shows that the state 0 is an absorption state.

Zero in the second line reflects the impossibility of transition from the state $1 \le i \le N - 1$ to the state $j > i + 1$, which results in the end of the service for one customer.

The expression for the transition probability in the third line is $\mathsf{P}\{\varsigma_1^{(\varepsilon)} \le t, \zeta^{(\varepsilon)}(t) = i - j + 1\}$, i.e., the probability that the time of the service for the first customer is less than or equal to t and that exactly $i - j + 1$ customers come in the system during this time.

Zero in the fourth line reflects the impossibility of transition from the state $1 \le i \le N - 1$ to the state $j = 1$, because the transition in the absorption state 0 occurs if i customers come in the system before the end of the service of the first customer.

The expression for the transition probability in the fifth line is $\mathsf{P}\{\kappa_1^{(\varepsilon)} + \cdots + \kappa_i^{(\varepsilon)} \leq \min(t, \varsigma_1^{(\varepsilon)})\}$, i.e., the probability that i customers come in the system before the end of the service of the first customer and before time t.

The expressions in the sixth and seventh lines reflect the fact that the waiting time for the first customer coming in the system, after the moment when the queue buffer becomes empty, has an exponential distribution.

As usual, we use the symbols $\mathsf{P}_i(A)$ and $\mathsf{E}_i\xi$ for the conditional probabilities and the expectations calculated under the condition that $\xi^{(\varepsilon)}(0) = \eta^{(\varepsilon)}(0) = i$.

An important property of the process $\xi^{(\varepsilon)}(t)$ and the imbedded semi-Markov process $\eta^{(\varepsilon)}(t)$ is that the first hitting time to the state 0, denoted above by $\mu^{(\varepsilon)}$, is the same for both processes.

Also the first hitting time to the state N, which results in the end of the service of a customer, denoted above by $\tau_1^{(\varepsilon)}$, is the same for both processes.

It follows from this observation that, for any initial state $1 \leq i \leq N$, the hitting probabilities $_0 f_{iN}^{(\varepsilon)}$ and the absorption probabilities $_N f_{i0}^{(\varepsilon)}$ defined for the imbedded semi-Markov processes $\eta^{(\varepsilon)}(t)$ coincide, respectively, with the hitting probabilities $\mathsf{P}_i\{\tau_1^{(\varepsilon)} \leq \mu^{(\varepsilon)}\}$ and the absorption probabilities $\mathsf{P}_i\{\mu^{(\varepsilon)} \leq \tau_1^{(\varepsilon)}\}$ defined for the process $\xi^{(\varepsilon)}(t)$.

Moreover, for any initial state $1 \leq i \leq N$, the distribution of hitting times, $_0 G_{iN}^{(\varepsilon)}(t)$, and the distribution of absorption times, $_N G_{i0}^{(\varepsilon)}(t)$, defined for the imbedded semi-Markov processes $\eta^{(\varepsilon)}(t)$ coincide, respectively, with the distributions $\mathsf{P}_i\{\tau_1^{(\varepsilon)} \leq t, \tau_1^{(\varepsilon)} \leq \mu^{(\varepsilon)}\}$ and the absorption probabilities $\mathsf{P}_i\{\mu^{(\varepsilon)} \leq t, \mu^{(\varepsilon)} < \tau_1^{(\varepsilon)}\}$ defined for the process $\xi^{(\varepsilon)}(t)$.

Let the initial state satisfy $\xi^{(\varepsilon)}(0) = i$. In terms of the characteristics used for regenerative processes in Chapter 3, the distribution function $_0 G_{NN}^{(\varepsilon)}(t)$ plays the role of the distribution function for the duration of the regeneration period $F^{(\varepsilon)}(t) = \mathsf{P}_N\{\tau_1^{(\varepsilon)} \leq t, \tau_1^{(\varepsilon)} \leq \mu^{(\varepsilon)}\}$; the absorption probability $_N f_{N0}^{(\varepsilon)}$ replaces the stopping probability for one regeneration cycle $f^{(\varepsilon)} = 1 - F^{(\varepsilon)}(\infty)$; the moment generating function $_0 \phi_{NN}^{(\varepsilon)}(\rho) = \int_0^\infty e^{\rho t} \, _0 G_{NN}^{(\varepsilon)}(dt)$ is now the moment generating function $\phi^{(\varepsilon)}(\rho)$; and the characteristic root $\rho^{(\varepsilon)}$ is the root of the equation $_0 \phi_{NN}^{(\varepsilon)}(\rho) = 1$. Also, the distribution function $_i G_{iN}^{(\varepsilon)}(t)$ replaces the distribution function for the duration of the transition period $\tilde{F}^{(\varepsilon)}(t) = \mathsf{P}_i\{\tau_1^{(\varepsilon)} \leq t, \tau_1^{(\varepsilon)} \leq \mu^{(\varepsilon)}\}$; the absorption probability $_i f_{N0}^{(\varepsilon)}$ is the stopping probability for the transition period $\tilde{f}^{(\varepsilon)} = 1 - \tilde{F}^{(\varepsilon)}(\infty)$; the moment generating function $_0 \phi_{iN}^{(\varepsilon)}(\rho) = \int_0^\infty e^{\rho t} \, _0 G_{iN}^{(\varepsilon)}(dt)$ becomes the moment generating function $\tilde{\phi}^{(\varepsilon)}(\rho)$; and the quantity $_0 \phi_{iN}^{(\varepsilon)}(\rho^{(\varepsilon)})$ replaces the quantity $\tilde{\phi}^{(\varepsilon)}$.

Now, we can apply the methods of asymptotic analysis for perturbed semi-Markov processes developed in Chapter 4 and 5 to asymptotic analysis of the corresponding characteristics related to regeneration cycles of the regenerative process $\xi^{(\varepsilon)}(t)$.

The coincidence for the corresponding absorption probabilities and the hitting distributions for the processes $\eta^{(\varepsilon)}(t)$ and $\xi^{(\varepsilon)}(t)$ should not lead to a misconception of the probabilities $\mathsf{P}_i\{\eta^{(\varepsilon)}(t) = j, \mu^{(\varepsilon)} > t\}$ and $\mathsf{P}_i\{\xi^{(\varepsilon)}(t) = j, \mu^{(\varepsilon)} > t\}$, which are the main object of our asymptotic analysis.

By the definition, the trajectories of the processes $\eta^{(\varepsilon)}(t)$ and $\xi^{(\varepsilon)}(t)$ in the interval $[0, \mu^{(\varepsilon)}]$ are synchronised at the moments $\alpha_n^{(\varepsilon)}, n \leq \nu^{(\varepsilon)}$. But trajectories of the process $\xi^{(\varepsilon)}(t)$ may deviate from the corresponding trajectories of the process $\eta^{(\varepsilon)}(t)$ between these moments due to new customers coming in the system. That is why the probabilities $\mathsf{P}_i\{\eta^{(\varepsilon)}(t) = j, \mu^{(\varepsilon)} > t\}$ and $\mathsf{P}_i\{\xi^{(\varepsilon)}(t) = j, \mu^{(\varepsilon)} > t\}$ may not coincide, as well as the corresponding quasi-stationary distributions and other related quasi-stationary quantities.

The process $\xi^{(\varepsilon)}(t), t \geq 0$, belongs to the class of so-called stochastic processes with semi-Markov modulation (switchings). We refer to books by Gikhman and Skorokhod (1971), Silvestrov (1980a), Koroliuk and Limnios (2005), and Anisimov (2008), where one can find a description of such processes.

6.2.3   Hitting and absorption probabilities for the imbedded semi-Markov process

Let us study relations between the conditions $\mathbf{T_5}$, $\mathbf{E_5'}$, $\mathbf{E_5''}$, and the conditions for the mixed ergodic and limit/large deviation theorems introduced in Chapters 3 and 4 for the imbedded semi-Markov process $\eta^{(\varepsilon)}(t), t \geq 0$.

It follows from formula (6.2.2) that the transition probabilities for the imbedded discrete time Markov chain $\eta_n^{(\varepsilon)}, n = 0, 1, \ldots$, corresponding to the semi-Markov process $\eta^{(\varepsilon)}(t), t \geq 0$, has the following form:

$$
p_{ij}^{(\varepsilon)} = \begin{cases}
0 & \text{if } i = 0, \\
0 & \text{if } 1 \leq i \leq N - 1, j > i + 1, \\
p_{i-j+1}^{(\varepsilon)}(\lambda) & \text{if } 1 \leq i \leq N - 1, 2 \leq j \leq i + 1, \\
0 & \text{if } 1 \leq i \leq N - 1, j = 1, \\
\bar{p}_i^{(\varepsilon)}(\lambda) & \text{if } 1 \leq i \leq N - 1, j = 0, \\
1 & \text{if } i = N, j = N - 1, \\
0 & \text{if } i = N, j \neq N - 1,
\end{cases}
\tag{6.2.3}
$$

where

$$
p_r^{(\varepsilon)} = p_r^{(\varepsilon)}(\lambda) = \int_0^\infty \frac{(\lambda s)^r}{r!} e^{-\lambda s} G^{(\varepsilon)}(ds), \quad r = 0, 1, \ldots,
\tag{6.2.4}
$$

and

$$\bar{p}_{r+1}^{(\varepsilon)} = \bar{p}_{r+1}^{(\varepsilon)}(\lambda) = \int_0^\infty \frac{\lambda^{r+1} s^r}{r!} e^{-\lambda s} (1 - G^{(\varepsilon)}(s))\, ds$$

$$= 1 - \sum_{l=0}^{r} p_l^{(\varepsilon)}(\lambda), \quad r = 0, 1, \ldots . \tag{6.2.5}$$

The corresponding matrix of transition probabilities has the following form:

$$\mathbf{P}^{(\varepsilon)} = \begin{bmatrix} 1 & 0 & 0 & 0 & \cdots & 0 & 0 \\ \bar{p}_1^{(\varepsilon)} & 0 & p_0^{(\varepsilon)} & 0 & \cdots & 0 & 0 \\ \bar{p}_2^{(\varepsilon)} & 0 & p_1^{(\varepsilon)} & p_0^{(\varepsilon)} & \cdots & 0 & 0 \\ \vdots & \vdots & \vdots & \vdots & \vdots & \vdots & \vdots \\ \bar{p}_{N-1}^{(\varepsilon)} & 0 & p_{N-2}^{(\varepsilon)} & p_{N-3}^{(\varepsilon)} & \cdots & p_1^{(\varepsilon)} & p_0^{(\varepsilon)} \\ 0 & 0 & 0 & 0 & \cdots & 1 & 0 \end{bmatrix}. \tag{6.2.6}$$

By the assumptions on the initial model, the distribution $G^{(\varepsilon)}(t)$ is assumed to be not concentrated in zero for every $\varepsilon > 0$. This assumption implies that **(a)** $p_k^{(\varepsilon)} > 0, k = 0, 1, \ldots$; **(b)** $\sum_{k=0}^{\infty} p_k^{(\varepsilon)} = 1$; **(c)** $0 < \bar{p}_k^{(\varepsilon)} < 1, k = 0, 1, \ldots$.

Condition $\mathbf{T_5}$ implies that condition $\mathbf{T_1}$ **(a)** holds, i.e., that the following relations hold for $i = 0, j \in X$:

$$p_{ij}^{(\varepsilon)} \to p_{ij}^{(0)} \quad \text{as } \varepsilon \to 0. \tag{6.2.7}$$

Indeed, formulas (6.2.2)–(6.2.5) and condition $\mathbf{T_5}$ imply that the following relations holds for $r = 0, \ldots$:

$$p_r^{(\varepsilon)} \to p_r^{(0)} \quad \text{as } \varepsilon \to 0, \tag{6.2.8}$$

and

$$\bar{p}_r^{(\varepsilon)} \to \bar{p}_r^{(0)} \quad \text{as } \varepsilon \to 0. \tag{6.2.9}$$

Relation (6.2.8) follows from continuity and boundedness of the integrands in the integrals which define the probabilities $p_n^{(\varepsilon)}$ and condition $\mathbf{T_5}$. Relation (6.2.9) is a direct corollary of relation (6.2.8).

Relations **(a)**–**(c)** imply that the hitting probabilities satisfy **(d)** $_0 f_{ij}^{(\varepsilon)} > 0, i \neq 0, j \neq 0, 1$; the hitting probabilities are zero, **(e)** $_0 f_{i1}^{(\varepsilon)} = 0, i \neq 0$; **(f)** the absorption probabilities are positive $_j f_{i0}^{(\varepsilon)} > 0, i, j \neq 0$.

Thus, in the prelimit case, i.e., for $\varepsilon > 0$, the class of recurrent-without-absorption states is the set $X_1^{(\varepsilon)} = \{2, \ldots, N\}$ if $N > 2$ or $X_1^{(\varepsilon)} = \{1, 2\}$ if $N = 2$, while the class of non-recurrent-without-absorption states is the set $X_2^{(\varepsilon)} = \{1\}$ if $N > 2$ or $X_2^{(\varepsilon)} = \varnothing$ if $N = 2$.

If condition $\mathbf{E}_5'$ holds, **(g)** $p_0^{(0)} = 1$ and $p_n^{(0)} = 0$, $n = 1, 2, \ldots$; **(h)** $\bar{p}_n^{(0)} = 0$, $n = 1, 2, \ldots$.

In this case, the corresponding matrix of the transition probabilities takes the following form:

$$\mathbf{P}^{(0)} = \begin{bmatrix} 1 & 0 & 0 & 0 & \cdots & 0 & 0 \\ 0 & 0 & 1 & 0 & \cdots & 0 & 0 \\ 0 & 0 & 0 & 1 & \cdots & 0 & 0 \\ \vdots & \vdots & \vdots & \vdots & \vdots & \vdots & \vdots \\ 0 & 0 & 0 & 0 & \cdots & 0 & 1 \\ 0 & 0 & 0 & 0 & \cdots & 1 & 0 \end{bmatrix}. \tag{6.2.10}$$

In this case, **(i)** the hitting probabilities satisfy $_0 f_{ij}^{(0)} = 1$ for any $1 \leq i < j \leq N$ and $i = N$, $j = N - 1, N$; **(j)** the hitting probabilities are zero, $_0 f_{ij}^{(0)} = 0$, for any $1 \leq j \leq i < N$ and $i = N$, $j < N - 1$; **(k)** the absorption probabilities are $_j f_{i0}^{(0)} = 0$ for any $i, j \neq 0$.

Thus, in this limit case, i.e., for $\varepsilon = 0$, the set $X_1^{(0)} = \{N - 1, N\}$ is the class of recurrent-without-absorption states while the set $X_2^{(0)} = \{1, \ldots, N - 2\}$ is the class of non-recurrent-without-absorption states.

If condition $\mathbf{E}_5''$ holds then **(l)** $p_n^{(0)} > 0$, $n = 0, 1, \ldots$; **(m)** $\sum_{n=0}^{\infty} p_n^{(0)} = 1$; **(n)** $0 < \bar{p}_n^{(0)} < 1$, $n = 1, 2, \ldots$. In this case, the corresponding matrix of transition probabilities has the same structure as in the prelimit case.

This implies that **(o)** $_0 f_{ij}^{(0)} > 0$, $i \neq 0$, $j \neq 0, 1$; **(p)** $_0 f_{i1}^{(0)} = 0$, $i \neq 0$; **(q)** $_j f_{i0}^{(0)} > 0$, $i, j \neq 0$.

Thus, in the limit case, i.e., for $\varepsilon = 0$, we have $X_1^{(0)} = \{2, \ldots, N\}$ and $X_2^{(0)} = \{1\}$ if $N > 2$, or $X_1^{(0)} = \{1, 2\}$ and $X_2^{(0)} = \varnothing$ if $N = 2$.

Condition $\mathbf{T}_5$ also implies that condition $\mathbf{T}_1$ **(b)** holds, i.e., that the following relation holds for $i \neq 0$, $j \in X$:

$$Q_{ij}^{(\varepsilon)}(\cdot) \Rightarrow Q_{ij}^{(0)}(\cdot) \text{ as } \varepsilon \to 0. \tag{6.2.11}$$

Relation (6.2.11) follows from continuity and boundedness of the integrands in the integrals which define the probabilities $Q_{ij}^{(\varepsilon)}(t)$ and condition $\mathbf{T}_5$.

It should be noted that in the case where condition $\mathbf{E}_5'$ holds, the distributions of sojourn times, $F_i^{(0)}(t)$, are concentrated in 0, i.e., $F_i^{(0)}(0) = 1$ for $1 \leq i \leq N - 1$. Thus, these states are instant for the limit process. However, the distribution function is $F_N^{(0)}(t) = 1 - e^{-\lambda t}$. This directly follows from formulas (6.2.2).

In the case where condition $\mathbf{E}_5''$ holds, the distributions of sojourn times, $F_i^{(0)}(t)$, are not concentrated in zero for any $i \neq 0$. This also follows from formulas (6.2.2).

6.2.4　Asymptotics of absorption probabilities in the pseudo-stationary case

Let us discuss in more details the question about the asymptotic behaviour of absorption probabilities in the pseudo-stationary case where conditions $\mathbf{T}_5$ and $\mathbf{E}_5'$ hold.

We say that positive functions $a_\varepsilon, \varepsilon > 0$, and $b_\varepsilon, \varepsilon > 0$, are *comparable*, if $a_\varepsilon/b_\varepsilon \to c$ as $\varepsilon \to 0$, where $0 \le c \le \infty$. We shall use the symbols $a_\varepsilon \prec b_\varepsilon$ if $c = 0$; $a_\varepsilon \preceq b_\varepsilon$ if $0 \le c < \infty$; and $a_\varepsilon \asymp b_\varepsilon$ if $0 < c < \infty$, respectively. Recall also that the notation $a_\varepsilon \sim b_\varepsilon$ was used in the case $c = 1$.

The following relations can be written for the absorption probabilities $_{i+1}f_{i0}^{(\varepsilon)}$, $i = 1, \ldots, N - 1$,

$$_{i+1}f_{i0}^{(\varepsilon)} = \bar{p}_i^{(\varepsilon)} + \sum_{k=1}^{i-1} p_{i-k}^{(\varepsilon)} {}_{i+1}f_{k+10}^{(\varepsilon)}. \tag{6.2.12}$$

Transitions from a state $1 \le i \le N - 1$ are possible with positive probabilities only to the states $j \le i + 1$. This obviously implies that, for any $1 \le i < r < j \le N$,

$$_0f_{ij}^{(\varepsilon)} = {}_0f_{ir}^{(\varepsilon)} {}_0f_{rj}^{(\varepsilon)}. \tag{6.2.13}$$

This relation implies the following formula for any $1 \le i < j \le N$:

$$_jf_{i0}^{(\varepsilon)} = {}_{i+1}f_{i0}^{(\varepsilon)} + \sum_{r=i+1}^{j-1} {}_0f_{ir}^{(\varepsilon)} {}_{r+1}f_{r0}^{(\varepsilon)}. \tag{6.2.14}$$

Using formula (6.2.14) on can rewrite equations (6.2.12) in the following form for $i = 1, \ldots, N - 1$:

$$_{i+1}f_{i0}^{(\varepsilon)} = \bar{p}_i^{(\varepsilon)} + \sum_{k=1}^{i-1} p_{i-k}^{(\varepsilon)} \big({}_{k+2}f_{k+10}^{(\varepsilon)} + \sum_{r=k+2}^{i} {}_0f_{k+1r}^{(\varepsilon)} {}_{r+1}f_{r0}^{(\varepsilon)} \big). \tag{6.2.15}$$

This relations can be rewritten in the following form for $i = 1, \ldots, N - 1$:

$$_{i+1}f_{i0}^{(\varepsilon)} = \frac{\bar{p}_i^{(\varepsilon)} + \sum_{k=1}^{i-2} p_{i-k}^{(\varepsilon)} \big({}_{k+2}f_{k+10}^{(\varepsilon)} + \sum_{r=k+2}^{i-1} {}_0f_{k+1r}^{(\varepsilon)} {}_{r+1}f_{r0}^{(\varepsilon)} \big)}{1 - p_1^{(\varepsilon)} - \sum_{k=1}^{i-2} p_{i-k}^{(\varepsilon)} {}_0f_{k+1i}^{(\varepsilon)}}. \tag{6.2.16}$$

If conditions $\mathbf{T}_5$ and $\mathbf{E}_5'$ hold, then $(\mathbf{r})$ $p_n^{(\varepsilon)} \to 0$ as $\varepsilon \to 0$ for $n \ge 1$; $(\mathbf{s})$ $_0f_{ij}^{(\varepsilon)} \to 1$ as $\varepsilon \to 0$ for $1 \le i < j \le N$; and $(\mathbf{t})$ $_jf_{i0}^{(\varepsilon)} \to 0$ as $\varepsilon \to 0$ for $1 \le i < j \le N$.

Using relations $(\mathbf{r})$–$(\mathbf{t})$, we get from (6.2.16) the following asymptotic relations for $i = 1, \ldots, N - 1$:

$$_{i+1}f_{i0}^{(\varepsilon)} \sim \bar{p}_i^{(\varepsilon)} + \sum_{k=1}^{i-2} p_{i-k}^{(\varepsilon)} \sum_{r=k+1}^{i-1} {}_{r+1}f_{r0}^{(\varepsilon)} \quad \text{as } \varepsilon \to 0. \tag{6.2.17}$$

By regrouping terms in the sum in the right-hand side of (6.2.17) we can rewrite this relation for $i = 1, \ldots, N - 1$ in the following equivalent form:

$$_{i+1}f_{i0}^{(\varepsilon)} \sim \bar{p}_i^{(\varepsilon)} + \sum_{k=2}^{i-1} {}_{k+1}f_{k0}^{(\varepsilon)} \sum_{r=i-k+1}^{i-1} p_r^{(\varepsilon)} \quad \text{as } \varepsilon \to 0. \tag{6.2.18}$$

In particular, the asymptotic relations (6.2.18) yield that $_2f_{10}^{(\varepsilon)} \sim \bar{p}_1^{(\varepsilon)}$; $_3f_{20}^{(\varepsilon)} \sim \bar{p}_2^{(\varepsilon)}$; $_4f_{30}^{(\varepsilon)} \sim \bar{p}_3^{(\varepsilon)} + {}_3f_{20}^{(\varepsilon)}p_2^{(\varepsilon)}$; $_5f_{40}^{(\varepsilon)} \sim \bar{p}_4^{(\varepsilon)} + {}_4f_{30}^{(\varepsilon)}(p_2^{(\varepsilon)} + p_3^{(\varepsilon)}) + {}_3f_{20}^{(\varepsilon)}p_3^{(\varepsilon)}$ as $\varepsilon \to 0$, etc.

The recurrence asymptotic relations (6.2.18) allow to make a desirable asymptotic analysis for absorption probabilities. Note that for the cyclic absorption probability, we have

$$_Nf_{N0}^{(\varepsilon)} = {}_Nf_{N-1\,0}^{(\varepsilon)}. \tag{6.2.19}$$

Let us introduce the following asymptotic order condition:

$\mathbf{O_{11}}$: $\bar{p}_k^{(\varepsilon)} p_r^{(\varepsilon)} \prec \bar{p}_i^{(\varepsilon)}$ for $2 \le k \le i - 1, i - k + 1 < r \le i - 1, 2 < i \le N - 1$.

Let us show that, under conditions $\mathbf{T_5}$, $\mathbf{E_5'}$, condition $\mathbf{O_{11}}$ is equivalent to the following asymptotic relations for absorption probabilities for $j = 1, \ldots, N - 1$:

$$_{j+1}f_{j0}^{(\varepsilon)} \sim \bar{p}_j^{(\varepsilon)} \quad \text{as } \varepsilon \to 0. \tag{6.2.20}$$

The asymptotic relations (6.2.18) directly imply that the asymptotic relation in (6.2.20) holds for $j = 1, 2$.

Let us first assume that condition $\mathbf{O_{11}}$ holds and that the asymptotic relation in (6.2.20) is true for $j = 1, \ldots, i - 1$, where $2 < i < N - 1$, and prove that, in this case, the asymptotic relation in (6.2.20) holds for $j = i$.

Indeed, the induction assumption made above and condition $\mathbf{O_{11}}$ permit to transform asymptotic relation (6.2.18) for $j = i$ to the following form:

$$_{i+1}f_{i0}^{(\varepsilon)} \sim \bar{p}_i^{(\varepsilon)} + \sum_{k=2}^{i-1} {}_{k+1}f_{k0}^{(\varepsilon)} \sum_{r=i-k+1}^{i-1} p_r^{(\varepsilon)}$$

$$\sim \bar{p}_i^{(\varepsilon)} + \sum_{k=2}^{i-1} \sum_{r=i-k+1}^{i-1} \bar{p}_k^{(\varepsilon)} p_r^{(\varepsilon)} \sim \bar{p}_i^{(\varepsilon)} \quad \text{as } \varepsilon \to 0. \tag{6.2.21}$$

Assume now that the asymptotic relation (6.2.20) holds for $j = 1, \ldots, N - 1$ and the asymptotic relation in condition $\mathbf{O_{11}}$ is verified for $i = 1, \ldots, j - 1$, where $2 < j < N - 1$, and prove that the asymptotic relation in condition $\mathbf{O_{11}}$ holds for $i = j$.

Indeed, the induction assumptions made above permit to transform the asymptotic relation (6.2.18) for $i = j$ to the following form:

$$\bar{p}_j^{(\varepsilon)} \sim {}_{j+1}f_{j0}^{(\varepsilon)} \sim \bar{p}_j^{(\varepsilon)} + \sum_{k=2}^{j-1} {}_{k+1}f_{k0}^{(\varepsilon)} \sum_{r=j-k+1}^{j-1} p_r^{(\varepsilon)}$$

$$\sim \bar{p}_j^{(\varepsilon)} + \sum_{k=2}^{j-1} \sum_{r=j-k+1}^{j-1} \bar{p}_k^{(\varepsilon)} p_r^{(\varepsilon)} \quad \text{as } \varepsilon \to 0, \tag{6.2.22}$$

and, therefore,

$$\sum_{k=2}^{j-1} \sum_{r=j-k+1}^{j-1} \bar{p}_k^{(\varepsilon)} p_r^{(\varepsilon)} \to 0 \quad \text{as } \varepsilon \to 0. \tag{6.2.23}$$

Relation (6.2.23) implies that the asymptotic relation in condition $\mathbf{O}_{11}$ holds for $i = j$.

Introduce the following asymptotic order condition:

$\mathbf{O}_{12}$: $\bar{p}_{i+1}^{(\varepsilon)} \prec \bar{p}_i^{(\varepsilon)}$, for $i = 1, \ldots, N - 1$.

Let us show that, under conditions $\mathbf{T}_5$, $\mathbf{E}_5'$, condition $\mathbf{O}_{12}$ is equivalent to the following condition:

$\mathbf{O}_{12}'$: $\bar{p}_i^{(\varepsilon)} \sim p_i^{(\varepsilon)}$ as $\varepsilon \to 0$, for $i = 1, \ldots, N - 1$.

Indeed, the identities $\bar{p}_i^{(\varepsilon)} = p_i^{(\varepsilon)} + \bar{p}_{i+1}^{(\varepsilon)}, i \geq 1$, condition $\mathbf{O}_{12}$ imply that $1 = p_i^{(\varepsilon)}/\bar{p}_i^{(\varepsilon)} + \bar{p}_{i+1}^{(\varepsilon)}/\bar{p}_i^{(\varepsilon)} \sim p_i^{(\varepsilon)}/\bar{p}_i^{(\varepsilon)}$ as $\varepsilon \to 0$, while condition $\mathbf{O}_{12}'$ implies that $1 = p_i^{(\varepsilon)}/\bar{p}_i^{(\varepsilon)} + \bar{p}_{i+1}^{(\varepsilon)}/\bar{p}_i^{(\varepsilon)} \sim 1 + \bar{p}_{i+1}^{(\varepsilon)}/\bar{p}_i^{(\varepsilon)}$ as $\varepsilon \to 0$ and, therefore, $\bar{p}_{i+1}^{(\varepsilon)}/\bar{p}_i^{(\varepsilon)} \to 0$ as $\varepsilon \to 0$.

Finally, under conditions $\mathbf{T}_5$, $\mathbf{E}_5'$, $\mathbf{O}_{11}$, and $\mathbf{O}_{12}$, we can rewrite the asymptotic relations (6.2.20) in the following simpler form for $i = 1, \ldots, N - 1$:

$$_{i+1}f_{i0}^{(\varepsilon)} \sim p_i^{(\varepsilon)} \quad \text{as } \varepsilon \to 0. \tag{6.2.24}$$

The following condition gives a typical example where both conditions $\mathbf{O}_{11}$ and $\mathbf{O}_{12}$ obviously hold:

$\mathbf{O}_{13}$: $\bar{p}_i^{(\varepsilon)} \sim \bar{p}_i \, \varepsilon^i$ as $\varepsilon \to 0$ for $i = 1, \ldots, N-1$, where $\bar{p}_i > 0$ for $i = 1, \ldots, N-1$.

Let us show that, under conditions $\mathbf{T}_5$, $\mathbf{E}_5'$, condition $\mathbf{O}_{13}$ is equivalent to the following asymptotic relations for absorption probabilities for $j = 1, \ldots, N - 1$:

$$_{j+1}f_{j0}^{(\varepsilon)} \sim \bar{p}_j \, \varepsilon^j \quad \text{as } \varepsilon \to 0, \tag{6.2.25}$$

where $\bar{p}_j > 0$, for $j = 1, \ldots, N - 1$.

The proofs is needed only for the converse statement that the asymptotic relations (6.2.25) for $j = 1, \ldots, N - 1$ imply condition $\mathbf{O}_{13}$.

Indeed, let us assume that the asymptotic relations (6.2.25) hold for $j = 1, \ldots, N - 1$, and that the asymptotic relation in condition $\mathbf{O}_{13}$ holds for $i = 1, \ldots, j - 1$, where $1 \leq j < N - 1$, and prove that the asymptotic relation in condition $\mathbf{O}_{13}$ holds for $i = j$.

The induction assumptions made above mean that $\bar{p}_i^{(\varepsilon)} \sim p_i^{(\varepsilon)} \sim \bar{p}_i \varepsilon^i$ as $\varepsilon \to 0$ for $i = 1, \ldots, j - 1$. This relations permits to transform the asymptotic relation (6.2.25) for $i = j$ to the following form:

$$\bar{p}_j \varepsilon^j \sim {}_{j+1}f_{j0}^{(\varepsilon)} \sim \bar{p}_j^{(\varepsilon)} + \sum_{k=2}^{j-1} {}_{k+1}f_{k0}^{(\varepsilon)} \sum_{r=j-k+1}^{j-1} p_r^{(\varepsilon)}$$

$$\sim \bar{p}_j^{(\varepsilon)} + \sum_{k=2}^{j-1} \sum_{r=j-k+1}^{j-1} \bar{p}_k \varepsilon^k \bar{p}_r \varepsilon^r \quad \text{as } \varepsilon \to 0. \tag{6.2.26}$$

Relation (6.2.26) obviously implies that

$$\bar{p}_j^{(\varepsilon)} \sim \bar{p}_j \varepsilon^j \quad \text{as } \varepsilon \to 0. \tag{6.2.27}$$

The following condition, which is more general than $\mathbf{O}_{13}$, also gives a typical example in which conditions $\mathbf{O}_{11}$ and $\mathbf{O}_{12}$ obviously hold:

$\mathbf{O}_{14}$: $\bar{p}_i^{(\varepsilon)} \sim \bar{p}_i \varepsilon^{h_i}$ as $\varepsilon \to 0$ for $i = 1, \ldots, N - 1$, where (a) $\bar{p}_i > 0$ for $i = 1, \ldots, N - 1$, and (b) $h_k + h_r > h_i$ if $k + r > i$ for $2 \leq k, r, i \leq N - 1$.

A proof similar to the one given above yields that, under conditions $\mathbf{T}_5$, $\mathbf{E}_5'$, condition $\mathbf{O}_{14}$ is equivalent to the following asymptotic relations for absorption probabilities for $j = 1, \ldots, N - 1$:

$$_{j+1}f_{j0}^{(\varepsilon)} \sim \bar{p}_j \varepsilon^{h_j} \quad \text{as } \varepsilon \to 0, \tag{6.2.28}$$

where $\bar{p}_j > 0$ for $j = 1, \ldots, N - 1$.

Let us consider a model in which conditions $\mathbf{T}_5$, $\mathbf{E}_5'$, $\mathbf{O}_{11}$, and $\mathbf{O}_{12}$ hold.

Assume that the following condition, which implies condition $\mathbf{T}_5$, holds:

$\mathbf{T}_6$: $G^{(\varepsilon)}(t) = G(t/\lambda^{(\varepsilon)})$, where (a) $G(t)$ is a proper distribution function concentrated on the interval $[0, \infty)$ and not concentrated in zero, and (b) $\lambda^{(\varepsilon)} > 0$ for $\varepsilon > 0$ and $\lambda^{(\varepsilon)} \to \lambda^{(0)} < \infty$ as $\varepsilon \to 0$.

There are two cases to consider. The first one corresponds to a model with asymptotically quick service, where the following condition holds:

$\mathbf{E}_6'$: $\lambda^{(0)} = 0$.

This is the case with pseudo-stationary asymptotics for the corresponding probabilities conditioned by non-absorption events.

The second case, which corresponds to a model with the usual service; here the following condition holds:

$\mathbf{E}_6''$: $\lambda^{(0)} > 0$.

The model described above can be interpreted in the following way. Let us change the time scale such that the time t is counted as the time $\tilde{t} = t/\lambda^{(\varepsilon)}$ in the new time scale. In the new time scale, the lifetime will take the value $\tilde{\mu}^{(\varepsilon)} = \mu^{(\varepsilon)}/\lambda^{(\varepsilon)}$. Correspondingly, the service time will have the distribution function $G(t)$ and the input flow of customers will be a Poisson flow with the parameter $\lambda\lambda^{(\varepsilon)}$. This flow has a lower intensity, if the parameter $\lambda^{(\varepsilon)}$ takes a small value. In particular the model can be interpreted as a queueing system with an asymptotically lower intensity of input flow if $\lambda^{(\varepsilon)} \to 0$ as $\varepsilon \to 0$.

In the model described above, the probabilities $p_i^{(\varepsilon)}$, $i = 0, 1, \ldots$, are given by the following formulas:

$$p_i^{(\varepsilon)} = \int_0^\infty \frac{(\lambda s)^i}{i!} e^{-\lambda s} G(ds/\lambda^{(\varepsilon)}) = (\lambda^{(\varepsilon)})^i \int_0^\infty \frac{(\lambda s)^i}{i!} e^{-\lambda\lambda^{(\varepsilon)}s} G(ds). \quad (6.2.29)$$

Let us consider the pseudo-stationary case, where condition $\mathbf{E}_6'$ holds. Denote

$$M_n = \int_0^\infty s^n G(ds), \ n = 1, 2, \ldots .$$

Let us also assume the following condition to hold:

$\mathbf{M}_{14}^{(N)}$: $M_N = \int_0^\infty s^N G(ds) < \infty$.

Conditions $\mathbf{T}_6$, $\mathbf{E}_6'$, and $\mathbf{M}_{14}^{(N)}$ imply that the following asymptotic relations hold for $i = 1, \ldots, N - 1$:

$$p_i^{(\varepsilon)} \sim (\lambda^{(\varepsilon)})^i \cdot \frac{\lambda^i M_i}{i!} \ \text{ as } \varepsilon \to 0. \quad (6.2.30)$$

Also, using a standard representation of the exponential function in the form of a power series, we get for $i = 1, \ldots, N - 1$ that

$$\begin{aligned}
\bar{p}_{i+1}^{(\varepsilon)} &= \int_0^\infty e^{-\lambda\lambda^{(\varepsilon)}s} \left(\frac{(\lambda\lambda^{(\varepsilon)}s)^{i+1}}{(i+1)!} + \frac{(\lambda\lambda^{(\varepsilon)}s)^{i+2}}{(i+2)!} + \cdots \right) G(ds) \\
&= \int_0^\infty \frac{(\lambda\lambda^{(\varepsilon)}s)^{i+1}}{(i+1)!} e^{-\lambda\lambda^{(\varepsilon)}s} \left(1 + \frac{(\lambda\lambda^{(\varepsilon)}s)}{(i+2)} + \cdots \right) G(ds) \\
&\leq \int_0^\infty \frac{(\lambda\lambda^{(\varepsilon)}s)^{i+1}}{(i+1)!} e^{-\lambda\lambda^{(\varepsilon)}s} \left(1 + \frac{(\lambda\lambda^{(\varepsilon)}s)}{1!} + \cdots \right) G(ds) \\
&= \int_0^\infty \frac{(\lambda\lambda^{(\varepsilon)}s)^{i+1}}{(i+1)!} G(ds) = (\lambda^{(\varepsilon)})^{i+1} \frac{\lambda^{i+1} M_{i+1}}{(i+1)!}. \quad (6.2.31)
\end{aligned}$$

Relations (6.2.30) and (6.2.31) imply in an obvious way that both conditions $\mathbf{O_{11}}$ and $\mathbf{O_{12}}$ hold.

Thus, under conditions $\mathbf{T_6}$, $\mathbf{E_6'}$, and $\mathbf{M_{14}^{(N)}}$, the following asymptotic relations hold for $i = 1, \ldots, N - 1$:

$$_{i+1}f_{i0}^{(\varepsilon)} \sim \bar{p}_i^{(\varepsilon)} \sim p_i^{(\varepsilon)} \sim (\lambda^{(\varepsilon)})^i \cdot \frac{\lambda^i M_i}{i!} \quad \text{as } \varepsilon \to 0. \tag{6.2.32}$$

Let us assume that, additionally to $\mathbf{T_6}$, $\mathbf{E_6'}$, and $\mathbf{M_{14}^{(N)}}$, the following condition holds:

$\mathbf{O_{15}}$: $\lambda^{(\varepsilon)} \sim \lambda_1 \varepsilon$ as $\varepsilon \to 0$, where $\lambda_1 > 0$.

In this case, the following asymptotic relations hold for $i = 1, \ldots, N - 1$:

$$_{i+1}f_{i0}^{(\varepsilon)} \sim \bar{p}_i^{(\varepsilon)} \sim (\lambda^{(\varepsilon)})^i \cdot \frac{\lambda^i M_i}{i!} \sim \frac{(\lambda_1 \lambda)^i M_i}{i!} \varepsilon^i \quad \text{as } \varepsilon \to 0. \tag{6.2.33}$$

Thus, condition $\mathbf{O_{13}}$ holds.

Let us now assume that, additionally to $\mathbf{T_6}$, $\mathbf{E_6'}$, and $\mathbf{M_{14}^{(N)}}$, the following condition holds:

$\mathbf{O_{16}}$: $\lambda^{(\varepsilon)} \sim \lambda_h \varepsilon^h \varepsilon$ as $\varepsilon \to 0$, where $\lambda_h > 0$ and h is a positive integer number.

In this case, the following asymptotic relations hold for $i = 1, \ldots, N - 1$:

$$_{i+1}f_{i0}^{(\varepsilon)} \sim \bar{p}_i^{(\varepsilon)} \sim (\lambda^{(\varepsilon)})^i \cdot \frac{\lambda^i M_i}{i!} \sim \frac{(\lambda_h \lambda)^i M_i}{i!} \varepsilon^{ih} \quad \text{as } \varepsilon \to 0. \tag{6.2.34}$$

Thus, condition $\mathbf{O_{14}}$ holds with the parameters $h_i = hi, i = 1, \ldots, N - 1$.

The following example shows that there are models in which the asymptotic order condition $\mathbf{O_{11}}$ holds but condition $\mathbf{O_{12}}$ does not.

Let us consider a model where the distribution $G^{(\varepsilon)}(t)$ has two atoms in the point 0 and $\lambda^{(\varepsilon)} > 0$ with the probabilities $1 - q^{(\varepsilon)}$ and $q^{(\varepsilon)} > 0$, respectively. We assume that (v) $q^{(\varepsilon)} \to q^{(0)} = 0$ as $\varepsilon \to 0$, and (w) $\lambda^{(\varepsilon)} \to \lambda^{(0)}$ as $\varepsilon \to 0$, where $0 \leq \lambda^{(0)} \leq \infty$. The assumption (v) obviously implies that conditions $\mathbf{T_5}$ and $\mathbf{E_5'}$ hold.

In this example, the probabilities $p_i^{(\varepsilon)}, i = 0, 1, \ldots$, are given by the following formulas:

$$p_i^{(\varepsilon)} = \frac{(\lambda \lambda^{(\varepsilon)})^i}{i!} e^{-\lambda \lambda^{(\varepsilon)}} q^{(\varepsilon)}. \tag{6.2.35}$$

Let $i < j + r, 1 \leq i, j, r \leq N - 1$. The following asymptotic estimate takes place:

$$\frac{\bar{p}_j^{(\varepsilon)} p_r^{(\varepsilon)}}{\bar{p}_i^{(\varepsilon)}} = \frac{\left(\sum_{k=j}^{i-1} \frac{(\lambda \lambda^{(\varepsilon)})^k}{k!} + \sum_{k=i}^{\infty} \frac{(\lambda \lambda^{(\varepsilon)})^k}{k!} \right) e^{-\lambda \lambda^{(\varepsilon)}} q^{(\varepsilon)} \cdot \frac{(\lambda \lambda^{(\varepsilon)})^r}{r!} e^{-\lambda \lambda^{(\varepsilon)}} q^{(\varepsilon)}}{\sum_{k=i}^{\infty} \frac{(\lambda \lambda^{(\varepsilon)})^k}{k!} e^{-\lambda \lambda^{(\varepsilon)}} q^{(\varepsilon)}}$$

$$\leq e^{-\lambda \lambda^{(\varepsilon)}} q^{(\varepsilon)} \lambda \lambda^{(\varepsilon)} \sum_{k=j}^{i-1} \frac{(\lambda \lambda^{(\varepsilon)})^{k+r-i-1} i!}{k! r!}$$

$$+ e^{-\lambda \lambda^{(\varepsilon)}} q^{(\varepsilon)} \frac{(\lambda \lambda^{(\varepsilon)})^r}{r!} \to 0 \quad \text{as } \varepsilon \to 0. \tag{6.2.36}$$

Indeed, the expression in the right-hand side of (6.2.36) tends to 0 as $\varepsilon \to 0$ whatever value the limit $0 \leq \lambda^{(0)} \leq \infty$ takes. In case $\lambda^{(0)} = 0$, this is true because $q^{(\varepsilon)} \to 0$ and $\lambda^{(\varepsilon)} \to 0$ as $\varepsilon \to 0$, while $e^{-\lambda\lambda^{(\varepsilon)}} \to 1$ as $\varepsilon \to 0$. Note also that all power indices in the sum in the right-hand side of (6.2.36) are $k + r - i - 1 \geq 0$ if $j \leq k \leq i - 1$. If $0 < \lambda^{(0)} < \infty$, this can be seen, since $q^{(\varepsilon)} \to 0$ and $\lambda^{(\varepsilon)} \to \lambda^{(0)}$ as $\varepsilon \to 0$. In case $\lambda^{(0)} = \infty$, this is true, since $q^{(\varepsilon)} \to 0$ and $e^{-\lambda\lambda^{(\varepsilon)}}(\lambda^{(\varepsilon)})^{k+r} \to 0$ as $\varepsilon \to 0$.

Relation (6.2.36) implies that condition $\mathbf{O}_{11}$ holds and, therefore, the following asymptotic relations take place for $i = 1, \ldots, N - 1$:

$$_{i+1}f_{i0}^{(\varepsilon)} \sim \bar{p}_i^{(\varepsilon)} \quad \text{as } \varepsilon \to 0. \tag{6.2.37}$$

If $\lambda^{(0)} = 0$, then for $i \geq 1$,

$$p_i^{(\varepsilon)} \sim (\lambda^{(\varepsilon)})^i q^{(\varepsilon)} \cdot \frac{\lambda^i}{i!} \quad \text{as } \varepsilon \to 0, \tag{6.2.38}$$

and

$$\begin{aligned}
\bar{p}_{i+1}^{(\varepsilon)} &= p_{i+1}^{(\varepsilon)} + p_{i+2}^{(\varepsilon)} + \cdots \\
&= (\lambda^{(\varepsilon)})^{i+1} q^{(\varepsilon)} \frac{\lambda^{i+1}}{(i+1)!}\left(1 + \frac{\lambda\lambda^{(\varepsilon)}}{i+2} + \cdots\right)e^{-\lambda\lambda^{(\varepsilon)}} \\
&\leq (\lambda^{(\varepsilon)})^{i+1} q^{(\varepsilon)} \frac{\lambda^{i+1}}{(i+1)!}.
\end{aligned} \tag{6.2.39}$$

Therefore, condition $\mathbf{O}_{12}$ also holds and the following asymptotic relation takes place for $i = 1, \ldots, N - 1$:

$$_{i+1}f_{i0}^{(\varepsilon)} \sim \bar{p}_i^{(\varepsilon)} \sim p_i^{(\varepsilon)} \sim (\lambda^{(\varepsilon)})^i q^{(\varepsilon)} \cdot \frac{\lambda^i}{i!} \quad \text{as } \varepsilon \to 0. \tag{6.2.40}$$

If $0 < \lambda^{(0)} < \infty$, then, for $i \geq 1$,

$$p_i^{(\varepsilon)} \sim q^{(\varepsilon)} \cdot \frac{(\lambda\lambda^{(0)})^i}{i!} e^{-\lambda\lambda^{(0)}} \quad \text{as } \varepsilon \to 0, \tag{6.2.41}$$

while

$$\bar{p}_i^{(\varepsilon)} = \sum_{j=i}^{\infty} \frac{(\lambda\lambda^{(\varepsilon)})^j}{j!} e^{-\lambda\lambda^{(\varepsilon)}} q^{(\varepsilon)} \sim q^{(\varepsilon)} \cdot M_i^{(0)} \quad \text{as } \varepsilon \to 0, \tag{6.2.42}$$

where

$$M_i^{(\varepsilon)} = \sum_{j=i}^{\infty} \frac{(\lambda\lambda^{(\varepsilon)})^j}{j!} e^{-\lambda\lambda^{(\varepsilon)}}.$$

Relation (6.2.42) implies that condition $\mathbf{O_{12}}$ does not hold, since all the probabilities $\bar{p}_i^{(\varepsilon)}$, $i \geq 1$, have the same asymptotic order. Therefore, the following asymptotic relations take place for $i = 1, \ldots, N - 1$:

$$_{i+1}f_{i0}^{(\varepsilon)} \sim \bar{p}_i^{(\varepsilon)} \sim q^{(\varepsilon)} \cdot M_i^{(0)} \quad \text{as } \varepsilon \to 0. \tag{6.2.43}$$

If $\lambda^{(0)} = \infty$, then, for $i \geq 1$,

$$p_i^{(\varepsilon)} \sim (\lambda^{(\varepsilon)})^i e^{-\lambda\lambda^{(\varepsilon)}} q^{(\varepsilon)} \cdot \frac{\lambda^i}{i!} \quad \text{as } \varepsilon \to 0, \tag{6.2.44}$$

while

$$\bar{p}_i^{(\varepsilon)} = \sum_{j=i}^{\infty} \frac{(\lambda\lambda^{(\varepsilon)})^j}{j!} e^{-\lambda\lambda^{(\varepsilon)}} q^{(\varepsilon)} \sim (\lambda^{(\varepsilon)})^i e^{-\lambda\lambda^{(\varepsilon)}} q^{(\varepsilon)} \bar{M}_i^{(\varepsilon)} \quad \text{as } \varepsilon \to 0, \tag{6.2.45}$$

where

$$\bar{M}_i^{(\varepsilon)} = \sum_{j=i}^{\infty} \frac{(\lambda\lambda^{(\varepsilon)})^{j-i}}{j!} \to \infty \quad \text{as } \varepsilon \to 0.$$

Relations (6.2.44) and (6.2.45) imply that condition $\mathbf{O_{12}}$ does not hold, because $p_i^{(\varepsilon)} \prec \bar{p}_i^{(\varepsilon)}$ for $i \geq 1$, since $p_i^{(\varepsilon)}/\bar{p}_i^{(\varepsilon)} \to 0$ as $\varepsilon \to 0$ for $i \geq 1$.

Moreover, in this case, for $i \geq 1$,

$$\bar{p}_i^{(\varepsilon)} \sim \bar{p}_{i+1}^{(\varepsilon)} \quad \text{as } \varepsilon \to 0. \tag{6.2.46}$$

Indeed, in this case the identities $\bar{p}_i^{(\varepsilon)} = p_i^{(\varepsilon)} + \bar{p}_{i+1}^{(\varepsilon)}$, $i \geq 1$, imply that $1 = p_i^{(\varepsilon)}/\bar{p}_i^{(\varepsilon)} + \bar{p}_{i+1}^{(\varepsilon)}/\bar{p}_i^{(\varepsilon)} \sim \bar{p}_{i+1}^{(\varepsilon)}/\bar{p}_i^{(\varepsilon)}$ as $\varepsilon \to 0$.

Thus all the probabilities $\bar{p}_i^{(\varepsilon)}$, $i \geq 1$, have the same asymptotic order, moreover, they are asymptotically equivalent. Therefore, the following asymptotic relations take place for $i = 1, \ldots, N - 1$:

$$_{i+1}f_{i0}^{(\varepsilon)} \sim \bar{p}_1^{(\varepsilon)} \sim (\lambda^{(\varepsilon)}) e^{-\lambda\lambda^{(\varepsilon)}} q^{(\varepsilon)} \bar{M}_1^{(\varepsilon)} \quad \text{as } \varepsilon \to 0. \tag{6.2.47}$$

6.2.5   Conditions for mixed ergodic and limit/large deviation theorems for imbedded semi-Markov processes

Let us assume that condition $\mathbf{T_5}$ and either condition $\mathbf{E_5'}$ or $\mathbf{E_5''}$ hold.

As was pointed above, these conditions imply that condition $\mathbf{T_1}$ holds for the imbedded semi-Markov process $\eta^{(\varepsilon)}(t)$, $t \geq 0$.

Conditions $\mathbf{T_5}$ and $\mathbf{E_5'}$ correspond to the pseudo-stationary case. In this case, the set of recurrent-without-absorption states is $X_1^{(0)} = \{2, \ldots, N\}$ and the set of non-recurrent-without-absorption states is $X_2^{(0)} = \{1\}$.

Conditions $\mathbf{T_5}$ and $\mathbf{E_5''}$ correspond to the quasi-stationary case. Here, the set of recurrent-without-absorption states is $X_1^{(0)} = \{1, \ldots, N\}$ and, for the set of non-recurrent-without-absorption states, we have $X_2^{(0)} = \varnothing$.

Condition $\mathbf{I_4}$ holds, since the distribution function $F_N^{(\varepsilon)}(t) = 1 - e^{-\lambda t}$ of sojourn time in state N is exponential with parameter λ for every $\varepsilon \geq 0$. Therefore, the distribution ${}_0 G_{NN}^{(\varepsilon)}(t)$ is not concentrated in zero. Moreover, it is obvious that ${}_0 G_{NN}^{(\varepsilon)}(0) = F_N^{(\varepsilon)}(0) {}_0 G_{N-1N}^{(\varepsilon)}(0) = 0$.

Conditions $\mathbf{N_1}$ and $\mathbf{N_2}$ also hold, since the distribution function $F_N^{(\varepsilon)}(t) = 1 - e^{-\lambda t}$ of sojourn time in state N is exponential with parameter λ for every $\varepsilon \geq 0$.

The power moments $p_{ij}^{(\varepsilon)}[n]$, $n = 0, 1, \ldots$, take the following form:

$$
p_{ij}^{(\varepsilon)}[n] = \int_0^\infty s^n Q_{ij}^{(\varepsilon)}(ds)
$$

$$
= \begin{cases}
p_{i-j+1}^{(\varepsilon)}[n] & \text{if } 1 \leq i \leq N-1, 2 \leq j \leq i+1, \\
\bar{p}_{i-j}^{(\varepsilon)}[n] & \text{if } 1 \leq i \leq N-1, j = 0, \\
n!/\lambda^n & \text{if } i = N, j = N-1, \\
0 & \text{otherwise,}
\end{cases}
\tag{6.2.48}
$$

where

$$
p_k^{(\varepsilon)}[n] = \int_0^\infty s^n \frac{(\lambda s)^k}{k!} e^{-\lambda s} G^{(\varepsilon)}(ds)
$$

$$
= \frac{(k+n)!}{\lambda^n k!} p_{k+n}^{(\varepsilon)}(\lambda) < \infty, \quad n = 0, 1, \ldots,
\tag{6.2.49}
$$

and

$$
\bar{p}_{k+1}^{(\varepsilon)}[n] = \int_0^\infty s^n \frac{\lambda^{k+1} s^k}{k!} e^{-\lambda s}(1 - G^{(\varepsilon)}(s))\, ds = \frac{(k+n)!}{\lambda^n k!} \bar{p}_{k+n+1}^{(\varepsilon)}(\lambda)
$$

$$
= \frac{(k+n)!}{\lambda^n k!} \left(1 - \sum_{r=0}^{k+n} p_r^{(\varepsilon)}(\lambda)\right) < \infty, \quad n = 0, 1, \ldots.
\tag{6.2.50}
$$

These formulas imply that moment conditions $\mathbf{M_{11}}$ holds and condition $\mathbf{M_{13}^{(k)}}$ also holds for every $k \geq 1$.

The mixed power-exponential moments $p_{ij}^{(\varepsilon)}[\rho, n]$, $n = 0, 1, \ldots$, take the following form for $\rho < \lambda$:

$$p_{ij}^{(\varepsilon)}[\rho, n] = \int_0^\infty s^n e^{s\rho} Q_{ij}^{(\varepsilon)}(ds)$$

$$= \begin{cases} p_{i-j+1}^{(\varepsilon)}[\rho, n] & \text{if } 1 \le i \le N-1, 2 \le j \le i+1, \\ \bar{p}_{i-j}^{(\varepsilon)}[\rho, n] & \text{if } 1 \le i \le N-1, j = 0, \\ n!\lambda/(\lambda - \rho)^{n+1} & \text{if } i = N, j = N-1, \\ 0 & \text{otherwise}, \end{cases} \qquad (6.2.51)$$

where

$$p_r^{(\varepsilon)}[\rho, n] = \int_0^\infty s^n e^{s\rho} \frac{(\lambda s)^r}{r!} e^{-\lambda s} G^{(\varepsilon)}(ds)$$

$$= \frac{\lambda^r (r+n)!}{(\lambda - \rho)^{r+n} r!} p_{r+n}^{(\varepsilon)}(\lambda - \rho) < \infty, \qquad r, n = 0, 1, \ldots, \qquad (6.2.52)$$

and

$$\bar{p}_{r+1}^{(\varepsilon)}[\rho, n] = \int_0^\infty s^n e^{s\rho} \frac{\lambda^{r+1} s^r}{r!} e^{-\lambda s} (1 - G^{(\varepsilon)}(s)) \, ds$$

$$= \frac{\lambda^{r+1}(r+n)!}{(\lambda - \rho)^{r+n+1} r!} \bar{p}_{r+n+1}^{(\varepsilon)}(\lambda - \rho)$$

$$= \frac{\lambda^{r+1}(r+n)!}{(\lambda - \rho)^{r+n+1} r!} \left(1 - \sum_{l=0}^{r+n} p_l^{(\varepsilon)}(\lambda - \rho)\right) < \infty, \qquad r, n = 0, 1, \ldots . \qquad (6.2.53)$$

As was mentioned above, conditions $\mathbf{T_5}$ and $\mathbf{E_5'}$ correspond to the pseudo-stationary case.

Formulas (6.2.51), (6.2.52), and (6.2.53) imply that condition $\mathbf{C_{12}}$ holds for any $0 < \delta < \lambda$. In this case, as follows from Lemma 4.5.8, condition $\mathbf{C_{13}}$ also holds.

Also, as was mentioned above, conditions $\mathbf{T_5}$ and $\mathbf{E_5''}$ correspond to the quasi-stationary case.

Let us show that conditions $\mathbf{G_1}$ and $\mathbf{G_2}$ hold with the parameter $\rho^* = \lambda$.

In the quasi-stationary case, the set of recurrent-without-absorption states is $X_1^{(0)} = \{2, \ldots, N\}$, and the set of non-recurrent-without-absorption states is $X_2^{(0)} = \{1\}$.

In this case, as follows from formula (6.2.52), **(a)** $p_{ij}^{(0)}[\rho, 0] = p_{i-j+1}^{(0)}[\rho, 0] < \infty$, $\rho < \lambda$ for $1 \le i \le N-1, 2 \le j \le i+1$; **(b)** $p_{NN-1}^{(0)}[\rho, 0] = \lambda/(\lambda - \rho) < \infty$, $\rho < \lambda$; **(c)** $p_{i0}^{(0)}[\rho, 0] = \bar{p}_i^{(0)}[\rho, 0] < \infty$ for $1 \le i \le N-1$; and **(d)** $p_{NN-1}^{(0)}[\rho, 0] = \lambda/(\lambda - \rho) \to \infty$ as $\rho < \lambda$, $\rho \to \lambda$. Relations **(a)**–**(d)** imply that condition $\mathbf{G_1}$ holds with the parameter $\rho^* = \lambda$.

Also, relations (6.2.51), (6.2.52), and (6.2.53) imply that **(e)** $p_{ij}^{(\varepsilon)}[\rho, 0] \to p_{ij}^{(0)}[\rho, 0]$ $< \infty$ as $\varepsilon \to 0$ for $\rho < \lambda$ and $i \neq 0$, $j \in X$. Relation **(e)** obviously implies that condition $\mathbf{G_2}$ **(a)** holds with the parameter $\rho^* = \lambda$. Since, in this case, $_0\varphi_{12}^{(0)}(\rho) = p_{12}^{(0)}[\rho, 0] < \infty$, condition $\mathbf{G_2}$ **(b)** also holds with the parameter $\rho^* = \lambda$.

As was shown in Lemma 4.5.10, condition $\mathbf{C_{13}}$ is implied by conditions $\mathbf{G_1}$ and $\mathbf{G_2}$. In this case, the characteristic root $0 < \rho^{(0)} < \rho^* = \lambda$.

Therefore, by Lemma 4.5.10, **(e)** conditions $\mathbf{C_{12}}$ and $\mathbf{C_{13}}$ hold for the state $i = N$ and some $0 < \beta_N = \beta < \delta < \rho^* = \lambda$ for the imbedded semi-Markov processes $\eta^{(\varepsilon)}(t)$, $t \geq 0$.

Also, by Lemma 4.5.10, **(f)** conditions $\mathbf{C_{14}}$–$\mathbf{C_{16}}$ hold for the state $i = N$ and some $0 < \beta_N = \beta < \delta < \rho^* = \lambda$ and $\varepsilon_7', \varepsilon_8', \varepsilon_{15}' > 0$ (the parameters used in these conditions) for the imbedded semi-Markov processes $\eta^{(\varepsilon)}(t)$, $t \geq 0$.

6.2.6   Conditions for mixed ergodic and limit/large deviation theorems for the regenerative processes $\xi^{(\varepsilon)}(t)$

As was pointed out in Subsection 6.2.2, the subsequent moments of jumps from the state $N - 1$ to the state N are regenerative moments for both processes $\eta^{(\varepsilon)}(t)$ and $\xi^{(\varepsilon)}(t)$.

This implies that the distribution functions $_0G_{NN}^{(\varepsilon)}(t)$ and $G_{NN}^{(\varepsilon)}(t)$, for the imbedded semi-Markov process $\eta^{(\varepsilon)}(t)$, coincide, respectively, with the distribution functions $F^{(\varepsilon)}(t)$ and $\hat{F}^{(\varepsilon)}(t)$ for the regenerative process $\xi^{(\varepsilon)}(t)$.

Also, if the initial state is $\xi^{(\varepsilon)}(t) = i \neq 0, N$, then the distribution functions $_0G_{iN}^{(\varepsilon)}(t)$ and $G_{iN}^{(\varepsilon)}(t)$ for the imbedded semi-Markov process $\eta^{(\varepsilon)}(t)$ coincide, respectively, with the distribution functions $\tilde{F}^{(\varepsilon)}(t)$ and $\hat{\tilde{F}}^{(\varepsilon)}(t)$ for the regenerative process $\xi^{(\varepsilon)}(t)$.

In Chapter 4 it was shown in what way the conditions used in Chapter 3 for regenerative processes are employed as the corresponding conditions used in Chapter 4 for semi-Markov processes when considering these processes as regenerative processes with return times in some recurrent state that is regarded as regeneration times.

By applying these results to the semi-Markov processes $\xi^{(\varepsilon)}(t)$ we get that $\mathbf{T_5}$ and $\mathbf{E_5'}$ imply that conditions $\mathbf{D_{13}}$, $\mathbf{M_8}$, and $\mathbf{D_{17}}$–$\mathbf{D_{20}}$ hold as well as conditions $\mathbf{C_3}$ and $\mathbf{C_8}$. Also, conditions $\mathbf{T_5}$ imply that conditions $\mathbf{D_{21}}$, $\mathbf{C_6}$ hold as well as conditions $\mathbf{D_{23}}$, $\mathbf{C_{10}}$, and $\mathbf{C_{11}}$.

The only conditions that require special considerations are $\mathbf{F_7}$–$\mathbf{F_9}$. This is because the forcing functions $\mathsf{P}_N\{\xi^{(\varepsilon)}(t) = j, \tau_1^{(\varepsilon)} \wedge \mu^{(\varepsilon)} > t\}$, $t \geq 0$, and $\mathsf{P}_N\{\eta^{(\varepsilon)}(t) = j, \tau_1^{(\varepsilon)} \wedge \mu^{(\varepsilon)} > t\}$, $t \geq 0$, for the corresponding renewal equations, which can be written for distributions of the regenerative process $\xi^{(\varepsilon)}(t)$ and the imbedded semi-Markov process $\eta^{(\varepsilon)}(t)$, are not the same.

For $i, j = 1, \ldots, N$, denote

$$q_{ij}^{(\varepsilon)}(t) = \mathsf{P}_i\{\xi^{(\varepsilon)}(t) = j, \tau_1^{(\varepsilon)} \wedge \mu^{(\varepsilon)} > t\}, \quad t \geq 0.$$

In order to check condition $\mathbf{F_7}$, we shall prove that **(a)** condition $\mathbf{T_5}$ implies the local uniform convergence of the functions $q_{ij}^{(\varepsilon)}(t)$ to the function $q_{ij}^{(0)}(t)$ as $\varepsilon \to 0$ almost everywhere with respect to the Lebesgue measure on $[0, \infty)$ for every $i, j = 1, \ldots, N$.

Let us use the random variables $\alpha_0^{(\varepsilon)} = 0$ and $\alpha_n^{(\varepsilon)}$, $n = 1, \ldots$, introduced in Subsection 6.2.2. Also define the random variables $\vartheta_n^{(\varepsilon)} = \alpha_1^{(\varepsilon)} + \cdots + \alpha_n^{(\varepsilon)}$, $n = 0, 1, \ldots$, and also let $v_{0N}^{(\varepsilon)} = \min(n \geq 1 : \xi^{(\varepsilon)}(\vartheta_n^{(\varepsilon)}) \in \{0, N\})$.

Using these random variables we can write the following formula for the probabilities $q_{ij}^{(\varepsilon)}(t)$ for $i = 1, \ldots, N$, $j = 1, \ldots, N-1$:

$$q_{ij}^{(\varepsilon)}(t) = \sum_{k=j}^{N-1} \sum_{n=0}^{\infty} \mathsf{P}_i\{v_{0N}^{(\varepsilon)} > n, \vartheta_n^{(\varepsilon)} \leq t, \xi^{(\varepsilon)}(\vartheta_n^{(\varepsilon)}) = k,$$

$$\vartheta_{n+1}^{(\varepsilon)} > t, \zeta^{(\varepsilon)}(t) - \zeta^{(\varepsilon)}(\vartheta_n^{(\varepsilon)}) = k - j\}, \quad t \geq 0. \quad (6.2.54)$$

Also, for $i = 1, \ldots, N$, $j = N$,

$$q_{iN}^{(\varepsilon)}(t) = \delta(i, N)e^{-\lambda t}, \quad t \geq 0. \quad (6.2.55)$$

Formulas (6.2.54) and (6.2.55) show that the probability $q_{ij}^{(\varepsilon)}(t)$, if regarded as a function of $t \geq 0$, has not more than a countable set $R_{ij}^{(\varepsilon)}$ of discontinuity points for every $i, j = 1, \ldots, N$.

According to the remarks made in Subsections 3.2.3 and 3.3.3, this implies that conditions $\mathbf{F_8}$ and $\mathbf{F_9}$ hold, since conditions $\mathbf{M_8}$ and $\mathbf{C_6}$ hold.

Let us prove that **(b)** condition $\mathbf{T_5}$ implies the local uniform convergence of the functions $q_{ij}^{(\varepsilon)}(t)$ to the function $q_{ij}^{(0)}(t)$ as $\varepsilon \to 0$, $t \in S_{ij}^{(0)}$, for every $i, j = 1, \ldots, N$, where $\bar{S}_{ij}^{(0)} = [0, \infty) \setminus S_{ij}^{(0)}$, $i, j = 1, \ldots, N$, are some at most countable subsets of $[0, \infty)$.

Since the sets $\bar{S}_{ij}^{(0)}$, $i, j = 1, \ldots, N$, are at most countable, statement **(b)** implies **(a)**.

Formula (6.2.55) obviously implies that statement **(b)** holds for $i = 1, \ldots, N$, $j = N$.

The following obvious formula also permits to reduce the proof of statement **(b)** for $i = N$, $j = 1, \ldots, N-1$ to the case where $i = N-1$, $j = 1, \ldots, N-1$,

$$q_{Nj}^{(\varepsilon)}(t) = \int_0^t q_{N-1j}^{(\varepsilon)}(t - s)\lambda e^{-\lambda s}\, ds, \quad t \geq 0. \quad (6.2.56)$$

Statement **(b)** for the functions $q_{N-1j}^{(\varepsilon)}(t)$, $t \geq 0$, means that there exists a set $S_{N-1j}^{(0)}$ such that its complement, $\bar{S}_{N-1j}^{(0)}$, is a countable set and, for any $t_\varepsilon \to t_0$ as $\varepsilon \to 0$, where $t_0 \in S_{N-1j}^{(0)}$, we have $q_{N-1j}^{(\varepsilon)}(t_\varepsilon) \to q_{N-1j}^{(0)}(t_0)$ as $\varepsilon \to 0$.

Let us choose $T > t_0$. There exists $\varepsilon_0 > 0$ such that $t_\varepsilon \leq T$ for $\varepsilon \leq \varepsilon_0$. Let us also define functions $q_{N-1\,j}^{(\varepsilon)}(s) = 0$ for $s < 0$.

Now, by applying formula (6.2.56), the Lebesgue theorem, and statement **(b)** to the functions $q_{N-1\,j}^{(\varepsilon)}(t)$, we get

$$q_{Nj}^{(\varepsilon)}(t_\varepsilon) = \int_0^T q_{N-1\,j}^{(\varepsilon)}(t_\varepsilon - s)\lambda e^{-\lambda s}\,ds$$

$$\rightarrow \int_0^T q_{N-1\,j}^{(0)}(t_0 - s)\lambda e^{-\lambda s}\,ds = q_{Nj}^{(0)}(t_0) \quad \text{as } \varepsilon \to 0. \tag{6.2.57}$$

Thus statement **(b)** also holds for the functions $q_{Nj}^{(\varepsilon)}(t)$, $t \geq 0$. The set $S_{Nj}^{(0)} = S_{N-1\,j}^{(0)}$ can be taken as a corresponding set of local uniform convergence.

Therefore, it only remains to consider the cases $i, j = 1, \dots, N - 1$.

Let us first consider a more simple case where conditions $\mathbf{T_5}$ and $\mathbf{E_5'}$ hold.

In this case, the following simple estimate can be used for $i, j = 1, \dots, N - 1$:

$$q_{ij}^{(\varepsilon)}(t) \leq \mathsf{P}_i\{\mu^{(\varepsilon)} \wedge \tau_1^{(\varepsilon)} > t\} = 1 - G_{iN}^{(\varepsilon)}(t), \quad t \geq 0. \tag{6.2.58}$$

As was pointed out in Subsection 6.2.2, conditions $\mathbf{T_5}$ and $\mathbf{E_5'}$ imply that all limit distributions of sojourn times, $F_j^{(0)}(t)$, $j = 1, \dots, N - 1$, are concentrated in 0. Therefore, the distributions $G_{iN}^{(\varepsilon)}(t)$, $i = 1, \dots, N - 1$, are also concentrated in 0, i.e., $1 - G_{iN}^{(\varepsilon)}(t) = 0$, $t \geq 0$, for $i = 1, \dots, N - 1$. On the other hand, by Lemma 4.2.3, $G_{iN}^{(\varepsilon)}(\cdot) \Rightarrow G_{iN}^{(0)}(\cdot)$ as $\varepsilon \to 0$ for $i = 1, \dots, N - 1$ that, in this case, is equivalent to the relation $1 - G_{iN}^{(\varepsilon)}(t) \to 1 - G_{iN}^{(0)}(t) = 0$ as $\varepsilon \to 0$ for $t > 0$ and $i = 1, \dots, N - 1$. Since the functions $1 - G_{iN}^{(\varepsilon)}(t)$ are non-increasing, the last relation implies that **(c)** the functions $1 - G_{iN}^{(\varepsilon)}(t)$ locally uniformly converge to the function $1 - G_{iN}^{(0)}(t) = 0$ as $\varepsilon \to 0$ for any $t > 0$ and $i = 1, \dots, N - 1$.

Due to estimate (6.2.58), statement **(c)** implies that **(d)** the functions $q_{ij}^{(\varepsilon)}(t)$ locally uniformly converge to the function $q_{ij}^{(0)}(t) = 0$ as $\varepsilon \to 0$ for any $t > 0$ and $i, j = 1, \dots, N - 1$. Thus statement **(b)** holds true with the sets $S_{ij}^{(0)} = (0, \infty)$, $i, j = 1, \dots, N - 1$.

Let us now consider a more complex case where conditions $\mathbf{T_5}$ and $\mathbf{E_5''}$ hold. Introduce Markov renewal functions for $i, j = 1, \dots, N - 1$ by

$$U_{ij}^{(\varepsilon)}(t) = \sum_{n=0}^{\infty} U_{ij}^{(\varepsilon)}(n, t), \quad t \geq 0,$$

where

$$U_{ij}^{(\varepsilon)}(n, t) = \mathsf{P}_i\{\nu_{0N}^{(\varepsilon)} > n, \vartheta_n^{(\varepsilon)} \leq t, \xi^{(\varepsilon)}(\vartheta_n^{(\varepsilon)}) = j\}, \quad t \geq 0.$$

By the definition, for $i, j = 1, \ldots, N - 1$,

$$U_{ij}^{(\varepsilon)}(0, t) = \delta(i, j), \quad U_{ij}^{(\varepsilon)}(1, t) = Q_{ij}^{(\varepsilon)}(t), \quad t \geq 0. \tag{6.2.59}$$

Also, the functions $U_{ij}^{(\varepsilon)}(n, t)$ satisfy the following recurrence convolution relations for $i, j = 1, \ldots, N - 1$ and $n = 1, 2, \ldots$:

$$U_{ij}^{(\varepsilon)}(n, t) = \sum_{k=1}^{N-1} \int_0^t U_{kj}^{(\varepsilon)}(n - 1, t - s) Q_{ik}^{(\varepsilon)}(ds), \quad t \geq 0. \tag{6.2.60}$$

Relations (6.2.11), (6.2.59), and (6.2.60) imply that the following weak convergence relations hold for the functions $U_{ij}^{(\varepsilon)}(n, t)$ (which are, in fact, improper distribution functions in the argument $t \geq 0$), for $n = 0, 1, \ldots, i, j = 1, \ldots, N - 1$:

$$U_{ij}^{(\varepsilon)}(n, \cdot) \Rightarrow U_{ij}^{(0)}(n, \cdot) \quad \text{as } \varepsilon \to 0. \tag{6.2.61}$$

Let us denote by $\nu^{(\varepsilon)}(t) = \max(n : \varsigma_1^{(\varepsilon)} + \cdots + \varsigma_n^{(\varepsilon)} \leq t), t \geq 0$, the renewal process with inter-renewal times $\varsigma_n^{(\varepsilon)}$, $n = 1, 2, \ldots$, which are subsequent service times.

Note that **(m)** in the case where the initial state is $\xi^{(\varepsilon)}(0) = i \neq 0, N$, the random event satisfies $\{\nu_{0N}^{(\varepsilon)} > n, \vartheta_n^{(\varepsilon)} \leq t\} \subseteq \{\varsigma_1^{(\varepsilon)} + \cdots + \varsigma_n^{(\varepsilon)} \leq t\} = \{\nu^{(\varepsilon)}(t) \geq n\}$ for $t \geq 0$ and $n = 0, 1, \ldots$.

The following asymptotic estimate follows from conditions $\mathbf{T_5}$ and $\mathbf{E_5''}$, the corresponding estimate (1.2.55) for moments of the renewal process, and relations **(m)** for $t \geq 0$ and $n = 0, 1, \ldots$:

$$\lim_{n \to \infty} \overline{\lim_{\varepsilon \to 0}} \sum_{k=n}^{\infty} U_{ij}^{(\varepsilon)}(n, t) \leq \lim_{n \to \infty} \overline{\lim_{\varepsilon \to 0}} \sum_{k=n}^{\infty} \mathsf{P}\{\nu^{(\varepsilon)}(t) \geq k\}$$

$$\leq \overline{\lim_{\varepsilon \to 0}} \, \mathsf{E}\nu^{(\varepsilon)}(t)^2 \lim_{n \to \infty} \sum_{k=n}^{\infty} 1/k^2 = 0. \tag{6.2.62}$$

Relations (6.2.61) and (6.2.62) imply that the renewal functions $U_{ij}^{(\varepsilon)}(t)$ generate, on the Borel σ-algebra of interval $[0, \infty)$, measures that take finite values on finite intervals, $U_{ij}^{(\varepsilon)}((u, v]) = U_{ij}^{(\varepsilon)}(v) - U_{ij}^{(\varepsilon)}(u) < \infty, 0 \leq u \leq v < \infty$, and these measures weakly converge for $i, j = 1, \ldots, N - 1$,

$$U_{ij}^{(\varepsilon)}(\cdot) \Rightarrow U_{ij}^{(0)}(\cdot) \quad \text{as } \varepsilon \to 0. \tag{6.2.63}$$

Now, using the independence of the service processes and the input flow, and the Markov property of the process $\xi^{(\varepsilon)}(t)$ at the moments $\vartheta_n^{(\varepsilon)}$, we can re-write formula (6.2.58) in the following form for $i, j = 1, \ldots, N - 1$:

$$q_{ij}^{(\varepsilon)}(t) = \sum_{k=j}^{N-1} \int_0^t \frac{(\lambda(t - s))^{k-j}}{(k - j)!} e^{-\lambda(t-s)} (1 - G^{(\varepsilon)}(t - s)) U_{ik}^{(\varepsilon)}(ds), \quad t \geq 0. \tag{6.2.64}$$

Let also $S_i^{(0)}$ be the set of all points $t_0 \in [0, \infty)$ such that (1) the measures $U_{ik}^{(0)}(A)$, $k = 1, \ldots, N - 1$, have no atom at the point t_0; (2) every point $t_0 - s$, where s is some point in which at least one of the measures $U_{ik}^{(0)}(A)$, $k = 1, \ldots, N-1$, has an atom, is a point of continuity of the function $1 - G^{(0)}(u)$. Obviously, the set $\bar{S}_i^{(0)} = [0, \infty) \setminus S_i^{(0)}$ is at most countable.

In order to prove statement **(b)** for the functions $q_{ij}^{(\varepsilon)}(t)$, we should prove that $q_{ij}^{(\varepsilon)}(t_\varepsilon) \to q_{ij}^{(0)}(t_0)$ as $\varepsilon \to 0$ for any $t_\varepsilon \to t_0$ as $\varepsilon \to 0$, where $t_0 \in S_i^{(0)}$.

Let us choose $T > t_0$ such that the point T is a point of continuity of all renewal functions $U_{ik}^{(0)}(t)$, $k = 1, \ldots, N - 1$. There exists $\varepsilon_0 > 0$ such that $t_\varepsilon \leq T$ for $\varepsilon \leq \varepsilon_0$.

Conditions $\mathbf{T_5}$ and $\mathbf{E_5''}$ imply that **(e)** $1 - G^{(\varepsilon)}(t_\varepsilon - s) \to 1 - G_{ij}^{(0)}(t_0 - s)$ as $\varepsilon \to 0$ for every s such that the point $t_0 - s$ is a continuity point of the limit function. Since the functions $1 - G^{(\varepsilon)}(t_\varepsilon - s)$ are non-increasing in the argument s, statement **(e)** implies that **(f)** the functions $1 - G^{(\varepsilon)}(t_\varepsilon - s)$ converge locally uniformly to the function $1 - G^{(0)}(t_0 - s)$ as $\varepsilon \to 0$ at every point of continuity of the limit function.

The assumptions that $t_0 \in S_i^{(0)}$ and T is a point of continuity of all renewal functions $U_{ik}^{(0)}(t)$, $k = 1, \ldots, N - 1$, the relation of weak convergence (6.2.63), and statement **(f)** permit to apply Lemma 1.2.2 to the measures $U_{ik}^{(\varepsilon)}(A \cap [0, T])$ and the functions $\frac{(\lambda(t_\varepsilon - s))^{k-j}}{(k-j)!} e^{-\lambda(t_\varepsilon - s)}(1 - G^{(\varepsilon)}(t_\varepsilon - s))\chi(s \leq t_\varepsilon)$ for every $i, k, j = 1, \ldots, N-1$. This yields the following relation:

$$q_{ij}^{(\varepsilon)}(t_\varepsilon) = \sum_{k=j}^{N-1} \int_0^T \frac{(\lambda(t_\varepsilon - s))^{k-j}}{(k-j)!}$$

$$\times\, e^{-\lambda(t_\varepsilon - s)}(1 - G^{(\varepsilon)}(t_\varepsilon - s))\chi(s \leq t_\varepsilon)U_{ik}^{(\varepsilon)}(ds)$$

$$\to q_{ij}^{(0)}(t_0) = \sum_{k=j}^{N-1} \int_0^T \frac{(\lambda(t_0 - s))^{k-j}}{(k-j)!}$$

$$\times\, e^{-\lambda(t_0 - s)}(1 - G^{(0)}(t_0 - s))\chi(s \leq t_0)U_{ik}^{(0)}(ds) \quad \text{as } \varepsilon \to 0. \quad (6.2.65)$$

Thus statement **(b)** holds for the function $q_{ij}^{(\varepsilon)}(t)$ with the set $S_{ij}^{(0)} = S_i^{(0)}$ for every $i, j = 1, \ldots, N - 1$.

6.2.7   Quasi-stationary distributions for the regenerative processes $\xi^{(\varepsilon)}(t)$

We will use the moment generation functions for $i = 1, \ldots, N$,

$$_0\phi_{iN}^{(\varepsilon)}(\rho) = \int_0^\infty e^{\rho t}\, _0G_{iN}^{(\varepsilon)}(dt), \quad \phi_{iN}^{(\varepsilon)}(\rho) = \int_0^\infty e^{\rho t}\, G_{iN}^{(\varepsilon)}(dt),$$

As was mentioned above, $\mathbf{T_5}$ implies that conditions $\mathbf{C_{12}}$–$\mathbf{C_{16}}$ hold for the imbedded semi-Markov processes $\eta^{(\varepsilon)}(t)$ with some $\beta < \delta < \lambda$ and, therefore, by Lemmas 4.5.6

and 4.5.7, there exists $\varepsilon_0 > 0$ such that, for $\varepsilon \le \varepsilon_0$, **(a)** $_0\phi_{NN}^{(\varepsilon)}(\beta) \in (1,\infty)$; **(b)**
$_0\phi_{iN}^{(\varepsilon)}(\beta) < \infty, i = 1,\ldots, N$; **(c)** $\phi_{iN}^{(\varepsilon)}(\beta) < \infty, i = 1,\ldots, N$.

Let us also introduce moment generating functions by

$$\omega_{iN}^{(\varepsilon)}(\rho) = \int_0^\infty e^{\rho t}(1 - G_{iN}^{(\varepsilon)}(t))\, dt, \quad \omega_{iNj}^{(\varepsilon)}(\rho) = \int_0^\infty e^{\rho t} q_{ij}^{(\varepsilon)}(t)\, dt.$$

Obviously,

$$\omega_{iN}^{(\varepsilon)}(\rho) = \sum_{j=1}^N \omega_{iNj}^{(\varepsilon)}(\rho). \tag{6.2.66}$$

Note that **(d)** $1 + \rho\omega_{iN}^{(\varepsilon)}(\rho) = \phi_{iN}^{(\varepsilon)}(\rho)$ for $i = 1,\ldots, N$. This relation implies that
(e) $\omega_{iN}^{(\varepsilon)}(\beta) < \infty, i = 1,\ldots, N$; **(f)** $\omega_{iNj}^{(\varepsilon)}(\beta) < \infty, i, j = 1,\ldots, N$.

Formula (3.3.11) found for the quasi-stationary distribution of regenerative process given in Subsection 3.3.3, if applied to the regenerative process $\xi^{(\varepsilon)}(t)$, gives the following formula for its quasi-stationary distribution:

$$\pi_j^{(\varepsilon)}(\rho^{(\varepsilon)}) = \frac{\omega_{NNj}^{(\varepsilon)}(\rho^{(\varepsilon)})}{\omega_{NN}^{(\varepsilon)}(\rho^{(\varepsilon)})}, \quad j = 1,\ldots, N, \tag{6.2.67}$$

where $\rho^{(\varepsilon)} < \beta$ is the unique root of the characteristic equation

$$_0\phi_{NN}^{(\varepsilon)}(\rho) = 1. \tag{6.2.68}$$

Formula (6.2.67) for the quasi-stationary distribution of the regenerative process $\xi^{(\varepsilon)}(t)$ is similar to the corresponding variant of formula (5.5.36) for the quasi-stationary distribution specified for the imbedded semi-Markov process $\eta^{(\varepsilon)}(t)$. In these formulas, $\omega_{NNj}^{(\varepsilon)}(\rho^{(\varepsilon)})$, $j \neq 0$, are, respectively, the mixed power-exponential moment of the generating functions for the times spend by the regenerative process $\xi^{(\varepsilon)}(t)$ or the imbedded semi-Markov process $\eta^{(\varepsilon)}(t)$ in a given state before hitting the state N or being absorbed.

Let us introduce the following mixed power-exponential moment generating functions for times spend by the process $\xi^{(\varepsilon)}(t)$ in a given state before hitting the state N or being absorbed for $i, j = 1,\ldots, N, n = 0, 1,\ldots$, and $\rho < \lambda$:

$$\varphi_{ij}^{(\varepsilon)}[\rho, n] = \mathsf{E}_i \int_0^{\alpha_1^{(\varepsilon)}} s^n e^{s\rho} \chi(\xi^{(\varepsilon)}(s) = j)\, ds$$

$$= \int_0^\infty s^n e^{s\rho} \mathsf{P}_i\{\xi^{(\varepsilon)}(s) = j, \alpha_1^{(\varepsilon)} > s\}\, ds.$$

The following formula takes place for $i, j = 1, \ldots, N$ and $n = 0, 1, \ldots$, and $n = 0, 1, \ldots$:

$$\varphi_{ij}^{(\varepsilon)}[\rho, n] =$$

$$\begin{cases} \dfrac{\lambda^{i-j}(i-j+n)!}{(\lambda-\rho)^{i-j+1}(i-j)!}\, \bar{p}_{i-j+n}^{(\varepsilon)}(\lambda - \rho) & \text{if } 1 \leq i \leq N - 1, 1 \leq j \leq i, \\[2mm] 0 & \text{if } 1 \leq i \leq N - 1, i + 1 \leq j \leq N, \\[2mm] 0 & \text{if } i = N, 1 \leq j \leq N - 1, \\[2mm] \dfrac{n!}{(\lambda-\rho)^{n+1}}, & \text{if } i = N, j = N. \end{cases} \qquad (6.2.69)$$

Indeed, formula (6.2.70) is obvious in the cases $i = N$ and $1 \leq i \leq N - 1$, $i + 1 \leq j \leq N$, while for the case $1 \leq i \leq N - 1, 1 \leq j \leq i$,

$$\varphi_{ij}^{(\varepsilon)}[\rho, n] = \int_0^\infty s^n e^{s\rho} \mathsf{P}_i\{\xi^{(\varepsilon)}(s) = j, \alpha_1^{(\varepsilon)} > s\}\, ds$$

$$= \int_0^\infty s^n \frac{(\lambda s)^{i-j}}{(i-j)!} e^{-(\lambda-\rho)s}(1 - G^{(\varepsilon)}(s))\, ds$$

$$= \frac{\lambda^{i-j}(i-j+n)!}{(\lambda-\rho)^{i-j+n+1}(i-j)!}\, \bar{p}_{i-j+n}^{(\varepsilon)}(\lambda - \rho). \qquad (6.2.70)$$

The following representation can be obtained for $j \neq 0$:

$$\int_0^{\mu^{(\varepsilon)} \wedge \tau_1^{(\varepsilon)}} e^{s\rho} \chi(\xi^{(\varepsilon)}(s) = j)\, ds$$

$$= \int_0^{\alpha_1^{(\varepsilon)}} e^{s\rho} \chi(\xi^{(\varepsilon)}(s) = j)\, ds + \sum_{i \neq 0, N} \chi(\xi^{(\varepsilon)}(\alpha_1^{(\varepsilon)}) = i)$$

$$\times e^{\rho \alpha_1^{(\varepsilon)}} \int_0^{\mu^{(\varepsilon)} \wedge \alpha_1^{(\varepsilon)} - \tau_1^{(\varepsilon)}} e^{s\rho} \chi(\xi^{(\varepsilon)}(\alpha_1^{(\varepsilon)} + s) = j)\, ds. \qquad (6.2.71)$$

Taking expectations on the left- and the right-hand sides of (6.2.71) we get the following system of linear equations for the moment generating functions $\omega_{iNj}^{(\varepsilon)}(\rho), i = 1, \ldots, N$, for every $\rho \leq \beta, \varepsilon \leq \varepsilon_0$, and $j = 1, \ldots, N$:

$$\omega_{iNj}^{(\varepsilon)}(\rho) = \varphi_{ij}^{(\varepsilon)}[\rho, 0] + \sum_{r \neq 0, N} p_{ir}^{(\varepsilon)}[\rho, 0]\omega_{rNj}^{(\varepsilon)}(\rho), \quad i \neq 0. \qquad (6.2.72)$$

This system of linear equations should be compared with the system of linear equations for mixed power-exponential moment generating functions for the times spend by the imbedded semi-Markov process $\eta^{(\varepsilon)}(t)$ in a given state before hitting the state N or an absorption.

Both systems have the same coefficient matrix $_N\mathbf{P}^{(\varepsilon)}(\rho)$ and differ only in the free terms. In the case of the imbedded semi-Markov process, the free terms $\varphi_{ij}^{(\varepsilon)}[\rho, 0]$, $i = 1, \ldots, N$, should be replaced with the free terms $\tilde{\varphi}_{ij}^{(\varepsilon)}[\rho, 0] = \delta(i, j) \sum_{r=1}^{i} \varphi_{ir}^{(\varepsilon)}[\rho, 0]$, $i = 1, \ldots, N$. This is because the imbedded semi-Markov process $\eta^{(\varepsilon)}(t)$ for the process $\xi^{(\varepsilon)}(t)$ is defined in such a way that $\mathsf{P}_i\{\eta^{(\varepsilon)}(s) = j, \alpha_1^{(\varepsilon)} > s\} = \delta(i, j) \cdot \sum_{r=1}^{i} \mathsf{P}_i\{\xi^{(\varepsilon)}(s) = r, \alpha_1^{(\varepsilon)} > s\}$ for every $s \geq 0$ and $i, j = 1, \ldots, N$. Both systems have unique solutions for every $\rho \leq \beta$ and $\varepsilon \leq \varepsilon_0$. A detailed discussion concerned the latter system can be found in Section 5.5.7, while the corresponding threshold value ε_0 is determined in Subsection 4.5.4.

In the quasi-stationary case, where conditions $\mathbf{T_5}$ and $\mathbf{E_5''}$ hold, the limit quasi-stationary probabilities $\pi_j^{(0)}(\rho^{(0)})$, $j = 1, \ldots, N$, are positive and can be found using formulas (6.2.66) and (6.2.67). The corresponding moment generating function $\omega_{NNj}^{(0)}(\rho^{(0)})$, $j = 1, \ldots, N$, can be found by solving the system of linear equations (6.2.72), which should be solved for the value $\rho = \rho^{(0)} > 0$ which is a solution of equation (6.2.68). Note also that, in this case, the transition quantities are positive, $_0\phi_{iN}^{(0)}(\rho^{(0)}) > 0$, $j = 1, \ldots, N - 1$.

In a simpler pseudo-stationary case, where conditions $\mathbf{T_5}$ and $\mathbf{E_5'}$ hold, the limit quasi-stationary probabilities satisfy $\pi_j^{(0)}(\rho^{(0)}) = \delta(j, N)$, $j = 1, \ldots, N$. In this case, $\rho^{(0)} = 0$ and, as follows from relations (6.2.55) and (6.2.56) and statement (I) formulated in Subsection 6.2.6, we have $q_{Nj}^{(0)}(t) = \delta(j, N)e^{-\lambda t}$, $t \geq 0$, for $j = 1, \ldots, N$. Note also that, in this case, the transition stopping probabilities satisfy $_0G_{iN}^{(0)}(\infty) = 1$, $j = 1, \ldots, N - 1$.

So, all conditions introduced in Chapter 3 are verified. All the results given in Chapter 3 are applicable to the regenerative process $\xi^{(\varepsilon)}(t)$, $t \geq 0$, which describes functioning of M/G queueing systems with quick service and a bounded queue buffer.

6.2.8   Mixed ergodic and limit/large deviation theorems

Conditions $\mathbf{T_5}$ and $\mathbf{E_5'}$ correspond to a pseudo-stationary model with the absorption probabilities asymptotically vanishing to zero. Under these conditions there are analogues of mixed ergodic and limit Theorems 3.2.1 and 3.2.5 and mixed ergodic and large deviation Theorems 3.3.1 and 3.3.8 covering the case of the pseudo-stationary asymptotics. $\mathbf{T_5}$ and $\mathbf{E_5''}$ correspond to a quasi-stationary model with absorption probabilities asymptotically separated from zero. Under these conditions, analogues of mixed ergodic and large deviation Theorems 3.3.2–3.3.4 and 3.3.7–3.3.12 covering the case of quasi-stationary asymptotics can be formulated, together with Theorem 3.3.5, 3.3.6, and 3.3.13, which give conditions for convergence of the corresponding quasi-stationary distributions. It is useful to note that the theorems listed above cover the pseudo-stationary case, where conditions $\mathbf{T_5}$ and $\mathbf{E_5''}$ hold as well.

The limit parameters, including the stationary and quasi-stationary distributions, can be calculated by using the corresponding algorithms and formulas given in Chapter 4. They should be specified for the semi-Markov processes $\eta^{(\varepsilon)}(t)$, $t \geq 0$, introduced in the current section.

6.3   Exponential asymptotics for M/G queueing systems with quick service

In this section, we continue to study M/G queueing systems with quick service and nonlinearly perturbed parameters.

6.3.1   Perturbation conditions in mixed ergodic and large deviation theorems

Let us formulate perturbation conditions that replace the corresponding perturbation conditions introduced in Chapters 3 and 5 in mixed ergodic and limit/large deviation theorems.

Let us introduce the following perturbation condition for $\rho < \lambda$:

$\mathbf{P}_{32}^{(\rho,k)}$: $p_n^{(\varepsilon)}(\lambda - \rho) = p_n^{(0)}(\lambda - \rho) + e_n[\rho, 1]\varepsilon + \cdots + e_n[\rho, k]\varepsilon^k + o(\varepsilon^k)$ for $n = 0, \ldots, N + k$, where $|e_n[\rho, l]| < \infty$, $n = 0, \ldots, N + k$, $l = 1, \ldots, k$.

It is also convenient to define $e_n[\rho, 0] = p_n^{(0)}(\lambda - \rho)$ for $n = 0, \ldots, N + k$.

Note that condition $\mathbf{P}_{32}^{(\rho,k)}$ implies that the following asymptotic expansions take place for $n = 0, \ldots, N + k$:

$$\bar{p}_{n+1}^{(\varepsilon)}(\lambda - \rho) = 1 - \sum_{m=0}^{n} p_m^{(\varepsilon)}(\lambda - \rho)$$

$$= \bar{p}_{n+1}^{(0)}(\lambda - \rho) + \bar{e}_{n+1}[\rho, 1]\varepsilon + \cdots + \bar{e}_{n+1}[\rho, k]\varepsilon^k + o(\varepsilon^k), \quad (6.3.1)$$

where $\bar{e}_n[\rho, 0] = \bar{p}_n^{(0)}(\lambda - \rho)$ for $n = 0, \ldots, N + k$, and, for $n = 0, \ldots, N + k$, $l = 0, \ldots, k$,

$$\bar{e}_{n+1}[\rho, l] = \begin{cases} 1 - \sum_{m=0}^{n} e_m[\rho, l] & \text{if } l = 0, \\ -\sum_{m=0}^{n} e_m[\rho, l] & \text{if } l = 1, \ldots, k. \end{cases} \quad (6.3.2)$$

These conditions should be used for the value $\rho = 0$ in the pseudo-stationary case, i.e., if conditions $\mathbf{T}_5$ and $\mathbf{E}_5'$ hold, and for the value $\rho = \rho^{(0)} > 0$, which is a solution of equation (6.2.68), in the quasi-stationary case, i.e., if conditions $\mathbf{T}_5$ and $\mathbf{E}_5''$ hold.

As an example, let us consider a model for which condition $\mathbf{T}_6$ holds. In this case, it is natural to replace the perturbation condition $\mathbf{P}_{32}^{(\rho,k)}$ with the following perturbation condition imposed on the parameter $\lambda^{(\varepsilon)}$:

$\mathbf{P}_{33}^{(k)}$: $\lambda^{(\varepsilon)} = \lambda^{(0)} + \lambda_1 \varepsilon + \cdots + \lambda_k \varepsilon^k + o(\varepsilon^k)$, where $|\lambda_r| < \infty, r = 1, \ldots, k$.

It is also convenient to define $\lambda_0 = \lambda^{(0)}$.

In this model, for $n = 0, 1, \ldots$,

$$p_n^{(\varepsilon)}(\lambda - \rho) = \int_0^\infty \frac{((\lambda - \rho)s)^n}{n!} e^{-(\lambda-\rho)s} G(ds/\lambda^{(\varepsilon)})$$

$$= (\lambda^{(\varepsilon)})^n \int_0^\infty \frac{((\lambda - \rho)s)^n}{n!} e^{-(\lambda-\rho)\lambda^{(\varepsilon)}s} G(ds). \tag{6.3.3}$$

Let us consider the pseudo-stationary case, where condition $\mathbf{E}_6'$ holds, i.e., $\lambda^{(0)} = 0$. Let us first consider the case where $\lambda^{(\varepsilon)} = O(\varepsilon)$, i.e., the following condition also holds:

$\mathbf{O}_{17}$: $\lambda_1 > 0$.

Let us show that, in this case, conditions $\mathbf{P}_{33}^{(k)}$, $\mathbf{E}_6'$, $\mathbf{O}_{17}$, and $\mathbf{M}_{14}^{(N+k)}$ imply that condition $\mathbf{P}_{32}^{(\rho,k)}$ holds for every $\rho < \lambda$ with the coefficients in the corresponding expansions,

$$e_n[\rho, l] = \begin{cases} 0 & \text{if } 0 \leq l \leq k, k < n \leq N + k, \\ 0 & \text{if } 0 \leq l < n, 0 \leq n \leq k, \\ \text{given in (6.3.9) and (6.3.10)} & \text{if } n \leq l \leq k, 0 \leq n \leq k. \end{cases} \tag{6.3.4}$$

The case $k < n \leq N + k$ is obvious, since, in this case, formula (6.3.3) implies that

$$p_n^{(\varepsilon)}(\lambda - \rho) \leq (\lambda^{(\varepsilon)})^n \int_0^\infty \frac{((\lambda - \rho)s)^n}{n!} G(ds)$$

$$= (\lambda^{(\varepsilon)})^n \frac{(\lambda - \rho)^n}{n!} M_n = o(\varepsilon^k). \tag{6.3.5}$$

Let us now assume that $0 \leq n \leq k$. In this case, formula (6.3.3) implies that

$$p_n^{(\varepsilon)}(\lambda - \rho) = (\lambda^{(\varepsilon)})^n \int_0^\infty \frac{((\lambda - \rho)s)^n}{n!} \left(\sum_{r=0}^{k-n} \frac{(-(\lambda - \rho)\lambda^{(\varepsilon)}s)^r}{r!} \right) G(ds)$$

$$+ (\lambda^{(\varepsilon)})^{k+1} \int_0^\infty \frac{((\lambda - \rho)s)^{k+1}}{n!}$$

$$\times \left(\sum_{r=k-n+1}^\infty \frac{(-(\lambda - \rho)\lambda^{(\varepsilon)}s)^{r-k+n-1}}{r!} \right) G(ds), \quad (6.3.6)$$

where $(-(\lambda - \rho)\lambda^{(\varepsilon)}s)^0$ should be counted as 1 if $(\lambda - \rho)\lambda^{(\varepsilon)}s = 0$.

Note that for $(\lambda - \rho)\lambda^{(\varepsilon)}s \geq 0$,

$$\left| \sum_{r=k-n+1}^{\infty} \frac{(-(\lambda - \rho)\lambda^{(\varepsilon)}s)^{r-k+n-1}}{r!} \right| \leq \frac{1}{(k-n+1)!} \leq 1. \qquad (6.3.7)$$

Thus, by continuing the estimate in (6.3.6) and using estimate (6.3.7), we can conclude that there exists $-1 \leq \theta^{(\varepsilon)} \leq 1$ such that

$$p_n^{(\varepsilon)}(\lambda - \rho) = \sum_{r=0}^{k-n} (\lambda^{(\varepsilon)})^{n+r} \frac{(\lambda - \rho)^{n+r}(-1)^r}{n!r!} M_{n+r}$$

$$+ (\lambda^{(\varepsilon)})^{k+1}\theta^{(\varepsilon)} \frac{(\lambda - \rho)^{k+1}}{n!} M_{k+1}$$

$$= e_n[\rho, n]\varepsilon^n + \cdots + e_n[\rho, k]\varepsilon^k + o(\varepsilon^k), \qquad (6.3.8)$$

where the coefficients $e_n[\rho, l], l = n, \ldots, k$, are given by the formulas

$$e_n[\rho, l] = \sum_{r=0}^{k-n} (-1)^r \frac{(\lambda - \rho)^{n+r}(n+r)!}{n!r!} M_{n+r} \cdot \sum_{n_1,\ldots,n_k \in D_{n+r,l}} \prod_{q=1}^{k} \frac{\lambda^{n_q}}{n_q!}, \qquad (6.3.9)$$

where $D_{n+r,l}$ is the set of all nonnegative integer solutions of the system

$$n_1 + \cdots + n_k = n + r, \quad n_1 + \cdots + kn_k = l. \qquad (6.3.10)$$

The coefficients $e_n[\rho, l]$ can be calculated by formulas (6.3.9) and (6.3.10) for any $l = n, n+1, \ldots$, in particular,

$$e_n[\rho, n] = \lambda_1^n \frac{(\lambda - \rho)^n}{n!} M_n,$$

$$e_n[\rho, n+1] = n\lambda_1^{n-1}\lambda_2 \frac{(\lambda - \rho)^n}{n!} M_n - \lambda_1^{n+1} \frac{(\lambda - \rho)^{n+1}}{n!} M_{n+1},$$

$$e_n[\rho, n+2] = n\lambda_1^{n-1}\lambda_3 \frac{(\lambda - \rho)^n}{n!} M_n - (n+1)\lambda_1^n\lambda_2 \frac{(\lambda - \rho)^{n+1}}{n!} M_{n+1}$$

$$+ \lambda_1^{n+2} \frac{(\lambda - \rho)^{n+2}}{2n!} M_{n+2}, \qquad (6.3.11)$$

and, therefore,

$$e_0[\rho, 0] = 1,$$

$$e_0[\rho, 1] = -\lambda_1(\lambda - \rho)M_1,$$

$$e_0[\rho, 2] = -\lambda_2(\lambda - \rho)M_1 + \lambda_1^2\frac{(\lambda - \rho)^2}{2}M_2,$$

$$e_1[\rho, 1] = \lambda_1(\lambda - \rho)M_1,$$

$$e_1[\rho, 2] = \lambda_2(\lambda - \rho)M_1 - \lambda_1^2(\lambda - \rho)^2 M_2,$$

$$e_1[\rho, 3] = \lambda_3(\lambda - \rho)M_1 - 2\lambda_1\lambda_2(\lambda - \rho)^2 M_2 + \lambda_1^3\frac{(\lambda - \rho)^3}{2}M_3,$$

$$e_2[\rho, 2] = \lambda_1^2\frac{(\lambda - \rho)^2}{2}M_2,$$

$$e_2[\rho, 3] = 2\lambda_1\lambda_2\frac{(\lambda - \rho)^2}{2}M_2 - \lambda_1^3\frac{(\lambda - \rho)^2}{2}M_3,$$

$$e_2[\rho, 4] = 2\lambda_1\lambda_3\frac{(\lambda - \rho)^2}{2}M_2 - 3\lambda_1^2\lambda_2\frac{(\lambda - \rho)^3}{2}M_3 + \lambda_1^4\frac{(\lambda - \rho)^4}{4}M_4. \quad (6.3.12)$$

Note that formulas (6.3.11) imply that $e_n[\rho, n] > 0$ for $n = 0, 1, \ldots$.

In this case, also the coefficients in the asymptotic expansions for the coefficients $\bar{p}_{n+1}^{(\varepsilon)}(\lambda - \rho)$, $n = 0, 1, \ldots$, can be expressed, by using formula (6.3.4), in a form simpler than in (6.3.2),

$$\bar{e}_{n+1}[\rho, l] = \begin{cases} 0 & \text{if } 0 \leq l < n + 1, \\ \sum_{m=n+1}^{l} e_m[\rho, l] & \text{if } n + 1 \leq l \leq k. \end{cases} \quad (6.3.13)$$

Formulas (6.3.9) and (6.3.10) are also valid in the case where the following condition, more general than $\mathbf{O_{17}}$, holds for some $1 \leq h \leq k$:

$\mathbf{O_{18}}$: $\lambda_l = 0$, $l < h$, $\lambda_h > 0$.

In this case, formulas (6.3.9) and (6.3.10) yield that $e_n[\rho, l] = 0$ for $l < nh$.

Now, let us consider the quasi-stationary case where condition $\mathbf{E_6''}$ holds, i.e., $\lambda^{(0)} > 0$.

Let us show that, in this case, conditions $\mathbf{P_{33}^{(k)}}$ and $\mathbf{E_6''}$ imply that condition $\mathbf{P_{32}^{(\rho, k)}}$ holds for every $\rho < \lambda$ with the coefficients in the corresponding expansions given below by formulas (6.3.18) and (6.3.19).

Denote $\lambda_1^{(\varepsilon)} = \lambda^{(\varepsilon)} - \lambda^{(0)}$. Since $\lambda^{(0)} > 0$ and $\lambda^{(\varepsilon)} \to \lambda^{(0)}$ as $\varepsilon \to 0$, there exists $\varepsilon' < 0$ such that $|\lambda_1^{(\varepsilon)}| \leq \frac{\lambda^{(0)}}{2}$ and, therefore, $\lambda^{(\varepsilon)} \geq \frac{\lambda^{(0)}}{2}$ for $\varepsilon \leq \varepsilon'$.

In this case, formula (6.3.3) implies that

$$p_n^{(\varepsilon)}(\lambda - \rho) = (\lambda^{(\varepsilon)})^n \int_0^\infty \frac{((\lambda - \rho)s)^n}{n!} e^{-(\lambda-\rho)\lambda^{(0)}s}$$

$$\times \left(\sum_{r=0}^k \frac{(-(\lambda - \rho)\lambda_1^{(\varepsilon)}s)^r}{r!} \right) G(ds)$$

$$+ (\lambda_1^{(\varepsilon)})^{k+1}(\lambda^{(\varepsilon)})^n \int_0^\infty \frac{((\lambda - \rho)s)^{n+k+1}}{n!} e^{-(\lambda-\rho)\lambda^{(0)}s}$$

$$\times \left(\sum_{r=k+1}^\infty \frac{(-(\lambda - \rho)\lambda_1^{(\varepsilon)}s)^{r-k-1}}{r!} \right) G(ds). \tag{6.3.14}$$

Note that

$$\left| \sum_{r=k+1}^\infty \frac{(-(\lambda - \rho)\lambda_1^{(\varepsilon)}s)^{r-k-1}}{r!} \right| \leq \sum_{r=k+1}^\infty \frac{((\lambda - \rho)|\lambda_1^{(\varepsilon)}|s)^{r-k-1}}{r!}$$

$$\leq e^{(\lambda-\rho)|\lambda_1^{(\varepsilon)}|s}, \tag{6.3.15}$$

and, for any $m = 0, 1, \ldots$ and $\delta > 0$,

$$M_m(\delta) = \int_0^\infty s^m e^{-\delta s} G(ds) < \infty. \tag{6.3.16}$$

Thus, by continuing the estimate in (6.3.6) and using estimate (6.3.15), we can conclude that there exists $-1 \leq \theta^{(\varepsilon)} \leq 1$ such that, for $\varepsilon \leq \varepsilon'$,

$$p_n^{(\varepsilon)}(\lambda - \rho) = (\lambda^{(\varepsilon)})^n \sum_{r=0}^k (\lambda_1^{(\varepsilon)})^r \frac{(\lambda - \rho)^{n+r}(-1)^r}{n!r!} M_{n+r}((\lambda - \rho)\lambda^{(0)})$$

$$+ (\lambda^{(\varepsilon)})^{k+1}\theta^{(\varepsilon)} \frac{(\lambda - \rho)^{n+k+1}}{n!} M_{n+k+1}\left((\lambda - \rho)\frac{\lambda^{(0)}}{2}\right)$$

$$= e_n[\rho, 0] + e_n[\rho, 1]\varepsilon + \cdots + e_n[\rho, k]\varepsilon^k + o(\varepsilon^k), \tag{6.3.17}$$

where the coefficients $e_n[\rho, l]$, $l = 0, \ldots, k$, are given by the formulas

$$e_n[\rho, l] = \sum_{m=0}^n \sum_{r=0}^k C_n^m (\lambda^{(0)})^{n-m}(-1)^r \frac{(\lambda - \rho)^{n+r}(n + r)!}{n!r!}$$

$$\times M_{n+r}((\lambda - \rho)\lambda^{(0)}) \cdot \sum_{n_1,\ldots,n_k \in D_{n+r,l}} \prod_{q=1}^k \frac{\lambda^{n_q}}{n_q!}, \tag{6.3.18}$$

where $D_{n+r,l}$ is the set of all nonnegative integer solutions of the system

$$n_1 + \cdots + n_k = n + r, \quad n_1 + \cdots + k n_k = l. \tag{6.3.19}$$

6.3.2   Asymptotic expansions for mixed power-exponential moments for hitting times for the imbedded semi-Markov processes

Consider the mixed power-exponential generation functions for $\rho \in \mathbb{R}_1$ and $i = 1, \ldots, N$,

$$_0\phi_{iN}^{(\varepsilon)}[\rho, n] = \int_0^\infty t^n e^{\rho t}\, _0G_{iN}^{(\varepsilon)}(dt).$$

As was shown in Subsection 5.3.5, relation $_0\phi_{iN}^{(\varepsilon)}(\beta) < \infty, i = 1, \ldots, N$, implies that **(a)** $_0\phi_{iN}^{(\varepsilon)}[\rho, n] < \infty, n = 0, 1, \ldots, i = 1, \ldots, N$, for $\rho < \beta$.

To construct asymptotic expansions for the characteristic root $\rho^{(\varepsilon)}$ of equation (6.2.68) we should, according to the results given in Chapters 3 and 5, use the asymptotic expansions for the mixed power-exponential generation functions $_0\phi_{iN}^{(\varepsilon)}[\rho, n]$, $i \neq 0, n = 0, \ldots, k$. These expansions are based on the perturbation condition $\mathbf{P}_{25}^{(\rho,k)}$, which in its turn is based on the asymptotic expansions for the mixed power-exponential generation functions $p_{ij}^{(\varepsilon)}[\rho, n], i, j \neq 0, n = 0, \ldots, k$. The corresponding algorithm is described in Section 5.2.

Now, the perturbation condition $\mathbf{P}_{32}^{(\rho,k)}$ and formula (6.2.52) imply that the following asymptotic expansions take place for $r = 1, \ldots, N, n = 0, \ldots, k$:

$$p_r^{(\varepsilon)}[\rho, n] = p_r^{(0)}[\rho, n] + e_r[\rho, 1, n]\varepsilon + \cdots + e_r[\rho, k, n]\varepsilon^k + o(\varepsilon^k), \qquad (6.3.20)$$

where $p_r^{(0)}[\rho, n] = e_r[\rho, 0, n]$ for $n = 0, \ldots, k$ and, for $r = 1, \ldots, N, n, l = 0, \ldots, k$,

$$e_r[\rho, l, n] = \frac{\lambda^r (r + n)!}{(\lambda - \rho)^{r+n} r!} e_{r+n}[\rho, l]. \qquad (6.3.21)$$

Formula (6.2.51) and the asymptotic relations (6.3.20) imply that the following asymptotic expansions take place for $i, j \neq 0, n = 0, \ldots, k$:

$$p_{ij}^{(\varepsilon)}[\rho, n] = p_{ij}^{(0)}[\rho, n] + e_{ij}[\rho, 1, n]\varepsilon + \cdots + e_{ij}[\rho, k, n]\varepsilon^k + o(\varepsilon^k), \qquad (6.3.22)$$

where $p_{ij}^{(0)}[\rho, n] = e_{ij}[\rho, 0, n]$, for $n = 0, \ldots, k$ and, for $i, j \neq 0, n, l = 0, \ldots, k$,

$$e_{ij}[\rho, l, n] = \begin{cases} e_{i-j+1}[\rho, l, n] & \text{if } 1 \leq i \leq N - 1, 2 \leq j \leq i + 1, \\ \delta(l, 0)\dfrac{n!}{(\lambda-\rho)^{n+1}} & \text{if } i = N, j = N - 1, \\ 0 & \text{otherwise.} \end{cases} \qquad (6.3.23)$$

Therefore, the perturbation condition $\mathbf{P}_{32}^{(\rho,k)}$ implies that the perturbation condition $\mathbf{P}_{25}^{(\rho,k)}$ holds, which, in this case, is represented by the asymptotic relations (6.3.22). As was mentioned above, this condition permits to construct the asymptotic expansions for the power moments $_0\phi_{iN}^{(\varepsilon)}[0,n]$, $i \neq 0$, $n = 0,\ldots,k$, which constitute the perturbation conditions $\mathbf{P}_9^{(k)}$ and $\mathbf{P}_{10}^{(\bar{k})}$, or to construct the asymptotic expansions for mixed power-exponential moments $_0\phi_{iN}^{(\varepsilon)}[\rho^{(0)},n]$, $i \neq 0$, $n = 0,\ldots,k$, which constitute the perturbation conditions $\mathbf{P}_{11}^{(k)}$ and $\mathbf{P}_{12}^{(\bar{k})}$.

6.3.3   Asymptotic expansions for mixed power-exponential moments for the time spent by the process $\xi^{(\varepsilon)}(t)$ in a given state before hitting a given state or an absorption

Let us return to the system of linear equations (6.2.72). As was pointed out in Subsection 6.2.7, **(a)** $\varphi_{ij}^{(\varepsilon)}[\rho,0] < \infty$ for $i,j \neq 0$, for $\rho < \lambda$ and $\varepsilon \geq 0$; and **(b)** $\omega_{iNj}^{(\varepsilon)}(\rho) < \infty$, $i,j \neq 0$ for $\rho < \beta$ and $\varepsilon \leq \varepsilon_0$.

Relations **(a)** and **(b)** imply that **(c)** $\varphi_{ij}^{(\varepsilon)}[\rho,n] < \infty$ for $i,j \neq 0$, $n = 0,1,\ldots$, and $\varepsilon \geq 0$; **(d)** $\omega_{iNj}^{(\varepsilon)}(\rho) < \infty$, $i,j \neq 0$, $n = 0,1,\ldots$, for $\rho < \beta$ and $\varepsilon \leq \varepsilon_0$, and that **(e)** we can differentiate (with respect to the variable ρ) equations of system (6.2.72) and get the following system of linear equation (which coincides with system (6.2.72) for $n = 0$) for every $\rho < \beta$, $\varepsilon \leq \varepsilon_0$, $j \neq 0$, and $n = 0,1,\ldots$:

$$\omega_{iNj}^{(\varepsilon)}[\rho,n] = \vartheta_{iNj}^{(\varepsilon)}[\rho,n] + \sum_{r \neq 0,N} p_{ir}^{(\varepsilon)}(\rho)\omega_{rNj}^{(\varepsilon)}[\rho,n], \ i \neq 0, \tag{6.3.24}$$

where $\omega_{iNj}^{(\varepsilon)}[\rho,0] = \omega_{iNj}^{(\varepsilon)}(\rho)$, $i,j \neq 0$ and, for $i,j \neq 0$, $n = 0,1,\ldots$,

$$\vartheta_{iNj}^{(\varepsilon)}[\rho,n] = \varphi_{ij}^{(\varepsilon)}[\rho,n] + \sum_{r \neq 0,N} \sum_{m=1}^{n} C_n^m p_{ir}^{(\varepsilon)}[\rho,m]\omega_{rNj}^{(\varepsilon)}[\rho,n-m]. \tag{6.3.25}$$

The systems of linear equation (6.3.24) can be rewritten in a matrix form. Introduce the vectors for $\rho < \beta$, $\varepsilon \leq \varepsilon_0$, $j \neq 0$, and $n = 0,1,\ldots$ by

$$\boldsymbol{\omega}_{\mathrm{N}j}^{(\varepsilon)}[\rho,n] = \begin{bmatrix} \omega_{1Nj}^{(\varepsilon)}[\rho,n] \\ \vdots \\ \omega_{NNj}^{(\varepsilon)}[\rho,n] \end{bmatrix}, \qquad \boldsymbol{\vartheta}_{\mathrm{N}j}^{(\varepsilon)}[\rho,n] = \begin{bmatrix} \vartheta_{1Nj}^{(\varepsilon)}[\rho,n] \\ \vdots \\ \vartheta_{NNj}^{(\varepsilon)}[\rho,n] \end{bmatrix},$$

and

$$\boldsymbol{\varphi}_{\mathrm{j}}^{(\varepsilon)}[\rho,n] = \begin{bmatrix} \varphi_{1j}^{(\varepsilon)}[\rho,n] \\ \vdots \\ \varphi_{Nj}^{(\varepsilon)}[\rho,n] \end{bmatrix}.$$

Then the system of linear equations (6.3.24) can be rewritten in the following form for every $\rho < \beta$, $\varepsilon \leq \varepsilon_0$, $j \neq 0$, and $n = 0,1,\ldots$:

$$\boldsymbol{\omega}_{\mathrm{N}j}^{(\varepsilon)}[\rho,n] = \boldsymbol{\vartheta}_{\mathrm{N}j}^{(\varepsilon)}[\rho,n] + {}_{\mathrm{N}}\mathbf{P}^{(\varepsilon)}[\rho,n]\,\boldsymbol{\omega}_{\mathrm{N}j}^{(\varepsilon)}[\rho,n], \tag{6.3.26}$$

where the matrices $_N\mathbf{P}^{(\varepsilon)}[\rho, n]$ have the following form for $n = 0, 1, \ldots$:

$$
_N\mathbf{P}^{(\varepsilon)}[\rho, n] =
\begin{bmatrix}
0 & p_0^{(\varepsilon)}[\rho, n] & 0 & \cdots & 0 & 0 \\
0 & p_1^{(\varepsilon)}[\rho, n] & p_0^{(\varepsilon)}[\rho, n] & \cdots & 0 & 0 \\
\vdots & \vdots & \vdots & \vdots & \vdots & \vdots \\
0 & p_{N-2}^{(\varepsilon)}[\rho, n] & p_{N-3}^{(\varepsilon)}[\rho, n] & \cdots & p_1^{(\varepsilon)}[\rho, n] & 0 \\
0 & 0 & 0 & \cdots & \dfrac{n!}{(\lambda-\rho)^{n+1}} & 0
\end{bmatrix}.
\tag{6.3.27}
$$

The systems of linear equations (6.3.26) are systems of ρ-hitting type with the same coefficient matrix $_N\mathbf{P}^{(\varepsilon)}[\rho, 0]$.

Under conditions $\mathbf{T_5}$, these systems have unique solutions for every $\varepsilon \le \varepsilon_0$, $\rho < \beta$, $j \ne 0$, and $n = 0, 1, \ldots$. These solutions have the following forms:

$$
\omega_{Nj}^{(\varepsilon)}[\rho, n] = [\mathbf{I} - {}_N\mathbf{P}^{(\varepsilon)}[\rho, 0]]^{-1} \vartheta_{Nj}^{(\varepsilon)}[\rho, n].
\tag{6.3.28}
$$

Systems (6.3.26) have, for a given $j \ne 0$, the same coefficient matrix $_N\mathbf{P}^{(\varepsilon)}(\rho)$ but different inhomogeneous terms $\vartheta_{Nj}^{(\varepsilon)}[\rho, n]$ for $n = 0, 1, \ldots$. These systems should be solved recursively.

First, system (6.3.26) should be solved for the case $n = 0$. Second, system (6.3.26) should be solved for the case $n = 1$. Note that the expressions for the inhomogeneous terms $\theta_{iNj}^{(\varepsilon)}[\rho, 1] = \varphi_{ij}^{(\varepsilon)}[\rho, 1] + \sum_{r \ne 0, N} p_{ir}^{(\varepsilon)}[\rho, 1]\omega_{rNj}^{(\varepsilon)}[\rho, 0]$, $i \ne 0$, given in (6.3.25) include the solutions $\omega_{iNj}^{(\varepsilon)}[\rho, 0]$, $i \ne 0$, of systems (6.3.26) for $n = 0$.

This recursive procedure should be repeated for $n = 1, 2, \ldots$. The expressions for the inhomogeneous terms $\vartheta_{ijl}^{(\varepsilon)}[\rho, n]$ given in (6.3.25) include the solutions $\omega_{ijl}^{(\varepsilon)}[\rho, m]$, $i \ne 0$, of systems (6.3.26) for $m = 0, 1, \ldots, n - 1$.

Formula (6.3.28) can be rewritten in the following vector form for every $\rho < \beta$, $\varepsilon \le \varepsilon_0$, and $n = 0, 1, \ldots, j \ne 0$:

$$
\vartheta_{Nj}^{(\varepsilon)}[\rho, n] = \varphi_j^{(\varepsilon)}[\rho, n] + \sum_{m=1}^{n} C_n^m \, _N\mathbf{P}^{(\varepsilon)}[\rho, m]\omega_{Nj}^{(\varepsilon)}[\rho, n - m].
\tag{6.3.29}
$$

Using formula (6.3.29) we can re-write formula (6.3.28) in the following form representing the recursive procedure for calculating solutions of the system of linear equations (6.3.26) for $\rho < \beta$, $\varepsilon \le \varepsilon_0$, and $n = 0, 1, \ldots, j \ne 0$,

$$
\omega_{Nj}^{(\varepsilon)}[\rho, n] = [\mathbf{I} - {}_N\mathbf{P}^{(\varepsilon)}(\rho)]^{-1} \vartheta_{Nj}^{(\varepsilon)}[\rho, n]
$$

$$
= [\mathbf{I} - {}_N\mathbf{P}^{(\varepsilon)}(\rho)]^{-1} \varphi_j^{(\varepsilon)}[\rho, n]
$$

$$
+ \sum_{m=1}^{n} C_n^m [\mathbf{I} - {}_N\mathbf{P}^{(\varepsilon)}(\rho)]^{-1} \, _N\mathbf{P}^{(\varepsilon)}[\rho, m]\omega_{Nj}^{(\varepsilon)}[\rho, n - m].
\tag{6.3.30}
$$

We are now in a position to give asymptotic expansions for the functions $\omega_{iNj}^{(\varepsilon)}[\rho, n]$, $i, j \neq 0, n = 0, 1, \ldots, k$.

Formula (6.2.69) and the perturbation condition $\mathbf{P}_{32}^{(\rho,k)}$ imply the following asymptotic expansion for the vectors $\boldsymbol{\varphi}_j^{(\varepsilon)}[\rho, n], j \neq 0, n = 0, \ldots, k, i, j \neq 0, n = 0, \ldots, k$:

$$\varphi_{ij}^{(\varepsilon)}[\rho, n] = \varphi_{ij}^{(0)}[\rho, n] + w_{ij}[\rho, 1, n]\varepsilon + \cdots + w_{ij}[\rho, k, n]\varepsilon^k + o(\varepsilon^k), \qquad (6.3.31)$$

where $\varphi_{ij}^{(0)}[\rho, n] = w_{ij}[\rho, 0, n]$ for $i, j \neq 0, n = 0, \ldots, k$, and, $w_{ij}[\rho, l, n]$ are given for $i, j \neq 0, n, l = 0, \ldots, k$ by the following formulas:

$$w_{ij}[\rho, l, n] =$$

$$\begin{cases} \frac{\lambda^{i-j}(i-j+n)!}{(\lambda-\rho)^{i-j+1}(i-j)!}\bar{e}_{i-j+n}(\lambda - \rho) & \text{if } 1 \leq i \leq N - 1, 1 \leq j \leq i, \\ 0 & \text{if } 1 \leq i \leq N - 1, i + 1 \leq j \leq N, \\ 0 & \text{if } i = N, 1 \leq j \leq N - 1, \\ \delta(l, 0)\frac{n!}{(\lambda-\rho)^{n+1}} & \text{if } i = N, j = N. \end{cases} \qquad (6.3.32)$$

Note also that the asymptotic relation (6.3.20) implies that the following matrix asymptotic expansion can be written for the matrices ${}_N\mathbf{P}^{(\varepsilon)}[\rho, n]$ for $n = 0, 1, \ldots$:

$$_N\mathbf{P}^{(\varepsilon)}[\rho, n] = {}_N\mathbf{P}^{(0)}[\rho, n] + {}_N\mathbf{E}[\rho, 1, n]\varepsilon + \cdots + {}_N\mathbf{E}[\rho, k, n]\varepsilon^k + \mathbf{o}(\varepsilon^k), \quad (6.3.33)$$

where ${}_N\mathbf{P}^{(0)}[\rho, n] = {}_N\mathbf{E}[\rho, 0, n]$ for $n = 0, \ldots, k$, and, for $l, n = 0, \ldots, k$,

$$_N\mathbf{E}^{(\varepsilon)}[\rho, l, n] =$$

$$\begin{bmatrix} 0 & e_0[\rho, l, n] & 0 & \cdots & 0 & 0 \\ 0 & e_1[\rho, l, n] & e_0[\rho, l, n] & \cdots & 0 & 0 \\ \vdots & \vdots & \vdots & \vdots & \vdots & \vdots \\ 0 & e_{N-2}[\rho, l, n] & e_{N-3}[\rho, ln] & \cdots & e_1[\rho, l, n] & 0 \\ 0 & 0 & 0 & \cdots & \delta(l, 0)\frac{n!}{(\lambda-\rho)^{n+1}} & 0 \end{bmatrix}. \qquad (6.3.34)$$

Let us introduce the vectors for $j \neq 0, l, n = 0, \ldots, k$,

$$\mathbf{c}_{Nj}[\rho, l, n] = \begin{bmatrix} c_{1Nj}[\rho, l, n] \\ \vdots \\ c_{NNj}[\rho, l, n] \end{bmatrix}, \qquad \mathbf{w}_j[\rho, l, n] = \begin{bmatrix} w_{1j}[\rho, l, n] \\ \vdots \\ w_{Nj}[\rho, l, n] \end{bmatrix}.$$

The following asymptotic relations are corollaries of Lemma 5.3.3 applied to the functions $\omega_{Nj}^{(\varepsilon)}[\rho, n]$ for $j \neq 0$ and $n = 0, 1, \ldots, k$. Proofs of these relations are analogous to those given in Lemmas 5.3.5–5.3.6 and 5.5.7. Thus, under conditions $\mathbf{T}_5$ and $\mathbf{P}_{32}^{(\rho,k)}$,

$$\omega_{Nj}^{(\varepsilon)}[\rho, n] = \omega_{Nj}^{(0)}[\rho, n] + \mathbf{c}_{Nj}[\rho, 1, n]\varepsilon + \cdots + \mathbf{c}_{Nj}[\rho, k, n]\varepsilon^k + \mathbf{o}(\varepsilon^k), \qquad (6.3.35)$$

where the vector coefficients $\mathbf{c}_{Nj}[\rho, r, n]$ are given by the recurrence formulas $\mathbf{c}_{Nj}[\rho, 0, 0] = \boldsymbol{\omega}_{Nj}^{(0)}[\rho, 0] = [\mathbf{I} - {}_{N}\mathbf{P}^{(0)}(\rho)]^{-1}\boldsymbol{\theta}_{Nj}^{(0)}[\rho, 0]$ and, in general, for $r = 0, \ldots, k$ (for a given n) and subsequently for $n = 0, \ldots, k$,

$$\mathbf{c}_{Nj}[\rho, r, n] = [\mathbf{I} - {}_{N}\mathbf{P}^{(0)}(\rho)]^{-1}(\mathbf{w}_{j}[\rho, r, n]$$

$$+ \sum_{m=1}^{n} C_{n}^{m} \sum_{q=0}^{r} {}_{N}\mathbf{E}[\rho, q, m]\mathbf{c}_{Nj}[\rho, r - q, n - m]$$

$$+ \sum_{p=1}^{r} \mathbf{E}_{N}[\rho, p, 0]\mathbf{c}_{Nj}[\rho, r - p, n]). \tag{6.3.36}$$

The asymptotic relation (6.3.35) allows to construct, in an obvious way, asymptotic expansions for mixed power-exponential moments for $A \subseteq X_0 = \{j \neq 0\}, i \neq 0$, $n = 0, 1, \ldots$,

$$\omega_{iNA}^{(\varepsilon)}[\rho, n] = \sum_{j \in A} \omega_{iNj}^{(\varepsilon)}[\rho, n],$$

which has the following form:

$$\omega_{iNA}^{(\varepsilon)}[\rho, n] = \omega_{iNA}^{(0)}[\rho, n] + c_{iNA}[\rho, 1, n]\varepsilon + \cdots + c_{iNA}[\rho, k, n]\varepsilon^{k} + o(\varepsilon^{k}), \tag{6.3.37}$$

where $\omega_{iNA}^{(0)}[\rho, n] = c_{iNA}[\rho, 0, n], n = 0, \ldots, k$, and, for $A \subseteq X_0 = \{j \neq 0\}, i \neq 0$, $l, n = 0, \ldots, k$,

$$c_{iNA}[\rho, l, n] = \sum_{j \in A} c_{iNj}[\rho, l, n].$$

Finally, we conclude that conditions $\mathbf{T}_5$, $\mathbf{P}_{32}^{(\rho, k)}$, and $\mathbf{E}_5'$ imply that the perturbation condition $\mathbf{P}_{16}^{(k)}$ holds. It is represented by the asymptotic expansion (6.3.37) that should be taken for $\rho = 0$. Also, conditions $\mathbf{T}_5$, $\mathbf{P}_{32}^{(\rho, k)}$, and $\mathbf{E}_5''$ imply that the perturbation condition $\mathbf{P}_{15}^{(k)}$ holds. It is represented by the asymptotic expansion (6.3.37) that should be taken for $\rho = \rho^{(0)} > 0$.

6.3.4   Exponential expansions in mixed ergodic and large deviation theorems

All results given in Chapter 3 can be applied to the regenerative processes $\xi^{(\varepsilon)}(t)$, $t \geq 0$, that describe functioning of queueing M/G type systems with quick service and a bounded queue buffer.

Conditions $\mathbf{T}_5$, $\mathbf{E}_5'$, $\mathbf{P}_{32}^{(k)}$ correspond to a pseudo-stationary model with the absorption probabilities asymptotically vanishing to zero. Under these conditions, analogues of Theorems 3.4.1–3.4.3 and 3.4.6–3.4.8, which give exponential asymptotic expansions in the corresponding mixed ergodic and large deviation theorems, can be formulated.

For example, if conditions $\mathbf{T_5}$, $\mathbf{E_5'}$, the perturbation condition $\mathbf{P_{32}^{(0,k)}}$, and the balancing condition $\mathbf{B_5^{(r)}}$ (for some $1 \leq r \leq k$) hold, the asymptotic relation given in Theorem 3.4.2 takes the following form for $i, j \neq 0$:

$$\frac{\mathsf{P}_i\{\xi^{(\varepsilon)}(t^{(\varepsilon)}) = j, \mu^{(\varepsilon)} > t^{(\varepsilon)}\}}{\exp\{-(a_1\varepsilon + \cdots + a_{r-1}\varepsilon^{r-1})t^{(\varepsilon)}\}} \to \delta(j, N)e^{-\lambda_r a_r} \quad \text{as } \varepsilon \to 0. \qquad (6.3.38)$$

If $k \geq N - 1$ and the order condition $\mathbf{O_{13}}$ holds, then, as follows from Theorem 3.4.2, $a_l = 0, l < N - 1$, while $a_{N-1} > 0$. In particular, condition $\mathbf{O_{13}}$ is implied by conditions $\mathbf{T_6}$ and $\mathbf{P_{33}^{(k)}}$.

Conditions $\mathbf{T_5}$, $\mathbf{E_5''}$, and $\mathbf{P_{32}^{(\rho^{(0)},k)}}$ correspond to a quasi-stationary model with the absorption probabilities asymptotically separated from zero. Under these conditions, analogues of Theorems 3.4.4–3.4.5 and 3.4.9–3.4.10, which give exponential asymptotic expansion in mixed ergodic and large deviation theorems, can be formulated. Also, analogues of Theorems 3.5.2–3.5.4 give higher order asymptotic expansions for quasi-stationary distributions. It is useful to note that the theorems listed above cover the pseudo-stationary case as well.

The limit parameters, including the stationary and the quasi-stationary distributions, and the coefficients in the corresponding asymptotic expansions, can be calculated by using the corresponding algorithms and formulas given in Chapters 3–5. They should be specified for the imbedded semi-Markov processes $\eta^{(\varepsilon)}(t)$, $t \geq 0$, and the regenerative processes $\xi^{(\varepsilon)}(t)$, $t \geq 0$, with the use of the corresponding formulas given in Sections 6.2 and 6.3.

6.4   Quasi-stationary phenomena in birth-and-death type stochastic systems

In this section, we investigate conditions for pseudo- and quasi-stationary asymptotics for stochastic systems birth-and-death type.

6.4.1   Semi-Markov processes of birth-and-death type

Let $\eta^{(\varepsilon)}(t)$, $t \geq 0$, for every $\varepsilon \geq 0$, be a semi-Markov process with the phase space $X = \{0, 1, \ldots, N\}$ and the transition probabilities

$$Q_{ij}^{(\varepsilon)}(t) = \begin{cases} p_{i,+}^{(\varepsilon)} F_i^{(\varepsilon)}(t) & \text{if } i = 1, \ldots, N-1, \, j = i+1 \text{ or } i = j = N, \\ p_{i,-}^{(\varepsilon)} F_i^{(\varepsilon)}(t) & \text{if } i = 1, \ldots, N-1, \, j = i-1, \\ 1 & \text{if } i = j = 0, \\ 0 & \text{otherwise.} \end{cases} \qquad (6.4.1)$$

The imbedded discrete time Markov chain $\eta_n^{(\varepsilon)}$, $n = 0, 1, \ldots$, for the semi-Markov process $\eta^{(\varepsilon)}(t)$, $t \geq 0$, has the matrix of transition probabilities,

$$
\mathbf{P}^{(\varepsilon)} =
\begin{bmatrix}
1 & 0 & 0 & \cdots & 0 & 0 & 0 \\
p_{1,-}^{(\varepsilon)} & 0 & p_{1,+}^{(\varepsilon)} & \cdots & 0 & 0 & 0 \\
\vdots & \vdots & \vdots & \vdots & \vdots & \vdots & \vdots \\
0 & 0 & 0 & \cdots & p_{N-1,-}^{(\varepsilon)} & 0 & p_{N-1,+}^{(\varepsilon)} \\
0 & 0 & 0 & \cdots & 0 & p_{N,-}^{(\varepsilon)} & p_{N,+}^{(\varepsilon)}
\end{bmatrix}.
\tag{6.4.2}
$$

As in Chapter 4, we denote by $\nu_j^{(\varepsilon)}$ and $\mu_j^{(\varepsilon)}$ the first hitting time to a state $j \in X$ for the imbedded Markov chain $\eta_n^{(\varepsilon)}$, $n = 0, 1, \ldots$, and the semi-Markov process $\eta^{(\varepsilon)}(t)$, $t \geq 0$, respectively.

The state 0 is an absorption state and, therefore, the random variable $\mu_0^{(\varepsilon)}$ is the absorption time that is an object of our study.

The condition $\mathbf{T_1}$ takes in this case the following form:

$\mathbf{T_7}$: (a) $p_{i,\pm}^{(\varepsilon)} \to p_{i,\pm}^{(0)}$ as $\varepsilon \to 0$, $i \neq 0$;

(b) $F_i^{(\varepsilon)}(\cdot) \Rightarrow F_i^{(0)}(\cdot)$ as $\varepsilon \to 0$, $i \neq 0$.

Condition $\mathbf{E_2}$ becomes:

$\mathbf{E_7}$: $p_{i,+}^{(0)} > 0$, $i \neq 0$.

Condition $\mathbf{E_7}$ can be realised in two variants. Let us introduce the set

$$
E = \{1 \leq i \leq N : p_{i,-}^{(0)} = 0\}.
$$

The following condition is the first variant for realisation of condition $\mathbf{E_7}$:

$\mathbf{E_7'}$: $p_{i,+}^{(0)} > 0$, $i = 1, \ldots, N$, and $E \neq \varnothing$.

In this case, denote

$$
i_+ = \max(i \geq 1 : p_{i,-}^{(0)} = 0), \quad i_- = \min(i \geq 1 : p_{i,-}^{(0)} = 0).
$$

This is a pseudo-stationary case. The hitting probabilities satisfy $_0 f_{ij}^{(0)} = 1$ for $i \geq i_-$, $j \geq i$, while $0 < {}_0 f_{ij}^{(0)} < 1$ for $1 \leq i < i_-$, $j \geq i$. Also, the absorption probabilities satisfy $_j f_{i0}^{(0)} = 0$ for $i \geq i_-$, $j \geq i$, while $0 < {}_j f_{i0}^{(0)} < 1$ for $1 \leq i < i_-$, $j \geq i$. The set of recurrent-without-absorption states is $X_1^{(0)} = \{i_+, \ldots, N\}$, while for the set of non-recurrent-without-absorption states, we have $X_2^{(0)} = \{1, \ldots, i_+ - 1\}$.

Conditions $\mathbf{T_7}$ and $\mathbf{E_7'}$ do not guarantee that $p_{i,-}^{(\varepsilon)} = 0$, $i \in E$ for ε small enough. In order to have this property, one should require, additionally to $\mathbf{T_7}$ and $\mathbf{E_7'}$, the following condition to hold:

$\mathbf{E_8}$: There exists a state $i^* \in E$ and $\varepsilon_1 > 0$ such that $p_{i,-}^{(\varepsilon)} > 0$, $i = i^* + 1, \ldots, N$, and $p_{i^*,-}^{(\varepsilon)} = 0$ for every $0 < \varepsilon \leq \varepsilon_1$.

Let conditions $\mathbf{T_7}$, $\mathbf{E_7'}$, and $\mathbf{E_8}$ hold. Conditions $\mathbf{T_7}$ and $\mathbf{E_7'}$ imply that there exists $\varepsilon_2 > 0$ such that $p_{i,+}^{(\varepsilon)} > 0$, $i = 1, \ldots, N$, and $p_{i,-}^{(\varepsilon)} > 0$, $i \notin E$, $i \neq 0$, for every $\varepsilon \leq \varepsilon_2$. Then, for $0 < \varepsilon \leq \varepsilon_1 \wedge \varepsilon_2$, the set of recurrent-without-absorption states is $X_1^{(\varepsilon)} = \{i^*, \ldots, N\}$, while the set of non-recurrent-without-absorption states equals $X_2^{(\varepsilon)} = \{1, \ldots, i^* - 1\}$.

Note also that conditions $\mathbf{T_7}$, $\mathbf{E_7'}$, and $\mathbf{E_8}$ imply that $p_{i,-}^{(0)} = 0$, i.e., $i^* \in E$. Thus, $i_- \leq i^* \leq i_+$ and, therefore, $X_1^{(0)} \subseteq X_1^{(\varepsilon)}$ for $\varepsilon \leq \varepsilon_1 \wedge \varepsilon_2$.

The following condition is the second variant for realisation of condition $\mathbf{E_7}$:

$\mathbf{E_7''}$: $q_{i,+}^{(0)} > 0$, $i = 1, \ldots, m$, and $E = \varnothing$.

This is the quasi-stationary case. The hitting probabilities satisfy $0 < {}_0 f_{ij}^{(0)} < 1$ for $i \neq 0$, $j \geq i$, and for the absorption probabilities, $0 < {}_j f_{i0}^{(0)} < 1$ for $1 \leq i \neq 0$, $j \geq i$. Thus, the set of recurrent-without-absorption states is $X_1^{(0)} = \{1, \ldots, N\}$, while for the set of non-recurrent-without-absorption states, we have $X_2^{(0)} = \varnothing$.

Conditions $\mathbf{T_7}$ and $\mathbf{E_7''}$ imply that there exists $\varepsilon_3 > 0$ such that $p_{i,\pm}^{(\varepsilon)} > 0$, $i \neq 0$, for $\varepsilon \leq \varepsilon_3$. Then, the set of recurrent-without-absorption states is $X_1^{(\varepsilon)} = \{1 \ldots, N\}$, while for the set of non-recurrent-without-absorption states, $X_2^{(\varepsilon)} = \varnothing$, for $0 < \varepsilon \leq \varepsilon_3$.

Condition $\mathbf{I_4}$ takes in this case the following form:

$\mathbf{I_7}$: $F_i^{(0)}(0) < 1$ for some state $i \in X_1^{(0)}$.

The non-periodicity conditions $\mathbf{N_1}$ and $\mathbf{N_2}$ have the same formulations as for the general semi-Markov processes. As was mentioned in Subsection 4.3.1, these conditions are not restrictive. For example, condition $\mathbf{N_1}$ holds if the distribution function $F_i^{(0)}(t)$ has an absolutely continuous component at least for some $i \neq 0$. In particular, this is the case for the standard birth-and-death type model. Note that, in this case, condition $\mathbf{I_6}$ holds as well.

6.4.2 Asymptotics of hitting and absorption probabilities

Note first of all that condition $\mathbf{T_8}$ (a) implies that, for $l \neq 0$, $i, j \in X$, $i \neq j$,

$$ {}_i f_{lj}^{(\varepsilon)} \to {}_i f_{lj}^{(0)} \quad \text{as } \varepsilon \to 0. \tag{6.4.3} $$

Let us consider, together with the usual hitting probabilities $_if_{lj}^{(\varepsilon)} = \mathsf{P}_l\{v_j^{(\varepsilon)} < v_i^{(\varepsilon)}\}$, the modified hitting probabilities $_ig_{lj}^{(0)}$, which take into account a possible hitting to the states i and j at the moment 0, and are connected with the former hitting probabilities by the following formulas for $i, j \in X, i \neq j, l \neq 0$:

$$_ig_{lj}^{(\varepsilon)} = \begin{cases} _if_{lj}^{(0)} & \text{if } l \neq i, j, \\ 0 & \text{if } l = i, \\ 1 & \text{if } l = j. \end{cases} \tag{6.4.4}$$

Let $0 \leq i < j \leq N, i \leq l \leq j$. The hitting probabilities $_ig_{lj}^{(0)}, i \leq l \leq j$ satisfy the following system of linear equations:

$$_ig_{lj}^{(\varepsilon)} = p_{l,-}^{(\varepsilon)} \, _ig_{l-1j}^{(\varepsilon)} + p_{l,+}^{(\varepsilon)} \, _ig_{l+1j}^{(\varepsilon)}, \quad i < l < j, \quad _ig_{ij}^{(\varepsilon)} = 0, \quad _ig_{jj}^{(\varepsilon)} = 1. \tag{6.4.5}$$

The equations in (6.4.5) can be rewritten in the following form for $i < l < j$:

$$\left(_ig_{l+1j}^{(\varepsilon)} - \, _ig_{lj}^{(\varepsilon)} \right) p_{l,+}^{(\varepsilon)} = p_{l,-}^{(\varepsilon)} \left(_ig_{lj}^{(\varepsilon)} - \, _ig_{l-1j}^{(\varepsilon)} \right). \tag{6.4.6}$$

This recurrence relation yields the following relation for $i < l < j$:

$$\left(_ig_{l+1j}^{(\varepsilon)} - \, _ig_{lj}^{(\varepsilon)} \right) = \left(_ig_{lj}^{(\varepsilon)} - \, _ig_{l-1j}^{(\varepsilon)} \right) \prod_{r=i+1}^{l} \frac{p_{r,-}^{(\varepsilon)}}{p_{r,+}^{(\varepsilon)}}. \tag{6.4.7}$$

By summing up from $l = i$ to $l = j - 1$ in (6.4.7) and taking into account equalities $_ig_{ij}^{(\varepsilon)} = 0$ and $_ig_{jj}^{(\varepsilon)} = 1$, we get the following equality:

$$1 = \sum_{l=i}^{j-1} \left(_ig_{l+1j}^{(\varepsilon)} - \, _ig_{lj}^{(\varepsilon)} \right) = \, _ig_{i+1j}^{(\varepsilon)} \, A_j^{(\varepsilon)}[j, j], \tag{6.4.8}$$

where, for $i \leq k \leq j$,

$$A_k^{(\varepsilon)}[i, j] = \sum_{l=i}^{k-1} \prod_{r=i+1}^{l} \frac{p_{r,-}^{(\varepsilon)}}{p_{r,+}^{(\varepsilon)}}.$$

Note that $A_i^{(\varepsilon)}[i, j] = 0$ and $A_k^{(\varepsilon)}[i, j] \geq 1$ for $i < k \leq j$ (by the definition, the sum and the product with the upper index less than the lower index should be taken equal to 0 and 1, respectively).

Finally, by calculating the sum from $l = i$ to $l = k - 1$ in (6.4.7) and taking into account equality (6.4.8), we get the following equalities for $0 \leq i < j \leq N$, $i \leq k \leq j$:

$$_ig_{kj}^{(\varepsilon)} = \frac{A_k^{(\varepsilon)}[i, j]}{A_j^{(\varepsilon)}[i, j]}. \tag{6.4.9}$$

Let us now use this formula and investigate the asymptotics of the cyclic absorption
probabilities

$$
Nf_{N0}^{(\varepsilon)} = p_{N,-}^{(\varepsilon)}\left(1 - {}_0g_{N-1N}^{(\varepsilon)}\right) = p_{N,-}^{(\varepsilon)}\left(1 - \frac{A_{N-1}^{(\varepsilon)}[0, N]}{A_N^{(\varepsilon)}[0, N]}\right)
$$

$$
= p_{N,-}^{(\varepsilon)}\frac{\prod_{r=1}^{N-1}\frac{p_{r,-}^{(\varepsilon)}}{p_{r,+}^{(\varepsilon)}}}{A_N^{(\varepsilon)}[0, N]} = p_{N,+}^{(\varepsilon)}\frac{\prod_{r=1}^{N}\frac{p_{r,-}^{(\varepsilon)}}{p_{r,+}^{(\varepsilon)}}}{A_N^{(\varepsilon)}[0, N]}. \tag{6.4.10}
$$

If conditions $\mathbf{T_7}$ and $\mathbf{E_7'}$ hold, then (a) $p_{r,-}^{(\varepsilon)} \to 0$ and $p_{r,+}^{(\varepsilon)} \to 1$ as $\varepsilon \to 0$ for
$r \in E$, while (b) $p_{r,\pm}^{(\varepsilon)} \to p_{r,\pm}^{(0)} \in (0, 1)$ as $\varepsilon \to 0$ for $i \notin E$. Also, (c) $A_{N-1}^{(\varepsilon)}[0, N] \to$
$A_{i_-}^{(0)}[0, N]$ as $\varepsilon \to 0$.

Relations (a)–(c) and formula (6.4.10) imply that

$$
Nf_{N0}^{(\varepsilon)} \sim \prod_{r\in E} p_{r,-}^{(\varepsilon)} \cdot p_{N,+}^{(0)} \frac{\prod_{r\notin E}\frac{p_{r,-}^{(0)}}{p_{r,+}^{(0)}}}{A_{i_-}^{(0)}[0, N]} \quad \text{as } \varepsilon \to 0. \tag{6.4.11}
$$

6.4.3 Moment conditions

Condition $\mathbf{M_{11}'}$ takes the form:

$\mathbf{M_{15}}$: $m_i^{(\varepsilon)} = \int_0^\infty t F_i^{(\varepsilon)}(dt) \to m_i^{(0)} = \int_0^\infty t F_i^{(0)}(dt) < \infty$ as $\varepsilon \to 0, i \neq 0$.

For $i \neq 0$ and $\rho \geq 0$, denote

$$
\psi_i^{(\varepsilon)}(\rho) = \int_0^\infty e^{\rho t} F_i^{(\varepsilon)}(dt).
$$

The following condition replaces condition $\mathbf{C_{12}}$:

$\mathbf{C_{19}}$: $\overline{\lim}_{0\leq\varepsilon\to 0}\, \psi_i^{(\varepsilon)}(\delta) < \infty, i \neq 0$, for some $\delta > 0$.

The pseudo-stationary case, where the characteristic root $\rho^{(0)} = 0$, corresponds
with conditions $\mathbf{T_7}$, $\mathbf{E_7'}$. In this case, as follows from Lemmas 4.5.8, conditions $\mathbf{T_7}$,
$\mathbf{E_7'}$, and $\mathbf{C_{19}}$ imply that condition $\mathbf{C_{13}}$ also holds.

The quasi-stationary case, where the characteristic root $\rho^{(0)} > 0$, corresponds with
conditions $\mathbf{T_7}$, $\mathbf{E_7''}$. As was mentioned above, condition $\mathbf{C_{19}}$ replaces condition $\mathbf{C_{12}}$.
Condition $\mathbf{C_{13}}$ requires a special consideration.

Let us also denote, for $l \neq 0, i, j \in X$ and $\rho \in \mathbb{R}_1$,

$$
{}_i\phi_{lj}^{(\varepsilon)}(\rho) = \int_0^\infty e^{\rho t}\, {}_iG_{lj}^{(\varepsilon)}(dt).
$$

Note first of all that, for $\rho \geq 0$,

$$_0\phi_{NN}^{(\varepsilon)}(\rho) = p_{N,+}^{(\varepsilon)}\psi_N^{(\varepsilon)}(\rho) + p_{N,-}^{(\varepsilon)}\psi_N^{(\varepsilon)}(\rho)\,_0\phi_{N-1N}^{(\varepsilon)}(\rho). \tag{6.4.12}$$

Also, the following recurrence relation takes place for $0 \leq i \leq l < j \leq N$ and $\rho \geq 0$:

$$_i\phi_{lj}^{(\varepsilon)}(\rho) = \prod_{r=l}^{j-1} {}_i\phi_{rr+1}^{(\varepsilon)}(\rho). \tag{6.4.13}$$

Let us note that conditions $\mathbf{T_7}$ and $\mathbf{E_7''}$ imply that there exists $\varepsilon' > 0$ such that $p_{r,\pm}^{(\varepsilon)} > 0$, $r \neq 0$, for all $\varepsilon \leq \varepsilon'$.

Taking into account relation (6.4.13) we can write the following system of equations for the moment generating functions $_0\phi_{rr+1}^{(\varepsilon)}(\rho)$, $r = 1, \ldots, N-1$, for every $\rho \geq 0$ and $\varepsilon \leq \varepsilon'$:

$$_0\phi_{12}^{(\varepsilon)}(\rho) = p_{1,+}^{(\varepsilon)}\psi_1^{(\varepsilon)}(\rho),$$

$$_0\phi_{rr+1}^{(\varepsilon)}(\rho) = p_{r,+}^{(\varepsilon)}\psi_r^{(\varepsilon)}(\rho)$$
$$+ p_{r,-}^{(\varepsilon)}\psi_r^{(\varepsilon)}(\rho)\,_0\phi_{r-1r}^{(\varepsilon)}(\rho)\,_0\phi_{rr+1}^{(\varepsilon)}(\rho), \quad 1 < r \leq N-1. \tag{6.4.14}$$

Note that the equations in (6.4.14) can take the form $\infty = \infty$ if the corresponding moment generation functions take the value ∞.

Let us assume that $0 < \beta < \delta$, where δ is taken from condition $\mathbf{C_{19}}$, and $\varepsilon \leq \varepsilon'$.

The equations (6.4.14) yield that **(a)** $_0\phi_{12}^{(\varepsilon)}(\beta) < \infty$ and that **(b)** $_0\phi_{rr+1}^{(\varepsilon)}(\beta) < \infty$ if and only if $p_{r,-}^{(\varepsilon)}\psi_r^{(\varepsilon)}(\beta)\,_0\phi_{r-1r}^{(\varepsilon)}(\beta) < 1$ for given $1 < r \leq N-1$. Also, equations (6.4.14) yield that **(c)** if $p_{r,-}^{(\varepsilon)}\psi_r^{(\varepsilon)}(\beta)\,_0\phi_{r-1r}^{(\varepsilon)}(\beta) \geq 1$ and, therefore, $_0\phi_{rr+1}^{(\varepsilon)}(\beta) = \infty$, hence $_0\phi_{kk+1}^{(\varepsilon)}(\beta) = \infty$ for every $r \leq k \leq N-1$, and **(d)** if $p_{r,-}^{(\varepsilon)}\psi_r^{(\varepsilon)}(\beta)\,_0\phi_{r-1r}^{(\varepsilon)}(\beta) < 1$, then $_0\phi_{rr+1}^{(\varepsilon)}(\beta) < \infty$ and so $_0\phi_{kk+1}^{(\varepsilon)}(\beta) < \infty$ for every $1 < k \leq r$. Also, **(e)** relation (6.4.12) implies that $_0\phi_{NN}^{(\varepsilon)}(\beta) < \infty$ if and only if $_0\phi_{N-1N}^{(\varepsilon)}(\beta) < \infty$.

Summarising the remarks above, we can conclude **(f)** $_0\phi_{NN}^{(\varepsilon)}(\beta) < \infty$ if and only if $p_{r,-}^{(\varepsilon)}\psi_r^{(\varepsilon)}(\beta)\,_0\phi_{r-1r}^{(\varepsilon)}(\beta) < 1$ for $1 < r \leq N-1$.

In this case, equations (6.4.14) can be solved for every $\varepsilon \leq \varepsilon'$ and $\rho \leq \beta$ recurrently for $r = 2, \ldots, N-1$,

$$_0\phi_{rr+1}^{(\varepsilon)}(\rho) = \frac{p_{r,+}^{(\varepsilon)}\psi_r^{(\varepsilon)}(\rho)}{1 - p_{r,-}^{(\varepsilon)}\psi_r^{(\varepsilon)}(\rho)\,_0\phi_{r-1r}^{(\varepsilon)}(\rho)}. \tag{6.4.15}$$

Let $p_{r,-}^{(\varepsilon)}\psi_r^{(\varepsilon)}(\beta)\,_0\phi_{r-1r}^{(\varepsilon)}(\beta) < 1$ for $1 < r \leq N-1$. In this case, equations (6.4.14) imply that **(g)** $_0\phi_{rr+1}^{(\varepsilon)}(\beta) > 1$ if and only if $p_{r,+}^{(\varepsilon)}\psi_r^{(\varepsilon)}(\beta) + p_{r,-}^{(\varepsilon)}\psi_r^{(\varepsilon)}(\beta)\,_0\phi_{r-1r}^{(\varepsilon)}(\beta) >$

1 for given $1 < r \leq N - 1$. Also, equations (6.4.14) yield that **(h)** if ${}_0\phi_{r-1r}^{(\varepsilon)}(\beta) > 1$, then ${}_0\phi_{kk+1}^{(\varepsilon)}(\beta) > 1$ for every $r \leq k \leq N - 1$, while relation (6.4.12) implies that **(i)** if ${}_0\phi_{N-1N}^{(\varepsilon)}(\beta) > 1$, then ${}_0\phi_{NN}^{(\varepsilon)}(\beta) > 1$.

Consider the condition:

$\mathbf{C_{20}}$: There exist $1 < r \leq N - 1$ and $0 < \beta < \delta$, for δ defined in condition $\mathbf{C_{19}}$, such that: **(a)** $p_{r,-}^{(0)} \psi_r^{(0)}(\beta) \, {}_0\phi_{r-1r}^{(0)}(\beta) < 1$, and **(b)** $p_{r,+}^{(0)} \psi_r^{(0)}(\beta) + p_{r,-}^{(0)} \psi_r^{(0)}(\beta)$ $\cdot \, {}_0\phi_{r-1r}^{(0)}(\beta) > 1$.

The quantities ${}_0\phi_{r-1r}^{(0)}(\beta)$, $1 < r \leq N - 1$, that enter condition $\mathbf{C_{20}}$ should be calculated recursively with the use of relations (6.4.15).

The remarks above imply that the conditions $\mathbf{C_{19}}$ and $\mathbf{C_{20}}$ are sufficient for condition $\mathbf{C_{13}}$ to hold.

Another variant of conditions that in quasi-stationary case are sufficient for condition $\mathbf{C_{13}}$ to hold are the corresponding variants of conditions $\mathbf{G_1}$ and $\mathbf{G_2}$.

For $i \neq 0$, denote

$$\rho_i^* = \sup(\rho \geq 0 : \psi_i^{(0)}(\rho) < \infty), \quad \rho^* = \min(\rho_i^*, i \neq 0).$$

Conditions $\mathbf{G_1}$ and $\mathbf{G_2}$ take in this case the following form:

$\mathbf{G_3}$: **(a)** $\rho^* > 0$;

   (b) $\psi_i^{(0)}(\rho^*) = \infty$ for some $i \neq 0$;

and

$\mathbf{G_4}$: $\overline{\lim}_{\varepsilon \to 0} \psi_i^{(\varepsilon)}(\rho) < \infty$, $i \neq 0$, for $\rho < \rho^*$.

Conditions $\mathbf{T_7}$, $\mathbf{E_7''}$ and $\mathbf{C_{19}}$–$\mathbf{C_{20}}$ or $\mathbf{G_3}$–$\mathbf{G_4}$ also imply that conditions $\mathbf{C_{14}}$ and $\mathbf{C_{15}}$ hold. Condition $\mathbf{C_{16}}$ is not needed, since, under conditions $\mathbf{T_7}$ and $\mathbf{E_7''}$, the set of non-recurrent-without absorption states is $X_2^{(0)} = \varnothing$.

6.4.4 Mixed ergodic and limit/large deviation theorems

Conditions $\mathbf{T_7}$, and $\mathbf{E_7'}$ correspond to the pseudo-stationary model with the absorption probabilities asymptotically vanishing to zero. Under these conditions, analogues for mixed ergodic and limit Theorems 4.4.1 and 4.4.2 and mixed ergodic and large deviation Theorems 4.6.1 and 4.6.2 covering the case of the pseudo-stationary asymptotics can be formulated. Also, the analogues of Theorems 4.4.3 and 4.4.5 give convergence conditions for the corresponding stationary distributions under the additional non-absorption condition $\mathbf{A_{10}}$.

Conditions $\mathbf{T_7}$, and $\mathbf{E_7''}$ correspond to a quasi-stationary model with the absorption probabilities asymptotically separated from zero. Under these conditions, analogues of mixed ergodic and large deviation Theorems 4.6.3, 4.6.5, 4.6.8, 4.6.7, and 4.6.11 covering the case of quasi-stationary asymptotics can be formulated as well as Theorem 4.6.6. They give conditions for convergence of the corresponding quasi-stationary distributions. Note that the theorems listed above can also be applied in the pseudo-stationary case as well.

The limit parameters, including the stationary and quasi-stationary distributions can be calculated by using the corresponding algorithms and formulas given in Chapter 4. They should be specified for the semi-Markov processes of birth-and-death type $\eta^{(\varepsilon)}(t), t \geq 0$, introduced in this section.

6.4.5   Perturbation conditions in mixed ergodic and large deviation theorems

Let us formulate perturbation conditions, which replace the perturbation conditions introduced in Chapter 5 in the theorems representing exponential asymptotic expansions in mixed ergodic and limit/large deviation theorems.

We shall assume that conditions $\mathbf{T_7}$, $\mathbf{E_7}$, and $\mathbf{C_{19}}$ hold.

The perturbation condition $\mathbf{P^{(k)}_{18}}$ take in this case the following form:

$\mathbf{P^{(k)}_{34}}$: $\quad p^{(\varepsilon)}_{i,\pm} = p^{(0)}_{i,\pm} + e_{i,\pm}[1]\varepsilon + \cdots + e_{i,\pm}[k]\varepsilon^k + o(\varepsilon^k)$ for $i \neq 0$, where $|e_{i,\pm}[l]| < \infty$,
$\quad\quad i \neq 0$.

It is convenient to define $e_{i,\pm}[0] = p^{(0)}_{i,\pm}$ for $i \neq 0$.

Note that condition $\mathbf{P^{(k)}_{34}}$ (for any $k = 0, 1, \ldots$) implies condition $\mathbf{T_7}$.

For $i \neq 0$ and $\rho \geq 0$, denote

$$\psi^{(\varepsilon)}_i[\rho, n] = \int_0^\infty t^n e^{\rho t} F^{(\varepsilon)}_i(dt).$$

Condition $\mathbf{C_{19}}$ guarantees that for every $0 < \beta < \delta$ there exists $\varepsilon' = \varepsilon(\beta) > 0$ such that for $n = 0, 1, \ldots, i \neq 0$, and $\varepsilon \leq \varepsilon'$, (a) $\psi^{(\varepsilon)}_i[\rho, n] < \infty$.

The perturbation condition $\mathbf{P^{(\rho,k)}_{26}}$ takes in this case the following form for $\rho < \beta$:

$\mathbf{P^{(\rho,k)}_{35}}$: $\quad \psi^{(\varepsilon)}_i[\rho, n] = \psi^{(0)}_i[\rho, n] + e_i[\rho, 1, n]\varepsilon + \cdots + e_i[\rho, k-n, n]\varepsilon^{k-n} + o(\varepsilon^{k-n})$ for
$\quad\quad n = 0, \ldots, k, i \neq 0$, where $|e_n[\rho, r, n]| < \infty, r = 1, \ldots, k-n, n = 0, \ldots, k$,
$\quad\quad i \neq 0$.

It is convenient to define $e_i[\rho, 0, n] = \psi^{(0)}_i[\rho, n]$, for $n = 0, \ldots, k, i \neq 0$.

Note that condition $\mathbf{P^{(\rho,k)}_{35}}$ should be used for the value $\rho = 0$ in the pseudo-stationary case, where conditions $\mathbf{T_7}$, $\mathbf{E'_7}$, and $\mathbf{C_{19}}$ hold. In this case, $\psi^{(\varepsilon)}_i[0, n] = \int_0^\infty t^n F^{(\varepsilon)}_i(dt), n = 1, \ldots$, are, in fact, power moments for the distribution functions $F^{(\varepsilon)}_i(t)$.

Condition $\mathbf{P}_{35}^{(\rho,k)}$ should be used for the value $\rho = \rho^{(0)} > 0$, where $\rho^{(0)}$ is a unique root of the characteristic equation ${}_0\phi_{NN}^{(\varepsilon)}(\rho) = 1$ in the quasi-stationary case, where conditions $\mathbf{T}_7$, $\mathbf{E}_7''$, and $\mathbf{C}_{19}$ hold.

6.4.6 Pivotal properties of asymptotic expansions for absorption probabilities

According to Lemma 5.1.5, condition $\mathbf{P}_{34}^{(k)}$ implies that the absorption probabilities have the following asymptotic expansion for $i \neq 0$:

$$_N f_{i0}^{(\varepsilon)} = {}_N f_{i0}^{(0)} + {}_N b_{i0}[1]\varepsilon + \cdots + {}_N b_{i0}[k]\varepsilon^k + o(\varepsilon^k). \tag{6.4.16}$$

In the quasi-stationary case, i.e., if conditions $\mathbf{P}_{34}^{(k)}$ and $\mathbf{E}_7''$ hold, the limit absorption probabilities satisfy ${}_N f_{i0}^{(0)} > 0$, $i \neq 0$, and, therefore, the asymptotic expansion (6.4.16) is $(0, k)$-pivotal.

Let us consider in more details a more interesting pseudo-stationary case, where conditions $\mathbf{P}_{34}^{(k)}$ and $\mathbf{E}_7'$ hold.

We restrict consideration to the most important case, where the following asymptotic order condition holds in addition to conditions $\mathbf{P}_{34}^{(k)}$:

$\mathbf{O}_{19}$: $e_{i,-}[0] = 0, e_{i,-}[1] > 0, i \neq 0$.

In this case, the transition probabilities satisfy $p_{i-}^{(\varepsilon)} = O(\varepsilon)$, $i \neq 0$, and $E = \{1, \ldots, N\}$.

Formula (6.4.9) implies, in this case, that, for $1 \leq i < N$,

$$_N f_{i0}^{(\varepsilon)} = 1 - \frac{A_i^{(\varepsilon)}[0, N]}{A_N^{(\varepsilon)}[0, N]} = \frac{\sum_{l=i}^{N-1} \prod_{r=1}^{l} \frac{p_{r,-}^{(\varepsilon)}}{p_{r,+}^{(\varepsilon)}}}{A_N^{(\varepsilon)}[0, N]}$$

$$\sim \prod_{r=1}^{i} p_{r,-}^{(\varepsilon)} \sim \left(\prod_{r=1}^{i} e_{r,-}[1]\right)\varepsilon^i \quad \text{as } \varepsilon \to 0. \tag{6.4.17}$$

Asymptotic relation (6.4.11) yields, since $E = \{1, \ldots, N\}$, that there is the pivotal asymptote given by relation (6.4.17) for the cyclic absorption probabilities ${}_N f_{N0}^{(\varepsilon)}$ as well, i.e.,

$$_N f_{N0}^{(\varepsilon)} = p_{N,-}^{(\varepsilon)}\left(1 - \frac{A_{N-1}^{(\varepsilon)}[0, N]}{A_N^{(\varepsilon)}[0, N]}\right) = p_{N,+}^{(\varepsilon)}\frac{\prod_{r=1}^{N} \frac{p_{r,-}^{(\varepsilon)}}{p_{r,+}^{(\varepsilon)}}}{A_N^{(\varepsilon)}[0, N]}$$

$$\sim \prod_{r \in E} p_{r,-}^{(\varepsilon)} \sim \left(\prod_{r=1}^{N} e_{r,-}[1]\right)\varepsilon^N \quad \text{as } \varepsilon \to 0. \tag{6.4.18}$$

Let us now consider a more general case, where the following asymptotic order condition holds additionally to conditions $\mathbf{E}_7'$ and $\mathbf{P}_{34}^{(k)}$:

$\mathbf{O}_{20}$: $e_{i,-}[l] = 0, l < k_i, e_{i,-}[k_i] > 0$ for $i \in E' \subseteq E$, where $1 \le k_i \le k, i \ne E'$ and $e_{i,-}[l] = 0, l \le k$, for $i \in E'' = E \setminus E'$.

In this case, the transition probabilities are $p_{i,-}^{(\varepsilon)} = O(\varepsilon^{k_i}), i \in E', p_{i,-}^{(\varepsilon)} = o(\varepsilon^k)$, $i \in E''$, while $p_{i,-}^{(\varepsilon)} \to p_{i,-}^{(0)} > 0, i \notin E$.

Let us also define $j(i) = \min(j \in E, j > i), h_i = \sum_{r \in E', r \le i} k_r, h = \sum_{r \in E'} k_r$, where $\tilde{h}_i$ is the number of states in the set $E'' \cap \{j \ne 0, j \le i\}$ and $\tilde{h}$ is the number of states in E''.

In this case, the absorption probabilities have the following asymptotics as $\varepsilon \to 0$:

$$N f_{i0}^{(\varepsilon)} = \frac{\sum_{l=i}^{N-1} \prod_{r=1}^{l} \frac{p_{r,-}^{(\varepsilon)}}{p_{r,+}^{(\varepsilon)}}}{A_N^{(\varepsilon)}[0, N]} \sim \tag{6.4.19}$$

$$\sim \begin{cases} \prod_{r \in E, r \le i} p_{r,-}^{(\varepsilon)} \cdot \left(\sum_{l=i}^{j(i)-1} \prod_{r \notin E, r \le l} \frac{p_{r,-}^{(0)}}{p_{r,+}^{(0)}} \right) & \text{if } 1 \le i < i_+, \\[2ex] \prod_{r \in E} p_{r,-}^{(\varepsilon)} \cdot \left(\sum_{l=i}^{N-1} \prod_{r \notin E, r \le l} \frac{p_{r,-}^{(0)}}{p_{r,+}^{(0)}} \right) & \text{if } i_+ \le i < N, \\[2ex] \prod_{r \in E} p_{r,-}^{(\varepsilon)} \cdot \left(p_{N,+}^{(0)} \prod_{r \notin E} \frac{p_{r,-}^{(0)}}{p_{r,+}^{(0)}} \right) & \text{if } i = N, \end{cases}$$

$$\sim \begin{cases} \varepsilon^{h_i} \left(\prod_{r \in E, r \le i} c_{r,-}[k_r] \sum_{l=i}^{j(i)-1} \prod_{r \notin E, r \le l} \frac{p_{r,-}^{(0)}}{p_{r,+}^{(0)}} \right) & \text{if } 1 \le i < i_+ \text{ and } \tilde{h}_i = 0, \\[2ex] \varepsilon^{h} \left(\prod_{r \in E} c_{r,-}[k_r] \sum_{l=i}^{N-1} \prod_{r \notin E, r \le l} \frac{p_{r,-}^{(0)}}{p_{r,+}^{(0)}} \right) & \text{if } i_+ \le i < N \text{ and } \tilde{h} = 0, \\[2ex] \varepsilon^{h} \left(\prod_{r \in E} c_{r,-}[k_r] p_{N,+}^{(0)} \prod_{r \notin E} \frac{p_{r,-}^{(0)}}{p_{r,+}^{(0)}} \right) & \text{if } i = N \text{ and } \tilde{h} = 0, \\[2ex] o(\varepsilon^{h_i + k\tilde{h}_i}) & \text{if } 1 \le i < i_+ \text{ and } \tilde{h}_i > 0, \\[2ex] o(\varepsilon^{h + k\tilde{h}}) & \text{if } i_+ \le i \le N \text{ and } \tilde{h} > 0. \end{cases}$$

6.4.7   Asymptotics of potential matrices

Asymptotic expansions for potential matrices play a key role in algorithms for calculating coefficients in asymptotic expansions for the characteristic roots $\rho^{(\varepsilon)}$ in mixed ergodic and large deviation theorems as well as in asymptotic expansions for stationary and quasi-stationary distributions.

Let us investigate in more details the asymptotics of potential matrices corresponding to the state N, which is an obvious natural choice for this model.

Let us introduce $N \times N$ matrices,

$$
{}_N\mathbf{P}^{(\varepsilon)} =
\begin{bmatrix}
0 & p_{1,+}^{(\varepsilon)} & 0 & \cdots & 0 & 0 & 0 \\
p_{2,-}^{(\varepsilon)} & 0 & p_{2,+}^{(\varepsilon)} & \cdots & 0 & 0 & 0 \\
\vdots & \vdots & \vdots & \vdots & \vdots & \vdots & \vdots \\
0 & 0 & 0 & \cdots & p_{N-1,-}^{(\varepsilon)} & 0 & 0 \\
0 & 0 & 0 & \cdots & 0 & p_{N,-}^{(\varepsilon)} & 0
\end{bmatrix},
\tag{6.4.20}
$$

and also $N \times N$ matrices ${}_N\mathbf{E}[0] = {}_N\mathbf{P}^{(0)}$ and, in general, for $n = 0, 1, \ldots$,

$$
{}_N\mathbf{E}[n] =
\begin{bmatrix}
0 & e_{1,+}[n] & 0 & \cdots & 0 & 0 & 0 \\
e_{2,-}[n] & 0 & e_{2,+}[n] & \cdots & 0 & 0 & 0 \\
\vdots & \vdots & \vdots & \vdots & \vdots & \vdots & \vdots \\
0 & 0 & 0 & \cdots & e_{N-1,-}[n] & 0 & 0 \\
0 & 0 & 0 & \cdots & 0 & e_{N,-}[n] & 0
\end{bmatrix}.
\tag{6.4.21}
$$

The perturbation condition $\mathbf{P}_{34}^{(k)}$ gives the following matrix expansion:

$$
{}_N\mathbf{P}^{(\varepsilon)} = {}_N\mathbf{P}^{(0)} + {}_N\mathbf{E}[1]\varepsilon + \cdots + {}_N\mathbf{E}[k]\varepsilon^k + \mathbf{o}(\varepsilon^k).
\tag{6.4.22}
$$

Let us also introduce the potential matrices ${}_N\mathbf{U}^{(\varepsilon)} = [\mathbf{I} - {}_N\mathbf{P}^{(\varepsilon)}]^{-1}$, and denote, for $n = 0, 1, \ldots$,

$$
{}_N\mathbf{U}^{(\varepsilon)} =
\begin{bmatrix}
{}_N u_{11}^{(\varepsilon)} & \cdots & {}_N u_{1N}^{(\varepsilon)} \\
\vdots & \vdots & \vdots \\
{}_N u_{N1}^{(\varepsilon)} & \cdots & {}_N u_{NN}^{(\varepsilon)}
\end{bmatrix}, \quad
{}_N\mathbf{U}[n] =
\begin{bmatrix}
{}_N u_{11}[n] & \cdots & {}_N u_{1N}[n] \\
\vdots & \vdots & \vdots \\
{}_N u_{N1}[n] & \cdots & {}_N u_{NN}[n]
\end{bmatrix}.
\tag{6.4.23}
$$

According to Lemma 5.1.1, conditions $\mathbf{P}_{34}^{(k)}$ and $\mathbf{E}_7$ imply that the following asymptotic matrix expansions takes place:

$$
{}_N\mathbf{U}^{(\varepsilon)} = {}_N\mathbf{U}^{(0)} + {}_N\mathbf{U}[1]\varepsilon + \cdots + {}_N\mathbf{U}[k]\varepsilon^k + \mathbf{o}(\varepsilon^k),
\tag{6.4.24}
$$

where the matrix coefficients ${}_N\mathbf{U}[r]$ are given by the recursion formulas ${}_N\mathbf{U}[0] = {}_N\mathbf{U}^{(0)} = [\mathbf{I} - {}_N\mathbf{P}^{(0)}]^{-1}$ and, for $r = 1, \ldots, k$,

$$
{}_N\mathbf{U}[r] = {}_N\mathbf{U}[0] \sum_{q=1}^{r} {}_N\mathbf{E}[q, 0]\, {}_N\mathbf{U}[r - q].
\tag{6.4.25}
$$

The most interesting is the pseudo-stationary case, where conditions $\mathbf{P}_{34}^{(k)}$ and $\mathbf{E}_7'$ hold.

Let us recall that elements of the potential matrix have the following probabilistic representation for $i, j \neq 0$:

$$_N u_{ij}^{(\varepsilon)} = \mathsf{E}_i \delta_{Nj}^{(\varepsilon)}, \tag{6.4.26}$$

where

$$\delta_{Nj}^{(\varepsilon)} = \sum_{n=1}^{\nu_N^{(\varepsilon)} \wedge \nu_0^{(\varepsilon)}} \delta(\eta_{n-1}^{(\varepsilon)}, j) = \sum_{n=1}^{\infty} \chi(\nu_N^{(\varepsilon)} \wedge \nu_0^{(\varepsilon)} > n - 1, \eta_{n-1}^{(\varepsilon)} = j).$$

The case $N = 1$ is trivial, since $_1 u_{11}^{(\varepsilon)} = 1$. Let $N > 1$.

If $j = N$ then, for $i \neq 0$,

$$_N u_{iN}^{(\varepsilon)} = \delta(N, i). \tag{6.4.27}$$

Consider the case where $1 \leq j < N$. The distributions of the random variable $\delta_{Nj}^{(\varepsilon)}, 1 \leq j < N$, can be expressed in terms of the hitting probabilities introduced in formulas (6.4.4) and (6.4.9). They take the following form for $n = 1, 2, \ldots$:

$$\mathsf{P}_i\{\delta_{Nj}^{(\varepsilon)} \geq n\} = \begin{cases} _0 g_{ij}^{(\varepsilon)}(g_j^{(\varepsilon)})^{n-1} & \text{if } 1 \leq i < j < N, \\ (g_j^{(\varepsilon)})^{n-1}, & \text{if } 1 \leq i = j < N, \\ (1 - {_j}g_{iN}^{(\varepsilon)})(g_j^{(\varepsilon)})^{n-1} & \text{if } 1 \leq j < i < N, \\ p_{N,-}^{(\varepsilon)}(1 - {_j}g_{N-1N}^{(\varepsilon)})(g_j^{(\varepsilon)})^{n-1} & \text{if } 1 \leq j < i = N, \end{cases} \tag{6.4.28}$$

where,

$$g_j^{(\varepsilon)} = p_{j,-}^{(\varepsilon)} {_0}g_{j-1j}^{(\varepsilon)} + p_{j,+}^{(\varepsilon)}(1 - {_j}g_{j+1N}^{(\varepsilon)}). \tag{6.4.29}$$

It follows from formula (6.4.28) that, for $1 \leq j < N$,

$$_N u_{ij}^{(\varepsilon)} = \mathsf{E}_i \delta_{Nj}^{(\varepsilon)} = \begin{cases} _0 g_{ij}^{(\varepsilon)}/(1 - g_j^{(\varepsilon)}) & \text{if } 1 \leq i \leq j < N, \\ 1/(1 - g_j^{(\varepsilon)}) & \text{if } 1 \leq i = j < N, \\ (1 - {_j}g_{iN}^{(\varepsilon)})/(1 - g_j^{(\varepsilon)}) & \text{if } 1 \leq j < i \leq N, \\ p_{N,-}^{(\varepsilon)}(1 - {_j}g_{N-1N}^{(\varepsilon)})/(1 - g_j^{(\varepsilon)}) & \text{if } 1 \leq j < i = N. \end{cases} \tag{6.4.30}$$

Let us restrict the considerations to the basic case where conditions $\mathbf{P}_{34}^{(k)}$ and $\mathbf{O}_{19}$ hold.

Formula (6.4.9) implies that, for $i < k \leq j$, **(a)** $A_k^{(\varepsilon)}[i, j] \to 1$ as $\varepsilon \to 0$, while **(b)** $A_j^{(\varepsilon)}[i, j] - A_k^{(\varepsilon)}[i, j] \sim (\prod_{r=i+1}^{k} e_{r,-}[1])\varepsilon^{k-i}$ as $\varepsilon \to 0$, and, therefore,

$$_i g_{kj}^{(\varepsilon)} = A_k^{(\varepsilon)}[i, j]/A_j^{(\varepsilon)}[i, j] \to 1 \text{ as } \varepsilon \to 0, \tag{6.4.31}$$

and

$$1 - {}_i g_{kj}^{(\varepsilon)} = (A_j^{(\varepsilon)}[i,j] - A_k^{(\varepsilon)}[i,j])/A_j^{(\varepsilon)}[i,j]$$

$$\sim \left(\prod_{r=i+1}^{k} e_{r,-}[1] \right) \varepsilon^{k-i} \quad \text{as } \varepsilon \to 0. \tag{6.4.32}$$

Relations (6.4.31) and (6.4.32) imply that

$$g_j^{(\varepsilon)} = \begin{cases} (e_{j,-}[1] + e_{j,-}[1])\varepsilon + o(\varepsilon) & \text{if } 1 < j < N, \\ e_{2,-}[1]\varepsilon + o(\varepsilon) & \text{if } j = 1. \end{cases} \tag{6.4.33}$$

Finally, using formula (6.4.30) and the asymptotic relations (6.4.31), (6.4.32) and (6.4.33) and applying the proposition **(iii)** from Lemma 8.1.1, we get the following asymptotic relation, which gives a complete description of pivotal properties of the potential matrices:

$$_N u_{ij}^{(\varepsilon)}$$

$$= \begin{cases} 1 + e_{2,-}[1]\varepsilon + o(\varepsilon) & \text{if } 1 = i = j, \\ 1 + (e_{j,-}[1] + e_{j+1,-}[1] - e_{1,-}[1])\varepsilon + o(\varepsilon) & \text{if } 1 = i < j < N, \\ 1 + (e_{j,-}[1] + e_{j+1,-}[1])\varepsilon + o(\varepsilon) & \text{if } 1 < i \le j < N, \\ 0 & \text{if } 1 \le i < j = N, \\ 1 & \text{if } i = j = N, \\ (\prod_{r=j+1}^{i} e_{r,-}[1])\varepsilon^{i-j} + o(\varepsilon^{i-j}) & \text{if } 1 \le j < i \le N. \end{cases} \tag{6.4.34}$$

The asymptotic relation (6.4.34) permits, in particular, to give explicit values for elements of the first two matrices in the asymptotic expansion (6.4.24) for potential matrices,

$$_N u_{ij}^{(0)} = {}_N u_{ij}[0] = \begin{cases} 1 & \text{if } 1 \le i \le j < N, \\ 0 & \text{if } 1 \le i < j = N, \\ 1 & \text{if } i = j = N, \\ 0 & \text{if } 1 \le j < i \le N, \end{cases} \tag{6.4.35}$$

and

$$_N u_{ij}[1] = \begin{cases} e_{2,-}[1] & \text{if } 1 = i = j, \\ e_{j,-}[1] + e_{j+1,-}[1] - e_{1,-}[1] & \text{if } 1 = i < j < N, \\ e_{j,-}[1] + e_{j+1,-}[1] & \text{if } 1 < i \le j < N, \\ 0 & \text{if } 1 \le i \le j = N, \\ e_{j+1,-}[1] & \text{if } 1 \le j = i-1 \le N-1, \\ 0 & \text{if } 1 \le j < i-1 \le N-1. \end{cases} \tag{6.4.36}$$

Two cases should be considered, **(1)** $k \geq N$ or **(2)** $k < N$.

In case **(1)**, if conditions $\mathbf{P}_{34}^{(k)}$, $\mathbf{E}_7'$, and $\mathbf{O}_{19}$ hold, the asymptotic relation (6.4.34) implies that $Z_0 = X_1^{(0)} = \{N\}$, $Z_r = \{N - r\}$, $r = 1, \ldots, N - 1$, and $Z_r = \varnothing$, $r = N, \ldots, k + 1$ (these sets were defined in Subsection 5.1.2). Also, $Y_{0,0} = \varnothing$, $Y_{0,1} = \{1\}$, $Y_{0,r} = \varnothing, r = 1, \ldots, k$, and $Y_{0,k+1} = \{2, \ldots, N\}$ (these sets were defined in Subsection 5.1.5). Consequently, $W_{0,r} = \varnothing, r = 0, \ldots, N - 1$, $W_{0,N} = \{1\}$, $W_{0,r} = \varnothing, r = N + 1, \ldots, k$, and $W_{0,k+1} = \{2, \ldots, N\}$. Condition $\mathbf{O}_4^{(N)}$ holds in this case.

By Theorem 5.4.1, the characteristic root $\rho^{(\varepsilon)}$ of the equation $_0\phi_{NN}^{(\varepsilon)}(\rho) = 1$ has, in this case, the following (N, k)-asymptotic pivotal expansion:

$$\rho^{(\varepsilon)} = a_N \varepsilon^N + \cdots + a_k \varepsilon^k + o(\varepsilon^k). \tag{6.4.37}$$

In case **(2)**, if conditions $\mathbf{P}_{34}^{(k)}$, $\mathbf{E}_7''$, and $\mathbf{O}_{19}$ hold, then the asymptotic relation (6.4.34) implies that $Z_0 = X_1^{(0)} = \{N\}$, $Z_r = \{N - r\}$, $r = 1, \ldots, k$, and $Z_{k+1} = \{1, \ldots, N - k - 1\}$ (these sets were defined in Subsection 5.1.2). Also $Y_{0,0} = \varnothing$, $Y_{0,1} = \{1\}$, $Y_{0,r} = \varnothing$, $r = 1, \ldots, k$, and $Y_{0,k+1} = \{2, \ldots, N\}$ (these sets were defined in Subsection 5.1.5). Hence, $W_{0,r} = \varnothing, r = 0, \ldots, k$, $W_{0,k+1} = \{1, \ldots, N\}$. Condition $\mathbf{O}_3^{(k)}$ holds in this case. By Theorem 5.4.1, the characteristic root of the equation $_0\phi_{NN}^{(\varepsilon)}(\rho) = 1$ has the asymptotics $\rho^{(\varepsilon)} = o(\varepsilon^k)$.

The case, where conditions $\mathbf{P}_{34}^{(k)}$ and $\mathbf{O}_{20}$ hold, can be treated in an analogous way.

The pivotal properties of the asymptotic expansion for the characteristic root $\rho^{(\varepsilon)}$ can be analysed with the use of the asymptotic relation (6.4.19). Indeed, $\rho^{(\varepsilon)}$ has a pivotal (h, k)-expansion for some $h \leq k$ if the cyclic absorption probabilities $_Nf_{N0}^{(\varepsilon)}$ have pivotal (h, k)-expansions, and $\rho^{(\varepsilon)} = o(\varepsilon^h)$ if $_Nf_{N0}^{(\varepsilon)} = o(\varepsilon^h)$ for some $h \leq k$. Thus, if the parameters h and $\tilde{h}$ that enter the asymptotic expansion (6.4.19) take values, respectively, less than or equal k and equal to 0, then $\rho^{(\varepsilon)}$ has a pivotal (h, k)-expansion; if $h + \tilde{h} \leq k$ and $\tilde{h} > 0$, then $\rho^{(\varepsilon)} = o(\varepsilon^{h+\tilde{h}})$; and if $h + \tilde{h} > k$, then $\rho^{(\varepsilon)} = o(\varepsilon^k)$.

6.4.8   Standard birth-and-death type processes

A standard birth-and-death type model corresponds to the case where the distribution functions $F_i^{(\varepsilon)}(t) = 1 - \exp\{-q_i^{(\varepsilon)}t\}$, $t \geq 0$, $i \neq 0$, are exponential with parameters $q_i^{(\varepsilon)} > 0$ for $i \neq 0$.

In the case of a standard birth-and-death type model, condition $\mathbf{T}_7$ takes the following form:

$\mathbf{T}_8$:   **(a)** $p_{i,\pm}^{(\varepsilon)} \to p_{i,\pm}^{(0)}$ as $\varepsilon \to 0, i \neq 0$;

   (b) $q_i^{(\varepsilon)} \to q_i^{(0)} > 0$ as $\varepsilon \to 0, i \neq 0$.

In this case, as follows from Lemma 4.5.10, conditions $\mathbf{T_8}$ and $\mathbf{E_7''}$ imply that conditions $\mathbf{G_3}$ and $\mathbf{G_4}$ automatically hold with $\rho^* = \min(q_i^{(0)}, i \neq 0)$.

In the case of a standard birth-and-death type model mentioned above, the perturbation condition $\mathbf{P_{35}^{(\rho,k)}}$ can be replaced with the following condition:

$\mathbf{P_{36}^{(k)}}$: $q_i^{(\varepsilon)} = q_i^{(0)} + q_i[1]\varepsilon + \cdots + q_i[k]\varepsilon^k + o(\varepsilon^k)$ for $i \neq 0$, where $|q_i[r]| < \infty$ for $i \neq 0$ and $r = 1, \ldots, k$.

It is convenient to define $q_i[0] = q_i^{(0)}$, for $i \neq 0$.

In this case the corresponding power moments and the moment generating functions are given by the following formulas for $i \neq 0$ and $n = 0, 1, \ldots$:

$$m_i^{(\varepsilon)}[n] = \int_0^\infty s^n q_i^{(\varepsilon)} \exp\{-q_i^{(\varepsilon)}s\}\, ds = n!(q_i^{(\varepsilon)})^{-n} < \infty, \tag{6.4.38}$$

and

$$\psi_i^{(\varepsilon)}[\rho, n] = \int_0^\infty s^n e^{\rho s} q_i^{(\varepsilon)} \exp\{-q_i^{(\varepsilon)}s\}\, ds$$

$$= \frac{n!q_i^{(\varepsilon)}}{(q_i^{(\varepsilon)} - \rho)^{n+1}} < \infty, \quad \rho < q_i^{(\varepsilon)}. \tag{6.4.39}$$

These expansions can be obtained with the use of Lemmas 5.6.3 and 5.6.5. Note that the only condition $\mathbf{P_{36}^{(k)}}$ is required.

In applications, transition probabilities $p_{i,\pm}^{(\varepsilon)}, i \neq 0$, and the parameters $q_i^{(\varepsilon)}, i \neq 0$, are expressed in terms of the corresponding transition intensities $q_{i,\pm}^{(\varepsilon)} = g_{iN,\pm} q_\pm^{(\varepsilon)}$, $i \neq 0$, by the following formulas:

$$q_i^{(\varepsilon)} = g_{iN,-} q_-^{(\varepsilon)} + g_{iN,+} q_+^{(\varepsilon)}, \qquad p_{i,\pm}^{(\varepsilon)} = g_{iN,\pm} q_\pm^{(\varepsilon)} / q_i^{(\varepsilon)}, \tag{6.4.40}$$

where **(a)** $g_{iN,\pm} > 0, i \neq 0$, and **(b)** $q_\pm^{(\varepsilon)} > 0$, for $\varepsilon > 0$.

In the case of a standard birth-and-death type model, condition $\mathbf{T_8}$ takes the following form:

$\mathbf{T_9}$: $q_\pm^{(\varepsilon)} \to q_\pm^{(0)} > 0$ as $\varepsilon \to 0$.

Conditions $\mathbf{E_7'}$ and $\mathbf{E_7''}$ that correspond, respectively, to pseudo- and quasi-stationary cases take the following form:

$\mathbf{E_9'}$: $q_+^{(0)} > 0, q_-^{(0)} = 0$;

and

$\mathbf{E_9''}$: $q_\pm^{(0)} > 0$.

In this case, the perturbation conditions $\mathbf{P}_{34}^{(k)}$ and $\mathbf{P}_{36}^{(k)}$ can be replaced with the following condition:

$\mathbf{P}_{37}^{(k)}$: $q_{\pm}^{(\varepsilon)} = q_{\pm}^{(0)} + q_{\pm}[1]\varepsilon + \cdots + q_{\pm}[k]\varepsilon^k + o(\varepsilon^k)$, where $|q_{\pm}[r]| < \infty$, for $r = 1, \ldots, k$.

It is convenient to define $q_{\pm}[0] = q_{\pm}^{(0)}$.

Condition $\mathbf{P}_{37}^{(k)}$ implies, by Lemma 8.1.1, that both conditions $\mathbf{P}_{36}^{(k)}$ and $\mathbf{P}_{34}^{(k)}$ hold with the coefficients in the corresponding asymptotic expansions given for $r = 0, \ldots, k$ and $i \neq 0$ by

$$q_i[r] = g_{iN,-}q_-[r] + g_{iN,+}q_+[r], \qquad (6.4.41)$$

and

$$p_{i,\pm}[r] = \left(g_{iN,\pm}q_{\pm}[r] - \sum_{m=0}^{r-1} p_{i,\pm}[m]q_i[r-m] \right) / q_i[0]. \qquad (6.4.42)$$

6.4.9   Exponential expansions in mixed ergodic and large deviation theorems

All results given in Chapter 5 can be used for semi-Markov processes of birth-and-death type.

Conditions $\mathbf{T}_7$, $\mathbf{E}_7'$, and $\mathbf{P}_{34}^{(k)}$–$\mathbf{P}_{37}^{(k)}$ correspond to a pseudo-stationary model with the absorption probabilities asymptotically vanishing to zero. Under these conditions, analogues of Theorems 5.4.1, 5.4.2, which give exponential asymptotic expansion in mixed ergodic and large deviation theorems, can be formulated. Also an analogue of Theorem 5.5.1 gives higher order asymptotic expansions for stationary distributions under the additional non-absorption condition $\mathbf{A}_8$.

Conditions $\mathbf{T}_7$, $\mathbf{E}_7''$, and $\mathbf{P}_{34}^{(k)}$–$\mathbf{P}_{37}^{(k)}$ correspond to a quasi-stationary model with the absorption probabilities asymptotically separated from zero. Under these conditions, analogues of Theorem 5.4.3 and 5.4.4, which give an exponential asymptotic expansion in mixed ergodic and large deviation theorems, can be formulated. Also, analogues of Theorems 5.5.4 and 5.5.5 give higher order asymptotic expansions for quasi-stationary distributions. Note that these theorems can also be applied in the pseudo-stationary case.

The limit parameters, including the stationary and the quasi-stationary distributions, and the coefficients in the corresponding asymptotic expansions, can be calculated by using the corresponding algorithms and formulas given in Chapters 4 and 5. They are used for the semi-Markov processes $\eta^{(\varepsilon)}(t)$, $t \geq 0$, introduced in this section.

6.4.10   Nonlinearly perturbed queueing systems of M/M type

There are many models of queueing systems of M/M type that can be described with the use of standard birth-and-death type Markov processes.

For example, let us consider a standard birth-and-death type Markov process with transition intensities given by the following formula for $i \neq 0$:

$$q_{i,-}^{(\varepsilon)} = i q_-^{(\varepsilon)}, \qquad q_{i,+}^{(\varepsilon)} = (N - i) q_+^{(\varepsilon)}. \qquad (6.4.43)$$

A queueing system with N servers working independently and possessing exponential working and repairing times with intensities $q_-^{(\varepsilon)}$ and $q_+^{(\varepsilon)}$, respectively, can be described with the use of a continuous time Markov chain $\eta^{(\varepsilon)}(t)$, $t \geq 0$, with transition intensities given by formula (6.4.43).

In this case, $\eta^{(\varepsilon)}(t)$ is the number of working servers at the moment t, while the lifetime is the first moment at which all the servers are being repaired, i.e., the process $\eta^{(\varepsilon)}(t)$ occurs in the state 0.

According to the definition given above, the intensity of the virtual transition $N \to N$ given by formula (6.4.43) equals 0 and, consequently, the corresponding transition probability is $p_{N,+}^{(\varepsilon)} = 0$. This contradicts the conditions $\mathbf{E}_7'$ and $\mathbf{E}_7''$, according to which the probabilities $p_{N,+}^{(\varepsilon)}$ should take positive values for all ε small enough. This side effect can be easily overcome by applying the regularisation procedure described in Subsection 5.6.2. The intensity $q_{N,+}^{(\varepsilon)} = 0$ can be replaced with a new value $\tilde{q}_{N,+}^{(\varepsilon)} = q > 0$. As was explained in Subsection 5.6.2, this regularisation procedure does affect neither the Markov chain $\eta^{(\varepsilon)}(t)$ nor the hitting time to a give state. In queueing terms, the additional virtual transitions $N \to N$ can be interpreted as a realisation of special test procedures which confirm that all servers are in the working position and do not require repairing.

Another example is a standard birth-and-death type Markov process with transition intensities given by the following formula for $i \neq 0$:

$$q_{i,-}^{(\varepsilon)} = i q_-^{(\varepsilon)}, \qquad q_{i,+}^{(\varepsilon)} = q_+^{(\varepsilon)}. \qquad (6.4.44)$$

These intensities also correspond to a model of queueing system with N servers working independently and possessing exponential working and repairing times with the intensities $q_-^{(\varepsilon)}$ and $q_+^{(\varepsilon)}$, respectively.

The difference with the former one is that there is only one service place for repairs in the system. The broken servers can be repaired only in a sequential order while, in the former model, simultaneous repairs of all broken servers were possible.

As in the first example, $\eta^{(\varepsilon)}(t)$ is the number of working servers at the moment t, while the lifetime is the first moment at which all servers occur in a repairing state, i.e., the process $\eta^{(\varepsilon)}(t)$ occurs in the state 0.

The regularisation procedure is not required for the model described above.

Many other modification of queueing systems of M/M type can be considered with the use of standard birth-and-death type Markov processes. The results presented in the current section permit to make an asymptotic analysis of pseudo- and quasi-stationary phenomena in the corresponding queueing systems with nonlinearly perturbed parameters.

6.4.11   Quasi-stationary phenomena in population and epidemic models

As is well known, standard birth-and-death type Markov processes can be also used for a description of various types of population and epidemic models.

In particular, the classical population model is described by the birth-and-death process with transition intensities given by the following formula, for $i \neq 0$:

$$q_{i,-}^{(\varepsilon)} = i q_-^{(\varepsilon)}, \qquad q_{i,+}^{(\varepsilon)} = i q_+^{(\varepsilon)}. \tag{6.4.45}$$

Here, $q_{i,+}^{(\varepsilon)}$ and $q_-^{(\varepsilon)}$ are, respectively, the individual birth and the death intensities.

In this case, $\eta^{(\varepsilon)}(t)$ is the number of individuals in the population at the moment t, while the lifetime is the first moment at which individuals die, i.e., the process $\eta^{(\varepsilon)}(t)$ occurs in the state 0. Note that we consider a model with a bounded maximal size N of population.

The regularisation procedure is not required for the model described above.

Another classical example is an epidemic model described by a birth-and-death process with the transition intensities given by the following formula for $i \neq 0$:

$$q_{i,-}^{(\varepsilon)} = i q_-^{(\varepsilon)}, \qquad q_{i,+}^{(\varepsilon)} = i(N - i) q_+^{(\varepsilon)}. \tag{6.4.46}$$

Here $q_{i,-}^{(\varepsilon)}$ and $q_+^{(\varepsilon)}$ are, respectively, individual recovering and contamination intensities.

In this case, $\eta^{(\varepsilon)}(t)$ is the number of ill individuals in a population at the moment t, while the epidemic lifetime is the first moment when all individuals recover, i.e., the process $\eta^{(\varepsilon)}(t)$ occurs in the state 0. Note that the model with bounded maximal size of population N is considered.

According to the definition given above, the intensity of the virtual transition $N \to N$ given by formula (6.4.46) equals 0 and, consequently, the corresponding transition probability is zero, $p_{N,+}^{(\varepsilon)} = 0$. This contradicts conditions $\mathbf{E}_7'$ and $\mathbf{E}_7''$, according to which the probabilities $p_{N,+}^{(\varepsilon)}$ should take positive values for all ε small enough. This side effect can be easily corrected by applying the regularisation procedure described in Subsection 5.6.2. The intensity $q_{N,+}^{(\varepsilon)} = 0$ can be replaced with a new value $\tilde{q}_{N,+}^{(\varepsilon)} = q > 0$. As was explained in Subsection 5.6.2, this regularisation procedure does not affect either the Markov chain $\eta^{(\varepsilon)}(t)$ or the time hitting a give state.

Many other modifications of population and epidemic models type can be considered with the use of standard birth-and-death type Markov processes. The results given in the current section permit to make an asymptotic analysis of pseudo- and quasi-stationary phenomena in the corresponding population and epidemic models with nonlinearly perturbed parameters.

6.5   Quasi-stationary phenomena in metapopulation models

Natural populations of most species have a spatial structure. Several habitat patches can support local populations. Such a set of local populations is called a *metapopulation*. Local populations in a metapopulation are connected by migration. A local population may go extinct due to a catastrophe. An empty patch may be colonised by migrants originating from other patches.

It is useful to note that there is a clear analogy between metapopulation models and epidemic models and that, in fact, the epidemic models are particular cases of metapopulation models. The analogy is actually takes place at different hierarchical levels. The most obvious analogy identifies patches with host individuals, colonisation with infection, and local extinction with recovery. Occupied patches correspond to infected, and empty ones to susceptible individuals. Another interpretation is to identify patches with groups of individuals, for instance, households or other groups of individuals with similar behaviour.

6.5.1   Metapopulation models

We consider a system consisting of m patches which at any time $t \geq 0$ can be either occupied or empty. The state of the metapopulation at time t is described by the vector $\eta^{(\varepsilon)}(t) = (\eta_1^{(\varepsilon)}(t), \ldots, \eta_m^{(\varepsilon)}(t))$, $t \geq 0$, where $\eta_i^{(\varepsilon)}(t)$ is either 1 or 0, according to whether the ith patch is occupied or empty at time t.

Now let us describe a stochastic process, which defines the metapopulation model, more precisely. We suppose that **(a)** in the absence of migration, the local population inhibiting a patch i will go extinct in a time interval of length Δt with probability $p_{ii}^{(\varepsilon)}(\Delta t) = g_{ii}^{(\varepsilon)} \Delta t + o(\Delta t)$, **(b)** the probability that an empty patch j will be colonised in a time interval of the length Δt by migrants originating from patch i is $p_{ij}^{(\varepsilon)}(\Delta t) = g_{ij}^{(\varepsilon)} \Delta t + o(\Delta t)$, **(c)** all interaction processes (extinction and colonisation) are independent.

The local dynamics is thus modelled by preassigning an $m \times m$ interaction matrix $G^{(\varepsilon)} = \|g_{ij}^{(\varepsilon)}\|$, where $0 \leq g_{ij}^{(\varepsilon)} < \infty$, $i, j = 1, \ldots, m$.

Having described the dynamics of a local patch we can deduce the law governing the time evolution of the process $\eta^{(\varepsilon)}(t)$. We suppose that the process $\eta^{(\varepsilon)}(t)$, $t \geq 0$ is a continuous from the right homogeneous Markov chain with absorption, which is a particular case of the model considered in Section 5.6.

The phase space of the random process $\eta^{(\varepsilon)}(t)$ is the m-dimensional binary hypercube $X = \{\bar{x} = (x_1, \ldots, x_m) : x_i \in \{0, 1\}\}$. The state $\bar{0} = (0, \ldots, 0)$ corresponding to metapopulation extinction is an absorption state, and the first time $\mu_0^{(\varepsilon)}$ the process hits this state can be interpreted as the extinction time of the metapopulation.

The local transition intensities are given by

$$
q_{\bar{x}\bar{y}}^{(\varepsilon)} = \begin{cases} g_{ii}^{(\varepsilon)} & \text{if } |\bar{x} - \bar{y}| = 1, x_i = 1, y_i = 0, \\ \sum_{i:x_i=1} g_{ij}^{(\varepsilon)} & \text{if } |\bar{x} - \bar{y}| = 1, x_j = 0, y_j = 1, \\ 0 & \text{otherwise.} \end{cases}
\tag{6.5.1}
$$

In this case, the parameter of the corresponding exponential distribution of sojourn time in a state $\bar{x} = (x_1, \ldots, x_m) \in X$ is given by the following formula:

$$
q_{\bar{x}}^{(\varepsilon)} = \sum_{\bar{y} \neq \bar{x}} q_{\bar{x}\bar{y}}^{(\varepsilon)} = \sum_{i:x_i=1} g_{ii}^{(\varepsilon)} + \sum_{j:x_j=0} \sum_{i:x_i=1} g_{ij}^{(\varepsilon)}.
\tag{6.5.2}
$$

Let $\tau^{(\varepsilon)}(n)$, $n = 0, 1, \ldots$, be subsequent moments of jumps for the Markov chain $\eta^{(\varepsilon)}(t)$, $t \geq 0$, and $\eta_n^{(\varepsilon)} = \eta^{(\varepsilon)}(\tau^{(\varepsilon)}(n))$, $n = 0, 1, \ldots$, be states of the Markov chain $\eta^{(\varepsilon)}(t)$ at the subsequent moments of jumps. By the definition, the random sequence $\eta_n^{(\varepsilon)}$, $n = 0, 1, \ldots$, is an imbedded discrete time Markov chain for the continuous time Markov chain $\eta^{(\varepsilon)}(t)$, $t \geq 0$.

Let us also define the hitting times in a state $\bar{y} \in X$ for the imbedded Markov chain $\eta_n^{(\varepsilon)}$ and the Markov chain $\eta^{(\varepsilon)}(t)$, respectively, by $v_{\bar{y}}^{(\varepsilon)} = \min(n \geq 1 : \eta_n^{(\varepsilon)} = \bar{y})$ and $\mu_{\bar{y}}^{(\varepsilon)} = \tau^{(\varepsilon)}(v_{\bar{y}}^{(\varepsilon)})$.

The first hitting time $\mu_{\bar{0}}^{(\varepsilon)}$ is the lifetime of the system. By the definition, $\mu_{\bar{0}}^{(\varepsilon)}$ is the first time at which all patches become empty.

The model described above is a particular case of the model considered in Section 5.6. The difference is in the notations used for the states of the corresponding Markov chain above and in Section 5.6 and can easily be removed by a natural re-numeration of the states, $(0, 0, \ldots, 0) \leftrightarrow 0$, $(0, 0, \ldots, 1) \leftrightarrow 1, \ldots, (1, 1, \ldots, 1) \leftrightarrow 2^m - 1$. Note that in this model the number of non-absorption states is $N = 2^m - 1$.

It should also be noted that the metapopulation model described above has an obvious similarity with the queueing system described in Section 6.1. The corresponding Markov chains has the same structure of the phase space and differ only in formulas for the corresponding transition intensities. Thus, one can expect a certain similarity in the conditions and the results related to the analysis of pseudo- and quasi-stationary asymptotics. In order to stress this similarity, we try to keep the presentation of conditions and the results in the current section close to that given in Section 6.1.

Let us use all the notations introduced in Chapter 4 for transition and hitting probabilities and moment generation functions with the only change in the notations of indices used for states. Instead of the indices $i, j, \ldots$, we shall use the indices $\bar{x}, \bar{y}, \ldots$, etc.

As was pointed out in Subsection 4.3.1, we need not assume that $\bar{0} = (0, \ldots, 0)$ is an absorption state, since the probabilities $\mathsf{P}_{\bar{x}}\{\eta^{(\varepsilon)}(t) = \bar{y}, \mu_{\bar{0}}^{(\varepsilon)} > t\}$ do not depend on the transition probabilities from the state $\bar{0}$.

Let us therefore see what form will the corresponding conditions formulated in Chapters 4 and 5 take. It is natural to re-formulate these conditions in terms of the interaction matrix $G^{(\varepsilon)} = \|g_{ij}^{(\varepsilon)}\|$.

We assume that the following condition holds, which permits us to interpret the models with $\varepsilon > 0$ as perturbed versions of the model with $\varepsilon = 0$,

$\mathbf{T_{10}}$:　$g_{ij}^{(\varepsilon)} \to g_{ij}^{(0)}$ as $\varepsilon \to 0$, $i, j = 1, \ldots, m$.

As far as the limit model is concerned, we assume that the following condition holds:

$\mathbf{E_{10}}$:　$g_{ij}^{(0)} > 0$, $i \neq j$.

Condition $\mathbf{E_{10}}$ can be realised in two variants. Condition $\mathbf{T_{10}}$ obviously implies that the limit intensities $g_{ii}^{(0)} \geq 0$, $i = 1, \ldots, m$. Let us introduce the set

$$E = \{1 \leq i \leq m : g_{ii}^{(0)} = 0\}.$$

The following condition is the "pseudo-stationary" variant for realisation of condition $\mathbf{E_{10}}$:

$\mathbf{E_{10}'}$:　$g_{ij}^{(0)} > 0$, $i \neq j$, and $E \neq \varnothing$.

In this case we assume, without loss of generality, that there exists $1 \leq l \leq m$ such that $E = \{1, \ldots, l\}$. We shall refer to patches with the indices $i \in E$ as "mainland" patches. Indeed, in the limit case $\varepsilon = 0$, the patches $i \in E$ act as mainland; its population will never go extinct.

It is obvious that conditions $\mathbf{T_{10}}$ and $\mathbf{E_{10}'}$ imply that the following condition holds:

$\mathbf{A_{10}}$:　$g_{ij}^{(\varepsilon)} > 0$, $i \neq j$, and $g_{ii}^{(\varepsilon)} > 0$, $i = l + 1, \ldots, m$, $\varepsilon \leq \varepsilon_1$ for some $\varepsilon_1 > 0$.

Condition $\mathbf{A_{10}}$ guarantees that, for every model with the parameter ε small enough, every patch with index $j = 1, \ldots, m$ can be colonised, i.e., the state 0 changes to the state 1 with the positive colonisation intensity $\sum_{i:x_i=1} g_{ij}^{(\varepsilon)}$, and that every patch with index $i = l + 1, \ldots, m$ can change the state 1 to the state 0 with the positive extinction intensity $g_{ii}^{(\varepsilon)}$. As far as patches with the indices $i = 1, \ldots, l$ are concerned, it is only certain, due to conditions $\mathbf{T_{10}}$ and $\mathbf{E_{10}'}$, that the corresponding extinction intensities are small for ε small enough and that these patches can not become empty for the unperturbed model ($\varepsilon = 0$).

As was mentioned above, conditions $\mathbf{T_{10}}$ and $\mathbf{E_{10}'}$ do not guarantee that the extinction intensities vanish, $g_{ii}^{(\varepsilon)} = 0$, $i = 1, \ldots, l$, for ε small enough. In order to guarantee this property, one needs to impose, additionally to $\mathbf{T_{10}}$ and $\mathbf{E'_{10}}$, the following condition:

$\mathbf{A_{11}}$: $g_{ii}^{(\varepsilon)} = 0$, $i \in E$, $\varepsilon \leq \varepsilon_2$ for some $\varepsilon_2 > 0$.

Condition $\mathbf{A_{10}}$ and $\mathbf{A_{11}}$ mean that, for every model with the parameter ε small enough, every patch with index $i = 1, \ldots, m$ can be colonised, every patch with index $i = 1, \ldots, l$ can not become empty, while every patch with index $i = l + 1, \ldots, m$ may become empty.

The following condition is the "quasi-stationary" variant for realisation of condition $\mathbf{E_{10}}$:

$\mathbf{E''_{10}}$: $g_{ij}^{(0)} > 0$, $i \neq j$, and $E = \emptyset$.

It is obvious that conditions $\mathbf{T_{10}}$ and $\mathbf{E''_{10}}$ imply that the following condition holds:

$\mathbf{A_{12}}$: $g_{ij}^{(\varepsilon)} > 0$, $i, j = 1, \ldots, m$, $\varepsilon \leq \varepsilon_3$ for some $\varepsilon_3 > 0$.

Condition $\mathbf{A_{12}}$ means that, for every model with the parameter ε small enough, every patch can become empty, i.e., a change from the state 1 to the state 0 is admitted, and every patch can be colonised, i.e., the processes can change the state 0 to the state 1.

Gyllenberg and Silvestrov (1994) considered such a metapopulation model in discrete time. For the analysis of quasi-stationary phenomena of a similar model, but without the limit procedure, we refer to Day and Possingham (1995). The introduction for the introduction contains a more detailed explanation of the difference between quasi- and pseudo-stationary phenomena.

Condition $\mathbf{E_{10}}$ guarantees that $q_{\bar{x}}^{(0)} > 0$ for every $\bar{x} \neq \bar{1}$, where $\bar{1} = (1, 1, \ldots, 1)$. The state $\bar{1}$ has a special status. For this state, the intensity satisfies $q_{\bar{1}}^{(0)} = \sum_{i=1}^{m} g_{ii}^{(0)}$. If all the intensities are zeros, $g_{ii}^{(0)} = 0$, $i = 1, \ldots, m$, then $q_{\bar{1}}^{(0)} = 0$ and, therefore, the state $\bar{1}$ is an absorption state. To change this situation, we shall use a regularisation transformation by adding an additional intensity $q > 0$ in this state for the virtual transition $\bar{1} \to \bar{1}$.

Thus, let $\tilde{\eta}^{(\varepsilon)}(t) = (\tilde{\eta}_1^{(\varepsilon)}(t), \ldots, \tilde{\eta}_m^{(\varepsilon)}(t))$, $t \geq 0$, be for every $\varepsilon \geq 0$ a semi-Markov process with a phase space X and the transition probabilities

$$Q_{\bar{x}\bar{y}}^{(\varepsilon)}(t) = \frac{\tilde{q}_{\bar{x}\bar{y}}^{(\varepsilon)}}{\tilde{q}_{\bar{x}}^{(\varepsilon)}} \left(1 - \exp\{-\tilde{q}_{\bar{x}}^{(\varepsilon)} t\}\right), \quad \bar{x}, \bar{y} \in X, \quad t \geq 0, \tag{6.5.3}$$

where

$$\tilde{q}_{\bar{x}\bar{y}}^{(\varepsilon)} = \begin{cases} g_{ii}^{(\varepsilon)} & \text{if } |\bar{x} - \bar{y}| = 1, x_i = 1, y_i = 0, \\ q & \text{if } \bar{x} = \bar{1}, \bar{x} = \bar{y}, \\ \sum_{i:x_i=1} g_{ij}^{(\varepsilon)} & \text{if } |\bar{x} - \bar{y}| = 1, x_j = 0, y_j = 1, \\ 0 & \text{otherwise,} \end{cases} \tag{6.5.4}$$

and

$$\tilde{q}_{\bar{x}}^{(\varepsilon)} = \sum_{\bar{y} \neq \bar{x}} q_{\bar{x}\bar{y}}^{(\varepsilon)} + q\delta(\bar{x}, \bar{1})$$

$$= \sum_{i:x_i=1} g_{ii}^{(\varepsilon)} + \sum_{j:x_j=0} \sum_{i:x_i=1} g_{ij}^{(\varepsilon)} + q\delta(\bar{x}, \bar{1}). \qquad (6.5.5)$$

As was mentioned in Section 5.6, the semi-Markov process $\tilde{\eta}^{(\varepsilon)}(t)$, $t \geq 0$, and the Markov chain $\eta^{(\varepsilon)}(t)$, $t \geq 0$, coincide. Consequently, the absorption times $\mu_{\bar{0}}^{(\varepsilon)}$ and $\tilde{\mu}_{\bar{0}}^{(\varepsilon)} = \inf(t > 0 : \tilde{\eta}^{(\varepsilon)}(t) = \bar{0})$ coincide and, therefore, the probabilities $\mathsf{P}_{\bar{x}}\{\eta^{(\varepsilon)}(t) = \bar{y}, \mu_{\bar{0}}^{(\varepsilon)} > t\}$ and $\mathsf{P}_{\bar{x}}\{\tilde{\eta}^{(\varepsilon)}(t) = \bar{y}, \tilde{\mu}_{\bar{0}}^{(\varepsilon)} > t\}$ are equal for any $\bar{x}, \bar{y} \neq \bar{0}$ and $t \geq 0$.

Thus, we can always consider a regularised semi-Markov process $\tilde{\eta}^{(\varepsilon)}(t)$ instead of the Markov chain $\eta^{(\varepsilon)}(t)$. To simplify the notations, we interpret $\tilde{\eta}^{(\varepsilon)}(t)$ as a regularised version of the Markov chain $\eta^{(\varepsilon)}(t)$ and use the initial notation $\eta^{(\varepsilon)}(t)$, $t \geq 0$, for the regularised version as well as the notations $q_{\bar{x}\bar{y}}^{(\varepsilon)}$ and $q_{\bar{x}}^{(\varepsilon)}$ for the corresponding transition intensities.

6.5.2　Conditions for mixed ergodic and limit/large deviation theorems

Let us study relations between conditions $\mathbf{T_{10}}$ and $\mathbf{E_{10}}$ and conditions for mixed ergodic and limit/large deviation theorems introduced in Chapter 4.

Denote

$$X_0 = \{\bar{x} \in X : \bar{x} \neq \bar{0}\},$$
$$X_1^{(0)} = \{\bar{x} \in X_0 : x_1, \ldots, x_l = 1\},$$
$$X_2^{(0)} = X_0 \setminus X_1^{(0)}, \qquad (6.5.6)$$

and also

$$U_0 = X_1^{(0)}, \quad U_r = \{\bar{x} \in X_2^{(0)} : \sum_{i=1}^{l} \delta(x_i, 0) = r\}, \quad r = 1, \ldots, l. \qquad (6.5.7)$$

By the definition,

$$\bigcup_{r=1}^{l} U_r = X_2^{(0)}.$$

As was mentioned above, conditions $\mathbf{T_{10}}$ and $\mathbf{E'_{10}}$ imply that condition $\mathbf{A_{10}}$ holds. According to this condition, for every $0 \leq \varepsilon \leq \varepsilon_1$, the hitting probabilities satisfy $_{\bar{0}}f_{\bar{x}\bar{y}}^{(\varepsilon)} > 0$ for any $\bar{x} \in X_0$, $\bar{y} \in X_1^{(0)}$.

Indeed, in this case, all colonisation and extinction intensities, which can provide a transition from the state $\bar{x}$ to the state $\bar{y}$ without visiting the state $\bar{0}$, are positive.

The property of hitting probabilities described above means that, for every $0 \leq \varepsilon \leq \varepsilon_1$, the Markov chain $\eta^{(\varepsilon)}(t)$ has one class of recurrent-without-absorption states $X_1^{(\varepsilon)} \supseteq X_1^{(0)}$.

As far as the absorption probabilities are concerned, condition $\mathbf{A_{10}}$ guarantees only that, for every $0 \leq \varepsilon \leq \varepsilon_1$, the absorption probabilities are positive, $_{\bar{y}}f_{\bar{x}\bar{0}}^{(\varepsilon)} > 0$, for any $\bar{x} \in U_l$, $\bar{y} \in X_1^{(0)}$.

Indeed, in this case, the class U_l consists of states of the form $(0, \ldots, 0, x_{l+1}, \ldots, x_m)$ such that $\sum_{k=l+1}^{m} \delta(x_k, 1) \geq 1$. All colonisation and extinction intensities, which can provide any transition within the class U_l, are positive. The one-step transition to the absorption state $\bar{0} = (0, \ldots, 0)$ is possible from the states in this class such that $\sum_{k=l+1}^{m} \delta(x_k, 1) = 1$.

Also, in the prelimit cases, i.e., if $0 < \varepsilon \leq \varepsilon_1$, the absorption probabilities satisfy $_{\bar{y}}f_{\bar{x}\bar{0}}^{(\varepsilon)} > 0$ for the states $\bar{x} \in X_0 \setminus U_l$, $\bar{y} \in X_1^{(0)}$ such that $x_i = 1$ if the corresponding extinction intensity is positive, $g_{ii}^{(\varepsilon)} > 0$, and $x_i = 0$ if the corresponding failure intensity is $g_{ii}^{(\varepsilon)} = 0$.

Indeed, in this case, it is possible that a transition to some state $\bar{z}$ from the class U_l occurs without visiting the class $X_1^{(0)}$ and then a transition from the state $\bar{z}$ to the state $\bar{0}$ occurs without leaving the class U_l.

The situation is different in the limit case where $\varepsilon = 0$. In this case, as was pointed out above, the hitting probabilities satisfy $_{\bar{z}}f_{\bar{x}\bar{y}}^{(0)} = 0$ for any $\bar{x}, \bar{z} \in X_1^{(0)}$, $\bar{y} \in X_2^{(0)}$.

This is so because all extinction intensities equal zero for patches with the indices $i \in E$.

Thus, in this case, $X_1^{(0)}$ is a closed set of recurrent-without absorption states, while $X_2^{(0)}$ is a class of non-recurrent-without absorption states.

The absorption probabilities satisfy $_{\bar{y}}f_{\bar{x}\bar{0}}^{(0)} = 0$ for any $\bar{x} \in X_0 \setminus U_l$, $\bar{y} \in X_1^{(0)}$ but $_{\bar{y}}f_{\bar{x}\bar{0}}^{(0)} \in (0, 1)$, for $\bar{x} \in U_l$, $\bar{y} \in X_1^{(0)}$.

This also is true because all extinction intensities equal zero for the servers with indices $i \in E$.

Thus, the case where conditions $\mathbf{T_{10}}$ and $\mathbf{E'_{10}}$ hold, corresponds to a pseudo-stationary model. Moreover, in this case, the class of recurrent-without-absorption states is $X_1^{(0)}$ and the class of non-recurrent-without-absorption states is $X_2^{(0)}$.

As was mentioned above, conditions $\mathbf{T_{10}}$ and $\mathbf{E''_{10}}$ imply condition $\mathbf{A_{12}}$ to hold. According to this condition, for every $0 \leq \varepsilon \leq \varepsilon_3$, the hitting probabilities are positive, $_{\bar{0}}f_{\bar{x}\bar{y}}^{(\varepsilon)} > 0$ for any $\bar{x}, \bar{y} \in X_0$. This means that the Markov chain $\eta^{(\varepsilon)}(t)$ has one class of recurrent-without-absorption states $X_1^{(\varepsilon)} = X_0$. Moreover, in this case, it can readily be seen that also $_{\bar{y}}f_{\bar{x}\bar{0}}^{(\varepsilon)} > 0$ for any $\bar{x}, \bar{y} \in X_0$.

This holds because the colonisation and extinction intensities are positive for all patches.

Thus, the case where conditions $\mathbf{T}_{10}$ and $\mathbf{E}''_{10}$ hold, corresponds to a quasi-stationary model. Moreover, in this case, the class of recurrent-without-absorption states is $X_1^{(0)} = X_0$.

Let us discuss how conditions introduced in Chapter 4 are realised for the model described above. We assume that conditions $\mathbf{T}_{10}$ and $\mathbf{E}_{10}$ hold.

Relations (6.5.1), (6.5.2), and condition $\mathbf{T}_{10}$ obviously imply that, for $\bar{x}, \bar{y} \in X$,

$$q^{(\varepsilon)}_{\bar{x}\bar{y}} \to q^{(0)}_{\bar{x}\bar{y}} \quad \text{as } \varepsilon \to 0, \tag{6.5.8}$$

and

$$q^{(\varepsilon)}_{\bar{x}} \to q^{(0)}_{\bar{x}} \quad \text{as } \varepsilon \to 0. \tag{6.5.9}$$

By formula (6.5.3), the distributions of sojourn times are exponential, for $\bar{x}, \bar{y} \in X$,

$$F^{(\varepsilon)}_{\bar{x}\bar{y}}(t) = F^{(\varepsilon)}_{\bar{x}}(t) = 1 - \exp\{-q^{(\varepsilon)}_{\bar{x}}t\}, \quad t \geq 0. \tag{6.5.10}$$

Relations (6.5.8) and (6.5.9), and formula (6.5.10) imply that condition $\mathbf{T}_3$ holds. Thus, condition $\mathbf{T}_1$ holds. Formulas (6.5.9) and (6.5.10) imply that condition $\mathbf{I}_3$ holds. Relation (6.5.9) and formula (6.5.10) also imply that conditions $\mathbf{M}_{11}$ and $\mathbf{M}_{12}$ hold.

The non-arithmetic conditions $\mathbf{N}_1$ and $\mathbf{N}_2$ for return times in states $\bar{x} \in X_1^{(0)}$ obviously hold since the distributions of sojourn times are exponential.

Let us define

$$\rho^* = \min(q^{(0)}_{\bar{x}}, \bar{x} \neq \bar{0}). \tag{6.5.11}$$

By formula (6.5.5), $\rho^* > 0$.

As was mentioned in Section 5.6, condition $\mathbf{C}_{12}$ holds for any $0 < \delta < \rho^*$.

Conditions $\mathbf{T}_{10}$ and $\mathbf{E}'_{10}$ correspond to the pseudo-stationary case. In this case, as follows from Lemma 4.5.8, condition $\mathbf{C}_{13}$ also holds.

Conditions $\mathbf{T}_{10}$ and $\mathbf{E}''_{10}$ correspond to the quasi-stationary case. In this case, condition $\mathbf{G}_1$ holds with the parameter ρ^* given in formula (6.5.11). Condition $\mathbf{G}_2$ also holds since in this case $X_2^{(0)} = \varnothing$. Therefore, as follows from Lemma 4.5.10, conditions $\mathbf{C}_{12}$ and $\mathbf{C}_{13}$ also hold.

Conditions $\mathbf{C}_{14}$ and $\mathbf{C}_{15}$ also hold as follows from the remarks made in Section 5.6. Condition $\mathbf{C}_{16}$ is not needed, since $X_2^{(0)} = \varnothing$ in the quasi-stationary case.

So, all conditions introduced in Chapter 4 are verified. All results given in Chapter 4 can also be applied to the Markov chain $\eta^{(\varepsilon)}(t) = (\eta_1^{(\varepsilon)}(t), \ldots, \eta_m^{(\varepsilon)}(t))$, $t \geq 0$ that describes stochastic dynamics of the metapopulation model introduced in the current section.

6.5.3   Mixed ergodic and limit/large deviation theorems

Conditions $\mathbf{T}_{10}$ and $\mathbf{E}'_{10}$ correspond to a pseudo-stationary model with the absorption probabilities asymptotically vanishing to zero. Under these conditions, we can obtain analogues of mixed ergodic and limit Theorems 4.4.1 and 4.4.2 and mixed ergodic and large deviation Theorems 4.6.1 and 4.6.2, covering the case of the pseudo-stationary asymptotics. Also, analogues of Theorems 4.4.3 and 4.4.5 give conditions for convergence of the corresponding stationary distributions under the additional non-absorption condition $\mathbf{A}_{11}$.

Conditions $\mathbf{T}_{10}$ and $\mathbf{E}''_{10}$ correspond to the quasi-stationary model with the absorption probabilities asymptotically separated from zero. Under these conditions, analogues of mixed ergodic and large deviation Theorems 4.6.3, 4.6.5, 4.6.7, and 4.6.11, covering the case quasi-stationary asymptotics can be formulated as well as Theorem 4.6.6, which give conditions for convergence of the corresponding quasi-stationary distributions. It is also useful to note that the theorems listed above can be applied in the pseudo-stationary case as well.

The limit parameters, including stationary and quasi-stationary distributions, can be calculated by using the corresponding algorithms and formulas given in Chapter 4. They should be specified for the semi-Markov processes $\eta^{(\varepsilon)}(t)$, $t \geq 0$, introduced in the current section.

6.5.4   Conditions for exponential expansions in mixed ergodic and large deviation theorems

Let us formulate conditions that replace the corresponding perturbation conditions introduced in Chapter 5 in the corresponding theorems representing exponential asymptotic expansions in mixed ergodic and limit/large deviation theorems.

The perturbation condition $\mathbf{P}_{30}^{(k)}$ can be replaced with the following condition:

$\mathbf{P}_{38}^{(k)}$: $g_{ij}^{(\varepsilon)} = g_{ij}^{(0)} + g_{ij}[1]\varepsilon + \cdots + g_{ij}[k]\varepsilon^k + o(\varepsilon^k)$ for $i, j = 1, \ldots, m$, where $|g_{ij}[r]| < \infty$, $r = 0, \ldots, k$, $i, j = 1, \ldots, m$.

It is also convenient to define $g_{ij}[0] = g_{ij}^{(0)}$, $i, j = 1, \ldots, m$.

Note that condition $\mathbf{P}_{38}^{(k)}$ obviously implies condition $\mathbf{T}_{10}$.

Condition $\mathbf{P}_{38}^{(k)}$ also implies that condition $\mathbf{P}_{30}^{(k)}$ and the corresponding expansions for the transition intensities $q_{\bar{x}\bar{y}}^{(\varepsilon)}$ and $q_{\bar{x}}^{(\varepsilon)}$ take the following form for $\bar{x} \neq \bar{0}, \bar{y} \in X$:

$$q_{\bar{x}\bar{y}}^{(\varepsilon)} = q_{\bar{x}\bar{y}}^{(0)} + q_{\bar{x}\bar{y}}[1]\varepsilon + \cdots + q_{\bar{x}\bar{y}}[k]\varepsilon^k + o(\varepsilon^k), \tag{6.5.12}$$

where, for $r = 1, \ldots, k$,

$$q_{\bar{x}\bar{y}}[r] = \begin{cases} g_{ii}[r] & \text{if } |\bar{x} - \bar{y}| = 1, x_i = 1, y_i = 0, \\ \sum_{i:x_i=1} g_{ij}[r] & \text{if } |\bar{x} - \bar{y}| = 1, x_j = 0, y_j = 1, \\ 0 & \text{otherwise}, \end{cases} \tag{6.5.13}$$

and

$$q_{\bar{x}}^{(\varepsilon)} = q_{\bar{x}}^{(0)} + q_{\bar{x}}[1]\varepsilon + \cdots + q_{\bar{x}}[k]\varepsilon^k + o(\varepsilon^k), \tag{6.5.14}$$

where, for $r = 1, \ldots, k$,

$$q_{\bar{x}}[r] = \sum_{i:x_i=1} g_{ii}^{(\varepsilon)}[r] + \sum_{j:x_j=0}\sum_{i:x_i=1} g_{ij}^{(\varepsilon)}[r]. \tag{6.5.15}$$

It is also convenient to define $q_{\bar{x}\bar{y}}[0] = q_{\bar{x}\bar{y}}^{(0)}$ and $q_{\bar{x}}[0] = q_{\bar{x}}^{(0)}$ for $\bar{x} \neq \bar{0}$, $\bar{y} \in X$.

As soon as the asymptotic expansions for the transition intensities $q_{\bar{x}\bar{y}}^{(\varepsilon)}$ and $q_{\bar{x}}^{(\varepsilon)}$ are obtained, one can get, for every $\bar{x} \in X_0$ and $n = 0, 1, \ldots$, asymptotic expansions for power moments

$$m_{\bar{x}\bar{y}}^{(\varepsilon)}[n] = m_{\bar{x}}^{(\varepsilon)}[n]$$

$$= \int_0^\infty s^n q_{\bar{x}}^{(\varepsilon)} \exp\{-q_{\bar{x}}^{(\varepsilon)}s\}\, ds = n!(q_{\bar{x}}^{(\varepsilon)})^{-n} < \infty, \tag{6.5.16}$$

and moment generating functions

$$\psi_{\bar{x}\bar{y}}^{(\varepsilon)}[\rho, n] = \psi_{\bar{x}}^{(\varepsilon)}[\rho, n] = \int_0^\infty s^n e^{\rho s} q_{\bar{x}}^{(\varepsilon)} \exp\{-q_{\bar{x}}^{(\varepsilon)}s\}\, ds$$

$$= \frac{n!q_{\bar{x}}^{(\varepsilon)}}{(q_{\bar{x}}^{(\varepsilon)} - \rho)^{n+1}} < \infty, \quad \rho < q_{\bar{x}}^{(\varepsilon)}. \tag{6.5.17}$$

These expansions can be obtained with the use of Lemmas 5.6.3 and 5.6.5. Note that here the only condition $\mathbf{P}_{38}^{(k)}$ is required.

Therefore, condition $\mathbf{P}_{38}^{(k)}$ implies that the perturbation conditions $\mathbf{P}_{18}^{(k)}$ and $\mathbf{P}_{20}^{(k)}$ hold and that the perturbation conditions $\mathbf{P}_{24}^{(\rho,k)}$ and $\mathbf{P}_{26}^{(\rho,k)}$ hold for $\rho < \rho^*$.

The only conditions that need an additional analysis are conditions $\mathbf{O}_3^{(h)}$, $\mathbf{O}_4^{(h)}$.

Recall that these conditions determine pivotal properties of asymptotic expansions for the characteristic root $\rho^{(\varepsilon)}$ of the equation $_{\bar{0}}\phi_{\bar{x}\bar{x}}^{(\varepsilon)}(\rho) = 1$ for $\bar{x} \in X_1^{(0)}$ in the most interesting pseudo-stationary case.

Let us first consider the case where the following condition, additional to conditions $\mathbf{E}_{10}'$ and $\mathbf{P}_{38}^{(k)}$, holds:

$\mathbf{O}_{21}$: $g_{ii}[1] > 0$, $i \in E$.

Condition $\mathbf{O}_{21}$ means that the extinction intensities for the main patches are of the order $O(\varepsilon)$.

Let us first clarify pivotal properties for the cyclic potential quantities,

$$_{\bar{x}}u_{\bar{x}\bar{y}}^{(\varepsilon)} = \mathsf{E}_{\bar{x}} \sum_{n=1}^{\nu_{\bar{x}}^{(\varepsilon)} \wedge \nu_{\bar{0}}^{(\varepsilon)}} \delta(\eta_{n-1}^{(\varepsilon)}, \bar{y}), \quad \bar{x} \in X_1^{(0)}, \quad \bar{y} \in X_0.$$

Two cases should be considered **(1)** $k \geq l$ and **(2)** $k < l$. Recall that $k \geq 0$ is the order of the asymptotic expansion in the perturbation condition $\mathbf{P}_{38}^{(k)}$, while $l \geq 1$ is the number of indices in the set E, i.e., the number of the main servers.

As shown in Lemma 5.1.1, the potential quantities $_{\bar{x}}u_{\bar{x}\bar{y}}^{(\varepsilon)}$ admit $(0,k)$-expansions,

$$_{\bar{x}}u_{\bar{x}\bar{y}}^{(\varepsilon)} = {}_{\bar{x}}u_{\bar{x}\bar{y}}[0] + {}_{\bar{x}}u_{\bar{x}\bar{y}}[1]\varepsilon + \cdots + {}_{\bar{x}}u_{\bar{x}\bar{y}}[k]\varepsilon^k + o(\varepsilon^k). \tag{6.5.18}$$

Let us show that, **(a)** in case **(1)**, where $k \geq l$, conditions $\mathbf{E}'_{10}$, $\mathbf{P}_{38}^{(k)}$, and $\mathbf{O}_{21}$ imply that $_{\bar{x}}u_{\bar{x}\bar{y}}[n] = 0$, $n < r$, while $_{\bar{x}}u_{\bar{x}\bar{y}}[r] > 0$ for $\bar{x} \in X_1^{(0)} = Z_0$, $\bar{y} \in U_r, 0 \leq r \leq l$ (the sets U_r were defined in relations (6.5.6) and (6.5.7)).

According to Lemma 5.1.2, statement **(a)** is equivalent to that **(b)** the global asymptotic order is $\hat{r}(\bar{x}, \bar{y}) = r$ for states $\bar{x} \in U_0$, $\bar{y} \in U_r, 0 \leq r \leq l$.

Statement **(b)** follows from an obvious observation that **(c)** the local asymptotic order $r(\bar{x}', \bar{x}'')$ equals 1 if $\bar{x}' \in U_n, \bar{x}'' \in U_{n+1}$, for some $n = 0, \ldots, l-1$, or 0 otherwise. Let $(\bar{x}_0, \ldots, \bar{x}_q)$ be an irreducible $(\bar{x}', \bar{x}'')$-chain of states from X_0, where $\bar{x}' = \bar{x}_0 \in U_0, \bar{x}'' = \bar{x}_q \in U_r$. Obviously, **(d)** the minimal number of pairs $(\bar{x}_n, \bar{x}_{n+1})$ with the local asymptotic order $r(\bar{x}_n, \bar{x}_{n+1}) = 1$ in such an irreducible $(\bar{x}', \bar{x}'')$-chain is r. These pairs correspond to transitions from some state in the class U_n to some state in the class U_{n+1} for $n = 0, \ldots, r-1$. This proves statement **(b)** and, consequently, statement **(a)**.

Note that the statement **(a)** implies that **(e)** $Z_r = U_r, r = 0, \ldots, l$ (the sets Z_r were defined in Section 5.1). Also, $Z_r = \varnothing, r = l+1, \ldots, k+1$, since $\bigcup_{n=0}^{l} U_n = X_0$.

Denote by U'_{l-1} the set of states $\bar{x} = (x_1, \ldots, x_m)$ such that $x_i = 1$ for some $1 \leq i \leq l$ and $x_j = 0$ for $j \neq i$, and by U''_l the set of states $\bar{x} = (x_1, \ldots, x_m)$ such that $x_i = 1$ for some $l+1 \leq i \leq m$ and $x_j = 0$ for $j \neq i$. By the definition, $U'_{l-1} \subseteq U_{l-1}$ and $U''_l \subseteq U_l$.

Formula (6.5.3) and conditions $\mathbf{E}'_{10}$ and $\mathbf{O}_{21}$ imply that **(f)** $q_{\bar{x}\bar{0}}^{(\varepsilon)} = 0$ if $\bar{x} \in \bigcup_{n=0}^{l-2} U_n$; **(g)** $q_{\bar{x}\bar{0}}^{(\varepsilon)} \sim g_{i,-}[1]\varepsilon$ as $\varepsilon \to 0$ (note that $g_{i,-}[1] > 0$) if $x_i = 1, \bar{x} \in U'_{l-1}$, for $i = 1, \ldots, l$, and $q_{\bar{x}\bar{0}}^{(\varepsilon)} = 0$ if $\bar{x} \in U_{l-1} \setminus U'_{l-1}$; **(h)** $q_{\bar{x}\bar{0}}^{(\varepsilon)} \sim g_{ii}^{(0)}$ as $\varepsilon \to 0$ (note that $g_{ii}^{(0)} > 0$) if $x_i = 1, \bar{x} \in U''_l$, for $i = l+1, \ldots, m$, and $q_{\bar{x}\bar{0}}^{(\varepsilon)} = 0$ if $\bar{x} \in U_l \setminus U''_l$.

Using relations **(f)–(h)**, the asymptotic expansions (6.5.12) and (6.5.14) and the quotient rule for the asymptotic expansions given in statement **(iv)** of Lemma 8.1.1, one can readily get properties similar to **(f)–(h)** for the one-step absorption transition probabilities $p_{\bar{x}\bar{0}}^{(\varepsilon)} = q_{\bar{x}\bar{0}}^{(\varepsilon)}/q_{\bar{x}}^{(\varepsilon)}$. These properties are: **(i)** the transition probabilities are zero, $p_{\bar{x}\bar{0}}^{(\varepsilon)} = 0$ if $\bar{x} \in \bigcup_{n=0}^{l-2} U_n$; **(j)** $p_{\bar{x}\bar{0}}^{(\varepsilon)} \sim g_{i,-}[1]\varepsilon/q_{\bar{x}}^{(0)}$ as $\varepsilon \to 0$ if $x_i = 1$, $\bar{x} \in U'_{l-1}$ for $i = 1, \ldots, l$, and $p_{\bar{x}\bar{0}}^{(\varepsilon)} = 0$ if $\bar{x} \in U_{l-1} \setminus U'_{l-1}$; **(k)** $p_{\bar{x}\bar{0}}^{(\varepsilon)} \sim g_{i,-}^{(0)}/q_{\bar{x}}^{(0)}$ as $\varepsilon \to 0$ if $x_i = 1, \bar{x} \in U''_l$ for $i = l+1, \ldots, m$, and $p_{\bar{x}\bar{0}}^{(\varepsilon)} = 0$ if $\bar{x} \in U_l \setminus U''_l$.

Statements **(i)–(k)** imply that **(l)** the set $Y_{\bar{0},r}$ (defined in Subsection 5.1.5) coincides with U''_l for $r = 0$, with U'_{l-1} for $r = 1$, with $\varnothing$ for $r = 2, \ldots, k$, and with $X_0 \setminus (U'_{l-1} \cup U''_l)$ for $r = k+1$.

Statements **(a)** and **(l)** imply in an obvious way that **(m)** the set $W_{\bar{0},r}$ (defined in Subsection 5.1.5) coincides with $\varnothing$ for $r = 0,\ldots,l-1$, with $U'_{l-1} \cup U''_l$ for $r = l$, with $\varnothing$ for $r = l+1,\ldots,k$, and with $X_0 \setminus (U'_{l-1} \cup U''_l)$ for $r = k+1$.

Therefore, **(n)** if conditions $\mathbf{E}'_{10}$ and $\mathbf{P}^{(k)}_{38}$, and $\mathbf{O}_{21}$ hold and the additional assumption **(1)** $k \geq l$ holds, then condition $\mathbf{O}^{(l)}_4$ holds.

In this case, by Lemmas 5.1.5 and 5.1.7, the cyclic absorption probabilities $_{\bar{x}}f^{(\varepsilon)}_{\bar{x}\bar{0}}$, $\bar{x} \in X^{(0)}_1$ have (l,k)-asymptotic pivotal expansions (with the coefficients $_{\bar{x}}\bar{b}_{\bar{x}\bar{0}}[l] > 0$),

$$_{\bar{x}}f^{(\varepsilon)}_{\bar{x}\bar{0}} = {}_{\bar{x}}\bar{b}_{\bar{x}\bar{0}}[l]\varepsilon^l + \cdots + {}_{\bar{x}}\bar{b}_{\bar{x}\bar{0}}[k]\varepsilon^k + o(\varepsilon^k). \tag{6.5.19}$$

Also, by Theorem 5.4.1, the characteristic root $\rho^{(\varepsilon)}$ of the equation $_{\bar{0}}\phi^{(\varepsilon)}_{\bar{x}\bar{x}}(\rho) = 1$ has the following (l,k)-asymptotic pivotal expansion:

$$\rho^{(\varepsilon)} = a_l\varepsilon^l + \cdots + a_k\varepsilon^k + o(\varepsilon^k). \tag{6.5.20}$$

Similarly, it can be shown that **(o)** in case **(2)**, where $k < l$, conditions $\mathbf{E}'_{10}$, $\mathbf{P}^{(k)}_{38}$, and $\mathbf{O}_{21}$ imply that the coefficients satisfy $_{\bar{x}}u_{\bar{x}\bar{y}}[n] = 0$, $n < r$, while $_{\bar{x}}u_{\bar{x}\bar{y}}[r] > 0$, for $\bar{x} \in X^{(0)}_1 = U_0$, $\bar{y} \in U_r$, $0 \leq r \leq k$ (these sets were defined in relations (6.1.6) and (6.1.7)), while **(p)** $_{\bar{x}}u_{\bar{x}\bar{y}}[n] = 0$, $n < k+1$ for $\bar{x} \in X^{(0)}_1 = U_0$, $\bar{y} \in \bigcup^l_{r=k+1} U_r$.

Statement **(o)** implies that **(q)** $Z_r = U_r$, $r = 0,\ldots,k$, and $Z_{k+1} = \bigcup^l_{r=k+1} U_r$. The sets $Y_{\bar{0},r}$ were introduced in statement **(l)**.

It follows from statements **(o)** and **(l)** that **(r)** the set $W_{\bar{0},r}$ coincides with $\varnothing$, for $r = 0,\ldots,k$ and $W_{\bar{0},k+1} = X_0$.

Therefore, **(s)** if conditions $\mathbf{E}'_{10}$ and $\mathbf{P}^{(k)}_{38}$, and $\mathbf{O}_{21}$ hold and the additional assumption **(1)** $k \leq l$ is true, then condition $\mathbf{O}^{(k)}_3$ holds.

By Lemmas 5.1.5 and 5.1.7, the cyclic absorption probabilities $_{\bar{x}}f^{(\varepsilon)}_{\bar{x}\bar{0}}$, $\bar{x} \in X^{(0)}_1$, satisfy the following relations: **(t)** $_{\bar{x}}f^{(\varepsilon)}_{\bar{x}\bar{0}} = o(\varepsilon^k)$, $\bar{x} \in X^{(0)}_1$. Also, by Theorem 5.4.1, the characteristic root $\rho^{(\varepsilon)}$ of the equation $_{\bar{0}}\phi^{(\varepsilon)}_{\bar{x}\bar{x}}(\rho) = 1$ satisfies the relation **(u)** $\rho^{(\varepsilon)} = o(\varepsilon^k)$.

The asymptotic relations and **(t)** and **(u)** are not surprising. One should not expect to get more precise asymptotics under the perturbation condition $\mathbf{P}^{(k)}_{38}$ that give exact asymptotics for transition intensities only up to the terms of the order ε^k.

Let us also consider a more complicated case where the following condition, additional to conditions $\mathbf{E}'_{10}$ and $\mathbf{P}^{(k)}_{38}$, holds:

$\mathbf{O}_{22}$: $g_{ii}[n] = 0$, $n < k_i$, $i \in E$ and $g_{ii}[k_i] > 0$, $i \in E$, for some $1 \leq k_i \leq k$, $\quad i = 1,\ldots,m$.

Condition $\mathbf{O}_{22}$ means that the extinction intensities for the main patches $i \in E$ may be of the different orders $O(\varepsilon^{k_i})$.

As above, two cases should be considered, **(3)** $k \geq k_1 + \cdots + k_l = \bar{k}_l$ and **(4)** $k < \bar{k}_l$.

By repeating the asymptotic analysis carried out above, it can be shown that, under conditions $\mathbf{E}'_{10}$, $\mathbf{P}^{(k)}_{38}$, and $\mathbf{O}_{22}$, condition $\mathbf{O}_3^{(\bar{k}_l)}$ holds in case **(3)**, or condition $\mathbf{O}_4^{(k)}$ holds in case **(4)**. Conditions $\mathbf{O}_5^{(\rho, h)}$, $\mathbf{O}_6^{(\rho, h)}$, as was mentioned in Subsection 5.3.7, can not be expressed so explicitly in terms of coefficients in the asymptotic expansions used in the perturbation condition $\mathbf{P}^{(k)}_{38}$. Here, we should use the coefficients of the corresponding expansions for the moment generating functions $_{\bar{0}}\phi^{(\varepsilon)}_{\bar{x}\bar{x}}(\rho)$, $x \in X_1^{(0)}$, given in Lemma 5.3.5 and specified for the semi-Markov processes considered in the current section.

6.5.5   Exponential expansions in mixed ergodic and large deviation theorems

All results given in Chapter 5 can be used for the semi-Markov processes $\eta^{(\varepsilon)}(t), t \geq 0$ that describe stochastic dynamics of metapopulation models.

Conditions $\mathbf{T}_{10}$, $\mathbf{E}'_{10}$, $\mathbf{P}^{(k)}_{38}$, and $\mathbf{O}_{21}$ or $\mathbf{O}_{22}$ correspond to a pseudo-stationary model with the absorption probabilities asymptotically vanishing to zero. Under these conditions, there are analogues of Theorems 5.4.1 and 5.4.2, which give exponential asymptotic expansions in mixed ergodic and large deviation theorems. Also an analogue of Theorem 5.5.1 gives higher order asymptotic expansions for stationary distributions under the additional non-absorption condition $\mathbf{A}_{11}$.

Conditions $\mathbf{T}_{10}$, $\mathbf{E}''_{10}$, and $\mathbf{P}^{(k)}_{38}$ correspond to a quasi-stationary model with the absorption probabilities asymptotically separated from zero. Under these conditions, one can obtain analogues of Theorems 5.4.3 and 5.4.4, which give exponential asymptotic expansion in mixed ergodic and large deviation theorems. Also, analogues of Theorems 5.5.4 and 5.5.5 give higher order asymptotic expansions for quasi-stationary distributions. It is useful to note that the theorems listed above also cover the pseudo-stationary case, i.e., they can also applied if conditions $\mathbf{T}_{10}$, $\mathbf{E}'_{10}$, and $\mathbf{P}^{(k)}_{38}$ hold.

The limit parameters, including the stationary and quasi-stationary distributions, and the coefficients in the corresponding asymptotic expansions, can be calculated by using the corresponding algorithms and formulas given in Chapters 4 and 5. They should be specified for the semi-Markov processes $\eta^{(\varepsilon)}(t), t \geq 0$, introduced in the current section.

Chapter 7

Nonlinearly perturbed risk processes

Chapter 7 contains results that extend the classical Cramér–Lundberg and diffusion approximations for the ruin probabilities to a model of nonlinearly perturbed risk processes. Both approximations are presented in a unified way using the techniques of perturbed renewal equations developed in Chapters 1 and 2. Correction terms are obtained for the Cramér–Lundberg and the diffusion type approximations, which provides a needed asymptotic behaviour of relative errors in the perturbed model. We study the dependence of these correction terms on the relations between the rate of perturbation and the rate of growth of the initial capital. We also give various variants of the diffusion type approximation, including the asymptotics for increments and derivatives of the ruin probabilities.

In the traditional risk theory, the *risk process*

$$\xi_u(t) = u + vt - \sum_{k=1}^{\eta(t)} \kappa_k, \quad t \geq 0, \tag{7.0.1}$$

is used to model functioning of an insurance company. In (7.0.1), u is a nonnegative constant that denotes an initial capital of a company, a positive constant v denotes the gross premium rate, $\eta(t)$ is the Poisson process (with a parameter λ) that counts the number of claims to the company in the time-interval $[0, t]$, κ_k, $k = 1, 2, \ldots$, is a sequence of nonnegative i.i.d. random variables (with a finite mean μ) which are independent for the process $\eta(t)$, and κ_k is the amount of the kth claim. An important object of the study is the *ruin probability* for a company that has an initial capital u,

$$P(u) = \mathsf{P}\{\inf_{t \geq 0} \xi_u(t) < 0\}, \quad u \geq 0.$$

Of a special interest is the asymptotics of the ruin probabilities for large values of the initial capital u.

The loading rate of claims to the company is characterized by a constant $\alpha = \frac{\lambda\mu}{v}$. Only the subcritical case with $\alpha < 1$ is nontrivial (if $\alpha \geq 1$, then the ruin probability $P(u) \equiv 1$). The classical result, known as the *Cramér–Lundberg approximation*, gives an asymptotics for the ruin probability under Cramér type conditions conditions on the claim distribution $G(t)$,

$$\frac{P(u)}{e^{-\rho u}} \to \pi \quad \text{as } u \to \infty, \tag{7.0.2}$$

where the constant ρ, known as the Lundberg exponent, and the limit constant π are determined from the constant α and the claim distribution $G(u)$.

Of a special interest also is the asymptotics of the ruin probabilities for large values of u and the values of α which are less than but close to 1. Here, the so-called diffusion approximation gives an answer. In the simplest case, it describes the asymptotic behaviour of the ruin probabilities when the initial capital is $u \to \infty$, and, simultaneously, the growth premium rate satisfies $v \downarrow \lambda\mu$. Under the condition that $u \to \infty$ and $\alpha \to 1$ such that $(1-\alpha)u \to \varrho$, the *diffusion approximation* yields an asymptotics for the ruin probability,

$$P(u) \to e^{-\varrho/a_1} \quad \text{as} \quad u \to \infty, \tag{7.0.3}$$

where the constant a_1 is determined by the claim distribution $G(t)$.

Taking the asymptotic relations (7.0.2) and (7.0.3) as a starting point, we formulate the problem in a more general way. We consider a family $\xi_u^{(\varepsilon)}(t)$, $t \geq 0$, of risk processes that depend on a small parameter $\varepsilon \geq 0$. The process $\xi_u^{(\varepsilon)}(t)$ is considered as a perturbation of the process $\xi_u^{(0)}(t)$ and, therefore, we assume some weak continuity conditions satisfied by the characteristic quantities $v = v^{(\varepsilon)}$, $\lambda = \lambda^{(\varepsilon)}$, and the distributions $G(u) = G^{(\varepsilon)}(u)$, considered as functions of ε, at the point $\varepsilon = 0$. Moreover, we allow for nonlinear perturbations, which means that the characteristic quantities of the perturbed risk processes are nonlinear functions of ε. However, we restrict our attention to the case of a smooth perturbation and, hence, assume that these functions can be expanded in a power series with respect to ε up to and including the order k. The object of our study is the asymptotic behaviour of the ruin probabilities $P^{(\varepsilon)}(u)$ for $u \to \infty$ and $\varepsilon \to 0$.

The relationship between the rates at which ε tends to zero and the initial capital u tends to infinity has an influence upon the obtained results. Without loss of generality it can be assumed that $u = u^{(\varepsilon)}$ is a function of the parameter ε. The relationship between the rate of perturbation and the rate of growth of the initial capital is characterized by the following balancing condition:

$$\varepsilon^r u^{(\varepsilon)} \to \varrho_r \quad \text{as} \quad \varepsilon \to 0 \tag{7.0.4}$$

for some positive integer r and $\varrho_r \in [0, \infty)$.

Under the assumptions described above and some natural Cramér type condition on the claim distributions $G^{(\varepsilon)}$, we obtain the asymptotic relation

$$\frac{P^{(\varepsilon)}(u^{(\varepsilon)})}{\exp\{-(\rho + a_1\varepsilon + \cdots + a_{r-1}\varepsilon^{r-1})u^{(\varepsilon)}\}} \to e^{-\varrho_r a_r}\pi \quad \text{as} \quad \varepsilon \to 0, \tag{7.0.5}$$

where ρ and π are the same as in (7.0.2) and (7.0.3), and we give an explicit recurrence algorithm for calculating the coefficients $a_1, \ldots, a_r$ as rational functions of the coefficients in the expansions for the characteristic quantities of the perturbed risk processes.

The ruin probability $P^{(\varepsilon)}(u^{(\varepsilon)}) = \mathsf{P}\{\inf_{t \geq 0} \xi_0^{(\varepsilon)}(t) < -u^{(\varepsilon)}\}$ can be interpreted as a tail probability for the infimum of the risk process $\xi_0^{(\varepsilon)}(t)$. With this interpretation, the asymptotic relation (7.0.5) takes the form of a large deviation theorem. Depending on whether ϱ_r equals zero or is positive, the balancing condition (7.0.4) states that either $u^{(\varepsilon)} = o(\varepsilon^{-r})$ or $u^{(\varepsilon)} = O(\varepsilon^{-r})$. Following the standard terminology of large deviation theory we refer to this asymptotic behaviour of $u^{(\varepsilon)}$ as different large deviation zones.

The main new element in our results is a high order exponential asymptotic expansion which extends in a unified way both the classical Cramér–Lundberg approximation and the diffusion approximation of ruin probabilities to a model of nonlinearly perturbed risk processes. The derived approximations have an optimal asymptotic behaviour of the relative errors. This behaviour depends on the balance between the order of the perturbation parameter ε and the order of growth of the initial capital u. The approximations are supplemented with an explicit algorithm for calculating the asymptotic corrections.

We pay an additional attention to the case of the diffusion approximation. In addition to asymptotics (7.0.5), we also give asymptotics for increments of the ruin probabilities and for densities of the non-ruin probabilities.

Theorem 7.2.1 and Lemmas 7.2.1–7.2.3 give various diffusion type asymptotics for the ruin probabilities. Theorems 7.2.2 and 7.2.3 give asymptotics for increments and densities of ruin probabilities for nonlinearly perturbed risk type processes in the model of diffusion approximation. Theorem 7.3.1 gives an approximation for ruin probabilities that covers both Cramér–Lundberg and diffusion approximations in the model where the gross premium rate, the parameter of the Poisson claim flow and the claim distribution are nonlinearly perturbed. Theorem 7.3.2 gives exponential expansions which improve the result of Theorem 7.3.1. Approximations for the ruin probabilities based on exponential asymptotic expansions of type (7.0.5) are given in Subsection 7.4.2. Theorems 7.5.1–7.5.3 give diffusion and Cramér–Lundberg type approximations for distribution of the capital surplus prior and at ruin for nonlinearly perturbed risk processes.

In conclusion, we would like to note that the results obtained in Chapter 7 also have interpretations in the queueing theory. As is known, the Cramér–Lundberg approximation describes the asymptotics of steady tail probabilities for waiting times of the classical M/G/1 queueing system in the subcritical ergodic case, while the diffusion approximation describes the corresponding asymptotics in the critical so-called heavy traffic case. In this context, the results obtained in the present section yield new large deviation asymptotics for waiting times of perturbed M/G/1 queueing system.

7.1   Cramér–Lundberg and diffusion approximations

In this section we give a short summary of classical results concerning the Cramér–Lundberg and the diffusion approximations for ruin probabilities.

7.1.1   Classical Cramér–Lundberg approximation for ruin probabilities

Let $\xi_u(t)$, $t \geq 0$ be a standard *risk process* defined in the following way:

$$\xi_u(t) = u + \upsilon t - \sum_{k=1}^{\eta(t)} \kappa_k, \quad t \geq 0, \tag{7.1.1}$$

where (**a**) u is a nonnegative constant, (**b**) υ is a positive constant, (**c**) κ_k, $k = 1, 2, \ldots$, is a sequence of nonnegative i.i.d. random variables with a distribution function $G(u)$ that has a finite expectation $\mu > 0$, (**d**) $\eta(t)$, $t \geq 0$ is a Poisson process with parameter $\lambda > 0$, (**e**) the sequence κ_k, $k = 1, 2, \ldots$, and the process $\eta(t)$, $t \geq 0$, are independent.

The risk process (7.1.1) has been used as a standard model of the business of an insurance company. The constant υ is the gross premium rate, the Poisson process $\eta(t)$ counts the number of claims to the company in the time interval $[0, t]$, and κ_k is the amount of the kth claim.

One of the classical objects of study is the *ruin probability* of the company that has an initial capital $u \geq 0$,

$$P(u) = \mathsf{P}\{\inf_{t \geq 0} \xi_u(t) < 0\}. \tag{7.1.2}$$

Of a special interest is the asymptotics of the ruin probability $P(u)$ when the initial capital u tends to infinity.

The following constant is a threshold parameter that determines the principal behaviour of the ruin probabilities:

$$\alpha = \frac{\lambda \mu}{\upsilon}.$$

It is known (see, for example, Feller (1966)) that the ruin probability $P(u)$, considered as a function in $u \geq 0$, satisfies, in the case where $\alpha \leq 1$, the renewal equation

$$P(u) = \alpha(1 - \bar{G}(u)) + \alpha \int_0^u P(u - s)\bar{G}(ds), \quad u \geq 0, \tag{7.1.3}$$

where

$$\bar{G}(u) = \frac{1}{\mu} \int_0^u (1 - G(s))\, ds.$$

The distribution function that generates this equation is $F(u) = \alpha \bar{G}(u)$. The total variation of the distribution function $F(u)$ is α.

Only the case $\alpha < 1$ is non-trivial, since $P(u) \equiv 1$ is a solution of equation (7.1.3) if $\alpha = 1$. Note also that in the case where $\alpha > 1$, the ruin probability is still $P(u) \equiv 1$, but it does not satisfy (7.1.3) in this case.

A standard condition under which the Cramér–Lundberg approximation is valid is the following:

$\mathbf{C_{21}}$: There exists $\delta > 0$ such that:

(a) $\int_0^\infty e^{\delta s} \bar{G}(ds) < \infty$;

(b) $\alpha \int_0^\infty e^{\delta s} \bar{G}(ds) > 1$.

Condition $\mathbf{C_{21}}$ guarantees, by Lemma 1.4.3, that the characteristic equation

$$\alpha \int_0^\infty e^{\rho s} \bar{G}(ds) = 1 \tag{7.1.4}$$

has a unique positive root ρ.

The root ρ is called the *Lundberg exponent*.

Multiplying equation (7.1.3) by $e^{\rho u}$ transforms it into a proper renewal equation for the function $P(u)e^{\rho u}$. Applying the renewal theorem (see, for example, Feller 1971) to this equation one obtains the following asymptotic relation known as the *Cramér–Lundberg approximation*:

$$\frac{P(u)}{e^{-\rho u}} \to \pi \ \text{ as } \ u \to \infty, \tag{7.1.5}$$

where

$$\pi = \frac{\int_0^\infty e^{\rho s}(1 - \bar{G}(s))\, ds}{\int_0^\infty s e^{\rho s} \bar{G}(ds)}. \tag{7.1.6}$$

7.1.2   Diffusion approximation for ruin probabilities

The Cramér–Lundberg approximation describes the asymptotics of the ruin probabilities for fixed values of $\alpha < 1$.

On the other hand, it does not give a detailed description of the asymptotics for the ruin probability $P(u)$ for large values of u and values of α that are less than but close to 1. For example, it does not describe the asymptotic behaviour of the ruin probabilities if the initial capital satisfies $u \to \infty$ and, simultaneously, the gross premium rate is $v \downarrow \lambda\mu$. Here, the so-called diffusion approximation gives an answer. The asymptotics depends on a balancing relation between the speeds at which v tends to $\lambda\mu$ from above and u tending to infinity. Typical conditions include a balancing condition of the type $(1 - \alpha)u \to \varrho$ as $\varepsilon \to 0$ plus a condition of finiteness of the second moment of the claim distributions. Under these condition, the *diffusion approximation* yields an asymptotics of the ruin probability in the form

$$P(u) \to e^{-\varrho/\bar{m}} \ \text{ as } \ u \to \infty, \tag{7.1.7}$$

where $\bar{m} = \beta/2\mu$ and $\beta = \mathsf{E}X_1^2$.

A traditional way to obtain a diffusion type asymptotics is based on approximating the risk processes by a Wiener process with a shift. Weak convergence of the risk processes, normalized in a proper way, to a Wiener process, and the corresponding functional limit theorems for the infinite interval are proved and then used to show that the ruin probabilities for the risk processes can be approximated by the corresponding ruin probabilities for the limit Wiener process with a shift; finally provide explicit formulas to calculate the corresponding ruin probabilities for the limit process are also used. The method described above gave the name for diffusion approximation.

7.2　Diffusion approximation for perturbed risk processes

In this section we present asymptotics results concerning the diffusion approximation of the ruin probabilities.

7.2.1　Diffusion approximation of the ruin probabilities for perturbed risk processes

Let $\xi_u^{(\varepsilon)}(t)$, $t \geq 0$, be a standard risk process defined for every $\varepsilon \geq 0$ in the following way:

$$\xi_u^{(\varepsilon)}(t) = u + \upsilon^{(\varepsilon)}t - \sum_{k=1}^{\eta^{(\varepsilon)}(t)} \kappa_k^{(\varepsilon)}, \quad t \geq 0, \tag{7.2.1}$$

where **(a)** u is a nonnegative constant, **(b)** $\upsilon^{(\varepsilon)}$ is a positive constant, **(c)** $\kappa_k^{(\varepsilon)}$, $k = 1, 2, \ldots$, is a sequence of nonnegative i.i.d. random variables with a distribution function $G^{(\varepsilon)}(u)$ that has a finite expectation $\mu^{(\varepsilon)} = \int_0^\infty u\,G^{(\varepsilon)}(du) > 0$, **(d)** $\eta^{(\varepsilon)}(t)$, $t \geq 0$, is a Poisson process with parameter $\lambda^{(\varepsilon)} > 0$, **(e)** the sequence of the random variables $\kappa_k^{(\varepsilon)}$, $k = 1, 2, \ldots$, and the process $\eta^{(\varepsilon)}(t)$, $t \geq 0$ are independent.

The object of our interest is the ruin probability

$$P^{(\varepsilon)}(u) = \mathsf{P}\{\inf_{t \geq 0} \xi_u^{(\varepsilon)}(t) < 0\}, \quad u \geq 0. \tag{7.2.2}$$

The ruin probability $P^{(\varepsilon)}(u)$ satisfies, for every $\varepsilon \geq 0$, the renewal equation (7.1.3) that takes the following form:

$$P^{(\varepsilon)}(u) = \alpha^{(\varepsilon)}(1 - \bar{G}^{(\varepsilon)}(u)) + \alpha^{(\varepsilon)} \int_0^u P^{(\varepsilon)}(u - s)\bar{G}^{(\varepsilon)}(ds), \quad u \geq 0, \tag{7.2.3}$$

where

$$\alpha^{(\varepsilon)} = \frac{\lambda^{(\varepsilon)}\mu^{(\varepsilon)}}{\upsilon^{(\varepsilon)}},$$

and

$$\bar{G}^{(\varepsilon)}(u) = \frac{1}{\mu^{(\varepsilon)}} \int_0^u (1 - G^{(\varepsilon)}(s))\,ds.$$

To ensure that for every $\varepsilon \geq 0$, $P^{(\varepsilon)}(u)$ is a solution of (7.2.3), we assume that the following condition is satisfied:

$\mathbf{H_1}$:   **(a)** $0 < v^{(\varepsilon)}, \lambda^{(\varepsilon)}, \mu^{(\varepsilon)} < \infty$ for every $\varepsilon \geq 0$;

   (b) $\alpha^{(\varepsilon)} \in (0, 1]$ for every $\varepsilon \geq 0$.

The mean value of the distribution function $\bar{G}^{(\varepsilon)}(u)$ can be expressed via the first two moments of the claim distribution function $G^{(\varepsilon)}(u)$,

$$m^{(\varepsilon)} = \int_0^\infty u \bar{G}^{(\varepsilon)}(du) = \frac{\beta^{(\varepsilon)}}{2\mu^{(\varepsilon)}}, \quad \beta^{(\varepsilon)} = \int_0^\infty u^2 G^{(\varepsilon)}(du). \tag{7.2.4}$$

Let us introduce the following condition:

$\mathbf{D_{28}}$: $\lim_{0 \leq t \to 0} \overline{\lim}_{\varepsilon \to 0} G^{(\varepsilon)}(t) < 1$.

Note that, according to the condition $\mathbf{E_1}$, $\mu^{(0)} > 0$, we have that $G^{(0)}(0) < 1$ as well.

We will also use the following condition:

$\mathbf{M_{16}}$: $\lim_{T \to \infty} \overline{\lim}_{0 \leq \varepsilon \to 0} \int_T^\infty u^2 G^{(\varepsilon)}(du) = 0$.

Note that condition $\mathbf{M_{16}}$ reduces to the condition $\beta^{(0)} < \infty$ if the claim distributions $G^{(\varepsilon)}(u) \equiv G^{(0)}(u)$ do not depend on the parameter ε.

Let $f^{(\varepsilon)} \geq 0$ and $g^{(\varepsilon)} > 0$ be functions of $\varepsilon \geq 0$. We use the notation $f^{(\varepsilon)} \sim g^{(\varepsilon)}$ if **(a)** $f^{(\varepsilon)}/g^{(\varepsilon)} \to 1$ as $\varepsilon \to 0$ and $f^{(\varepsilon)} \sim O(g^{(\varepsilon)})$ if **(b)** $\underline{\lim}_{\varepsilon \to 0} f^{(\varepsilon)}/g^{(\varepsilon)} > 0$, and **(c)** $\overline{\lim}_{\varepsilon \to 0} f^{(\varepsilon)}/g^{(\varepsilon)} < \infty$.

The following condition is a variant of the balancing condition:

$\mathbf{B_9}$: $u^{(\varepsilon)} \to \infty$ and $\alpha^{(\varepsilon)} \to 1$ in such a way that $(1 - \alpha^{(\varepsilon)})u^{(\varepsilon)} \sim O(1)$ as $\varepsilon \to 0$.

The main result of this section is the following theorem.

Theorem 7.2.1. *Let conditions* $\mathbf{H_1}$, $\mathbf{D_{28}}$, $\mathbf{M_{16}}$, *and* $\mathbf{B_9}$ *hold. Then*

$$\frac{P^{(\varepsilon)}(u^{(\varepsilon)})}{e^{-(1-\alpha^{(\varepsilon)})u^{(\varepsilon)}/m^{(\varepsilon)}}} \to 1 \quad as \ \varepsilon \to 0. \tag{7.2.5}$$

Proof. We prove several lemmas that are weak variants of Theorem 7.2.1. These lemmas have their own value and divide the proof of Theorem 7.2.1 into several logical steps.

The first lemma is based on the following conditions:

$\mathbf{D_{29}}$: $\bar{G}^{(\varepsilon)}(\cdot) \Rightarrow \bar{G}(\cdot)$ as $\varepsilon \to 0$, where $\bar{G}(u)$ is a non-arithmetic distribution function;

and

$\mathbf{M_{17}}$: $\bar{m}^{(\varepsilon)} \to \bar{m} = \int_0^\infty u\bar{G}(du) < \infty$ as $\varepsilon \to 0$.

Remark 7.2.1. As was shown in Remark 1.2.2, condition $\mathbf{M_{17}}$ is equivalent, under condition $\mathbf{D_{29}}$, to the following relation:

$$\lim_{T \to \infty} \overline{\lim_{0 \le \varepsilon \to 0}} \int_T^\infty u\bar{G}^{(\varepsilon)}(du) = 0. \tag{7.2.6}$$

Let us also use the following variant of the balancing condition:

$\mathbf{B_{10}}$: $u^{(\varepsilon)} \to \infty$ and $\alpha^{(\varepsilon)} \to 1$ in such a way that $(1 - \alpha^{(\varepsilon)})u^{(\varepsilon)} \to \varrho \in [0, \infty]$ as $\varepsilon \to 0$.

Lemma 7.2.1. *Let conditions* $\mathbf{H_1}$, $\mathbf{D_{29}}$, $\mathbf{M_{17}}$, *and* $\mathbf{B_{10}}$ *hold. Then*

$$P^{(\varepsilon)}(u^{(\varepsilon)}) \to e^{-\varrho/\bar{m}} \quad as \ \varepsilon \to 0. \tag{7.2.7}$$

Proof. We are going to apply Theorem 1.3.1 to the renewal equation (7.2.3).

Conditions $\mathbf{B_{10}}$ and $\mathbf{D_{29}}$ imply that condition $\mathbf{D_6}$ holds for the distribution functions $F^{(\varepsilon)}(u) = \alpha^{(\varepsilon)}\bar{G}^{(\varepsilon)}(u)$ with the limit distribution function $F^{(0)}(u) = \bar{G}(u)$.

Conditions $\mathbf{B_{10}}$ and $\mathbf{M_{17}}$ imply that condition $\mathbf{M_6}$ holds for the distribution functions $F^{(\varepsilon)}(u)$ with $\bar{m}$ being a limit value of the expectation.

In this case, the free term in equation (7.2.3) is $q^{(\varepsilon)}(u) = \alpha^{(\varepsilon)}(1 - \bar{G}^{(\varepsilon)}(u))$. Monotonicity of this function and conditions $\mathbf{D_{29}}$ and $\mathbf{B_{10}}$ imply in an obvious way that condition $\mathbf{F_2}$ **(a)** holds. The set of points of continuity of the limit function $q_0(u) = 1 - \bar{G}(u)$ can serve as a set of local uniform convergence in condition $\mathbf{F_2}$ **(a)**.

Condition $\mathbf{F_2}$ **(b)** holds, since $0 \le q^{(\varepsilon)}(u) \le 1$.

Condition $\mathbf{F_2}$ **(c)** follows from (7.2.6) and the following estimate:

$$h \sum_{r \ge T/h} \sup_{rh \le u \le (r+1)h} |q^{(\varepsilon)}(u)| = h \sum_{r \ge T/h} \alpha^{(\varepsilon)}(1 - \bar{G}^{(\varepsilon)}(rh))$$

$$\le \int_{T-h}^\infty (1 - \bar{G}^{(\varepsilon)}(s))\, ds$$

$$\le \int_{T-h}^\infty s\bar{G}^{(\varepsilon)}(ds). \tag{7.2.8}$$

Condition $\mathbf{B_{10}}$ coincides with $\mathbf{B_1}$.

Finally, the limit constant $x^{(0)}(\infty)$ takes the following form:

$$x^{(0)}(\infty) = \frac{\int_0^\infty (1 - \bar{G}(u))\, du}{\int_0^\infty u\bar{G}(du)} = 1. \tag{7.2.9}$$

The proof can be completed by applying Theorem 1.3.1 to the renewal equation (7.2.3). $\qquad\square$

We can now improve conditions of Lemma 7.2.1 by using the conditions formulated directly in terms of the claim distributions $G^{(\varepsilon)}(u)$. Let us assume that the following condition is satisfied:

$\mathbf{D_{30}}$: $G^{(\varepsilon)}(\cdot) \Rightarrow G^{(0)}(\cdot)$ as $\varepsilon \to 0$.

Note that condition $\mathbf{H_1}$ implies that the distribution function $G^{(0)}(u)$ is not concentrated in zero.

Lemma 7.2.2. *Let conditions* $\mathbf{H_1}$, $\mathbf{D_{30}}$, $\mathbf{M_{16}}$ *and* $\mathbf{B_{10}}$ *hold. Then*

$$P^{(\varepsilon)}(u^{(\varepsilon)}) \to e^{-\varrho/\bar{m}^{(0)}} \quad \text{as } \varepsilon \to 0, \tag{7.2.10}$$

where $\bar{m}^{(0)} = \beta^{(0)}/2\mu^{(0)}$.

Proof. Conditions $\mathbf{D_{30}}$ and $\mathbf{M_{16}}$ obviously imply that

$$\mu^{(\varepsilon)} \to \mu^{(0)} \quad \text{as } \varepsilon \to 0. \tag{7.2.11}$$

Condition $\mathbf{H_1}$ implies that $\mu^{(0)} > 0$ and, therefore, $G^{(0)}(0) < 1$. This relation and condition $\mathbf{D_{30}}$ imply that condition $\mathbf{D_{28}}$ holds.

Condition $\mathbf{D_{30}}$ and relation (7.2.11) imply that

$$\bar{G}^{(\varepsilon)}(\cdot) \Rightarrow \bar{G}^{(0)}(\cdot) \quad \text{as } \varepsilon \to 0. \tag{7.2.12}$$

Since the distribution function $G^{(0)}(u)$ is not concentrated in zero, the distribution function $\bar{G}^{(0)}(u)$ has a non-zero absolutely continuous component and, hence, it is non-arithmetic.

Relation (7.2.12) and the remark above imply that condition $\mathbf{D_{29}}$ holds for the limit distribution function $\bar{G}(u) = \bar{G}^{(0)}(u)$.

With $\mathbf{D_{30}}$, condition $\mathbf{M_{16}}$ is equivalent to the following relation:

$$\beta^{(\varepsilon)} \to \beta^{(0)} = \int_0^\infty u^2 G^{(0)}(du) < \infty \quad \text{as } \varepsilon \to 0. \tag{7.2.13}$$

Relations (7.2.11) and (7.2.13) imply that

$$m^{(\varepsilon)} \to m^{(0)} = \beta^{(0)}/2\mu^{(0)} \quad \text{as } \varepsilon \to 0. \tag{7.2.14}$$

Relation (7.2.14) implies that condition $\mathbf{M_{17}}$ holds with the limit constant $\bar{m} = m^{(0)}$.

The proof can now be completed by applying Lemma 7.2.1. $\square$

We can now weaken the conditions of Lemma 7.2.2 by omitting the conditions of weak convergence of the corresponding distributions used in this lemma. Let us now assume the following condition:

$\mathbf{M_{18}}$: $\bar{m}^{(\varepsilon)} \to \bar{m}$ as $\varepsilon \to 0$.

Note that conditions $\mathbf{D_{28}}$ and $\mathbf{M_{16}}$ imply that $\bar{m} \in (0, \infty)$ in condition $\mathbf{M_{18}}$.

Lemma 7.2.3. *Let conditions* $\mathbf{H_1}$, $\mathbf{D_{28}}$, $\mathbf{M_{16}}$, $\mathbf{B_{10}}$, *and* $\mathbf{M_{18}}$ *hold. Then*

$$P^{(\varepsilon)}(u^{(\varepsilon)}) \to e^{-\varrho/\bar{m}} \quad \text{as } \varepsilon \to 0. \tag{7.2.15}$$

Proof. Let $\varepsilon_n > 0$ be an arbitrary subsequence such that $\varepsilon_n \to 0$ as $n \to \infty$. One can always select a subsequence $\varepsilon'_k = \varepsilon_{n_k} \to 0$ as $k \to \infty$ from the subsequence ε_n such that

$$G^{(\varepsilon'_k)}(\cdot) \Rightarrow G(\cdot) \quad \text{as } k \to \infty, \tag{7.2.16}$$

where $G(u)$ is a distribution function, possibly improper or concentrated in zero.

Conditions $\mathbf{D_{28}}$ and $\mathbf{M_{16}}$ imply, however, that the limit distribution function $G(u)$ is proper and not concentrated in zero.

Moreover, relation (7.2.16) and condition $\mathbf{M_{16}}$ imply that

$$\mu^{(\varepsilon'_k)} \to \mu = \int_0^\infty u\, G(du) \quad \text{as } k \to \infty, \tag{7.2.17}$$

and

$$\beta^{(\varepsilon'_k)} \to \beta = \int_0^\infty u^2 G(du) \quad \text{as } k \to \infty. \tag{7.2.18}$$

Relations (7.2.17) and (7.2.18), formula (7.2.4), and condition $\mathbf{M_{18}}$ imply that

$$\bar{m}^{(\varepsilon'_k)}(\cdot) \to \bar{m} = \beta/2\mu \quad \text{as } k \to \infty. \tag{7.2.19}$$

Note that the limit distribution $G(u)$ in (7.2.16), as well as the constants μ and β in (7.2.17) and (7.2.18), could depend on the choice of the subsequences ε_n and ε'_k, but the constant $\beta/2\mu$ must be equal to $\bar{m}$, due to condition $\mathbf{M_{18}}$ and, therefore, it does not depend on the choice of the subsequences ε_n and ε'_k.

Now we can apply Lemma 7.2.2 to the risk processes $\xi_u^{(\varepsilon'_k)}(t)$, $t \geq 0$, and get the following relation:

$$P^{(\varepsilon'_k)}(u^{(\varepsilon'_k)}) \to e^{-\varrho/\bar{m}} \quad \text{as } k \to \infty. \tag{7.2.20}$$

Since the choice of the initial subsequence ε_n was arbitrary, relation (7.2.20) implies that

$$P^{(\varepsilon)}(u^{(\varepsilon)}) \to e^{-\varrho/\bar{m}} \quad \text{as } \varepsilon \to 0. \tag{7.2.21}$$

The proof is complete. $\qquad\qquad\square$

Now we can finish the proof of Theorem 7.2.1. The last step is to weaken conditions $\mathbf{B_{10}}$ and $\mathbf{M_{18}}$.

So, we assume that conditions $\mathbf{H_1}$, $\mathbf{D_{28}}$, $\mathbf{M_{16}}$, and $\mathbf{B_9}$ hold.

Let $\varepsilon_n > 0$ be an arbitrary subsequence such that $\varepsilon_n \to 0$ as $n \to \infty$. One can always select a subsequence $\varepsilon'_k = \varepsilon_{n_k} \to 0$ as $k \to \infty$ from the subsequence ε_n such that

$$\bar{m}^{(\varepsilon'_k)} \to \bar{m} \quad \text{as} \quad \varepsilon \to 0, \tag{7.2.22}$$

where $\bar{m}$ is a constant from the interval $[0, \infty]$.

Conditions $\mathbf{D_{28}}$ and $\mathbf{M_{16}}$ imply, however, that the limit constant satisfies $\bar{m} \in (0, \infty)$.

Now one can select a subsequence $\varepsilon''_r = \varepsilon'_{k_r} \to 0$ as $r \to \infty$ from the subsequence ε'_k such that

$$(1 - \alpha^{(\varepsilon''_r)})u^{(\varepsilon''_r)} \to \varrho \quad \text{as} \quad \varepsilon \to 0, \tag{7.2.23}$$

where ϱ is a constant from the interval $[0, \infty]$.

Condition $\mathbf{B_9}$ implies, however, that $\varrho \in (0, \infty)$.

Now we can apply Lemma 7.2.3 to the risk processes $\xi_u^{(\varepsilon''_r)}(t)$, $t \geq 0$, and get the following relation:

$$P^{(\varepsilon''_r)}(u^{(\varepsilon''_r)}) \to e^{-\varrho/\bar{m}} \quad \text{as} \quad r \to \infty. \tag{7.2.24}$$

Due to (7.2.22) and (7.2.23), relation (7.2.24) can be rewritten in an equivalent form,

$$\frac{P^{(\varepsilon''_r)}(u^{(\varepsilon''_r)})}{e^{-\varrho/\bar{m}}} \sim \frac{P^{(\varepsilon''_r)}(u^{(\varepsilon''_r)})}{e^{-(1-\alpha^{(\varepsilon''_r)})u^{(\varepsilon''_r)}/\bar{m}^{(\varepsilon''_r)}}} \to 1 \quad \text{as} \quad r \to \infty. \tag{7.2.25}$$

Since the initial subsequence ε_n was chosen in an arbitrary way, relation (7.2.25) is equivalent to the relation

$$\frac{P^{(\varepsilon)}(u^{(\varepsilon)})}{e^{-(1-\alpha^{(\varepsilon)})u^{(\varepsilon)}/\bar{m}^{(\varepsilon)}}} \to 1 \quad \text{as} \quad \varepsilon \to 0. \tag{7.2.26}$$

The proof is complete. $\square$

It should be noted that the conditions of Theorem 7.2.1 do not require convergence of the parameters $\upsilon^{(\varepsilon)}$ and $\lambda^{(\varepsilon)}$, and the weak convergence of the claim distributions $G^{(\varepsilon)}(\cdot)$. In the case where the parameters $\upsilon^{(\varepsilon)}$ and $\lambda^{(\varepsilon)}$ converge to some limit values $\upsilon^{(0)}$ and $\lambda^{(0)}$, respectively, and condition $\mathbf{D_{30}}$ holds, one can consider the process $\xi_u^{(\varepsilon)}(t)$, $t \geq 0$, as a perturbation of the limit risk process $\xi_u^{(0)}(t)$, $t \geq 0$, with the gross premium rate $\upsilon^{(0)}$, the intensity of the claim flow $\lambda^{(0)}$, and the claim distribution function $G^{(0)}(u)$.

7.2.2　Asymptotics for increments of ruin probabilities

Let us now study the asymptotics of increments of the ruin probabilities $P^{(\varepsilon)}(u_\varepsilon + t) - P^{(\varepsilon)}(u_\varepsilon + t + h)$, where $t \in \mathbb{R}_1$, $h > 0$. As usual we assume that the ruin probability is $P_\varepsilon(u) = 0$ for $u < 0$.

The following theorem supplements Theorem 7.2.1.

Theorem 7.2.2. *Let conditions* $\mathbf{H_1}$, $\mathbf{D_{28}}$, $\mathbf{M_{16}}$, *and* $\mathbf{B_9}$ *hold. Then, for any* $t \in \mathbb{R}_1$, $h > 0$,

$$\frac{P^{(\varepsilon)}(u^{(\varepsilon)} + t) - P^{(\varepsilon)}(u^{(\varepsilon)} + t + h)}{(1 - \alpha^{(\varepsilon)})e^{-(1-\alpha^{(\varepsilon)})u^{(\varepsilon)}/\bar{m}^{(\varepsilon)}}\dfrac{h}{\bar{m}^{(\varepsilon)}}} \to 1 \quad as\ \varepsilon \to 0. \tag{7.2.27}$$

Proof. Introduce the renewal function generated by the distribution function $\alpha^{(\varepsilon)}\bar{G}_n^{(\varepsilon)}(u)$,

$$U^{(\varepsilon)}(u) = \sum_{n=0}^{\infty}(\alpha^{(\varepsilon)})^n \bar{G}_n^{(\varepsilon)}(u), \quad u \geq 0,$$

where $\bar{G}_n^{(\varepsilon)}(u)$ is the n-fold convolution of the distribution function $\bar{G}^{(\varepsilon)}(u)$.

Let also $U^{(\varepsilon)}(A)$ be a renewal measure on the Borel σ-algebra of $[0, \infty)$, which is determined by its values on the intervals $U^{(\varepsilon)}((a, b]) = U^{(\varepsilon)}(b) - U^{(\varepsilon)}(a)$, $0 \leq a \leq b < \infty$. Note that this measure has the atom 1 in the point 0, since the 0-fold convolution $\bar{G}_0^{(\varepsilon)}(s)$, by the definition, is a distribution function that has a unit jump in the point 0.

The renewal equation (7.2.3) has a unique solution given that is given by the same formula as for the proper renewal equation,

$$P^{(\varepsilon)}(u) = \int_0^u \alpha^{(\varepsilon)}(1 - \bar{G}^{(\varepsilon)}(u - s))U^{(\varepsilon)}(ds), \quad u \geq 0. \tag{7.2.28}$$

Standard calculations permit to transform this formula into the following form:

$$P^{(\varepsilon)}(u) = \sum_{n=0}^{\infty} \int_0^u \alpha^{(\varepsilon)}(1 - \bar{G}^{(\varepsilon)}(u - s))(\alpha^{(\varepsilon)})^n \bar{G}_n^{(\varepsilon)}(ds)$$

$$= \alpha^{(\varepsilon)} - \sum_{n=1}^{\infty}(1 - \alpha^{(\varepsilon)})(\alpha^{(\varepsilon)})^n \bar{G}_n^{(\varepsilon)}(u) = 1 - (1 - \alpha^{(\varepsilon)})U^{(\varepsilon)}(u).$$

This formula can be rewritten in the form,

$$1 - P^{(\varepsilon)}(u) = (1 - \alpha^{(\varepsilon)})U^{(\varepsilon)}(u). \tag{7.2.29}$$

Formula (7.2.29) is the famous Khintchine–Pollaczek formula. It permits to reduce studies of the asymptotics for the increments of ruin probabilities to studies of an asymptotics for the increments of the renewal function.

Let us consider the renewal equation with the forcing function $q^{(\varepsilon)}(u) = \chi_{[t-h,t]}(u)$, $u \geq 0$,

$$x^{(\varepsilon)}(u) = \chi_{[t-h,t]}(u) + \alpha^{(\varepsilon)} \int_0^u x^{(\varepsilon)}(u-s)\bar{G}^{(\varepsilon)}(ds), \quad u \geq 0. \tag{7.2.30}$$

Taking into account the indicator form of the forcing function and formula (7.2.29), we can represent the solution of this equation in the following form:

$$x^{(\varepsilon)}(u) = \int_0^u \chi_{[t-h,t]}(u-s))U^{(\varepsilon)}(ds) = U^{(\varepsilon)}(u-t+h) - U^{(\varepsilon)}(u-t)$$

$$= (P^{(\varepsilon)}(u-t) - P^{(\varepsilon)}(u-t+h))/(1-\alpha^{(\varepsilon)}), \quad u \geq 0. \tag{7.2.31}$$

Theorem 7.2.2 is an analogue of Theorem 7.2.1, in which the functions $x^{(\varepsilon)}(u)$ replace the ruin probabilities $P^{(\varepsilon)}(u)$.

The proof of Theorem 7.2.2 repeats the main steps of the proof of Theorem 7.2.1. The renewal equations (7.2.30) and (7.2.3) have, respectively, solutions $x^{(\varepsilon)}(u)$ and $P^{(\varepsilon)}(u)$. These equations have the same generating distribution function $\alpha^{(\varepsilon)}\bar{G}^{(\varepsilon)}(s)$ and differ only in the forcing functions. These are, respectively, the functions $\int_0^u \chi_{[t-h,t]}(u)$ and $\alpha^{(\varepsilon)}(1 - \bar{G}^{(\varepsilon)}(u))$. This permits to formulate and prove lemmas similar to Lemmas 7.2.1, 7.2.2, and 7.2.3.

For the lemmas formulated below, we replace $U^{(\varepsilon)}(u)$ with $(1-P^{(\varepsilon)}(u))/(1-\alpha^{(\varepsilon)})$ in the formula defining that defines the function $x^{(\varepsilon)}(u)$. Also, we use the argument $u^{(\varepsilon)} + 2t$ instead of $u^{(\varepsilon)}$. This does not change the corresponding balancing conditions $\mathbf{B_9}$ and $\mathbf{B_{10}}$.

Lemma 7.2.4. *Let conditions* $\mathbf{H_1}$, $\mathbf{D_{29}}$, $\mathbf{M_{17}}$, *and* $\mathbf{B_{10}}$ *hold. Then, for any* $t \in \mathbb{R}_1$, $h > 0$,

$$\frac{P^{(\varepsilon)}(u^{(\varepsilon)}+t) - P^{(\varepsilon)}(u^{(\varepsilon)}+t+h)}{(1-\alpha^{(\varepsilon)})} \to e^{-\varrho/\bar{m}}\frac{h}{\bar{m}} \quad as\ \varepsilon \to 0. \tag{7.2.32}$$

The proof is similar to the proof of Lemma 7.2.1. The difference is only in that the forcing function $\alpha_\varepsilon(1-\bar{G}^{(\varepsilon)}(u))$ has to be replaced with a simpler function $\chi_{[t-h,t]}(u)$. This function obviously satisfies condition $\mathbf{F_2}$.

Lemma 7.2.5. *Let conditions* $\mathbf{H_1}$, $\mathbf{D_{30}}$, $\mathbf{M_{16}}$ *and* $\mathbf{B_{10}}$ *hold. Then, for any* $t \in \mathbb{R}_1$, $h > 0$,

$$\frac{P^{(\varepsilon)}(u^{(\varepsilon)}+t) - P^{(\varepsilon)}(u^{(\varepsilon)}+t+h)}{(1-\alpha^{(\varepsilon)})} \to e^{-\varrho/\bar{m}_0}\frac{h}{\bar{m}_0} \quad as\ \varepsilon \to 0. \tag{7.2.33}$$

Lemma 7.2.6. *Let conditions* $\mathbf{H_1}$, $\mathbf{D_{28}}$, $\mathbf{M_{16}}$, $\mathbf{B_{10}}$ *hold. Then, for any* $t \in \mathbb{R}_1$, $h > 0$,

$$\frac{P^{(\varepsilon)}(u^{(\varepsilon)}+t) - P^{(\varepsilon)}(u^{(\varepsilon)}+t+h)}{(1-\alpha^{(\varepsilon)})} \to e^{-\varrho/\bar{m}}\frac{h}{\bar{m}} \quad as\ \varepsilon \to 0. \tag{7.2.34}$$

The proofs of Lemmas 7.2.5 and 7.2.6 are absolutely analogous to the proofs of Lemmas 7.2.2 and 7.2.3. Using Lemma 7.2.6 we can also complete the proof of Theorem 7.2.2 by repeating the selection arguments that were used in the proof of Theorem 7.2.1.
$\qquad\qquad\qquad\qquad\qquad\qquad\qquad\qquad\qquad\qquad\qquad\qquad\qquad\qquad\qquad$ □

7.2.3   Asymptotics for densities of non-ruin probabilities

We will also give asymptotics for densities of non-ruin probabilities.

Note that the non-ruin probability $1 - P^{(\varepsilon)}(u)$ is a distribution function on $[0, \infty)$ with the jump $1 - \alpha^{(\varepsilon)}$ in zero. The equation (7.2.3) can be rewritten as

$$1 - P^{(\varepsilon)}(u) = 1 - \alpha^{(\varepsilon)}$$

$$+ \alpha^{(\varepsilon)} \int_0^u (1 - P^{(\varepsilon)}(u - s)) \frac{1 - G^{(\varepsilon)}(s)}{\mu^{(\varepsilon)}} \, ds, \quad u \geq 0. \quad (7.2.35)$$

If $1 - G^{(\varepsilon)}(s)$ were a continuous function we could obtain, by "formally" differentiating (7.2.35), that $1 - P^{(\varepsilon)}(u)$ has a derivative $-p^{(\varepsilon)}(u)$ that satisfies the following equation:

$$-p^{(\varepsilon)}(u) = (1 - \alpha^{(\varepsilon)})\alpha^{(\varepsilon)} \frac{1 - G^{(\varepsilon)}(u)}{\mu^{(\varepsilon)}}$$

$$+ \alpha^{(\varepsilon)} \int_0^u -p^{(\varepsilon)}(u - s) \frac{1 - G^{(\varepsilon)}(s)}{\mu^{(\varepsilon)}} \, ds, \quad u \geq 0. \quad (7.2.36)$$

The renewal equation (7.2.36) has a unique solution even if the function $1 - G^{(\varepsilon)}(s)$ is not assumed to be continuous. One can easily check that the function $1 - \alpha^{(\varepsilon)} + \int_0^u -p^{(\varepsilon)}(s) \, ds$, $u \geq 0$, satisfies equation (7.2.35) and, hence, $-p^{(\varepsilon)}(u)$ is indeed a density of the distribution function $1 - P^{(\varepsilon)}(u)$ (without the the assumption that $1 - G^{(\varepsilon)}(s)$ is continuous), and

$$1 - P^{(\varepsilon)}(u) = 1 - \alpha^{(\varepsilon)} + \int_0^u -p^{(\varepsilon)}(s) \, ds, \quad u \geq 0. \quad (7.2.37)$$

It follows from (7.2.36) that $-p_\varepsilon(u)$ is a càdlàg function. It is given by the formula

$$-p^{(\varepsilon)}(u) = (1 - \alpha^{(\varepsilon)})\alpha^{(\varepsilon)} \int_0^u \frac{1 - G^{(\varepsilon)}(u - s)}{\mu^{(\varepsilon)}} U^{(\varepsilon)}(ds), \quad u \geq 0. \quad (7.2.38)$$

Formula (7.2.38) implies that $-p^{(\varepsilon)}(u)$ is a non-negative function.

As follows from Lemma 1.1.1 and Remark 1.1.1, conditions $\mathbf{H_1}$ and $\mathbf{D_{28}}$ imply that there exists $\varepsilon_0 > 0$ such that for any $h > 0$,

$$\sup_{\varepsilon \leq \varepsilon_0} \sup_{0 \leq u \leq v \leq u + h} (U^{(\varepsilon)}(v) - U^{(\varepsilon)}(u)) = K_h < \infty. \quad (7.2.39)$$

Indeed, condition $\mathbf{M_{16}}$ implies that

$$\varlimsup_{\varepsilon \to 0} \beta^{(\varepsilon)} = B < \infty. \tag{7.2.40}$$

Using formula (7.2.38) and relations (7.2.39) and (7.2.40) we get the following estimate for $h > 0$:

$$\varlimsup_{\varepsilon \to 0} \sup_{u \geq 0} \frac{-p^{(\varepsilon)}(u)}{1 - \alpha^{(\varepsilon)}} \leq \varlimsup_{\varepsilon \to 0} \sup_{u \geq 0} \sum_{k \leq [u/h]+1} (1 - G^{(\varepsilon)}(kh))$$

$$\times \, (U^{(\varepsilon)}(u - (k-1)h) - U^{(\varepsilon)}((u - kh) \wedge 0))$$

$$\leq \varlimsup_{\varepsilon \to 0} \sup_{u \geq 0} \sum_{k \leq [u/h]+1} \frac{\beta^{(\varepsilon)}}{(kh)^2} K_h \leq \frac{BK_h}{(kh)^2} < \infty. \tag{7.2.41}$$

Estimate (7.2.41) means that $\frac{-p^{(\varepsilon)}(u)}{1-\alpha^{(\varepsilon)}}$ are asymptotically uniformly bounded functions, as $\varepsilon \to 0$.

The following theorem gives asymptotics for the normalised densities $\frac{-p^{(\varepsilon)}(u)}{1-\alpha^{(\varepsilon)}}$ and supplements Theorem 7.2.2.

Theorem 7.2.3. *Let conditions* $\mathbf{H_1}$, $\mathbf{D_{28}}$, $\mathbf{M_{16}}$, *and* $\mathbf{B_9}$ *hold. Then*

$$\frac{-p^{(\varepsilon)}(u^{(\varepsilon)})}{(1 - \alpha^{(\varepsilon)})e^{-(1-\alpha^{(\varepsilon)})u^{(\varepsilon)}/\bar{m}^{(\varepsilon)}}\frac{1}{\bar{m}^{(\varepsilon)}}} \to 1 \;\; as \;\; \varepsilon \to 0. \tag{7.2.42}$$

Proof. Equation (7.2.36) can be rewritten in the following form, which is fitted for applying Theorem 1.3.1:

$$\frac{-p^{(\varepsilon)}(u)}{1 - \alpha^{(\varepsilon)}} = \alpha^{(\varepsilon)}\frac{1 - G^{(\varepsilon)}(u)}{\mu^{(\varepsilon)}}$$

$$+ \, \alpha^{(\varepsilon)} \int_0^u \frac{-p^{(\varepsilon)}(u - s)}{1 - \alpha^{(\varepsilon)}} \bar{G}^{(\varepsilon)}(ds), \quad u \geq 0. \tag{7.2.43}$$

Now, for the functions $\frac{-p^{(\varepsilon)}(u)}{1-\alpha^{(\varepsilon)}}$, we can prove an analogue of Theorem 7.2.1 by repeating the proof of this theorem. Thus we only formulate analogues of the corresponding lemmas, which are also useful by themselves.

Lemma 7.2.7. *Let conditions* $\mathbf{H_1}$, $\mathbf{D_{28}}$, $\mathbf{M_{16}}$, $\mathbf{D_{30}}$, *and* $\mathbf{B_{10}}$ *hold. Then*

$$\frac{-p^{(\varepsilon)}(u^{(\varepsilon)})}{(1 - \alpha^{(\varepsilon)})} \to e^{-\varrho/\bar{m}_0}\frac{1}{\bar{m}_0} \;\; as \;\; \varepsilon \to 0. \tag{7.2.44}$$

Proof. The proof combines elements of the proofs of Lemmas 7.2.1 and 7.2.2.

In the proof of Lemma 7.2.2 it was shown that conditions $\mathbf{M_{16}}$ and $\mathbf{D_{30}}$ imply that condition $\mathbf{D_{29}}$ holds for the distribution functions $\bar{G}^{(\varepsilon)}(u)$ with the limit distribution function $\bar{G}^{(0)}(u)$.

In place of the forcing function $\alpha^{(\varepsilon)}(1 - \bar{G}^{(\varepsilon)}(u))$, we have a simpler forcing function, $\alpha^{(\varepsilon)}\frac{1-G^{(\varepsilon)}(u)}{\mu^{(\varepsilon)}}$. Condition $\mathbf{D_{30}}$ and relation (7.2.11), obtained in the proof of Lemma 7.2.2, imply due to monotonicity of these forcing functions that they satisfy condition $\mathbf{F_2}$ **(a)**. The set of points of continuity of the limit function $\frac{1-G^{(0)}(u)}{\mu^{(0)}}$ can serve as a set of convergence in condition $\mathbf{F_2}$ **(a)**. Condition $\mathbf{F_2}$ **(b)** holds, since $0 \le \alpha^{(\varepsilon)}\frac{1-G^{(\varepsilon)}(u)}{\mu^{(\varepsilon)}} \le \frac{1}{\mu^{(\varepsilon)}}$. Condition $\mathbf{F_2}$ **(c)** follows from the following estimate:

$$
h \sum_{r \ge T/h} \sup_{rh \le u \le (r+1)h} \alpha^{(\varepsilon)}\frac{1 - G^{(\varepsilon)}(u)}{\mu^{(\varepsilon)}} = h \sum_{r \ge T/h} \alpha^{(\varepsilon)}\frac{(1 - G^{(\varepsilon)}(rh))}{\mu^{(\varepsilon)}}
$$

$$
\le \frac{1}{\mu^{(\varepsilon)}} \int_{T-h}^{\infty} (1 - G^{(\varepsilon)}(s))\, ds
$$

$$
\le 1 - \bar{G}^{(\varepsilon)}(T - h) \le \frac{\bar{m}^{(\varepsilon)}}{T - h}. \qquad (7.2.45)
$$

Condition $\mathbf{B_{10}}$ coincides with $\mathbf{B_1}$.

Finally, the corresponding renewal limit takes the following form:

$$
x^{(0)}(\infty) = \frac{1}{\bar{m}^{(0)}} \int_0^{\infty} \frac{1 - G^{(0)}(u)}{\mu^{(0)}}\, du = \frac{1}{\bar{m}^{(0)}}. \qquad (7.2.46)
$$

The proof can be completed by applying Theorem 1.3.1 to equation (7.2.36). $\qquad \square$

Lemma 7.2.8. *Let conditions* $\mathbf{H_1}$, $\mathbf{D_{28}}$, $\mathbf{M_{16}}$, $\mathbf{B_{10}}$, *and* $\mathbf{M_{17}}$ *hold. Then*

$$
\frac{-p^{(\varepsilon)}(u^{(\varepsilon)})}{(1 - \alpha^{(\varepsilon)})} \to e^{-\varrho/\bar{m}}\frac{1}{\bar{m}} \quad as \ \varepsilon \to 0. \qquad (7.2.47)
$$

The proof of Lemma 7.2.8 is absolutely similar to the proof of Lemma 7.2.3. Using Lemma 7.2.8 we can also complete the proof of Theorem 7.2.3 by repeating the selection arguments used in the proof of Theorem 7.2.1. $\qquad \square$

7.3 Asymptotic expansions in Cramér–Lundberg approximation for perturbed risk processes

In this section we give approximation for perturbed risk processes which generalises the classical Cramér–Lundberg and diffusion approximations and exponential asymptotic expansions for ruin probabilities for this approximation.

7.3.1   Generalised Cramér–Lundberg and diffusion approximations for perturbed risk processes

Let us assume that conditions $\mathbf{H_1}$ and $\mathbf{D_{30}}$ hold. Let us also introduce conditions:

$\mathbf{D_{31}}$: $v^{(\varepsilon)} \to v^{(0)}$ as $\varepsilon \to 0$,

and

$\mathbf{D_{32}}$: $\lambda^{(\varepsilon)} \to \lambda^{(0)}$ as $\varepsilon \to 0$.

Consider the following moment generating function:

$$\varphi^{(\varepsilon)}(\rho) = \int_0^\infty e^{\rho s}(1 - G^{(\varepsilon)}(s))\,ds = \mu^{(\varepsilon)} \int_0^\infty e^{\rho s} \bar{G}^{(\varepsilon)}(ds).$$

The following condition is an analogue of the Cramér type condition $\mathbf{C_2}$:

$\mathbf{C_{22}}$:   There exists $\delta > 0$ such that:

(a) $\overline{\lim}_{0 \leq \varepsilon \to 0}\, \varphi^{(\varepsilon)}(\delta) < \infty$;

(b) $\frac{\lambda^{(0)}}{c^{(0)}} \varphi^{(0)}(\delta) = \alpha^{(0)} \int_0^\infty e^{\delta s} \bar{G}^{(0)}(ds) \in (1, \infty)$.

Condition $\mathbf{C_{22}}$ obviously implies condition $\mathbf{M_{16}}$. As was pointed out in the proof of Lemma 7.2.2, conditions $\mathbf{H_1}$, $\mathbf{D_{30}}$ and $\mathbf{M_{16}}$ imply that (a) $\mu^{(\varepsilon)} \to \mu^{(0)}$ as $\varepsilon \to 0$; (b) $\bar{G}^{(\varepsilon)}(\cdot) \Rightarrow \bar{G}^{(0)}(\cdot)$ as $\varepsilon \to 0$; and that (c) $\bar{G}^{(0)}(s)$ is a non-arithmetic distribution function.

Relation (b) and conditions $\mathbf{H_1}$ and $\mathbf{C_{22}}$ imply that (d) $\varphi^{(\varepsilon)}(\rho) \to \varphi^{(0)}(\rho)$ as $\varepsilon \to 0$, for $\rho < \delta$. Statement (a) and conditions $\mathbf{D_{31}}$ and $\mathbf{D_{32}}$ obviously imply that (e) $\alpha^{(\varepsilon)} \to \alpha^{(0)}$ as $\varepsilon \to 0$.

Let us consider the following characteristic equation:

$$\frac{\lambda^{(\varepsilon)}}{c^{(\varepsilon)}} \int_0^\infty e^{\rho s}(1 - G^{(\varepsilon)}(s))\,ds = 1. \tag{7.3.1}$$

Relations (b) and (e), and condition $\mathbf{C_{22}}$, with a use of Lemma 1.4.3, imply that (f) there exists $\varepsilon_1 > 0$ such that the characteristic equation (7.3.1) for every $\varepsilon \leq \varepsilon_1$ has a unique non-negative root $\rho^{(\varepsilon)}$, and (g) $\rho^{(\varepsilon)} \to \rho^{(0)}$ as $\varepsilon \to 0$.

For $n = 0, 1, \ldots$, let us also introduce the mixed power-exponential moment generating functions

$$\varphi^{(\varepsilon)}[\rho, n] = \int_0^\infty s^n e^{\rho s}(1 - G^{(\varepsilon)}(s))\,ds = \mu^{(\varepsilon)} \int_0^\infty s^n e^{\rho s} \bar{G}^{(\varepsilon)}(ds).$$

By the definition, $\varphi^{(\varepsilon)}[\rho, 0] = \varphi^{(\varepsilon)}(\rho)$.

Let us choose an arbitrary $\rho^{(0)} < \beta < \delta$. Condition $\mathbf{C_{22}}$ **(a)** implies that **(h)** there exists $\varepsilon_2 = \varepsilon_2(\beta) > 0$ such that, for $\varepsilon \leq \varepsilon_2$ and $n = 0, 1, \ldots,$

$$\varphi^{(\varepsilon)}[\beta, n] \leq c_n \int_0^\infty e^{\delta s}(1 - G^{(\varepsilon)}(s))\, ds < \infty, \qquad (7.3.2)$$

where $c_n = c_n(\delta, \beta) = \sup_{s \geq 0} s^n e^{-(\delta - \beta)s} < \infty$.

Note also that $\varphi^{(\varepsilon)}[\rho, n]$, for $\rho \leq \beta$, is the derivative of order n of the function $\varphi^{(\varepsilon)}(\rho)$.

Denote

$$\pi^{(\varepsilon)}(\rho^{(\varepsilon)}) = \frac{\int_0^\infty e^{\rho^{(\varepsilon)}s}(1 - \bar{G}^{(\varepsilon)}(s))\, ds}{\int_0^\infty s e^{\rho^{(\varepsilon)}s}\bar{G}^{(\varepsilon)}(ds)}$$

$$= \frac{\int_0^\infty e^{\rho^{(\varepsilon)}s}(\int_s^\infty (1 - G^{(\varepsilon)}(u))\, du)\, ds}{\int_0^\infty s e^{\rho^{(\varepsilon)}s}(1 - G^{(\varepsilon)}(s)\, ds}. \qquad (7.3.3)$$

Relation **(g)** implies that **(i)** there exists $\varepsilon_3 = \varepsilon_3(\beta) > 0$ such that $\rho^{(\varepsilon)} < \beta$. Define

$$\varepsilon_0 = \min(\varepsilon_1, \varepsilon_2, \varepsilon_3). \qquad (7.3.4)$$

Relations **(b)** and **(g)**, and conditions $\mathbf{H_1}$ and $\mathbf{C_{22}}$ imply that **(j)** $\int_0^\infty e^{\rho^{(\varepsilon)}s}(1 - \bar{G}^{(\varepsilon)}(s))\, ds < \infty$ and **(k)** $\int_0^\infty s e^{\rho^{(\varepsilon)}s}\bar{G}^{(\varepsilon)}(ds) < \infty$ for $\varepsilon \leq \varepsilon_3$. Therefore, the quantity $\pi^{(\varepsilon)}(\rho^{(\varepsilon)})$ is well defined for all $\varepsilon \leq \varepsilon_3$.

The following theorem generalises both the classical approximations for ruin probabilities, namely the Cramér–Lundberg approximation given by the asymptotic relation (7.1.5), and the diffusion approximation given by the asymptotic relation (7.1.7).

Theorem 7.3.1. *Let conditions* $\mathbf{H_1}$, $\mathbf{D_{30}}$–$\mathbf{D_{32}}$, *and* $\mathbf{C_{22}}$ *hold. Then for any* $0 \leq u^{(\varepsilon)} \to \infty$ *as* $\varepsilon \to 0$ *the following asymptotic relation holds:*

$$\frac{P^{(\varepsilon)}(u^{(\varepsilon)})}{\exp\{-\rho^{(\varepsilon)}u^{(\varepsilon)}\}} \to \pi^{(0)}(\rho^{(0)}) \;\; \text{as} \;\; \varepsilon \to 0. \qquad (7.3.5)$$

Proof. We can apply Theorem 1.4.2 to the renewal equation (7.2.3). In this case, the forcing function is $q^{(\varepsilon)}(u) = \alpha^{(\varepsilon)}(1 - \bar{G}^{(\varepsilon)}(u))$ and the distribution function that generates this equation is $F^{(\varepsilon)}(u) = \alpha^{(\varepsilon)}\bar{G}^{(\varepsilon)}(u)$.

Conditions $\mathbf{H_1}$ and $\mathbf{D_{30}}$–$\mathbf{D_{32}}$ imply, as was mentioned in Subsection 7.3.1, that **(b)** $\bar{G}^{(\varepsilon)}(\cdot) \Rightarrow \bar{G}^{(0)}(\cdot)$ as $\varepsilon \to 0$, **(c)** the limit distribution function $\bar{G}^{(0)}(u)$ is non-arithmetic, and **(e)** $\alpha^{(\varepsilon)} \to \alpha^{(0)}$ as $\varepsilon \to 0$. Relations **(b)**, **(c)**, and **(e)** imply that condition $\mathbf{D_7}$ holds. Note that, according condition $\mathbf{H_1}$, the parameter $f^{(\varepsilon)} = 1 - \alpha^{(\varepsilon)} \in [0, 1)$.

Let first consider the case, where $\rho^{(0)} > 0$. Note that **(l)** $\rho^{(0)} > 0$ if and only if $\alpha^{(0)} < 1$,

Condition $\mathbf{C_2}$ takes in this case the form of condition $\mathbf{C_{22}}$.

Monotonicity of the function $q^{(\varepsilon)}(u) = \alpha^{(\varepsilon)}(1 - \bar{G}^{(\varepsilon)}(u))$ and relations **(b)**, **(e)**, and **(g)** $\rho^{(\varepsilon)} \to \rho^{(0)}$ as $\varepsilon \to 0$ imply in an obvious way that condition $\mathbf{F_6}$ **(a)** holds. The set of points of continuity of the limit function $q^{(0)}(u) = \alpha^{(0)}(1 - \bar{G}^{(0)}(u))$ can serve as a set of local uniform convergence in condition $\mathbf{F_6}$ **(a)**.

Condition $\mathbf{F_6}$ **(b)** holds, since $0 \leq q^{(\varepsilon)}(u) \leq 1$.

As was mention above, conditions $\mathbf{H_1}$, $\mathbf{D_{30}}$–$\mathbf{D_{32}}$, and $\mathbf{C_{22}}$ imply that

$$\lim_{\varepsilon \to 0} \int_0^\infty e^{\beta s}(1 - \bar{G}^{(\varepsilon)}(s))\, ds < \infty. \tag{7.3.6}$$

Using relation (7.3.6) we get for any $\rho^{(0)} < \beta \leq \delta$ and $\gamma \leq \beta - \rho^{(0)}$ that

$$\lim_{\varepsilon \to 0} h \sum_{r \geq T/h} \sup_{rh \leq u \leq (r+1)h} e^{(\rho^{(0)}+\gamma)u}\,|\,q^{(\varepsilon)}(u)\,|$$

$$\leq \lim_{\varepsilon \to 0} h \sum_{r \geq T/h} e^{(\rho^{(0)}+\gamma)(r+1)h}\alpha^{(\varepsilon)}(1 - \bar{G}^{(\varepsilon)}(rh))$$

$$\leq e^{\beta h} \lim_{\varepsilon \to 0} \int_{T-h}^\infty e^{\beta s}(1 - \bar{G}^{(\varepsilon)}(s))\, ds \to 0 \quad \text{as } T \to \infty. \tag{7.3.7}$$

Relation (7.3.7) implies that condition $\mathbf{F_6}$ **(c)** also holds.

Let now consider the case, where $\rho^{(0)} = 0$. Note that **(m)** $\rho^{(0)} = 0$ if and only if $\alpha^{(0)} = 1$.

Condition $\mathbf{C_1}$ takes the form of condition $\mathbf{C_{22}}$ **(a)** if $\rho^{(0)} = 0$.

Condition $\mathbf{F_5}$ is verified in the same way as it was done above for condition $\mathbf{F_6}$.

Finally, the renewal limit $\tilde{x}^{(0)}(\infty)$ takes the form of the quantity $\pi^{(0)}(\rho^{(0)})$ in formula (7.3.3).

The proof can be completed by applying Theorem 1.4.2 to the renewal equation (7.2.3). The statements **(i)** or **(ii)** of this theorem should be used in the cases $\rho^{(0)} > 0$ and $\rho^{(0)} = 0$, respectively. $\qquad\qquad\square$

As was mentioned in the proof, **(l)** $\rho^{(0)} > 0$ if and only if $\alpha^{(0)} < 1$, while **(m)** $\rho^{(0)} = 0$ if and only if $\alpha^{(0)} = 1$. The case **(l)** $\rho^{(0)} > 0$ corresponds to a model of the Cramér–Lundberg approximation, while the case **(m)** $\rho^{(0)} = 0$ corresponds to a model of the diffusion approximation.

In the case **(m)** $\rho^{(0)} = 0$, Theorem 7.3.1 impose stronger condition on the claim distributions then Lemma 7.2.2, which is, in fact, a direct analogue of the classical diffusion approximation in the case of perturbed risk processes. Indeed, conditions $\mathbf{C_{22}}$ and $\mathbf{M_{16}}$ are used in Theorem 7.3.1 and Lemma 7.2.2, respectively.

However, no balancing condition that would restrict the rate of growth of the time parameter $u^{(\varepsilon)}$ to ∞ is assumed in Theorem 7.3.1, while the balancing condition $\mathbf{B_{10}}$ is assumed in Lemma 7.2.2.

Note also that, in the case **(m)** $\rho^{(0)} = 0$, conditions $\mathbf{H_1}$, $\mathbf{D_{30}}$ and $\mathbf{C_{22}}$ imply, by Lemma 1.4.2 and formula (7.2.4), the following asymptotic relation

$$\rho^{(\varepsilon)} \sim (1 - \alpha^{(\varepsilon)}) \cdot \frac{2\mu^{(\varepsilon)}}{\beta^{(\varepsilon)}} \quad \text{as } \varepsilon \to 0. \tag{7.3.8}$$

7.3.2   Perturbation conditions for risk processes

Let us formulate and clarify a connection between different forms of perturbation conditions for the risk processes.

We assume that condition $\mathbf{H_1}$ holds.

The perturbation conditions for the parameters $\upsilon^{(\varepsilon)}$ and $\lambda^{(\varepsilon)}$ are formulated in the following way:

$\mathbf{P_{39}^{(k)}}$: $\upsilon^{(\varepsilon)} = \upsilon^{(0)} + d_1\varepsilon + \cdots + d_k\varepsilon^k + o(\varepsilon^k)$, where $|d_l| < \infty, l = 1, \ldots, k$,

and

$\mathbf{P_{40}^{(k)}}$: $\lambda^{(\varepsilon)} = \lambda^{(0)} + e_1\varepsilon + \cdots + e_k\varepsilon^k + o(\varepsilon^k)$, where $|e_l| < \infty, l = 1, \ldots, k$.

It is also convenient to define $d_0 = \upsilon^{(0)}$ and $e_0 = \lambda^{(0)}$. Note that, according to condition $\mathbf{E_1}$, we have **(a)** $\upsilon^{(0)}, \lambda^{(0)} > 0$.

Also, conditions $\mathbf{P_{39}^{(k)}}$ and $\mathbf{P_{40}^{(k)}}$ imply that conditions $\mathbf{D_{31}}$ and $\mathbf{D_{32}}$ hold.

Let us also introduce the following perturbation condition for $\rho < \beta$:

$\mathbf{P_{41}^{(\rho,k)}}$: $\varphi^{(\varepsilon)}[\rho, n] = \varphi^{(0)}[\rho, n] + v[\rho, 1, n]\varepsilon + \cdots + v[\rho, k - n, n]\varepsilon^{k-n} + o(\varepsilon^{k-n})$,
where $|v[\rho, r, n]| < \infty$ for $r = 1, \ldots, k - n, n = 0, \ldots, k$.

It is also convenient to set $v[\rho, 0, n] = \varphi^{(0)}[\rho, n], n = 0, \ldots, k$.

The distribution function, which generates the renewal equation (7.2.3), is $F^{(\varepsilon)}(u) = \alpha^{(\varepsilon)}\bar{G}^{(\varepsilon)}(u)$. In order to apply theorems given in Chapter 1, we should also obtain asymptotic expansions for the corresponding mixed power-exponential generating functions,

$$\tilde{\varphi}^{(\varepsilon)}[\rho, n] = \alpha^{(\varepsilon)} \int_0^\infty s^n e^{\rho s} \bar{G}^{(\varepsilon)}(ds) = \frac{\lambda^{(\varepsilon)}}{\upsilon^{(\varepsilon)}} \varphi^{(\varepsilon)}[\rho, n]. \tag{7.3.9}$$

Lemma 7.3.1. *Let conditions* $\mathbf{H_1}$, $\mathbf{C_{22}}$, $\mathbf{P_{39}^{(k)}}$, $\mathbf{P_{40}^{(k)}}$ *hold and* $\mathbf{P_{41}^{(\rho,k)}}$ *be satisfied for some* $\rho \leq \beta$. *Then the following asymptotic expansion takes place:*

$$\tilde{\varphi}^{(\varepsilon)}[\rho, n] = \tilde{\varphi}^{(0)}[\rho, n] + b[\rho, 1, n]\varepsilon + \cdots + b[\rho, k - n, n]\varepsilon^{k-n} + o(\varepsilon^{k-n}), \tag{7.3.10}$$

where the coefficients $b[\rho, r, n]$ *are given by the recurrence formulas* $b[\rho, 0, 0] = \frac{\lambda^{(0)}\varphi^{(0)}[\rho,n]}{\upsilon^{(0)}} = \frac{e_0 v[\rho,0,n]}{d_0}$ *and, for* $r = 0, \ldots, k - n$ *and given* $n \leq k$ *and sequentially for,* $n = 0, \ldots, k$,

$$b[\rho, r, n] = \sum_{m=0}^r h_m v[\rho, r - m, n], \tag{7.3.11}$$

where

$$h_m = d_0^{-1}\left(e_m - \sum_{q=1}^{m} d_q h_{m-q}\right). \tag{7.3.12}$$

Proof. The proof follows from formula (7.3.9) and statements **(v)** and **(iii)** of Lemma 8.1.1, which should be applied first to the quotient $\frac{\lambda^{(\varepsilon)}}{v^{(\varepsilon)}}$ and then to the product $\frac{\lambda^{(\varepsilon)}}{v^{(\varepsilon)}}\cdot\varphi^{(\varepsilon)}[\rho,n]$. $\qquad\square$

Remark 7.3.1. The first coefficients $b[\rho,r,n]$, $r = 0,1,2$ and $n = 0,1,\ldots$, which are used below with $n = 0,1,2$ for calculating the coefficients a_1 and a_2 in the corresponding expansions for characteristic roots $\rho^{(\varepsilon)}$, are given by the following formulas:

$$b[\rho,0,n] = \frac{e_0}{d_0}v[\rho,0,n],$$

$$b[\rho,1,n] = \frac{e_0}{d_0}v[\rho,1,n] + \frac{e_1 - d_1 h_0}{d_0}v[\rho,0,n]$$

$$= \frac{e_0}{d_0}v[\rho,1,n] + \frac{d_0 e_1 - d_1 e_0}{d_0^2}v[\rho,0,n],$$

$$b[\rho,2,n] = \frac{e_0}{d_0}v[\rho,2,n] + \frac{e_1 - d_1 h_0}{d_0}v[\rho,1,n]$$

$$+ \frac{e_2 - d_1 h_1 - d_2 h_0}{d_0}v[\rho,2,n]. \tag{7.3.13}$$

Finally, let us introduce the following balancing condition:

$\mathbf{B}_{11}^{(r)}$: $0 \le u^{(\varepsilon)} \to \infty$ in such a way that $\varepsilon^r u^{(\varepsilon)} \to \varrho_r$, where $0 \le \varrho_r < \infty$.

7.3.3   Exponential expansions for ruin probabilities

Now we are in a position to apply theorems of Chapter 1 to the perturbed renewal equation (7.2.3).

Theorem 7.3.2. *Let conditions* $\mathbf{H}_1$, $\mathbf{D}_{30}$, $\mathbf{C}_{22}$, $\mathbf{P}_{39}^{(k)}$, $\mathbf{P}_{40}^{(k)}$, *and* $\mathbf{P}_{41}^{(\rho^{(0)},k)}$ *hold. Then*

(i) *The root* $\rho^{(\varepsilon)}$ *of equation (7.3.1) has the asymptotic expansion*

$$\rho^{(\varepsilon)} = \rho^{(0)} + a_1\varepsilon + \cdots + a_k\varepsilon^k + o(\varepsilon^k), \tag{7.3.14}$$

where the coefficients a_n are given by the recurrence formulas $a_1 = -\dfrac{b[\rho^{(0)},1,0]}{b[\rho^{(0)},0,1]}$

and in general for $n = 1, \ldots, k$,

$$a_n = -b[\rho^{(0)}, 0, 1]^{-1} \left(b[\rho^{(0)}, n, 0] + \sum_{q=1}^{n-1} b[\rho^{(0)}, n-q, 1] a_q \right.$$

$$\left. + \sum_{2 \le m \le n} \sum_{q=m}^{n} b[\rho^{(0)}, n-q, m] \cdot \sum_{n_1, \ldots, n_{q-1} \in D_{m,q}} \prod_{p=1}^{q-1} a_p^{n_p} / n_p! \right), \quad (7.3.15)$$

where the coefficients $b[\rho^{(0)}, r, n]$ are given by (7.3.11) and $D_{m,q}$, for every $2 \le m \le q < \infty$, is the set of all nonnegative, integer solutions of the system

$$n_1 + \cdots + n_{q-1} = m, \quad n_1 + \cdots + (q-1)n_{q-1} = q. \quad (7.3.16)$$

(ii) *If $b[\rho^{(0)}, l, 0] = 0$, $l = 1, \ldots, r$, for some $1 \le r \le k$, then $a_1, \ldots, a_r = 0$. If $b[\rho^{(0)}, l, 0] = 0$, $l = 1, \ldots, r-1$, but $b[\rho^{(0)}, r, 0] < 0$ for some $1 \le r \le k$, then $a_1, \ldots, a_{r-1} = 0$ but $a_r > 0$.*

(iii) *If, additionally, condition $\mathbf{B}_{11}^{(r)}$ is satisfied for some $1 \le r \le k$, then the following asymptotic relation holds:*

$$\frac{P^{(\varepsilon)}(u^{(\varepsilon)})}{\exp\{-(\rho^{(0)} + a_1\varepsilon + \cdots + a_{r-1}\varepsilon^{r-1})u^{(\varepsilon)}\}}$$

$$\to e^{-\varrho_r a_r} \pi^{(0)}(\rho^{(0)}) \quad as \ \varepsilon \to 0. \quad (7.3.17)$$

Proof. We can apply Theorem 2.2.1 to the renewal equation (7.2.3). In this case, the forcing function is $q^{(\varepsilon)}(u) = \alpha^{(\varepsilon)}(1 - \bar{G}^{(\varepsilon)}(u))$ and the distribution function that generates this equation is $F^{(\varepsilon)}(u) = \alpha^{(\varepsilon)}\bar{G}^{(\varepsilon)}(u)$.

As was mentioned above, conditions $\mathbf{H}_1$, $\mathbf{D}_{30}$, and the perturbation conditions $\mathbf{P}_{39}^{(k)}$, $\mathbf{P}_{40}^{(k)}$, and $\mathbf{P}_{41}^{(\rho^{(0)},k)}$ imply that conditions $\mathbf{D}_{31}$ and $\mathbf{D}_{32}$ hold. Therefore, all conditions of Theorem 7.3.1 hold. As was shown in the proof of this theorem, condition $\mathbf{D}_7$ holds. This condition coincides in this case with condition $\mathbf{D}_{11}$.

Also, it was show in the proof of Theorem 7.3.1 that condition $\mathbf{F}_6$ holds.

Condition $\mathbf{C}_2$ takes in this case the form of condition $\mathbf{C}_{22}$.

Condition $\mathbf{P}_3^{(k)}$ is implied by conditions $\mathbf{P}_{39}^{(k)}$, $\mathbf{P}_{40}^{(k)}$, and $\mathbf{P}_{41}^{(\rho^{(0)},k)}$ that follows from Lemma 7.3.1. Formulas for the coefficients in the corresponding asymptotic expansion are given by formula (7.3.11).

The balancing condition $\mathbf{B}_3^{(r)}$ takes in this case the form of condition $\mathbf{B}_{11}^{(r)}$.

Finally, the renewal limit $x^{(0)}(\infty)$ takes the form of the quantity $\pi^{(0)}(\rho^{(0)})$ in formula (7.3.3).

The proof of the theorem can be completed by applying Theorem 2.2.1 to the renewal equation (7.2.3). $\qquad\square$

Remark 7.3.2. The coefficients a_1 and a_2 can be calculated with the use of formulas (2.2.6) given in Subsection 2.2.1, where the coefficients $b_{r,n}$ should be replaced with the coefficients $b[\rho^{(0)}, r, n]$ calculated with the use of formulas (7.3.13) given in Subsection 7.3.1, i.e.,

$$a_1 = -\frac{b[\rho^{(0)}, 1, 0]}{b[\rho^{(0)}, 0, 1]},$$

$$a_2 = -\frac{1}{b[\rho^{(0)}, 0, 1]}\left(b[\rho^{(0)}, 2, 0] + b[\rho^{(0)}, 1, 1]a_1 + \frac{1}{2}b[\rho^{(0)}, 0, 2]a_1^2\right)$$

$$= -\frac{b[\rho^{(0)}, 2, 0]}{b[\rho^{(0)}, 0, 1]} + \frac{b[\rho^{(0)}, 1, 1]b[\rho^{(0)}, 1, 0]}{b[\rho^{(0)}, 0, 1]^2} - \frac{b[\rho^{(0)}, 0, 2]b[\rho^{(0)}, 1, 0]^2}{2b[\rho^{(0)}, 0, 1]^3},$$

$$a_3 = -\frac{1}{b[\rho^{(0)}, 0, 1]}\left(b[\rho^{(0)}, 3, 0] + b[\rho^{(0)}, 2, 1]a_1 + b[\rho^{(0)}, 1, 1]a_2\right.$$

$$\left. + \frac{1}{2}b[\rho^{(0)}, 1, 2]a_1^2 + b[\rho^{(0)}, 0, 2]a_1a_2 + \frac{1}{6}b[\rho^{(0)}, 0, 3]a_1^3\right). \tag{7.3.18}$$

Let us also formulate an analogue of Theorem 2.2.2 with regard to the renewal equation (7.2.3).

Introduce the integer parameters $1 \leq h \leq k_0$ and $k_r \geq 1, r = 1, \ldots,$ and define the vector parameter $\bar{k} = (h, k_0, \ldots, k_{\tilde{n}})$, where $\tilde{n} = \max(n : nh \leq k_0)$.

Consider the following perturbation condition:

$\mathbf{P}_{42}^{(\rho, \bar{k})}$: **(a)** $\varphi^{(\varepsilon)}[\rho, 0] = \varphi^{(0)}[\rho, 0] + v[\rho, h, 0]\varepsilon^h + \cdots + v[\rho, k_0, 0]\varepsilon^{k_0} + o(\varepsilon^{k_0})$,

where $|v[\rho, r, 0]| < \infty, r = 1, \ldots, k_0$;

(b) $\varphi^{(\varepsilon)}[\rho, n] = \varphi^{(0)}[\rho, n] + v[\rho, 1, n]\varepsilon + \cdots + v[\rho, k_n, n]\varepsilon^{k_n} + o(\varepsilon^{k_n})$,

where $|v[\rho, r, n]| < \infty, r = 1, \ldots, k_n, n = 1, \ldots, \tilde{n}$.

As above, it will also be convenient to set $v[\rho, 0, n] = \varphi^{(0)}[\rho, n], n = 0, \ldots, k$.

Condition $\mathbf{P}_{42}^{(\rho, k)}$ is more general than condition $\mathbf{P}_{41}^{(k)}$, and the former reduces to the latter if $h = 1, k_0 = k$ (in this case, $\tilde{n} = k$) and $k_n = k - n, r = 1, \ldots, \tilde{n}$.

The following theorem generalises Theorems 7.3.2. The proof is the same as the one given for Theorems 7.3.2 and is based on applying Theorem 2.2.2 to the renewal equation (7.2.3).

Theorem 7.3.3. *Let conditions* $\mathbf{H}_1, \mathbf{D}_{30}, \mathbf{C}_{22}, \mathbf{P}_{39}^{(k)}, \mathbf{P}_{40}^{(k)},$ *and* $\mathbf{P}_{42}^{(\rho^{(0)}, \tilde{k})}$ *hold. Then*

(i) *For all ε small enough there exists a unique nonnegative root $\rho^{(\varepsilon)}$ of equation (7.3.1) having the following asymptotic expansion:*

$$\rho^{(\varepsilon)} = \rho^{(0)} + a_h\varepsilon^h + \cdots + a_k\varepsilon^k + o(\varepsilon^k), \tag{7.3.19}$$

where $k = \min(k_0, k_1 + h, \ldots, k_{\tilde{n}} + h\tilde{n})$ and the coefficients $a_n, n = h, \ldots, k$, are given by the recurrence formulas $a_h = -b[\rho^{(0)}, h, 0]/b[\rho^{(0)}, 0, 1]$ and, for $n = h, \ldots, k$,

$$a_n = -b[\rho^{(0)}, 0, 1]^{-1}\left(b[\rho^{(0)}, n, 0] + \sum_{i=h}^{n-1} b[\rho^{(0)}, n - i, 1]a_i \right.$$

$$\left. + \sum_{r=2}^{[n/h]} \sum_{j=rh}^{n} b[\rho^{(0)}, n - j, r] \cdot \sum_{n_h, \ldots, n_{j-1} \in D_{h,r,j}} \prod_{i=h}^{j-1} a_i^{n_i}/n_i! \right), \quad (7.3.20)$$

where the coefficients $b[\rho^{(0)}, r, n]$ are given by formula (7.3.11), and $D_{h,r,j}$, for every $h \geq 1$ and $2 \leq r \leq j < \infty$, is the set of all nonnegative, integer solutions of the system

$$n_h + \cdots + n_{j-1} = r, \quad hn_h + \cdots + (j - 1)n_{j-1} = j. \quad (7.3.21)$$

(ii) *If $b[\rho^{(0)}, l, 0] = 0, l = h, \ldots, r$ for some $h \leq r \leq k$, then $a_1, \ldots, a_r = 0$. If $b[\rho^{(0)}, l, 0] = 0, l = h, \ldots, r - 1$, but $b[\rho^{(0)}, r, 0] < 0$ for some $h \leq r \leq k$, then $a_h, \ldots, a_{r-1} = 0$ but $a_r > 0$.*

(iii) *If, additionally, conditions $\mathbf{B}_{11}^{(r)}$ hold for some $h \leq r \leq k$, then*

$$\frac{P^{(\varepsilon)}(u^{(\varepsilon)})}{\exp\{-(\rho^{(0)} + a_h\varepsilon + \cdots + a_{r-1}\varepsilon^{r-1})u^{(\varepsilon)}\}}$$

$$\to e^{-\varrho_r a_r} \pi^{(0)}(\rho^{(0)}) \quad \text{as} \quad \varepsilon \to 0. \quad (7.3.22)$$

7.3.4 Modifications of perturbation conditions

In this subsection, we consider modifications of the perturbation and other conditions introduced in Subsection 7.3.2.

Let us first consider conditions $\mathbf{P}_{39}^{(k)}$ and $\mathbf{P}_{40}^{(k)}$.

One of possible interpretations for these conditions is based on an assumption that the intensity of the claim flow, $\lambda = \lambda(\upsilon)$, is a function of the gross premium rate. Let us assume that **(a)** $\lambda(\upsilon)$ has the continuous derivative of an order k, **(b)** $0 < \lambda(0+) \leq \infty$, **(c)** $\overline{\lim}_{\upsilon \to \infty} \frac{\lambda(\upsilon)\mu^{(0)}}{\upsilon^{(0)}} < \alpha^{(0)} \leq 1$. Under these assumptions, there exists the greatest a largest root of the equation

$$\frac{\lambda(\upsilon^{(0)})\mu^{(0)}}{\upsilon^{(0)}} = \alpha^{(0)}. \quad (7.3.23)$$

The balance relation connected to the diffusion approximation suggests that the difference $\upsilon - \upsilon^{(0)}$ can be regarded as the parameter ε. Note that, due to condition $\mathbf{H}_1$, we need to consider only the values $\upsilon \geq \upsilon^{(0)}$.

The expansion in condition $\mathbf{P}_{39}^{(k)}$ takes the following form:

$$v^{(\varepsilon)} = v^{(0)} + \varepsilon. \tag{7.3.24}$$

The expansion in condition $\mathbf{P}_{40}^{(k)}$ for the function $\lambda^{(\varepsilon)} = \lambda(v^{(\varepsilon)})$ becomes

$$\lambda(v^{(\varepsilon)}) = \lambda(v^{(0)}) + \lambda_1(v^{(0)})\varepsilon + \cdots + \frac{1}{k!}\lambda_k(v^{(0)})\varepsilon^k + o(\varepsilon^k), \tag{7.3.25}$$

where $\lambda_r(v)$ is the derivative of order r of the function $\lambda(v)$ for $r = 1, \ldots, k$.

Now, let us consider the perturbation condition $\mathbf{P}_{42}^{(\rho^{(0)},\tilde{k})}$.

Introduce the moment generating functions

$$\psi^{(\varepsilon)}(\rho) = \int_0^\infty e^{\rho s} G^{(\varepsilon)}(ds)$$

and the mixed power-exponential moment generating functions for $n = 0, 1, \ldots,$

$$\psi^{(\varepsilon)}[\rho, n] = \int_0^\infty s^n e^{\rho s} G^{(\varepsilon)}(ds).$$

By the definition, $\psi^{(\varepsilon)}[\rho, 0] = \psi^{(\varepsilon)}(\rho)$.

The following relation links the moment generating functions $\varphi^{(\varepsilon)}(\rho)$ and $\psi^{(\varepsilon)}(\rho)$,

$$\varphi^{(\varepsilon)}(\rho) = \int_0^\infty e^{\rho s}(1 - G^{(\varepsilon)}(s))\,ds = \mathsf{E}\int_0^{\kappa_1^{(\varepsilon)}} e^{\rho s}\,ds$$

$$= \begin{cases} (\psi^{(\varepsilon)}[\rho, 0] - 1)/\rho & \text{for } \rho \neq 0, \\ \psi^{(\varepsilon)}[0, 1] & \text{for } \rho = 0. \end{cases} \tag{7.3.26}$$

Relation (7.3.26) implies that the asymptotic inequality in condition $\mathbf{C}_{22}$ **(a)** is equivalent to the following relation:

$$\varlimsup_{0 \leq \varepsilon \to 0} \psi^{(\varepsilon)}(\delta) < \infty. \tag{7.3.27}$$

Relations (7.3.26) and (7.3.27) imply that, for $\varepsilon \leq \varepsilon_2$ and $n = 0, 1, \ldots,$

$$\psi^{(\varepsilon)}[\beta, n] \leq c_n \int_0^\infty e^{\delta s} G^{(\varepsilon)}(ds) = \psi^{(\varepsilon)}(\beta) < \infty. \tag{7.3.28}$$

Note also that $\psi^{(\varepsilon)}[\rho, n]$ is the derivative of order n of the function $\psi^{(\varepsilon)}(\rho)$ for $\rho \leq \beta$.

The following relation connects the derivatives of the functions $\varphi^{(\varepsilon)}[\rho, n]$ and $\psi^{(\varepsilon)}[\rho, n]$ for $n = 1, 2, \ldots, \varepsilon \leq \varepsilon_2$, and $\rho \leq \beta$:

$$\varphi^{(\varepsilon)}[\rho, n] = \begin{cases} (\psi^{(\varepsilon)}[\rho, n] - n\varphi^{(\varepsilon)}[\rho, n - 1])/\rho & \text{for } \rho \neq 0, \rho \leq \beta, \\ \psi^{(\varepsilon)}[0, n + 1]/(n + 1) & \text{for } \rho = 0. \end{cases} \tag{7.3.29}$$

Relation (7.3.29) can be obtained similarly to relation (5.5.43). Relation **(a)** $\psi^{(\varepsilon)}[\rho, 0] = \varphi^{(\varepsilon)}[\rho, 0]\rho + 1$, $\rho \leq \beta$, follows from formula (7.3.26). Differentiating **(a)** yields the following relation: **(b)** $\psi^{(\varepsilon)}[\rho, n] = \varphi^{(\varepsilon)}[\rho, n]\rho + n\varphi^{(\varepsilon)}[\rho, n - 1]$, $n = 1, 2, \ldots, \rho \leq \beta$. In particular, **(b)** yields the following relation for $\rho = 0$: **(c)** $\psi^{(\varepsilon)}[0, n] = n\varphi^{(\varepsilon)}[0, n - 1]$, $n = 1, 2, \ldots, \rho \leq \beta$. Relations **(a)**–**(c)** imply relation (7.3.29).

Let us formulate a perturbation condition that would imply condition $\mathbf{P}_{41}^{(\rho,k)}$,

$\mathbf{P}_{43}^{(\rho,k+1)}$: $\psi^{(\varepsilon)}[\rho, n] = \psi^{(0)}[\rho, n] + u[\rho, 1, n]\varepsilon + \cdots + u[\rho, k + 1 - n, n]\varepsilon^{k+1-n} + o(\varepsilon^{k+1-n})$, where $|u[\rho, r, n]| < \infty, r = 1, \ldots, k+1-n, n = 0, \ldots, k+1$.

It also is convenient to define $u[\rho, 0, n] = \psi^{(0)}[\rho, n], n = 0, \ldots, k$.

The following lemma is an obvious corollary of relations (7.3.26) and (7.3.29).

Lemma 7.3.2. *Let condition* $\mathbf{H}_1$, $\mathbf{C}_{22}$ *hold, and condition* $\mathbf{P}_{43}^{(\rho,k+1)}$ *be satisfied for some* $\rho \leq \beta$. *Then condition* $\mathbf{P}_{41}^{(\rho,k)}$ *holds for* $\rho \leq \beta$, *and the asymptotic expansions in this condition take the following form for* $n = 0, \ldots, k$:

$$\varphi^{(\varepsilon)}[\rho, n] = \varphi^{(0)}[\rho, n] + v[\rho, 1, n]\varepsilon + \cdots + v[\rho, k - n, n]\varepsilon^{k-n} + o(\varepsilon^{k-n}), \tag{7.3.30}$$

where $v[\rho, r, n]$, $r = 0, \ldots, k - n$, $n = 0, \ldots, k$, *for* $r = 0, \ldots, k$, $n = 0$, *are given by the formulas*

$$v[\rho, r, 0] = \begin{cases} (u[\rho, r, 0] - \delta(r, 0))/\rho & \text{if } \rho \neq 0, \\ u[0, r, 1] & \text{if } \rho = 0, \end{cases} \tag{7.3.31}$$

and, for $r = 0, \ldots, k - n$, $n = 1, \ldots, k$,

$$v[\rho, r, n] = \begin{cases} (u[\rho, r, n] - nv[\rho, r, n - 1])/\rho & \text{if } \rho \neq 0, \\ u[0, r, n + 1]/(n + 1) & \text{if } \rho = 0. \end{cases} \tag{7.3.32}$$

Remark 7.3.3. Note that condition $\mathbf{P}_{43}^{(\rho,k+1)}$ can be replaced with condition $\mathbf{P}_{41}^{(\rho,k)}$ in Lemma 7.3.2 if $\rho > 0$ and the corresponding asymptotic expansion in condition $\mathbf{P}_{43}^{(\rho,k+1)}$ can be omitted in Lemma 7.3.2 for $n = 0$ if $\rho = 0$.

Remark 7.3.4. According to Lemma 7.3.2, condition $\mathbf{P}_{41}^{(\rho,k)}$ can be replaced in Theorem 7.3.2 with condition $\mathbf{P}_{43}^{(\rho,k+1)}$.

Condition $\mathbf{P}_{43}^{(\rho,k+1)}$ is more convenient to work with in applications, since it is formulated for the claim distributions in terms of the usual moment generating functions $\psi^{(\varepsilon)}(\rho)$ and its derivatives $\psi^{(\varepsilon)}[\rho,n]$.

In many cases there are explicit formulas for $\psi^{(\varepsilon)}(\rho)$. They can effectively be used if the perturbation conditions are given in terms of the parameters of these distribution.

For example, let $G^{(\varepsilon)}(u)$ be a gamma-distribution with parameters $\sigma^{(\varepsilon)}$ and v, i.e.,

$$\psi^{(\varepsilon)}(\rho) = \frac{(\sigma^{(\varepsilon)})^{v}}{(\sigma^{(\varepsilon)} - \rho)^{v}}, \qquad \rho < \sigma^{(\varepsilon)}, \tag{7.3.33}$$

and

$$\psi^{(\varepsilon)}[\rho,n] = a(v,n)\frac{(\sigma^{(\varepsilon)})^{v}}{(\sigma^{(\varepsilon)} - \rho)^{v+n}}, \qquad \rho < \sigma^{(\varepsilon)}, \tag{7.3.34}$$

where $a(v,n) = \prod_{l=1}^{n}(v + n - l)$.

Let us assume that the following perturbation condition holds:

$\mathbf{P}_{44}^{(k+1)}$: $\sigma^{(\varepsilon)} = \sigma^{(0)} + \sigma_{1}\varepsilon + \cdots + \sigma_{k+1}\varepsilon^{k+1} + o(\varepsilon^{k+1})$, where $0 < \sigma^{(0)} < \infty$ and $|\sigma_{r}| < \infty, r = 1,\ldots,k+1$.

As usual we will denote $\sigma_{0} = \sigma^{(0)}$.

Let us fix some $0 < \beta < \sigma^{(0)}$. It is obvious that every function $\psi^{(\varepsilon)}[\rho,n]$, $n = 0, 1, \ldots$, has an asymptotic $(0, k+1)$-expansion.

We will not try to get a general formula but just give explicit expressions for the first two terms in the corresponding asymptotic expansions for the non-trivial case $v \neq 1$.

Thus, let $k \geq 1$. By expanding the function $(\sigma^{(\varepsilon)})^{v}$ in a power series at the point $\varepsilon = 0$, we get that **(a)** $(\sigma^{(\varepsilon)})^{v} = (\sigma_{0})^{v} + \varepsilon v(\sigma_{0})^{v-1}\sigma_{1} + \varepsilon^{2}(\frac{1}{2}v(v-1)\sigma_{0}^{v-2}(\sigma_{1})^{2} + v(\sigma_{0})^{v-1}\sigma_{2}) + o(\varepsilon^{2})$. Similarly, by expanding the function $(\sigma^{(\varepsilon)} - \rho)^{v+n}$ in a power series at the point $\varepsilon = 0$, we obtain **(b)** $(\sigma^{(\varepsilon)} - \rho)^{v+n} = (\sigma_{0} - \rho)^{v+n} + \varepsilon(v+n)(\sigma_{0} - \rho)^{v+n-1}\sigma_{1} + \varepsilon^{2}(\frac{1}{2}(v+n)(v+n-1)(\sigma_{0} - \rho)^{v+n-2}(\sigma_{1})^{2} + v(\sigma_{0} - \rho)^{v-1}\sigma_{2}) + o(\varepsilon^{2})$. Now the corresponding asymptotic $(0, 2)$-expansions for the functions $\psi^{(\varepsilon)}[\rho,n]$ can be obtained by gathering and equating the coefficients at powers of ε in the asymptotic identity

$$(u[\rho, 0, n] + \varepsilon u[\rho, 1, n] + \varepsilon^{2}u[\rho, 2, n] + o(\varepsilon^{2}))$$
$$\times ((\sigma_{0} - \rho)^{v+n} + \varepsilon(v + n)(\sigma_{0} - \rho)^{v+n-1}\sigma_{1}$$
$$+ \varepsilon^{2}\Big(\frac{1}{2}(v + n)(v + n - 1)(\sigma_{0} - \rho)^{v+n-2}(\sigma_{1})^{2}$$
$$+ v(\sigma_{0} - \rho)^{v-1}\sigma_{2}\Big) + o(\varepsilon^{2}))$$
$$= a(v,n)((\sigma_{0})^{v} + \varepsilon v(\sigma_{0})^{v-1}\sigma_{1}$$
$$+ \varepsilon^{2}\Big(\frac{1}{2}v(v - 1)\sigma_{0}^{v-2}(\sigma_{1})^{2} + v(\sigma_{0})^{v-1}\sigma_{2}\Big) + o(\varepsilon^{2})). \tag{7.3.35}$$

This algorithm yields the following formulas:

$$u[\rho, 0, n] = \frac{1}{(\sigma_0 - \rho)^{\nu+n}} a(\nu, n)(\sigma_0)^\nu,$$

$$u[\rho, 1, n] = \frac{1}{(\sigma_0 - \rho)^{\nu+n}} \nu(\sigma_0)^{\nu-1}\sigma_1 - u[\rho, 0, n](\nu + n)(\sigma_0 - \rho)^{\nu+n-1}\sigma_1,$$

$$u[\rho, 2, n] = \frac{1}{(\sigma_0 - \rho)^{\nu+n}} \left(\frac{1}{2}\nu(\nu - 1)\sigma_0^{\nu-2}(\sigma_1)^2 + \nu(\sigma_0)^{\nu-1}\sigma_2 \right.$$

$$- u[\rho, 0, n]\left(\frac{1}{2}(\nu + n)(\nu + n - 1)(\sigma_0 - \rho)^{\nu+n-2}(\sigma_1)^2 \right.$$

$$\left. + \nu(\sigma_0 - \rho)^{\nu-1}\sigma_2 \right) - u[\rho, 1, n](\nu + n)(\sigma_0 - \rho)^{\nu+n-1}\sigma_1 \bigg). \quad (7.3.36)$$

Conditions of Theorems 7.3.2 and 7.3.3 become simpler in a model where the claim distribution $G^{(\varepsilon)}(u) \equiv G^{(0)}(u)$ does not depend on the perturbation parameter ε.

In this case, condition $\mathbf{C_{22}}$ reduces to condition $\mathbf{C_{21}}$. Condition $\mathbf{D_{30}}$ automatically holds.

Conditions $\mathbf{P_{41}^{(\rho,k)}}$ and $\mathbf{P_{42}^{(\rho,\tilde{k})}}$ also automatically hold for any $k = 0, 1, \ldots$. Moreover, in this case, $v[\rho, 0, n] = \psi^{(0)}[\rho, n]$ and $v[\rho, r, n] = 0$ for $r = 1, 2, \ldots, n = 0, 1, \ldots$. This permits to write formulas (7.3.11) for the coefficients $b[\rho, r, n]$, given in Lemma 7.3.1, in a simpler form for $r, n = 0, 1, \ldots$,

$$b[\rho, r, n] = h_r v[\rho, 0, n], \quad (7.3.37)$$

where, as in Lemma 7.3.1,

$$h_m = d_0^{-1}\left(e_m - \sum_{q=1}^{m} e_q h_{m-q} \right). \quad (7.3.38)$$

In some cases, it is convenient to express the condition for the claim distributions directly in terms of the distributions $G^{(\varepsilon)}(u)$. An example of this is a model of contaminated claim distributions, $G^{(\varepsilon)}(u) = (1 - \varepsilon)G^{(0)}(u) + \varepsilon\tilde{G}(u)$. Here ε serves as the probability of contamination and $\tilde{G}(u)$ is the contamination distribution. The contamination formula for $G^{(\varepsilon)}(u)$ can be rewritten as $G^{(\varepsilon)}(u) = G^{(0)}(u) + \varepsilon(\tilde{G}(u) - G^{(0)}(u))$. This representation suggests the following form of the perturbation condition for the claim distributions:

$\mathbf{P_{45}^{(k)}}$: $G^{(\varepsilon)}(u) = G^{(0)}(u) + G_1(u)\varepsilon + \cdots + G_k(u)\varepsilon^k + G_{k+1}^{(\varepsilon)}(u)o(\varepsilon^k)$, where $G_l(u)$,

$l = 1, \ldots, k$, and $G_{k+1}^{(\varepsilon)}(u)$ are functions (in u) that are right continuous and bounded by a constant that does not depend on ε.

The following condition replaces condition $\mathbf{C_{22}}$:

$\mathbf{C_{23}}$: There exists $\delta > 0$ such that:

(a) $\overline{\lim}_{0 \le \varepsilon \to 0} \varphi^{(\varepsilon)}(\delta) < \infty$;

(b) $\frac{\lambda^{(0)}}{c^{(0)}} \varphi^{(0)}(\delta) = \alpha^{(0)} \int_0^\infty e^{\delta s} \bar{G}^{(0)}(ds) \in (1, \infty)$;

(c) $\int_0^\infty e^{\delta s} |G_l(s)|\, ds < \infty,\ l = 1, \ldots, k$;

(d) $\overline{\lim}_{\varepsilon \to 0} \int_0^\infty e^{\delta s} |G_{k+1}^{(\varepsilon)}(s)|\, ds < \infty$.

It is readily seen that conditions $\mathbf{C_{23}}$ and $\mathbf{P_{45}^{(k)}}$ imply that condition $\mathbf{C_{22}}$ holds and conditions $\mathbf{P_{42}^{(\rho,k)}}$ is satisfied for any $\rho < \delta$.

The coefficients in the expansion given in condition $\mathbf{P_{45}^{(k)}}$ can be calculated with the use of the following formulas:

$$v[\rho, 0, n] = \int_0^\infty s^n e^{\rho s} (1 - G^{(0)}(s))\, ds, \quad n = 0, 1, \ldots, \tag{7.3.39}$$

and

$$v[\rho, r, n] = -\int_0^\infty s^n e^{\rho s} G_l(s)\, ds, \quad r = 1, 2, \ldots, \quad n = 0, 1, \ldots. \tag{7.3.40}$$

7.4 Approximations for ruin probabilities based on asymptotic expansions

In this section we give approximations for ruin probabilities based on asymptotic expansions.

7.4.1 Asymptotic expansions for ruin renewal limits

One can weaken the balancing condition $\mathbf{B_{11}^{(r)}}$ used in the Theorem 7.3.2 and replace it with the following:

$\mathbf{B_{12}^{(r)}}$: $u^{(\varepsilon)} \to \infty$ in such a way that $\overline{\lim}_{\varepsilon \to 0} \varepsilon^r u^{(\varepsilon)} = \varrho_r \in [0, \infty)$.

Under conditions $\mathbf{H_1}, \mathbf{D_{30}}, \mathbf{C_{22}}, \mathbf{P_{39}^{(k)}}, \mathbf{P_{40}^{(k)}}, \mathbf{P_{41}^{(\rho^{(0)},k)}}$, and $\mathbf{B_{12}^{(r)}}$, we have the following asymptotic relation that is equivalent to the asymptotic relation (7.3.17):

$$\frac{P^{(\varepsilon)}(u^{(\varepsilon)})}{e^{-\rho_r^{(\varepsilon)} u^{(\varepsilon)}} \pi^{(0)}(\rho^{(0)})} \to 1 \quad \text{as } \varepsilon \to 0, \tag{7.4.1}$$

where

$$\rho_r^{(\varepsilon)} = \rho^{(0)} + a_1 \varepsilon + \cdots + a_r \varepsilon^r, \tag{7.4.2}$$

and $\pi^{(\varepsilon)}(\rho^{(\varepsilon)})$ is the renewal limit that corresponds to the renewal equation (7.2.3) and is given by formula (7.3.3), i.e.,

$$\pi^{(\varepsilon)}(\rho^{(\varepsilon)}) = \frac{\int_0^\infty e^{\rho^{(\varepsilon)}s}(1 - \bar{G}^{(\varepsilon)}(s))\,ds}{\int_0^\infty s e^{\rho^{(\varepsilon)}s}\bar{G}^{(\varepsilon)}(ds)}$$

$$= \frac{\int_0^\infty e^{\rho^{(\varepsilon)}s}(\int_s^\infty (1 - G^{(\varepsilon)}(u))\,du)\,ds}{\int_0^\infty s e^{\rho^{(\varepsilon)}s}(1 - G^{(\varepsilon)}(s)\,ds}.$$

The asymptotic relation (7.4.1) suggests that we can use the expression $e^{-\rho_r^{(\varepsilon)}u^{(\varepsilon)}} \cdot \pi^{(0)}(\rho^{(0)})$ as an approximation for the ruin probability $P^{(\varepsilon)}(u^{(\varepsilon)})$.

The approximation $e^{-\rho_r^{(\varepsilon)}u^{(\varepsilon)}}\pi^{(0)}(\rho^{(0)})$ is obtained by replacing the Lundberg exponent $\rho^{(\varepsilon)}$ with its approximation $\rho_r^{(\varepsilon)}$ calculated using from the asymptotic expansion resulted from applying Theorem 7.3.2 to the renewal equation (7.2.3),

$$\rho^{(\varepsilon)} = \rho^{(0)} + a_1\varepsilon + \cdots + a_k\varepsilon^k + o(\varepsilon^k). \tag{7.4.3}$$

The approximation $e^{-\rho_r^{(\varepsilon)}u^{(\varepsilon)}}\pi^{(0)}(\rho^{(0)})$ also contains the limit constant $\pi^{(0)}(\rho^{(0)})$ given by formula (7.3.3).

Theorem 1.4.4, applied to the renewal equation (7.2.3), shows that

$$\pi^{(\varepsilon)}(\rho^{(\varepsilon)}) \to \pi^{(0)}(\rho^{(0)}) \quad \text{as} \ \varepsilon \to 0. \tag{7.4.4}$$

Indeed, as was shown in the proof of Theorem 7.3.2, conditions $\mathbf{H}_1$, $\mathbf{D}_{30}$ and $\mathbf{C}_{22}$, $\mathbf{P}_{39}^{(k)}$, $\mathbf{P}_{40}^{(k)}$, and $\mathbf{P}_{41}^{(\rho^{(0)},k)}$ imply that conditions $\mathbf{D}_7$, $\mathbf{C}_2$ and $\mathbf{F}_6$ hold in the case where $\rho^{(0)} > 0$ or $\mathbf{D}_6$, $\mathbf{C}_1$, and $\mathbf{F}_5$ if $\rho^{(0)} = 0$. Thus, Theorem 1.4.4 can be applied to the renewal equation (7.2.3) to give relation (7.4.4).

On the other hand, conditions $\mathbf{H}_1$, $\mathbf{D}_{30}$, and $\mathbf{C}_{22}$ imply that there exists $\varepsilon_0 > 0$ such that (a) condition $\mathbf{C}_{21}$ holds for the risk process $\xi_u^{(\varepsilon)}(t)$, $t \geq 0$, for every $\varepsilon \leq \varepsilon_0$. Thus, for every $\varepsilon \leq \varepsilon_0$, the classical Cramér–Lundberg approximation for ruin probabilities takes place,

$$P^{(\varepsilon)}(u) \sim e^{-\rho^{(\varepsilon)}u}\pi^{(\varepsilon)}(\rho^{(\varepsilon)}) \quad \text{as} \ u \to \infty. \tag{7.4.5}$$

This shows that the approximation $e^{-\rho_r^{(\varepsilon)}u^{(\varepsilon)}}\pi^{(0)}(\rho^{(0)})$ could be improved by replacing the coefficient $\pi^{(0)}(\rho^{(0)})$ with $\pi^{(\varepsilon)}(\rho^{(\varepsilon)})$. Moreover, it also seems logical to replace $\pi^{(\varepsilon)}(\rho^{(\varepsilon)})$ with some approximation using the expansion of this coefficient in an asymptotic power series with respect to the perturbation parameter ε. Such expansions can be obtained by applying Theorem 2.3.2 to the renewal equation (7.2.3).

Introduce the mixed power-exponential moment generating functions by

$$\omega^{(\varepsilon)}(\rho) = \int_0^\infty e^{\rho s}\left(\int_s^\infty (1 - G^{(\varepsilon)}(u))\,du\right)ds$$

$$= \mu^{(\varepsilon)}\int_0^\infty e^{\rho s}(1 - \bar{G}^{(\varepsilon)}(s))\,ds,$$

and the mixed power-exponential moment generating functions for $n = 0, 1, \ldots$,

$$\omega^{(\varepsilon)}[\rho, n] = \int_0^\infty s^n e^{\rho s} \left(\int_s^\infty (1 - G^{(\varepsilon)}(u))\, du \right) ds$$

$$= \mu^{(\varepsilon)} \int_0^\infty s^n e^{\rho s} (1 - \bar{G}^{(\varepsilon)}(s))\, ds.$$

By the definition, $\omega^{(\varepsilon)}[\rho, 0] = \omega^{(\varepsilon)}(\rho)$.

Choose an arbitrary $\rho^{(0)} < \beta < \delta$.

Condition $\mathbf{C_{22}}$ **(a)** implies that **(b)** $\varphi^{(\varepsilon)}(\beta) < \infty$ for $\varepsilon \leq \varepsilon_2$. On the other hand, integration by parts shows that **(c)** $\omega^{(\varepsilon)}(\beta) = \beta^{-1}(\varphi^{(\varepsilon)}(\beta) - 1)$. Using **(c)**, we get for $\varepsilon \leq \varepsilon_2$ and $n = 0, 1, \ldots$ that

$$\omega^{(\varepsilon)}[\beta, n] \leq c_n \omega^{(\varepsilon)}(\beta) < \infty, \tag{7.4.6}$$

where $c_n = c_n(\delta, \beta) = \sup_{s \geq 0} s^n e^{-(\delta - \beta)s} < \infty$.

Also note that for $\rho \leq \beta$, $\omega^{(\varepsilon)}[\rho, n]$ is the n-order derivative of order n of the function $\omega^{(\varepsilon)}(\rho)$.

Let $\bar{\kappa}_1^{(\varepsilon)}$ be a random variable with the distribution function $\bar{G}^{(\varepsilon)}(s)$.

The following relation links the moment generating functions $\omega^{(\varepsilon)}(\rho)$ and $\varphi^{(\varepsilon)}(\rho)$:

$$\omega^{(\varepsilon)}(\rho) = \mu^{(\varepsilon)} \int_0^\infty e^{\rho s}(1 - \bar{G}^{(\varepsilon)}(s))\, ds = \mu^{(\varepsilon)} \mathsf{E} \int_0^{\bar{\kappa}_1^{(\varepsilon)}} e^{\rho s}\, ds$$

$$= \begin{cases} \mu^{(\varepsilon)}(\mathsf{E}e^{\rho \bar{\kappa}_1^{(\varepsilon)}} - 1)/\rho & \text{for } \rho \neq 0, \\ \mu^{(\varepsilon)} \mathsf{E}\bar{\kappa}_1^{(\varepsilon)} & \text{for } \rho = 0, \end{cases}$$

$$= \begin{cases} (\varphi^{(\varepsilon)}[\rho, 0] - \varphi^{(\varepsilon)}[0, 0])/\rho & \text{for } \rho \neq 0, \\ \varphi^{(\varepsilon)}[0, 1] & \text{for } \rho = 0. \end{cases} \tag{7.4.7}$$

Note that we have used the identity $\mu^{(\varepsilon)} = \varphi^{(\varepsilon)}[0, 0]$ in relation (7.4.7).

Relation (7.4.7) yields the following recurrence relations for $n = 1, 2, \ldots, \varepsilon \leq \varepsilon_2$, and $\rho \leq \beta$:

$$\omega^{(\varepsilon)}[\rho, n] = \begin{cases} (\varphi^{(\varepsilon)}[\rho, n] - n\omega^{(\varepsilon)}[\rho, n-1])/\rho & \text{for } \rho \neq 0, \rho \leq \beta, \\ \varphi^{(\varepsilon)}[0, n+1]/(n+1) & \text{for } \rho = 0. \end{cases} \tag{7.4.8}$$

Relation (7.4.8) can be obtained similarly to relation (7.3.29)). Relation **(a)** $\varphi^{(\varepsilon)}[\rho, 0] = \omega^{(\varepsilon)}[\rho, 0]\rho + \mu^{(\varepsilon)}$, $\rho \leq \beta$, follows from formula (7.4.7). By differentiating **(a)** we get the following relation: **(b)** $\varphi^{(\varepsilon)}[\rho, n] = \omega^{(\varepsilon)}[\rho, n]\rho + n\omega^{(\varepsilon)}[\rho, n-1]$, $n = 1, 2, \ldots, \rho \leq \beta$. In particular, **(b)** yields the following relation for $\rho = 0$: **(c)** $\varphi^{(\varepsilon)}[0, n] = n\omega^{(\varepsilon)}[0, n-1]$, $n = 1, 2, \ldots, \rho \leq \beta$. Relations **(a)**–**(c)** imply relation (7.4.8).

Let us formulate a perturbation condition implied by condition $\mathbf{P}_{41}^{(\rho,k+1)}$,

$\mathbf{P}_{46}^{(\rho,k)}$: $\omega^{(\varepsilon)}[\rho, n] = \omega^{(0)}[\rho, n] + w[\rho, 1, n]\varepsilon + \cdots + w[\rho, k - n, n]\varepsilon^{k-n} + o(\varepsilon^{k-n})$,
where $|w_r[\rho, n]| < \infty$, $r = 1, \ldots, k - n$, $n = 0, \ldots, k$.

It also is convenient to define $w[\rho, 0, n] = \omega^{(0)}[\rho, n]$, $n = 0, \ldots, k$.

The following lemma is an obvious corollary of relations (7.4.7) and (7.4.8).

Lemma 7.4.1. *Let condition* $\mathbf{H}_1$, $\mathbf{C}_{22}$ *hold, and condition* $\mathbf{P}_{41}^{(\rho,k+1)}$ *be satisfied for some* $\rho \leq \beta$. *Then condition* $\mathbf{P}_{46}^{(\rho,k)}$ *holds for* $\rho \leq \beta$, *and the asymptotic expansions in this condition take the following form for* $n = 0, \ldots, k$:

$$\omega^{(\varepsilon)}[\rho, n] = \omega^{(0)}[\rho, n] + w[\rho, 1, n]\varepsilon + \cdots + w[\rho, k - n, n]\varepsilon^{k-n} + o(\varepsilon^{k-n}), \quad (7.4.9)$$

where the coefficients $w[\rho, r, n]$, $r = 0, \ldots, k - n$, $n = 0, \ldots, k$, *are given for* $r = 0, \ldots, k$, $n = 0$ *by*

$$w[\rho, r, 0] = \begin{cases} (v[\rho, r, 0] - v[0, r, 0])/\rho & \text{if } \rho \neq 0, \\ v[0, r, 1] & \text{if } \rho = 0, \end{cases} \quad (7.4.10)$$

and, for $r = 0, \ldots, k - n$, $n = 1, \ldots, k$, *by*

$$w[\rho, r, n] = \begin{cases} (v[\rho, r, n] - nw[\rho, r, n - 1])/\rho & \text{if } \rho \neq 0, \\ v[0, r, n + 1]/(n + 1) & \text{if } \rho = 0. \end{cases} \quad (7.4.11)$$

Remark 7.4.1. Note that condition $\mathbf{P}_{41}^{(\rho,k+1)}$ can be replaced in Lemma 7.4.1 with condition $\mathbf{P}_{32}^{(\rho,k)}$ if $\rho > 0$, and the corresponding asymptotic expansion in condition $\mathbf{P}_{41}^{(\rho,k+1)}$ can be omitted in Lemma 7.3.2 for $n = 0$ if $\rho = 0$.

Remark 7.4.2. According to Lemma 7.3.2, condition $\mathbf{P}_{41}^{(\rho,k+1)}$ is implied by condition $\mathbf{P}_{43}^{(\rho,k+2)}$.

The following theorem gives asymptotic expansions for the renewal limits $\pi^{(\varepsilon)}(\rho^{(\varepsilon)})$. This is a corollary of Theorem 2.3.2 applied to the renewal equation (7.2.3).

Note that we use condition $\mathbf{P}_{43}^{(\rho^{(0)},k+2)}$ which, according to Lemmas 7.3.2 and 7.4.1, implies that both conditions $\mathbf{P}_{46}^{(\rho^{(0)},k+1)}$ and $\mathbf{P}_{41}^{(\rho^{(0)},k)}$ hold.

Theorem 7.4.1. *Let conditions* $\mathbf{H}_1$ $\mathbf{D}_{30}$, $\mathbf{C}_{22}$, $\mathbf{P}_{39}^{(k)}$, $\mathbf{P}_{40}^{(k)}$, *and* $\mathbf{P}_{43}^{(\rho^{(0)},k+2)}$ *hold. Then the functional* $\pi^{(\varepsilon)}$ *has the asymptotic expansions*

$$\begin{aligned} \pi^{(\varepsilon)}(\rho^{(\varepsilon)}) &= \frac{\omega^{(0)}[\rho^{(0)}, 0] + f_1'\varepsilon + \cdots + f_k'\varepsilon^k + o(\varepsilon^k)}{\varphi^{(0)}[\rho^{(0)}, 1] + f_1''\varepsilon + \cdots + f_k''\varepsilon^k + o(\varepsilon^k)} \\ &= \pi^{(0)}(\rho^{(0)}) + f_1\varepsilon + \cdots + f_k\varepsilon^k + o(\varepsilon^k), \end{aligned} \quad (7.4.12)$$

where the coefficients f'_n, f''_n are given by the formulas $f'_0 = \omega^{(0)}[\rho^{(0)}, 0] = w[\rho^{(0)}, 0, 0]$, $f'_1 = w[\rho^{(0)}, 1, 0] + w[\rho^{(0)}, 0, 1]a_1$, $f''_0 = \varphi^{(0)}[\rho^{(0)}, 1] = v[\rho^{(0)}, 0, 1]$, $f''_1 = v[\rho^{(0)}, 1, 1] + v[\rho^{(0)}, 0, 2]a_1$, and, for $n = 0, \ldots, k$,

$$f'_n = w[\rho^{(0)}, n, 0] + \sum_{q=1}^{n} w[\rho^{(0)}, n-q, 1]a_q$$

$$+ \sum_{2 \leq m \leq n} \sum_{q=m}^{n} w[\rho^{(0)}, n-q, m] \cdot \sum_{n_1, \ldots, n_{q-1} \in D_{m,q}} \prod_{p=1}^{q-1} a_p^{n_p} / n_p!, \qquad (7.4.13)$$

and

$$f''_n = v[\rho^{(0)}, n, 1] + \sum_{q=1}^{n} v[\rho^{(0)}, n-q, 2]a_q$$

$$+ \sum_{2 \leq m \leq n} \sum_{q=m}^{n} v[\rho^{(0)}, n-q, m+1] \cdot \sum_{n_1, \ldots, n_{q-1} \in D_{m,q}} \prod_{p=1}^{q-1} a_p^{n_p} / n_p!, \quad (7.4.14)$$

and the coefficients f_n are given by the recurrence formulas $f_0 = \pi^{(0)}(\rho^{(0)}) = f'_0 / f''_0$ and, for $n = 0, \ldots, k$,

$$f_n = \left(f'_n - \sum_{q=0}^{n-1} f''_{n-q} f_q \right) / f''_0. \qquad (7.4.15)$$

Proof. We shall apply Theorem 2.3.2 to the renewal equation (7.2.3).

As usual, we can take $\rho^{(0)} < \beta < \delta$.

Conditions $\mathbf{H_1}$, $\mathbf{D_{30}}$ and $\mathbf{C_{22}}$, $\mathbf{P_{39}^{(k)}}$, $\mathbf{P_{40}^{(k)}}$, and $\mathbf{P_{41}^{(\rho^{(0)}, k)}}$ obviously imply that conditions $\mathbf{D_{11}}$, $\mathbf{C_2}$ hold. Conditions $\mathbf{E_1}$, $\mathbf{C_{22}}$, $\mathbf{P_{43}^{(\rho^{(0)}, k+2)}}$ imply that conditions $\mathbf{P_{41}^{(\rho^{(0)}, k+1)}}$ and $\mathbf{P_{46}^{(\rho^{(0)}, k)}}$ hold. This follows from Lemmas 7.3.2 and 7.4.1. Condition $\mathbf{P_3^{(k+1)}}$ takes the form of condition $\mathbf{P_{41}^{(\rho^{(0)}, k+1)}}$, and condition $\mathbf{P_7^{(k)}}$ becomes condition $\mathbf{P_{46}^{(\rho^{(0)}, k)}}$. Finally, as was shown in the proof of Theorem 7.3.2, conditions $\mathbf{H_1}$, $\mathbf{D_{30}}$ and $\mathbf{C_{22}}$, $\mathbf{P_{39}^{(k)}}$, $\mathbf{P_{40}^{(k)}}$, and $\mathbf{P_{41}^{(\rho^{(0)}, k)}}$ imply that condition $\mathbf{F_6}$ also holds.

Thus, we apply Theorem 2.3.2 to the renewal equation (7.2.3) and obtain the asymptotic relations (7.4.12). $\qquad \square$

Remark 7.4.3. The coefficients f'_1, f''_1, f_1 and f'_2, f''_2, f_2 can be calculated with the use of formulas (2.3.24) given in Subsection 2.2.1, where the coefficients $b_{r,n}$ and $c_{r,n}$ should be replaced, respectively, with the coefficients $v[\rho^{(0)}, r, n]$ and $w[\rho^{(0)}, r, n]$, i.e.,

$$f_0' = w[\rho^{(0)}, 0, 0],$$

$$f_1' = w[\rho^{(0)}, 1, 0] + w[\rho^{(0)}, 0, 1]a_1$$

$$= w[\rho^{(0)}, 1, 0] - \frac{w[\rho^{(0)}, 0, 1]v[\rho^{(0)}, 1, 0]}{v[\rho^{(0)}, 0, 1]},$$

$$f_2' = w[\rho^{(0)}, 2, 0] + w[\rho^{(0)}, 0, 1]a_2 + w[\rho^{(0)}, 1, 1]a_1 + \frac{w[\rho^{(0)}, 0, 2]a_1^2}{2},$$

$$f_0'' = v[\rho^{(0)}, 0, 1],$$

$$f_1'' = v[\rho^{(0)}, 1, 1] + v[\rho^{(0)}, 0, 2]a_1$$

$$= v[\rho^{(0)}, 1, 1] - \frac{v[\rho^{(0)}, 0, 2]v[\rho^{(0)}, 1, 0]}{v[\rho^{(0)}, 0, 1]},$$

$$f_2'' = v[\rho^{(0)}, 2, 1] + v[\rho^{(0)}, 0, 2]a_2 + v[\rho^{(0)}, 1, 2]a_1 + \frac{v[\rho^{(0)}, 0, 3]a_1^2}{2},$$

$$f_0 = \frac{f_0'}{f_0''} = \frac{w[\rho^{(0)}, 0, 0]}{v[\rho^{(0)}, 0, 1]},$$

$$f_1 = \frac{1}{f_0''}(f_1' - f_0 f_1'') = \frac{w[\rho^{(0)}, 1, 0]}{v[\rho^{(0)}, 0, 1]} - \frac{w[\rho^{(0)}, 0, 1]v[\rho^{(0)}, 1, 0]}{v[\rho^{(0)}, 0, 1]^2},$$

$$- \frac{w[\rho^{(0)}, 0, 0]v[\rho^{(0)}, 1, 1]}{v[\rho^{(0)}, 0, 1]^2} + \frac{w[\rho^{(0)}, 0, 0]v[\rho^{(0)}, 0, 2]v[\rho^{(0)}, 1, 0]}{v[\rho^{(0)}, 0, 1]^3},$$

$$f_2 = \frac{1}{f_0''}(f_2' - f_0 f_2'' - f_1 f_1''). \tag{7.4.16}$$

7.4.2 Approximations for ruin probabilities

The asymptotic expansions (7.4.12) can be used to construct approximations for the coefficients $\pi^{(\varepsilon)}(\rho^{(\varepsilon)})$. We can use the following approximation formula for $\pi^{(\varepsilon)}(\rho^{(\varepsilon)})$, based on the asymptotic expansion (7.4.12):

$$\pi_l^{(\varepsilon)} = \pi^{(0)}(\rho^{(0)}) + f_1\varepsilon + \cdots + f_l\varepsilon^l. \tag{7.4.17}$$

Approximations $\rho_r^{(\varepsilon)} = \rho^{(0)} + a_1\varepsilon + \cdots + a_r\varepsilon^r$ and $\pi_l^{(\varepsilon)} = \pi^{(0)}(\rho^{(0)}) + f_1\varepsilon + \cdots + f_l\varepsilon^l$ for the Lundberg exponent $\rho^{(\varepsilon)}$ and the coefficients $\pi^{(\varepsilon)}(\rho^{(\varepsilon)})$, respectively, leads to the following approximation for the ruin probabilities:

$$P_{r,l}^{(\varepsilon)}(u) = e^{-\rho_r^{(\varepsilon)}u}\pi_l^{(\varepsilon)}. \tag{7.4.18}$$

By using the parameters r and l one can control the highest order of moments of the claim distribution involved in the approximation formulas.

Approximation $P_{r,l}^{(\varepsilon)}(u)$ is asymptotically equivalent, under conditions of Theorem 7.4.1 and condition $\mathbf{B}_{12}^{(r)}$, to the initial approximation given by the Cramér–Lundberg approximation for the ruin probability $P^{(\varepsilon)}(u)$ by the quantity $\tilde{P}^{(\varepsilon)}(u) = e^{-\rho^{(\varepsilon)}u} \cdot \pi^{(\varepsilon)}(\rho^{(\varepsilon)})$ or by its variant $\hat{P}^{(\varepsilon)}(u) = e^{-\rho^{(\varepsilon)}u}\pi^{(0)}(\rho^{(0)})$, i.e.,

$$\frac{P^{(\varepsilon)}(u^{(\varepsilon)})}{P_{r,l}^{(\varepsilon)}(u^{(\varepsilon)})} \sim \frac{P^{(\varepsilon)}(u^{(\varepsilon)})}{\tilde{P}^{(\varepsilon)}(u^{(\varepsilon)})} \sim \frac{P^{(\varepsilon)}(u^{(\varepsilon)})}{\hat{P}^{(\varepsilon)}(u^{(\varepsilon)})} \to 1 \text{ as } \varepsilon \to 0. \tag{7.4.19}$$

All the approximation considered above give a zero asymptotic relative error for the ruin probabilities both in models where the limit Lundberg exponent is $\rho^{(0)} > 0$, which corresponds to the case of the Cramér–Lundberg approximation for perturbed risk processes, and in a model where $\rho^{(0)} = 0$, which corresponds to the case the diffusion approximation.

7.5   Asymptotic expansions for the distribution of the surplus prior to and at the time of a ruin

In this section, we consider an additional example that shows how to apply the methods of asymptotic analysis of a perturbed renewal equation to perturbed risk processes. We obtain asymptotic expansions for the distribution of the surplus prior to and at the time of a ruin.

7.5.1   The surplus prior to and at the time of a ruin

Let us introduce, for $u \geq 0$, a random variable, which is the time of a ruin,

$$\tau_u^{(\varepsilon)} = \inf(t \geq 0 : \xi_u^{(\varepsilon)}(t) < 0).$$

This random variable takes values in the interval $(0, \infty]$. The ruin probability can be expressed with the use of this random variable as $P^{(\varepsilon)}(u) = \mathsf{P}\{\tau_u^{(\varepsilon)} < \infty\}$.

Let us now introduce, for $u, x, y \geq 0$, the random variables $\xi_u^{(\varepsilon)}(\tau_u^{(\varepsilon)} - 0)$ and $-\xi_u^{(\varepsilon)}(\tau_u^{(\varepsilon)})$ that are, respectively, the surplus prior to and at the time of the ruin. These random variables are well defined if $\tau_u^{(\varepsilon)} < \infty$. Let us assign them value $+\infty$ if $\tau_u^{(\varepsilon)} = \infty$.

Let us now introduce the surplus probabilities that define the joint distribution of the surpluses prior to and at the time of the ruin,

$$P_{x,y}^{(\varepsilon)}(u) = \mathsf{P}\{\tau_u^{(\varepsilon)} < \infty, -\xi_u^{(\varepsilon)}(\tau_u^{(\varepsilon)}) > x, \xi_u^{(\varepsilon)}(\tau_u^{(\varepsilon)} - 0) > y\}, \quad x, y \geq 0.$$

A starting point in our asymptotic analysis is the following renewal equation for the surplus probabilities $P_{x,y}^{(\varepsilon)}(u), u \geq 0$ given in Schmidli (1999):

$$P_{x,y}^{(\varepsilon)}(u) = \alpha^{(\varepsilon)}(1 - \bar{G}^{(\varepsilon)}(u \vee y + x))$$

$$+ \alpha^{(\varepsilon)} \int_0^u P_{x,y}^{(\varepsilon)}(u - s)\bar{G}^{(\varepsilon)}(ds), \quad u \geq 0. \qquad (7.5.1)$$

This renewal equation should be compared with the renewal equation (7.2.3) satisfied by the ruin probabilities $P^{(\varepsilon)}(u), u \geq 0$. Both equations have the same generating distribution $F^{(\varepsilon)}(s) = \alpha^{(\varepsilon)}\bar{G}^{(\varepsilon)}(s)$ and differ only in the forcing functions. In the case of the renewal equation (7.5.1), the forcing function is $q^{(\varepsilon)}(s) = \alpha^{(\varepsilon)}(1 - \bar{G}^{(\varepsilon)}(u \vee y + x))$, while in the case of the renewal equation (7.2.3), the forcing function is given by $q^{(\varepsilon)}(s) = \alpha^{(\varepsilon)}(1 - \bar{G}^{(\varepsilon)}(u))$.

Note that, by the definition, both random variables, the surplus prior to the ruin, $\xi_u^{(\varepsilon)}(\tau_u^{(\varepsilon)} - 0)$, and at the time of the ruin, $-\xi_u^{(\varepsilon)}(\tau_u^{(\varepsilon)})$, are positive random variables with probability 1. This implies that $P_{0,0}^{(\varepsilon)}(u) = P^{(\varepsilon)}(u), u \geq 0$.

7.5.2 Diffusion approximation for the surplus prior to and at the time of a ruin for perturbed risk processes

For $x, y \geq 0$, denote

$$\pi_{x,y}^{(\varepsilon)} = \frac{\int_0^\infty (1 - \bar{G}^{(\varepsilon)}(u \vee y + x))\, du}{\int_0^\infty u\bar{G}^{(\varepsilon)}(du)}. \qquad (7.5.2)$$

Conditions $\mathbf{H_1}$ and $\mathbf{M_{16}}$ imply that both integrals are finite, $\int_0^\infty (1 - \bar{G}^{(\varepsilon)}(u \vee y + x))\, du < \infty$ and $\int_0^\infty u\bar{G}^{(\varepsilon)}(du) < \infty$, and therefore, the quantity $\pi_{x,y}^{(\varepsilon)}$ is well defined for all ε small enough.

By the definition, **(a)** $P_{x,y}^{(\varepsilon)} \in [0, 1]$, **(b)** $P_{x,y}^{(\varepsilon)}(u)$ is a non-increasing function in $x \geq 0$ and $y \geq 0$, and **(c)** $P_{0,0}^{(\varepsilon)}(u) = P^{(\varepsilon)}(u)$.

The following theorem is a variant of the diffusion approximation for perturbed risk processes given in Lemma 7.2.2.

Theorem 7.5.1. *Let conditions* $\mathbf{H_1}$, $\mathbf{D_{30}}$, $\mathbf{M_{16}}$, *and* $\mathbf{B_{10}}$ *be satisfied. Then, for* $x, y \geq 0$,

$$P_{x,y}^{(\varepsilon)}(u^{(\varepsilon)}) \to \pi_{x,y}^{(0)}\, e^{-\varrho/\bar{m}^{(0)}} \quad \text{as } \varepsilon \to 0, \qquad (7.5.3)$$

where $\bar{m}^{(0)} = \beta^{(0)}/2\mu^{(0)}$.

Proof. The proof is absolutely analogous to the proof of Lemma 7.2.2. The only difference is in the formula for the corresponding renewal limit. In the case of the ruin probabilities $P_\varepsilon(u)$ treated in Lemma 7.2.2, the corresponding renewal limit, given in (7.2.9), equals 1. In the case of the quantities $P_{x,y}^{(\varepsilon)}(u)$, the corresponding renewal limit is $P_{x,y}^{(0)}$, which explains the difference between the limits in the asymptotic relations (7.2.10) and (7.5.3). $\qquad \square$

7.5.3   Generalised Cramér–Lundberg and diffusion approximations for the surplus prior to and at the time of a ruin for perturbed risk processes

Denote

$$
\begin{aligned}
\pi_{x,y}^{(\varepsilon)}(\rho^{(\varepsilon)}) &= \frac{\int_0^\infty e^{\rho^{(\varepsilon)}s}(1 - \bar{G}^{(\varepsilon)}(s \vee y + x))\, ds}{\int_0^\infty s e^{\rho^{(\varepsilon)}s} \bar{G}^{(\varepsilon)}(ds)} \\
&= \frac{\int_0^\infty e^{\rho^{(\varepsilon)}s}(\int_{s \vee y + x}^\infty (1 - G^{(\varepsilon)}(u))\, du)\, ds}{\int_0^\infty s e^{\rho^{(\varepsilon)}s}(1 - G^{(\varepsilon)}(s))\, ds}.
\end{aligned}
\tag{7.5.4}
$$

Conditions $\mathbf{H_1}$, $\mathbf{D_{30}}$–$\mathbf{D_{32}}$, and $\mathbf{C_{22}}$ imply that $\int_0^\infty e^{\rho^{(\varepsilon)}}s(1 - \bar{G}^{(\varepsilon)}(u \vee y + x))\, du < \infty$ and $\int_0^\infty u e^{\rho^{(\varepsilon)}u} \bar{G}^{(\varepsilon)}(du) < \infty$ and, therefore, the quantity $\pi_{x,y}^{(\varepsilon)}(\rho^{(\varepsilon)})$ is well defined for all ε small enough.

The following theorem is an analogue of Theorem 7.3.1.

Theorem 7.5.2. *Let conditions* $\mathbf{H_1}$, $\mathbf{D_{30}}$–$\mathbf{D_{32}}$, *and* $\mathbf{C_{22}}$ *hold. Then, for any* $0 \leq u^{(\varepsilon)} \to \infty$ *as* $\varepsilon \to 0$, *the following asymptotic relation holds for* $x, y \geq 0$:

$$
\frac{P_{x,y}^{(\varepsilon)}(u^{(\varepsilon)})}{\exp\{-\rho^{(\varepsilon)}u^{(\varepsilon)}\}} \to \pi_{x,y}^{(0)}(\rho^{(0)}) \ \text{ as } \ \varepsilon \to 0.
\tag{7.5.5}
$$

Proof. The proof repeats the proof of Theorem 7.3.1. The only difference is in the formula for the corresponding renewal limit. In the case of the ruin probabilities $P_\varepsilon(u)$ considered in Theorem 7.3.1, the corresponding renewal limit given in (7.3.3) is $\pi^{(0)}(\rho^{(0)})$. In the case of the quantities $P_{x,y}^{(\varepsilon)}(u)$, the corresponding renewal limit is $\pi_{x,y}^{(0)}(\rho^{(0)})$, which explains the difference between the limits in the asymptotic relations (7.3.5) and (7.5.5). $\square$

7.5.4   Exponential expansion for the distribution of the surplus prior to and at the time of a ruin

Finally, let us formulate an analogue of Theorem 7.3.2 for for the distribution of the surplus prior to and at the time of a ruin. Statements **(i)** and **(ii)** do not change. Thus we only give an analogue of statement **(iii)** of the theorem.

Theorem 7.5.3. *Let conditions* $\mathbf{H_1}$, $\mathbf{D_{30}}$, $\mathbf{C_{22}}$, $\mathbf{P_{39}^{(k)}}$, $\mathbf{P_{40}^{(k)}}$, *and* $\mathbf{P_{41}^{(\rho^{(0)},k)}}$ *hold and condition* $\mathbf{B_{11}^{(r)}}$ *be satisfied for some* $1 \leq r \leq k$. *Then the following asymptotic relation holds for* $x, y \geq 0$:

$$
\frac{P_{x,y}^{(\varepsilon)}(u^{(\varepsilon)})}{\exp\{-(\rho^{(0)} + a_1\varepsilon + \cdots + a_{r-1}\varepsilon^{r-1})u^{(\varepsilon)}\}} \to e^{-\varrho_r a_r} \pi_{x,y}^{(0)}(\rho^{(0)}) \ \text{ as } \ \varepsilon \to 0,
\tag{7.5.6}
$$

where the coefficients a_n, $n = 1, \ldots, k$, *are given by formulas (7.3.15).*

Proof. The proof repeats the proof of Theorem 7.3.2. Again, the only difference is in the formula for the corresponding renewal limit. In the case of the ruin probabilities $P_\varepsilon(u)$ treated in Theorem 7.3.2, the corresponding renewal limit given in (7.3.3) is $\pi^{(0)}(\rho^{(0)})$. In the case of the quantities $P_{x,y}^{(\varepsilon)}(u)$, the corresponding renewal limit is $\pi_{x,y}^{(0)}(\rho^{(0)})$, which explains the difference between the limits in the asymptotic relations (7.3.17) and (7.5.6). $\qquad\square$

Chapter 8

Supplements

The last Chapter 8 contains three supplements. The first one gives some basic arithmetic operation formulas for scalar and matrix asymptotic expansions.

In the second supplement, we discuss some new prospective directions for future research studies in the area and make comments on some already available results. Let us point out some of them. In the book, we study quasi-stationary phenomena for nonlinearly perturbed stochastic systems with continuous time. However, it is clear that similar results can also be obtained for discrete time models which are important by themselves and in applications. Another direction for future studies is the exponential asymptotic expansions, which are based on non-polynomial systems of infinitesimals, in mixed ergodic and large-deviation theorems for nonlinearly perturbed regenerative, Markov, and semi-Markov processes. In the book, we treat models of nonlinearly perturbed Markov chains and semi-Markov processes with a finite phase space. It would be important to realise an analogous program of studies for Markov chains and semi-Markov type processes with countable and general phase spaces. Another direction for advancing the studies carried out in the book is connected with models for nonlinearly perturbed Markov chains and semi-Markov processes with uncoupled phase spaces. Explicit estimates for the remainder terms in the exponential asymptotic expansions for nonlinearly perturbed stochastic systems also make a natural object for future studies. Studies of pseudo- and quasi-stationary phenomena for nonlinearly perturbed queueing and reliability systems, various epidemic models, models of population dynamics, and other biological systems are very important and prospective directions for the future research. Examples given in the book make just the first step in exploring this unlimited area.

Quasi-stationary phenomena and related problems were a subject of intensive studies during several decades. The bibliography contained in the book has more than 1000 references. The last supplement contains brief bibliographical remarks on the works in he related areas dealt with in this comprehensive bibliography.

8.1 Arithmetics of asymptotic expansions

In this section we give formulas for working with asymptotic expansions. These formulas are used in the book on numerous occasions. They are known and can be found in the existing literature. We give them here to make the book more self-contained.

8.1.1 Asymptotic expansions

Let $A(\varepsilon)$ be a function that is defined on an interval $0 < \varepsilon \le \varepsilon_0$ and given on this interval by an *asymptotic expansion*,

$$A(\varepsilon) = a_h \varepsilon^h + \cdots + a_k \varepsilon^k + o(\varepsilon^k), \tag{8.1.1}$$

where **(a)** $0 \le h \le k < \infty$ are integers, and **(b)** the coefficients $a_h, \dots, a_k$ are real numbers. We call such an asymptotic expansion an (h, k)-*expansion*.

In fact, representation (8.1.1) means that $A(\varepsilon)$ is a sum of a polynomial function of ε of the order k with the first $h - 1$ coefficients equal to zero, and a function $o(\varepsilon^k)$ satisfying $o(\varepsilon^k)/\varepsilon^k \to 0$ as $\varepsilon \to 0$. In particular, $(0, 0)$-expansion has the form $A(\varepsilon) = a_0 + o(1)$, where $o(1) \to 0$ as $\varepsilon \to 0$.

We say that an (h, k)-expansion $A(\varepsilon)$ is *pivotal* if it is known that $a_h \ne 0$.

Consider four asymptotic expansions, $A(\varepsilon) = a_{h_A} \varepsilon^{h_A} + \cdots + a_{k_A} \varepsilon^{k_A} + o(\varepsilon^{k_A})$, $B(\varepsilon) = b_{h_B} \varepsilon^{h_B} + \cdots + b_{k_B} \varepsilon^{k_B} + o(\varepsilon^{k_B})$, $C(\varepsilon) = c_{h_C} \varepsilon^{h_C} + \cdots + c_{k_C} \varepsilon^{k_C} + o(\varepsilon^{k_C})$, and $D(\varepsilon) = d_{h_D} \varepsilon^{h_D} + \cdots + d_{k_D} \varepsilon^{k_D} + o(\varepsilon^{k_D})$.

Lemma 8.1.1. *The above asymptotic expansions have the following "operational" rules:*

    (i) *If $A(\varepsilon)$ is an (h_A, k_A)-expansion and d is a constant, then $dA(\varepsilon)$ is an (h_A, k_A)-expansion with the corresponding coefficients $da_r, r = h_A, \dots, k_A$. This expansion is pivotal if and only if $da_{h_A} \ne 0$.*

    (ii) *If $A(\varepsilon)$ is an (h_A, k_A)-expansion and $B(\varepsilon)$ is an (h_B, k_B)-expansion, then $C(\varepsilon) = A(\varepsilon) + B(\varepsilon)$ is an (h_C, k_C)-expansion with $h_C = h_A \wedge h_B$, $k_C = k_A \wedge k_B$, and the coefficients satisfy $c_{h_C + r} = a_{h_C + r} + b_{h_C + r}, r = 0, \dots, k_C - h_C$, where $a_{h_C + r} = 0$ for $0 \le r < h_A - h_C$ and $b_{h_C + r} = 0$ for $0 \le r < h_B - h_C$. The expansion $C(\varepsilon)$ is pivotal if and only if $c_{h_C} = a_{h_C} + b_{h_C} \ne 0$.*

    (iii) *If $A(\varepsilon)$ is an (h_A, k_A)-expansion and $B(\varepsilon)$ is an (h_B, k_B)-expansion, then $C(\varepsilon) = A(\varepsilon) \times B(\varepsilon)$ is an (h_C, k_C)-expansion with $h_C = h_A + h_B$, $k_C = (h_A + k_B) \wedge (k_A + h_B)$, and the coefficients are given by $c_{h_C + r} = \sum_{0 \le i \le r} a_{h_A + i} b_{h_B + r - i}, \ r = 0, \dots, k_C - h_C$. The expansion $C(\varepsilon)$ is pivotal if and only if $c_{h_C} = a_{h_A} b_{h_B} \ne 0$.*

    (iv) *If $B(\varepsilon)$ is a pivotal (h_B, k_B)-expansion, then $C(\varepsilon) = \varepsilon^{h_B}/B(\varepsilon)$ is a pivotal $(0, k_C)$-expansion with $k_C = k_B - h_B$ and the coefficients satisfy $c_0 = b_{h_B}^{-1}$ and $c_r = -b_{h_B}^{-1} \sum_{1 \le i \le r} b_{h_B + i} c_{r - i}, r = 1, \dots, k_C$.*

    (v) *If $A(\varepsilon)$ is an (h_A, k_A)-expansion, $B(\varepsilon)$ is a pivotal (h_B, k_B)-expansion, then $D(\varepsilon) = \varepsilon^{h_B} A(\varepsilon)/B(\varepsilon)$ is an (h_D, k_D)-expansion with $h_D = h_A, k_D = k_A \wedge (h_A + k_B - h_B)$, and $d_{h_D + r} = \sum_{0 \le i \le r} c_i a_{h_A + r - i}, r = 0, \dots, k_D - h_D$, where $c_j, j = 0, \dots, k_C$ are coefficients of the $(0, k_C)$-expansion $C(\varepsilon) = \varepsilon^{h_B}/B(\varepsilon)$*

given in **(iv)**. *The expansion $D(\varepsilon)$ is pivotal if and only if $d_{h_D} = a_{h_A} c_0 = a_{h_A}/b_{h_B} \neq 0$.*

(vi) *If $A(\varepsilon)$ is an (h_A, k_A)-expansion, $B(\varepsilon)$ is a pivotal (h_B, k_B)-expansion, then $D(\varepsilon) = \varepsilon^{h_B} A(\varepsilon)/B(\varepsilon)$ is an (h_D, k_D)-expansion with $h_D = h_A$, $k_D = k_A \wedge (h_A + k_B - h_B)$, and the coefficients satisfying $d_{h_D+r} = b_{h_B}^{-1}(a_{h_A+r} - \sum_{1 \le i \le r} b_{h_B+i} d_{h_D+r-i})$, $r = 0, \ldots, k_D - h_D$. The expansion $D(\varepsilon)$ is pivotal if and only if $d_{h_D} = a_{h_A}/b_{h_B} \neq 0$.*

8.1.2 Asymptotic matrix expansions

Lemma 8.1.1 can be generalised to a model of asymptotic matrix expansions. Let $A(\varepsilon) = \|A_{ij}(\varepsilon)\|$ be an $m \times n$ matrix-valued function that is defined on an interval $0 < \varepsilon \le \varepsilon_0$ and represented on this interval by the *asymptotic matrix expansion*

$$\mathbf{A}(\varepsilon) = \mathbf{A}_h \varepsilon^h + \cdots + \mathbf{A}_k \varepsilon^k + \mathbf{o}(\varepsilon^k), \tag{8.1.2}$$

where **(a)** $0 \le h \le k < \infty$ are integers, and **(b)** the matrix coefficients $\mathbf{A}_h = \|a_{h,ij}\|, \ldots, \mathbf{A}_k = \|a_{k,ij}\|$ are $m \times n$ matrices with real-valued elements, and **(c)** $\mathbf{o}(\varepsilon^k) = \|o_{ij}(\varepsilon^k)\|$ is an $m \times n$ matrix-valued function of ε such that $o_{ij}(\varepsilon^k)/\varepsilon^k \to 0$ as $\varepsilon \to 0$, $i = 1, \ldots, m$, $j = 1, \ldots, n$. We call such an asymptotic expansion an *(m, n)-matrix (h, k)-expansion*.

Let us now consider four asymptotic matrix expansions $\mathbf{A}(\varepsilon) = \mathbf{A}_{h_A} \varepsilon^{h_A} + \cdots + \mathbf{A}_{k_A} \varepsilon^{k_A} + \mathbf{o}(\varepsilon^{k_A})$, $\mathbf{B}(\varepsilon) = \mathbf{B}_{h_B} \varepsilon^{h_B} + \cdots + \mathbf{B}_{k_B} \varepsilon^{k_B} + \mathbf{o}(\varepsilon^{k_B})$, $\mathbf{C}(\varepsilon) = \mathbf{C}_{h_C} \varepsilon^{h_C} + \cdots + \mathbf{C}_{k_C} \varepsilon^{k_C} + \mathbf{o}(\varepsilon^{k_C})$, and $\mathbf{D}(\varepsilon) = \mathbf{D}_{h_D} \varepsilon^{h_D} + \cdots + \mathbf{D}_{k_D} \varepsilon^{k_D} + \mathbf{o}(\varepsilon^{k_D})$. The number of lines and columns in the corresponding matrices will be specified in each particular case.

Lemma 8.1.2. *The above asymptotic matrix expansions have the following have the following "operational" rules:*

(i) *If $\mathbf{A}(\varepsilon)$ is an (m, n)-matrix (h_A, k_A)-expansion and d is a constant, then $d\,\mathbf{A}(\varepsilon)$ is an (m, n)-matrix (h_A, k_A)-expansion with the corresponding matrix coefficients $d\,\mathbf{A}_r$, $r = h_A, \ldots, k_A$.*

(ii) *If $\mathbf{A}(\varepsilon)$ is an (m, n)-matrix (h_A, k_A)-expansion, $\mathbf{B}(\varepsilon)$ is an (m, n)-matrix (h_B, k_B)-expansion, then $\mathbf{C}(\varepsilon) = \mathbf{A}(\varepsilon) + \mathbf{B}(\varepsilon)$ is an (m, n)-matrix (h_C, k_C)-expansion with $h_C = h_A \wedge h_B$, $k_C = k_A \wedge k_B$, $\mathbf{C}_{h_C+r} = \mathbf{A}_{h_C+r} + \mathbf{B}_{h_C+r}$, $r = 0, \ldots, k_C - h_C$, where $\mathbf{A}_{h_C+r} = \mathbf{0}$ for $0 \le r < h_A - h_C$ and $\mathbf{B}_{h_C+r} = \mathbf{0}$ for $0 \le r < h_B - h_C$, where $\mathbf{0}$ is a matrix with all elements equal to zero.*

(iii) *If $\mathbf{A}(\varepsilon)$ is an (m, l)-matrix (h_A, k_A)-expansion and $\mathbf{B}(\varepsilon)$ is an (l, n)-matrix (h_B, k_B)-expansion, then $\mathbf{C}(\varepsilon) = \mathbf{A}(\varepsilon) \times \mathbf{B}(\varepsilon)$ is an (m, n)-matrix (h_C, k_C)-expansion with $h_C = h_A + h_B$, $k_C = (h_A + k_B) \wedge (k_A + h_B)$, and the coefficients are given by $\mathbf{C}_{h_C+r} = \sum_{0 \le i \le r} \mathbf{A}_{h_A+i} \mathbf{B}_{h_B+r-i}$, $r = 0, \ldots, k_C - h_C$.*

(iv) *If* $\mathbf{B}(\varepsilon)$ *is an* (n,n)-*matrix* (h_B,k_B)-*expansion and* $\det(\mathbf{B}_h) \neq 0$, *then* $\mathbf{C}(\varepsilon) = \varepsilon^{h_B}\mathbf{B}(\varepsilon)^{-1}$ *is an* (n,n)-*matrix* $(0,k_C)$-*expansion with* $k_C = k_B - h_B$, *and* $\mathbf{C}_0 = \mathbf{B}_{h_B}^{-1}$ *and* $\mathbf{C}_r = -\mathbf{B}_{h_B}^{-1}\sum_{1\leq i \leq r}\mathbf{B}_{h_B+i}\mathbf{C}_{r-i}$, $r = 1,\ldots,k_C$.

(v) *If* $\mathbf{A}(\varepsilon)$ *is an* (n,m)-*matrix* (h_A,k_A)-*expansion,* $\mathbf{B}(\varepsilon)$ *is an* (n,n)-*matrix* (h_B,k_B)-*expansion, and* $\det(\mathbf{B}_{h_B}) \neq 0$, *then* $\mathbf{D}(\varepsilon) = \varepsilon^{h_B}\mathbf{B}(\varepsilon)^{-1}\mathbf{A}(\varepsilon)$ *is an* (n,m)-*matrix* (h_C,k_C)-*expansion with* $h_D = h_A$, $k_D = k_A \wedge (h_A + k_B - h_B)$, $\mathbf{D}_{h_D+r} = \sum_{0\leq i \leq r}\mathbf{C}_i \cdot \mathbf{A}_{h_A+r-i}$, $r = 0,\ldots,k_D - h_D$, *where* $\mathbf{C}_j, j = 0,\ldots,k_C$, *are coefficients of the* (n,n)-*matrix* $(0,k_C)$-*expansion* $\mathbf{C}(\varepsilon) = \varepsilon^{h_B}\mathbf{B}(\varepsilon)^{-1}$ *given in* **(iv)**.

(vi) *If* $\mathbf{A}(\varepsilon)$ *is an* (n,m)-*matrix* (h_A,k_A)-*expansion,* $\mathbf{B}(\varepsilon)$ *is an* (n,n)-*matrix* (h_B,k_B)-*expansion,* $\det(\mathbf{B}_{h_B}) \neq 0$, *then* $\mathbf{D}(\varepsilon) = \varepsilon^{h_B}\mathbf{B}(\varepsilon)^{-1}\mathbf{A}(\varepsilon)$ *is an* (n,m)-*matrix* (h_D,k_D)-*expansion with* $h_D = h_A$, $k_D = k_A \wedge (h_A + k_B - h_B)$, *and* $\mathbf{D}_{h_D+r} = \mathbf{B}_{h_B}^{-1}(\mathbf{A}_{h_A+r} - \sum_{1\leq i \leq r}\mathbf{B}_{h_B+i}\mathbf{D}_{h_d+r-i})$, $r = 0,\ldots,k_D - h_D$.

8.1.3 Nonlinearly perturbed systems of linear equations

Let $\mathbf{B}(\varepsilon) = ||B_{ij}(\varepsilon)||$ be an $n \times n$ matrix-valued function and $\mathbf{a}(\varepsilon) = ||a_i(\varepsilon)||$ be an $n \times 1$ matrix-valued (column vector-valued) function with real-valued elements that are defined in an interval $0 < \varepsilon \leq \varepsilon_0$. We consider a system of linear equations

$$\mathbf{B}(\varepsilon)\mathbf{x}(\varepsilon) = \mathbf{a}(\varepsilon). \tag{8.1.3}$$

A solution of the system $\mathbf{x}(\varepsilon) = ||x_i(\varepsilon)||$ is sought for in the class of $(n \times 1)$-matrix (column vector-valued) functions with real-valued elements that are defined on an interval $0 < \varepsilon \leq \varepsilon_0'$, where $0 < \varepsilon_0' < \varepsilon_0$.

We assume that $\mathbf{a}(\varepsilon)$ and $\mathbf{B}(\varepsilon)$ can be represented in the form of asymptotic matrix expansions, $\mathbf{a}(\varepsilon) = \mathbf{a}_{h_a}\varepsilon^{h_a} + \cdots + \mathbf{a}_{k_a}\varepsilon^{k_a} + \mathbf{o}(\varepsilon^{k_a})$ with $n \times 1$ matrix-valued (vector-valued) coefficients $\mathbf{a}_r = ||a_{r,i}||, r = h_a,\ldots,k_a$, and $\mathbf{B}(\varepsilon) = \mathbf{B}_{h_B}\varepsilon^{h_B} + \cdots + \mathbf{B}_{k_B}\varepsilon^{k_B} + \mathbf{o}(\varepsilon^{k_B})$ with $n \times n$ matrix-valued coefficients $\mathbf{B}_r = ||b_{r,ij}||, r = h_B,\ldots,k_B$.

The following lemma gives conditions for system (8.1.3) to have a unique solution that can be represented in the form of an asymptotic expansion.

Lemma 8.1.3. *Let* $\mathbf{a}(\varepsilon)$ *be an* $(n, 1)$-*matrix* (h_a,k_a)-*expansion,* $\mathbf{B}(\varepsilon)$ *an* (n,n)-*matrix* (h_B,k_B)-*expansion, and* $\det(\mathbf{B}_{h_B}) \neq 0$. *Then we have the following:*

(i) $\tilde{\mathbf{x}}(\varepsilon) = \varepsilon^{h_B}\mathbf{B}(\varepsilon)^{-1}\mathbf{a}(\varepsilon)$ *is an* $(n, 1)$-*matrix (vector)* (h_x,k_x)-*expansion with* $h_x = h_a, k_x = k_a \wedge (h_a + k_B - h_B)$, *and*

$$\tilde{\mathbf{x}}_{h_x+r} = \mathbf{B}_{h_B}^{-1}(\mathbf{a}_{h_a+r} - \sum_{1\leq i \leq r}\mathbf{B}_{h_B+i}\tilde{\mathbf{x}}_{h_x+r-i}), \quad r = 0,\ldots,k_x - h_x.$$

(ii) $\tilde{\mathbf{x}}(\varepsilon) = \varepsilon^{h_B} \mathbf{B}(\varepsilon)^{-1} \mathbf{a}(\varepsilon)$ *is an* $(n, 1)$-*matrix (vector)* (h_x, k_x)-*expansion with* $h_x = h_a, k_x = k_a \wedge (h_a + k_B - h_B)$, *and* $\tilde{\mathbf{x}}_{h_x+r} = \sum_{0 \le i \le r} \mathbf{C}_i \mathbf{a}_{h_a+r-i}, r = 0, \ldots, k_x - h_x$, *where* $\mathbf{C}_j, j = 0, \ldots, k_C$, *are coefficients of the* (n, n)-*matrix* $(0, k_C)$-*expansion* $\mathbf{C}(\varepsilon) = \varepsilon^{h_B} \mathbf{B}(\varepsilon)^{-1}$ *given in item* **(iv)** *of Lemma 8.1.2, i.e., defined by the recurrence formulas* $\mathbf{C}_0 = \mathbf{B}_{h_B}^{-1}$ *and*

$$\mathbf{C}_r = -\mathbf{B}_{h_B}^{-1} \sum_{1 \le i \le r} \mathbf{B}_{h_B+i} \mathbf{C}_{r-i}, \quad r = 1, \ldots, k_C.$$

Proof. Obviously, $\varepsilon^{-h_B} \mathbf{B}(\varepsilon) \to \mathbf{B}_{h_B}$ as $\varepsilon \to 0$. Therefore, the matrix $\mathbf{B}(\varepsilon)$ has nonzero determinant for all positive ε small enough, say $0 < \varepsilon \le \varepsilon_0' < \varepsilon_0$. Thus, for such ε, there exists a unique solution of the system of linear equations (8.1.3), which is given by the formula $\mathbf{x}(\varepsilon) = \mathbf{B}(\varepsilon)^{-1} \mathbf{a}(\varepsilon)$. Now, the proof follows from parts **(v)** and **(vi)** of Lemma 8.1.2. $\square$

Remark 8.1.1. It is useful to note that $\tilde{\mathbf{x}}(\varepsilon) = \varepsilon^{h_B} \mathbf{x}(\varepsilon)$, where $\mathbf{x}(\varepsilon) = \mathbf{B}(\varepsilon)^{-1} \mathbf{a}(\varepsilon)$ is a unique solution of the system of linear equations (8.1.3).

Let $\mathbf{B}(\varepsilon) = ||B_{ij}(\varepsilon)||$ be an $n \times n$ matrix-valued function and $\mathbf{a}(\varepsilon) = ||a_i(\varepsilon)||$ an $n \times 1$ matrix-valued (column vector-valued) function with real-valued elements that are defined on an interval $0 < \varepsilon \le \varepsilon_0$. We consider a system of linear equations given in a form that often appears in applications to stochastic systems,

$$\mathbf{x}(\varepsilon) = \mathbf{a}(\varepsilon) + \mathbf{A}(\varepsilon)\mathbf{x}(\varepsilon). \tag{8.1.4}$$

As before, we assume that $\mathbf{a}(\varepsilon)$ and $\mathbf{A}(\varepsilon)$ can be represented in the form of asymptotic matrix expansions, $\mathbf{a}(\varepsilon) = \mathbf{a}_{h_a} \varepsilon^{h_a} + \cdots + \mathbf{a}_{k_a} \varepsilon^{k_a} + \mathbf{o}(\varepsilon^{k_a})$ with $n \times 1$ matrix-valued (vector-valued) coefficients $\mathbf{a}_r = ||a_{r,i}||, r = h_a, \ldots, k_a$, and $\mathbf{A}(\varepsilon) = \mathbf{A}_{h_A} \varepsilon^{h_A} + \cdots + \mathbf{A}_{k_A} \varepsilon^{k_A} + \mathbf{o}(\varepsilon^{k_A})$ with $n \times n$ matrix-valued coefficients $\mathbf{A}_r = ||a_{r,ij}||, r = h_A, \ldots, k_A$.

Below, we denote by $\mathbf{0}$ the $n \times n$ matrix with all zero elements and by $\mathbf{I} = ||\delta_{i,j}||$ a unit $n \times n$ matrix with zero non-diagonal and unit diagonal elements. Let us also introduce a norm of an $n \times n$ matrix $\mathbf{A}$ by

$$|\mathbf{A}| = \max_{1 \le i \le n} \sum_{1 \le j \le n} |a_{ij}|.$$

Lemma 8.1.4. *Let* $\mathbf{a}(\varepsilon)$ *be an* $(n, 1)$-*matrix* (h_a, k_a)-*expansion,* $\mathbf{A}(\varepsilon)$ *an* (n, n)-*matrix* $(0, k_A)$-*expansion, and* $|\mathbf{A}_0^n| \to 0$ *as* $n \to \infty$. *Then the following holds*:

(i) $\mathbf{x}(\varepsilon) = [\mathbf{I} - \mathbf{A}(\varepsilon)]^{-1} \mathbf{a}(\varepsilon)$ *is an* $(n, 1)$-*matrix* (h_x, k_x)-*expansion with* $k_a \wedge h_x = h_a, k_x = (h_a + k_A)$, *and* $\mathbf{x}_{h_x+r} = [\mathbf{I} - \mathbf{A}_0]^{-1} (\mathbf{a}_{h_a+r} + \sum_{1 \le i \le r} \mathbf{A}_i \mathbf{x}_{h_x+r-i})$, $r = 0, \ldots, k_x - h_x$.

(ii) $\mathbf{x}(\varepsilon) = [\mathbf{I} - \mathbf{A}(\varepsilon)]^{-1} \mathbf{a}(\varepsilon)$ *is an* $(n, 1)$-*matrix* (h_x, k_x)-*expansion with* $h_x = h_a, k_x = k_a \wedge (h_a + k_A)$, *and* $\mathbf{x}_{h_x+r} = \sum_{0 \le i \le r} \mathbf{C}_i \mathbf{a}_{h_a+r-i}, r = 0, \ldots,$

$k_x - h_x$, where $\mathbf{C}_j$, $j = 0, \ldots, k_C = k_A$, are coefficients of the (n, n)-matrix $(0, k_C)$-expansion $\mathbf{C}(\varepsilon) = [\mathbf{I} - \mathbf{A}(\varepsilon)]^{-1}$ defined by the recurrence formulas $\mathbf{C}_0 = [\mathbf{I} - \mathbf{A}_0]^{-1}$ and $\mathbf{C}_r = [\mathbf{I} - \mathbf{A}_0]^{-1} \sum_{1 \leq i \leq r} \mathbf{C}_i \mathbf{A}_{r-i}, r = 1, \ldots, k_C$.

Proof. We employ two basic inequalities for the matrix norms, $|\mathbf{AB}| \leq |\mathbf{A}||\mathbf{B}|$ and $|\mathbf{A} + \mathbf{B}| \leq |\mathbf{A}| + |\mathbf{B}|$. Since $|\mathbf{A}_0^n| \to 0$ as $n \to \infty$, there exists m such that $|\mathbf{A}_0^m| < a < 1$. Let also $b = \max(|\mathbf{A}_0|, \ldots, |\mathbf{A}_0^{m-1}|)$. Obviously, $\mathbf{A}(\varepsilon) \to \mathbf{A}(0)$ as $\varepsilon \to 0$ and, therefore, $|\mathbf{A}(\varepsilon)^m| \to |\mathbf{A}(0)^m|$ as $\varepsilon \to 0$, and $b_\varepsilon = \max(|\mathbf{A}(\varepsilon)|, \ldots, |\mathbf{A}(\varepsilon)^{m-1}|) \to b$ as $\varepsilon \to 0$. Thus, $|\mathbf{A}(\varepsilon)|^n \leq (b + 1)a^{[n/m]}$, $n = 0, 1, \ldots$, for all positive ε small enough, say $0 < \varepsilon < \varepsilon_0'$. This implies that $\sup_{\varepsilon \leq \varepsilon_0'} |\mathbf{A}(\varepsilon)|^n \to 0$ as $n \to \infty$. The last asymptotic relation implies that there exists an inverse matrix, $(\mathbf{I} - \mathbf{A}(\varepsilon))^{-1} = \mathbf{I} + \mathbf{A}(\varepsilon) + \mathbf{A}(\varepsilon)^2 + \cdots$, for every $0 < \varepsilon < \varepsilon_0'$. For such ε, there exists a unique solution of the system of linear equations (8.1.3), which is given by the formula $\mathbf{x}(\varepsilon) = (\mathbf{I} - \mathbf{A}(\varepsilon))^{-1} \mathbf{a}(\varepsilon)$. Now, the proof follows from Lemma 8.1.3 applied to the case where $\mathbf{B}(\varepsilon) = (\mathbf{I} - \mathbf{A}(\varepsilon))$. $\quad\square$

Lemma 8.1.5. *Let $\mathbf{a}(\varepsilon)$ be an $(n, 1)$-matrix (h_a, k_a)-expansion, $\mathbf{A}(\varepsilon)$ an (n, n)-matrix (h_A, k_A)-expansion, and $h_A > 0$. Then*

  (i) $\mathbf{x}(\varepsilon) = [\mathbf{I} - \mathbf{A}(\varepsilon)]^{-1} \mathbf{a}(\varepsilon)$ *is an $(n, 1)$-matrix (h_x, k_x)-expansion with $h_x = h_a$, $k_x = (h_a + k_A) \wedge k_a$, and $\mathbf{x}_{h_x+r} = [\mathbf{I} - \mathbf{A}_0]^{-1}(\mathbf{a}_{h_a+r} + \sum_{h_A \leq i \leq r} \mathbf{A}_i \mathbf{x}_{h_x+r-i})$, $r = 0, \ldots, k_x - h_x$.*

  (ii) $\mathbf{x}(\varepsilon) = [\mathbf{I} - \mathbf{A}(\varepsilon)]^{-1} \mathbf{a}(\varepsilon)$ *is an $(n, 1)$-matrix (h_x, k_x)-expansion with $h_x = h_a$, $k_x = (h_a + k_A) \wedge k_a$, and $\mathbf{x}_{h_x+r} = \sum_{0 \leq i \leq r} \mathbf{C}_i \mathbf{a}_{h_a+r-i}, r = 0, \ldots, k_x - h_x$, where $\mathbf{C}_j$, $j = 0, \ldots, k_C$ are coefficients of the (n, n)-matrix $(0, k_C)$-expansion $\mathbf{C}(\varepsilon) = [\mathbf{I} - \mathbf{A}(\varepsilon)]^{-1}$ defined by the recurrence formulas $\mathbf{C}_0 = [\mathbf{I} - \mathbf{A}_0]^{-1}$ and $\mathbf{C}_r = [\mathbf{I} - \mathbf{A}_0]^{-1} \sum_{h_A \leq i \leq r} \mathbf{C}_i \mathbf{A}_{r-i}, r = 1, \ldots, k_C$.*

Proof. Lemma 8.1.5 is a corollary of Lemma 8.1.4. Indeed, the matrix (h_x, k_x)-expansion $\mathbf{A}(\varepsilon) = \mathbf{A}_{h_A} \varepsilon^{h_A} + \cdots + \mathbf{A}_{k_A} \varepsilon^{k_A} + \mathbf{o}(\varepsilon^{k_A})$ can be represented in the form of a matrix $(0, k_x)$-expansion as $\mathbf{A}(\varepsilon) = \mathbf{0} + \mathbf{0}\varepsilon + \cdots + \mathbf{0}\varepsilon^{h_A-1} + \mathbf{A}_{h_A} \varepsilon^{h_A} + \cdots + \mathbf{A}_{k_A} \varepsilon^{k_A} + \mathbf{o}(\varepsilon^{k_A})$. In this case, $\mathbf{A}_0 = \mathbf{0}$, thus condition, $\mathbf{A}_0^n \to \mathbf{0}$ as $\varepsilon \to 0$, holds. The asymptotic matrix expansions for $\mathbf{x}(\varepsilon)$ given in Lemma 8.1.4 reduce to the corresponding expansions given in Lemma 8.1.5. $\quad\square$

8.2   New directions for the research

This book is devoted to studies of quasi-stationary phenomena in nonlinearly perturbed stochastic processes and systems. Asymptotic results presented in the book include the following: (i) mixed ergodic theorems (for the state of the process or the stochastic system) and limit theorems (for lifetimes) which describe transition phenomena; (ii) mixed ergodic and large deviation theorems that describe pseudo- and quasi-stationary phenomena; (iii) exponential expansions in mixed ergodic and large deviation theorems; (iv) theorems on convergence of quasi-stationary distributions; and (v) asymptotic expansions for quasi-stationary distributions.

The used methods are based on exponential asymptotics for a nonlinearly perturbed renewal equation. The classes of processes for which this program is realised include nonlinearly perturbed regenerative processes, semi-Markov processes, and continuous time Markov chains with absorption. Applications to nonlinearly perturbed queueing systems, population dynamics and epidemic models, and risk processes are considered.

In this section, we discuss and comment on some new directions in the research concerned pseudo- and quasi-stationary phenomena for perturbed stochastic systems that relate to the theory developed in this book. We hope that this discussion will stimulate a further research in the area.

8.2.1   Nonlinearly perturbed stochastic systems with discrete time

There is no doubt that all results on continuous time stochastic processes and systems presented in the book should have their analogues for discrete time stochastic processes and systems. It should be noted that discrete time models are interesting by themselves and have important applications.

Some results concerned asymptotic expansions in mixed ergodic and large deviation theorems for nonlinearly perturbed regenerative processes with discrete time can be found in Englund and Silvestrov (1997) and Englund (2000, 2001), and in Silvestrov (2000a) for discrete time Markov chains with absorption. However, these results give discrete time analogues just for a small part of the results from the theory developed in the present book for continuous time processes.

A development of a similar complete theory for nonlinearly perturbed stochastic processes and system with discrete time requires an additional comprehensive research.

8.2.2   Asymptotic expansions based on non-polynomial systems of infinitesimals

Asymptotic expansions in mixed ergodic and limit/large deviation theorems and asymptotic expansions for quasi-stationary distributions can also be obtained for nonlinearly perturbed stochastic processes and systems where the expansions in the initial perturbation conditions are based on different systems of infinitesimals.

In this book, all expansions are based on the integer powers $\varphi_n(\varepsilon) = \varepsilon^n$, $n = 0, 1, \ldots$, for the simplest infinitesimal $\varphi_1(\varepsilon) = \varepsilon$. This is a very natural system of infinitesimals that uses Taylor type expansions for nonlinearly perturbed characteristics of the corresponding processes and systems.

A natural generalisation of the theory based on polynomial type infinitesimals can be developed for a model where the corresponding expansions include products of integer powers $\varphi_{\bar{n}}(\varepsilon) = \prod_{i=1}^{m} \varepsilon^{n_i \omega_i}$, $\bar{n} = (n_1, \ldots, n_m)$, $n_1, \ldots, n_m = 0, 1, \ldots$, of infinitesimals taken from some *base set* $\{\varphi_i(\varepsilon) = \varepsilon^{\omega_i}, i = 1,\ldots,m\}$, where ω_i, $i = 1,\ldots,m$, are positive real numbers. The difference in the corresponding expansions should mainly be caused by new forms in arithmetics for asymptotic expansions, i.e., in formulas for coefficients in sums, products, quotients, as well as in the power, exponential, and other functions in the asymptotic expansions. The importance of such expansions is that they permit to obtain a more dense net of infinitesimals in the corresponding expansions.

We have a conjecture that analogous asymptotic expansions can also be constructed for a model where the corresponding expansions include products of integer powers $\varphi_{\bar{n}}(\varepsilon) = \prod_{i=1}^{m}(\varphi_i(\varepsilon))^{n_i}, \bar{n} = (n_1,\ldots,n_m), n_1,\ldots,n_m = 0,1,\ldots, m = 1,2,\ldots$, of infinitesimals taken from a finite or countable base set $\{\varphi_i(\varepsilon), i = 1,2,\ldots\}$. The only condition should be imposed on these infinitesimals that any two infinitesimals of the product form given above, $\varphi'_{\bar{n}'}(\varepsilon)$ and $\varphi''_{\bar{n}''}(\varepsilon)$, should be asymptotically comparable, i.e., there should be a constant $0 \leq c \leq \infty$ (determined by these infinitesimals) such that $\varphi'_{\bar{n}'}(\varepsilon)/\varphi''_{\bar{n}'}(\varepsilon) \to c$ as $\varepsilon \to 0$.

An example of such expansions in mixed ergodic and large deviation theorems for regenerative processes and their applications to queueing systems can be found in Englund and Silvestrov (1997) and Englund (1999, 2000, 2001). In these works, nonlinear perturbations with the base set of the infinitesimals $\{\varepsilon, e^{-a/\varepsilon}\}$ was considered.

8.2.3 Asymptotic expansions in mixed ergodic and large deviation theorems for semi-Markov type processes with countable and general phase spaces

The asymptotic results obtained in Chapters 4 and 5 of this book relate to nonlinearly perturbed Markov chains and semi-Markov processes with finite phase spaces. These processes possess a regeneration property at return moments in a fixed state. This makes it possible to use the asymptotic results for nonlinearly perturbed regenerative processes presented in Chapter 3.

Markov chains and semi-Markov processes with countable phase spaces possess similar regeneration properties. Moreover, the method of artificial regeneration developed in the works of Kovalenko (1977), Nummelin (1978a), Athreya and Ney (1978a), permits to construct regeneration moments for Markov and semi-Markov processes with a general phase space. In this way, the results concerning the asymptotic analysis of pseudo- and quasi-stationary phenomena for nonlinearly perturbed regenerative processes can be applied to nonlinearly perturbed Markov chains and semi-Markov processes with countable and general phase spaces.

There exists a large number of works devoted to ergodic theorems, limit theorems for random functionals of hitting time types, as well as large deviation theorems for such functionals in non-mixed or mixed forms (with ergodic theorems). These works relate to Markov and semi-Markov type processes with finite, countable, and general phase spaces. The results presented in Chapter 4 are related to this direction. We list and make comments on the corresponding references in the bibliographical remarks.

It should be noted that the results given in Chapter 4 play only a preparatory role with respect to the results related to exponential asymptotic expansions in mixed ergodic and large deviation theorem, and asymptotic expansions for quasi-stationary distributions given in Chapter 5. These theorems also require many auxiliary results that are important by themselves. Such results, for example, include asymptotic cyclic solidarity properties for the corresponding processes and asymptotic expansions for absorption and hitting probabilities, power and mixed power-exponential moments for the first hitting times and related functionals, and others. We believe that, altogether, these results make one of the main new authors' contributions to the study of quasi-stationary phenomena for perturbed stochastic processes and systems. Most of these results have no analogues for Markov and semi-Markov type processes with countable and general phase spaces, and constitute a new direction for the future research. As usual, operator analogues for matrix techniques used in the models with finite phase spaces should be employed for models with countable and general phase spaces, which significantly complicates the problem under consideration.

8.2.4 Mixed ergodic and large deviation theorems for semi-Markov type processes with asymptotically uncoupled phase space

Results presented in Chapters 4 and 5 are related to a basic model of nonlinearly perturbed semi-Markov processes with one limit class of recurrent-without-absorption states. It would be very interesting to extend the theory to models in which the set of non-absorption states for perturbed processes asymptotically splits into several classes of recurrent-without-absorption states which do not communicate with each others.

We give corresponding references in the bibliographical remarks. It should be noted that most of the results for models with an asymptotically uncoupled phase space are related to weak convergence limit theorems for distributions of hitting times for pseudo-stationary models mixed with the corresponding ergodic theorems and related theorems on asymptotic aggregation of Markov and semi-Markov type processes. Also a few results concerned asymptotic expansions for linearly perturbed Markov and semi-Markov type processes are known.

At the same time, a study of large deviation theorems for hitting times, as well as exponential asymptotic expansions in mixed ergodic and large deviation theorems for hitting times and asymptotic expansions for quasi-stationary distributions in models with asymptotically uncoupled sets of recurrent-without-absorption states for linearly and all the more nonlinearly perturbed and Markov and semi-Markov type processes, have not been conducted before. Such theorems certainly constitute an additional prospective direction for the future research.

In principle, the method based on applying asymptotic results to perturbed regenerative processes developed in Chapter 3 can be used in combination with the method of recurrence asymptotic analysis for power and mixed power-exponential moments of hitting times developed in Chapter 5. However, one should expect that pseudo-absorption effects caused by asymptotic vanishing of communication between different recurrent-without-absorption classes of states may complicate the asymptotic analysis for power and mixed power-exponential moments of hitting times for models with asymptotically uncoupled phase space.

8.2.5 Mixed ergodic and large deviation theorems for stochastic processes with semi-Markov modulation (switchings)

Methods of the asymptotic analysis of pseudo- and quasi-stationary phenomena in nonlinearly perturbed Markov chains and semi-Markov processes also can be applied to more general processes of Markov and semi-Markov types. In order to be able to apply the methods of asymptotic analysis developed for perturbed renewal equations, such processes should possess appropriate regeneration properties. In particular, these are so-called stochastic processes with semi-Markov modulation (switchings), which possess imbedded semi-Markov processes. In the Markov case, these processes are also known as Markov processes with discrete components. We refer to the book of Gikhman and Skorokhod (1975), Silvestrov (1980a), Koroliuk and Limnios (2005), and Anisimov (2008), where one can find descriptions of these classes of stochastic processes.

Sections 6.2 and 6.3 contain a detailed description of a typical example of M/G queueing systems with quick service and a bounded queue buffer, which can be described with the use of processes with semi-Markov modulation. This example shows in which way the methods developed in the book can be applied to the asymptotic analysis of pseudo- and quasi-stationary phenomena in nonlinearly perturbed processes with semi-Markov modulation. The main point in the corresponding asymptotic analysis is to express the perturbation and other conditions in terms of local characteristics related to the individual semi-Markov regeneration cycles.

We have no doubts that the main results of the theory presented in this book can be extended to this class of processes that are more general than regenerative and semi-Markov processes.

8.2.6 Double asymptotic expansions in limit and large deviation theorems for lifetimes in nonlinearly perturbed stochastic processes and systems

The expansions obtained in Chapter 5, applied to marginal distributions of absorption times, have the form $\mathsf{P}\{\varepsilon\mu^{(\varepsilon)} > t/\varepsilon^{r-1}\}/\pi e^{-(\rho_0 + a_1\varepsilon + \cdots + a_r\varepsilon^r)t/\varepsilon^r} = 1 + o(1)$ for $r = 1, \ldots, k$. These expansions describe the behaviour of relative errors in large deviation zones of different orders, when approximating the distribution of absorption times with exponential type distributions that have specially fitted parameters.

Our conjecture is that it should also be possible to expand the residual term $o(1)$ in the asymptotic relation given above. Our hope is based on the results given in Korolyuk, Penev and Turbin, (1972), Poliščuk and Turbin (1973), and Silvestrov and Abadov (1991, 1993), where such expansions have been obtained in pseudo-stationary model ($\rho_0 = 0$), respectively, for the case $r = 1$ in the first two papers and for the cases $r = 1$ and $r = 2$ in the last two papers. Note that the case $r = 1$ corresponds to asymptotic expansions in usual weak limit theorem while the case $r = 2$ corresponds asymptotic expansions large deviation theorems for the zone of large deviation of the order $O(\varepsilon^{-1})$. In the cases $r > 2$, the question about such "double" expansions is an open problem. These remarks can be also related to exponential expansions in mixed ergodic and large deviation theorems.

8.2.7   Explicit upper bounds for the remainder terms in asymptotic expansions for nonlinearly perturbed stochastic systems

The problem of finding explicit upper bounds for remainder terms in exponential asymptotic expansions in mixed ergodic and limit/large deviation theorems for nonlinearly perturbed stochastic processes and systems is of a special importance. For example, such explicit upper bounds for the remainder terms are obtained in the asymptotic expansions given in the papers Silvestrov and Abadov (1991, 1993) mentioned above.

Of course, the research of the rates of convergence in limit theorems for lifetime functionals for regenerative processes and Markov- and semi-Markov type processes (which correspond to the problem of finding explicit upper bounds for the remainder terms in zero-order expansions) continue to be actual. The corresponding references related to these problems are given in the bibliographical remarks.

Explicit upper bounds for the remainder terms in asymptotic expansions of stationary and quasi-stationary distributions are known for linearly perturbed Markov- and semi-Markov processes. Similar problems related to explicit upper bounds for the remainder terms in asymptotic expansions of stationary and quasi-stationary distributions for nonlinearly perturbed stochastic processes and systems have not been conducted so far.

8.2.8   Quasi-stationary phenomena in nonlinearly perturbed queueing and reliability systems

Asymptotic analysis of pseudo- and quasi-stationary phenomena in nonlinearly perturbed queueing systems and networks of M/M, M/G, and G/M types, which can be described with the use of Markov chains, semi-Markov processes, and stochastic processes with semi-Markov modulation, is an unlimited area for applications of the theory developed in the book. Examples of such applications are given in Sections 6.1–6.4. These examples show that, in such applications, the main problems translate to technical and sometimes non-trivial and interesting calculations connected with a reformulation of the perturbation conditions and rewriting formulas for the coefficients of the asymptotic expansions and quasi-stationary distributions in terms of local characteristics used to define the corresponding queueing systems.

It would be interesting to develop an asymptotic analysis of pseudo- and quasi-stationary phenomena for nonlinearly perturbed queueing systems with a more general structure of input flows of customers and service processes, e.g., systems with input flows of group income of customers and group service, different models of vacations, systems with parameters in input flows and service processes depending on a queue, systems with modulated input flows and switching service regimes, systems with several kinds of customers, systems with several types of servers, with and without reservation, etc.

As well known, queueing systems of G/G type can also be described with the use of regenerative processes. Here the moments when the queue becomes empty play, as a rule, the role of regeneration times. Thus, we can use the methods of asymptotic analysis of pseudo- and quasi-stationary phenomena for perturbed regenerative processes. However, the analysis of regeneration cycles and a reduction of the conditions and formulas that are based on the global characteristics connected with the regeneration cycles to, correspondingly, conditions and formulas that would be based on local characteristics used to define the corresponding systems is expected to be much more difficult for such systems.

8.2.9 Quasi-stationary phenomena in nonlinearly perturbed models of biological type systems

Another unlimited area for applications of the asymptotic results dealing with pseudo- and quasi-stationary phenomena in stochastic systems are nonlinearly perturbed epidemic and population dynamics models which can be described with the use of Markov chains, semi-Markov processes with finite phase spaces. Examples of such applications are given in Sections 6.4 and 6.5. Similarly to the examples from the queueing theory, the main problems are to reformulate the perturbation conditions and to express the formulas for the coefficients in the asymptotic expansions and quasi-stationary distributions in terms of the local characteristic used to define the corresponding epidemic or population dynamics models.

Nonlinearly perturbed epidemic and population dynamics models with different types, e.g., sex, age, health status, etc., of individuals in the population, forms of interactions between individuals, environmental modulation, complex spatial structure, etc., make interesting objects for asymptotic analysis of pseudo- and quasi-stationary phenomena.

8.2.10 Asymptotics of probabilities and other characteristics connected with ruins for perturbed risk type processes

Chapter 7 of the book is devoted to studies of quasi-stationary phenomena for nonlinearly perturbed risk processes. Asymptotic results for ruin probabilities, densities of non-ruin probabilities, and distributions of the capital surplus prior to and at the time of a ruin are given. However, there are many other functional characteristics connected with risk processes that satisfy renewal equations. In such cases, analogous asymptotic results can be obtained with the use of asymptotic methods developed for perturbed renewal equations in Chapters 1 and 2.

More general risk type processes may also supply models with functional characteristics satisfying renewal equations. For example, such are risk processes with investment components described by Brownian motions, more general Lévy type risk processes, risk processes with modulated claim flows, etc.

8.2.11   Markov and semi-Markov processes describing dynamics of credit ratings

Finally, we would like also to mention a possible prospective area of financial applications for asymptotic methods developed in the book. The methods of asymptotic analysis of pseudo- and quasi-stationary characteristics may be also applied to studies of default type distributions for Markov and semi-Markov processes with absorption which become common to use for description of dynamics of credit ratings.

8.2.12   Numerical studies connected with asymptotic expansions describing quasi-stationary phenomena in nonlinearly perturbed stochastic systems

Numerical and simulation studies make a natural supplement to analytical results, especially such as asymptotic expansions for functional characteristics of perturbed stochastic processes and systems. For example, asymptotic results presented in Section 7.4 yield formulas for approximations of ruin probabilities based on higher order moments of the claim distributions. These approximations are asymptotically optimal in the sense that the corresponding relative errors tend to zero. It would be interesting to compare them with other known approximations. This, however, would require to carry out additional comprehensive analytical, numerical, and simulation studies that are beyond the frame of the present book. We hope that these experimental studies will be realised in a future for this model as well as for other perturbed models of stochastic processes and systems.

8.2.13   Asymptotics for perturbed equations of renewal type

The program for studies of pseudo- and stationary phenomena in nonlinearly perturbed stochastic processes and systems realised in the book is based on the method of the asymptotic analysis for perturbed renewal equations presented in Chapters 1 and 2. We have no doubts that the main asymptotic results for the perturbed renewal equations presented in these chapters should hold for more general equations of renewal type. In many cases, such results can be achieved by a suitable reduction to the case of classical renewal equation treated in the book. For example, Shurenkov (1980a, 1980b, 1980c) gave such a generalisation of some results presented in Chapter 1 to the case of a matrix renewal equation. Chapter 4 contains a more through and extended presentation of matrix version of asymptotic results for perturbed renewal equation given in Chapter 1 for the model case of perturbed semi-Markov processes with absorption. Chapter 5 essentially improves these asymptotic results to the much more advanced form of the corresponding asymptotic expansions. In fact, this chapter contains a matrix generalisation of the results of Chapter 2 to the model case of nonlinearly perturbed semi-Markov processes with absorption.

There exist other versions of renewal type equations, for example, the so-called renewal equations with waits, operator renewal type equations, renewal equations on a whole line etc. It would be interesting to extend the theory presented in the present book to such equations.

There are also other areas, beyond ergodic, pseudo- and quasi-stationary problems, where the renewal equation and the corresponding asymptotic results play an essential role. For example, these are asymptotic problems for moment functionals appearing in the theory of branching processes and many others. We hope that the book will stimulate future works directed to new applications of the theory presented in this book.

8.3   Bibliographical remarks

Our bibliographical remarks are structured by chapters according to the contents of the book. We first refer to works that contain results given in the corresponding chapters and indicate new results included in the book and published for the first time. Then we refer to some works that contain results directly connected with the results given in the corresponding chapters. We also collect and give references to works in related areas, dealing with ergodic and quasi-ergodic theorems, stability theorems, limit and large deviation theorems for lifetime-type functionals and asymptotic aggregation theorems for regenerative, Markov, and semi-Markov type processes, as well as applications of such theorems to queueing systems, models of population dynamics, epidemic models, and other stochastic systems.

Chapters 1. Chapter 1 presents results related to a generalisation of the classical renewal theory to a model of the perturbed renewal equation.

We begin with the references concerning the classical renewal theorem. The final form of this important theorem (see, Theorem 1.1.1), describing the asymptotics of a solution of the renewal equation, was given in Feller (1966).

Chapter 1 is based on the results obtained in Silvestrov (1976, 1978, 1979a), where a generalisation of the renewal theorem to a model of the perturbed renewal equation was given. The main results contained in these papers are given in Theorems 1.2.1 and 1.3.1, which are direct analogues of the classical renewal theorems for a model of a perturbed renewal equation, and also in Theorems 1.4.1 and 1.4.2, which describe the asymptotics of a solution of a perturbed renewal equation under the Cramér type moment conditions imposed on the distributions that generate the corresponding renewal equations.

Shurenkov (1980a, 1980b, 1980c) has extended some of the results mentioned above to a model of perturbed matrix renewal equations. Englund and Silvestrov (1997) and Englund (2000, 2001) gave variants of the corresponding results for a discrete time perturbed renewal equation.

References to some selected works related to asymptotic renewal theory are Feller (1941, 1949, 1950, 1961, 1966), Täcklind (1945), Blackwell (1948, 1953), Erdös, Feller and Pollard (1949), Cox and Smith (1953), Smith (1954, 1955b, 1958, 1961a, 1962), Dynkin (1955), Feller and Orey (1961), Farrell (1962), Garsia and Lamperti (1962/1963), Garsia (1963), Borovkov (1964), Stone (1965a, 1965b, 1966), Heyde (1967b), Stone and Wainger (1967), Teugels (1967, 1968a, 1970, 1990), Rogozin (1973, 1976), Sevast'yanov (1974), Zubkov (1975), Silvestrov (1976, 1978, 1979a, 1980c, 1980e, 1983a, 1984a, 1984b, 1994), Arjas, Nummelin and Tweedie (1978), Kalasnikov (1978b), Kartashov (1978, 1979a, 1982a, 1982b), Silvestrov and Banakh (1981), Thorisson (1987), Bertoin (1999), Grey (2001), and Yin Chuancun and Zhao Junsheng (2006).

We would also like to refer to works, where asymptotic results in the renewal theory were extended to more general renewal type models and equations, Karlin (1955), Smith (1955a, 1960, 1961b, 1969), Hatori (1959, 1960), Farrell (1964, 1966a, 1966b), Gerber (1970), Su Yeong Tzay (1993), Woodroofe (1982), Gut (1983, 1988), Janson (1983a), Alsmeyer (1986, 1987, 1988a, 1988b, 1988c, 1991a, 1991b, 1992, 1994a, 1994b, 1995, 1996), Alsmeyer and Irle (1986), Lam and Lehoczky (1991), Schmidli (1997), Su Yeong-Tzay (1998), Su Yeong Tzay and Wang Chi Tshung (1998), Kim Dong Yun and Woodroofe (2003), Sgibnev (2003, 2006), and Kartashov and Stroev (2005).

Results presented in Chapter 1 form a basis for studying a nonlinearly perturbed renewal equation, which is done in Chapter 2.

Chapters 2. In this chapter, we present exponential expansions for solutions of nonlinearly perturbed renewal equations. These expansions essentially improve the asymptotics of solutions of perturbed renewal equations given in Chapter 1. Chapter 2 is based on the results obtained in the papers Silvestrov (1995) and Gyllenberg and Silvestrov (1998, 1999a, 2000a), and on extensions of these results to a model of a perturbed renewal equation satisfying more general perturbation conditions.

In Silvestrov (1995), nonlinear perturbation conditions were imposed on the moments of the distribution functions that generate the corresponding renewal equation. Explicit expansions for the corresponding characteristic roots were given, and the corresponding exponential expansions were obtained for solutions of nonlinearly perturbed renewal equations. These results are given in Theorems 2.1.1 and 2.1.2. Theorems mentioned above cover the case of an asymptotically proper renewal equation. The general case, where the limit renewal equation can be proper or improper, was considered in Gyllenberg and Silvestrov (1998, 1999a, 2000a). The corresponding result is formulated in Theorem 2.2.1. The latter paper also contains asymptotical expansions for the renewal limit given in Theorem 2.3.2. Theorems 2.1.3 and 2.2.2 present new results that generalise Theorems 2.1.1, 2.1.2, and 2.2.1 mentioned above to models with more general perturbation conditions.

The results mentioned above are related to a model of a continuous time renewal equation. The case of a discrete time renewal equation is interesting on its own right. Some results are given in the papers Englund and Silvestrov (1997) and Englund (2000, 2001). Also, exponential expansions, which are based on non-polynomial perturbation conditions, were considered for discrete and continuous time perturbed renewal equations in Englund and Silvestrov (1997), and Englund (1999a, 1999b, 2001).

The results presented in Chapters 1 and 2 make a basis for our studies of pseudo- and quasi-stationary phenomena in nonlinearly perturbed regenerative processes.

Chapter 3. In Chapter 3, we present applications of the asymptotic results obtained in Chapters 1 and 2 to a model of nonlinearly perturbed regenerative processes. Chapter 3 is based on the results obtained in the papers Silvestrov (1995) and Gyllenberg, and Silvestrov (1998, 1999a, 2000a), as well as on extensions of these results to a model of perturbed regenerative processes with more general perturbation conditions and regenerative processes with transition period.

We formulate mixed ergodic and limit/large deviation theorems for joint distributions of positions of perturbed regenerative processes and regeneration stopping times. This is done in Theorems 3.2.1–3.3.6, which describe pseudo-stationary and quasi-stationary phenomena for perturbed regenerative processes. In fact, these theorems are direct corollaries of the renewal theorem for perturbed renewal equations proved in Silvestrov (1976, 1978, 1979a) and given in Chapter 1. These results have analogues in the literature. In particular, some slight modifications were given in the papers Shurenkov (1980a, 1980b, 1980c). Some similar results related for discrete time nonlinearly perturbed regenerative processes as well as some results related to mixed ergodic and limit/large deviation theorems for nonlinearly perturbed regenerative processes with non-polynomial perturbations can be found in Englund and Silvestrov (1997) and Englund (2000, 2001).

Our main new achievements are Theorems 3.4.2 and 3.4.4 that are based on the results of the papers Silvestrov (1995) and Gyllenberg and Silvestrov (2000a). The main new elements are nonlinear perturbations and higher order exponential expansions, which cover large deviation theorems for regenerative stopping times in various large deviation zones that are determined from the balancing conditions imposed on the rate of growth of time and the rate of perturbation. Also, a new element is a mixed ergodic and large deviation form of the corresponding theorems that describe asymptotics for joint distribution of the position of regenerative process and the regeneration stopping time. Theorems 3.4.3 and 3.4.5 present new results that generalise Theorems 3.4.2 and 3.4.4 mentioned above, covering the cases with more general perturbation conditions, while Theorems 3.4.6–3.4.10 extend all the above theorems to a model of nonlinearly perturbed regenerative processes with transition period. Theorems 3.5.1, 3.5.2, and 3.5.4 also contain new results on asymptotic expansions for quasi-stationary distributions of nonlinearly perturbed regenerative processes.

We mentioned above some papers which contain results related to mixed ergodic and limit/large deviation theorems for regenerative processes. These theorems describe asymptotics of the joint distribution of the position of regenerative process and the regeneration stopping time.

In a marginal form, these theorems reduce either to individual ergodic theorems for perturbed regenerative processes, which describe asymptotics of distributions of regenerative processes, or to limit and large deviation theorems for regeneration stopping times.

Let us mention some works on ergodic theorems for regenerative processes as well as rates of convergence in these theorems and stability of stationary characteristics. The stochastic process setting and use of some constructive probabilistic methods such as coupling slightly distingue these results of the corresponding asymptotic results in renewal theory. These works are Smith (1955a, 1958), Lamperti (1961, 1962), Feller (1966), Lindvall (1977, 1979a, 1982, 1983, 1986, 1992a, 1992b), Silvestrov (1976, 1978, 1979a, 1980c, 1980e, 1983a, 1984a, 1984b, 1994), Kalashnikov (1978b, 1981a, 1990, 1994b, 1994c), Banakh (1981, 1982a, 1982b), Thorisson (1983, 1985a, 1992, 1998, 2000), Anichkin (1985), Lindvall and Rogers (1996), Silvestrov and Stenflo (1998), Stenflo (1998), and Glynn and Thorisson (2004).

A regeneration stopping time can be represented in a form of a geometric random sum with a reminder term, as was shown in Section 3.1. That is why, limit theorems for regeneration stopping times are closely connected with limit theorems for geometric random sums that were a subject of thorough studies in Rényi (1956), Kovalenko (1965), Keilson (1966a, 1966b, 1978), Gnedenko and Kovalenko (1964), Gnedenko and Frayer (1969), Silvestrov (1969, 1980a), Masol (1973, 1974, 1975a, 1975b, 1976a, 1977), Solov'ev (1971), Korolyuk, D. (1982), Kruglov and Korolev (1990), Klebanov, Melamed, and Umarov (1991), Gnedenko and Korolev (1996), Silvestrov (2004a), Silvestrov and Teugels (2004), and Silvestrov and Drozdenko (2005, 2006). We refer to the book Silvestrov (2004a) for a detailed list of references.

The rates of convergence and large deviation theorems in limit theorems for regeneration stopping times and geometric sums and related results were studied in Feller (1966), Silvestrov (1969a, 1969c, 1980a, 1995), Freyer (1970), Solov'ev (1971), Szynal (1976), Kovalenko (1980, 1994), Kalashnikov (1983, 1986, 1993, 1994b, 1997), Khusanbayev (1983, 1984), Kalashnikov and Vsekhsvyatski (1985, 1989), Vsekhsvyatski and Kalashnikov (1985a), Grishechkin (1986), Korolev (1987, 1989a, 1989b), Melamed (1987, 1989), Kovalenko and Kuznetsov (1988), Brown (1990), Hsu Guanghui and He Qiming (1990), Kruglov and Korolev (1990), Sudakova (1990, 1993), Kartashov (1991, 1996b), Willekens and Teugels (1992), Gnedenko and Korolev (1996), Kalashnikov and Konstantinidis (1996), Asmussen and Kalashnikov (1999), Bon and Kalashnikov (1999, 2001), Kalashnikov and Tsitsiashvili (1999a, 1999b, 2001), Bening and Korolev (2002), and Glynn and Thorisson (2004).

It should be noted that the works listed above give the usual asymptotics in weak limit theorems and the corresponding estimates for the rate of convergence in the pseudo-stationary case, where the stopping probabilities for one regeneration cycle (or success probability for the geometric random index) tend to zero, while results concerning large deviation asymptotics mainly relate to the quasi-stationary case, where the stopping probabilities for one regeneration cycle (or success probability for geometric the random index) are separated from zero. Both the models with light and heavy tail distributions of summands were considered.

The results on weak convergence and large deviation asymptotics for regeneration stopping times given in Theorems 3.2.1–3.3.6 deal with the case of asymptotically ergodic regenerative processes with light tail distributions of regeneration moments, but give these results in a united form covering both the pseudo- and quasi-stationary cases.

The exponential expansions in the mixed ergodic and limit or large deviation theorems for absorption times and the asymptotic expansions of the quasi-stationary distributions for nonlinearly perturbed regenerative processes, which are given in Theorems 3.4.2–3.5.4, have no analogues in the works listed above.

We would like also to mention some basic papers and books on renewal theory and regenerative processes and related problems that are Smith (1955b), Cox (1962), Kingman (1972), Silvestrov (1980a), Shedler (1987, 1993), Gut (1988), Bingham, Goldie, and Teugels (1989), Shurenkov (1989), Alsmeyer, (1991b), Kalashnikov (1994b, 1997), and Thorisson (2000).

The results presented in Chapter 3 make a basis for our studies of pseudo- and quasi-stationary phenomena in nonlinearly perturbed semi-Markov processes.

Chapter 4. We make here a very detailed analysis of pseudo- and quasi-stationary phenomena for perturbed semi-Markov processes with a finite set of states and absorption. Results presented in Chapter 4 extend the corresponding results obtained in Gyllenberg and Silvestrov (1999a, 2000a), and Silvestrov (2000a).

There are two classes of perturbed models that differ in the asymptotical properties of the absorption probabilities. The first class includes models with small local absorption probabilities. More precisely, these probabilities are functions of a perturbation parameter ε and tend to zero if ε tends to zero. In such models, the absorption times $\mu^{(\varepsilon)}$ tend in probability to infinity as $\varepsilon \to 0$. The second class includes models with some of the local absorption probabilities separated from zero. In such models, the absorption times $\mu^{(\varepsilon)}$ are asymptotically stochastically bounded random variables as $\varepsilon \to 0$. To distinguish these two cases, we introduce the term *pseudo-stationary* for mixed limit/large deviation theorems in the first case, while the term *quasi-stationary* is traditionally used in the second case.

In Chapter 4 we consider mixed ergodic and limit theorems and mixed ergodic and large deviation theorems for perturbed semi-Markov processes in the usual triangular array models, where the perturbation conditions are given in the simplest form that assumes convergence of local transition probabilities of the perturbed processes to the corresponding characteristics of the limit processes as the perturbation parameter $\varepsilon \to 0$. The main results are Theorems 4.4.1 and 4.4.2, which represent mixed ergodic and limit theorems for perturbed semi-Markov processes under minimal moment conditions on the sojourn times, while Theorems 4.6.1–4.6.2 and 4.6.3–4.6.4, which represent mixed ergodic and large deviation theorems for perturbed semi-Markov processes under Cramér type moment conditions imposed on the sojourn times, respectively, for pseudo-stationary and quasi-stationary models. Also, Theorems 4.6.5 and 4.6.9 give conditions for convergence of conditional distributions of the processes, assuming a non-absorption condition, to the corresponding quasi-stationary distributions, while Theorems 4.6.6 and 4.6.10 give conditions for convergence of quasi-stationary distributions for perturbed semi-Markov processes.

Results given in Shurenkov (1980a, 1980b, 1980c), and Alimov and Shurenkov (1990a, 1990b), Shurenkov and Degtyar (1994), and Eleĭko and Shurenkov (1995b) are the most close with the results presented in Chapter 4. In the earlier works these authors used the method based on the use of regeneration properties of semi-Markov processes and reduction of the analysis to the case of scalar perturbed renewal equation, then switched to methods that are based on a direct asymptotic analysis of matrix or general Markov perturbed renewal equations. We do prefer to use the former regeneration based method. This method has a significant advantage, since it allows to not only get the corresponding mixed ergodic and limit/large deviation theorems but also improve them to a form of exponential asymptotic expansions as well as get asymptotic expansions for quasi-stationary distributions. Such expansions are hardly expected to be obtained with a use of direct matrix calculations that takes an awkward form. We would like also to refer to works Silvestrov (1981), Abadov (1984), and Silvestrov and Abadov (1984, 1991, 1993), where large deviation theorems and the corresponding asymptotic expansions for distributions of hitting times for perturbed Markov chains were given under pseudo-stationary conditions. These results stimulated ours studies related to exponential asymptotics in large deviation theorems for perturbed semi-Markov processes.

There exist a huge number of works in related areas. We first would like to refer to papers that are related to pseudo-stationary asymptotics. The model with finite phase space was most extensively investigated. Limit theorems for random functionals, known as first hitting times, first passage times, first record times, etc., and closely related to problems of asymptotic aggregation for Markov type processes, have been investigated by many researchers. Together with the limit theorems, the corresponding estimates for the rate of convergence in the models with coupled and nearly uncoupled phase space have been intensively studied. The characteristic feature of these results is that some explicit algebraic algorithms can be given for calculating the parameters of limit law, constants in the corresponding estimates, etc.

Limit theorems, rates of convergence for distributions of hitting times and related problems were studied in Harris (1952), Meshalkin (1958), Keilson (1966b, 1974,

1978, 1998), Solov'ev (1966), Darroch and Morris (1967), Korolyuk (1969), Silvestrov (1969c, 1970, 1971a, 1971b, 1972c, 1974, 1980a), Anisimov (1970, 1971a, 1971b), Korolyuk and Turbin (1970, 1976a), Holst (1971, 1977, 1979, 2001), Turbin (1971, 1972), Masol and Silvestrov (1972), Zakusilo (1972a, 1972b), Kovalenko (1973, 1975), Zubkov, and Mihaĭlov (1974), Masol (1976b), Brown (1983), Brown and Guang Ping (1984), Gaver, Jacobs, and Latouche (1984), Gut and Holst (1984), Latouche, Jacobs, and Gaver (1984), Aldous (1986), Brown and Shao (1987), Korolyuk, D. (1990), Latouche (1991), Asmussen (1994), Gyllenberg and Silvestrov (1994, 1999a, 2000a), Chen Dayue, Feng Jianfeng, and Qian Minping (1996), Stewart (1997, 1998, 2001, 2003), Ferrari and Garcia (1998), Keilson and Vasicek (1998), Yin and Zhang (1998), and Zhang and Yin (2004), and Silvestrov and Drozdenko (2005, 2006).

Related problems of asymptotic aggregation for Markov chains and semi-Markov processes with finite phase space were studied in Meshalkin (1958), Simon and Ando (1961), Hanen (1963a, 1963b, 1963c, 1963d), Anisimov (1972, 1973), Courtois (1975, 1977), Gaitsgory and Pervozvansky (1975), Kalashnikov (1978a), Anisimov and Sityuk (1980), Delebecque and Quadrat (1981a, 1981b), Coderch, Willsky, Sastry, and Castanon (1983a, 1983b), Delebecque (1983a, 1983b), Stewart (1983, 1984, 1991, 1993, 1998, 2001), Delebecque, Quadrat, and Kokotović (1984), Schweitzer (1984, 1991), Cao and Stewart (1985), Schweitzer, Puterman, and Kindle (1985), Schweitzer and Kindle (1986a, 1986b), Rohlicek (1987), Rohlicek and Willsky (1988a, 1988b), Stewart and Sun (1990), Kim and Smith (1991), Stewart and Zhang (1991a, 1991b), Sumita and Rieders (1991), Ledoux, Rubino, and Sericola (1994), Stewart, G., Stewart, W., and McAllister (1994), Stewart (1998, 2001), Yin and Zhang (1998), Yin, Zhang, and Badowski (2000a, 2000b, 2000c), Badowski, Yin, and Zhang (2003), and Yin and Zhang (2003).

Similar problems for Markov chains and Markov type processes with countable and general phase space, in particular, limit theorems for hitting times and rates of convergence were obtained in Gaver (1964, 1976), Heyde (1964, 1966, 1967a), Silvestrov (1969b, 1969c, 1972a, 1972b, 1972c, 1974, 1979b, 1980a, 1981, 1995, 2000a), Rosén (1970, 1978), Gani and Lehoczky (1971), Basu (1972, 1978), Gani (1972), Solov'ev (1972), Chow and Hsiung (1976), Ahmad, and Basu (1983), Gut (1974a, 1974b, 1975, 1983, 1988, 1990a, 1990b), Mileshina (1975, 1977), Korolyuk and Turbin (1976b, 1978), Kalashnikov (1978b, 1986, 1994a, 1994b), Voĭna (1978), Kaplan (1979a, 1979b, 1980), Korolyuk, Turbin, and Tomusjak (1979), Zubkov (1979), Chow, Hsiung, and Yu, (1980), Wang Zi Kun (1980), Aldous (1982, 1989, 1991), Janson (1983b), Korolyuk, D. and Silvestrov (1983, 1984), Vsekhsvyatski and Kalashnikov (1985b), Lalley (1984), Kartashov (1986, 1987, 1996b), Martin-Löf (1986a), Anisimov (1988), Borovkov (1998b, 1998c), Silvestrov and Velikii (1988), Kostava and Shurenkov (1988), Todorovic and Gani (1989), Korolyuk, D. (1990), Aldous and Brown (1992, 1993), Ferrari, Galves, and Landim (1994), Ferrari, Galves, and Liggett (1995), Sevast'yanov (1995, 1998, 2000, 2006), Motsa and Silvestrov (1996),

Walker (1998), Gasanenko (1999, 2001), Kesten and Maller (1999), Korolyuk (1999), Korolyuk and Limnios (2005), Puchała and Rolski (2005), and Asselah and Ferrari (2006).

The asymptotic aggregation and related problems were studied in Gusak and Korolyuk (1971), Anisimov (1975, 1980, 1984, 1986, 1988, 2008), Korolyuk and Turbin (1976a, 1978, 1982), Voĭna (1978, 1979), Arias and Korolyuk (1980), Kalashnikov (1981a), Anisimov, Voĭna, and Lebedev (1983), Bobrova (1983), Korolyuk, V.V. (1983), Korolyuk, V.S and Korolyuk, V.V. (1999), Kartashov (1986, 1987, 1996b), Korolyuk (1986, 1993, 1997), Korolyuk and Limnios (1998, 1999, 2000, 2002, 2004a, 2004b, 2005), Korolyuk and Svishchuk (1990, 1992, 1995), Al-Eideh (1995), Kovalenko, Birolini, Kuznetsov, and Shenbukher (1998), and Korolyuk and Mamonova (2003).

Ergodic theorems and the corresponding rates of convergence were obtained for discrete and continuous time Markov chains in Markov (1906), Kolmogorov (1931, 1937), Doeblin (1936, 1937, 1938, 1940), Lévy (1951), Harris (1956), Kendall (1959), Kingman (1963), Vere-Jones (1962, 1963, 1964, 1967, 1968), Kesten and Spitzer (1963), Kesten (1963, 1970), Cheong (1967), Teugels (1968c, 1972), Orey (1971), Callaert and Keilson (1973), Pitman (1974), Griffeath (1974/75, 1976, 1978), Revuz (1975), Nummelin and Tweedie (1976), Kovalenko (1977), Athreya and Ney (1978b), Keilson (1978, 1998), Nummelin (1978a, 1984), Seneta (1979, 1984), Borovkov (1980, 1998a), Brockwell, Gani, and Resnick (1982, 1983), Nummelin and Tuominen (1982), Tan Choon Peng (1982, 1983), Tweedie (1983a, 1983b, 2001), Kartashov (1984, 1985d, 1996b, 1996c), Gani and Tin Pyke (1985), van Doorn (1985, 2001, 2003), Chen (1991, 1996, 2005), Semal (1991), Zeĭfman (1991b), Borovkov and Foss (1992, 1994), Lindvall (1992a), Meyn and Tweedie (1992, 1993a, 1993b, 1993c, 1994a, 1994b), Spieksma and Tweedie (1994), Tuominen and Tweedie (1994), Down, Meyn, and Tweedie (1995), Kesten (1995), Rosenthal (1995), Zeĭfman (1995), Hoppensteadt, Salehi, and Skorokhod (1996a, 1996b), Lund and Tweedie (1996), Lund, Meyn, and Tweedie (1996), Mengersen and Tweedie (1996), Roberts and Tweedie (1996, 1999, 2001), Scott and Tweedie (1996), Granovskiĭ and Zeĭfman (1997, 2000, 2004), Roberts and Rosenthal (1997), Keilson and Vasicek (1998), Corcoran and Tweedie (2001), Guglielmi and Tweedie (2001), Jarner and Tweedie (2001, 2002, 2003), Zhang, Chen, Lin and Hou (2001), Fort and Moulines (2003), Borovkov and Hordijk (2004), Elesin, Kuznetsov, and Zeĭfman (2004), Andrieu and Moulines (2006), Mao (2006), Roberts, Rosenthal, Segers, and Sousa (2006), Zeĭfman, Leorato, Orsingher, Satin, and Shilova (2006), Roberts and Rosenthal (2007), and Sirl, Zhang Hanjun, and Pollett (2007).

Similar problems for more general semi-Markov and related processes were studied in Nagaev (1965, 1982), Teugels (1968b, 1970), Çinlar (1969a, 1969b, 1974a, 1974b, 1975), Shur (1967, 1976, 1977, 1984, 1985, 2002, 2005), Keilson (1969), Jacod (1970, 1971), Cheong and Teugels (1973), Kesten (1974), Gikhman and Skorokhod (1975), Ezhov and Shurenkov (1976), Athreya and Ney (1978b), Athreya, McDonald, and

Ney (1978), Nummelin (1978b), Eleĭko (1980, 1990, 1998), Silvestrov (1980a, 1980b, 1983c, 1994), Shurenkov (1981, 1983, 1989), Silvestrov and Pezhinska (1984), Kartashov (1985c, 1996a, 1997, 2000, 2005), Anderson and Athreya (1987), Aldous (1988), Ethier and Kurtz (1988), Meyn (1989), Alimov and Shurenkov (1990b), Meyn and Caines (1990), Thorisson (1990, 1998, 2000), Asmussen (1992), Meyn and Guo Lei (1993), Alimov (1994a, 1994b), Alsmeyer (1994c, 1997), Eleĭko and Shurenkov (1995a), Sgibnev (1996), Stenflo (1996, 1998), Hoppensteadt, Salehi, and Skorokhod (1997), Iosifescu (1999), Konstantopoulos and Last (1999), Balaji and Meyn (2000), Roberts and Tweedie (2000), Alsmeyer and Hoefs (2001, 2002, 2004), Fuh Cheng-Der and Lai Tze Leung (2001), Klüppelberg and Pergamenchtchikov (2003), Kontoyiannis and Meyn (2003, 2005), Fuh Cheng-Der (2004), and Baxendale (2005).

Stability problems, rates of approximation for the perturbation of stationary distributions and related characteristics for Markov chains and semi-Markov type processes were studied in Seneta (1967, 1968, 1973, 1980, 1981, 1988, 1991, 1993a, 1991, 1993b, 1993c), Schweitzer (1968), Kurtz (1971), Tweedie (1971, 1975a, 1975b, 1980), Golub and Seneta (1974), Courtois (1975, 1977), Courtois and Louchard (1976), Allen, Anderssen, and Seneta (1977), Borovkov (1977), Kalashnikov (1978b, 1994a), Kalashnikov and Anichkin (1981), Aissani and Kartashov (1983b), Cocozza-Thivent, Kipnis, and Roussignol (1983), Courtois and Semal (1984a, 1984b, 1984c 1991), Haviv and Rothblum (1984), Haviv and Van der Heyden (1984), Kartashov (1985b, 1996b), Hunter (1986), Haviv (1986, 1988, 1992, 1999), Haviv and Ritov (1986, 1993), Burnley (1987), Gibson and Seneta (1987), Johnson and Isaacson (1988), Zeĭfman (1989a, 1989b, 1991a), Semal and Courtois (1991), Stewart (1991b), Borovkov, Fayolle, and Korshunov (1992), Hassin and Haviv (1992), Haviv, Ritov, and Rothblum (1992), Barlow (1993a, 1993b), Meyn and Tweedie (1992, 1993a, 1993b, 1993c, 1998), O'Cinneide (1993), Lasserre (1994), Zeĭfman and Isaacson (1994), Eleĭko and Shurenkov (1996), Glynn and Meyn (1996), Borovkov (1998a), Roberts, Rosenthal, and Schwartz (1998), Tweedie (1998), Andreev, Elesin, Krylov, Kuznetsov, and Zeĭfman (2002), Avrachenkov and Haviv (2003), Craven (2003), Zhang and Yin (2004), Hunter (2005, 2006), and Yin George and Zhang Hanqin (2007).

We would also like to refer to works related to studies of convergence for rareficated (thinned) stochastic flows. These are Rényi (1956), Belyaev (1963), Kovalenko (1965), Mogyoródi (1971, 1972a, 1972b), Szantai (1971a, 1971b), Råde (1972a, 1972b), Zakusilo (1972a, 1972b), Jagers (1974), Jagers and Lindvall (1974), Tomko (1974), Kallenberg (1975), Lindvall (1976), Serfozo (1976, 1980, 1984a, 1984b), Anisimov and Gasanenko (1978), Gasanenko (1980, 1983, 1991, 1998a, 1998b), Grandell (1980), Anisimov and Chernyak (1982, 1983), Böker and Serfozo (1983), Anisimov (1986, 2000), Manor (1998), and Silvestrov and Drozdenko (2006).

As far as quasi-stationary distributions and related problems are concerned, the only case of non-perturbed processes was mainly studied. Models of Markov and semi-Markov type processes with discrete phase space were studied in Vere-Jones (1962, 1969), Kingman (1963), Darroch and Seneta (1965, 1967), Seneta (1966a,

1966b, 1973, 1981), Seneta and Vere-Jones (1966), Daley (1968), Buiculescu (1972, 1975), Cheong (1972), Flaspohler and Holmes (1972), Pakes (1973, 1978a, 1978b, 1978/79, 1995), Flaspohler (1974), Green (1976), Pakes and Tavaré (1981), Anisimov and Pushkin (1985), Pollett (1986, 1988a, 1990, 1991, 1995, 1999, 2001a), Abate and Whitt (1989), Pollett and Roberts (1990), Kijima (1992a, 1993a, 1997), Pollett and Vere-Jones (1992), Nair and Pollett (1993), Pollett and Stewart (1994), Ferrari, Kesten, Martinez, and Picco (1995), Jacka and Roberts (1995), van Doorn and Schrijner (1996), Ferrari, Kesten, and Martinez (1996), Hart and Pollett (1996), Coolen-Schrijner, Hart, and Pollett (1999), Darlington and Pollett (2000), Elmes, Pollett, and Walker (2000), Gyllenberg and Silvestrov (1999a, 2000a), Hart and Pollett (2000), Silvestrov (2000a), Dickman and Vidigal (2002), Asselah and Castell (2003), Latouche, Remiche, and Taylor (2003), Coolen-Schrijner and van Doorn (2006a), and Ferrari, and Marić (2007).

Quasi-stationary distributions for Markov and semi-Markov type processes with general phase space and related problems were studied in Iglehart (1974a, 1974b, 1975), Tweedie (1974a, 1974b, 1974c), Pollard and Tweedie (1975, 1976), Bolthausen (1976), Nummelin (1976, 1977, 1979, 1984), Nummelin and Tweedie (1976, 1978), Arjas and Nummelin (1977/78), Durrett (1978, 1980), Tuominen and Tweedie (1979a, 1994), Arjas, Nummelin, and Tweedie (1980/81), Pinsky (1985), Seneta and Tweedie (1985), Abadov (1984), Silvestrov and Abadov (1984, 1991, 1993), Bertoin and Doney (1994a), Ferrari, Martinez, and San Martin (1997), Klebaner, Lazar, and Zeitouni (1998), Martinez, Picco, and San Martin (1998), Breyer and Hart (2000), Lladser and San Martin (2000), Moler, Plo, and San Miguel (2000), Mei Qi Xiang and Lin Xiang (2001), Glynn and Thorisson (2001, 2002), Jacka and Warren (2002), Jacka, Lazic, and Warren (2005a, 2005b), and Kyprianou and Palmowski (2006).

Chapter 4 also contains a number of lemmas. Lemmas 4.1.1–4.4.1 have a standard preparatory character. Lemmas 4.5.1–4.5.10 present a series of solidarity statements concerning existence conditions and convergence properties for moment generating functions of hitting times for perturbed semi-Markov processes. They are interesting by themselves and could have applications beyond the scope of quasi-stationary studies. Other lemmas given in Chapter 4 are of a standard character.

In connection with these results, we would like to mention the works which deal with moment solidarity properties conditions of existence and upper bounds for power and exponential moments of hitting times, Kolmogorov (1937), Chung (1953, 1954, 1960), Hodges and Rosenblatt (1953), Loève (1955), Pyke and Schanflie (1964), Teugels (1968b), Cheong and Teugels (1972), Kalashnikov (1973, 1978b), Silvestrov (1970, 1974, 1980a, 1983a, 1983b, 1996, 2004b), Silvestrov and Poleščuk (1974), Poleščuk (1977), Nummelin and Tuominen (1983), Tweedie (1983b), Gaver, Jacobs and Latouche (1984), Sbignev (1996), Pyke (1999), and Silvestrov and Drozdenko (2006).

We would also like to mention some first papers on semi-Markov processes, which are Takács (1954), Lévy (1954/56), Smith (1954/56), Pyke (1961a, 1961b), Janssen

(1964, 1965), and some surveys and important books on Markov, semi-Markov and related processes, Romanovskiĭ (1949), Doob (1953), Chung Kai Lai (1960), Dynkin (1963), Gikhman and Skorokhod (1965, 1971, 1973, 1975), Çinlar (1975), Kemeny, Snell, and Knapp (1966, 1976), Çinlar (1969a), Foguel (1969), Orey (1971), Rosenblatt (1971), Korolyuk, Brodi, and Turbin (1974), Revuz (1975), Koroljuk an Turbin (1976a, 1978, 1982a), Teugels (1976, 1986), Silvestrov (1980a), Borovkov (1980, 1998a), Kovalenko, Kuznetsov, and Shurenkov (1983), Nummelin (1984), Ethier and Kurtz (1986), Anderson (1991), Korolyuk and Svishchuk (1992, 1995), Wang Zi Kun and Yang Xiang Qun (1992), Meyn and Tweedie (1993c), Kijima (1997), Ramaswami (1997), Latouche and Ramaswami (1999), Limnios, and Oprişan (2001), Bini, Latouche, and Meini (2005), Koroliuk and Limnios (2005), and Janssen and Manca (2006, 2007), Anisimov (2008), and Harlamov (2008).

The results obtained in Chapter 4 play a preparatory role and are needed for constructing asymptotic expansions for nonlinearly perturbed semi-Markov processes with finite phase space and absorption, which are given in Chapter 5.

Chapter 5. This Chapter plays a key role in the book. Our main results presented in this chapter include exponential expansions in mixed ergodic and large deviation theorems for nonlinearly perturbed semi-Markov processes and the corresponding asymptotic expansions for quasi-stationary distributions. The chapter is based in part on the results obtained in the papers Gyllenberg and Silvestrov (1998, 1999a, 2000a) and Silvestrov (2000a, 2007a) and extends these results in several directions.

We think that the method used for obtaining the expansions mentioned above has its own value and great potential for future studies. The first step is a standard one. It is an imbedding of the corresponding semi-Markov process with absorption to a model of regenerative process, where the regeneration times are subsequent return moments to a fixed state $i \neq 0$, where 0 is the absorption state.

This imbedding permits to separate the second step, connected with a construction of asymptotic expansions for power or mixed power-exponential moments of regeneration (return) times, and the third step in the corresponding algorithm, connected with a construction of asymptotic expansions for the characteristic roots $\rho^{(\varepsilon)}$. This separation significantly simplifies the asymptotic analysis.

The second step is to construct asymptotic expansions for higher order moments of the hitting-without-absorption times. These moments satisfy recurrence systems of linear equations with the same perturbed coefficient matrix and the free terms connected by special recurrence systems of relations in which the free terms for the moments of a given order are given as polynomial functions of moments of lower orders. This permits to build an effective recurrence algorithm for constructing the corresponding asymptotic expansions. Each sub-step in this recurrence algorithm is a *matrix* but linear-type problem, where the solution of the system of linear equations with nonlinearly perturbed coefficients and free terms should be expanded in asymptotic series.

The third step is to construct an asymptotic expansion for the root $\rho^{(\varepsilon)}$ of the nonlinear but *scalar* equation ${}_0\phi_{ii}^{(\varepsilon)}(\rho) = 1$, where ${}_0\phi_{ii}^{(\varepsilon)}(\rho)$ is a moment generating function for the corresponding return-without-absorption time. Here, the asymptotic expansions constructed at the second step are also essentially used.

As soon as an asymptotic expansion for the characteristic root $\rho^{(\varepsilon)}$ is constructed, one can use it directly to give corresponding exponential expansions in the mixed ergodic and large deviation theorems.

Asymptotic expansions for quasi-stationary distributions require two more steps, which are needed for constructing asymptotic expansions at the point $\rho^{(\varepsilon)}$ for the corresponding moment generation functions, giving expressions for quasi-stationary probabilities in the quotient form, and for transforming the corresponding asymptotic quotient expressions to the form of Taylor type asymptotic expansions.

As a result, we get an effective algorithm for a construction of the needed asymptotic expansions.

Theorems 5.4.1 and 5.4.2 give exponential expansions in mixed ergodic and large deviation theorems for nonlinearly perturbed semi-Markov processes in the pseudo-stationary case, while Theorems 5.4.3 and 5.4.4 give the corresponding expansions in the quasi-stationary case. Theorems 5.5.2–5.5.5 give the corresponding asymptotic expansions of quasi-stationary distributions for nonlinearly perturbed semi-Markov processes. These expansions may be compared with analogous asymptotic expansions for the usual stationary distributions given in Theorem 5.5.1. It should be noted that the construction of the asymptotic expansions for quasi-stationary distributions is much more complicated than for ordinary stationary distributions. This is so, since the quasi-stationary probabilities are functions of the characteristic roots $\rho^{(\varepsilon)}$ and, therefore, are very nonlinear functionals of the corresponding moments of the transition probabilities. While ordinary stationary probabilities are simple rational functions of the corresponding stationary probabilities of imbedded Markov chains and expectations for the sojourn times.

We also specify perturbation conditions in all the theorems mentioned above for an important case of nonlinearly perturbed Markov chains with absorption.

The main new elements, as compared with the results listed above and given in Gyllenberg and Silvestrov (1998, 1999a, 2000a), consist in the following. Firstly, we use a more advanced and compact matrix form of the corresponding asymptotic expansions. Secondly, we also consider models with a non-empty class of non-recurrent-without-absorption states, which is important for applications. Thirdly, we give a detailed analysis of pivotal properties of the corresponding asymptotic expansions. Also we would like to mention Lemmas 5.1.1–5.3.9 that give asymptotic expansions for hitting probabilities, power- and mixed power-exponential moments of hitting times for nonlinearly perturbed semi-Markov processes. These results have their own values and possible applications beyond the problems studied in the book.

It should be mentioned that some results similar to the results obtained in Chapter 5 were given for nonlinearly perturbed discrete time Markov chains in Silvestrov (2000a).

A list of other works, where asymptotic expansions for some characteristics related to expectations of hitting times and stationary distributions were studied, include the papers Poliščuk and Turbin (1973), Korolyuk and Turbin (1976a, 1978), Latouche and Louchard (1978), Louchard and Latouche (1982), Aissani and Kartashov (1983b), Abadov (1984), Kartashov (1985d, 1985e, 1996b), Latouche (1991), Hassin and Haviv (1992), Haviv, Ritov, and Rothblum (1992), Haviv and Ritov (1993), Englund and Silvestrov (1997), Cao (1998), Yin and Zhang (1998, 2003), Avrachenkov, and Lasserre (1999), Englund (1999a, 1999b, 2001), Li, Yin, G., Yin, K., and Zhang (1999), Silvestrov (2000a), Avrachenkov, Haviv, and Howlett (2001), Yin, G., Zhang Yang, and Yin, K. (2001), and Avrachenkov and Haviv (2004).

Asymptotic expansions for distributions of hitting times and related functionals were studied in Korolyuk, Penev, and Turbin, (1972, 1981), Turbin (1972), Poliščuk and Turbin (1973), Korolyuk and Turbin (1976a, 1978), Abadov (1984), Silvestrov and Abadov (1984, 1991, 1993), Korolyuk and Tadzhiev (1986), Latouche (1988), Courtois and Semal (1991), Kartashov (1985d, 1985e, 1996b), Englund and Silvestrov (1997), Englund (1999a, 1999b, 2001), Silvestrov (2000a), and Koroliuk and Limnios (2005).

Papers that contains asymptotic expansions for other types of functionals for Markov type processes are Nagaev (1957, 1961), Leadbetter (1963), Gray and Pinsky (1983), Liao Ming (1988), Stewart and Sun Ji Guang (1990), Khasminskii, Yin, and Zhang (1996a, 1996b), Yin and Zhang (1998), Fuh Cheng-Der and Lai Tze Leung (2001), Gasanenko (2003), Samoĭlenko (2006a, 2006b), and Fuh Cheng-Der (2007).

It should be noted that the works listed above contain neither exponential expansions in mixed ergodic and large deviation theorems for nonlinearly perturbed semi-Markov processes, except Silvestrov and Abadov (1984, 1991, 1993), nor the corresponding asymptotic expansions for quasi-stationary distributions.

We also would like to mention some works that concern with recurrence relations for different order moments for hitting time type functionals essentially used in our studies. Such recurrence relations have been obtained for Markov chains in Chung (1953, 1954, 1960). A further development was achieved in Hodges and Rosenblatt (1953), Lamperty (1960), Kemeny and Snell (1961a, 1961b), Pitman (1974, 1976, 1977), Kalashnikov (1978b), Silvestrov (1980a, 1980d, 1983b, 1996, 2004b), Nummelin and Tuominen (1983), and Tweedie (1983b).

Some relevant books concerned perturbation problems for stochastic processes studied in Chapter 5 are Erdélyi (1956), Kato (1966), Cole (1968), Korolyuk and Turbin (1976a, 1978), Wentzell and Freidlin (1979), Kevorkian and Cole (1981, 1996), Ho and Cao (1991), Kartashov (1996b), Yin and Zhang (1998), and Stewart (1998, 2001), Latouche and Ramaswami (1999), and Koroliuk and Limnios (2005).

Chapters 1–5 present a theory that can be applied to study pseudo- and quasi-stationary phenomena in nonlinearly perturbed stochastic systems, and nonlinearly perturbed risk processes. These applications are presented, respectively, in Chapters 6 and 7.

Chapter 6. In this chapter, we give several typical examples that show how the results of Chapters 1–5 can be applied to an analysis of pseudo- and quasi-stationary phenomena in nonlinearly perturbed stochastic systems. This chapter is partly based on the results of the papers Gyllenberg and Silvestrov (1994, 1998a, 2000a). In these papers, applications to the analysis of quasi-stationary phenomena in nonlinearly perturbed multi-server highly reliable queueing systems and metapopulation dynamic models were considered. We also consider new applications to nonlinearly perturbed M/G queueing systems with quick service and stochastic systems of birth-death type, including epidemic and population dynamics models.

In all the cases, we first give conditions for mixed ergodic and limit/large deviation theorems. These results have many analogues in the prior works. Second, we formulate nonlinear perturbation conditions that permit to construct exponential expansions in mixed ergodic and limit/large deviation theorems, as well as asymptotic expansions of the corresponding quasi-stationary distributions. These results have no analogues in the preceding works, except for the papers Englund (1999a, 1999b, 2000, 2001), where analogues of some of our results were given for nonlinearly perturbed M/M/m queueing systems and Bernoulli random walks.

We give references to works on limit theorems for different lifetime functionals such as occupation times or waiting times in queueing systems, lifetime in reliability models, ruin times for insurance models, etc. In this connection, we need to mention the works Kingman, (1961, 1962), Gnedenko (1964a, 1964b, 1067a, 1967b), Gnedenko and Kovalenko (1964), Solov'ev (1964, 1966, 1970, 1971, 1983), Çinlar (1967), Pakes (1971, 1972), Masol (1972), Gnedenko and Solov'ev (1974, 1975), Kovalenko and Musaev (1974), Brown (1975, 1984), Keilson (1975, 1986), Kovalenko (1975, 1977, 1980, 1994, 1995, 2000), Zajtsev and Solov'ev (1975), Ovchinnikov (1976), Solov'ev and Sahobov (1976, 1977), Sahobov and Solov'ev (1977), Ivchenko, Kashtanov, and Kovalenko (1979), Walker (1980), Kovalenko and Kuznetsov (1981, 1988), Keilson and Sumita (1983), Anisimov and Zaĭnutdinov (1985), Butko and Korolyuk (1984), Voĭna and Pushkin (1984), Anisimov, Zakusilo, and Donchenko (1987), Assmussen (1987, 2003), Brown and Shao (1987), Rohlicek (1987), Kalashnikov and Rachev (1988), Anisimov and Sztrik (1989) Sztrik and Anisimov (1989), Konstantinidis and Solov'ev (1991), Solov'ev and Konstant (1991), Willekens and Teugels (1992), Kalashnikov (1994c), Zhang Hanqin, Hsu Guang-Hui, and Wang Rongxin (1995), Roy (1996), Kovalenko, Kuznetsov, Pegg (1997), Gyllenberg and Silvestrov (1998, 2000a), Kolowrocki (1998, 2000), Kovalenko, Birolini, Kuznetsov, and Shenbukher (1998), Englund (1999a, 1999b, 2001), Korolyuk (1999), Anisimov (2000, 2008), Anisimov and Kurtulush (2001), Frostig (2002), Lebedev (2002), and Korolyuk and Limnios (2005).

Quasi-stationary distributions in queueing systems and related problems were studied in Pakes (1975), Thorisson (1985b), Abate and Whitt (1988, 1997, 1999), Abate, Kijima and Whitt (1991), Kijima and Makimoto (1992), Willekens and Teugels (1992), Kijima (1993b, 1995), Abate, Choudhury, and Whitt (1994a, 1994b, 1995), Choudhury and Whitt (1994), Glynn and Whitt (1994), van Doorn (1981), Gyllenberg and Silvestrov (1998, 2000a), Englund (1999a, 1999b, 2001), Gaver and Jacobs (1999) Kijima and Makimoto (1999), and Whitt (2000), while the problems of stability of stationary distributions, ergodic theorems, and some asymptotic expansions for characteristics of perturbed queueing systems were studied in Neuts and Teugels (1969), Kyprianou (1972a, 1972b), Zolotarev (1975, 1976, 1977), Kartashov (1979b, 1981, 1984, 1985a, 1985b), Tuominen and Tweedie (1979b), Borovkov (1980), Aissani and Kartashov, (1983a, 1983b), Brandt, Franken, and Lisek (1984), Franken, Kirstein, and Streller (1984), Kalashnikov and Rachev (1984, 1985, 1986, 1987), Latouche (1988), Kalashnikov and Rachev (1988), Rachev (1991), Meyn and Down (1994), Down and Meyn (1995), Asmussen and Teugels (1996), Boucherie (1997), Asmussen (1999), Anisimov and Artalejo (2002), Asmussen and O'Cinneide (2002), and Andreev, Elesin, Krylov, Kuznetsov, and Zeĭfman (2003).

We would also like to mention some books and surveys related to asymptotic problems for queueing systems; these are Khintchine, (1955, 1963), Cox and Smith (1961), Borovkov (1972, 1980), Neuts (1981, 1989), Franken, König, Arndt, and Schmidt (1982), Kalashnikov and Rachev (1988), Kovalenko and Kuznetsov (1988), Kalashnikov (1994c), Kovalenko (1994), Kovalenko, Kuznetsov, and Pegg (1997), Limnios and Oprişan (2001), Whitt (2002), Hassin and Haviv (2003), and Anisimov (2008).

Quasi-stationary distributions and related problems for birth-and-death processes were studied in Good (1968), Solov'ev (1972), Callaert and Keilson (1973), Cavender (1978), Lindvall (1979b), Keilson and Ramaswamy (1984, 1986), Brockwell (1985), van Doorn (1987, 1991), Latouche (1987), Pakes and Pollett (1989), Ferrari, Martinez, and Picco (1991, 1992), Kijima and Seneta (1991), Kijima (1992b), Martinez (1993), Roberts and Jacka (1994), Jacka and Roberts (1995), Martinez and Vares (1995), van Doorn and Schrijner (1995a, 1995b), Bean, Bright, Latouche, Pearce, Pollett, and Taylor (1997), Kijima, Nair, Pollett, and van Doorn (1997), Roberts, Jacka, and Pollett (1997), Schrijner and van Doorn (1997), Bean, Pollett, and Taylor (1998, 2000), Cairns and Pollett (2004) van Doorn and Zeifman (2005), Coolen-Schrijner and van Doorn (2006b), van Doorn (2006), and Pollett, Zhang Hanjun, and Cairns (2007).

As was mentioned above, the chapter contains an analysis of pseudo- and quasi-stationary phenomena for a continuous-time stochastic metapopulation model. As in other examples, the main point in our results is obtaining corresponding exponential expansions in mixed ergodic and large deviation theorems. It seems that some comments concerned the term *metapopulation* may be useful. Natural populations of most species have a spatial structure. Several habitat patches can support local populations. Such a set of local populations is called a *metapopulation*. Local populations in a metapopulation are connected by migration. A local population may go extinct due to a catastrophe. An empty patch may be colonised by migrants originating from other patches. Due to a fragmentation of natural habitats, mathematical metapopulation models have attracted an immense attention in the past few years. It is also useful to note that there is a clear analogy between metapopulation models and epidemic models and that, in fact, the epidemic models are particular cases of metapopulation models. The analogy actually takes place at different hierarchical levels. The most obvious analogy identifies patches with host individuals, colonisation with infection, and local extinction with recovery. Occupied patches correspond to infected, and empty ones to susceptible individuals.

For an account of the current state of the subject as well as for many specific models and references to the literature, we refer to the works Gani (1971), Hethcote and Yorke (1984), Gyllenberg and Hanski (1992), Gyllenberg and Silvestrov (1994, 1998a, 2000a), Hanski and Gilpin (1997), Daley and Gani (1999), and Hanski (1999).

In connection with our metapopulation example, we mention the papers on asymptotics of distributions of the extinction times, and quasi-stationary phenomena in population dynamic and epidemic models, as well as quasi-stationary phenomena in stochastic models of chemical reactions and other non-technical stochastic systems, which can be modelled within the framework of continuous time Markov chains or continuous time Markov processes with absorption. We refer here to the papers Weiss and Dishon (1971), Barbour (1976), Oppenheim, Shuler and Weiss (1977), Turner and Malek-Mansour (1978), Cohen (1979), Pakes (1984), Parsons and Pollett (1987), Pakes (1988a, 1988b, 1989a, 1989b,1989c, 1990), Pollett (1988b, 2001b), Kryscio and Lefèvre (1989), Durrett and Neuhauser (1991), Heyde (1981, 1884), Nåsell (1991, 1996, 1999a, 1999b, 2001, 2002a, 2002b), Al-Towaiq and Al-Eideh (1994), Gilpin and Taylor (1994), Gyllenberg and Silvestrov (1994, 1998, 2000a), Day and Possingham (1995), Kijima (1995), Hanski and Gilpin (1997), Högnäs (1997), Martin-Löf (1998), Ramanan and Zeitouni (1999), Clancy, O'Neill, and Pollett (2001, 2003), Bobrowski (2004), Cairns and Pollett (2004), and Ross and Pollett (2006).

Chapter 7. In this chapter, we present results that extend classical diffusion and Cramér–Lundberg approximations for ruin probabilities, as well as some other ruin based functionals, to a model of nonlinearly perturbed risk processes. The chapter is based on the results obtained in the papers Gyllenberg and Silvestrov (1999b, 2000b) and Silvestrov (2000b), and Silvestrov (2007b).

Our approach is based on the well-known fact that the ruin probability for classical risk process satisfies a renewal equation, if considered as a function of the initial capital. Thus, the asymptotic results for perturbed renewal equation, obtained in Chapters 1 and 2, can be applied. Using this approach we get an extension of the classical diffusion and Cramér–Lundberg approximations for ruin probabilities to a model of nonlinearly perturbed risk processes. We introduce correction terms for the diffusion and Cramér–Lundberg type approximations, which provide the right asymptotic behaviour of relative errors in the perturbed model. The dependence of these correction terms on the relations between the rate of the perturbation and the speed of growth of the initial capital is investigated. Various types of perturbations of risk processes are discussed.

Theorems 7.2.1, 7.2.2, and 7.2.3 present new variants of the diffusion approximation, respectively, for the ruin probabilities, increments of the ruin probabilities, and density of the ruin probability for perturbed risk processes. These results were obtained in Silvestrov (2000b). Theorem 7.3.1 presents a new generalised unified variant of the diffusion and Cramér-Lundberg approximations for the perturbed risk processes. Theorems 7.3.2 and 7.3.3 present exponential asymptotic expansions of the ruin probabilities for nonlinearly perturbed risk processes. These results were obtained in Gyllenberg and Silvestrov (1999b, 2000b). It is useful to note that these theorems cover, in a unified form, both the diffusion approximation case, which corresponds, in terms used in the book, to a pseudo-stationary model, as well as the Cramér–Lundberg approximation case, which corresponds to a quasi-stationary model. In Section 7.4, we also give variants of approximations for the ruin probabilities by using the corresponding asymptotic expansions. These approximations involve higher order moments of the claim distributions and have zero asymptotic relative errors. Numerical studies show that the corresponding approximations are comparable with other approximations known in the literature. Finally, we also give the diffusion and the Cramér–Lundberg type approximations for the distribution of the surplus prior to and at the time of the ruin for nonlinearly perturbed risk processes in Theorems 7.5.1, 7.5.2, and 7.5.3. These are new results given in Silvestrov (2007b).

Our results are connected with the results related to various sufficient conditions for the diffusion and the Cramér–Lundberg type approximations and approximations for the ruin probabilities given in Lundberg (1903, 1909, 1926), Cramér (1926, 1930, 1955), Esscher (1932), Segerdahl (1942, 1948), Iglehart (1969), Grandell and Segerdahl (1971), Beekman (1974), Siegmund (1975), Harrison (1977), Grandell (1977, 1978/79 1991), Thorin (1977, 1982), de Vylder (1978, 1996), Gerber (1979), Janssen and Delfosse (1982), Wikstad (1983), Asmussen (1985, 1987, 1989, 2000), Glynn (1990), Sundt (1991), Schmidli (1992, 1994a, 1994b, 1995, 1997, 1999), Embrechts and Klüppelberg (1993), Asmussen and Rolski (1994), Barndorff-Nielsen and Schmidli (1995), Klüppelberg and Mikosch (1995), Asmussen and Klüppelberg (1996), Beirlant, Teugels and Vynckier (1996), Kalashnikov and Konstantinidis (1996, 2000), de Vylder and Marceau (1996), Embrechts, Klüpenberg, and Mikosch (1997), Asmussen, Schmidli, and Schmidt (1999), Gyllenberg and Silvestrov (1999b, 2000b), Kalashnikov (1999), Kartashov (1999), Silvestrov (2000b, 2000c), Enikeeva, Kalashnikov, and Rusaityte (2001), Hipp and Schmidli (2004), Kartashov and Stroev (2005), and Silvestrov and Drozdenko (2006).

We would also like to mention results related to asymptotics of the ruin probabilities for risk processes with heavy tailed claim distributions and more general risk type processes given in Täcklind (1942), Andersen, E. Sparre (1957), von Bahr (1975), Thorin and Wikstad (1976), Embrechts, Goldie and Veraverbeke (1979), Embrechts and Veraverbeke (1982), Martin-Löf (1986b), Klüppelberg (1989), Embrechts and Klüppelberg (1993), Asmussen, Fle, and Klüppelberg (1994), Bertoin and Doney (1994b), Assmussen (1995, 1996, 2000, 2003), Klüppelberg and Mikosch (1995, 1997), Asmussen and Teugels (1996), Grandell (1997), Klüppelberg and Stadmular (1998), Zinchenko (1999), Bening, Korolev, and Kudryavtsev (2001), Kalashnikov and Norberg (2002), Klüppelberg, Kyprianou, and Maller (2004), Klüppelberg and Kyprianou (2006), Ming Ruixing, Modibo Diarra, and Hu Yijun (2007).

The surplus prior to and at the time of the ruin and its asymptotics have been studied for non-perturbed risk processes in Gerber, Goovarerts, and Kaas (1987), Dufresne and Gerber (1988), Dickson (1992, 2005), Dickson, Dos Reis, and Waters (1995), Willmot and Lin (1998), Schmidli (1999), Rolski, Schmidli, Schmidt, and Teugels (1999), and Badescu, Breuer, Drekic, Latouche, and Stanford (2005).

Some basic books on risk theory and related topics are Cramér (1930, 1955), Gerber (1979), Bingham, Goldie, and Teugels (1989), Grandell (1991), Sundt (1991), Daykin, Pentikäinen, and Pesonen (1994), de Vylder (1996), Embrechts, Klüppelberg, and Mikosch (1997), Rolski, Schmidli, Schmidt, and Teugels (1999), Asmussen (2000), Bening and Korolev (2002), Mikosch (2004), and Dickson (2005).

Bibliography

[1] Abadov, Z. A. (1984) Asymptotical Expansions with Explicit Estimation of Constants for Exponential Moments of Sums of Random Variables Defined on a Markov Chain and their Applications to Limit Theorems for First Hitting Times. Candidate of Science dissertation, Kiev State University.

[2] Abate, J., Choudhury, G. L., Whitt, W. (1994a) Asymptotics for steady-state tail probabilities in structured Markov queueing models. *Comm. Statist. Stoch. Models*, **10**, no. 1, 99–143.

[3] Abate, J., Choudhury, G. L., Whitt, W. (1994b) Waiting-time tail probabilities in queues with long-tail service-time distributions. *Queueing Systems Theory Appl.*, **16**, no. 3–4, 311–338.

[4] Abate, J., Choudhury, G. L., Whitt, W. (1995) Exponential approximations for tail probabilities in queues. I. Waiting times. *Oper. Res.*, **43**, no. 5, 885–901.

[5] Abate, J., Kijima, M., Whitt, W. (1991) Decompositions of the $M/M/1$ transition function. *Queueing Systems Theory Appl.*, **9**, no. 3, 323–336.

[6] Abate, J., Whitt, W. (1988) Simple spectral representations for the $M/M/1$ queue. *Queueing Systems Theory Appl.*, **3**, no. 4, 321–345.

[7] Abate, J., Whitt, W. (1989) Spectral theory for skip-free Markov chains. *Prob. Eng. Inf. Sci.*, **3**, 77–88.

[8] Abate, J., Whitt, W. (1997) Asymptotics for $M/G/1$ low-priority waiting-time tail probabilities. *Queueing Systems Theory Appl.* **25**, no. 1–4, 173–233.

[9] Abate, J., Whitt, W. (1999) Explicit $M/G/1$ waiting-time distributions for a class of long-tail service-time distributions. *Oper. Res. Lett.*, **25**, no. 1, 25–31.

[10] Ahmad, I. A., Basu, A. K. (1983) Equivalence of rates of convergence in the CLT between maximum partial sums and first passage times of random walks. *Scand. Actuar. J.*, no. 2, 89–105.

[11] Aissani, D., Kartashov, N. V. (1983a) Strong stability of an imbedded Markov chain in an $M/G/1$ system. *Teor. Veroyatn. Mat. Stat.*, **29**, 3–7 (English translation in *Theory Probab. Math. Statist.*, **29**, 1–5).

[12] Aissani, D., Kartashov, N. V. (1983b) Ergodicity and stability of Markov chains with respect to operator topologies in a space of transition kernels. *Dokl. Akad. Nauk Ukr. SSR*, Ser. A, no. 11, 3–5.

[13] Aldous, D. J. (1982) Markov chains with almost exponential hitting times. *Stoch. Process. Appl.*, **13**, 305–310.

[14] Aldous, D. J. (1986) Self-intersections of 1-dimensional random walks. *Probab. Theory Related Fields*, **72**, no. 4, 559–587.

[15] Aldous, D. J. (1988) Finite-time implications of relaxation times for stochastically monotone processes. *Probab. Theory Related Fields*, **77**, no. 1, 137–145.

[16] Aldous, D. J. (1989) Hitting times for random walks on vertex-transitive graphs. *Math. Proc. Cambridge Philos. Soc.*, **106**, no. 1, 179–191.

[17] Aldous, D. J. (1991) Threshold limits for cover times. *J. Theor. Probab.*, **4**, 197–211.

[18] Aldous, D. J., Brown, M. (1992) Inequalities for rare events in time-reversible Markov chains. I. In: Shaked, M, Tong. Y. L (eds) *Stochastic Inequalities*. AMS-IMS-SIAM Joint Summer Research Conference held in Seattle, Washington, 1991. Institute of Mathematical Statistics Lecture Notes, Monograph Series, **22**, Hayward, CA, 1–16.

[19] Aldous, D. J., Brown, M. (1993) Inequalities for rare events in time-reversible Markov chains. II. *Stoch. Process. Appl.*, **44**, 15–25.

[20] Al-Eideh, B.M. (1995) Quasi-stationary distributions in Markov chains with absorbing subchains. *J. Inform. Optim. Sci.*, **16**, no. 2, 281–286.

[21] Alimov, D. (1994a) Markov functionals of an ergodic Markov process. *Teor. Veroyatn. Primen.*, **39**, 618–626 (English translation in *Theory Probab. Appl.*, **39**, 504–512).

[22] Alimov, D. (1994b) Limit-ergodic Markov functionals of an ergodic process. *Teor. Veroyatn. Primen.*, **39**, 657–668 (English translation in *Theory Probab. Appl.* **39**, 537–546).

[23] Alimov, D., Shurenkov, V. M. (1990a) Markov renewal theorems in triangular array model. *Ukr. Mat. Zh.*, **42**, 1443–1448 (English translation in *Ukr. Math. J.*, **42**, 1283–1288).

[24] Alimov, D., Shurenkov, V. M. (1990b) Asymptotic behavior of terminating Markov processes that are close to ergodic. *Ukr. Mat. Zh.*, **42**, 1701–1703 (English translation in *Ukr. Math. J.*, **42**, 1535–1538).

[25] Allen, B., Anderssen, R. S., Seneta, E. (1977) Computation of stationary measures for infinite Markov chains. In: *Algorithmic Methods in Probability*. Studies in the Management Sci., **7**, North-Holland, Amsterdam, 13–23.

[26] Alsmeyer, G. (1986) Parameter-dependent renewal theorems with applications. *Z. Oper. Res.*, Ser. A-B **30**, no. 3, A111–A134.

[27] Alsmeyer, G. (1987) On the moments of certain first passage times for linear growth processes. *Stoch. Process. Appl.*, **25**, 109–136.

[28] Alsmeyer, G. (1988a) Second-order approximations for certain stopped sums in extended renewal theory. *Adv. Appl. Probab.*, **20**, 391–410.

[29] Alsmeyer, G. (1988b) On first and last exit times for curved differentiable boundaries. *Sequent. Anal.*, **7**, no. 4, 345–362.

[30] Alsmeyer, G. (1988c) On the variance of first passage times in the exponential case. In: *Transactions of the Tenth Prague Conference on Information Theory, Statistical Decision Functions, Random Processes*, Vol. A, Prague, 1986. Reidel, Dordrecht, 183–192.

[31] Alsmeyer, G. (1991a) Some relations between harmonic renewal measures and certain first passage times. *Statist. Probab. Lett.*, **12**, no. 1, 19–27.

[32] Alsmeyer, G. (1991b) *Erneuerungstheorie. Analyse stochastischer Regenerationsschemata.* Teubner Skripten zur Mathematischen Stochastik, B. G. Teubner, Stuttgart.

[33] Alsmeyer, G. (1992) On generalized renewal measures and certain first passage times. *Ann. Probab.*, **20**, 1229–1247.

[34] Alsmeyer, G. (1994a) Blackwell's renewal theorem for certain linear submartingales and coupling. *Acta Appl. Math.*, **34**, no. 1–2, 135–150.

[35] Alsmeyer, G. (1994b) Random walks with stochastically bounded increments: renewal theory via Fourier analysis. *Yokohama Math. J.*, **42**, no. 1, 1–21.

[36] Alsmeyer, G. (1994c) On the Markov renewal theorem. *Stoch. Process. Appl.*, **50**, 37–56.

[37] Alsmeyer, G. (1995) Random walks with stochastically bounded increments: renewal theory. *Math. Nachr.*, **175**, 13–31.

[38] Alsmeyer, G. (1996) Superposed continuous renewal processes: a Markov renewal approach. *Stoch. Process. Appl.*, **61**, 311–322.

[39] Alsmeyer, G. (1997) The Markov renewal theorem and related results. *Markov Process. Related Fields*, **3**, no. 1, 103–127.

[40] Alsmeyer, G., Hoefs, V. (2001) Markov renewal theory for stationary m-block factors. *Markov Proces. Related Fields*, **7**, no. 2, 325–348.

[41] Alsmeyer, G., Hoefs, V. (2002) Markov renewal theory for stationary $(m + 1)$-block factors: convergence rate results. *Stoch. Process. Appl.*, **98**, 77–112.

[42] Alsmeyer, G., Hoefs, V. (2004) Markov renewal theory for stationary $(m + 1)$-block factors: first passage time and overshoot. In: *Semi-Markov Processes: Theory and Applications. Comm. Statist. Theory Methods*, **33**, no. 3, 545–568.

[43] Alsmeyer, G., Irle, A. (1986) Asymptotic expansions for the variance of stopping times in nonlinear renewal theory. *Stoch. Process. Appl.*, **23**, 235–258.

[44] Al-Towaiq, M. H., Al-Eideh, B. M. (1994) Quasi-stationary distribution in absorbing Markov chains with r absorbing states. *J. Inform. Optim. Sci.*, **15**, no. 1, 89–95.

[45] Andersen, E. Sparre (1957) On the collective theory of risk in the case of contagion between the claims. In: *Transactions of the XVth International Congress of Actuaries*, Vol. II, New York, 1957, 219–229.

[46] Anderson, K. K., Athreya, K. B. (1987) A renewal theorem in the infinite mean case. *Ann. Probab.*, **15**, 388–393.

[47] Anderson, W. J. (1991) *Continuous-Time Markov Chains: An Applications-Oriented Approach.* Probability and Its Applications, Springer, New York.

[48] Andreev, D. B., Elesin, M. A., Krylov, E. A., Kuznetsov, A. V., Zeĭfman, A. I. (2002) On ergodicity and stability estimates for some nonhomogeneous Markov chains. In: *Proceedings of the Seminar on Stability Problems for Stochastic Models*, Part III, Eger, 2001. *J. Math. Sci.*, **112**, no. 2, 4111–4118.

[49] Andreev, D. B., Elesin, M. A., Krylov, E. A., Kuznetsov, A. V., Zeĭfman, A. I. (2003) Ergodicity and stability of nonstationary queueing systems. *Teor. Ĭmovīr. Mat. Stat.*, **68**, 1–10 (English translation in *Theory Probab. Math. Statist.*, **68**, 1–10).

[50] Andrieu, C., Moulines, E. (2006) On the ergodicity properties of some adaptive MCMC algorithms. *Ann. Appl. Probab.*, **16**, 1462–1505.

[51] Anichkin, S. A. (1985) Rate of convergence estimates in Blackwell's theorem. In: *Stability Problems for Stochastic Models*, Varna, 1985. VNIISI, Moscow, 95–102 (English translation in *J. Soviet Math.*, **40**, no. 4, 449–453).

[52] Anisimov, V. V. (1970a) Limit distributions for functionals on semi-Markov processes defined on fixed set of states up to the exit time. *Dokl. Acad. Nauk SSSR*, **193**, 743–745 (English translation in *Soviet Math. Dokl.*, **11**, 1002–1004).

[53] Anisimov, V. V. (1971a) Limit theorems for sums of random variables on a Markov chain, connected with the exit from a set that forms a single class in the limit. *Teor. Veroyatn. Mat. Stat.*, **4**, 3–17 (English translation in *Theory Probab. Math. Statist.*, **4**, 1–13).

[54] Anisimov, V. V. (1971b) Limit theorems for sums of random variables in array of sequences defined on a subset of states of a Markov chain up to the exit time. *Teor. Veroyatn. Mat. Stat.*, **4**, 18–26 (English translation in *Theory Probab. Math. Statist.*, **4**, 15–22).

[55] Anisimov, V. V. (1972) On limiting behaviour of a semi-Markov process with splitting set of states. *Dokl. Acad. Nauk SSSR*, **206**, 777–779 (English translation in *Soviet Math. Dokl.*, **13**, 1276–1279).

[56] Anisimov, V. V. (1973) Asymptotical consolidation of states for random processes. *Kibernetika*, no. 3, 109–117.

[57] Anisimov, V. V. (1975) Limit Theorems for Random Processes and their Application to Processes with Discrete Component. Doctor of Science dissertation, Kiev State University.

[58] Anisimov, V. V. (1980) Limit theorems for processes admitting the asymptotical consolidation of states. *Theor. Veroyatn. Mat. Stat.*, **22**, 3–15 (English translation in *Theory Probab. Math. Statist.*, **22**, 1–13).

[59] Anisimov, V. V. (1984) *Asymptotic Methods of Analysis for Stochastic Systems*. Mitsnerba, Tbilisi.

[60] Anisimov, V. V. (1986) Approximation of Markov processes that can be asymptotically lumped. *Teor. Veroyatn. Mat. Stat.*, **34**, 1–12 (English translation in *Theory Probab. Math. Statist.*, **34**, 1–11).

[61] Anisimov, V. V. (1988) *Random Processes with Discrete Components*. Vysshaya Shkola and Izdatel'stvo Kievskogo Universiteta, Kiev.

[62] Anisimov, V. V. (2000) Asymptotic analysis of reliability for switching systems in light and heavy traffic conditions. In: Limnios, N., Nikulin, M. (eds) *Recent Advances in Reliability Theory: Methodology, Practice and Inference*. Birkhäuser, Boston, 119–133.

[63] Anisimov, V. V. (2008) *Switching Processes in Queuing Models*. Applied Stochastic Methods Series. ISTE and Wiley, London.

[64] Anisimov, V. V., Artalejo, J. R. (2002) Approximation of multiserver retrial queues by means of generalized truncated models. *Top*, **10**, no. 1, 51–66.

[65] Anisimov, V. V., Chernyak, A. V. (1982) Limit theorems for certain rare functionals on Markov chains and semi-Markov processes. *Teor. Veroyatn. Mat. Stat.*, **26**, 3–8 (English translation in *Theory Probab. Math. Statist.*, **26**, 1–6).

[66] Anisimov, V. V., Chernyak, A. V. (1983) Asymptotic behavior of nonhomogeneous flows of rare events. *Izv. Akad. Nauk SSSR*, Tekhn. Kibernet., no. 6, 73–80 (English translation in *Engrg. Cybernetics*, **21**, no. 6, 127–135).

[67] Anisimov, V. V., Gasanenko, V. A. (1978) Controllable flows subject to rarefaction. *Kibernetika*, no. 1, 119–122 (English translation in *Cybernetics*, **14**, no. 1, 110–124).

[68] Anisimov, V. V., Kurtulush, M. (2001) Some Markov retrial queueing systems under light traffic conditions. *Kibernet. Sistem. Anal.*, no. 6, 110–126 (English translation in *Cybernet. Systems Anal.*, **37**, no. 6, 876–887).

[69] Anisimov, V. V., Pushkin, S. G. (1985) Limiting conditional distribution for a homogeneous Markov chain with a set of strongly communicating states. *Dokl. Akad. Nauk Ukr. SSR*, Ser. A, no. 12, 61–63.

[70] Anisimov, V. V., Sityuk, V. N. (1980) On the convergence of a nonhomogeneous Markov chain to a process with independent values controlled by a nonhomogeneous Markov process. *Teor. Veroyatn. Mat. Stat.*, **23**, 3–10 (English translation in *Theory Probab. Math. Statist.*, **23**, 1–8).

[71] Anisimov, V. V., Sztrik, J. (1989) Asymptotic analysis of some complex renewable systems operating in random environments. *Eur. J. Oper. Res.*, **41**, no. 2, 162–168.

[72] Anisimov, V. V., Voĭna, A. A., Lebedev, E. A. (1983) Asymptotic estimation of integral functionals and consolidation of stochastic systems. *Vestnik Kiev. Univ.*, Model. Optim. Slozhn. Sist., no. 2, 41–50.

[73] Anisimov, V. V., Zaĭnutdinov, G. F. (1985) Limit distributions for the time of the first loss of a customer in controllable Markov queueing systems with priorities. *Vestnik Kiev. Univ.*, Model. Optim. Slozhn. Sist., no. 4, 3–9.

[74] Anisimov, V. V., Zakusilo, O. K., Donchenko, V. S. (1987) *Elements of Queueing and Asymptotical Analysis of Systems*. Lybid', Kiev.

[75] Arjas, E., Korolyuk, V. S. (1980) Stationary phase lumpability of Markov renewal processes. *Dokl. Akad. Nauk Ukr. SSR*, Ser. A, no. 8, 3–6.

[76] Arjas, E., Nummelin, E. (1977/78) Semi-Markov processes and α-invariant distributions. *Stoch. Process. Appl.*, **6**, 53–64.

[77] Arjas, E., Nummelin, E., Tweedie, R. L. (1978) Uniform limit theorems for nonsingular renewal and Markov renewal processes. *J. Appl. Probab.*, **15**, 112–125.

[78] Arjas, E., Nummelin, E., Tweedie, R. L. (1980/81) Semi-Markov processes on a general state space: α-theory and quasistationarity. *J. Austral. Math. Soc.*, Ser. A **30**, no. 2, 187–200.

[79] Asmussen, S. (1985) Conjugate processes and the simulation of ruin problems. *Stoch. Process. Appl.*, **20**, 213–229.

[80] Asmussen, S. (1987, 2003) *Applied Probability and Queues*. Wiley Series in Probability and Mathematical Statistics, Wiley, New York and Applications of Mathematics, **51**, Springer, New York.

[81] Asmussen, S. (1989) Risk theory in a Markovian environment. *Scand. Actuar. J.*, no. 2, 69–100.

[82] Asmussen, S. (1992) On coupling and weak convergence to stationarity. *Ann. Appl. Probab.*, **2**, 739–751.

[83] Asmussen, S. (1994) Busy period analysis, rare events and transient behavior in fluid flow models. *J. Appl. Math. Stoch. Anal.*, **7**, no. 3, 269–299.

[84] Asmussen, S. (1996) Rare events in the presence of heavy tails. In: *Stochastic Networks*. New York, 1995. Lecture Notes in Statistics, **117**, Springer, New York, 197–214.

[85] Asmussen, S. (1999) Semi-Markov queues with heavy tails. In: Janssen, J., Limnios, N. (eds) *Semi-Markov Models and Applications*. Kluwer, Dordrecht, 269–284.

[86] Asmussen, S. (2000) *Ruin Probabilities*. Advanced Series on Statistical Science & Applied Probability, **2**, World Scientific, Singapore.

[87] Asmussen, S., Fløe, H.L., Klüppelberg, C. (1994) Large claims approximations for risk processes in a Markovian environment. *Stoch. Process. Appl.*, **54**, 29–43.

[88] Asmussen, S., Kalashnikov, V. (1999) Failure rates of regenerative systems with heavy tails. In: *Stability Problems for Stochastic Models*, Part III, Hajdúszoboszló, 1997. *J. Math. Sci.*, **93**, no. 4, 486–500.

[89] Asmussen, S., Klüppelberg, C. (1996) Large deviation results for subexponential tails, with applications to insurance risk. *Stoch. Process. Appl.*, **64**, 103–125.

[90] Asmussen, S., O'Cinneide, C. (2002) On the tail of the waiting time in a Markov-modulated $M/G/1$ queue. *Oper. Res.*, **50**, no. 3, 559–565.

[91] Asmussen, S., Rolski, T. (1994) Risk theory in a periodic environment: the Cramér-Lundberg approximation and Lundberg's inequality. *Math. Oper. Res.*, **19**, no. 2, 410–433.

[92] Asmussen, S., Schmidli, H., Schmidt, V. (1999) Tail probabilities for non-standard risk and queueing processes with subexponential jumps. *Adv. Appl. Probab.*, **31**, 422–447.

[93] Asmussen, S., Teugels, J. L. (1996) Convergence rates for $M/G/1$ queues and ruin problems with heavy tails. *J. Appl. Probab.*, **33**, 1181–1190.

[94] Asselah, A, Castell, F. (2003) Existence of quasi-stationary measures for asymmetric attractive particle systems on $\mathbb{Z}^d$. *Ann. Appl. Probab.*, **13**, 1569–1590.

[95] Asselah, A., Ferrari, P. A. (2006) Hitting times for independent random walks on $\mathbb{Z}^d$. *Ann. Probab.*, **34**, 1296–1338.

[96] Athreya, K. B., McDonald, D., Ney, P. (1978) Limit theorems for semi-Markov processes and renewal theory for Markov chains. *Ann. Probab.*, **6**, 788–797.

[97] Athreya, K. B., Ney, P. (1978a) A new approach to the limit theory of recurrent Markov chains. *Trans. Amer. Math. Soc.*, **245**, 493–501.

[98] Athreya, K. B., Ney, P. E. (1978b) Limit theorems for semi-Markov processes. *Bull. Austral. Math. Soc.*, **19**, no. 2, 283–294.

[99] Avrachenkov, K. E., Haviv, M. (2003) Perturbation of null spaces with application to the eigenvalue problem and generalized inverses. *Linear Algebra Appl.*, **369**, 1–25.

[100] Avrachenkov, K. E., Haviv, M. (2004) The first Laurent series coefficients for singularly perturbed stochastic matrices. *Linear Algebra Appl.*, **386**, 243–259.

[101] Avrachenkov, K. E., Haviv, M., Howlett, P. G. (2001) Inversion of analytic matrix functions that are singular at the origin. *SIAM J. Matrix Anal. Appl.*, **22**, no. 4, 1175–1189.

[102] Avrachenkov, K. E., Lasserre, J. B. (1999) The fundamental matrix of singularly perturbed Markov chains. *Adv. Appl. Probab.*, **31**, 679–697.

[103] Badescu, A. L., Breuer, L., Drekic, S., Latouche, G., Stanford, D. A. (2005) The surplus prior to ruin and the deficit at ruin for a correlated risk process. *Scand. Actuar. J.*, no. 6, 433–445.

[104] Badowski, G., Yin, G., Zhang, Q. (2003) Near-optimal controls of discrete-time dynamic systems driven by singularly-perturbed Markov chains. *J. Optim. Theory Appl.*, **116**, no. 1, 131–166.

[105] Balaji, S., Meyn, S. P. (2000) Multiplicative ergodicity and large deviations for an irreducible Markov chain. *Stoch. Process. Appl.*, **90**, 123–144.

[106] Banakh, D. (1981) Explicit estimates of the rate of convergence in a discrete renewal theorem. *Dokl. Akad. Nauk Ukr. SSR*, Ser. A, no. 11, 3–4.

[107] Banakh, D. (1982a) Estimates of the rate of convergence in a discrete renewal theorem. *Teor. Veroyatn. Mat. Stat.*, **26**, 9–16 (English translation in *Theory Probab. Math. Statist.*, **26**, 7–14).

[108] Banakh, D. (1982b) Explicit Estimates of the Rate of Convergence for Regenerative Processes with Discrete Time. Candidate of Science dissertation, Kiev State University.

[109] Barbour, A. D. (1976) Quasi-stationary distributions in Markov population processes. *Adv. Appl. Probab.*, **8**, 296–314.

[110] Barlow, J. L. (1993a) Perturbation results for nearly uncoupled Markov chains with applications to iterative methods. *Numer. Math.*, **65**, 51–62.

[111] Barlow, J. L. (1993b) Error bounds for the computation of null vectors with applications to Markov chains. *SIAM J. Matrix Anal. Appl.*, **14**, no. 3, 798–812.

[112] Barndorff-Nielsen, O. E., Schmidli, H. (1995) Saddlepoint approximations for the probability of ruin in finite time. *Scand. Actuar. J.*, no. 2, 169–186.

[113] Basu, A. K. (1972) Invariance theorems for first passage time variables. *Canad. Math. Bull.*, **15**, 171–176.

[114] Basu, A. K. (1978) On the rates of convergence in limit theorems for one-sided stopping rules. *Scand. Actuar. J.*, no. 3, 169–174.

[115] Baxendale, P. H. (2005) Renewal theory and computable convergence rates for geo-metrically ergodic Markov chains. *Ann. Appl. Probab.*, **15**, no. 1B, 700–738.

[116] Bean, N. G., Bright, L., Latouche, G., Pearce, C. E. M., Pollett, P. K., Taylor, P. G. (1997) The quasi-stationary behavior of quasi-birth-and-death processes. *Ann. Appl. Probab.*, **7**, 134–155.

[117] Bean, N., Pollett, P., Taylor, P. (1998) The quasistationary distributions of level-independent quasi-birth-and-death processes. In: *Special Issue in Honor of Marcel F. Neuts. Comm. Statist. Stoch. Models*, **14**, no. 1–2, 389–406.

[118] Bean, N., Pollett, P., Taylor, P. (2000) Quasistationary distributions for level-dependent quasi-birth-and-death processes. *Comm. Statist. Stoch. Models*, **16**, no. 5, 511–541.

[119] Beekman, J. A. (1974) *Two Stochastic Processes*. Almqvist & Wiksell, Stockholm and Halsted Press, New York.

[120] Beirlant, J., Teugels, J. L., Vynckier, P. (1996) *Practical Analysis of Extreme Values*. Leuven University Press, Leuven.

[121] Belyaev, Yu. K. (1963) Limit theorems for dissipative flows. *Teor. Veroyatn. Primen.*, **8**, 175–184 (English translation in *Theory Probab. Appl.*, **8**, 165–173).

[122] Bening, V. E., Korolev, V. Yu. (2002) *Generalized Poisson Models and their Applications in Insurance and Finance*. Modern Probability and Statistics, VSP, Utrecht.

[123] Bening, V. E., Korolev, V. Yu., Kudryavtsev, A. A. (2001) On the computation of confidence intervals for the ruin probability in generalized risk. *Vestnik Moskov. Univ.*, Ser. XV Vychisl. Mat. Kibernet., no. 3, 32–39 (English translation in *Moscow Univ. Comput. Math. Cybernet.*, no. 3, 29–37).

[124] Berman, S. M. (1992) *Sojourns and Extremes of Stochastic Processes*. The Wadsworth & Brooks/Cole Statistics/Probability Series, Wadsworth & Brooks/Cole, Pacific Grove, CA.

[125] Bertoin, J. (1999) Renewal theory for embedded regenerative sets. *Ann. Probab.*, **27**, 1523–1535.

[126] Bertoin, J., Doney, R. A. (1994a) On conditioning a random walk to stay nonnegative. *Ann. Probab.*, **22**, 2152–2167.

[127] Bertoin, J., Doney, R. A. (1994b) Some asymptotic results for transient random walks. *Adv. Appl. Probab.*, **28**, 207–226.

[128] Bingham, N. H., Goldie, C. M., Teugels, J. L. (1989) *Regular Variation*. Encyclopedia of Mathematics and its Applications, **27**, Cambridge University Press, Cambridge.

[129] Bini, D. A., Latouche, G., Meini, B. (2005) *Numerical Methods for Structured Markov Chains*. Numerical Mathematics and Scientific Computation, Oxford Science Publications, Oxford University Press, New York.

[130] Blackwell, D. (1948) A renewal theorem. *Duke Math. J.*, **15**, 145–150.

[131] Blackwell, D. (1953) Extension of a renewal theorem. *Pacific J. Math.*, **3**, 315–320.

[132] Bobrova, A. F. (1983) Estimates of accuracy of an asymptotic consolidation of countable Markov chains. In *Stability Problems for Stochastic Models*. Trudy Seminara, VNIISI, Moscow, 16–24.

[133] Bobrowski, A. (2004) Quasi-stationary distributions of a pair of Markov chains related to time evolution of a DNA locus. *Adv. Appl. Probab.*, **36**, 57–77.

[134] Böker, F., Serfozo, R. (1983) Ordered thinnings of point processes and random measures. *Stoch. Process. Appl.*, **15**, 113–132.

[135] Bolthausen, E. (1976) On a functional central limit theorem for random walks conditioned to stay positive. *Ann. Probab.*, **4**, 480–485.

[136] Bon, J.-L., Kalashnikov, V. (1999) Bounds for geometric sums used for evaluation of reliability of regenerative models. In: *Proceedings of the 18th Seminar on Stability Problems for Stochastic Models*, Part III, Hajdúszoboszló, 1997. *J. Math. Sci.*, **93**, no. 4, 501–510.

[137] Bon, J.-L., Kalashnikov, V. (2001) Some estimates of geometric sums. *Statist. Probab. Lett.*, **55**, no. 1, 89–97.

[138] Borovkov, A. A. (1964) Remarks on the theorems of Wiener and Blackwell. *Teor. Veroyatn. Primen.*, **9**, 331–343 (English translation in *Theory Probab. Appl.*, **9**, 303–312).

[139] Borovkov, A. A. (1972) *Probabilistic Processes in Queueing Theory*. Probability Theory and Mathematical Statistics, Nauka, Moscow (English edition: *Stochastic Processes in Queueing Theory*. Applications of Mathematics, **4**, Springer, New York (1976)).

[140] Borovkov, A. A. (1977) Some estimates for the rate of convergence in stability theorems. *Teor. Veroyatn. Primen.*, **22**, 689–699 (English translation in *Theory Probab. Appl.*, **22**, 668–678).

[141] Borovkov, A. A. (1980) *Asymptotic Methods in Queueing Theory*. Probability Theory and Mathematical Statistics, Nauka, Moscow (English edition: Wiley Series in Probability and Mathematical Statistics: Applied Probability and Statistics, Wiley, New York (1984)).

[142] Borovkov, A. A. (1998a) *Ergodicity and Stability of Stochastic Processes*. Wiley Series in Probability and Statistics: Probability and Statistics, Wiley, New York.

[143] Borovkov, A. A. (1998b) On the asymptotics of distributions of first-passage times. I. *Mat. Zametki*, **75**, no. 1, 24–39 (English translation in *Math. Notes*, **75**, no. 1–2, 23–37).

[144] Borovkov, A. A. (1998c) On the asymptotics of distributions of first-passage times. II. *Mat. Zametki*, **75**, no. 3, 350–359 (English translation in *Math. Notes*, **75**, no. 3–4, 322–330).

[145] Borovkov, A. A., Fayolle, G., Korshunov, D. A. (1992) Transient phenomena for Markov chains and applications. *Adv. Appl. Probab.*, **24**, 322–342.

[146] Borovkov, A. A., Foss, S. G. (1992) Stochastically recursive sequences and their generalizations. *Sib. Adv. Math.*, **2**, no. 1, 16–81.

[147] Borovkov, A. A., Foss, S. G. (1994) Two ergodicity criteria for stochastically recursive sequences. *Acta Appl. Math.*, **34**, no. 1–2, 125–134.

[148] Borovkov, A. A., Hordijk, A. (2004) Characterization and sufficient conditions for normed ergodicity of Markov chains. *Adv. Appl. Probab.*, **36**, 227–242.

[149] Boucherie, R. J. (1997) On the quasi-stationary distribution for queueing networks with defective routing. *J. Austral. Math. Soc.*, Ser. B **38**, no. 4, 454–463.

[150] Brandt, A., Franken, P., Lisek, B. (1984) Ergodicity and steady state existence. Continuity of stationary distributions of queueing characteristics. In: *Modelling and Performance Evaluation Methodology*, Paris, 1983. Lecture Notes in Control and Inform. Sci., **60**, Springer, Berlin, 275–296.

[151] Breyer, L. A., Hart, A. G. (2000) Approximations of quasi-stationary distributions for Markov chains. In: *Stochastic Models in Engineering, Technology, and Management*, Gold Coast, 1996. *Math. Comput. Modell.*, **31**, no. 10–12, 69–79.

[152] Brockwell, P. J. (1985) The extinction time of a birth, death and catastrophe process and of a related diffusion model. *Adv. Appl. Probab.*, **17**, 42–52.

[153] Brockwell, P. J., Gani, J., Resnick, S. I. (1982) Birth, immigration and catastrophe processes. *Adv. Appl. Probab.*, **14**, 709–731.

[154] Brockwell, P. J., Gani, J., Resnick, S. I. (1982) Catastrophe processes with continuous state-space. *Austral. J. Statist.*, **25**, no. 2, 208–226.

[155] Brown, M. (1975) The first passage time distribution for a parallel exponential system with repair. In: *Reliability and Fault Tree Analysis*, Univ. Berkeley, CA, 1974. Soc. Indust. Appl. Math., Philadelphia, 365–396.

[156] Brown, M. (1983) Approximating IMRL distributions by exponential distributions, with applications to first passage times. *Ann. Probab.*, **11**, 419–427.

[157] Brown, M. (1984) On the reliability of repairable systems. *Oper. Res.*, **32**, no. 3, 607–615.

[158] Brown, M. (1990) Error bounds for exponential approximations of geometric convolutions. *Ann. Probab.*, **18**, 1388–1402.

[159] Brown, M. G., Guang Ping (1984) On the waiting time for the first occurrence of a pattern. In: *Reliability Theory and Models*, Charlotte, N.C., 1983. Notes Rep. Comput. Sci. Appl. Math., **10**, Academic Press, Orlando, FL, 267–272.

[160] Brown, M., Shao, Y. (1987) Identifying coefficients in spectral representation for first passage-time distributions. *Prob. Eng. Inf. Sci.*, **1**, 69–74.

[161] Buiculescu, M. (1972) Quasi-stationary distributions for continuous-time Markov processes with a denumerable set of states. *Rev. Roum. Math. Pures Appl.*, **17**, 1013–1023.

[162] Buiculescu, M. (1975) On quasi-stationary distributions for multi-type Galton-Watson processes. *J. Appl. Probab.*, **12**, 60–68.

[163] Burnley, C. (1987) Perturbation of Markov chains. *Math. Mag.*, **60**, no. 1, 21–30.

[164] Butko, T. K., Korolyuk, V. S. (1984) The potential method in the investigation of a queueing system of type $GI|M|1|\infty$. In: *Analytic Methods in Problems of Probability Theory.* Akad. Nauk Ukr. SSR, Inst. Mat., Kiev, 16–27.

[165] Cairns, B., Pollett, P. K. (2004) Extinction times for a general birth, death and catastrophe process. *J. Appl. Probab.,* **41**, 1211–1218.

[166] Callaert, H., Keilson, J. (1973) On exponential ergodicity and spectral structure for birth-death processes. I, II. *Stoch. Process. Appl.,* **1**, 187–216, 217–235.

[167] Cao, W.-L., Stewart, W. J. (1985) Iterative aggregation/disaggregation techniques for nearly uncoupled Markov chains. *J. ACM,* **32**, 702–719.

[168] Cao, Xi-Ren (1998) The Malaurin series for performance functions of Markov chains. *Adv. Appl. Prob.,* **30**, 676–692.

[169] Cavender, J. A. (1978) Quasistationary distributions for birth-death processes. *Adv. Appl. Probab.,* **10**, 570–586.

[170] Chen, M.-F. (1991) Exponential L^2-convergence and L^2-spectral gap for Markov processes. *Acta. Math. Sinica,* New Ser., **7**, 19–37.

[171] Chen, M.-F. (1996) Estimation of spectral gap for Markov chains. *Acta. Math. Sinica,* New Ser., **12**, 337–360.

[172] Chen, M.-F. (2005) *Eigenvalues, Inequalities, and Ergodic Theory.* Probability and its Applications, Springer, London.

[173] Chen, Dayue; Feng, Jianfeng; Qian, Minping (1996) Metastability of exponentially perturbed Markov chains. *Sci. China,* Ser. A **39**, no. 1, 7–28.

[174] Cheong, C. K. (1967) Geometric convergence of semi-Markov transition probabilities. *Z. Wahrsch. Verw. Gebiete,* **7**, 122–130.

[175] Cheong, C. K. (1972) Quasi-stationary distributions for the continuous-time Galton-Watson process. *Bull. Soc. Math. Belg.,* **24**, 343–350.

[176] Cheong, C. K., Teugels, J. L. (1972) General solidarity theorems for semi-Markov processes. *J. Appl. Probab.,* **9**, 789–802.

[177] Cheong, C. K., Teugels, J. L. (1973) On a semi-Markov generalization of the random walk. *Stoch. Process. Appl.,* **1**, 53–66.

[178] Choquet, G., Deny, J. (1960) Sur l'équation de convolution $\mu = \mu * \sigma$. *C.R. Acad. Sci. Paris,* **250**, 799–801.

[179] Choudhury, G. L., Whitt, W. (1994) Heavy-traffic asymptotic expansions for the asymptotic decay rates in the $BMAP/G/1$ queue. *Comm. Statist. Stoch. Models,* **10**, no. 2, 453–498.

[180] Chow, Y. S., Hsiung, C. A. (1976) Limiting behavior of $\max_{j \le n} S_j / j^{-\alpha}$ and first passage time in random walk with positive drift. *Bull. Inst. Math. Acad. Sinica,* **4**, 35–44.

[181] Chow, Y. S., Hsiung, C. A., Yu, K. F. (1980) Limit theorems for a positively drifting process and its related first passage times. *Bull. Inst. Math. Acad. Sinica,* **8**, 141–172.

[182] Chung, Kai Lai (1953) Contributions to the theory of Markov chains. *J. Research Nat. Bur. Standards*, **50**, 203–208.

[183] Chung, Kai Lai (1954) Contributions to the theory of Markov chains II. *Trans. Amer. Math. Soc.*, **76**, 397–419.

[184] Chung, Kai Lai (1960, 1967) *Markov Chains with Stationary Transition Probabilities.* Fundamental Principles of Mathematical Sciences, **104**, Springer, Berlin.

[185] Çinlar, E. (1967) Queues with semi-Markovian arrivals. *J. Appl. Probab.*, **4**, 365–379.

[186] Çinlar, E. (1969a) Markov renewal theory. *Adv. Appl. Probab.*, **1**, 123–187.

[187] Çinlar, E. (1969b) On semi-Markov processes on arbitrary spaces. *Proc. Cambridge Philos. Soc.*, **66**, 381–392.

[188] Çinlar, E. (1974a) Periodicity in Markov renewal theory. *Adv. Appl. Probab.*, **6**, 61–78.

[189] Çinlar, E. (1974b) Markov renewal theory: a survey. *Manag. Sci.*, **21**, no. 7, 727–752.

[190] Çinlar, E. (1975) *Introduction to Stochastic Processes.* Prentice-Hall, Englewood Cliffs, N.J.

[191] Clancy, D., O'Neill, P. D., Pollett, P. K. (2001) Approximations for the long-term behavior of an open-population epidemic model. *Methodol. Comput. Appl. Probab.*, **3**, no. 1, 75–95.

[192] Clancy, D., Pollett, P. K. (2003) A note on quasi-stationary distributions of birth-death processes and the SIS logistic epidemic. *J. Appl. Probab.*, **40**, 821–825.

[193] Cocozza-Thivent, C., Kipnis, C., Roussignol, M. (1983) Stabilité de la récurrence nulle pour certaines chanes de Markov perturbées. *J. Appl. Probab.*, **20**, 482–504.

[194] Coderch, M., Willsky, A. S., Sastry, S. S., Castanon, D. A. (1983a) Hierarchical aggregation of singulary perturbed finite state Markov processes. *Stochastics*, **8**, 259–289.

[195] Coderch, M., Willsky, A. S., Sastry, S. S., Castanon, D. A. (1983b) Hierarchical aggregation of linear systems with multiple time scales. *IEEE Trans. Autom. Contr.*, AC **28**, 1017–1030.

[196] Cohen, J. E. (1979) Ergodic theorems in demography. *Bull. Amer. Math. Soc. (N.S.)*, **1**, no. 2, 275–295.

[197] Cole, J. D. (1968) *Perturbation Methods in Applied Mathematics.* Blaisdell, Waltham, Mass.

[198] Collet, P., Martinez, S., Schmitt, B. (1996) Quasi-stationary distribution and Gibbs measure of expanding systems. In: *Instabilities and Nonequilibrium Structures*, Santiago, 1993. Nonlinear Phenom. Complex Systems, **1**, Kluwer, Dordrecht, 205–219.

[199] Coolen-Schrijner, P., Hart, A., Pollett, P. (1999) Quasi-stationarity of discrete-time Markov chains with drift to infinity. *Methodol. Comput. Appl. Probab.*, **1**, no. 1, 81–96.

[200] Coolen-Schrijner, P., Hart, A., Pollett, P. (2000) Quasistationarity of continuous-time Markov chains with positive drift. *J. Austral. Math. Soc.*, Ser. B **41**, no. 4, 423–441.

[201] Coolen-Schrijner, P., van Doorn, E. A. (2006a) Quasi-stationary distributions for a class of discrete-time Markov chains. *Methodol. Comput. Appl. Probab.*, **8**, no. 4, 449–465.

[202] Coolen-Schrijner, P., van Doorn, E. A. (2006b) Quasi-stationary distributions for birth-death processes with killing. *J. Appl. Math. Stoch. Anal.*, Art. ID 84640, 15 pp.

[203] Corcoran, J. N., Tweedie, R. L. (2001) Perfect sampling of ergodic Harris chains. *Ann. Appl. Probab.*, **11**, 438–451.

[204] Courtois, P. J. (1975) Error analysis in nearly-completely decomposable stochastic systems. *Econometrica*, **43**, no. 4, 691–709.

[205] Courtois, P. J. (1977) *Decomposibility: Queueing and Computer System Applications.* ACM Monograph Series, Academic Press, New York.

[206] Courtois, P. J., Louchard, G. (1976) Approximation of eigencharacteristics in nearly-completely decomposable stochastic systems. *Stoch. Process. Appl.*, **4**, 283–296.

[207] Courtois, P. J., Semal, P. (1984a) Error bounds for the analysis by decomposition of non-negative matrices. In: Iazeolla, G., Courtois, P. J., Hordijk, A. (eds) *Mathematical Computer Performance and Reliability.* North-Holland, Amsterdam, 209–224.

[208] Courtois, P. J., Semal, P. (1984b) Block decomposition and iteration in stochastic matrices. *Philips J. Res.*, **39**, no. 4–5, 178–194.

[209] Courtois, P. J., Semal, P. (1984c) Bounds for the positive eigenvectors of nonnegative matrices and for their approximations by decomposition. *J. Assoc. Comput. Mach..* **31**, no. 4, 804–825.

[210] Courtois, P. J., Semal, P. (1991) Bounds for transient characteristics of large or infinite Markov chains. In: Stewart, W. J. (ed) *Numerical Solution of Markov Chains.* Probability: Pure and Applied, **8**, Marcel Dekker, New York, 413–434.

[211] Cox, D. R. (1962) *Renewal Theory.* Methuen, London and Wiley, New York.

[212] Cox, D. R., Smith, W. L. (1953) A direct proof of a fundamental theorem of renewal theory. *Skand. Aktuarietidskr.*, **36**, 139–150.

[213] Cox, D. R., Smith, W. L. (1961) *Queues.* Methuen's Monographs on Statistical Subjects, Methuen, London and Wiley, New York.

[214] Cramér, H. (1926) Review of F. Lundberg. *Skand. Aktuarietidskr.*, **9**, 223–245.

[215] Cramér, H. (1930) *On the Mathematical Theory of Risk.* Skandia Jubilee Volume, Stockholm.

[216] Cramér, H. (1955) *Collective Risk Theory.* Skandia Jubilee Volume, Stockholm.

[217] Craven, B. D. (2003) Perturbed Markov processes. *Stoch. Models*, **19**, no. 2, 269–285.

[218] Daley, D. (1968) Quasi-stationary behaviour of a left-continuous random walk. *Ann. Math. Statist.*, **40**, 532–539.

[219] Daley, D. J., Gani, J. (1999) *Epidemic Modelling: an Introduction.* Cambridge Studies in Mathematical Biology, **15**, Cambridge University Press, Cambridge.

[220] Darlington, S. J., Pollett, P. K. (2000) Quasistationarity in continuous-time Markov chains where absorption is not certain. *J. Appl. Probab.*, **37**, 598–600.

[221] Darroch, J. N., Morris, K. W. (1967) Some passage-time generating functions for discrete-time and continuous-time finite Markov chains. *J. Appl. Probab.*, **4**, 496–507.

[222] Darroch, J., Seneta, E. (1965) On quasi-stationary distributions in absorbing discrete-time finite Markov chains. *J. Appl. Probab.*, **2**, 88–100.

[223] Darroch, J., Seneta, E. (1967) On quasi-stationary distributions in absorbing continuous-time finite Markov chains. *J. Appl. Probab.*, **4**, 192–196.

[224] Day, J. R., Possingham, H. P. (1995) A stochastic metapopulation model with variability in patch size and position. *Theor. Popul. Biol.*, **48**, 333–360.

[225] Daykin, C. D., Pentikäinen, T., Pesonen, M. (1994) *Practical Risk Theory for Actuaries.* Monographs on Statistics and Applied Probability, **53**, Chapman & Hall, London.

[226] Delebecque, F. (1983a) Perturbation d'une chane de Markov. In: *Mathematical Tools and Models for Control, Systems Analysis and Signal Processing*, Vol. 3, Toulouse/Paris, 1981/1982. Travaux Rech. Coop. Programme, **567**, CNRS, Paris, 241–258.

[227] Delebecque, F. (1983b) A reduction process for perturbed Markov chains. *SIAM J. Appl. Math.*, **43**, 325–330.

[228] Delebecque, F., Quadrat J. P. (1981a) The optimal cost expansion of finite controls finite states Markov chains with weak and strong interactions. In: *Analysis and Optimization of Systems*, Versailles, 1980. Lecture Notes in Control and Information Sci., **28**, Springer, Berlin-New York, 322–337.

[229] Delebecque, F., Quadrat J. P. (1981b) Optimal control of Markov chains admitting strong and weak interactions. *Automatika*, **17**, 281–296.

[230] Delebecque, F., Quadrat, J. P., Kokotović, P. V. (1984) A unified view of aggregation and coherency in networks and Markov chains. *Inter. J. Control*, **40**, 939–952.

[231] de Vylder, F. (1978) A practical solution to the problem of ultimate ruin probability. *Scand. Actuar. J.*, 114–119.

[232] de Vylder, F. (1996) *Advanced Risk Theory. A Self-Contained Introduction.* Editions de l'Université de Bruxelles and Swiss Association of Actuaries, Zürich.

[233] de Vylder, F., Marceau, E. (1996) Classical numerical ruin probabilities. *Scand. Actuar. J.*, no. 2, 109–123.

[234] Dickman, R., Vidigal, R. (2002) Quasi-stationary distributions for stochastic processes with an absorbing state. *J. Phys.*, A **35**, no. 5, 1147–1166.

[235] Dickson, D. C. M. (1992) On the distribution of the surplus prior to ruin. *Insur. Math. Econom.*, **11**, no. 3, 191–207.

[236] Dickson, D. C. M. (2005) *Insurance Risk and Ruin.* International Series on Actuarial Science, Cambridge University Press, Cambridge.

[237] Dickson, D. C. M., Dos Reis, A. D. E., Waters, H. R. (1995) Some stable algorithms in ruin theory and their applications. *ASTIN Bull.*, **25**, 153–175.

[238] Diekmann, O., Heesterbeek, J. A. P. (2000) *Mathematical Epidemiology of Infectious Diseases: Model Building, Analysis and Interpretation.* Wiley Series in Mathematical and Computational Biology, Wiley, Chichester.

[239] Doeblin, W. (1936) Sur les chaînes de Markoff. *C. R. Acad. Sci. Paris*, **203**, no. 1, 24–26.

[240] Doeblin, W. (1937) Sur deux propiétés asymptotiques de mouvement régis par certains tupes de chaînes simples. *Bull. Math. Soc. Raum. Sci.*, **39**, no. 1, 57–115, no. 2, 3–61.

[241] Doeblin, W. (1938) Sur deux problems de Kolmogoroff concentrat les chaînes décombrables. *Bull. Soc. Math. France*, **66**, 210–220.

[242] Doeblin, W. (1940) Éléments d'une théorie générale des chaînes simples constantes de Markoff. *Ann. École Norm.*, (3) **57**, 61–111.

[243] Doob, J. L. (1953, 1990) *Stochastic Processes.* Wiley, New York; Chapman & Hall, London and Wiley Classics Library, A Wiley-Interscience Publication, Wiley, New York.

[244] Down, D., Meyn, S. P. (1995) Stability of acyclic multiclass queueing networks. *IEEE Trans. Automat. Control*, **40**, no. 5, 916–919.

[245] Down, D., Meyn, S. P., Tweedie, R. L. (1995) Exponential and uniform ergodicity of Markov processes. *Ann. Probab.*, **23**, 1671–1691.

[246] Dufresne, F., Gerber, H. U. (1988) The surpluses immediately before and at ruin, and the amount of the claim causing ruin. *Insur. Math. Econom.*, **7**, no. 2, 193–199.

[247] Durrett, R. (1978) Conditioned limit theorems for some null recurrent Markov processes. *Ann. Probab.*, **6**, 798–828.

[248] Durrett, R. (1980) Conditioned limit theorems for random walks with negative drift. *Z. Wahrsch. Verw. Gebiete*, **52**, 277–287.

[249] Durrett, R., Neuhauser, C. (1991) Epidemics with recovery in $D = 2$. *Ann. Appl. Probab.*, **1**, 189–206.

[250] Dynkin, E. B. (1955) Some limit theorems for sums of independent random variables with infinite mathematical expectations. *Izv. Acad. Nauk SSSR*, Ser. Mat., **19**, 247–266 (English translation in *Select. Transl. Math. Statist. Probab.*, **1**, 171–189 (1961)).

[251] Dynkin, E. B. (1963) *Markov Processes.* Fizmatgiz, Moscow (English edition: *Markov Processes*, Vols. I, II. Academic Press, New York and Springer, Berlin).

[252] Eleĭko, Ya.I. (1980) Limit distributions for a semi-Markov process with arbitrary phase space. *Teor. Veroyatn. Mat. Stat.*, **23**, 51–58 (English translation in *Theory Probab. Math. Statist.*, **23**, 55–61).

[253] Eleĭko, Ya. I. (1990) Limiting distributions of time averages of additive functionals defined on a semi-Markov process. *Ukr. Mat. Zh.*, **42**, 843–847 (English translation in *Ukr. Math. J.*, **42**, 744–747).

[254] Eleĭko, Ya. I. (1998) Some refinements of limit theorems for semi-Markov processes with a general phase space. *Teor. Ĭmorvirn. Mat. Stat.*, **59**, 57–65 (English translation in *Theory Probab. Math. Statist.*, **59**, 57–65).

[255] Eleĭko, Ya. I., Shurenkov, V. M. (1995a) Some properties of random evolutions. *Ukr. Mat. Zh.*, **47**, 1333–1337 (English translation in *Ukr. Math. J.*, **47**, 1519–1525).

[256] Eleĭko, Ya. I., Shurenkov, V. M. (1995b) Transient phenomena in a class of matrix-valued stochastic evolutions. *Teor. Ĭmorvirn. Mat. Stat.*, **52**, 72–76 (English translation in *Theory Probab. Math. Statist.*, **52**, 75–79).

[257] Eleĭko, Ya. I., Shurenkov, V. M. (1996) On the asymptotic representation of the Perron root of a matrix-valued stochastic evolution. *Ukr. Mat. Zh.*, **48**, 35–43 (English translation in *Ukr. Math. J.*, **48**, 38–47).

[258] Elesin, M. A., Kuznetsov, A. V., Zeĭfman, A. I. (2004) On the rate of convergence for some birth and death processes. *J. Math. Sci.*, **122**, no. 4, 3359–3364.

[259] Elmes, S., Pollett, P., Walker, D. (2000) Further results on the relationship between μ-invariant measures and quasi-stationary distributions for absorbing continuous-time Markov chains. In: *Stochastic Models in Engineering, Technology, and Management*, Gold Coast, 1996. *Math. Comput. Modell.*, **31**, no. 10–12, 107–113.

[260] Embrechts, P., Goldie, C. M., Veraverbeke, N. (1979) Subexponentiality and infinite divisibility. *Z. Wahrsch. Verw. Gebiete*, **49**, 335–347.

[261] Embrechts, P., Klüppelberg, C. (1993) Some aspects of insurance mathematics. *Teor. Veroyatn. Primen.*, **38**, 374–416 (Also in *Theory Probab. Appl.*, **38**, 262–295).

[262] Embrechts, P., Klüppelberg, C., Mikosch, T. (1997) *Modelling Extremal Events for Insurance and Finance*. Applications of Mathematics, **33**, Springer, Berlin.

[263] Embrechts, P., Veraverbeke, N. (1982) Estimates for the probability of ruin with special emphasis on the possibility of large claims. *Insur. Math. Econom.*, **1**, no. 1, 55–72.

[264] Englund, E. (1999a) Perturbed renewal equations with application to M/M queueing systems. 1. *Teor. Ĭmovirn. Mat. Stat.*, **60**, 31–37 (Also in *Theory Probab. Math. Statist.*, **60**, 35–42).

[265] Englund, E. (1999b) Perturbed renewal equations with application to M/M queueing systems. 2. *Teor. Ĭmovirn. Mat. Stat.*, **61**, 21–32 (Also in *Theory Probab. Math. Statist.*, **61**, 21–32).

[266] Englund, E. (2000) Nonlinearly perturbed renewal equations with applications to a random walk. In: Silvestrov, D., Yadrenko, M., Olenko A., Zinchenko, N. (eds) *Proceedings of the Third International School on Applied Statistics, Financial and Actuarial Mathematics*, Feodosiya, 2000. *Theory Stoch. Process.*, **6(22)**, no. 3–4, 33–60.

[267] Englund, E. (2001) Nonlinearly Perturbed Renewal Equations with Applications. Ph.D. Thesis, Umeå University.

[268] Englund, E., Silvestrov, D. S. (1997) Mixed large deviation and ergodic theorems for regenerative processes with discrete time. In: Jagers, P., Kulldorff, G., Portenko, N., Silvestrov, D. (eds) *Proceedings of the Second Scandinavian–Ukrainian Conference in Mathematical Statistics*, Vol. I, Umeå, 1997. *Theory Stoch. Process.*, **3(19)**, no. 1–2, 164–176.

[269] Enikeeva, F., Kalashnikov, V., Rusaityte, D. (2001) Continuity estimates for ruin probabilities. *Scand. Actuar. J.*, no. 1, 18–39.

[270] Erdélyi, A. (1956) *Asymptotic Expansions*. Dover, New York.

[271] Erdös, W., Feller, W., Pollard, H. (1949) A theorem on power series. *Bul. Amer. Math. Soc.*, **55**, 201–204.

[272] Esscher, F. (1932) On the probability function in the collective theory of risk. *Skand. Aktuarietidskr.*, **15**, 175–195.

[273] Ethier, S. N., Kurtz, T. G. (1986) *Markov Processes. Characterization and Convergence*. Wiley Series in Probability and Mathematical Statistics: Probability and Mathematical Statistics, Wiley, New York.

[274] Ethier, S. N., Kurtz, T. G. (1988) Coupling and ergodic theorems for Fleming-Viot processes. *Ann. Probab.*, **26**, 533–561.

[275] Ezhov, I. I., Shurenkov, V. M. (1976) Ergodic theorems connected with the Markov property of random processes. *Teor. Veroyatn. Primen.*, **21**, 635–639 (English translation in *Theory Probab. Appl.*, **21**, 620–623).

[276] Farrell, R. H. (1962) Asymptotic renewal theorems in the absolutely continuous case. *Duke Math. J.*, **29**, 33–40.

[277] Farrell, R. H. (1964) Limit theorems for stopped random walks. *Ann. Math. Statist.*, **35**, 1332–1343.

[278] Farrell, R. H. (1966a) Limit theorems for stopped random walks. II. *Ann. Math. Statist.*, **37**, 860–865.

[279] Farrell, R. H. (1966b) Limit theorems for stopped random walks. III. *Ann. Math. Statist.*, **37**, 1510–1527.

[280] Feller, W. (1941) On the integral equation of renewal theory. *Ann. Math. Statist.*, **12**, 243–267.

[281] Feller, W. (1949) Fluctuation theory of recurrent events. *Trans. Amer. Math. Soc.*, **67**, 98–119.

[282] Feller, W. (1950, 1957, 1968) *An Introduction to Probability Theory and Its Applications*, Vol. I. Wiley Series in Probability and Statistics, Wiley, New York.

[283] Feller, W. (1961) A simple proof for renewal theorems. *Comm. Pure Appl. Math.*, **14**, 285–293.

[284] Feller, W. (1966, 1971) *An Introduction to Probability Theory and Its Applications*, Vol. II. Wiley Series in Probability and Statistics, Wiley, New York.

[285] Feller, W., Orey, S. (1961) A renewal theorem. *J. Math. Mech.*, **10**, 619–624.

[286] Ferrari, P. A., Galves, A., Landim, C. (1994) Exponential waiting time for a big gap in a one-dimensional zero-range process. *Ann. Probab.*, **22**, 284–288.

[287] Ferrari, P. A., Galves, A., Liggett, T. M. (1995) Exponential waiting time for filling a large interval in the symmetric simple exclusion process. *Ann. Inst. H. Poincaré Probab. Statist.*, **31**, no. 1, 155–175.

[288] Ferrari, P. A., Garcia, N. L. (1998) One-dimensional loss networks and conditioned $M/G/\infty$ queues. *J. Appl. Probab.*, **35**, 963–975.

[289] Ferrari, P. A., Kesten, H., Martinez, S. (1996) R-positivity, quasi-stationary distributions and ratioo limit theorems for a class of probabilistic automata. *Ann. Appl. Probab.*, **6**, 577–616.

[290] Ferrari, P. A., Kesten, H., Martinez, S., Picco, P. (1995) Existence of quasi-stationary distributions. A renewal dynamical approach. *Ann. Probab.*, **23**, 501–521.

[291] Ferrari, P. A., Marić, N. (2007) Quasi stationary distributions and Fleming-Viot processes in countable spaces. *Electron. J. Probab.*, **12**, no. 24, 684–702.

[292] Ferrari, P. A., Martinez, S., Picco, P. (1991) Some properties of quasi-stationary distributions in the birth and death chains: a dynamical approach. In: *Instabilities and Nonequilibrium Structures*, III, Valparaiso, 1989. Mathematics and its Applications, **64**, Kluwer, Dordrecht, 177–187.

[293] Ferrari, P. A., Martinez, S., Picco, P. (1992) Existence of non-trivial quasi-stationary distributions in the birth-death chain. *Adv. Appl. Probab.*, **25**, 82–102.

[294] Ferrari, P. A., Martinez, S., San Martin, J. (1997) Phase transition for absorbed Brownian motion with drift. *J. Statist. Phys.*, **86**, no. 1–2, 213–231.

[295] Flaspohler, D. C. (1974) Quasi-stationary distributions for absorbing continuous-time denumerable Markov chains. *Ann. Inst. Statist. Math.*, **26**, 351–356.

[296] Flaspohler, D. C., Holmes, P. T. (1972) Additional quasi-stationary distributions for semi-Markov processes. *J. Appl. Probab.*, **9**, 671–676.

[297] Foguel, S. R. (1969) *The Ergodic Theory of Markov Processes*. Van Nostrand Mathematical Studies, **21**, Van Nostrand, New York.

[298] Fort, G., Moulines, E. (2003) Polynomial ergodicity of Markov transition kernels. *Stoch. Process. Appl.*, **103**, 57–99.

[299] Franken, P., Kirstein, B.-M., Streller, A. (1984) Reliability analysis of complex systems with repair. *Elektron. Infor. Kybern.*, **20**, no. 7–9, 407–422.

[300] Franken, P., König, D., Arndt, U., Schmidt, V. (1982) *Queues and Point Processes*. Wiley Series in Probability and Mathematical Statistics: Applied Probability and Statistics, Wiley, Chichester.

[301] Freyer, B. (1970) Die Klasse der Grenzverteilungen von Summen gleichverteilter Zufallsgrößen mit einer zufälligen Anzahl von Summanden. *Math. Nachr.*, **44**, 341–350.

[302] Frostig, E. (2002) Comparison between future lifetime distribution and its approximations. *Noth. Amer. Actuar. J.*, **6**, no. 2, 11–17.

[303] Fuh, Cheng-Der (2004) Uniform Markov renewal theory and ruin probabilities in Markov random walks. *Ann. Appl. Probab.*, **14**, 1202–1241.

[304] Fuh, Cheng-Der (2007) Asymptotic expansions on moments of the first ladder height in Markov random walks with small drift. *Adv. Appl. Probab.*, **39**, 826–852.

[305] Fuh, Cheng-Der; Lai, Tze Leung (2001) Asymptotic expansions in multidimensional Markov renewal theory and first passage times for Markov random walks. *Adv. Appl. Probab.*, **33**, 652–673.

[306] Gaitsgory, V. G., Pervozvansky, A. A. (1975) Agregation of systems in Markov chains with weak interactions. *Kibernetika*, no. 11, 91–98.

[307] Gani, J., Lehoczky, J. (1971) An asymptotic result in traffic theory. *J. Appl. Probab.*, **8**, 815–820.

[308] Gani, J. (1971) Point processes in epidemiology. In: *Stochastic Point Processes: Statistical Analysis, Theory, and Applications*, Conf., IBM Res. Center, Yorktown Heights, New York, 1971, 756–773.

[309] Gani, J. (1972) First emptiness problems in queueing, storage, and traffic theory. In: *Proceedings of the Sixth Berkeley Symposium on Mathematical Statistics and Probability*, Vol. III, Berkeley, CA, 1970/1971. Univ. California Press, Berkeley, CA, 515–532.

[310] Gani, J.; Tin, Pyke (1985) On a class of continuous-time Markov processes. *J. Appl. Probab.*, **22**, 804–815.

[311] Garsia, A. M. (1963) Some tauberian therems and the asymptotic behavior of probabilities of reccurent events. *J. Math. Analysis Appl.*, **7**, no. 1, 146–162.

[312] Garsia, A., Lamperti, J. (1962/1963) A discrete renewal theorem with infinite mean. *Comment. Math. Helv.*, **37**, 221–234.

[313] Gasanenko, V. A. (1980) On thinning a semi-Markov process with a countable set of states. *Teor. Veroyatn. Mat. Stat.*, **22**, 25–29 (English translation in *Theory Probab. Math. Statist.*, **22**, 25–29).

[314] Gasanenko, V. A. (1983) Thinning processes. *Ukr. Mat. Zh.*, **35**, 27–30 (English translation in *Ukr. Math. J.*, **35**, 24–27).

[315] Gasanenko, V. A. (1991) Applications of thinning processes. *Ann. Univ. Sci. Budapest*, Sect. Comput., **11**, 35–51.

[316] Gasanenko, V. A. (1998a) A limit theorem for thinning processes with mixing. I. *Ukr. Mat. Zh.*, **50**, 471–475 (English translation in *Ukr. Math. J.*, **50**, 533–538).

[317] Gasanenko, V. A. (1998b) A limit theorem for thinning processes with mixing. II. *Ukr. Mat. Zh.*, **50**, 603–612 (English translation in *Ukr. Math. J.*, **50**, 685–696).

[318] Gasanenko, V. A. (1999) Normed Poisson process in one-sided curvelinear boundary domain. In: *Proceedings of the Third Ukrainian-Scandinavian Conference in Probability Theory and Mathematical Statistics*, Kiev, 1999. *Theory Stoch. Process.*, **5(21)**, no. 3–4, 64–71.

[319] Gasanenko, V. A. (2001) On the asymptotics of the sojourn probability of a Poisson process between two divergent nonlinear boundaries. *Ukr. Mat. Zh.*, **53**, 14–22 (English translation in *Ukr. Math. J.*, **53**, 15–25).

[320] Gasanenko, V. A. (2003) On asymptotic independence of the exit moment and position from a small domain for diffusion processes. *Cent. Eur. J. Math.*, **1**, no. 1, 86–96.

[321] Gaver, D. P., Jr. (1964) An absorption probability problem. *J. Math. Anal. Appl.*, **9**, 384–393.

[322] Gaver, D. P. (1976) Random record models. *J. Appl. Probab.*, **13**, 538–547.

[323] Gaver, D. P., Jacobs, P. A. (1999) Waiting times when service times are stable laws: tamed and wild. In: *Applied Probability and Stochastic Processes*. Internat. Ser. Oper. Res. Management Sci., **19**, Kluwer, Boston, 219–229.

[324] Gaver, D. P., Jacobs, P. A., Latouche, G. (1984) Finite birth-and-death models in randomly changing environments. *Adv. Appl. Probab.*. **16**, 715–731.

[325] Gerber, H. U. (1970) An extension of the renewal equation and its application in the collective theory of risk. *Skand. Aktuarietidskr.*, 205–210.

[326] Gerber, H. U. (1979) *An Introduction to Mathematical Risk Theory*. Huebner Foundation Monographs, Philadelphia.

[327] Gerber, H. U., Goovarerts, M. J., Kaas, R. (1987) On the probability of severity of ruin. *ASTIN Bull.*, **17**, 151–163.

[328] Gibson, D., Seneta, E. (1987) Augmented truncations of infinite stochastic matrices. *J. Appl. Probab.*, **24**, 600–608.

[329] Gikhman, I. I., Skorokhod, A. V. (1965) *Introduction to the Theory of Random Processes*. Probability Theory and Mathematical Statistics, Nauka, Moscow (English editions: W. B. Saunders Co., Philadelphia (1969) and Dover, Mineola, NY (1996)).

[330] Gikhman, I. I., Skorokhod, A. V. (1971) *Theory of Random Processes*. 1. Probability Theory and Mathematical Statistics, Nauka, Moscow (English editions: *The Theory of Stochastic Processes*. I. Classics in Mathematics, Springer, Berlin (2004) and Fundamental Principles of Mathematical Sciences, **210**, Springer, New York (1974) and Berlin (1980)).

[331] Gikhman, I. I., Skorokhod, A. V. (1973) *Theory of Random Processes*. 2. Probability Theory and Mathematical Statistics, Nauka, Moscow (English editions: *The Theory of Stochastic Processes*. II. Classics in Mathematics, Springer, Berlin (2004) and Fundamental Principles of Mathematical Sciences, **218**, Springer, New York (1975)).

[332] Gikhman, I. I., Skorokhod, A. V. (1975) *Theory of Random Processes*. 3. Probability Theory and Mathematical Statistics, Nauka, Moscow (English editions: *The Theory of Stochastic Processes*. III. Classics in Mathematics, Springer, Berlin (2007) and Fundamental Principles of Mathematical Sciences, **232**, Springer, Berlin (1979)).

[333] Gilpin, M., Taylor, B. L. (1994) Reduced dimensional population transition matrices: Extinction distributions from Markovian dynamics. *Theor. Popul. Biol.*, **46**, 121–130.

[334] Glynn, P. W. (1990) Diffusion approximations. In: Heyman, D. P., Sobel, M. J. (eds) *Stochastic Models*. Handbooks Oper. Res. Management Sci., Vol. 2, North-Holland, Amsterdam, 145–198.

[335] Glynn, P. W., Meyn, S. P. (1996) A Liapounov bound for solutions of the Poisson equation. *Ann. Probab.*, **24**, 916–931.

[336] Glynn, P. W., Thorisson, H. (2001) Two-sided taboo limits for Markov processes and associated perfect simulation. *Stoch. Process. Appl.*, **91**, 1–20.

[337] Glynn, P. W., Thorisson, H. (2002) Structural characterization of taboo-stationarity for general processes in two-sided time. *Stoch. Process. Appl.*, **91**, 311–318.

[338] Glynn, P. W., Thorisson, H. (2004) Limit theory for taboo-regenerative processes. *Queueing Syst.*, **46**, no. 3–4, 271–294.

[339] Glynn, P. W., Whitt, W. (1994) Logarithmic asymptotics for steady-state tail probabilities in a single-server queue. In: Galambos, J., Gani, J. (eds) *Studies in Applied Probability. Essays in Honor of Lajos Takács. J. Appl. Probab.*, **31A**, 131–156.

[340] Gnedenko, B. V. (1964a) On non-loaded duplication. *Izv. Akad. Nauk SSSR*, Tekhn. Kibernet., no. 4, 3–12.

[341] Gnedenko, B. V. (1964b) Doubling with renewal. *Izv. Akad. Nauk SSSR*, Tekhn. Kibernet., no. 5, 111–118.

[342] Gnedenko, B. V. (1967a) Some theorems on standbys. In: *Proceedings Fifth Berkeley Sympos. Mathematical Statistics and Probability*, Vol. III, Berkeley, CA, 1965/66. Univ. California Press, Berkeley, CA, 285–291.

[343] Gnedenko, B. V. (1967b) The connection between the theory of summation of independent random variables and the problems of queueing theory and reliability theory. *Rev. Roum. Math. Pures Appl.*, **12**, 1243–1253.

[344] Gnedenko, B. V., Frayer, B. (1969) A few remarks on a result by I.N. Kovalenko. *Litov. Mat. Sbornik*, **9**, 463–470.

[345] Gnedenko, B. V., Korolev, V. Yu. (1996) *Random Summation. Limit Theorems and Applications*. CRC Press, Boca Raton, FL.

[346] Gnedenko, B. V., Kovalenko, I. N. (1966, 1987) *Introduction to Queueing Theory*. Nauka, Moscow (English editions: Israel Program for Scientific Translations, Jerusalem, Daniel Davey & Co., Inc., Hartford, Conn. (1968) and Mathematical Modeling, **5**, Birkhäuser, Boston (1989)).

[347] Gnedenko, D. B., Solov'ev, A. D. (1974) A general model for standby with renewal. *Izv. Akad. Nauk SSSR*, Tekhn. Kibernet., no. 6, 113–118 (English translation in *Engrg. Cybernetics*, **12**, no. 6, 82–86).

[348] Gnedenko, D. B., Solov'ev, A. D. (1975) Estimation of reliability for complex recovering systems. *Izv. Akad. Nauk SSSR*, Tekhn. Kibernet., no. 3, 121–128 (English translation in *Engrg. Cybernetics*, **13**, no. 3, 89–96).

[349] Golub, G. H., Seneta, E. (1974) Computation of the stationary distribution of an infinite stochastic matrix of special form. *Bull. Austral. Math. Soc.*, **10**, 255–261.

[350] Good, P. (1968) The limiting behaviour of transient birth-death processes conditioned on survival. *J. Austral. Math. Soc.*, **8**, 716–722.

[351] Grandell, J. (1977) A class of approximations of ruin probabilities. *Scand. Actuar. J.*, Suppl., 37–52.

[352] Grandell, J. (1978/79) Empirical bounds for ruin probabilities. *Stoch. Process. Appl.*, **8**, 243–255.

[353] Grandell, J. (1980) Approximate waiting times in thinned point processes. *Litov. Mat. Sbornik*, **20**, no. 4, 29–47.

[354] Grandell, J. (1991) *Aspects of Risk Theory*. Probability and Its Applications, Springer, New York.

[355] Grandell, J. (1997) *Mixed Poisson Processes*. Chapman & Hall, London.

[356] Grandell, J., Segerdahl, C.O. (1971) A comparison of some approximations of ruin probabilities. *Skand. Aktuarietidskr.*, 114–158.

[357] Granovskiĭ, B. L., Zeĭfman, A. I. (1997) The decay function of nonhomogeneous birth-death processes, with application to mean-field models. *Stoch. Process. Appl.*, **72**, 105–120.

[358] Granovskiĭ, B. L., Zeĭfman, A. I. (2000) Nonstationary Markovian queues. In: *Proceedings of the 19th Seminar on Stability Problems for Stochastic Models*, Part II, Vologda, 1998. *J. Math. Sci.*, **99**, no. 4, 1415–1438.

[359] Granovskiĭ, B. L., Zeĭfman, A. I. (2004) A lower bound on the spectrum for some mean-field models. *Teor. Veroyatn. Primen.*, **49**, 164–171 (English translation in *Theory Probab. Appl.*, **49**, 148–155).

[360] Gray, A., Pinsky, M. A. (1983) The mean exit time from a small geodesic ball in a Riemannian manifold. *Bull. Sci. Math.* (2) **107**, no. 4, 345–370.

[361] Green, P. J. (1976) The maximum and time to absorption of a left-continuous random walk. *J. Appl. Probab.*, **13**, 444–454.

[362] Grey, D. R. (2001) Renewal theory. In: *Stochastic Processes: Theory and Methods*. Handbook of Statistics, **19**, North-Holland, Amsterdam, 413–441.

[363] Griffeath, D. (1974/75) A maximal coupling for Markov chains. *Z. Wahrsch. Verw. Gebiete*, **31**, 95–106.

[364] Griffeath, D. (1976) Coupling Methods for Markov Processes. Ph.D. Thesis, Cornell University, Ithaca.

[365] Griffeath, D. (1978) Coupling methods for Markov processes. In: *Studies in Probability and Ergodic Theory*. Adv. in Math. Suppl. Stud., **2**, Academic Press, New York-London, 1–43.

[366] Grishechkin, S. A. (1986) On rate of convergence of random sum to non-normal limit distribution. *Vestnik Mosk. Univ.*, Ser. Mat. Meh., no. 6, 64–67.

[367] Guglielmi, A., Tweedie, R. L. (2001) Markov chain Monte Carlo estimation of the law of the mean of a Dirichlet process. *Bernoulli*, **7**, no. 4, 573–592.

[368] Gusak, D. V., Korolyuk, V. S. (1971) Asymptotic behaviour of semi-Markov processes with a decomposable set of states. *Teor. Veroyatn. Mat. Stat.*, **5**, 43–50 (English translation in *Theory Probab. Math. Statist.*, **5**, 43–51).

[369] Gut, A. (1974a) On the moments and limit distributions of some first passage times. *Ann. Probab.*, **2**, 277–308.

[370] Gut, A. (1974b) On convergence in r-mean of some first passage times and randomly indexed partial sums. *Ann. Probab.*, **2**, 321–323.

[371] Gut, A. (1975) Weak convergence and first passage times. *J. Appl. Probab.*, **12**, 324–332.

[372] Gut, A. (1983) Renewal theory and ladder variables. In: *Probability and Mathematical Statistics*. Uppsala University, Uppsala, 25–39.

[373] Gut, A. (1988) *Stopped Random Walks. Limit Theorems and Applications*. Applied Probability. A Series of the Applied Probability Trust, **5**, Springer, New York.

[374] Gut, A. (1990a) Convergence rates for record times and the associated counting process. *Stoch. Process. Appl.*, **36**, 135–151.

[375] Gut, A. (1990b) Limit theorems for record times. In: *Probability Theory and Mathematical Statistics*, Vol. I, Vilnius, 1989. Mokslas, Vilnius, 490–503.

[376] Gut, A., Holst, L. (1984) On the waiting time in a generalized roulette game. *Statist. Probab. Lett.*, **2**, no. 4, 229–239.

[377] Gyllenberg, M., Hanski, I. (1992) Single-species metapopulation dynamics: a structured model. *Theor. Popul. Biol.*, **42**, no. 1, 35–61.

[378] Gyllenberg, M., Silvestrov, D. S. (1994) Quasi-stationary distributions of a stochastic metapopulation model. *J. Math. Biol.*, **33**, 35–70.

[379] Gyllenberg, M., Silvestrov, D. S. (1998) Quasi-stationary phenomena in semi-Markov models. In: Janssen, J., Limnios, N. (eds) *Proceedings of the Second International Symposium on Semi-Markov Models: Theory and Applications*, Compiègne, 1998, 87–93.

[380] Gyllenberg, M., Silvestrov, D. S. (1999a) Quasi-stationary phenomena for semi-Markov processes. In: Janssen, J., Limnios, N. (eds) *Semi-Markov Models and Applications*. Kluwer, Dordrecht, 33–60.

[381] Gyllenberg, M., Silvestrov, D. S. (1999b) Cramér–Lundberg and diffusion approximations for nonlinearly perturbed risk processes including numerical computation of ruin probabilities. In: Silvestrov, D. S., Yadrenko, M., Borisenko, O., Zinchenko, N. (eds) *Proceedings of the Second International School on Actuarial and Financial Mathematics*, Kiev, 1999. *Theory Stoch. Process.*, **5(21)**, no. 1–2, 6–21.

[382] Gyllenberg, M., Silvestrov, D. S. (2000a) Nonlinearly perturbed regenerative processes and pseudo-stationary phenomena for stochastic systems. *Stoch. Process. Appl.*, **86**, 1–27.

[383] Gyllenberg, M., Silvestrov, D. S. (2000b) Cramér–Lundberg approximation for nonlinearly perturbed risk processes. *Insur. Math. Econom.*, **26**, no. 1, 75–90.

[384] Hanen, A. (1963a) Probème central limite dans le cas Markovien fini. La matrice limite n'a qu'une seule classe ergodique et pas d'état transitoire. *C. R. Acad. Sci. Paris*, **256**, 68–70.

[385] Hanen, A. (1963b) Problème central limite dans le cas Markovien fini. II. La matrice limite a plusieurs classes ergodiques et pas d'états transitoires. *C. R. Acad. Sci. Paris*, **256**, 362–364.

[386] Hanen, A. (1963c) Problème central limite dans le cas Markovien fini. Cas général. *C. R. Acad. Sci. Paris*, **256**, 575–577.

[387] Hanen, A. (1963d) Théorèmes limites pour une suite de chaînes de Markov. *Ann. Inst. H. Poincaré*, **18**, 197–301.

[388] Hanski, I. (1999) *Metapopulation Ecology*. Oxford Series in Ecology and Evolution, Oxford University Press, Oxford.

[389] Hanski, I., Gilpin, M. E. (1997) *Metapopulation Biology: Ecology, Genetics, and Evolution*. Academic Press, London.

[390] Harlamov, B. (2008) *Conditinuous Semi-Markov Processes*. Applied Stochastic Methods Series. ISTE and Wiley, London.

[391] Harris, T. E. (1952) First passage and recurrence distributions. *Trans. Amer. Math. Soc.*, **73**, 471–486.

[392] Harris, T. E. (1956) The existence of stationary measures for certain Markov processes. In: *Proceedings of the Third Berkeley Symposium on Mathematical Statistics and Probability*, Vol. II, Berkeley, CA, 1954–1955. Univ. California Press, Berkeley and Los Angeles, 113–124.

[393] Harrison, J. M. (1977) Ruin problems with compound assets. *Stoch. Process. Appl.*, **5**, 67–79.

[394] Hart, A. G., Pollett, P. K. (1996) Direct analytical methods for determining quasistationary distributions for continuous-time Markov chains. In: *Athens Conference on Applied Probability and Time Series Analysis*, Vol. I, Athens, 1995. Lecture Notes in Statistics, **114**, Springer, New York, 116–126.

[395] Hart, A. G., Pollett, P. K. (2000) New methods for determining quasi-stationary distributions for Markov chains. In: *Stochastic Models in Engineering, Technology, and Management*, Gold Coast, 1996. *Math. Comput. Modell.*, **31**, no. 10–12, 143–150.

[396] Hassin, R., Haviv, M. (1992) Mean passage times and nearly uncoupled Markov chains. *SIAM J. Disc. Math.*, **5**, 386–397.

[397] Hassin, R., Haviv, M. (2003) *To Queue or not to Queue: Equilibrium Behavior in Queueing Systems*. Internat. Ser. Oper. Res. Management Sci., **59**, Kluwer, Boston.

[398] Hatori, H. (1959) Some theorems in an extended renewal theory. I. *Kōdai Math. Semin. Repts.*, **11**, no. 3, 139–146.

[399] Hatori, H. (1960) Some theorems in an extended renewal theory. II. *Kōdai Math. Semin. Repts.*, **12**, no. 1, 21–27.

[400] Haviv, M. (1986) An approximation to the stationary distribution of a nearly completely decomposable Markov chain and its error analysis. *SIAM J. Algeb. Discr. Methods*, **7**, no. 4, 589–593.

[401] Haviv, M. (1988) Error bounds on an approximation to the dominant eigenvector of a nonnegative matrix. *Linear Multilin. Algebra*, **23**, no. 2, 159–163.

[402] Haviv, M. (1992) An aggregation/disaggregation algorithm for computing the stationary distribution of a large Markov chain. *Comm. Statist. Stoch. Models*, **8**, no. 3, 565–575.

[403] Haviv, M. (1999) On censored Markov chains, best augmentations and aggregation/disaggregation procedures. In: *Aggregation and Disaggregation in Operations Research. Comput. Oper. Res.*, **26**, no. 10–11, 1125–1132.

[404] Haviv, M., Ritov, Y. (1986) An approximation to the stationary distribution of a nearly completely decomposable Markov chain and its error bound. *SIAM J. Algebr. Discr. Methods*, **7**, no. 4, 583–588.

[405] Haviv, M., Ritov, Y. (1993) On series expansions and stochastic matrices. *SIAM J. Matrix Anal. Appl.*, **14**, no. 3, 670–676.

[406] Haviv, M., Ritov, Y., Rothblum, U. G. (1992) Taylor expansions of eigenvalues of perturbed matrices with applications to spectral radii of nonnegative matrices. *Linear Algebra Appl.*, **168**, 159–188.

[407] Haviv, M., Rothblum, U. G. (1984) Bounds on distances between eigenvalues. *Linear Algebra Appl.*, **63**, 101–118.

[408] Haviv, M., Van der Heyden, L. (1984) Perturbation bounds for the stationary probabilities of a finite Markov chain. *Adv. Appl. Probab.*, **16**, 804–818.

[409] Hethcote, H. W., Yorke, J. A (1984) *Gonorrhea Transmission Dynamics and Control.* Lecture Notes in Biomathematics, **56**, Springer, Berlin.

[410] Heyde, C. C. (1964) Two probability theorems and their application to some first passage problems. *J. Austral. Math. Soc.*, **4**, 214–222.

[411] Heyde, C. C. (1966) Some renewal theorems with application to a first passage problem. *Ann. Math. Statist.*, **37**, 699–710.

[412] Heyde, C. C. (1967a) A limit theorem for random walks with drift. *J. Appl. Probab.*, **4**, 144–150.

[413] Heyde, C. C. (1967b) Asymptotic renewal results for a natural generalization of classical renewal theory. *J. Royal Statist. Soc.*, Ser. B **29**, 141–150.

[414] Heyde, C. C. (1981) On the survival of a gene represented in a founder population. *J. Math. Biol.*, **12**, no. 1, 91–99.

[415] Heyde, C. C. (1984) On limit theorems for gene survival. In: *Limit Theorems in Probability and Statistics*, Vol. I, II, Veszprém, 1982. Colloq. Math. Soc. János Bolyai, **36**, North-Holland, Amsterdam, 573–586.

[416] Hipp, C., Schmidli, H. (2004) Asymptotics of ruin probabilities for controlled risk processes in the small claims case. *Scand. Actuar. J.*, no. 5, 321–335.

[417] Hodges, J. L., Jr., Rosenblatt, M. (1953) Recurrence-time moments in random walks. *Pacific J. Math.*, **3**, 127–136.

[418] Ho, Y. C., Cao, X. R. (1991) *Perturbation Analysis of Discrete Event Dynamic Systems.* Internat. Ser. Engineering and Computer Science, Kluwer, Boston.

[419] Högnäs, G. (1997) On the quasi-stationary distribution of a stochastic Ricker model. *Stoch. Process. Appl.*, **70**, 243–263.

[420] Holst, L. (1971) Limit theorems for some occupancy and sequential occupancy problems. *Ann. Math. Statist.*, **42**, 1671–1680.

[421] Holst, L. (1977) Some asymptotic results for occupancy problems. *Ann. Probab.*, **5**, 1028–1035.

[422] Holst, L. (1979) A unified approach to limit theorems for urn models. *J. Appl. Probab.*, **16**, 154–162.

[423] Holst, L. (2001) Extreme value distributions for random coupon collector and birthday problems. *Extremes*, **4**, no. 2, 129–145.

[424] Hoppensteadt, F., Salehi, H., Skorokhod, A. (1996a) On the asymptotic behavior of Markov chains with small random perturbations of transition probabilities. In: *Multidimensional Statistical Analysis and Theory of Random Matrices*, Bowling Green, OH, 1996. VSP, Utrecht, 93–100.

[425] Hoppensteadt, F., Salehi, H., Skorokhod, A. (1996b) Markov chain with small random perturbations with applications to bacterial genetics. *Random Oper. Stoch. Equat.*, **4**, no. 3, 205–227.

[426] Hoppensteadt, F., Salehi, H., Skorokhod, A. (1997) Discrete time semigroup transformations with random perturbations. *J. Dynam. Dif. Equat.*, **9**, no. 3, 463–505.

[427] Hsu, Guanghui; He, Qiming (1990) The residual life for alternating renewal processes. *Asia-Pac. J. Oper. Res.*, **7**, 76–81.

[428] Hunter, J. J. (1986) Stationary distributions of perturbed Markov chains. *Linear Algebra Appl.*, **82**, 201–214.

[429] Hunter, J. J. (2005) Stationary distributions and mean first passage times of perturbed Markov chains. *Linear Algebra Appl.*, **410**, 217–243.

[430] Hunter, J. J. (2006) Mixing times with applications to perturbed Markov chains. *Linear Algebra Appl.*, **417**, 108–123.

[431] Iglehart, D. L. (1969) Diffusion approximations in collective risk theory. *J. Appl. Probab.*, **6**, 285–292.

[432] Iglehart, D. L. (1974a) Random walks with negative drift conditioned to stay positive. *J. Appl. Probab.*, **11**, 742–751.

[433] Iglehart, D. L. (1974b) Functional central limit theorem for random walks conditioned to stay positive. *Ann. Probab.*, **2**, 608–619.

[434] Iglehart, D. L. (1975) Conditioned limit theorems for random walks. In: *Stochastic Processes and Related Topics (Proceedings of the Summer Research Institute on Statistical Inference for Stochastic Processes, Vol. 1; dedicated to Jerzy Neyman)*, Indiana Univ., Bloomington, Ind., 1974. Academic Press, New York, 167–194.

[435] Iosifescu, M. (1999) A generalisation of semi-Markov processes. In: Janssen, J., Limnios, N. (eds) *Semi-Markov Models and Applications*. Kluwer, Dordrecht, 23–32.

[436] Ivchenko, G. I., Kashtanov, V. A., Kovalenko, I. N. (1979) *Theory of Queueing*. Vysshaya Shkola, Moscow.

[437] Jacka, S. D., Lazic, Z., Warren, J. (2005a) Conditioning an additive functional of a Markov chain to stay nonnegative. I. Survival for a long time. *Adv. Appl. Probab.*, **37**, 1015–1034.

[438] Jacka, S. D., Lazic, Z., Warren, J. (2005b) Conditioning an additive functional of a Markov chain to stay nonnegative. II. Hitting a high level. *Adv. Appl. Probab.*, **37**, 1035–1055.

[439] Jacka, S. D., Roberts, G. O. (1995) Weak convergence of conditional processes on a countable state space. *J. Appl. Probab.*, **32**, 902–916.

[440] Jacka, S., Warren, J. (2002) Examples of convergence and non-convergence of Markov chains conditioned not to die. *Electron. J. Probab.*, **7**, no. 1, 22 pp.

[441] Jacod, J. (1970) Un theoréme de renouvellement pour les chaînes semi-Markoviennes. *C. R. Acad. Sci. Paris*, **270**, no. 4, 255–258.

[442] Jacod, J. (1971) Theoréme de renouvellement et classification pour pour les chaînes semi-Markoviennes. *Ann. Inst. H. Poincaré*, Sect. B (N.S.), **7**, 83–129.

[443] Jagers, P. (1974) Aspects of random measures and point processes. In: *Advances in Probability and Related Topics*, Vol. 3. Dekker, New York, 179–239.

[444] Jagers, P., Lindvall, T. (1974) Thinning and rare events in point processes. *Z. Wahrsch. Verw. Gebiete*, **28**, 89–98.

[445] Janson, S. (1983a) Renewal theory for M-dependent variables. *Ann. Probab.*, **11**, 558–568.

[446] Janson, S. (1983b) On waiting times in games with disasters. In: *Contributions to Probability and Statistics in Honour of Gunnar Blom*, University of Lund, Lund, 1985, 195–204.

[447] Janssen, J. (1964) Processus de renouvellemenst markoviens et processus semi-markoviens. I. *Cahiers Centre Études Recherche Opér.*, **6**, 81–105.

[448] Janssen, J. (1965) Processus de renouvellements markoviens et processus semi-markoviens. II. Stationnarité et application à un problme d'invalidité. *Cahiers Centre Études Recherche Opér.*, **7**, 126–141.

[449] Janssen, J., Delfosse, Ph. (1982) Some numerical aspects in transient risk theory. *Astin Bull.*, **13**, no. 2, 99–113.

[450] Janssen, J., Manca, R. (2006) *Applied Semi-Markov Processes*. Springer, New York.

[451] Janssen, J., Manca, R. (2007) *Semi-Markov Risk Models for Finance, Insurance and Reliability*. Springer, New York.

[452] Jarner, S. F., Tweedie, R. L. (2001) Locally contracting iterated functions and stability of Markov chains. *J. Appl. Probab.*, **38**, 494–507.

[453] Jarner, S. F., Tweedie, R. L. (2002) Convergence rates and moments of Markov chains associated with the mean of Dirichlet processes. *Stoch. Process. Appl.*, **101**, 257–271.

[454] Jarner, S. F., Tweedie, R. L. (2003) Necessary conditions for geometric and polynomial ergodicity of random-walk-type Markov chains. *Bernoulli*, **9**, no. 4, 559–578.

[455] Johnson, J., Isaacson, D. (1988) Conditions for strong ergodicity using intensity matrices. *J. Appl. Probab.*, **25**, 34–42.

[456] Kalashnikov, V. V. (1973) Property of γ-recurrence for Markov sequences. *Dokl. Acad. Nauk SSSR*, **213**, 1243–1246 (English translation in *Soviet Math. Dokl.*, **14**, 1869–1873).

[457] Kalashnikov, V. V. (1978a) Solution of the problem of approximating a denumerable Markov chain. *Engrg. Cybernetics*, no. 3, 92–95.

[458] Kalashnikov, V. V. (1978b) *Qualitative Analysis of the Behaviour of Complex Systems by the Method of Test Functions*. Series in Theory and Methods of Systems Analysis, Nauka, Moscow.

[459] Kalashnikov, V. V. (1981a) Estimations of convergence rate and stability for regenerative and renovative processes. In: *Point Processes and Queuing Problems*, Keszthely, 1978. Colloq. Math. Soc. János Bolyai, **24**, North-Holland, Amsterdam, 163–180.

[460] Kalashnikov, V. V. (1981b) Methods of estimation of continuity in problems of state consolidation. In: *Stability Problems for Stochastic Models*, Panevezhis, 1980. VNIISI, Moscow, 48–53.

[461] Kalashnikov, V. V. (1983) Estimation of the rate of convergence in Rényi's theorem. In: *Stability Problems for Stochastic Models*. Trudy Seminara, VNIISI, Moscow, 48–51.

[462] Kalashnikov, V. V. (1986) Approximation of some stochastic models. In: *Semi-Markov Models*, Brussels, 1984. Plenum, New York, 319–335.

[463] Kalashnikov, V. V. (1990) Regenerative queueing processes and their qualitative and quantitative analysis. *Queueing Systems Theory Appl.*, **6**, no. 2, 113–136.

[464] Kalashnikov, V. V. (1993) Two-side estimates of geometric convolutions. In: Kalashnikov, V. V., Zolotarev, V. M. (eds) *Stability Problems for Stochastic Models*, Suzdal, 1991. Lecture Notes in Mathematics, **1546**, Springer, Berlin, 76–88.

[465] Kalashnikov, V. V. (1994a) Regeneration and general Markov chains. *J. Appl. Math. Stoch. Anal.*, **7**, no. 3, 357–371.

[466] Kalashnikov, V. V. (1994b) *Topics on Regenerative Processes*. CRC Press, Boca Raton, FL.

[467] Kalashnikov, V. V. (1994c) *Mathematical Methods in Queuing Theory*. Mathematics and its Applications, **271**, Kluwer, Dordrecht.

[468] Kalashnikov, V. V. (1997) *Geometric Sums: Bounds for Rare Events with Applications*. Mathematics and its Applications, **413**, Kluwer, Dordrecht.

[469] Kalashnikov, V. V. (1999) Bounds for ruin probabilities in the presence of large claims and their comparison. With a discussion by John A. Beekman and a reply by the author. *Noth. Amer. Actuar. J.*, **3**, no. 2, 116–129.

[470] Kalashnikov, V. V., Anichkin, S.A. (1981) Continuity of random sequences and approximation of Markov chains. *Adv. Appl. Probab.*, **13**, 402–414.

[471] Kalashnikov, V. V., Konstantinidis, D. (1996) Ruin probability. In: *Research Papers Dedicated to the Memory of B. V. Gnedenko*. Fundam. Prikl. Mat., **2**, no. 4, 1055–1100.

[472] Kalashnikov, V. V., Konstantinides, D. (2000) Ruin under interest force and subexponential claims: a simple treatment. *Insur. Math. Econom.*, **27**, no. 1, 145–149.

[473] Kalashnikov, V. V., Norberg, R. (2002) Power tailed ruin probabilities in the presence of risky investments. *Stoch. Process. Appl.*, **98**, 211–228.

[474] Kalashnikov, V. V., Rachev, S. T. (1984) Characterization of queueing models and its stability. In: *Stability Problems for Stochastic Models*. Trudy Seminara, VNIISI, Moscow, 61–89.

[475] Kalashnikov, V. V., Rachev, S. T. (1985) Characterization problems in queueing and their stability. *Adv. Appl. Probab.*, **17**, 868–886.

[476] Kalashnikov, V. V., Rachev, S. T. (1986) Characterizations of inverse problems in queueing and their stability. *J. Appl. Probab.*, **23**, 459–473.

[477] Kalashnikov, V. V., Rachev, S. T. (1987) Characterization of queues and its stability estimates. In: *Probability Theory and Mathematical Statistics*, Vol. II, Vilnius, 1985. VNU Sci. Press, Utrecht, 37–53.

[478] Kalashnikov, V. V., Rachev, S. T. (1988) *Mathematical Methods for Construction of Queueing* Models. Nauka, Moscow (English edition: The Wadsworth & Brooks/Cole Operations Research Series, Wadsworth & Brooks/Cole, Pacific Crove, CA (1990)).

[479] Kalashnikov, V. V., Tsitsiashvili, G. (1999a) Tails of waiting times and their bounds. *Queueing Systems Theory Appl.*, **32**, no. 1–3, 257–283.

[480] Kalashnikov, V. V., Tsitsiashvili, G. (1999b) Two-sided estimates for random sums with subexponential summands. *Probl. Peredachi Infor.*, **35**, no. 3, 67–79 (English translation in *Probl. Inform. Transmis.*, **35**, no. 3, 248–258).

[481] Kalashnikov, V. V., Tsitsiashvili, G. (2001) Asymptotically correct bounds of geometric convolutions with subexponential components. In: *Stability Problems for Stochastic Models*, Part III, Nałęczow, 1999. *J. Math. Sci.*, **106**, no. 2, 2806–2819.

[482] Kalashnikov, V. V., Vsekhsvyatski, S. Yu. (1985) Metric estimates of the first occurrence time in regenerative processes. In: Kalashnikov, V. V., Zolotarev, V. M. (eds) *Stability Problems for Stochastic Models*, Uzhgorod, 1984. Lecture Notes in Mathematics, **1155**, Springer, Berlin, 102–130.

[483] Kalashnikov, V. V., Vsekhsvyatski, S. Yu. (1989) On the connection of Rényi's theorem and renewal theory. In: Zolotarev, V. M., Kalashnikov, V. V. (eds) *Stability Problems for Stochastic Models*, Sukhumi, 1987. Lecture Notes in Mathematics, **1412**, Springer, Berlin, 83–102.

[484] Kallenberg, O. (1975) Limits of compound and thinned point processes. *J. Appl. Probab.*, **12**, 269–278.

[485] Kaplan, E. I. (1979a) Limit distribution for exit times of nonstationary random sequences. *Dokl. Akad. Nauk Ukr. SSR*, Ser. A, no. 11, 900–902.

[486] Kaplan, E. I. (1979b) Limit theorems for exit times of random sequences with mixing. *Teor. Veroyatn. Mat. Stat.*, **21**, 53–59 (English translation in *Theory Probab. Math. Statist.*, **21**, 59–65).

[487] Kaplan, E. I. (1980) Limit Theorems for Sum of Switching Random Variables with an Arbitrary Phase Space of Switching Component. Candidate of Science dissertation, Kiev State University.

[488] Karlin, S. (1955) On the renewal equation. *Pacific J. Math.*, **5**, 229–257.

[489] Kartashov, N. V. (1978) On explicit estimates of the rate of convergence in the renewal theorem. *Teor. Veroyatn. Mat. Stat.*, **18**, 74–79 (English translation in *Theory Probab. Math. Statist.*, **18**, 77–82).

[490] Kartashov, N. V. (1979a) Power estimates for the rate of convergence in the renewal theorem. *Teor. Veroyatn. Primen.*, **24**, 600–607 (English translation in *Theory Probab. Math. Statist.*, **24**, 606–612).

[491] Kartashov, N. V. (1979b) Quantitative estimates of ergodicity in the system $GI/GI/1$. *Teor. Veroyatn. Mat. Stat.*, **21**, 59–65 (English translation in *Theory Probab. Math. Statist.*, **21**, 67–74).

[492] Kartashov, N. V. (1981) An estimate of ergodicity in the system $M/M/1$. *Teor. Veroyatn. Mat. Stat.*, **24**, 53–59 (English translation in *Theory Probab. Math. Statist.*, **24**, 59–66).

[493] Kartashov, N. V. (1982a) A generalization of Stone's representation and necessary conditions for uniform convergence in the renewal theorem. *Teor. Veroyatn. Mat. Stat.*, **26**, 49–62 (English translation in *Theory Probab. Math. Statist.*, **26**, 53–67).

[494] Kartashov, N. V. (1982b) Equivalence of uniform renewal theorems and their criteria. *Teor. Veroyatn. Mat. Stat.*, **27**, 51–60 (English translation in *Theory Probab. Math. Statist.*, **27**, 55–64).

[495] Kartashov, N. V. (1984) Criteria for uniform ergodicity and strong stability of Markov chains with a common phase space. *Teor. Veroyatn. Mat. Stat.*, **30**, 65–81 (English translation in *Theory Probab. Math. Statist.*, **30**, 71–89).

[496] Kartashov, N. V. (1985a) Inequalities in stability and ergodicity theorems for Markov chains with a general phase space. I. *Teor. Veroyatn. Primen.*, **30**, 230–240 (English translation in *Theory Probab. Appl.*, **30**, 247–259).

[497] Kartashov, N. V. (1985b) Inequalities in stability and ergodicity theorems for Markov chains with a general phase space. II. *Teor. Veroyatn. Primen.*, **30**, 478–485 (English translation in *Theory Probab. Appl.*, **30**, 507–515).

[498] Kartashov, N. V. (1985c) A refinement of estimates for the ergodicity of renewal processes and Markov chains. *Teor. Veroyatn. Mat. Stat.*, **32**, 27–33 (English translation in *Theory Probab. Math. Statist.*, **32**, 27–34).

[499] Kartashov, N. V. (1985d) Asymptotic representations in an ergodic theorem for general Markov chains and their applications. *Teor. Veroyatn. Mat. Stat.*, **32**, 113–121 (English translation in *Theory Probab. Math. Statist.*, **32**, 131–139).

[500] Kartashov, N. V. (1985e) Asymptotic expansions and inequalities in stability theorems for general Markov chains under relatively bounded perturbations. In: *Stability Problems for Stochastic Models*, Varna, 1985. VNIISI, Moscow, 75–85 (English translation in *J. Soviet Math.*, **40**, no. 4, 509–518).

[501] Kartashov, N. V. (1986) Inequalities in theorems of consolidation of Markov chains. *Theor. Veroyatn. Mat. Stat.*, **34**, 62–73 (English translation in *Theory Probab. Math. Statist.*, **34**, 67–80).

[502] Kartashov, N. V. (1987) Estimates for the geometric asymptotics of Markov times on homogeneous chains. *Teor. Veroyatn. Mat. Stat.*, **37**, 66–77 (English translation in *Theory Probab. Math. Statist.*, **37**, 75–88).

[503] Kartashov, N. V. (1991) Inequalities in Rénei's theorem. *Teor. Ĭmovirn. Mat. Stat.*, **45**, 27–33 (English translation in *Theory Probab. Math. Statist.*, **45**, 23–28).

[504] Kartashov, N. V. (1996a) Uniformly ergodic jump Markov processes with bounded intensities. *Teor. Ĭmovirn. Mat. Stat.*, **52**, 91–103 (English translation in *Theory Probab. Math. Statist.*, **52**, 86–98).

[505] Kartashov, M. V. (1996b) *Strong Stable Markov Chains*. VSP, Utrecht and TBiMC, Kiev.

[506] Kartashov, M. V. (1996c) Computation and estimation of the exponential ergodicity exponent for general Markov processes and chains with recurrent kernels. *Teor. Ĭmovirn. Mat. Stat.*, **54**, 47–57 (English translation in *Theory Probab. Math. Statist.*, **54**, 49–60).

[507] Kartashov, M. V. (1997) Calculation of the exponential ergodicity exponent for birth and death processes. *Teor. Ĭmovirn. Mat. Stat.*, **57**, 51–59 (English translation in *Theory Probab. Math. Statist.*, **57**, 53–60).

[508] Kartashov, M. V. (1999) On ruin probabilities for a risk process with bounded reserves. *Teor. Ĭmovirn. Mat. Stat.*, **60**, 46–58 (English translation in *Theory Probab. Math. Statist.*, **60**, 53–65).

[509] Kartashov, M. V. (2000) Calculation of the spectral ergodicity exponent for the birth and death process. *Ukr. Mat. Zh.*, **52**, 889–897 (English translation in *Ukr. Math. J.*, **52**, 1018–1028).

[510] Kartashov, M. V. (2005) Ergodicity and stability of quasihomogeneous Markov semigroups of operators. *Teor. Ĭmovirn. Mat. Stat.*, **72**, 54–62 (English translation in *Theory Probab. Math. Statist.*, **72**, 59–68).

[511] Kartashov, M. V., Stroev, O. M. (2005) Lundberg approximation for the risk function in an almost homogeneous environment. *Teor. Ĭmovirn. Mat. Stat.*, **73**, 63–71 (English translation in *Teory Probab. Math. Statist.*, **73**, 71–79).

[512] Kato, T. (1966) *Perturbation Theory for Linear Operators*. Springer, Berlin.

[513] Keener, R. W. (1992) Limit theorems for random walks conditioned to stay positive. *Ann. Probab.*, **20**, 801–824.

[514] Keilson, J. (1966a) A limit theorem for passage times in ergodic regenerative processes. *Ann. Math. Statist.*, **37**, 866–870.

[515] Keilson, J. (1966b) A technique for discussing the passage time distribution for stable systems. *J. Royal Statist. Soc.*, Ser. B **28**, 477–486.

[516] Keilson, J. (1969) On the matrix renewal function for Markov renewal processes. *Ann. Math. Statist.*, **40**, 1901–1907.

[517] Keilson, J. (1974) Sojourn times, exit times and jitter in multivariate Markov processes. *Adv. Appl. Probab.*, **6**, 747–756.

[518] Keilson, J. (1975) Systems of independent Markov components and their transient behavior. In: *Reliability and Fault Tree Analysis*, Conf., Univ. California, Berkeley, CA, 1974. Soc. Indust. Appl. Math., Philadelphia, 351–364.

[519] Keilson, J. (1978) *Markov Chain Models – Rarity and Exponentiality*. Applied Mathematical Sciences, **28**, Springer, New York.

[520] Keilson, J. (1986) Stochastic models in reliability theory. In: *Theory of Reliability*, Varenna, 1984. Proceedings of the International School of Physics "Enrico Fermi", XCIV, North-Holland, Amsterdam, 23–54.

[521] Keilson, J. (1998) Covariance and relaxation time in finite Markov chains. *J. Appl. Math. Stoch. Anal.*, **11**, 391–396.

[522] Keilson, J., Ramaswamy, R. (1984) Convergence of quasistationary distributions in birth-death processes. *Stoch. Process. Appl.*, **18**, 301–312.

[523] Keilson, J., Ramaswamy, R. (1986) The bivariate maximum process and quasistationary structure of birth-death processes. *Stoch. Process. Appl.*, **22**, 27–36.

[524] Keilson, J., Sumita, U. (1983) The depletion time for $M/G/1$ systems and a related limit theorem. *Adv. Appl. Probab.*, **15**, 420–443.

[525] Keilson, J., Vasicek, O. A. (1998) Monotone measures of ergodicity for Markov chains. *J. Appl. Math. Stoch. Anal.*, **11**, 283–288.

[526] Kemeny, J. G., Snell, J. L. (1961a) Potentials for denumerable Markov chains. *J. Math. Anal. Appl.*, **6**, 196–260.

[527] Kemeny, J. G., Snell, J. L. (1961b) Finite continuous time Markov chains. *Theor. Veroyatn. Primen.*, **6**, 110–115 (Also in *Theory Probab. Appl.*, **6**, 101–105).

[528] Kemeny, J. G., Snell, J. L., Knapp, A. W. (1966, 1976) *Denumerable Markov Chains*. Van Nostrand, Princeton and Springer, New York.

[529] Kendall, D. G. (1959) Unitary dilations of Markov transition operators, and the corresponding integral representations for transition-probability matrices. In: Grenander, U. (ed) *Probability and Statistics: The Harald Cramér volume*. Almqvist & Wiksell, Stockholm and Wiley, New York, 139–161.

[530] Kesten, H. (1963) Ratio theorems for random walks. II. *J. Analyse Math.*, **11**, 323–379.

[531] Kesten, H. (1974) Renewal theory for functionals of a Markov chain with general state space. *Ann. Probab.*, **2**, 355–386.

[532] Kesten, H. (1970) A ratio limit theorem for symmetric random walk. *J. Analyse Math.*, **23**, 199–213.

[533] Kesten, H. (1995) A ratio limit theorem for (sub) Markov chains on $\{1, 2, \ldots\}$ with bounded jumps. *Adv. Appl. Probab.*, **27**, 652–691.

[534] Kesten, H., Maller, R. A. (1999) Stability and other limit laws for exit times of random walks from a strip or a halfplane. *Ann. Inst. H. Poincaré Probab. Statist.*, **35**, no. 6, 685–734.

[535] Kesten, H., Spitzer, F. (1963) Ratio theorems for random walks. I. *J. Analyse Math.*, **11**, 285–322.

[536] Kevorkian, J., Cole, J. D. (1981) *Perturbation Methods in Applied Mathematics.* Applied Mathematical Sciences, **34**, Springer, New York.

[537] Kevorkian, J., Cole, J. D. (1996) *Multiple Scale and Singular Perturbation Methods.* Applied Mathematical Sciences, **114**, Springer, New York.

[538] Khasminskii, R. Z., Yin, G., Zhang, Q. (1996a) Singularly perturbed Markov chains: quasi-stationary distribution and asymptotic expansion. In: *Proceedings of Dynamic Systems and Applications*, Vol. 2, Atlanta, GA, 1995. Dynamic, Atlanta, GA, 301–308.

[539] Khasminskii, R. Z., Yin, G., Zhang, Q. (1996b) Asymptotic expansions of singularly perturbed systems involving rapidly fluctuating Markov chains. *SIAM J. Appl. Math.*, **56**, no. 1, 277–293.

[540] Khintchine, A. Y. (1955) *Mathematical Methods in the Theory of Queueing.* Trudy Mat. Inst. Steklov. **49** (English edition: Griffin's Statistical Monographs & Courses, **7**, Hafner, New York (1960, 1969)).

[541] Khintchine, A. Y. (1963) *Works on Theory of Queueing Systems.* Fizmatgiz, Moscow.

[542] Khusanbayev, Ya. M. (1983) Limit Theorems for Processes with Conditionally Independent Increments and Functionals of Additive Type for Regenerative Processes. Candidate of Science dissertation, Kiev State University.

[543] Khusanbayev, Ya. M. (1984) Conditions of convergence for stochastic processes in moment of under-jump. *Ukr. Mat. Zh.*, **36**, 126–129 (English translation in *Ukr. Math. J.*, **36**, 116–119).

[544] Kijima, M. (1992a) On the existence of quasi-stationary distributions in denumerable R-recurrent Markov chains. *J. Appl. Probab.*, **29**, 21–36.

[545] Kijima, M. (1992b) Evaluation of the decay parameter for some specialized birth-death processes. *J. Appl. Probab.*, **29**, 781–791.

[546] Kijima, M. (1993a) Quasi-limiting distributions of Markov chains that are skip-free to the left in continuous time. *J. Appl. Probab.*, **30**, 509–517.

[547] Kijima, M. (1993b) Quasi-stationary distributions of single-server phase-type queues. *Math. Oper. Res.*, **18**, no. 2, 423–437.

[548] Kijima, M. (1995) Bounds for the quasi-stationary distribution of some specialized Markov chains. In: *Stochastic Models in Engineering, Technology and Management*, Gold Coast, 1994. *Math. Comput. Modell.*, **22**, no. 10–12, 141–147.

[549] Kijima, M. (1997) *Markov Processes for Stochastic Modelling.* Stochastic Modeling Series, Chapman & Hall, London.

[550] Kijima, M., Makimoto, N. (1992) Computation of the quasi-stationary distributions in $M(n)/GI/1/K$ and $GI/M(n)/1/K$ queues. *Queueing Systems Theory Appl.*, **11**, no. 3, 255–272.

[551] Kijima, M., Makimoto, N. (1999) Quasi-stationary distributions of Markov chains arising from queueing processes: a survey. In: Shanthikumar, J.G., Sumita, U. (eds) *Applied Probability and Stochastic Processes*. Internat. Ser. Oper. Res. Management Sci., **19**, Kluwer, Boston, 277–311.

[552] Kijima, M., Nair, M. G., Pollett, P. K., van Doorn, E. A. (1997) Limiting conditional distributions for birth-death processes. *Adv. Appl. Probab.*, **29**, 185–204.

[553] Kijima, M., Seneta, E. (1991) Some results for quasi-stationary distributions of birth-death processes. *J. Appl. Probab.*, **28**, 503–511.

[554] Kim, D. S., Smith, R. L. (1991) An exact aggregation-disaggregation algorithm for mandatory set decomposable Markov chains. In: Stewart, W. J. (ed) *Numerical Solution of Markov Chains*. Probability: Pure and Applied, **8**. Marcel Dekker, New York, 89–103.

[555] Kim, Dong Yun, Woodroofe, M. (2003) Non-linear renewal theory with stationary perturbations. *Sequent. Anal.*, **22**, no. 1–2, 55–74.

[556] Kingman, J. F. C. (1961) The single server queue in heavy traffic. *Proc. Cambridge Philos. Soc.*, **57**, 902–904.

[557] Kingman, J. F. C. (1962) On queues in heavy traffic. *J. Royal Statist. Soc.*, Ser. B **24**, 383–392.

[558] Kingman, J. F. C. (1963) The exponential decay of Markovian transition probabilities. *Proc. London Math. Soc.*, **13**, 337–358.

[559] Kingman, J. F. C. (1972) *Regenerative Phenomena*. Wiley Series in Probability and Mathematical Statistics, Wiley, London.

[560] Klebaner, F. C., Lazar, J., Zeitouni, O. (1998) On the quasi-stationary distribution for some randomly perturbed transformations of an interval. *Ann. Appl. Probab.*, **8**, 300–315.

[561] Klebanov, L. B., Melamed, I. A., Umarov, A. Yu. (1991) An analogue of A.N. Kolmogorov's uniform limit theorem under summation of a geometric number of random variables. *Soobshch. Akad. Nauk Gruzii*, **143**, no. 2, 129–131.

[562] Klüppelberg, C. (1989) Estimation of ruin probabilities by means of hazard rates. *Insur. Math. Econom.*, **8**, no. 4, 279–285.

[563] Klüppelberg, C., Kyprianou, A. E. (2006) On extreme ruinous behaviour of Lévy insurance risk processes. *J. Appl. Probab.*, **43**, 594–598.

[564] Klüppelberg, C., Kyprianou, A. E., Maller, R. A. (2004) Ruin probabilities and overshoots for general Lévy insurance risk processes. *Ann. Appl. Probab.*, **14**, 1766–1801.

[565] Klüppelberg, C., Mikosch, T. (1995) Delay in claim settlement and ruin probability approximations. *Scand. Actuar. J.*, no. 2, 154–168.

[566] Klüppelberg, C., Mikosch, T. (1997) Large deviations of heavy-tailed random sums with applications in insurance and finance. *J. Appl. Probab.*, **34**, 293–308.

[567] Klüppelberg, C., Pergamenchtchikov, S. (2003) Renewal theory for functionals of a Markov chain with compact state space. *Ann. Probab.*, **31**, 2270–2300.

[568] Klüppelberg, C., Stadtmuller, U. (1998) Ruin probability in the presence of heavy-tails and interest rates. *Scand. Actuar. J.*, no. 1, 49–58.

[569] Kolmogorov, A. N. (1931) Über die analytischen Methoden in der Wahrscheinlichkeit-srechnung. *Math. Ann.*, **104**, 415–458 (Russian translation in *Uspehi Mat. Nauk*, **5**, 5–41).

[570] Kolmogorov, A. N. (1937) Markov chains with countable number of states. *Bul. MGU*, Sec. A, Mat. Mekh., **1**, no. 3, 1–16.

[571] Kolowrocki, K. (1998) On limit reliability functions of large systems. In: Ionescu, D. C., Limnios, N. (eds) *Statistical and Probabilistic Models in Reliability*. Birkhäuser, Boston, 153–183.

[572] Kolowrocki, K. (2000) An asymptotic approach to multivariate system reliability eva-lution In: Limnios, N., Nikulin, M. (eds) *Recent Advances in Reliability Theory: Methodology, Practice and Inference*. Birkhäuser, Boston, 163–180.

[573] Konstantinidis, D. G., Solov'ev, A. D. (1991) A uniform estimate for the reliability of a complex repairable system with an unlimited number of repair units. *Vestnik Moskov. Univ.*, Ser. I, Mat. Mekh., no. 3, 21–26 (English translation in *Moscow Univ. Math. Bull.*, **46**, no. 3, 21–24).

[574] Konstantopoulos, T., Last, G. (1999) On the use of Lyapunov function methods in renewal theory. *Stoch. Process. Appl.*, **79**, 165–178.

[575] Kontoyiannis, I., Meyn, S. P. (2003) Spectral theory and limit theorems for geometri-cally ergodic Markov processes. *Ann. Appl. Probab.*, **13**, 304–362.

[576] Kontoyiannis, I., Meyn, S. P. (2005) Large deviations asymptotics and the spectral theory of multiplicatively regular Markov processes. *Electron. J. Probab.*, **10**, no. 3, 61–123.

[577] Korolev, V. Yu. (1987) On approximation of densities of random sums of independent random variables by densities of mixtures of normal laws. In: *Stochastic Analysis*. Izdatel'stvo MGU, Moscow, 50–64.

[578] Korolev, V. Yu. (1989a) The asymptotic distributions of random sums. In: Kalashnikov, V. V., Zolotarev, V. M., (eds) *Stability Problems for Stochastic Models*, Sukhumi, 1987. Lecture Notes in Mathematics, **1412**, Springer, Berlin, 110–123.

[579] Korolev, V. Yu. (1989b) Approximation of distributions of of random sums of inde-pendent random variables by densities of mixtures of normal laws. *Teor. Veroyatn. Primen.*, **34**, 581–588 (English translation in *Theory Probab. Appl.*, **34**, 523–531).

[580] Korolyuk, D. V. (1982) Limit distributions of additive functionals of regenerative pro-cesses up to the time of reaching asymptotically receding regions. *Dokl. Akad. Nauk Ukr. SSR*, Ser. A, no. 8, 16–18.

[581] Korolyuk, D. V. (1983) Limit Theorems for Hitting Time Type Functionals Defined on Processes with Semi-Markov Switchings. Candidate of Science dissertation, Kiev State University.

[582] Korolyuk, D. V. (1990) Asymptotics of the time of attaining a distant level by a random walk with delay. *Teor. Veroyatn. Mat. Stat.*, **42**, 64–67 (English translation in *Theory Probab. Math. Statist.*, **42**, 73–76).

[583] Korolyuk, D. V., Silvestrov D. S. (1983) Entry times into asymptotically receding domains for ergodic Markov chains. *Teor. Veroyatn. Primen.*, **28**, 410–420 (English translation in *Theory Probab. Appl.*, **28**, 432–442).

[584] Korolyuk, D. V., Silvestrov D. S. (1984) Entry times into asymptotically receding regions for processes with semi-Markov switchings. *Teor. Veroyatn. Primen.*, **29**, 539–544 (English translation in *Theory Probab. Appl.*, **29**, 558–563).

[585] Korolyuk, V. S. (1969) On asymptotical estimate for time of a semi-Markov process being in the set of states. *Ukr. Mat. Zh.*, **21**, 842–845 (English translation in *Ukr. Math. J.*, **21**, 705–710).

[586] Korolyuk, V. S. (1986) Markov renewal processes in reliability analysis. In: *Semi-Markov Models*, Brussels, 1984. Plenum, New York, 217–230.

[587] Korolyuk, V. S. (1993) *Stochastic Models of Systems*. Libid', Kiev.

[588] Korolyuk, V. S. (1997) Singularly perturbed stochastic systems. *Ukr. Mat. Zh.*, **49**, 25–34 (English translation in *Ukr. Math. J.*, **49**, 25–35).

[589] Korolyuk, V. S. (1999) Semi-Markov random walks. In: Janssen, J., Limnios, N. (eds) *Semi-Markov Models and Applications*. Kluwer, Dordrecht, 61–75.

[590] Korolyuk, V. S., Brodi, S. M., Turbin, A. F. (1974) Semi-Markov processes and their application. In: *Itogi Nauki i Tehniki*, Ser. Teor. Veroyatn. Mat. Stat. Teor. Kibernet., **11**, VINITI, Moscow, 47–97 (English translation in *J. Math. Sci.*, **4**, no. 3, 244–280).

[591] Korolyuk, V. S., Korolyuk, V. V. (1999) *Stochastic Models of Systems*. Mathematics and its Applications, **469**, Kluwer, Dordrecht.

[592] Korolyuk, V. S., Limnios, N. (1998) Diffusion approximation of integral functionals in merging and averaging scheme. *Teor. Ĭmovirn. Mat. Stat.*, **59**, 99–105 (English translation in *Theory Probab. Math. Statist.*, **59**, 101–107).

[593] Korolyuk, V. S., Limnios, N. (1999) Diffusion approximation for integral functionals in the double merging and averaging scheme. *Teor. Ĭmovirn. Mat. Stat.*, **60**, 77–84 (English translation in *Theory Probab. Math. Statist.*, **60**, 87–94).

[594] Korolyuk, V. S., Limnios, N. (2000) Evolutionary systems in an asymptotic split phase space. In: Limnios, N., Nikulin, M. (eds) *Recent Advances in Reliability Theory: Methodology, Practice and Inference*. Birkhäuser, Boston, 145–161.

[595] Korolyuk, V. S., Limnios, N. (2002) Markov additive processes in a phase merging scheme. In: Korolyuk, V., Prokhorov, Yu., Khokhlov, V., Klesov, O. (eds) *Proceedings of the Conference Dedicated to the 90th Anniversary of Boris Vladimirovich Gnedenko*, Kiev, 2002. *Theory Stoch. Process.*, **8**, no. 3–4, 213–225.

[596] Korolyuk, V. S., Limnios, N. (2004a) Average and diffusion approximation of stochastic evolutionary systems in an asymptotic split state space. *Ann. Appl. Probab.*, **14**, 489–516.

[597] Korolyuk, V. S., Limnios, N. (2004b) Diffusion approximation of evolutionary systems with equilibrium in asymptotic split phase space. *Teor. Ĭmovirn. Mat. Stat.*, **70**, 63–73 (English translation in *Theory Probab. Math. Statist.*, **70**, 71–82).

[598] Koroliuk, V. S., Limnios, N. (2005) *Stochastic Systems in Merging Phase Space*. World Scientific, Singapore.

[599] Korolyuk, V. S., Mamonova, G. V. (2003) Stationary phase consolidation of the superposition of two renewal processes. *Dopov. Nats. Akad. Nauk Ukr.*, Mat. Prirodozn. Tekh. Nauki, no. 5, 21–25.

[600] Korolyuk, V. S., Penev, I. P., Turbin, A. F. (1972) The asymptotic behavior of the distribution of the absorption time of a Markov chain. *Kibernetika*, no. 2, 20–22.

[601] Korolyuk, V. S., Penev, I. P., Turbin, A. F. (1981) Asymptotic expansion for the distribution of the absorption time of a weakly inhomogeneous Markov chain. In: Korolyuk, V. S. (ed) *Analytic Methods of Investigation in Probability Theory*. Akad. Nauk Ukr. SSR, Inst. Mat., Kiev, 97–105.

[602] Korolyuk, V. S., Svishchuk, A. V. (1990) Weak convergence of semi-Markov random evolutions in an averaging scheme. *Teor. Ĭmovirn. Mat. Stat.*, **42**, 53–64 (English translation in *Theory Probab. Math. Statist.*, **42**, 61–71).

[603] Korolyuk, V. S., Svishchuk, A. V. (1992) *Semi-Markov Random Evolutions*. Naukova Dumka, Kiev (English edition: Mathematics and its Applications, **308**, Kluwer Academic Publishers, Dordrecht (1995)).

[604] Korolyuk, V. S., Svishchuk, A. V. (1995) *Evolution of Systems in Random Media*. CRC Press, Boca Raton, FL.

[605] Korolyuk, V. V., Tadzhiev, A. (1986) Asymptotic behavior of Markov evolutions prior to the time of absorption. *Ukr. Mat. Zh.*, **38**, 248–251 (English translation in *Ukr. Math. J.*, **38**, 219–222).

[606] Korolyuk, V. S., Turbin, A. F. (1970) On the asymptotic behaviour of the occupation time of a semi-Markov process in a reducible subset of states. *Teor. Veroyatn. Mat. Stat.*, **2**, 133–143 (English translation in *Theory Probab. Math. Statist.*, **2**, 133–143).

[607] Korolyuk, V. S., Turbin, A. F. (1976a) *Semi-Markov Processes and its Applications*. Naukova Dumka, Kiev.

[608] Korolyuk, V. S., Turbin, A. F. (1976b) Asymptotic enlarging of semi-Markov processes with an arbitrary state space. In: *Proceedings of the Third Japan-USSR Symposium on Probability Theory*, Tashkent, 1975. Lecture Notes in Mathematics, **550**, Springer, Berlin, 297–315.

[609] Korolyuk, V. S., Turbin, A. F. (1978) *Mathematical Foundations of the State Lumping of Large Systems*. Naukova Dumka, Kiev (English edition: Mathematics and its Applications, **264**, Kluwer, Dordrecht (1993)).

[610] Korolyuk, V. S., Turbin, A. F. (1982) *Markov Renewal Processes in Problems of System Reliability*. Naukova Dumka, Kiev.

[611] Korolyuk, V. S., Turbin, A. F., Tomusjak, A. A. (1979) Sojourn time of a semi-Markov process in a decomposing set of states. In: *Analytical Methods of Probability Theory*. Naukova Dumka, Kiev, 69–79.

[612] Korolyuk, V. V. (1983) Homogeneous processes with independent increments with semi-Markov switch-overs that admit asymptotic phase lumping. *Dokl. Akad. Nauk Ukr. SSR*, Ser. A, no. 10, 8–11.

[613] Kostava, B. A., Shurenkov, V. M. (1988) Limit distributions that are connected with the exit of a random walk from a half-space. In: *Selected Problems in the Modern Theory of Random Processes*. Akad. Nauk Ukr. SSR, Inst. Mat., Kiev, 96–101.

[614] Kovalenko, I. N. (1965) On the class of limit distributions for thinning flows of homogeneous events. *Litov. Mat. Sbornik*, **5**, 569–573.

[615] Kovalenko, I. N. (1973) An algorithm of asymptotic analysis of a sojourn time of Markov chain in a set of states. *Dokl. Acad. Nauk Ukr.* SSR, Ser. A, no. 6, 422–426.

[616] Kovalenko, I. N. (1975) *Studies in the Analysis of Reliability of Complex Systems*. Naukova Dumka, Kiev.

[617] Kovalenko, I. N. (1977) Limit theorems of reliability theory. *Kibernetika*, no. 6, 106–116 (English translation in *Cybernetics*, **13**, 902–914).

[618] Kovalenko, I. N. (1980) *Analysis of Rare Events in Estimation of Efficiency and Reliability of Systems*. Sovetskoe Radio, Moscow.

[619] Kovalenko, I. N. (1994) Rare events in queuing theory – a survey. *Queuing Systems Theory Appl.*, **16**, no. 1–2, 1–49.

[620] Kovalenko, I. N. (1995) Approximation of queues via small parameter method. In: *Advances in Queueing*. Probab. Stochastics Ser., CRC, Boca Raton, FL, 481–506.

[621] Kovalenko, I. N. (2000) Estimation of the intensity of a flow on nonmonotone failures of the queueing system $(\leq \lambda)/G/m$. *Ukr. Mat. Zh.*, **52**, 1219–1225 (English translation in *Ukr. Math. J.*, **52**, 1396–1402).

[622] Kovalenko, I. N., Birolini, A., Kuznetsov, N. Yu., Shenbukher, M. (1998) On the reduction of the phase space of a random process in analyzing the reliability of systems. *Kibernet. Sistem. Anal.*, no. 4, 134–144 (English translation in *Cybernet. Systems Anal.*, **34**, no. 4, 588–595).

[623] Kovalenko, I. N., Kuznetsov, N. Yu. (1981) Renewal process and rare events limit theorems for essentially multidimensional queueing processes. *Math. Operat. Statist.*, Ser. Statist., **12**, no. 2, 211–224.

[624] Kovalenko, I. N., Kuznetsov, N. Yu. (1988) *Methods for Calculating Highly Reliable Systems*. Radio and Svyaz', Moscow.

[625] Kovalenko, I. N., Kuznetsov, N. Yu., Pegg, P. A. (1997) *Mathematical Theory of Reliability of Time Dependent Systems with Practical Applications*. Wiley Series in Probability and Statistics, Wiley, New York.

[626] Kovalenko, I. N., Kuznetsov, N. Yu., Shurenkov, V. M. (1983) *Models of Random Processes. A Handbook for Mathematicians and Engineers*. Naukova Dumka, Kiev (English edition: CRC Press, Boca Raton, FL (1996)).

[627] Kovalenko, I. N., Musaev, E. M. (1974) The reliability of complex systems that are under periodic perturbations. *Kibernetika*, no. 3, 89–93 (English translation in *Cybernetics*, **10**, no. 3, 477–482).

[628] Kruglov, V. M., Korolev, V. Yu. (1990) *Limit Theorems for Random Sums*. Izdatel'stvo Moskovskogo Universiteta, Moscow.

[629] Kryscio, R. J., Lefèvre, C. (1989) On extinction of the S-I-S stochastic logistic epidemic. *J. Appl. Probab.*, **27**, 685–694.

[630] Kurtz, T. G. (1971) Comparison of semi-Markov and Markov processes. *Ann. Math. Statist.*, **42**, 991–1002.

[631] Kyprianou, E. K. (1972a) On the quasi-stationary distributions of the $GI/M/1$ queue. *J. Appl. Probab.*, **9**, 117–128.

[632] Kyprianou, E. K. (1972b) The quasi-stationary distributions of queues in heavy traffic. *J. Appl. Probab.*, **9**, 821–831.

[633] Kyprianou, A. E., Palmowski, Z. (2006) Quasi-stationary distributions for Lévy processes. *Bernoulli*, **12**, no. 4, 571–581.

[634] Lalley, S. P. (1984) Limit theorems for first-passage times in linear and nonlinear renewal theory. *Adv. Appl. Probab.*, **16**, 766–803.

[635] Lam, C. Y. T., Lehoczky, J. P. (1991) Superposition of renewal processes. *Adv. Appl. Probab.*, **23**, 64–85.

[636] Lamperti, J. (1960) The first-passage moments and the invariant measure of a Markov chain. *Ann. Math. Statist.*, **31**, 515–517.

[637] Lamperti, J. (1961) A contribution to renewal theory. *Proc. Amer. Math. Soc.*, **12**, 724–731.

[638] Lamperti, J. (1962) An invariance principle in renewal theory. *Ann. Math. Statist.*, **33**, 685–696.

[639] Lasserre, J. B. (1994) A formula for singular perturbations of Markov chains. *J. Appl. Probab.*, **31**, 829–833.

[640] Latouche, G. (1987) A note on two matrices occurring in the solution of quasi-birth-and-death processes. *Comm. Statist. Stoch. Models*, **3**, no. 2, 251–257.

[641] Latouche, G. (1988) Perturbation analysis of a phase-type queue with weakly correlated arrivals. *Adv. Appl. Probab.*, **20**, 896–912.

[642] Latouche, G. (1991) First passage times in nearly decomposable Markov chains. In: Stewart, W. J. (ed) *Numerical Solution of Markov Chains*. Probability: Pure and Applied, **8**. Marcel Dekker, New York, 401–411.

[643] Latouche, G., Jacobs, P. A., Gaver, D. P. (1984) Finite Markov chain models skip-free in one direction. *Naval Res. Logist. Quart.*, **31**, no. 4, 571–588.

[644] Latouche, G., Louchard, G. (1978) Return times in nearly decomposible stochastic processes. *J. Appl. Probab.*, **15**, 251–267.

[645] Latouche, G., Ramaswami, V. (1999) *Introduction to Matrix Analytic Methods in Stochastic Modeling*. ASA-SIAM Series on Statistics and Applied Probability, Society for Industrial and Applied Mathematics (SIAM), Philadelphia, PA and American Statistical Association, Alexandria, VA.

[646] Latouche, G., Remiche, M. A., Taylor, P. (2003) Transient Markov arrival processes. *Ann. Appl. Probab.*, **13**, 628–640.

[647] Lebedev, E. A. (2002) On the first passage time of removing level for retrial queues. *Dopov. Nats. Akad. Nauk Ukr.*, Mat. Prirodozn. Tekh. Nauki, no. 3, 47–50.

[648] Lévy, P. (1951) Systemes markoviens et stationnares. Gas denombrable. *Ann. Sci. Ecole Norm. Sup.*, **68**, 327–381.

[649] Lévy, P. (1954/56) Processus semi-markoviens. In: Erven P., Noordhoff N. V. (eds) *Proceedings of the International Congress of Mathematicians*, Vol. III, Amsterdam, 1954. North-Holland, Amsterdam, 416–426.

[650] Leadbetter, M. R. (1963) On series expansion for the renewal moments. *Biometrika*, **50**, no. 1–2, 75–80.

[651] Leadbetter, M. R., Lindgren, G., Rootzén, H. (1983) *Extremes and Related Properties of Random Sequences and Processes*. Probability and Its Applications, Springer, New York.

[652] Ledoux, J., Rubino, G., Sericola, B. (1994) Exact aggregation of absorbing Markov processes using the quasi-stationary distribution. *J. Appl. Probab.*, **31**, 626–634.

[653] Li, X., Yin, G., Yin, K., Zhang, Q. (1999) A numerical study of singularly perturbed Markov chains: quasi-equilibrium distributions and scaled occupation measures. In: *Differential Equations and Dynamical Systems*, Waterloo, ON, 1997. *Dynam. Contin. Discrete Impuls. Systems*, **5**, no. 1–4, 295–304.

[654] Liao, Ming (1988) Hitting distributions of small geodesic spheres. *Ann. Probab.*, **16**, 1039–1050.

[655] Limnios, N., Oprişan, G. (2001) *Semi-Markov Processes and Reliability*. Statistics for Industry and Technology, Birkhäuser, Boston.

[656] Lindvall, T. (1976) An invariance principle for thinned random measures. *Studia Sci. Math. Hung.*, **11**, no. 3–4, 269–275.

[657] Lindvall, T. (1977) A probabilistic proof of Blackwell's renewal theorem. *Ann. Probab.*, **5**, 482–485.

[658] Lindvall, T. (1979a) On coupling of discrete renewal processes. Z. *Wahrsch. Verw. Gebiete*, **48**, 57–70.

[659] Lindvall, T. (1979b) A note on coupling of birth and death processes. *J. Appl. Probab.*, **16**, 505–512.

[660] Lindvall, T. (1982) On coupling for continuous time renewal process, *J. Appl. Probab.*, **19**, 82–89.

[661] Lindvall, T. (1983) On coupling of diffusion processes. *J. Appl. Probab.*, **20**, 82–93.

[662] Lindvall, T. (1986) On coupling of renewal processes with use of failure rates. *Stoch. Process. Appl.*, **22**, 1–15.

[663] Lindvall, T. (1992a, 2002) *Lecture on the Coupling Method*. Wiley Series in Probability and Mathematical Statistics: Probability and Mathematical Statistics, Wiley, New York and Dover, Mineola, NY.

[664] Lindvall, T. (1992b) A simple coupling of renewal processes. *Adv. Appl. Probab.*, **24**, 1010–1011.

[665] Lindvall, T., Rogers, L. C. G. (1996) On coupling of random walks and renewal processes. *J. Appl. Probab.*, **33**, 122–126.

[666] Lladser, M., San Martin, J. (2000) Domain of attraction of the quasi-stationary distributions for the Ornstein-Uhlenbeck process. *J. Appl. Probab.*, **37**, 511–520.

[667] Loève, M. (1955, 1963) *Probability Theory*. Van Nostrand, Toronto and Princeton.

[668] Louchard, G., Latouche, G. (1982) Random times in nearly-completely decomposable, transient Markov chains. *Cahiers Centre Études Rech. Opér.*, **24**, no. 2–4, 321–352.

[669] Lund, R. B., Tweedie, R. L. (1996) Geometric convergence rates for stochastically ordered Markov chains. *Math. Oper. Res.*, **21**, no. 1, 182–194.

[670] Lund, R. B., Meyn, S. P., Tweedie, R. L. (1996) Computable exponential convergence rates for stochastically ordered Markov processes. *Ann. Appl. Probab.*, **6**, 218–237.

[671] Lundberg, F. (1903) I. *Approximerad framställning av sannolikhetsfunktionen.* II. *Återförsäkring av kollektivrisker.* Almqvist & Wiksell, Uppsala.

[672] Lundberg, F. (1909) Über die der Theorie Rückversicherung. In: VI *Internationaler Kongress für Versicherungswissenschaft.* Bd. 1, Vien, 1909, 877–955.

[673] Lundberg, F. (1926) *Försäkringsteknisk riskutjämning.* F. Englunds boktryckeri AB, Stockholm.

[674] Manor, O. (1998) Bernoulli thinning of Markov renewal process. *Appl. Stoch. Models Data Anal.*, **14**, no. 3, 229–240.

[675] Mao, Y.-H. (2006) Convergence rates in strong ergodicity for Markov processes. *Stoch. Process. Appl.*, **116**, 1964–1976.

[676] Markov, A. A. (1906) Generalisation of the law of large numbers on dependent trials. *Izv. Kazan. Fiz.-Mat. Obsch.*, **15**, no. 4, 135–156.

[677] Martinez, S. (1993) Quasi-stationary distributions for birth-death chains. Convergence radii and Yaglom limit. In: *Cellular Automata and Cooperative Systems*, Les Houches, 1992. NATO Adv. Sci. Inst. Ser. C Math. Phys. Sci., **396**, Kluwer, Dordrecht, 491–505.

[678] Martinez, S., Picco, P., San Martin, J. (1998) Domain of attraction of quasi-stationary distributions for the Brownian motion with drift. *Adv. Appl. Probab.*, **30**, 385–408.

[679] Martinez, S., Vares, M. E. (1995) A Markov chain associated with the minimal quasi-stationary distribution of birth-death chains. *J. Appl. Probab.*, **32**, 25–38.

[680] Martin-Löf, A. (1986a) Entropy estimates for the first passage time of a random walk to a time dependent barrier. *Scand. J. Statist.*, **13**, no. 3, 221–229.

[681] Martin-Lof, A. (1986b) Entropy, a useful concept in risk theory. *Scand. Actuar. J.*, no. 3–4, 223–235.

[682] Martin-Löf, A. (1998) The final size of a nearly critical epidemic, and the first passage time of a Wiener process to a parabolic barrier. *J. Appl. Probab.*, **35**, 671–682.

[683] Masol, V. I. (1972) Asymptotic behavior of the time of failure-free performance in a certain queueing system with constraints. *Kibernetika*, no. 3, 108–111.

[684] Masol, V. I. (1973) On the asymptotic behaviour of the time of arrival at a high level by a regenerating stable process with independent increments. *Teor. Veroyatn. Mat. Stat.*, **9**, 131–136 (English translation in *Theory Probab. Math. Statist.*, **9**, 139–144).

[685] Masol, V. I. (1974) Limit Distributions for Functionals of Additive Type of Regenerative Processes. Candidate of Science dissertation, Kiev State University.

[686] Masol, V. I. (1975a) Limit distributions of the time of the first passage through a high level by certain regenerating random processes with independent increments. *Dop. Akad. Nauk Ukr. RSR*, Ser. A, no. 12, 1080–1083.

[687] Masol, V. I. (1975b) The limit distribution of functionals of the trajectory of a random process of diffusion type. *Teor. Veroyatn. Mat. Stat.*, **13**, 100–106 (English translation in *Theory Probab. Math. Statist.*, **13**, 109–115).

[688] Masol, V. I. (1976a) Limit theorems for functionals of additive type of regenerative diffusion processes. *Dokl. Akad. Nauk Ukr. SSR*, Ser. A, no. 1, 15–18.

[689] Masol, V. I. (1976b) Limit theorems for regenerating random processes that are controlled by a finite Markov chain. *Dokl. Akad. Nauk Ukr. SSR*, Ser. A, no. 3, 206–209.

[690] Masol, V. I. (1977) Limit distributions of functionals of additive type on diffusion processes. *Teor. Veroyatn. Mat. Stat.*, **16**, 59–64 (English translation in *Theory Probab. Math. Statist.*, **16**, 63–68).

[691] Masol, V. I., Silvestrov, D. S. (1972) Record values of the occupation time of a semi-Markov process. *Visn. Kiïv. Univ.*, Ser. Mat. Meh., **14**, 81–89.

[692] Mei, Qi Xiang; Lin, Xiang (2001) The relationship between μ-invariant distribution and quasi-stationary distributions for continuous-time Markov chains. *Math. Theory Appl.*, **21**, no. 2, 26–29.

[693] Melamed, I. A. (1987) A local limit theorem for densities and large deviations of sums of a random number of random variables. In: *Stability Problems for Stochastic Models*. Trudy Seminara, VNIISI, Moscow 70–75 (English translation in *J. Soviet Math.*, **47**, no. 5, 2738–2742).

[694] Melamed, J. A. (1989) Limit theorems in the set-up of summation of a random number of independent identically distributed random variables. In: Kalashnikov, V. V., Zolotarev, V. M., (eds) *Stability Problems for Stochastic Models*, Sukhumi, 1987. Lecture Notes in Mathematics, **1412**, Springer, Berlin, 194–228.

[695] Mengersen, K. L., Tweedie, R. L. (1996) Rates of convergence of the Hastings and Metropolis algorithms. *Ann. Statist.*, **24**, 101–121.

[696] Meshalkin, L. D. (1958) Limit theorems for Markov chains with a finite number of states. *Teor. Veroyatn. Primen.*, **3**, 361–385 (English translation in *Theory Probab. Appl.*, **3**, 335–357).

[697] Meyn, S. P. (1989) Ergodic theorems for discrete time stochastic systems using a stochastic Lyapunov function. *SIAM J. Control Optim.*, **27**, no. 6, 1409–1439.

[698] Meyn, S. P., Caines, P. E. (1990) Mean exit times and rates of convergence for stochastic systems with applications to adaptive control. In: *Adaptive Systems in Control and Signal Processing*, Glasgow, 1989. IFAC Sympos. Ser., no. 1, Pergamon, Oxford, 177–182.

[699] Meyn, S. P., Down, D. (1994) Stability of generalized Jackson networks. *Ann. Appl. Probab.*, **4**, 124–148.

[700] Meyn, S. P.; Guo, Lei (1993) Geometric ergodicity of a doubly stochastic time series model. *J. Time Series Anal.*, **14**, no. 1, 93–108.

[701] Meyn, S. P., Tweedie, R. L. (1992) Stability of Markovian processes. I. Criteria for discrete-time chains. *Adv. Appl. Probab.*, **24**, 542–574.

[702] Meyn, S. P., Tweedie, R. L. (1993a) Stability of Markovian processes. II. Continuous-time processes and sampled chains. *Adv. Appl. Probab.*, **25**, 487–517.

[703] Meyn, S. P., Tweedie, R. L. (1993b) Stability of Markovian processes. III. Foster–Lyapunov criteria for continuous-time processes. *Adv. Appl. Probab.*, **25**, 518–548.

[704] Meyn, S. P., Tweedie, R. L. (1993c) *Markov Chains and Stochastic Stability*. Communications and Control Engineering Series, Springer, London.

[705] Meyn, S. P., Tweedie, R. L. (1994a) State-dependent criteria for convergence of Markov chains. *Ann. Appl. Probab.*, **4**, 149–168.

[706] Meyn, S. P., Tweedie, R. L. (1994b) Computable bounds for geometric convergence rates of Markov chains. *Ann. Appl. Probab.*, **4**, 981–1011.

[707] Mikosch, T. (2004) *Non-life Insurance Mathematics. An Introduction with Stochastic Processes*. Springer, Berlin.

[708] Mileshina, R. I. (1975) Limit Theorems for Controlled Random Walks. Candidate of Science dissertation, Kiev State University.

[709] Mileshina, R. I. (1977) Limit theorems for sums of random variables defined on a Bernoulli walk in a scheme of arrays and stopped at the moment of reaching a removed level. *Visn. Kiïv. Univ.*, Ser. Mat. Meh., no. 19, 98–103.

[710] Ming, Ruixing; Modibo, Diarra; Hu, Yijun (2007) A local asymptotic behavior for ruin probability in the renewal risk model. *Wuhan Univ. J. Nat. Sci.*, **12**, no. 3, 407–411.

[711] Mogyoródi, J. (1971) Some remarks on the rarefaction of the renewal processes. *Litov. Mat. Sbornik*, **11**, 303–315.

[712] Mogyoródi, J. (1972a) On the rarefaction of renewal processes, I, II. *Studia Sci. Math. Hung.*, **7**, 258–291, 293–305.

[713] Mogyoródi, J. (1972b) On the rarefaction of renewal processes, III, IV, V, VI. *Studia Sci. Math. Hung.*, **8**, 21–28, 29–38, 193–198, 199–209.

[714] Moler, J. A., Plo, F., San Miguel, M. (2000) Minimal quasi-stationary distribution under null R-recurrence. *Test*, **9**, 455–470.

[715] Motsa, A. I., Silvestrov, D. S. (1996) Asymptotics of extremal statistics and functionals of additive type for Markov chains. In: Klesov, O., Korolyuk, V., Kulldorff, G., Silvestrov, D. (eds) *Proceedings of the First Ukrainian–Scandinavian Conference on Stochastic Dynamical Systems*, Uzhgorod, 1995. *Theory Stoch. Process.*, **2(18)**, no. 1–2, 217–224.

[716] Nagaev, S. V. (1957) Some limit theorems for stationary Markov chains. *Teor. Veroyatn. Primen.*, **2**, 389–416 (English translation in *Theory Probab. Appl.*, **2**, 378–406).

[717] Nagaev, S. V. (1961) A refinement of limit theorems for homogeneous Markov chains. *Teor. Veroyatn. Primen.*, **6**, 67-86 (English translation in *Theory Probab. Appl.*, **6**, 62–81).

[718] Nagaev, S. V. (1965) On ergodic theorem for Markov processes with discrete time. *Sib. Mat. Zh.*, **6**, no. 2, 413–432.

[719] Nagaev, S. V. (1982) On ergodic theorem for for homogeneous Markov chains. *Dokl. Akad. Nauk SSSR*, **263**, 27-30 (English translation in *Soviet Math. Dokl.*, **25**, 281–284).

[720] Nair, M. G., Pollett, P. K. (1993) On relationship between μ-invariant measures and quasi-stationary distributions for continuous-time Markov chains. *Adv. Appl. Probab.*, **25**, 82–102.

[721] Nåsell, I. (1991) On the quasi-stationary distribution of the Ross malaria model. *Math. Biosci.*, **107**, 187–208.

[722] Nåsell, I. (1996) The quasi-stationary distribution of the closed endemic SIS model. *Adv. Appl. Probab.*, **28**, 895–932.

[723] Nåsell, I. (1999a) On the time to extinction in recurrent epidemics. *J. Royal Statist. Soc.*, Ser. B **61**, 309–330.

[724] Nåsell, I. (1999b) On the quasi-stationary distribution of the stochastic logistic epidemic. In: *Epidemiology, Cellular Automata, and Evolution*, Sofia, 1997. *Math. Biosci.*, **156**, no. 1–2, 21–40.

[725] Nåsell, I. (2001) Extinction and quasi-stationarity in the Verhulst logistic model. *J. Theoret. Biol.*, **211**, 11–27.

[726] Nåsell, I. (2002a) Stochastic models of some endemic infections. *Math. Biosci.*, **179**, no. 1, 1–19.

[727] Nåsell, I. (2002b) Endemicity, persistence, and quasi-stationarity. In: *Mathematical Approaches for Emerging and Reemerging Infectious Diseases: An Introduction*, Minneapolis, MN, 1999. The IMA Volumes in Mathematics and its Applications, **125**, Springer, New York, 199–227.

[728] Neuts, M. F. (1981) *Matrix-geometric Solutions in Stochastic Models. An Algorithmic Approach*. Johns Hopkins Series in the Mathematical Sciences, **2**, Johns Hopkins University Press, Baltimore.

[729] Neuts, M. F. (1989) *Structured Stochastic Matrices of $M/G/1$ Type and their Applications*. Probability: Pure and Applied, **5**, Marcel Dekker, New York.

[730] Neuts, M. F., Teugels, J. L. (1969) Exponential ergodicity of the $M/G/1$ queue. *SIAM J. Appl. Math.*, **17**, 921–929.

[731] Nummelin, E. (1976) Limit theorems for α-recurrent semi-Markov processes. *Adv. Appl. Probab.*, **8**, 531–547.

[732] Nummelin, E. (1977) On the concepts of α-recurrence and α-transience for Markov renewal processes. *Stoch. Process. Appl.*, **5**, 1–19.

[733] Nummelin, E. (1978a) A splitting techniques for certain Markov chains and its applications. *Z. Wahrsch. Verw. Gebiete*, **43**, 309–318.

[734] Nummelin, E. (1978b) Uniform and ratio limit theorems for Markov renewal and semi-regenerative processes on a general state space. *Ann. Inst. H. Poincaré*, Sect. B (N.S.) **14**, no. 2, 119–143.

[735] Nummelin, E. (1979) Strong ratio limit theorems for φ-recurrent Markov chains. *Ann. Probab.*, **7**, 639–650.

[736] Nummelin, E. (1984) *General Irreducible Markov Chains and Nonnegative Operators.* Cambridge Tracts in Mathematics, **83**, Cambridge University Press, Cambridge.

[737] Nummelin, E., Tuominen, P. (1982) Geometric ergodicity of Harris recurrent Markov chains with applications to renewal theory. *Stoch. Process. Appl.*, **12**, 187–202.

[738] Nummelin, E., Tuominen, P. (1983) The rate of convergence in Orey's theorem for Harris recurrent Markov chains with applications to renewal theory. *Stoch. Process. Appl.*, **15**, 295–311.

[739] Nummelin, E., Tweedie, R. L. (1976) Geometric ergodicity for a class of Markov chains. In: *École d'Été de Calcul des Probabilités de Saint-Flour*, Saint-Flour, 1976. *Ann. Sci. Univ. Clermont*, **61**, 145–154.

[740] Nummelin, E., Tweedie, R. L. (1978) Geometric ergodicity and R-positivity for general Markov chains. *Ann. Probab.*, **6**, 404–420.

[741] O'Cinneide, C. A. (1993) Entrywise perturbation theory and error analysis for Markov chains. *Numer. Math.*, **65**, 109–120.

[742] Oppenheim, I., Shuler, K. E., Weiss, G. H. (1977) Stochastic theory of nonlinear rate processes with multiple stationary states. *Phisica*, **88A**, 191–214.

[743] Orey, S. (1971) *Lecture Notes on Limit Theorems for Markov Chain Transition Probabilities.* Van Nostrand, London.

[744] Ovchinnikov, V. N. (1976) Asymptotical behaviour of the first falure time in model of non-homogeneous reservation with quick repearing. *Izv. Akad. Nauk SSSR*, Tekhn. Kibernet., no. 2, 82–90.

[745] Pakes, A. G. (1971) The serial correlation coefficients of waiting times in the stationary $GI/M/1$ queue. *Ann. Math. Statist.*, **42**, 1727–1734.

[746] Pakes, A. G. (1972) A $GI/M/1$ queue with a modified service mechanism. *Ann. Inst. Statist. Math.*, **24**, 589–597.

[747] Pakes, A. G. (1973) Conditional limit theorems for a left-continuous random walk. *J. Appl. Probab.*, **10**, 39–53.

[748] Pakes, A. G. (1975) On the tails of waiting-time distributions. *J. Appl. Probab.*, **12**, 555–564.

[749] Pakes, A. G. (1978a) On the age distribution of a Markov chain. *J. Appl. Probab.*, **15**, 65–77.

[750] Pakes, A. G. (1978b) On the maximum and absorption time of left-continuous random walk. *J. Appl. Probab.*, **15**, 292–299.

[751] Pakes, A. G. (1978/79) The age of a Markov process. *Stoch. Process. Appl.*, **8**, 277–303.

[752] Pakes, A. G. (1984) The Galton-Watson process with killing. *Math. Biosci.*, **69**, 171–188.

[753] Pakes, A. G. (1988a) The supercritical birth, death and catastrophe process: limit theorems on the set of nonextinction. *J. Math. Biol.*, **26**, 405–420.

[754] Pakes, A. G. (1988b) The Markov branching process with density-independent catastrophes. I. Behaviour of extinction probabilities. *Math. Proc. Cambridge Philos. Soc.*, **103**, no. 2, 351–366.

[755] Pakes, A. G. (1989a) On the asymptotic behaviour of the extinction time of the simple branching process. *Adv. Appl. Probab.*, **21**, 470–472.

[756] Pakes, A. G. (1989b) Asymptotic results for the extinction time of Markov branching processes allowing emigration. I. Random walk decrements. *Adv. Appl. Probab.*, **21**, 243–269.

[757] Pakes, A. G. (1989c) The Markov branching process with density-independent catastrophes. II. The subcritical and critical cases. *Math. Proc. Cambridge Philos. Soc.*, **106**, no. 2, 369–383.

[758] Pakes, A. G. (1990) The Markov branching process with density-independent catastrophes. III. The supercritical case. *Math. Proc. Cambridge Philos. Soc.*, **107**, no. 1, 177–192.

[759] Pakes, A. G. (1995) Quasi-stationary laws for Markov processes: examples of an alwas proximate absorbing state. *Adv. Appl. Probab.*, **27**, 120–145.

[760] Pakes, A. G., Pollett, P. K. (1989) The supercritical birth, death and catastrophe process: limit theorems on the set of extinction. *Stoch. Process. Appl.*, **32**, 161–170.

[761] Pakes, A. G., Tavaré, S. (1981) Comments on the age distribution of Markov processes. *Adv. Appl. Probab.*, **13**, 681–703.

[762] Parsons, R. W., Pollett, P. K. (1987) Quasistationary distributions for some autocatalytic reaction schemes. *J. Statist. Phys.*, **46**, 249–254.

[763] Pinsky, R. G. (1985) On the convergence of diffusion processes conditioned to remain in a bounded region for large time to limiting positive recurrent diffusion processes. *Ann. Probab.*, **13**, 363–378.

[764] Pitman, J. W. (1974) Uniform rates of convergence for Markov chain transition probabilities. *Z. Wahrsch. Verw. Gebiete*, **29**, 193–227.

[765] Pitman, J. W. (1976) An identity for stopping times of a Markov process. In: Williams, E. J. (ed) *Studies in Probability and Statistics (papers in honour of Edwin J. G. Pitman)*. North-Holland, Amsterdam, 41–57.

[766] Pitman, J. W. (1977) Occupation measures for Markov chains. *Adv. Appl. Probab.*, **9**, 69–86.

[767] Poliščuk, L. I., Turbin, A. F. (1973) Asymptotic expansions for certain characteristics of semi-Markov processes. *Teor. Veroyatn. Mat. Stat.*, **8** 122–127 (English translation in *Theory Probab. Math. Statist.*, **8**, 121–126).

[768] Poleščuk V. S. (1977) Limit Theorems for Processes of Step Sums of Random Variables Defined on Recurrent Markov Chains in Scheme of Series. Candidate of Science dissertation, Kiev State University.

[769] Pollard, D. B., Tweedie, R. L. (1975) R-theory for Markov chains on a topological state space. I. *J. London Math. Soc.* (2) **10**, no. 4, 389–400.

[770] Pollard, D. B., Tweedie, R. L. (1976) R-theory for Markov chains on a topological state space. II. *Z. Wahrsch. Verw. Gebiete*, **34**, 269–278.

[771] Pollett, P. K. (1986) On the equivalence of μ-invariant measures for the minimal process and its q-matrix. *Stoch. Process. Appl.*, **22**, 203–221.

[772] Pollett, P. K. (1988a) Reversibility, invariance and μ-invariance. *Adv. Appl. Probab.*, **20**, 600–621.

[773] Pollett, P. K. (1988b) On the problem of evaluating quasistationary distributions for open reaction schemes. *J. Statist. Phys.*, **53**, 1207–1215.

[774] Pollett, P. K. (1990) A note on the classification of Q-processes when Q is not regular. *J. Appl. Probab.*, **27**, 278–290.

[775] Pollett, P. K. (1991) On the construction problem for single-exit Markov chains. *Bull. Austral. Math. Soc.*, **43**, no. 3, 439–450.

[776] Pollett, P. K. (1995) The determination of quasistationary distributions directly from the transition rates of an absorbing Markov chain. In: *Stochastic Models in Engineering, Technology and Management*, Gold Coast, 1994. *Math. Comput. Modell.*, **22**, no. 10–12, 279–287.

[777] Pollett, P. K. (1999) Quasistationary distributions for continuous time Markov chains when absorption is not certain. *J. Appl. Probab.*, **36**, 268–272.

[778] Pollett, P. K. (2001a) Similar Markov chains. In: *Probability, Statistics and Seismology. J. Appl. Probab.*, **38A**, 53–65.

[779] Pollett, P. K. (2001b) Quasi-stationarity in populations that are subject to large-scale mortality or emigration. *Environ. Internat.*, **27**, 231–236.

[780] Pollett, P. K., Roberts, A. J. (1990) A description of the long-term behaviour of absorbing continuous-time Markov chains using a centre manifold. *Adv. Appl. Probab.*, **22**, 111–128.

[781] Pollett, P. K., Stewart, D. E. (1994) An efficient procedure for computing quasistationary distributions of Markov chains with sparse transition structure. *Adv. Appl. Probab.*, **26**, 68–79.

[782] Pollett, P. K., Vere-Jones, D. (1992) A note on evanescent processes. *Austral. J. Statist.*, **34**, no. 3, 531–536.

[783] Pollett, P.; Zhang, Hanjun; Cairns, B. J. (2007) A note on extinction times for the general birth, death and catastrophe process. *J. Appl. Probab.*, **44**, 566–569.

[784] Puchała, Z., Rolski, T. (2005) The exact asymptotic of the time to collision. *Electron. J. Probab.*, **10**, no. 40, 1359–1380.

[785] Pyke, R. (1961a) Markov renewal processes: definitions and preliminary properties. *Ann. Math. Statist.*, **32**, 1231–1242.

[786] Pyke, R. (1961b) Markov renewal processes with finitely many states. *Ann. Math. Statist.*, **32**, 1243–1259.

[787] Pyke, R. (1999) The solidarity of Markov renewal processes. In: Janssen, J., Limnios, N. (eds) *Semi-Markov Models and Applications*. Kluwer, Dordrecht, 3–21.

[788] Pyke, R., Schanflie, R. (1964) Limit theorems for Markov renewal processes. *Ann. Math. Statist.*, **35**, 1746–1764.

[789] Rachev, S. T. (1991) *Probability Metrics and the Stability of Stochastic Models*. Wiley Series in Probability and Mathematical Statistics: Applied Probability and Statistics, Wiley, Chichester.

[790] Råde, L. (1972a) Limit theorems for thinning of renewal point processes. *J. Appl. Probab.*, **9**, 847–851.

[791] Råde, L. (1972b) *Thinning of Renewal Point Processes. A Flow Graph Study*. Matematisk Statistik AB, Göteborg.

[792] Ramanan, K., Zeitouni, O. (1999) The quasi-stationary distribution for small random perturbations of certain one-dimensional maps. *Stoch. Process. Appl.*, **84**, 25–51.

[793] Ramaswami, V. (1997) Matrix analytic methods: a tutorial overview with some extensions and new results. In: *Matrix-analytic Methods in Stochastic Models*. Lecture Notes in Pure and Applied Mathematics, **183**, Dekker, New York, 261–296.

[794] Rényi, A. (1956) A characterization of Poisson processes. *Magyar. Tud. Akad. Mat. Kutató Int. Közl.*, **1**, 519–527.

[795] Revuz, D. (1975, 1984) *Markov Chains*. North-Holland Mathematical Library, **11**, North-Holland, Amsterdam and Elsevier, New York.

[796] Roberts, G. O., Jacka, S. D. (1994) Weak convergence of conditioned birth and death processes. *J. Appl. Probab.*, **31**, 90–100.

[797] Roberts, G. O., Jacka, S. D., Pollett, P. K. (1997) Non-explosivity of limits of conditioned birth and death processes. *J. Appl. Probab.*, **34**, 35–45.

[798] Roberts, G. O., Rosenthal, J. S. (1997) Shift-coupling and convergence rates of ergodic averages. *Comm. Statist. Stoch. Models*, **13**, no. 1, 147–165.

[799] Roberts, G. O., Rosenthal, J. S. (2007) Coupling and ergodicity of adaptive Markov chain Monte Carlo algorithms. *J. Appl. Probab.*, **44**, 458–475.

[800] Roberts, G. O., Rosenthal, J. S., Schwartz, P. O. (1998) Convergence properties of perturbed Markov chains. *J. Appl. Probab.*, **35**, 1–11.

[801] Roberts, G. O., Rosenthal, J. S., Segers, J., Sousa, B. (2006) Extremal indices, geometric ergodicity of Markov chains, and MCMC. *Extremes*, **9**, no. 3–4, 213–229.

[802] Roberts, G. O., Tweedie, R. L. (1996) Geometric convergence and central limit theorems for multidimensional Hastings and Metropolis algorithms. *Biometrika*, **83**, no. 1, 95–110.

[803] Roberts, G. O., Tweedie, R. L. (1999) Bounds on regeneration times and convergence rates for Markov chains. *Stoch. Process. Appl.*, **80**, 211–229.

[804] Roberts, G. O., Tweedie, R. L. (2000) Rates of convergence of stochastically monotone and continuous time Markov models. *J. Appl. Probab.*, **37**, 359–373.

[805] Roberts, G. O., Tweedie, R. L. (2001) Geometric L^2 and L^1 convergence are equivalent for reversible Markov chains. In: *Probability, Statistics and Seismology. J. Appl. Probab.*, **38A**, 37–41.

[806] Rogozin, B. A. (1973) An estimate of the remainder term in limit theorems of renewal theory. *Teor. Veroyatn. Primen.*, **18**, 703–717 (English translation in *Theory Probab. Appl.*, **18**, 662–677).

[807] Rogozin, B. A. (1976) Asymptotic analysis of the renewal function. *Teor. Veroyatn. Primen.*, **21**, 689–706 (English translation in *Theory Probab. Appl.*, **21**, 669–686).

[808] Rohlicek, J. R. (1987) Aggregation and Time Scale Analysis of Perturbed Markov Systems. Ph.D. dissertation. Mass. Inst. Technol., Cambridge, Mass.

[809] Rohlicek, J. R., Willsky, A. S. (1988a) The reduction of perturbed Markov generators: an algorithm exposing the role of transient states. *J. Ass. Comp. Mach.*, **35**, 675–696.

[810] Rohlicek, J. R., Willsky, A. S. (1988b) Multiple time scale decomposition of descrete time Markov chains. *Systems Control Lett.*, **11**, 309–314.

[811] Rolski, T., Schmidli, H., Schmidt, V., Teugels, J. (1999) *Stochastic Processes for Insurance and Finance.* Wiley Series in Probability and Statistics, Wiley, New York.

[812] Romanovskiĭ, V. I. (1949) *Discrete Markov Chains.* Gostehizdat, Moscow (English edition: Wolters-Noordhoff Publishing, Groningen (1970)).

[813] Rootzén, H. (1988) Maxima and exceedances of stationary Markov chains. *Adv. Appl. Probab.*, **20**, 371–390.

[814] Rosén, B. (1970) On the coupon collector's waiting time. *Ann. Math. Statist.*, **41**, 1952–1969.

[815] Rosén, B. (1978) A tool for derivation of asymptotic results for sampling and occupancy problems, illustrated by its application to some general capture-recapture sampling problems. In: *Proceedings of the Second Prague Symposium on Asymptotic Statistics*, Hradec Králové, 1978. North-Holland, Amsterdam, 67–79.

[816] Rosenblatt, M. (1971) *Markov Processes. Structure and Asymptotic Behavior.* Die Grundlehren der mathematischen Wissenschaften, **184**, Springer, Heidelberg.

[817] Rosenthal, J. S. (1995) Minorization conditions and convergence rates for Markov chain Monte Carlo. *J. Amer. Statist. Assoc.*, **90**, 558–566.

[818] Ross, J. V., Pollett, P. K. (2006) Extinction times for a birth-death process with two phases. *Math. Biosci.*, **202**, no. 2, 310–322.

[819] Roy, A. (1996) Transient Analysis of Queueing Systems. Ph.D. Thesis, University of Rochester, NY.

[820] Sahobov, O., Solov'ev, A. D., (1977) Two-sided estimates of reliability in a general standby model with one renewal unit. *Izv. Akad. Nauk SSSR*, Tekhn. Kibernet., no. 4, 94–99 (English translation in *Engrg. Cybernetics*, **15**, no. 4, 58–63).

[821] Samoĭlenko, Ī. V. (2006a) Asymptotic expansion of Markov random evolution. *Ukr. Mat. Visn.*, **3**, no. 3, 394–407 (English translation in *Ukr. Math. Bull.*, **3**, no. 3, 381–394).

[822] Samoĭlenko, Ī. V. (2006b) Asymptotic expansion of semi-Markov random evolution. *Ukr. Mat. Zh.*, **58**, 1234–1248 (English translation in *Ukr. Math. J.*, **58**, 1396–1414).

[823] Schmidli, H. (1992) A General Insurance Risk Model. Ph.D. Thesis, ETH, Zurich.

[824] Schmidli, H. (1994a) Corrected diffusion approximations for a risk process with the possibility of borrowing and investment. *Schweiz. Verein. Versicherungsmath. Mitt.*, no. 1, 71–82.

[825] Schmidli, H. (1994b) Diffusion approximations for a risk process with the possibility of borrowing and investment. *Comm. Statist. Stoch. Models*, **10**, 365–388.

[826] Schmidli, H. (1995) Cramér-Lundberg approximations for ruin probabilities of risk processes perturbed by diffusion. *Insur. Math. Econom.*, **16**, no. 2, 135–149.

[827] Schmidli, H. (1997) An extension to the renewal theorem and an application to risk theory. *Ann. Appl. Probab.*, **7**, 121–133.

[828] Schmidli, H. (1999) On the distribution of the surplus prior and at ruin. *Astin Bull.*, **29**, no. 2, 227–244.

[829] Schrijner, P., van Doorn, E. A. (1997) Weak convergence of conditioned birth-death processes in discrete time. *J. Appl. Probab.*, **34**, 46–53.

[830] Schweitzer, P. J. (1968) Perturbation theory and finite Markov chains. *J. Appl. Probab.*, **5**, 401–413.

[831] Schweitzer, P. J. (1984) Aggregation methods for large Markov chains. In: *Mathematical Computer Performance and Reliability*, Pisa, 1983. North-Holland, Amsterdam, 275–286.

[832] Schweitzer, P. J. (1991) A survey of aggregation-disaggregation in large Markov chains. In: Stewart, W. J. (ed) *Numerical Solution of Markov Chains*. Probability: Pure and Applied, **8**, Marcel Dekker, New York, 63–88.

[833] Schweitzer, P. J., Kindle, K. W. (1986a) An iterative aggregation-disaggregation algorithm for solving linear equations. *Appl. Math. Comput.*, **18**, no. 4, 313–353.

[834] Schweitzer, P. J., Kindle, K. W. (1986b) Iterative aggregation for solving undiscounted semi-Markovian reward processes. *Comm. Statist. Stoch. Models*, **2**, no. 1, 1–41.

[835] Schweitzer, P. J., Puterman, M. L., Kindle, K. W. (1985) Iterative aggregation-disaggregation procedures for discounted semi-Markov reward processes. *Oper. Res.*, **33**, no. 3, 589–605.

[836] Scott, D. J., Tweedie, R. L. (1996) Explicit rates of convergence of stochastically ordered Markov chains. In: *Athens Conference on Applied Probability and Time Series Analysis*, Vol. I, Athens, 1995. Lecture Notes in Statistics, **114**, Springer, New York, 176–191.

[837] Segerdahl, C. O. (1942) Übn einige risi-kotheoretische fragestellungen. *Skand. Aktuarietidskr.*, **25**, 43–83.

[838] Segerdahl, C. O. (1948) Some problems of the ruin functions in the collective theory of risk. *Skand. Aktuarietidskr.*, **31**, 46–87.

[839] Semal, P. (1991) Iterative algorithms for large stochastic matrices. *Linear Algebra Appl.*, **154/156**, 65–103.

[840] Semal, P., Courtois, P. J. (1991) Computing conditional distributions of queuing network matrices. In: Stewart, W. J. (ed) *Numerical Solution of Markov Chains*. Probability: Pure and Applied, **8**, Marcel Dekker, New York, 329–356.

[841] Seneta, E. (1966a) Quasi-stationary distributions and time-reversion in genetics. *J. Royal Statist. Soc.*, Ser. B **28**, 253–277.

[842] Seneta, E. (1966b) Quasi-stationary behaviour in the random walk with continuous time. *Austral. J. Statist.*, **8**, 92–98.

[843] Seneta, E. (1967) Finite approximations to infinite non-negative matrices. *Proc. Cambridge Philos. Soc.*, **63**, 983–992.

[844] Seneta, E. (1968) Finite approximations to infinite non-negative matrices. II. Refinements and applications. *Proc. Cambridge Philos. Soc.*, **64**, 465–470.

[845] Seneta, E. (1973) *Non-negative Matrices. An Introduction to Theory and Applications*. Halsted Press, New York.

[846] Seneta, E. (1979) Coefficients of ergodicity: structure and applications. *Adv. Appl. Probab.*, **11**, 576–590.

[847] Seneta, E. (1980) Computing the stationary distribution for infinite Markov chains. *Linear Algebra Appl.*, **34**, 259–267.

[848] Seneta, E. (1981, 2006) *Non-negative Matrices and Markov Chains*. Springer Series in Statistics, Springer, New-York.

[849] Seneta, E. (1984) Explicit forms for ergodicity coefficients and spectrum localization. *Linear Algebra Appl.*, **60**, 187–197.

[850] Seneta, E. (1988) Perturbation of the stationary distribution measured by ergodicity coefficients. *Adv. Appl. Probab.*, **20**, 228–230.

[851] Seneta, E. (1991) Sensitivity analysis, ergodicity coefficients, and rank-one updates for finite Markov chains. In: Stewart, W. J. (ed) *Numerical Solution of Markov Chains*. Probability: Pure and Applied, **8**, Marcel Dekker, New York, 121–129.

[852] Seneta, E. (1993a) Sensitivity of finite Markov chains under perturbation. *Statist. Probab. Lett.*, **17**, no. 2, 163–168.

[853] Seneta, E. (1993b) Applications of ergodicity coefficients to homogeneous Markov chains. In: *Doeblin and Modern Probability*, Blaubeuren, 1991. Contemporary Mathematics, **149**, Amer. Math. Soc., Providence, RI, 189–199.

[854] Seneta, E. (1993c) Explicit forms for ergodicity coefficients of stochastic matrices. *Linear Algebra Appl.*, **191**, 245–252.

[855] Seneta, E., Tweedie, R. L. (1985) Moments for stationary and quasistationary distributions of Markov chains. *J. Appl. Probab.*, **22**, 148–155.

[856] Seneta, E., Vere-Jones, D. (1966) On quasi-stationary distributions in discrete-time Markov chains with a denumerable infinity of states. *J. Appl. Probab.*, **3**, 403–434.

[857] Serfozo, R. F. (1976) Compositions, inverses and thinnings of random measures. *Z. Wahrsch. Verw. Gebiete*, **37**, 253–265.

[858] Serfozo, R. F. (1980) High-level exceedances of regenerative and semistationary processes. *J. Appl. Probab.*, **17**, 423–431.

[859] Serfozo, R. F. (1984a) Rarefications of compound point processes. *J. Appl. Probab.*, **21**, 710–719.

[860] Serfozo, R. F. (1984b) Thinning of cluster processes: Convergence of sums of thinned point processes. *Math. Oper. Res.*, **9**, 522–533.

[861] Sevast'yanov, B. A. (1974) Renewal Theory. In: *Itogi Mauki i Tehniki*. Ser. Teor. Veroyatn. Mat. Stat. Teor. Kibernet., **11**, VINITI, Moscow, 99–128 (English translation in *J. Math. Sci.*, **4**, no. 3, 281–302).

[862] Sevast'yanov, B. A. (1995) Extinction probabilities of branching processes bounded from below. *Teor. Veroyatn. Primen.*, **40**, 556–564 (English translation in *Theory Probab. Appl.*, **40**, 495–502).

[863] Sevast'yanov, B. A. (1998) Asymptotic behavior of extinction probabilities of stopped branching processes. *Teor. Veroyatn. Primen.*, **43**, 390–397 (English translation in *Theory Probab. Appl.*, **43**, 315–322).

[864] Sevast'yanov, B. A. (2000) Sojourn times in a finite set of states of Markov branching processes and the probabilities of extinction of a modified Galton-Watson process. *Diskret. Mat.*, **12**, no. 4, 39–45 (English translation in *Discrete Math. Appl.*, **10**, no. 6, 535–541).

[865] Sevast'yanov, B. A. (2006) Asymptotics of the survival probability of a bounded from below Markov critical process with continuous time and infinite variance. *Diskret. Mat.*, **18**, no. 1, 3–8 (English translation in *Discrete Math. Appl.*, **16**, no. 1, 1–5).

[866] Sgibnev, M. S. (1996) An infinite variance solidarity theorem for Markov renewal functions. *J. Appl. Probab.*, **33**, 434–438.

[867] Sgibnev, M. S. (2003) Systems of renewal equations on the line. *J. Math. Sci. Univ. Tokyo*, **10**, no. 3, 495–517.

[868] Sgibnev, M. S. (2006) Exact asymptotics of solutions for multidimensional renewal equations. *Izv. Ross. Akad. Nauk*, Ser. Mat. **70**, no. 2, 159–178 (English translation in *Izv. Math.*, **70**, no. 2, 363–383).

[869] Shedler, G. S. (1987) *Regeneration and Networks of Queues*. Probability and Its Applications, Springer, New York.

[870] Shedler, G. S. (1993) *Regenerative Stochastic Simulation*. Statistical Modeling and Decision Science, Academic Press, Boston.

[871] Shur, M. G. (1967) Ergodic theorems for a class of Markov processes. *Teor. Veroyatn. Primen.*, **12**, 493–505 (English translation in *Theory Probab. Appl.*, **12**, 443-454).

[872] Shur, M. G. (1976) An ergodic theorem for Markov processes. I. *Teor. Veroyatn. Primen.*, **21**, 410–416 (English translation in *Theory Probab. Appl.*, **21**, 400–406).

[873] Shur, M. G. (1977) An ergodic theorem for Markov processes. II. *Teor. Veroyatn. Primen.*, **22**, 712–728 (English translation in *Theory Probab. Appl.*, **22**, 692–707).

[874] Shur, M. G. (1984) Asymptotic properties of the powers of positive operators. I. *Teor. Veroyatn. Primen.*, **29**, 692–702 (English translation in *Theory Probab. Appl.*, **29**, 719–730).

[875] Shur, M. G. (1985) Asymptotic properties of the powers of positive operators. II. *Teor. Veroyatn. Primen.*, **30**, 241–251 (English translation in *Theory Probab. Appl.*, **30**, 656–667).

[876] Shur, M. G. (2002) New ratio limit theorems for Markov chains. In: *Analytic Methods in Applied Probability*. Amer. Math. Soc. Transl., Ser. 2, **207**, Amer. Math. Soc., Providence, RI, 203–212.

[877] Shur, M. G. (2005) The uniform integrability condition in strong ratio limit theorems. I. *Teor. Veroyatn. Primen.*, **50**, 517–532 (English translation in *Theory Probab. Appl.*, **50**, 436–447).

[878] Shurenkov, V. M. (1980a) Transient phenomena of renewal theory in asymptotic problems of the theory of random processes. In: *Probabilistic Methods of Infinite-Dimensional Analysis*. Akad. Nauk Ukr. SSR, Inst. Mat., Kiev, 133–168.

[879] Shurenkov, V. M. (1980b) Transition phenomena of the renewal theory in asymptotical problems of theory of random processes. I. *Mat. Sbornik*, **112(154)**, no. 1(5), 115–132 (English translation in *Math. USSR-Sbornik*, **40**, no. 1, 107–123).

[880] Shurenkov, V. M. (1980c) Transition phenomena of the renewal theory in asymptotical problems of theory of random processes. II. *Mat. Sbornik*, **112(154)**, no. 2(6), 226–241 (English translation in *Math. USSR-Sbornik*, **40**, no. 2, 211–225).

[881] Shurenkov, V. M. (1981) *Ergodic Theorems and Related Problems*. Naukova Dumka, Kiev (English edition: VSP, Utrecht, 1998).

[882] Shurenkov, V. M. (1983) Final probabilities of ergodic Markov processes. In: Itô, K., Prokhorov, Yu. V. (eds) *Probability Theory and Mathematical Statistics. Proceedings of the Fourth USSR–Japan Symposium*, Tbilisi, 1982. Lecture Notes in Mathematics, **1021**, Springer, Berlin, 655–665.

[883] Shurenkov, V. M. (1989) *Ergodic Markov Processes*. Probability Theory and Mathematical Statistics, Nauka, Moscow.

[884] Shurenkov, V. M., Degtyar, S. V. (1994) Markov renewal theorems in a scheme of arrays. In: *Asymptotic Analysis of Random Evolutions*. Akad. Nauk Ukr., Inst. Mat., Kiev, 270–305.

[885] Siegmund, D. (1975) The time until ruin in collective risk theory. *Mitt. Verein. Schweiz. Versicherungsmath.*, **75**, no. 2, 157–165.

[886] Silvestrov, D. S. (1969a) Limit theorems for a non-recurrent walk connected with a Markov chain. *Ukr. Mat. Zh.*, **21**, 790–804 (English translation in *Ukr. Math. J.*, **21**, 657–669).

[887] Silvestrov, D. S. (1969b) Asymptotic behaviour of the first passage time for sums of random variables controlled by a regular semi-Markov process. *Dokl. Akad. Nauk SSSR*, **189**, 949–951 (English translation in *Soviet Math. Dokl.*, **10**, 1541–1544).

[888] Silvestrov, D. S. (1969c) Limit Theorems for Semi-Markov Processes and their Applications to Random Walks. Candidate of Science dissertation, Kiev State University.

[889] Silvestrov, D. S. (1970) Limit theorems for semi-Markov processes and their applications. 1, 2. *Teor. Veroyatn. Mat. Stat.*, **3**, 155–172, 173–194 (English translation in *Theory Probab. Math. Statist.*, **3**, 159–176, 177–198).

[890] Silvestrov, D. S. (1971a) Limit theorems for semi-Markov summation schemes. 1. *Teor. Veroyatn. Mat. Stat.*, **4**, 153–170 (English translation in *Theory Probab. Math. Statist.*, **4**, 141–157).

[891] Silvestrov, D. S. (1971b) Uniform estimates of the rate of convergencxe for sums of random variables defined on a finite homogeneous Markov chain with absorption. *Theor. Veroyatn. Mat. Stat.*, **5**, 116–127 (English translation in *Theory Probab. Math. Statist.*, **5**, 123–135).

[892] Silvestrov, D. S. (1972a) Limit distributions for sums of random variables that are defined on a countable Markov chain with absorption. *Dokl. Acad. Nauk Ukr. SSR*, Ser. A, no. 4, 337–340.

[893] Silvestrov, D. S. (1972b) Estimation of the rate of convergence for sums of random variables that are defined on a countable Markov chain with absorption. *Dokl. Acad. Nauk Ukr. SSR*, Ser. A, no. 5, 436–438.

[894] Silvestrov, D. S. (1972c) Limit Theorems for Composite Random Functions. Doctor of Science dissertation, Kiev State University.

[895] Silvestrov, D. S. (1974) *Limit Theorems for Composite Random Functions*. Vysshaya Shkola and Izdatel'stvo Kievskogo Universiteta, Kiev.

[896] Silvestrov, D. S. (1976) A generalization of the renewal theorem. *Dokl. Akad. Nauk Ukr. SSR*, Ser. A, no. 11, 978–982.

[897] Silvestrov, D. S. (1978) The renewal theorem in a series scheme. 1. *Teor. Veroyatn. Mat. Stat.*, **18**, 144–161 (English translation in *Theory Probab. Math. Statist.*, **18**, 155–172).

[898] Silvestrov, D. S. (1979a) The renewal theorem in a series scheme 2. *Teor. Veroyatn. Mat. Stat.*, **20**, 97–116 (English translation in *Theory Probab. Math. Statist.*, **20**, 113–130).

[899] Silvestrov, D. S. (1979b) A remark on limit distributions for times of attainment for asymptotically recurrent Markov chains. *Theor. Sluch. Proces.*, **7**, 106–109.

[900] Silvestrov, D. S. (1980a) *Semi-Markov Processes with a Discrete State Space*. Library for an Engineer in Reliability, Sovetskoe Radio, Moscow.

[901] Silvestrov, D. S. (1980b) Remarks on the strong law of large numbers for accumulation processes. *Teor. Veroyatn. Mat. Stat.*, **22**, 118–130 (English translation in *Theory Probab. Math. Statist.*, **22**, 131–143).

[902] Silvestrov, D. S. (1980c) Synchronized regenerative processes and explicit estimates for the rate of convergence in ergodic theorems. *Dokl. Acad. Nauk Ukr. SSR*, Ser. A, no. 11, 22–25.

[903] Silvestrov, D. S. (1980d) Mean hitting times for semi-Markov processes, and queueing networks. *Elektron. Infor. Kybern.*, **16**, 399–415.

[904] Silvestrov, D. S. (1980e) Explicit estimates in ergodic theorems for regenerative processes. *Elektron. Infor. Kybern.*, **16**, 461–463.

[905] Silvestrov, D. S. (1981) Theorems of large deviations type for entry times of a sequence with mixing. *Teor. Veroyatn. Mat. Stat.*, **24**, 129–135 (English translation in *Theory Probab. Math. Statist.*, **24**, 145–151).

[906] Silvestrov, D. S. (1983a) Method of a single probability space in ergodic theorems for regenerative processes 1. *Math. Operat. Statist.*, Ser. Optim., **14**, 285–299.

[907] Silvestrov, D. S. (1983b) Invariance principle for the processes with semi-Markov switch-overs with an arbitrary state space. In: Itô, K., Prokhorov, Yu. V. (eds) *Probability Theory and Mathematical Statistics. Proceedings of the Fourth USSR–Japan Symposium*, Tbilisi, 1982. Lecture Notes in Mathematics, **1021**, Springer, Berlin, 617–628.

[908] Silvestrov, D. S. (1983c) Resolvent conditions for uniform mixing and stability of frequency characteristics for recurrent Markov chains with an arbitrary phase space. *Dokl. Akad. Nauk Ukr. SSR*, Ser. A, no. 10, 18–21.

[909] Silvestrov, D. S. (1984a) Method of a single probability space in ergodic theorems for regenerative processes 2. *Math. Operat. Statist.*, Ser. Optim., **15**, 601–612.

[910] Silvestrov, D. S. (1984b) Method of a single probability space in ergodic theorems for regenerative processes 3. *Math. Operat. Statist.*, Ser. Optim., **15**, 613–622.

[911] Silvestrov, D. S. (1994) Coupling for Markov renewal processes and the rate of convergence in ergodic theorems for processes with semi-Markov switchings. *Acta Appl. Math.*, **34**, 109–124.

[912] Silvestrov, D. S. (1995) Exponential asymptotic for perturbed renewal equations. *Teor. Ĭmovirn. Mat. Stat.*, **52**, 143–153 (English translation in *Theory Probab. Math. Statist.*, **52**, 153–162).

[913] Silvestrov, D. S. (1996) Recurrence relations for generalised hitting times for semi-Markov processes. *Ann. Appl. Probab.*, **6**, 617–649.

[914] Silvestrov, D. S. (2000a) Nonlinearly perturbed Markov chains and large deviations for lifetime functionals. In: Limnios, N., Nikulin, M. (eds) *Recent Advances in Reliability Theory: Methodology, Practice and Inference*. Birkhäuser, Boston, 135–144.

[915] Silvestrov, D. S. (2000b) Perturbed renewal equation and diffusion type approximation for risk processes. *Teor. Ĭmovirn. Mat. Stat.*, **62**, 134–144 (English translation in *Theory Probab. Math. Statist.*, **62**, 145–156).

[916] Silvestrov, D. S. (2000c) Generalized exceeding times, renewal and risk processes. In: Silvestrov, D., Yadrenko, M., Olenko, A., Zinchenko, N. (eds) *Proceedings of the Third International School on Applied Statistics, Financial and Actuarial Mathematics*, Feodosiya, 2000. *Theory Stoch. Process.*, **6(22)**, no. 3–4, 125–182.

[917] Silvestrov D. S. (2004a) *Limit Theorems for Randomly Stopped Stochastic Processes.* Probability and Its Applications, Springer, London.

[918] Silvestrov, D. S. (2004b) Upper bounds for exponential moments of hitting times for semi-Markov processes. In: *Semi-Markov Processes: Theory and Applications. Comm. Statist. Theory Methods*, **33**, no. 3, 533–544.

[919] Silvestrov, D. S. (2007a) Asymptotic expansions for quasi-stationary distributions of nonlinearly perturbed semi-Markov processes. *Theory Stoch. Process.*, **13(29)**, no. 1–2, 267–271.

[920] Silvestrov, D. S. (2007b) Asymptotic expansions for distributions of the surplus prior and at the time of ruin. In: Silvestrov, D., Borisenko, O. (eds) *Proceedings of the International Summer School "Insurance and Finance: Science, Practice and Education"*, Foros, 2007. *Theory Stoch. Process.*, **13(29)**, no. 4, 183–188.

[921] Silvestrov, D. S., Abadov, Z. A. (1984) Asymptotic behaviour for exponential moments of sums of random variables defined on exponentially ergodic Markov chains. *Dokl. Acad. Nauk Ukr. SSR*, Ser. A, no. 4, 23–25.

[922] Silvestrov, D. S., Abadov, Z. A. (1991) Uniform asymptotic expansions for exponential moments of sums of random variables defined on a Markov chain and distributions of passage times. 1. *Teor. Veroyatn. Mat. Stat.*, **45**, 108–127 (English translation in *Theory Probab. Math. Statist.*, **45**, 105–120).

[923] Silvestrov, D. S., Abadov, Z. A. (1993) Uniform representations of exponential moments of sums of random variables defined on a Markov chain, and of distributions of passage times. 2. *Teor. Veroyatn. Mat. Stat.*, **48**, 175–183 (English translation in *Theory Probab. Math. Statist.*, **48**, 125–130).

[924] Silvestrov, D. S., Banakh, D. (1981) Uniform mean ergodic theorems for accumulation processes. *Dokl. Acad. Nauk Ukr. SSR*, Ser. A, no. 2, 36–38.

[925] Silvestrov, D. S., Drozdenko, M. O. (2005) Necessary and sufficient conditions for the weak convergence of the first-rare-event times for semi-Markov processes. *Dopov. Nac. Akad. Nauk Ukr.*, Mat. Prirodozn. Tekh. Nauki, no. 11, 25–28.

[926] Silvestrov, D. S., Drozdenko, M. O. (2006) Necessary and sufficient conditions for weak convergence of first-rare-event times for semi-Markov processes. I, II. In: Silvestrov, D., Borisenko, O. (eds) *Proceedings of the International Summer School "Insurance and Finance: Science, Practice and Education"*, Foros, 2006. *Theory Stoch. Process.*, **12 (28)**, no. 3–4, 161–199, 200–216.

[927] Silvestrov, D. S., Pezhinska, G. (1984) Explicit estimates for the rate of convergence in ergodic theorems for random processes with semi-Markov switchings. In: *Random Analysis and Asymptotical Problems of Probability Theory and Mathematical Statistics*. Mitsnerba, Tbilisi, 72–78.

[928] Silvestrov, D. S., Poleščuk, V. S. (1974) Cyclic conditions for the convergence of sums of random variables defined on a recurrent Markov chain. *Dokl. Acad. Nauk Ukr. SSR,* Ser. A, no. 9, 790–792.

[929] Silvestrov, D. S., Stenflo, Ö. (1998) Ergodic theorems for iterated function systems controlled by regenerative sequences. *J. Theor. Probab.,* **11**, 589–608.

[930] Silvestrov, D.S., Teugels, J. L. (2004) Limit theorems for mixed max-sum processes with renewal stopping. *Ann. Appl. Probab.,* **14**, 1838–1868.

[931] Silvestrov, D. S., Velikii, Yu. A. (1988) Necessary and sufficient conditions for convergence of attainment times. In: Zolotarev, V. M., Kalashnikov, V. V. (eds) *Stability Problems for Stochastic Models.* Trudy Seminara, VNIISI, Moscow, 129–137 (English translation in *J. Soviet. Math.,* **57**, no. 4, 3317–3324).

[932] Simon, H. A., Ando, A. (1961) Aggregation of variables in dynamic systems. *Econometrica,* **29**, 111–138.

[933] Sirl, D.; Zhang, Hanjun; Pollett, P. (2007) Computable bounds for the decay parameter of a birth-death process. *J. Appl. Probab.,* **44**, 476–491.

[934] Skorokhod, A. V. (1956) Limit theorems for stochastic processes. *Teor. Veroyatn. Primen.,* **1**, 289–319 (English translation in *Theory Probab. Appl.,* **1**, 261–290).

[935] Smith, W. L. (1954) Asymptotic renewal theorems. *Proc. Roy. Soc. Edinburgh.,* Sect. A. **64**, 9–48.

[936] Smith, W. L. (1955a) Extensions of a renewal theorem. *Proc. Cambridge Philos. Soc.,* **51**, 629–638.

[937] Smith, W. L. (1955b) Regenerative stochastic processes. *Proc. Royal Soc. London,* Ser. A **232**, 6–31.

[938] Smith, W. L. (1954/56) Regenerative stochastic processes. In: Erven P., Noordhoff N. V. (eds) *Proceedings of the International Congress of Mathematicians,* Vol. II, Amsterdam, 1954. North-Holland, Amsterdam, 304–305.

[939] Smith, W. L. (1957) On renewal theory, counter problems, and quasi-Poisson processes. *Proc. Cambridge Philos. Soc.,* **53**, 175–193.

[940] Smith, W. L. (1958) Renewal theory and its ramifications. *J. Royal Statist. Soc.,* Ser. B **20**, 243–302.

[941] Smith, W. L. (1960) Infinitesimal renewal processes. In: *Contributions to Probability and Statistics.* Stanford Univ. Press, Stanford, CA, 396–413.

[942] Smith, W. L. (1961a) A note on the renewal theorem when the mean renewal lifetime is infinite. *J. Royal Statist. Soc.,* Ser. B **23**, 230–237.

[943] Smith, W. L. (1961b) On some general renewal thorems for nonidentically distributed variables. In: *Procedings of 4-th Berkeley Symposium. Mathematical Statisticsand Probability,* Vol. II, Berkeley, CA, 1960. Univ. California Press, Berkeley, CA, 467–514.

[944] Smith, W. L. (1962) On necessary and sufficient conditions for the convergence of the renewal density. *Trans. Amer. Math. Soc.,* **104**, no. 1, 79–100.

[945] Smith, W. L. (1969) Remarks on renewal theory when the quality of renewals varies. *Bull. Inst. Internat. Statist.*, **42**, 1049–1061.

[946] Solov'ev, A. D. (1964) Asymptotical distribution of lifetime of a doubled element. *Izv. Akad. Nauk SSSR*, Tekhn. Kibernet., no. 5, 199–121.

[947] Solov'ev, A. D. (1966) A combinatorial identity and its application to the problem on the first occurrence of a rare event. *Teor. Veroyatn. Primen.*, **11**, 313–320 (English translation in *Theory Probab. Appl.*, **11**, 276–282).

[948] Solov'ev, A. D. (1970) Standby with rapid renewal. *Izv. Akad. Nauk SSSR*, Tekhn. Kibernet., no. 1, 56–71 (English translation in *Engrg. Cybernetics.*, **8**, no. 1, 49–64).

[949] Solov'ev, A. D. (1971) Asymptotical behaviour of the first occurrence time of a rare event in a regenerative process. *Izv. Akad. Nauk SSSR*, Tekhn. Kibernet., no. 6, 79–89 (English translation in *Engrg. Cybernetics.*, **9**, no. 6, 1038–1048).

[950] Solov'ev, A. D. (1972) Asymptotic distribution of the moment of first crossing of a high level by a birth and death process. In: *Proceedings of the Sixth Berkeley Symposium on Mathematical Statistics and Probability*, Berkeley, CA, 1970/1971, Vol. III: Probability theory. Univ. California Press, Berkeley, CA, 71–86.

[951] Solov'ev, A. D. (1983) Analytical methods for computing and estimating reliability. In: Gnedenko, B. V. (ed) *Problems of Mathematical Theory of Reliability*. Radio i Svyaz', Moscow, 9–112.

[952] Solov'ev, A. D., Konstant, D. G. (1991) Reliability estimation of a complex renewable system with an unbounded number of repair units. *J. Appl. Probab.*, **28**, 833–842.

[953] Solov'ev, A. D., Sahobov, O. (1976) Two-sided estimates of the reliability of repairable systems. *Izv. Akad. Nauk UzSSR*, Ser. Fiz.-Mat., no. 5, 28–33.

[954] Solov'ev, A. D., Sahobov, O. (1977) Two-sided estimates for the probability of the failure of a system over a single period of regeneration. *Izv. Akad. Nauk UzSSR*, Ser. Fiz.-Mat. Nauk, no. 2, 41–46.

[955] Solov'yev, A. D., Zaytsev, V. A. (1975) Standby with incomplete renewal. *Izv. Akad. Nauk SSSR*, Tekhn. Kibernet., no. 1, 72–81 (English translation in *Engrg. Cybernetics*, **13**, no. 1, 58–62).

[956] Spieksma, F. M., Tweedie, R. L. (1994) Strengthening ergodicity to geometric ergodicity for Markov chains. *Comm. Statist. Stoch. Models*, **10**, no. 1, 45–74.

[957] Stenflo, Ö. (1996) Iterated function systems controlled by a semi-Markov chain. In: Klesov, O., Korolyuk, V., Kulldorff, G., Silvestrov, D. (eds) *Proceedings of the First Ukrainian–Scandinavian Conference on Stochastic Dynamical Systems*, Uzhgorod, 1995. *Theory Stoch. Process.*, **2(18)**, no. 1–2, 305–315.

[958] Stenflo, Ö. (1998) Ergodic Theorems for Iterated Function Systems Controlled by Stochastic Sequences. Ph.D. Thesis, Umeå University.

[959] Stewart, G. W. (1983) Computable error bounds for aggregated Markov chains. *J. Assoc. Comput. Mach.*, **30**, no. 2, 271–285.

[960] Stewart, G. W. (1984) On the structure of nearly uncoupled Markov chains. In: *Mathematical Computer Performance and Reliability*, Pisa, 1983. North-Holland, Amsterdam, 287–302.

[961] Stewart, G. W. (1991) On the sensitivity of nearly uncoupled Markov chains. In: Stewart W. J. (ed) *Numerical Solution of Markov Chains*. Probability: Pure and Applied, **8**, Marcel Dekker, New York, 105–119.

[962] Stewart, G. W. (1993) On the perturbation of Markov chains with nearly transient states. *Numer. Math.*, **65**, no. 1, 135–141.

[963] Stewart, G. W. (1997) On Markov chains with sluggish transients. *Comm. Statist. Stoch. Models*, **13**, no. 1, 85–94.

[964] Stewart, G. W. (1998) *Matrix algorithms. Vol. I. Basic Decompositions*. Society for Industrial and Applied Mathematics, Philadelphia, PA.

[965] Stewart, G. W. (2001) *Matrix Algorithms. Vol. II. Eigensystems*. Society for Industrial and Applied Mathematics, Philadelphia, PA.

[966] Stewart, G. W. (2003) On the powers of a matrix with perturbations. *Numer. Math.*, **96**, no. 2, 363–376.

[967] Stewart, G. W., Stewart, W. J., McAllister, D. F. (1994) A two-stage iteration for solving nearly completely decomposable Markov chains. In: *Recent Advances in Iterative Methods*. IMA Vol. Math. Appl., **60**, Springer, New York, 201–216.

[968] Stewart, G. W.; Sun, Ji Guang (1990) *Matrix Perturbation Theory*. Computer Science and Scientific Computing, Academic Press, Boston.

[969] Stewart, G. W., Zhang, G. (1991a) On a direct method for the solution of nearly uncoupled Markov chains. *Numer. Math.*, **59**, no. 1, 1–11.

[970] Stewart, G. W., Zhang, G. (1991b) Eigenvalues of graded matrices and the condition numbers of a multiple eigenvalue. *Numer. Math.*, **58**, no. 7, 703–712.

[971] Stone, C. (1965a) On moment generating functions and renewal theory. *Ann. Math. Statist.*, **36**, 1298–1301.

[972] Stone, C. (1965b) On characteristic functions and renewal theory. *Trans. Amer. Math. Soc.*, **120**, 327–342.

[973] Stone, C. (1966) On absolutely continuous components and renewal theory. *Ann. Math. Statist.*, **37**, 271–275.

[974] Stone, C., Wainger, S. (1967) One-sided error estimates in renewal theory. *J. Analys. Math.*, **20**, 325–352.

[975] Su, Yeong Tzay (1993) Limit theorems for first-passage times in nonlinear Markov renewal theory. *Sequent. Anal.*, **12**, no. 3–4, 235–245.

[976] Su, Yeong Tzay (1998) A renewal theory for perturbed Markov random walks with applications to sequential analysis. *Statist. Sinica*, **8**, no. 2, 429–443.

[977] Su, Yeong Tzay; Wang, Chi Tshung (1998) Nonlinear renewal theory for lattice Markov random walk. *Sequent. Anal.*, **12**, no. 3–4, 267–277.

[978] Sudakova, E. V. (1990) Exponential asymptotics of sums of geometrical number of independent random variables. *Theor. Ĭmovirn. Mat. Stat.*, **42**, 161–166 (English translation in *Theory Probab. Math. Statist.*, **42**, 135–139).

[979] Sudakova, E. V. (1993) Estimation with weigt for the characterization of asymptotics of sums of independent random variables. *Theor. Ĭmovirn. Mat. Stat.*, **48**, 185–190 (English translation in *Theory Probab. Math. Statist.*, **42**, 131–134).

[980] Sumita, U., Rieders, M. (1991) Numerical comparison of the replacement process approach with the aggregation-disaggregation algorithm for row-continuous Markov chains. In: Stewart W. J. (ed) *Numerical Solution of Markov Chains*. Probability: Pure and Applied, **8**, Marcel Dekker, New York, 287–302.

[981] Sundt, B. (1991) *An Introduction to Non-Life Insurance Mathematics*. Versicherungswirtschaft, Karlsruhe.

[982] Szantai, T. (1971a) On limiting distributions for sums of random variables concerning the rarefaction of recurrent processes. *Studia Sci. Math, Hung.*, **6**, 443–452.

[983] Szantai, T. (1971b) On an invariance problem related to different rarefactions of recurrent processes. *Studia Sci. Math, Hung.*, **6**, 453–456.

[984] Sztrik, J., Anisimov, V. V. (1989) Reliability analysis of a complex renewable system with fast repair. *J. Inform. Process. Cybernet.*, **25**, no. 11–12, 573–580.

[985] Szynal, D. (1976) On limit distribution theorems for sums of a random number of random variables appearing in the study of rarefaction of a recurrent process. *Zastosow. Mat.*, **15**, 277–288.

[986] Täcklind, S. (1942) Sur le risque du ruine dans des jeux inéqutables. *Skand. Aktuarietidskr.*, **25**, 1–42.

[987] Täcklind, S. (1945) Fourieranalytische Behandlung vom Erneuerungsproblem. *Skand. Aktuarietidskr.*, **28**, 68–105.

[988] Takács, L. (1954) Some investigations concerning recurrent stochastic processes of a certain type. *Magyar. Tud. Akad. Mat. Kutató Int. Közl.*, **3**, 114–128.

[989] Tan, Choon Peng (1982) A functional form for a particular coefficient of ergodicity. *J. Appl. Probab.*, **19**, 858–863.

[990] Tan, Choon Peng (1983) Coefficients of ergodicity with respect to vector norms. *J. Appl. Probab.*, **20**, no. 2, 277–287.

[991] Teugels, J. L. (1967) Exponential decay in renewal theorems. *Bull. Soc. Math. Belg.*, **19**, 259–276.

[992] Teugels, J. L. (1968a) Renewal theorems when the first or the second moment is infinite. *Ann. Math. Statist.*, **39**, 1210–1219.

[993] Teugels, J. L. (1968b) Exponential ergodicity in Markov renewal processes. *J. Appl. Probab.*, **33**, 434–438.

[994] Teugels, J. L. (1968c) Exponential ergodicity in derived Markov chains. *J. Appl. Probab.*, **5**, 669–678.

[995] Teugels, J. L. (1970) Regular variation of Markov renewal functions. *J. London Math. Soc.* (2), **2**, 179–190.

[996] Teugels, J. L. (1972) An example on geometric ergodicity of a finite Markov chain. *J. Appl. Probab.*, **9**, 466–469.

[997] Teugels, J. L. (1976) A bibliography on semi-Markov processes. *J. Comput. Appl. Math.*, **2**, 125–144.

[998] Teugels, J. L. (1986) A second bibliography on semi-Markov processes. In: *Semi-Markov models*, Brussels, 1984. Plenum, New York, 507–584.

[999] Teugels, J. L. (1990) Renewal theory and applications in engineering. In: *Bericht über die Wissenschaftliche Jahrestagung der GAMM*, Karlsruhe, 1989. *Z. Angew. Math. Mech.*, **70**, no. 6, T513–T518.

[1000] Thorin, O. (1977) Ruin probabilities prepared for numerical calculations. *Scand. Actuar. J.*, Suppl., 7–17.

[1001] Thorin, O. (1982) Probabilities of ruins. *Scand. Actuar. J.*, no. 2, 65–102.

[1002] Thorin, O., Wikstad, N. (1976) Calculation of ruin probabilities when the claim distributions is lognormal. *Astin Bulletin*, **IX**, 231–246.

[1003] Thorisson, H. (1983) The coupling of regenerative processes. *Adv. Appl. Probab.*, **15**, 531–561.

[1004] Thorisson, H. (1985a) Periodic regeneration. *Stoch. Process. Appl.*, **20**, 85–104.

[1005] Thorisson, H. (1985b) On regenerative and ergodic properties of the k-server queue with nonstationary Poisson arrivals. *J. Appl. Probab.*, **22**, 893–902.

[1006] Thorisson, H. (1987) A complete coupling proof of Blackwell's renewal theorem. *Stoch. Process. Appl.*, **26**, 87–97.

[1007] Thorisson, H. (1990) The classical coupling, a refinement. *Teor. Veroyatn. Primen.*, **35**, 809–817 (Also in *Theory Probab. Appl.*, **35**, 810–819).

[1008] Thorisson, H. (1992) The coupling method and regenerative processes. In: *Analysis, Algebra, and Computers in Mathematical Research*, Luleå, 1992. Lecture Notes in Pure and Applied Mathematics, **156**, Dekker, New York, 347–363.

[1009] Thorisson, H. (1998) Coupling. In: *Probability Towards 2000*, New York, 1995. Lecture Notes in Statistics, **128**, Springer, New York, 319–339.

[1010] Thorisson, H. (2000) *Coupling, Stationarity and Regeneration*. Probability and Its Applications, Springer, New York.

[1011] Todorovic, P., Gani, J. (1989) A Markov renewal process imbedded in a Markov chain. *Stoch. Anal. Appl.*, **7**, no. 3, 339–353.

[1012] Tomko, J. (1974) On rerafication of multivariate point processes. In: *Progress in Statistics*, Vol. II, European Meeting Statisticians, Budapest, 1972. Colloq. Math. Soc. Janos Bolyai, **9**, North-Holland, Amsterdam, 843–868.

[1013] Tuominen, P., Tweedie, R. L. (1979a) Exponential decay and ergodicity of general Markov processes and their discrete skeletons. *Adv. Appl. Probab.*, **11**, 784–803.

[1014] Tuominen, P., Tweedie, R. L. (1979b) Exponential ergodicity in Markovian queueing and dam models. *J. Appl. Probab.*, **16**, 867–880.

[1015] Tuominen, P., Tweedie, R. L. (1994) Subgeometric rates of convergence of f-ergodic Markov chains. *Adv. Appl. Probab.*, **26**, 775–798.

[1016] Turbin, A. F. (1971) On asymptotic behavior of time of a semi-Markov process being in a reducible set of states. Linear case. *Teor. Veroyatn. Mat. Stat.*, **4**, 179–194 (English translation in *Theory Probab. Math. Statist.*, **4**, 167–182).

[1017] Turbin, A. F. (1972) An application of the theory of perturbations of linear operators to the solution of certain problems that are connected with Markov chains and semi-Markov processes. *Teor. Veroyatn. Mat. Stat.*, **6**, 118–128 (English translation in *Theory Probab. Math. Statist.*, **6**, 119–130).

[1018] Turner, J. W., Malek-Mansour, M. (1978) On absorbing zero boundary problem in birth death processes. *Phisica*, **93A**, 517–525.

[1019] Tweedie, R. L. (1971) Truncation procedures for non-negative matrices. *J. Appl. Probab.*, **8**, 311–320.

[1020] Tweedie, R. L. (1974a) R-theory for Markov chains on a general state space. I: Solidarity properties and r-recurrent chains. *Ann. Probab.*, **2**, 840–864.

[1021] Tweedie, R. L. (1974b) R-theory for Markov chains on a general state space. II: r-subvariant measures for r-recurrent chains. *Ann. Probab.*, **2**, 865–878.

[1022] Tweedie, R. L. (1974c) Quasi-stationary distributions for Markov chains on a general state space. *J. Appl. Probab.*, **11**, 726–741.

[1023] Tweedie, R. L. (1975a) The robustness of positive recurrence and recurrence of Markov chains under perturbations of the transition probabilities. *J. Appl. Probab.*, **12**, 744–752.

[1024] Tweedie, R. L. (1975b) Truncation approximation of the limit probabilities for denumerable semi-Markov processes. *J. Appl. Probab.*, **12**, 161–163.

[1025] Tweedie, R. L. (1980) Perturbations of countable Markov chains and processes. *Ann. Inst. Statist. Math.*, **32**, no. 2, 283–290.

[1026] Tweedie, R. L. (1981) Criteria for ergodicity, exponential ergodicity and strong ergodicity of Markov processes. *J. Appl. Probab.*, **18**, 122–130.

[1027] Tweedie, R. L. (1983a) The existence of moments for stationary Markov chains. *J. Appl. Probab.*, **20**, 191–196.

[1028] Tweedie, R. L. (1983b) Criteria for rates of convergence of Markov chains, with application to queueing and storage theory. In: *Probability, Statistics and Analysis*. London Math. Soc. Lecture Notes, **79**, Cambridge Univ. Press, Cambridge-New York, 260–276.

[1029] Tweedie, R. L. (1998) Truncation approximations of invariant measures for Markov chains. *J. Appl. Probab.*, **35**, 517–536.

[1030] Tweedie, R. L. (2001) Markov chains: structure and applications. In: *Stochastic Processes: Theory and Methods*. Handbook of Statistics, **19**, North-Holland, Amsterdam, 817–851.

[1031] van Doorn, E. A. (1981) The transient state probabilities for a queueing model where potential customers are discouraged by queue length. *J. Appl. Probab.*, **18**, 499–506.

[1032] van Doorn, E. A. (1985) Conditions for exponential ergodicity and bounds for the decay parameter of a birth-death process. *Adv. Appl. Probab.*, **17**, 514–530.

[1033] van Doorn, E. A. (1987) The indeterminate rate problem for birth-death processes. *Pacific J. Math.*, **130**, no. 2, 379–393.

[1034] van Doorn, E. A. (1991) Quasi-stationary distributions and convergence to quasi-stationarity of birth-death processes. *Adv. Appl. Probab.*, **23**, 683–700.

[1035] van Doorn, E. A. (2001) Representations for the rate of convergence of birth-death processes. *Teor. Ĭmovirn. Mat. Stat.*, **65**, 33–38 (Also in *Theory Probab. Math. Statist.*, **65**, 37–43).

[1036] van Doorn, E. A. (2003) Birth-death processes and associated polynomials. In: *Proceedings of the Sixth International Symposium on Orthogonal Polynomials, Special Functions and their Applications*, Rome, 2001. *J. Comput. Appl. Math.*, **153**, no. 1–2, 497–506.

[1037] van Doorn, E. A. (2006) On the α-classification of birth-death and quasi-birth-death processes. *Stoch. Models*, **22**, no. 3, 411–421.

[1038] van Doorn, E. A., Schrijner, P. (1995a) Ratio limits and limiting conditional distributions for discrete-time birth-death processes. *J. Math. Anal. Appl.*, **190**, no. 1, 263–284.

[1039] van Doorn, E. A., Schrijner, P. (1995b) Geometric ergodicity and quasi-stationarity in discrete-time birth-death processes. *J. Austral. Math. Soc.*, Ser. B **37**, no. 2, 121–144.

[1040] van Doorn, E. A., Schrijner, P. (1996) Limit theorems for discrete-time Markov chains on the nonnegative integers conditioned on recurrence to zero. *Comm. Statist. Stoch. Models*, **12**, no. 1, 77–102.

[1041] van Doorn, E. A., Zeĭfman, A. I. (2005) Extinction probability in a birth-death process with killing. *J. Appl. Probab.*, **42**, 185–198.

[1042] Vere-Jones, D. (1962) Geometric ergodicity in denumerable Markov chains. *Quart. J. Math.*, **13**, 7–28.

[1043] Vere-Jones, D. (1963) On the spectra of some linear operators associated with queueing systems. *Z. Wahrsch. Verw. Gebiete*, **2**, 12–21.

[1044] Vere-Jones, D. (1964) A rate of convergence problem in the theory of queues. *Teor. Veroyatn. Primen.*, **9**, 104–112 (Also in *Theory Probab. Appl.*, **9**, 94–103).

[1045] Vere-Jones, D. (1967) Ergodic properties of nonnegative matrices. I. *Pacific J. Math.*, **22**, 361–386.

[1046] Vere-Jones, D. (1968) Ergodic properties of nonnegative matrices. II. *Pacific J. Math.*, **26**, 601–620.

[1047] Vere-Jones, D. (1969) Some limit theorems for evanescent processes. *Austral. J. Statist.*, **11**, 67–78.

[1048] Voĭna, A. A. (1978) Limit Theorems for Shemes of Summation on Processes with Arbitrary Sets of States and their Applications. Candidate of Science dissertation, Kiev State University.

[1049] Voĭna, A. A. (1979) Sojourn time in a fixed subset and the consolidation of states of random processes with an arbitrary phase space. *Theor. Veroyatn. Mat. Stat.*, **20**, 30–39 (English translation in *Theory Probab. Math. Statist.*, **20**, 35–44).

[1050] Voĭna, A. A., Pushkin, S.G. (1984) Limit behavior of the emptying time of a queueing system. *Dokl. Akad. Nauk Ukr. SSR*, Ser. A, no. 3, 63–66.

[1051] von Bahr, B. (1975) Asymptotic ruin probabilities when exponential moments do not exist. *Scand. Actuar. J.*, 6–10.

[1052] Vsekhsvyatski, S. Yu., Kalashnikov, V. V. (1985a) Estimates for the occurrence times of rare events in regenerative processes. *Teor. Veroyatn. Primen.*, **30**, 580–583 (English translation in *Theory Probab. Appl.*, **30**, 618–621).

[1053] Vsekhsvyatski, S. Yu., Kalashnikov, V. V. (1985b) Estimates of the instant of the first occurrence of a rare event in birth and death processes. *Izv. Akad. Nauk SSSR*, Tekhn. Kibernet., no. 2, 214–216 (English translation in *Soviet J. Comput. Systems Sci.*, **23**, no. 5, 145–147).

[1054] Walker, B. K. (1980) A semi-Markov approach to quantifying foult-tolerant system perfomance. Ph.D. Thesis. Mass. Inst. Technol., Cambridge, Mass.

[1055] Walker, D. M. (1998) The expected time until absorption when absorption is not certain. *J. Appl. Probab.*, **35**, 812–823.

[1056] Wang, Zi Kun (1980) Sojourn times and first passage times for birth and death processes. *Sci. Sinica*, **23**, no. 3, 269–279.

[1057] Wang, Zi Kun; Yang, Xiang Qun (1992) *Birth and Death Processes and Markov Chains*. Springer, Berlin and Science Press, Beijing.

[1058] Weiss, G. H., Dishon, M. (1971) On asymptotic behaviour of the stochastic and deterministic models of epidemic. *Math. Biosci.*, **11**, 261–265.

[1059] Wentzell, A. D., Freidlin, M. I. (1979) *Fluctuations in Dynamical Systems Subject to Small Random Perturbations*. Probability Theory and Mathematical Statistics, Nauka, Moscow (English edition: *Random Perturbations of Dynamical Systems*. Fundamental Principles of Mathematical Sciences, **260**, Springer, New York (1998)).

[1060] Whitt, W. (2000) The impact of a heavy-tailed service-time distribution upon the $M/GI/s$ waiting-time distribution. *Queueing Systems Theory Appl.*, **36**, no. 1–3, 71–87.

[1061] Whitt, W. (2002) *Stochastic-Process Limits. An Introduction to Stochastic-Process Limits and their Application to Queues*. Springer Series in Operations Research, Springer, New York.

[1062] Wikstad, N. (1983) A numerical illustration of differences between ruin probabilities originated in the ordinary and in stationary cases. *Scand. Actuar. J.*, no. 1, 47–48.

[1063] Willekens, E., Teugels, J. L. (1992) Asymptotic expansions for waiting time probabilities in an $M/G/1$ queue with long-tailed service time. *Queueing Systems Theory Appl.*, **10**, no. 4, 295–311.

[1064] Willmot, G. E., Lin, X. S. (1998) Exact and approximate properties of the distribution of surplus before and after ruin. *Insur. Math. Econom.*, **23**, no. 1, 91–110.

[1065] Woodroofe, M. (1982) *Nonlinear Renewal Theory in Sequential Analysis.* CBMS-NSF Regional Conference Series in Applied Mathematics, **39**, Soc. Industr. Appl. Math., Philadelphia.

[1066] Yin, George; Zhang, Hanqin (2007) Singularly perturbed Markov chains: limit results and applications. *Ann. Appl. Probab.*, **17**, 207–229.

[1067] Yin, Chuancun; Zhao, Junsheng (2006) Nonexponential asymptotics for the solutions of renewal equations, with applications. *J. Appl. Probab.*, **43**, 815–824.

[1068] Yin, G. G., Zhang, Q. (1998) *Continuous-time Markov Chains and Applications. A Singular Perturbation Approach.* Applications of Mathematics, **37**, Springer, New York.

[1069] Yin, G., Zhang, Q. (2003) Discrete-time singularly perturbed Markov chains. In: *Stochastic Modeling and Optimization.* Springer, New York, 1–42.

[1070] Yin, G., Zhang, Q., Badowski, G. (2000a) Asymptotic properties of a singularly perturbed Markov chain with inclusion of transient states. *Ann. Appl. Probab.*, **10**, 549–572.

[1071] Yin, G., Zhang, Q., Badowski, G. (2000b) Occupation measures of singularly perturbed Markov chains with absorbing states. *Acta Math. Sin.*, **16**, no. 1, 161–180.

[1072] Yin, G., Zhang, Q., Badowski, G. (2000c) Singularly perturbed Markov chains: convergence and aggregation. *J. Multivar. Anal.*, **72**, no. 2, 208–229.

[1073] Yin, G., Zhang, Q., Liu, Q. G. (2000) Error bounds for occupation measure of singularly perturbed Markov chains including transient states. *Probab. Engrg. Inform. Sci.*, **14**, no. 4, 511–531.

[1074] Yin, G., Zhang, Q., Yang, H., Yin, K. (2001) Discrete-time dynamic systems arising from singularly perturbed Markov chains. In: *Proceedings of the Third World Congress of Nonlinear Analysts*, Part 7, Catania, 2000. *Nonlinear Anal.*, **47**, no. 7, 4763–4774.

[1075] Zajtsev, V. A., Solov'ev, A. D. (1975) Redundancy of complex systems. *Izv. Akad. Nauk SSSR*, Tekhn. Kibernet., no. 4, 83–92 (English translation in *Engrg. Cybernetics*, **13**, no. 4, 66–75).

[1076] Zakusilo, O. K. (1972a) Thinning semi-Markov processes. *Teor. Veroyatn. Mat. Stat.*, **6**, 54–59 (English translation in *Theory Probab. Math. Statist.*, **6**, 53–58).

[1077] Zakusilo, O. K. (1972b) Necessary conditions for convergence of semi-Markov processes that thin. *Teor. Veroyatn. Mat. Stat.*, **7**, 65–69 (English translation in *Theory Probab. Math. Statist.*, **7**, 63–66).

[1078] Zeĭfman, A. I. (1989a) Quasi-ergodicity and stability of some nonhomogeneous Markov processes. *Sib. Mat. Zh.*, **30**, no. 2, 85–89 (English translation in *Sib. Math. J.*, **30**, no. 2, 236–239).

[1079] Zeĭfman, A. I. (1989b) Quasi-ergodicity for nonhomogeneous continuous-time Markov chains. *J. Appl. Probab.*, **26**, 643–648.

[1080] Zeĭfman, A. I. (1991a) Stability of nonhomogeneous Markov chains with continuous time. *Izv. Vyssh. Uchebn. Zaved. Mat.*, no. 7, 33–39 (English translation in *Soviet Math. (Iz. VUZ)*, **35**, no. 7, 29–34).

[1081] Zeĭfman, A. I. (1991b) Some estimates of the rate of convergence for birth and death processes. *J. Appl. Probab.*, **28**, 268–277.

[1082] Zeĭfman, A. I. (1995) Upper and lower bounds on the rate of convergence for nonhomogeneous birth and death processes. *Stoch. Process. Appl.*, **59**, 157–173.

[1083] Zeĭfman, A. I., Isaacson, D. L. (1994) On strong ergodicity for nonhomogeneous continuous-time Markov chains. *Stoch. Process. Appl.*, **50**, 263–273.

[1084] Zeĭfman, A., Leorato, S., Orsingher, E., Satin, Ya., Shilova, G. (2006) Some universal limits for nonhomogeneous birth and death processes. *Queueing Syst.*, **52**, no. 2, 139–151.

[1085] Zhang, H., Chen, A., Lin, X., Hou, Z. (2001) Strong ergodicity of monotone transition functions. *Statist. Prob. Lett.*, **55**, 63–69.

[1086] Zhang, Hanqin; Hsu, Guang-Hui; Wang, Rongxin (1995) Heavy traffic limit theorems for a sequence of shortest queueing systems. *Queueing Systems Theory Appl*, **21**, no. 1–2, 217–238.

[1087] Zhang, Q., Yin, G. (2004) Exponential bounds for discrete-time singularly perturbed Markov chains. *J. Math. Anal. Appl.*, **293**, no. 2, 645–662.

[1088] Zinchenko, N. (1999) Risk processes in the presence of large claims. In: Silvestrov, D., Yadrenko, M., Borysenko, O., Zinchenko, N. (eds) *Proceedings of the Second International School on Applied Statistics, Financial and Actuarial Mathematics*, Kyiv, 1999. *Theory Stoch. Process.*, **5(21)**, no. 1–2, 200–211.

[1089] Zolotarev, V. M. (1975) Quantitative estimates in problems on the continuity of queueing systems. *Teor. Veroyatn. Primen.*, **20**, 215–218 (English translation in *Theory Probab. Appl.*, **20**, 211–213).

[1090] Zolotarev, V. M. (1976) On the stochastic continuity of queueing systems of the type $G \mid G \mid 1$. *Teor. Veroyatn. Primen.*, **21**, 260–279 (English translation in *Theory Probab. Appl.*, **21**, 250–269).

[1091] Zolotarev, V. M. (1977) Quantitative estimates for the continuity property of queueing systems of type $G/G/\infty$. *Teor. Veroyatn. Primen.*, **22**, 700–711 (English translation in *Theory Probab. Appl.*, **22**, 679–691).

[1092] Zubkov, A. M. (1975) The rate of convergence of a renewal density. *Mat. Sbornik*, **98(140)**, no. 1(9), 143–156 (English translation in *Math. USSR-Sbornik*, **27**, no. 1, 131–142).

[1093] Zubkov, A. M. (1979) Inequalities for restricted transition probabilities and their applications. *Mat. Sbornik*, **109(151)**, no. 4, 491–532 (English translation in *Math. USSR-Sbornik*, **37**, no. 4, 451–488).

[1094] Zubkov, A. M., Mihaĭlov, V. G. (1974) Limit distributions of random variables that are connected with long duplications in a sequence of independent trials. *Teor. Veroyatn. Primen.*, **19**, 173–181 (English translation in *Theory Probab. Appl.*, **19**, 172–179).

Index